THE BIOLOGY OF NEMATODES

This book is dedicated to Alan F. Bird and Kenneth A. Wright in memory of their great contributions to the study of nematodes. They were an inspiration to others and their enthusiasm was unabated up to the time of their deaths.

THE BIOLOGY OF NEMATODES

Edited by

Donald L. Lee

School of Biology
University of Leeds, UK

London and New York

First published 2002
by Taylor & Francis
11 New Fetter Lane, London EC4P 4EE

Simultaneously published in the USA and Canada
by Taylor & Francis Inc
29 West 35th Street, New York, NY 10001

Taylor & Francis is an imprint of the Taylor & Francis Group

© 2002 Taylor & Francis

Typeset in Malaysia by Expo Holdings
Printed and bound in Singapore by Fabulous Printers Pte Ltd

British Library Cataloguing in Publication Data
A catalogue record for this book is available from the British Library

Library of Congress Cataloging in Publication Data
A catalog record for this book has been requested.

ISBN 0-415-27211-4

CONTENTS

FOREWORD
by Bridget M Ogilvie, DBE, ScD

To most people, nematodes seem dull if not repulsive and their mainly cylindrical, thread like shape does not help their image. Their main uniting feature seems to be their life cycle, an egg followed by 5 stages with moults between. But beneath these rather banal features lies an infinite degree of complexity and variation. This is true both of those which live at least part of their lives as parasites of plants or animals and of the majority that are free living, and are found in all manner of environments, including many that are extreme.

To scientists, the most famous worm today is almost certainly *Caenorhabditis elegans*, used as a model for detailed cellular and molecular studies of a multicellular animal for the past 3 decades, culminating with the publication of the complete sequence of its genome in 1998. These studies have enabled *C. elegans* to be used for the investigation of a variety of biological questions relevant to many species including our own.

The extraordinary depth and detail of our knowledge of *C. elegans* has only just begun to illuminate our understanding of the huge numbers of species that are classified as nematodes. This multi-authored volume edited by Donald Lee is the culmination of his long term commitment to furthering our understanding of these organisms which are ubiquitous in nature and often important parasites of plants and animals. Our understanding of the huge nematode family at the genetic/systematics level does not permit us yet to relate the complexities of these creatures to their environment or their genetic distance from each other. Instead, in this book the vast body of knowledge about nematodes is organised into themes and within each theme, the available knowledge is discussed across the whole arena of nematode variation.

It is 20 years since I too was engaged in research into parasitic nematodes. I am delighted to be asked to write this foreword and thus have the opportunity to see the work of many old friends and even more new scientists engaged in the study of this complex group of animals.

It is particularly pleasing to read reviews that incorporate work done not only at the end of the 20th Century but in some cases, as long ago as the 19th Century. This is a tribute to the scholarship of the participating authors and thus, a commendation of this interesting volume.

PREFACE

The objective of this multi-author book is to bring together various aspects of the biology of free-living, plant-parasitic and animal-parasitic nematodes in a form that will be suitable for research scientists, academics involved in the teaching of nematodes and some aspects of parasitology, and post-graduates and undergraduates who wish to have the diverse literature available in readily accessible reviews.

Nematodes have tended to be studied because of their medical and veterinary importance and because they are major pests of crops. There is an extensive literature on their structure, systematics, biochemistry, physiology, immunology and molecular biology. However, large numbers of free-living nematodes inhabit marine or freshwater mud and the soil, with several million being found per square metre in the top few centimetres of sub-littoral mud and cultivated soil, and they are being increasingly studied.

The place of nematodes in the early cytological literature is not generally realised. The first description of the formation of polar bodies by sub-division of the nucleus of the ovum; the discovery that both parents contribute equal numbers of chromosomes to the offspring; and the discovery that the cells which eventually form certain organs, such as the reproductive system or the intestine, could be identified as early as the fourth cleavage of the egg, were by scientists who worked on the large ascarid *Parascaris equorum* (syn. *Ascaris megalocephala*) from horses during the 19th century (see Chitwood and Chitwood (1950) *An Introduction to Nematology*, Baltimore: Monumental Printing Co.). Nematodes have again become prominent as model organisms. In 1965 Sydney Brenner chose the free-living nematode *Caenorhabditis elegans* to study the developmental biology, genetics, molecular biology and behaviour of a single organism. Since then its complete cell lineage has been described with every cell being identified and assigned a unique label and the connectivity of its nervous system has been worked out. There is now an extensive literature on all aspects of the biology of this nematode. Information obtained from *C. elegans* is being used in the study of other nematodes and this is reflected in several chapters in this book.

"Larva" instead of "juvenile", "pharynx" instead of "oesophagus" and "epidermis" instead of "hypodermis" have been used throughout the book. The reasons for this are given in the relevant sections. No attempt has been made to standardise the classification used in the various chapters as this is currently undergoing proposed modification (see Chapter 1).

Acknowledgements

I would like to thank the contributors to the book for the generous allocation of their valuable time to the writing of their chapters and for their patience during the long gestation period. Many people have encouraged and supported me in the compilation of this book. I am particularly grateful to Howard Atkinson, Jean Bird, Ian Hope, John Jones, Rolo Perry, David Wharton and Urs Wyss for their constructive advice and encouragement. Also to Jean Bird, John Jones, Warwick Nicholas and Jörg Ott who supplied original diagrams, photographs and electron micrographs. The staff at the publishers have been a constant support.

I particularly wish to thank my wife, Shirley, who has been so very supportive throughout my career and especially during the compilation and writing of this book.

LIST OF CONTRIBUTORS

Alexander, R. McN.
School of Biology
University of Leeds
Leeds LS2 9JT
UK

Atkinson, H.J.
School of Biology
University of Leeds
Leeds LS2 9JT
UK

Behm, C.A.
Division of Biochemistry and Molecular Biology
Faculty of Science
Australian National University
Canberra ACT 0200
AUSTRALIA

Blaxter, M.L.
Institute of Cell, Animal and Population Biology
Ashworth Laboratories
University of Edinburgh
Edinburgh EH9 3JT
UK

Bundy, D.A.P.
World Bank
1818 HStreet NW
Washington DC 20433
USA

Conder, G.A.
Global Research and Development
Pfizer Inc.
Eastern Point Road
Groton CT 06340
USA

De Ley, P.
Department of Nematology
University of California at Riverside
Riverside CA 92521
USA

Dobson, R.J.
CSIRO Livestock Industries
McMaster Laboratory
Blacktown NSW 2148
AUSTRALIA

Geary, T. G.
Animal Health Discovery Research
Pharmacia and Upjohn
301 Henrietta Street
Kalamazoo MI 49007
USA

Gems, D.
Department of Biology
University College London
4 Stephenson Way
London NW1 2HE
UK

Gibbons, L.M.
Pathology and Infectious Diseases
The Royal Veterinary College
University of London
Herts. Hatfield AL9 7TA
UK

Guyatt, H.L.
The Wellcome Trust Centre for the
   Epidemiology of Infectious Disease
University of Oxford
South Parks Road
Oxford OX1 3FY
UK

Hominick, W.M.
CABI Bioscience
UK Centre (Egham)
Egham
Surrey TW20 9TY
UK

Hope, I.A.
School of Biology
University of Leeds
Leeds LS2 9JT
UK

Ishibashi, N.
Department of Applied Biological
 Sciences
Saga University
Saga 840-8502
JAPAN

Jones, J.
Unit of Mycology, Bacteriology and Nematology
Scottish Crop Research Institute
Invergowrie
Dundee DD2 5DA
UK

Justine, J-L.
Lab. de Biologie Parasitaire, Protistologie, Helminthologie
Muséum National d'Histoire Naturelle
61 rue Buffon
75231 Paris cédex 05
FRANCE

Kerry, B.R.
Department of Entomology and Nematology
IACR-Rothamsted
Harpenden
Herts. AL5 2JQ
UK

Lee, D.L.
School of Biology
University of Leeds
Leeds LS2 9JT
UK

Martin, R.J.
Department of Biomedical Sciences
College of Veterinary Medicine
Iowa State University
Ames IA 50011-1250
USA

Maruyama, H.
Department of Medical Zoology
Nagoya City University Medical School
Kawasumi-1, Mizuho
Nagoya 467-8601
JAPAN

Michael, E.
The Wellcome Trust Centre for the
    Epidemiology of Infectious Disease
University of Oxford
Oxford OX1 3FY
UK

Munn, E.A.
Department of Immunology
Babraham Institute
Cambridge CB2 4AT
UK

Munn, P.D.
Department of Immunology
Babraham Institute
Cambridge CB2 4AT
UK

Nawa, Y.
Department of Parasitology
Miyazaki Medical College
Kuyotake
Miyazaki 889-1692
JAPAN

Perry, R.N.
Department of Entomology and Nematology
IACR-Rothamsted
Harpenden
Herts. AL5 2JQ
UK

Purcell, J.
Department of Preclinical Veterinary
 Sciences, R.(D.)S.V.S.
University of Edinburgh
Summerhall
Edinburgh EH9 1QH
UK

Robertson, A.P.
Department of Preclinical Veterinary
 Sciences, R.(D.)S.V.S.
University of Edinburgh
Summerhall
Edinburgh EH9 1QH
UK

Sangster, N.C.
Department of Veterinary Anatomy and Pathology
University of Sydney
Sydney 2006
AUSTRALIA

Thompson, D.P.
Animal Health Discovery Research
Pharmacia and Upjohn
301 Henrietta Street
Kalamazoo MI 49007
USA

Valkanov, M.A.
Department of Preclinical Veterinary
Sciences, R.(D.)S.V.S.
University of Edinburgh
Summerhall
Edinburgh EH9 1QH
UK

Wharton, D.A.
Department of Zoology
University of Otago
PO Box 56
Dunedin
NEW ZEALAND

Wyss, U.
Institute für Phytopathologie
Universität Kiel
Hermann-Rodewald-Straße 9
D-24118 Kiel
GERMANY

1. Systematic Position and Phylogeny

Paul De Ley*
Vakgroep Biologie, Universiteit Gent, Ledeganckstraat 35, B-9000 Gent, Belgium

Mark Blaxter
Institute of Cell, Animal and Population Biology, King's Buildings, University of Edinburgh, West Mains Road, Edinburgh EH9 3JT, UK

Keywords

Classification, morphology, sequence analysis, Ecdysozoa, invertebrates, parasitism

Introduction

Nematodes are a highly diverse and very important group of multicellular animals, but their systematics have always been volatile and are currently entering a new phase of turbulence. At the moment of writing this chapter, molecular methods and phylogenetic models are bringing new insights that require significant changes in nematode systematics. To provide an overview at this particular time is therefore both an exciting and impossible challenge, since we can roughly predict the extent of the changes to come, but not yet stipulate their precise form.

Several authors have reviewed the complex history of nematode systematics over the past 150 years, and the many individuals that contributed to it (e.g. Micoletzky, 1921; Andrássy, 1976; Lorenzen, 1983; Inglis, 1983). In this chapter, we therefore do not attempt to present a comprehensive overview of the many different classifications of nematodes proposed in the past. Rather, we will sketch the history of past phylogenetic frameworks for nematodes, briefly discuss how these phylogenies were translated into classifications, and compare them with the first results from molecular phylogenetic analyses. We will also elaborate a first tentative approximation to a classification based on these molecular analyses, combined with morphological data.

Fundamental questions of organismal classification and phylogenetic inference are continuously being debated in ever-growing numbers of books and journals. These debates are becoming ever more specialised and complex, suggesting that a consensus will not be reached any time soon. Our purpose here is not to add our own voice to the chorus, neither by attempting to establish any purported 'supremacy' of molecular phylogenies over morphological data (cf. Moritz and Hillis, 1996 for a sensible assessment), nor by providing the 'one true and definitive' classification for nematodes. Irrespective of all debates about the best methods and data, we only wish to demonstrate three practical points:

(i) There is a substantial degree of overall congruence between morphological and molecular phylogenetic analyses of nematodes.
(ii) Nematode phylogeny is *not* an intractable problem — although certainly not an easy one either.
(iii) Nematologists must incorporate the latest methods of phylogenetic inference into their repertoire as an integral part of nematode systematics.

We believe the latter point is an essential prerequisite for the continued survival and renewed growth of nematology in the new century (cf. Ferris, 1994). Nematode taxonomy will not be able to assume its proper place within biological systematics unless it catches up with the newest developments of systematic theory and practice, and unless it proves able to estimate nematode diversity, recover nematode relationships, and hierarchically organise nematode classification on a genealogical basis.

The Higher Classification

Nemata or Nematoda?

Taxonomic nomenclature provides the basis for recording and communicating classifications. It is therefore an essential biological tool, but not a fundamental axiom underpinning all inferences and assumptions. We consider it of great importance that systematic hierarchies are proposed, compared and rejected, based on the theoretical and relational framework provided by phylogenetic inference. We consider it far less important which names exactly are assigned to the different taxa occurring within such hierarchies, as long as the great majority of users of any given system will understand which taxon is represented by which name.

A good example of a long-lasting nomenclatorial debate about taxon names in nematodes, is the name of the encompassing taxon itself: Nemata or Nematoda? At present, it is nearly universally agreed that nematodes should indeed be ranked as a phylum,

*Current address: Department of Nematology, University of California, Riverside CA 92521, USA.

especially now that the older encompassing taxa Aschelminth(e)s or Nemathelminth(e)s are more and more being questioned (see page 3). Adherents to the name Nemata argue that this was the name used in the first proposal (by Cobb, 1919) of phylum rank for nematodes, and that it should therefore be upheld over the corrupted name Nematoda (Chitwood, 1957, 1958). The latter was a modification of Nematodes as first used at family level by Burmeister (1837), and ultimately derives from Nematoidea, an order-level name originally proposed by Rudolphi (1808). On the other hand (see Steiner, 1960), Nematoda has been in use longer, and the original proposal of the phylum by Cobb (1919) actually introduced the name Nemates, which was only later amended to Nemata by Chitwood (1958).

The International Code for Zoological Nomenclature does not rule on taxon names above family level, and for decisions on name validity at lower levels it applies a combination of both chronological priority (arguably favoring Nematoda) with rank-specificity (arguably favoring Nemata). Hence, there is no obvious official guideline or precedent to be followed — which undoubtedly contributes to the further life of the issue. Since there is no convincing solution in sight, and an ever-growing list of scientifically more interesting questions remain to be answered, we do not wish to devote more time and space to the matter than necessary. We leave it to the reader to apply his or her preferred choice of phylum name as he or she sees fit. In this chapter, we have opted to use the name 'Nematoda' only for pragmatic purposes, i.e. due to its older and somewhat more widespread usage.

New Data and Old Questions: The Uncertain Position of Nematodes in Metazoa

The settings

In the last few decades, molecular approaches to biological systematics have pervaded the field, providing fresh perspectives on a wealth of new and old issues. This transformation is most evident in traditionally popular groups such as vertebrates (Mallat and Sullivan, 1998; Alvarez et al., 1999), vascular plants (Hoot, Magallon and Crane, 1999; Qiu and Palmer, 1999) and arthropods (Eernisse, 1997; Spears and Abele, 1997; Yeates and Wiegmann, 1999), but the practical and theoretical implications and applications of molecular evolution have also drastically redirected research on many other biota (Stenroos and De Priest, 1998; Lang et al., 1999; Sogin and Silbermann, 1998; Brinkmann and Philippe, 1999).

Taxonomical studies on microscopic organisms such as nematodes traditionally receive little attention, and have often generated considerable frustration for both their scientific producers and users alike. This is partly due to the severe practical difficulties of obtaining a representative sampling of all relevant taxa. An even greater constraint is formed by observational restrictions on the number of characters available for diagnosis and classification within such groups. In the case of nematodes, this has typically led to textbook

phrases referring to the 'great morphological uniformity of nematodes'. This persistent myth was based on a mere handful of well studied model species and largely ignores the numerous bizarre and perplexing morphologies, ecologies and ontogenies described in specialised literature for over a century (see De Ley, 2000). Nematode morphology is actually rich in potentially useful characters, but an ideal instrument for observation is still missing: light microscopy does not provide enough resolution and requires substantial experience, while electron microscopy (particularly TEM) is far too costly in time and equipment for effective use on a routine basis.

By comparison, modern molecular tools can quickly and affordably provide a wealth of characters, using standardised basic methods applicable in almost any taxon. The systematics of 'morphologically uniform' groups therefore stands to benefit tremendously from these new approaches. In microbiology, for example, prokaryote taxonomy has been revolutionised and is now firmly based on DNA sequence data and the application of the latest methods for computer assisted character analysis (Woese, 1994, 1996; Brinkmann and Philippe, 1999). Sequence analyses of nematode relationships have also begun to appear at a rapidly increasing pace, addressing such issues as the position of nematodes within Metazoa (Sidow and Thomas, 1994; Vanfleteren et al., 1994; Aguinaldo et al., 1997; Aleshin et al., 1998c), relationships among higher-level taxa within nematodes (Vanfleteren et al., 1994; Kampfer, Sturmbauer and Ott, 1998; Blaxter et al., 1998; Aleshin et al., 1998a, b), and relationships within and among previously intractable families and genera (see Fitch, Bugaj-gaweda and Emmons, 1995; Al-banna, Williamson and Gardner, 1997; Nadler and Hudspeth, 1998; Adams, 1998a, b; De Ley et al., 1999).

Another potentially vast source of taxonomic characters is developmental lineage mapping. Like Transmission Electron Microscopy, this technique remains more limited in routine applicability and cost-effectiveness, but lineaging and TEM can nevertheless provide far superior character resolution than light microscopy of preserved material. They are especially powerful when combined with phylogenetic frameworks provided by sequence analyses (cf. Baldwin et al. 1997a, b; Goldstein, Frisse and Thomas, 1998; Sommer et al., 1999).

Apart from increasing character resolution as such, alternatives to light-microscopical morphology are of fundamental importance for a second reason: character diversity and divergence in nematode taxa is often quite different depending on the character suites in question. Thus, divergence in ribosomal DNA sequence is remarkably high within certain nematode clades with fairly low morphological diversity, and remarkably low in other clades with much higher morphological diversity (Blaxter et al., 1998). **There is no obvious correlation between rates of morphological evolution and rates of fixation of mutations**. In a developmental analog, nematodes can exhibit substantial diversity in the mechanisms that specify cell fates, even in the absence of any evident differences in ultimate cell fates themselves, i.e. differences in final morphology (see review by Félix, 1999).

Any classification that claims to represent 'natural relationships' must therefore attempt to combine morphology with other character suites — and this even when relatedness is not directly equated to phylogeny. In this chapter, we assume that classification *should* be based primarily on phylogenetic relationships, even though we are aware that phylogenetic analysis of biological organisms (like any other scientific methodology) is not an infallible approach guaranteed to produce definite answers to all pertinent problems. Rather, our point is that phylogeny provides (*a*) a neutral ground for analysing, comparing and combining any set of characters reflecting evolutionary history and (*b*) a theoretical framework for translating that evolutionary history into classification along less arbitrary lines than those followed by individual intuition. Within a phylogenetic framework, new character suites such as molecular data do not replace morphological data, but rather they reduce the number of inconclusive or deficient analyses and thus improve the overall outcome (Patterson, Williams and Humphries, 1993).

Understanding the results obtained with molecular and embryological methodologies requires skills and knowledge that were hardly relevant to nematological systematics before, but which are rapidly becoming essential for both the users and suppliers of nematode classifications. Thus, a new phase is reached in the continuing process of taxonomic revision as driven by advances in accuracy and resolution of characters. Some excellent examples of the concomitant advantages, pitfalls and continuities are provided by recent developments in studies on the phylogenetic position of nematodes within Metazoa, and we will first focus on this issue.

New life for an old debate

The use of molecular datasets for the analysis of metazoan phylogeny has a short and spectacular history (Field *et al.*, 1988; Christen *et al.*, 1991; Adoutte and Philippe, 1993; Sidow and Thomas, 1994; Halanych *et al.*, 1995; Wray, Levington and Shapiro, 1996; Aguinaldo *et al.*, 1997; Adoutte *et al.*, 1999). The placement of phyla within the radiation of Metazoa has been revised significantly compared to some morphological views, but concurs in many details with re-evaluations based on extant morphology (Nielsen, 1995) and on early Cambrian 'missing link' fossils (Conway Morris, 1993, 1994). In general, few genes have been used for reconstruction of the Metazoan radiation, and particular emphasis is placed on the SSU rDNA gene.

Adoutte *et al.* (1999) point out that the Cambrian radiation may have taken place in a very short period (geologically speaking) with the origin of all modern phyla occurring between 600 My and 550 My BP (Philippe, Chenuil and Adoutte, 1994; but see Conway Morris, 1997). The nature of molecular evolution suggests that very little phylogenetic signal from this radiation may be left in animal genomes, as any gene evolving at a suitable rate at that time may well have diverged so far by now, that any remaining signal has been obscured by the noise of homoplasy and back-mutation. Some analyses of metazoan evolution using molecular clock models place the divergence of the phyla back to > 1000 My (Raff, Marshall and Turbeville, 1994; Vanfleteren *et al.*, 1994; Doolittle *et al.*, 1996; Wray *et al.*, 1996; Feng, Cho and Doolittle, 1997). While the application of modern molecular clock measures to ancient lineages (and clocks based mainly on the deuterostome Chordata to all phyla) is open to serious question, there is a remarkable consistency in the estimates of the radiation of the phyla (circa 1000 My to 700 My). The metazoan phyla had already diverged at the onset of the Cambrian, as evidenced by the Lagerstatte assemblages of the Burgess Shale and related faunas (Gould, 1989; Foot *et al.*, 1992; Conway Morris, 1993; Conway Morris, 1994). In the absence of informative fossils, the debate must therefore continue. Although nematode fossils are known (mostly preserved in amber), few are older than 30 My and none are informative with regards to the origin of Nematoda (Poinar, 1992; Poinar, Acra and Acra, 1994; Manum *et al.*, 1994). The fossil record is therefore particularly unhelpful in reconstruction of the relationships of the Nematoda (Conway Morris, 1981).

The molecular placement of the Nematoda was initially based on the SSU rDNA sequence from *Caenorhabditis elegans* (Ellis, Sulston and Coulson, 1986). The anomalous behaviour of this sequence (and, ironically, that of the other major protostome model organism, *Drosophila melanogaster*) in molecular phylogenetic reconstruction led to the explicit exclusion of these species from analyses. An example is found in Philippe, Chenuil and Adoutte (1994), where *Drosophila* and *Caenorhabditis* can be robustly placed as sister taxa at the base of the protostomes (while other insect sequences cluster correctly with arthropods deep within the Protostomia). For the arthropods, other SSU rDNA sequences were available that 'behaved better' under various models of sequence evolution and could be used to place Arthropoda within the Metazoa. By contrast, the *C. elegans* SSU rDNA sequence was until recently the only one available for the Nematoda, and the acquisition of SSU rDNA sequences from other rhabditids and from strongylid or strongyloidid parasites (Putland *et al.*, 1993; Zarlenga, Lichtenfels and Stringfellow, 1994; Zarlenga *et al.*, 1994; Fitch, Bugajgaweda and Emmons, 1995; Fitch and Thomas, 1997) failed to significantly improve inferred branch lengths or resolution. To avoid branch length artefacts, nematodes were often simply left out of reconstructions (Lake, 1990; Philippe, Chenuil and Adoutte 1994; Winnepenninckx, Backeljau and De Wachter, 1995).

In the last few years, the search for less anomalous SSU rDNA sequences was extended to the whole phylum Nematoda, and species from the zooparasitic Spirurida and Trichocephalida proved to have less extreme branch lengths (Halanych *et al.*, 1995; Aguinaldo *et al.*, 1997). Analysis of these sequences with a dataset encompassing the rest of the Metazoa yields a hypothesis of Metazoan evolution substantially different from the accepted models: nematodes are united with Kinorhyncha, Priapulida, Nematomorpha, Tardigrada, Onychophora and Arthropoda into a clade of moulting animals, the Ecdysozoa (Aguinaldo *et al.*, 1997). Reassessment of other phyla using SSU rDNA has yielded a hypothesis of metazoan relationships where the Protostomia can be split into two major groups: the Ecdysozoa and the Lophotrochozoa. There are several morphological and ontologenetic characters/suites of characters that can be adduced as evidence for these clades (Nielsen, 1995).

Although the proposition of the Ecdysozoa is supported by other characters such as sequence analysis of EF-1α (McHugh, 1997; Garey and Schmidt-Rhaesa, 1998), moulting, loss of locomotory cilia, trilayered cuticle structure and microvillar epicuticle formation (Nielsen, 1995; Schmidt-Rhaesa *et al.*, 1998), acceptance of such a radical departure from current mainstream thought is slow to accrue. Much of this mainstream opinion derives from Grobben's (1910) 'Aschelminthes' concept, and is actually less entrenched in taxonomical tradition than its adherents might assume. Thus, a wide range of different scenarios was proposed by Grobben's predecessors and contemporaries, and one of the competing schools of thought specifically argued for close relationship between nematodes, tardigrades and/or insects. Such proposals included the 'Chitinophores' concept of Perrier (1897), as well as the extensive argument by Seurat (1920) for common ancestry of arthropods and nematodes (see review in Chitwood, 1974). Perrier's view was based on the assumption that nematodes originated as a parasitic lineage deriving from arthropod ancestry, and rapidly lost credibility as the anatomy and diversity of free-living nematodes became more evident. Seurat's case for a shared origin was less easily discounted, however: it seems to have been mostly ignored rather than refuted, and should now be given due credit as a largely forgotten forerunner of the Ecdysozoa concept.

Future directions

SSU rDNA is a very useful phylogenetic marker, but it is neither the only appropriate choice, nor a consistently reliable one (Hasegawa and Hashimoto, 1993; Sidow and Thomas, 1994), and its support for the Ecdysozoan clade has been questioned by Wagele *et al.* (1999). A wider route to confirmation or rejection of Ecdysozoa is provided by the use of independent sequence datasets for analysis. There is as yet no consensus on any single gene useful for deep analysis of Metazoa in parallel with SSU and LSU rDNA, not least because analyses of proteins and protein genes are also subject to various biases, both at the level of individual sequences (Sidow and Thomas, 1994; Foster and Hickey, 1999) and in comparisons across multiple loci (Guigo, Muchnik and Smith, 1996). Nevertheless, the technologies of sequencing, sequence analysis and sequence database management are developing at such a pace, that it is rapidly becoming possible to analyse multiple loci per organism and thus evaluate the phylogenetic qualities of each locus (see e.g. O'Grady, 1999; Curole and Kocher, 1999).

In this manner, Mushegian *et al.* (1998) performed an extensive analysis of quartets of 42 orthologous proteins encoded by the *C. elegans*, human and *Drosophila melanogaster* genomes, rooted by orthologues from the yeast *Saccharomyces cerevisiae*. They found that 24 different proteins supported a closer relationship between flies and humans (with high mean bootstrap support, >80%), while only 11 supported an arthropod-nematode relationship (with lower overall bootstrap support, mean 61%). This is in apparent conflict with the results derived by Aguinaldo *et al.* (1997) using SSU rDNA, and argues against the 'Ecdysozoa' as a natural group. However, Mushegian *et al.* (1998) pointed out that a larger proportion of slowly evolving genes (where fewer substitutions are inferred in the lineages leading from the respective common ancestors to yeast versus fly, or to yeast versus nematode) supported a nematode + arthropod clade (8 out of 18 protein quartets). In addition, the *C. elegans* protein dataset was shown to have a greater overall substitution rate than the fly dataset, with 25 out of 36 *C. elegans* proteins being more distant from yeast than their fly orthologues (Mushegian *et al.*, 1998). The consensus result, a placement of nematodes basal to a fly + human clade, may thus once again be an artefact of the use of rapidly evolving sequences from *C. elegans*, mirroring the earlier SSU rDNA analyses.

This kind of analysis will undoubtedly benefit from sequence datasets derived from other nematode taxa, particularly those in clades displaying lower substitution rates in SSU rDNA. The inception of EST-based genome projects on nematodes of economic, health and research interest will hopefully yield data sets for such analysis in the near future (Blaxter *et al.*, 1996; Blaxter *et al.*, 1999; Daub *et al.*, 2000; Johnston *et al.*, 1999). An alternative molecular system for phylogenetics is the analysis of gene families that are involved in setting up or executing the body plan of Metazoa. These genes may retain phylogenetic signatures in their sequence and expression patterns that reveal deep relationships. One such family is the homeodomain-containing genes of the HOX cluster (de Rosa *et al.*, 1999). This cluster of paralogous genes, which each contain a short domain involved in binding DNA (the homeodomain), is closely linked in the genomes of many species, and is involved in the core process of specifying and executing anterior-posterior identity during development. Importantly, in most taxa the genes are arranged along the chromosome in the same order as their corresponding sites of activity in specifying tissue/organ fate on the A-P axis.

First discovered in *Drosophila*, the HOX cluster has also been found in mammals, where the general genome duplications evident in the chordate lineages have produced four paralogous HOX clusters. HOX gene datasets, and in some cases demonstration of linkage on the chromosome, have now been acquired for additional taxa, spanning many phyla (de Rosa *et al.*, 1999). The *C. elegans* HOX cluster was described through classical developmental genetics coupled with gene cloning, and is significantly different from both fly and mammal clusters (Costa *et al.*, 1988; Wang *et al.*, 1993; Kenyon, 1994; Salser and Kenyon, 1994). There appear to be fewer genes in the cluster, which is arranged in a different order and interspersed with non-HOX genes over a large region of chromosome III. These features suggested to early reviewers that the *C. elegans* cluster might be a relic of an ancestral cluster from which the fly and mammal ones were derived. This was based on the assumption that nematodes were basal to arthropods and deuterostomes, an assumption that remains popular (Vanfleteren and Vierstraete, 1999) albeit with increasingly cautious support (see Fitch and Thomas, 1997).

Completion of the *C. elegans* genome project, which revealed additional HOX genes in the cluster, and the proposal of Ecdysozoa have led to reappraisal of the *C. elegans* HOX cluster as a very derived pattern. Analysis of HOX genes from additional phyla from Ecdysozoa and Lophotrochozoa led de Rosa *et al.* (1999) to propose that there is an Ecdysozoan-specific feature of HOX clusters of Priapulida, Arthropoda, Onychophora and Nematoda: the presence of a gene similar to arthropod *Abdominal-B* (AbdB). The arthropod gene *Ultrabithorax* (Ubx) could also be defined in Onychophora and Priapulida, but a nematode (i.e. *C. elegans*) orthologue of Ubx is difficult to define (de Rosa *et al.*, 1999). This model of the metazoan HOX clusters suggests that the Ur-metazoan had nine or ten genes, and that the *C. elegans* cluster has diverged very significantly from this archetypal arrangement. Again, analysis of HOX gene clusters from other nematode taxa will be very informative, particularly from clades known to be less derived in morphology and to have lower substitution rates for the loci in question.

Early Explorations of Relationships within Nematoda

First discoveries of taxa and characters

With the first taxonomic and anatomical studies of parasitic and free-living nematodes, zoologists of the nineteenth and early twentieth century were given just an inkling of the diversity and complexity of this group. Thus, the earliest taxonomic systems were by necessity limited to very incomplete representation and to excessive emphasis on very few characters. Convergence could not yet be recognised as such, due to limitations of the available microscope optics, and because of the absence of an established framework for accurate description and interpretation. Famous examples of approaches that overemphasized a single character suite and therefore resulted in largely artificial classifications, include the system of Schneider (1866) based on somatic musculature, and the proposals of Cobb (1919) based on stoma armature.

The nematologists and helminthologists of the time were well aware of the exploratory nature of their efforts. At such an early stage in the exploration of the diversity of nematodes, attempts to establish a comprehensive phylogenetic overview were generally recognised as being largely speculative (see de Man, 1876; Micoletzky, 1922). Most classifications of that time were therefore proposed purely on typological

and ecological grounds. Also, although the numbers of known parasitic and free-living species were very limited, synthetic approaches were hindered from the outset by difficulties of comparing the seemingly simple anatomy of microscopic species with the more obviously complex characters of large zooparasites. Combined with the different backgrounds and purposes of different authors, this set a trend for scientific specialisation and fragmentation which was to plague nematology and helminthology throughout their further history. Nevertheless, some attempts were made to integrate all known Nematoda into a single classification, e.g. by simply allocating all zooparasites to a different higher taxon from all other nematodes (Perrier, 1897; Stiles and Hassall, 1926). In effect, this represented an ecological pendant to the morphologically simplistic systems of Schneider and Cobb.

Other authors realised that the affinities between nematode parasites of vertebrates and other nematodes were more complex and required a subtler approach. Thus, Baylis and Daubney (1926) distinguished five encompassing orders within the Nematoda (Ascaroidea, Strongyloidea, Filarioidea, Dioctophymoidea and Trichinelloidea) that contained both zooparasitic and interstitial representatives. Likewise, Filipjev (1929, 1934) rejected a basic division into two groups based on presence versus absence of zooparasitism and proposed no less than eleven orders: Enoplata, Chromadorata, Desmoscolecata, Monhysterata, Anguillulata, Oxyurata, Ascaridata, Spirurata, Filariata, Dioctophymata and Trichurata. Two of these orders, Enoplata and Anguillulata, contained both zooparasitic and non-zooparasitic representatives. Filipjev (1934) also argued strongly in favour of marine nematodes as representing the earliest forms of nematodes. In many respects, his system was highly influential, especially on the subsequent classification of Chitwood (1937). Most of Filipjev's proposed orders remain in use today, albeit with changed endings and with some changes in contents. Filipjev even considered an overall division along the lines of Chitwood's soon-to-follow subclasses Phasmidia/Aphasmidia, but preferred not to group his orders in higher taxa (p. 6 in Filipjev, 1934).

An early phylogeny: Micoletzky (1922)

Those few authors who actually dared to speculate on nematode interrelationships around the turn of the century, were often very cautious about their own phylogenetic and taxonomic inferences. An interesting example of an early system is provided by Micoletzky (1922), who tentatively suggested highly resolved evolutionary trees for 76 genera (see condensed tree in Figure 1.1) of the 167 non-zooparasitic genera known to him at the time. Clearly influenced by Cobb (1919) and Marcinowski (1909), Micoletzky considered stoma structure to be of primary importance and therefore grouped genera into five families based on stomatal characters. This resulted e.g. in the inclusion of all nematodes with teeth in one family (Odontopharyngidae), while all nematodes with a stylet-like organ were united in one other family (Tylenchidae). Both these

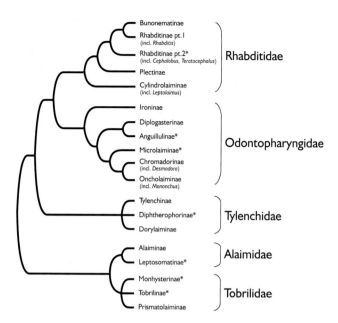

Figure 1.1. Overview of the phylogenetic relationships among non-zooparasitic nematodes, as cautiously proposed by Micoletzky (1922) on the basis of buccal morphology. Taxa marked with an asterisk were assumed to be paraphyletic. Compiled and modified from figures RW in Micoletzky (1922); see text for details.

groupings later turned out to be extremely artificial. By comparison, his arrangement of plectids, teratocephalids, cephalobids and rhabditids *sensu stricto* (united in one family Rhabditidae) withstood the test of time much better, presaging both the branching order and derived position found for these taxa in much more recent phylogenies (cf. Figures 1.4; 1.6).

Bipartite Systems and Comprehensive Classifications

The origin of Chitwood's Adenophorea/Secernentea system

Probably the single most important event in the history of nematode systematics, was the realisation that nematodes with phasmids share numerous other characters and therefore represent a single, highly diversified and yet highly coherent unit (Chitwood and Chitwood, 1933; Chitwood, 1937). As a result of his extensive experience with parasitic and non-parasitic nematodes alike, Chitwood succeeded in cutting across many of the previous simplistic classifications, bringing together in one class most of the important zooparasitic taxa with one of the main free-living groups (rhabditids *sensu lato*), as well as the largest taxon of predominantly phyto- and mycoparasitic nematodes (tylenchids). Sixty years of subsequent research have proven the diagnostic and phylogenetic soundness of this grouping, and Chitwood's system was generally adhered to by many nematologists for over forty years.

As was common practice in pre-Hennigian taxonomy, the remainder of nematodes was also grouped by Chitwood (1937), even though no unifying characters existed for it. At first, he respectively

proposed the names Phasmidia and Aphasmidia for the two groups, but subsequently (Chitwood, 1958) suggested replacement with Secernentea (='secretors', referring to presence of an excretory system with lateral canals) and Adenophorea (='gland-bearers', referring to presence of caudal glands) as modifications of terms coined by von Linstow (1905), in order to avoid confusion with the scientific name already in use for stick-insects.

Chitwood and Chitwood (1933) assumed that bacterial-feeding forms with tubular stoma and valvate basal bulb were closest to the common ancestor of all nematodes. In particular, plectids were presumed to have preserved the morphology of the earliest Adenophorea, while rhabditids were thought to represent the earliest branch within Secernentea. As explained next, these assumptions were later modified or rejected by other authors, and in particular Adenophorea would subsequently be recognised as a paraphyletic group. Most of the major changes to classification proposed after Chitwood (1937) therefore consist of attempts to correct for the lack of phylogenetic and morphological unity of Adenophorea. Thus, our own proposed system (see pages 12 and 13) includes neither the name Adenophorea nor Secernentea as valid taxon, and we will therefore use both names between parentheses throughout the remainder of this chapter.

Uprooting the bipartite system: Maggenti (1963–1983)

Maggenti (1963) elaborated on Filipjev's and Chitwood's observations, focusing particularly on aspects of pharynx structure and excretory system as markers of evolutionary relationships. This led him to the realisation that the pharynx structure of plectids and rhabditids could not arguably be close to that of the common ancestor of all nematodes. Basically, he considered a cylindrical pharynx to be ancestral, suggesting that enoplids and monhysterids were morphologically much nearer to the origin of nematodes than plectids and rhabditids. Furthermore, the origin of rhabditids (and by extension all 'Secernentea') was inferred to lie close to the origin of chromadorids and leptolaimids, implying that 'Adenophorea' were paraphyletic (Figure 1.2).

These observations provided the basis for several later rejections or qualifications of 'Adenophorea' by other authors (see pages 7–9), and thus lie at the root of the eventual replacement of Chitwood's bipartite system by various tripartite systems. However, Maggenti (1963, 1983, 1991) never rejected Adenophorea himself. Although he later stated that classification should be based on phylogeny (Maggenti, 1983), like many evolutionary taxonomists he adhered to the more permissive monophyly criterion of Simpson (1961), where valid taxa should have exclusive ancestry in a single taxon of lower rank, rather than in a single species. He therefore did not apply paraphyly as sufficient cause to invalidate taxa. Even in 1983, Maggenti still maintained 'Secernentea' and 'Adenophorea' as sister taxa (Figure 1.3).

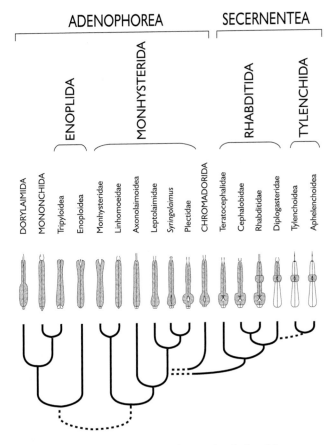

Figure 1.2. Overview of the phylogenetic relationships among non-zooparasitic nematodes, as proposed by Maggenti (1963) on the basis of pharyngeal structure and excretory system. Dotted lines represent more tenuous hypotheses of relationship. Enoplida, Monhysterida and Rhabditida were assumed to be paraphyletic. Modified from Maggenti (1963); see text for details.

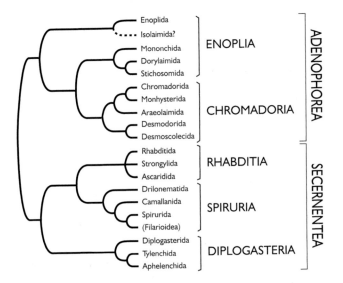

Figure 1.3. Overview of the phylogenetic relationships among all nematodes, as proposed by Maggenti (1983). All taxa were apparently assumed to be monophyletic. Modified from Maggenti (1983); see text for details.

Maggenti rejected close relationships between plectids and rhabditids, arguing that the valve structure in the basal bulb of larger *Plectus* species must be convergent with the 'butterfly valves' of rhabditids, rather than homologous. He may have deemed this necessary to strengthen his case for a basic error in Chitwood's rooting of the nematode tree. However, many of the smaller Plectidae do have butterfly valves (as noted by Lorenzen, 1981, 1994), and a close relationship between plectids and rhabditids is actually quite compatible with a different placement of the root for nematodes. In retrospect, it is therefore ironic that Maggenti actually discarded a strong morphological argument *in favour* of a revised polarisation of pharyngeal characters in nematodes.

Towards Tripartite Systems: Critical Scrutiny and Fragmented Classifications

More explicit phylogenies and departures from bipartite systems

The Adenophorea/Secernentea dichotomy found its way into many textbooks and classifications, but it was never universally accepted. Even while maintaining 'Adenophorea' as a valid taxon, many authors overtly or covertly recognised its lack of monophyly or homogeneity compared to 'Secernentea' (Maggenti, 1963; De Coninck, 1965; Gadéa, 1973; Lorenzen, 1981, 1994; Malakhov, Ryzhikov and Sonin, 1982; Inglis, 1983; Adamson, 1987). More radically, Goodey (1963) simply rejected both classes, *i.a.* on the grounds of the intermediate position of teratocephalids, and instead maintained a multitude of orders as the highest-ranked subphyletic taxa. This proposal did not gain wider acceptance, but more influential were the works of Maggenti (1963) and De Coninck (1965). The latter promulgated the use of the subclasses Chromadoria and Enoplia within 'Adenophorea' to indicate their status as distinct major groups, while the former demonstrated the more derived status of 'Secernentea' and thus spurred later proposals to upgrade the enoplid and chromadorid subclasses to the same rank as their secernentean counterpart (Andrássy, 1974, 1976; Malakhov, Ryzhikov and Sonin, 1982; Inglis, 1983; Adamson, 1987). In some cases these proposals for a tripartite system were grounded in first applications of cladistics, in other cases the inference methods remained intuitive while emphasizing the need for diagnostic coherence of higher taxa. Analysis, depiction and argumentation of evolutionary relationships gained prominence, and it became more important to compile sets of supporting characters, rather than to propose just-so scenarios.

The system of Andrássy (1974, 1976): typology and paraphyly

The earliest exponent of a subdivision into three major nematode groups, the phylogeny and system of Andrássy (1974, 1976) provide a highly individual mixture of an explicitly typological outlook combined with largely implicit evolutionary assumptions and deductions. While his phylogenetic framework followed Maggenti (1963) in depicting Adenophorea as paraphyletic (Figure 1.4), he altogether omitted inclusion of the major zooparasitic groups and attached no noticeable significance to monophyly as a criterion of taxon validity. Thus, he retained both Rhabditida and Araeolaimida as valid orders, despite their perceived paraphyly, and was not at all troubled by discrepancies of rank between presumed ancestral and descendant taxa (he assumed that the Order Rhabditida arose from within the Family Plectoidea, and the Order Tylenchida from Suborder Diplogastrina).

Instead, Andrássy rejected Adenophorea on the basis of an unweighted count of diagnostic characters: as a group, they could only be defined by 4 'constant characters', versus no less than 17 for Secernentea. He therefore concluded 'that Adenophorea ... cannot possibly be a unified group' (p. 48 in Andrássy, 1976). By comparison, the enoplid and chromadorid groups were much more homogeneous in their own rights (with resp. 16 and 14 'constant characters') and therefore merited equivalent status to Secernentea. Andrássy also made unweighted tabular counts of diagnostic characters at lower ranks, to justify taxon validity and separation within his three subclasses at order and suborder level. This primary concern for diagnosis became all the more evident in his subsequently published encyclopaedic key to all known soil and freshwater nematode species (Andrássy, 1984).

Also, since he assumed that Nematoda and Nematomorpha jointly constituted a phylum, Nematoda itself could not be ranked higher than a class, and the three subdivisions within it should therefore be ranked as subclasses. Instead of using subclass names derived from species names, he preferred names referring to characters and therefore proposed Torquentia and Penetrantia (referring to amphidial structure) as replacements for Chromadoria and Enoplia, while modifying the character-based name Secernentea (referring to the excretory system) to Secernentia, so as to match subclass endings.

Despite this heavy emphasis on diagnosis and typology, Andrássy (1976) did provide a reconstruction of ancestral characters and morphologies, and he also presented a resolved phylogeny of all non-zooparasitic nematode suborders, apparently inferred from an intuitive combination of character counts and reconstruction of so-called 'evolutionary series' of species. His typological/gradistic approach to phylogeny and classification was deemed theoretically and factually flimsy and received extensive criticism (Coomans, 1977; Lorenzen, 1981, 1983, 1994; Maggenti, 1983, 1991). Nevertheless, compared to recent phylogenies as obtained by ribosomal DNA sequencing and a whole battery of highly advanced tree-construction computer algorithms (Figure 1.6), Andrássy's evolutionary framework was certainly no worse than that of many contemporary schemes claiming a sounder theoretical basis. In many ways, his works epitomize the pragmatic, encyclopaedic, prolific and authoritatively idiosyncratic character of nematode taxonomy in the past century. Although often controversial, the direct connection between diagnostic characters and classification did enable him to formulate identification keys for all groups covered (later extended to species level in his book of 1984), a feat that is unlikely to ever again be repeated. Because of the need for identification tools of many users of nematode taxonomy, his system has become quite influential and will probably remain so for decades to come.

The system of Lorenzen (1981, 1994): intensiveness, cladistics and monophyly

Lorenzen (1981, 1994) provided the first phylogenetic system with emphatic focus on the 'Adenophorea' and on rigorously cladistic principles (Figure 1.5). His analysis included intense scrutiny of character polarity, and sought to establish monophyly of taxa wherever possible. This yielded many instances of nominal taxa that lacked a defining synapomorphy and were therefore not demonstrably monophyletic. For each such case, Lorenzen amalgamated the groups in question into a single potentially paraphyletic taxon awaiting further resolution. Indeed, absence of evidence does not equate evidence of absence, and it would have been premature for him to reject such taxa outright. However, the unfortunate corollary of his approach was that many separate and distantly related families were effectively being lumped into potentially paraphyletic 'bucket groups'. For instance, within Chromadorida he

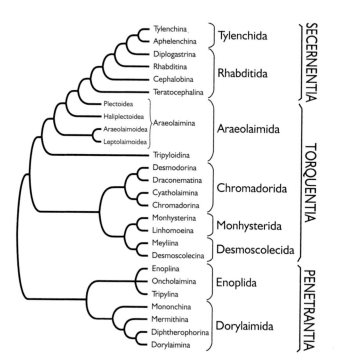

Figure 1.4. Overview of the phylogenetic relationships among non-zooparasitic nematodes, as proposed by Andrássy (1976) on the basis of counts of diagnostic characters. Torquentia, Araeolaimida and Rhabditida are assumed to be paraphyletic. Compiled and modified from figures 21–32 in Andrássy (1976); see text for details.

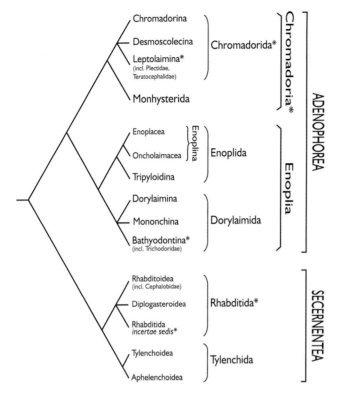

Figure 1.5. Overview of the phylogenetic relationships among non-zooparasitic nematodes, as proposed by Lorenzen (1981, 1994) on the basis of cladistic analysis of morphological characters. Taxa marked with asterisk are assumed to be paraphyletic. Modified from Lorenzen (1994); see text for details.

included no less than 18 families in the suborder Leptolaimina, among which such disparate taxa as Teratocephalidae, Ceramonematidae and Rhabdolaimidae. In hindsight, this approach led to a paradoxical situation: Lorenzen's analysis most forcefully demonstrated the complete lack of support for 'Adenophorea' as monophyletic taxon, but his classification nevertheless maintained it, for lack of alternative resolution. Thus, he simultaneously demonstrated that cladistics could indeed be applied to such a scientific quagmire as nematode phylogeny — but also that useful morphological characters were too few in number to get sufficient resolution for a well-resolved classification.

A shortage of synapomorphies not only precluded positive proof of paraphyly for 'Adenophorea', but also prevented certain homoplasies from being detected. For example, Lorenzen interpreted the posterior position of the dorsal pharyngeal gland opening in taxa such as trichodorids, alaimids and dorylaimids as synapomorphic, and therefore united the former two with some other taxa in the 'bucket suborder' Bathyodontina. However, more intuitive morphological inferences (Riemann, 1972; Siddiqi, 1974, 1983a, b; Decraemer, 1995) as well as ribosomal data (Blaxter *et al.*, 1997; Vanfleteren and De Ley, unpublished data) show that alaimids and trichodorids belong in or near Enoplida. A posterior displacement of the gland opening may well

be functionally linked to the hydrodynamics of feeding through a very narrow spear and/or mouth opening. If so, the character may easily have arisen repeatedly in unrelated groups. A case in point are the aphelenchids, which are certainly not related to dorylaims and thus show that the posteriad shift has occurred at the very least twice in Nematoda.

A last general critique of Lorenzen's approach is that he placed the origin of Nematoda in between 'Adenophorea' and 'Secernentea', an assumption that was largely outdated by 1981. Although he correctly criticized Maggenti (1963) for having exaggerated the degrees of difference between the basal bulb in plectids and rhabditids, Lorenzen (1981, 1983, 1994) did not himself address the question of character polarity of the rhabditid/plectid basal bulb. He also did not consider its possible validity as a synapomorphy for a clade comprising of teratocephalids, plectids and 'Secernentea'. Thus, while Lorenzen pointed out that Maggenti's error had actually deprived him of an argument in favour of his tree of pharyngeal character evolution, Lorenzen in turn failed to pick up on a character that could have settled his own uncertainty as to the paraphyly of 'Adenophorea'.

With the wisdom of hindsight, these comments are easily made, but the importance of Lorenzen's efforts should not be underestimated. His work placed a long-overdue emphasis on the taxonomy and phylogeny of marine nematodes, and was therefore extremely influential on subsequent research into ecology and taxonomy of marine nematodes. He also directly influenced the successive papers by Malakhov, Ryzhikov and Sonin (1982), Inglis (1983) and Adamson (1987), in which the names 'Adenophorea' and 'Secernentea' were abandoned altogether. Inglis (1983) proposed the names Enoplea, Chromadorea and Rhabditea for classes equivalent to the subclasses Enoplia, Chromadoria and Rhabditia in Lorenzen (1981) and Malakhov, Ryzhikov and Sonin (1982), and expressed the view that Chromadorea and Rhabditea were closer to one another than to Enoplea. Unlike Lorenzen, Inglis assumed an overtly non-cladistic philosophical stance, arguing that monophyly for nematode taxa (and even for the entire phylum Nematoda) was too difficult to ascertain, and therefore of limited relevance to their classification. Partly following and partly disagreeing with Inglis (1983), Adamson (1987) subsequently presented a cladistic analysis at subclass level and proposed that Rhabditea should also include all taxa grouped in Chromadorea by Inglis (1983). Thus, he reverted to a bipartite system, but one that partitioned the nematode orders quite differently from that of Chitwood (1937).

Developing an updated Linnean Classification

The latest system is never the last one

Nematode systematics is inherently prone to controversy and instability. This is probably even more true now than ever before. On the one hand, it is eminently necessary to incorporate the new molecular phylogenies into a comprehensive classification, not least because this will allow us to reunite parasitic and non-parasitic taxa once again into a comprehensive single framework. On the other hand, molecular sampling of taxonomic diversity within Nematoda is still very limited, especially within 'Adenophorea' and at lower levels of classification. Here too, the greater ease of collection and culture of 'secernentean' species has led to more extensive and intensive analysis, and concomitant taxonomic changes can therefore be proposed with greater confidence and detail for 'Secernentea' than for 'Adenophorea'. Furthermore, phylogenetic analysis is becoming an ever more sophisticated and complex discipline in itself, which nevertheless remains inherently prone to sources of bias and uncertainty, despite the many theoretical and methodological advances of the last decades. Above all, we have hardly begun the collection and comparison of results obtained from multiple character suites, such as those provided by different molecular loci, or by molecular versus morphological versus developmental data.

We may therefore rest assured that any system formulated today will be outdated almost immediately. Nevertheless, some of the phylogenetic relationships supported by molecular analyses are sufficiently robust or surprising to warrant formal representation in classification. At the very least, this will provide a new set of hypotheses for corroboration, and with a little luck it may contribute to more accurate application and interpretation of comparative studies on nematode genetics, biochemistry, development, morphology, ecology and control. With these provisos, a hybrid classification is presented below, derived partly from more recent morphological systems (e.g. Lorenzen, 1981, 1994; Inglis, 1983; Malakhov, 1994) and partly from results obtained with SSU rDNA sequences (Figure 1.6). As a rule of thumb, the latter were given primacy whenever they provided strong bootstrap support (above 85%), while the former were applied in cases where bootstrap values of molecular phylogenies were insignificant (< 65%), or in groups that were not yet included in sequence analyses[1]. In cases of moderate bootstrap support (65–85%), we attempted to reach a consensus between molecular and morphological patterns of relationships, if possible.

Major changes in perspective

Undoubtedly, our system will receive a mixed welcome, and numerous modifications may still be required before it can eventually be deemed compatible with an accurate phylogenetic framework. Irrespective of the precise nomenclatorial choices made below, we are confident of one important change: a major shift of balance is required, in order to combine parasitic and non-parasitic taxa within a single phylogeny-based hierarchy. Many taxa formerly placed within 'Adenophorea' need to be retained at, or upgraded to

comparatively high rank, while many parasitic taxa within 'Secernentea' must be downgraded to lower rank than has become accepted in recent years. This balancing act is actually well overdue, even on morphological grounds alone, since it follows inevitably from the combination of two earlier hypotheses: the paraphyly of 'Adenophorea' with respect to 'Secernentea', and the assumption that all parasitic nematode taxa derive from free-living ancestry.

Both hypotheses were previously proposed on the grounds of morphology (Lorenzen, 1981, 1994) and life cycle data (Inglis, 1983; Anderson, 1984), and both are now supported strongly by sequence analysis. Lorenzen (1981, 1994) preserved 'Secernentea' as a class, placing the root of Nematoda between 'Secernentea' and 'Adenophorea'. However, he emphasized that synapomorphies were lacking for 'Adenophorea' and several of its constituent taxa, which were therefore possibly paraphyletic. Sequence data now confirm his suspicions of paraphyly, and place the root of Nematoda somewhere between chromadorids, enoplids and dorylaimids. Thus, although our system differs significantly from Lorenzen's in ranks and names (especially for many chromadorid/'secernentean' taxa), there is nevertheless more agreement in terms of branching orders than meets the eye.

In hindsight, it is evident that previous systems have, by default, overemphasized morphological differences in ranking decisions, at the expense of probable phylogenetic nestedness. This has led to the current tendency to even out morphological ranks across very different nematode groups: individual nematode taxonomists have always tended to either split or lump taxa rather evenly across the board, throughout the entire phylum. Averaging out these trends, this can now be perceived as overemphasis on higher ranks in morphologically highly divergent parasitic taxa such as strongylids and tylenchs, and underemphasis in morphologically cryptic free-living taxa such as enoplids and rhabditids. In this way, molecular phylogenies cut across morphological splitting and lumping approaches, in a way that neither entirely supports nor completely rejects either. Apart from perceived morphological support, numbers of currently known species may provide another source for discomfort with our system. However, in a phylogenetic system species numbers are at best a minor concern[2] in ranking decisions — another respect in which our system diverges from the usual splitting/lumping disputes.

We also introduce a series of infraorders for certain 'secernentean' parasitic taxa, in order to preserve maximum stability at family rank and below. Pearse (1936; 1942) provided no ending for the infraorder rank, but it is fairly often used in e.g. insect systematics, where the most common ending is

---

[1] These threshold values of 65% and 85% are approximations derived from the simulation studies by Hillis and Bull (1993) of the relationship between bootstrap and probability of correct tree topology.
[2] The comparatively high ranks for 'adenophorean' groups are actually matched by the likelihood that these groups contain a much greater proportion of nematode species richness than currently known. Thus, even if species richness were relevant to taxonomic rank, then it must be noted that future explorations can be expected to make up more than adequately for the relatively few species currently distinguished within 'adenophorean' groups.

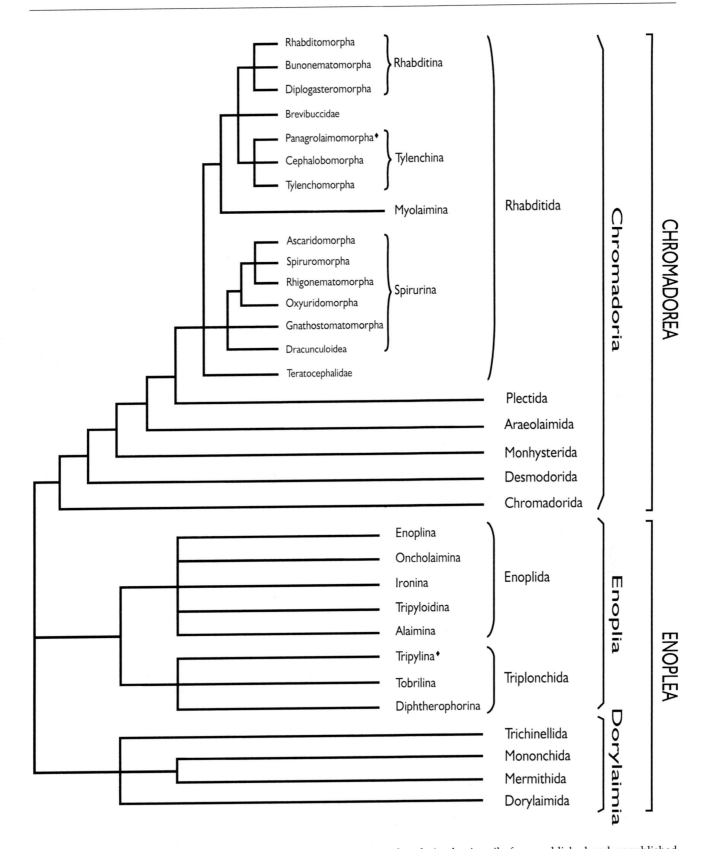

Figure 1.6. Overview of the phylogenetic relationships among nematodes, derived primarily from published and unpublished analyses of SSU rDNA sequence data (see text for details). Taxa marked with a diamond are positioned on the basis of other criteria (axis determination in the case of Panagrolaimomorpha, spicular musculature in the case of Tripylina). Only those taxa are shown for which SSU rDNA sequence data are known. The differing levels of resolution of the tree primarily reflect the extent to which taxa have been sampled for SSU data, and not the known or suspected diversity of the taxa in question.

-*omorpha* (Barnard, 1999; Yeates and Wiegmann, 1999). In most cases (our suborders Spirurina and Tylenchina) the insertion of infraorders in our nematode classification largely allows preservation of the currently accepted superfamilies (as respectively listed in Schmidt, Roberts and Janovy, 1995 and Maggenti *et al.*, 1987). In the case of our suborder Rhabditina, however, the inclusion of an embedded strongylid group would require an additional ranking level to preserve superfamilies. Rather than introducing a more exotic rank in between infraorder and superfamily (e.g. suprafamily), for which there are few or no precedents in other animal phyla, we have lowered the superfamilies of the former Strongylida (as listed in Schmidt, Roberts and Janovy, 1995) back to family rank, and acted likewise with the similar case of the former Aphelenchina (as listed in Hunt, 1992). We assume that frequent usage of the rank of tribe will be required in future, in order to properly resolve strongylid and aphelench classification below family rank. Analogy can again be made with insects, where the use of tribe rank is quite common in highly diverse and derived taxa such as Diptera (Harbach and Kitching, 1998) and Hymenoptera (cf. Gauld and Bolton, 1988), which typically also contain major parasitic clades.

Finally, it must be noted that some authors argue that Linnean classifications should be abandoned altogether, in favor of truly phylogenetic systems without predetermined taxon ranks or type subtaxa (cf. de Quieroz and Gauthier, 1992, 1994; Ax, 1996). Such an approach would be impractical to introduce here and now, not least because of the firmly Linnean premises of all existing nematode classifications, and the fact that this radical departure from accepted practice still needs to be considered and ratified by the taxonomical community at large (see the preface by Minelli and Kraus in ICZN, 1999). Nevertheless, it does illustrate the extent to which systematics has evolved for other taxa than nematodes, and how radically different all mainstream classifications might eventually be formulated — even for nematodes.

A Tentative Classification Emphasizing Monophyletic Taxa

The system presented below is a linnean classification derived from phylogenetic analyses and considerations, i.e. it is a hierarchical system of nested, presumably monophyletic taxa assigned to the various ranks commonly applied in zoological classification. A few clarifications are required with regard to form and contents:

– Although derived from trees, this classification does *not* aim to represent a tree in its own right. Thus, the sequential order of subtaxa of equal rank is *not* intended to strictly reflect phylogenetic relationships within a given encompassing taxon, e.g. the first order listed within a subclass is not necessarily the one presumed to have arisen first within that subclass. As a specific example, the

order Desmoscolecida is listed first within subclass Chromadoria because its position with respect to other orders of Chromadoria remains unresolved, *not* because we assume that it is closest to order Chromadorida. Taxa that are truly *incertae sedis* (i.e. not only of uncertain position among taxa with rank equal to theirs, but also among several taxa of higher rank) are always listed first, preceding the various higher taxa to which they might belong with equal uncertainty.

– Families are marked with [†] when SSU rDNA sequence has been obtained from at least one of its species. This provides some idea of the taxonomic extent of sampling obtained so far in our molecular analyses: in most groups, the majority of families have not yet been included. For our classification, the available molecular data are therefore primarily relevant to higher taxa, while the distinction and placement of families remains largely based on existing (morphological) systems.

– Wherever possible, we have attempted to use taxa that are presumably monophyletic on the basis of current data (morphological and/or molecular). However, complete resolution is not always possible, mainly because of the as yet very incomplete molecular sampling. A few taxa are therefore maintained despite strong indications that they are paraphyletic and may in future be greatly restricted in species and genus content. Such families are marked with *.

– It was not always possible to verify the correct taxonomic author for suprageneric taxa. Some older publications were not available to us, while others often list a different taxonomic author for the same taxon. For order- and class-group taxon names, many older sources do not follow the taxon endings proposed by Pearse (1936, 1942). In those cases, we consider it important that a taxon of a given rank was proposed, but not whether the name of that taxon ended exactly as in the Pearse system. For the taxonomic author of a given order- or class-group taxon, we list the name of the earliest proponent of that taxon name with the relevant rank and etymological stem, regardless of the exact ending of that name as originally spelled. Thus, we list Hyman (1951) as the author of order Mermithida, even though her work actually proposed the order name Mermithoidea.

– For family-group taxa, we have followed the principle of coordination as defined by the International Code of Zoological Nomenclature (Art. 36.1. in ICZN, 1999), where the correct taxonomic author for **all** family-group names based on a given stem, is the earliest author to propose **any** single taxon of family-group rank (subfamily, family or superfamily) with that particular stem. Thus, we now propose a family Mesorhabditidae and superfamily Mesorhabditoidea, but the taxonomic author for both is Andrássy (1976) because he first proposed the subfamily Mesorhabditinae.

PHYLUM NEMATODA Potts, 1932
Incertae sedis:
ORDER BENTHIMERMITHIDA Tchesunov, 1995
 Family Benthimermithidae Petter, 1980
Incertae sedis:
ORDER RHAPTOTHYREIDA Tchesunov, 1995
 Family Rhaptothyreidae Hope and Murphy, 1969

CLASS ENOPLEA Inglis, 1983

SUBCLASS ENOPLIA Pearse, 1942

ORDER ENOPLIDA Filipjev, 1929
 Incertae sedis: Family Andrassyidae Tchesunov and Gagarin, 1999

 Suborder Enoplina Chitwood and Chitwood, 1937
  Superfamily Enoploidea Dujardin, 1845
   Family Enoplidae[†] Dujardin, 1845
   Family Thoracostomopsidae Filipjev, 1927
   Family Anoplostomatidae Gerlach and Riemann, 1974
   Family Phanodermatidae Filipjev, 1927
   Family Anticomidae Filipjev, 1918

 Suborder Oncholaimina De Coninck, 1965
  Superfamily Oncholaimoidea Filipjev, 1916
   Family Oncholaimidae[†] Filipjev, 1916
   Family Enchelidiidae[†] Filipjev, 1918

 Suborder Ironina Siddiqi, 1983
  Superfamily Ironoidea de Man, 1876
   Family Ironidae[†] de Man, 1876
   Family Leptosomatidae Filipjev, 1916
   Family Oxystominidae Chitwood, 1935

 Suborder Tripyloidina De Coninck, 1965
  Superfamily Tripyloidoidea Filipjev, 1928
   Family Tripyloididae[†] Filipjev, 1928

 Suborder Alaimina Clark, 1961
  Superfamily Alaimoidea Micoletzky, 1922
   Family Alaimidae[†] Micoletzky, 1922

ORDER TRIPLONCHIDA Cobb, 1920

 Suborder Diphtherophorina Coomans and Loof, 1970
  Superfamily Diphtherophoroidea Micoletzky, 1922
   Family Diphtherophoridae[†] Micoletzky, 1922
   Family Trichodoridae[†] Thorne, 1935

 Suborder Tobrilina Tsalolikhin, 1976
  Superfamily Tobriloidea De Coninck, 1965
   Family Tobrilidae[†] De Coninck, 1965
   Family Triodontolaimidae De Coninck, 1965
   Family Rhabdodemaniidae Filipjev, 1934
   Family Pandolaimidae Belogurov, 1980
  Superfamily Prismatolaimoidea Micoletzky, 1922
   Family Prismatolaimidae[†] Micoletzky, 1922

 Suborder Tripylina Andrássy, 1974
  Superfamily Tripyloidea de Man, 1876
   Family Tripylidae[†] de Man, 1876
   Family Onchulidae Andrássy, 1963

ORDER TREFUSIIDA Lorenzen, 1981
 Superfamily Trefusioidea Gerlach, 1966
  Family Simpliconematidae Blome and Schrage, 1985
  Family Trefusiidae Gerlach, 1966
  Family Laurathonematidae Gerlach, 1953
  Family Xenellidae De Coninck, 1965

SUBCLASS DORYLAIMIA Inglis, 1983

ORDER DORYLAIMIDA Pearse, 1942

 Suborder Dorylaimina Pearse, 1942
  Superfamily Dorylaimoidea de Man, 1876
   Family Dorylaimidae[†] de Man, 1876
   Family Aporcelaimidae[†] Heyns, 1965
   Family Qudsianematidae[†] Jairajpuri, 1965
   Family Nordiidae[†] Jairajpuri and Siddiqi, 1964
   Family Longidoridae[†] Thorne, 1935
   Family Actinolaimidae[†] Thorne, 1939
  Superfamily Belondiroidea Throne, 1939
   Family Belondiridae Throne, 1939
  Superfamily Tylencholaimoidea Filipjev, 1934
   Family Leptonchidae Thorne, 1935
   Family Tylencholaimidae[†] Filipjev, 1934
   Family Aulolaimoididae Jairajpuri, 1964
   Family Mydonomidae Thorne, 1964

 Suborder Nygolaimina Thorne, 1935
  Superfamily Nygolaimoidea Thorne, 1935
   Family Nygolaimidae Thorne, 1935
   Family Nygellidae Andrássy, 1958
   Family Aetholaimidae Jairajpuri, 1965
   Family Nygolaimellidae Clark, 1961

 Suborder Campydorina Jairajpuri, 1983
  Superfamily Campydoroidea Thorne, 1935
   Family Campydoridae Thorne, 1935

ORDER MONONCHIDA Jairajpuri, 1969

 Suborder Bathyodontina Siddiqi, 1983
  Superfamily Cryptonchoidea Chitwood, 1937
   Family Bathyodontidae Clark, 1961
   Family Cryptonchidae Chitwood, 1937
  Superfamily Mononchuloidea De Coninck, 1965
   Family Mononchulidae De Coninck, 1965

 Suborder Mononchina Kirjanova and Krall, 1969
  Superfamily Anatonchoidea Jairajpuri, 1969
   Family Anatonchidae[†] Jairajpuri, 1969
  Superfamily Mononchoidea Chitwood, 1937
   Family Mononchidae[†] Chitwood, 1937
   Family Mylonchulidae[†] Jairajpuri, 1969

ORDER ISOLAIMIDA Cobb, 1920
    Superfamily Isolaimoidea Timm, 1969
      Family Isolaimiidae Timm, 1969

ORDER DIOCTOPHYMATIDA Baylis and Daubney, 1926

    Suborder Dioctophymatina Skrjabin, 1927
      Family Dioctophymatidae Castellani and
      Chalmers, 1910
      Family Soboliphymatidae Petrov, 1930

ORDER MUSPICEIDA Bain and Chabaud, 1959

    Suborder Muspiceina Bain and Chabaud, 1959
      Family Muspiceidae Sambon, 1925
      Family Robertdollfusiidae Chabaud and
      Campana, 1950

ORDER MARIMERMITHIDA Rubtzov, 1980
    Family Marimermithidae Rubtzov and
    Platonova, 1974

ORDER MERMITHIDA Hyman, 1951

    Suborder Mermithina Andrássy, 1974
    Superfamily Mermithoidea Braun, 1883
      Family Mermithidae[†] Braun, 1883
      Family Tetradonematidae Cobb, 1919

ORDER TRICHINELLIDA Hall, 1916
    Superfamily Trichinelloidea Ward, 1907
      Family Anatrichosomatidae Yamaguti, 1961
      Family Capillariidae Railliet, 1915
      Family Cystoopsidae Skrjabin, 1923
      Family Trichinellidae[†] Ward, 1907
      Family Trichosomoididae Hall, 1916
      Family Trichuridae[†] Ransom, 1911

CLASS CHROMADOREA Inglis, 1983

SUBCLASS CHROMADORIA Pearse, 1942

ORDER DESMOSCOLECIDA Filipjev, 1929

    Suborder Desmoscolecina Filipjev, 1934
    Superfamily Desmoscolecoidea Shipley, 1896
      Family Desmoscolecidae Shipley, 1896
      Family Meyliidae De Coninck, 1965
      Family Cyartonematidae Tchesunov, 1990

ORDER CHROMADORIDA Chitwood, 1933
    Suborder Chromadorina Filipjev, 1929
    Superfamily Chromadoroidea Filipjev, 1917
      Family Chromadoridae[†] Filipjev, 1917
      Family Ethmolaimidae Filipjev and Schuurmans
      Stekhoven, 1941
      Family Neotonchidae Wieser and Hopper, 1966
      Family Achromadoridae Gerlach and Riemann,
      1973
      Family Cyatholaimidae[†] Filipjev, 1918

ORDER DESMODORIDA De Coninck, 1965

    Suborder Desmodorina De Coninck, 1965
    Superfamily Desmodoroidea Filipjev, 1922
      Family Desmodoridae[†] Filipjev, 1922
      Family Epsilonematidae Steiner, 1927
      Family Draconematidae Filipjev, 1918
    Superfamily Microlaimoidea Micoletzky, 1922
      Family Microlaimidae Micoletzky, 1922
      Family Aponchiidae Gerlach, 1963
      Family Monoposthiidae Filipjev, 1934

ORDER MONHYSTERIDA Filipjev, 1929

    Suborder Monhysterina De Coninck and
    Schuurmans Stekhoven, 1933
    Superfamily Monhysteroidea de Man, 1876
      Family Monhysteridae[†] de Man, 1876
    Superfamily Sphaerolaimoidea Filipjev, 1918
      Family Xyalidae[†] Chitwood, 1951
      Family Sphaerolaimidae Filipjev, 1918

    Suborder Linhomoeina Andrássy, 1974
    Superfamily Siphonolaimoidea Filipjev, 1918
      Family Siphonolaimidae Filipjev, 1918
      Family Linhomoeidae Filipjev, 1922
      Family Fusivermidae Tchesunov, 1996

ORDER ARAEOLAIMIDA De Coninck and Schuurmans
Stekhoven, 1933
    Superfamily Axonolaimoidea Filipjev, 1918
      Family Axonolaimidae[†] Filipjev, 1918
      Family Comesomatidae Filipjev, 1918
      Family Diplopeltidae[†] Filipjev, 1918
      Family Coninckiidae Lorenzen, 1981

ORDER PLECTIDA Malakhov, 1982
    Superfamily Leptolaimoidea Örley, 1880
      Family Leptolaimidae Örley, 1880
      Family Rhadinematidae Lorenzen, 1981
      Family Aegialoalaimidae Lorenzen, 1981
      Family Diplopeltoididae Tchesunov, 1990
      Family Paramicrolaimidae Lorenzen, 1981
      Family Ohridiidae Andrássy, 1976
      Family Bastianiidae De Coninck, 1935
      Family Odontolaimidae Gerlach and Riemann,
      1974
      Family Rhabdolaimidae Chitwood, 1951
    Superfamily Ceramonematoidea Cobb, 1933
      Family Tarvaiidae Lorenzen, 1981
      Family Ceramonematidae Cobb, 1933
      Family Tubolaimoididae Lorenzen, 1981
    Superfamily Plectoidea Örley, 1880
      Family Plectidae[†] Örley, 1880
      Family Chronogasteridae Gagarin, 1975
      Family Metateratocephalidae Eroshenko, 1973
    Superfamily Haliplectoidea Chitwood, 1951
      Family Peresianidae Vitiello and De Coninck, 1968
      Family Haliplectidae Chitwood, 1951
      Family Aulolaimidae Jairajpuri and Hooper, 1968

ORDER RHABDITIDA Chitwood, 1933
Incertae sedis: Family Teratocephalidae† Andrássy, 1958
Incertae sedis: Family Chambersiellidae Thorne, 1937
Incertae sedis: Family Brevibuccidae† Paramonov, 1956

Suborder Spirurina
Incertae sedis: Superfamily Dracunculoidea Stiles, 1907
    Family Dracunculidae Stiles, 1907
    Family Philometridae† Baylis and Daubney, 1926
    Family Phlyctainophoridae Roman, 1965
    Family Skrjabillanidae Schigin and Schigina, 1958
    Family Anguillicolidae Yamaguti, 1935
    Family Guyanemidae Petter, 1975
    Family Micropleuridae Baylis and Daubney, 1926

INFRAORDER GNATHOSTOMATOMORPHA n. infraord.
Superfamily Gnathostomatoidea Railliet, 1895
    Family Gnathostomatidae† Railliet, 1895

INFRAORDER OXYURIDOMORPHA n. infraord.
Superfamily Thelastomatoidea Travassos, 1929
    Family Thelastomatidae Travassos, 1929
    Family Travassosinematidae Rao, 1958
    Family Hystrignathidae Travassos, 1919
    Family Protrelloididae Chitwood, 1932
Superfamily Oxyuroidea Cobbold, 1864
    Family Oxyuridae Cobbold, 1864
    Family Pharyngodonidae Travassos, 1919
    Family Heteroxynematidae† Skrjabin and Shikhobalova, 1948

INFRAORDER RHIGONEMATOMORPHA n. infraord.
Superfamily Rhigonematoidea Artigas, 1930
    Family Rhigonematidae Artigas, 1930
    Family Ichthyocephalidae Travassos and Kloss, 1958
Superfamily Ransomnematoidea Travassos, 1930
    Family Ransomnematidae† Travassos, 1930
    Family Carnoyidae Filipjev, 1934
    Family Hethidae Skrjabin and Shikhobalova, 1951

INFRAORDER SPIRUROMORPHA n. infraord.
Superfamily Camallanoidea Railliet and Henry, 1915
    Family Camallanidae Railliet and Henry, 1915
Superfamily Physalopteroidea Railliet, 1893
    Family Physalopteridae Railliet, 1893
Superfamily Rictularoidea Hall, 1915
    Family Rictulariidae Hall, 1915
Superfamily Thelazoidea Skrjabin, 1915
    Family Thelaziidae Skrjabin, 1915
    Family Rhabdochonidae Travassos, Artigas and Pereira, 1928
    Family Pneumospiruridae Wu and Hu, 1938
Superfamily Spiruroidea Örley, 1885
    Family Gongylonematidae Hall, 1916
    Family Spiruridae Örley, 1885

    Family Spirocercidae Chitwood and Wehr, 1932
    Family Hartertiidae Quentin, 1970
Superfamily Habronematoidea Chitwood and Wehr, 1932
    Family Hedruridae Railliet, 1916
    Family Habronematidae Chitwood and Wehr, 1932
    Family Tetrameridae Travassos, 1914
    Family Cystidicolidae Skrjabin, 1946
Superfamily Acuarioidea Railliet, Henry and Sisoff, 1912
    Family Acuariidae Railliet, Henry and Sisoff, 1912
Superfamily Filarioidea Weinland, 1858
    Family Filariidae Weinland, 1858
    Family Onchocercidae† Leiper, 1911
Superfamily Aproctoidea Yorke and Maplestone, 1926
    Family Aproctidae Yorke and Maplestone, 1926
    Family Desmidocercidae Cram, 1927
Superfamily Diplotriaenoidea Skrjabin, 1916
    Family Diplotriaenidae Skrjabin, 1916
    Family Oswaldofilariidae Chabaud and Choquet, 1953

INFRAORDER ASCARIDOMORPHA n. infraord.
Superfamily Ascaridoidea Baird, 1853
    Family Heterocheilidae† Railliet and Henry, 1912
    Family Ascarididae† Baird, 1853
    Family Raphidascarididae† Hartwich, 1954
    Family Anisakidae† Railliet and Henry, 1912
Superfamily Cosmocercoidea Skrjabin and Schikhobalova, 1951
    Family Cosmocercidae Railliet, 1916
    Family Atractidae Railliet, 1917
    Family Kathlaniidae† Lane, 1914
Superfamily Heterakoidea Railliet and Henry, 1914
    Family Heterakidae Railliet and Henry, 1912
    Family Aspidoderidae Skrjabin and Schikhobalova, 1947
    Family Ascaridiidae† Travassos, 1919
Superfamily Subuluroidea Travassos, 1914
    Family Subuluridae Travassos, 1914
    Family Maupasinidae Lopez-Neyra, 1945
Superfamily Seuratoidea Hall, 1916
    Family Seuratidae Hall, 1916
    Family Cucullanidae† Cobbold, 1864
    Family Quimperiidae Gendre, 1928
    Family Chitwoodchabaudiidae Puylaert, 1970
    Family Schneidernematidae Freitas, 1956

Suborder Myolaimina Inglis, 1983

Superfamily Myolaimoidea Andrássy, 1958
    Family Myolaimidae† Andrássy, 1958

Suborder Tylenchina Thorne, 1949

INFRAORDER PANAGROLAIMOMORPHA n. infraord.
Superfamily Panagrolaimoidea Thorne, 1937
    Family Panagrolaimidae† Thorne, 1937

Superfamily Strongyloidoidea Chitwood and McIntosh, 1934 n.superfam.
    Family Steinernematidae[†] Filipjev, 1934
    Family Strongyloididae[†] Chitwood and McIntosh, 1934
    Family Rhabdiasidae[†] Railliet, 1916

INFRAORDER CEPHALOBOMORPHA n. infraord.
    Superfamily Cephaloboidea Filipjev, 1934
        Family Cephalobidae[†] Filipjev, 1934
        Family Elaphonematidae Heyns, 1962
        Family Osstellidae Heyns, 1962
        Family Alirhabditidae Suryawanshi, 1971
        Family Bicirronematidae Andrássy, 1978

INFRAORDER TYLENCHOMORPHA n. infraord.
    Superfamily Aphelenchoidea Fuchs, 1937
        Family Aphelenchidae[†] Fuchs, 1937
        Family Aphelenchoididae[†] Skarbilovich, 1947
    Superfamily Criconematoidea Taylor 1936
        Family Criconematidae[†] Taylor, 1936
        Family Hemicycliophoridae Skarbilovich, 1959
        Family Tylenchulidae[†] Skarbilovich, 1947
    Superfamily Sphaerularioidea Lubbock, 1861
        Family Anguinidae[†] Nicoll, 1935
        Family Sphaerulariidae Lubbock, 1861
        Family Neotylenchidae Thorne, 1941
        Family Iotonchidae Goodey, 1935
    Superfamily Tylenchoidea Örley, 1880
        Family Hoplolaimidae[†] Filipjev, 1934
        Family Meloidogynidae[†] Skarbilovich, 1959
        Family Tylenchidae Örley, 1880
        Family Belonolaimidae*[†] Whitehead, 1959
        Family Pratylenchidae*[†] Thorne, 1949
    Superfamily Myenchoidea Pereira, 1931

INFRAORDER DRILONEMATOMORPHA n. infraord.
    Superfamily Drilonematoidea Pierantoni, 1916
        Family Drilonematidae Pierantoni, 1916
        Family Ungellidae Chitwood, 1950
        Family Homungellidae Timm, 1966
        Family Pharyngonematidae Chitwood, 1950
        Family Creagrocercidae Baylis, 1943

Suborder Rhabditina Chitwood, 1933

INFRAORDER BUNONEMATOMORPHA n. infraord.
    Superfamily Bunonematoidea Micoletzky, 1922
        Family Bunonematidae[†] Micoletzky, 1922
        Family Pterygorhabditidae Goodey, 1963

INFRAORDER DIPLOGASTEROMORPHA n. infraord.
    Superfamily Cylindrocorporoidea Goodey, 1939
        Family Cylindrocorporidae[†] Goodey, 1939
    Superfamily Odontopharyngoidea Micoletzky, 1922
        Family Odontopharyngidae Micoletzky, 1922
    Superfamily Diplogasteroidea Micoletzky, 1922
        Family Pseudodiplogasteroididae[†] Körner, 1954
        Family Diplogasteroididae[†] Filipjev and Schuurmans Stekhoven, 1941

        Family Diplogasteridae[†] Micoletzky, 1922
        Family Neodiplogasteridae[†] Paramonov, 1952
        Family Mehdinematidae Farooqui, 1967
        Family Cephalobiidae Filipjev, 1934

INFRAORDER RHABDITOMORPHA n. infraord.
Incertae sedis: Family Carabonematidae Stammer and Wachek, 1952
Incertae sedis: Family Agfidae Dougherty, 1955

    Superfamily Mesorhabditoidea Andrássy, 1976 n. superfam.
        Family Mesorhabditidae[†] Andrássy, 1976 n. superfam.
        Family Peloderidae[†] Andrássy, 1976 n.fam.
    Superfamily Rhabditoidea* Örley, 1880
        Family Diploscapteridae[†] Micoletzky, 1922
        Family Rhabditidae*[†] Örley, 1880
    Superfamily Strongyloidea Baird, 1853
        Family Heterorhabditidae[†] Poinar, 1975
        Family Strongylidae[†] Baird, 1853
        Family Ancylostomatidae[†] Looss, 1905
        Family Trichostrongylidae Witenberg, 1925
        Family Metastrongylidae Leiper, 1908

Brief characteristics of major groups:

The following paragraphs expand on various taxa listed above, and particularly on those for which the diagnosis and/or contents are changed substantially in our classification compared to previous systems.

– *CLASS ENOPLEA:* SSU rDNA analysis strongly supports distinction of three basal clades in Nematoda: dorylaims, enoplids and chromadorids (resp. clade I, II and C+S in Blaxter *et al.*, 1998). Relationships between these clades cannot be resolved clearly at present, but some analyses do support sister taxon status of dorylaims and enoplids. Since this agrees with all previous systems, we therefore retain dorylaims and enoplids for the present within the encompassing class Enoplea.

– **SUBCLASS ENOPLIA:** Lorenzen (1981, 1994) distinguished two orders within Enoplia: Enoplida and Trefusiida. He deduced monophyly of Enoplida from the presence of metanemes. Analysis of the few available SSU rDNA sequences does not resolve higher taxa within Enoplia very well, but disagrees significantly with Lorenzen's system in at least one respect: alaimids, prismatolaimids and diphtherophorids are firmly placed within the enoplid clade, even though they are not known to have metanemes, and the latter two groups are found to be surprisingly close to tobrilids. In order to combine patterns of metaneme presence and SSU rDNA affinity, we therefore transfer the order Triplonchida to Enoplia and expand its contents. We assume for the time being that metanemes do indeed represent a synapomorphy for Enoplia, but also that these structures were lost secondarily in several of its member taxa.

– **ORDER ENOPLIDA:** Lorenzen (1981, 1994) distinguished two suborders within the order: Enoplina and Tripyloidina. The latter taxon was presumed to be paraphyletic, while the former was considered monophyletic on the basis of one synapomorphy (caudal glands usually extending well anterior to anus). SSU rDNA data do not clearly resolve relationships among various relevant taxa, and in particular do not support a single clade for taxa with preanal caudal gland extensions. However, sequence data do strongly suggest that Enoplida *sensu* Lorenzen (1981) are paraphyletic with respect to Triplonchida *sensu* Siddiqi (1983b). By transferring tobrilids and relatives from Enoplida to Triplonchida (see below) we eliminate the obvious paraphyly of Enoplida — at least for the time being: SSU rDNA analysis still does not positively group the remaining member taxa of Enoplida into a monophyletic clade. More extensive sampling is definitely required, and may well lead to further changes in rank and/or composition of this group.

– **Suborder Oncholaimina:** Lorenzen (1981) classified oncholaims within Enoplina[3] on the basis of a single character: the shared presence of preanal caudal gland extensions. However, he noted that this character is variable in at least two other taxa of Enoplina: Enoploidea and Ironidae. Such variability indicates that the extent of the caudal glands must have been subject to homoplastic evolution, through multiple independent origins of the character and/or secondary reversal of the derived condition to its ancestral state. In the absence of other synapomorphies for Enoplina *sensu* Lorenzen, it is impossible to distinguish between these two types of homoplasy, and the monophyly of the taxon is therefore dubious. SSU rDNA analysis does not confirm exclusive common origin of extended glands in oncholaims and enoplids, but instead yields a moderately supported clade grouping oncholaims with the suborder Tripyloidina. Additional morphological synapomorphies need to be recognised and more unequivocal molecular analyses need to be obtained using a wider range of taxa. For the time being we therefore classify oncholaims at equal rank with both Enoplina and Tripyloidina, i.e. as a separate suborder.

– **Suborder Ironina:** SSU rDNA data firmly place ironids in Enoplia, but do not provide clues on their affinities to other member taxa of this subclass. Superfamily Ironoidea *sensu* Lorenzen (1981) includes both species with and without preanal extension of the caudal glands, and also species with or without metanemes. Lorenzen (1981) placed this taxon within his rankless taxon Enoplacea, on the basis of synapomorphies with respect to cervical gland position and muscular connections between the anterior pharynx and the body wall. Both characters are variable within Ironidae, and the muscular connections in Enoplidae

are not obviously homologous, neither structurally nor positionally, to the retractor muscles occurring in some Ironidae (e.g. compare Figure 2 in Inglis, 1964 with Figure 1A in Van Der Heiden, 1974). The affinities of ironids within Enoplida therefore remain unclear, and we place them as a separate suborder until more data become available.

– **Suborder Alaimina:** SSU rDNA analysis unequivocally places alaimids within Enoplia, and not in Dorylaimida, where Lorenzen (1981) placed them on the basis of the posteriorly located dorsal gland opening. As discussed more fully in the text, this character may well be homoplastic. We therefore include alaims in Enoplida and rank them as a separate suborder, since they appear to lack caudal glands and current SSU data do not further resolve their position.

– **ORDER TRIPLONCHIDA:** Trichodorids and their relatives were traditionally placed with dorylaims because of the shared presence of a protrusible, piercing feeding structure. This resemblance was later found to be rather superficial, and several authors suggested closer affinity with enoplids (see review in Decraemer, 1995). Lorenzen (1981, 1994) maintained them within Dorylaimida on the basis of the posteriorly located pharyngeal gland openings. In view of the strong support of SSU rDNA data for inclusion in Enoplia, we assume that this character is convergent (see page 9).

Sequence analysis very robustly positions a clade of nematodes without protrusible spear (tobrilids and prismatolaimids) as sister taxon to triplonchs. We therefore expand the contents of Triplonchida to include the free-living taxa in question, thereby removing a component from Enoplida which would otherwise render the latter order paraphyletic with respect to Triplonchida. This action is not entirely without morphological grounds. Some free-living Enoplia resemble Trichodoridae in several respects (Siddiqi, 1983b). Most notably, the spicule protractor muscles are modified into capsule-like structures that surround the anterior half of each spicule and which appear to squeeze out the spicules rather than simply pulling them backwards. Although this character needs further study, it may represent a synapomorphy for Triplonchida as expanded here.

– **Suborder Tobrilina:** SSU rDNA sequence emphatically unites Tobrilidae and Prismatolaimidae into a separate clade that represents a sister taxon to trichodorids and relatives. A close relationship between these free-living and plant-parasitic taxa is not evident from any morphological character, and Prismatolaimidae were placed in Chromadoria rather than Enoplia by Lorenzen (1981). Given the robustness of the sequence data, we unite tobrilids with prismatolaimids into a suborder

---

[3] Within Enoplida, Lorenzen (1981) proposed the taxa Enoplacea and Oncholaimacea at unspecified ranks in between order and superfamily. Inglis (1983) criticized this proposal because he considered it unnecessary to introduce an additional ranking level in classification, and because the — acea ending of both names was reserved for the rank of tribe in the nomenclatorial system of Pearse (1936; 1942).

Tobrilina, and presume that this suborder also includes several families that have not yet been included in molecular studies. Distinct spicule protrusion capsules do not occur in any of these families, but the anterior end of the spicules is embedded in muscle tissue in at least some *Prismatolaimus* (cf. Figure 3J,K in Coomans and Raski, 1988) and *Tobrilus* species (personal observation), suggesting that this group may contain muscle arrangements intermediate between completely separate protractors *versus* true capsules. Ultrastructural studies are needed to prove or disprove this hypothesis.

– Suborder Tripylina: Tripylids and tobrilids are traditionally classified closely to one another, and Lorenzen (1981) classified them as families within the suborder Tripyloidina. Only a single sequence has been obtained so far from Tripylidae *sensu* Lorenzen (1981), and this sequence provides no resolution other than placement in Enoplia. Despite the absence of any definite support from sequence data, we propose to include Tripylidae in Triplonchida, on the strength of the prominent spicule protrusion capsules found in this family. Another family with prominent capsules is Onchulidae, and we therefore classify it with Tripylidae in suborder Tripylina.

– SUBCLASS DORYLAIMIA: Dorylaims and relatives are traditionally classified in the same order, subclass or class as enoplids and relatives (De Coninck, 1965; Andrássy, 1976; Lorenzen, 1981; Inglis, 1983; Maggenti, 1983; Adamson, 1986; Malakhov, 1994). SSU rDNA does not resolve the position of dorylaims with respect to chromadorids and enoplids *sensu lato*. This unresolved position hints at the intriguing possibility that dorylaims and relatives might actually represent a full-blown third class of nematodes. This possibility has not been considered seriously in previous systems, nor do we consider it appropriate until definite resolution is obtained at the base of the nematode tree.

Nevertheless, when we consider that none of the free-living taxa within Dorylaimia occupy truly marine habitats, an interesting point is raised: a putative position of the dorylaimid clade as sister taxon to Enoplia + Chromadorea might require us to reconsider the ecology of the common ancestor of all nematodes. Was this ancestor really a marine organism, as proposed by Filipjev (1929, 1934) and now generally assumed, or could it instead have been an inhabitant of freshwater sediments — or even terrestrial deposits? At present, this still seems unlikely and indirectly pleads against a separate class for dorylaims.

In defining Dorylaimia and allocating taxa to it, Lorenzen (1981, 1994) placed great emphasis on posteriorly placed pharyngeal gland openings as an autapomorphy. As discussed on page 9, we strongly suspect this derived character to have arisen more frequently in

nematodes than allowed for in Lorenzen's system. Sequence analysis strongly supports a less comprehensive dorylaim clade, agreeing much more with some of the arguments of Siddiqi (1983b) in that Triplonchida and Alaimina are excluded. Because this clade is well-supported and so clearly separated from the enoplid clade, we agree with Inglis (1983) in allocating to it the rank of subclass.

– ORDER DORYLAIMIDA: The order Dorylaimida may well constitute the most diverse order of soil and freshwater nematodes, and particularly so in the tropics. Could this diversity be due to very early origins, rather than/in addition to rapid speciation? The presence of a protrusible spear in dorylaims undoubtedly represents an apomorphy, presumably resulting from a complex evolutionary pathway. However, SSU rDNA divergence within the suborder Dorylaimina is minimal, suggesting either that this locus has evolved at a very slow rate, or that the origins of Dorylaimina are relatively recent. Several key taxa with less derived buccal morphology were not yet sampled (Nygolaimina, Campydorina) and these issues must therefore remain speculative for the time being.

- ORDER DIOCTOPHYMATIDA: No dioctophymatids have yet been sampled for sequence data, but all modern authors agree on a placement within the dorylaim clade (Anderson and Bain, 1982; Skrjabin, 1991).

- ORDER TRICHINELLIDA[4]: Only two taxa, from two different families, have been sampled through sequencing thus far, and these place the Trichinellida at or near the base of the Dorylaimia (Blaxter *et al.*, 1998). This placement may be due to long-branch attraction artefacts, and additional sequences from other taxa, particularly capillarid and cystoopsid species may in future resolve both the placement of this group and its relation to the other zooparasitic Dorylaimia, the Mermithida and Dioctophymatida (Anderson and Bain, 1982; Skrjabin, 1991).

– *CLASS CHROMADOREA*: In our classification, Chromadorea contains the bulk of taxa within Nematoda, entirely including the 'Secernentea', as strongly supported by molecular data. 'Secernentea' is not a sister taxon of Chromadorea (as allowed by Inglis, 1983, where 'Rhabditea' replaces 'Secernentea') and must therefore receive a lower rank than its encompassing clade. Unlike Adamson (1986), we prefer to use the name Chromadorea for this encompassing clade, because it has line precedence over Rhabditea in Inglis (1983: p. 244) and because it better represents the fundamental split between the enoplid and chromadorid lineages.

---

[4] To our knowledge, Hall (1916 in Baylis and Daubney, 1926) was the first to propose an order-level taxon for trichurids, trichinellids and their relatives. He used the name Trichinelloidea for this order, preceding by at least ten years the order names Trichocephalida, Trichosyringida or Trichurida, each of which subsequently became popular in different parts of the world. We emend his order name here to Trichinellida and consider this name to have priority over the other three.

Furthermore, SSU rDNA strongly indicates that the stem of the chromadorid clade consists of a comb-like sequence of monophyletic groups, culminating in the 'secernentean' radiation. As a result, 'Secernentea' cannot just be downgraded to subclass, unless we either maintain a paraphyletic subclass for all stem taxa within Chromadoria, or propose a separate subclass for each of these stem taxa. Neither option is acceptable, and we therefore downgrade 'Secernentea' to ordinal rank. There are numerous order names available within the former 'Secernentea' and we have chosen the one with chronological priority, *i.e.* Rhabditida (see below).

– **SUBCLASS CHROMADORIA:** This subclass already existed in previous systems. We therefore preserve it here, but it is only of nominal importance since we do not recognize any other subclasses within Chromadorea.

– **ORDER DESMOSCOLECIDA:** This group of unusual nematodes has long puzzled taxonomists, and no sequence data have as yet been obtained. We therefore follow Lorenzen (1981, 1994) in classifying it within Chromadoria, but as an order rather than a suborder, because of the higher ranks accorded to other components of Chromadoria.

- **ORDER CHROMADORIDA:** Unlike Lorenzen's analysis, SSU rDNA phylogenies strongly support a monophyletic clade of chromadorids *sensu stricto*, as sister taxon to all other hitherto sequenced Chromadoria. In order to allow a reasonably high rank for all other clades within Chromadoria, this taxon must therefore receive ordinal rank (it corresponds to the superfamily Chromadoroidea in Lorenzen, 1981, 1994!).

- **ORDER DESMODORIDA:** Within Chromadoria, this group does not share exclusive common ancestry with chromadorids *sensu stricto*. It is instead the next separate lineage to diverge from the stem of Chromadorea, and must therefore also receive order rank.

- **ORDER MONHYSTERIDA:** Likewise, monhysterids do not share exclusive common ancestry with Chromadorida *sensu* Lorenzen (1981, 1994), but rather represent the next lineage in a nested series within Chromadorea. Also, their position does not agree with the assumption that their morphology represents an ancestral grade (cf. Maggenti, 1963). Lorenzen (1981) defined Monhysterida on the basis of the presence of the outstretched female reproductive branch(es), a credible apomorphy compared with the reflexed condition found in Chromadorida and Desmodorida. On the basis of this character, he therefore included the superfamilies Monhysteroidea, Siphonolaimoidea and Axonolaimoidea within Monhysterida. However, SSU rDNA data clearly show that Axonolaimoidea are sister taxon to a clade of nematodes with reflexed female reproductive branches, suggesting that this character has reverted at least once within Chromadoria. We therefore classify Axonolaimoidea within the order Araeolaimida instead of Monhysterida, similar to the systems of Malakhov,

Ryzhikov and Sonin (1982), Inglis (1983) and Maggenti (1983). Lorenzen noted that autapomorphies could neither be found for Axonolaimoidea, nor for Siphonolaimoidea, and the latter have not yet been sampled for SSU rDNA sequence data. For the time being we maintain the equivalent of Lorenzen's Siphonolaimoidea as suborder Linhomoeina within Monhysterida.

- **ORDER ARAEOLAIMIDA:** Ever since its proposal by De Coninck and Schuurmans Stekhoven (1933), this order has been used as a 'bucket taxon' for numerous families of uncertain affinities. In the system of Lorenzen (1981, 1994) it was rejected and its components were subdivided over the Monhysterida and Chromadorida: Leptolaimina, which constituted the single largest and least resolved suborder in his analysis. Although only two SSU rDNA sequences are available to date from superfamily Axonolaimoidea *sensu* Lorenzen (Félix *et al.*, 2000; Vanfleteren and Moens, unpublished data), these data show that the superfamily clearly does not belong in Monhysterida (contrary to Lorenzen, 1981, 1994), but rather that it constitutes the sister taxon of a clade including Plectidae and all 'Secernentea'. We therefore consider Axonolaimoidea to be a part of an order separate from both monhysterids and plectids. The correct name for this order is Araeolaimida, because *Araeolaimus* is currently classified in Axonolaimoidea (Lorenzen, 1981, 1994). The wider contents of Araeolaimida remain entirely unclear for the present, since known morphological data do not include clear apomorphies shared with other putative relatives of axonolaimids, and more sequence data are not yet available (see below). Araeolaimida *sensu nobis* therefore largely matches the contents and contents specified for the order by Malakhov, Ryzhikov and Sonin (1982), with the important addition of an outstretched female reproductive system as diagnostic character shared with Monhysterida.

- **ORDER PLECTIDA:** The position of the family Plectidae was not resolved in the analyses of Blaxter *et al.* (1998), where it sometimes grouped with Spiruromorpha *sensu nobis*, while ending up as sister taxon to all 'Secernentea' at other times. The latter position is much more in agreement with morphological characters, and is also given stronger support by analyses including additional taxa (Félix *et al.*, 2000; Vanfleteren and De Ley, unpublished). This implies that Plectidae require ranking equivalent to that of Araeolaimida and Rhabditida *sensu nobis*, consistent with the proposal of the order Plectida by Malakhov, Ryzhikov and Sonin (1982). However, a large number of families sharing characters with Araeolaimida and/ or Plectidae remain excluded from SSU rDNA analysis. Most of these families correspond with Lorenzen's suborder Leptolaimina, which he characterised as a presumably paraphyletic taxon lacking resolution of internal relationships (Lorenzen, 1981, 1994).

We consider it likely that this unresolved group constitutes a mixture of paraphyletic and/or misplaced families, and provisionally attempt to classify them by combining the systems of Malakhov, Ryzhikov and Sonin (1982) and Lorenzen (1981, 1994). Thus, the equivalent of Lorenzen's Leptolaimina is placed here as superfamily Leptolaimoidea within the order Plectida. Unlike Lorenzen (1981, 1994), we maintain a separate superfamily Ceramonematoidea, including in it the putatively related but problematic families Tarvaiidae and Tubolaimoididae, and unlike Malakhov, Ryzhikov and Sonin (1982) we provisionally place this group within Plectida rather than Desmodorida.

- **ORDER RHABDITIDA:** In our system the order Rhabditida is equivalent with all of 'Secernentea' in most previous classifications, a major change that will undoubtedly meet with ample criticism. Nevertheless, this is a logical monophyly-based alternative to current covertly or overtly paraphyletic systems, and it follows inexorably from the application of Linnean ranks to the strongly supported pattern of relationships provided by SSU rDNA. Morphology has not (yet) yielded sufficient synapomorphies to recognise the series of monophyletic clades preceding 'Secernentea' in evolution, so that past emphasis on diagnosis and typology has allowed an escalation of increasingly higher ranks, without any correction for probable evolutionary relationships. We believe that phylogeny must be the basis for classification, and therefore propose corrections compatible with sound phylogenetic resolution.

– Family Teratocephalidae: The genus *Teratocephalus* is morphologically intermediate between 'Adenophorea' and 'Secernentea' and has received a different position in nearly every classification. SSU rDNA sequence analysis also does not clearly resolve its position, variously placing it near Plectidae and/or Spiruromorpha. Until better resolution is obtained, we consider the genus to be the only known member of a family *incertae sedis*. The family Metateratocephalidae used to be classified together with *Teratocephalus* in one superfamily (cf. Andrássy, 1984) but it exhibits characters that are much closer to those of Plectida (cf. Karegar, De Ley and Geraert, 1997) and sequence data have not yet been obtained.

– Family Chambersiellidae: No member of this family has as yet been included in sequence analysis. Chambersiellid species are characterised by presence of liplets, setiform labial sensilla and relatively posteriad amphids. All these character states may well be plesiomorphic. As far as known, no species have offset spermathecae. Andrássy (1984) included chambersiellids in his suborder Teratocephalina, but since the supporting characters may be symplesiomorphic we do not consider it appropriate to unite the two relevant families.

– Family Brevibuccidae: SSU rDNA sequence analysis does not resolve the position of *Brevibucca* sp., placing it variously with panagrolaims, rhabditids or near the base of Rhabditida *sensu nobis* (Félix *et al.*, 2000). Morphological data on *Brevibucca* and related genera do not suffice to assign it to any suborder or infraorder, still requiring e.g. ultrastructural studies of the buccal cavity to verify presence and structure of the stegostom.

– **Suborder Spirurina:** The molecular analyses of Blaxter *et al.* (1998) suggest that the four zooparasitic orders Spirurida, Ascaridida, Oxyurida and Rhigonematida are very closely related and comprise a clade (clade III) nested within Rhabditida but distinct from other groups included therein. In some analyses, Plectida are the sister group of this clade III within Rhabditida, but, as explained above, a basal position of Plectida is favoured. Linking of Ascaridida with Spirurida and Oxyurida with Rhigonematida has been suggested previously by several authors, but clade III has emerged only from molecular analyses. To conform to the schema presented here, the orders have been reclassified as infraorders. In analysis of SSU rDNA sequences, the Dracunculoidea is placed either basal to these groups, or has its origin in an unresolved polytomy (Blaxter, unpublished). We therefore consider the superfamily as of uncertain position within the suborder. As with the Gnathostomatoidea discussed below, Dracunculoidea may need to be elevated in status (e.g. to infraorder).

– INFRAORDER GNATHOSTOMATOMORPHA: The traditional order Spirurida is not monophyletic by molecular criteria, with Gnathostomatoidea being placed basal to clade III, and thus outside the derived group of other spirurids, Oxyuridomorpha, Rhigonematomorpha and Ascaridomorpha (Blaxter, unpublished). They are therefore elevated to infraorder rank.

– INFRAORDER OXYURIDOMORPHA: The Oxyuridomorpha is represented by a single heteroxurid taxon in molecular analyses. This is placed robustly in clade III, but additional taxa would improve understanding, particularly from insect-parasitic Thelastomatoidea. The taxonomy of Oxyuroidea is taken from Petter (1976).

– INFRAORDER RHIGONEMATOMORPHA: Classification within Rhigonematomorpha basically follows Hunt (1996). Only one species of this infraorder has so far been included in sequence analyses, and it is therefore not yet possible to address the (in)validity of classifying Rhigonematoidea and Ransomnematoidea as sister taxa.

– INFRAORDER SPIRUROMORPHA: Sequence data is thus far available for only one superfamily within the Spiruromorpha, the Filaroidea, and this group is monophyletic. In the absence of conflicting evidence we maintain the remaining superfamilies within Spiruromorpha, although Anderson, Chabaud and Willmont *et al.* (1974) and Chabaud and Bain (1994) suggest that the Physalopteroidea and Camallanoidea may be similarly distanced from the 'crown spirurids'. Classification follows Chabaud (1975a, 1975b), Anderson (1976) and Anderson and Bain (1976).

– INFRAORDER ASCARIDOMORPHA: This taxon is monophyletic by molecular and morphological criteria (Nadler, 1995; Blaxter *et al.*, 1998; Nadler and Hudspeth, 1998). Classification is based on Hartwich (1974) and Chabaud (1978), except for classification of Ascaridoidea which is based on the morphological analysis of Fagerholm (1991) and compatible with the molecular phylogenies obtained by Nadler and Hudspeth (1998).

– **Suborder Myolaimina:** This suborder was first proposed by Inglis (1983) for the families Myolaimidae, Carabonematidae, Agfidae and Chitwoodiellidae. The family Myolaimidae only contains the genus *Myolaimus*, of which all known species are free-living and are characterized by a detached outer cuticular layer, complete absence of spicules and presence of a bursa-like structure on the posterior end of the male body. The other three families all contain very poorly known zooparasitic species with normal cuticle and of which the males have spicules and a more typical rhabditid bursa. SSU rDNA data suggests that *Myolaimus* represents a sister taxon to the combined clade of suborders Tylenchina and Rhabditina *sensu nobis*, and must therefore be placed in a separate suborder. However, we do not agree with Inglis (1983) that the three zooparasitic taxa are demonstrably related to *Myolaimus*. The few data available suggest to us that these three families belong somewhere within Rhabditina (see below), and we place them there as *incertae sedis*.

– **Suborder Tylenchina:** SSU rDNA strongly supports exclusive common ancestry for the morphologically disjunct tylenchs and cephalobs, confirming the varied arguments by Siddiqi (1980 and 1985) against a derivation of tylenchs from diplogasterid ancestry. Blaxter *et al.* (1998) reported moderate support for the inclusion of panagrolaims, steinernematids and strongyloidids in one clade with tylenchs and cephalobs, but this support dwindles with the addition of more sequences (Félix *et al.*, 2000). Nevertheless, an independent synapomorphy supporting inclusion was identified by Goldstein, Frisse and Thomas (1998), who found a similar pattern of embryonic axis specification in early development in these five taxa. We therefore place them together for the time being.

Some important characters of the hypothetical most recent common ancestor of Tylenchina *sensu nobis* include: stegostom without epithelial interradial cells or glottoid part, median bulb absent, mono-prodelphy, bursa absent. All these could well be ancestral to the entire order Rhabditida as defined here. In other words: no obvious morphological synapomorphies for the suborder are known at present.

– INFRAORDER PANAGROLAIMOMORPHA: Most taxa sequenced from this clade exhibit relatively long branch lengths in SSU rDNA phylogenies, as well as AT-rich SSU sequences. Both of these features are symptomatic of potential artefacts in phylogenetic resolution of the sequence data. Thus, both the composition and resolution of this infraorder are much more dubious than that of its presumed sister taxon Cephalobomorpha + Tylenchomorpha. Within Panagrolaimomorpha, we propose a grouping of steinernematids and strongyloidids in a superfamily Strongyloidoidea, because these taxa share didelphy, zooparasitism/zoopathogenicity and presence of a discrete Dauer stage in the life cycle as putative synapomorphies within Tylenchina. We also include Rhabdiasidae within this group, based on sequence analysis of *Rhabdias bufonis* (Blaxter and Dorris, unpublished) and shared characters as above. However, the position of *Steinernema* does not resolve clearly with SSU sequence data, and the above character polarisations are therefore tentative.

– INFRAORDER CEPHALOBOMORPHA: All cephalobid SSU rDNA sequences determined so far (Blaxter *et al.*, 1998; Goldstein, Frisse and Thomas, 1998; Félix *et al.*, 2000; Vanfleteren and De Ley, unpublished data) are derived from within the family Cephalobidae. This taxon is both morphologically and genetically homogeneous compared to other families within the superfamily Cephaloboidea, which have not yet been included in molecular studies. It is therefore possible that one or more of these will later turn out to be basal taxa within Tylenchina.

– INFRAORDER TYLENCHOMORPHA: Sequence analysis within this taxon is as yet largely restricted to the superfamily Tylenchoidea *sensu* Maggenti *et al.* (1987). It is evident from these preliminary analyses that the families Belonolaimidae and Pratylenchidae are not demonstrably monophyletic, and that cyst nematodes are closely related to Hoplolaimidae while excluding root-knot nematodes (Szalanski, Adams and Powers, 1997; De Ley *et al.*, unpublished). Considering the morphological support for an affiliation of cyst nematodes with Hoplolaimidae and root-knot nematodes with Pratylenchidae (Geraert, 1997), we consider it appropriate to include Heteroderinae as a subfamily within Hoplolaimidae, and conversely to classify Meloidogininae as a fully separate family.

Blaxter *et al.* (1998) included *Aphelenchus avenae* and *Bursaphelenchus* sp. in their analysis, and found conflicting results in terms of resolution of the aphelenchs: *Aphelenchus* was consistently placed as sister taxon to all tylenchs, while *Bursaphelenchus* was usually placed at the base of the clade named here as Panagrolaimomorpha. Additional tylench or panagrolaim sequences do not change this pattern (unpublished data), and our attempts to sequence more Aphelenchoidoidea *sensu* Hunt (1992) have failed so far. We assume that the positioning of *Bursaphelenchus* sp. in Blaxter *et al.* (1998) is an artefact of branch length and/or elevated AT-contents, and that it is closer to *Aphelenchus avenae* than suggested by current SSU data.

Both genera are traditionally placed in separate superfamilies (resp. Aphelenchoidoidea and Aphelenchoidea) within a suborder Aphelenchina or an order Aphelenchida, which is characterised i.a. by aspects of the median bulb structure that are undoubtedly apomorphic (most notably: the median bulb contains the dorsal gland opening). We maintain this clade, but as a superfamily instead of a (sub)order, and comprising of only two families: Aphelenchidae (presumed autapomorphy: males with bursa with rays) and Aphelenchoididae (presumed autapomorphy: isthmus strongly or completely reduced). For the sake of overall rank balancing, the families previously classified within Aphelenchoidoidea *sensu* Hunt (1992) are better classified as subfamilies within Aphelenchoididae *sensu nobis*, and previously recognised subfamilies (Hunt, 1992) can instead be ranked as tribes.

– INFRAORDER DRILONEMATOMORPHA: Drilonematids share the presence of an offset spermatheca with Cephaloboidea, and some genera are particularly similar to the cephalob family Osstellidae in pharyngeal and stomatal characters. Neither drilonematids nor osstellids have as yet been included in molecular studies, and the precise position of both remains unclear. For the time being, we therefore classify the former Drilonematida as an infraorder within Tylenchina, on the as yet unverified assumption that they constitute a sister taxon to Cephalobomorpha or to Cephalobomorpha + Tylenchomorpha.

– **Suborder Rhabditina:** Compared to other changes proposed in the classification of taxa formerly grouped in Secernentea, the suborder Rhabditina more or less retains the composition allocated to it by Andrássy (1984), i.e. it includes diplogasterids (contrary to Maggenti, 1983 and Inglis, 1983) and bunonematids (contrary to Inglis, 1983). However, a minor change is the transfer of Alloionematidae to Panagrolaimomorpha, and a truly major change is the inclusion of all strongylids in Rhabditina. This inclusion of strongylids is not only supported very emphatically by SSU rDNA sequence data (Baldwin *et al.* in Fitch and Thomas, 1997; Blaxter *et al.* 1998; Aleshin *et al.*, 1998a; Sudhaus and Fitch, in press) but also consistent with two morphological character complexes: (*a*) muscular part of stegostom short, consisting of less than four muscle sets in at least the juvenile stages; (*b*) presence of a bursa with rays in the males. Both character complexes are credible synapomorphies when compared to the corresponding conditions in Teratocephalidae as outgroup for the suborders Myolaimina, Tylenchina and Rhabditina *sensu nobis* (De Ley *et al.*, unpublished).

– INFRAORDER BUNONEMATOMORPHA: Bunonematids are enigmatic creatures with a strongly modified symmetry of all external structures. This modified symmetry is distinctly autapomorphic for the taxon, and therefore does not provide clues on their relationship with other Rhabditida. However, males do have one bursal wing with rays, and the extent of muscularisation of the stegostom looks basically similar to that of Rhabditomorpha. Contrary to the unresolved position obtained by Blaxter *et al.* (1998) with a single *Bunonema* SSU rDNA sequence, inclusion of more species results

in placement basally within Rhabditina (Fitch *et al.*, unpublished).

– INFRAORDER DIPLOGASTEROMORPHA: Diplogasterids are extremely diverse in buccal morphology, but only one species has been studied in sufficient detail to resolve the arrangement of the musculature of the stegostom (cf. Baldwin *et al.*, 1997a). This study supports placement of diplogasterids as sister taxon to, or basal taxon within Rhabditina. Also, most species of diplogasterids have an at least partially developed bursa with rays. SSU rDNA sequence analysis supports a basal position of this taxon within Rhabditina (Baldwin *et al.* in Fitch and Thomas, 1997; Blaxter *et al.*, 1998; Sudhaus and Fitch, in press) and the study with most extensive representation of Rhabditina suggests a close affinity between *Rhabditoides* species and diplogasterids (Sudhaus and Fitch, in press). Molecular relationships within diplogasterids are not yet clearly resolved, but preliminary analyses by Luong *et al.* (1999) do not lend particular support to the gradistic classification of Andrássy (1984). For lack of a well-resolved alternative, we maintain for the present most families recognised by Andrássy (1984), while adding the zooparasitic taxa Cephalobiidae and Mehdinematidae (for the latter, see Luong *et al.* 1999).

– INFRAORDER RHABDITOMORPHA: Excluding some *Rhabditoides* species, this infraorder more or less combines the subfamily Rhabditinae *sensu* Sudhaus (1976) or superfamily Rhabditoidea *sensu* Andrássy (1976, 1983, 1984) with Heterorhabditidae and Strongylida. Andrássy (1976) proposed subfamilies within Rhabditidae on the basis of a combination of the number of female reproductive branches and the extent of the bursa on the tail. Considering subsequent developmental and molecular data (Sommer and Sternberg, 1994; Fitch, Bugaj-gaweda and Emmons, 1995; Fitch, 1997), both characters were presumably subject to significant amounts of parallellism within the group. Nevertheless, a clade including modified versions of Andrássy's subfamilies Mesorhabditinae and Peloderinae is supported by SSU rDNA analysis (Sudhaus and Fitch, in press) and is retained here as a separate superfamily Mesorhabditoidea. Its sister taxon is the superfamily Rhabditoidea *sensu nobis*, which is distinguished from Mesorhabditoidea by the autapomorphy of having the male phasmid posterior to all other bursal rays (Kiontke and Sudhaus, in press). As listed above, Rhabditoidea is undoubtedly paraphyletic, since strongylids and heterorhabditids are actually embedded within this taxon, despite their being classified here as a separate superfamily (see below). We resort to this paraphyletic status as a temporary measure pending further phylogenetic resolution within Rhabditoidea, and to avoid further lowering of the rank of the strongylid taxon.

– Superfamily Strongyloidea: As with tylenchs and aphelenchs, balancing ranks across the entire phylum also requires significantly lower ranks for the suprageneric taxa of the former order Strongylida. The entire order must be reallocated to superfamily level, the superfamilies within Strongylida are brought back to family level, and families or subfamilies of previous systems may need to be accorded subfamily and tribe ranks, respectively. This will undoubtedly meet with criticism from specialists of strongylids, but it is a necessary consequence of the derived position of strongylids within the wide radiation of free-living rhabditids. Once again it needs to be emphasized that taxonomic ranks are not proportional to species richness. Instead, they are determined by degrees of nestedness within a given system.

In nematodes, many free-living taxa are very poorly known in terms of character resolution, and current classification of these taxa therefore requires relatively few ranks. By comparison, zooparasitic taxa provide more readily accessible characters, and the nestedness of zooparasitic taxa within radiations of free-living taxa therefore requires more ranks at lower levels for the former than for the latter. The strongylids are particularly well studied (Lichtenfels, 1980; Beveridge and Durette-Desset, 1994; Durette-Desset *et al.*, 1994; Hoberg and Lichtenfels, 1994; Ben Slimane *et al.*, 1996; Durette-Desset *et al.*, 1999), and the richness of characters available for morphological analysis has led to the proposal of strong and apparently robust phylogenetic hypotheses for the group.

Within Strongyloidea there is, in contrast to the morphological diversity, little diversity in SSU rRNA sequence (Zarlenga *et al.*, 1994a,b; Blaxter *et al.*, 1998; Dorris, De Ley and Blaxter, 1999), and thus other genes, particularly the ribosomal cistron internal transcribed spacer region (ITS) are being used to reveal phylogenetic patterns (Campbell, Gasser and Chilton, 1995; Chilton, Gasser and Beveridge, 1995, 1997; Chilton, Beveridge and Andrews, 1997; Chilton *et al.*, 1997, 1998; Hoste *et al.*, 1995; Stevenson, Chilton and Gasser, 1995; Stevenson, Gasser and Chilton, 1996; Hung *et al.*, 1996, 1997; Romstad *et al.*, 1997a,b, 1998; Gasser *et al.*, 1998; Monti *et al.*, 1998; Newton *et al.*, 1998). For clarity's sake, we should specify that we do not consider measures of SSU rRNA sequence divergence relevant to decisions on appropriate ranking of taxa: our lowering of the strongylid taxon to superfamily level is *only* based on nestedness of the obtained phylogenetic tree, and *not* on the uniformity of SSU rDNA sequences among strongylids.

The Lower Classification

Species Concepts in Nematodes

Typology and trade-offs

Because of the medical, veterinary or agricultural importance of many parasitic nematodes, one of the main tasks of nematode taxonomists is the identification of potentially harmful species. In practice, such identifications used to rely largely on light microscopical observation of fixed specimens, and were often based on characters that are barely visible at the limits of optical resolution, are prone to individual variation and fixation artefacts, and which often

require subjective interpretation skills based on years of experience. Furthermore, it was realised that autotokous reproduction occurs widely in nematodes, and that different reproductive strategies can occur in related species, or even within one species (Triantaphyllou and Hirschmann, 1964 and 1980; Poinar and Hansen, 1983). As a result, decisions on validity and diagnosis of nematode species have traditionally been constrained to largely typological methods, and nematode taxonomists have rarely been able or even willing to participate in theoretical debates about species concepts, or in experimental studies of population dynamics, dispersal and vicariance (but see Ferris, 1983).

Instead, debates about taxonomy of nematode species usually focus mainly on issues of splitting vs. lumping of populations into species, and of species into higher taxa. Arguments typically revolve around questions such as the number and extent of differences required for considering two populations as separate species, preferable size versus ease of use of genera and families, diagnostic relevance of variability studies, etc. Those concerned with accurate characterisation of individual species are often required to devote all available time to disentangling overlapping patterns of variability (Anderson and Hooper, 1970; Fortuner and Quénéhervé, 1980) and chasing the relevant type material (De Ley, Siddiqi and Boström, 1993; Karssen and Van Hoenselaar, 1998) while those who dare to attempt constructing larger overviews (Andrássy, 1984; Siddiqi, 1985; Jairajpuri and Ahmad, 1992) must of necessity trade specific detail for taxonomic scope, resulting in typological keys and classifications bearing little connection with actual patterns of relatedness and variation.

Morphology does not suffice — even for morphospecies

At present, there is ample evidence to suggest that light microscopical evidence alone simply cannot provide the diagnostic resolution required for consistent species identification in many of the larger genera and families of nematodes. On the one hand, it is known (but not necessarily acknowledged) that individual species may vary considerably, to such an extent that one species may include character combinations supposedly differentiating between multiple species or even genera (Anderson and Hooper, 1970; Fortuner and Quénéhervé, 1980). On the other hand, few or no morphological differences may occur between nematode populations with clearly different reproductive, developmental and/or molecular characters (Sudhaus, 1978; Butler *et al.*, 1981; Sommer *et al.*, 1996; Karssen, 1996; De Ley *et al.*, 1999).

Furthermore, intraspecific polymorphism in diagnostically important characters occurs in certain groups of nematodes, and offspring of heteromorphic parents can display character states different from either parent (Hirschmann, 1950). Finally, and perhaps most frustratingly, some of the commonest nematoda genera contain large numbers of poorly known nominal species, of which the only available data are those found in the original description, which often lacks many characters that are considered essential by present standards. For instance, Hunt (1993) listed 138 'valid' species for the genus *Aphelenchoides*, but immediately cautioned that many are inadequately characterized for reliable recognition. And in a similar vein, the rapidly deteriorating situation in *Xiphinema* led Heyns (1983) to comment that, paradoxically, morphological characters are insufficient to distinguish morphospecies in the *X. americanum* group!

In view of the already abundant methodological complications of defining species, very little is known about the frequency of hybridisation in nematodes. However, it has been suggested that hybridisation has occurred naturally in the genus *Meloidogyne* (Triantaphyllou, 1985), and evidence was found recently for at least two independent hybrid lineages (Hugall, Stanton and Moritz, 1999): conflicting patterns of sequence diversity in the Internal Transcribed Spacer (ITS) region and in mitochondrial DNA (mtDNA) suggest reticulate origins for certain populations that were identified morphologically and electrophoretically as *M. hapla*, or as one of the species belonging to the *M. incognita/ M.arenaria/M. javanica* complex. While it is at present impossible to evaluate the overall frequency of hybridisation events in nematode phylogeny, we should clearly keep in mind that reticulate evolution probably does indeed occur in nematodes.

Beyond Morphospecies: Integration of Diverse Kinds of Data

New tools

While the light microscope remains the primary instrument for most nematode taxonomists, the past three decades have brought us a panoply of new tools capable of providing many new characters for the diagnosis of species. Some of these require availability of live specimens, while others can be applied to material fixed and preserved by traditional means. Thus, it has finally become possible to make meaningful departures from purely typological approaches, and to adopt and adapt new methods and theoretical frameworks.

In many groups of nematodes, SEM observations have quickly become an essential requirement for accurate interpretation and description of external features, and especially those found on or near the anterior end (Corbett and Clark, 1983; Hirschmann, 1985; Sauer and Annells, 1985; Stewart and Nicholas, 1994; Neuhaus *et al.*, 1997; Hunt and Moore, 1999). Methods have also been developed to study sclerotized internal structures such as stylets (Eisenback, 1993), buccal cavities (Borgonie, Van Driessche and Coomans, 1995) and male reproductive organs (Nguyen and Smart, 1997).

A wide range of molecular and biochemical methods has been adapted for use in nematodes. Because of its apparently uniform structure, the intestine is largely ignored in traditional procedures for description and diagnosis. However, immunofluorescence techniques and vital stains reveal substantial diversity in biochemical properties of intestinal cells of bacterial-feeding

Rhabditida (Borgonie *et al.*, 1995), suggesting that taxonomic applications could be developed with relative ease. Allozyme analysis through electrophoresis is now a common approach to the characterisation of problematic taxa, often allowing separation between morphologically cryptic species (Jagdale, Gordon and Vrain, 1996; Mattiucci *et al.*, 1998; Tastet *et al.*, 1999). PCR-RFLP characterisation of selected loci (usually the ITS region) has been applied extensively (Gasser *et al.*, 1996; Powers *et al.*, 1997), particularly in some of the most challenging taxa such as *Xiphinema* (Vrain, 1993), Heteroderidae (Ferris *et al.*, 1995; Subbotin *et al.*, 1999) and the entomopathogenic genera *Steinernema* and *Heterorhabditis* (Reid, Hominick and Briscoe, 1997). Sequence analysis of selected variable loci provides a potential plethora of diagnostic characters (Thomas *et al.*, 1997), and also allows phylogenetic analysis of at least some of the relationships between analysed species (Adams, Burnell and Powers, 1998; De Ley *et al.*, 1999). Sequence divergence among related species depends greatly on both the locus and species in question, and it is therefore important to choose the right locus for a given taxon. Thus, ribosomal loci such as the ITS region or the D2/D3 expansion segment of the LSU rDNA gene generally work well to distinguish among congeneric species, but in some taxa such as strongylid nematodes it is better to compare more rapidly evolving loci such as the mitochondrial ND4 gene (Anderson, Blouin and Beech, 1998; Blouin *et al.*, 1998)

Some highly specialised tools were developed for embryological studies of *Caenorhabditis elegans*, and these can be applied with relative ease to any other nematode with sufficiently rapid development and transparent embryo. Four-dimensional microscopy can be used to record and analyse development of the entire cell lineage during early development (Schnabel *et al.*, 1997). Microscopy with Nomarski/DIC optics can also yield information on postembryonic characters such as male tail development (Fitch, 1997; Nguyen *et al.*, 1999) and vulva development (Sommer *et al.*, 1999). These studies often reveal substantial differences between closely related species, which are not only directly relevant to taxonomy as new character suites, but also indirectly important in the evaluation of the final morphological characters. For example, within the exceedingly confusing genus *Acrobeloides*, one species displays an autapomorphy with respect to chirality of the second division of the AB blastomere (Félix *et al.*, 1996; De Ley *et al.*, 1999), while another species (or species complex) appears to be autapomorphic with respect to the cell fate of ventral cord cells P9.p and P10.p (Félix *et al.*, 2000). Thus, developmental characters provide a potentially vast source of characters, and this not only in patterns of cell lineages in normal development, but also with respect to artificially induced alterations due to cell ablation, dislocation or other types of manipulations.

New concepts

Not surprisingly, such a wealth of new data also allows for the formulation and application of more sophisticated species concepts. In a recent essay that looks set to become highly influential, Adams (1998) advocated a combination of diagnostic procedures borrowed from the Phylogenetic Species Concept of Rosen (1978), and the theoretical concept of 'largest integrating lineages' adapted from the Evolutionary Species Concept of Simpson (1961). He proposed to use the presence *versus* absence of autapomorphies as a basis for deciding on the status as valid species of related populations. Although Adams did not specifically restrict the relevant source of characters to sequence analysis, it has first been applied to sequence data (Adams, Burnell and Powers, 1998; De Ley *et al.*, 1999). At least a minimal amount of phylogenetic analysis is required to demonstrate whether a given character is autapomorphic or not, and in nematodes sequence data are often more directly amenable to such analysis. Furthermore, molecular phylogenies can also be used to polarise other characters (Fitch, 1997; Goldstein *et al.*, 1998; Nadler and Hudspeth, 1998; Sommer *et al.*, 1999; Félix *et al.*, 2000) and thus to pinpoint non-molecular autapomorphies indirectly. In the coming years, we may expect more (and more extensive) datasets to be constructed for developmental and morphological characters alike, which will undoubtedly allow wider experimentation with phylogeny-based species concepts such as that of Adams.

Population Dynamics, Gene Flow and Genetic Structure of Nematode Species

From diagnosis to dynamics

Almost nothing is known about the movements of mutations and alleles between conspecific populations of free-living or phytoparasitic nematodes. Population dynamics in natural conditions remain unknown even for *Caenorhabditis elegans*, probably the single best-known metazoan organism in terms of DNA sequence, embryology and neurology. Plant parasites have received greater attention (Vrain *et al.*, 1997) but these studies tend to be strictly limited to monitoring of nematode densities. Some data are available on their molecular population biology (Hyman, 1996; Hyman and Whipple, 1996), but no comprehensive studies have as yet been published. This major gap in our knowledge has direct implications for many fundamental assumptions about speciation and other evolutionary processes in nematodes. Until we know how related populations behave through time and space, no sound inferences are possible about e.g. the taxonomic significance of observed differences between those populations. Even at higher levels of classification, it is important to have quantitative and qualitative insights into population dynamics, for example, in order to interpret observed differences in genetic divergence within and between related higher taxa (see pages 2 and 23).

From dynamics to diversity

By comparison, studies of zooparasitic nematodes have already produced interesting precedents for applications throughout the phylum. For example, there is evidence for extensive gene flow between populations of certain nematode trichostrongylid species parasitising domestic animals (see review by Anderson, Blouin and Beech, 1998), suggesting that human transportation of these animals has basically removed all geographical barriers between the parasite populations, and that it encourages rapid spreading of resistance to anthelminthics among these parasites. In contrast, populations of the deer parasite *Mazamastrongylus odocoilei* appear to be more strongly isolated from one another (Blouin *et al.*, 1995).

Furthermore, strong correlations with effective population size and life cycle have been detected. Thus, the low effective infrapopulation sizes of the amphimictic pig parasite *Ascaris suum* may explain why greater genetic differentiation occurs among infrapopulations of this species than in trichostrongylids, despite intensive movement of the hosts by humans (Nadler *et al.*, 1995; Nadler, 1996). And as another example, the alternately hermaphroditic and amphimictic entomopathogen *Heterorhabditis marelatus* displays a combination of low genetic diversity with clearly subdivided population structure (Blouin, Liu and Berry, 1999). In this case, population sizes and gene flow are presumably both quite low, and the species may exemplify dynamics similar to those of specialised microhabitat-colonizers among free-living nematodes.

These studies illustrate that it is now possible — at last — to quantify previously intractable properties of nematode populations, and open up prospects for applications in measurements of dispersal and species turnover. Thus, the coming decades may not only provide us with a factual understanding of population dynamics and genetics, but also with the first objective methods for estimating and comparing nematode species diversity on local, regional and global scales.

Acknowledgements

We are greatly indebted to Pierre Baujard, Michael Blouin, Sven Boström, Marie-Anne Félix, David Fitch, Steve Nadler, Reyes Peña Santiago, Nikolai Petrov, Björn Sohlenius and Walter Sudhaus, for many stimulating comments. While their suggestions and remarks were instrumental in producing this manuscript, we take full responsibility for its contents and failings — no doubt there are more of the latter than our own expertise and experience allowed for. We are also grateful to Wilfrida Decraemer, David Fitch, Manuel Mundo-Ocampo, Steve Nadler and Leonid Rusin for helping us track down and/or translate some important papers.

References

Adams, B.J. (1998) Species concepts and the evolutionary paradigm in modern nematology. *Journal of Nematology*, **30**, 1–21.

Adams, B.J., Burnell, A.M. and Powers, T.O. (1998) A phylogenetic analysis of *Heterorhabditis* (Nemata: Rhabditidae) based on Internal Transcribed Spacer 1 DNA sequence data. *Journal of Nematology*, **30**, 22–39.

Adamson, M.L. (1987) Phylogenetic analysis of the higher classification of the Nematoda. *Canadian Journal of Zoology*, **65**, 1478–1482.

Adoutte, A., Balavoine, G., Lartillot, N. and de Rosa, R. (1999) Animal evolution. The end of the intermediate taxa? *Trends in Genetics*, **15**, 104–108.

Adoutte, A. and Philippe, H. (1993) The major lines of metazoan evolution: summary of traditional evidence and lessons from ribosomal RNA sequence analysis. In *Comparative Molecular Neurobiology* edited by Y. Pichon, pp. 1–30. Basel: Birkhäuser.

Aguinaldo, A.M.A., Turbeville, J. M., Linford, L.S., Rivera, M.C., Garey, J. R., Raff, R.A. and Lake, J. A. (1997) Evidence for a clade of nematodes, arthropods and other moulting animals. *Nature*, 387, 489–493.

Al-Banna, L., Williamson, V. and Gardner, S.L. (1997) Phylogenetic analysis of nematodes of the genus *Pratylenchus* using nuclear 26S rDNA. *Molecular Phylogenetics and Evolution*, 7, 94–102.

Aleshin, V.V., Kedrova, O.S., Milyutina, I.A., Vladychenskaya, N.S. and Petrov, N.B. (1998a) Secondary structure of some elements of 18S rRNA suggests that strongylid and a part of rhabditid nematodes are monophyletic. *FEBS Letters*, 429, 4–8.

Aleshin, V.V., Kedrova, O.S., Milyutina, I.A., Vladychenskaya, N.S. and Petrov, N.B. (1998b) Relationships among nematodes based on the analysis of 18S rRNA gene sequences: molecular evidence for monophyly of chromadorian and secernentian nematodes. *Russian Journal of Nematology*, 6, 175–184.

Aleshin, V.V., Milyutina, I.A., Kedrova, O.S., Vladychenskaya, N.S. and Petrov, N.B. (1998c) Phylogeny of Nematoda and Cephalorhyncha derived from 18S rDNA. *Journal of Molecular Evolution*, 47, 597–605.

Alvarez ,Y., Juste, J., Tabares, E., Garrido-Pertierra, A., Ibanez, C. and Bautista, J.M. (1999) Molecular phylogeny and morphological homoplasy in fruitbats. *Molecular Biology and Evolution*, 16, 1061–1067.

Anderson, R.C. (1984) The origins of zooparasitic nematodes. *Canadian Journal of Zoology*, 62, 317–328.

Anderson, R.C. and Bain, O. (1976) *Keys to the genera of the Order Spirurida part 3. Diplotriaenoidea, Aproctoidea and Filaroidea.* Farnham Royal, UK: Commonwealth Agricultural Bureaux.

Anderson, R.C. and Bain, O. (1982) *Keys to genera of the superfamilies Rhabditoidea, Dioctophymatoidea, Trichinelloidea and Muspiceoidea.* Farnham Royal, UK: Commonwealth Agricultural Bureaux.

Anderson, R.C., Chabaud, A.G. and Willmont, S. (1974) *General introduction (Vol 1 of CIH Keys to the Nematode Parasites of Vertebrates).* London: CAB International.

Anderson, R.V. and Hooper, D.J. (1970) A neotype for *Cephalobus persegnis* Bastian, 1865, redescription of the species, and observations on the variability in taxonomic characters. *Canadian Journal of Zoology*, 48, 457–469.

Anderson, T.J.C., Blouin, M.S. and Beech, R.N. (1998) Population biology of parasitic nematodes: applications of genetic markers. *Advances in Parasitology*, 41, 219–283.

Andrássy, I. (1974) A nematodák evolúciója és reszenderése (The evolution and systematization of Nematoda). *Magyar Tudomanyos Aka demia Biologiai Tudomanyok Osztalyanak Kozlemenyei*, 17, 13–58.

Andrássy, I. (1976) *Evolution as a basis for the systematization of nematodes,* 287 pp. London: Pitman Publishing.

Andrássy, I. (1983) *A taxonomic review of the suborder Rhabditina (Nematoda: Secernentia),* 241 pp. Paris: ORSTOM.

Andrássy, I. (1984) *Klasse Nematoda,* 509 pp. Stuttgart: Gustav Fischer Verlag.

Ax, P. (1996) *Multicellular Animals, Volume 1: A New Approach to the Phylogenetic Order in Nature,* 225 pp. Berlin: Springer.

Baldwin, J.G., Giblin-Davis, R.M., Eddleman, C.D., Williams, D.S., Vida, J.T. and Thomas, W.K. (1997a) The buccal capsule of *Aduncospiculum halicti* (Nemata: Diplogasterina): an ultrastructural and molecular phylogenetic study. *Canadian Journal of Zoology*, 75, 407–423.

Baldwin, J.G., Frisse, L.M., Vida, J.T., Eddleman, C.D. and Thomas, W.K. (1997b) An evolutionary framework for the study of developmental evolution in a set of nematodes related to *Caenorhabditis elegans*. *Molecular Phylogenetics and Evolution*, 8, 249–259.

Barnard, P.C. (1999) *Identifying British insects and arachnids*, 353 pp. Cambridge: Cambridge University Press.

Baylis, H.A. and Daubney, R. (1926) *A synopsis of the families and genera of Nematoda,* 277 pp. London: The British Museum (Natural History).

Ben Slimane, B., Chabaud, A.G. and Durette-Desset, M.-C. (1996) Les nématodes Trichostrongylina parasites d'amphibiens et de reptiles: problèmes taxinomiques, phylétiques et biogéographiques. *Systematic Parasitology*, 35, 179–206.

Beveridge, I. and Durette-Desset, M.C. (1994) Comparative ultrastructure of the cuticle of trichostrongyle nematodes. *International Journal for Parasitology*, 24, 887–98.

Blaxter, M.L., Aslett, M., Daub, J., Guiliano, D. and The Filarial Genome Project (1999) Parasitic helminth genomics. *Parasitology*, 118, S39–S51.

Blaxter, M.L., De Ley, P., Garey, J., Liu, L.X., Scheldeman, P., Vierstraete, A., *et al.* (1998) A molecular evolutionary framework for the phylum Nematoda. *Nature*, 392, 71–75.

Blaxter, M.L., Raghavan, N., Ghosh, I., Guiliano, D., Lu, W., Williams, S.A., *et al.* (1996) Genes expressed in *Brugia malayi* infective third stage larvae. *Molecular and Biochemical Parasitology*, 77, 77–96.

Blouin, M.S., Liu, J. and Berry, R.E. (1999) Life cycle variation and the genetic structure of nematode populations. *Heredity*, 83, 253–259.

Blouin, M.S., Yowell, C.A., Courtney, C.H. and Dame, J.B. (1995) Host movement and the genetic structure of populations of parasitic nematodes. *Genetics*, 141, 1007–1014.

Blouin, M.S., Yowell, C.A., Courtney, C.H. and Dame, J.B. (1998) Substitution bias, rapid saturation, and the use of mtDNA for nematode systematics. *Molecular Biology and Evolution*, 15, 1719–1727.

Borgonie, G., Claeys, M., De Waele, D. and Coomans, A. (1995) *In vivo* and *in vitro* characterization of the intestine of fifteen bacteriophagous nematodes (Nematoda: Rhabditida) *Fundamental and Applied Nematology*, 18, 115–122.

Borgonie, G., Van Driessche, R. and Coomans, A. (1995) Scanning electron microscopy of the outer and inner surface of the buccal cavity of some Mononchida. *Fundamental and Applied Nematology*, 18, 1–10.

Brinkmann, H. and Philippe, H. (1999) Archaea sister group of bacteria? Indications from tree reconstruction artifacts in ancient phylogenies. *Molecular Biology and Evolution*, 16, 817–825.

Burmeister, H. (1837) *Handbuch der Naturgeschichte. 2 Abt.: Zoologie.* pp. 369–858. Berlin.

Butler, M.H., Wall, S.M., Luehrsen, K.R., Fox, G.E. and Hecht, R.M. (1981) Molecular relationships between closely related strains and species of nematodes. *Journal of Molecular Evolution*, 18, 18–23.

Campbell, A.J.D., Gasser, R.B. and Chilton, N.B. (1995) Differences in a ribosomal DNA sequence of *Strongylus* species allows identification of single eggs. *International Journal for Parasitology*, 25, 359–365.

Chabaud, A.G. (1975) *Keys to the genera of the Order Spirurida part 1. Camallanoidea, Dracunculoidea, Gnathostomatoidea, Physalopteroidea, Rictularoidea and Thelazoidea.* pp. 1–27. Farnham Royal, UK: Commonwealth Agricultural Bureaux.

Chabaud, A.G. (1975) *Keys to the genera of the Order Spirurida part 2. Spiruroidea, Habronematoidea and Acuaroidea.* pp. 28–58. Farnham Royal, UK: Commonwealth Agricultural Bureaux.

Chabaud, A.G. (1978) *Keys to genera of the Superfamilies Cosmocercoidea, Seuratoidea, Heterakoidea and Subuluroidea.* pp. 1–71. Farnham Royal, UK: Commonwealth Agricultural Bureaux.

Chabaud, A.G. and Bain, O. (1994) The evolutionary expansion of the Spirurida. *International Journal for Parasitology*, 24, 1179–201.

Chilton, N.B., Beveridge, I. and Andrews, R.H. (1997) An electrophoretic analysis of patterns of speciation in *Cloacina clarkae, C. communis, C. petrogale* and *C. similis* (Nematoda: Strongyloidea) from macropodid marsupials. *International Journal for Parasitology*, 27, 483–93.

Chilton, N.B., Beveridge, I., Hoste, H. and Gasser, R.B. (1997) Evidence for hybridisation between *Paramacropostrongylus iugalis* and *P. typicus* (Nematoda: Strongyloidea) in grey kangaroos, *Macropus fuliginosus* and *M. giganteus*, in a zone of sympatry in eastern Australia. *International Journal for Parasitology*, 27, 475–82.

Chilton, N.B., Gasser, R.B. and Beveridge, I. (1995) Differences in ribosomal DNA sequence of morphologically indistinguishable species within the *Hypodontus macropi* complex (Nematoda: Strongyloidea) *International Journal for Parasitology*, 25, 647–651.

Chilton, N.B., Gasser, R.B. and Beveridge, I. (1997) Phylogenetic relationships of Australian strongyloid nematodes inferred from ribosomal DNA sequence data. *International Journal for Parasitology*, 27, 1481–94.

Chilton, N.B., Hoste, H., Newton, L.A., Beveridge, I. and Gasser, R.B. (1998) Common secondary structures for the second internal transcribed spacer pre-rRNA of two subfamilies of trichostrongylid nematodes. *International Journal for Parasitology*, 28, 1765–73.

Chitwood, B.G. (1937) A revised classification of the Nematoda. In *Papers on helminthology, 30 year jubileum K.J. Skrjabin*, edited by Anon., pp. 67–79. Moscow: All-Union Lenin Academy of Agricultural Sciences.

Chitwood, B.G. (1957) The english word 'Nema' revisited. *Systematic Zoology* 6, 184–186.

Chitwood, B.G. (1958) The designation of official names for higher taxa of invertebrates. *Bulletin of Zoological Nomenclature* 15, 860–895.

Chitwood, B.G. and Chitwood, M.B. (1933) The characters of a protonematode. *Journal of Parasitology*, 20, 130.

Chitwood, B.G. and Chitwood, M.B. (1974) *Introduction to nematology.* Consolidated edition. 334 pp. Baltimore: University Park Press.

Christen, R., Ratto, A., Baroin, A., Perasso, R., Grell, K.G. and Adoutte, A. (1991) An analysis of the origin of metazoans, using comparisons of partial sequences of the 28S RNA, reveals an early emergence of triploblasts. *European Molecular Biology Organisation Journal,* **10,** 499–503.

Cobb, N.A. (1919) The orders and classes of nemas — Contributions to a science of nematology, 8. In *Contributions to a science of nematology, 1914–1935* by N.A.Cobb. pp. 212–216. Baltimore: Waverly Press.

Conway Morris, S. (1981) Parasites and the fossil record. *Parasitology,* **82,** 489–509.

Conway Morris, S. (1993) The fossil record and the early evolution of the Metazoa. *Nature,* **361,** 219–225.

Conway Morris, S. (1994) Why molecular biology needs palaeontology. *Development* Supplement, 1–13.

Conway Morris, S. (1997) Molecular clocks: defusing the Cambrian 'explosion'? *Current Biology,* **7,** R71–R74.

Coomans, A. (1977) Evolution as a basis for the systematization of nematodes — a critical review and exposé. *Nematologica* **23,** 129–136.

Coomans, A. and Raski, D.J. (1988) Two new species of *Prismatolaimus* de Man, 1880 (Nemata: Prismatolaimidae) in southern Chile. *Journal of Nematology,* **20,** 288–303.

Corbett, D.C.M. and Clark, S.A. (1983) Surface features in the taxonomy of *Pratylenchus* species. *Revue de Nématologie* **6,** 85–98.

Costa, M., Weir, M., Coulson, A., Sulston, J. and Kenyon, C. (1988) Posterior pattern formation in *C. elegans* involves position-specific expression of a gene containing a homeobox. *Cell* **55,** 747–756.

Daub, J., Loukas, A., Pritchard, D.I. and Blaxter, M.L. (2000) A survey of genes expressed in adults of the human hookworm, *Necator americanus. Parasitology,* **120,** 171–184.

De Coninck, L.A.P. (1965) Systématique des nématodes. In *Traité de Zoologie: Anatomie, Systématique, Biologie,* 4 edited by P.P. Grassé. pp. 586–731. Paris: Masson et Cie.

De Coninck, L.A.P. and Schuurmans Stekhoven, J.H. (1933) *The freeliving marine nemas of the belgian coast. II.* Verhandelingen van het Koninklijk Natuurhistorisch Museum van België, 58, Brussel, 163 pp.

Decraemer, W. (1995) *The family Trichodoridae: stubby root and virus vector nematodes.* 360 pp. Dordrecht: Kluwer Academic Publishers.

De Ley, P. (2000). Lost in worm space: phylogeny and morphology as road maps to nematode diversity. *Nematology,* **2,** 9–16.

De Ley, P., Siddiqi, M.R. and Boström, S. (1993) A revision of the genus *Pseudacrobeles* Steiner, 1938 (Nematoda, Cephalobidae) 2. Subgenus *Bunobus* subgen. n., problematical species, discussion and key. *Fundamental and Applied Nematology,* **16,** 289–308.

De Ley, P., Félix, M.-A., Frisse, L.M., Nadler, S.A., Sternberg, P.W. and Thomas, W.K. (1999) Molecular and morphological characterisation of two reproductively isolated species with mirror-image anatomy (Nematoda: Cephalobidae). *Nematology,* **6,** 591–612.

de Man, J.G. (1876) Onderzoekingen over vrij in de aarde levende nematoden. *Tijdschrift der Nederlandsche Dierkundige Vereeniging,* **2,** 78–196.

de Rosa, R., Grenier, J.K., Andreeva, T., Cook, C.E., Adoutte, A., Akam, M., Carroll, S.B. and Balavoine, G. (1999) Hox genes in brachiopods and priapulids and protostome evolution. *Nature,* **399,** 772–6.

Dolinski, C., Borgonie, G., Schnabel, R. and Baldwin, J.G. (1998) Buccal capsule development as a consideration for phylogenetic analysis of Rhabditida (Nemata) *Development, Genes and Evolution,* **208,** 495–503.

Doolittle, R.F., Feng, D.-F., Tsang, S., Cho, G. and Little, E. (1996) Determining divergence times of the major kingdoms of living organisms with a protein clock. *Science* **271,** 470–477.

Dorris, M., De Ley, P. and Blaxter, M. (1999) Molecular analysis of nematode diversity. *Parasitology Today* **15,** 188–193.

Durette-Desset, M., Gasser, R.B. and Beveridge, I. (1994) The origins and evolutionary expansion of the Strongylida (Nematoda) *International Journal for Parasitology,* **24,** 1139–1166.

Durette-Desset, M.C., Hugot, J.P., Darlu, P. and Chabaud, A.G. (1999) A cladistic analysis of the Trichostrongyloidea (Nematoda) *International Journal for Parasitology,* **29,** 1065–1086.

Curole, A.P. and Kocher, T.D. (1999) Mitogenomics: digging deeper with complete mitochondrial genomes. *Trends in Ecology and Evolution,* **14,** 394–398.

Eernisse, D.J. (1997) Arthropod and annelid relationships re-examined. In *Arthropod Relationships,* edited by R.A. Fortey and R.H. Thomas. pp. 43–56.

The Systematics Association Special Volume Series 55. London: Chapman and Hall.

Eisenback, J.D. (1993) Morphological comparisons of females, males, and 2nd-stage juveniles of cytological race-a and race-b of *Meloidogyne hapla* Chitwood, 1949. *Fundamental and Applied Nematology,* **16,** 259–271.

Ellis, R.E., Sulston, J.E. and Coulson, A.R. (1986) The rDNA of *C. elegans*: sequence and structure. *Nucleic Acids Research* **14,** 2345–2364.

Fagerholm, H.P. (1991) Systematic implications of male caudal morphology in ascaridoid nematode parasites. *Systematic Parasitology* **19,** 215–228.

Félix, M.-A. (1999) Evolution of developmental mechanisms in nematodes. *Journal of Experimental Zoology (Mol. Dev. Evol.),* **285,** 3–18.

Félix, M.-A., De Ley, P., Sommer, R.J., Frisse, L., Nadler, S.A., Thomas, W.K., Vanfleteren, J. and Sternberg, P.W. (2000) Evolution of vulva development in the Cephalobina (Nematoda). *Developmental Biology,* **221,** 68–86.

Félix, M.-A., Sternberg, P. and De Ley, P. (1996) Sinistral nematode population. *Nature,* **381,** 122.

Feng, D.-F., Cho, G. and Doolittle, R.F. (1997) Determining divergence times with a protein clock: Update and reevaluation. *Proceedings of the National Academy of Sciences, USA,* **94,** 13028–13033.

Ferris, V.R. (1983) Phylogeny, historical biogeography and the species concept in soil nematodes. In *Concepts in nematode systematics,* edited by A.R. Stone, H.M. Platt, and L.F. Khalil. pp. 143–161. The Systematics Association special volume no. 22. London: Academic Press.

Ferris, V.R. (1994) The future of nematode systematics. *Fundamental and Applied Nematology,* **17,** 97–101.

Ferris V.R., Miller, L.I., Faghihi, J. and Ferris, J.M. (1995) Ribosomal DNA comparisons of *Globodera* from 2 continents. *Journal of Nematology,* **27,** 273–283.

Field, K.G., Olsen, G.J., Lane, D.J., Giovannoni, S.J., Ghiselin, M.T., Raff, E.C. *et al.* (1988) Molecular phylogeny of the animal kingdom. *Science* **239,** 748–753.

Filipjev, I.N. (1929) Classification of freeliving Nematoda and their relations to parasitic forms. *Journal of Parasitology,* **15,** 281–282.

Filipjev, I.N. (1934) *The classification of the freeliving nematodes and their relation to the parasitic nematodes.* Smithsonian Miscellanous Collections **89** (6), 63 pp.

Fitch, D.H.A. (1997) Evolution of male tail development in rhabditid nematodes related to *Caenorhabditis elegans. Systematic Biology,* **46,** 145–179.

Fitch, D.H.A., Bugaj-gaweda, B. and Emmons, S.W. (1995) 18S ribosomal gene phylogeny for some Rhabditidae related to *Caenorhabditis elegans. Molecular Biology and Evolution,* **12,** 346–358.

Fitch, D.H.A. and Thomas, W.K. (1997) Evolution. In *C. elegans II,* edited by D.L. Riddle, T. Blumenthal, B.J. Meyer and J.R. Priess, pp. 815–850. Cold Spring Harbor, New York: Cold Spring Harbor Laboratory Press.

Foot, M., Gould, S.J., Lee, M.S.Y., Briggs, D.E.G., Fortey, R.A. and Wills, M.A. (1992) Cambrian and recent morphological diversity. *Science,* **258,** 1816–1818.

Fortuner, R. and Quénéhervé, P. (1980) Morphometrical variability in *Helicotylenchus* Steiner, 1954. 2: Influence of the host on *H. dihystera* (Cobb, 1893) Sher, 1961. *Revue de Nématologie,* **3,** 291–296.

Foster, P.G. and Hickey, D.A.(1999) Compositional bias may affect both DNA-based and protein-based phylogenetic reconstructions. *Journal of Molecular Evolution,* **48,** 284–290.

Gadéa, E. (1973) Sobre la filogenia interna de los Nematodos. *Publicaciones del Instituto de Biología Aplicada,* **54,** 87–92.

Garey, J.R. and Schmidt-Rhaesa, A. (1998) The essential role of 'minor' phyla in molecular studies of animal evolution. *American Zoologist,* **38,** 907–917.

Gasser, R.B., Monti, J.R., Bao-Zhen, Q., Polderman, A.M., Nansen, P. and Chilton, N.B. (1998) A mutation scanning approach for the identification of hookworm species and analysis of population variation. *Molecular and Biochemical Parasitology,* **92,** 303–12.

Gasser, R.B., Stevenson, L.A., Chilton, N.B., Nansen, P., Bucknell, D.G. and Beveridge, I. (1996) Species markers for equine strongyles detected in intergenic rDNA by PCR-RFLP. *Molecular and Cellular Probes,* **10,** 371–378.

Gauld, I. and Bolton, B. (1988) *The Hymenoptera.* 332 pp. Oxford: Oxford University Press.

Geraert, E. (1997) Comparison of the head patterns in the Tylenchoidea (Nematoda). *Nematologica,* **43,** 283–294.

Goldstein, B., Frisse, L.M. and Thomas, W.K. (1998) Embryonic axis specification in nematodes: evolution of the first step in development. *Current Biology,* **8,** 157–160.

Goodey, J.B. (1963) *Soil and Freshwater Nematodes.* Second edition. 544 pp. London: Methuen.

Gould, S.J. (1989) *Wonderful Life. The Burgess shale and the nature of history*. London: Hutchinson Radius.

Grobben, K. (1908) Die systematische Einteilung des Tierreiches. *Verhandlungen K.K. Zoologischer Botanischer Gesellschaft in Wien*, **58**, 491–511.

Guigo, R., Muchnik, I. and Smith, T.F. (1996) Reconstruction of ancient molecular phylogeny. *Molecular Phylogenetics and Evolution*, **6**, 189–213.

Halanych, K.M., Bacheller, J.D., Aguinaldo, A.M.A., Liva, S.M., Mills, D.H. and Lake, J.A. (1995) Evidence from 18S ribosomal DNA that the lophophorates are protostome animals. *Science*, **267**, 1641–1643.

Harbach, R.E. and Kitching, I.J. (1998) Phylogeny and classification of the Culicidae (Diptera). *Systematic Entomology*, **23**, 327–370.

Hartwich, G. (1974) *Keys to genera of the Ascaridoidea*. Farnham Royal, UK: Commonwealth Agricultural Bureaux.

Hasegawa, M. and Hashimoto, T. (1993) Ribosomal-RNA trees misleading? *Nature*, **361**, 23.

Heyns, J. Problems of species delimitation in the genus *Xiphinema*, with special reference to monosexual species. In *Concepts in nematode systematics*, edited by A.R. Stone, H.M. Platt, and L.F. Khalil. pp. 163–174. The Systematics Association special volume no. 22. London: Academic Press.

Hillis, D.M. and Bull, J.J. (1993) An empirical test of bootstrapping as a method for assessing confidence in phylogenetic analysis. *Systematic Biology*, **42**, 182–192.

Hirschmann, H. (1950) Über das vorkommen zweier Mundhöhletypen bei *Diplogaster lheritieri* Maupas und *Diplogaster biformis* n.sp. und die Entstehung dieser hermaphroditischen Art aus *Diplogaster lheritieri*. *Zoologischer Jahrbücher (Systematik)*, **80**, 132–170.

Hirschmann, H. (1985) The genus *Meloidogyne* and morphological characters differentiating its species. In *An advanced treatise on Meloidogyne, Vol. 1.Biology and Control*, edited by J.N. Sasser and C.C. Carter. pp. 79–93. Raleigh: North Carolina State University/ USAID.

Hoberg, E.P. and Lichtenfels, J.R. (1994) Phylogenetic systematic analysis of the Trichostrongylidae (Nematoda), with an initial assessment of coevolution and biogeography. *Journal of Parasitology*, **80**, 976–996.

Hoot, S.B., Magallon, S. and Crane, P.R. (1999) Phylogeny of basal eudicots based on three molecular data sets: *atpB*, *rbcL*, and 18S nuclear ribosomal DNA sequences. *Annals of the Missouri Botanical Garden*, **86**, 1–32.

Hoste, H., Chilton, N.B., Gasser, R.B. and Beveridge, I. (1995) Differences in the second internal transcribed spacer (ribosomal DNA) between five species of *Trichostrongylus* (Nematoda: Trichostrongylidae). *International Journal for Parasitology*, **25**, 75–80.

Hugall, A., Stanton, J. and Moritz, C. (1999) Reticulate evolution and the origins of ribosomal Internal Transcribed Spacer diversity in apomictic *Meloidogyne*. *Molecular Biology and Evolution*, **16**, 157–164.

Hung, G.-C., Jacobs, D.E., Krecek, R.C., Gasser, R.B. and Chilton, N.B. (1996) *Strongylus asini* (Nematoda: Strongyloidea): genetic relationships with other *Strongylus* species determined by ribosomal DNA. *International Journal for Parasitology*, **26**, 1407–1411.

Hung, G.C., Chilton, N.B., Beveridge, I., McDonnell, A., Lichtenfels, J.R. and Gasser, R.B. (1997) Molecular delineation of Cylicocyclus nassatus and C. ashworthi (Nematoda:Strongylidae). *International Journal for Parasitology*, **27**, 601–5.

Hunt, D.J. (1993) *Aphelenchida, Longidoridae and Trichodoridae: their systematics and bionomics*. 352 pp. Wallingford: CAB International.

Hunt, D.J. (1996) A synopsis of the Rhigonematidae (Nematoda), with an outline classification of the Rhigonematida. *Afro-Asian Journal of Nematology*, **6**, 137–150.

Hunt, D.J. and Moore, D. Rhigonematida from New Britain diplopods. 2. The genera *Rhigonema* Cobb, 1898 and *Zalophora* Hunt, 1994 (Rhigonematoidea: Rhigonematidae) with descriptions of three new species. *Nematology*, **1**, 225–242.

Hyman, B.C. (1996) Molecular systematics and population biology of phytonematodes: some unifying principles. *Fundamental and Applied Nematology*, **4**, 309–313.

Hyman, B.C. and Whipple, L.E. (1996) Application of mitochondrial DNA polymorphism to *Meloidogyne* molecular population biology. *Journal of Nematology*, **28**, 268–276.

Hyman, L.H. (1951) The invertebrates. Vol. III: Acanthocephala, Aschelminthes and Entoprocta — The pseudocoelomate Bilateria. 572 pp. New York: McGraw-Hill.

Inglis, W.G. (1964) The marine Enoplida (Nematoda): a comparative study of the head. *Bulletin of the British Museum (Natural History)*, **11**, 263–376.

Inglis, W.G. (1983) An outline classification of the phylum Nematoda. *Australian Journal of Zoology*, **31**, 243–255.

ICZN — International Commission on Zoological Nomenclature (1999) *International Code of Zoological Nomenclature*. 4th edition. 336 pp. London: International Trust for Zoological Nomenclature.

Jagdale, G.B., Gordon, R. and Vrain, T.C. (1996) Use of cellulose acetate electrophoresis in the taxonomy of steinernematids (Rhabditida, Nematoda). *Journal of Nematology*, **28**, 301–309.

Jairajpuri, M.S. and Ahmad, W. (1992) *Dorylaimida — Free-living, predaceous and plant-parasitic nematodes*. 458 pp. Leiden: Brill.

Johnston, D.A., Blaxter, M.L., Degrave, W.M., Foster, J., Ivens, A.C. and Melville, S.E. (1999) Genomics and the biology of parasites. *BioEssays*, **21**, 131–147.

Kampfer, S., Sturmbauer, C. and Ott, J.A. (1998) Phylogenetic analysis of rDNA sequences from adenophorean nematodes and implications for the Adenophorea-Secernentea controversy. *Invertebrate Biology*, **117**, 29–36.

Karegar, A., De Ley, P. and Geraert, E. (1997) A detailed morphological study of *Acromoldavicus skrjabini* (Nesterov and Lisetskaya, 1965) Nesterov, 1970 (Nematoda:Cephaloboidea) from Iran and Spain. *Fundamental and Applied Nematology*, **20**, 277–283.

Karssen, G. (1996) Description of *Meloidogyne fallax* n. sp. (Nematoda: Heteroderidae), a root-knot nematode from the Netherlands. *Fundamental and Applied Nematology*, **19**, 593–599.

Karssen, G. and Van Hoenselaar, T. (1998) Revision of the genus *Meloidogyne Goldi*, 1892 (Nematoda : Heteroderidae) in Europe. *Nematologica*, **44**, 713–788.

Kenyon, C. (1994) If birds can fly, why can't we? Homeotic genes and evolution. *Cell*, **78**, 175–180.

Kiontke, K. and Sudhaus, W. (in press) Phasmids in male Rhabditida and other secernentean nematodes. *Journal of Nematode Morphology and Systematics*, — .

Lake, J.A. (1990) Origin of the Metazoa. *Proceedings of the National Academy of Sciences USA*, **86**, 763–766.

Lang, B.F., Seif, E., Gray, M.W., O'Kelly, C.J. and Burger, G. (1999) Comparative genomics approach to the evolution of eukaryotes and their mitochondria. *Journal of Eukaryotic Microbiology*, **46**, 320–326.

Lichtenfels, J.R. (1980) *Keys to the genera of the Superfamilies Ancylostomatoidea and Diaphanocephaloidea*. Farnham Royal, UK: Commonwealth Agricultural Bureaux.

Lorenzen, S. (1981) Entwurf eines phylogenetischen Systems der freilebenden Nematoden. *Feröffentlichungen des Institut für Meeresforschungen Bremerhaven, Supplement*, **7**, 472 pp.

Lorenzen, S. (1983) Phylogenetic systematics: problems, achievements and its application to the Nematoda. In *Concepts in nematode systematics*, edited by A.R. Stone, H.M. Platt, and L.F. Khalil. pp. 11–23. The Systematics Association special volume no. 22. London: Academic Press.

Lorenzen, S. (1994) *The phylogenetic systematics of freeliving nematodes*. 383 pp. London: The Ray Society.

Luong, L.T., Platzer, E.G., De Ley, P. and Thomas, W.K. (1999) Morphological, molecular and biological characterization of *Mehdinema alii* (Nematoda: Diplogasterida) from the decorated cricket (*Gryllodes sigillatus*). *Journal of Parasitology*, **85**, 1053–1064.

Maggenti, A.R. (1963) Comparative morphology in nemic phylogeny. In *The lower Metazoa, comparative biology and phylogeny*, edited by E.C. Dougherty. pp. 273–282. Berkeley: University of California Press.

Maggenti, A.R. (1983) Nematode higher classification as influenced by species and family concepts. In *Concepts in nematode systematics*, edited by A.R. Stone, H.M. Platt, and L.F. Khalil. pp. 25–40. The Systematics Association special volume no. 22. London: Academic Press.

Maggenti, A.R. (1991) Nemata: higher classification. In *Manual of Agricultural Nematology*, edited by W.R. Nickle. pp. 147–187. New York: Marcel Dekker Inc.

Maggenti, A.R., Luc, M., Raski, D.J., Fortuner, R. and Geraert, E. (1987) A reappraisal of Tylenchina (Nemata) 2. Classification of the suborder Tylenchina (Nemata: Diplogasteria). *Revue de Nématologie* **10**, 135–142.

Malakhov, V.V., Ryzhikov, K. and Sonin, M. (1982) The system of large taxa of nematodes: subclasses, orders, suborders. *Zoologicheskij Zhurnal*, **61**, 1125–1134.

Malakhov, V.V. (1994) *Nematodes — Structure, development, classification and phylogeny*. 286 pp. Washington: Smithsonian Institution Press.

Mallat, J. and Sullivan, J. (1998) 28S and 18S rDNA sequences support the monophyly of lampreys and hagfishes. *Molecular Biology and Evolution*, **15**, 1706–1718.

Manum, S.B., Bose, M.N., Sayer, R.T. and Boström, S. (1994) A nematode (*Captivonema cretacea* gen. et sp.n.) preserved in a clitellate cocoon wall from the early cretaceous. *Zoologica Scripta*, 23, 27–31.

Marcinowski, K. (1909) Parasitisch und semiparasitisch Pflanzen lebende Nematoden. *Arbeiten aus der Kaiserlichen Biologischen Anstalt für Land- und Forstwirtschaft*, 7, 1–192.

Mattiucci, S., Paggi, L., Nascetti, G., Ishikura, H., Kikuchi, K., Sato, N. *et al.* (1998) Allozyme and morphological identification of *Anisakis*, *Contracaecum* and *Pseudoterranova* from Japanese waters (Nematoda, Ascaridoidea). *Systematic Parasitology*, 40, 81–92.

McHugh, D. (1997) Molecular evidence that echiurans and pogonophorans are derived annelids. *Proceedings of the National Academy of Sciences, USA*, 94, 8006–8009.

Micoletzky, H. (1922) Die freilebenden Erd-Nematoden, mit besonderer Berücksichtigung der Steiermark und der Bukowina, zugleich mit einer Revision sämtlicher nicht mariner, freilebender Nematoden in Form von Genus-Beschreibungen und Bestimmungsschlüsseln. *Archiv für Naturgeschichte, Abteilung A*, 87, 1–650.

Monti, J.R., Chilton, N.B., Qian, B.Z. and Gasser, R.B. (1998) Specific amplification of *Necator americanus* or *Ancylostoma duodenale* DNA by PCR using markers in ITS-1 rDNA, and its implications. *Molecular Cell Probes*, 12, 71–8.

Moritz, C. and Hillis, D.M. (1996) Molecular systematics: context and controversies. In *Molecular systematics*, edited by D.M. Hillis, C. Moritz, and B.K. Mable, pp. 1–13. Sunderland: Sinauer Associates.

Mushegian, A.R., Garey, J.R., Martin, J. and Liu, L.X. (1998) Large-scale taxonomic profiling of eukaryotic model organisms: a comparison of orthologous proteins encoded by the human, fly, nematode and yeast genomes. *Genome Research*, 8, 590–598.

Nadler, S.A. (1995) Advantages and disadvantages of molecular phylogenetics: a case study of ascaridoid nematodes. *Journal of Nematology*, 27, 423–432.

Nadler, S. (1996) Microevolutionary patterns and molecular markers: the genetics of geographic variation in *Ascaris suum*. *Journal of Nematology*, 28, 277–285.

Nadler, S. and Hudspeth, D.S.S. (1998) Ribosomal DNA and phylogeny of the Ascaridoidea (Nemata: Secernentea): implications for morphological evolution and classification. *Molecular Phylogenetics and Evolution*, 10, 221–236.

Nadler, S., Lindquist, R.L. and Near, T.J. (1995) Genetic structure of midwestern *Ascaris suum* populations: a comparison of isoenzyme and RAPD markers. *Journal of Parasitology*, 81, 385–394.

Neuhaus, B., Bresciani, J., Christensen, C.M. and Sommer, C. (1997) Morphological variation of the corona radiata in *Oesophagostomum dentatum*, *O. quadrispinulatum*, and *O. radiatum* (Nematoda: Strongyloidea) *Journal of the Helminthological Society of Washington*, 64, 128–136.

Newton, L.A., Chilton, N.B., Beveridge, I. and Gasser, R.B. (1998) Systematic relationships of some members of the genera *Oesophagostomum* and *Chabertia* (Nematoda : Chabertiidae) based on ribosomal DNA sequence data. *International Journal for Parasitology*, 28, 1781–1789.

Nielsen, C. (1995) *Animal evolution. Interrelationships of the living phyla.* Oxford: Oxford University Press.

Nguyen, C.Q., Hall, D.H., Yang, Y., Fitch, D.H.A. (1999) Morphogenesis of the *Caenorhabditis elegans* male tail tip. *Developmental Biology*, 207, 86–106.

Nguyen, K.B. and Smart, G.C. Jr. (1997) Scanning electron microscope studies of spicules and gubernacula of *Steinernema* spp. (Nemata: Steinernematidae) *Nematologica*, 43, 465–480.

Nielsen, C. (1995) *Animal evolution — interrelationships of the living phyla.* 467 pp. Oxford: University Press.

O'Grady, P.M. (1999) Reevaluation of phylogeny in the *Drosophila obscura* species group based on combined analysis of nucleotide sequences. *Molecular Phylogenetics and Evolution*, 12, 124–139.

Patterson, C., Williams, D.M. and Humphries, C.J. (1993) Congruence between molecular and morphological phylogenies. *Annual Review of Ecology and Systematics*, 24, 153–188.

Pearse, A.S. (1936) *Zoological names, a list of phyla, classes and orders, Section F.* 24 pp. Durham: American Association for the Advancement of Science, Duke University Press.

Pearse, A.S. (1942) *An introduction to parasitology.* 375 pp. Springfield: Charles C. Thomas.

Perrier, E. (1897) *Traité de Zoologie, Fascicule 4.* pp. 1345–2136. Paris: Savy.

Petter, A.J. (1976) *Keys to the genera of the Oxyuroidea.* pp. 1–30. Farnham Royal, UK: Commonwealth Agricultural Bureaux.

Philippe, H., Chenuil, A. and Adoutte, A. (1994) Can the cambrian explosion be inferred through molecular phylogeny? *Development* Supplement, 15–25.

Poinar, G.O. (1992) *Life in amber.* 350 pp. Stanford: Stanford University Press.

Poinar, G.O., Acra, A. and Acra, F. (1994) Earliest fossil nematode (Mermithidae) in cretaceous Lebanese amber. *Fundamental And Applied Nematology*, 17, 475–477.

Poinar, G.O. Jr. and Hansen, E. (1983) Sex and reproductive modifications in nematodes. *Helminthological Abstracts — Series B*, 52, 145–163.

Powers, T.O., Todd, T.C., Burnell, A.M., Murray, P.C.B., Fleming, C.C., Szalanski, A.L., Adams, B.A. and Harris, T.S. (1997) The rDNA Internal Transcribed Spacer region as a taxonomic marker for nematodes. *Journal of Nematology*, 29, 441–450.

Putland, R.A., Thomas, S.M., Grove, D.I. and Johnson, A.M. (1993) Analysis of the 18S ribosomal RNA gene of *Strongyloides stercoralis*. *International Journal of Parasitology*, 23, 149–51.

Qiu, Y.L. and Palmer, J.D. (1999) Phylogeny of early land plants: insights from genes and genomes. *Trends in Plant Science*, 4, 26–30.

de Quieroz, K. and Gauthier, J. (1992) Phylogenetic taxonomy. *Annual Review of Ecology and Systematics*, 23, 449–480.

de Quieroz, K. and Gauthier, J. (1994) Toward a phylogenetic system of biological nomenclature. *Trends in Ecology and Evolution*, 9, 27–31.

Raff, R.A., Marshall, C.R. and Turbeville, J.M. (1994) Using DNA sequences to unravel the Cambrian radiation of the animal phyla. *Annual Reviews of Ecology and Systematics*, 25, 351–375.

Reid, A.P., Hominick, W.M. and Briscoe, B.R. (1997) Molecular taxonomy and phylogeny of entomopathogenic nematode species (Rhabditida: Steinernematidae) by RFLP analysis of the ITS region of the ribosomal DNA repeat unit. *Systematic Parasitology*, 37, 187–193.

Riemann, F. (1972). *Kinonchulus sattleri* n.g., n.sp. (Enoplida, Tripyloidea), an aberrant free-living nematode from the lower Amazonas. *Veroffentlichungen der Institut für Meeresforschungen Bremerhaven*, 13, 317–326.

Romstad, A., Gasser, R.B., Monti, J.R., Polderman, A.M., Nansen, P., Pit, D.S. and Chilton, N.B. (1997a) Differentiation of *Oesophagostomum bifurcum* from *Necator americanus* by PCR using genetic markers in spacer ribosomal DNA. *Molecular Cell Probes*, 11, 169–76.

Romstad, A., Gasser, R.B., Nansen, P., Polderman, A.M. and Chilton, N.B. (1998) *Necator americanus* (Nematoda: Ancylostomatidae) from Africa and Malaysia have different ITS-2 rDNA sequences. *International Journal for Parasitology*, 28, 611–5.

Romstad, A., Gasser, R.B., Nansen, P., Polderman, A.M., Monti, J.R. and Chilton, N.B. (1997b) Characterization of *Oesophagostomum bifurcum* and *Necator americanus* by PCR-RFLP of rDNA. *Journal of Parasitology*, 83, 963–6.

Rudolphi, C.A. (1808) *Entozoorum sive vermium intestinalium historia naturalis, vol. 1.* 527 pp. Amstelaedami.

Salser, S.J. and Kenyon, C. (1994) Patterning *C. elegans*: homeotic cluster genes, cell fates and cell migrations. *Trends in Genetics*, 10, 159–164.

Sauer, M.R. and Annells, C.M. (1985) Lip region structure in Acrobelinae (Nematoda: Cephalobidae) *Nematologica*, 30, 140–150.

Schmidt, G.D., Roberts, L.S. and Janovy, J.Jr. (1995) *Foundations of Parasitology.* pp. . Boston: McGraw Hill.

Schmidt-Rhaesa, A., Bartolomaeus, T., Lemburg, C., Ehlers, U. and Garey, J. (1998) The position of the Arthropoda in the phylogenetic system. *Journal of Morphology*, 238, 263–285.

Schnabel, R., Hutter, H., Moerman, D. and Schnabel, H (1997) Assessing normal embryogenesis in *Caenorhabditis elegans* using a 4D microscope: Variability of development and regional specification. *Developmental Biology*, 184, 234–265.

Schneider, A. (1866) *Monographie der Nematoden.* 357 pp. Berlin: Georg Reimer.

Seurat, L.G. (1920) *Histoire naturelle des nématodes de la Berbérie. Première partie. Morphologie, développement, éthologie et affinités des nématodes.* 221 pp. Alger.

Siddiqi, M.R. (1974) Systematics of the genus *Trichodorus* Cobb, 1913 (Nematoda: Dorylaimida) with descriptions of three new species. *Nematologica*, 9, 259–278.

Siddiqi, M.R. (1980) The origin and phylogeny of the nematode orders Tylenchida Thorne, 1949 and Aphelenchida, n. ord. *Helminthological Abstracts — Series B*, 49, 143–170.

Siddiqi, M.R. (1983a) Evolution of plant parasitism in nematodes. In *Concepts in nematode systematics*, edited by A.R. Stone, H.M. Platt, and L.F. Khalil. pp. 113–129. The Systematics Association special volume no. 22. London: Academic Press.

Siddiqi, M.R. (1983b) Phylogenetic relationships of the soil nematode orders Dorylaimida, Mononchida, Triplonchida and Alaimida, with a revised classification of the subclass Enoplia. *Pakistan Journal of Nematology*, **1**, 79–110.

Siddiqi, M.R. (1985) *Tylenchida — Parasites of plants and insects.* 645 pp. Farnham Royal, UK: Commonwealth Agricultural Bureaux.

Sidow, A. and Thomas, W.K. (1994) A molecular evolutionary framework for eukaryotic model organisms. *Current Biology*, **4**, 596–603.

Simpson, G.G. (1961) *Principles of animal taxonomy.* 247 pp. New York: Columbia University Press.

Skrjabin, K.I., Ed. (1991) *Key to parasitic Nematodes.* Moscow: Izdatel'stvo Akademii Nauk SSSR.

Sogin, M.L. and Silberman J.D. (1998) Evolution of the protists and protistan parasites from the perspective of molecular systematics. *International Journal for Parasitology*, **28**, 11–20.

Sommer, R.J. and Sternberg, P.W. (1994). Changes of induction and competence during the evolution of vulva development in nematodes. *Science*, **265**, 114–118.

Sommer, R.J., Carta, L.K., Kim, S.-Y. and Sternberg, P.W. (1996) Morphological, genetic and molecular description of *Pristionchus pacificus* sp. n. (Nematoda: Neodiplogasteridae). *Fundamental and Applied Nematology*, **19**, 511–521.

Sommer, R.J., Sigrist, C.B., Grandien, K., Jungblut, B., Eizinger, A., Adamis, H. and Schlak, I. (1999) A phylogenetic interpretation of nematode vulval variations. *Invertebrate Reproduction and Development*, **36**, 57–65.

Spears, T. and Abele, L.G. (1997) Crustacean phylogeny inferred from 18S rDNA. In *Arthropod Relationships*, edited by R.A. Fortey and R.H. Thomas. pp. 169–187. The Systematics Association Special Volume Series 55. London: Chapman and Hall.

Steiner, G. (1960) The Nematoda as a taxonomic category and their relationship to other animals. In *Nematology*, edited by J.N. Sasser and W.R. Jenkins. pp. 12–18. Chapel Hill: The University of North Carolina Press.

Stenroos, S.K. and DePriest, P.T. (1998) SSU rDNA phylogeny of cladoniiform lichens. *American Journal of Botany*, **85**, 1548–1559.

Stevenson, L.A., Chilton, N.B. and Gasser, R.B. (1995) Differentiation of *Haemonchus placei* from *H. contortus* (Nematoda: Trichostrongylidae) by the ribosomal DNA second internal transcribed spacer. *International Journal for Parasitology*, **25**, 483–488.

Stevenson, L.A., Gasser, R.B. and Chilton, N.B. (1996) The ITS-2 rDNA of *Teladorsagia circumcincta*, *T. trifurcata* and *T. davtiani* (Nematoda: Trichostrongylidae) indicates these taxa are one species. *International Journal for Parasitology*, **26**, 1123–1126.

Stewart, A.C. and Nicholas, W.L. (1994) New species of Xyalidae (Nematoda: Monhysterida) from Australian ocean beaches. *Invertebrate Taxonomy*, **8**, 91–115.

Stiles, C.W. and Hassall, A. (1926) Key-catalogue of the worms reported for man. *Hygiene Laboratory of the U.S. Public Health Service — Bulletin*, **142**, 69–196.

Subbotin, S.A., Waeyenberge, L., Molokanova, I.A. and Moens, M. (1999) Identification of *Heterodera avenae* group species by morphometrics and rDNA-RFLPs. *Nematology*, **1**, 195–207.

Sudhaus, W. (1978) Der 'Kryptopsezies'-Begriff zur Kennzeichnung genetisch isolierter allopatrischer Populationen einer Morpho- und Ökospezies am Beispiel van *Rhabditis spiculigera* (Nematoda) *Zeitschrift für zoologische Systematik und Evolutionsforschung*, **16**, 102–107.

Sudhaus, W. and Fitch, D.A. (in press) *Comparative studies on the phylogeny and systematics of the Rhabditidae (Nematoda).* Society of Nematologists.

Szalanski, A.L., Adams, B.J. and Powers, T.O. (1997) Molecular systematics of Tylenchida using 18S ribosomal DNA. *Journal of Nematology*, **29**, 607–608.

Tastet, C; Bossis, M., Gauthier, J.-P., Renault, L. and Mugniéry, D. (1999) *Meloidogyne chitwoodi* and *M. fallax* protein variation assessed by two-dimensional electropherogram computed analysis. *Nematology*, **1**, 301–314.

Thomas, W.K., Vida, J.T., Frisse, L.M., Mundo, M. and Baldwin, J.G. (1997) DNA sequences from formalin-fixed nematodes: integrating molecular and morphological approaches to taxonomy. *Journal of Nematology*, **29**, 250–254.

Triantaphyllou, A.C. (1985) Cytogenetics, cytotaxonomy and phylogeny of root-knot nematodes. In *An advanced treatise on Meloidogyne, Vol. 2. Methodology*, edited by K.R. Barker, C.C. Carter, and J.N. Sasser. pp. 113–126. Raleigh: North Carolina State University/ USAID.

Triantaphyllou, A.C. and Hirschmann, H. (1964) Reproduction in plant and soil nematodes. *Annual Review of Phytopathology*, **2**, 57–80.

Triantaphyllou, A.C., and Hirschmann, H. (1980) Cytogenetics and morphology in relation to evolution and speciation of plant-parasitic nematodes. *Annual Review of Phytopathology*, **18**, 333–359.

Van der Heiden, A. (1974) The structure of the anterior feeding apparatus in members of the Ironidae (Nematoda: Enoplida). *Nematologica*, **20**, 419–436.

Vanfleteren, J.R., Blaxter, M.L., Tweedie, S.A.R., Trotman, C., Lu, L., Van Heuwaert, M.-L. and Moens, L. (1994) Molecular genealogy of some nematode taxa as based on cytochrome *c* and globin amino acid sequences. *Molecular Phylogenetics and Evolution*, **3**, 92–101.

Vanfleteren, J.R. and Vierstraete, A.R. (1999) Insertional RNA editing in metazoan mitochondria: the cytochrome *b* gene in the nematode *Teratocephalus lirellus*. *RNA*, **5**, 622–624.

Vrain, T.C. (1993) Restriction Fragment Length Polymorphism separates species of the *Xiphinema americanum* group. *Journal of Nematology*, **25**, 361–364.

Vrain, T.C., Forge, T.A. and De Young, R. (1997) Population dynamics of *Pratylenchus penetrans* parasitizing raspberry. *Fundamental and Applied Nematology*, **20**, 29–36.

Wagele, J.W., Erikson, T., Lockhart, P. and Misof, B. (1999) The Ecdysozoa: Artifact or monophylum? *Journal of Zoological Systematics and Evolutionary Research*, **37**, 211–223.

Wang, B.B., Müller-Immergluck, M.M., Austin, J., Robinson, N.T., Chisholm, A. and Kenyon, C. (1993) A homeotic gene cluster patterns the anterioposterior body axis of *C. elegans*. *Cell*, **74**, 29–42.

Winnepenninckx, B., Backeljau, T. and De Wachter, R. (1995) Phylogeny of protostome worms derived from 18S rRNA sequences. *Molecular Biology and Evolution*, **12**, 641–649.

Woese, C.R. (1994) Microbiology In Transition. *Proceedings of the National Academy of Science, USA*, **91**, 1601–1603.

Woese, C.R. (1996) Phylogenetic trees: Whither microbiology? *Current Biology*, **6**, 1060–1063.

Wray, G.A., Levinton, J.S. and Shapiro, L.H. (1996) Molecular evidence for deep precambrian divergences among metazoan phyla. *Science*, **274**, 568–573.

von Linstow, O. (1905) Neue Helminthen. *Archiv für Naturgeschichte*, **71**, 267–276.

Yeates, D.K. and Wiegmann, B.M. (1999) Congruence and controversy: Toward a higher-level phylogeny of diptera. *Annual Review of Entomology*, **44**, 397–428.

Zarlenga, D.S., Lichtenfels, J.R. and Stringfellow, F. (1994) Cloning and sequence analysis of the small subunit ribosomal RNA gene from *Nematodirus battus*. *Journal of Parasitology*, **80**, 342–344.

Zarlenga, D.S., Stringfellow, F., Nobary, M. and Lichtenfels, J.R. (1994) Cloning and characterisation of ribosomal RNA genes from three species of *Haemonchus* (Nematoda: Trichostrongyloidea) and identification of PCR primers for rapid differentiation. *Experimental Parasitology*, **78**, 28–36

2. General Organisation

Lynda M. Gibbons

Department of Pathology and Infectious Diseases, The Royal Veterinary College, University of London, Hertfordshire AL9 7TA, United Kingdom

Keywords

Nematoda Organisation Morphology

Introduction

Nematodes are colloquially referred to as roundworms but this is rarely reflected in their gross appearance and refers mainly to their shape in cross section. The complete worms are often thread-like, cylindrical, generally fusiform, more rarely sac-like (e.g. female *Tetrameres* and *Meloidogyne*). They vary in length from microscopic (e.g. the females of *Aphelenchoides bicaudatus* which only reach up to 0.47 mm long) to several metres in length (e.g. the females of *Placentonema gigantissima* which have been recorded as over 8 metres long). The body is covered with a cuticle which may or may not exhibit any number of variations of complex ridges, spines or hooks. The body has no internal segmentation and beneath the cuticle the wall is composed of an epidermis (hypodermis) and a single layer of muscle cells.

The mouth or oral opening, which is generally terminal, may or may not be surrounded by lips and sensory organs. The mouth opens into a mouth cavity or stoma, which is more commonly referred to as the buccal cavity. This is followed by the pharynx (also referred to as the oesophagus), intestine and rectum which opens to the exterior via an anus in the female and larvae and a cloaca formed by the vas deferens joining the rectum in the male. The anus and cloaca may be terminal or subterminal. If subterminal then they always open ventrally.

The majority of nematodes exhibit sexual dimorphism. The male reproductive system opening to the exterior via the cloaca situated towards the posterior end of the body and the female system opening via a pore called the vulva which is always found in the ventral body wall. The vulva may be situated at any position between the anterior and posterior ends of the body.

Nematodes have a secretory-excretory system and a complex nervous system but no circulatory system.

The Nematoda are split into two subclasses the Adenophorea and Secernentea (Chabaud, 1974) (but see chapter 1). The Adenophorea are characterized by having few or no male caudal papillae, a secretory-excretory system which has no lateral canals or terminal ducts lined with cuticle, the absence of phasmids, the presence of epidermal glands, a cylindrical pharynx or pharyngeal glands free in the psuedocoelom forming a stichosome or trophosome, unsegmented eggs with a plug at either of the poles which may hatch *in utero* and first stage larvae which may have a stylet and are usually infective to the final host. The Secernentea are characterized by having numerous male caudal papillae, and a secretory-excretory system with lateral canals and a terminal duct lined with cuticle, the presence of phasmids, the absence of epidermal glands, a fully developed pharynx, eggs generally without polar plugs occasionally operculate at one pole and third stage larvae infective to the final host.

The organisation of nematodes is both varied and complex throughout the free-living and parasitic groups and is generally related to feeding mechanisms (see Chapter 8) and habitat. The general body structure, digestive tract and reproductive systems are reviewed and more detailed studies of particular structures which may be consulted are indicated where appropriate.

General Body Structure

Cuticle

The cuticle forms a flexible exoskeleton which invaginates at the mouth opening, amphids, phasmids, the secretory-excretory pore, vulva, cloaca and rectum. The complex structure of the cuticle itself will be dealt with in Chapter 7. The cuticle however has many features derived from the modification of its various layers and these will be briefly described.

Punctations are seen in marine and freshwater nematodes (Adenophoreans) and are small round areas beneath the cuticle. Pores seen in the Leptolaimina are found along the body but pores restricted to the anterior end are mainly lateral or sublateral in distribution and those restricted to further along the body are dorsal or ventral in distribution (Bird and Bird, 1991). These appear to be absent in the Secernentea.

Transverse striations or striae take many forms and are found throughout the nematodes groups. They may be easily distinguishable along the body, but sometimes they are fine and difficult to see or only clearly visible in particular regions. The distance between two striations is referred to as the interstitial region. Annulations are formed from deep striations and the distance between is referred to as the annule. Willmott (1974) defines striations as fine transverse grooves occurring at regular intervals and annulations as deep transverse grooves occurring at regular intervals giving the body a segmented appearance. These are common in the nematode parasites of vertebrates (*Dentostomella*, Figure 2.1) and also in the plant-parasitic nematodes particularly the Criconematidae as well as the marine nematodes. Muller (1979) has demonstrated how these features can be used in the identification of female *Onchocerca* spp.

Although longitudinal cuticular ridges are present in several groups including the Filarioidea (*Dipetalonema repens*, Figure 2.2) they are a particular feature of bursate nematodes belonging to the superfamilies Trichostrongyloidea, Molineoidea and Heligmosomoidea (*Paraheligmonina*, Figure 2.3). In these superfamilies the longitudinal cuticular ridges generally arise posterior to the cephalic region and extend along the body for part or all of its length. They may form a continuous or discontinuous line. Lee (1965) and Lee and Nicholls (1983) examined *Nippostrongylus brasiliensis* and demonstrated that proteinaceous struts were present along these ridges within the fluid layer of the cuticle which enabled support without loss of mobility. Durette-Desset (1971) described the arrangement of these ridges in cross section as the synlophe and in her studies of the Heligmosomoidea she identified four different types of longitudinal cuticular ridge. The four types are defined as follows: a ridge with an internal support which looks spine-like in cross section is termed an arête; a ridge formed by the fusion of two

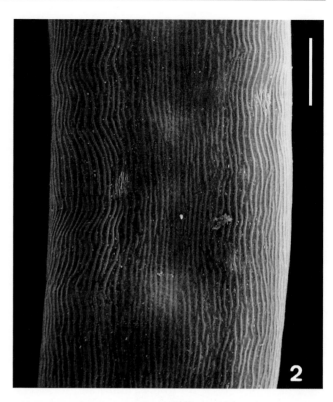

Figure 2.2. *Dirofilaria repens*, Filarioidea, showing discontinuous longitudinal ridges or striations on the body wall. Scale bar = 100 μm. Reproduced courtesy of CABI Biosciences.

Figure 2.1. *Dentostomella*, Oxyuroidea, anterior end showing marked cuticular striations. Scale bar = 10 μm. Reproduced courtesy of CABI Biosciences.

Figure 2.3. *Paraheligmonina*, Heligmosomoidea, longitudinal cuticular ridges extending along the body posterior to the cephalic region. Scale bar = 100 μm. Reproduced courtesy of CABI Biosciences.

or more arêtes is termed a comarête; a ridge without any internal support is termed a crête and a large, wide ridge found in the left lateral field which incorporates

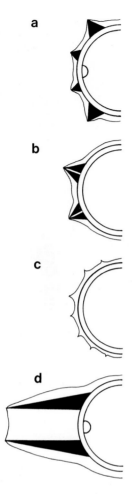

Figure 2.4. Diagram of the four types of longitudinal cuticular ridge: a. arête; b. comarête; c. crête; d. carène.

two enlarged arêtes is termed a carène (Figure 2.4). The ridges may be orientated perpendicularly or obliquely to the body surface and may function as a means of attachment to the intestinal villi, orientation of the nematode body, aid to locomotion or as a mechanism for scraping the host mucosa.

Alae are thickened wings of cuticle which are found laterally or sublaterally on the body. Cervical alae by definition are restricted to the cervical region and occur only in certain groups of animal-parasitic nematodes. There may be one (*Toxocara canis*, Figure 2.5) or three (*Physocephalus sexalatus*, Figure 2.6). Lateral alae occur in both sexes and extend along the length of the body overlying the lateral field (*Paraorientatractis semiannulata*, Figure 2.7). Caudal alae are seen on the posterior end of the male (*Abbreviata paradoxa*, Figure 2.8). The alae may be narrow or broad and in some groups are so developed they form a copulatory bursa (*Haemonchus longistipes*, Figure 2.9). The caudal alae will be discussed in more detail in the section on the male reproductive system.

Spines are projections on the cuticle and take many forms. They may be restricted to a particular part of the body such as the cephalic region (*Carnoya fimbriata*, Figure 2.10) or the tip of the female tail (*Ancylostoma*

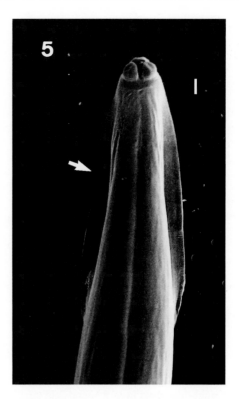

Figure 2.5. *Toxocara canis*, Ascaridoidea, anterior region showing a single cervical ala in each lateral field (arrowed). Scale bar = 100 μm. Reproduced from Gibbons (1986) with kind permission from CABI Publishing.

Figure 2.6. *Physocephalus sexalatus*, Spiruroidea, anterior region showing three cervical alae in each lateral field (arrowed). Scale bar = 100 μm. Reproduced from Gibbons (1986) with kind permission from CABI Publishing.

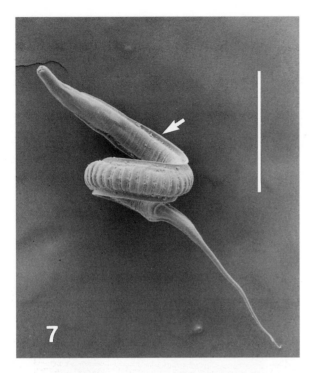

Figure 2.7. *Paraorientatractis semiannulata*, Cosmocercoidea, showing a single lateral ala in each lateral field (arrowed). Scale bar = 231 μm. Reproduced courstesy of CABI Biosciences.

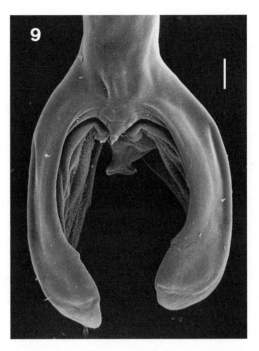

Figure 2.9. *Haemonchus longistipes*, Trichostrongyloidea, posterior end of male with a well developed copulatory bursa. Scale bar = 100 μm. Reproduced from Gibbons (1986) with kind permission of CABI Publishing.

Figure 2.8. *Abbreviata paradoxa*, Physalopteroidea, posterior end of male showing wide caudal alae (arrowed). Scale bar = 100 μm. Reproduced from Gibbons (1986) with kind permission of CABI Publishing.

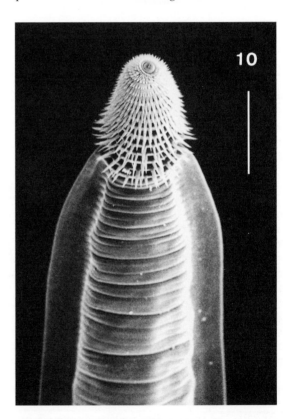

Figure 2.10. *Carnoya fimbriata*, Rhigonematoidea, anterior end showing spines restricted to the cephalic region and the beginning of the lateral alae. Scale bar = 100 μm. Reproduced from Hunt and Sutherland (1984) *Systematic Parasitology* **6**, plate 1A page 233 with kind permission of Kluwer Academic Publishers.

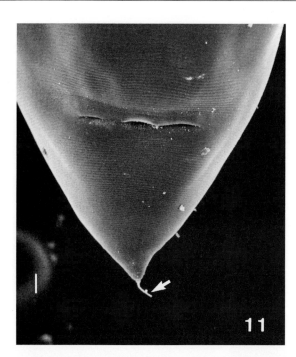

Figure 2.11. *Ancylostoma duodenale*, Ancylostomatoidea, posterior end of female showing a spine at the distal tip of the tail (arrowed). Scale bar = 10 μm. Reproduced courtesy of CABI Biosciences.

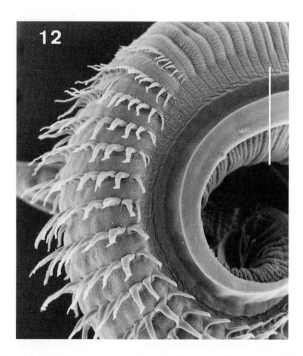

Figure 2.12. *Podocnematractis colombiaensis*, Cosmocercoidea, showing spines arranged in numerous rows on the dorsal surface of the body, each spine has more than one point at its distal tip. Scale bar = 120 μm. Reproduced courtesy of CABI Biosciences.

duodenale, Figure 2.11), arranged in one, two or numerous rows (*Podocnematractis colombiaensis*, Figure 2.12) respectively or arranged in different regions of the body

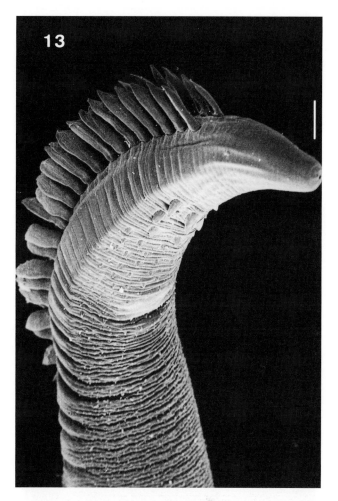

Figure 2.13. *Buckleyatractis marinkelli*, Cosmocercoidea, showing spines arranged in three rows along the median dorsal surface and a group of spines on the ventral surface. Scale bar = 100 μm. Reproduced from Khalil and Gibbons (1988) *Systematic Parasitology* 12, figure 27 page 196 with kind permission from Kluwer Academic Publishers.

(*Buckleyatractis marinkelli*, Figure 2.13). The spines themselves may be simple points, large and hook-like (*Pterygodermatites (Mesopectines) alphi*, Figure 2.14) or may be divided into more than one point at their distal tips (*Podocnematractis colombiaensis*, Figure 2.12).

Cuticular bosses and protuberences are present in some species. Their function is unknown and they are sometimes referred to as excrescences. These are clearly demonstrated in the cone-shaped protuberences of *Nilonema gymnarchi* (Figure 2.15) and the more modest bosses typical of *Loa loa*.

Rugae, sometime referred to as transverse ridges, have the appearance of wrinkled folds of cuticle or raised annulations. They are interrupted in the lateral fields and form incomplete rings around the body. These are typical of the females of the genus *Onchocerca* (Filarioidea) (*Onchocerca armillata*, Figure 2.16) and their variation in the different species is fully described by Muller (1979) and Bain (1981).

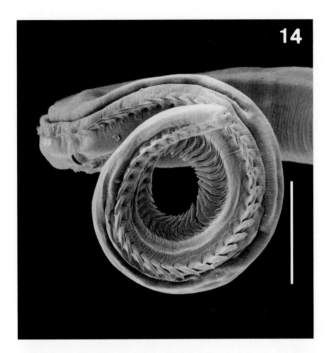

Figure 2.14. *Pterygodermatites* (*Mesopectines*) *alphi*, Rictularioidea, showing large hook-like spines arranged in two rows along the body. Scale bar = 430 μm. Reproduced courtesy of CABI Biosciences.

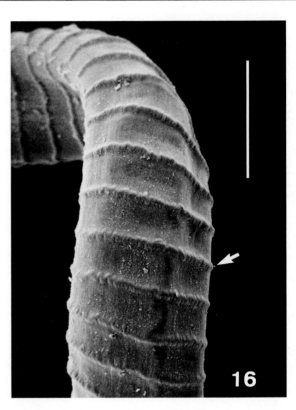

Figure 2.16. *Onchocerca armillata*, Filarioidea, part of a female showing the rugae on the body wall (arrowed). Scale bar = 100 μm. Reproduced courtesy of CABI Biosciences.

Epidermis

The epidermis, also referred to as the hypodermis, is covered by the zones of the cuticle and may be syncytial or cellular. This layer protrudes into the pseudocoelom in the lateral, mid-dorsal and mid-ventral regions and these protrusions are referred to as the epidermal or lateral cords. The lateral cords are the most prominent and incorporate the excretory canals if present.

In the Adenophorea the epidermis is interrupted by the presence of epidermal glands. Bird and Bird (1991) gave a detailed description of these glands in free-living marine nematodes. In the genera *Trichuris*, *Capillaria* and *Trichinella* (see Kim and Ledbetter, 1980) these glands, modifications of the lateral and median epidermal cords, are arranged in longitudinal rows referred to as bacillary bands. In the genus *Trichuris* there is a single lateral bacillary band which extends from behind the cephalic region to near the level of the posterior end of the pharynx (*Trichuris parvispicularis*, Figure 2.17). It starts as a narrow band and increases in width posteriorly. The bacillary band is replaced posteriorly by scattered epidermal glands in the lateral field. The genus *Capillaria* has bacillary bands in the ventral and lateral regions which extend the length of the body. The bacillary bands contain both unicellular epidermal gland cells and nonglandular cells. Each epidermal gland cell is associated with a bipolar nerve cell and opens through a cuticular pore which may be filled with a plug giving the band a honeycomb texture.

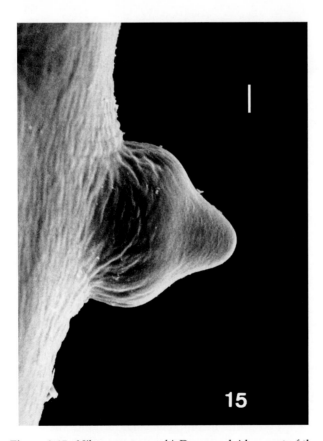

Figure 2.15. *Nilonema gymnarchi*, Dracunculoidea, part of the body wall showing one of the cone-shaped excrescences projecting above the surface. Scale bar = 10 μm. Reproduced courtesy of CABI Biosciences.

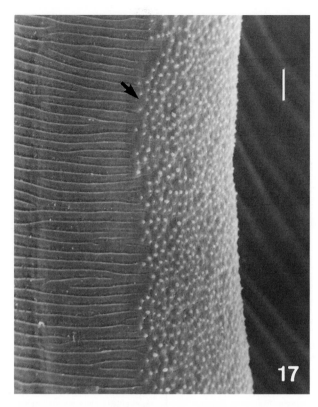

Figure 2.17. *Trichuris parvispicularis*, Trichinelloidea, part of the anterior end showing the bacillary band (arrowed). Scale bar = 10 μm. Reproduced courtesy of CABI Biosciences.

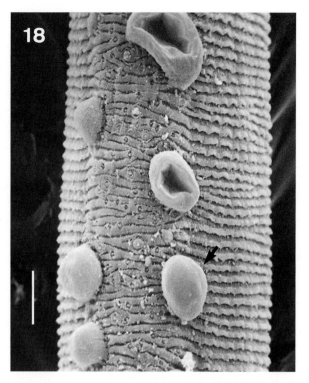

Figure 2.18. *Trichuris parvispicularis*, Trichinelloidea, showing the anterior end of the bacillary band bordered with the cuticular expansions called cephalic glands (arrowed). Scale bar = 10 μm. Reproduced courtesy of CABI Biosciences.

For a more detailed discussion of the structure and function of these epidermal glands Sheffield (1963), Wright (1963, 1968), Wright and Chan (1973) and Jenkins (1969, 1970) should be consulted. Along the borders near the anterior end of the bacillary bands an additional structure may be present in the form of cuticular expansions or 'vesicles'. These are referred to as cephalic glands (*Trichuris parvispicularis*, Figure 2.18).

Muscles

The neuromuscular system will be dealt with in Chapter 12 and only a brief outline of the organisation of the musculature is given here.

Somatic muscle

The somatic muscle cells are found grouped between the epidermal cords. The number of rows of muscle cells between the cords varies. One to two rows are referred to as holomyarian, two to five rows as meromyarian and a larger number of rows as polymyarian. The muscle cells are spindle-shaped, longitudinally orientated and divided into a contractile and non-contractile portion. The shape of the contractile portion groups the cells into three types, namely, platymyarian with a wide shallow contractile portion close to the epidermis, coelomyarian with a contractile portion extending up the lateral sides of the cell (Dioctophymatoidea, Figure 2.19) and circomyarian where the striated muscle fibres envelop the non-

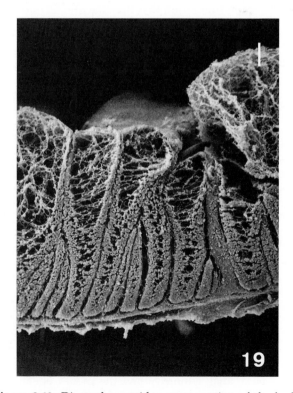

Figure 2.19. Dioctophymatoidea, cross section of the body wall showing the coelomyarian cell type with the contractile part of the cell extending up the lateral walls. Scale bar = 10 μm. Reproduced courtesy of CABI Biosciences.

contractile portion. Chitwood and Chitwood (1974) suggested each muscle cell has an 'innervation' process. The process extends from a thickening of the sarco-plasm in the region of the nucleus and communicates with the nerve ring or one of the dorsal, ventral and submedian nerves. Bird and Bird (1991) referred to these as sarcoplasmic connections and indicated that there are muscle-to-muscle connections within and between quadrants of muscle cells.

Specialised muscle

Specialised muscle cells are mainly associated with the digestive system and the male and female reproductive tracts. The pharynx is composed of a thick muscular layer. In plant- parasitic nematodes and the few animal-parasitic nematodes which have a stylet some of the muscles of the pharynx and cephalic region have become modified to form stylet protractor muscles. Other specialized muscles include the somato-intest-inal, copulatory, bursal, spicular, gubernacular and vulval muscles.

Pseudocoelom

The pseudocoelom is the space posterior to the nerve ring between the somatic muscles and the digestive tract. The space is fluid-filled which coats the digestive tract and reproductive organs suspended in it. Coelo-mocytes occur in the pseudocoelom, these are large amoeboid cells which vary in number, size and shape. The function of the coelomocytes is unknown although there is some suggestion that they may be secretory or phagocytic (Bird and Bird, 1991).

Digestive Tract

Bird and Bird (1991) suggested the division of the digestive tract into three main regions: the stomodeum which includes the mouth, lips, buccal cavity and pharynx; intestine and proctodeum which includes the female rectum and male cloaca. This proposal is based on the function of the constituent parts i.e. pharyngeal glands open into the stomodeum, rectal glands into the proctodeum and the different embryological origin of the three regions. Here the constituent parts of each of these regions are dealt with individually.

Mouth

The mouth and lips are covered in cuticle similar to that of the rest of the body (Bird and Bird, 1991). The orientation of the mouth opening in the majority of the nematodes is terminal but may be displaced antero-dorsally to dorsally as in the species of the Ancylostomatoidea and Rictular-ioidea (*Pterygodermatites (Mesopectines) alphi*, Figure 2.20) or ventrally (*Ichthyocephaloides dasyacanthus*, Figure 2.21). The ancestral nematode described by De Coninck (1965) has a hexaradial symmetry and the mouth opening is surrounded by six distinct lips or labia. This hexaradial symmetry is still common in the Aphelenchida (Hunt, 1993). Fusion of these lips results in several groups, such as the Heterakoidea and Ascaridoidea (*Ascaridia dissimilis*,

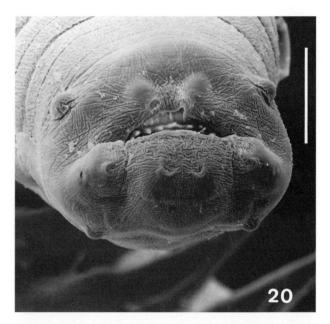

Figure 2.20. *Pterygodermatites (Mesopectines) alphi*, Rictular-ioidea, anterior end showing the mouth displaced antero-dorsally. Scale bar = 38 μm. Reproduced courtesy of CABI Biosciences.

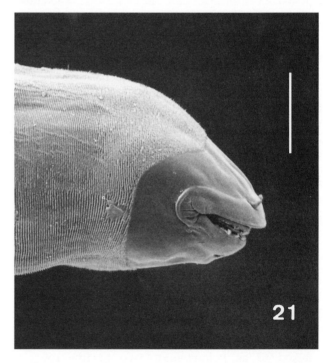

Figure 2.21. *Icthyocephaloides dasyacanthus*, Rhigonematoidea, anterior end showing mouth opening ventrally. Scale bar = 100 μm. Reproduced from Hunt and Sutherland (1984) *Systematic Parasitology*, 6, figure 2C page 144 with kind permission of Kluwer Academic Publishers.

Figure 2.22), having three lips; each lip may be divided into an apical and basal region. This division is clearly seen in *Parascaris equorum* and species of *Falcaustra* (*Falcaustra tikasinghi*, Figure 2.23). A transverse groove

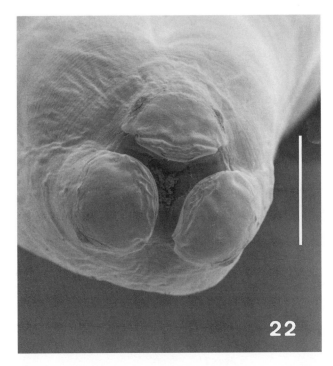

Figure 2.22. *Ascaridia dissimilis*, Heterakoidea, cephalic region with three lips. Scale bar = 100 μm. Reproduced courtesy of CABI Biosciences.

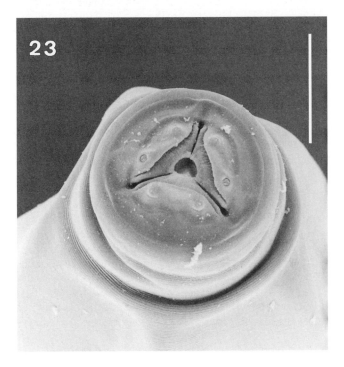

Figure 2.23. *Falcaustra tikasinghi*, Cosmocercoidea, cephalic region with three lips each divided into an apical and basal region. Scale bar = 75 μm. Reproduced courtesy of CABI Biosciences.

mented by an anterior cuticular flange and many of the ascarid nematodes such as *Ascaris*, *Porrocaecum* and *Amplicaecum* have dentigerous ridges at the outer margin of the internal surface of each lip (*Amplicaecum africanum*, Figure 2.24). Each dentigerous ridge is formed of a single row of denticles, the ridge separated from the outer surface by a groove which Angel, Thibaut and Saez (1974) refer to as a peribuccal groove. Rarely the denticles of the ridges are enlarged such as in *Terranova caballeroi* where the posterior most denticle on each subventral lip is enlarged to form a cusp. Cuticular outgrowths may arise between the lips or labia and these are referred to as interlabia. Rarely surrounding the base of the lips there may be fringe-like structures referred to as fimbriae (*Crossophorus collaris*, Figure 2.25). In some species lateral cuticular outgrowths develop over the primary lips and replace them. These are referred to as false lips or pseudolabia and generally take the form of two large lateral lips (*Physaloptera gemina*, Figure 2.26). Ornamentation may be associated with the pseudolabia such as the denticulate posterior border of *Histiocephalus bucorvi* which also has a cervical collar with longitudinal ridges or cephalic shields (*Parabronema indicum*, Figure 2.27).

The pseudolabia may be set off from the rest of the body in a cephalic bulb. This is a characteristic of the family Gnathostomatidae (*Gnathostoma doloresi*, Figure 2.28). There may be a single swelling as in *Gnathostoma* or the bulb may be divided into lobes as in *Tanqua tiara*. Gibbons and Keymer (1991) describe the cephalic bulb of *Tanqua tiara* using light and scanning electron microscopy and found that the cephalic bulb of

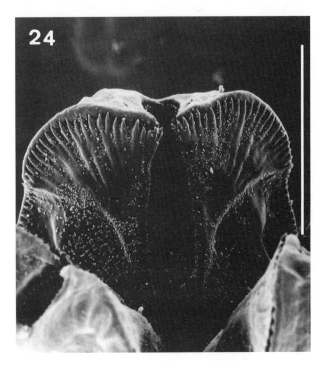

Figure 2.24. *Amplicaecum africanum*, Ascaridoidea, internal surface of one of the three lips showing the dentigerous ridge. Scale bar = 100 μm. Reproduced from Gibbons (1986) with kind permission of CABI Publishing.

or labial sinus divides the lips of these species into two clear regions. The lips can be modified by the presence of additional structures. The lips of *Strongyluris* are aug-

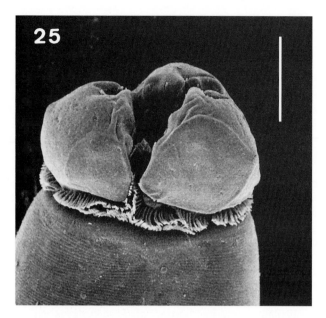

Figure 2.25. *Crossophorus collaris*, Ascaridoidea, cephalic region with fimbriae surrounding the base of the three lips. Scale bar = 100 μm. Reproduced courtesy of CABI Biosciences.

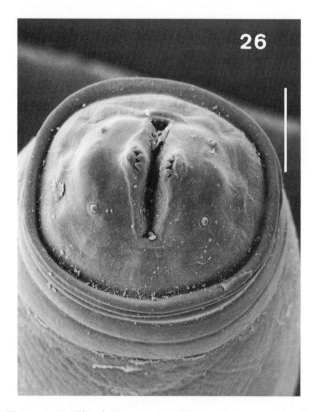

Figure 2.26. *Physaloptera gemina*, Physalopteroidea, cephalic region showing the two large lateral false lips or pseudolabia and odontia. Scale bar = 100 μm. Reproduced courtesy of CABI Biosciences.

this species is divided into four segments each containing a walled cavity termed a ballonet. Each ballonet is associated with an elongate, blindly ending sac which extends posteriorly from the cephalic region and whose

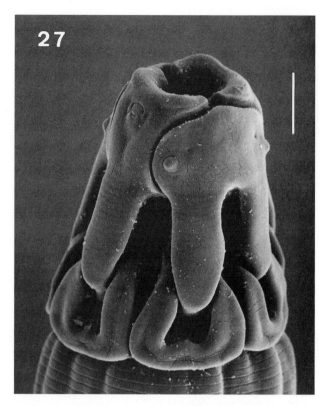

Figure 2.27. *Parabronema indicum*, Habronematoidea, cephalic region showing the looped cephalic shields posterior to the pseudolabia. Scale bar = 10 μm. Reproduced courtesy of CABI Biosciences.

surface has spirally arranged fibrils. The work of Baylis and Lane (1920) and Baylis (1939) suggests that the contraction of the fibrils changes the shape of the cervical sacs putting pressure on the contents of the ballonets enabling the cephalic bulb to change shape which assists in holding the parasite in position within the host gut. In other members of the family the cephalic bulb may be replaced by long ramified appendages (*Ancyracanthus*, Figure 2.29). Elaborate head structures are not restricted to the parasites of vertebrates. *Travassosinema thyropygi* (Figure 2.30) from spirobolid millipedes has an elaborate umbraculum with elements radiating from the lip and head region.

A further modification of the mouth opening is the presence of leaf crowns or *corona radiata*. These are leaf-like elements which surround the mouth opening of many species, particularly in the superfamily Strongyloidea (*Oesophagostomum columbianum*, Figures 2.31, 2.32). Chitwood and Chitwood (1974) suggested that these are the divided apical region of the ancestral lips. Generally two circles of elements are present, referred to as external and internal leaf crowns respectively. The point of attachment of the internal leaf crown may be associated with the rim or wall of the buccal capsule (Lichtenfels, 1975). Rarely the two leaf crowns may be fused (*Chabertia ovina*), reduced to a single leaf crown (*Cylicodontostomum purvisi*) or reduced to hook-like elements (*Agriostomum*) The number of elements in the

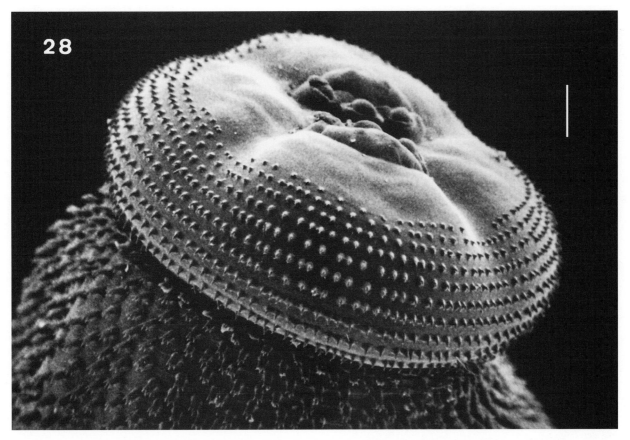

Figure 2.28. *Gnathostoma doloresi*, Gnathostomatoidea, pseudolabia incorporated within a cephalic bulb separated from the cervical region. Scale bar = 100 μm. Reproduced from Gibbons (1986) with kind permission of CABI Publishing.

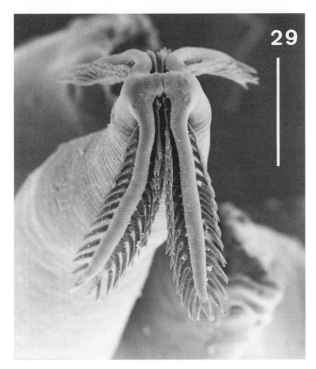

Figure 2.29. *Ancyracanthus*, Gnathostomatoidea, cephalic region with ramified appendages. Scale bar = 300 μm. Reproduced courtesy of CABI Biosciences.

Figure 2.30. *Travassosinema thyropygi*, Travassosinematidae, *en face* view showing the elements of the umbraculum. Scale bar = 25 μm. Reproduced courtesy of Dr David Hunt.

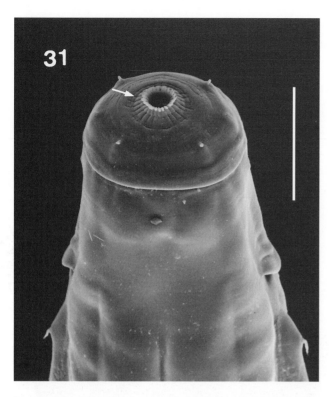

Figure 2.31. *Oesophagostomum columbianum*, Strongyloidea, anterior end showing the leaf crown or *corona radiata* (arrowed). Scale bar = 86 μm. Reproduced courtesy of CABI Biosciences.

leaf crowns can vary from six (*Deletrocephalus dimidiatus*) to 80 (*Strongylus equinus*).

Many nematodes have no lips or additional structures and the mouth opening is surrounded by a simple circumoral membrane. In the absence of lips the mouth opening can vary in shape. In the superfamily Diaphanocephaloidea the mouth is dorso-laterally flattened and in *Allodapa suctoria* the opening is hexagonal. A thickening of the cuticle surrounding the mouth, separated by a transverse groove from the body, is referred to as the oral collar; this can be inflated or flattened and is a typical feature of many species in the superfamily Strongyloidea.

Cuticular thickenings extending from the mouth opening, called cordons, are a feature primarily found in the Acuarioidea. These cordons can be elaborate and may extend posteriorly in anastomosing loops (*Desportesius*, Figure 2.33) or form a collarette (*Stegophorus*).

Cephalic pores, lateral to the mouth opening, associated with protrusible trident-like structures are seen in the subfamily Diplotriaeninae of the superfamily Diplotriaenoidea. Ornamentation of the cephalic region may take the form of cuticular thickenings forming striations which encompass the cephalic sensory organs (*Striatofilaria*) or form a peribuccal ring which is a raised area surrounding the mouth and this may also incorporate cuticular elevations (*Setaria*) or spines (*Stephanofilaria*, Figure 2.34).

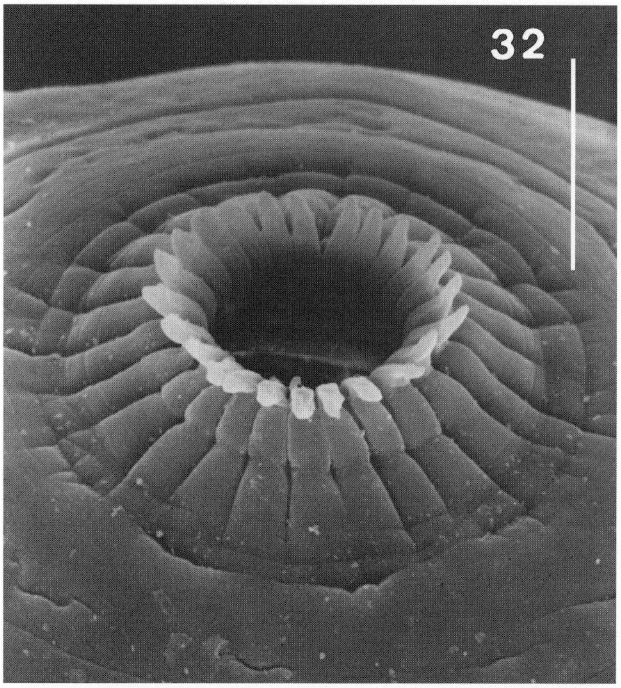

Figure 2.32. *Oesophagostomum columbianum*, Strongyloidea, higher magnification of the leaf crown showing the external leaf crown. Scale bar = 18 μm. Reproduced courtesy of CABI Biosciences.

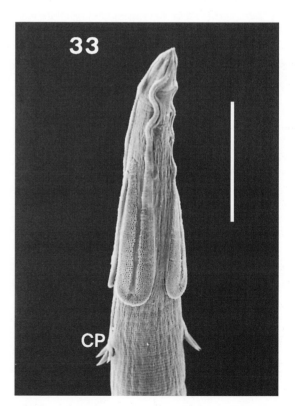

Figure 2.33. *Desportesius* sp., Acuarioidea, with cordons expanding posteriorly changing from raised ridges to rows of spines. CP = Cervical papillae or deirids. Scale bar = 152 μm Reproduced courtesy of CABI Biosciences.

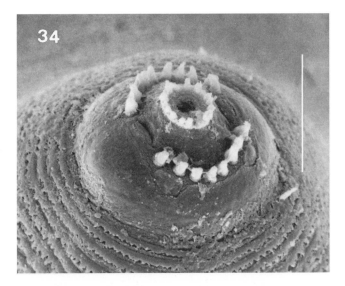

Figure 2.34. *Stephanofilaria* sp., Filarioidea, cephalic region ornamented by a circle of spines. Scale bar = 10 μm. Reproduced courtesy of CABI Biosciences.

Buccal Capsule

The buccal capsule is the part of the digestive tract between the mouth opening and the proximal end of the pharynx. The space enclosed is referred to as the buccal cavity or stoma. Chitwood and Chitwood (1974) defined three regions within the cavity, namely, the cheilostom or lip cavity, the protostom or cylindrical part and telostom or end cavity. The walls adjoining these regions are therefore known as the cheilorhabdions, protorhabdions and telorhabdions respectively. Later Bird and Bird (1991) divided the buccal cavity into five regions, namely, the cheilostom, prostom, mesostom, metastom and telostom. For a more detailed discussion of these regions of the buccal cavity Bird and Bird (1991) should be consulted. The development of the buccal capsule and cavity reflects the method of feeding (Bird and Bird, 1991) It varies in shape and form as well as in the presence or absence of stylets, teeth, subventral lancets, cutting plates, dorsal gutter and dorsal cone.

In secernentean plant-parasitic nematodes, some entomophilic nematodes and predatory nematodes, the buccal cavity may be transformed to incorporate a protrusible spear or stomatostyle (*Meloidogyne*, Figure 2.35). This hollow structure consists of three parts, a cone, a shaft and a posterior part which swells to form three basal knobs which act as attachments for the muscles (Hirschmann, 1983, Eisenback *et al.*, 1980 and Eisenback and Hirschmann, 1981). Adenophorean plant-parasitic nematodes have a stylet made up of only two parts, an anterior odonstostyle and a posterior odontophore (Bird and Bird, 1991).

The buccal capsule varies in shape from rudimentary in the majority of the Trichostrongyloidea, narrow in the Rhabditida to well developed in the Strongyloidea and from cylindrical (*Oesophagostomum*) to globular (*Strongylus*). Generally the buccal capsule is symme-

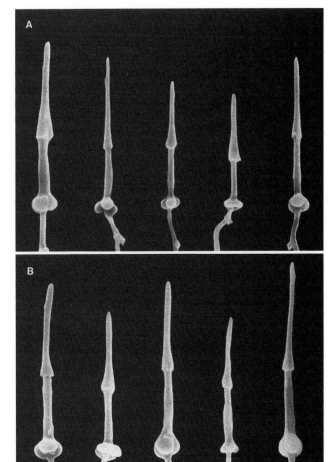

Figure 2.35. *Meloidogyne* sp., Hoplolaimoidea, stylets extracted from male root-knot nematodes: A. *M. arenaria, M. carolinensis, M. exigua, M. graminicola, M. hapla*; B. *M. incognita, M. javanica, M. megatyla, M. naasi, M. nataliei*. Reproduced courtesy of Professor J.D. Eisenback.

trical but can be asymmetrical as in the hookworm *Ancylostoma*, the strongyle *Chabertia ovina* and the members of the superfamily Rictularioidea.

In some members of the Spiruroidea the lining of the buccal capsule projects on to the body surface forming a cuticular circumoral elevation. This forms a distinct rim around the mouth opening (*Spirura*).

Members of the Ancylostomatoidea have semilunar cutting plates on the outer margin of the buccal capsule. Generally two large ventro-lateral plates are visible through the mouth opening and in some species two very small dorso-lateral cutting plates are also present (*Grammocephalus clathratus*, Figure 2.36).

Teeth

Inglis (1966) and Bird (1971) divided the teeth into two groups depending on their point of attachment. Teeth which originate in the labial or cheilostom region are referred to as odontia (*Physaloptera* Figure 2.26, *Abbreviata, Cyathospirura* and *Ancylostoma*) and those which

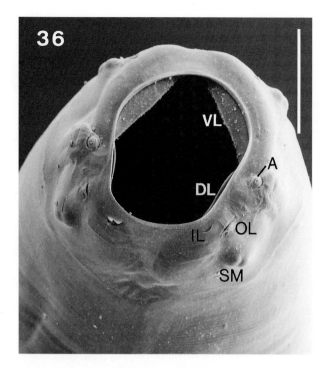

Figure 2.36. *Grammocephalus clathratus*, Ancylostomatoidea, cephalic region showing the two ventro- and two dorso-lateral cutting plates attached to the margin of the buccal capsule. A = amphid; DL = Dorso-lateral cutting plate; IL = inner labial sensillum; OL = outer labial sensillum; SM = Submedian cephalic papillae; VL = Ventro-lateral cutting plate. Scale bar = 100 μm. Reproduced courtesy of CABI Biosciences.

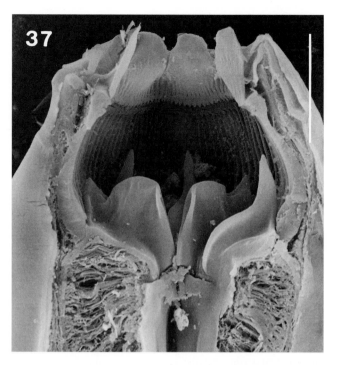

Figure 2.37. *Paradeletrocephalus minor*, Strongyloidea, showing the buccal capsule cut open to reveal some of the nine teeth, three with large semilunar bases and short points and six with small semilunar bases and large points. Scale bar = 43 μm. Reproduced courtesy of CABI Biosciences.

originate posteriorly, including those associated with the pharynx, are referred to as onchia (*Haemonchus*, *Triodontophorus*, *Ternidens deminutus*, female *Oxyuris* and *Strongylus*).

The number and form of the teeth may also vary (*Streptopharagus pigmentatus* has six, *Ternidens deminutus* has three, one dorsal, two subventral and *Haemonchus* has a single dorsal tooth). Teeth may be simple pointed structures or more elaborate with ridges (*Strongylus asini*), each tooth may have three sharp points (*Triodontophorus tenuicollis*), be folded (*Ternidens deminutus*), rounded (*Strongylus vulgaris*) or different shape and sizes (*Paradeletrocephalus minor*, Figure 2.37).

Subventral lancets are sharply pointed teeth near the base of the buccal cavity in the Ancylostomatoidea. *Oxyuris equi* also has finely branched bristles associated with large teeth which have small denticles on their outer proximal margins in the female and bristles without teeth in the male. The pharyngeal funnel may have teeth which do not reach into the buccal capsule (*Gyalocephalus capitatus*).

More rarely, complex structures are associated with the buccal capsule. In *Labiduris* the lining of the buccal cavity is developed into pectinate lobes which unfold and project through the ventral part of the mouth opening (Figure 2.38).

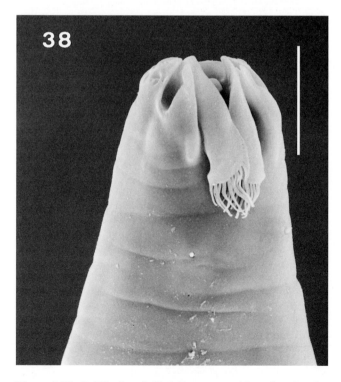

Figure 2.38. *Labiduris zschokkei*, Cosmocercoidea, showing the two fringed pectinate lobes, formed from the extension of the lining of the buccal cavity, protruding through the mouth opening. Scale bar = 38 μm. Reproduced courtesy of CABI Biosciences.

Dorsal Gutter

The dorsal gutter is the name given to a channel which may extend for part or all the length of the wall of the buccal capsule. It contains the duct of the dorsal pharyngeal gland which is generally completely enclosed within the channel and opens at the distal end (*Strongylus equinus*, Figure 2.39). Mobarak and Ryan (1998) have given a detailed account of the ultrastructure of the dorsal gutter of *Strongylus vulgaris*.

The dorsal cone is a modification of the dorsal gutter which takes the form of a tooth-like projection into the buccal cavity supporting the duct of the dorsal pharyngeal gland (*Bunostomum*) (Wilfred and Lee, 1981).

Pharyngeal Funnel

The posterior boundary of the buccal capsule joins the next region of the alimentary canal, the pharynx or oesophagus. The buccal capsule-pharynx junction is referred to by Chitwood and Chitwood (1974) as the telostom. The buccal cavity opens into the lumen of the pharynx which may be dilated forming another cavity which is referred to as the pharyngeal funnel or oesophagostom. Additional features such as teeth, bristles or pectinate blades may be present or absent within the funnel. *Crossocephalus viviparus* has three pairs of pectinate blades, one pair associated with each of the three lobes of the pharynx. The anterior part of the pharynx is capable of everting through the mouth opening extending these pectinate blades (Figure 2.40).

Pharynx

The pharynx is an elongate structure between the buccal capsule and the remainder of the gut. It is lined with cuticle, has a triradiate lumen and generally is divided into an anterior muscular part and a posterior glandular part. The pharynx may be composed of three regions, namely, the corpus, isthmus and bulb with a tricuspid valve and throughout the nematode groups one or all these regions are present. The oxyurid pharynx, for example, has all three regions present, namely, the corpus, isthmus and bulb. The rhabditoid pharynx has an additional bulb at the posterior margin of the corpus dividing this region into the anterior procorpus which is more or less uniform in width and a posterior metacorpus or median bulb. The strongylid pharynx, however, is claviform or club-shaped without a clear demarcation of the regions (*Paradeletrocephalus minor*, Figure 2.41). There are also many ranges of organisation of the pharynx within the various groups. An example of this is the genus *Labiduris* which has a typical oxyurid pharynx with a corpus, isthmus and posterior bulb incorporating a tricuspid valve but also has an additional small bulb without a valve just anterior to the posterior bulb.

The pharynx has three glands associated with it, one dorsal which opens near the anterior end of the

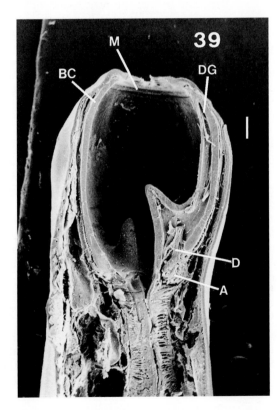

Figure 2.39. *Strongylus equinus*, Strongyloidea, showing the buccal capsule cut open to reveal the dorsal gutter, the ampulla of the dorsal pharyngeal gland and a portion of the dorsal pharyngeal gland duct. A = Ampulla of the dorsal pharyngeal gland; BC = Buccal capsule; D = Duct of dorsal pharyngeal gland; DG = Dorsal gutter; M = Mouth. Scale bar = 100 μm. Reproduced from Gibbons (1986) with kind permission of CABI Publishing.

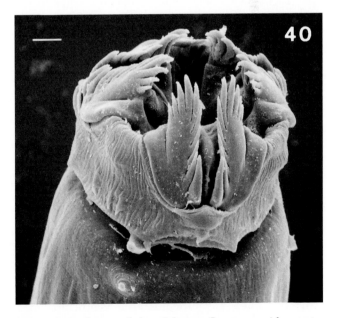

Figure 2.40. *Crossocephalus viviparus*, Cosmocercoidea, anterior lobes of pharynx everted through the mouth opening revealing three pairs of armed pectinate blades. Scale bar = 10 μm. Reproduced from Gibbons (1986) with kind permission of CABI Publishing.

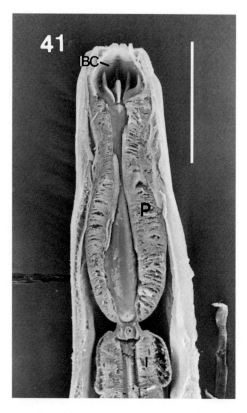

Figure 2.41. *Paradeletrocephalus minor*, Strongyloidea, showing the anterior body cut open longitudinally to reveal the clavate pharynx. BC = Buccal cavity; I = Intestine; P = Pharynx. Scale bar = 200 μm. Reproduced courtesy of CABI Biosciences.

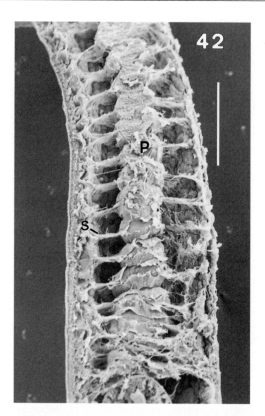

Figure 2.42. *Trichuris parvispicularis*, Trichinelloidea, showing part of the anterior portion cut open longitudinally to show the narrow pharynx and the thin cell walls of the stichocytes forming the stichosome. P = Pharynx; S = Wall of stichocyte. Scale bar = 100 μm Reproduced courtesy of CABI Biosciences.

pharynx or may extend into the buccal capsule in the dorsal gutter or dorsal cone and two subventral glands which open more posteriorly. It is rare for all three to open into the stoma. In the superfamily Trichinelloidea the glands have multiplied into a column of cells called stichocytes which form a stichosome outside the reduced muscle tissue of the pharynx (*Trichuris parvispicularis*, Figure 2.42). Extensive illustration of the variation in pharynx formation can be seen in Chitwood and Chitwood (1974).

Pharyngeo-intestinal Valve

The pharynx connects to the intestine through a short structure termed the pharyngeo-intestinal valve which may project into the intestine. It is lined with cuticle, triradiate and dorso-ventrally or laterally flattened. Lichtenfels *et al.* (1988) used the length of this valve in the systematics of the ostertagid nematodes of domesticated animals. In some groups (Anisakidae, Ascaridoidea) the distal end of the pharynx is modified to form a separate thick-walled spherical or cylindrical structure called a ventriculus. The ventriculus may have an additional solid appendage which is directed posteriorly, dorsal to the gut and is referred to as the ventricular appendage (*Contracaecum*). For the variations of these features in the Ascaridoidea see Hartwich (1974).

Intestine

The intestine is composed of a single layer of epithelial cells whose internal surface is usually covered by microvilli. These cells are often covered by irregular muscle fibres and a connective tissue sheath. The intestine generally forms a straight tube but in some species there is a blind pouch at the proximal end near the junction of the pharynx or ventriculus and referred to as the intestinal caecum (*Porrocaecum*). For a detailed description of the ultrastructure of this region see Bird and Bird (1991).

Rectum and cloaca

At the junction of the intestine and rectum there is an intestino-rectal sphincter or valve formed from intestinal tissue. The rectum is lined with cuticle and there are three glands which open into the rectum, one dorsal and two subventral. In the female the rectum opens through a simple pore called the anus. In some species the rectum and anus are non functional and the intestine terminates in a blind ending sac. In the male the vas deferens opens into the rectum prior to the external opening of the gut and the combined duct which opens to the exterior is referred to as the cloaca. Further modifications of this region are also seen in the male and will be discussed in association with the male reproductive system. The female anus and the male cloaca both open ventrally to the exterior of the body.

Nervous System

The nerve ring encircles the pharynx and is associated with longitudinal nerves which extend anteriorly and posteriorly. Six nerves extend anteriorly from the nerve ring, each with three main distal branches which are associated with the cephalic sensory organs. (See Chapter 12 for the details of neuromuscular transmission). Bird and Bird (1991) considered that the most suitable term for nematode sensory organs is sensillum. These sensory organs or sensilla are either mechanoreceptors or have a chemosensory function. The structure and function of these sensory organs will be dealt with in Chapter 14. De Coninck (1965) proposed a hypothesis for the number and arrangement of cephalic papillae in ancestral nematodes based on a hexaradial symmetry. His hypothesis proposed that two concentric rings of six papillae, referred to as internal and external rings, surround the mouth opening which has six lips. Each ring of sensory organs has a single papilla on the anterior margin of each of the six lips, referred to as inner labial papillae, and the posterior margin, referred to as the outer labial papillae. He also proposed that external to these two rings of sensory organs there is a third ring of four papillae or setae referred to as the submedian cephalic papillae or setae (Figure 2.43). The variations in number and arrangement of these sensory organs from the ancestral form described by De

Coninck are reviewed by Chitwood and Chitwood (1974) and Grassé (1965a and b). Associated with the submedian cephalic papillae is a pair of lateral sensory organs which Cobb (1923) referred to as the amphids. The full number of cephalic sensory organs is generally present in marine forms but are reduced in the animal-parasitic forms (*Brugia pahangi*, Figure 2.44).

The cephalic region incorporating these sensory organs may be inflated to form a cephalic vesicle. Wright (1975), following his studies on *Nippostrongylus*, suggested that this is the result of the median zone of the cuticle being swollen with fluid. Wright suggested that stimuli on the vesicle from environmental contact may be transmitted to the sensory organs. The vesicle may be large and divided into a wide anterior region and narrower posterior region (*Cooperia*) or be restricted to the anterior end (*Ostertagia*). This swelling of the anterior end should not be confused with the dilated cephalic region seen in the genera *Brugia* and *Wuchereria*.

Posterior to the nerve ring there are dorsal, ventral, four submedian and 1–3 pairs of lateral nerves in the epidermis. A variable number of commissures are present between the longitudinal nerves. The lateral nerve contains a few ganglia, the ventral nerve a chain of ganglia while the dorsal and submedian nerves have no ganglia. Processes which Chitwood and Chitwood (1974) referred to as 'innervation processes' extend from the somatic muscle cells to the nerve ring or dorsal, ventral and submedian nerves. The nervous

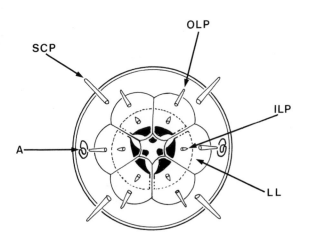

Figure 2.43. Diagram showing the arrangement of cephalic sensory papillae in an ancestral nematode. A = Amphid; ILP = Inner labial papilla; LL = Lateral lip; OLP = Outer labial papilla; SCP = Submedian cephalic papilla. Redrawn after De Coninck (1965).

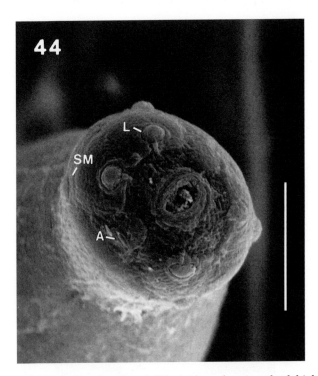

Figure 2.44. *Brugia pahangi*, Filarioidea, showing the labial and submedian cephalic papillae with one of the amphids. A = Amphid; L = Labial papilla; SM = Submedian cephalic papilla. Scale bar = 10 μm. Reproduced from Gibbons (1986) with kind permission of CABI Publishing.

system serves a scattered number of tactile papillae or sensory organs on the body surface. One pair is confined to the cervical region, innervated by the lateral nerve and is referred to as the cervical papillae or deirids. Morphologically these papillae appear in many different forms ranging from an almost indistinct rounded structure (*Trichostrongylus*) to an enlarged conical process, extending beyond the body surface, which may be either bluntly rounded (*Haemonchus similis*, Figure 2.45) or needle-like (*Oesophagostomum columbianum*, Figure 2.46). They may also form a very elaborate structure with several prongs (*Desportesius*, Figure 2.47). Lichtenfels *et al.* (1995) reported the work of Chalfie and White (1988) and Ward *et al.* (1975) on the nematode *Caenorhabditis elegans*. In these studies a second pair of deirids was reported near the junction of the middle and posterior thirds of the body. Lichtenfels *et al.* (1995) adopted the terms anterior deirids for the cervical papillae and posterior deirids for the second pair of papillae. The remaining body papillae may be restricted in the male to the lateral and ventral part of the posterior end of the body forming the male caudal papillae. The arrangement of these will be dealt with in the discussion of the male reproductive system. Additionally a pair of postanal papillae, similar to the amphids and referred to as the phasmids, may be present. In plant-parasitic nematodes multiple phasmids may be present and their arrangement has been used as a taxonomic character in the genus *Hoplolaimus*.

More recently Lichtenfels *et al.* (1995) described a new epidermal gland in female nematodes in the family Trichostrongylidae which they described as the perivulval pores because of their close association with the vulva. In *Haemonchus* they are located slightly dorsal to

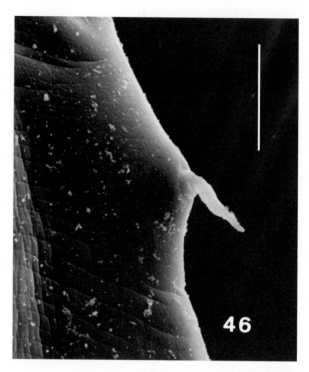

Figure 2.46. *Oesophagostomum columbianum*, Strongyloidea, showing one of the needle-like cervical papillae. Scale bar = 14 μm. Reproduced courtesy of CABI Biosciences.

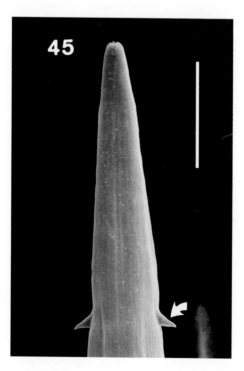

Figure 2.45. *Haemonchus similis*, Trichostrongyloidea, anterior end showing the conical cervical papillae projecting above the body surface (arrowed). Scale bar = 100 μm. Reproduced courtesy of CABI Biosciences.

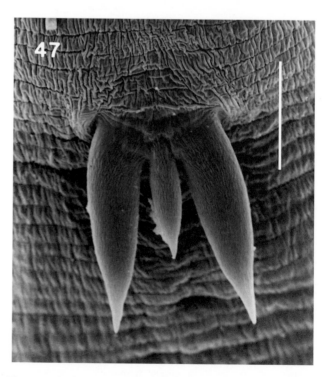

Figure 2.47. *Desportesius* sp., Acuarioidea, showing one of the cervical papillae with three prongs. Scale bar = 15 μm. Reproduced courtesy of CABI Biosciences.

each lateral line in the region of the infundibulum of the posterior ovejector (Figure 2.48). Their function is at present unknown but is being investigated (Lichtenfels *pers. comm.*). Lichtenfels *et al.* (1995) suggested that they appear to have some similarities to the phasmids. The hemizonid is ventrally situated near the secretory-excretory pore and represents a ventro-lateral commissure of the nervous system. The hemizonid is approximately 2.5 times the normal annular striations in width and lies between the cuticle and epidermis. Bird (1971) suggested they are a modified cuticular sense organ but their function is still to be determined. Hunt (1993) described the hemizonid in the Aphelenchida as generally located posterior to the secretory-excretory pore which opens on the 'ventro-median line behind the median bulb'. Other sense organs include hemizonions, cephalids, caudalids and ocelli or eye spots.

Secretory-Excretory System

Externally the presence of this system is identified by what is termed in the majority of descriptions as the excretory pore. The pore is found on the ventral surface in the region of the pharynx at any level from the cephalic region to just posterior to the pharyngeo-intestinal junction (*Torynurus convolutus*, Figure 2.49). Chitwood and Chitwood (1974) described two basic types of excretory system within which there is a wide variety of form.

The first type consists of two canals which are found in the lateral cords and are connected anteriorly and ventrally by an excretory sinus. The two lateral canals

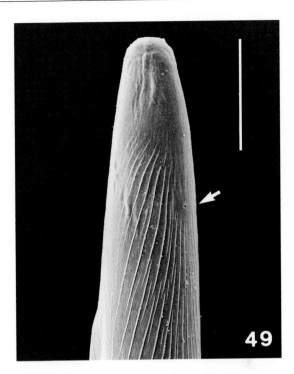

Figure 2.49. *Torynurus convolutus,* Metastrongyloidea, anterior end showing the position of the secretory-excretory pore (arrowed). Scale bar = 136 μm. Reproduced courtesy of CABI Biosciences.

may extend forward forming the H-shaped system. Variations in this system include the presence or absence of two subventral secretory-excretory glands situated in the body cavity and the canals being confined to one lateral cord. If there are no extensions forward the arrangement is referred to as the inverted U-shaped system. If only one lateral canal is present the system is termed asymmetric. The canals link to the secretory-excretory pore via a median duct. Paired subventral glands open into the transverse canal. They have a granular cytoplasm and are thought to be secretory. These types of excretory system are found in the Secernentea i.e. the nematodes which have phasmids.

The second type of excretory system consists of a single, ventral, gland cell or renette situated in the body cavity or pseudocoelom which connects directly to the ventral secretory-excretory pore. The duct connecting to the pore is lined with cuticle at its distal end only. In some species the duct is looped within the tissue of the ventral gland. At the point of connection there may be a dilation or ampulla. This type is found in the Adenophorea i.e. the nematodes without phasmids which includes many free-living, marine and freshwater species.

More recently Bird (1971) and Bird and Bird (1991) questioned the validity of referring to this system as solely an excretory system. They considered that this is primarily based on morphological grounds and the evidence is insufficiently clear-cut for this name to reflect the function of the system. Several examples of secretion in nematode species are cited and they adopted the term

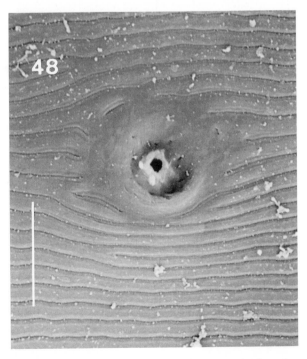

Figure 2.48. *Haemonchus similis,* Trichostrongyloidea, showing the perivulval pore posterior to the vulva. Scale bar = 8 μm. Reproduced courtesy of CABI Biosciences.

secretory-excretory system which they consider more appropriate to its function and this terminology is now gaining acceptance in the literature. Thus the external pore is referred to as the secretory-excretory pore.

Reproductive System

Female

The female has one to two tubular ovaries, rarely more, although in some animal-parasitic nematodes there may be up to four (*Polydelphis*), six (*Hexametra*) or thirty-two (*Placentonema gigantissima*). The ovary, which is a blind ending sac, is connected to a uterus which terminates in a vaginal opening, the vulva, on the ventral surface of the body. The position of this opening may be anywhere along the length of the body from near the anterior end (*Serratospiculum seurati*, Figure 2.50) at the end of a vulvar flap (*Haemonchus similis*, Figure 2.51) near the posterior end (*Oesophagostomum columbianum*, Figure 2.52) or protruding from the body surface near the tail (*Pseudalius inflexus*, Figure 2.53). In telogonic forms the ovary consists of two zones, namely, the germinal zone and the growth zone. The germinal zone is a region of rapid division of small cells. The growth zone is a region where these cells gradually increase in size. This latter zone may form the larger part of the ovary and be exceptionally long in parasitic species. The vulva in free-living forms is a simple, often slit-like, opening with a transverse vagina leading into a pair of opposed uteri without any highly developed musculature at their distal ends. The

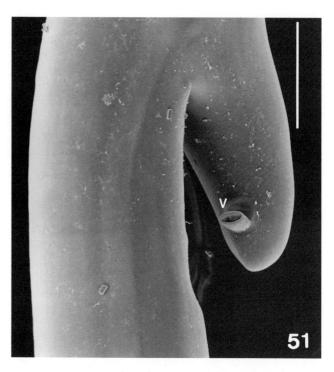

Figure 2.51. *Haemonchus similis*, Trichostrongyloidea, showing vulvar opening near the tip of a vulva flap. V = Vulva. Scale bar = 120 μm. Reproduced courtesy of CABI Biosciences.

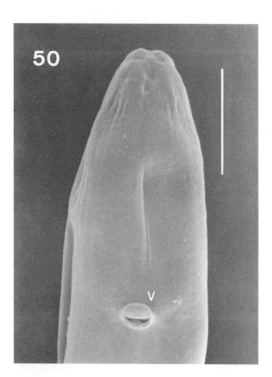

Figure 2.50. *Serratospiculum seurati*, Filarioidea, showing the vulvar opening near the anterior end. V = Vulva. Scale bar = 380 μm. Reproduced courtesy of CABI Biosciences.

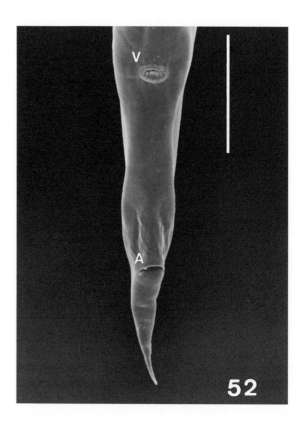

Figure 2.52. *Oesophagostomum columbianum*, Strongyloidea, showing the vulvar opening near the posterior end anterior to the anus. A = Anus; V = Vulva. Scale bar = 270 μm. Reproduced courtesy of CABI Biosciences.

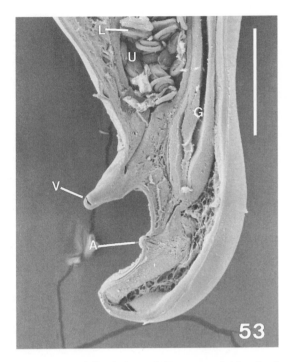

Figure 2.53. *Pseudalius inflexus*, Metastrongyloidea, body cut open to see the uterus with larvae and the vulva opening on a projection just anterior to the anus near the tail. A = Anus; G = Alimentary tract; L = Larva; U = Uterus; V = Vulva. Scale bar = 231 μm. Reproduced courtesy of CABI Biosciences.

uteri lead into short, sometimes little differentiated, oviducts and short tubular ovaries. A seminal receptacle may be present as a small swelling of each uterus or the ovarial end of the uterus may take on this function. The reproductive system may recurve at the junction of the ovary and uterus or ovary and oviduct if present.

In animal-parasitic forms the entire reproductive tract increases in length, which enables an increase in egg production, and the uterus and vagina have a highly developed musculature designed for the efficient ejection of eggs and referred to as the ovejector. The uteri are connected to the vagina which often develops into an elongated muscular tube lined with cuticle, called the vagina vera. Seurat (1912) named the components of the ovejector as the vestibule, sphincter and trunk or trompe. The latter component is now referred to as the infundibulum (Durette-Desset, 1983). When Lichtenfels (1980) described the development of this structure for the Strongyloidea he divided the morphology of the ovejectors into two types, namely, 'ovejector I' which is usually Y-shaped (*Triodontophorus serratus*) and 'ovejector II' which is generally J-shaped (*Oesophagostomum dentatum*). A seminal receptacle may be present as an outpocketing of the vagina vera or as a modification of the distal ends of the uteri while in monodelphic forms the seminal receptacle may be formed as a sac considered to be the vestigial remains of the second reproductive system. The oviduct represents a thick-walled region between the uterus and ovary. Bladder-like structures called

vaginal glands may be present near the distal part of the vagina formed from muscle cell bodies pressed out of the muscle cell walls. Sphincter muscles may be present at or near the vulval opening.

Chitwood and Chitwood (1974) followed the terminology of Seurat in defining the three variations in arrangement of the uteri, namely, amphidelphic when the uteri are opposed at their distal ends, opisthodelphic when the uteri are parallel and directed posteriorly and prodelphic when the uteri are directed anteriorly. Other terms accepted by Chitwood and Chitwood (1974) and which are also widely accepted are monodelphic for a single set of female reproductive organs, didelphic for two complete sets of reproductive organs, tetradelphic for four complete sets of reproductive organs and polydelphic for more than four sets of reproductive organs (Figure 2.54).

Male

The male reproductive system consists of a testis, seminal vesicle and vas deferens. In some forms a vas efferens is present and separates the testis from the seminal vesicle (Figure 2.55). A single set of reproductive organs is referred to as monorchic and a double set is referred to as diorchic. The posterior end of the vas deferens, if provided with a muscle layer, may act as an ejaculatory duct. In addition ejaculatory glands, which open into the posterior end of the vas deferens, may be present. In telogonic forms the testis has a germinal zone and a growth zone. In hologonic forms germ cell production can occur along any part of the testis and there are no clear divisions into zones. The vas deferens enters the rectum from the ventral side in all species except the trichinelloids where it enters on the dorso-lateral side. In the phasmidian nematode groups the junction of the vas deferens with the hind gut is very close to the intestino-rectal valve. In these species there is almost no rectum and the whole of the hind gut forms part of the cloaca. In the adenophorean groups, particularly the Enoplida, the vas deferens may join the rectum more posteriorly so that there is a rectum and a cloaca.

Additional features of the male reproductive system are the spicules (*Gazellostrongylus lerouxi*, Figure 2.56), gubernaculum, telamon and spicular pouch. The spicules generally enter the cloaca from the dorsal side. In free-living nematodes the spicular pouch is often very delicate while in parasitic forms the spicular pouch is distinct. The spicules have protractor muscles forming a complete longitudinal layer on the pouch surface in parasitic forms and on the dorso-lateral sides in free-living forms. In the majority of the species each spicule is provided with two retractor and two protractor muscles attached to the proximal end of the spicule. The retractor muscles extend anteriorly towards the lateral chord where they attach to the body wall. The protractor muscles may be inserted post anally to the body wall or dorsal side of the spicular pouch. If only one spicule is present the retractor muscle extends to both right and left body walls. In the trichurids the cloacal lining is eversible forming a sheath covering part of the spicule which may be spinous. The sheath is continuous with the external cuticle and is sometimes

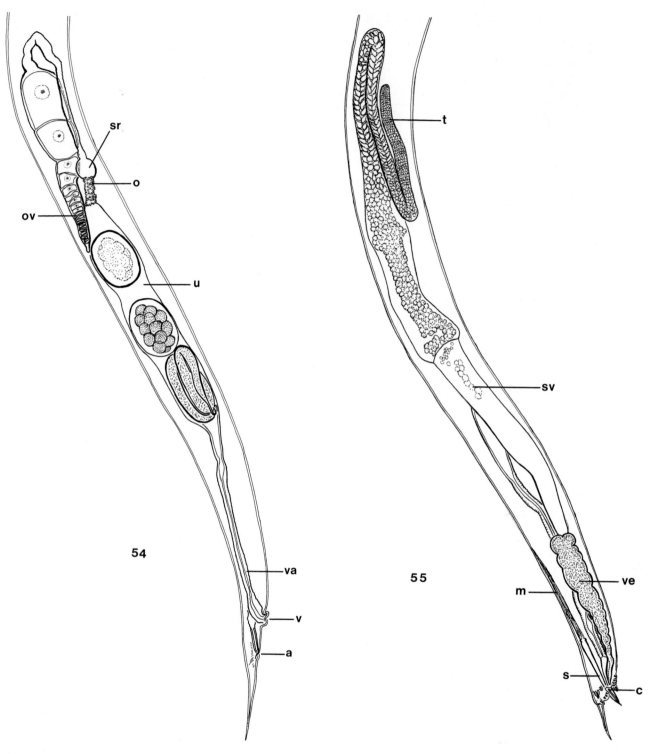

Figure 2.54. Monodelphic female reproductive system of the genus *Labiduris*. a = anus; o = oviduct; ov = ovary; sr = seminal receptacle; u = uterus; v = vulva; va = vagina. Original.

Figure 2.55. Monorchic male reproductive system of the genus *Labiduris*. c = cloaca; m = muscles; s = spicules (only one of the pair shown complete); sv = seminal vesicle; t = testis; ve = vas deferens and ejaculatory gland. Original.

referred to as a cirrus (*Trichuris suis*, Figure 2.57). There are generally two spicules but there are species with only one and some where this organ is absent. Each spicule consists of a tube covered by sclerotized cuticle containing

a central cytoplasmic core. Lee (1973) observed nerve axons in the cytoplasmic core of the spicules of *Heterakis gallinarum* and *Nippostrongylus brasiliensis*. Chitwood and Chitwood (1974) divided each spicule into three regions,

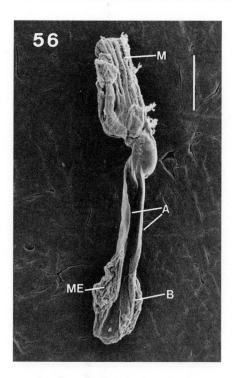

Figure 2.56. *Gazellostrongylus lerouxi*, Trichostrongyloidea, one of the two spicules dissected out of the body showing the muscles, branch , alae and additional membranes. A = Alae; B = Branch; M = Muscles; ME = Membrane. Scale bar = 100 μm. Reproduced from Gibbons (1986) with kind permission of CABI Publishing.

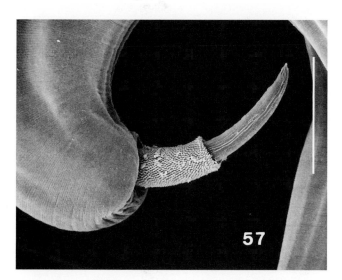

Figure 2.57. *Trichuris suis*, Trichinelloidea, posterior end of male showing the single spicule protruding through the cloaca and part of the spinous sheath or cirrus. Scale bar = 100 μm. Reproduced from Gibbons (1986) with kind permission of CABI Publishing.

namely, the head or capitulum, the shaft or calomus and the blade or lamina which may have one or more flanges. Free-living nematodes generally have equal, similar spicules which are short in relation to body length. Parasitic forms may also have equal and similar spicules

but some species may have unequal spicules which are different in shape and form. In species which have unequal spicules the left is generally longer than the right. In addition the spicules of many parasitic forms are branched, alate or ornamented (Andreeva, 1957).

The gubernaculum is formed from a cuticular thickening of the dorsal wall of the spicular pouch. The posterior end may project into the lumen of the spicular pouch separating the spicules. It is essentially a curved plate which forms a groove in which the spicules can move. In the superfamily Metastrongyloidea the gubernaculum forms three distinct regions, namely, the capitulum, corpus and crura (Anderson, 1978).

The term telamon was proposed by Hall (1921) for the supporting structures formed in the ventral and lateral walls of the cloaca of *Hyostrongylus rubidus* and which enable the spicules to be turned posteriorly when they protrude from the spicular pouch. This structure is only required in the bursate nematodes where the spicular pouch is not directly opposite the cloaca.

Sensory organs on the posterior end of the male tail of the Adenophorea are few (*Trichuris trichiura*, Figure 2.58) but in the Secernentea they are numerous and their arrangement is characteristic of each of the orders. Chabaud and Petter (1961a) considered the ancestral form to have 21 caudal papillae. These papillae are arranged in various ways and sometimes reduced. The arrangement within the orders can be briefly described as follows: Rhabditida with a simple bursa and numerous papillae (*Rhabditis*, Figure 2.59); Oxyurida reduced number of papillae surrounding the cloaca and a small cluster of papillae in the mid tail region (*Passalurus ambiguus*,

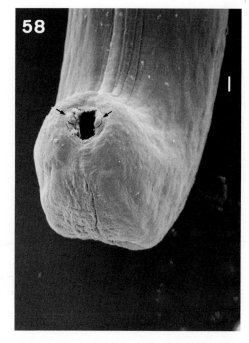

Figure 2.58. *Trichuris trichiura*, Trichinelloidea, posterior end of male showing the two large papillae either side of the cloaca (arrowed). Scale bar = 10 μm. Reproduced from Gibbons (1986) with kind permission of CABI Publishing.

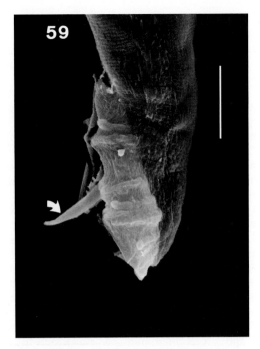

Figure 2.59. *Rhabditis* sp., Rhabditoidea, showing lateral view of posterior end of male with a simple bursa, spicule (arrowed) protruding through cloaca. Scale bar = 10 μm. Reproduced from Gibbons (1986) with kind permission of CABI Publishing.

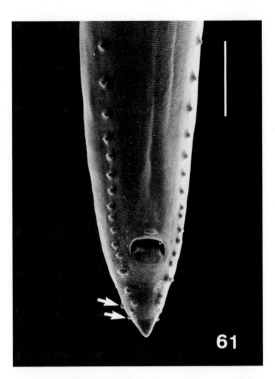

Figure 2.61. *Toxascaris leonina*, Ascaridoidea, posterior end of male with numerous papillae, some dorso-lateral in position (arrowed). Scale bar = 100 μm. Reproduced from Gibbons (1986) with kind permission of CABI Publishing.

Figure 2.60); Ascaridida with numerous papillae of which two or three of the pairs are dorso-lateral in position (*Toxascaris leonina*, Figure 2.61); the Spirurida with ventral or ventro-lateral papillae (*Procyrnea mansioni*, Figure 2.62) and the Strongylida in which the papillae have expanded into lateral and dorsal bursal lobes (*Haemonchus long-*

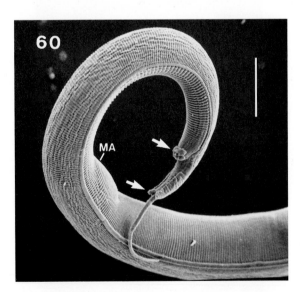

Figure 2.60. *Passalurus ambiguus*, Oxyuroidea, showing posterior end of male with a reduced number of papillae (arrowed) and showing one of the mamelons on the ventral surface. MA = Mamelon. Scale bar = 100 μm. Reproduced courtesy of CABI Biosciences.

Figure 2.62. *Procyrnea mansioni*, Habronematoidea, posterior end of male showing the papillae ventral in position (arrowed), spicule protruding through cloaca. S = Spicule. Scale bar = 100 μm. Reproduced from Gibbons (1986) with kind permission of CABI Publishing.

istipes, Figure 2.9). The terminology applied to the papillae is dependant on their position in relation to the cloaca. Papillae anterior to the cloaca are preanal or precloacal, papillae lateral and level with the cloaca are adanal or adcloacal and papillae posterior to the cloaca are postanal or postcloacal. Osche (1958) first numbered the papillae of the simplified bursa of *Rhabditis* and later Chabaud and Petter (1961b) expanded this numerical system to all the groups of the Secernentea from vertebrates so that the arrangement of the papillae can be directly compared. Chabaud *et al.* (1970) demonstrated the homology between the papillae of the Rhabditida and the Strongylida and produced a numerical system for naming the bursal rays which is now widely used in descriptions of species. The terminology is listed below:

Ray or papillae number	Description
0	ventral papilla or raylet
1	Prebursal papilla
2	Antero-ventral or ventro-ventral ray
3	Postero-ventral or latero-ventral ray
4	Antero-lateral or externo-lateral ray
5	Medio-lateral ray
6	Postero-lateral ray
7	Rays dorsal to the cloaca, often associated with an accessory bursal membrane, referred to as dorsal papillae or raylets
8	Externo-dorsal ray
9 and 10	Dorsal ray

Durette-Desset (1983) grouped the bursal rays of the superfamily Trichostrongyloidea so that the pattern of the rays could be used in descriptions and keys. Thus with the two ventrals rays close together, the antero-lateral separate from the other laterals and the medio-lateral and postero-lateral rays close together the bursa was designated the 2–1–2 type. Other variations included the 2–2–1 type bursa with the ventrals close together, antero-lateral and medio-lateral rays close together and the postero-lateral ray separated and the 1–3–1 type bursa with the antero-ventral and postero-lateral rays separate and the postero-ventral, antero-lateral and medio-lateral rays close together. The complete range of variations are given in Durette-Desset (1983) (Figure 2.63).

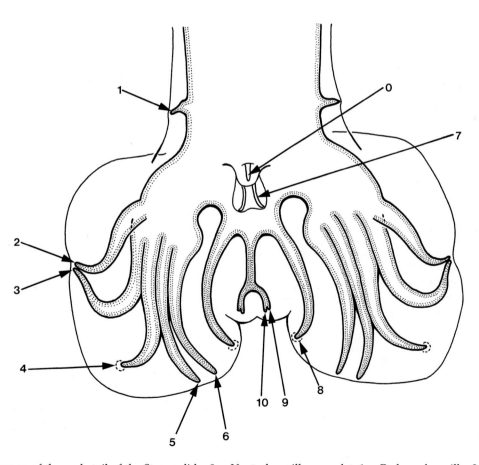

Figure 2.63. Diagram of the male tail of the Strongylida. 0 = Ventral papilla or raylet; 1 = Prebursal papilla; 2 = Antero-ventral ray; 3 = Postero-ventral ray; 4 = Antero-lateral ray; 5 = Medio-lateral ray; 6 = Postero-lateral ray; 7 = Dorsal papillae or raylets; 8 = Externo-dorsal ray; 9 and 10 = Dorsal ray. Reproduced from Gibbons and Khalil (1982) *Journal of Helminthology*, 56, figure 1–2.1 page 187 with kind permission of CABI Publishing.

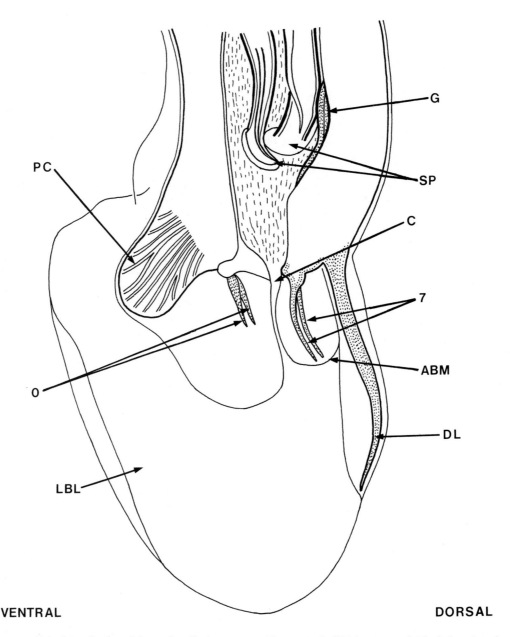

Figure 2.64. Diagram of the lateral view of the male tail of an ostertagid nematode (Trichostrongyloidea) showing the components of the genital cone. ABM = Accessory bursal membrane; C = Cloaca; DL = Dorsal lobe and ray; G = Gubernaculcum; LBL = Lateral bursal lobe; PC = proconus; SP = Spicules; 0 = Ventral raylets ; 7 = Dorsal raylets. Reproduced from Gibbons and Khalil (1983) with kind permission of the Systematics Association.

The bursate nematodes have an additional group of structures which are formed by the shortening of the tail and the expansion of the cuticle of the lateral and dorsal walls. The name given to this group of structures is the genital cone and they are completely surrounded by the lobes of the bursa (Figure 2.64). Ventral to the cloaca is the region called the proconus, which may be expanded (*Ostertagia*) or not. Between the proconus and the cloaca there is a papilla(e) which may be single (*Trichostrongylus colubriformis*), double (*Ostertagia ostertagi*) or triple (*Parostertagia heterospiculum*) surrounded by a simple membrane and is referred to as papilla(e) 0 by Chabaud *et al.* (1970), ventral rays by Andreeva

(1957) and ventral raylets by Stringfellow (1970, 1972). Dorsal to the cloaca there is a pair of papillae associated with an accessory bursal membrane referred to as papillae 7 by Chabaud *et al.* (1970), dorsal membrane and rays by Andreeva (1957) and dorsal raylets by Stringfellow (1971) (*Oesophagostomum columbianum*, Figure 2.65). Sjöberg in 1926 (in Dróżdż, 1965) first recognised a modification of the dorsal raylets and accessory bursal membrane in *Teladorsagia trifurcata* (=*Ostertagia trifurcata*). Dróżdż (1965) called this structure the Sjöberg organ and suggested it was formed by the secondary joining of an extension of the dorsal wall of the genital cone with the accessory bursal membrane.

Figure 2.65. *Oesophagostomum columbianum*, Strongyloidea, showing the ventral view of the genital cone with a single papilla 0 or ventral raylet, no proconus and the pair of papillae 7 or dorsal raylets with a reduced dorsal accessory bursal membrane. O = Ventral papilla or raylet; 7 = Dorsal papilla or raylet. Scale bar = 30 μm. Reproduced courtesy of CABI Biosciences.

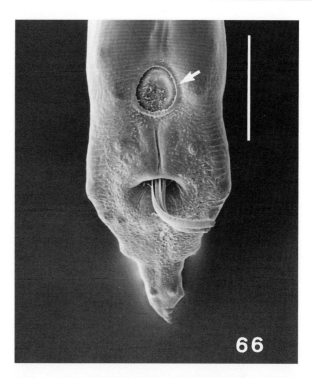

Figure 2.66. *Ascaridia galli*, Heterakoidea, male tail with precloacal sucker (arrowed) on its ventral surface. Scale bar = 380 um. Reproduced courtesy of CABI Biosciences.

Additional structures may be present lateral to the cloaca and between the papillae and the bursal lobes. These are referred to as genital appendages (*Ashworthius pattoni*) and cuticular appendages (*Strongylus equinus*). For a review of the structures of the genital cone see Gibbons and Khalil (1983).

In the superfamilies Cosmocercoidea, Heterakoidea, Oxyuroidea, Seuratoidea and Subuluroidea a structure called the precloacal sucker may be present; the number present is variable. This may be simple, forming a small depression or more complex with a sclerotized supporting ring (*Ascaridia galli*, Figure 2.66). The most complex form is seen in *Hoplodontophorus flagellum* in which the sucker is horseshoe-shaped and is strengthened by parallel struts.

Ornamentation of the posterior end of the male takes various forms including venation (*Haemonchus*) and spines (*Paracooperia*) on the ventral surface of bursal lobes, short discontinuous longitudinal ridges (*Cyathospirura seurati*) and a complete covering of spines (*Gnathostoma*). In the Oxyuroidea (*Passalurus ambiguus*, Figure 2.60, *Syphacia obvelata*, Figure 2.67) structures termed 'mobile gripping organs' by Dick and Wright (1974), more generally known as mamelons, are present. Two to four of these oval cushion-like structures may be present and they are formed of enlarged annules with central rows of spines. In the Filarioidea a region composed of transverse bands of ornamentation,

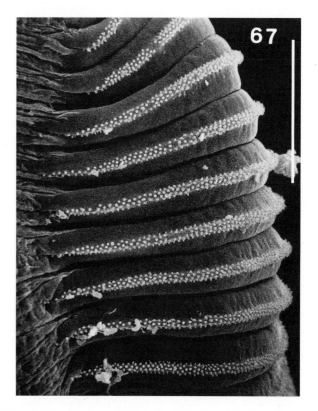

Figure 2.67. *Syphacia obvelata*, Oxyuroidea, a single mamelon or gripping organ found on the ventral surface in the posterior half of the body. Scale bar = 10 μm. Reproduced courtesy of CABI Biosciences.

Figure 2.68. *Dipetalonema gracile*, Filarioidea, showing the ornamentation on the ventral surface of the male just anterior to the cloaca. Scale bar = 10 μm. Reproduced from Gibbons (1986) with kind permission of CABI Publishing.

referred to as the *area rugosa* (*Dipetalonema gracile*, Figure 2.68), is seen anterior to the cloaca.

Further reading

For details of Life cycles see Chapter 3; the male and female gametes and fertilisation see Chapter 4, and Embryology see Chapter 5. For more detailed studies of the structures of nematodes Chitwood and Chitwood (1974), Grassé (1965a and 1965b), Bird and Bird (1991), Wright (1991) and Maggenti (1991) are recommended.

Acknowledgements

I would like to express my thanks to Professor J.D. Eisenbeck, Department of Plant Pathology, Physiology and Weed Science, College of Agriculture and Life Sciences, Virginia Polytechnic and State University for permission to use Figure 2.35; Dr D.J. Hunt, CABI Biosciences for Figures 2.10, 2.21, 2.30; Dr L.F. Khalil and Mr R. Tranfield, formerly of CAB International, for making specimens available and assisting in their preparation; CABI Biosciences for giving permission to reproduce the non published scanning electron micrographs as indicated in the text; CABI Publishing and Kluwer Academic Publishers for permission to use the previously published figures indicated in the text; Professor D.E. Jacobs of The Royal Veterinary College and Dr L.F. Khalil for their constructive comments on the draft manuscript which assisted in the preparation of this chapter.

References

Anderson, R.C. (1978) Key to genera of the superfamily Metastrongyloidea. In *CIH Keys to the Nematode Parasites of Vertebrates*, edited by R.C. Anderson, A.G. Chabaud and S. Willmott, No. 5 pp. 1–40. Farnham Royal, Slough: Commonwealth Agricultural Bureaux.

Andreeva, N.K. (1957) [*Atlas of helminths (Strongylata) of domesticated and wild ruminants in Kazakh SSR*] pp. 215, plates 98. Institut Veterinarii Kazakhskogo Filiala Vashnil, Tasjkent.

Ansel, M., Thibaut, M. and Saez, H. (1974) Scanning electron microscopy on *Parascaris equorum* (Goeze, 1782) Yorke and Maplestone, 1926. *International Journal for Parasitology* **4**, 17–23.

Bain, O. (1981) Le genre *Onchocerca*: hypothèses sur son évolution et clé dichotomique des espèces. *Annales de Parasitologie Humaine et Comparée* **56**, 503–526.

Bird, A.F. (1971) *The structure of nematodes*. pp. xi + 318. New York & London: Academic Press.

Bird, A.F. and Bird, J. (1991) *The structure of nematodes*. 2nd Edition, pp. xviii + 316. San Diego: Academic Press.

Baylis, H.A. and Lane, C. (1920) A revision of the nematode family Gnathostomidae. *Proceedings of the Zoological Society of London*, **18–21**, 245–310.

Baylis, H.A. (1939) *The Fauna of British India, including Ceylon and Burma. Nematoda, Volume II (Filarioidea, Dioctophymatoidea and Trichinelloidea)* pp. 198–200. London: Taylor and Francis.

Chabaud, A.G. (1974) Keys to subclasses, orders and superfamilies. In *CIH Keys to the Nematode Parasites of Vertebrates*, edited by R.C. Anderson, A.G. Chabaud and S. Willmott, No. 1 pp. 6–17. Farnham Royal, Slough: Commonwealth Agricultural Bureaux.

Chabaud, A.G. and Petter, A. (1961a) Évolution et valeur systématique des papilles cloacales chez les nématodes phasmidiens parasites de vertébrés. *Comptes rendus des séances de l'Académie des Sciences*, **252**, 1684–1686.

Chabaud, A.G. and Petter, A.J. (1961b) Remarques sur l'evolution des papilles cloacales chez les nematodes phasmidiens parasites de vertebrés. *Parassitologia*, **3**, 51–70.

Chabaud, A.G., Puylaert, F., Bain, O., Petter, A.J. and Durette-Desset, M.-C. (1970) Remarques sur l'homologie entre les papilles cloacales des Rhabditides et les côtes dorsales des Strongylida. *Comptes rendus des séances de l'Académie des Sciences*, **271**, 1771–1774.

Chalfie, M. and White, J. (1988) The nervous system. In *The nematode Caenorhabditis elegans*, edited by W.B. Wood, pp. 337–391. Cold Spring Harbor: Cold Spring Harbor Laboratory Press.

Chitwood, B.G. and Chitwood, M.G. (1974) *Introduction to Nematology*. Revised edition reprinted in memory of Benjamin G. Chitwood, 1907–1972. pp. v + 334. Baltimore, Maryland: University Park Press.

Cobb, N.A. (1923) Interesting features in the anatomy of nemas. *Journal of Parasitology*, **9**, 242–244.

De Coninck, L.A.P. (1965) *Classe de Nématodes*. In *Traité de Zoologie. Anatomie, Sytematique, Biologie*, edited by P.P. Grassé, Volume 4 (2) Némathelminthes, pp. 1–217. Paris: Masson et Cie.

Dick, T.A. and Wright, K.A. (1974) The ultrastructure of the cuticle of the nematode *Syphacia obvelata* (Rudolphi, 1802). III Cuticle associated with the male reproductive structures. *Canadian Journal of Zoology*, **52**, 178–182.

Dróżdż, J. (1965) Studies on helminths and helminthiases in Cervidae. I. Revision of the subfamily Ostertagiinae Sarwar, 1956 and an attempt to explain the phylogenesis of its representatives. *Acta Parasitologica Polonica*, **13**, 445–481.

Durette-Desset, M.-C. (1971) Essai de classification des nématodes héligmosomes. Correlations avec la paleobiogeographie des hôtes. *Mémoires du Muséum National d'Histoire Naturelle Nouvelle Série A Zoologie*, **69**, 1–126.

Durette-Desset, M.-C. (1983) Keys to genera of the superfamily Trichostrongyloidea. In *CIH Keys to the Nematode Parasites of Vertebrates*, edited by R.C. Anderson and A.G. Chabaud, No. 10 pp. 1–86. Farnham Royal, Slough: Commonwealth Agricultural Bureaux.

Eisenback, J.D., Hirschmann, H. and Triantaphyllou, A.C. (1980) Morphological comparison of *Meloidogyne* female head structures, perineal patterns and stylets. *Journal of Nematology* **12**, 300–313.

Eisenback, J.D. and Hirschmann, H. (1981) Identification of *Meloidogyne* species on the basis of head shape and stylet morphology of the male. *Journal of Nematology* **13**, 513–521.

Gibbons, L.M. (1986) SEM guide to the morphology of nematode parasites of vertebrates. pp. 199. CAB International, Farnham Royal, Slough, U.K.

Gibbons, L.M. and Keymer, I.F. (1991) Redescription of *Tanqua tiara* (Nematoda, Gnathostomidae), and associated lesions in the stomach of the Nile monitor Lizard (*Varanus niloticus*). *Zoologica Scripta*, **20**, 7–14.

Gibbons, L.M. and Khalil, L.F. (1982) A key for the identification of the genera of the nematode family Trichostrongylidae Leiper, 1912. *Journal of Helminthology*, **56**, 185–233.

Gibbons, L.M. and Khalil, L.F. (1983) Morphology of the genital cone in the nematode family Trichostrongylidae and its value as a taxonomic character. Systematics Association Special Volume No. 22, *Concepts in Nematode Systematics*, edited by A.R. Stone, H.M. Platt and L.F. Khalil] pp. 261–271. London and New York: Academic Press.

Gibbons, L.M. and Khalil, L.F. (1988) Two nematodes, *Paratractis hystrix* (Diesing, 1851) and *Buckleyatractis marinkelli* n. g., n. sp. (Atractidae: Cosmocercoidea) from *Podocnemis* spp. in Colombia. *Systematic Parasitology*, **12**, 187–198.

Grassé, P.P. [Editor] (1965a) *Traité de Zoologie, Anatomie, Systématique, Biologie, Tome IV, Fasc 2 Némathelminthes (Nématodes)*, pp. 731. Paris: Masson et Cie.

Grassé, P.P. [Editor] (1965b) *Traité de Zoologie Anatomie, Systématique, Biologie. Tome IV, Fasc 3 Némathelminthes (Nématodes–Gordiacés) rotifères, gastrotriches, kinorhynches*, pp. 764. Paris: Masson et Cie.

Hall, M.C. (1921)* Two new genera of nematodes with a note on a neglected nematode structure. *Proceedings of the United States National Museum*, **59**, 541–546.

Hartwich, G. (1974) Keys to the genera of the Ascaridoidea. In *CIH Keys to the Nematode Parasites of Vertebrates*, edited by R.C. Anderson, A.G. Chabaud and S. Willmott, No. 2 pp. iv + 15. Farnham Royal, Slough: Commonwealth Agricultural Bureaux.

Hirschmann, H. (1983) Scanning electron microscopy as a tool in nematode taxonomy. Systematics Association Special Volume No. 22, *Concepts in Nematode Systematics*. edited by Stone, H.M. Platt and L.F. Khalil] pp. 95–111. London and New York: Academic Press.

Hunt, D.J. (1993) *Aphelenchida, Longidoridae and Trichodoridae: Their systematics and bionomics*. pp. xx + 352. Wallingford: CAB International.

Hunt, D.J. and Sutherland, J.A. (1984a) *Ichthyocephaloides dasyacanthus* n. g., n. sp. (Nematoda: Rhigonematoidea) from millipede from Papua New Guinea. *Systematic Parasitology*, **6**, 141–146.

Hunt, D.J. and Sutherland, J. (1984b) Three new species of *Carnoya* Gilson, 1898 (Nematoda: Rhigonematidae) from Papua New Guinea. *Systematic Parasitology*, **6**, 229–236.

Inglis, W.G. (1966) The origin and function of the cheilostom complex in the nematode *Falcaustra stewarti*. *Proceedings of the Linnean Society of London*, **177**, 55–62.

Jenkins, T. (1969) Electron microscopy observations of the body wall of *Trichuris suis* (Shrank, 1788) (Nematoda: Trichuroidea) 1. The cuticle and bacillary band. *Zeitschrift für Parasitenkunde*, **32**, 374–387.

Jenkins, T. (1970) A morphological and histochemical study of *Trichuris suis* (Shrank, 1788) with special reference to the host-parasite relationship. *Parasitology*, **61**, 357–374.

Kim, C. and Ledbetter, M.C. (1980) Surface morphology of *Trichinella spiralis* by scanning electron microscopy. *Journal of Parasitology*, **66**, 75–81.

Lee, D.L. (1965) The cuticle of adult *Nippostrongylus brasiliensis*. *Parasitology* **55**, 173–181.

Lee, D.L. (1973) Evidence for a sensory function for the copulatory spicules of nematodes. *Journal of Zoology*, **169**, 281–285.

Lichtenfels, J.R. (1975) Helminths of domestic equids. Illustrated key to genera and species with emphasis on North American forms. *Proceedings of the Helminthological Society of Washington*, **42** (special volume), pp. v + 92.

Lichtenfels, J.R. (1980) Keys to the superfamily Strongyloidea. In *CIH Keys to the Nematode Parasites of Vertebrates*, edited by R.C. Anderson, A.G.

Chabaud and S.M. Willmott, No. 7. pp. 1–41. Farnham Royal, Slough: Commonwealth Agricultural Bureaux.

Lichtenfels, J.R., Pilitt, P.A. and Lancaster, M.B. (1988) Systematics of the nematodes that cause ostertagiasis in cattle, sheep and goats in North America. *Veterinary Parasitology* **27**, 3–12.

Lichtenfels, J.R., Wergin, W.P., Murphy, C. and Pilitt, P.A. (1995) Bilateral, perivulval cuticular pores in trichostrongylid nematodes. *Journal of Parasitology*, **81**, 633–636.

Maggenti, A.R. (1991) General nematode morphology. In *Manual of Agricultural Nematology*, edited by W.R. Nickle pp. 3–46. New York: Marcel Decker.

Mobarak, M.S. and Ryan, M.F. (1998) Ultrastructure of the buccal capsule of the equine nematode *Strongylus vulgaris* with special reference to the dorsal gutter. *Journal of Helminthology*, **72**, 167–177.

Muller, R. (1979) Identification of *Onchocerca*. In *Problems in the Identification of parasites and their vectors*, edited by A.E.R. Taylor and R.L. Muller. Symposium of the British Society for Parasitology, London 27 October, 1978, Volume 17, pp. 175–206. Oxford, U.K: Blackwell Scientific Publications.

Osche, G. (1958) Die Bursa und Schwazstruckturen und ihre Aberrationen bei den Strongylina (Nematoda). Morphologische Studien zum Problem der Pluri- und Paripotenzerscheinungen. *Zeitschrift für Morphologie und Ökologie der Tiere*, **46**, 571–635.

Seurat, L.G. (1912) Sur la morphologie de l'ovijecteur de quelques nématodes. *Compte Rendu de séances de la Société de Biologie*, **72**, 778–781

Sheffield, H.G. (1963) Electron microscopy of the bacillary band and stichosome of *Trichuris muris* and *T. vulpis*. *Journal of Parasitology*, **49**, 998–1009.

Stringfellow, F. (1970) Comparative morphology of the genital cones of *Cooperia* (Nematoda, Trichostrongylidae) from cattle and sheep in the United States with a key to the common species. *Journal of Parasitology*, **56**, 1189–1198.

Stringfellow, F. (1971) Functional morphology and histochemistry of structural proteins of the genital cone of *Ostertagia ostertagi*, with a comparison of the genital cones of other *Ostertagia* common in cattle in the United States. *Journal of Parasitology*, **57**, 823–831.

Stringfellow, F. (1972) Comparative morphology of the genital cones of *Ostertagia* from sheep in the United States. *Journal of Parasitology*, **58**, 265–270.

Ward, S., Thomson, N., White, J.G. and Brenner, S. (1975) Electron microscopical reconstruction of the anterior sensory anatomy of the nematode *Caenorhabditis elegans*. *Journal of Comparative Neurology*, **160**, 313–338.

Wilfred, M. and Lee, D.L. (1981) Observations on the buccal capsule and associated glands of adult *Bunostomum trigonocephalum* (Nematoda). *International Journal for Parasitology*, **11**, 485–492.

Willmott, S. (1974) Glossary of terms. In *CIH Keys to the Nematode Parasites of Vertebrates*, edited by R.C. Anderson, A.G. Chabaud and S.M. Willmott] No. 1, pp. 1–5. Farnham Royal, Slough: Commonwealth Agricultural Bureaux.

Wright, K.A. (1963) Cytology of the bacillary bands of the nematode *Capillaria hepatica* (Bancroft, 1863). *Journal of Morphology*, **112**, 233–259.

Wright, K.A. (1968) Structure of the bacillary band of *Trichuris myocastoris*. *Journal of Parasitology*, **54**, 1106–1110.

Wright, K.A. (1975) Cephalic sense organs of the rat hookworm *Nippostrongylus brasiliensis*: form and function. *Canadian Journal of Zoology*, **53**, 1131–1146.

Wright, K.A. and Chan, J. (1973) Sense receptors in the bacillary band of trichuroid nematodes. *Tissue and Cell*, **5**, 373–380.

Wright, K.A. (1991) Nematoda. In *Microscopic Anatomy of Invertebrates*, edited by F.W. Harrison and E.E. Rupert Volume 4, pp. 111–195. New York: John Wiley and Sons, Inc.

3.   Life Cycles

Donald L. Lee

School of Biology, Leeds University, Leeds LS2 9JT, U.K

Patterns of Life History

The life history of all nematodes consists of an egg, four larval stages (L1, L2, L3, L4) and a fifth, adult, stage in which the reproductive and associated structures become fully developed and functional; they all moult four times during development (Figure 3.1).

The immature stages of nematodes were initially called larvae, however, the term 'juvenile' is now used by some authors and by some nematological journals for the immature stages of free-living and plant-parasitic nematodes. The synonymous term 'larva' is extensively used to describe the immature stages of animal-parasitic nematodes and in the literature on *Caenorhabditis*. Bird and Bird (1991) give reasoned arguments in favour of retaining the term 'larva' instead of the synonym 'juvenile' and 'larva' will be used throughout this book.

Many species of nematode possess an obligatory larval stage (e.g. trichostrongyles) which becomes arrested in development at the second moult as a specially adapted L3. It is non-feeding, is usually ensheathed in the cuticle of the L2 which has not been ecdysed (refered to as ensheathed larvae), is more resistant to environmental factors than other stages in the life cycle, has a different behaviour, and is adapted to dispersal. Fuchs (1915) coined the term 'Dauerlarven' (enduring larvae) for the larvae of *Diplogaster* and *Rhabditolaimus* found associated with some species of bark-beetles and this German term has continued to be used, as 'dauer larvae', to describe the resistant L3 of many free-living and insect-parasitic nematodes (Bovien, 1937). The most extensively studied dauer larva is that of *Caenorhabditis elegans* (Riddle, 1988; Riddle and Albert, 1997). One of the earliest descriptions of dauer larvae is by Maupas (1899) who found larvae of *Diplogaster* being carried under the elytra of coprophagous and wood-inhabiting beetles.

The sheath of ensheathed larvae can be important in the survival of the L3 until further development is able to take place. The dauer larvae of many species possess a sheath but some, such as *Caenorhabditis elegans*, do not although they do have structural differences from normal L3 (Riddle, 1988; Riddle and Albert, 1997). Dauer larvae, ensheathed larvae, larvae that are retained in an intermediate host, and larvae that do not hatch until ingested by a host, have an important role in the life cycle of nematodes as they become arrested in their development until they are stimulated by suitable environmental triggers to continue their development. With free-living nematodes, such as *C. elegans*, specific chemo-sensory cues initiate both entry into and exit from the dauer stage; the cues inform the larva if it is in an environment that contains sufficient food to support its reproduction (Riddle, 1988; Riddle and Albert, 1997). The L3 of trichostrongyles that infect the alimentary tract of ruminants require physiological triggers, such as appropriate levels of CO_2, temperature and pH to initiate exsheathment of the ensheathed larvae in the most suitable region of the alimentary tract for further development to occur (Rogers, 1962). Fully developed eggs of some nematodes, such as *Ascaris suum*, require similar stimuli to initiate hatching and this enhances the probability of the larvae emerging in the most suitable region of the host (see Chapter 6). The L4 of some trichostrongyles (*Ostertagia ostertagi*, *Haemonchus contortus*) exhibit a type of arrested development, called hypobiosis; this may be induced by environmental factors and/or the genetic makeup of the strain of nematode (see below and Chapter 15). Alterations to behaviour result in the dauer larva of some species (*C. elegans*) (Riddle, 1988; Riddle and Albert, 1997) and the ensheathed larva of others (*Nippostrongylus brasiliensis*, *Ancylostoma* spp.) mounting projections and waving their bodies in three-dimensional spirals and loops (nictation) whilst anchored by their tails (Croll and Matthews, 1977). This behaviour enhances the possibility that the dauer larvae of *C. elegans* (Riddle and Albert, 1997) and the dung-inhabiting nematodes *Diplogaster* spp. and *Rhabditis coarctata* (Bovien, 1937) make contact with, and attach to, an insect which may transport them (phoresis) from an environment that is no longer suitable for feeding and reproduction to a more suitable environment. It also improves the chances of skin-penetrating larvae, such as *N. brasiliensis* and *A. duodenale*, making contact with the surface of a suitable host prior to infecting it. The behaviour of ensheathed larvae of some trichostrongyles (*H. contortus*, *O. ostertagi*, *Trichostrongylus colubriformis*) differs from that of the L1 and L2 resulting in migration from the dung, in which they have developed to the ensheathed L3 stage, onto the surrounding herbage. Here they ascend blades of grass and are thus readily available to be ingested by grazing herbivores. The L3 of the cattle lungworm, *Dictyocaulus viviparus*, do not readily migrate from the dung, yet they become widely dispersed on the surrounding herbage. This dispersal is brought about by the fungus *Pilobolus*. The L3 move to the surface of the dung, which is usually heavily colonized by the fungus, and ascend the sporangiophores. *Pilobolus* discharges its sporangia violently in response to changes in illumination and it catapults these L3, together with the sporangia, onto adjacent pasture where both the larvae and the fungal spores are more readily available to grazing animals (Robinson, 1962).

Free-Living Nematodes

Nematodes occupy many different habitats and their life cycle strategies usually reflect this. Most species of nematode are free-living with enormous numbers living in the soil and in marine and freshwater sediments. Whilst most of these free-living nematodes feed mainly upon microorganisms some are predators, however, they all have a similar life cycle (Figure 3.1). The oocytes of *Panagrellus redivivus*, a microbivorous nematode, after fertilization by sperm from the male, continue to develop to the L2 stage in the uterus and the female gives birth to active larvae (ovoviviparity). These rapidly develop to the adult through another three moults and under optimal conditions the entire life cycle is complete in 69 hours (Dropkin, 1980). *Caenorhabditis elegans* is the most thoroughly studied nematode as a result of the extensive and detailed research that has been done over the past 35 years using this free-living nematode, selected in 1965 by Sydney Brenner as a model animal to study animal development, genetics and behaviour (Brenner, 1973, 1974, 1988; Riddle *et al.*, 1997; Wood, 1988a). It is a microbivorous soil-dwelling nematode, found in many parts of the world, with a life cycle of about 3 days and a normal life-span of about 2 weeks under optimal conditions (Wood, 1988b; Riddle and Albert, 1997). There are two sexes, hermaphrodites and males. The adult hermaphrodite is structurally a female and can reproduce by self-fertilization but hermaphrodites are unable to fertilize each other (Wood, 1988b). Males arise spontaneously and at low frequency and can fertilize hermaphrodites. Hermaphrodites are protandrous, with sperm they have produced being stored in the spermatheca of the adult hermaphrodite. The hermaphrodite then uses these sperm to fertilize its own oocytes. A hermaphrodite that has self-fertilized lays about 300 eggs in its reproductive lifetime (Wood, 1988b) but it produces many more oocytes than sperm so with self-fertilization the number of offspring is limited by the number of sperm inside the adult. However, the adult hermaphrodite has the potential to produce more than 1000 offspring after mating with a male (Riddle *et al.*, 1997). The fertilized egg develops rapidly at 25°C with embryogenesis being completed in 14 hours, followed by hatching of the L1 and development through four larval stages to the adult in 36 hours in the presence of abundant food and suitable environmental conditions (Riddle, 1988; Riddle *et al.*, 1997; Riddle and Albert, 1997). With increased population density and an increasingly limited food supply, that is when environmental conditions become unsuitable for successful reproduction, development is arrested at the second moult with the L3 being a dauer larva (Figure 3.1).

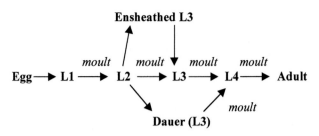

Figure 3.1. Basic life cycle of nematodes. Some species produce a modified L3 that is a resistant dauer larva; some others produce an obligatory ensheathed L3. Both of these types are arrested in their development and require suitable stimuli to continue development.

Plant-Parasitic Nematodes

Plant-parasitic nematodes live both in soil and in or on plants and details of their life cycle vary among species. However, the life cycles of plant-parasitic nematodes are mostly simple and direct but usually take longer than free-living nematodes. The L1 develops within the egg and the first moult is within the egg. The L2 hatches from the egg and then undergoes a further three moults to become adult. Surface feeding nematodes, such as *Trichodorus* and *Paratylenchus*, move freely over and feed on the roots of plants (see Chapter 9). Their life cycle progresses as outlined above, with eggs being laid in the soil and, under suitable conditions, there can be a rapid build up of the population. Ectoparasitic feeders, such as *Longidorus* and *Xiphinema*, also lay their eggs in the soil but they may take up to 2 years to complete their life cycle. Several pairs of *Anguina tritici*, which causes 'ear cockles' in wheat, induce the formation of a gall (cockle) that replaces the grain in an infected floret (Jones and Jones, 1974). This gall falls to the ground at harvesting and remains in the soil until the following growing season when larvae escape from the gall, seek out wheat seedlings at up to 30 cm distance, and invade the seedling. The larvae make their way to the growing point of the seedling and are carried up with the ear as the plant grows. They become adult within the ear of wheat where several pairs mate and lay thousands of eggs inside the ensuing gall. These eggs soon hatch to release L1, which moult to the L2 stage. The L2, which is the infective stage, is very resistant to desiccation and can survive in the desiccated state inside the gall for years. *Ditylenchus dipsaci*, the stem and bulb nematode, is able to invade, and develop on, a number of different plant hosts such as oats, clover, potato, onion and narcissi (Jones and Jones, 1974). Infective-stage larvae (L4) enter the tissues of the host plant and move about actively thus causing extensive damage in nonvascular tissues of the stem, leaves or cotyledons. The life cycle is simple and is complete in 19–23 days on onions at room temperature but is slower under field conditions. L4 within the soil, or planted together with the bulbs of onion or narcissi, invade the tissues of the host. The many eggs produced by the adults hatch within the plant and the released larvae feed upon it. They develop to the L4 stage and in onion and narcissi the

L4 become a form of dauer larva. These L4 congregate around the base of the bulb in large numbers in aggregations referred to as 'nematode wool'. The individuals on the surface die and form a protective shield around the individuals inside the mass of nematodes. The individuals within the 'wool' can survive drying of the infected bulbs for long periods. They revive in the presence of moisture and move through the soil to infect the next crop.

In species that are more highly adapted to parasitism the female tends to become sedentary, swollen, loses the power of movement and concentrates on feeding and egg laying. Adult females of endoparasitic species, such as *Heterodera, Globodera* and *Meloidogyne*, are adapted to a sessile life within their host plants; they become swollen and remain in one position whereas the males remain active. The life cycles of *Globodera* and *Heterodera* are very similar (Figure 3.2; see also Chapter 9). Eggs hatch, usually in response to a stimulus from the host (see Chapter 6), to release a second-stage larva, which is attracted to the roots of host plants. These L2 invade the plant intracellularly just behind the root tip, or near the emergence of lateral roots, and move into the cortex causing considerable cell death (Trudgill, 1997) (see also Chapter 9). The L3 of the male lies just beneath the epidermis of the root or has ruptured the cortex. The male L4 elongates within the cuticle of the L3 and after the fourth moult the male escapes into the soil where it attempts to seek out females. The L3 of the female migrates further into the root and eventually settles with its head near the stele. The saliva of the female larva induces the formation of multinucleate syncytia, which have characteristics similar to those of normal transfer cells, upon which they feed (see Chapters 8 and 9). The sex of larvae of these nematodes is not predetermined, the sex becoming obvious at the L3 stage. Larvae that produce syncytia that are restricted in size by competition, by a resistant reaction

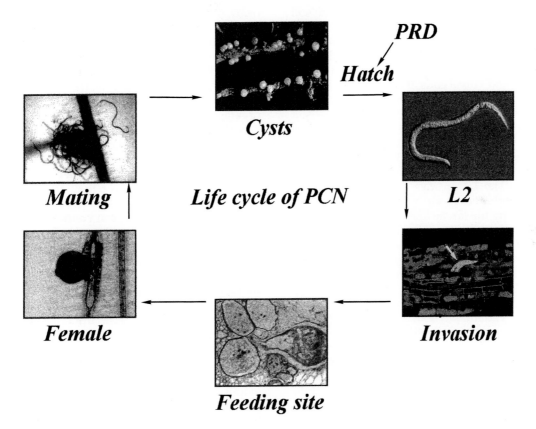

Life cycle of PCN

Cysts

Hatch *PRD*

Mating   **L2**

Female   **Invasion**

Feeding site

Figure 3.2. Life cycle of a cyst nematode (PCN). Bold type refers to nematode life stages represented in the figure above. Cysts, each containing up to 300 eggs lie dormant in the soil until diffusates from the roots of a suitable host plant are detected. Each egg contains a single L2 (the moult from L1 to L2 having occurred in the egg) which responds to root diffusate (PRD) from the host plant by taking up water and hatching. The L2 migrate through the soil until they locate the host plant roots, which they invade just behind the growing root tip. Following invasion the nematodes set up a multinucleate feeding site on which they depend for the nutrients required for development through a further three moults to the adult stage. [Note that the feeding site image shown is that of a root knot nematode (*Meloidogyne*) rather than of a cyst nematode.] Females grow to an extent that they burst out of the roots and adopt a spherical appearance while the males retain the vermiform shape of the L2. After mating the female dies and her body wall forms the cyst which protects the unhatched L2 in the soil until another host is detected. All images obtained from the Nemapix CD ROM with the exception of the invading nematode, which was obtained from Dr A. Kumar (J. Jones: Scottish Crop Research Institute). The images have been converted into black and white from colour images. The image of the L2 is courtesy of Prof. M. McClure, the feeding site courtesy of Prof. Lopez and the images of the female, mating nematodes and cysts courtesy of Dr U. Zunke. In all cases the original authors retain copyright of the images.

of the host, or because they were confined in a lateral root, become males. Larvae that induce the formation of larger syncytia become females (Trudgill, 1997). L4 females enlarge and become saccate and eventually rupture their cortex so that most of their body lies on the outside of the root but their head remains buried in the root, being held in place by a sticky secretion or cement. The ovaries become greatly enlarged in the young female and a gelatinous egg sac is produced. Fertilization takes place at this stage. The female produces up to 500 or more eggs, some of which may be deposited in the egg sac, but most are retained within the body of the female. After the female dies her body wall hardens (see Chapter 7) and forms a protective envelope or cyst for the enclosed eggs. Depending on the species this cyst may remain white or turn cream, golden yellow or brown (Ellenby, 1946). The cyst drops off the root into the soil where the L2 develop within the eggs inside the cyst. Between crops most of the larvae remain dormant within the eggs inside the cyst thus enabling them to persist in the soil for many years. Some larvae hatch under suitable environmental conditions but exudates (hatching factor) from appropriate host plants stimulate hatching of the L2 which escape from the cyst through apertures at the former head and vulval ends of the dead female (see Chapter 6). The number of generations per year is governed by environmental conditions, such as soil temperature and soil moisture, and the length of time the host plant occupies the ground (Jones and Jones, 1974). The life cycle of *Meloidogyne* is similar to that of *Heterodera* and *Globodera* but the enlarged females do not form cysts. The L2 invades roots of host plants intercellularly and undergoes a quick succession of moults. Feeding by the larvae and the adult causes the cells at the feeding site to enlarge and their nuclei repeatedly divide to form multinucleate giant cells (see Chapter 9). Several species of *Meloidogyne* reproduce sexually (Santos, 1972) and the sex is determined as described above for *Heterodera*, but some species (e.g. *Meloidogyne incognita* and *M. javanica*) are mitotically parthenogenetic (Triantaphyllou, 1979). The eggs are laid in the egg sac and both the female and the egg sac remain buried within a gall containing large multinucleate giant cells that have been induced by the feeding nematode. Eggs of *Meloidogyne* readily hatch in water and root exudates play a much less important role in hatching than in *Heterodera* and *Globodera*. For more detailed information on hatching see Perry (Chapter 6) and for more on life cycles see Jones and Jones (1974).

Entomogenous Nematodes

Mermithids are parasitic as larvae in invertebrates whereas the adults are free-living and non-feeding. Adult females of *Mermis subnigrescens* live coiled up in the soil over winter; during warm showery weather they move onto the surface of the soil and ascend the stems of plants where they lay their eggs on the foliage. If the eggs are eaten by a phytophagous arthropod, such as an earwig or grasshopper, the larva hatches in

the alimentary tract of the insect. With the aid of a mouth stylet, it pierces the gut wall and passes into the haemocoele where it feeds on the fat body and viscera of the insect, grows and moults. It completes its growth in 8 to 10 weeks, after which it emerges from the insect as a large L4, killing the insect as it does so, and migrates into the soil where the final moult occurs the following spring. Copulation takes place in the soil and the female becomes filled with eggs but she does not ascend to lay these until the following summer (Christie, 1937).

A number of nematodes belonging to the Steinernematidae (e.g. *Steinernema feltiae, S. glaseri*) and Heterorhabditidae (e.g. *Heterorhabditis bacteriophora*) are obligate parasites of insects and some are now being used as biological control agents (see Chapter 19). In the case of *Heterorhabditis* L3 in the soil penetrate through the cuticle of the host insect and migrate to the haemocoele, whereas L3 of *Steinernema* spp. in the soil invade the haemocoele through natural body openings or by penetrating the gut after ingestion by the insect. These L3 transport bacteria belonging to the genus *Xenorhabdus* into the insect, inside which the bacteria cause septicaemia and ultimately the death of the insect. The bacteria establish conditions suitable for the growth and reproduction of the nematodes and also inhibit colonization of the cadaver by other microorganisms (Poinar, 1979; see also Chapter 19). The nematodes pass through several generations in the dying insect and its cadaver then, presumably when food resources dwindle, thousands of dauer larvae in the L3 stage are formed and they migrate from the cadaver into the soil where they attempt to contact a new host (see Chapter 20). Another entomogenous nematode used in biological control is *Beddingia* (syn. *Deladenus*) *siricidicola* (see Chapter 19). This nematode is a facultative parasite of the woodwasp (*Sirex noctilio*) and has both free-living (mycophagous) and parasitic generations. The female insect introduces the spores of the fungus, *Amylostereum areolatum*, together with toxic mucus, into a tree when it inserts its ovipositor into the tree during egg laying. The insect larvae feed on the fungus. Some female *Sirex* that are infected with the nematode contain sterile eggs, each containing up to 200 nematode larvae. These larvae escape from the eggs inside the tree, or escape from the insect during oviposition, and begin to feed on the fungus. They pass through several generations feeding on the fungus but if a *Sirex* larva is present then the nematode differentiates into a morphologically distinct female that does not feed on the fungus. After mating these females enter the haemocoel of the insect larva, but they do not develop a reproductive system until the insect larva pupates. Larval nematodes within the female hatch, but do not leave, the female nematode until the later stages of pupation of the insect when they emerge into the haemocoele of the insect and migrate to the reproductive organs. Here they enter the developing eggs of the insect and pass into another tree at oviposition (Bedding, 1984; Chapter 19).

Several insects act as hosts for certain species of

oxyurid nematode, such as *Leidynema appendiculata* in cockroaches. The life cycle of these nematodes is direct with the adults inhabiting the hind-gut of the insect and eggs passed in the faeces develop to an infective-stage larva within the egg. The cycle continues when the egg is eaten by the insect. These nematodes appear to cause no harm to the insect.

For more information on the life cycles of entomogenous nematodes see Chapters 19 and 20.

Animal-Parasitic Nematodes

Several major groups of animal-parasitic nematode are monoxenous in that they have a direct life cycle involving only one host. In primary monoxeny (Anderson 1988, 1992) it is probable that no intermediate host was involved in transmission at any stage in the evolution of parasitism. Monoxenous nematodes can infect the host through ingestion of an egg containing the infective larva (*Enterobius vermicularis* in humans; *Leidynema appendiculata* in cockroaches) or through ingestion of an infective-stage larva (*Haemonchus contortus* and *Trichostrongylus colubriformis* in ruminants) and are not involved in tissue migration (Figure 3.3). Eggs of trichostrongyle nematodes, such as *Trichostrongylus colubriformis* and *H. contortus*, are passed to the exterior in the faeces of their host where they hatch as an L1. The L1 and L2 are microbivorous but at the second moult the L3, which is a non-feeding stage that is arrested in its development, retains the cuticle of the L2 as a sheath (Figure 3.1). The ensheathed L3 migrates from the faeces and ascends herbage. If eaten by a suitable host the L3 is stimulated to exsheath in an appropriate part of the alimentary tract (rumen for *H. contortus*, abomasum for *T. colubriformis*) then passes into the next region of the alimentary tract where it moults into the L4 and then to the adult nematode. The larvae of some species (*H. contortus, Ostertagia ostertagi*) enter the wall of the alimentary tract and undergo further development there before returning to the surface of the mucosa where they become adult. Other monoxenous nematodes infect the host through the percutaneous route as free-living infective-stage larvae (*Necator americanus, Strongyloides* spp., *Nippostrongylus brasiliensis*). These then undergo a tissue migration, usually involving migration through the lymph and/or the blood vessels, the heart and the lungs before being coughed up and swallowed to become established as adults in their final location in the alimentary tract (Figure 3.3). The infective-stage larvae of some species can infect the host orally as well as through the skin. For example, the larvae of *Ancylostoma duodenale* that enter the host through the skin then undergo tissue migration before establishing as adults in the alimentary tract, whereas those that are ingested orally do not undergo a tissue migration but develop directly into adults in the alimentary tract (Schad, Nawalinski and Kochar, 1983) (Figure 3.3). Transmammary transmission can occur in the infective-stage larvae of some of the nematodes that enter the mother percutaneously (*Strongyloides* sp.) (Brown and Girardeau, 1977) and this is an important route of

infection for *Uncinaria lucasi* in northern fur seals (*Callorhinus ursinus*) and northern sea lions (*Eumetopias jubata*) (Olsen and Lyons, 1965). Secondary monoxeny (Anderson, 1988, 1992) applies to those nematodes which some authors suggest have lost an intermediate host during the course of evolution and have reverted to direct transmission, thus the final host serves as its own intermediate host. This is thought to occur in *Ascaris lumbricoides* and *A. suum* which hatch in the alimentary tract and then undergo a tissue and organ migration before returning to the alimentary tract where the final moult takes place (Anderson, 1992) (Figure 3.3). Similar tissue migration occurs with *Toxocara canis* and can result in transplacental transmission with the result that the unborn pups are infected prior to birth. It should be noted that basically the ascaridoids are heterogenous (see below).

The life cycle of *Trichinella spiralis* is unusual in that it uses the final definitive host as the intermediate host for the next generation (Figure 3.3). Adults live in the cytoplasm of the epithelial cells of the intestine of mammals (Dunn and Wright, 1985) where the female gives birth to about 1000 L1 over 5 days (Despommier, 1998). These newborn larvae penetrate the lamina propria of the intestine and migrate via the blood stream and lymphatic system and become distributed to all tissues of the host. The larvae move from the capillaries to surrounding cells which they penetrate, resulting in the death of the cell, unless it is a striated skeletal muscle fibre. Only skeletal muscle fibres provide the conditions necessary to support the growth and development of the larva (Despommier, 1998). The larva is not surrounded by a host membrane but lies within the cytoplasm (Lee and Shivers, 1987) of the muscle fibre where it moves, feeds and grows (Jasmer, 1995; Despommier, 1998). Secretions from the larva induce the formation of hypertrophic nuclei, loss of the myofibrils and other characteristic organelles resulting in the formation of a nurse cell (Purkeson and Despommier, 1974; Despommier, 1990, 1998; Ko *et al.*, 1994). This nurse cell becomes encapsulated in collagen fibres and surrounded by a network of capillaries (Wright, *et al.*, 1988; Baruch and Despommier, 1991). Apparently the larva controls the structure and function of the altered muscle fibre through these secretions (Despommier, 1990; Jasmer, 1990, 1993, 1995; Ko *et al.*, 1994; Mak and Ko, 1999) resulting in remodeling of the muscle fibre into the nurse cell. The nematode induces the infected muscle cell to re-enter the cell cycle and to replicate its DNA. The expression of the structural and regulatory genes of the muscle fibre are downregulated at the transcriptional level resulting in the muscle differentiation programme becoming inactivated in the nurse cell. The muscle cell becomes chronically suspended in a cell cycle phase consistent with G2/M, with a 4N DNA complement in the nuclei of the nurse cell (Jasmer, 1993, 1995). Single and double-stranded nucleases have been demonstrated in the excretions/secretions of *Trichinella* and these may be involved in the arrest of the infected muscle at the G2/M phase (Mak and Ko, 1999). This seizure of control of a

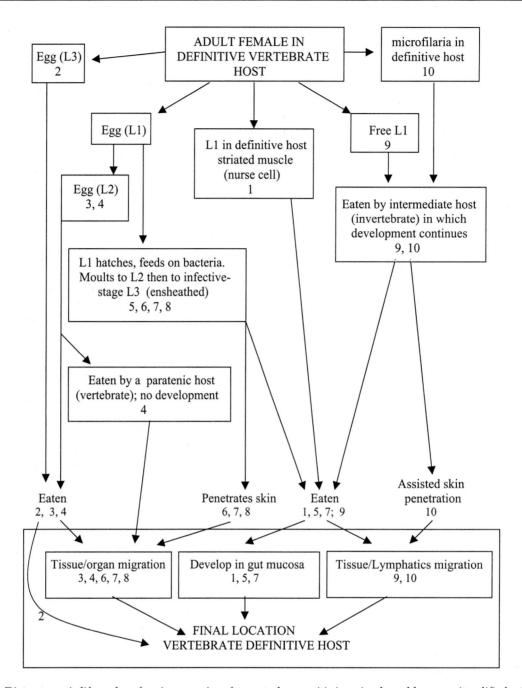

Figure 3.3. Diagrammatic life cycles of various species of nematode parasitic in animals and humans (modified with permission from Crompton and Joyner, 1980. *Parasitic worms.* London: Wykeham Publications (London) Ltd.). *1. Trichinella spiralis. 2. Enterobius vermicularis. 3. Ascaris lumbricoides. 4. Toxocara cati. 5. Haemonchus contortus. 6. Nippostrongylus brasiliensis. 7. Ancylostoma duodenale. 8. Necator americanus. 9. Dracunculus medinensis. 10. Wuchereria bancrofti and Onchocerca volvulus.*

metazoan cell in one organism by another metazoan organism is unusual. The plant-parasitic nematode *Globodera rostochiensis* does something similar when it induces the formation of a giant cell in the root of its host (see chapter 9). As well as several hypertrophic nuclei the nurse cell contains extensive rough and smooth endoplasmic reticulum, and Golgi complexes. Vesicles that are budded from the Golgi complexes pass through the cytoplasm to empty into the cavity in

which the nematode lies (Lee and Shivers, 1987). Presumably the contents of the vesicles sustain the larva. The L1 develops once the host animal is eaten by another suitable host when the L1 escapes from the nurse cell, burrows into the cytoplasm of the epithelial cells of the intestinal wall (Dunn and Wright, 1985; Wright, 1979) and undergoes four quick moults before developing into the adult male or female. *T. spiralis* and *T. pseudospiralis* larvae can lead to major reductions in power output, fatigue resistance and the ability of infected muscles to do work, thus probably increasing the possibility of predation and thereby increasing the opportunity for the life cycle to be completed (Winter et al., 1994; Harwood et al., 1996).

Heterogenous nematodes use one or more intermediate hosts in their life cycle and heteroxeny is common in nematode parasites of vertebrates. In some life cycles the nematode develops to the third-stage larva in the intermediate host and the final host becomes infected when it eats the intermediate host containing these infective-stage larvae. With these life cycles the intermediate hosts can be terrestrial snails, slugs or earthworms (Metastrongyloidea), aquatic and terrestrial arthropods (Spirurida, Dracunculoidea), vertebrates (Ascaridoidea), or oligochaetes (Dioctophymatoidea) (Anderson, 1988, 1992) (Figure 3.3). In other life cycles involving an intermediate host the nematode develops in a vector, such as a biting insect or tick, which then transmits the larva to the next host (Filarioidea) (Figure 3.3). Filarial nematodes, such as *Wuchereria*, *Onchocerca* and *Brugia* are usually found, as adults, in the lymphatic system, subcutaneous connective tissue or other connective tissues of their definitive hosts. The mature females are ovoviviparous or viviparous and release microfilariae (embryos which develop within a thin 'shell' that is retained as a sheath in some species) that are found in the blood or skin. With some species (*Wuchereria*) the microfilariae exhibit diurnal periodicity which brings them to the peripheral circulation when the blood-feeding vector/intermediate host is likely to be seeking a blood meal. When ingested by a suitable arthropod, such as a mosquito or blackfly, they pass into the mid-gut where sheathed microfilariae lose their sheath. The larvae then penetrate the mid-gut wall, cross the haemocoel and enter the thoracic flight muscles. Here they develop into short thick 'sausage' larvae, grow and moult twice to become filariform infective-stage L3. The L3 migrate from the thoracic muscles to the head region of the arthropod and are released onto the skin of the host from where they enter the wound made by the mouthparts when the arthropod feeds. The larvae migrate within the definitive host, moulting twice more, before reaching their final location. The adult nematodes are usually long-lived (up to several years in some species) and the time taken for microfilariae to appear may be up to several months (Muller and Wakelin, 1998).

Some of the advantages of using an intermediate host in the life cycle are thought to be: (1) an increase in the spatial distribution of the nematode; (2) a possible increase in longevity of the infective stage; (3) the intermediate host enables the larva to grow to such an extent that it can influence the maturation of the nematode in the final host; (4) protection of the larva from adverse effects of the external environment; (5) increased chances of transmission to the final host, either because the intermediate host is a food item or is a biting arthropod that serves as a vector (Anderson, 1992; Chabaud, 1955; Inglis, 1965). Some animal-parasitic nematodes employ paratenic (Beaver, 1969)

or transport hosts in their life cycle. A paratenic host is 'an organism which serves to transfer a larval stage or stages from one host to another but in which little or no development takes place' (Anderson, 1988, 1992). The paratenic host is usually a food item in the diet of the final host and, as such, increases the chances of successful transmission. For example, migration and encystment of infective-stage larvae of *Toxocara cati* in paratenic hosts such as mice, or of *Heterakis gallinarum* in earthworms, exploits the predator-prey relationship (Figure 3.3). In some nematode parasites of animals development proceeds beyond the L3 to the L4 or even to the sub-adult stage in the intermediate host. This is called precocity and is defined as 'growth and/or development beyond the expected in the intermediate host' (Anderson, 1988, 1992). Precocious development in intermediate hosts can be reflected in development of genital primordia in larvae. This may have benefits for the parasite if it accelerates gamete production in final hosts in which possible transmission of the next generation is reduced in time and/or space due to the behaviour of the final host (Anderson, 1988). For example, adult *Hedrurus androphora* live in the intestine of salamanders and larvae develop to the sub-adult stage in the haemocoele of the aquatic isopod *Asellus aquaticus*. The salamanders are mainly terrestrial but enter water to spawn. If isopods that became infected through ingesting eggs deposited by salamanders when they entered the water for spawning the previous year, and which now contain sub-adult stages of the nematode, are ingested by salamanders when they are in the water to spawn, then the nematodes will mature rapidly and will produce eggs before the salamander returns to land (Anderson, 1992). Precocity can also be reflected in growth without moulting resulting in enlarged larvae, as in *Anisakis* sp. and *Pseudoterranova decipiens* found in the body cavity and musculature of fish. Some parasitic nematodes, having invaded the final host, are able to persist for several generations in the host through autoinfection. Eggs may hatch in the host and some of the larvae develop, sometimes involving tissue migration, to the adult stage within the same individual (*Strongyloides stercoralis*, *Ollulanus tricuspis*, *Probstmayria vivipara*) (Anderson, 1992). Alternation of parasitic hermaphroditic or parthenogenetic females with free-living larvae and adults (males and females), referred to as alternation of homogonic and heterogonic generations, occurs in the life cycle of some animal-parasitic nematodes (*Strongyloides stercoralis*).

Arrested Larval Development (Hypobiosis)

Arrested larval development (hypobiosis) is important in the life cycle of many trichostrongyles and occurs at a particular stage in development of the nematode (Armour and Duncan, 1982). Arrested larval development of *Ostertagia ostertagi* in cattle has been the subject of much interest as large numbers can accumulate in infected animals during the grazing season resulting in outbreaks of disease when recent infection with the parasites could not have occurred (Michel, 1974, 1976; Smeal and Donald, 1981, 1982a,b; Armour and Duncan,

1982; Eysker, 1997; Fernandez, Fiel and Steffan, 1999). Eggs of *O. ostertagi* passed in the faeces of infected cattle hatch to release free-living microbivorous larvae which develop in the faecal pat to the ensheathed, infective stage, L3. If ingested by a suitable host the L3 exsheath in the abomasum where they invade the gastric pits and glands, initiate the formation of nodules and undergo the third moult. After a few days L4 leave the nodules and enter the lumen of the abomasum where they mature. However, *Ostertagia* has the propensity to become arrested as an early L4 in the wall of the abomasum, subsequently continuing development a few months later (this can result in Type II Ostertagiasis). This arrested larval development, which resembles diapause, appears to be triggered by cold conditioning of the free-living stages of the larvae in the Northern Hemisphere, with the L4 continuing their development within the host the following spring. The effect of this is to delay the production of eggs of the parasite over the winter and to make them available on the pasture the following spring when their chances of survival and development to the L3 are greatly enhanced. A similar phenomenon occurs with the L4 of *Haemonchus contortus* and *Cooperia* spp. (Armour and Duncan, 1982). In some countries that have a hot dry season, inhibition of trichostrongyle L4 appears to be associated with survival during the dry season (Jacquiet *et al.*, 1995). In Australia, by using a population of *O. ostertagi* from a coastal region of New South Wales, which has an all year-round mild climate, and another population from the cooler Northern Tablelands, which has cold, dry winters and mild summers, it was originally suggested that the propensity of *O. ostertagi* to become arrested is genetically determined and is independent of cold conditioning (Smeal and Donald, 1981). However, it is now thought that the genetically determined differences described by Smeal and Donald (1981) may be accounted for by different cattle production systems (Smeal and Donald, 1982a). There is still some controversy about the effect of cold temperatures on inhibition of development of *O. ostertagi* in Australia (Smeal and Donald, 1982b). Armour and Duncan (1987) concluded that when arrested larval development occurs it is a heritable characteristic of the larva because different isolates differ in their propensity for arrested larval development in response to environmental pressures. Other factors probably have a role in initiating arrested development. Light, as well as temperature, may be a factor in inducing L3 of *O. ostertagi* to become arrested in cattle in some regions of the world (Fernandez, Fiel and Steffan, 1999).

For more detailed information on the life cycles of animal-parasitic nematodes see the relevant chapters in Cox, Kreier and Wakelin (1998) and Anderson (1992).

Dauer Larva

The most thoroughly studied dauer larva is that of *Caenorhabditis elegans*. This dauer larva appears when environmental conditions are no longer suitable for successful reproduction through increased competition, a limited food supply and an increase in a dauer-inducing pheromone in the environment (Riddle, 1988; Riddle and Albert, 1997). With increased population density and an increasingly limited food supply, environmental conditions become unsuitable for successful reproduction, thus development is arrested at the second inter-moult resulting in the formation of a dauer larva (Figure 3.1). A pheromone is produced throughout the life cycle of the nematode and formation of the dauer larva requires continuous exposure to this pheromone after the middle of the L1 stage (Riddle, 1988). The dauer larva differs structurally and behaviourally from all other stages of the life cycle, has a non-functional alimentary tract, does not feed, can survive for much longer than the normal 2–week life span of the adult hermaphrodite of *Caenorhabditis* and is specialized for dispersal and long-term survival (Riddle, 1988; Riddle *et al.*, 1997; Riddle and Albert, 1997). Population density is apparently measured by an increase in a dauer-inducing pheromone that is produced by both *C. elegans* and by *C. briggsae* but pheromones produced by other soil-inhabiting rhabditid nematodes are not active on *C. elegans* (Fodor *et al.*, 1983; Golden and Riddle, 1982,1984; Riddle, 1988; Riddle and Albert, 1997). When favourable conditions, such as a suitable and adequate food supply, become available the dauer larva begins to feed after 2-3 hours, moults to the L4 after a further 8-10 hours and then to the adult (Riddle, 1988). Several other species of free-living, plant-parasitic and entomogenous nematodes also produce a dauer-like larval stage which enables them to resist adverse conditions and which may also help in dispersal of the nematode.

Ensheathed Larvae

The free-living L3 of many animal-parasitic nematodes are similar to the dauer larvae described above in that they enter a form of diapause, retain the apolysed, but not ecdysed, cuticle of the L2, are non-feeding, are resistant to environmental stresses, require specific stimuli to continue their development and are important in dispersal. However, unlike the dauer larvae of free-living nematodes the ensheathed L3 of animal-parasitic nematodes are an obligatory stage of the life cycle and must gain access to a suitable host to continue their development. Eggs of *Ancylostoma duodenale* deposited in the faeces of infected humans hatch to release free-living microbivorous L1 which moult into a microbivorous L2. The ensheathed infective-stage L3 arises from the incomplete moult of the L2 and moves to the surface of the faecal/soil mixture where it attaches to the substratum by its tail and waves in the air (nictating) (Croll and Matthews, 1977). No further development takes place until it makes contact with the skin of the human host when it leaves the sheath, penetrates the skin, migrates to the heart, lungs, bronchi, trachea and then to the alimentary tract where it establishes in the jejunum. It moults to the L4 and to the adult stage during this journey (Figure 3.3). The L3 of *Necator americanus*, which is parasitic in humans, and the L3 of *Nippostrongylus brasiliensis*, which is parasitic in rats, have a similar migration to that of *Ancylostoma*

(Figure 3.3). However, if the L3 of *Ancylostoma duodenale* is ingested then it does not undergo a tissue/organ migration but develops to the L4 and then to the adult stage in the alimentary tract (Figure 3.3).

The ensheathed infective-stage L3 of *Haemonchus contortus* and *Trichostrongylus colubriformis*, which develop from free-living microbivorous L2 (Figure 3.3), require specific stimuli to exsheath with the nature of the stimuli determining in which part of the alimentary tract of their hosts they will develop into adults (Rogers, 1962; Rogers and Sommerville, 1968). Thus the L3 of *Haemonchus* exsheaths best at 37°C, in fluid at a neutral pH and at high concentrations of CO_2, such as are found in the rumen of ruminants. The larval nematodes then pass through to the abomasum where the L4 develop in glands of the abomasal wall. The adults reside in the lumen of the abomasum where they use a tooth to lacerate blood vessels to feed on blood. The L3 of *T. colubriformis* exsheath best under acid conditions and in the presence of lower concentrations of CO_2, although the latter is not essential, and normally begin exsheathment in the abomasum or stomach and develop into adults in the small intestine (Rogers, 1962; Rogers and Sommerville, 1968).

Life-Cycle Strategies

r- and K-selection

MacArthur and Wilson (1967) coined the terms 'r-selection' and 'K-selection' where r refers to the maximal intrinsic rate of natural increase [r_{max}] of an organism and K refers to the carrying capacity of the habitat, that is, the maximum number of organisms the habitat will support. With organisms that exhibit r-selection, density and competition effects in the environment are minimal and fecundity (progeny) can be maximized (selection favours high productivity and rapid development of offspring at the expense of survival of the adult). With K-selection, competition and/or predation in the environment is high and the emphasis is on survival of adults which put more energy into each of fewer offspring (selection favours lower fecundity and slower development of well provisioned offspring) (Pianka, 1970; Jennings and Calow, 1975; Calow, 1983b; Wharton, 1986). Theoretically r and K features should not be selected for simultaneously because organisms are supposed to trade-off high fecundity with low post-reproductive survival. However, no organism is completely 'r-selected' or completely 'K-selected' but must come to some compromise between these two extremes (Pianka, 1970). When the soil-inhabiting nematode *Caenorhabditis* first colonizes a new habitat it will tend to exhibit r strategy as it has a small body size, has a short period of repeated breeding, relatively high fecundity, rapid development of the offspring, and short adult survival periods. This will enable it to rapidly colonise a substrate. However, as the population increases, competition will also increase resulting in the formation

of dauer larvae that have longer survival periods, which is a K characteristic. *Plectus palustris*, a microbivorous benthic-inhabiting nematode, lives in an environment where there is lower food abundance but higher predictability of food supply in a stable environment and it is likely to experience greater competition (Schiemer, 1982a, b; 1983; Schiemer, Duncan and Klekowski, 1980). It exhibits certain K-strategies as it has longer developmental periods, delayed reproduction, longer adult survival and lower fecundity (Wharton, 1986). Animal-parasitic nematodes tend to show a mixture of r- and K-strategies in that they have high fecundity (r-strategy), made possible because of the stable, nutrient-rich environment provided by the host, but with low larval survival, yet the adults have relatively long survival periods and comparatively long periods of development (K-strategy) (Calow, 1983a, b; Wharton, 1986). The traditional explanation for this is that the parasites have to compensate for a massive mortality during the transmission phases of their life cycles that is not experienced by free-living species (Calow, 1983b). However, this explanation implies that free-living species could be more fecund if there had been appropriate selection pressures and is apparently contra-Darwinian because, under that logic, reproductive output should always be maximized (Calow, 1983b). The neo-Darwinian hypothesis is that natural selection will favour traits that maximize fecundity (n) and survival (s) and minimize generation time (t). The intrinsic rate of increase (r) should increase as n increases yet this is not a universal feature of organisms. It is probable that this is because there is interaction between s, n and t (Calow, 1983b; Calow and Sibly, 1983; Sibly and Calow, 1983), with a trade-off between n and s_a (where $_a$ is the period between breeding seasons in the adult phase) in which fecundity is supposed to have a negative effect on the survival of the parent. In principle, selection should push reproductive output to the physiological limit of the system but, in practice, this is not possible because of the trade-off between S_a and n (Calow, 1983b) (see below).

Reproductive potential

The intrinsic rate of natural increase r_m of a population can be calculated from the Lotka equation for a population with a stable age-distribution and growing in an unlimited environment (Vranken and Heip, 1983; Wharton, 1986):

$$\sum_{x=0}^{max\ age} e_m^{-rx}\, l_x m_x = 1 \qquad \text{Equation 3.1}$$

or

$$\sum_0^\infty l_x m_x \exp\left(-r_m x\right) = 1 \qquad \text{Equation 3.2}$$

where l_x is age-specific survival and m_x age-specific fecundity which are summarised in life tables.

The reproductive potential of a population can be calculated from the average number of female offspring produced per female entering the population (R_0 — the net reproductive rate) and the instantaneous growth rate of the population, where there is a stable age distribution and where the conditions of space and resources are not limiting (Wharton, 1986). However, some authors use r_m as the rate of increase whether or not the population has a stable age distribution (Vranken and Heip, 1983):

$$dN/dt = r_m N \qquad \text{Equation 3.3}$$

where t is time and N is the number of individuals.

The r_m values of the bacteriophagous nematodes *Caenorhabditis briggsae* and *Mesodiplogaster lheriteri*, which have short generation times, are among the highest recorded for animals in this size range whereas the predacious nematodes *Labronema vulvapillatum* and *Oncholaimus oxyuris* and the herbivorous nematode *Eudiplogaster paramatus*, which have relatively long developmental periods, have lower r_m values (Heip, Smol and Absillis, 1978; Romeyn, Bouwman and Admiraal, 1983; Wharton, 1986).

The capacity for increase of a population is dependent upon the fecundity of females, the generation time, and adult and larval survival. The reproductive potential of an organism can therefore be increased by increasing fecundity of the females, by increasing survival of the larvae (e.g. delayed hatching, dauer larvae, ensheathed L3 of trichostrongyles, hypobiosis) or adults, and by decreasing the generation time (Wharton, 1986). Animal-parasitic nematodes usually have much higher fecundity (*Haemonchus contortus* can lay up to 10,000 eggs per day at a rate of 1 egg every 10 seconds) but have longer generation times (for *Haemonchus* the prepatent period, i.e. the time taken for an egg to develop into an egg-laying female, is 1 month) than free-living nematodes which may produce fewer than 20 eggs in a lifetime but have a generation time of just a few days (Crofton, 1966). Thus, the reproductive potential of animal-parasitic nematodes may be less than that of free-living nematodes which produce many fewer eggs per female but may have a much quicker generation time. Theoretically one fertilized female *Haemonchus contortus* laying 10,000 eggs per day, 5000 of which are potential females, and with a generation time of one month could produce about 10^{11} egg-laying females in three months whereas the insect-parasitic/saprophagous nematode *Steinernema* (syn. *Neoaplectana glaseri*), which produces 15 eggs, 8 of which may produce females, in a lifetime but has a generation time of 5 to 7 days, could in three months produce about 10^{13} egg-producing females (Crofton, 1966). However, the full reproductive potential of these, and other nematodes, is never achieved because of environmental constraints, such as temperature, accident, disease, predation and/or the defensive responses of the host. Whilst one female *H. contortus* under laboratory conditions can theoretically give rise to 10^{11} egg-laying females in three generations, under field conditions

there is likely to be only one or two generations per year, large numbers of the eggs will not develop to the L3 stage and of these a relatively small number will manage to infect a suitable host. This high fecundity is possible because adult *Haemonchus* have access to a plentiful food supply. In this situation a population that has a plentiful supply of resources might invest more in fecundity, even with lower survival of larvae as occurs with *Haemonchus* and many other animal-parasitic nematodes, and individuals may survive longer than a population with poor resource availability. Animal-parasitic nematodes do appear to have higher levels of growth and reproduction and survive longer than their free-living relatives and this may be due to better resource availability (see Calow, 1983b for further details). The high r_m of free-living nematodes, such as *Caenorhabditis*, which have a short generation time, enables them to rapidly colonize a food source. Shortening the life cycle can be achieved by employing ovoviviparity, by releasing eggs at an advanced stage of development, and/or by avoiding the need to spend time finding a mate as exhibited by parthenogenetic and self-fertilising hermaphrodites (Wharton, 1986).

The intrinsic rate of increase of an organism should increase as fecundity increases but the maximisation of fecundity does not appear to be a universal feature of organisms. The most probable reason for this, as mentioned above, is that there is interaction between survival, fecundity and generation time, with there being a tradeoff between fecundity and survival of the adult because fecundity is supposed to have a negative effect on the survival of the parent. In organisms that have a finite amount of resources available for their metabolism then investment of these resources in reproduction reduces the resources available for maintenance of the organism and may affect survival of the adult. Extrinsic factors, such as predation, disease, accident and, in parasitic organisms, the defensive responses of the host, are likely to have an adverse effect on survival of the adult (Calow, 1983a, b). However, a population experiencing a plentiful and constant food supply, such as occurs with adult *Haemonchus* and adult *Ascaris* (the female can produce 200,000 eggs/day), is able to invest more in the production of gametes, because its basic metabolic needs are constantly being met, than a population that has limited access to a food supply. This may partly explain why animal-parasitic nematodes tend to be longer-lived and produce more eggs than free-living nematodes. However, relatively few of these offspring survive to continue the life cycle, which is one reason why so many eggs are produced.

Acknowledgments

I am grateful to Professor Peter Calow and Dr David Wharton who read and commented on earlier drafts of this chapter. Whilst their comments were helpful in producing this chapter, I take full responsibility for its contents.

References

Anderson, R.C. (1988) Nematode transmission patterns. *Journal of Parasitology*, **74**, 30–45.

Anderson, R.C. (1992) *Nematode parasites of vertebrates. Their development and transmission*. Wallingford: CAB International.

Armour, J. and Duncan, M. (1982) Arrested larval development in cattle nematodes. *Parasitology Today*, **3**, 171–176.

Baruch, A.M. and Despommier, D.D. (1991) Blood vessels in *Trichinella spiralis* infections: a study using vascular casts. *Journal of Parasitology*, **77**, 99–103.

Beaver, P.C. (1969) The nature of visceral larva migrans. *Journal of Parasitology*, **55**, 3–12.

Bedding, R.A. (1984) Nematode parasites of Hymenoptera. In *Plant and insect nematodes*, edited by W.R. Nickle, pp. 755–794. New York: Marcel Dekker.

Bird, A.F. and Bird, J. (1991) *The Structure of Nematodes*, 2nd edn. San Diego: Academic Press.

Bovien, P. (1937) Some types of association between nematodes and insects. *Videnskkabelige Meddelelser fra Dansk Naturhistorik Forening I Kjøbenhavn*, **101**, 1–114

Brenner, S. (1973) The genetics of behaviour. *British Medical Bulletin*, **29**, 269–271.

Brenner, S. (1974) The genetics of *Caenorhabditis elegans*. *Genetics*, **77**, 71–94.

Brenner, S. (1988) The molecular evolution of genes and proteins: a tale of two serines. *Nature*, **334**, 528–530.

Brown, R.C. and Girardeau, M.H.F. (1977) Transmammary passage of *Strongyloides* sp. larvae in the human host. *American Journal of Tropical Medicine and Hygiene*, **26**, 215–219.

Calow, P. (1983a) Energetics of reproduction and its evolutionary implications. *Biological Journal of the Linnaean Society*, **20**, 153–165.

Calow, P. (1983b) Pattern and paradox in parasite reproduction. In *The reproduction biology of parasites*, edited by P.J. Whitfield. *Symposia of the British Society for Parasitology*, **20**. *Parasitology*, **86**, 197–207.

Calow, P. and Sibly, R.M. (1983) Physiological trade-offs and the evolution of life cycles. *Science Progress*, **68**, 177–188.

Chabaud, A.G. (1955) Essai d'interprétation phylétique des cycles évolutifs chez les Nématodes parasites de vertébrés. *Annales de Parasitologie Humaine et Comparée*, **30**, 83–126.

Christie, J.R. (1937) *Mermis subnigrescens*, a nematode parasite of grasshoppers. *Journal of Agricultural Research*, **55**, 253–263.

Cox, F.E.G., Kreier, J.P. and Wakelin, D. (1998) *Parasitology*, Volume 5. In *Topley and Wilson's Microbiology and Microbial Infections*, 9th edn., edited by L. Collier, A. Balows and M. Sussman. London: Arnold.

Crofton, H.D. (1966) *Nematodes*. London: Hutchinson.

Croll, N.A. and Matthews, B.E. (1977) *Biology of Nematodes*. Glasgow: Blackie.

Despommier, D.D. (1990) The worm that would be virus. *Parasitology Today*, **6**, 193–195.

Despommier, D.D. (1998) *Trichinella* and *Toxocara*. In Volume 5, *Parasitology*, edited by F.E.G. Cox, J.P. Kreier and D. Wakelin, pp. 597–607. *Topley and Wilson's Microbiology and Microbial Infections*, 9th edn., edited by L. Collier, A. Balows and M. Sussman. London: Arnold.

Dropkin, V.H. (1980) *Introduction to Plant Nematology*. New York: Wiley.

Dunn, I.J. and Wright, K.A. (1985) Cell injury caused by *Trichinella spiralis* in the mucosal epithelium of B10A mice. *Journal of Parasitology*, **71**, 757–766.

Ellenby, C. (1946) Nature of the cyst wall of the potato-root eelworm *Heterodera rostochiensis*, Wollenweber, and its permeability to water. *Nature*, **157**, 302.

Eysker, M. (1997) Some aspects of inhibited development of trichostrongylids in ruminants. *Veterinary Parasitology*, **72**, 265–272.

Fernandez, A.S., Fiel, C.A. and Steffan, P.E. (1999) Study on the inductive factors of hypobiosis of *Ostertagia ostertagi* in cattle. *Veterinary Parasitology*, **81**, 295–307.

Fodor, A., Riddle, D.L., Nelson, F.K. and Golden, J.W. (1983) Comparison of a new wild-type *Caenorhabditis briggsae* with laboratory strains of *Caenorhabditis briggsae* and *Caenorhabditis elegans*. *Nematologica*, **29**, 203–217.

Fuchs, G. (1915) Die Naturgeschichte der Nematoden und eineger anderer Parasiten. 1. Des *Ips typographus*, 2. Des *Hylobius abietis*. *Zoologische Jahrbücher Systematik*, **38**, 109–227.

Golden, J.W. and Riddle, D.L. (1982) A pheromone influences larval development in the nematode *Caenorhabditis elegans*. *Science* **218**, 578–580.

Golden, J.W. and Riddle, D.L. (1984) The *Caenorhabditis elegans* dauer larva: Developmental effects of pheromone, food, and temperature. *Developmental Biology*, **102**, 368–378.

Harwood, C.L., Young, I.S., Lee, D.L. and Altringham, J.D. (1996) The effect of *Trichinella spiralis* infection on the mechanical properties of the mammalian diaphragm. *Parasitology*, **113**, 535–543.

Heip, C., Smol, N. and Absillis, V. (1978) Influence of temperature on the reproductive potential of *Oncholaimus oxyuris* (Nematoda: Oncholaimidae). *Marine Biology*, **45**, 255–260.

Inglis, W.G. (1965) Patterns of evolution in parasitic nematodes. In *Evolution of Parasites*, edited by A.E.R. Taylor, pp. 79–124. Third Symposium of the British Society for Parasitology. Oxford: Blackwell.

Jacquiet, P., Cabaret, J., Cheikh, D. and Thiam, A. (1995) Experimental study of survival strategy of *Haemonchus contortus* in sheep during the dry season in desert areas of the Mauritania. *Journal of Parasitology*, **81**, 1013–1015.

Jasmer, D.P. (1990) *Trichinella spiralis*: Altered expression of muscle proteins in trichinosis. *Experimental Parasitology*, **72**, 321–331.

Jasmer, D.P. (1993) *Trichinella spiralis* infected skeletal muscle cells arrest in G_2/M and cease muscle gene expression. *Journal of Cell Biology*, **121**, 785–793.

Jasmer, D.P. (1995) *Trichinella spiralis*: Subversion of differentiated mammalian skeletal muscle cells. *Parasitology Today*, **11**, 185–188.

Jennings, J.B. and Calow, P. (1975) The relationship between high fecundity and the evolution of parasitism. *Oecologia*, **21**, 109–115.

Jones, F.G.W. and Jones, M.G. (1974) *Pests of field crops*. 2nd edn. London: Edward Arnold.

Ko, R.C., Fan, L., Lee, D.L. and Compton, H. (1994) Changes in host muscles induced by excretory/secretory products of larval *Trichinella spiralis* and *Trichinella pseudospiralis*. *Parasitology*, **108**, 195–205.

Lee, D.L. and Shivers, R.R. (1987) A freeze-fracture study of muscle fibres infected with *Trichinella spiralis*. *Tissue and Cell*, **19**, 665–671.

MacArthur, R.H. and Wilson, E.O. (1967) *The theory of island geography*. Princeton: Princeton University Press.

Mak, C. and Ko, R.C. (1999) Characterization of endonuclease activity from excretory/secretory products of a parasitic nematode, *Trichinella spiralis*. *European Journal of Biochemistry*, **260**, 477–481.

Maupas, E. (1899) La mue et l'enkystement chez les nématodes. *Archives de Zoologie et Expérimentale*, **7**, 563–628.

Michel, J.F. (1974) Arrested development of nematodes and some related phenomena. *Advances in Parasitology*, **13**, 279–366.

Michel, J.F. (1976) The epidemiology and control of some nematode infections in grazing animals. *Advances in Parasitology*, **14**, 355–397.

Muller, R. and Wakelin, D. (1998) Lymphatic filariasis. In Volume 5, *Parasitology*, edited by F.E.G. Cox, J.P. Kreier and D. Wakelin, pp. 609–619. *Topley and Wilson's Microbiology and Microbial Infections*, 9th edn., edited by L. Collier, A. Balows and M. Sussman. London: Arnold.

Olsen, O.W. and Lyons, E.T. (1965) Life cycle of *Uncinaria lucasi* Stiles, 1901 (Nematoda: Ancylostomatidae) of fur seals, *Callirhinus ursinus* Linn., on the Pribilof Islands, Alaska. *Journal of Parasitology*, **51**, 689–700.

Pianka, E.R. (1970) On r- and K-selection. *American Naturalist*, **104**, 592–597.

Poinar, G.O. (1979) *Nematodes for biological control of insects*. Florida, Boca Raton: CRC Press Inc.

Purkeson, M. and Despommier, D.D. (1974) Fine structure of the muscle phase of *Trichinella spiralis* in the mouse. In *Trichinellosis*, edited by C.W. Kim, pp. 7–24. New York: Intext Educational Publishers.

Riddle, D.L. (1988) The dauer larva. In *The nematode* Caenorhabditis elegans, edited by W.B. Wood, pp. 393–412. Cold Spring Harbor: Cold Spring Harbor Laboratory Press.

Riddle, D.L. and Albert, P.S. (1997) Genetic and environmental regulation of dauer larva development. In *C. elegans II*, edited by D.L. Riddle, T. Blumenthal, B.J. Meyer and J.R. Preiss, pp. 739–768. Cold Spring Harbor: Cold Spring Harbor Laboratory Press.

Riddle, D.L., Blumenthal, T., Meyer, B.J. and Priess, J.R. (1997) Introduction to *C. elegans*. In *C. elegans II*, edited by D.L. Riddle, T. Blumenthal, B.J. Meyer and J.R. Preiss, pp. 1–22. Cold Spring Harbor: Cold Spring Harbor Laboratory Press.

Robinson, J. (1962) *Pilobolus* spp. and the translation of infective larvae of *Dictyocaulus viviparus* from faeces to pasture. *Nature*, **193**, 353–354.

Rogers, W.P. (1962). *The nature of parasitism. The relationship of some metazoan parasites to their hosts*. Academic Press: New York.

Rogers, W.P. and Sommerville, R.I. (1968) The infectious process, and its relation to the development of early parasitic stages of nematodes. *Advances in Parasitology*, **6**, 327–348.

Romeyn, K., Bouwman, L.A. and Admiraal, W. (1983) Ecology and cultivation of the herbivorous brackish-water nematode *Eudiplogaster paramatus*. *Marine Ecology Progress Series*, **12**, 145–153.

Santos, M.S.N.de A. (1972) Production of male *Meloidogyne* spp. and attraction to their females. *Nematologica*, **18**, 291.

Schad, G.A., Nawalinski, T.A. and Kochar, V. (1983) Human ecology and the distribution and abundance of hookworm populations. In *Human Ecology and Infectious Diseases*, pp. 187–223. London: Academic Press.

Schiemer, F. (1982a) Food dependence and energetics of free-living nematodes. I. Respiration, growth and reproduction of *Caenorhabditis briggsae* (Nematoda) at different levels of food supply. *Oecologia* **54**, 108–121.

Schiemer, F. (1982b) Food dependence and energetics of free-living nematodes. II. Life history parameters of *Caenorhabditis briggsae* (Nematoda) at different levels of food supply. *Oecologia* **54**, 122–128.

Schiemer, F. (1983) Comparative aspects of food dependence and energetics of free-living nematodes. *Oikos*, **41**, 32–42.

Schiemer, F., Duncan, A. and Klekowski, R.Z. (1980) A bioenergentic study of a benthic nematode, *Plectus palustris* de Mann 1880, throughout its life cycle. II. Growth, fecundity and energy budgets at different densities of bacterial food and general ecological considerations. *Oecologia*, **44**, 205–212.

Sibly, R. and Calow, P. (1983) An integrated approach to life cycle evolution using selective landscapes. *Journal of Theoretical Biology*, **102**, 527–547.

Smeal, M.G. and Donald, A.D. (1981) Effects of inhibition of development of the transfer of *Ostertagia ostertagi* between geographical regions of Australia. *Parasitology*, **82**, 389–399.

Smeal, M.G. and Donald, A.D. (1982a) Inhibited development of *Ostertagia ostertagi* in relation to production systems in cattle. *Parasitology*, **85**, 21–25.

Smeal, M.G. and Donald, A.D. (1982b) Inhibition of development of *Ostertagia ostertagi* — effect of temperature on the infective larval stage. *Parasitology*, **85**, 27–32.

Triantaphyllou, A.C. (1979) Cytogenetics of root-knot nematodes. In *Root Knot Nematodes* Meloidogyne *species; Systematics, Biology and Control*, edited by F. Lamberti and C.E. Taylor, pp. 85–109. London: Academic Press.

Trudgill, D.L. (1997) Parthenogenetic root-knot nematodes (*Meloidogyne* spp.); how can these biotrophic endoparasites have such an enormous host range? *Plant Pathology*, **46**, 26–32.

Vranken, G. and Heip, C. (1983) Calculation of the intrinsic rate of natural increase, r_m, with *Rhabditis marina* Bastian 1865 (Nematoda). *Nematologica*, **29**, 468–477.

Wharton, D. A. (1986) *A functional biology of nematodes*. London: Croom Helm.

Winter, M.D., Ball, M.L., Altringham, J.D. and Lee, D.L. (1994). The effect of *Trichinella spiralis* and *Trichinella pseudospiralis* on the mechanical properties of mammalian diaphragm muscle. *Parasitology* **109**, 129–134.

Wood, W.B. (1988a) *The nematode* Caenorhabditis elegans. Cold Spring Harbor: Cold Spring Harbor Laboratory Press.

Wood, W.B. (1988b). Introduction to *C. elegans* biology. In *The nematode* Caenorhabditis elegans, edited by W.B. Wood, pp. 1–16. Cold Spring Harbor: Cold Spring Harbor Laboratory Press.

Wright, K.A. (1979) *Trichinella spiralis*: an intracellular parasite in the intestinal phase. *Journal of Parasitology*, **65**, 441–445.

Wright, K.A., Matta, I., Hong, H.P. and Flood, N. (1988) *Trichinella* larvae and the vasculature of the murine diaphragm. In *Proceedings of the Seventh International Conference on Trichinellosis*, edited by C.E. Tanner, A.R. Martinez-Fernandez and F. Bolas-Fernandez, pp. 70–75. Madrid: Consejo Superior de Investigaciones Cientificas Press.

4. Male and Female Gametes and Fertilisation

Jean-Lou Justine

Laboratoire de Biologie Parasitaire, Protistologie, Helminthologie, Muséum National d'Histoire Naturelle, 61 rue Buffon, 75231 Paris cédex 05, France

Introduction

Nematodes are fundamentally bisexual, but certain species are parthenogenetic, hermaphroditic or are pseudogamenous, in which development of the ovum is stimulated by a male gamete the nucleus of which does not fuse with that of the ovum and contributes nothing to the hereditary composition of the embryo. The nematode testis can be in the form of the widespread telogonic testis, in which the germ cells are produced from a single terminal cell, or the less common hologonic testis, in which proliferation of the germ cells takes place along the whole length of the testis. The male reproductive system generally comprises a testis (rarely two) that runs along most of the length of the nematode. The testis has a distal zone of germ-cell formation containing the spermatogonia, a zone containing the spermatocytes and spermatids, a seminal vesicle and a vas deferens that contains the sperm (see diagram in Gibbons, Chapter 2). Maturation of spermatids into spermatozoa appears to be triggered by mating and is thought to take place within the vas deferens (Scott, 1996). A muscularised region of the vas deferens forms an ejaculatory duct. The male copulatory organs generally include two innervated cuticular spicules (sometimes one), which are probably sensory and act as guide to sperm during copulation with the female. Some species posses a gubernaculum which guides the spicules during their eversion. A copulatory bursa is present in many species and aids in copulation by grasping the genital region of the female. The male copulatory apparatus is well endowed with sensillae. The female reproductive system comprises one or two ovaries, rarely more, a seminal receptacle, one or two uteri, a vagina vera and a vulva (see Chapter 2 on General Organization). There appear to be no sense organs associated with the female genitalia. Internal fertilization through female genital ducts is the general rule in nematodes. Hermaphrodites produce sperm that are used to fertilise the eggs of the same individual, as in *Caenorhabditis elegans*, but some hermaphroditic species occasionally produce a male which is capable of fertilising the hermaphroditic female, as in *C. elegans*. A few cases of traumatic insemination have been reported.

Our knowledge of nematode sperm is somewhat unbalanced. On the one hand, the ultrastructure of sperm has been studied in a relatively large number of species (now 56 genera, 73 species (Justine and Jamieson, 1999)). However, this may be considered a small representative sample for such a large phylum. Information is completely lacking, or restricted to a single species, for several large groups. Conversely, the highest current levels of technology and sophistication have been used to study the motile cellular system of nematode sperm of only two species, *Caenorhabditis elegans* and *Ascaris suum*. A review of nematode sperm should encompass observations made with a variety of techniques, from basic light microscope observations to those on the three-dimensional structure of proteins of nematode spermatozoa.

Our knowledge of oocyte ultrastructure is limited to a smaller number of species. Recent studies have concentrated on the genes responsible for oocyte maturation in *Caenorhabditis elegans*. Adaptation to parasitism has resulted in some species having a large egg output, for example the female of *Ascaris lumbricoides*, can produce 200,000 eggs per day.

General Characters of Male Gametes

Morphological Characters of Male Gametes of Nematodes

The main characteristics of nematode sperm are:

1. Constant absence of cilia or flagella.
2. Development of pseudopodia, at least *in utero*, and amoeboid motility.
3. Presence of a special protein, major-sperm-protein, associated with amoeboid motility in most sperm studied. This is correlated with the presence of 'fibrous bodies' in early stages, and a fibrillar cytoskeleton in mature spermatozoa. Exceptions may be because of insufficient data.
4. Presence of unusual membranous organelles in mature spermatozoa of most nematodes studied.
5. Presence of modified centriole(s) with triplets replaced by singlets.
6. Absence of a nuclear envelope at maturity (except in the Enoplea).
7. A variable form, from rounded to elongate.

A brief overview of some of these characteristics follows.

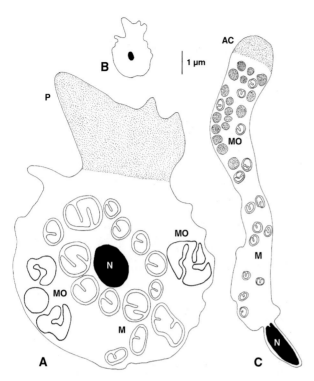

Figure 4.1. Variety of form and size of nematode spermatozoa. Spermatozoa contain mitochondria (M), membranous organelles (MO) and a fibrous cytoplasm containing MSP in the pseudopod (P) or anterior cap (AC). The nucleus (N) is generally devoid of nuclear envelope. A, *Caenorhabditis elegans*: small, rounded spermatozoon. B, diagram at the same magnification as the other species. C, *Heligmosomoides polygyrus*: large, elongate spermatozoon. Scale Bar 1 μm valid for B, C. A, B, redrawn from Ward, S., Hogan, E. and Nelson, G.A. (1983) *Developmental Biology*, **98**, 70–79; C, redrawn from Mansir, A. and Justine, J.-L. (1998) *Molecular Reproduction and Development*, **49**, 150–167.

Form

Nematode spermatozoa always lack a flagellum and do not have a 'typical' head and flagellated tail. Foor (1970b) proposed a classification into four types, but subsequent discovery of several types of sperm which do not fit into this classification means that this is no longer applicable. Nematode sperm can be broadly divided into round and elongate types (Figure 4.1).

Round-type spermatozoa occur in the ascarids, such as *Ascaris*, (corresponding to Foor's ascaroid type, with a typical refringent body) and the dioctophymatoids (corresponding to Foor's dioctophymatoid type, in which membranous organelles are absent). Rhabditids, such as *Caenorhabditis elegans*, monhysterids (*Sphaerolaimus*) which have giant round sperm cells, and dorylaimids (*Xiphinema*) which produce numerous finger-like pseudopodia belong to this type.

Elongate spermatozoa have an anterior part that bears a pseudopod and a posterior part trailing behind. This type of sperm is found in the strongylids (corresponding with Foor's strongyloid type), in which there is a large anterior cytoplasmic 'head-like' mass

that produces pseudopods and a posterior 'tail-like' structure that contains a conoid nucleus (as in *Heligmosomoides*, *Nippostrongylus* and *Nematodirus*); in the mermithid *Gastromermis*, which has sperm resembling the strongyloid type, but with an additional filopod containing microtubules and a different location of the centriole; in *Aspiculuris* (corresponding to Foor's oxyuroid type) in which the sperm have a 'tail' containing bundles of microtubules and the nuclear material and a 'head' which produces filopods; in *Capillaria*; and in the Enoplida (*Enoplus*). Many nematode spermatozoa undergo changes in their morphology in the male genital tract and later in the female tract during and after copulation (e.g. the elongate spermatozoon of strongyloids becomes rounded and produces filopods in the female). Spermatozoa considered as being of the round type before production of pseudopods can be regarded as bipolar when a pseudopod is produced. It would appear that the value of morphological classification of spermatozoa is probably low and some early descriptions should probably be re-evaluated.

Centrioles

Centrioles in nematode spermatozoa do not conform to the usual nine-triplet pattern found in the cells of other organisms but are generally made up of nine singlets (Figure 4.2A). Centrioles composed of doublets have been described in the spermatids of *Gastromermis* (Poinar and Hess-Poinar, 1993) (it is not known whether these are further simplified into singlets in mature sperm) and in *Heligmosomoides polygyrus* the centrioles are composed of ten singlets (Mansir and Justine, 1995, 1998) (Figure 4.2B).

Nuclear envelope

The spermatid and spermatozoon of most species of nematode are devoid of a nuclear envelope. This envelope apparently disappears at an earlier stage in some plant-parasitic species, as in the spermatocytes of *Heterodera* (Shepherd, Clark and Kempton, 1973) and in the secondary spermatocytes of *Ekphymatodera* (Cares and Baldwin, 1994b) and *Verutus* (Cares and Baldwin, 1994a). The only known exceptions are in the Enoplida, in which the nuclear envelope continues to be present in the mature spermatozoon. However, the nuclear envelope is not 'typical' in the Enoplida, because it produces protrusions that may extend as far as the peripheral cytoplasm.

The nucleus of nematode spermatozoa is usually surrounded by a halo of electron-dense material which contains a protein and is disorganised in mutants of *Caenorhabditis elegans spe-11* (Browning and Strome, 1995; Roush, 1996); see additional details in the systematic section for *C. elegans*.

Membranous organelles

Membranous organelles (Figure 4.3) are intimately related to the cytoskeleton and the major sperm protein in those nematodes that possess them. The nomenclature of these membranous organelles is confused in the literature. With the exception of the oxyurid *Aspiculuris*

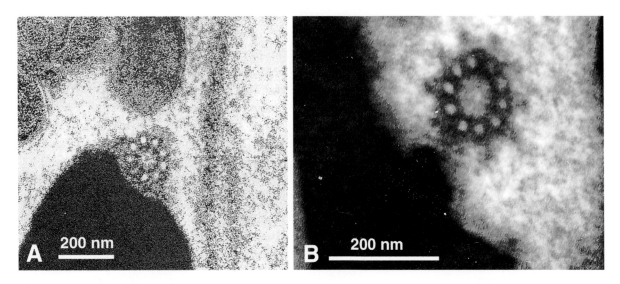

Figure 4.2. Centrioles in nematode spermatozoa are usually located near the nucleus. A, *Ancylostoma caninum*, centriole made up of 9 singlets, the usual configuration in nematode sperm, here in a spermatid; B, *Heligmosomoides polygyrus*, centriole with 10 singlets in spermatozoon. A, from Ugwunna, S.C. and Foor, W.E. (1982) *Journal of Parasitology*, **68**, 817–823. B, from Mansir, A. and Justine, J.-L. (1995) *Mémoires du Muséum National d'Histoire Naturelle*, **166**, 119–128, reproduced with permission.

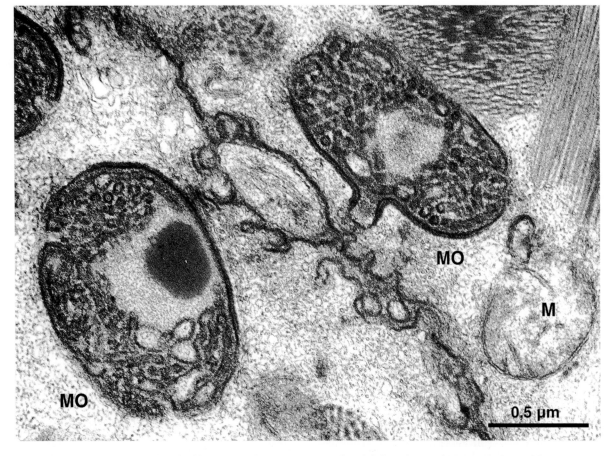

Figure 4.3. Membranous organelles (MO) in two adjacent spermatids of *Sphaerolaimus hirsutus*. Left, membranous organelle containing amorphous body; right, membranous organelle with protrusion directed toward the cell membrane. M, mitochondrion. Original.

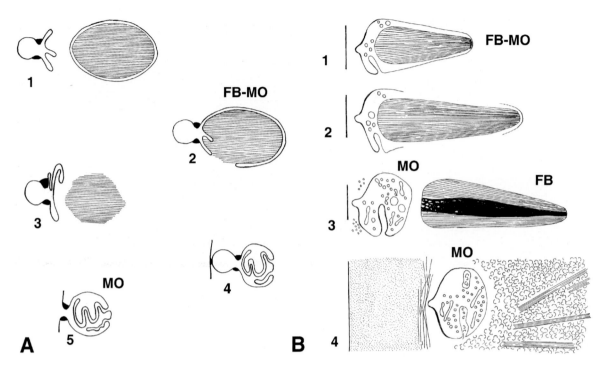

Figure 4.4. Fate of the fibrous body-membranous organelle (FB-MO) complex in the spermatids of two species of nematodes. A, *Caenorhabditis elegans*, steps 1–5, with final fusion of the membranous organelle (MO) with the plasma membrane of the spermatozoon (step 5); B, *Sphaerolaimus hirsutus*, steps 1–4, with fibrous body (FB) disappearing to produce a fibrillar cytoplasm. A, redrawn from Wolf, N., Hirsh, D. and McIntosh, J.R. (1978), *Journal of Ultrastructure Research*, **63**, 155–169. B, redrawn from Noury-Sraïri, N., Gourbault, N. and Justine, J.-L. (1993) *Biology of the Cell*, **79**, 231–241.

tetraptera (Lee and Anya, 1967), the dioctophymid *Dioctophyma renale*, dorylaimids (Shepherd, 1981), and non-aphelenchin tylenchids (Shepherd, 1981), membranous organelles seem to be a consistent feature of nematode sperm. They are membrane bound vesicles that contain internal projections of the membrane, which, in some respects, resemble cristae of mitochondria. They were at one time thought to be specialized mitochondria (Foor, 1970b; Foor, Johnson and Beaver, 1971), and were ambivalently called 'mitochondrion-like inclusions' and 'specialized mitochondria' (Jamuar, 1966). However, the membranous organelles are now known to be Golgi-derived structures in *Ascaris megalocephala* (Favard, 1961); *Heterakis gallinarum* (Lee, 1971); *Rhabditis pellio* (Beams and Sekhon, 1972); *Panagrellus silusiae* (Pasternak and Samoiloff, 1972); *Deontostoma californicum* (Wright, Hope and Jones, 1973); *Dipetalonema viteae* (McLaren, 1973a); *Caenorhabditis elegans* (Wolf, Hirsh and McIntosh, 1978); *Ancylostoma caninum* (Ugwunna and Foor, 1982a); *Sphaerolaimus hirsutus* (Noury-Sraïri, Gourbault and Justine, 1993); *Heligmosomoides polygyrus* (Mansir and Justine, 1996), to name some of the examples.

The contents of the membranous organelles are PAS and acid phosphatase positive (Clark, Moretti and Thomson, 1967; Jamuar, 1966) as in typical acrosomes. Because of this and also from morphological considerations it was suggested that the membranous organelles function as acrosomes and as a result they were called proacrosomal granules (Clark, Moretti and Thomson,

1967; Favard, 1961); a function accepted by Beams and Sekhon (1972). Secretions released by the membranous organelles at the onset of fertilization in *Dipetalonema viteae* (McLaren, 1973a, b) were thought to be responsible for removing the surface coat of the oocyte. However, Foor (1968, 1970b) suggested that physical non-lytic activity of the pseudopodia brought about removal of the surface coat of the oocyte and disputed an acrosomal function. Discontinuity of the oocyte membrane in the region of contact with filopodia of the sperm in *Globodera* (syn. *Heterodera*) *rostochiensis*, the sperm of which lack membranous organelles, led Shepherd, Clark and Kempton (1973) to suggest the possibility of an acrosomal function for the filopodia. Several workers (Burghardt and Foor, 1978; Foor, 1968, 1970b; Lee, 1971; Ugwunna and Foor, 1982a) have argued against an acrosomal action by the membranous organelles because the organelles maintain a posterior position during penetration of the oocyte by the sperm, having already exocytosed their contents; undergo no apparent morphological change immediately prior to contact of sperm and oocyte; and remain intact for some time in the zygote cytoplasm.

Immunocytochemical studies using antibodies against the major sperm protein (MSP), the protein which is characteristic of fibrous bodies, have demonstrated that MSP stored in the fibrous bodies is released as fibres which eventually make up the cytoskeleton of the pseudopodia (see Systematic Review of Nematode Sperm Ultrastructure). Membranous organelles and

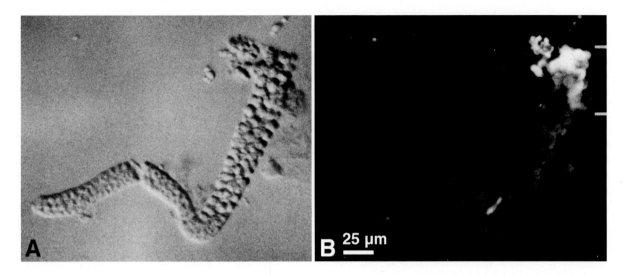

Figure 4.5. Immunolocalization of MSP in testis of *Caenorhabditis elegans*. Dissected testis with the distal end containing spermatogonia at the left. A, interference contrast; B, immunolabelling of MSP. Spermatogenesis is initiated at the bend in the lower right and spermatocytes mature toward the top; note that fluorescence, revealing MSP, is limited to the primary and secondary spermatocytes. From Ward, S. and Klass, M. (1982) *Developmental Biology*, **92**, 203–208, reproduced with permission.

fibrous bodies share a common origin in several species (Noury-Sraïri, Gourbault and Justine, 1993); the structure at the origin of these two structures is known as the fibrous body-membranous organelle (FB-MO) complex (Figure 4.4).

The function of the membranous organelles cannot be considered without reference to the MSP (see next section).

The Sperm Cytoskeleton

Nematodes are characterized by a unique cytoskeleton, including a protein, the major sperm protein (MSP), which is found only in nematodes and only in male germ cells (Figure 4.5).

Major sperm protein

In most amoeboid cells, such as leucocytes, motility is based on actin and its associated proteins. In nematode sperm, which are amoeboid, major sperm protein (MSP) forms an extensive filament system that appears to take the place and role of actin in other amoeboid cells (King *et al.*, 1992). Most work on the MSP has been done on *Caenorhabditis elegans* and on *Ascaris suum*. *C. elegans* has been preferred for genetic studies on the MSP genes because the genetics of this organism is one of the best known and the genome has now been sequenced (*C. elegans* Sequencing Consortium, 1998). *A. suum* has been used for studies on the protein itself, because the large size of the worm makes large quantities of sperm, and therefore of the protein, available. Other species have also been studied, but to a lesser extent.

a. Biochemical characteristics of MSP. MSP of *C. elegans* represents 15% of the total sperm proteins (Ward *et al.*, 1988) and has a molecular weight of 15.5–15.6 kDa (Burke and Ward, 1983; Klass and Hirsh, 1981). Three

isomers constitute respectively 67%, 31% and 2% of the MSP and their isoelectric pHs are 8.8, 8.7, and 7.1. The calculated molecular weight from the sequences of three variants is 14,063 Da, 14,078 Da and 14,093 Da. MSP of *Ascaris* sperm represents 15% of the total sperm proteins (Nelson and Ward, 1981) and has a molecular weight of 14 kDa (Abbas and Cain, 1981; Sepsenwol, Ris and Roberts, 1989). It is a dimeric molecule that polymerizes to form non-polar filaments that are constructed from two helical subfilaments that wind around each other (Haaf *et al.*, 1996; Haaf *et al.*, 1998). The cytoskeleton of the sperm contains two isoforms of MSP, alpha-MSP and beta-MSP. The calculated molecular weights from the sequences of the isoforms are 14,317 Da (alpha-MSP) and 14,175 Da (beta-MSP). Neither of the isoforms shares sequence homology with other cytoskeletal proteins (King *et al.*, 1992).

Some data are available on other species of nematodes. Molecular studies of the MSP of the sperm of *Onchocerca volvulus* (Scott *et al.*, 1989b), *Globodera rostochiensis* (Novitski *et al.*, 1993) and *Dictyocaulus viviparus* (Schnieder, 1993) indicate that the calculated molecular weight in these species falls within the same range as *C. elegans* and *A. suum*. In *Onchocerca volvulus* and *Brugia malayi* there are also two isoforms of MSP and they are encoded on two separate genes (Scott *et al.*, 1989b; Scott, 1996).

b. Sequences and genes of MSP. There are 126 amino acids in the MSP of *Caenorhabditis elegans*, *Ascaris suum* and *Onchocerca volvulus*, but 125 in *Globodera rostochiensis*. Between species the amino acid sequences of the MSPs are 80% to 90% identical (Scott, 1996), however, comparison of MSP amino acid sequences suggests that *C. elegans*, *A. suum* and *O. volvulus* may be more closely related to each other than they are to *G. rostochiensis* (Novitski *et al.*, 1993).

Numerous MSP genes are expressed in roughly equal amounts in *C. elegans*, but they produce only three variants, which are distinguishable in terms of pHi (Klass, Ammons and Ward, 1988). The number of gene copies has been estimated as more than 50 (Ward *et al.*, 1988). About half of the genes are pseudogenes interspersed among the functional MSP genes (Scott *et al.*, 1989a). Of the 40 identified genes, 37, including all those known to be transcribed, are organized in six clusters each comprising 3 to 13 genes. The clusters are located in three loci, one in the right arm of chromosome 2, and two in the median region of chromosome 4, where many germ cell specific genes are located (Burke and Ward, 1983; Klass, Kinsley and Lopez, 1984; Ward *et al.*, 1988). However, Wilson *et al.* (1994), who completed the nucleotide sequence of a contiguous 2 181 032 base pairs in the central gene cluster of chromosome 3, found a sequence similar to the MSP gene on that chromosome. In each cluster, the genes are not in tandem but are separated by at least several thousands bases (Ward *et al.*, 1988). Most of the MSP genes are transcribed, with each gene contributing 1–3% of the total poly(A) + RNA (Klass, Ammons and Ward, 1988). The MSP genes are transcriptionally regulated and coordinated, with the corresponding mRNA being found only in primary spermatocytes (Klass, Dow and Herndon, 1982). This coordinate expression of dispersed genes is made possible by common cis-acting regulatory sequences that are found in the highly conserved 5′ flanking region of each gene (Klass, Ammons and Ward, 1988). It has been proposed that the MSP gene family is the target for the ELT-1 protein of *C. elegans*, a homologue of the vertebrate GATA transcription factor family (Shim, 1999).

Two genes of MSP are known in *Ascaris* and two isoforms, alpha and beta, have been described. The two isoforms each contain 126 amino acids but differ at four residues (positions 14, 15, 54 and 67) resulting in alpha-MSP being 142 Da larger and 0.6 pH unit more basic than beta-MSP; the beta-MSP is five times more abundant than the alpha-MSP (King *et al.*, 1992).

In *Onchocerca volvulus*, two MSP genes have been cloned and sequenced and the two known sequences differ in five amino acids in positions 14, 32, 91, 109 and 110 (Scott *et al.*, 1989b). The genes contain 99- and 282-base coding regions in the first and second exons, respectively, interrupted by a single intron of 153 nucleotides located between the codons for amino acids 33 and 34. The coding regions show a high (95%) similarity, but the introns have a lower similarity (79%). The 5′ flanking region of the two genes of *Onchocerca* are 40% similar, whereas there is a high degree of similarity in the 5′ flanking sequences for different MSP genes of *C. elegans* (Klass, Ammons and Ward, 1988).

In *Globodera rostochiensis*, sequences of msp1 and msp2 differ in four ammoniac's in positions 6, 13, 32 and 38, and msp3 differs from msp1 in only one amino acid (position 83) (Novitski *et al.*, 1993). Three MSP genes have been cloned and sequenced. The genes contain 96- and 282-base coding regions in the first and

second exons, respectively, interrupted by a single intron of 57 nucleotides located between the codons for amino acids 32 and 33. The intron sequences are less conserved than the codon region (Novitski *et al.*, 1993).

The Dv3-14 gene fragment of *Dictyocaulus viviparus* is 471 bp, of which about 90% is translated and the molecular mass of the translation product is 15.5 kDa. DNA sequence homologies were found with MSPs from *A. suum*, *C. elegans* and *Onchocerca*. The sequence of 62 amino acids coded by the gene fragment Dv 3-14 shows a similarity of 84% with the MSPs of *Ascaris suum* and of *C. elegans* (Schnieder, 1993) and 80% homology with other nematodes (Setterquist and Fox, 1995). These results suggest that the protein encoded by Dv3-14 is an MSP of *D. viviparus* (Schnieder, 1993).

Several gene fragments have been sequenced in *Pratylenchus* spp., four in *P. scribneri* and six in *P. penetrans*. A highly variable segment in an otherwise conserved gene sequence of MSP allowed these two nematodes to be distinguished from each other (Setterquist *et al.*, 1996).

The two isoforms of the MSP of *A. suum* exhibit no sequence homology with any other cytoskeletal protein such as actin, myosin and intermediate filament proteins (King *et al.*, 1992); this includes the lack of reputed nucleotide and cation binding sites. However, these authors pointed out that the lack of sequence similarity does not contraindicate a possible three-dimensional resemblance. Proteins showing sequence homologies with nematode MSP have been detected in human neuromuscular tissues (Auffray *et al.*, 1995), and in the central nervous system and gill of the mollusc *Aplysia* (Skehel *et al.*, 1995), but these sequence similarities do not reflect any functional homology. Results to date indicate that MSP is a protein characteristic of nematode sperm.

Ward *et al.* (1988) interpreted the large number of MSP gene copies of *C. elegans* as a mechanism that allows rapid synthesis of MSP in the developing spermatocyte. However, although *Ascaris* and the other nematodes studied also need an important synthesis of MSP in germ cells, the number of gene copies is apparently smaller in these nematodes.

The MSP of *C. elegans* shows neither glycosylation nor phosphorylation (Ward *et al.*, 1988), but the two isoforms of MSP of *Ascaris* are acetylated on the N-terminal alanine (King *et al.*, 1992). No data about possible post-translational modifications of MSP have been reported for other species.

c. Structure and assembly of the MSP protein. Several authors have reported differing diameters for the fibres of MSP when they are packed in the fibrous bodies (see Figure 4.6) (Beams and Sekhon, 1972; Sepsenwol, Ris and Roberts, 1989; Wolf, Hirsh and McIntosh, 1978), however, there now seems to be agreement that the diameter of MSP fibres seen in pseudopods of nematode sperm is 2 nm (Abbas and Cain, 1984; Burghardt and Foor, 1978; Foor, 1974; Foor, Johnson and Beaver, 1971; Noury-Sraïri, Gourbault and Justine, 1993; Roberts, Pavalko and Ward, 1986; Ward and Klass, 1982; Wright and Sommerville, 1984, 1985b).

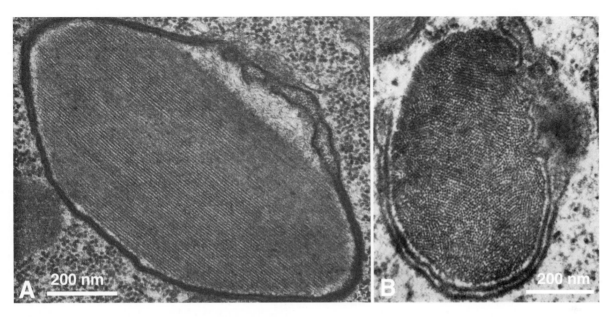

Figure 4.6. MSP fibres in fibrous bodies. A, *Rhabditis pellio*, longitudinal section; B, *Heterakis gallinarum*, cross section. A, from Beams, H.W. and Sekhon, S.S. (1972) *Journal of Ultrastructure Research*, **38**, 511–527; B, from Lee, D.L. (1971) *Journal of Zoology (London)*, **164**, 181–187, reproduced with permission.

Three-dimensional reconstruction of negatively stained MSP filaments of *Ascaris suum* showed that they have dots spaced at roughly 11 nm intervals and that they are constructed from a hierarchy of helices (Figure 4.7). Helical subunits are arranged into two subfilament strands, which are based on left-handed helices of pitch 9 nm. Each subfilament is wrapped around the other in a right-handed helix of pitch about 22.5 nm to form an individual MSP filament which is 10 nm in diameter (Stewart, King and Roberts, 1994). Further coiling of filaments about themselves produces fibre complexes *in vivo* and helical macrofibres *in vitro* in the absence of accessory proteins (King, Stewart and Roberts, 1994; King *et al.*, 1992). The MSP filaments are

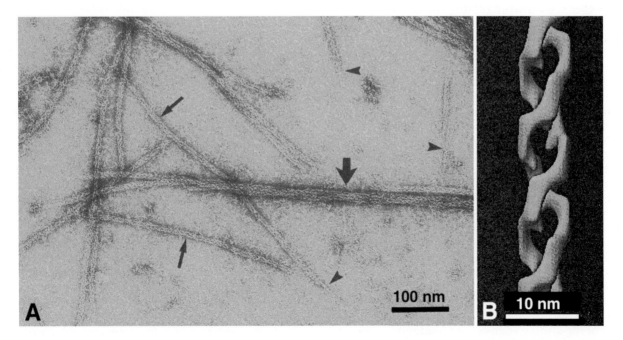

Figure 4.7. Structure of MSP filaments. A, MSP filaments of *Ascaris suum* negatively stained with uranyl acetate. Filaments (small arrows) have a characteristic axial pattern of dots repeated every 11 nm. Sometimes subfilaments (arrowheads) are seen projecting from filaments or in the background. Filaments bundle to form macrofibres (large arrows) commonly containing 3 filaments. Bar = 100 nm. B, Three dimensional reconstruction of negatively stained MSP filaments. Bar = 10 nm. From Stewart, M., King, K.L. and Roberts, T.M. (1994) *Journal of Molecular Biology*, **243**, 60–71, reproduced with permission.

arranged into long, branched fibre complexes that span the length of the pseudopod. Fibres radiating from these complexes become anchored in a dense layer of material that lines the inside of the pseudopod membrane. Where the fibre complexes attach to the pseudopod membrane the membrane protrudes to form villipodia. As the sperm move the fibre complexes flow centripetally toward the cell body due to filament assembly at the leading edge and disassembly at the rear of the pseudopod. This results in the fibre complexes moving rearwards at 10-50 μm/min, which is the same rate as the sperm crawls forward (Roberts and King, 1991; Sepsenwol, Ris and Roberts, 1989). Vectorial filament assembling and bundling of MSP are the two phenomena associated with amoeboid locomotion. The pitch of the subfilaments in the MSP filament is out of phase, with the consequence that interaction will occur between only a fraction of the subunits in different strands of an MSP filament, in contrast to actin where the interactions are stronger (Stewart, King and Roberts, 1994).

In ethanol, or in other water-miscible alcohols, at pH ranging from 5.7 to 9.5 and temperature from 2 to 39°C, each of the two isoforms of *Ascaris* MSP assembles into filaments 10 nm wide and with a characteristic substructure repeating axially at 9 nm. These filaments were found to be indistinguishable from fibres isolated from detergent-lysed sperm (King *et al.*, 1992). Italiano *et al.* (1996) developed an *in vitro* motility system for the sperm of *Ascaris* and found that addition of ATP to sperm extracts induced the formation of fibres that displayed the key features of the MSP cytoskeleton *in vivo*. These fibres were about 2 μm in diameter and consisted of a meshwork of MSP filaments. Each fibre had at one end a vesicle derived from the plasma membrane of the leading edge of the sperm. Fibre growth was due to filament assembly at the vesicle resulting in the vesicles being transported along the bundle. This demonstrated that localised polymerisation and bundling of filaments could move membranes. Polymerisation along the leading edge of the sperm appears to drive protrusion. *Ascaris* sperm are able to regulate the state of MSP polymerisation by varying intracellular pH (King *et al.*, 1994). During spermatogenesis there is a cycle of MSP assembly-disassembly-reassembly that coincides with changes in internal pH. In spermatocytes, in which the MSP is contained in the paracrystalline fibrous bodies, the internal pH was 6.8; this was 0.6 units higher than in the spermatids in which the fibrous bodies disassemble and contain no assemblies of MSP filaments. Activation of the spermatids to complete their development resulted in a rapid increase in internal pH to 6.4 and the reappearance of filaments in the cytoplasm. The sperm establish a pH gradient in the pseudopod with the pH being 0.15 units higher at the leading edge of the pseudopod, where the fibre complexes assemble, than at the base of the pseudopod, where disassembly of the fibre complexes occurs (King *et al.*, 1994). Italiano, Stewart and Roberts (1999) found that the behaviour of sperm tethered to the substrate showed that an additional force is required to pull the cell body forward. They used pH to uncouple cytoskeletal polymerisation and depolymerisation of *Ascaris* sperm and found that at pH 6.75 protrusion of the leading edge slowed dramatically whereas both cytoskeletal disassembly at the base of the pseudopodium and cell body retraction continued. At pH 6.35 the cytoskeleton pulled away from the leading edge and receded through the pseudopodium whilst undergoing disassembly at the cell body. At pH 5.5 the cytoskeleton quickly disassembled but when the cells were washed with physiological buffer the cytoskeleton quickly reassembled. Cytoskeletal reassembly took place at the pseudopodial margin resulting in membrane protrusion, but the cell body did not move until the cytoskeleton was rebuilt and depolymerisation resumed. It appears that cell body retraction is mediated by tension in the cytoskeleton, correlated with depolymerisation of the MSP at the base of the pseudopodium.

Fibre growth occurs by the addition of subunits at the ends of filaments, but is also regulated by pressure-sensitive factors (Roberts, Salmon and Stewart, 1998). In solution MSP, unlike actin, exists as a symmetrical dimeric molecule that polymerises to form non-polar filaments constructed from two helical subfilaments that wind round one another (Haaf *et al.*, 1996; Haaf *et al.*, 1998). The MSP filaments can interact with one another to form higher-order assemblies without requiring the range of accessory proteins usually employed in actin-based systems (Haaf *et al.*, 1998). The rate of MSP polymerisation associated with the movement of vesicles in an *in vitro* motility assay is enhanced by magnesium and manganese ions (Figure 4.8). These ions appear to be associated with the assembly of subfilaments into filaments and their subsequent aggregation into bundles (Haaf *et al.*, 1998). The structure of *Ascaris* MSP subfilament helices has been determined by X-ray crystallography. MSP helices are constructed from dimers (Figure 4.9) and have no overall polarity; this suggests that no molecular motors play a direct role in the generation of movement (Bullock *et al.*, 1998). Protein-protein interactions of MSP have been evaluated by studying dimer formation of various MSP molecules experimentally inserted into bacteria (Smith and Ward, 1998).

Although associated proteins are claimed to be uninvolved in motility linked to MSP, several proteins have been found with the same localization. Protein-10 and Protein-11 are located in the fibrous body complex of *C. elegans* sperm (Ward, 1986). Particles, 30 nm in diameter, are present in triton X-100 protein extracts of *Ascaris* pseudopods, and are composed of proteins with molecular weights of 33 kDa, 42 kDa, and a group between 52 and 56 kDa (Holliday and Roberts, 1995). Also, immunofluorescence labelling showed that particle-associated proteins are present in the pseudopods of *Ascaris* sperm, which indicates that the particles are elements of the motility apparatus of the sperm (Holliday and Roberts, 1995).

Mansir and Justine (1996) showed by immunofluorescence that MSP co-localizes with actin in the spermatids (see Figure 4.11), but not the sperm, of

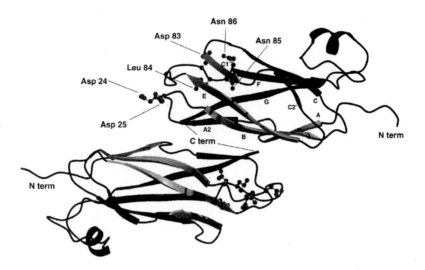

Figure 4.8. Global fold of the polypeptide chain in a MSP dimer of *Ascaris suum*. The six residues identified as being involved in the Mn2+ / Mg2+ binding sites are shown in ball and stick representation. From Haaf, A., LeClaire, L.I., Roberts, G., Kent, H.M., Roberts, T.M., Stewart, M. *et al.* (1998) *Journal of Molecular Biology*, **284**, 1611–1624, reproduced with permission.

Heligmosomoides polygyrus and *Graphidium strigosum* (Mansir and Justine, 1996, 1999), and with tropomyosin in *H. polygyrus* (Mansir and Justine, 1996). They suggested that the actin is not involved in cell motility but may have a role in shaping of the cell and in the arrangement of its organelles during spermiogenesis (Mansir and Justine, 1996).

d. MSP and movement. As described above, motility in most, if not all, nematode sperm is dependent on the formation of MSP fibre complexes and the formation of pseudopodia (Scott, 1996). Large fibre complexes (see above) can be seen in the pseudopods of live, crawling sperm of *Ascaris* where, with a continuous flow of membrane specialisations, they form the villipodia, at the leading edge of the sperm pseudopod (Sepsenwol, Ris and Roberts, 1989). Villipodia are projections of the surface of the pseudopod formed at the leading edge of the sperm and they act as the initial substrate adhesion sites. These sites then expand to form contact zones. There is rapid elongation of the pseudopod (within 9–12 min) from the cell body of immotile spermatids of *A. suum* and fusion of the membranous organelles with the plasma membrane when the sperm are treated with an extract from the glandular vas deferens of the male under strict anaerobic conditions (Sepsenwol and Taft, 1990; Scott, 1996). During forward movement the contact zones at the leading edge of the pseudopod remain fixed as the rest of the cell moves over them; this is analogous to the close contacts between the pseudo-pods and their substrate described underneath other crawling eukaryotic cells. Forward movement ceases if the sperm are exposed to air. Even before the pseudopod attaches to the substratum, the entire cytoskeleton and villipodia move continuously in unison in a posterior direction toward the cell body of the sperm. During crawling the complexes and villo-podia disappear at the junction between pseudopod and cell body. In this region there is high turnover of membrane and the terminations of branched refringent fibres, which extend the length of the pseudopod,

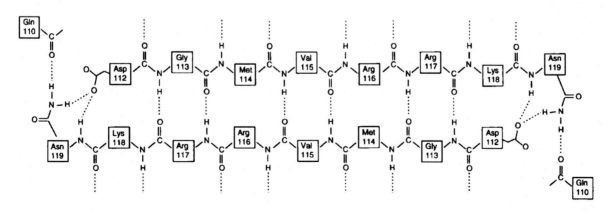

Figure 4.9. Schematic illustration of the putative hydrogen bonds formed at the interface between successive MSP dimers along the helice. From Bullock, T.L., McCoy, A.J., Kent, H.M., Roberts, T.M. and Stewart, M. (1998) *Nature Structural Biology*, **5**, 184–189, reproduced with permission.

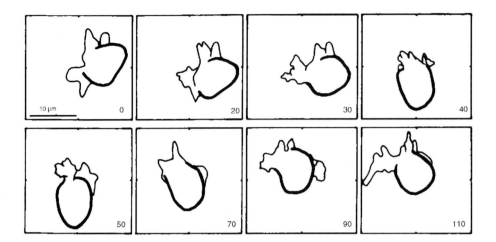

Figure 4.10. Movement of nematode spermatozoa. Series of tracings from videotapes of *Ascaris lumbricoides* spermatozoon translocating on glass. Cell body is outlined with heavy line. Time in seconds. From Nelson, G.A. and Ward, S. (1981) *Experimental Cell Research*, **131**, 149–160, reproduced with permission.

occurs here (Sepsenwol, Ris and Roberts, 1989; Sepsenwol and Taft, 1990; Scott, 1996).

Nematode sperm move comparatively quickly, but the speed varies from species to species and also depends on the conditions of observation. Forward movement of spermatozoa on glass averages 9.4 μm/min for *C. elegans* sperm (Nelson, Roberts and Ward, 1982; Roberts and Streitmatter, 1984), 12.3 μm/min for *Nippostrongylus brasiliensis* (Wright and Sommerville,

1984), 7.3 μm/min for *Heligmosomoides polygyrus* (Wright and Sommerville, 1977), and 11 μm/min on glass slides (Nelson and Ward, 1981) for *Ascaris lumbricoides* (Figure 4.10), and a rapid 70 μm/min, or more, for *Ascaris suum* spermatozoa (Sepsenwol and Taft, 1990).

It is difficult to generalise on the movement of the amoeboid sperm of nematodes, as there appear to be significant differences among the families of nema-

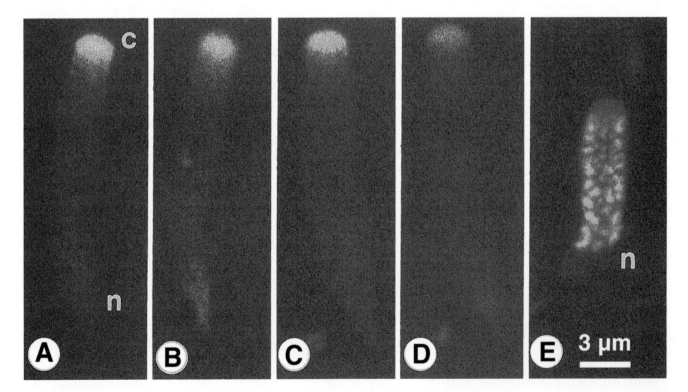

Figure 4.11. Colocalization of MSP and actin in the anterior cap of spermatozoa of *Heligmosomoides polygyrus*. A-B and C-D, respectively, double labelling of actin and MSP with specific antibodies revealed by second antibodies conjugated with FITC (A, D) and TRITC (B, C). E, free spermatid. The dots are labelled by the anti-MSP antibody. N, position of nucleus. From Mansir, A. and Justine, J.-L. (1996) Actin and major sperm protein in spermatids and spermatozoa of the parasitic nematode *Heligmosomoides polygyrus*. *Molecular Reproduction and Development*, **45**, 332–341, Copyright © 1996 Wiley-Liss, Inc. Reprinted by permission of Wiley-Liss, Inc., a subsidiary of John Wiley & Sons, Inc.

todes. The spermatozoa of *Nippostrongylus brasiliensis* creep by means of transverse waves of constriction passing rearwards over their short pseudopods rather than by forming individual villipodia (Wright and Sommerville, 1984). Pseudopods are the source of movement in the spermatozoa of *C. elegans* and by changing their contours they control the direction of movement. It is probable that the progressive formation of new attachment sites on the pseudopod sets the new direction. In contrast to free-living amoebae and leucocytes there is apparently no extension/retraction cycle of the leading margin of the pseudopods and no cytoplasmic streaming in moving sperm of *C. elegans*. Observations on mutant sperm of *C. elegans* support the view that projections are required for movement of the sperm of this nematode (Nelson, Roberts and Ward, 1982). As described above, independent processes of continuous extension at the leading edge and continuous shortening at the base of the pseudopod act to propel the sperm of *Ascaris* forward (Sepsenwol and Taft, 1990).

It is interesting to compare actin and MSP, as they are both proteins known to be associated with amoeboid movement. Both are globular proteins that polymerize to make fibrous polymers, and produce movement by vectorial assembling and filament bundling. However, there are many differences between them. The molecular weight of MSP is three times smaller than actin; there are no similarities of sequence between the two proteins; actin and MSP filaments both polymerize as right-handed helices, but the pitch is 70 nm in actin whereas it is 20 nm in MSP (King, Stewart and Roberts, 1994; King *et al.*, 1992). In solution, MSP exists as symmetrical dimers (Haaf *et al.*, 1996), in contrast to actin. Also, in contrast to MSP, polymerization of actin is modulated by bivalent cations and implicates ATP (King *et al.*, 1992). Accessory proteins (actin-binding proteins) are required for actin polymerization and motility but they are not required for polymerization and motility of MSP, at least *in vitro* (King, Stewart and Roberts, 1994; King *et al.*, 1992). Therefore, similarities between actin and MSP are related more to the mechanism by which these two proteins generate motility than to their molecular structure and assembly dynamics.

Actin and other proteins

Earlier workers looked for the presence of actin in nematode sperm because actin is associated with amoeboid movement in most cells. By using an immunofluorescent technique, Nelson and Ward (1981) found actin in the pseudopodia of spermatozoa of *Ascaris lumbricoides*. However, several studies have indicated that actin does not have a role in the movement of nematode sperm. Gel electrophoresis has been used to demonstrate that actin is only 0.5% of the cell's protein (Nelson and Ward, 1981). Also, sperm movement is not affected by cytochalasin D or by DNAase I. However, Foor (1983b) showed that DNAase-I-FITC, which binds specifically to globular actin, produced a 'floral-like arrangement' (by labelling

the fibrous bodies) in a crescent, probably corresponding to the pseudopod, in spermatozoa (termed spermatocytes) of *Ascaris*.

Later work on *C. elegans* and *Ascaris suum* demonstrated the role of MSP in the movement of nematode sperm (see above). The presence of actin was therefore not considered of functional significance by most authors studying the MSP cytoskeleton (reviews in Roberts and Stewart, 1995; Roberts and Stewart, 1997; Scott, 1996; Theriot, 1996).

Revived interest in the presence of actin in nematode sperm has resulted in the detection, by immunolabelling, of actin in the fibrous bodies of the large spermatids of *Sphaerolaimus hirsutus*, but this study did not include results for fully mature spermatozoa (Noury-Sraïri, Gourbault and Justine, 1993). Immunolabelling also demonstrated the presence of actin in the spermatozoa of *C. elegans* but the small size of these spermatozoa did not make them good models for immunocytochemical work on cell organelles (Noury-Sraïri, Gourbault and Justine, 1993). Spermatozoa of trichostrongyle nematodes have large, elongate spermatids and spermatozoa (about 15–20 μm in length) which are clearly polarised as they have an anterior pseudopod that is easily recognised. Several anti-actin antibodies have been used to demonstrate the presence of actin in male germ cells of *Heligmosomoides polygyrus*. Actin is present in the fibrous bodies (which are known to contain MSP in an inactive state) of spermatids and in the anterior cap of the spermatozoon, which also contains MSP in an active state (Mansir and Justine, 1996) (Figure 4.11). This co-localization of actin and MSP in all stages is intriguing. Actin has also been detected in the anterior cap of the spermatozoon of *Trichostrongylus colubriformis* and *Teladorsagia* (syn. *Ostertagia*) *circumcincta* (Mansir *et al.*, 1997), and co-localization of actin and MSP was demonstrated in *Graphidium strigosum* by Mansir and Justine (1999). Tropomyosin, one of the numerous actin-associated proteins, has also been demonstrated in the anterior cap of *H. polygyrus* (Mansir and Justine, 1996; Mansir *et al.*, 1997).

Although MSP appears to be the sole protein involved in movement of the sperm of *C. elegans* and *A. suum*, recent studies on trichostrongyle nematodes suggest that actin may have a role in spermatozoa in these and some other nematodes. For example, it has been suggested that actin may have a role in shaping the cell and also in the arrangement of the organelles during spermiogenesis in nematodes (Mansir and Justine, 1996).

Tubulin

Tubulin is not a major cytoskeletal protein in the spermatozoa of nematodes because the spermatozoa have no flagella. However, although *C. elegans* and *A. suum* have no tubulin in mature sperm and in spermatids, this is not the case in all nematodes. Microtubules are present under the plasma membrane of the sperm of some tylenchids and mermithids and in the 'tail' of *Aspiculuris tetraptera* (Lee and Anya, 1967) but nothing is known about tubulin in these species, apart from the ultrastructural localisation of the microtubules. Tubulin is a component of the cytoskeleton in spermatids of trichostrongyle nematodes, but disappears at the end of spermiogenesis, and is almost absent in mature spermatozoa. A conspicuous, but transient, microtubular system is present in cells at various stages during spermatogenesis in *Heligmosomoides polygyrus* (Mansir and Justine, 1998) (Figure 4.12; see also Figure 4.26). The presence or absence of certain post-translational modifications of tubulin has been studied in this nematode but so far in no other nematode (Mansir and Justine, 1998). See the systematics section for further details.

Nematode Sperm Ultrastructure

The systematic order used is that proposed in DeLey and Blaxter in this volume (see also Blaxter *et al.* (1998)). In this system, there are two classes, Enoplea, with subclasses Enoplia and Dorylaimia, and Chromadorea, with a single subclass, Chromadoria, which includes most parasitic nematodes. The order Rhabditida, which includes most parasites, is a terminal group within the Chromadoria. The Strongyloidea are close to *Caenorhabditis elegans*.

This section of the chapter is based only on ultrastructural descriptions of nematode sperm and spermatogenesis.

Class Enoplea, Subclass Enoplia, Order Enoplida

Baccetti *et al.* (1983) described the spermatozoon of *Mesacanthion hirsutum* (Figure 4.13) and, on the basis of the presence of a nuclear envelope, which is absent in all other groups of nematode, the authors separated the Enoplida from all other nematodes. These authors regarded the spermatozoon of the Enoplida as being the most primitive type in the Nematoda. The spermatozoon of *M. hirsutum* is lenticular, with a central nucleus limited by a typical nuclear envelope, surrounded by 'spheroidal membranous vesicles', which appear to correspond to the membranous organelles of other nematodes, and a few mitochondria.

The spermatocytes of *Enoplus anisospiculus* contain Golgi bodies, rough endoplasmic reticulum, ribosomes, mitochondria and membranous organelles in their cytoplasm (Yushin and Malakhov, 1998). Apparently development of the membranous organelles proceeds along two parallel lines that are characteristic of two different types of spermatocyte. In the first type, the membranous organelles begin as a system of cisternae, whereas in the second type the membranous organelles first appear as large vesicles filled with electron-dense

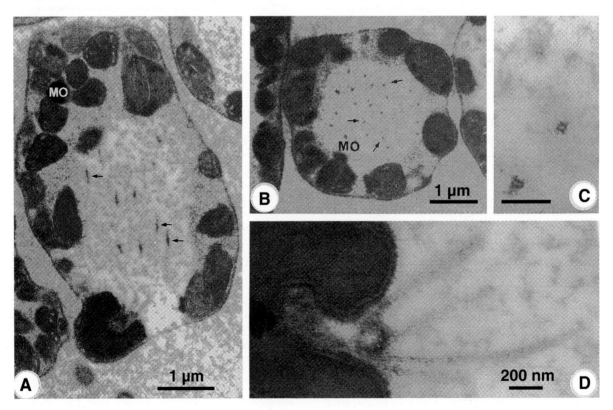

Figure 4.12. Tubulin in spermatids of *Heligmosomoides polygyrus*. A, early elongate spermatid, longitudinal section. Longitudinal microtubules (arrows) diverging from the centrioles in the nuclear fossa; B, elongate spermatid, transverse section. Microtubules (arrows) occupy the center of the cell which is devoid of other organelles. Membranous organelles (MO) occupy the periphery of the cell. C, microtubules, as in B; D, microtubules diverging from centriole in nuclear fossa. From Mansir, A. and Justine, J.-L. (1998) The microtubular and posttranslationally modified tubulin during spermatogenesis in a parasitic nematode with amoeboid and aflagellate spermatozoa. *Molecular Reproduction and Development*, **49**, 150–167, Copyright © 1998 Wiley-Liss, Inc. Reprinted by permission of Wiley-Liss, Inc., a subsidiary of John Wiley & Sons, Inc.

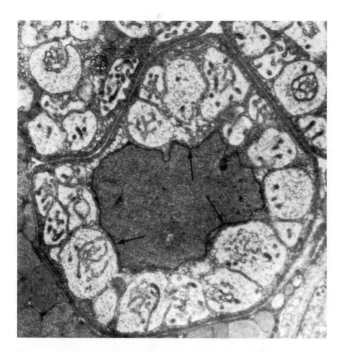

Figure 4.13. Spermatozoon of *Mesacanthion hirsutum*. Note the presence of a nuclear envelope (arrows), which is characteristic of enoplids and not found in other groups of nematodes. From Baccetti, B., Dallai, R., Grimaldi de Zio, S. and Marinari, A. (1983) The evolution of the Nematode spermatozoon. *Gamete Research*, **8**, 309–323, Copyright © 1983 Wiley-Liss, Inc. Reprinted by permission of Wiley-Liss, Inc., a subsidiary of John Wiley & Sons, Inc.

material. Fibrous bodies are apparently absent in the spermatocytes and young spermatids and no evidence of formation of membranous organelles/fibrous bodies was found at any stage of spermiogenesis. If these findings are confirmed then the development of two types of membranous organelles should be considered unique for the Nematoda. The nuclei of spermatids have a distinct nuclear envelope. Mitochondria in older spermatids form a layer at the future anterior end of the nucleus, all membranous organelles become positioned posteriorly, and fibrous bodies with a marked radial orientation appear first between the anterior layer of the mitochondria and the nucleus. This cluster of organelles is retained in the immature spermatozoon after detachment of the residual body (Figure 4.14).

Spermatozoa in the uteri of female *Enoplus demani* and *Enoplus anisospiculus* are bipolar and those taken from the females of both species, as well as those activated *in vitro*, showed amoeboid motility at the broad anterior (pseudopod) end of the sperm. The posterior region of the sperm is cone-shaped and has a more constant form. Live spermatozoa taken from female *E. demani* measure 18–20 μm and those from female *E. anisospiculus* measure 24–28 μm in length and can crawl across the surface of glass (Yushin and Malakhov, 1994). The nucleus of the sperm of both species has a distinct nuclear envelope made up of two superimposed membranes, with the outer membrane forming projections into the cytoplasm. Numerous

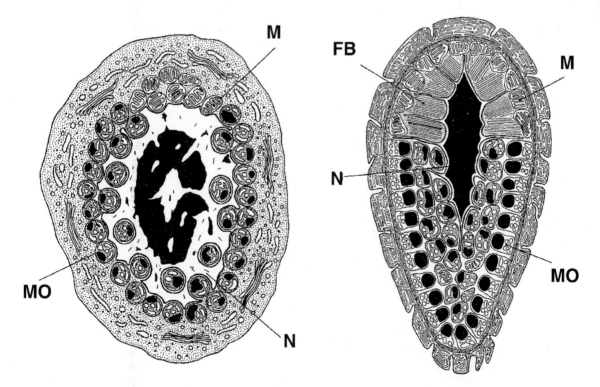

Figure 4.14. Diagram of the spermatid and mature spermatozoon in *Enoplus anisospiculus*. Left, young spermatid. The mitochondria (M) are grouped in the anterior region; right, immature spermatozoon. The anterior part contains radially oriented fibrous bodies (FB) and mitochondria (M), membranous organelles (MO) are grouped in the posterior part. N, nucleus. From Yushin, V.V. and Malakhov, V.V. (1998) *Fundamental and Applied Nematology*, **21**, 213–225, reproduced with permission.

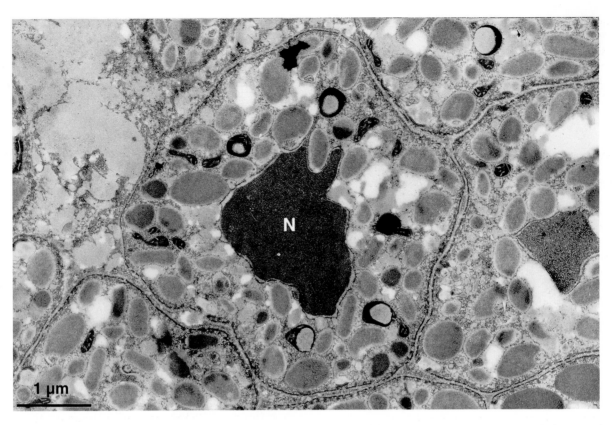

Figure 4.15. Spermatozoon of *Enoploides delamarei*. Note the presence of a nuclear envelope around the nucleus (N). From Noury-Sraïri (1994), Thèse d'État, Université Cadi Ayyad, Marrakech, Morocco, reproduced with permission.

round or polygonal mitochondria are dispersed throughout the cone part of the cell although there is a concentration of the mitochondria anterior to the nucleus and at the base of the pseudopod. Sperm in the uteri of the two species contain two types of membranous organelles. Spherical organelles are located centrally, are not connected with the plasma membrane and contain microvilli that project into an electron-dense material. Electron lucent membranous organelles are concentrated beneath the plasma membrane, contain an elaborate system of internal microvilli and open to the exterior via pores. The anterior pseudopod contains a fibrillar network of fine fibres that attach to the plasma membrane. This is reminiscent of the MSP cytoskeleton described in *C. elegans* and other species (Yushin and Malakhov, 1994). These authors claim that the 'basic and primitive type of sperm' in the nematodes should be considered as bipolar, as in *Enoplus*, and not globular as suggested by Baccetti *et al.* (1983).

Spermatogenesis and the spermatids of *Enoploides delamarei* (Figure 4.15) were described by Noury-Sraïri (1994) (see also Justine and Jamieson, 1999). The spermatocytes have numerous dictyosomes that produce dense bodies. The nuclei of the spermatids have a nuclear envelope made up of two membranes, as in other enoplids. There are large fibrous bodies, which have electron-dense contents, in the cytoplasm and

peripheral membranous organelles are in contact with the plasma membrane. The plasma membrane is lined internally by an electron-dense layer, as described by Baccetti *et al.* (1983) in *Mesacanthion*. Spermatids in male *Enoploides* are round cells whereas mature spermatozoa in the female genital tract of *Enoplus* are bipolar cells (Yushin and Malakhow, 1994). The spermatids of *Enoplus anisospiculus* are slightly elongated cells (Yushin and Malakhov 1998). These differences between *Enoploides* and *Enoplus* may be attributed to differences in the stages of development of the male cells, which probably undergo morphological changes in the female.

The sperm of *Deontostoma californica* in the male and female reproductive tracts are bluntly rod-shaped and are about $8 \times 18 \ \mu$m (Wright, Hope and Jones, 1973). There is a central, irregular-shaped, nuclear mass, about $3 \times 15 \ \mu$m, which consists of electron-dense fibrous material. According to the authors there is no nuclear membrane, however, the boundary of the nuclear mass resembles the nuclear envelope found in other enoplids. Membranous organelles, about 0.55 μm in diameter, possess internal microvillus-like projections and are connected to the plasma membrane by a narrow neck surrounded by a dense collar. The volume of the contents of the membranous organelles, which are presumably released at the cell surface once the sperm enter the uterus, was estimated at 100 μm^3 or one seventh of the total sperm volume. The ectoplasmic zone of sperm adjacent to the uterine epithelium was found to possess hemidesmosome-like attachments to the epithelium and rarely contained the open membranous organelles found elsewhere. Dense bodies, averaging 0.64 μm in diameter, contain microvillus-like internal projections and are interspersed between the membranous organelles. It is probable that they are membranous organelles that have yet to connect to the plasma membrane as some of them have neck-like projections. Some apparently degenerate granular mitochondria are present. Groups of rarely found cross-sectioned tubules were thought possibly to be rudiments of centrioles. Absence of true centrioles and the appearance of the membranous organelles were considered by Wright, Hope and Jones (1973) to be of interest in view of the suggestion by Sprent (1962) that ascaridoid nematodes may have evolved from members of the marine order Enoplida. This hypothesis should be rejected because the absence of centrioles in an ultrastructural study without exhaustive study of serial sections is dubious, and because other 'enoplid' characteristics can be found not only in ascaridoids but also in other nematodes. Although the authors claim that the nucleus of *D. californica* lacks an envelope, the electron micrographs in their article show a nuclear envelope similar to that of other enoplid sperm. It appears that the sperm of all members of the Enoplida possess a nuclear envelope and this differentiates them from all other nematodes. Another general characteristic of enoplid sperm is the presence of a dense layer lining the plasma membrane internally.

The oxystominid *Halalaimus dimorphus* is unique in having both egg and sperm dimorphism; the posterior and anterior testes produce sperm of different sizes (about 10 μm and 4 μm). Turpeenniemi (1998) described the ultrastructure of spermiogenesis and the drop-shaped spermatozoa. The nucleus is located in the rounded end of the spermatozoon and has a nuclear envelope. The cytoplasm contains many mitochondria and an extensive network of small tubules (not microtubules); in the center there is a star-shaped cavity made up of membranous elements. The later looks like a single giant membranous organelle.

Electron micrographs and brief descriptions of spermatozoa of the oncholaimids *Adoncholaimus fuscus* and *Metoncholaimus denticaudatus* published by Calcoen and Dekegel (1979) indicate that sperm, in the demanian organ of the female (see below) and in the ductus ejaculatorius of the male are amoeboid. The sperm are apparently not divided into two regions. They have an almost central nucleus and some additional smaller dense masses which may be chromatin, some mitochondria (at least in *M. denticaudatus*) and vesiculated cytoplasm. The presence of typical membranous organelles cannot be established from their electron micrographs. A nuclear envelope is not visible in the micrographs, but the presence of an electron-lucent area around the nucleus is reminiscent of that seen in enoplids which possess a nuclear envelope. Females of oncholaimids possess a demanian

organ, which has been interpreted as an adaptation to traumatic insemination (Chabaud *et al.*, 1983). It would be of interest to study this unusual aspect of reproduction further, not least to determine if it is correlated with adapted spermatozoal morphology. Interestingly, *Adoncholaimus* is grouped with the Trichocephalida in Blaxter's analysis (Blaxter *et al.*, 1998) and not with the other enoplids.

Class Enoplea, Subclass Enoplia, Order Triplonchida

Sperm structure is apparently relatively varied, as shown by a light microscopy study of *Trichodorus* spp. by Decraemer (1988) (Figure 4.16), but no electron microscopy study is available.

Class Enoplea, Subclass Dorylaimia, Order Dorylaimida

The spermatozoa of *Xiphinema diversicaudatum* are rounded cells that have many peripheral evaginations (Baccetti *et al.*, 1983). There is a central nucleus that is surrounded by numerous mitochondria and filaments (10 nm in diameter) lie between the mitochondria. Many microtubules lie beneath the cell membrane. There is no nuclear envelope and the chromatin is organized in dense clumps.

Kruger (1991) has described spermatogenesis and the ultrastructure of the spermatozoon of *Xiphinema theresiae* (Figure 4.17). The mature spermatozoon is slightly elongate and produces numerous filopods that give it a 'woolly' appearance. The central nucleus lacks an envelope and is surrounded by regularly arranged mitochondria situated in dense cytoplasm. The peripheral cytoplasm contains dense bundles of filaments, which probably correspond to the fibrous bodies of other nematodes, and numerous peripheral filopods. During spermiogenesis the spermatid shows two very distinct stages, with an early stage which lacks filopods and a later stage which possesses filopods.

Van de Velde *et al.* (1991) have described spermatozoa found in the female genital tract of *X. theresiae* and *X. pinoides*. The spermatozoa in the uterus have abundant peripheral protrusions that were thought to be associated with amoeboid movement of the sperm during their migration up the uterus. These peripheral protrusions are lost when the sperm reach the oviduct. This loss of filopods also occurs when the spermatozoa reach the oocyte region of *X. diversicaudatum* (Bleve-Zacheo, Melillo and Zacheo, 1993). Thus, the filopods of sperm in the female tract of *Xiphinema* actively participate in movement of the sperm along the uterus but are then lost prior to fertilization.

Class Enoplea, Subclass Dorylaimia, Order Mononchida

The spermatozoon of *Mylonchulus nainitalensis* is 4.5 μm in length and 1.5 μm in width. It has a nuclear region, which contains a conical nucleus lacking an envelope but which is surrounded by a dense halo, and an amoeboid cytoplasmic region that contains the organelles and produces filopodia . There are numerous microtubules at the periphery of the cell. No centrioles

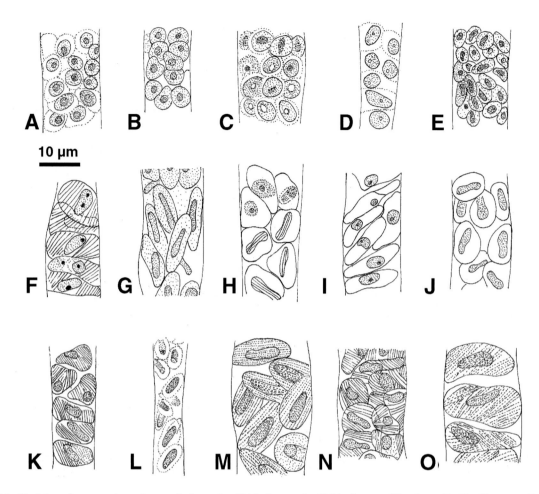

Figure 4.16. Variety of spermatozoal morphology in *Trichodorus*. A, *Trichodorus californicus*; B, *T. dilatatus*; C, *T. elegans*; D, *T. intermedius*; E, *T. proximus*; F, *T. borneoensis*; G, *T. coomansi*; I, *T. cottieri*; J, *T. cylindricus*; J, *T. eburnens*; K, *T. primitivus*; L, *T. similis*; M, *T. taylori*; N, *T. cedarus*; O, *T. variopapillatus*. Redrawn from Decraemer, W. (1988) *Bulletin de l'Institut Royal des Sciences Naturelles de Belgique, Biologie*, **58**, 29–44.

or mitochondria were found. The amoeboid region with pseudopodia should be called 'anterior' rather than 'basal' as in Baccetti's brief description (Baccetti *et al.*, 1983).

Class Enoplea, Subclass Dorylaimia, Order Dioctophymatida

The immature sperm in the seminal vesicle of *Dioctophyme renale* are rodlike (Foor, 1970b). They lack membranous organelles but the surface of the cell is highly modified. There are two parallel membranes, of which the outer (the plasma membrane) is separated from the inner by 20 nm, and there are dense, often beaded material, between these two layers. The cytoplasm contains large clumps of chromatin, centrioles and whorls or sheets of membranous elements that in places disrupt the surface membrane arrangement. More mature sperm in the seminal vesicle, or *in utero*, are more round in appearance, lack the inner surface membrane and centrioles were not detected. An unusual feature is the absence of mitochondria, which is in striking contrast with the sperm of other

nematodes. Chromatin in the mature sperm is homogeneous and fills most of the cell. The membranous elements of the mature sperm are compact, are associated with tortuous channels that open to the exterior, and appear to be different from those of ascaroid sperm. Pseudopodia are formed after insemination and presumably are associated with movement of the sperm, but no other changes were noted.

Class Enoplea, Subclass Dorylaimia, Order Mermithida

Poinar and Hess-Poinar (1993) have described the ultrastructure of spermatogenesis and the spermatozoa of the mermithid *Gastromermis* sp. The spermatozoon is 35–40 µm in length, has a broad anterior extremity bearing pseudopods, and a posterior cone-shaped region. The spermatozoon resembles those found in trichonstrongylids such as *Nippostrongylus brasiliensis*, *Heligmosomoides polygyrus* and *Nematodirus battus* (Jamuar, 1966; Wright and Sommerville, 1977; Martin and Lee, 1980). However, the nucleus in *Gastromermis* is

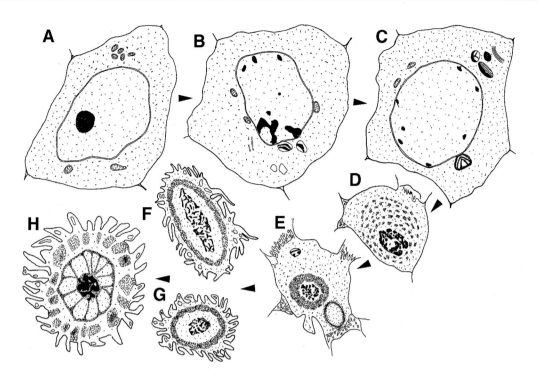

Figure 4.17. Diagram of spermiogenesis in *Xiphinema theresiae*. A-C, spermatocytes; D, E, young spermatids; F, G, older spermatids, H, early spermatozoon. From Kruger, J.C.d.W. (1991) Ultrastructure of sperm development in the plant parasitic nematode *Xiphinema theresiaie*. *Journal of Morphology*, **210**, 163–174, Copyright © 1991 Wiley-Liss, Inc. Reprinted by permission of Wiley-Liss, Inc., a subsidiary of John Wiley & Sons, Inc.

elongate and occupies most of the length of the cone-shaped region, whereas it is a shorter cone with almost no associated cytoplasm associated with it in *Heligmosomoides*. The filopods bear microspikes that contain microtubules and the spermatids lack the fibrous bodies which are associated with the production of MSP in other nematodes (this is another difference from the trichostrongylids). In the cone region, the mitochondria are arranged in parallel rows and form an elongate sheath around the nucleus. Longitudinally oriented microtubules lie beneath the cell membrane and between the nucleus and the mitochondrial sheath. The centrioles, which consist of 9 doublets in spermatids, are situated at the tip of the nucleus, in contrast to trichostrongylids in which they are at the junction between the nucleus and the cytoplasmic region of the sperm body. Dense tubular structures, of unknown composition and function, are intercalated between the mitochondria and have not been recorded in other nematode sperm. Membranous organelles are present in mature sperm but they apparently arise by a different process from that in other nematodes, in which they are usually associated with fibrous bodies but which are absent in *Gastromermis*. The sperm becomes motile in conditions that have been used to study MSP-based motility in other species of nematode, and the filopods were seen vibrating, elongating, shortening and coalescing. As these sperm lack the fibrous bodies that are associated with motility in other nematodes it would be of interest to ascertain the nature of the protein that is associated with movement

of filopods in the sperm of *Gastromermis*. For example, are tubulin, actin or MSP involved?

The sperm of *Gastromermis* thus has several unusual features. These include the presence of dense tubular structures, longitudinal microtubules in the sperm body, microtubules in spikes originating from the pseudopods, membranous organelles that lack fibrous bodies, and the probable absence of MSP.

The sperm of other mermithids appear to be similar to those of *Gastromermis*. Light micrographs of spermatozoa of *Pheromermis cornicularis* and of *Hexamermis brevis* show them to be elongate spermatozoa, with, in the first species, an elongate nucleus (Kaiser, 1991). An electron micrograph of the spermatozoa of *Amphimermis elegans* shows globular mitochondria and the same dense tubular structures as found in the spermatozoa of *Gastromermis* (Kaiser, 1991).

Class Enoplea, Subclass Dorylaimia, Order Trichinellida

Trichurids and the dioctophymatids differ from the vast majority of nematodes in having hologonic testes, with spermatogenesis occurring in centripetal sequence along the testis. Other nematodes are telogonic, with spermatogenesis proceeding in sequence from the tip distally.

The mature spermatozoon of *Capillaria hepatica* is about 14×2 μm and shows a separation by a slight constriction into a blunter 'head' region and a more tapering 'tail' (Neill and Wright, 1973; Wright, 1991). The trilobed nucleus lies in the head region close to the middle of the sperm and lacks a nuclear envelope. The late spermatids possess two centrioles that lie close to the nucleus and each centriole has nine singlet microtubules. Mitochondria encircle the nucleus in the spermatids but become more widely distributed throughout the cytoplasm in the mature spermatozoon. Microtubules lie at the periphery of the cell beneath the plasma membrane. In late spermatids and spermatozoa the cytoplasm contains an anastomosing network of material which, especially near the cell periphery, appears tubular. This network of material does not seem to be smooth ER and the authors suggested that it may be a fixation pattern of the cytoplasmic matrix. In the vas deferens, pseudopodia were seen to project irregularly from both the 'head' and 'tail', but predominantly the 'head', of the sperm. The plasma membrane of the primary spermatocytes develops simple or multiple evaginations and these ultimately form characteristic membrane-stacks. These evaginations persist into the secondary spermatocytes but do not appear in the spermatids. The membrane specializations do not resemble those found in other types of nematode sperm and they also appear in a way that is different from the mode of formation described for other nematodes. The young spermatid contains many tubules, 85–125 nm in diameter, some of which appear to collapse to form concave vesicles with an intermembrane space of 26–38 nm. These in turn appear to close upon themselves and fuse with adjacent units to give double membrane loops enclosing a portion of cytoplasm. These membrane specializations come to lie just below the cell membrane around the 'tail' portion of the mature sperm (Neill and Wright, 1973). The origin of the tubules is uncertain; it is possible that they represent a transformation of the irregularly expanded endoplasmic reticulum of spermatocytes that is no longer recognisable in spermatids. Neill and Wright (1973) drew attention to the similarity of these membrane specializations to the membrane specializations in the peripheral cytoplasm of tick spermatocytes, which ultimately fuse to form a central vacuole and the motile processes of the tick's sperm. Sperm in the uterus of female *Capillaria hepatica* resemble those in the vas deferens of the male (Neill and Wright, 1973).

The spermatogonia and the spermatocytes of *Trichuris muris* have been described by Jenkins, Larkman and Funnell (1979) and by Jenkins and Larkman (1981) respectively. The centrioles of the spermatogonia and the spermatocytes have nine doublets; this is interesting as the centrioles found in most other nematodes consist of nine singlets. According to Jenkins and Larkman (1981) some of the subsurface vesicles of the spermatocyte appear to be derived from the Golgi apparatus, which indicates that there may be some homology of trichurid membrane specializations with the membranous organelles other nematodes. The mature sperm of *Trichuris vulpis* has an amoeboid shape and possesses elongate multilamellar bodies in a row beneath the plasma membrane. The amorphous cytoplasmic matrix is less dense than in earlier stages of development and contains small mitochondria. No Golgi apparatus or

endoplasmic reticulum is present in the mature sperm. One or two electron dense masses of chromatin are present in the centre of the sperm and the centrioles are situated close to them (Nakashima, 1972).

Spermatogenesis in male *Trichinella spiralis* has been studied by Slomianny *et al.* (1981) and by Takahashi, Goto and Kita (1994) and spermatozoa in the female reproductive tract have been described by Takahashi, Goto and Kita (1995). The spermatocytes have a well-developed Golgi system and cup-shaped vesicles lie at the periphery of the cell, but fibrous bodies are apparently not present in the spermatocytes or the spermatids. The spermatozoon has mitochondria arranged around the nucleus and membranous organelles are present at the periphery of the cell. These membranous organelles show a strong antigenicity related to phosphorylcholine (Takahashi, Homan and Pak Leong Lim, 1993). Two centrioles, each composed of 9 singlets, lie close to the nucleus. Spermatozoa are present as packets in the male tract and also in the uterus, where they were seen to be still maturing (Foor, 1983b).

Descriptions of spermiogenesis in males (Hulinská and Shaikenov, 1983) and of spermatozoa in the female tract (Hulinská and Shaikenov, 1982) of *Trichinella nativa, T. pseudospiralis* and *T. spiralis* indicate that there may be variations in the number of mitochondria and in the shape of membranous organelles between these species. However, the variation in the shape of the membranous organelles is probably attributable to different stages of differentiation rather than to species-specific differences.

Class Chromadorea, Subclass Chromadoria, Order Monhysterida

The only case of a spermatophore being present in a nematode is that described in *Prorhynchonema warwicki*. Females were found with one or two spermatophores, each about 10 μm in diameter, and containing numerous spermatozoa (Gourbault and Renaud-Mornant, 1983). Our knowledge of sperm structure in these marine forms is rudimentary in comparison to the variation that probably occurs and deserves further study.

In the Monhysterina, spermiogenesis has been described in detail in the marine nematode *Sphaerolaimus hirsutus* by Noury-Sraïri (1994) and by Noury-Sraïri, Gourbault and Justine (1993); see also Justine and Jamieson (1999) and Figure 4.18. The spermatocytes are large cells and have a large nucleus bearing a normal nuclear envelope, but the spermatids have no nuclear envelope. Fibrous bodies and membranous organelles in the spermatids have a common origin in fibrous body-membranous organelle (FB-MO) complexes. The spermatozoa are large lens-shaped cells, about 15 μm in diameter, and have a central nucleus that has no envelope. By following maturation of the cytoplasmic organelles it has been shown that dictyosomes produce membrane-bound vesicles in the spermatocytes. Each vesicle contains smaller vesicles, about 50–200 nm in

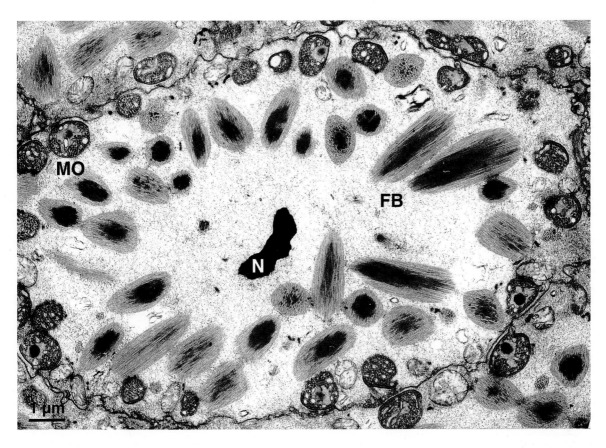

Figure 4.18. Spermatid of *Sphaerolaimus hirsutus*. Note nucleus (N) devoid of nuclear envelope, elongate fibrous bodies (FB) with dense core, and peripheral membranous organelles (MO). From Noury-Sraïri (1994), Thèse d'État, Université Cadi Ayyad, Marrakech, Morocco, reproduced with permission.

diameter, an amorphous body about 300 nm in diameter, and a fibrous body. In the spermatids, the fibrous bodies are arranged radially like the spokes of a wheel. The large size of these cells and their organelles enables them to be observed with the light microscope and labelling of the cell with anti-actin antibody clearly reveals the fibrous bodies. In mature spermatids, the fibrous body becomes separated from the rest of the vesicle. The membranous part of the vesicle forms the membranous organelle, which lies close to the cell membrane, and the fibrous body, devoid of membrane, remains in a more central position. At this stage, transverse sections of the fibrous body show a central dense region and a peripheral region containing wavy lines. In the spermatozoon the membranous organelles become arranged in a ring in the central region of the cell and the contents of the fibrous bodies disperse to form the cytoskeleton. The peripheral part of the cell contains thin fibres that are probably MSP. The fibrous bodies contain MSP (Justine and Jamieson, 1999), as do the fibrous bodies of all nematodes, but they also exhibited strong labelling by anti-actin antibodies. Noury-Sraïri, Gourbault and Justine (1993) claimed that immunocytochemistry carried out on the large spermatocytes, spermatids and sperm of *Sphaerolaimus* produces images more informative than for *C. elegans*, which has small germ cells in which organelles cannot

be recognized by light microscopy. The spermatozoa described in *Sphaerolaimus* have their membranous organelles situated away from the cell membrane and are never associated with it. They do not, therefore, show the final stage of differentiation seen in other nematodes, where the mature spermatozoon has membranous organelles associated with, and in continuity with, the cell membrane. It is probable that, in this species, this final stage of differentiation occurs in the female tract.

Mature sperm of *Daptonema* (Figure 4.19) in the male show a stage that has not been found in *Sphaerolaimus* (Noury-Sraïri, 1994; Justine and Jamieson, 1999). The spermatozoa are round shaped cells and contain a central nucleus which has no envelope. The cytoplasm contains fibrous bodies and membranous organelles, with these latter being associated with the cell membrane. Fibres, probably of MSP, are present in the pseudopods.

The spermatids of the xyalid nematode *Gonionchus australis* are round and have a central nuclear mass of chromatin which is encircled by mitochondria and, more peripherally, by membranous organelles (Nicholas and Stewart, 1997). The spermatozoa are more elongate than the spermatids and contain a dense elongate nucleus, without a nuclear envelope, mitochondria and four dense fibrous bodies.

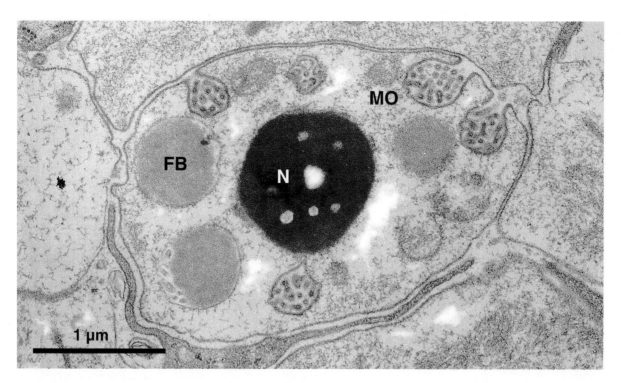

Figure 4.19. Spermatozoon of *Daptonema* sp. Note nucleus (N) devoid of nuclear envelope, fibrous bodies (FB), and membranous organelles (MO). From Noury-Sraïri (1994), Thèse d'État, Université Cadi Ayyad, Marrakech, Morocco, reproduced with permission.

Class Chromadorea, Subclass Chromadoria, Order Rhabditida, Suborder Spirurina, Infraorder Gnathostomatomorpha

Foor (1970b) has briefly described the sperm of *Gnathostoma* sp.. The main part of the cell is bluntly conoid, has peripheral membranous organelles, some of which open to the surface, has at the periphery 'lipid-like droplets' which are inconsistent in their presence, more centrally situated mitochondria, and central chromatin masses which are not membrane-bound. The remainder of the cell is pseudopodial in appearance and does not have these organelles. The 'lipid-like droplets', because of their location within the cell and their aspect are probably homologous to the MSP-containing fibrous bodies that are present in the sperm of other nematodes.

Class Chromadorea, Subclass Chromadoria, Order Rhabditida, Suborder Spirurina, Infraorder Oxyuridomorpha

Aspiculuris tetraptera (Figure 4.20) differs from the sperm of all other nematodes so far described (Lee and Anya, 1967). It has a distinct 'head', that is 4–6 μm long, and a distinct 'tail', about 30 μm long. A single large mitochondrion extends centrally from near the tip of the head to near the tip of the tail. No other mitochondria are present in the spermatozoon. This single mitochondrion appears to be formed by the merging of several smaller mitochondria during spermiogenesis. Two bundles of microtubules, oriented parallel to the longitudinal axis, also extend along the length of the 'tail' on each side of the long mitochondrion and where the mitochondrion terminates near the

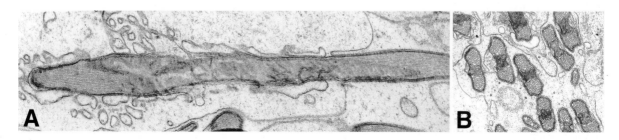

Figure 4.20. Spermatozoon of *Aspicularis tetraptera*. A, ''head'' of spermatozoon, longitudinal section; B, transverse section. This region of the spermatozoon contains the mitochondrion and peripheral microtubules. Original micrographs, courtesy of D.L. Lee and A.O. Anya.

tip of the 'tail' the two bundles of microtubules merge to form a single bundle. The two bundles also extend into the 'head' of the spermatozoon. The microtubules are arranged in parallel in the bundles and each bundle is partly surrounded by an electron-dense sheath. Feulgen staining and acridine orange fluorescent microscopy demonstrated the presence of DNA in the electron-dense sheath or associated with the longitudinal bundles of microtubules. Two centrioles, each with 9 singlet tubules, are located among the microtubules. The sperm is capable of amoeboid movement; the 'head' puts out pseudopodia which, it is suggested, may be involved in locomotion but the tail has not been observed to move *in vitro*. Whilst the authors speculated that the microtubules may be involved in motility it is more probable that they have a structural function.

There are no refringent bodies. Of particular interest is the absence of anything resembling membranous organelles or fibrous bodies at any stage of spermatogenesis, including in the spermatozoon. Whilst no electron micrograph of the anterior part of the spermatozoon was published by Lee and Anya (1967) unpublished electron micrographs of the head of the spermatozoon reveal the presence of an extensive system of large-diameter tubules situated in the cytoplasm around the central mitochondrion and its associated bundles of microtubules (Figure 4.20), but there is no evidence of membranous organelles or fibre bundles (Lee and Anya, unpublished) .This system of tubules is somewhat reminiscent of the large-diameter tubules found in the spermatozoon of *Capillaria hepatica* (Neil and Wright, 1973) but unlike the sperm of *Capillaria* there is no evidence that the tubules collapse to form membrane loops. This requires further study because the absence of an MSP cytoskeleton would represent a major difference between the Oxyurida and other nematodes, with the possible exception of some tylenchids and *Capillaria*. It is a puzzle as to how the pseudopodia of the sperm that lack membrane organelles and fibre bodies, and thus presumably MSP, are produced and bring about motility in the sperm.

In several oxyurids, such as *Passalurus ambiguus* (Hugot, Bain and Cassone 1982) and *Auchenacantha* spp. (Hugot, 1984), light microscope observations have led to the hypothesis that insemination is traumatic, with the presence of a epidermal pocket beneath the cuticle. This latter structure is apparently unknown in other nematodes and does not occur in all oxyurids. Spermatozoa, which appear to have a globular part and an elongate part, have been seen within the epidermal pocket of *P. ambiguus* (Hugot, Bain and Cassone, 1983). This peculiar method of fertilization apparently also occurs in some oncholaimids (Chabaud *et al.*, 1983). Traumatic insemination is unlikely to be a general occurrence in the oxyurids. It would be interesting to know if it is correlated with the unusual sperm structure of this group, but unfortunately this is known at the ultrastructural level only in *Aspiculuris* and there is no evidence that traumatic insemination occurs in this species.

Class Chromadorea, Subclass Chromadoria, Order Rhabditida, Suborder Spirurina, Infraorder Rhigonematomorpha

In the reproductive tract of male *Rhigonema madecassum*, the spermatozoon is elongate and has a nucleus made up of several discrete elements without a nuclear envelope. One extremity probably corresponds to the anterior pseudopod-producing end and has a granular cytoplasm whilst the rest of the cell contains dense, probably fibrous, bodies and peripheral membranous organelles (Van Waerebeke, Noury-Sraïri and Justine, 1990). Mitochondria were originally thought to be absent but Noury-Sraïri (1994) mentioned the presence of a few mitochondria in these spermatozoa. Observations at the light microscope level have shown that profound morphological changes occur when the spermatozoa are in the female genital tract of rhigonematids (Van Waerebeke, 1984, 1985).

Class Chromadorea, Subclass Chromadoria, Order Rhabditida, Suborder Spirurina, Infraorder Spiruromorpha

The sperm of spirurids were considered by Foor (1970b) to be the ascaroid type. They are similar to ascaroid sperm in being amoeboid, having a conoid form in most examples, having membranous organelles with internal microvillus-like projections at the periphery of the cell, having numerous mitochondria and having no nuclear envelope. They differ from the ascaroid type of sperm in having no reconstruction of the nucleus after meiosis, in the persistence of separate compact chromosomes in the spermatozoon and having no single, large refringent body. Lipid-like inclusions, which are probably fibrous bodies (see below), have been described in the sperm of filariins.

Spermatozoa in the female genital tract of *Tetrameres columbicola* are round, 3.5–4 μm in diameter, with a nucleus, 'dense granules' that probably correspond to fibrous bodies, mitochondria, and membranous organelles that open widely on the plasma membrane. Electron-dense cytoplasm around the nucleus of a spermatozoon in an oocyte was interpreted as a proliferation of ribosomes (Simpson, Carlisle and Conti, 1984).

Spermiogenesis and the sperm of *Brugia malayi* have been described by Scott (1996). The spermatids are rounded cells as they bud out from the residual body, but the fully developed spermatids take on a rod-like appearance. The nuclear material in the spermatid is highly condensed and is organized into two or three discrete non-membrane-bound bodies. The spermatids also contain the membrane-bound organelles that are typically found in most, but not all, of the nematode species studied to date. These organelles include mitochondria, a large refringent body and membranous organelles. Fibrous arrays are not present in rounded spermatids but appear later in development, as the spermatids become elongate. They extend to the tip of the cell, but do not appear to be attached or anchored to the tip. These arrays are positioned along the margins of the cell just beneath the cell membrane and probably contribute to the rod-shaped appearance of the spermatids. Maturation of spermatids to spermatozoa appears to be triggered by mating and can also be triggered *in vitro* by proteases or by a rapid increase in intracellular pH. The pseudopod contains a filamentous material that is distributed throughout the pseudopod and becomes organized into a cytoskeleton composed of fibres arranged radially into long, branched fibre complexes. This probably corresponds to the MSP cytoskeleton as there are two isoforms of MSP in *B. malayi*. The surface of the pseudopod has numerous small projections (villipodia) that are substrate adhesion sites.

Burghardt and Foor (1975) have described changes in sperm morphology in the uterus of *Brugia pahangi*. In recently mated females the proximal portion of the uterus contains sperm that range in morphology from a rigid, non-motile form to mature amoeboid sperm. The immotile sperm are identical in appearance to those found in the male seminal vesicle. Transformation to mature amoeboid sperm occurs within 1 hr in the uterus. They form a membranous system within minutes, the membranous organelles become spherical, fuse with the plasma membrane and release their electron-dense contents into the lumen of the uterus. The filamentous rod-like elements of MSP in the peripheral cytoplasm disintegrate, presumably to form the cytoskeleton, and pseudopods are formed.

Spermatozoa in the uterus of *Dirofilaria immitis* are bipolar. They have an anterior region that contains no organelles, which is probably the pseudopod containing an MSP cytoskeleton, and a posterior region that contains 'lipid-like inclusions' which are probably fibrous bodies, and membranous organelles that have openings to the exterior of the sperm (Foor, 1970b).

After penetration of the oocyte the sperm nucleus was not observed, but mitochondria and membranous organelles were still seen (Foor, 1970b). Maeda (1968) and Maeda *et al.* (1970), both quoted by McLaren (1973a), reported that the nuclear membrane is not reconstructed in this species and that the male gametes contain either 5 or 6 chromosomes. These authors referred to the membranous organelles as ovoid bodies. Delves, Howells and Post (1986) have carried out a light microscopy study of spermiogenesis in this species. Spermatozoa of *Dirofilaria immitis* in the uterus have also been described by Lee (1975), who termed the membranous organelles as 'mesosomelike vesicles'.

Spermiogenesis in *Acanthocheilonema (Dipetalonema) viteae* has been described by McLaren (1973a). Spermatogonia are developed in association with a syncytial rachis, from which the gametes are detached when they become secondary spermatocytes. The nuclear membrane is not reformed in the secondary spermatocytes. Typical membranous organelles appear in the primary spermatocytes and are present in the secondary spermatocytes and the spermatids. The centrioles consist of nine singlets. The spermatid becomes an elongate cell, up to 7.4 μm in length, and contains a few mitochondria, membranous organelles and chromatin material, but no nuclear envelope. The membranous organelles release their contents as fibre bundles into the cytoplasm and a pseudopod is formed at the anterior end of the cell. The membranous organelles later connect to the outer membrane of the cell by means of short channels. Spermatids removed from the reproductive tract of males and suspended in physiological saline showed no signs of motility. The mature spermatozoon in the male tract possesses an anterior pseudopod, a few mitochondria, discrete masses of chromatin, and a few membranous organelles situated

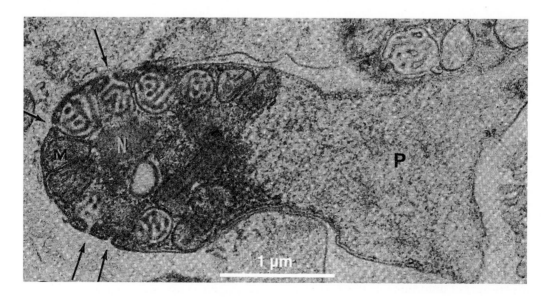

Figure 4.21. Spermatozoon of *Acanthocheilonema (Dipetalonema) viteae* which has been present in the female reproductive tract for a period of 6 days or less. The membranous organelles have fused (arrows) with the cell membrane. M, mitochondrion; N, nucleus; P, pseudopod. From Foor, W.E., Johnson, M.H. and Beaver, P.C. (1971) *Journal of Parasitology*, **57**, 1163–1169, reproduced with permission.

just beneath the plasma membrane of the cell and communicate with it through narrow channels (McLaren 1973a). According to McLaren (1973a) the late spermatids and spermatozoa in male *A. setariosum*, *Dirofilaria immitis* and *Litomosoides sigmodontis* (syn. *carinii*) are similar to those of *Acanthocheilonema viteae*, although the sperm of *L. carinii* are about half the size of the other species. Spermatozoa within the fertilization chamber of female *A. viteae* are almost identical to those within the male, except that the cell is more conical in shape and the membranous organelles are now spherical, all open to the exterior and have no obvious contents (McLaren, 1973a). Foor (1970a) and Foor, Johnson and Beaver (1971) have also studied differences between sperm found in the male and in the uterus of female *A. viteae* (Figure 4.21). Spermatozoa in the seminal vesicle are 1.5–2 μm in diameter and 7–7.5 μm in length. Within 24 hr after insemination, they become rounded, possess a large pseudopod and large amounts of endoplasmic reticulum appear in the cytoplasm. The membranous organelles migrate to the periphery of the cell within 6 days *in utero*, fuse with the plasma membrane and discharge their electron-dense contents.

Marcaillou and Szöllösi (1980), in a study of the testis wall of *Molinema (Dipetalonema) dessetae*, published electron micrographs of male germ cells that clearly show fibrous bodies with parallel fibres of MSP within the cells.

A spermatozoon (called a 'spermatid', although it was present in the seminal receptacle) of *Loa loa*, has been illustrated in a publication on the female reproductive tract (Weber, 1987). Membranous organelles are present at the periphery of the spermatozoon.

The spermatozoon present in a male *Physaloptera* sp. is illustrated in a micrograph published by Foor (1970b). Large fibrous bodies are associated with developing membranous organelles in the spermatocyte and there is an indication that it is of the ascaridoid type although, as in other spirurids, it lacks a refringent body.

Class Chromadorea, Subclass Chromadoria, Order Rhabditida, Suborder Spirurina, Infraorder Ascaridomorpha

The spermatozoon of ascarid nematodes has been extensively studied for many years, including by light microscopy (Fauré-Frémiet, 1913; Tuzet and Sanchez, 1951; Van Beneden and Julin, 1884), by early (Favard, 1961) and more recent (Foor, 1970b; Goldstein, 1977) electron microscopy, and by analysis of the protein content (Abbas and Cain, 1981, 1984; Clark, Moretti and Thomson, 1982; Panijel, 1950). There is an extensive literature dealing with sperm movement (Nelson and Ward, 1981; Sepsenwol and Taft, 1990) and with molecular studies of MSP (Bullock *et al.*, 1996; Italiano *et al.*, 1996; Royal *et al.*, 1995; Sepsenwol, Ris and Roberts, 1989; Stewart, King and Roberts, 1994). The 'ascaroid type' of nematode sperm was defined by Foor (1970b) for *Parascaris equorum* (syn. *Ascaris megalocephala*) (Favard, 1961) and *A. suum* (Abbas and Cain, 1979; Burghardt and Foor, 1978; Foor, 1970b; Foor and McMahon, 1973; Goldstein, 1977), and also for *Poly-*

delphis sp. (Foor, 1970b, 1983a) and *Toxocara canis* (Foor, 1970b).

The mature spermatozoon (Figure 4.22) in the uterus of female *A. suum* has a clear anterior region, which forms a pseudopod, and a denser posterior region which contains a round, dense nucleus which lacks a nuclear membrane, has a characteristic refringent body, mitochondria and membranous organelles. The cytoplasm of the pseudopod possesses fibrillar elements, consisting of the active form of MSP, and forms pseudopodial processes. The posterior region contains a large 'refringent body' (described below) which has a glistening appearance under the light microscope. Refringent bodies appear to be present only in ascarid nematodes (Foor, 1970b, 1983). Membranous organelles containing microvillus-like processes lie adjacent to, and many are continuous with, the plasma membrane of the posterior region.

In the seminal vesicle of male *A. suum*, the spermatozoon, called the spermatid by Burghardt and Foor (1978), is round and radially symmetrical. It has a central mass of chromatin, which lacks a nuclear membrane, surrounded by a granular area, mitochondria and several spherical refringent bodies. Membranous organelles are located near the periphery of the cell. This difference in shape of the spermatozoon in the male reproductive tract and in the female illustrates the difficulty in attempting to classify nematode sperm on morphological criteria. Sperm described as 'round' or 'globular' in the male may take on a bipolar shape, with well-defined pseudopods, in the female.

The refringent body forms a large 'refringent cone' (Goldstein, 1977) or 'cône réfringent' (Favard, 1961) that occupies most of the conoid region of *Ascaris* sperm. It is formed in the young spermatozoon by fusion of many smaller refringent bodies and varies in appearance; *in utero* it may be no longer apparent or may be appear as poorly defined electron dense material. The refringent body of *A. suum* sperm consists of only 2 to 6% lipid (Abbas and Cain, 1981), thus the description of it as 'lipid-like' (Foor, 1970b; Foor and McMahon, 1973) is inappropriate. The non-lipid fraction, which forms an insoluble fibrillar network, contains three polypeptides (the refringent granule proteins, RGP) of small molecular weight (7,000–14,000) in addition to an 18,400 Da polypeptide that occurs in other fractions of the sperm. Abbas and Cain (1984) elucidated the amino acid and lipid composition of refringent granules isolated from *Ascaris* sperm. The RGP proteins are rich in aspartic acid and asparagine, are PAS positive, are sensitive to tryptic hydrolysis and have solubility properties similar to the protein that Panijel (1950) recognized in the granules and called 'ascaridine', a term anglicised as ascaridin. The 14 kDa polypeptide is predominant and is probably identifiable with ascaridin, which is probably synonymous with MSP. The refringent granules correspond to the fibrous bodies as found in other nematodes, as they contain MSP in an inactive state. In ascarids the fibrous bodies coalescence into the large refringent body, which finally releases its contents as MSP fibres in the pseudopod when the sperm becomes motile.

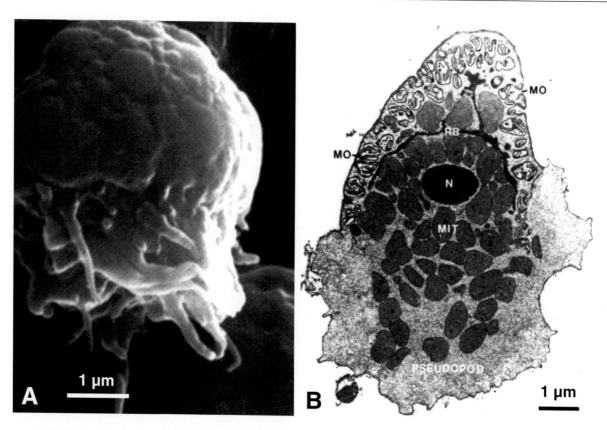

Figure 4.22. Spermatozoon of *Ascaris* spp. A, *Ascaris suum*, scanning electron microscopy of spermatozoon, after addition of sperm activating substances extracted from the male accessory glands. Note filipodia at the anterior end. B. Mature spermatozoon of *Ascaris lumbricoides* dissected from the uterine-oviduct junction of a female, transmission electron microscopy. MO, membranous organelle; N, nucleus; MIT, mitochondria; RB, residual body. A, From Abbas, M. and Cain, G.D. (1979) *Cell & Tissue Research*, **200**, 273–284; B, From Nelson, G.A. and Ward, S. (1981) *Experimental Cell Research*, **131**, 149–160, reproduced with permission.

Although Foor (1970b) did not mention a centriole in the mature sperm of *Ascaris lumbricoides*, Goldstein (1977) described a centriole that is present throughout spermatogenesis in *A. suum*. It is embedded in ribonucleoprotein, is 0.18 μm in diameter and 0.27 μm in length, has a central lumen 79 nm wide and has nine single microtubules connected by a dense, continuous ring but it lacks a central cartwheel.

Mitochondria are said to be exclusively or predominantly maternal in most Eukaryotes. During fertilization of the ovum in *Ascaris* the whole sperm, including the many mitochondria, enter the ovum raising the possibility that transmission of paternal mitochondria to the offspring may occur. However, Anderson *et al.* (1995) used mitochondrial DNA restriction patterns and reported the absence of paternal transmission of mitochondrial DNA in *Ascaris*. They suggested that the mitochondria of *Ascaris* sperm are actively destroyed or outcompeted by maternal mitochondria in the zygote.

Actin has been detected in *Ascaris* sperm by labelling them with DNAaseI-FITC, which binds specifically with the globular form, G-actin (Foor, 1983b). Studies on sperm movement have demonstrated the presence

of actin in *Ascaris* sperm (Nelson and Ward, 1981). However, Roberts and Stewart (1995, 1997), Scott (1996) and Theriot (1996) claim that actin is not involved in amoeboid movement of *Ascaris* sperm, but that movement depends exclusively upon the MSP cytoskeleton. See the section on MSP and cytoskeletal proteins for more details.

The spermatozoon of *Polydelphis* sp. possesses a very large disc-like refringent body that almost fills the spermatozoon (Foor, 1970b). Except for a small pseudopodial outgrowth, the entire circular margin of the spermatozoon is fringed by membranous organelles. According to Foor this sperm is similar to that of *Ascaris*. Later, Foor (1983b) showed the parallel arrangement of MSP fibres in the fibrous bodies of the sperm of *Polydelphis*.

MacKinnon and Burt (1992) published an electron micrograph of a spermatozoon in the female genital ducts of *Pseudoterranova decipiens*. The cell is bipolar, has a large pseudopod devoid of organelles, and a region with membranous organelles around the periphery.

Spermatocytes united by a cytophore in *Ascaridia* sp. have been briefly illustrated and described by Foor (1983b).

Lee (1971) classified the sperm of *Heterakis gallinarum* as the ascaroid type, although it has no refringent body. The early spermatocytes are in cytoplasmic continuity with a cytoplasmic, but anucleate, rachis. The nucleus of the spermatocyte possesses a nuclear membrane (Lee, personal communication). As the spermatocytes descend the testis there is an increase in number of Golgi complexes and the cisternae of the endoplasmic reticulum becomes more swollen. Both the Golgi complexes and vesicles that originate from the Golgi complexes gave strong positive results for acid phosphatase activity. In the spermatid the chromatin, which is not surrounded by a membrane, is centrally placed and is surrounded by mitochondria and, at the periphery, by membranous organelles (Lee called these 'alpha bodies'). Microtubules are present in the cytoplasm around the nucleus and the centrioles, which have nine peripheral singlets, are closely associated with these microtubules at the periphery of the chromatin. One relatively small area of the spermatid comes to contain abundant rough endoplasmic reticulum and ribosomes and subsequently a cell membrane is formed and cuts off this area from the remainder of the spermatid. The fibrous material of the membranous organelles increases in length and thus elongates the organelle. A cross section of a fibrous body clearly demonstrates the parallel fibres of inactive MSP (Figure 4.6B). More mature stages in the seminal vesicle are regarded as immature spermatozoa. In these the electron dense component of the membranous organelles decreases in amount, the microvillus-like invaginations become larger, and the fibrous component is released to become a loose fibrillar component in the outer zone of the spermatozoon with a tendency to accumulate at one pole. The mitochondria decrease in number and enlarge. The membranous organelles continue to give a strong reaction for acid phosphatase. Small membrane bound vesicles appear in the dense cytoplasm amongst the mitochondria and membranous organelles. After fertilisation the sperm become stored in the seminal receptacle of the female, the dense nucleus becomes surrounded by a halo of less dense chromatin and also by the mitochondria. The membranous organelles have become vesiculate and contain many microvillus-like extensions of the outer plasma membrane and some open to the surface of the spermatozoon. The sperm have a much more amoeboid appearance, pseudopodial like extensions of many of them being anchored in cells of the seminal receptacle wall. At fertilisation, the membrane of the spermatozoon fuses with the oolemma of the egg and the contents of the spermatozoon, including the mitochondria, enter the egg, leaving the membrane of the sperm as part of the outer egg membrane. Immediately the spermatozoon becomes part of the fertilised ovum there is a rapid production of ribosomes by the spermatozoon and the nucleus becomes less compact; the mitochondria migrate away from the centre of the sperm but the membranous organelles remain central, become more vacuolated and persist for some time (Lee and Lestan, 1971).

Class Chromadorea, Subclass Chromadoria, Order Rhabditida, Suborder Tylenchina, Infraorder Panagrolaimomorpha

Hess and Poinar (1989) used light and electron microscopy, and immunocytochemistry to study spermatogenesis and sperm in the insect-parasitic nematode *Steinernema (Neoaplectana) intermedium* (Steinernematidae). Spermatogenesis occurs as in other nematodes, but the spermatids are often associated as tetrads. The fibrous bodies have a fibrillar structure typical of MSP, and an anti-MSP antibody, raised against *C. elegans* MSP, specifically stained the fibrous part of the fibrous bodies. The spermatozoon is bipolar, with a main cell body containing the organelles and nuclear material at one end and pseudopodia without organelles, but containing filaments 6 to 8 nm in diameter, at the other end. Membranous organelles have fused with the plasma membrane and some microtubules are associated with these organelles. A cortical layer of microtubules is present beneath the cell membrane in the main cell body, but is absent in the pseudopod. The nucleus is composed of many discrete electron-dense patches. Spermatozoa form chains with junctions between the cells. Hess and Poinar (1989) found that different fixatives produced very different pictures of spermatozoa ultrastructure. For example, microtubules and microfilaments were not seen in specimens fixed with phosphate-buffered fixatives followed by alcohol dehydration. Because of this, caution should be used in the use of characters such as the presence of microtubules for a phylogenetic appraisal of the nematodes (see also Cares and Baldwin (1994a)).

Class Chromadorea, Subclass Chromadoria, Order Rhabditida, Suborder Tylenchina, Infraorder Cephalobomorpha

Baccetti *et al.* (1983) gave a brief description of the sperm of *Cephalobus* cf. *quinilineatus* (Cephalobidae). The rounded spermatozoa contain a nucleus in a peripheral position, membranous organelles, and bundles of microtubules running parallel to the cell. Contacts between sperm cells, called 'desmosome like junctions', were described.

The sperm of *Panagrellus silusiae* are round cells with a diameter of about 5μm and have membranous organelles (vesicular organelles' or 'v-bodies') around the periphery of the cell (Pasternak and Samoiloff, 1972). The nucleus of the mature sperm lacks a nuclear membrane, and the spermatozoon also lacks centrioles, microtubules, Golgi material, and a refringent body. The nucleus is represented by several discrete condensed clumps of chromatin in the central region. 'Crystalline bodies', that are transient in the spermatocyte, are probably the equivalent of the fibrous bodies that contain MSP in other nematodes. It is said that 'they subsequently degenerate and no trace of their fibrillar material is observed in the spermatozoon' (Pasternak and Samoiloff, 1972). This could be explained by a lack of sections through the pseudopod region.

Class Chromadorea, Subclass Chromadoria, Order Rhabditida, Suborder Tylenchina, Infraorder Tylenchomorpha

Superfamily Aphelenchoidea

The mature spermatozoa of *Aphelenchoides blas-tophthorus* lie closely packed in a region of the reproductive tract called the seminal vesicle, are about 6 to 8 μm wide and 10 to 12 μm long and produce many pseudopodia (Shepherd and Clark, 1976). The nucleus of the sperm is an eccentric mass of condensed chromatin and lacks a nuclear membrane. The sperm contains membrane specializations, similar to normal membranous organelles, randomly distributed throughout the cytoplasm at this stage. Fibrillar material appears in the spermatocytes as fibrillar bodies that are associated with cup-shaped membranes and together they form the membranous organelles. This fibrillar material loses its association with the membrane specializations in the spermatozoon and is scattered in bundles of irregular shape and size in the cytoplasm. Microtubules, which extend throughout parts of the cytoplasm of the spermatid not occupied by membranous organelles, occur in some of the spermatozoa. There is a narrow zone of cytoplasm free of organelles around the periphery of the cell. Once in the female reproductive tract the membranous organelles take up a peripheral position and open to the surface. This late migration to the periphery of the cell, together with the presence of electron lucent vesicles which appear in the spermatocyte and persist in the sperm in the female, are unusual features compared with the sperm of other nematodes. After insemination, the sperm differentiates into an amoeboid region containing the remnants of the fibrillar material and a smaller non-amoeboid region containing the membranous organelles, mitochondria, lucent vesicles and the still eccentric nucleus.

Superfamily Tylenchoidea

Tylenchids are, with a few possible exceptions such as *Aspiculuris* and *Capillaria*, unusual in lacking membranous organelles, as found in most other nematodes, in their spermatids or spermatozoa. Sperm ultrastructure has been described in several species of *Heterodera* and *Globodera* (Shepherd, Clark and Kempton, 1973) and of *Meloidogyne* (Shepherd and Clark, 1983). The spermatozoa of these genera are very similar in structure but they show considerable interspecific and some intersubspecific variation (Shepherd and Clark, 1983; Shepherd, Clark and Kempton, 1973). Sperm of *Meloidogyne arenaria*, *M. hapla* and *M. incognita incognita* are, from early maturity, polarized into a nuclear region and a large pseudopodium. However, Goldstein and Triantaphyllou (1980) found the sperm of *M. hapla* to be globular in the facultatively parthenogenetic Race A, which can undergo oocytic meiosis, and in Race B, which is obligatorily parthenogenetic and undergoes only oocytic mitosis. In contrast, the sperm of *M. graminicola*, *M. incognita wartellei* and *M. oryzae* are not globular. The sperm of *M. acronea* are intermediate

in polarization as there is some indication of polarity in early sperm but the mature sperm are amoeboid (Shepherd and Clark, 1983). The sperm of *Meloidogyne*, *Heterodera* and *Globodera* have cortical microtubules in the pseudopodia and associated with the nuclear material, but microtubules are absent elsewhere in the cytoplasm. The sperm of *Globodera* (syn. *Heterodera*) *rostochiensis* have an outermost electron dense, rather fuzzy, 10 nm layer beneath which is a fine 5 nm electron-dense layer. The plasma membrane of the cell lies beneath the electron-dense layer and a layer of microtubules, 25 nm in diameter and 15 nm apart, adhere to its inner surface. The microtubules form a parallel array just under the surface of the sperm but they are spirally arranged on the pseudopod. In the sperm receptacle and oviduct of the female the chromatin changes from a compact dense mass to a coarsely stranded mass consisting of a branching network, to which are attached large, 0.4 to 0.6 μm wide, spongy spherical bodies. The mitochondria migrate from the peripheral cytoplasm to become clustered around the nuclear material with the result that most of the cytoplasm becomes devoid of organelles, other than the cortical microtubules (Shepherd, Clark and Kempton, 1973). It was suggested that the cortical microtubules have a skeletal rather than a contractile function. In developing sperm, fibrillar material in the cytoplasm becomes rather diffuse or disperses entirely as the sperm matures. The sperm of *Meloidogyne* (Figure 4.23) contain dense bodies, which Goldstein and Triantaphyllou (1980) called 'nucleolar bodies' because they appeared to be RNA-positive in *M. hapla*. These so-called 'nuclear bodies' are absent from the sperm of cyst nematodes (Shepherd and Clark, 1983).

According to Shepherd and Clark (1983) condensation of the sperm nucleus changes several times during spermiogenesis and insemination in the round-cyst nematode *Globodera* (*Heterodera*). The chromatin passes through a compact homogeneous phase, followed by a finely stranded phase, then another compact phase and finally a coarsely stranded phase. Such changes are unknown in other nematode sperm and do not occur in all *Heterodera*/*Globodera*.

The spermatozoon of *Tylenchulus semipenetrans* is small (2 μm) and round with a central rounded nucleus (Baccetti *et al.*, 1983). A pair of centrioles lies close to the nucleus. The cytoplasm of the spermatozoon has 'tubular invaginations' that later disappear in spermatozoa observed in the male genital tract. Small spheroidal bodies, which contain a dense granular material arranged in tortuous laminae but which are not membrane-bound, lie close to the nucleus. The authors found it difficult to relate these structures to the membranous organelles seen in other nematodes.

In a richly illustrated study, Endo, Zunke and Wergin (1997, 1998) have described spermatogenesis and the structure of the spermatozoon of *Pratylenchus penetrans*. Spermatocytes, linearly arranged in the testis, contain synaptonemal complexes characteristic of the pachytene stage of meiosis. The cytoplasm of spermatocytes

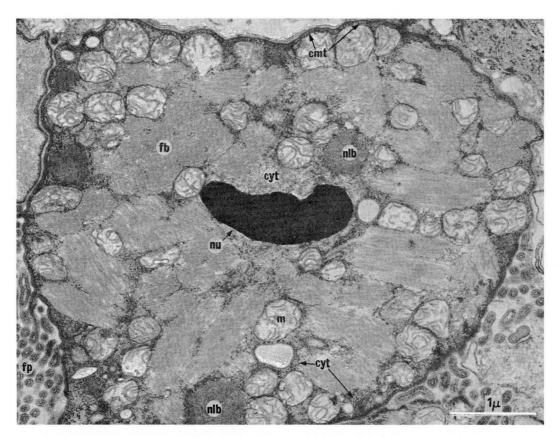

Figure 4.23. Testicular spermatozoon of *Meloidogyne incognita incognita*. Note numerous fibrous bodies (fb), central nucleus (nu) devoid of envelope, and cortical microtubules (cmt) lining the cell membrane. fp, filipodia; m, mitochondria; nlb, nucleolar bodies; cyt, tubular elements throughout cytoplasm. From Shepherd, A.M. and Clark, S.A. (1983) *Revue de Nématologie*, **6**, 17–32.

contain rough endoplasmic reticulum, free ribosomes, mitochondria, dictyosomes, and electron opaque accumulations. The spermatocytes are associated with a central rachis. In spermatids, the nuclear envelope disappears and the chromatin aggregates in an electron-opaque spherical unit. Fibrous bodies and an MSP cytoskeleton are present in the spermatid. Invaginations of the plasma membrane of the spermatid appear to be lined internally by dense material. Microtubules under the cell membrane were mentioned in spermatids and spermatozoa. Spermatozoa in the male and in the female possess fibrillar bundles of MSP free in the cytoplasm, together with mitochondria and a central nucleus. Membrane forms filopods, although sperm near the terminus of the vas deferens occasionally lack filopods. The present author believes that these spermatozoa may not represent the final stage in the development of the male gamete, because the MSP cytoskeleton appears to be incompletely matured.

Cares and Baldwin (1994a) described the spermatozoon of *Verutus volvingentis* (Figure 4.24B) and *Meloidodera floridensis* and found that in both species the spermatozoon is globular with many filopodia but the sperm of the two species are, nonetheless, highly divergent. The spermatozoon of *V. volvingentis* are about twice the size (4 to 8 μm in diameter) of those of

M. floridensis (2 × 2.5 to 3 × 4 μm) and can also be distinguished by the asymmetrical distribution of their filopodia and the relative lack of change in condensation of the chromatin after insemination. The fibres found in the cytoplasm of mature sperm of both species were called 'fibrous bodies' but this is incorrect, as they are not membrane-bound. These fibres probably correspond to a stage in the release of MSP from fibrous bodies, limited by a membrane, that are possibly present at an earlier stage in development. The so-called 'fibrous bodies' of the spermatozoa of *V. volvingentis* are less persistent than in the spermatozoa of *M. floridensis*. Unlike the sperm of some other Heteroderinae, the sperm of these two nematodes have no cortical microtubules.

The spermatozoon of *Ekphymatodera thomasoni* is also round in shape and has a centrally placed nucleus (Cares and Baldwin, 1994b). The spermatozoa originate from germ cells connected to a central rachis, a character they share with *Globodera* but not with other Heteroderinae apparently. Sperm in the spermatheca of the female are compactly stored. A layer of cortical microtubules lies just beneath the surface of the plasma membrane and it is claimed that there is a layer of spiral surface elevations on the filopodia, but these are not easily visible in the published micrograph. So-called

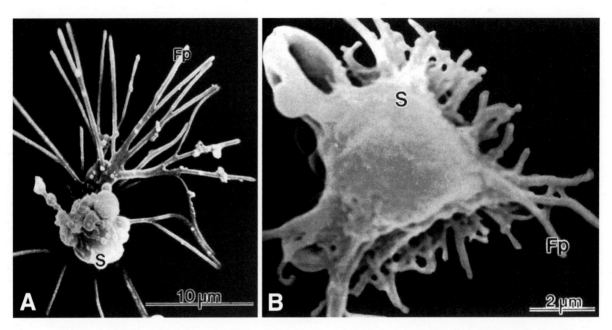

Figure 4.24. Scanning electron microscopy of tylenchid spermatozoa. A, Spermatozoon of *Heterodera schachti*. B, Mature testicular spermatozoon of *Verutus volvingentis*. S, sperm body; Fp, filipodia. A, from Cares, J.E. and Baldwin, J.G. (1995) *Canadian Journal of Zoology*, **73**, 309–320; B, from Cares, J.E. and Baldwin, J.G. (1994) *Canadian Journal of Zoology*, **72**, 1481–1491, reproduced with permission.

'fibrous bodies', as described for the sperm of *Verutus* and *Meloidodera*, are abundant in the spermatids but, unlike in these other two species, they do not persist in the sperm of *Ekphymatodera* (Cares and Baldwin, 1994a, b).

Punctodera chalcoensis has much smaller sperm (2 to 5 μm in diameter) than those of *Heterodera* and *Globodera* and they produce fewer filopodia (Cares and Baldwin, 1995). The so-called 'fibrous bodies' are of short persistence after spermiogenesis, as occurs in *Ekphymatodera*, with fibrous bodies being absent in mature sperm. There is a single layer of cortical microtubules under the plasma membrane and there are many mitochondria, which are not polarized in distribution. Chromatin remains unchanged with respect to condensation during development of the sperm. The spermatozoon of *Heterodera schachtii* has a large number of filopods (Cares and Baldwin, 1995) (Figure 4.24A).

Cares and Baldwin (1995) illustrated the spermatozoon of *Globodera tabacum* in a single scanning electron micrograph. This showed the main body of the sperm to be globular, about 4 to 7 μm in diameter, and to have 12 or more long pseudopodia that were concentrated on one side of the body. The authors mentioned in the text that microtubules occurred under the sperm membrane.

Lo (1993), in an unpublished report, described the ultrastructure of the spermatozoa of two plant-associated nematodes from West Africa. The spermatozoon of *Scutellonema cavenessi* is amoeboid, measures 3 × 4 μm, and has a main body and a pseudopod. The nucleus has an irregular shape and has no nuclear envelope. The cytoplasm contains many bundles of fibres and mitochondria. In the male the spermatozoa possess filopods and fibres. Spermatozoa in the

spermatheca are more globular, are 2 μm in diameter, are still polarized, the cytoplasm contains fewer bundles of fibres and membranous organelles lie close to the plasma membrane of the cell. The nucleus of the spermatozoon of *Helicotylenchus multicinctus* also lacks a nuclear envelope, and also possesses numerous bundles of fibres, mitochondria and pseudopods. The nucleus of the spermatozoon is more condensed in the spermatheca and the less abundant bundles of fibres gather in the pseudopod. Centrioles and microtubules were not seen in these nematodes and fibrous bodies were not shown in spermatocytes and spermatids.

Cares and Baldwin (1995) suggested that patterns of development of sperm may be useful for testing hypotheses of the phylogeny of Heteroderinae, but the diversity is so great that character coding will be required for a large number of representative species.

True membranous organelles are absent from all tylenchins, including, in addition to all genera cited above, *Ditylenchus dipsaci* and *Paratylenchus penetrans* (Shepherd and Clark, 1976, 1983). The absence of membranous organelles may be considered a characteristic of the tylenchids, as is the presence of filopodia. However, none of these characteristics is limited to this group, as nematodes belonging to other groups, such as *Aspiculuris* and *Capillaria*, appear to lack membranous organelles in their developing and mature sperm.

Class Chromadorea, Subclass Chromadoria, Order Rhabditida, Suborder Rhabditina, Infraorder Rhabditomorpha

Rhabditids are paraphyletic in the system of Blaxter *et al.* (1998), with *Caenorhabditis elegans* grouped with the strongylids and diplogasterids, and other families, such as the Steinernematidae and Cephalobidae, grouped with the Aphelenchida and Tylenchida. However, no ultrastructural resemblance can be found between the spermatozoa of *C. elegans* (round cells) and the strongylids (large elongate cells).

Superfamily Rhabditoidea

Spermatogenesis in *Caenorhabditis elegans* has been the subject of detailed ultrastructural studies (Ward, Argon and Nelson, 1981; Wolf, Hirsh and McIntosh, 1978). The normal sperm (Figure 4.25A) in the male is bipolar with organelles confined to a hemispherical region of the cell that is separated from an irregularly shaped pseudopod by an accumulation of laminar membranes. Fibrous bodies that are associated with developing membranous organelles in the early spermatid have dispersed by the late spermatid stage. Membranous organelles present in the spermatozoon have already fused with the plasma membrane and have released their fibrous contents to the exterior (Ward, Argon and Nelson, 1981). Wolf, Hirsh, and McIntosh (1978) described microvillus-like internal projections of the walls of the membranous organelle and referred to them as 'tortuous bits of membrane-bound cytoplasm'. The centriole they described consists of a ring surrounded by nine single microtubules. The pseudopod and the cell body contain an amorphous granular cytoplasm and are devoid of microtubules or microfilaments. The nucleus is irregular in shape, lacks a nuclear membrane and is surrounded by cristate mitochondria. Several other publications, which mainly concentrate on functional aspects of the cytoskeleton, have contributed to the description of *C. elegans* sperm (Nelson, Roberts and Ward, 1982; Pavalko and Roberts, 1987; Roberts and Streitmatter, 1984; Roberts and Ward, 1982a, b; Shakes and Ward, 1989b; Ward, Hogan and Nelson, 1983; Ward and Klass, 1982). Machaca, DeFelice and L'Hernault, 1996) described separation of the spermatocytes from an anuclear cytoplasmic core and their subsequent division. During the second division, intracellular organelles segregate specifically to the spermatids as they bud from the residual body. The authors applied patch-clamp techniques to study the distribution of membrane proteins during the divisions and found that membrane components follow a specific distribution pattern during development of the sperm. They detected several voltage-sensitive ion channel activities in the spermatocytes but only a single-channel type was detected in the spermatids, which they took to indicate that other channel activities are excluded from, or inactivated in, these cells as they form. The channel that was detected in the spermatids is an inward-rectifying chloride channel. Treatment of spermatids with chloride channel blockers induced them to differentiate into spermatozoa.

C. elegans has proved to be an excellent system for genetical studies, and mutations affecting spermatogenesis have been the subject of numerous studies, making this species the most thoroughly understood nematode for the genetics of sperm production and their genetic defects. Several genes that affect spermatogenesis and sperm structure have been identified. A series of mutations affects the fibrous bodies-membranous

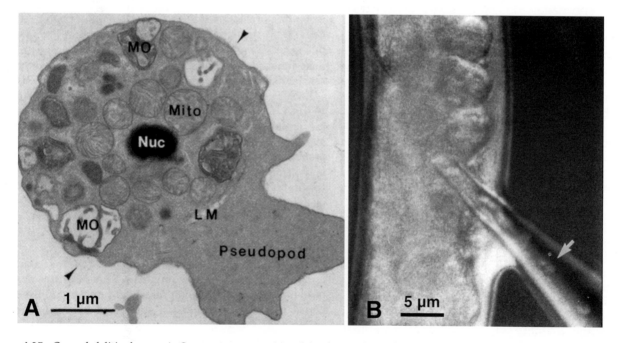

Figure 4.25. *Caenorhabditis elegans*. A, Spermatozoon activated *in vitro* with triethanolamine. MO, membranous organelle; Nuc, nucleus; Mito, mitochondrion; LM, laminar membranes. B, artificial insemination o*f an h*ermaphrodite. Light micrograph of an injection needle passed obliquely through the vulva; the spermatids have been injected, but some have flowed back in the needle and are visible (arrow). A, from Ward, S., Hogan, E. and Nelson, G.A. (1983) *Caenorhabditis elegans. Developmental Biology*, **98**, 70–79; B, from LaMunyon, C.W. and Ward, S. (1994) *Genetics*, **138**, 689–692, reproduced with permission.

organelles (FB-MO) complexes: mutations of *spe-10* (Shakes and Ward, 1989b), *spe-4* (L'Hernault and Arduengo, 1992) and *spe-5* (Machaca and L'Hernault, 1997) genes produce abnormal FB-MO complexes; mutations in the *spe-17* gene show abnormal presence of ribosomes on the FB-MO membrane (L'Hernault, Benian, and Emmons 1993); membranous organelles do not fuse with the sperm membrane in mutants of the *fer-1* gene (Achanzar and Ward, 1997; Argon and Ward, 1980; L'Hernault, Shakes and Ward, 1988; Ward and Miwa, 1978); MSP fails to assemble into fibrous bodies in mutations of the *spe-6* gene resulting in unassembled MSP becoming distributed throughout the cytoplasm (Varkey *et al.*, 1993); fibrous bodies are retained and pseudopods are defective in *fer-6* mutants (Ward, Argon and Nelson, 1981). Mutations in *spe-8*, *spe-12* and *spe-27* genes result in the production of abnormal pseudopods, after protease activation (L'Hernault, Shakes and Ward, 1988; LaMunyon and Ward, 1994; Minniti, Sadler and Ward, 1996; Shakes and Ward, 1989a); in addition, the protein encoded by *spe-12* is localized to the spermatid cell surface and plays a role in promoting spermiogenesis (Nance, Minniti, Sadler and Ward, 1999); mutations in the *spe-26* gene result in incomplete meiosis and the spermatocytes do not form haploid spermatids, possibly because of the defect of an actin-binding protein (Varkey *et al.*, 1995); fertilization-defective spermatozoa are produced by *spe-9* mutants (L'Hernault, Shakes and Ward, 1988; Singson, Mercer and L'Hernault, 1998); mutations in *spe-15* result in crawling sperm being unable to bring about fertilization (L'Hernault, Shakes and Ward, 1988); mutations in *fer-7* gene result in non-motile sperm (Argon and Ward, 1980); spermatids do not become activated into spermatozoa in *fer-15* mutants (Roberts and Ward, 1982b); and mutations of *fer-2* and *fer-4* genes produce non-functional, non-motile sperm with aberrant pseudopods and perinuclear tubules (Ward, Argon and Nelson, 1981). Mutations in the *spe-9* gene produce spermatozoa with wild-type morphology and motility but they cannot fertilize oocytes, even after contact between gametes, which suggests that disruption of *spe-9* function affects gamete recognition, adhesion, signalling and/or fusion (Singson, Mercer and L'Hernault, 1998). Mutants which have spermatids without nucleus produce spermatozoa which can induce both normal egg activation and anterior-posterior polarity in the one-cell *C. elegans* embryo, therefore demonstrating that neither the sperm chromatin mass nor a sperm nucleus is required for spermiogenesis, egg activation, of the induction or anterior-posterior polarity (Sadler and Shakes, 2000). LaMunyon and Ward (1994) have artificially inseminated *C. elegans* (Figure 4.25B) successfully with both male and with hermaphrodite sperm and been able to assess the viability of mutant and manipulated sperm. This, together with the sequencing of the complete 97-megabase genome of *C. elegans* (*C. elegans* sequencing consortium, 1998; Blaxter 1998), should enable more progress in the study of sperm of this model animal.

A mutation of the *spe-26* gene that reduces sperm production increased the lifespan by about 65% in both mated males and hermaphrodites. This suggests that spermiogenesis is a major factor reducing the lifespan of *C. elegans* (Van Voorhies, 1992).

C. elegans is also important in sperm competition studies. When male and hermaphrodite mate, spermatozoa of the male outcompete the sperm of the hermaphrodite and fertilize a majority of the offspring (LaMunyon and Ward 1995; Ward and Carrel 1979). Also, larger sperm outcompete smaller ones because larger sperm crawl faster and physically displace smaller sperm to take fertilization priority (LaMunyon and Ward, 1998, 1999).

Mature sperm in female *Rhabditis pellio* have an orientated layer of microtubules beneath the well-defined plasma membrane of the cell. The anterior amoeboid end, which presumably contains reorganised fibrous bodies, is sharply delimited from the organelle-containing posterior part of the sperm (Beams and Sekhon, 1972). The posterior end possesses numerous 'pockets' (membranous organelles) which contain microvillus-like structures, a dense nucleus that lacks a nuclear membrane, mitochondria clustered around the nucleus, and scattered microtubules. Ribosomes, Golgi cisternae and endoplasmic reticulum, which were present earlier, are eliminated with cytoplasm cast off from the spermatid.

Superfamily Strongyloidea

Spermatogenesis has been described in the insect-parasitic nematode *Heterorhabditis bacteriophora* (Poinar and Hess, 1985). This species has both hermaphroditic and male individuals but whereas *C. elegans* shows a succession of generations of hermaphrodites, with rare males, *Heterorhabditis* shows an alternation of generations, first hermaphroditic, then gonochoric. Spermatogenesis is similar in the two generations. Spermatozoa in the male are globular, have a small nucleus surrounded by a perinuclear halo but no nuclear envelope, and a pseudopod. Membranous organelles ('membranous specialisation') fuse with the plasma membrane at the completion of spermiogenesis. Study of the behaviour of spermatozoa in the genital tract of hermaphroditic and amphimictic females has shown that spermatozoa accumulate in the seminal receptacle, between the uterus and the oviduct. Three types of associations of the spermatozoa with the wall of the genital duct were described, and were interpreted as corresponding to sperm movement, attachment, and engulfment by the wall cells (Hess and Poinar, 1986).

The strongyloid, or strongylid (see Justine and Jamieson, 1999) sperm, as defined by Foor (1970b) (this does not include the heterorhabditids described above), has a distinct 'head' and nucleated 'tail' when in the seminal vesicle of the male. For convenience, Foor (1970b) included the closely related metastrongyle *Angiostrongylus cantonensis* in this group although its sperm do not possess the characteristic appearance.

Sperm in male *Ancylostoma caninum* possess a large anterior ('head') region that contains filamentous elements, numerous dense membranous organelles and mitochondria in the cytoplasm (Foor, 1970b; Ugwunna, 1986, 1990; Ugwunna and Foor, 1982a, b). The 'tail' region contains the nucleus, which consists of numerous dense filaments in a spiral-like arrangement (Ugwunna, 1990). The are two mutually perpendicular centrioles, each of which consists of nine peripheral tubules, at the junction of the nucleus with the anterior region. Cytoplasmic microtubules were found to be conspicuous in the secondary spermatocyte but were difficult to see in the sperm ('spermatid') in the seminal vesicle. Microtubules in developing spermatozoa of *A. caninum* seem to be involved in nuclear elongation because they invest the spermatid nucleus throughout the process of nuclear elongation and disappear once this is accomplished (Ugwunna and Foor, 1982b; Foor, 1983a). The spermatid and spermatozoa possess an abundant smooth endoplasmic reticulum (Ugwunna, 1986). Typical fibrous bodies are present in the cytoplasm of the spermatid (Ugwunna, 1986, 1990). After deposition in the female, the sperm of *A. caninum* become very polymorphic or amoeboid. Mixing of the nuclear and cytoplasmic regions occurs and the membranous organelles become confluent with the outer plasma membrane and lose their electron dense matrix. The cytoplasmic microtubules and centrioles apparently disappear.

Spermatozoa (called 'spermatids' although they were observed in the seminal vesicle) of *Protostrongylus rufescens* have been briefly described by Acosta-Garcia *et al.* (1985). They are round cells, possess a central nucleus and have membranous organelles at the periphery. Filaments that probably correspond to the MSP cytoskeleton are irregularly distributed in the cytoplasm. A single centriole of 9 singlets was seen close to the nucleus.

The spermatozoon of *Angiostrongylus cantonensis*, which has been described by Foor (1970b) and provisionally assigned by him to the strongyloid type of nematode sperm, differs from the sperm of other strongylids in being spherical and having a central nucleus when in the seminal vesicle. The nucleus is round and has dense filaments similar to those found in the sperm nucleus of *Ancylostoma*. The cytoplasm contains filaments, mitochondria and peripheral membranous organelles. After insemination of the female the spermatozoon in the uterus assumes an amoeboid appearance, somewhat similar to the sperm of ascarids. The membranous organelles fuse with the plasma membrane, the mitochondria appear larger than in the seminal vesicle and a mixing of organelles, similar to that in seen in sperm in the uterus of *Ancylostoma*, occurs.

Currey (1975) gave a brief description of the spermatozoa of *Metastrongylus apri*. They are globular, 7 μm in diameter, and contain fibrous bodies and membranous vesicles. Mitochondria and the other organelles surround a central nucleus.

As protostrongyle and metastrongyle spermatozoa have a globular shape their inclusion in Foor's strongylid type would appear to be unjustified.

Spermiogenesis and sperm of *Heligmosomoides polygyrus* (syn. *Nematospiroides dubius*) have been studied in some detail, including by scanning electron microscopy, transmission electron microscopy, and immunocytochemistry of the cytoskeletal proteins actin, tubulin and MSP (Mansir and Justine, 1995, 1996, 1998). These studies have confirmed the earlier descriptions by Wright and Sommerville (1985b). *H. polygyrus*, which is easily maintained in laboratory mice, is the strongylid in which spermatogenesis and sperm have been most thoroughly studied, and is therefore described in detail as a model to which other species will be compared.

The spermatocytes of *H. polygyrus* are round cells and contain fibrous bodies, membranous organelles and peripheral microtubules under the cell membrane. Early spermatids are globular, but late spermatids are relatively elongated cells with a pyriform (pear-shaped) nucleus. Spermatozoa are thinner and more elongate, have an anterior region that is longer and wider than the rest of the cell and a shorter conoid posterior region that contains the nucleus. The anterior end produces pseudopods. Both spermatids and spermatozoa possess a folded and pitted surface, except at the posterior end (Mansir and Justine, 1996; Wright and Sommerville, 1977). Five stages, mainly based on variation of the microtubular system, can be distinguished during elongation of the spermatid (Mansir and Justine, 1998) (Figure 4.26). In stage 1, round spermatids are linked to a cytoplasmic rachis and have a round nucleus that has a fossa or cleft in the nucleus. Tubulin is present as a perinuclear sheath, with intense labelling for tubulin in the fossa, and in a few microtubules that radiate in all directions from the sheath. In stage 2, referred to as the spermatids with a droplet stage, the spermatid has a slightly more elongate nucleus and a small bundle of parallel microtubules are associated with the nuclear fossa. The cytoplasm linked to the rachis, at the opposite end of the nucleus to the fossa, has elongated to form a cytoplasmic droplet, and microtubules that originate from the perinuclear sheath are present in this droplet. In stage 3, referred to as the elongate spermatids with a droplet stage, the bundle of microtubules that originated from the fossa area has elongated considerably while the microtubules around the nucleus have progressively disappeared. The cytoplasmic droplet at the opposite end to the bundle of microtubules contains tubulin, as revealed by intense labelling. The nucleus has become slightly more elongate at this stage. Tropomyosin is present as a cylinder around the microtubule bundle at this stage (Mansir and Justine, 1996). In stage 4, referred to as the elongate spermatids without a droplet stage, the microtubule bundle has reached its maximum length and the microtubules around the nucleus have disappeared. The nucleus is now pyriform (pear-shaped). In stage 4 the spermatids are no longer linked to the rachis and have lost the cytoplasmic droplet. At the end of this stage, the microtubular bundles disappear gradually, commencing at its apical extremity. In stage 5 mature spermatozoon in the male apparently have no

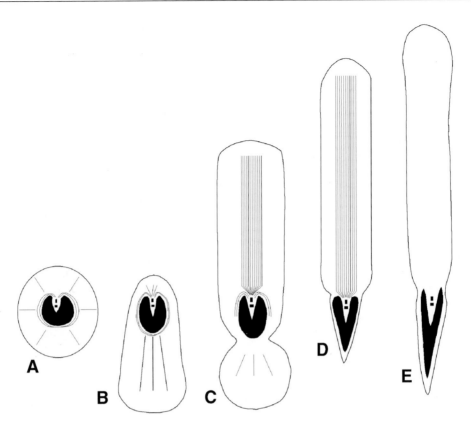

Figure 4.26. Diagram of the microtubular system during spermiogenesis of *Heligmosomoides polygyrus*. A, stage 1, round spermatid; B, stage 2, spermatid with droplet; C, stage 3, elongate spermatid with droplet; D, stage 4, elongate spermatid without droplet; E, stage 5, mature spermatozoon. From Mansir, A. and Justine, J.-L. (1998) *Molecular Reproduction and Development*, **49**, 150–167, reproduced with permission.

tubulin as none was detected by immunocytochemistry or by immunoblotting. The germ cells of this nematode thus possess a conspicuous but transient microtubular system as a cytoplasmic network in the spermatocyte, in the meiotic spindle, as a perinuclear network and as a longitudinal bundle of microtubules in the spermatid. Several post-translational modifications of tubulin have been studied in this transient system. Detyrosination, tyrosination and polyglutamylation were detected, but acetylation and polyglycylation were not; the absence of acetylation, which is generally related to stable microtubules, is consistent with the transient characters of the microtubular system of *H. polygyrus* (Mansir and Justine, 1998).

The MSP system in *H. polygyrus* is similar to that found in most other nematodes. Fibrous bodies, which contain MSP as parallel fibres, appear in the spermatocyte (Mansir and Justine, 1996; Wright and Sommerville, 1985b). In the elongating spermatid, these fibrous bodies are located in the peripheral region of the spermatid body, while the microtubule bundle occupies the central region. In the spermatozoon, MSP is released from the fibrous bodies and becomes located in the anterior crescent-shaped region or 'cap' (at the opposite end of the cell from the nucleus), where it forms a network of filaments.

The localisation of actin and MSP, studied by indirect immunofluorescence, is similar in male germ cells of

H. polygyrus. Spermatids show rows of dots, which correspond to the position of the fibrous bodies, around an unlabelled central longitudinal core, but spermatozoa are labelled only in the anterior cap. Phalloidin, which specifically labels F-actin, labels the spermatids, but not the anterior cap of the spermatozoon. This co-localization of actin and MSP appears to be incompatible with the theory that the MSP cytoskeleton of nematode germ cells operates in the absence of actin. Co-localization of tropomyosin, an actin-associated protein, in the MSP and in the actin-containing cap of the spermatozoon confirms the presence of actin in the spermatid and spermatozoon. However, the absence of F-actin in the mature spermatozoon indicates that actin does not have an active role in the cell at this stage of its development. It is possible that actin has a role in the shaping of the spermatid, but is not involved in sperm motility, which depends upon MSP (Mansir and Justine, 1996).

The mature spermatozoon of *H. polygyrus* is an elongate cell, with the nucleus in the cone-shaped posterior extremity and the filaments corresponding to the active MSP cytoskeleton in an anterior cap. Tubulin has not been detected in the spermatozoon but electron microscopy has revealed two centrioles with an unusual structure of 10 singlets (Mansir and Justine, 1995, 1998). These centrioles are unusual not because they are made up of singlets (which are widely found in nematode sperm) but because they have a 10-fold radial symmetry. Despite their unusual structure, the centrioles seem to act normally as microtubule-organising centres in the spermatid.

In the uterus of female *H. polygyrus*, the elongate spermatozoon of (16–18 μm long) becomes rounder in diameter (5–10 μm), with the cone-shaped nucleus projecting from the spherical sperm body (Sommerville and Weinstein, 1964; Wright and Sommerville, 1977). The spermatozoa move at about 7.3 μm/min, and exceptionally can move up to 23 μm/min, on glass covered with a layer of egg albumin (Wright and Sommerville, 1977).

Spermiogenesis of *Nippostrongylus brasiliensis* has been described by Jamuar (1966) and by Wright and Sommerville (1985a) and changes in morphology of the spermatozoon after insemination have been described by Wright, Cox and Dwarte (1980) and by Wright and Sommerville (1984). The spermatozoon resembles that of *Heligmosomoides polygyrus*, but there are some differences in dimensions (the anterior cytoplasmic region is 7 μm long and the posterior 'tail' is 11 μm long). Also, instead of two perpendicular centrioles in a nuclear fossa, as in *H. polygyrus*, the spermatozoon of *N. brasiliensis* has two parallel centrioles facing, but lying outside of, the anterior fossa of the nucleus. Each centriole has been described by Jamuar, (1966) as 'a central ring of dense material surrounded by 18 filaments'. However, close examination of Fig. 16 in Jamuar's paper makes it difficult to ascertain whether there are 18 microtubules or 9 singlet microtubules separated by spaces of the same dimension. The presence of microtubules around the two centrioles makes clarification more difficult. According to Wright and Sommerville (1985a) centrioles in the spermatocyte and in the spermatid of *N. brasiliensis* have 9 singlets so it is probable that the centrioles in the spermatozoon also consist of 9 singlets. The cytoplasm of the spermatid contains typical fibrous bodies and membranous organelles. In mature spermatozoa the membranous organelles connect to the plasma membrane of the cell (Wright and Sommerville, 1984). The presence of bundles of microtubules and of fibrous bodies in the cell described by Jamuar probably indicates that it was a late spermatid, especially as microtubules are absent in mature spermatozoa (Wright and Sommerville, 1984). If *H. polygyrus* represents a valid model for the trichostrongyles, then the presence of a bundle of microtubules in a male germ cell indicates a spermatid. Spermatozoa in the female of *N. brasiliensis* do not change shape as profoundly as do the sperm of *H. polygyrus* but they do become more elongate (Wright, Cox and Dwarte, 1980). The speed of these spermatozoa on glass coated with albumin reached 12.3 μm/min (Wright and Sommerville, 1984), which is much faster than the speed of ascent *in utero*, estimated to be 5.6 μm/min (Phillipson, 1969). Movement of the sperm was not stopped by cytochalasin (Wright and Sommerville, 1984), which is an actin inhibitor, and this is further evidence that actin in nematode sperm is not involved in motility.

Spermatogenesis in *Nematodirus battus* (Martin and Lee, 1980) follows the same pattern as in other trichostrongyles. Spermatogonia and spermatocytes are arranged upon a central cytoplasmic rachis (as in *Heligmosomoides polygyrus*) and possess 'cup-shaped membranous organelles' that probably correspond to the fibrous bodies and membranous organelles of the other species. A single centriole was found in the fossa of the nucleus. Spermatozoa in the male tract are elongate with an anterior cap and a posterior 'tail' containing the nucleus whereas spermatozoa in the seminal receptacle of the female are round but still with the nucleus in the stubby 'tail', as in *H. polygyrus*.

Spermatozoa in male trichostrongyles seem to have a similar morphology, with an elongate cytoplasmic region containing membranous organelles, fibrous bodies and with a cap at the anterior end, and a conical nucleus at the posterior end, because this morphology has been found in several other species. It occurs in *Teladorsagia* (syn. *Ostertagia*) *circumcincta* and in *Trichostrongylus colubriformis*, in which species actin was found to be located only in the anterior cap in mature spermatozoa (Mansir *et al.*, 1997), and in *Graphidium strigosum*, in which actin and MSP were found to be located in the cap (Mansir and Justine, 1999).

Phylogeny and Nematode Sperm Ultrastructure

There have been four major attempts to construct a classification of nematode sperm types. Anya (1976) and Foor (1983b) used morphological characteristics to form a classification or phylogenetic hierarchy respectively. Shepherd (1981) used the method of extrusion of the residual body (cytophore) and recognised two groups: tylenchoid and other (ascaroid, rhabditoid, aphelenchoid, etc.). Baccetti *et al.* (1983) proposed a classification which represents nematode sperm in three levels or 'steps' with the enoplids as the first step, followed by many types, and then ending in more simplified types.

Justine and Jamieson (1999) considered that reconstructing phylogeny for a group which contains more than 30,000 species from spermatological data based on about 70 species, many of which have not been thoroughly described, is a challenge that should probably not be attempted. A simple classification based on general sperm morphology is hampered by the fact that spermatozoa change shape in the genital ducts of males and/or females. However, these authors attempted some generalisations. Typical membranous organelles occur in most groups, suggesting that they are a plesiomorphic characteristic of the Nematoda, with their absence, if verified, being an apomorphy in certain groups. The Tylenchida are exceptional in lacking membranous organelles and in having cortical microtubules and superficial filopodia. However, cortical microtubules are also found in the mermithids. Absence of membranous organelles and mitochondria in the Dioctophymatida are presumably apomorphies of dioctophymatids. Typical fibrous bodies, showing evidence of the presence of MSP, a protein associated with sperm motility, have been found in most groups of nematodes, with some possible exceptions, such as the oxyurid, *Aspiculuris tetraptera*; thus, the presence of MSP should probably be considered a major synapomorphy for the Nematoda. Another observation of phylogenetic use is that the sperm of the Enoplida have a nuclear envelope, which is absent in all other groups; the absence of a nuclear envelope should be considered a synapomorphy of a taxon including all nematodes except the Enoplida. This is coherent with an interpretation of nematode relationships based on morphology (Adamson, 1987), and would perhaps help to resolve the polytomy found in a tree based on 18S DNA (Blaxter *et al.*, 1998). The classification used in this book, according to De Ley and Blaxter, and, for convenience to the reader, in this chapter, does recognize a subclass Enoplia, in which sperm ultrastructure (without nuclear envelope) is known only for the order Enoplida but not for the orders Triplonchida and Trefusiida; however, this subclass is grouped with the Subclass Dorylaimia within the Class Enoplea, but the Dorylaimia do have a nuclear envelope.

General Characters of Female Gametes

Morphological Characters of Female Gametes

The female genital system consists of ovary and gonoduct. There are two types of ovary in nematodes.

Telogonic ovaries, which are found in the majority of species, are usually didelphic and consist of a short germinal zone, which contains the oogonia, and a long growth zone, which contains the primary oocytes. Telogonic ovaries generally contain a rachis, a central cytoplasmic axis to which early germ cells are attached. In hologonic ovaries, which are found in the Trichinellida (*Trichuris*, *Capillaria*) and Dioctophymatida (*Dioctophyma*), germ cell formation occurs throughout the length of the ovary. The gonoduct consists of the oviduct, which may be small, the uterus, and vagina. A spermatheca (for storing sperm) occurs at the junction of the oviduct and the uterus. The hermaphroditic reproductive system of *C. elegans* is telogonic and amphidelphic and has two tubular ovotestes, which share a common uterus. This section of the review focuses on the ultrastructure of female gametes of nematodes. The structure of the eggshell will not be considered here, nor will the ovary wall and other ducts in which the oocytes pass during their development. Studies about the number of chromosomes in germ cells will also not be considered. There are not many reviews of oocyte and oogenesis ultrastructure, the latest being those by Anya (1976), Foor (1983b) and, more briefly, by Bird and Bird (1991).

Nematode oocytes, as in most phyla, display less variation than spermatozoa. Nematode oocytes are generally round cells, although spectacular variations of form may be found in certain species, such as *Ascaris*

Table 4.1. Classification of cytoplasmic inclusions of nematode oocytes, after Wharton (1979)

Proposed name	Names given in literature	Chemical nature	Function
Glycogen	Glycogen	Glycogen	Food reserve
	Glycogen reserves		Chitinous layer formation
Lipid droplet	Lipid droplet	Lipid	Lipid yolk
	Lipid yolk		
Dense granule	Dense granule	Protein	Protein yolk
	Fat globule		
	Yolk granule		
	Refringent protein body		
Hyaline granule	Lipoprotein granule	Protein and lipid	Lipoprotein yolk
Refringent granule	Refringent granule	Lipid	Formation of lipid layer
Shell granule	Refringent sphere	Protein?	Formation of non-chitin fraction of chitinous layer
	Large protein granule	Bound polyphenols?	
	Reticulate granule		
	Large sphere		
	Hyaline sphere		

(Wu and Foor, 1980). The nucleus is large with numerous nuclear pores. The cytoplasm contains mitochondria, ribosomes, endoplasmic reticulum, and a variety of cytoplasmic inclusions. These have been reviewed and classified by Wharton (1979), according to their chemical nature and function (Table 4.1).

In addition to the usual problem of visualisation of lipid inclusion in tissues dehydrated in lipid solvents when being prepared for transmission electron microscopy, it seems to the author of this review that the problems of fixation of nematode tissues, already mentioned in the sperm section, have hampered the correct description of the cytoplasmic inclusions in certain species. A comparative study of oocyte inclusions is therefore difficult.

Genetic Control of Oocyte Maturation in *Caenorhabditis elegans*

In *Caenorhabditis elegans* the cell-lineage pattern is known with precision. The gonad has two arms, each containing about 1000 cells. Only two germ cells are present at hatching. In hermaphroditic species such as *C. elegans*, the germ cells therefore have three possible fates: mitosis, meiosis and spermatogenesis, and meiosis and oogenesis (review in McLaren, 1993). In an adult hermaphrodite, these three fates are in temporal succession since the young adult is first male then female; the germ cells first produce 160 spermatozoa, then they produce oocytes. The study of genetic control of the fates of germ cells has produced an abundant literature, which is here only partially summarized (see also Hope, Chapter 5). Control of mitosis is made by the gene *glp-1*, which codes for membrane receptors specific to a signal delivered by the distal cap cell. The translation of this gene is temporally and spatially regulated in the embryo (Evans *et al.*, 1994). The gene *gld-2* is required for normal progression through meiotic prophase, and promotes entry into meiosis (Kadyk and Kimble, 1998). Mutants for *mog* genes (*m*asculinisation *of* the *g*erm line) make sperm continuously and do not switch into oogenesis (Graham and Kimble, 1993; Graham, Schedl and Kimble, 1993). Mutants of *fog-3* (for *f*eminisation *of* germ line) do not produce spermatozoa (Ellis and Kimble, 1995). This gene was predicted to encode for a protein of 263 aminoacids, and the mutation which produces the mutant phenotype of fog-3 was attributed to a frame shift at the sixth codon; this protein shows similarities with the tob family, which are involved in proliferation and differentiation (Chen, Singal, Kimble and Ellis, 2000). The gene *fem-1* is required for spermatogenesis in males and hermaphrodites (Doniach and Hodgkin, 1984). The gene *fem-3* directs spermatogenesis and the negative control of this gene, which allows switch to oogenesis, depends upon its 3′ untranslated region (Ahringer and Kimble, 1991). The *mog* genes repress the *fem-3* gene (Gallegos *et al.*, 1998). The gene *emo-1* is necessary for oocyte development and ovulation; it probably encodes a protein necessary for the transport of secreted and transmembrane proteins in oocytes (Iwasaki *et al.*, 1996). The gene *gld-1*

encodes a putative RNA binding protein essential for oocyte development; mutants have tumorous-like ovaries in which female germ cells undergo mitosis instead of meiosis (Francis *et al.*, 1995; Francis, Maine and Schedl, 1995; Jones, Francis and Schedl, 1996; Jones and Schedl, 1995).

In addition to genetic factors, the maturation of oocytes is also influenced by cells of the somatic sheath and spermathecal lineages (McCarter *et al.*, 1997). Defects of the *ceh-18* gene, a crucial determinant of sheath cell differentiation, causes abnormal meiosis (Rose *et al.*, 1997). Insemination also has a role and induces oocyte maturation (McCarter *et al.*, 1999).

Systematic Review of Nematode Female Germ Cells Ultrastructure

No ultrastructural study of female germ cells exists for the Order Enoplida, which are different from all other nematodes by the presence of the nuclear envelope in spermatozoa.

Class Enoplea, Subclass Dorylaimia, Order Dorylaimida

In *Xiphinema theresiae*, oocytes, 5 to 7 μm in diameter, are spherical and form large bulges that correspond to a respective cove in the neighbouring cell (Van de Velde and Coomans, 1988). Cytoplasmic processes of epithelial, non-germinative, cells can be seen between the oocytes. Some of these enlarged cytoplasmic processes form a 'central cytoplasmic mass' in the centre of the ovary. The oocytes develop villi in their region in contact to this structure. The cytoplasm of the oocyte is moderately electron-dense and contains remarkably few cell organelles: only free ribosomes, mitochondria and endoplasmic reticulum, an occasional dictyosome, and annulate lamellae are found. A pair of centrioles is situated close to the nuclear envelope and some rare microtubules are found in the cytoplasm. The absence of cytoplasmic inclusions (no lipid, no yolk) led the authors (Van de Velde and Coomans, 1988) to assume that only pre-vitellogenic oocytes were studied. According to Van de Velde and Coomans (1988), the most striking character of the ovary in this species is the absence of a rachis and the presence of a central cytoplasmic mass formed by the peripheral non-germinative cells.

In *Xiphinema diversicaudatum*, oogonia contain many free ribosomes, polysomes, dictyosomes and mitochondria (Bleve-Zacheo, Melillo and Zacheo, 1993). Oocytes gradually increase in size up to 19 μm in diameter. The cell membrane has a thick surface coat. They have a large nucleus, surrounded by a fibrillar zone and an electron-lucent, granular zone in the cytoplasm. The cytoplasm, which is rich in organelles, is more electron-dense than in the growth zone. Three types of cytoplasmic granules are observed. 'Type 1' granules, believed to contain lipoproteins, are spherical, solid in appearance and of different sizes. 'Type 2' granules, containing lipids, are almost spherical and arranged in groups. These two first types are supposed to participate in eggshell formation. 'Type 3' granules are dense spheres, probably yolk bodies. No glycogen was seen.

Xiphinema americanum has a reproductive system different from other *Xiphinema* species by the presence of symbiotic bacteria in the epithelial wall (Coomans and Claeys, 1998). Irregularly shaped oocytes have a large nucleus.

Class Enoplea, Subclass Dorylaimia, Order Trichinellida

The ovary of trichurids is hologonic. In *Trichinella spiralis*, oogenesis is subdivided into two phases, a nucleolar phase and a cytoplasmic phase. Inclusions in the cytoplasm were claimed to be only glycogen and lipids, without any protein yolk (Herbaut *et al.*, 1980). No difference was found in a comparative study between this species and *Trichinella nativa* (Hulinská and Shaikenov, 1982). In both species, the large oocytes, 9 μm in diameter, have microvilli, and contain large, transparent, membrane-bound structures and lipid droplets. Mature oocytes contain glycogen, large lipid droplets, and dense membrane-bound granules 2–3 μm in diameter. Later studies of oogenesis in the same species (Takahashi, Goto and Kita, 1995) added some details about the presence of numerous microvilli at the surface of oocytes (improperly termed 'ova') and also the timing of oocyte development (Xu, Nagano and Takahashi, 1997). A curious association between the oocytes and the intestine was described in *T. spiralis* (Herbaut *et al.*, 1979). The anterior ovary has no epithelial cells and is limited only by a basal lamina. Locally, the basal lamina of the intestine and ovary fuse and there is no layer of muscular peri-intestinal cells at this place. Near this fusion zone the endoplasmic reticulum of the oocyte makes a complex system of vacuoles. The plasma membrane of the oocyte is in contact with the intestine; several zones with this structure can be found in a single oocyte. This was interpreted as a case of accelerated cellular nutrition (Herbaut *et al.*, 1979).

In *Trichuris muris*, the hologonic arrangement of the germinal cells in the ovary results in a gradation of three morphologically distinct cell types which are visible in a single section: oogonia, vitellogenic primary oocytes, and late vitellogenic oocytes (Preston and Jenkins, 1983). Oogonia and young oocytes, 18 μm in diameter, have a large spherical nucleus and a prominent nucleolus. Their cytoplasm contains ribosomes and mitochondria. Synaptonemal complexes were observed. The previtellogenic primary oocytes have a nucleus with an irregular outline. Their cytoplasm contains dictyosomes, GER and ribosomes. Early vitellogenic oocytes are characterised by an increase of the lipid contents and have microvilli. Two types of lipid inclusions were described, spherical droplets and elongate clefts, and two types of granules were seen, homogeneous electron-dense granules, possibly lipid or lipoproteins (0.6 μm in diameter) and heterogeneous granules composed of one or more dense cores. Late oocytes have no more microvilli and

show increasing amounts of beta glycogen. The oocyte in the oviduct acquires an electron-dense layer on the external surface, which is suggested to arise from small units secreted from the oocytes, which subsequently polymerize. The presence of glycogen and a partially discharged reticulate granule in the oocyte have also been described in a study focussing on egg polar plugs (Preston and Jenkins, 1984).

Class Chromadorea, Subclass Chromadoria, Order Rhabditida, Suborder Spirurina, Infraorder Oxyuridomorpha

In *Aspiculuris tetraptera*, oogonia are arranged around a central rachis which contains numerous mitochondria (Wharton, 1979). Oocytes separate from the rachis and increase in size from 6 to 60 μm. The oocyte contains glycogen, numerous hyaline granules and electron-dense shell granules.

Gyrinicola batrachiensis (Figure 4.27) is an interesting species in which individual females produce two types of eggs: thin-shelled auto-infective eggs produced in the ventral horn of the ovary, and thick-shelled eggs produced in the dorsal horns and which serve for transmission. Oogenesis was studied in the two horns (Adamson, 1983). Oogonia, which are rich in ribosomes, early oocytes, and the rachis are similar in both horns. In the dorsal horn, the oocytes contain much rough endoplasmic reticulum, glycogen, numerous lipid droplets, and three types of granules. (1) Spherical granules, 1μm in diameter, which have a coarse granular core surrounded by an electron-dense outer zone, and which is sometimes membrane-bound. (2) Granules similar to, and apparently formed by, the coalescence of several smaller electron-dense granules, and which are associated with membranous elements. (3) Granules 3 to 5 μm in diameter, which have fine granular contents and one or more dense inclusions, and which have been interpreted as yolk granules. In the ventral horn, the oocytes contain many multivesicular bodies, large amounts of glycogen, occasional lipid droplets and multivesicular bodies, and granules 3 to 5 μm in diameter that are similar to those found in the dorsal horn. Therefore, the major difference between the two types of oocyte was in the amount of glycogen (high in the ventral horn and in the auto-infective eggs) and lipids (high in the dorsal horn and in the thick-shelled eggs).

Class Chromadorea, Subclass Chromadoria, Order Rhabditida, Suborder Spirurina, Infraorder Spiruromorpha

In *Acanthocheilonema (Dipetalonema) viteae*, the rachis is a central axial structure in the germinal region but is very branched in the growth region (McLaren, 1973b). Oocytes are attached to the rachis until they have become mature, and acquire a surface coat whilst in the growth zone. The oogonia, 3.5 × 4.2 μm in size, contain a large nucleus, ribosomes, small mitochondria and a few dictyosomes. The early oocytes contain the same elements and a few lamellar bodies. Large oocytes, 11.5 μm × 5.6 μm, have a large nucleus with a dense

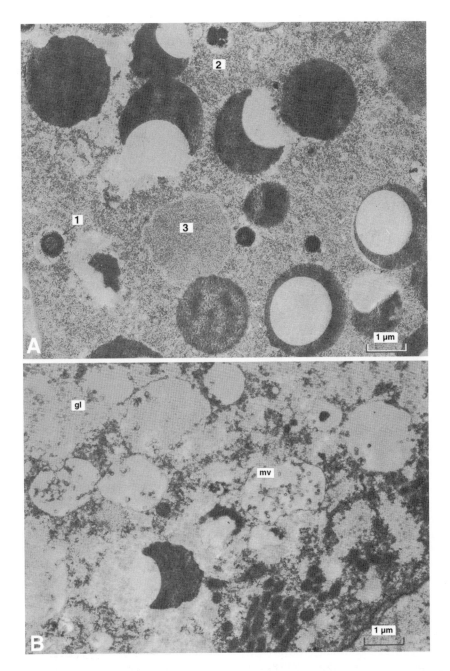

Figure 4.27. Mature oocytes of *Gyrinicola batrachiensis*. A, oocyte in the dorsal horn, showing three types of cytoplasmic granules (1, 2, 3). B, oocyte in the ventral horn, showing glycogen (gl) and multivesicular bodies (mv). From Adamson, M.L. (1983) *Parasitology*, **86**, 489–499.

nucleolus. Their cytoplasm contains small dense granules and occasional bundles of fibres.

In *Dirofilaria immitis*, the oogonia are confined to the distal part of the ovary and a terminal cell cap was not observed (Lee, 1975). Oogonia contain numerous dense bodies and lipid droplets; the endoplasmic reticulum is absent and dictyosomes are found only in the fully-grown oogonia. Dense bodies of the oogonia, elongate or oval, 0.5–1 × 1–3 μm in size, exhibit a positive reaction to mitochondrial enzyme. Primary oocytes, 5 × 12 μm in size, located in the middle part of the ovary, are arranged around a central rachis; they are large and

elongate, and have mitochondria, endoplasmic reticulum, dictyosomes, and very few dense bodies. The central rachis contains no organelles except a few dense granules. Late oocytes are larger, 10–15 μm, contain small dense granules, and are not connected to the rachis.

Some information about the ultrastructure of female germ cells of spirurid nematodes may also be found in various papers that do not concentrate on this subject. An image of the oocyte nucleus of *Dirofilaria immitis* shows synaptonemal complexes (Delves, Howells and Post, 1986). Oocytes of *Tetrameres columbicola* are apparently not membrane-bound in the germinal zone of the ovary; in the growth zone, they contain ribosomes and lipid vacuoles, and bear pseudopods; a rachis is present (Simpson, Carlisle and Conti, 1984). Oocytes of *Onchocerca volvulus* were claimed to contain numerous bacteria (Franz and Büttner, 1983), but this needs further study.

In certain filariae, the larvae, or microfilariae, are enclosed in a sheath when released by the female. This sheath is considered to be a modified eggshell and it has been shown that transcription of the gene coding for the major protein (gp22) of the sheath of *Litomosoides carinii* occurs mainly in oocytes (Conraths *et al.*, 1993).

In *Dirofilaria immitis*, exogenous ecdysone induces meiotic reinitiation of the oocytes (Barker *et al.*, 1991), thus supporting the theory that ecdysteroids play a hormonal role in nematodes, at least in filariids.

Class Chromadorea, Subclass Chromadoria, Order Rhabditida, Suborder Spirurina, Infraorder Ascaridomorpha

Oocyte development has been studied in detail in *Ascaris lumbricoides* (Foor, 1967). Oogonia and the rachis in the region of early oocytes contain numerous lamellar bodies. The oocytes are attached to the rachis by cytoplasmic bridges, which contain many microtubules. The oocytes contain lipid droplets, which arise by coalescence of small vesicles, numerous small granules, and large membrane-bound refringent granules. The refringent granules are 1.5–3 μm in young oocytes and 4–5 μm in old oocytes. Small vesicles are often found close to the early refringent granules. The mature refringent granules contain a homogeneous material, moderately electron-dense, and a dense material in their centre. They contain ascaroside esters and proteins, and contribute to the formation of one of the layers of the eggshell.

Oocytes of *Ascaris suum*, studied by scanning electron microscopy, display outstanding shapes (Wu and Foor, 1980). The immature oocytes are elongate, bipolar cells with an atypical region, which contains numerous lobate cytoplasmic processes, and a slender opposite pole corresponding to the basal region previously attached to the rachis; later, they gradually round up.

In *Heterakis gallinarum*, the oogonia and oocytes are arranged around a rachis. Oogonia contain glycogen and a few lipid droplets. The young oocytes contain lipid droplets, membrane-bound refringent granules, 2 to 3 μm in diameter, which give rise to the ascaroside layer of the eggshell, and 1 μm granules, probably yolk material (Lee and Lestan, 1971). The refringent granules are formed by the coalescence of several granules, which have an electron-dense centre surrounded by granular material. The yolk granules contain an electron-dense material with strands of electron-lucid material. The mature oocyte contains much glycogen and numerous lipid droplets, and has a coat 100 nm thick.

In *Pseudoterranova decipiens*, the oogonia contain small granules, large electron-dense granules and neutral-density granules (MacKinnon and Burt, 1992).

An unusual association of organelles, consisting of membranous whorls located in a moon-shaped indentation of the larger electron-dense granules, and glycogen, have been described. Oogonia then separate from the rachis. The oocytes contain few large dense granules, numerous small dense granules, and an increasing number of neutral-density granules, and they have numerous microvilli. Oocytes passing into the oviduct in a single file lose their microvilli.

In *Toxocara canis*, oogonia contain electron-lucent vesicles (Brunanská, 1993b). Young oocytes contain lipid droplets, small dense granules, and glycogen. Older oocytes contain dense granules, 0.3–0.7 μm in diameter, shell granules that are heterogeneous and 4 μm in diameter, and lipid droplets. The surface of the oocyte forms numerous invaginations and is covered by a dense material. The rachis contains neither organelles nor cytoplasmic inclusions in the germinal zone (Brunanská, 1994). Bridges between the rachis and oogonia contain microfilaments parallel to the long axis of the oocytes. At the beginning of the ovarian growth zone, the rachis contains lipid droplets and small dense granules, but in further portions it contains the same inclusions as the adjoining oocytes, i.e. lipids, dark dense granules, small shell granules and glycogen. The last portion of the ovary has no rachis.

Chen and Zhang (1984) have published some information about the chromatin of oocytes of *Meteterakis govindi*. Histochemical observations have been performed about lipid changes during oogenesis in *Ascaridia galli* (Parshad and Guraya, 1982).

Class Chromadorea, Subclass Chromadoria, Order Rhabditida, Suborder Tylenchina, Infraorder Tylenchomorpha

In *Meloidogyne javanica*, oogonia are arranged around a central rachis to which they are attached by cytoplasmic bridges (McClure and Bird, 1976). The rachis is anucleate and contains lipid droplets and refringent protein granules. Lipid droplets and refringent protein granules are also found in the cytoplasm of the oogonia. The oocytes, which detach from the rachis as they grow, contain lipid droplets, refringent protein granules and glycogen. The oocytes change in shape from round to oval as they pass through a region containing a valve or sphincter that lies between the oviduct and the spermatheca (McClure and Bird, 1976; Bird and Bird, 1991).

Endo, Zunke and Wergin (1997, 1999) have provided an extremely richly illustrated study of oogenesis (Figures 4.28, 4.29) in *Pratylenchus penetrans*. The germinal cells (oogonia) in the distal portion of the ovary have spheroid nuclei, numerous polyribosomes, and high concentrations of rough endoplasmic reticulum, mitochondria, dictyosomes, and electron-dense granules (Figure 4.28). Most nuclei of oocytes in the ovary are at the pachytene stage (first prophase of meiosis). In the midsection of the ovary, oocytes enlarge and accumulate in a single row (Figure 4.29). These cells are similar in their size and organelles to the oogonia. In the ovary, contact between germ cells and epithelial cells appears to be minimal. Oocytes prior to their entry

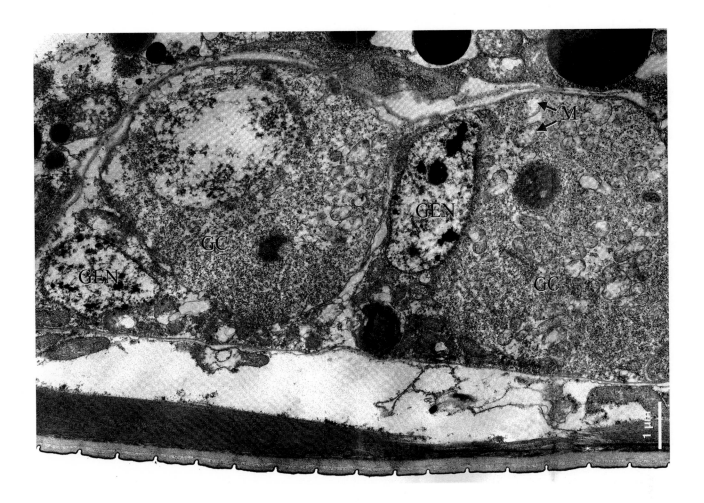

Figure 4.28. Distal region of the ovary of *Pratylenchus penetrans*, showing three distal cells. Germinal cells (GC) are precursor to oocyte development. GEN, gonad epithelium nucleus; M, mitochondria. From Endo, B.Y., Zunke, U. and Wergin, W.P. (1999) *Journal of the Helminthological Society of Washington*, **66**, 155–174, reproduced with permission.

into the oviduct show relatively few cytoplasmic inclusions, but oocytes in the spermatheca have numerous, large lipid droplets and protein granules.

Class Chromadorea, Subclass Chromadoria, Order Rhabditida, Suborder Rhabditina, Infraorder Rhabditomorpha

Superfamily rhabditoidea

The genital apparatus of *Caenorhabditis elegans* has been described by Hirsh, Oppenheim and Klass (1976) who

included some ultrastructural observations on the oocytes in their publication.

The nucleus of the oocytes of *C. elegans* has been the subject of detailed ultrastructural observations of the peri-nuclear and peri-nucleolar zones (Abirached and Brun, 1978) and of the synaptonemal complexes (Abirached and Brun, 1979). After treatment with DMSO (dimethyl suphoxide), synaptonemal complexes were found to be absent in the oocytes of *C. elegans* and abnormal gametes were produced (Goldstein and

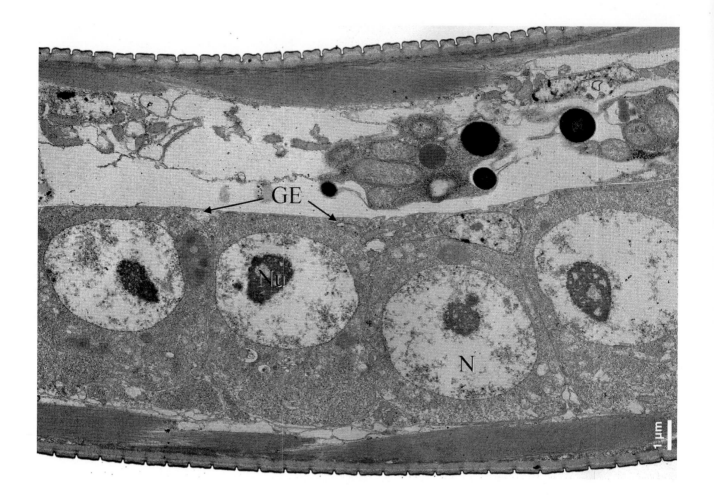

Figure 4.29. Proximal region of the ovary of *Pratylenchus penetrans*, showing growth stage of oocytes; GE, gonad epithelium; N, nucleus; Nu, nucleolus. From Endo, B.Y., Zunke, U. and Wergin, W.P. (1999) *Journal of the Helminthological Society of Washington*, **66**, 155–174, reproduced with permission.

Magnano, 1988). An ultrastructural and ultrastructural radioautographic study of RNA synthesis during oogenesis showed that metabolites are transferred from the germ cells to the rachis; the rachis was thus assigned a trophic role, as a common nutritive pool for the female germ cells (Gibert, Starck and Beguet, 1984). The meiotic segregation of the holocentric chromosomes has been followed in detail by transmission electron microscopy and by fluorescence immunomicroscopy (Albertson and Thomson, 1993). Visualization of actin

by phalloidin has shown that actin microfilaments are located in the periphery of oocytes (Strome, 1986).

Yolk proteins are known more precisely in *C. elegans* than in other species. They include four abundant glycoproteins which are specific to adult hermaphrodites: yp170A and yp170B, which are related, and yp115 and yp88, which are unrelated. Synthesis of these four proteins is correlated with the onset of the oogenetic phase (Starck, 1984). Products of RNA translated *in vitro* produced three proteins only, two corresponding to the yp170A/B and the third, with a molecular weight of 180 kDa, corresponding to a precursor of yp115 and yp88 (Kimble and Ward, 1988). These proteins are synthesized in the intestine; cleavage of the yp115/yp88 precursor probably occurs in the body cavity of the worm. Genes of vitellogenins include a family of five genes (*vit-1* to *vit-5*) which encode the yp170A/B proteins, and the gene *vit-6* which encodes the yp115/yp88 precursor (Kimble and Ward, 1988). The coding regions of the *vit* genes are highly conserved.

Three glycosylated proteins, VT1 (175 kDa), VT2 (107 kDa) and VT3 (82 kDa) have been found in *Dolichorhabditis* sp. (Winter, 1992).

Superfamily Strongyloida

In *Heligmosomoides polygyrus*, the oocytes are arranged around a rachis. Early oocytes have an elongate nucleus, 3.2×17 μm, which measures 13 μm in diameter in the fully developed (60×70 μm) oocyte, and has numerous nuclear pores. Three types of granule are found in the oocyte. The first granules to appear measure 1 to 1.5 μm, are amorphous and solid, are irregular in outline, are not electron-dense, and have been interpreted as lipoproteins. The second type of granule, which measures 1 to 8 μm, appears as large flocculent inclusions and is probably a lipid granule. The third type of granule, which measures 0.3 to 1 μm, is electron-dense, spherical and probably consists of protein or lipoprotein yolk. There is a delicate flocculent coat on the plasma membrane of the oocyte. RNA, detected by histochemistry, is abundant in the nucleolus and cytoplasm in the oogonia and oocytes (MacKinnon, 1987). The neutral lipid content of developing female gametes of *Heligmosomoides polygyrus* is higher in developing oocytes than in oogonia. It was shown to change according to the age of the female, rising to a maximum up to day 40 post-infection and then declining after 140 days (MacKinnon and Lee, 1988).

In *Syngamus trachea*, the germinal zone of the ovary forms the terminal part of the gonad (Brunanská, 1991). The cap cell contains fibrils, oriented vertically to the surface and lining the basal part of the cell, and vacuoles of varying size. The oogonia, 5 to 6 μm in diameter, are spherical. Their cytoplasm contains mitochondria, ribosomes, vacuoles of varying size, glycogen and lipids. The rachis contains lipid droplets. The growth zone of the ovary is subdivided into a shorter portion with a rachis and a longer portion without it (Brunanská, 1992). Young oocytes contain lipids and dense granules, which range from 0.1 μm to 1.7 μm in diameter, and which are probably proteins or lipoproteins. Large granules, 3 μm in size, are present in the maturing oocyte and contribute to the eggshell. The cell membrane of the oocyte has inconspicuous invaginations. In the zone without the rachis, numerous small dense granules, 0.3–0.5 μm in diameter, and large amounts of glycogen appear. Nuclei are about 10 μm in diameter. The cytoplasm of fertilized oocytes contains lipids, dense granules and shell granules (Brunanská, 1993a).

In *Dictyocaulus viviparus*, the germinal zone of the ovary is a multinuclear syncytium. A rachis is visible in the centre in the form of clustered tubules (Gutteková and Brunanská, 1988). The growth zone of the ovary was separated into six regions, the first four being associated with the rachis and the last two without an association with the rachis (Brunanská and Gutteková, 1989). In zone 1, the oocytes contain saturated lipids and small heterogeneous granules; in zone 2, the oocytes contain lipid droplets, small (0.2–0.5 μm) dense granules, and heterogeneous granules 4–5 μm in diameter; in zone 3, the oocytes are distinctly separated and contain small dense granules, heterogeneous granules near the nucleus, lipid droplets and glycogen; in zone 4, the number of dense granules and the amount of glycogen increases; in zone 5, the oocytes are larger and heterogeneous granules show signs of disintegration; in zone 6, the oocytes are 50×30 μm in size and have an eccentrically located nucleus. The surface of the oocyte possesses a dense coat, here termed the oolemma, which is sometimes interrupted (Brunanská and Gutteková, 1989). It was claimed that proteins from the intestinal cells make their way into the pseudocoelomic fluid and hence by micropinocytosis through the gonadal wall into the oocytes (Gutteková and Brunanská, 1990).

Fertilisation

The ultrastructural aspects of fertilisation have been studied in a very small number of species: *Ascaris lumbricoides* (Foor, 1968, 1970), *Acanthocheilonema (Dipetalonema) viteae* (McLaren, 1973b) and *Xiphinema diversicaudatum* (Bleve-Zacheo, Melillo and Zacheo, 1993).

As already stated, the nematode sperm has no acrosome and no flagellum and the process of fertilisation thus differs from that of other, more usual, animals.

In *Ascaris lumbricoides*, pseudopods emanating from the anterior cytoplasm make first contact with the primary oocyte (Foor, 1968, 1970b). The gamete membranes then interdigitate and finally fuse. After fusion of the gamete membranes, the sperm nucleus becomes disorganised and the refringent body, a special structure typical of *Ascaris* sperm, disappears when the sperm reaches the centre of the egg.

In *Acanthocheilonema (Dipetalonema) viteae*, polyspermy frequently occurs, and up to six spermatozoa have been observed either penetrating or within the oocyte (McLaren, 1973b). The anterior region of the spermatozoon is the first to contact the oocyte and causes a deep invagination of the oocyte membrane. No pseudopods were observed. The surface coat around the oocyte terminates where the membranes of the two gametes are in contact.

In *Dirofilaria immitis*, oocytes are mingled with spermatozoa in the seminal receptacle, where fertilisation occurs (Lee, 1975). Sperm penetration of oocytes was not observed, but a micrograph of a stage just posterior shows an opening in the oolemma and sperm contents still distinguishable from the oocyte cytoplasm. Later, the chromatin material disintegrates and the mitochondria and membranous organelles of the spermatozoon are decomposed into membranous scrolls.

In *Xiphinema diversicaudatum*, fertilisation occurs at a preferential pole of the oocyte (Bleve-Zacheo, Melillo and Zacheo, 1993). The cytoplasm of the oocyte in that area is devoid of granules, but light and dense vesicles are present. Two or more spermatozoa can be in contact with the oocyte but only one penetrates. Sperm penetration induces an invagination of the oocyte membrane. The spermatozoon can be recognized intact in the oocyte cytoplasm, but later it releases its nuclear contents.

In *Caenorhabditis elegans*, detailed observations of the fertilisation process have been done by light microscopy, with limited use of ultrastructural techniques. Fertilisation triggers active movement of the cytoplasmic granules of the oocyte. Sperm penetration is not required for this activation because sperm from fertilisation-defective animals also trigger activation without penetrating the oocyte (Ward and Carrel, 1979). Studies of mutants have shown that the presence of sperm chromatin mass nor a sperm nucleus is required for egg activation, or the induction of anterior-posterior polarity (Sadler and Shakes, 2000)

In *C. elegans*, Browning and Strome (1995) have demonstrated the first paternally contributed factor to be genetically identified and molecularly characterised (see also Roush, 1996). The protein encoded by the gene *spe-11* is expressed during spermiogenesis, is found in mature sperm, but is not expressed during oogenesis. Mutants that lack expression of spe-11 show normal sperm morphology, except that the electron-dense halo surrounding the nucleus and the pair of centrioles is unevenly distributed and contains additional granular material. Embryos produced by sperm from homozygous *spe-11* animals fail to complete meiosis of the oocyte and finally develop into non-viable multinucleate, single-cell embryos. Genetically engineered oocytes, expressing the *spe-11* gene, undergo a normal development when fertilised with spermatozoa lacking *spe-11* expression. These results show that the protein *spe-11* is not required during spermatogenesis, but is a sperm-supplied factor that participates directly in the development of the early embryo. After fertilisation, the egg undergoes successive developmental stages (see Hope, Chapter 5).

Acknowledgements

Dr Nezha Noury-Sraïri and Dr Aïcha Mansir, former students of J.-L. Justine, are thanked for documents provided. Dr Nicole Gourbault helped in providing specimens and discussions. Roselyne Tcheprakoff helped for the line drawings.

References

Abbas, M. and Cain, G.D. (1979) *In vitro* activation and behavior of the amoeboid sperm of *Ascaris suum* (Nematoda). *Cell & Tissue Research*, **200**, 273–284.

Abbas, M.K. and Cain, G.D. (1981) Subcellular fractions and the refringent granules of the spermatozoa of *Ascaris suum* (Nematoda). *Cell & Tissue Research*, **221**, 125–136.

Abbas, M.K. and Cain, G.D. (1984) Amino acid and lipid composition of refringent granules from the ameboid sperm of *Ascaris suum* (Nematoda). *Histochemistry*, **81**, 59–65.

Abirached, M. and Brun, J. (1979) L'évolution du complexe synaptonématique dans le noyau ovocytaire en prophase méiotique de *Caenorhabditis elegans* (Nematoda). *Comptes Rendus de l'Académie des Sciences, Paris, Série D*, **288**, 425–428.

Abirached, M. and Brun, J.-L. (1978) Ultrastructural changes in the nuclear and perinuclear regions of the oogonia and primary oocytes of *Caenorhabditis elegans*, Bergerac strain. *Revue de Nématologie*, **1**, 63–72.

Achanzar, W.E. and Ward, S. (1997) A nematode gene required for sperm vesicle fusion. *Journal of Cell Science*, **110**, 1073–1081.

Acosta-Garcia, I., Hernandez-Rodriguez, S., Gutierrez-Palomino, P. and Navarrete, I. (1985) Estudio al microscopio electronico de las espermatidas de *Protostrongylus rufescens* (Nematoda: Metastrongyloidea). *Revista Ibérica de Parasitologia*, **45**, 313–320.

Adamson, M.L. (1983) Ultrastructural observations on oogenesis and shell formation in *Gyrinicola batrachiensis* (Walton, 1929) (Nematoda: Oxyurida). *Parasitology*, **86**, 489–499.

Adamson, M.L. (1987) Phylogenetic analysis of the higher classification of the Nematoda. *Canadian Journal of Zoology*, **65**, 1478–1482.

Ahringer, J. and Kimble, J. (1991) Control of the sperm-oocyte switch in *Caenorhabditis elegans* hermaphrodites by the *fem-3* 3′ untranslated region. *Nature*, **349**, 346–348.

Albertson, D.G. and Thomson, J.N. (1993) Segregation of holocentric chromosomes at meiosis in the nematode, *Caenorhabditis elegans*. *Chromosome Research*, **1**, 15–26.

Anderson, T.J.C., Komuniecki, R., Komuniecki, P.R. and Jaenike, J. (1995) Are mitochondria inherited paternally in *Ascaris*? *International Journal for Parasitology*, **25**, 1001–1004.

Anya, A.O. (1976) Physiological aspects of reproduction in nematodes. *Advances in Parasitology*, **14**, 267–351.

Argon, Y. and Ward, S. (1980) *Caenorhabditis elegans* fertilization-defective mutants with abnormal sperm. *Genetics*, **96**, 413–33.

Auffray, C., Behar, G., Bois, F., Bouchier, C., Da Silva, C., Devignes, M.-D. *et al.* (1995) IMAGE: intégration au niveau moléculaire de l'analyse du génome humain et de son expression. *Comptes Rendus de l'Académie des Sciences Paris*, **318**, 263–272.

Baccetti, B., Dallai, R., Grimaldi de Zio, S. and Marinari, A. (1983) The evolution of the Nematode spermatozoon. *Gamete Research*, **8**, 309–323.

Barker, G.C., Mercer, J.G., Rees, H.H. and Howells, R.E. (1991) The effect of ecdysteroids on the microfilarial production of *Brugia pahangi* and the control of meiotic reinitiation in the oocytes of *Dirofilaria immitis*. *Parasitology Research*, **77**, 65–71.

Beams, H.W. and Sekhon, S.S. (1972) Cytodifferentiation during spermiogenesis in *Rhabditis pellio*. *Journal of Ultrastructure Research*, **38**, 511–527.

Bird, A.F. and Bird, J. (1991) *The structure of nematodes*, 2nd edn. San Diego: Academic Press.

Blaxter, M. (1998) *Caenorhabditis elegans* is a nematode. *Science*, **282**, 2041–2046.

Blaxter, M.L., De Ley, P., Garey, J.R., Liu, L.X., Scheldeman, P., Vierstraete, A. *et al.* (1998) A molecular evolutionary framework for the phylum Nematoda. *Nature*, **392**, 71–75.

Bleve-Zacheo, T., Melillo, M.T. and Zacheo, G. (1993) Ultrastructural studies on the nematode *Xiphinema diversicaudatum*: oogenesis and fertilization. *Tissue & Cell*, **25**, 375–388.

Browning, H. and Strome, S. (1995) A sperm-supplied factor required for embryogenesis in *C. elegans*. *Development*, **122**, 391–404.

Brunanská, M. (1991) An ultrastructural study on the germinal zone and rachis of the ovaries in *Syngamus trachea*. *Helminthologia*, **28**, 165–171.

Brunanská, M. (1992) The ultrastructure of the growth zone of the ovaries of *Syngamus trachea*. *Helminthologia*, **29**, 7–12.

Brunanská, M. (1993a) The structure and formation of the *Syngamus trachea* egg-shell (Nematoda: Syngamidae). *Folia Parasitologica*, **40**, 135–140.

Brunanská, M. (1993b) *Toxocara canis* (Nematoda, Ascarididae): the fine structure of the oogonia and oocytes. *Helminthologia*, **30**, 9–13.

Brunanská, M. (1994) *Toxocara canis* (Nematoda, Ascarididae): ultrastructure of the rachis and the ovarian wall. *Folia Parasitologica*, **41**, 149–153.

Brunanská, M. and Gutteková, A. (1989) The ultrastructure of the female reproductive organs in *Dictyocaulus viviparus*. II. The growth zone of the ovaries. *Helminthologia*, **26**, 129–136.

Bullock, T.L., McCoy, A.J., Kent, H.M., Roberts, T.M. and Stewart, M. (1998) Structural basis for amoeboid motility in nematode sperm. *Nature Structural Biology*, 5, 184–189.

Bullock, T.L., Parthasarathy, G., King, K.L., Kent, H.M., Roberts, T.M. and Stewart, M. (1996) New crystal forms of the motile major sperm protein (MSP) of *Ascaris suum*. *Journal of Structural Biology*, 116, 432–437.

Burghardt, R.C. and Foor, W.E. (1975) Rapid morphological transformations of spermatozoa in the uterus of *Brugia pahangi* (Nematoda, Filarioidea). *Journal of Parasitology*, 61, 343–350.

Burghardt, R.C. and Foor, W.E. (1978) Membrane fusion during spermiogenesis in *Ascaris*. *Journal of Ultrastructure Research*, 62, 190–202.

Burke, D.J. and Ward, S. (1983) Identification of a large multigene family encoding the major sperm protein of *Caenorhabditis elegans*. *Journal of Molecular Biology*, 171, 1–29.

Calcoen, J.A. and Dekegel, D. (1979) Spermatozoa in the demanian organ of female *Adoncholaimus fuscus* (Bastian, 1865) (Nematoda). *Netherlands Journal of Zoology*, 29, 142–143.

Cares, J.E. and Baldwin, J.G. (1994a) Comparative fine structure of sperm of *Verutus volvingentis* and *Meloidodera floridensis* (Heteroderinae, Nematoda). *Canadian Journal of Zoology*, 72, 1481–1491.

Cares, J.E. and Baldwin, J.G. (1994b) Fine structure of sperm of *Ekphymatodera thomasoni* (Heteroderinae, Nemata). *Journal of Nematology*, 26, 375–383.

Cares, J.E. and Baldwin, J.G. (1995) Comparative fine structure of sperm of *Heterodera schachtii* and *Punctodera chalcoensis*, with phylogenetic implications for Heteroderinae (Nemata: Heteroderidae). *Canadian Journal of Zoology*, 73, 309–320.

C. elegans Sequencing Consortium (1998) Genome sequence of the nematode *C. elegans*: A platform for investigating biology. *Science*, 282, 2012–2018.

Chabaud, A., Bain, O., Hugot, J.-P., Raush, R.L. and Raush, V.R. (1983) Organe de Monsieur de Man et insémination traumatique. *Revue de Nématologie*, 6, 127–131.

Chen, H. and Zhang, A. (1984) [Observations on the ultrastructure of intestinal cells and genital cells in nematodes (*Meteterakis govindi*)]. *Acta Zoologica Sinica*, 30, 148–152 + plate.

Chen, P.J., Singal, A., Kimble, J. and Ellis, R.E. (2000) A novel member of the tob family of proteins controls sexual fate in *Caenorhabditis elegans* germ cells. *Developmental Biology*, 217, 77–90.

Clark, W.H.J., Moretti, R.L. and Thomson, W.W. (1967) Electron microscopic evidence for the presence of an acrosomal reaction in *Ascaris lumbricoides* var. *suum*. *Experimental Cell Research*, 47, 643–647.

Clark, W.H.J., Moretti, R.L. and Thomson, W.W. (1982) Histochemical and ultracytochemical studies of the spermatids and sperm of *Ascaris lumbricoides* var. *suum*. *Biology of Reproduction*, 7, 145–159.

Conraths, F.J., Schützle, B., Schares, G., Christ, H., Hobom, G. and Zahner, H. (1993) The gene coding for the major sheath protein of *Litomosoides carinii* microfilariae, gp22, is transcribed in oocytes and embryonic cells. *Molecular and Biochemical Parasitology*, 60, 111–120.

Coomans, A. and Claeys, M. (1998) Structure of the female reproductive system of *Xiphinema americanum* (Nematoda: Longidoridae). *Fundamental and Applied Nematology*, 21, 569–580.

Currey, H.M. (1975) Ultrastructural studies of the male reproductive system of *Metastrongylus apri*. *Parasitology*, 71, xxii.

Decraemer, W. (1988) Morphometric variability and value of the characters used for specific identification in *Trichodorus* Cobb, 1913. *Bulletin de l'Institut Royal des Sciences Naturelles de Belgique, Biologie*, 58, 29–44.

Delves, C.J., Howells, R.E. and Post, R.J. (1986) Gametogenesis and fertilization in *Dirofilaria immitis* (Nematoda: Filarioidea). *Parasitology*, 92, 181–197.

Doniach, T. and Hodgkin, J. (1984) A sex-determining gene, *fem-1*, required for both male and hermaphrodite development in *Caenorhabditis elegans*. *Developmental Biology*, 106, 223–235.

Ellis, R.E. and Kimble, J. (1995) The fog-3 gene and regulation of cell fate in the germ line of *Caenorhabditis elegans*. *Genetics*, 1995, 561–577.

Endo, B.Y., Zunke, U. and Wergin, W.P. (1997) Ultrastructure of the lesion nematode, *Pratylenchus penetrans* (Nemata: Pratylenchidae). *Journal of the Helminthological Society of Washington*, 64, 59–95.

Endo, B.Y., Zunke, U. and Wergin, W.P. (1998) Ultrastructure of the male gonad and spermatogenesis in the lesion nematode, *Pratylenchus penetrans* (Nemata: Pratylenchidae). *Journal of the Helminthological Society of Washington*, 65, 227–242.

Endo, B.Y., Zunke, U. and Wergin, W.P. (1999) Ultrastructure of the female reproductive system of the lesion nematode, *Pratylenchus penetrans* (Nemata: Pratylenchidae). *Journal of the Helminthological Society of Washington*, 66, 155–174.

Evans, T.C., Crittenden, S.L., Kodoyianni, V. and Kimble, J. (1994) Translational control of maternal glp-1 mRNA establishes an asymmetry in the *C. elegans* embryo. *Cell*, 77, 183–184.

Fauré-Frémiet, E. (1913) Le cycle germinatif chez l'*Ascaris megalocephala*. *Archives d'Anatomie microscopique*, 15, 435–758 + Pl. XII–XIV.

Favard, P. (1961) Évolution des ultrastructures cellulaires au cours de la spermatogenèse de l'ascaris (*Ascaris megalocephala*, Schrank = *Parascaris equorum*, Goerze). *Annales des Sciences Naturelles, Zoologie et Biologie Animale, 12° Série*, 3, 52–152.

Foor, W.E. (1967) Ultrastructural aspects of oocyte development and shell formation in *Ascaris lumbricoides*. *Journal of Parasitology*, 53, 1245–1261.

Foor, W.E. (1968) Zygote formation in *Ascaris lumbricoides* (Nematoda). *Journal of Cell Biology*, 39, 119–134.

Foor, W.E. (1970a) Morphological changes of in utero *Dipetalonema viteae*. *The Journal of Parasitology*, 56, 103.

Foor, W.E. (1970b) Spermatozoan morphology and zygote formation in Nematodes. *Biology of Reproduction*, 2, 177–202.

Foor, W.E. (1974) Morphological changes of spermatozoa in the uterus and glandular vas deferens of *Brugia pahangi*. *Journal of Parasitology*, 60, 125–133.

Foor, W.E. (1983a) Nematoda. In *Reproductive Biology of Invertebrates, Volume I: Oogenesis, Oviposition, and Oosorption*, edited by K.G. Adiyodi and R.G. Adiyodi, pp. 223–256. Chichester: Wiley.

Foor, W.E. (1983b) Nematoda. In *Reproductive Biology of Invertebrates, Volume II: Spermatogenesis and Sperm Function*, edited by K.G. Adiyodi and R.G. Adiyodi, pp. 221–256. Chichester: Wiley.

Foor, W.E., Johnson, M.H. and Beaver, P.C. (1971) Morphological changes in the spermatozoa of *Dipetalonema viteae* in utero. *Journal of Parasitology*, 57, 1163–1169.

Foor, W.E. and McMahon, J.T. (1973) Role of the glandular vas deferens in the development of *Ascaris* spermatozoa. *Journal of Parasitology*, 59, 753–758.

Francis, R., Barton, M.K., Kimble, J. and Schedl, T. (1995) gld-1, a tumor suppressor gene required for oocyte development in *Caenorhabditis elegans*. *Genetics*, 139, 579–606.

Francis, R., Maine, E. and Schedl, T. (1995) Analysis of the multiple roles of gld-1 in germline development: interactions with the sex determination cascade and the glp-1 signaling pathway. *Genetics*, 139, 607–630.

Franz, M. and Büttner, D.W. (1983) The fine structure of adult *Onchocerca volvulus*. V. The digestive tract and the reproductive system of the female worm. *Tropenmedizin und Parasitologie*, 34, 155–161.

Gallegos, M., Ahringer, J., Crittenden, S. and Kimble, J. (1998) Repression by the 3′ UTR of fem-3, a sex-determining gene, relies on a ubiquitous mog-dependent control in *Caenorhabditis elegans*. *EMBO Journal*, 17, 6337–6347.

Gibert, M.-A., Starck, J. and Beguet, B. (1984) Role of the gonad cytoplasmic core during oogenesis in the nematode *Caenorhabditis elegans*. *Biology of the Cell*, 50, 77–86.

Goldstein, P. (1977) Spermatogenesis and spermiogenesis in *Ascaris lumbricoides* var. *suum*. *Journal of Morphology*, 154, 317–338.

Goldstein, P. and Magnano, L. (1988) Effects of dimethyl sulphoxide on early gametogenesis in *Caenorhabditis elegans*: ultrastructural aberrations and loss of synaptonemal complexes from pachytene nuclei. *Cytobios*, 56, 45–57.

Goldstein, P. and Triantaphyllou, A.C. (1980) The ultrastructure of sperm development in the plant-parasitic nematode *Meloidogyne hapla*. *Journal of Ultrastructure Research*, 71, 143–153.

Gourbault, N. and Renaud-Mornant, J. (1983) Système reproducteur d'un nématode marin à fécondation par spermatophore. *Revue de Nématologie*, 6, 51–56.

Graham, P.L. and Kimble, J. (1993) The mog-1 gene is required for the switch from spermatogenesis to oogenesis in *Caenorhabditis elegans*. *Genetics*, 133, 919–931.

Graham, P.L., Schedl, T. and Kimble, J. (1993) More mog genes that influence the switch from spermatogenesis to oogenesis in the hermaphrodite germ line of *Caenorhabditis elegans*. *Developmental Genetics*, 14, 471–484.

Gutteková, A. and Brunanská, M. (1988) The ultrastructure of the female reproductive organs in *Dictyocaulus viviparus*. I. Germinal zone of the ovaries. *Helminthologia*, 25, 235–243.

Gutteková, A. and Brunanská, M. (1990) *Dictyocaulus viviparus*: ultrastructure of the wall of the ovary, oviduct, vagina and vulva. *Helminthologia*, 27, 239–247.

Haaf, A., Butler, P.J., Kent, H.M., Fearnley, I.M., Roberts, T.M., Neuhaus, D. et al. (1996) The motile major sperm protein (MSP) from *Ascaris suum* is a symmetric dimer in solution. *Journal of Molecular Biology*, **260**, 251–260.

Haaf, A., LeClaire, L., Roberts, G., Kent, H.M., Roberts, T.M., Stewart, M. and Neuhaus, D. (1998) Solution structure of the motile sperm protein (MSP) of *Ascaris suum* — Evidence for two manganese binding sites and the possible role of divalent cations in filament formation. *Journal of Molecular Biology*, **284**, 1611–1624.

Herbaut, C., Slomianny, C., Vernes, A. and Biguet, J. (1979) Spécialisation du réticulum endoplasmique au contact de l'intestin postérieur, chez les ovocytes de *Trichinella spiralis* (Nématode, Trichuroïde). *Annales de Parasitologie Humaine et Comparée (Paris)*, **54**, 237–242.

Herbaut, C., Slomianny, C., Vernes, A. and Biguet, J. (1980) Étude ultrastructurale de l'ovogénèse chez *Trichinella spiralis* (Nématode Trichuroïde). *Annales de Parasitologie Humaine et Comparée (Paris)*, **55**, 679–685.

Hess, R. and Poinar, G.O.J. (1986) Ultrastructure of the genital ducts and sperm behavior in the insect parasitic nematode, *Heterorhabditis bacteriophora* Poinar (Heterorhabditidae: Rhabditida). *Revue de Nématologie*, **9**, 141–152.

Hess, R.T. and Poinar, G.O.J. (1989) Sperm development in the nematode *Neoaplectana intermedia* (Steinernematidae: Rhabditida). *Journal of Submicroscopic Cytology and Pathology*, **21**, 543–555.

Hirsh, D., Oppenheim, D. and Klass, M. (1976) Development of the reproductive system of *Caenorhabditis elegans*. *Developmental Biology*, **49**, 200–219.

Holliday, L.S. and Roberts, T.M. (1995) Isolation and characterization of 30-nm protein particles present in pseudopods of the amoeboid sperm of *Ascaris suum*. *Biochemical and Biophysical Research Communications*, **208**, 1073–1079.

Hugot, J.-P. (1984) L'insémination traumatique chez les oxyures de Dermpotères [sic] et de Léporidés. Étude morphologique comparée. considérations sur la phylogénèse. *Annales de Parasitologie Humaine et Comparée (Paris)*, **59**, 379–385.

Hugot, J.-P., Bain, O. and Cassone, J. (1982) Insémination traumatique et tube de ponte chez l'Oxyure parasite du Lapin domestique. *Comptes Rendus de l'Académie des Sciences, Paris*, **294, sér III**, 707–709.

Hugot, J.-P., Bain, O. and Cassone, J. (1983) Sur le genre *Passalurus* (Oxyuridae: Nematoda) parasite de Leporidés. *Systematic Parasitology*, **5**, 305–316.

Hulinská, D. and Shaikenov, B. (1982) The ultrastructure of the reproductive system, and the oogenesis of two day-old, fertilized females of *Trichinella spiralis* and *T. nativa*. *Folia Parasitologica (Praha)*, **29**, 39–44.

Hulinská, D. and Shaikenov, B. (1983) A comparative study on the development and the structure of sperms of *Trichinella nativa*, *T. pseudospiralis* and *T. spiralis* (Nematoda). *Folia Parasitologica (Praha)*, **30**, 31–36.

Italiano, J.E., Roberts, T.M., Stewart, M. and Fontana, C.A. (1996) Reconstitution *in vitro* of the motile apparatus from the amoeboid sperm of *Ascaris* shows that filament assembly and bundling move membranes. *Cell*, **84**, 105–114.

Iwasaki, K., McCarter, J., Francis, R. and Schedl, T. (1996) emo-1, a *Caenorhabditis elegans* Sec61p gamma homologue, is required for oocyte development and ovulation. *Journal of Cell Biology*, **134**, 699–714.

Jamuar, M.P. (1966) Studies of spermiogenesis in a nematode, *Nippostrongylus brasiliensis*. *Journal of Cell Biology*, **31**, 381–396.

Jenkins, T. and Larkman, A. (1981) Spermatogenesis in a trichuroid nematode, *Trichuris muris*. II. Fine structure of primary spermatocyte and first meiotic division. *International Journal for Invertebrate Reproduction*, **3**, 257–273.

Jenkins, T., Larkman, A. and Funnell, M. (1979) Spermatogenesis in a trichuroid nematode, *Trichuris muris* — I. Fine structure of spermatogonia. *International Journal for Invertebrate Reproduction*, **1**, 371–385.

Jones, A.R., Francis, R. and Schedl, T. (1996) GLD-1 a cytoplasmic protein essential for oocyte differentiation, shows stage- and sex-specific expression during *Caenorhabditis elegans* germline development. *Developmental Biology*, **180**, 165–183.

Jones, A.R. and Schedl, T. (1995) Mutations in gld-1, a female germ cell-specific tumor suppressor gene in *Caenorhabditis elegans*, affect a conserved domain also found in Src-associated protein Sam68. *Genes & Development*, **9**, 1491–1504.

Justine, J.-L. and Jamieson, B.G.M. (1999) Nematoda. In *Progress in Male Gamete Ultrastructure and Phylogeny*, edited by B.G.M. Jamieson.

Reproductive Biology of Invertebrates, edited by K.G. Adiyodi and G. Adiyodi, Volume IX, Part B. Oxford and IBH Publishing Co. Pvt. Lmd, New Delhi, Calcutta.

Kadyk, L.C. and Kimble, J. (1998) Genetic regulation of entry into meiosis in *Caenorhabditis elegans*. *Development*, **125**, 1803–1813.

Kaiser, H. (1991) Terrestrial and semiterrestrial Mermithidae. In *Manual of agricultural nematology*, edited by W.R. Nickle, pp.899–965. New York: Dekker.

Kimble, J. and Ward, S. (1988) Germ-line development and fertilization. In *The Nematode Caenorhabditis elegans*, edited by W.B. Wood, pp. 191–313. Cold Spring Harbor: Cold Spring Harbor Laboratory.

King, K.L., Stewart, M. and Roberts, T.M. (1994) Supramolecular assemblies of the *Ascaris suum* major sperm protein (MSP) associated with amoeboid cell motility. *Journal of Cell Science*, **107**, 2941–2949.

King, K.L., Stewart, M., Roberts, T.M. and Seavy, M. (1992) Structure and macromolecular assembly of two isoforms of the major sperm protein (MSP) from the amoeboid sperm of the nematode, *Ascaris suum*. *Journal of Cell Science*, **101**, 847–857.

Klass, M., Ammons, D. and Ward, S. (1988) Conservation of the 5' flanking sequence of transcribed members of the *Caenorhabditis elegans* major sperm protein gene family. *Journal of Molecular Biology*, **199**, 15–22.

Klass, M., Dow, B. and Herndon, M. (1982) Cell-specific transcriptional regulation of the major sperm protein in *Caenorhabditis elegans*. *Developmental Biology*, **93**, 152–164.

Klass, M.R. and Hirsh, D. (1981) Sperm isolation and biochemical analysis of the major sperm protein from *Caenorhabditis elegans*. *Developmental Biology*, **84**, 299–312.

Klass, M.R., Kinsley, S. and Lopez, L.C. (1984) Isolation and characterization of a sperm-specific gene family in the nematode *Caenorhabditis elegans*. *Molecular Cell Biology*, **4**, 529–537.

Kruger, J.C.d.W. (1991) Ultrastructure of sperm development in the plant parasitic nematode *Xiphinema theresiae*. *Journal of Morphology*, **210**, 163–174.

LaMunyon, C.W. and Ward, S. (1994) Assessing the viability of mutant and manipulated sperm by artificial insemination of *Caenorhabditis elegans*. *Genetics*, **138**, 689–692.

LaMunyon, C.W. and Ward, S. (1995) Sperm precedence in a hermaphroditic nematode (*Caenorhabditis elegans*) is due to competitive superiority of male sperm. *Experientia*, **51**, 817–823.

LaMunyon, C.W. and Ward, S. (1998) Larger sperm outcompete smaller sperm in the nematode *Caenorhabditis elegans*. *Proceedings of the Royal Society of London, B*, **265**, 1997–2002.

LaMunyon, C.W. and Ward, S. (1999) Evolution of sperm size in nematodes: sperm competition favours larger sperm. *Proceedings of the Royal Society of London, B*, **266**, 263–267.

Lee, C.-C. (1975) *Dirofilaria immitis*: ultrastructural aspects of oocyte development and zygote formation. *Experimental Parasitology*, **37**, 449–468.

Lee, D.L. (1971) The structure and development of the spermatozoon of *Heterakis gallinarum* (Nematoda). *Journal of Zoology (London)*, **164**, 181–187.

Lee, D.L. and Anya, A.O. (1967) The structure and development of the spermatozoon of *Aspiculuris tetraptera* (Nematoda). *Journal of Cell Science*, **2**, 537–544.

Lee, D.L. and Lestan, P. (1971) Oogenesis and egg shell formation in *Heterakis gallinarum* (Nematoda). *Journal of Zoology*, **164**, 189–196.

L'Hernault, S.W. and Arduengo, P.M. (1992) Mutation of a putative sperm membrane protein in *Caenorhabditis elegans* prevents sperm differentiation but not its associated meiotic divisions. *Journal of Cell Biology*, **119**, 55–68.

L'Hernault, S.W., Benian, G.M. and Emmons, R.B. (1993) Genetic and molecular characterization of the *Caenorhabditis elegans* spermatogenesis-defective gene spe-17. *Genetics*, **134**, 769–80.

L'Hernault, S.W., Shakes, D.C. and Ward, S. (1988) Developmental genetics of chromosome I spermatogenesis-defective mutants in the nematode *Caenorhabditis elegans*. *Genetics*, **120**, 435–452.

Lo, M. (1993) Étude structurale et ultrastructurale du mâle et de son système reproducteur chez les Tylenchina (Nemata: Diplogasteria). *Mémoire de Diplôme d'Études Approfondies de Biologie Animale*, University Cheikh Anta Diop, Faculté des Sciences et Techniques, Dakar, Sénégal.

Machaca, K., DeFelice, L.J. and L'Hernault, S.W. (1996) A novel chloride channel localizes to *Caenorhabditis elegans* spermatids and chloride channel blockers induce spermatid differentiation. *Developmental Biology*, **176**, 1–16.

Machaca, K. and L'Hernault, S.W. (1997) The *Caenorhabditis elegans* spe-5 gene is required for morphogenesis of a sperm-specific organelle and is associated with an inherent cold-sensitive phenotype. *Genetics*, **146**, 567–581.

MacKinnon, B.M. (1987) An ultrastructural and histochemical study of oogenesis in the trichostrongylid nematode *Heligmosomoides polygyrus. Journal of Parasitology*, **73**, 390–399.

MacKinnon, B.M. and Burt, M.D.B. (1992) Functional morphology of the female reproductive tract of *Pseudoterranova decipiens* (Nematoda) raised *in vivo* and *in vitro. Zoomorphology*, **112**, 237–245.

MacKinnon, B.M. and Lee, D.L. (1988) Age-related changes in *Heligmosomoides polygyrus* (Nematoda): neutral lipid content in developing oocytes. *Canadian Journal of Zoology*, **66**, 2791–2796.

Maeda, T. (1968) Electron microscopic studies on spermatogenesis in *Dirofilaria immitis. Kagoshima Igaku Zasshi*, **20**, 146–165. [Not seen in original].

Maeda, T., Harada, R., Nakashima, A., Sadakata, Y., Ando, M., Yonamine, K., Otsuji, Y. and Sato, H. (1970). Electron microscopic studies on spermatogensis in *Dirofilaria immitis*. In *Recent Advances in Researches on Filariasis and Schistosomiasis in Japan*, edited by M. Sasa, pp. 73–79. Tokyo: University of Tokyo Press. [Not seen in original].

Mansir, A. and Justine, J.-L. (1995) Centrioles with ten singlets in spermatozoa of the parasitic nematode *Heligmosomoides polygyrus*. In *Advances in spermatozoal phylogeny and taxonomy* edited by B.G.M. Jamieson, J. Ausio, and J.-L. Justine, *Mémoires du Muséum National d'Histoire Naturelle*, **166**, 119–128.

Mansir, A. and Justine, J.-L. (1996) Actin and major sperm protein in spermatids and spermatozoa of the parasitic nematode *Heligmosomoides polygyrus. Molecular Reproduction and Development*, **45**, 332–341.

Mansir, A. and Justine, J.-L. (1998) The microtubular system and posttranslationally modified tubulin during spermatogenesis in a parasitic nematode with amoeboid and aflagellate spermatozoa. *Molecular Reproduction and Development*, **49**, 150–167.

Mansir, A. and Justine, J.-L. (1999) Actin and major sperm protein in spermatozoa of a nematode, *Graphidium strigosum* (Strongylida: Trichostrongylidae). *Folia Parasitologica*, **46**, 47–51.

Mansir, A., Noury-Sraïri, N., Cabaret, J., Kerboeuf, D., Escalier, D., Durette-Desset, M.-C. *et al.* (1997) Actin in spermatids and spermatozoa of *Teladorsagia circumcincta* and *Trichostrongylus colubriformis* (Nematoda, Trichostrongylida). *Parasite*, **4**, 373–376.

Marcaillou, C. and Szöllösi, A. (1980) The 'blood-testis' barrier in a nematode and a fish: a generalizable concept. *Journal of Ultrastructure Research*, **70**, 128–136.

Martin, J. and Lee, D.L. (1980) Observations on the structure of the male reproductive system and spermatogenesis of *Nematodirus battus. Parasitology*, **81**, 579–586.

McCarter, J., Bartlett, B., Dang, T. and Schedl, T. (1997) Soma-germ cell interactions in *Caenorhabditis elegans*: multiple events of hermaphrodite germline development require the somatic sheath and spermathecal lineages. *Developmental Biology*, **181**, 121–143.

McCarter, J., Bartlett, B., Dang, T. and Schedl, T. (1999) On the control of oocyte meiotic maturation and ovulation in *Caenorhabditis elegans. Developmental Biology*, **205**, 111–128.

McClure, M.A. and Bird, A.F. (1976) The tylenchid (Nematoda) egg shell: formation of the egg shell in *Meloidogyne javanica. Parasitology*, **72**, 29–39.

McLaren, A. (1993) Germ cell sex determination. *Seminars in Developmental Biology*, **4**, 171–177.

McLaren, D.J. (1973a) The structure and development of the spermatozoon of *Dipetalonema viteae* (Nematoda: Filarioidea). *Parasitology*, **66**, 447–463.

McLaren, D.J. (1973b) Oogenesis and fertilization in *Dipetalonema viteae* (Nematoda: Filarioidea). *Parasitology*, **66**, 465–472.

Minniti, A.N., Sadler, C. and Ward, S. (1996) Genetic and molecular analysis of *spe-27*, a gene required for spermiogenesis in *Caenorhabditis elegans* hermaphrodites. *Genetics*, **143**, 213–223.

Nakashima, A. (1972) Electron microscopy of spermatogenesis in *Trichuris vulpis* (Floelich, 1785). *Acta Medicinensis Universitatis Kagoshima*, **14**, 211–228.

Nance, J., Minniti, A.N., Sadler, C. and Ward, S. (1999) *spe-12* encodes a sperm cell surface protein that promotes spermiogenesis in *Caenorhabditis elegans. Genetics*, **152**, 209–220.

Neill, B.W. and Wright, K.A. (1973) Spermatogenesis in the hologonic testis of the trichuroid nematode, *Capillaria hepatica* (Bancroft, 1893). *Journal of Ultrastructure Research*, **44**, 210–234.

Nelson, G.A., Roberts, T.M. and Ward, S. (1982) *Caenorhabditis elegans* spermatozoan locomotion: amoeboid movement with almost no actin. *Journal of Cell Biology*, **92**, 121–131.

Nelson, G.A. and Ward, S. (1981) Amoeboid motility and actin in *Ascaris lumbricoides* sperm. *Experimental Cell Research*, **131**, 149–160.

Nicholas, W.L. and Stewart, A.C. (1997) Ultrastructure of *Gonionchus australis* (Xyalida, Nematoda). *Journal of Nematology*, **29**, 133–143.

Noury-Sraïri, N. (1994) Cytosquelette des spermatozoïdes de Nématodes libres et parasites (étude ultrastructurale et immunocytochimique). *Thèse d'État*, Université Cadi Ayyad, Marrakech, Morocco.

Noury-Sraïri, N., Gourbault, N. and Justine, J.-L. (1993) The development and evolution of actin-containing organelles during spermiogenesis of a primitive nematode. *Biology of the Cell*, **79**, 231–241.

Novitski, C.E., Brown, S., Chen, R., Corner, A.S., Atkinson, H.J. and McPherson, M.J. (1993) Major sperm protein genes from *Globodera rostochiensis. Journal of Nematology*, **25**, 548–554.

Panijel, J. (1950) Recherches sur la nature et la signification de la protéine gram du gamète mâle d'*Ascaris megalocephala. Biochimica Biophysica Acta*, **6**, 79–93.

Parshad, V.R. and Guraya, S.S. (1982) Histochemical observations on lipid changes during oogenesis in the poultry nematode, *Ascaridia galli. International Journal of Invertebrate Reproduction*, **4**, 337–341.

Pasternak, J. and Samoiloff, M.R. (1972) Cytoplasmic organelles present during spermatogenesis in the free-living nematode *Panagrellus silusiae. Canadian Journal of Zoology*, **50**, 147–151.

Pavalko, F.M. and Roberts, T.M. (1987) *Caenorhabditis elegans* spermatozoa assemble proteins onto the surface at the tips of pseudopodial projections. *Cell Motility and the Cytoskeleton*, **7**, 169–177.

Phillipson, R.F. (1969) Reproduction of *Nippostrongylus brasiliensis* in the rat intestine. *Parasitology*, **59**, 961–971.

Poinar, G.O.J. and Hess, R. (1985) Spermatogenesis in the insect-parasitic nematode, *Heterorhabditis bacteriophora* Poinar (Heterorhabditidae: Rhabditida). *Revue de Nématologie*, **8**, 357–367.

Poinar, G.O.J. and Hess-Poinar, R.T. (1993) The fine structure of *Gastromermis* sp. (Nematoda: Mermithidae) sperm. *Journal of Submicroscopic Cytology and Pathology*, **25**, 417–431.

Preston, C.M. and Jenkins, T. (1983) Ultrastructural studies of early stages of oogenesis in a trichuroid nematode, *Trichuris muris. International Journal of Invertebrate Reproduction*, **6**, 77–91.

Preston, C.M. and Jenkins, T. (1984) *Trichuris muris*: structure and formation of the egg-shell. *Parasitology*, **89**, 263–273.

Roberts, T.M. and King, K.L. (1991) Centripetal flow and directed reassembly of the Major Sperm Protein (MSP) cytoskeleton in the amoeboid sperm of the nematode, *Ascaris suum. Cell Motility and the Cytoskeleton*, **20**, 228–241.

Roberts, T.M., Pavalko, F.M. and Ward, S. (1986) Membrane and cytoplasmic proteins transported in the same organelle complex during nematode spermatogenesis. *Journal of Cell Biology*, **102**, 1787–1796.

Roberts, T.M., Salmon, E.D. and Stewart, M. (1998) Hydrostatic pressure shows that lamellipodial motility in *Ascaris* requires membrane-associated major sperm protein filament nucleation and elongation. *Journal of Cell Biology*, **140**, 367–375.

Roberts, T.M. and Stewart, M. (1995) Nematode sperm locomotion. *Current Opinion in Cell Biology*, **7**, 13–17.

Roberts, T.M. and Stewart, M. (1997) Nematode sperm: amoeboid movement without actin. *Trends in Cell Biology*, **7**, 368–373.

Roberts, T.M. and Streitmatter, G. (1984) Membrane-substrate contact under the spermatozoon of *Caenorhabditis elegans*, a crawling cell that lacks filamentous actin. *Journal of Cell Science*, **69**, 117–126.

Roberts, T.M. and Ward, S. (1982a) Centripetal flow of pseudopodial surface components could propel the amoeboid movement of *Caenorhabditis elegans* spermatozoa. *Journal of Cell Biology*, **92**, 132–138.

Roberts, T.M. and Ward, S.1. (1982b) Membrane flow during nematode spermiogenesis. *Journal of Cell Biology*, **92**, 113–120.

Rose, K.L., Winfrey, V.P., Hoffman, L.H., Hall, D.H., Furuta, T. and Greenstein, D. (1997) The POU gene *ceh-18* promotes gonadal sheath cell differentiation and function required for meiotic maturation and ovulation in *Caenorhabditis elegans. Developmental Biology*, **192**, 59–77.

Roush, W. (1996) Sperm protein makes its mark upon the worm embryo. *Science*, **271**, 33.

Royal, D., Royal, M., Italiano, J., Roberts, T. and Soll, D.R. (1995) In *Ascaris* sperm pseudopods, MSP fibers move proximally at a constant rate regardless of the forward rate of cellular translocation. *Cell Motility and the Cytoskeleton*, **31**, 241–253.

Sadler, P.L. and Shakes, D.C. (2000) Anucleate *Caenorhabditis elegans* sperm can crawl, fertilize oocytes and direct anterior-posterior polarization of the 1-cell embryo. *Development*, **127**, 355–366.

Schnieder, T. (1993) The diagnostic antigen encoded by gene fragment Dv3-14: a major sperm protein of *Dictyocaulus viviparus. International Journal for Parasitology*, **23**, 383–389.

Scott, A.L. (1996) Nematode sperm. *Parasitology Today*, **12**, 425–430.

Scott, A.L., Dinman, J., Sussman, D.J. and Ward, S. (1989a) Major sperm protein and actin genes in free-living and parasitic nematodes. *Parasitology*, **98**, 471–478.

Scott, A.L., Dinman, J., Sussman, D.J., Yenbutr, P. and Ward, S. (1989b) Major sperm protein genes from *Onchocerca volvulus*. *Molecular and Biochemical Parasitology*, **36**, 119–126.

Sepsenwol, S., Ris, H. and Roberts, T.M. (1989) A unique cytoskeleton associated with crawling in the amoeboid sperm of the nematode, *Ascaris suum. Journal of Cell Biology*, **108**, 55–66.

Sepsenwol, S. and Taft, S.J. (1990) *In vitro* induction of crawling in the amoeboid sperm of the nematode parasite, *Ascaris suum. Cell Motility and the Cytoskeleton*, **15**, 99–110.

Setterquist, R.A. and Fox, G.E. (1995) *Dictyocaulus viviparus*: nucleotide sequence of Dv3–14. *International Journal for Parasitology*, **25**, 137–138.

Setterquist, R.A., Smith, G.K., Jones, R. and Fox, G.E. (1996) Diagnostic probes targeting the major sperm protein gene that may be useful in the molecular identification of nematodes. *Journal of Nematology*, **28**, 414–421.

Shakes, D.C. and Ward, S. (1989a) Initiation of spermiogenesis in *C. elegans*: a pharmacological and genetic analysis. *Developmental Biology*, **134**,

Shakes, D.C. and Ward, S. (1989b) Mutations that disrupt the morphogenesis and localization of a sperm-specific organelle in *Caenorhabditis elegans. Developmental Biology*, **134**, 307–316.

Shepherd, A.M. (1981) Interpretation of sperm development in nematodes. *Nematologica*, **27**, 122–125.

Shepherd, A.M. and Clark, S.A. (1976) Spermatogenesis and the ultrastructure of sperm and of the reproductive tract of *Aphelenchoides blastophthorus* (Nematoda: Tylenchida, Aphelenchina). *Nematologica*, **22**, 1–9.

Shepherd, A.M. and Clark, S.A. (1983) Spermatogenesis and sperm structure in some *Meloidogyne* species (Heteroderoidea, Meloidogynidae) and a comparison with those in some cyst nematodes (Heteroderoidea, Heteroderidae). *Revue de Nématologie*, **6**, 17–32.

Shepherd, A.M., Clark, S.A. and Kempton, A. (1973) Spermatogenesis and sperm ultrastructure in some cyst nematodes, *Heterodera* spp. *Nematologica*, **19**, 551–560.

Shim, Y.H. (1999) elt-1, a gene encoding a *Caenorhabditis elegans* GATA transcription factor, is highly expressed in the germ lines with msp genes as the potential targets. *Molecular Cells*, **9**, 535–541.

Simpson, C.F., Carlisle, J.W. and Conti, J.A. (1984) *Tetrameres columbicola* (Nematoda: Spiruridae) infection of pigeons: ultrastructure of the gravid female in glands of the proventriculus. *American Journal of Veterinary Research*, **45**, 1184–1192.

Singson, A., Mercer, K.B. and L'Hernault, S.W. (1998) The *C. elegans* spe-9 gene encodes a sperm transmembrane protein that contains EGF-like repeats and is required for fertilization. *Cell*, **93**, 71–79.

Skehel, P.A., Martin, K.C., Kandel, E.R. and Bartsch, D. (1995) A VAMP-binding protein from *Aplysia* required for neurotransmitter release. *Science*, **269**, 1580–1583.

Slomianny, C., Herbaut, C., Vernes, A. and Biguet, J. (1981) Étude ultrastructurale de la spermatogenèse chez *Trichinella spiralis* Owen 1835 (Nématode trichuroïde). *Zeitschrift für Parasitenkunde*, **64**, 207–215.

Smith, H.E. and Ward, S. (1998) Identification of protein-protein interactions of the major sperm protein (MSP) of *Caenorhabditis elegans. Journal of Molecular Biology*, **279**, 605–619.

Sommerville, R.I. and Weinstein, P.P. (1964) Reproductive behavior of *Nematospiroides dubius* in vivo and in vitro. *Journal of Parasitology*, **50**, 401–409.

Sprent, J.F.A. (1962) The evolution of the Ascaridoidea. *Journal of Parasitology*, **48**, 818–824.

Starck, J. (1984) Synthesis of oogenesis specific proteins in *Caenorhabditis elegans*: an approach to the study of vitellogenesis in a nematode. *International Journal of Invertebrate Reproduction and Development*, **7**, 149–160.

Stewart, M., King, K.L. and Roberts, T.M. (1994) The motile major sperm protein (MSP) of *Ascaris suum* forms filaments constructed from two helical subfilaments. *Journal of Molecular Biology*, **243**, 60–71.

Strome, S. (1986) Fluorescence visualization of the distribution of microfilaments in gonads and early embryos of the nematode *Caenorhabditis elegans. Journal of Cell Biology*, **103**, 2241–2252.

Takahashi, Y., Goto, C. and Kita, K.K. (1994) Ultrastructural study of *Trichinella spiralis* with emphasis on adult male reproductive organs. *Journal of Helminthology*, **68**, 353–358.

Takahashi, Y., Goto, C. and Kita, K.K. (1995) Ultrastructural study of *Trichinella spiralis* with emphasis on adult female reproductive organs. *Journal of Helminthology*, **69**, 247–252.

Takahashi, Y., Homan, W. and Pak Leong Lim, P.L. (1993) Immunocytochemical localization of the phosphorylcholine-associated antigen in *Trichinella spiralis. Journal of Parasitology*, **79**, 604–609.

Theriot, J.A. (1996) Worm sperm and advances in cell locomotion. *Cell*, **84**, 1–4.

Turpeenniemi, T.A. (1998) Ultrastructure of spermatozoa in the nematode *Halalaimus dimorphus* (Nemata, Oxystominidae). *Journal of Nematology*, **30**, 391–403.

Tuzet, O. and Sanchez, S. (1951) Sur l'acrosome du spermatozoïde de l'Ascaris. *Archives de Zoologie Expérimentale et Générale*, **88**, 142–148.

Ugwunna, S.C. (1986) The origin and some functions of smooth endoplasmic reticulum in *Ancylostoma caninum* sperm cells. *International Journal for Parasitology*, **16**, 289–296.

Ugwunna, S.C. (1990) Extrusion of the residual body in spermatids of *Ancylostoma caninum* (Nematoda, Strongyloidea). *Journal of Morphology*, **203**, 283–292.

Ugwunna, S.C. and Foor, W.E. (1982a) Development and fate of the membranous organelles in spermatozoa of *Ancylostoma caninum. Journal of Parasitology*, **68**, 834–844.

Ugwunna, S.C. and Foor, W.E. (1982b) The function of microtubules during spermatogenesis of *Ancylostoma caninum. Journal of Parasitology*, **68**, 817–823.

Van Beneden, E. and Julin, C. (1884) La spermatogénèse chez l'ascaride mégalocéphale. *Bulletin de l'Académie Royale de Belgique*, **7**, 312–342.

Van de Velde, M.C. and Coomans, A. (1988) Electron microscopy of the germ cells and the ovarian wall in *Xiphinema* (Nematoda). *Tissue & Cell*, **20**, 881–890.

Van de Velde, M.C., Coomans, A., Van Ranst, L., Kruger, J.C.D. and Claeys, M. (1991) Ultrastructure of sperm cells in the female gonoduct of *Xiphinema. Tissue & Cell*, **23**, 881–891.

Van Voorhies, W.A. (1992) Production of sperm reduces nematode lifespan. *Nature*, **360**, 456–458.

Van Waerebeke, D. (1984) *Rhigonema madecassum* n. sp. (Rhigonematidae; Nematoda), parasite de Diplopode à Madagascar: description et étude de la spermiogenèse. *Revue de Nématologie*, **7**, 271–276.

Van Waerebeke, D. (1985) Trois nouvelles espèces de *Glomerinema* Van Waerebeke, 1985 (Rhigonematidae, Nematoda) parasites de Sphaeroteroidea (Glomerida, Diplopoda) à Madagascar. *Revue de Nématologie*, **8**, 229–239.

Van Waerebeke, D., Noury-Sraïri, N. and Justine, J.-L. (1990) Spermatozoa of rhigonematid nematodes: morphology of 25 species and ultrastructure of *Rhigonema madecassum. International Journal for Parasitology*, **20**, 779–784.

Varkey, J.P., Jansma, P.L., Minniti, A.N. and Ward, S. (1993) The *Caenorhabditis elegans* spe-6 gene is required for major sperm protein assembly and shows second site non-complementation with an unlinked deficiency. *Genetics*, **133**, 79–86.

Varkey, J.P., Muhlrad, P.J., Minniti, A.N., Do, B. and Ward, S. (1995) The *Caenorhabditis elegans* spe-26 gene is necessary to form spermatids and encodes a protein similar to the actin-associated proteins kelch and scruin. *Genes and Development*, **9**, 1074–1086.

Ward, S. (1986) Asymmetric localization of gene products during the development of *Caenorhabditis elegans* spermatozoa. In *Gametogenesis and the early embryo*, edited by J.G. Gall, pp. 55–75. New York: Alan R. Liss.

Ward, S., Argon, Y. and Nelson, G.A. (1981) Sperm morphogenesis in wild-type and fertilization-defective mutants of *Caenorhabditis elegans. Journal of Cell Biology*, **91**, 26–44.

Ward, S., Burke, D.J., Sulston, J.E., Coulson, A.R., Albertson, D.G., Ammons, D. *et al.* (1988) Genomic organization of major sperm protein genes and pseudogenes in the nematode *Caenorhabditis elegans. Journal of Molecular Biology*, **199**, 1–13.

Ward, S. and Carrel, J.S. (1979) Fertilization and sperm competition in the nematode *Caenorhabditis elegans. Developmental Biology*, **73**, 304–321.

Ward, S., Hogan, E. and Nelson, G.A. (1983) The initiation of spermiogenesis in the nematode, *Caenorhabditis elegans. Developmental Biology*, **98**, 70–79.

Ward, S. and Klass, M. (1982) The location of the major sperm protein in *Caenorhabditis elegans* sperm and spermatocytes. *Developmental Biology*, **92**, 203–208.

Ward, S. and Miwa, J. (1978) Characterization of temperature-sensitive, fertilization-defective mutants of the nematode *Caenorhabditis elegans. Genetics*, **88**, 285–303.

Weber, P. (1987) The fine structure of the female reproductive tract of adult *Loa loa. International Journal for Parasitology*, **17**, 927–934.

Wharton, D.A. (1979) Oogenesis and egg-shell formation in *Aspiculuris tetraptera* Schulz (Nematoda: Oxyuroidea). *Parasitology*, **78**, 131–143.

Wilson, R., Ainscough, R., Anderson, K., Baynes, C., Berks, M., Bonfield, J. et al. (1994) 2.2 Mb of contiguous nucleotide sequence from chromosome III of *C. elegans*. *Nature*, **368**, 32–38.

Winter, C.E. (1992) The yolk polypeptides of a free-living rhabditid nematode. *Comparative Biochemistry and Physiology*, **103B**, 189–196.

Wolf, N., Hirsh, D. and McIntosh, J.R. (1978) Spermatogenesis in males of the free-living nematode, *Caenorhabditis elegans*. *Journal of Ultrastructure Research*, **63**, 155–169.

Wright, E.J., Cox, G. and Dwarte, D. (1980) Capacitation-like changes in the amoeboid sperm of a nematode. *Micron*, **11**, 377–378.

Wright, E.J. and Sommerville, R.I. (1977) Movement of a non-flagellate spermatozoon: a study of the male gamete of *Nematospiroides dubius* (Nematoda). *International Journal for Parasitology*, **7**, 353–359.

Wright, E.J. and Sommerville, R.I. (1984) Postinsemination changes in the amoeboid sperm of a nematode, *Nippostrongylus brasiliensis*. *Gamete Research*, **10**, 397–413.

Wright, E.J. and Sommerville, R.I. (1985a) Spermatogenesis in a nematode *Nippostrongylus brasiliensis*. *International Journal for Parasitology*, **15**, 283–299.

Wright, E.J. and Sommerville, R.I. (1985b) Structure and development of the spermatozoon of the parasitic nematode, *Nematospiroides dubius*. *Parasitology*, **90**, 179–192.

Wright, K.A. (1991) Nematoda. In *Microscopic Anatomy of Invertebrates*, edited by F.W. Harrison and E.E. Ruppert, pp. 111–195. New York: Wiley-Liss.

Wright, K.A., Hope, W.D. and Jones, N.O. (1973) The ultrastructure of the sperm of *Deontostoma californicum*, a free living marine nematode. *Proceedings of the Helminthological Society of Washington*, **40**, 30–36.

Wu, Y.-J. and Foor, W.E. (1980) *Ascaris* oocytes: ultrastructural and immunocytochemical changes during passage through the oviduct. *Journal of Parasitology*, **66**, 439–447.

Xu, D., Nagano, I. and Takahashi, Y. (1997) Electron microscopic observations on the normal development of *Trichinella spiralis* from muscle larvae to adult worms in BALB/c mice with emphasis on the body wall, genital organs and gastrointestinal organs. *Journal of Electron Microscopy*, **46**, 347–352.

Yushin, V.V. and Malakhov, V.V. (1994) Ultrastructure of sperm cells in the female gonoduct of free-living marine nematodes from genus *Enoplus* (Nematoda: Enoplida). *Fundamental and Applied Nematology*, **17**, 513–519.

Yushin, V.V. and Malakhov, V.V. (1998) Ultrastructure of sperm development in the free-living marine nematode *Enoplus anisospiculus* (Enoplida: Enoplidae). *Fundamental and Applied Nematology*, **21**, 213–225.

5. Embryology, Developmental Biology and the Genome

Ian A. Hope

School of Biology, The University of Leeds, Leeds, LS2 9JT, UK

Keywords

Embryology, Developmental Biology, Developmental Mechanisms, Genome, Genetics, *Caenorhabditis elegans*.

Introduction

Nematodes have a huge significance in our investigations of animal development. The transparency and availability of nematode embryos made them attractive subjects to study and classically, embryonic development was described for many nematode species, leading to several key concepts in developmental biology. Genetic tractability has made the nematode *Caenorhabditis elegans* one of the handful of model systems now dominating modern biological research. Molecular analyses have revealed just how extensively developmental mechanisms are evolutionarily conserved confirming the validity of *C. elegans* as an appropriate model through which to understand development across the animal kingdom. Completion of the sequencing of the *C. elegans* genome, a first for a multicellular organism, has emphasised the pre-eminence of this model system and stimulated molecular analysis in other nematode species.

Embryology

C. elegans is the only metazoan for which development has been described completely at the cellular level (Sulston and Horvitz, 1977, Sulston *et al.*, 1983; Deppe *et al.*, 1978, Kimble and Hirsh, 1979). However, descriptions of embryonic development for other species across the nematode phylum, to various levels of detail and accuracy, are scattered through the scientific literature over more than a century. It is perhaps of value to compare these accounts, both to each other and to the standard of *C. elegans*, with consideration to the current view of nematode evolutionary relationships as they have emerged, recently, from molecular analysis (Figure 5.1) (Blaxter *et al.*, 1998). Indeed, these accounts do appear to fit better with this current view than with traditional nematode phylogenetic trees (Voronov, Panchin and Spiridinov, 1998).

Clade V

Caenorhabditis elegans

Development is initiated by fertilisation of the mature oocyte in the spermatheca of the adult hermaphrodite. The sperm may have been produced either by the hermaphrodite herself, from the first 300 or so germ line cells to mature, or by a male, with transfer to the hermaphrodite's gonad during mating. The male sperm have an advantage over hermaphrodite sperm for oocyte fertilisation, although the basis of this advantage is not known. In well-fed hermaphrodites, embryos develop for 2 or 3 hours before laying but are retained, as long as until hatching, in adverse conditions such as starvation.

Within an hour of fertilisation, at $20°C$, the meiotic divisions of the oocyte are completed, the male and female pronuclei meet and fuse, the egg shell is produced and the first mitotic cell division occurs. A delay of meiosis of the ovum until after fertilisation and the absence of a clear animal vegetal axis, with yolk evenly distributed in the egg, are typical for nematodes (Malakhov, 1994). The first cell division is asymmetric, producing an anterior cell, AB, approximately twice the volume of the posterior cell, P_1. AB divides 18 minutes later, to produce two daughters of equal size. It appears as if AB attempts to divide perpendicularly to the long axis of the egg while the physical confines of the egg shell forces the spindle to adopt an oblique angle. P_1 divides slightly later than AB, but along the long axis of the egg, to generate a smaller, posterior cell, P_2, and a larger cell, EMS. The skewing of the AB division apparently causes EMS to take up, what is thereby defined as, the ventral position, with the four cell stage arranged as a rhombus, all four cells in the same plane (Figure 5.2).

The AB daughters divide next, 34 minutes after the first cleavage, the division axes being parallel to each other, across the egg, perpendicular to both the two previous divisions. Once again the confines of the egg shell apparently causes the spindles to be twisted from being completely perpendicular. The four AB granddaughters are of equal size but the two on the left are displaced slightly towards the anterior as compared to the two on the right. The six cell embryo then has a handedness that fixes the handedness of all subsequent development.

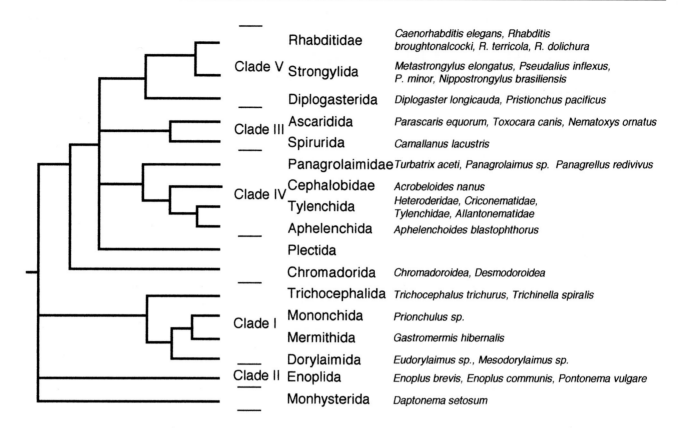

Rhabditidae — *Caenorhabditis elegans, Rhabditis broughtonalcocki, R. terricola, R. dolichura*

Clade V Strongylida — *Metastrongylus elongatus, Pseudalius inflexus, P. minor, Nippostrongylus brasiliensis*

Diplogasterida — *Diplogaster longicauda, Pristionchus pacificus*

Clade III Ascaridida — *Parascaris equorum, Toxocara canis, Nematoxys ornatus*

Spirurida — *Camallanus lacustris*

Panagrolaimidae — *Turbatrix aceti, Panagrolaimus sp. Panagrellus redivivus*

Clade IV Cephalobidae — *Acrobeloides nanus*

Tylenchida — *Heteroderidae, Criconematidae, Tylenchidae, Allantonematidae*

Aphelenchida — *Aphelenchoides blastophthorus*

Plectida

Chromadorida — *Chromadoroidea, Desmodoroidea*

Trichocephalida — *Trichocephalus trichurus, Trichinella spiralis*

Clade I Mononchida — *Prionchulus sp.*

Mermithida — *Gastromermis hibernalis*

Dorylaimida — *Eudorylaimus sp., Mesodorylaimus sp.*

Clade II Enoplida — *Enoplus brevis, Enoplus communis, Pontonema vulgare*

Monhysterida — *Daptonema setosum*

Figure 5.1. A current view of nematode phylogenetic relationships as deduced from molecular analyses (Blaxter *et al.*, 1998). This figure reflects the organisation of this chapter and the species/families referred to are listed on the right.

EMS divides, giving MS and E and producing a seven cell embryo, at 40 minutes of development. MS is anterior to and slightly larger than E. Division of P_2 at 44 minutes generates C and the smaller P_3 which together occupy the posterior surface of the eight cell embryo, C lying dorsal to P_3. There are subsequent divisions in the AB cell lineage, of MS, of E and of C and in the AB cell lineage again, before P_3 divides at 78 minutes to give D and P_4. Formation of D, dorsal to P_4, completes the generation of the founder cells. Cell divisions occur approximately synchronously within each founder cell lineage with the regular divisions in the founder cell lineages ceasing around 300 minutes after the first cell division of the embryo.

The almost completely invariant cell lineage has been followed entirely such that each cell born during *C. elegans* development has a unique identity and all have been given unique names (Sulston and Horvitz, 1977; Sulston *et al.*, 1983; Deppe *et al.*, 1978; Kimble and Hirsh, 1979). Key cells, such as the founder cells, have names of one or two capital letters or numbers. Progeny from these key cells are named with the key cell's name followed by a string of lower case letters describing the direction of the cell divisions through which the cell was generated from that key cell [a (anterior), p (posterior), d (dorsal), v (ventral), l (left) or r (right)]. The entire cell lineage, embryonic and postembryonic, with details of times and orientation of each cell division and each cell's fate, has been portrayed in a cell lineage

diagram and provides a crucial term of reference for developmental research on *C. elegans* (Sulston, Horvitz and Kimble, 1988). This diagram also indicates the few examples of variance in the cell lineage.

Gastrulation occurs during the middle third of the cell proliferation stage. First the daughters of E move from the ventral surface into the interior where they will give rise to the intestine. Then MS descendants, which form part of the pharynx and the myoderm, and the germ-line progenitor, P_4, follow the E lineage through the ventral cleft. Later cells of the D lineage and those descendants of C that also give rise to body wall muscle complete the cell movements that constitute gastrulation. Most of the so-called founder cells give rise to various cell types and not a single cell type as envisaged in earlier studies of embryonic development in other nematode species (see below).

Morphogenesis occurs after the cell proliferation stage, from 300 to 550 minutes after first cleavage. The cells change shape as a result of circumferential constriction around the whole embryo to bring about elongation into the worm form (Priess and Hirsh, 1986). During this stage the cells, including muscle cells and nerve cells, differentiate and the embryo begins to writhe within the confines of the egg shell. Developmental events during the final third of embryogenesis are not as obvious as at earlier stages, involving more localised maturation events, and hatching occurs at 800 minutes.

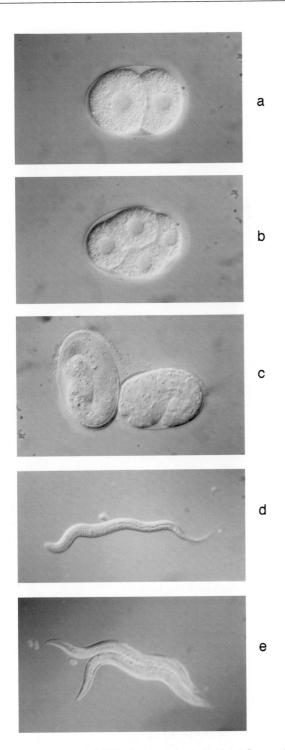

The hatchling, L1 larva, has no specialised surface structures. The cuticle runs smoothly from the oral opening at the tip of the blunt head to the finely pointed tail, the rectal opening being in a ventral position close to the tail. The underlying epidermis is a single cell layer, mostly syncytial, with 85 nuclei. This is separated from the intestinal tract, which runs directly between the oral and rectal openings, by the fluid-filled pseudocoelom. The wall of the intestine is also a single cell layer, containing 20 uninucleate cells. The more complex, tri-radiate pharynx runs about a third of the length of the L1, from the oral opening to the intestine, and contains muscle cells (20 cells, 37 nuclei), neurons (20), epithelial cells (9), gland cells (5) and marginal cells (7 cells, 9 nuclei). The body wall musculature lies beneath the epidermis and consists of 81 uninucleate cells. The nervous system, concentrated mainly in a ring around the pharynx and in the ventral nerve cord, consists of 202 neurons in the hermaphrodite and 204 in the male, plus 40 supporting cells and 6 glial cells. There are in addition 4 coelomocytes, 4 cells of the excretory system, a head mesodermal cell, one mesoblast, 30 further cells in the alimentary tract and 4 gonadal cells. This accounts for all the 558 and 560 cell nuclei of the L1 hermaphrodite and male, respectively. Embryonic development generates the L1 individual that is capable of feeding for growth and attainment of sexual maturity.

Postembryonic development involves proportional growth leading to a four fold increase in length, through four cuticular moults, over three days. There are 32 blast cells that continue to divide after hatching to generate the 959 and 1031 somatic nuclei of the adult hermaphrodite and male, respectively. Overall structure is retained. Internally the gonad develops and the germ-line proliferates. Externally the hermaphrodite vulva provides a mid-ventral opening from the uterus to the outside and the complex tail, to which the gonad is connected in the male, is elaborated for copulation.

This summarises the complete description of the development of this species that has been provided to the cellular level. Further subcellular details, such as intercellular connections, have also been determined from light and electron microscope observations. Furthermore, a considerable understanding of the mechanisms by which *C. elegans* development is achieved has been revealed (see below). This level of knowledge has not been produced for any other nematode species.

Other species in the family Rhabditidae

Embryonic development has been examined in three other soil-dwelling species of the family *Rhabditidae* to which *C. elegans* belongs, *Rhabditis broughtonalcocki*, *R. terricola* and *R. dolichura* (Skiba and Schierenberg, 1992). Embryogenesis for these three species is longer at 21°C, taking 22, 40 or 21 hours respectively, as compared to the 14 hours for *C. elegans*. The patterns of asymmetric cell divisions by which the founder cells are generated are identical for all four species and the rates of division within the founder cell lineages are

Figure 5.2. *Caenorhabditis elegans*. Photographs taken using Differential Interference Contrast microscopy. Anterior is to the left throughout. a. Two-cell embryo, viewed at 1000×. AB is at the anterior and P_1 is at the posterior. *C. elegans* eggs are 50 μm long. b. Four-cell embryo viewed at 1000×. Dorsal is to the top. ABa is anterior, ABp is dorsal, EMS is ventral and P_2 is posterior. c. The embryo on the left is fully elongated and that on the right has elongated to the $1\frac{1}{2}$ fold stage. These embryos are viewed at 1000×. For the embryo on the right, anterior is left and dorsal is to the top. d. L1 larva viewed at 400×. e. Adult hermaphrodites and three eggs viewed at 100×.

approximately proportional to the duration of embryogenesis for each species. However variation is seen in the relative timing of the germ-line divisions. P_3 divides proportionately earlier, relative to *C. elegans*, in *R. broughtonalcocki* and *R. terricola* and both P_2 and P_3 divide proportionately earlier, relative to *C. elegans*, in *R. dolichura*. Thus the arrangement of cells at the four cell stage is the same for all species, but older embryos of the different species could be distinguished. By the 28 cell stage, when gastrulation starts the cells in the embryos for these four species appear to be arranged identically and subsequent embryonic development of these *Rhabditis* species appears to be as in *C. elegans*. It has been suggested that advancement of germ-line divisions in more slowly developing embryos may have a role in protection of the germ-line quality. Cell number at the start of morphogenesis was found to be approximately 540 for *R. broughtonalcocki*, 590 for *R. terricola* and 570 for *R. dolichura* (all +/− 5%) which correspond closely to the 560 for *C. elegans* (Skiba and Schierenberg, 1992).

Postembryonic stages are very similar for these four species. The *R. dolichura* adult is slightly smaller than the other three (0.9 mm vs. 1.3 mm). *C. elegans* and *R. dolichura* are hermaphrodite with rare males whereas *R. broughtonalcocki* and *R. terricola* have male and female sexes in equal proportion. This is a relatively minor difference as the hermaphrodites are considered to be morphological females for which the first gametes mature as sperm. All female/hermaphrodite adults have very similar morphology, with bi-lobed reflexed gonads possessing a single mid-ventral opening.

Strongylida and Diplogasterida

Species in the Strongylida for which embryonic development has been studied include *Metastrongylus elongatus*, *Pseudalius inflexus* and *Pseudalius minor* (Martini, 1907; Chitwood and Chitwood, 1974). The initial cell divisions of embryos for this group appear to be very similar to those described above for *Rhabditis* (Malakhov, 1994). After the longitudinal division of the zygote, AB divides transversely. Subsequently, as P_1 divides, the AB daughters move such that the planar rhombus arrangement of the four-cell stage is formed. *Nippostrongylus brasiliensis*, a trichostrongyle, also shows this four-cell, rhombus arrangement (Figure 5.3).

Within the Diplogasterida, *Diplogaster longicauda* has been studied classically (Chitwood and Chitwood, 1974) and embryonic development was also considered similar to that of the rhabditids. Recently, *Pristionchus pacificus* has received considerable attention as a species with which to employ a comparative developmental biology approach (Sommer *et al.*, 1996). *P. pacificus* was selected as having an appropriate degree of evolutionary relatedness to the exceedingly well-characterised *C. elegans* for comparisons between these two species to be valuable in illuminating developmental and evolutionary mechanisms. In addition, *P. pacificus* shares key characteristics with *C. elegans*, such as a male plus self-fertilising hermaphrodite sexual system, that facilitates the genetic approach. Work on *P. pacificus* has

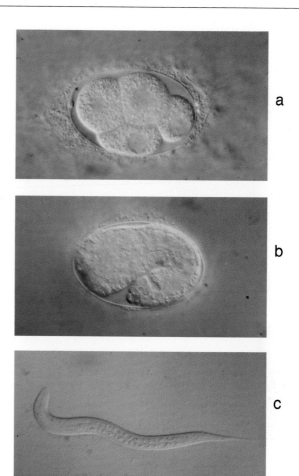

Figure 5.3. *Nippostrongylus brasiliensis*. Photographs taken using Differential Interference Contrast microscopy. Anterior is to the left throughout. Note the similarity to panels b, c and d of Figure 5.1. a. Four-cell embryo viewed at 1000×. Dorsal is to the top. *N. brasiliensis* eggs are 60 μm long. b. An embryo elongated to the $1\frac{1}{2}$ fold stage viewed at 1000×. Dorsal is to the top. c. L1 larva viewed at 400×.

concentrated, so far, on the post-embryonic developmental lineages and morphogenesis by which the vulva is formed (Sommer and Sternberg, 1996) (see below).

Clade III

Ascaradida

The ascarid *Parascaris equorum*, an intestinal parasite of horses, was a subject of many classical embryology studies by Boveri, zur Strassen and others (zur Strassen, 1896; Boveri, 1909), although this species was then known as *Ascaris megalocephala*. Nineteenth century studies on *P. equorum* revealed the halving of the number of chromosomes that occurs upon extrusion of the polar bodies at meiosis and the restoration of the diploid state upon fusion of the pronuclei at fertilisation. Chromatin diminution, the loss of a considerable proportion of the genetic material, that occurs early in the somatic lineages of ascarids although not in other nematodes, provided a marker that allowed a distinction between the soma and germ-line through development to be established.

As with *C. elegans*, the first division is transverse to produce a larger anterior blastomere, AB, and a smaller posterior blastomere, P_1 (Figure 5.4). The AB division is transverse, while P_1 divides longitudinally, to produce a symmetric, T-shape, four cell stage. This then rearranges to produce the planar rhombus arrangement as in the clade V species, described above. The difference in the initial arrangement of the four cell stage could be just a consequence of the physical characteristics of the egg shell. The extra space in the ascarid eggs might allow the default, T-shape form to be generated as an automatic consequence of the AB spindle aligning perpendicular to, and the P_1 spindle aligning with, the axis of the first cell division. Additional mechanisms, perhaps involving changes in intercellular adhesion, would be required to form the rhombus arrangement. Cell adhesion alone seems unlikely to be sufficient to maintain the planar (as distinct from tetrahedral) relationship amongst the cells. These additional mechanisms may be absent from the *Rhabditis* species, in which the confines of the egg shell might mean they are not required. In *C. elegans* eggs removed from the egg shell and the perivitelline membrane, the P_2 cell and its descendants may remain unassociated with the AB descendants (Edgar, 1995).

Subsequent embryonic cell divisions of *P. equorum* appear to be as for *C. elegans*. The cell lineage nomenclature originally developed for *P. equorum* has formed the basis of the cell lineage nomenclature for nematodes in general. The germ-line cell names (P_x, P for parental germinal cell) and most of the names of the founder cells were retained in the *C. elegans* nomenclature. However, an alternative nomenclature has been used in describing embryonic development in some nematode species (Figure 5.5). Successive somatic daughters of the germ-line have been called S_x, S_1 being the sister of P_1, S_2 the sister of P_2, etc. Daughters of subsequent somatic cell divisions are distinguished by roman versus greek letters or roman numerals or arabic numerals or apostrophes, depending on the orientations of the cell divisions. The nomenclature developed for *C. elegans* is much simpler.

The fates of the cell lineages were generally defined and, as such, were broadly the same as determined almost a century later for *C. elegans*. AB gives rise to ectoderm and nerve cells, E generates the intestine, MSt was considered to produce muscle cells, pharynx and stomodeal structures, C gives epidermis and muscle and D, just muscle. The cell number at the elongation stage was estimated as approximately 800 cells, but this was based on the assumption that the cell number doubled whenever mitosis was observed and so may in fact be nearer to the 560 cells known to be found in *C. elegans* embryos at this stage.

Other ascarids have also been studied including *Toxocara canis* (Walton, 1918) and *Nematoxys ornatus* (Martini, 1907).

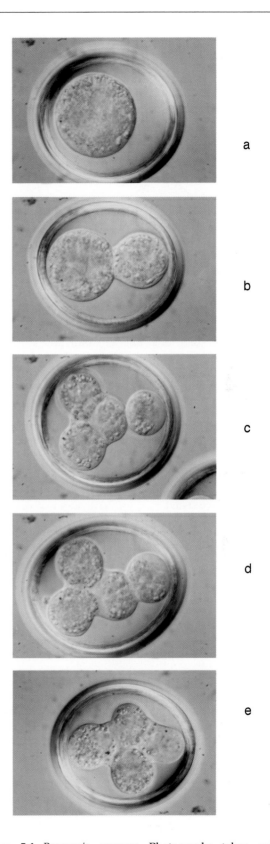

Figure 5.4. *Parascaris equorum.* Photographs taken using Differential Interference Contrast microscopy at 1000× magnification. a. One-cell embryo. *P. equorum* eggs are 90 μm long. b. Two-cell embryo. c, d, e. Four-cell embryos showing the symmetrical T-shape arrangement through to the rhombus arrangement.

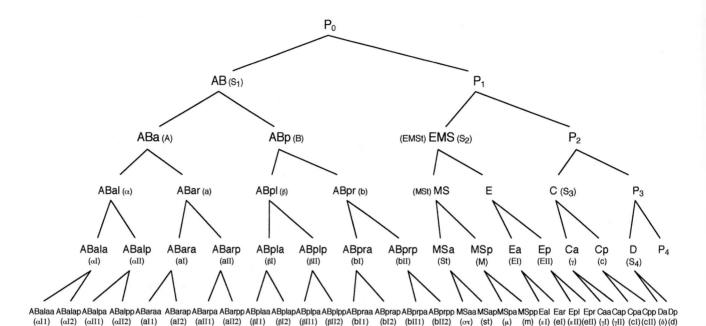

Figure 5.5. A summary of the nomenclatures used to describe nematode cell lineages, mainly referring to species in clades III, IV and V, and derived from Figure 151 of Chitwood and Chitwood (1974). The cell names are as used for *C. elegans*. Alternative nomenclature, used previously for other species, is given in brackets. In all the nematode cell lineage nomenclatures, the germ line cells are called P_x, where x refers to the number of divisions from the zygote. For *C. elegans* key blast cells have names of 1 or 2 capital letters. The names of other cells consist of the name of the blast cell from which that cell originated followed by a series of letters describing the orientations of the cell divisions by which the cell is produced (a anterior, p posterior, l left, r right, d dorsal, v ventral). The sisters of the germ line cells have been called S_x, where x refers to the number of divisions from the zygote. The classical nomenclature for cells of the somatic lineages is rather complex. AB (or S_1) divides to give A and B, with A and B dividing to give a and α, b and β, respectively. The second somatic founder cell, EMSt (for Endoderm, Mesoderm and Stomadeum, or S_2, called EMS for *C. elegans*) divides to give MSt (MS for *C. elegans*) and E, MSt dividing to give M and St. M then divides to give m and μ, St to give st and στ. C (or S_3) divides to give c and γ, D (or S_4) giving d and δ. The names of the daughters of these cells become even more complicated with a, α, b, β, m, μ, st, στ, c, γ, d and δ dividing to give aI and aII, αI and αII, etc. The daughters of the subsequent round of division are named by adding the numbers 1 or 2 to the ends of these names. In the E lineage, E gives EI and EII, which then divide to give eI and εI, eII and εII, respectively. The Roman versus Greek letters refers to daughters towards the right and left sides, respectively. Use of Roman numerals identifies anterior/posterior divisions. Arabic numerals and apostrophes, ' or '', allow additional daughter distinctions. In this figure the horizontal and vertical axes are arbitrary and do not have the spatial or temporal significance of other cell lineage diagrams.

Spirurida

The embryonic development of *Camallanus lacustris* has been described in considerable detail by Martini (1906) although the species was then called *Cucullanus elegans*. The initial divisions generate the rhomboid four cell stage immediately, without going through the T-shape form, and so the start of development is more like that in *Caenorhabditis elegans* than in the ascarids. The fates of the founder cells also appear to match those of *Caenorhabditis elegans*. Gastrulation, however, appears to be relatively retarded and doesn't start until 354 cells are present. Possibly as a consequence, the embryo prior to gastrulation appears flattened along the dorsal ventral axis, as compared to the previously described embryos, with the E cells still at the surface. Nevertheless, elongation and morphogenesis then generate a larva of typical nematode proportions.

Clade IV

Panagrolaimidae

Turbatrix aceti, or the vinegar eel worm, has also been an important species in embryological study (Pai, 1927, 1928), although originally under the name *Anguillula aceti*. The relative timing and orientation of the earliest cell divisions are remarkably similar to those of *Caenorhabditis elegans*. The degree of similarity meant that some significant differences in the fates of a few key cells between observations for *C. elegans* and the classical descriptions of *T. aceti* were surprising and led Sulston *et al.* (1983) to re-examine *T. aceti*. The C founder cells of *T. aceti* give rise to muscle cells and ectoderm, while the D founder cell gives rise only to muscle cells, exactly as for *C. elegans*, and not to exclusively ectoderm (C) or rectum (D) as in the earlier accounts.

Another key discrepancy in the literature has concerned the origin of the somatic gonad. At the end of embryogenesis both the *C. elegans* and *T. aceti* gonad primordia consist of 4 cells surrounded by a basement membrane. The adult germ-line arises from the central two cells while the somatic cells of the gonad arise from the two, peripheral cells. Pai asserted (Pai, 1927, 1928) that P_4 divided to give another somatic blast cell (S_5) and another germ-line cell (P_5), with S_5 and P_5 dividing once more to give the somatic cells and the germ-line, respectively, in the gonad primordium. To add to the confusion, classical descriptions of embryonic development of ascarids had claimed P_4, P_5 or even a P_6 cell as the primordial germ cell, on the basis of following the chromatin diminution events of the successive rounds of somatic cell formation. Several key texts (Croll and Matthews, 1977; Chitwood and Chitwood, 1974; Bird, 1971 (although not in the 2nd edn. (Bird and Bird, 1991))) have accepted the S_5/P_5 origin of the nematode gonad in general. Sulston *et al.* (1983) examined this specifically in *T. aceti*, finding that the somatic gonad was derived from the MS lineage and P_4 divided once to give the germ-line progenitors, precisely as in *C. elegans* and not as in earlier accounts. Subtle differences were found in the E cell lineage although E still produced all of and only intestine in both species (18 cells in *T. aceti*, 20 cells in *C. elegans*.) There could be similarly small differences in the AB and MS founder cell lineages. Nevertheless, the embryonic cell lineages are clearly very strongly conserved from *C. elegans* to *T. aceti*.

In one study of embryonic development in an undefined *Panagrolaimus* species (Skiba and Schierenberg, 1992) the generation of the founder cells was again found to be precisely as for *C. elegans*. One difference observed was in the timing of the parental germ-line divisions, with P_1, P_2 and P_3 dividing relatively early, as compared to the AB, MS, E and C founder cell lineages. This appears to reflect the same plasticity in the timing of the parental germinal cell divisions as was observed in comparisons within the *Rhabditidae* family (Skiba and Schierenberg, 1992). The timing of cell divisions in this *Panagrolaimus* species seemed to match those for *Rhabditis dolichura* except that P_1 also divides early, even before AB. By gastrulation the cells have the same organisation as for *C. elegans*.

The cell lineage of *Panagrellus redivivus* has been studied extensively (Sulston *et al.*, 1983; Sternberg and Horvitz, 1981, 1982). In this species, the cell divisions by which the founder cells are generated during embryogenesis, and their fates, are as for *C. elegans* and *T. aceti*. Not surprisingly, given their evolutionary relationships, the small difference found in the E lineage of *T. aceti* as compared to *C. elegans* was also found for *P. redivivus* (Sulston *et al.*, 1983). Post-embryonic development of *P. redivivus* is also very similar to *C. elegans*. In *P. redivivus* the vulva is formed more posteriorly from P5p, P6p, P7p and P8p, rather than just P5p, P6p and P7p. In addition, the *C. elegans* nerve cell PVR, one of the two, arguably-odd, nerve cell fates generated amongst the otherwise epidermal and muscle cell fates of the C lineage, is a post-embryonic blast cell in *P. redivivus* (Sulston *et al.*, 1983).

Cephalobidae

Skiba and Schierenberg (1992) also studied the embryonic lineage in *Acrobeloides nanus* (then described as a *Cephalobus* species). *A. nanus* is a soil-dwelling, microbivorous nematode like *C. elegans*. The two species are morphologically quite similar, although the *A. nanus* female is half the size of the *C. elegans* hermaphrodite and has a single lobed gonad opening more posteriorly. The eggs are of very similar dimensions but produced parthenogenetically. The initial asymmetric cell divisions of the embryo produce AB, MS, E, C, D and P_4 blastomeres with the same general fates as *C. elegans*. However, the parental germ-line divisions are even more precocious as compared to *C. elegans* than are those of *Panagrolaimus*. Both P_1 and P_2 divide before either of the somatic blastomeres, AB or EMS, and P_3 divides immediately after the first somatic cell division, of AB (Figure 5.6).

The timing of these cell divisions has a curious implication for the developmental process. The P_2 division generates P_3 and C side by side, across the long axis of the egg, with AB at the anterior and EMS centrally located. Then AB appears to attempt to divide transversely and, as in *C. elegans*, the confines of the egg causes the axis to tilt to an oblique angle, apparently pushing EMS to take up what is thereby defined as the ventral surface. However, the prior division of P_2 means that in the 5 cell embryo EMS may be, as a consequence, adjacent to either C or P_3. i.e. The dorsal/ventral arrangement of C/P_3 can be in either orientation. Despite this variation, by the 28 cell stage, as gastrulation starts, the cells of the *A. nanus* embryo have re-arranged and subsequent embryonic development occurs apparently as in *C. elegans*.

Tylenchida and Aphelenchida

Early embryonic development has been described for many species of Tylenchida perhaps because of their economic significance as parasites of plants. For many species of this order the first two divisions are both aligned to the long axis of the egg generating a four cell stage in which the cells have a linear arrangement (Figure 5.7). Perhaps the elongated shape of the egg restricts the spindles of the first cell divisions to the long axis. This blastomere organisation has been described for several species of the *Tylenchida*: *Radopholus similis* (van Weerdt, 1960), *Nacobbus serendipiticus* (Clark, 1967), *Rotylenchulus parvus* (Dasgupta and Raski, 1968), *Pratylenchus penetrans* and *Pratylenchus zeae* (Hung and Jenkins, 1969a, 1969b) and an *Hoplolaimus* sp., an *Helicotylenchus* sp. and a *Tylenchorhynchus* sp. (Drozdovskiy, 1968) of the *Heteroderidae*; and for *Criconemoides xenoplax* (Seshradi, 1964) and *Hemicriconemoides chitwoodi* (Fassuliotis, 1962) of the *Criconematidae*. This blastomere arrangement was also observed in *Seinura demani* and *Seinura steineri* of the *Aphelenchida* (Drozdovskiy, 1968).

An intermediate linear three cell stage may be formed when either the anterior blastomere (*Radopholus similis*, *N. serendipiticus*, *Rotylenchulus parvus*, *C. xenoplax* and the

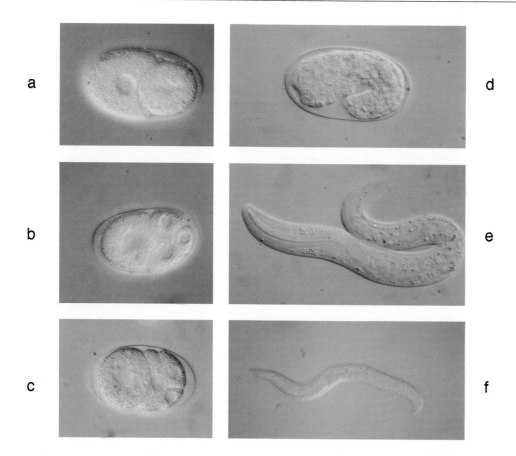

Figure 5.6. *Acrobeloides nanus.* Photographs taken using Differential Interference Contrast microscopy, all at 1000× magnification apart from panel f. Anterior is to the left throughout. a. Two-cell embryo. *A. nanus* eggs are 50 μm long. b. Four cell embryo with dorsal already fixed and to the top. AB is anterior, EMS is ventral, C is dorsal and P₃ is posterior. c. The alternative four cell stage in which the dorsal/ventral axis is not fixed until the subsequent division of AB (see text) (Skiba and Schierenberg, 1992). EMS is in the way of any contact between AB to the anterior and C and P₃ to the posterior. d. Embryo elongated to the $1\frac{1}{2}$ fold stage. Dorsal is to the top. e. L1 larva. f. Adult viewed at 200×.

Seinura sp.) or the posterior blastomere (*Hoplolaimus* sp., *Helicotylenchus* sp. and *Tylenchorhynchus* sp.) of the two cell stage divides first. In the four cell embryo the cells are not always perfectly linear, sometimes forming a zig-zag arrangement either as a result of cell rearrangement or of spindle axes being slightly out of alignment. For the *Pratylenchus* sp., *N. serendipiticus*, *Rotylenchulus parvus*, *Tylenchorhynchus* sp. and *Radopholus similis* the central, anterior blastomere moves to a dorsal position and the central, posterior blastomere takes up the ventral position. Thus, a version of the rhomboid form, as described above, was generated in these species, although for *Pratylenchus* and possibly *Rotylenchulus parvus*, the anterior cell has already divided before this rearrangement occurs. As if to catch up with these species, in *Radopholus similis* and *N. serendipiticus* the anterior cell of the four cell stage subsequently goes through two rounds of rapid cell divisions. For *C. xenoplax* the anterior three blastomeres of the four cell stage divide slightly ahead of the posterior blastomere and a rearrangement to a rhomboid form was not apparently observed, perhaps a distinction between the *Heteroderidae* and the *Criconematidae*.

Early embryonic cell divisions in *Ditylenchus dipsaci* (Yuksel, 1960) of the family *Tylenchidae* is apparently distinct from that in the other species of the order Tylenchida, addressed above. The initial division does generate a larger anterior cell that then divides to give a linear, three-cell form, but the most anterior cell then undergoes two rounds of rapid division to give a cluster of four cells at the anterior and the linear four cell stage is not observed. Subsequently, the posterior cell divides leaving the central cell (equivalent to *C. elegans* ABp) undivided. It is not apparent from the published description whether rearrangement of the central two cells (equivalent to *C. elegans* ABp and EMS) of the seven cell stage occurs to give a form equivalent to the rhombus. Perhaps this pattern of cleavage is a more extreme version of that for the *Pratylenchus* species and *Rotylenchulus parvus* with the anterior blastomere (equivalent to *C. elegans* ABa) dividing even earlier. Nevertheless, elongation and morphogenesis do subsequently occur to produce an apparently typical, first-stage, nematode larva.

Another tylenchid, *Bradynema rigidum* of the family *Allantonematidae* was also the subject of classical study

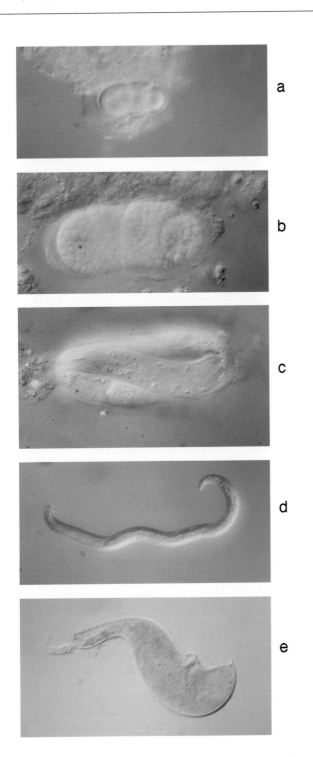

a

b

c

d

e

Figure 5.7. *Rotylenchulus reniformis*. Photographs taken using Differential Interference Contrast microscopy. Anterior is to the left throughout. a, b. A four cell embryo viewed at 400× and 1000× showing the linear arrangement of blastomeres and elongated shape of the egg. *R. reniformis* eggs are 75 μm long. c. An egg containing a fully elongated worm with the head in focus, viewed at 1000×. d. A juvenile viewed at 400×. e. An adult viewed at 200×. The adult swells from the typical nematode form but the vulva is visible along the upper surface in the figure and the tail is apparent to the right. The anterior is specialised for interfacing with the plant host.

(zur Strassen, 1892). It was claimed (zur Strassen, 1892) that in *Bradynema rigidum*, P_4 underwent an extra division to generate another somatic founder cell, S_5, the precursor of the somatic gonad, in the same way as is discussed above for *Turbatrix aceti*. Sulston *et al.* (1983) examined this issue specifically using *Aphelenchoides blastophthorus* and found that P_4 of *A. blastophthorus* doesn't divide in the embryo. Although *B. rigidum* is in the *Tylenchida* and *A. blastophthorus* is in the related order *Aphelenchida* (Blaxter *et al.*, 1998), the somatic gonad of this latter species at least, and therefore possibly those of the Tylenchida including *B. rigidum*, must originate from elsewhere, possibly as in *T. aceti* and *C. elegans*. Discrepancies in cell lineage descriptions between classical and more contemporary work may result from the classical work being dependent on observation of fixed material whereas more modern work has followed development by Nomarski optics in live embryos. Recent developments in 4D microscopy make the challenge of following embryonic lineages much easier now. The embryonic cell lineage could be fundamentally identical for all the species described so far.

Plectida and Chromadorida

The Plectida appear to be a separate clade, as distantly related to *C. elegans* as all the other species mentioned above apart from those within clade V. The Chromadorida are probably another distinct clade but more distantly related. Embryonic development in these orders (and the remaining orders described below) has received little attention, with the original literature, containing the most complete accounts, in Russian. This material has been summarized by Malakhov (1994).

Plectus species show one of two modes of cleavage (Malakhov, 1994). The first division generates a larger anterior AB blastomere with a smaller posterior P_1 cell. Transverse division of AB and longitudinal division of P_1 generates the T shape form. For some species this resolves to a planar rhombus arrangement, as for the Ascaradida, but for other species a tetrahedron organisation of the four cell stage is produced. EMS occupies the ventral surface and P_2, the posterior surface. Depending on the arrangement at the four cell stage for any particular species, the daughters of AB may either occupy left and right or anterior and dorsal surfaces.

Embryonic development in the Chromadorida has been described with reference to *Hypodontolaimus inaequalis* and *Chromadora nudicapitata* of the Chromadoroidea and for *Desmodora serpentulus* and *Spirina parasitifera* of the Desmodoroidea (Malakhov, 1981, 1994). For the Chromadoroidea, the very earliest stages are quite unlike those of any species described above. Between cell divisions, at the 1, 2 and 4 cell stages, the cells lose their uniform rounded shape and move within the egg. In addition, the spindles axes for the second round of cell division appear to adopt any relative orientation. Then, the smallest cell of the four cell stage (equivalent of *C. elegans* P_2) divides asymmetrically to give *C. elegans* P_3 and C equivalents. The

5 cells now present stop moving, adhere closely together and cell division pauses. When development starts again, the embryo behaves much more like that of the other nematode species described so far. The blastomeres no longer move around. Division of all cells except P_3 and C generates an 8 cell stage with C, P_3 and their cousins (E and MS) occupying the ventral surface, posterior to anterior, the other cells (equivalent to AB granddaughters) occupying the dorsal surface. The subsequent fates of these cells appear to be typical for nematodes, as described above.

For the *Desmodoroidea*, the striking cytoplasmic movements observed for the *Chromadoroidea* are again apparent at the 1 cell stage, but development appears to settle down faster with the four cell stage arranged in a T-shape that rearranges to the rhombus. Again, at the 8 cell stage, C, P_3, E and MS occupy the ventral surface, posterior to anterior and AB granddaughters occupy the dorsal surface. For the *Chromadoroidea* the AB daughters are left/right, where as for the *Desmodoroidea* the AB daughters are anterior/posterior, a difference reminiscent of a distinction, described above, between *Plectus* species. The left/right relationship between the immediate daughters of AB has not been encountered in any of the other species described so far.

Clade I

Embryonic development has been described for a few species representing the orders Mermithida, Dorylaimida, Mononchida and Trichocephalida of clade I. The detailed descriptions of early embryonic development of *Gastromermis hibernalis* (Mermithida) (Malakhov and Spiridonov, 1981), a *Eudorylaimus* sp. and a *Mesodorylaimus* sp. (Dorylaimida) (Drozdovskiy, 1975) and a *Prionculus* sp. (Mononchida) (Drozdovskiy, 1968) are very similar (Malakhov, 1994). The first division is, as for all nematodes, transverse, with the two blastomeres formed of very similar size for the Dorylaimida and Mononchida species and the anterior daughter larger for *Gastromermis*. The subsequent division of the two cells for all four species creates a four cell stage of the planar rhombus form.

Strikingly, the potentialities of the four cells of the rhombus form do not match those for the other nematode species described so far. The endoderm is derived from one of the daughters of the anterior blastomere of the two cell stage, rather than the posterior blastomere. Furthermore, primary ectoderm is derived, not only from the other daughter of the anterior blastomere, as expected, but also from one of the daughters of the posterior blastomere. This posterior blastomere daughter also gives rise to primary mesoderm, which is more consistent with lineage fates of the other nematode species.

The last of the four cells, one derived from the posterior blastomere, is a parental germ line cell called P_2 and equivalent to *C. elegans* P_2. P_2 goes through two more divisions to generate blastomeres C and D and the germ-line progenitor P_4. C and D are thought to give rise to posterior ectoderm rather than posterior ecto-

derm and mesoderm, but this was also a claim made in initial studies for other nematodes as well.

The labelling of the early blastomeres reflects these fates. The first cleavage generates P_1 and AESt or AFESt. (The F refers to frontal ectoderm while other letters match with the designations described above for the Ascaradida.) P_1 divides to give P_2 and BM while AESt (or AFESt) divides to give A and ESt (or FESt).

The early cleavages for *Trichocephalus trichurus* of the Trichocephalida are somewhat different (Malakhov, 1994). At the first division the posterior daughter, P_1, is larger than the anterior daughter AESt. P_1 then divides to give a linear three cell stage, with a centrally located BM and a posterior P_2. All three cells then divide together, AESt and P_2 dividing dorsal to ventral and BM dividing left to right, to give a unique six cell arrangement. The six blastomeres are labelled A, ESt, bm, $\beta\mu$, C and P_3. Despite the different pattern of cell divisions for this Trichocephalida embryo, the fates of these early blastomeres, as reflected in their names, do match those for the Mononchida, Mermithida and Dorylaimida, the other orders of this clade, distinguishing this group from other nematode species. Figure 5.8 provides micrograph images of another trichocephalid, *Trichinella spiralis*.

The striking difference between the early blastomere fates (e.g. origin of the intestine) for these clade I species as compared to the other nematode species discussed so far, raises the question as to how these fates could have been exchanged during nematode evolution without destroying embryogenesis. How can a characteristic nematode larva develop through two quite distinct, highly determinate cell lineages? The answer to this question has probably been provided in studies of embryonic development in the last major nematode clade, which includes the Enoplida.

Clade II

For the Enoplida (Figure 5.9), embryonic development has been examined principally in *Enoplus brevis* (Malakhov and Akimushkina, 1976; Voronov and Panchin, 1995a) and *Pontonema vulgare* (Malakhov and Cherdantzev, 1975) and appears to be similar in the two species (Malakhov, 1994). The early divisions are equal and approximately synchronous, giving blastomeres of the same size, displaying strong cytoplasmic activity that distorts their shapes. The axes of cell divisions appear to vary between embryos with the blastomeres moving within the egg such that all possible arrangements of the four cell stage are seen, even for embryos from a single female. At the fourth round of cell division, one of the cells starts to behave differently. Division of this cell is delayed and so a 15 cell stage is formed. The odd cell, now larger than the rest, starts to move into the interior, initiating gastrulation, and is the endodermal precursor. This cell has been labelled E and is equivalent to E in embryos of other nematodes such as *C. elegans*. Segregation of endodermal fate occurs at the eight cell stage in most nematodes including the Enoplida. This pattern of early cleavage means that even the posterior-anterior axis of the embryo is not

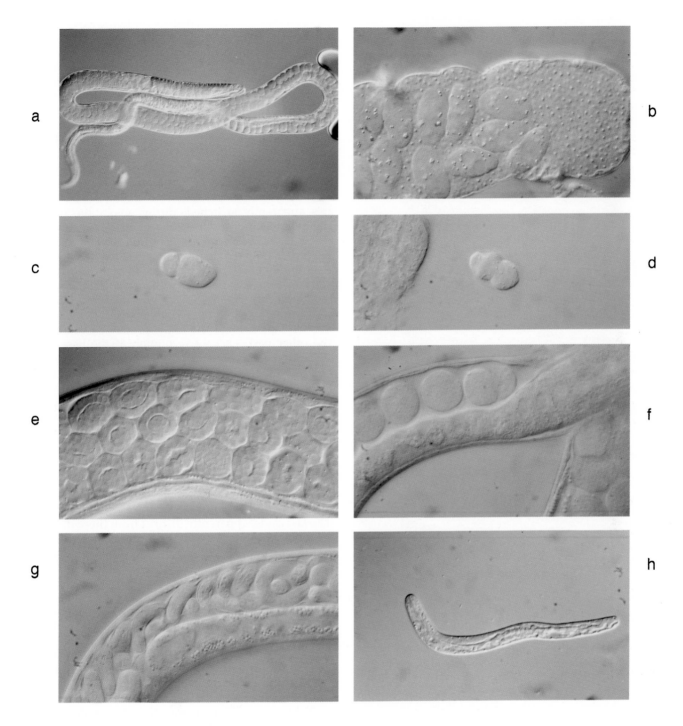

Figure 5.8. *Trichinella spiralis*. Photographs taken using Differential Interference Contrast microscopy. a. Adult female, day 4 post-infection, viewed at 200×. Embryos are retained within the mother until hatching and the egg coverings are much weaker than in the other nematode species photographed for this chapter. The gonad appears wider, containing more embryos in older females. Also the embryonic development appears to involve growth for this nematode species. The head is to the bottom left of this panel. b. Very early embryos in the distal end of a gonad dissected from a female, day 7 post-infection, viewed at 1000×. c, d. Two-cell and three-cell embryos, dissected from a female, day 3 post-infection, viewed at 1000×. Although development of *T. spiralis* has not been followed, from analogy with the descriptions for *Trichocephalus trichurus* (see text), anterior would be to the left, with AESt and P_1 blastomeres in panel c, AESt, BM and P_2 blastomeres in panel d, from left to right. *T. spiralis* eggs are 25 μm long. e. Gastrulating embryos in an adult female, day 7 post-infection, viewed at 1000×. f. Embryos starting to elongate within an adult female, day 4 post-infection, viewed at 1000×. The central embryo, in ventral view, is at the $1\frac{1}{2}$ fold stage, with the anterior to the left. Just to the right of this embryo, unfertilized oocytes can also be seen. g. Fully elongated embryos within an adult female, day 7 post-infection, viewed at 1000×. h. An invasive larval stage, day 7 post-infection, viewed at 1000×.

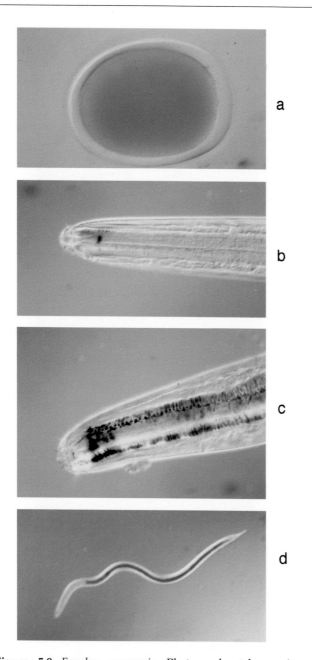

Figure 5.9. *Enoplus communis.* Photographs taken using Differential Interference Contrast microscopy. a. Egg, 180 μm long, viewed at 400×. Note the different magnification from the views of embryos in other figures in this chapter. b, c. Head of late larva and adult viewed at 200×. d. Late larva viewed at 25×.

apparent until gastrulation is well underway. Subsequent development, including gastrulation, morphogenesis and elongation, occurs to give a characteristic nematode larva in a manner apparently very similar to that observed for other species.

Other than E, the early blastomeres are not only morphologically indistinguishable but have also been shown to be undetermined (Voronov and Panchin, 1998, Voronov *et al.*, 1986, Voronov and Panchin, 1995b). To explore their fates the early blastomeres

were labelled by direct injection of marker dyes. While injection of one cell at the eight cell stage, presumably that destined to be E, resulted in all of and only the intestine being labelled, injection of other blastomeres never gave a consistent pattern from one embryo to another. It appears as if any of the blastomeres, with the exception of E, can give rise to any cell, other than an intestinal cell, in the fully formed larva. Such an indeterminate mode of development has not been found in any nematode order, other than the Enoplida.

Existence of an indeterminate mode of development in the nematodes may answer the question posed above as to how the determination for the intestinal cell fate has apparently switched, during evolution, between the anterior and posterior blastomeres of the two cell stage. If the indeterminate mode of development was ancestral, then originally the intestinal fate was not determined to either of the blastomeres at the two cell stage. Through evolution of the determinate mode of development, specification of intestinal fate was fixed to the anterior blastomere in one part of the nematode evolutionary lineage and to the posterior blastomere in the other part. Lack of resolution at the base of the nematode lineage means that whether evolution of the determinate mode of development occurred once with subsequent schism to the two forms or occurred independently twice cannot be resolved as yet. Perhaps in the original nematode even the intestine wasn't lineally specified and acquisition of intestinal lineage specification was the first step along the evolutionary pathway towards an almost fully determinate cell lineage, as observed for *C. elegans*.

Monhysterida

The last clade, apparently distinct from those described so far, is the Monhysterida and for this order embryonic development has been described for *Daptonema setosum* (Malakhov, 1994). The initial division gives a larger anterior AB cell and smaller posterior P_1. The second cell divisions are perpendicular to each other and a tetrahedron arrangement is formed. The endoderm is derived from the ventral daughter of P_1 and the AB daughters occupy left and right sides of the dorsal surface. The lack of cytoplasmic activity means that at least with regards early development, the Monhysterida are most similar to certain families of Plectida and are distinct from the *Chromadoroidea* that also share the same blastomere fate organization. These observations of the mode of early development could help clarify the evolutionary relationships for the Monhysterida, which have not yet been resolved by the molecular analyses.

Developmental Mechanisms

Notwithstanding the detailed descriptions of embryonic development summarised above, nematodes had severe limitations for classical investigations of the mechanisms of animal development. The transparency of the egg and the ready availability of the embryos made nematodes ideal subjects for direct observation, but small size and fragility when removed from the egg shell made investigation of development through experimental manipulation of nematode embryos very difficult. Isolation of developmentally competent cells or tissues from embryos, key to experiments of classical importance performed with, for example, amphibians and sea urchins (Gilbert, 1997), was impossible for nematodes until relatively recently. Apart from cell ablation experiments (see below), investigators had to rely upon observation of natural perturbations of nematode development. Rare, spontaneous fusion of eggs or cytoplasmic loss (von Ehrenstein and Schierenberg, 1980) showed that atypically sized eggs can nevertheless develop properly, giving rise to viable postembryonic stages. The mechanisms directing nematode development can compensate for changes in absolute dimensions.

The invariant cell lineage could be taken to suggest that nematode development was determinate. Indeed, nematodes were considered perfect examples of the mosaic mode of development with each cell apparently being very-specifically, developmentally determined at birth. However, invariant development could still result from an indeterminate mode of development regulated by intercellular communication if the signalling events were completely reproducible from individual to individual within a species. Demonstration that the latter is indeed the situation for nematodes has only been achieved relatively recently in studies with *C. elegans*. Even more recently, observation of the variable lineage of the Enoplida has revealed that nematodes can develop in a highly indeterminate manner (Voronov and Panchin, 1998).

Observation of Cell Autonomy

The first cell division is unequal for all nematodes, except the Enoplida, and appears to segregate discreet developmental potential to the two daughter cells. Originally, Boveri reached such conclusions from classical experiments on *Parascaris* eggs (Boveri, 1910). He observed the consequences of disturbance to the normal pattern of first cleavage either upon centrifugation or in dispermic eggs. The distinct patterns of cell division (equal for AB, unequal for P_1) and the phenomenon of chromatin diminution (which occurs in somatic cells but not the germ line of ascarids) could be used to follow cell fates in the abnormal embryos. Boveri concluded that a cytoplasmic determinant segregated to the posterior cell at the first cell cleavage to specify P_1 identity and absence of this determinant from the anterior cell resulted in AB identity.

The distinct determined fates of the two cells at the two cell stage was further demonstrated by cell ablation and cell isolation experiments. If either of the two cells of a *Parascaris* embryo was injured by UV irradiation without damaging the remaining cell, that remaining cell continued to develop, apparently as normal, demonstrating the distinct characteristics mentioned above. Similar results were observed if the cells were separated by ligature or if all but one cell of a four cell stage were destroyed.

Much more recently, establishment of techniques for isolation of viable blastomeres from *C. elegans* embryos allowed the development of individual cells to be followed, in the complete absence of the other cells of the embryo (Laufer, Bazzicalupo and Wood, 1980). The capacity for unequal cleavage that characterizes the germ-line, P_0-P_3, appears to be cell autonomously specified, with determinants segregating at each unequal cell division, in that isolated P_1, P_2 and P_3 continue to divide unequally. The orientations of these early cell divisions are also retained in these isolated blastomeres. Isolated AB divides to give a spherical clump of equally sized cells through a simple helical pattern of cleavage that may result from the orientations of successive cell divisions being perpendicular to the orientations of the preceding divisions but confined by adhesion between the cells. In contrast, isolated P_1 divides to give a clump of EMS derived cells linked, through P_4 and then D, to a smaller, C derived clump of cells, revealing the retention of germ line cell division organization, including the polarity reversal that occurs between P_1 and P_2 (Schierenberg, 1987). This reversal of germ line division polarity is not a universal nematode feature, being observed in several *Rhabditis* sp. and a *Panagrolaimus* sp., but not the Cephalobid *Acrobeloides nanus* (Skiba and Schierenberg, 1992).

Laser ablation of individual cells in late embryos revealed a high degree of cell autonomy in the later stages of embryonic development (Sulston *et al.*, 1983), as well as post-embryonic development (Sulston and White, 1980), of *C. elegans*, although a few examples of inductive signalling were revealed.

Observation of Inductive Signalling

Postembryonic signalling events

The complete cell lineage has permitted a thorough investigation of cell autonomy in the development of *C. elegans*. The *C. elegans* cell lineage is almost, but not quite, completely invariant. Postembryonically, there are just 11 pairs of cells (1 of which is hermaphrodite specific and 5 of which are male specific) for which one cell has one developmental fate and the other cell has another fate, but which cell follows which fate varies from individual to individual (Sulston and Horvitz, 1977). The decision, as to which cell will take which of the two mutually distinct fates followed by any particular cell pair, depends on intercellular signalling.

Laser ablation of cells revealed additional signalling events that are required postembryonically to direct certain cell fates even though normally the development of these cells shows no variation from individual to individual (Sulston and White, 1980). These inductions also only affect development of a few cells; two in the gonad, two in the ventral epidermis, six in formation of the hermaphrodite vulva and six in formation of the male tail.

The development of the vulva has received considerable attention. Six cells, called P3p, P4p, P5p, P6p, P7p, and P8p, forming the ventral epidermis of the *C. elegans*

L1 belong to an equivalence group, each with the potential to contribute to formation of the vulva. Normally, P6p forms the centre of the vulva (primary fate), P5p and P7p form the anterior and posterior of the vulva (secondary fate), while P3p, P4p and P8p divide once and then fuse with the main epidermal syncytium, hyp-7 (tertiary fate). The cells distinguish their fates through complex intercellular signalling that was characterised by genetic observations (see below) and laser ablation experiments. The anchor cell, in the gonad, provides the primary signal for vulva formation directing P6p to take the primary fate and P5p and P7p to follow the secondary fate. In addition hyp-7 provides an inhibitory signal, directing P3p, P4p and P8p to pursue the tertiary fate. Finally, P6p inhibits P5p and P7p from following the primary fate, ensuring that they follow the secondary fate.

The arrangement of the gonad and consequently the positioning of the vulva vary considerably between nematode species. Species with a central vulva usually have two ovaries (didelphic) symmetrically placed about the vulva, whereas species with a more posterior vulva have a single ovary (monodelphic). Comparative developmental biology studies of vulva formation have been undertaken to explore how development is modified during the course of evolution.

In all species from the same family as *C. elegans*, the Rhabditidae, that were examined, the vulva forms from P5p, P6p and P7p (Sommer and Sternberg, 1996). Although there are small differences in the number of cell divisions in the tertiary and secondary fates, the lineage of the primary fate is fully conserved for those Rhabditidae species that form the vulva in the central body region (*Pelodera, Oscheius*). However, P5p, P6p and P7p migrate to the appropriate position in those Rhabditidae species with a posterior vulva and formation of the vulva may be anchor cell dependent (*Cruznema*), as in *C. elegans*, or independent (*Mesorhabditis, Teratorhabditis*). For the latter the vulva fates may be followed autonomously.

Vulva formation has been examined in seven different genera of the Diplogasteridae, concentrating mostly on *Pristionchus pacificus*, and all show the same developmental pattern (Sommer, 1997). The vulva is once again formed from P5p, P6p and P7p, and P8p fuses with the epidermal syncytium, but P3p and P4p undergo programmed cell death like other Pnp cells of the ventral epidermis in these species. Interestingly despite the developmental pattern and morphology being so conserved the developmental mechanisms controlling vulva development in the Diplogasteridae vary significantly.

In *Panagrellus redivivus* (Sommer, 1997) and other members of the Panagrolaimidae (Felix and Sternberg, 1997) the cell lineage of vulva formation is somewhat different. The vulval equivalence group now extends to seven cells and includes P9p. P6p and P7p follow the primary fate, P5p and P8p follow the secondary fate and P3p, P4p and P9p follow the tertiary fate.

These studies on vulval development have suggested that developmental morphology can be conserved while developmental mechanisms change through the course of evolution. This would be facilitated by the redundancy in regulatory systems as observed in detail for vulva formation in *C. elegans*. The variation in vulval precursor cell lineages, shown by some species, may exemplify the evolutionary step required between two invariant modes of development. Observations on the variation in this very late step in nematode development may prove to be of considerable value for our understanding of evolution.

Embryonic signalling events

Embryonic development in *C. elegans* is completely invariant and so the cell lineage did not automatically reveal instances of inductive interaction during embryogenesis (Sulston *et al.*, 1983). Laser ablation of cells in the late embryo did reveal two further pairs of cells with distinct fates but equivalent developmental potential (Sulston *et al.*, 1983) to add to those described above, for postembryonic development. The fates of these cells again depend on intercellular signalling. Clearly, the vast majority of cells generated during *C. elegans* development appeared to develop cell autonomously, although the difficulty of performing laser ablations in the early embryo meant that additional inductions remained to be revealed.

Inductions affecting the AB lineage

The first evidence for major, intercellular signalling events directing development in the early *C. elegans* embryo came from an experiment to explore the distinction in the fates for the daughters of AB (Priess and Thomson, 1987). Using a blunt microneedle to push on the outside of the egg the positions of ABa and ABp were reversed in an embryo as P_1 was dividing to generate the four cell stage. Now ABp was in the anterior position, ABa was in what would have been the ventral position and EMS occupied what would have been the dorsal position, while P_2 was still in the normal posterior position. Surprisingly, it was found that all these manipulated embryos developed into perfectly normal worms. Therefore, ABa and ABp have equivalent developmental potential and their distinct fates, generating two thirds of the cells of the embryo (389 of 558), are dependent on intercellular signalling events from P_1 descendants. Considerable work has subsequently demonstrated convincingly that it is P_2 which signals to ABp directly to break the ABa/ABp equivalence (Moskowitz *et al.*, 1994, Mello, Draper and Priess, 1994; Mango *et al.*, 1994; Hutter and Schnabel, 1994; Bowerman *et al.*, 1992b).

The equivalency of the daughters of AB is more suggestively apparent in the embryos of for example the ascarids than in *C. elegans*. In the latter the cells of the AB lineage appear constrained to take nonequivalent positions by the confines of the egg casing, whereas in *Parascaris equorum*, the initial T-shape configuration of the four cell stage suggests that the AB daughters are equivalent. Rearrangement of the *P. equorum* four cell stage to the planar rhombus form places only one of the AB daughters in contact with P_2

and this would allow signals from P_2 to distinguish ABp from ABa fate.

Left-right axis specification is also dependent on inductive interactions. Despite the inversion of the dorsal/ventral axis, the *C. elegans* embryos manipulated to exchange ABa and ABp developed into worms with the normal, invariant, left-right anatomical handedness (Priess and Thomson, 1987). This suggested that the left-right axis, like the dorsal/ventral axis, was also not specified prior to the four cell stage. In the normal embryo, handedness first becomes apparent at the six cell stage, when the divisions of ABa and ABp, occurring across the egg, skew such that the daughters on the left side are slightly anterior to those on the right. By applying pressure to the egg shell above ABal with a blunt microneedle, both ABal and ABpl, the left daughters can be pushed towards the posterior such that the right daughters now take up an abnormal position slightly anterior to the left daughters (Wood, 1991). Once again these manipulated embryos develop perfectly normally with the one exception that the anatomical handedness is completely reversed. ABal and ABar share equivalent developmental potential, as do ABpl and ABpr. Most of the left/right differences result from an inductive signal from the MS founder cell to ABara and ABalp at the 12 cell stage (Hutter and Schnabel, 1994; Mango *et al.*, 1994). There are two additional cellular interactions that specify minor left/right differences in the AB lineage later in embryogenesis (Hutter and Schnabel, 1994, 1995). These observations revealed that the left and right sides are not intrinsically specified by determinants segregating at the first left/right cell division and the distinctions in fate between the blastomeres on the left and right sides are dependent on intercellular signalling.

Not all nematode species show invariant handedness like *C. elegans*. For *P. equorum*, 2.5% of individuals have reversed left-right asymmetry (zur Strassen, 1896) while *Bradynema rigidum* shows no preference in anatomical handedness (Wood, 1998). Indeed *C. elegans* embryos lacking the egg shell also develop with variable handedness (Wood and Kershaw, 1991). This variability would be consistent with left/right asymmetries being specified by intercellular signalling and would be harder to reconcile with a determinate, mosaic mode of development.

The pattern of cell divisions in the AB lineage is in marked contrast to that of the determinate divisions of the germ line. The divisions to generate the daughters and granddaughters of AB are oriented across, rather than along, the long axis of the egg and divisions within the AB lineage are synchronous and give cells of equal size. Distinction of fates within the AB lineage is heavily dependent on intercellular communication, in contrast to the autonomous specification of the germ line.

Inductions concerning E

EMS appears, at first glance, more like the autonomously specified germ line than AB. EMS divides along the long axis of the egg, EMS daughters are slightly different in size, and cells in the EMS lineage do not divide synchronously, the E and MS founder cell lineages having their own cell division tempos. Furthermore, apart from the germ line progenitor P_4, E is the only cell to give rise to all of and only one tissue, the intestine, and does so in all nematode species examined, including the highly indeterminate enoplids. Despite these observations making E a prime candidate for autonomous specification, E fate does indeed also depend on an inductive interaction in *C. elegans*.

The first suggestion that E fate may be dependent on an inductive interaction from P_2 came from observations on the development of four-cell *C. elegans* embryos from which P_2 was extruded (Schierenberg, 1987). In embryos from which P_2 was extruded, EMS went through the typical, unequal, cell division but; the nucleus of E did not take up it's characteristic peripheral position in the cell (usually adjacent to P_2), the E cell lineage continued to divide at the same rate as MS rather than slowing down and gut granules, characteristic of gut differentiation, were not seen. Similar observations were made when EMS was isolated from a four cell embryo (Goldstein, 1995; Goldstein, 1992; Goldstein, 1993). In contrast, E behaves more as it would in an intact embryo when generated from an isolated P_1, in which EMS and P_2 remain in contact, or if P_2 is placed back up against EMS after their separation.

E founder cell specification has also been investigated by cell extrusion experiments in the cephalobid *Acrobeloides nanus* (Wiegner and Schierenberg, 1998). The cells of the intestine in this species also appear to develop exclusively from an E founder cell generated by the same pattern of cell divisions as in *C. elegans*. However, P_2 was found to be unnecessary for EMS descendants to differentiate as gut cells. In fact, gut forming potential in *A. nanus* was apparently cell autonomous, and present in all three cells of the three cell stage; i.e. in AB and P_2 as well as in EMS! Once again intercellular communication is involved in gut specification, although in *A. nanus* the process appears to be inhibitory, to limit gut forming potential to EMS. A similar mechanism would be expected to exist in the enoplid embryo to restrict gut forming potential to a single cell of the eight cell stage, in the absence of specification of the remaining seven cells.

The distinction between the invariant mode of development for nematodes such as *C. elegans* or *A. nanus* and the indeterminate mode of the Enoplids may not be so significant. The level of conservation of developmental processes across the phylum could be further emphasized as molecular knowledge gained for *C. elegans* is extended to other nematode species.

Molecular Understanding of Developmental Mechanisms

Caenorhabditis elegans was originally selected as a model system with which to study animal development because of the suitability of this species for genetic analysis (Brenner, 1974). Genetics is a powerful approach with which to study complex biological topics such as development. With an appropriate screen of a mutagenised population, mutations can be rapidly isolated that will identify key components of a developmental process. Cloning of the gene affected by a mutation leads quickly onto an understanding of that developmental process at the molecular level. Molecular techniques could then allow homologous genes to be cloned from other nematode species that are not so suitable for genetic analysis, and this may provide a molecular understanding of developmental processes across the phylum and beyond.

Initial polarity

The orientation of the first embryonic cell division, along the anterior/posterior axis, is a characteristic of nematodes. Indeed, those initial, asymmetric divisions are fundamental to nematode embryogenesis segregating the germ line away from the soma. Mutations were isolated which affected the asymmetry of those first divisions in *C. elegans* and thereby identified the *par* genes (<u>par</u>titioning defective) (Kemphues and Strome, 1997).

Mutations in all the *par* genes, except *par-4*, result in the first division generating daughter cells of the same size which subsequently divide simultaneously, or nearly so. Both the daughter cells of that first division divide either transversely (like AB in a normal embryo) for *par-1, par-2, par-4* and *par-5* mutants or longitudinally (like P_1 in a normal embryo) for *par-3* and *par-6* mutants. Mutations in the *par* genes are maternal effect; i.e. the mother must be homozygous for a *par* gene mutation for the embryo to be affected and the *par* genes are required in the mother for proper development of her eggs. Antibodies raised to proteins generated from the cloned genes revealed that PAR-3 and PAR-6 proteins are localized to the anterior periphery of P_0 and PAR-1 and PAR-2 proteins are localized to the posterior periphery of P_0 (Kemphues and Strome, 1997; Hung and Kemphues, 1999). This localization of PAR-1 and PAR-2 to the side of the cell that will remain in the germ line, with PAR-3 and PAR-6 on the side of the cell that will be relegated to the soma, continues in the subsequent germ line cell divisions. In the somatic blastomeres, PAR-3 and PAR-6 remain evenly distributed around the cell periphery. PAR-3 and PAR-6 are required for the proper localisation of PAR-1 and PAR-2, and PAR-2 is required for the proper localisation of PAR-1, PAR-3 and PAR-6. PAR-1 is a putative serine/threonine kinase and PAR-3 and PAR-6 contain PDZ domains that appear to be involved in sub-membranous protein-protein associations. PAR-2 is a novel protein. An atypical protein kinase C (PKC-3) has been found to co-localise with PAR-3 and PAR-6, and is also required for correct localisation of PAR proteins (Tabuse *et al.*, 1998). Clearly these proteins have a major role in the initial asymmetric cell divisions and the specification of germ line and somatic identity. However, it is not yet fully understood how these proteins become localised in this fashion or how they lead to the distinct germ-line and somatic fates for the cells into which they are partitioned. Identification of *par* gene homologues in

insects and mammals has shown, however, that this mechanism is used extensively across the animal kingdom in asymmetric cell divisions.

The PAR system is involved in localisation of germ line determinants during the asymmetric divisions of the early embryo. Cytoplasmic streaming, visible in the light microscope, is thought to carry these determinants to the posterior of P_0 prior to the first division. P granules, observable with an electron microscope (Wolf, Priess and Hirsh, 1983) or more conveniently by immunofluorescence microscopy (Strome and Wood, 1982), co-localise to the germ line throughout development. P granules are ribonucleoprotein complexes and contain GLH-1 (a putative RNA helicase), MEX-3 (with a putative RNA binding domain) and PGL-1 (a novel protein) as well as poly(A)$^+$ RNAs, but evidence on the nature of the germ line determinants is still lacking.

The PAR system also has a role in directing the orientations of early cell divisions. During the first cell division the centrosomes are at opposite ends of the spindle aligned along the long axis of P_0. When the nuclei of AB and P_1 form, the centrosomes are on the anterior side of the nucleus in AB and on the posterior side in P_1. Upon replication of the centrosomes they move away from each other around the nuclei such that they are now arranged perpendicular to the longitudinal axis. In AB the spindle now forms between the two centrosomes, across the egg, and leads to a transverse cell division. However in P_1 the centrosomes are realigned so that the spindle forms along the long axis once again and a longitudinal division ensues. The centrosomes act as microtubule organising centres, and in P_1 a site at the anterior cortex of the cell captures the aster microtubules (Hyman, 1989; Hyman and White, 1987). One of the centrosomes, apparently either is equally likely, is then pulled towards this anterior site twisting the nucleus through 90° to attain the necessary alignment for a longitudinal division. This realignment also occurs at subsequent germ line divisions.

The orientation of the anterior/posterior axis appears to be established by the point of sperm entry (Goldstein and Hird, 1996). In *C. elegans* the sperm usually fuses with the oocyte on the side of the oocyte that first enters the spermatheca. By manipulating the conditions for fertilisation, the point of fusion was made more variable, but nevertheless the sperm pronucleus always marked the posterior of the embryo. Observations on ten other species of Rhabditidae, including species where the point of sperm entry varies naturally, suggest that the point of sperm entry also defines the posterior pole in other species of nematode. The asters of the sperm pronucleus may initiate cytoplasmic reorganisation, upon entry into the oocyte cytoplasm, which leads to the anterior/posterior polarity. Cytoplasmic streaming toward the sperm pronucleus can be seen and this may drive the sperm pronucleus into either of the poles of the oocyte. Curiously, in *Turbatrix aceti* and *Panagrellus redivivus*, of the Panagrolaimidae, sperm entry appears to specify the posterior pole, rather than the anterior pole (Goldstein, Hird and White, 1993). It is hard to see how such a reversal of the main embryonic axis could have occurred during evolution except via a form where the axis was not so specified and there are parthenogenetic nematode species which have no sperm with which to establish the anterior/posterior axis.

The primary signal for anterior/posterior axis specification does appear to vary across the nematode phylum. Molecular details of how that axis is interpreted are only starting to become clear for *C. elegans*. Extension of this analysis to other species in the phylum, starting from the cloned *C. elegans* genes and using standard molecular biology techniques, would reveal how this crucial developmental process, that initiates the entire developmental cycle, has been modified through evolution.

Specification of blastomere identity

The initial asymmetric cleavage provides the distinction between the AB founder cell and P_1, but genetic analysis has also identified some of the key regulators of specification of the other founder cells in *C. elegans*.

The fundamental distinction between soma and germ line may involve *pie-1* (pharynx, intestine in excess) (Mello *et al.*, 1992). The *pie-1* gene was identified with mutations that appeared to transform the germ line blastomere P_2 into an EMS-like cell, EMS being the somatic sister of P_2. There appears to be a general transcriptional repression in the germ line blastomeres and this is lost in *pie-1* mutants (Seydoux *et al.*, 1996). Many genes have been found which are normally transcribed in all somatic cells, but not the germ line cells, of the early embryo, with transcripts of some of these genes detectable as early as the four cell stage. Transcripts for these genes can be detected by *in situ* hybridisation in the germ-line blastomeres in *pie-1* mutants. Lack of such repression over an EMS promoting factor (e.g. *skn-1*, see below) in P_2 of a *pie-1* mutant could be the cause of the P_2 to EMS fate transformation. PIE-1 protein is only detected in the nuclei of the germ-line blastomeres P_1, P_2, P_3 and P_4 of the early embryo (Mello *et al.*, 1996). Segregation of PIE-1 to the germ line during the early cell divisions appears to be mediated by centrosome association. The lack of sequence homology for PIE-1, however, provides no clues as to how this protein mediates generalised, transcriptional repression in the germ line.

EMS founder cell identity may be specified by *skn-1* (skin (epidermis) in excess) (Bowerman *et al.*, 1992a). In *skn-1* mutants both founder cell daughters of EMS are affected; MS doesn't produce pharyngeal cells and E only generates intestinal cells in a fifth of the mutant embryos. Instead extra epidermal and muscle cells are present suggesting a transformation to the C founder cell fate. The *skn-1* mRNA appears uniformly distributed in 2 and 4 cell embryos (Seydoux and Fire, 1994) but post-transcriptional control leads to SKN-1 protein being most abundant in the nuclei of P_2 and EMS (Bowerman *et al.*, 1993). SKN-1 is a DNA binding protein and presumably controls expression of downstream genes. The lower SKN-1 protein levels in AB daughters and general, germ line, transcriptional repression in P_2 (see above) may mean that SKN-1 only activates transcription of those downstream genes in EMS.

Specification of the C and D founder cell fates may involve the *pal-1* gene (posterior alae) (Waring and Kenyon, 1991). This gene was originally identified from a postembryonic function but maternal loss of *pal-1* activity leads to the absence of C and D derived muscle cells (Hunter and Kenyon, 1996). PAL-1 protein is nuclear-localised in EMS and P_2, and their daughters. This observation and homology to Caudal, a homeobox protein involved in specifying posterior identity in *Drosophila melanogaster*, suggest PAL-1 also directs development by regulation of downstream genes.

Distinction of developmental fates within a founder cell lineage

While specification of the EMS blastomere has been addressed in the previous section, that of the E and MS cells has not. Recent molecular work may raise questions about the mechanistic significance of the designation of E and MS as founder cells, which was based on the classically perceived distinct developmental potential of the early blastomeres. The molecular mechanism that distinguishes E and MS in fact also distinguishes between daughters within other founder cell lineages. Mechanistically, perhaps AB, EMS, C and D, the immediate somatic daughters of all four asymmetric germ-line blastomere divisions, ought to be considered the key progenitor cells.

Distinction of the MS cell fate from that of E crucially involves *pop-1* (posterior pharynx defective). Mutations in the *pop-1* gene were originally identified from causing MS, which normally generates the posterior of the pharynx, to be transformed into an additional E cell (Lin, Thompson and Priess, 1995). However, POP-1 is also required in the anterior daughters, ABxxa cells, from the third round of cell divisions in the AB cell lineage (Lin, Hill and Priess, 1998). Immediately after cell divisions, both in the early embryo and postembryonically, POP-1 protein was detected at higher levels in anterior daughters and this distinction was only apparent in cell divisions aligned with the anterior/posterior axis. The POP-1 levels are reset before the next division. Cells appear to 'remember' the series of anterior versus posterior cell divisions by which they are generated, as recorded in the changes in POP-1 levels, in order to 'deduce' their final fates. After division of EMS, POP-1 is more abundant in MS than E and this is required for the MS fate. The POP-1 protein contains a HMG box suggesting possession of DNA binding activity and may, like SKN-1, PAL-1 and PIE-1, control development through regulation of expression of downstream genes.

The asymmetry of POP-1 is achieved by two distinct mechanisms (Meneghini *et al.*, 1999). POP-1 is down-regulated in posterior daughters, including E, by a protein kinase pathway. Components of this pathway are encoded by *mom-4* (more of MS) (a mitogen-activated protein kinase kinase kinase (MAP3K) homologue) and *lit-1* (loss of intestine) (a MAP kinase homologue), both identified genetically. Further components, WRM-1 (an armadillo/β-catenin homologue) and TAP-1 (a TAB1 like protein), have been identified through molecular approaches. POP-1 is also downregulated in E specifically by a Wnt signalling system emanating from P$_2$. Components of this system are encoded by *mom-1* (a porcupine homologue) and *mom-2* (a wingless/Wnt homologue), and include WRM-1 again. Components of this Wnt signalling system appear to influence cytoskeletal polarity in early blastomeres. Both these signalling systems are used extensively across the animal kingdom to control developmental processes.

The EMS blastomere is distinct from the AB, C and D blastomeres, the other somatic daughters of the germ line blastomeres, in that it divides along the anterior/posterior axis, a result of a specific realignment in response to a signal from P$_2$. Without this realignment, the POP-1 asymmetry that distinguishes E and MS would not be created and subsequent divisions within the EMS lineage would then be initially synchronous as in the AB, C and D lineages, the daughters of EMS having equivalent developmental potential. This conclusion then questions the evolutionary consistency, which might have been assumed from classical embryological descriptions, of the mechanism by which the single E blastomere, the intestinal progenitor, is specified across the nematode phylum.

Several of the proteins involved in other intercellular signalling events that distinguish cell fates in the AB lineage have also been identified through genetic analysis. The *glp-1* gene (germ line proliferation defective) (Priess, Schnabel and Schnabel, 1987; Austin and Kimble, 1987) encodes the receptor for the P$_2$ to ABp inductive signal and the MS to ABara and ABalp inductive signal (both described above) (Hutter and Schnabel, 1994; Mello *et al.*, 1994; Moskowitz *et al.*, 1994). The GLP-1 protein is also a receptor for other signalling events including some later in embryogenesis, where it acts redundantly with the homologous receptor LIN-12 (Christensen *et al.*, 1996), and one postembryonically, which maintains division of the germ line in the gonad, the function through which this gene was originally identified (Priess, Schnabel and Schnabel, 1987; Austin and Kimble, 1987). The *glp-1* transcript is detectable in all cells of the early embryo but is only translated in cells of the AB lineage, one consequence of the action of the *par* genes (Evans *et al.*, 1994, Crittenden *et al.*, 1997). The GLP-1 protein is localised around the entire plasma membrane surface of all cells of the AB lineage in the early embryo and therefore is appropriately positioned to receive the inductive signals from wherever they may originate (Evans *et al.*, 1994). Transduction of the signal to control gene expression appears to be mediated entirely by *lag-1* (*lin-12* and *glp-1*) (a homologue of CBF1 and Supressor of hairless) identified genetically from a mutant phenotype identical to that of *lin-12, glp-1* double mutants (Christensen *et al.*, 1996).

The *apx-1* gene (anterior pharynx in excess) encodes the signal for the P$_2$ to ABp induction, but not the MS to ABara/ABalp induction for which the signal has yet to be identified. Mutations in *apx-1* were identified as causing an ABp to ABa fate transformation (Mango *et al.*, 1994; Mello *et al.*, 1994). APX-1 protein is only produced in P$_2$ and is localised to the surface of P$_2$, adjacent to ABp, where it interacts directly with the GLP-1 receptor (Mickey *et al.*, 1996). The GLP-1/APX-1 interaction is one example of a type of homologous receptor/ligand interactions that includes the Notch/Delta interaction in *Drosophila melanogaster* and controls diverse aspects of development across the animal kingdom.

Other embryonically expressed transcription factors

Regulation of gene expression is a key aspect of developmental control and many other transcription factors have been identified that are either required for, or at least expressed during, later stages of *C. elegans* embryogenesis. Some have been identified through mutations (Way and Chalfie, 1988), some through expression pattern screens (Hope, 1994), some through sequence homology (Burglin *et al.*, 1989), some through purification of a specific DNA binding activity (Hawkins and McGhee, 1995). Presumably these also direct development through yet more gene expression regulation.

The E founder cell fate would appear to be the simplest founder cell fate to specify, generating just intestinal cells, but E founder cell fate specification has proven somewhat more complicated than may have been expected. A gut specific esterase, encoded by *ges-1*, is expressed specifically in the early E-lineage when the E lineage consists of just 4 cells (Edgar and McGhee, 1986). Analysis of the promoter region of *ges-1* identified elements, containing the sequence GATA, which could drive E lineage specific transcription (Egan *et al.*, 1995). A search for proteins that could bind to these promoter elements, identified the gene *elt-2*, that encodes a member of the GATA transcription factor family which were so named because they bind to GATA containing promoter elements (Hawkins and McGhee, 1995). ELT-2 is expressed at the 2 E cell stage, can be sufficient to drive expression of *ges-1* and is essential for development of the intestine (Fukushige *et al.*, 1998). However, *elt-2* is not essential for expression of *ges-2*, suggesting some redundancy in the regulators of this gene.

A genetic screen identified an 'endoderm-determining region', distinct from *elt-2*, required for E lineage development (Zhu *et al.*, 1997). GATA factors are frequently involved in gut specification in animals and yet another GATA transcription factor gene, *end-1*, was found within this region. *end-1* transcripts are detectable in the E founder cell and END-1 protein activates transcription of *elt-2*. However, *end-1* was found to be sufficient but not essential for direction of intestinal development (Zhu *et al.*, 1998) and the 'endoderm-determining region' appears to contain another gene that is redundant with *end-1* for E lineage specification. Redundancy may be a common characteristic of transcriptional regulators and makes it difficult to unravel the complexities of developmental control.

It has been suggested that some transcription factors may specify entire organs with apparently little regard to the cell lineage or tissue type (Mango, Lambie and Kimble, 1994; Schnabel and Schnabel, 1990; Labouesse and Mango, 1999). The gene *pha-4* (pharynx development abnormal) is thought to specify the cells of the pharynx. This gene was identified both genetically as a gene required for pharynx development (Mango, Lambie and Kimble, 1994) and as *Ce-fkh-1* through a screen for genes homologous to the Forkhead/HNF3β family of transcription factors (Azzaria *et al.*, 1996). That *pha-4* and *Ce-fkh-1* were the same gene was discovered subsequently (Horner *et al.*, 1998, Kalb *et al.*, 1998). PHA-4 acts through *pha-1*, which encodes a b-ZIP transcription factor, and *ceh-22*, which encodes a homeobox transcription factor, as well as other routes, to direct pharyngeal-specific gene expression. *pha-1* and *ceh-22* provide yet another example of redundancy in gene function (Okkema *et al.*, 1997).

Some transcription factors appear to be involved in specification of tissue types. The *hlh-1* gene (helix-loop-helix class of transcription factor), the *C. elegans* homologue of the MyoD family of vertebrate myogenin genes that promote muscle differentiation, was cloned (Krause *et al.*, 1990). This gene is indeed expressed in the embryonic cells that only give rise to body wall muscle cells suggesting that the *C. elegans* gene also has a role in muscle cell specification. However, HLH-1, although essential for viability, is not required for muscle cell differentiation or the expression of several muscle specific genes (Chen *et al.*, 1992). Once again, the developmental control appears more complicated than might, naively, have been expected.

There is a cluster of homeobox transcription factor genes, identified through genetic and molecular approaches, that appear to be involved in positional specification in *C. elegans* (Salser and Kenyon, 1994; Wang *et al.*, 1993; Kenyon and Wang, 1991; Burglin *et al.*, 1991, Ruvkun and Hobert, 1998). The homologous gene clusters in insects and vertebrates have a major role in positional specification down the anterior/posterior axis. The expression patterns and the mutant phenotypes described for the *C. elegans Hox* cluster genes suggests an equivalent role in *C. elegans*. A *Hox* cluster must have been involved in specification of anterior/posterior identity in the common ancestor of arthropods, chordates, nematodes and most other animal phyla, but the *C. elegans* version appears to be degenerate as compared to the ancestral *Hox* cluster. *C. elegans Hox* gene expression, in discrete overlapping domains down the anterior/posterior axis, starts during elongation of the embryo and continues through postembryonic development.

Many other transcription factors, beyond those discussed here, have been shown to be involved in embryonic development. Many more are likely to be involved (Ruvkun and Hobert, 1998).

Vulva development

Although little attempt has been made, so far, to extend our molecular understanding of *C. elegans* embryonic development to other nematode species, postembryonic development, of the vulva, has been explored, at a molecular genetic level, in *Pristionchus pacificus*. The vulva is not essential for viability in these two species; eggs, generated by self-fertilisation, can develop within the hermaphrodite gonad, without being laid, and the larva eventually escape from the dying mother. Viability of strains homozygous for mutations preventing formation of the vulva has made development of the vulva an excellent target for genetic analysis. Genetic analysis, along with cell ablation studies (see above), has provided a detailed understanding of the regulatory mechanisms by which vulva development in *C. elegans* is controlled (Kornfeld, 1997) and this has formed the basis for extension of the work to *P. pacificus* (Eizinger, Jungblut and Sommer, 1999).

Our molecular understanding of the development of the vulva in *C. elegans* mainly concerns the components of the three, signal transduction systems (summarised in (Kornfeld, 1997)). First, transduction of the anchor cell signal from the gonad to the vulval precursor cells (VPCs) is known in detail and includes, in order of action: the signal, encoded by *lin-3*, an epidermal growth factor homologue; the receptor, a tyrosine kinase, encoded by *let-23*; an SH2/SH3 adapter, encoded by *sem-5*; a RAS guanine nucleotide exchange factor, encoded by *let-60*; a kinase cascade, encoded by *lin-45*, *mek-2* and *mpk-1*; and transcription factors, encoded by *lin-1* and *lin-31*. This type of signal transduction system is used extensively across the animal kingdom. Second, the inhibitory signal from the syncytial epidermis consists of two parallel, redundant pathways, a situation difficult to approach genetically. However, such is the power of the genetics for this subject, that genetic analysis has provided the genes for many of the components of this signalling system, e.g. *lin-8*, *lin-9*, *lin-15*. And finally, the lateral inhibitory signal, from the VPC following the primary fate to those following the secondary fate, involves the receptor encoded by *lin-12*. This signal transduction pathway is thought to be homologous to that involving *glp-1* in the early embryo (see above).

The cells of the vulval equivalence group, cells with the potential to give rise to the vulva, are arranged along the anterior/posterior axis of the ventral epidermis and the *Hox* cluster genes, *mab-5* and *lin-39*, have key roles in patterning their developmental fates in both *C. elegans* and *P. pacificus* (Eizinger and Sommer, 1997; Clark, Chisholm and Horvitz, 1993). In *lin-39* mutants of both species, the vulva is not formed; all the VPCs follow the same fate as the cells of the vulval equivalence group that do not normally contribute to formation of the vulva, fusion with the epidermal syncytium in *C. elegans* and programmed cell death in *P. pacificus*. However, in *C. elegans lin-39* also has a later role in formation of the vulva in response to the inductive signal from the anchor cell (Maloof and

Kenyon, 1998). This role is not apparent in *P. pacificus* as in this species *lin-39* is not required, in the absence of programmed cell death, for vulva formation (Sommer *et al.*, 1998). In both species, *mab-5* is required for correct development of P8p, the cell of the vulval equivalence group immediately posterior to the VPCs, but this may be at a single step in *C. elegans* and two steps in *P. pacificus* (Eizinger, Jungblut and Sommer, 1999). This work is starting to reveal the complexity of change that occurs at the genetic level for even subtle evolutionary changes in developmental morphology.

Current Comprehension of Nematode Developmental Mechanisms

Although this is a brief summary of a large body of work, there are still relatively few genes so far identified as involved in *C. elegans* embryonic development. Most of the creative developmental processes are operating during early embryogenesis and many more genes than this would be expected to be involved. Gaps remaining in our knowledge include the immediate, mechanistic environment in which many of these proteins work and how the activities of the gene products, often transcriptional regulators, lead subsequently to the complex morphological development followed by the founder cell lineages. In addition, few homologues for these genes have as yet been characterised from other nematode species, other than the closely related *Caenorhabditis briggsae* (see below), so the generality of these processes across the nematode phylum is not known, although the homology to insects and vertebrates make it very likely.

The Genome Project

Molecular genetic analysis in *C. elegans* has been revolutionised first by the generation of the physical map of the *C. elegans* genome (Coulson *et al.*, 1988) and now by the completion of the sequencing of the genome (The *C. elegans* sequencing consortium, 1998). The resulting stimulation of progress in molecular analysis in *C. elegans* will facilitate extrapolation of our understanding of development across the nematode phylum. Furthermore, genome projects for other nematode species have spun off from the *C. elegans* genome project.

The Physical Map

While genetic analysis has always been rapid for *C. elegans*, in the initial phase of the *C. elegans* model system cloning of the genes defined by interesting mutations was very laborious. Usually, the cloning required extended chromosomal walks from the nearest cloned genetic marker. Successive hybridisations to genomic DNA libraries yielded a series of clones, each to be checked for the ability to reveal a chromosomal rearrangement likely to be responsible for the mutant phenotype. These chromosomal walks were carried out independently by many individual researchers. Construction of the physical map for the *C. elegans* genome meant that usually chromosome walking was no longer necessary.

The physical map is a fully catalogued set of overlapping genomic DNA clones for which relative positions in the genome have been determined (Coulson *et al.*, 1986). Techniques were developed so that clones with overlapping inserts could be rapidly and reliably identified from a shotgun, genomic DNA library. This allowed assembly of 'contigs', sets of clones covering a contiguous portion of the genome. The orientation and relative position of each contig also had to be determined, in terms of the pre-existing genetic map. Some contigs included previously cloned genetic markers and so could be placed in context immediately. Other contigs were positioned through chromosomal *in situ* hybridisation. Juxtaposition of some contigs revealed overlaps previously missed because of their small size. But there were still many gaps between contigs apparently due to segments of the genome that could not be cloned in the bacterial cosmid vectors used to construct the genomic DNA libraries. The advent of Yeast Artificial Chromosome (YAC) vectors, with larger inserts and eukaryotic host, allowed bridging of almost all of these gaps (Coulson *et al.*, 1988). Both the clones and the data making up the physical map were made freely available to other laboratories studying *C. elegans*.

Cloning a gene defined by mutation now involves genetic mapping with respect to connections between the physical and genetic map to identify an interval of the physical map where the gene must lie. Such connections are constantly being added both as more genes are cloned and, in particular, through characterisation of molecular polymorphisms. The more connections, the easier it is to locate a gene. Clones from that region of the physical map identified through genetic mapping are then assayed for the presence of the gene. Development of techniques for *C. elegans* transformation (Fire, 1986; Stinchcomb *et al.*, 1985) means that now clones from the physical map are frequently shown to contain a gene by their ability to complement a mutation.

After a gene had been cloned it would be sequenced and this, like the cloning, had also been done independently by individual researchers. In the same way as the physical map improved the efficiency of gene cloning within the field so the genome sequencing project has improved the efficiency of the sequencing of *C. elegans* genes.

Sequencing the *C. elegans* Genome

The *C. elegans* physical map was being completed as discussion about the possibility of sequencing the entire human genome was climaxing. The prior existence of the physical map, considered by many a crucial foundation for a genome sequencing attempt, has made the *C. elegans* genome project a key model for the range of genome projects initiated over the past decade. Completion of the 97Mbp *C. elegans* genomic sequence on schedule (The *C. elegans* sequencing consortium, 1998; Sulston *et al.*, 1992; Wilson *et al.*, 1994), the first genome of a multicellular organism to be sequenced, with an error rate of less than 10^{-4} has set quite a standard to follow.

Genomic DNA cosmid clones with minimal overlap were selected, from the physical map, for sequencing. Each clone was sequenced with a shotgun phase first, in which random sub-clones were sequenced, followed by a more directed phase to close gaps. Regions not covered by cosmids were sequenced using fosmid genomic DNA clones, using long range PCR (polymerase chain reaction) products or using YAC clones. Although a few difficult gaps in the sequence do remain these are considered unlikely to have much biological significance.

The genomic sequence has been subjected to preliminary computer analysis as generated, but full analysis will take many years. Repetitive DNA and tRNA genes are identified. Predictions of protein coding genes (19,099 in total), including the intron/exon structure, are built around splice site consensus sequences, codon usage bias, EST (Expressed Sequence Tag) data and sequence homology. Over 70 percent of the genes in the *C. elegans* genome show homology to genes identified in other organisms. Nearly half of these appear to show homology only to genes found in other nematodes and, although whole genome analysis is only just beginning, this does suggest a significant proportion of the genome may be nematode specific.

A *C. elegans* Database (ACeDB)

The sequence data, like the physical map data and the genetic map data, have always been made freely available and databases have been developed to make these data accessible. The current version, ACeDB (Thierry-Mieg, Thierry-Mieg and Stein, 1999; Eeckman and Durbin, 1995), is vital for working with the huge body of data that has been generated for this single species. At the heart of ACeDB are the three different views of the *C. elegans* genome; genetic loci related by recombination distance (the genetic map), genomic DNA clones with overlaps measured in *Hind*III restriction enzyme sites (the physical map) and the genomic sequence with scale in nucleotides (the sequence map). Each view of the genome includes related data, such as chromosomal rearrangements in the genetic map window and gene structure predictions in the sequence map window. Interconnections between these views, where identified, for example through a cloned genetic locus, are integral to the database. Annotation is extensive and additional links are provided from objects within these views to windows with further information. ACeDB is an essential resource, integrating the whole body of knowledge about this species and constantly referred to by laboratories dedicated to working with *C. elegans*.

Recently, steps have been taken to make ACeDB available over the internet (Thierry-Mieg, Thierry-Mieg and Stein, 1999) although initially some of the power of the locally maintained versions proved difficult to transfer to the web. Current sites for accessing ACeDB are:

http://wormbase.sanger.ac.uk/
http://alpha.crbm.cnrs-mop.fr/acedb/
http://www.wormbase.org/

While these addresses may change, access to a version of the *C. elegans* database is likely to be found through The Sanger Centre (*http://www.sanger.ac.uk/*) for many years to come.

Global Analysis

The availability of the genome sequence data has opened up entirely new and powerful avenues of biological study. First there is more detailed computer analysis of the data, which may focus on gene families and biological function (Clarke and Berg, 1998; Ruvkun and Hobert, 1998; Bargmann, 1998) or interspecies comparisons (Chervitz *et al.*, 1998; Blaxter, 1998). Second, the complete genomic sequence permits entire families of genes and their products to become subjects of study. The problem of overlapping activities, which can complicate genetic or biochemical studies, may be alleviated when the genome sequence is available and the full complement of a gene family present in an organism's genome is known. Third, techniques are being developed to examine the activity of the entire genome, specifically gene expression.

Expression pattern data for most, if not all, of the genes in the *C. elegans* genome may be one aspect of the knowledge base through which a full comprehension of this organism's biology will be achieved. Experiments to characterise the developmental expression patterns of genes predicted in the genome sequence data, with no prior consideration of biological function, have used reporter gene fusion (Lynch, Briggs and Hope, 1995) or *in situ* hybridisation (Tabara, Motohashi and Kohara, 1996; Birchall, Fishpool and Albertson, 1995) approaches. Steps are being taken to integrate gene expression pattern data into the *C. elegans* database ACeDB (Hope *et al.*, 1996).

Microarray technology is a new and powerful tool for simultaneously examining the expression levels of an entire gene complement (Shalon, Smith and Brown, 1996) and microarrays have been developed for *C. elegans* (*http://cmgm.stanford.edu/~kimlab/*). This technology, although presently lacking the capacity to provide the spatial aspect of developmental gene expression, should yield volumes of valuable data on gene regulation.

Genome Projects for other Nematode Species

A genome project has been initiated for another nematode of the same genus as *C. elegans*, *Caenorhabditis briggsae* (*http://genome.wustl.edu/gsc/Projects/briggsae.shtml*) and more than 8 per cent of the genome has been sequenced so far. The *C. briggsae* data is valuable for comparison with that for *C. elegans* as the evolutionary distance between these two species is considered sufficient that only functional elements are conserved. Intron/exon structure predictions may be confirmed by such comparisons (Clark *et al.*, 1995) and gene regulatory elements may be revealed (Gilleard, Barry and Johnstone, 1997). A genome project for the more distantly related *Pristionchus pacificus* has also been initiated purely for comparative purposes (Blaxter, 1998).

In addition to these two species, genome projects exist for several parasitic nematodes, primarily because of their medical or agricultural significance. Their distribution across the phylum, however, will also make them valuable for comparative purposes. The most advanced is for *Brugia malayi*, of the Spirurida, which causes human elephantiasis (*http://www.nematodes.org*) (Williams *et al.*, 1999), but genome projects, mostly EST projects, are underway for members of the Strongylida, Tylenchida, Ascaradida and Trichocephalida (Blaxter, 1998). The sequence data generated may prove to be of value in comprehension of developmental mechanisms across the nematode phylum.

Conclusion

The concept of Körpergrundgestalt or phylotypic stage (summarized in Slack *et al.*, 1993) suggests that there is a stage of development which all members of a phylum pass through and at which members of a phylum show maximum similarity. For different species within a phylum, development both before and after this stage may show considerable variation. The phylotypic stage may be a consequence of the constraint imposed on evolution by developmental mechanisms. Developmental processes at the beginning and end of development may be easier to modify during evolution.

For nematodes the phylotypic stage appears to be after completion of embryonic cell divisions and as elongation is commencing. Accounts of nematode embryonic development have tended to focus on the early cleavage events. This is, after all, the start of development and the fewer cells present make early development much easier to follow than post-proliferation development. Early cleavage does show considerable variation across the phylum from the invariant rhabditids to the indeterminate enoplids. Postembryonically, there is even greater variation in development as the adult nematode acquires the characteristics that allow a species to occupy a specific ecological niche. But the L1 stages appear very similar and this reflects the even greater similarity that exists amongst mid to late embryonic stages for species across the phylum.

Comprehension of nematode development to the molecular level is best for very early events (founder cell specification) and very late events (vulva development) in *C. elegans*. Perhaps we should not expect these processes to be strictly conserved across the phylum but their study beyond *C. elegans* is likely be of considerable value in understanding how development is modified through evolution. However, there are many transcription factors, including the *Hox* genes, identified in *C. elegans*, which are expressed around the phylotypic stage. Perhaps the functions of most of these genes will be strictly conserved, possibly beyond the phylum as for the *Hox* genes. An investigation of these genes in nematodes other than *C. elegans* may help to reveal the intricacies of that intractable phylotypic stage which appears so crucial to animal development.

Acknowledgements

I wish to thank the following for providing me with nematode species photographed in this chapter: Mandy Fearnehough, Fritz Muller, Stuart Pickersgill, Einhard Schierenberg, Derek Wakelin and Richard Warwick.

References

Austin, J. and Kimble, J. (1987) *glp-1* is required in the germ line for regulation of the decision between mitosis and meiosis in *C. elegans. Cell*, **51**, 589–599.

Azzaria, M., Goszczynski, B., Chung, M.A., Kalb, J.M. and McGhee, J.D. (1996) A *forkhead*/HNF-3 homolog expressed in the pharynx and intestine of the *C. elegans* embryo. *Developmental Biology*, **178**, 289–303.

Bargmann, C.I. (1998) Neurobiology of the *Caenorhabditis elegans* genome. *Science*, **282**, 2028–2033.

Birchall, P.S., Fishpool, R.M. and Albertson, D.G. (1995) Expression patterns of predicted genes from the *C. elegans* genome sequence visualized by FISH in whole organisms. *Nature Genetics*, **11**, 314–320.

Bird, A.F. (1971) *The structure of nematodes*. New York: Academic Press.

Bird, A.F. and Bird, J. (1991) *The structure of nematodes*. San Diego: Academic Press.

Blaxter, M. (1998) *Caenorhabditis elegans* is a nematode. *Science*, **282**, 2041–2046.

Blaxter, M.L., De Ley, P., Garey, J.R., Liu, L.X., Scheldman, P., Vierstraete, A., *et al.* (1998) A molecular evolutionary framework for the phylum Nematoda. *Nature*, **392**, 71–75.

Boveri, T. (1909) Die blastomerenkerne von *Ascaris megalocephala* und die theorie der chromosomenindividualität. *Archiv für experimentelle Zellforschung*, **3**, 181–268.

Boveri, T. (1910) Uber die Teilung zentrifugierter Eier von *Ascaris megalocephala. Archiv für Entwicklungsmechanik der Organismen.*, **30**, 101–125.

Bowerman, B., Draper, B.W., Mello, C.C. and Priess, J.R. (1993) The maternal gene *skn-1* encodes a protein that is distributed unequally in early *C. elegans* embryos. *Cell*, **74**, 443–452.

Bowerman, B., Eaton, B.A. and Priess, J.R. (1992a) *skn-1*, a maternally expressed gene required to specify the fate of ventral blastomeres in the early *C. elegans* embryo. *Cell*, **68**, 1061–1075.

Bowerman, B., Tax, F.E., Thomas, J.H. and Priess, J.R. (1992b) Cell interactions involved in development of the bilaterally symmetrical intestinal valve cells during embryogenesis in *Caenorhabditis elegans. Development*, **116**, 1113–1122.

Brenner, S. (1974) The genetics of *Caenorhabditis elegans. Genetics*, **77**, 71–94.

Burglin, T., Ruvkun, G., Coulson, A., Hawkins, N.C., McGhee, J.D., Schaller, D., *et al.* (1991) Nematode homeobox cluster. *Nature*, **351**, 703.

Burglin, T.R., Finney, M., Coulson, A. and Ruvkun, G. (1989) *Caenorhabditis elegans* has scores of homeobox-containing genes. *Nature*, **341**, 239–243.

Chen, L., Krause, M., Draper, B., Weintraub, H. and Fire, A. (1992) Body-wall muscle formation in *Caenorhabditis elegans* embryos that lack the MyoD homologue *hlh-1. Science*, **256**, 240–243.

Chervitz, S.A., Aravind, A., Sherlock, G., Balld, C.A., Koonin, E.V., Dwight, S.S., *et al.* (1998) Comparison of the complete protein sets of worm and yeast: orthology and divergence. *Science*, **282**, 2022–2028.

Chitwood, B.G. and Chitwood, M.B. (1974) *Introduction to Nematology*. Baltimore: University Park Press.

Christensen, S., Kodoyianni, V., Bosenberg, M., Friedman, L. and Kimble, J. (1996) *lag-1*, a gene required for *lin-12* and *glp-1* signaling in *Caenorhabditis elegans*, is homologous to human CBF1 and *Drosophila* Su(H). *Development*, **122**, 1373–1383.

Clark, D.V., Suleman, D.S., Beckenbach, K.A., Gilchrist, E.J. and Baillie, D.L. (1995) Molecular-cloning and characterization of the *dpy-20* gene of *Caenorhabditis elegans. Molecular & General Genetics*, **247**, 367–378.

Clark, S.A. (1967) The development and life history of the false root-knot nematode, *Nacobbus serendipiticus. Nematologica*, **13**, 91–101.

Clark, S.G., Chisholm, A.D. and Horvitz, H.R. (1993) Control of cell fates in the central body region of *C. elegans* by the homeobox gene *lin-39. Cell*, **74**, 43–55.

Clarke, N.D. and Berg, J.M. (1998) Zinc fingers in *Caenorhabditis elegans*: finding families and probing pathways. *Science*, **282**, 2018–2022.

Coulson, A., Sulston, J., Brenner, S. and Karn, J. (1986) Towards a physical map of the genome of the nematode *Caenorhabditis elegans*. *Proceedings of the National Academy of Sciences of the United States of America*, **83**, 7821–7825.

Coulson, A., Waterston, R., Kiff, J., Sulston, J. and Kohara, Y. (1988) Genome linking with yeast artificial chromosomes. *Nature*, **335**, 184–186.

Crittenden, S.L., Rudel, D., Binder, J., Evans, T.C. and Kimble, J. (1997) Genes required for GLP-1 asymmetry in the early *Caenorhabditis elegans* embryo. *Developmental Biology*, **181**, 36–46.

Croll, N.A. and Matthews, B.E. (1977) *Biology of Nematodes*. Glasgow: Blackie.

Dasgupta, D.R. and Raski, D.J. (1968) The biology of *Rotylenchulus parvus*. *Nematologica*, **14**, 429–440.

Deppe, U., Schierenberg, E., Cole, T., Krieg, C., Schmitt, D., Yoder, B. and von Ehrenstein, G. (1978) Cell lineages of the embryo of the nematode *Caenorhabditis elegans*. *Proceedings of the National Academy of Sciences of the United States of America*, **75**, 376–380.

Drozdovskiy, E.M. (1968) Contribution to the comparative study of the initial stages of ovum cleavage in nematodes. *Doklady Akademi Nauk SSSR*, **180**, 750–753.

Drozdovskiy, E.M. (1975) Ovum division among species of *Eudorylaimus* and *Mesodorylaimus* (Nematoda; Dorylaimida) and the role of cleavage in determining the composition of subclasses of nematode. *Doklady Akademii Nauk SSSR*, **222**, 1005–1008.

Edgar, L.G., and McGhee, J.D. (1986) Embryonic expression of a gut-specific esterase in *Caenorhabditis elegans*. *Developmental Biology*, **114**, 109–118.

Edgar, L.G. (1995) Blastomere Culture and Analysis. In *Caenorhabditis elegans: Modern biological analysis of an organism.*, Vol. 48 , edited by H.F. Epstein and D. C. Shakes, pp. 303–321. San Diego: Academic Press.

Eeckman, F.H. and Durbin, R. (1995) ACeDB and Macace. In *Caenorhabditis elegans Modern biological analysis of an organism*, Vol. 48 , edited by H.F. Epstein and D.C. Shakes, pp. 584–607. San Diego: Academic Press.

Egan, C.R., Chung, M.A., Allen, F.L., Heschl, M.F.P., van Buskirk, C.L. and McGhee, J.D. (1995) A gut-to-pharynx/tail switch in embryonic expression of the *Caenorhabditis elegans ges-1* gene centers on two GATA sequences. *Developmental Biology*, **170**, 397–419.

Eizinger, A., Jungblut, B. and Sommer, R.J. (1999) Evolutionary change in the functional specificity of genes. *Trends in Genetics*, **15**, 197–202.

Eizinger, A. and Sommer, R.J. (1997) The homeotic gene *lin-39* and the evolution of nematode epidermal cell fates. *Science*, **278**, 452–455.

Evans, T.C., Crittenden, S.L., Kodoyianni, V. and Kimble, J. (1994) Translational control of maternal *glp-1* mRNA establishes an asymmetry in the *C. elegans* embryo. *Cell*, **77**, 183–194.

Fassuliotis, G. (1962) Life history of *Hemicriconemoides chitwoodi* Esser. *Nematologica*, **8**, 110–116.

Felix, M.A. and Sternberg, P.W. (1997) Two nested gonadal inductions of the vulva in nematodes. *Development*, **124**, 253–259.

Fire, A. (1986) Integrative transformation of *Caenorhabditis elegans*. *European Molecular Biology Organization Journal*, **5**, 2673–2680.

Fukushige, T., Hawkins, M.G. and McGhee, J.D. (1998) The GATA-factor ELT-2 is essential for formation of the *Caenorhabditis elegans* intestine. *Developmental Biology*, **198**, 286–302.

Gilbert, S.F. (1997) Specification of cell fate by progressive cell-cell interactions. In *Developmental Biology* Chapter 15, pp. 591–633. Sunderland, Massachusetts: Sinauer Associates Inc.

Gilleard, J.S., Barry, J.D. and Johnstone, I.L. (1997) *cis* regulatory requirements for epidermal cell-specific expression of the *Caenorhabditis elegans* cuticle collagen gene *dpy-7*. *Molecular and Cellular Biology*, **17**, 2301–2311.

Goldstein, B. (1992) Induction of gut in *C. elegans* embryos. *Nature.*, **357**, 255–257.

Goldstein, B. (1993) Establishment of gut fate in the E lineage of *C. elegans*: the roles of lineage-dependent mechanisms and cell interactions. *Development*, **118**, 1267–1277.

Goldstein, B. (1995) An analysis of the response to gut induction in the *C. elegans* embryo. *Development*, **121**, 1227–1236.

Goldstein, B. and Hird, S.N. (1996) Specification of anteroposterior axis in *Caenorhabditis elegans*. *Development*, **122**, 1467–1474.

Goldstein, B., Hird, S.N. and White, J.G. (1993) Cell polarity in early *C. elegans* development. *Development*, **Supplement**, 279–287.

Hawkins, M.G. and McGhee, J.D. (1995) *elt-2*, a second GATA factor from the nematode *Caenorhabditis elegans*. *Journal of Biological Chemistry*, **270**, 14666–14671.

Hope, I.A. (1994) PES-1 is expressed during early embryogenesis in *Caenorhabditis elegans* and has homology to the fork head family of transcription factors. *Development*, **120**, 505–514.

Hope, I.A., Albertson, D.G., Martinelli, S.D., Lynch, A.S., Sonnhammer, E. and Durbin, R. (1996) The *C. elegans* expression pattern database: a beginning. *Trends in Genetics*, **12**, 370–371.

Horner, M.A., Quintin, S., Domeier, M.E., Kimble, J., Labouesse, M. and Mango, S. E. (1998) *pha-4*, an HNF-3 homolog, specifies pharyngeal organ identity in *Caenorhabditis elegans*. *Genes and Development*, **12**, 1947–1952.

Hung, C.-L. and Jenkins, W.R. (1969a) Comparative embryology of two species of *Pratylenchus*. *Journal of Nematology*, **1**, 11.

Hung, C.-L. and Jenkins, W.R. (1969b) Oogenesis and embryology of two plant-parasitic nematodes, *Pratylenchus penetrans* and *P. zeae*. *Journal of Nematology*, **1**, 352–356.

Hung, T.-J. and Kemphues, K.J. (1999) PAR-6 is a conserved PDZ domain-containing protein that colocalizes with PAR-3 in *Caenorhabditis elegans*. *Development*, **126**, 127–135.

Hunter, C.P. and Kenyon, C. (1996) Spatial and temporal controls target *pal-1* blastomere-specification activity to a single blastomere lineage in *C. elegans* embryos. *Cell*, **87**, 217–226.

Hutter, H. and Schnabel, R. (1994) *glp-1* and inductions establishing embryonic axes in *C. elegans*. *Development*, **120**, 2051–2064.

Hutter, H. and Schnabel, R. (1995) Establishment of left-right asymmetry in the *C. elegans* embryo: A multi-step process involving a series of inductive events. *Development*, **121**, 3417–3424.

Hyman, A.A. (1989) Centrosome movement in the early divisions of *Caenorhabditis elegans*: A cortical site determining centrosome position. *Journal of Cell Biology*, **109**, 1185–1193.

Hyman, A.A. and White, J.G. (1987) Determination of cell division axes in the early embryogenesis of *Caenorhabditis elegans*. *Journal of Cell Biology*, **105**, 2123–2135.

Kalb, J.M., Lau, K.K., Goszczynski, B., Fukushige, T., Moons, D., Okkema, P.G. and McGhee, J.D. (1998) *pha-4* is *Ce-fkh-1*, a Forkhead/HNF-3α, β, γ homolog that functions in organogenesis of the *C. elegans* pharynx. *Development*, **125**, 2171–2180.

Kemphues, K.J. and Strome, S. (1997) Fertilization and establishment of polarity in the embryo. In *C. elegans II*, edited by Riddle, D.L., Blumenthal, T., Meyer, B. J. and Priess, J.R., pp. 335–359. New York: Cold Spring Harbor Laboratory Press.

Kenyon, C. and Wang, B. (1991) A cluster of Antennapedia-class homeobox genes in a non-segmented animal. *Science*, **253**, 516–517.

Kimble, J. and Hirsh, D. (1979) The postembryonic cell lineages of the hermaphrodite and male gonads in *Caenorhabditis elegans*. *Developmental Biology*, **70**, 396–417.

Kornfeld, K. (1997) Vulval development in *C. elegans*. *Trends in Genetics*, **13**, 55–61.

Krause, M., Fire, A., Harrison, S.W., Priess, J. and Weintraub, H. (1990) CeMyoD accumulation defines the body wall muscle cell fate during *C. elegans* embryogenesis. *Cell*, **63**, 907–919.

Labouesse, M. and Mango, S.E. (1999) Patterning the *C. elegans* embryo; moving beyond the cell lineage. *Trends in Genetics*, **15**, 307–313.

Laufer, J.S., Bazzicalupo, P. and Wood, W.B. (1980) Segregation of developmental potential in early embryos of *Caenorhabditis elegans*. *Cell*, **19**, 569–577.

Lin, R., Hill, R.J. and Priess, J.R. (1998) POP-1 and anterior and posterior fate decisions in *C. elegans* embryos. *Cell*, **92**, 229–239.

Lin, R., Thompson, S. and Priess, J.R. (1995) *pop-1* encodes an HMG box protein required for the specification of a mesoderm precursor in early *C. elegans* embryos. *Cell*, **83**, 599–609.

Lynch, A.S., Briggs, D. and Hope, I.A. (1995) Developmental expression pattern screen for genes predicted in the *C. elegans* genome sequencing project. *Nature Genetics*, **11**, 309–313.

Malakhov, V.V. (1981) Embryogenesis of free-living marine nematodes from the orders Chromadorida and Desmodorida. *Zoologicheskii Zhurnal*, **60**, 485–495.

Malakhov, V.V. (1994) *Nematodes: structure, development, classification, and phylogeny*. Washington: Smithsonian Institution Press.

Malakhov, V.V. and Akimushkina, M.I. (1976) Embryogenesis of a free-living nematode *Enoplus brevis*. *Zoologicheskii Zhurnal*, **50**, 1788–1799.

Malakhov, V.V. and Cherdantzev, V.G. (1975) Embryogenesis of a free-living marine nematode, *Pontonema vulgare*. *Zoologicheskii Zhurnal*, **54**, 165–174.

Malakhov, V.V. and Spiridonov, S.E. (1981) Embryonic development of *Gastromermis* (Nematoda, Mermithida). *Zoologicheskii Zhurnal*, **60**, 1574–1577.

Maloof, J.N. and Kenyon, C. (1998) The Hox gene *lin-39* is required during *C. elegans* vulval induction to select the outcome of RAS signalling. *Development*, **125**, 181–190.

Mango, S.E., Lambie, E.J. and Kimble, J. (1994) The *pha-4* gene is required to generate the pharyngeal primordium of *Caenorhabditis elegans*. *Development*, **120**, 3019–3031.

Mango, S.E., Thorpe, C.J., Martin, P.R., Chamberlain, S.H. and Bowerman, B. (1994) Two maternal genes, *apx-1* and *pie-1* are required to distinguish the fates of equivalent blastomeres in the early *Caenorhabditis elegans* embryo. *Development*, **120**, 2305–2315.

Martini, E. (1906) Uber Subcuticula und Seitenfelder einiger Nematoden. I. *Zeitschrift für wissenschaftliche Zoologie*, **81**, 699–766.

Martini, E. (1907) Uber Subcuticula und Seitenfelder einiger Nematoden. II. *Zeitschrift für wissenschaftliche Zoologie*, **86**, 1–54.

Mello, C.C., Draper, B.W., Krause, M., Weintraub, H. and Priess, J.R. (1992) The *pie-1* and *mex-1* genes and maternal control of blastomere identity in early *C. elegans* embryos. *Cell*, **70**, 163–176.

Mello, C.C., Draper, B.W. and Priess, J.R. (1994) The maternal genes *apx-1* and *glp-1* and establishment of dorsal-ventral polarity in the early *C. elegans* embryo. *Cell*, **77**, 95–106.

Mello, C.C., Schubert, C., Draper, B., Zhang, W., Lobel, R. and Priess, J.R. (1996) The PIE-1 protein and germline specification in *C. elegans* embryos. *Nature*, **382**, 710–712.

Meneghini, M.D., Ishitani, T., Carter, J.C., Hisamoto, N., Ninomiya-Tsuji, J., Thorpe, C. *et al.* (1999) MAP kinase and Wnt pathways converge to downregulate an HMG-domain repressor in *Caenorhabditis elegans*. *Nature*, **399**, 793–797.

Mickey, K.M., Mello, C.C., Montgomery, M.K., Fire, A. and Priess, J.R. (1996) An inductive interaction in 4-cell stage *C. elegans* embryos involves APX-1 expression in the signalling cell. *Development*, **122**, 1791–1798.

Moskowitz, I.P., Gendreau, S.B. and Rothman, J.H. (1994) Combinatorial specification of blastomere identity by *glp-1*-dependent cellular interactions in the nematode *Caenorhabditis elegans*. *Development*, **120**, 771–790.

Okkema, P.G., Ha, E., Haun, C., Chen, W. and Fire, A. (1997) The *Caenorhabditis elegans* NK-2 homoebox gene *ceh-22* activates pharyngeal muscle gene expression in combination with *pha-1* and is required for normal pharyngeal development. *Development*, **124**, 3965–3973.

Pai, S. (1927) Lebenszyklus der *Anguillula aceti* Ehrbg. *Zoologische Anzeiger*, **74**, 257–270.

Pai, S. (1928) Die phasen des lebenscyclus der *Anguillula aceti* Ehrbg. und ihre expermentell-morphologische Beeinflussung. *Zeitschrift für wissenschaftliche Zoologie*, **131**, 293–344.

Priess, J.R. and Hirsh, D.I. (1986) *Caenorhabditis elegans* morphogenesis: The role of the cytoskeleton in elongation of the embryo. *Developmental Biology*, **117**, 156–173.

Priess, J.R., Schnabel, H. and Schnabel, R. (1987) The *glp-1* locus and cellular interactions in early *C. elegans* embryos. *Cell*, **51**, 601–611.

Priess, J.R. and Thomson, J.N. (1987) Cellular interactions in early *C. elegans* embryos. *Cell*, **48**, 241–250.

Ruvkun, G. and Hobert, O. (1998) The taxonomy of developmental control in *Caenorhabditis elegans*. *Science*, **282**, 2033–2041.

Salser, S.J. and Kenyon, C. (1994) Patterning *C. elegans*: homeotic cluster genes, cell fates and cell migrations. *Trends in Genetics*, **10**, 159–164.

Schierenberg, E. (1987) Reversal of cellular polarity and early cell-cell interaction in the embryo of *Caenorhabditis elegans*. *Developmental Biology*, **122**, 452–463.

Schnabel, H. and Schnabel, R. (1990) An organ-specific differentiation gene, *pha-1*, from *Caenorhabditis elegans*. *Science*, **250**, 686–688.

Seshradi, A.R. (1964) Investigations on the biology and life-cycle of *Criconemoides xenoplax*. *Nematologica*, **10**, 540–562.

Seydoux, G. and Fire, A. (1994) Soma-germline asymmetry in the distributions of embryonic RNAs in *Caenorhabditis elegans*. *Development*, **120**, 2823–2834.

Seydoux, G., Mello, C.C., Pettit, J., Wood, W.B., Priess, J. and Fire, A. (1996) Repression of gene expression in the embryonic germ lineage of *C. elegans*. *Nature*, **382**, 713–716.

Shalon, D., Smith, S.J. and Brown, P.O. (1996) A DNA microarray system for analyzing complex DNA samples using two-color fluorescent probe hybridization. *Genome Research*, **6**, 639–645.

Skiba, F. and Schierenberg, E. (1992) Cell lineages, developmental timing and spatial pattern formation in embryos of free-living soil nematodes. *Developmental Biology*, **151**, 597–610.

Slack, J.M.W., Holland, P.W.H. and Graham, C.F. (1993) The zootype and the phylotypic stage. *Nature*, **361**, 490–492.

Sommer, R.J. (1997) Evolutionary changes of developmental mechanisms in the absence of cell lineage alterations during vulva formation in the Diplogastridae (Nematoda). *Development*, **124**, 243–251.

Sommer, R.J., Carta, L.K., Kim, S.Y. and Sternberg, P.W. (1996) Morphological, genetic and molecular description of *Pristionchus pacificus* sp. n. (NEMATODA: *Neodiplogastridae*). *Fundamental and Applied Nematology*, **19**, 511–521.

Sommer, R.J., Eizinger, A., Lee, K.Z., Jungblut, B., Bubeck, A. and Schlak, I. (1998) The *Pristionchus* HOX gene *Ppa-lin-39* inhibits programmed cell death to specify the vulva equivalence group and is not required during vulval induction. *Development*, **125**, 3865–3873.

Sommer, R.J. and Sternberg, P.W. (1996) Evolution of nematode vulval fate patterning. *Developmental Biology*, **173**, 396–407.

Sternberg, P.W. and Horvitz, H.R. (1981) Gonadal cell lineages of the nematode *Panagrellus redivivus* and implications for evolution by the modification of cell lineage. *Developmental Biology*, **88**, 147–166.

Sternberg, P.W. and Horvitz, H.R. (1982) Postembryonic nongonadal cell lineages of the nematode *Panagrellus redivivus*: Description and comparison with those of *Caenorhabditis elegans*. *Developmental Biology*, **93**, 181–205.

Stinchcomb, D.T., Shaw, J.E., Carr, S.H. and Hirsh, D. (1985) Extrachromosomal DNA transformation of *Caenorhabditis elegans*. *Molecular and Cellular Biology*, **5**, 3484–3496.

Strome, S. and Wood, W.B. (1982) Immunofluorescence visualization of germ-line-specific cytoplasmic granules in embryos, larvae and adults of *Caenorhabditis elegans*. *Proceedings of the National Academy of Sciences of the United States of America*., **79**, 1558–1562.

Sulston, J., Du, Z., Thomas, K., Wilson, R., Hillier, L., Staden, R., *et al.* (1992) The *C. elegans* genome sequencing project: a beginning. *Nature*, **356**, 37–41.

Sulston, J., Horvitz, H.R. and Kimble, J. (1988) Appendix 3. Cell lineage. Part A. In *The nematode Caenorhabditis elegans*, edited by W.B. Wood, pp. 457–478. New York: Cold Spring Harbor Laboratory Press.

Sulston, J.E. and Horvitz, H.R. (1977) Post-embryonic cell lineages of the nematode, *Caenorhabditis elegans*. *Developmental Biology*, **56**, 110–156.

Sulston, J.E., Schierenberg, E., White, J.G. and Thomson, J.N. (1983) The embryonic cell lineage of the nematode *Caenorhabditis elegans*. *Developmental Biology*, **100**, 64–119.

Sulston, J.E. and White, J.G. (1980) Regulation and cell autonomy during postembryonic development of *Caenorhabditis elegans*. *Developmental Biology*, **78**, 577–597.

Tabara, H., Motohashi, T. and Kohara, Y. (1996) A multiwell version of *in situ* hybridization on whole mount embryos of *Caenorhabditis elegans*. *Nucleic Acids Research*, **24**, 2119–2124.

Tabuse, Y., Izumi, Y., Piano, F., Kemphues, K.J., Miwa, J. and Ohno, S. (1998) Atypical protein kinase C cooperates with PAR-3 to establish embryonic polarity in *Caenorhabditis elegans*. *Development*, **125**, 3607–3614.

The *C. elegans* sequencing consortium. (1998) Genome sequence of the nematode *C. elegans* A platform for investigating biology. *Science*, **282**, 2012–2018.

Thierry-Mieg, J., Thierry-Mieg, D. and Stein, L. (1999) *C. elegans* and the Web. In *C. elegans: A Practical Approach*, edited by I.A. Hope, Chapter 3. Oxford: Oxford University Press.

van Weerdt, L.G. (1960) Studies on the biology of *Radophulus similis*. *Nematologica*, **5**, 43–51.

von Ehrenstein, G. and Schierenberg, E. (1980) Cell lineages and development of *Caenorhabditis elegans* and other nematodes. In *Nematodes as biological models*, Vol. 1, edited by B.M. Zuckerman, pp. 2–71. New York: Academic Press.

Voronov, D.A., Makarenkova, E.P., Nezlin, L.P., Panchin, Y.V. and Spiridonov, S. E. (1986) The investigation of embryonic development of free-living marine nematode *Enoplus brevis* (Enoplida) by the method of blastomere labelling. *Doklady Akademii Nauk SSSR*, **286**, 201–204.

Voronov, D.A. and Panchin, Y.V. (1995a) The early stages of the cleavage in the free-living marine nematode *Enoplus brevis* (Enoplida, Enoplidae) in the normal and experimental conditions. *Zoologicheskii Zhurnal*, **74**, 31–38.

Voronov, D.A. and Panchin, Y.V. (1995b) Gastrulation in the free-living marine nematode *Enoplus brevis* (Enoplida, Enoplidae) and the localization of enthodermal material at the stage of two bastomeres in the nematodes of the order Enoplida. *Zoologicheskii Zhurnal*, **74**, 10–18.

Voronov, D.A. and Panchin, Y.V. (1998) Cell lineage in marine nematode *Enoplus brevis*. *Development*, **125**, 143–150.

Voronov, D.A., Panchin, Yu. V. and Spiridonov, S.E. (1998) Nematode phylogeny and embryology. *Nature*, **395**, 28.

Walton, A.C. (1918) The oogenesis and early embryology of *Ascaris canis* Werner. *Journal of Morphology*, **30**, 527–530.

Wang, B.B., Muller-Immergluck, M.M., Austin, J., Robinson, N.T., Chisholm, A. and Kenyon, C. (1993) A homeotic gene cluster patterns the anteroposterior body axis of *C. elegans*. *Cell*, **74**, 29–42.

Waring, D.A. and Kenyon, C. (1991) Regulation of cellular responsiveness to inductive signals in the developing *C. elegans* nervous system. *Nature*, **350**, 712–715.

Way, J.C. and Chalfie, M. (1988) *mec-3*, a homeobox-containing gene that specifies differentiation of the touch receptor neurons in *C. elegans*. *Cell*, **54**, 5–16.

Wiegner, O. and Schierenberg, E. (1998) Specification of gut cell fate differs significantly between the nematodes *Acrobeloides nanus* and *Caenorhabditis elegans*. *Developmental Biology*, **204**, 3–14.

Williams, S.A., Laney, S.J., Bierwert, L.A., LizotteWaniewski, M., Saunders, L., Lu, W.H., *et al.* (1999) Deep within the filarial genome: Progress of The Filarial Genome Project. *Parasitology Today*, **15**, 219–224.

Wilson, R. Ainscough, R., Anderson, K., Baynes, C., Berks, M., *et al.* (1994) 2.2 Mb of contiguous nucleotide sequence from chromosome III of *C. elegans*. *Nature*, **368**, 32–38.

Wolf, N., Priess, J. and Hirsh, D. (1983) Segregation of germline granules in early embryos of *Caenorhabditis elegans*: an electron microscope analysis. *Journal of Embryology Experimental Morphology*, **73**, 297–306.

Wood, W.B. (1991) Evidence from reversal of handedness in *C. elegans* embryos for early cell interactions determining cell fates. *Nature*, **349**, 536–538.

Wood, W.B. (1998) Handed asymmetry in nematodes. *Seminars in Cell and Developmental Biology*, **9**, 53–60.

Wood, W.B. and Kershaw, D. (1991) Handed asymmetry, handedness reversal and mechanisms of cell fate determination in *Caenorhabditis elegans*. *Ciba Foundation Symposium*, **162**, 143–159.

Yuksel, H.S. (1960) Observations on the life cycle of *Ditylenchus dipsaci* on onion seedlings. *Nematologica*, **5**, 289–296.

Zhu, J.W., Fukushige, T., McGhee, J.D. and Rothman, J.H. (1998) Reprogramming of early embryonic blastomeres into endodermal progenitors by a *Caenorhabditis elegans* GATA factor. *Genes and Development*, **12**, 3809–3814.

Zhu, J.W., Hill, R.J., Heid, P.J., Fukuyama, M., Sugimoto, A., Priess, J.R. and Rothman, J.H. (1997) *end-1* encodes an apparent GATA factor that specifies the endoderm precursor in *Caenorhabditis elegans*. *Genes and Development*, **11**, 2882–2896.

zur Strassen, O. (1892) *Bradynema rigidum* v. Sieb. *Zeitschrift für wissenschaftliche Zoologie*, **54**, 655–747.

zur Strassen, O. (1896) Embryonalentwickelung der *Ascaris megalocephala*. *Archiv für Entwicklungsmechanik der Organismen*, **3**, 27–105.

6. Hatching

Roland N. Perry

Entomology and Nematology Department, IACR-Rothamsted, Harpenden, Hertfordshire, AL5 2JQ, UK

Keywords

Eclosion, eggshell, enzymes, hatching, larva, nematodes.

Introduction

Although species of some nematode genera, such as *Panagrellus*, *Trichinella* and *Turbatrix*, are oviparous, where the larvae hatch within the female uterus and subsequently emerge, these are exceptions and the majority of larvae hatch from eggs which are laid by the adult female. Each egg contains a single larva and, despite the vast difference in size between adults of various species of nematodes, the majority of eggs are of similar size and morphology. In many species, it is the first-stage larva which hatches but in most plant-parasitic nematodes the larva moults within the egg and the resulting second-stage larva hatches. ('Juvenile' is the term of choice by Plant Nematologists; however, to conform with usage in this book, rather than personal preference, the term 'larva' will be used here.) In some animal-parasitic species, there is a further moult in the egg and it is the third-stage larva which hatches. Hatched larvae are very vulnerable to environmental stresses and, in plant endoparasitic species, they are viable in the soil for only a short period, usually days or weeks. For example, larvae of *Globodera rostochiensis* and *G. pallida* can survive for less than two weeks without feeding (Robinson, Atkinson and Perry, 1987a). In many species the egg is the main survival stage of the life cycle and, in species such as *G. rostochiensis* and *Ascaris suum*, the unhatched larva can remain viable for many years. The eggshell affords protection to the enclosed larva and hatch only occurs when environmental conditions are favourable. Physiological adaptations which enhance survival, such as quiescence and diapause, are frequently associated with the unhatched larva (Perry, 1989; Wharton, this volume).

Eggs of most species of nematodes are laid individually and exist without any additional protection. The best known exceptions are the plant-parasitic cyst nematodes (*Globodera* spp. and *Heterodera* spp.) and root-knot nematodes (*Meloidogyne* spp.). After fertilisation and death of the saccate female cyst nematode, the cuticle becomes tanned to form a tough, brown, round (*Globodera*) or lemon-shaped (*Heterodera*) cyst containing several hundred eggs. Each egg develops to contain a tightly coiled infective larva. There are small openings at the neck and the vulval ends of the cyst through which the hatched larvae escape into the soil. Females of the root-knot nematodes lay eggs into a gelatinous matrix, secreted by the six rectal glands, consisting of an irregular meshwork of glycoprotein material. The gelatinous matrix shrinks and hardens when dried (Bird and Soeffky, 1972), thus exerting mechanical pressure on the eggs to inhibit hatching during drought conditions.

This chapter will examine the stimuli which elicit hatching, the differences between species in hatching patterns and the hatching process. There are considerable variations between species of nematodes in the sequence of events during the hatching process which obviate presentation of a 'model' nematode hatching mechanism. Although the descriptions of hatching processes are based loosely on the chronological sequence of events from stimulation of hatch to eclosion (or movement out of the egg), the order of events, the overlap of individual responses and their significance may differ between species; where possible, differences and similarities are highlighted. Essentially, the hatching process can be divided into three phases: changes in the eggshell, activation of the larva, and eclosion. In many species, activation of the larva appears to precede, and may even cause, changes in eggshell structure; in others, alteration of eggshell permeability characteristics appears a necessary pre-requisite for metabolic, and consequent locomotory, changes in the larva.

Hatching Stimuli

In general, given suitable environmental conditions, such as appropriate temperature, oxygen availability and, for soil-dwelling forms, soil moisture levels, and an absence of physiological barriers, such as diapause, hatch of most species occurs without requiring specific cues emanating from the food source. Soil type is an important abiotic factor for soil dwelling species or life cycle stages of nematodes. In general, coarse-textured soils favour hatching and subsequent invasion of roots by plant-parasitic nematodes, providing suitable conditions for aeration and nematode migration. Maximum hatch of plant-parasitic nematodes usually occurs when soil water content is at field capacity, whilst drought and water-logging inhibit hatching.

Nematodes are poikilotherms and so temperature is an important influence in modifying their hatching activity. Clearly, hatching of parasitic nematodes in homoiothermic hosts occurs in a relatively stable temperature regime compared with the seasonal and diurnal fluctuations of temperature experienced by species of nematodes which hatch outside the host. The majority of research on the effects of temperature on the hatch of plant-parasitic nematodes has examined the *in vitro* hatch during exposure to constant temperatures. Such conditions do not relate to those experienced by the nematode *in vivo* and an exhaustive survey of research giving the various optimum temperatures for hatch of different species is of limited use. Perhaps unsurprisingly, low optimum temperatures for hatching are characteristic of those species of plant-parasitic nematodes that invade during winter or early spring, such as *Ditylenchus dipsaci* and *Heterodera cruciferae*, whereas nematodes that are adapted to warmer climates, such as *H. zeae*, *M. hapla* and *M. javanica*, exhibit higher temperature optima. Frequently, the optimum temperature for *in vitro* hatch is markedly different from that likely to be experienced *in vivo*. For example, the optimum temperature for *in vitro* hatch of *G. rostochiensis* is 20°C (Robinson, Atkinson and Perry, 1987b), a temperature that cysts at root depth in the soil are highly unlikely to experience during the normal field hatch time of April in the soils of Europe where this species is a major pest. Variation in the *range* of temperatures at which hatching can occur is an important adaptive characteristic. For example, the hatching of *G. pallida* is adapted to lower temperatures than that of *G. rostochiensis* (Robinson, Atkinson and Perry, 1987b), a factor perhaps associated with the dominance of *G. pallida* in northern parts of the UK.

Temperature extremes are also relevant to hatching biology. High temperatures can suppress hatch or are lethal to nematodes. For example, Stoyanov and Trifonova (1995) reported 30°C as the upper limit for hatch of *G. rostochiensis*, whilst no hatch was observed from *H. zeae* at 40°C (Hashmi and Krusberg, 1995). Dormancy can occur at most stages of the nematode's life cycle and, where the unhatched larva is the affected stage, it is of importance in mediating the hatching response. The transition from a fluctuating external environment to a stable increased temperature of 37°C can often be the stimulus for cessation of dormancy. Substantial hatching of *A. lumbricoides* and *Aspiculuris tetraptera* only occurs between 30°C and 40°C and they do not respond to temperatures below 30°C (Fairbairn, 1961; Anya, 1966). Temperature is an important environmental cue for the termination of diapause. *Nematodirus battus* produces an infective third-stage larva within the egg that overwinters on pasture. The autumn/winter period of low temperatures is required before a rise in temperature during the spring results in termination of diapause and hatch when host lambs are available for infection (Christie, 1962). The induction and termination of diapause in relation to hatching have been discussed by Evans and Perry (1976) and

Jones, Tylka and Perry (1998), and are examined in the context of nematode survival strategies by Wharton (Chapter 16).

Host-derived Hatching Stimulants

The life cycle of parasitic nematodes is frequently synchronised with that of their hosts to optimise the chances of successful invasion (Perry and Clarke, 1981). This synchrony is often centred on the stimulus for hatching being provided by the host itself. In some animal-parasitic and insect-parasitic nematodes, eggs are ingested by the host and the host gut provides the conditions for hatching. In some species of plant-parasitic nematodes, the stimulus for hatching emanates from host roots.

Much of the work to define conditions for the hatching of nematode parasites in the alimentary tract of homoiothermic hosts was done in the 1960s. Pioneering work by Fairbairn (1961) demonstrated that *Ascaris lumbricoides* was stimulated to hatch *in vitro* by treatment with carbon dioxide in the presence of a reducing agent at neutral pH and 37°C; these conditions are probably similar to those present within the intestine. Subsequently, Hass and Todd (1962) used Fairbairn's method and were able to hatch seven species of ascarids, *Heterakis gallinae*, *Trichuris suis* and *T. ovis in vitro*. Anya (1966) found that eggs of *Aspiculuris tetraptera* hatched best under air at pH 7–8, and that reducing conditions had little effect. However, for most species of animal-parasitic nematodes that hatch inside the host, reducing conditions are required for *in vitro* hatch. Lee and Atkinson (1976) suggested that the variation between species in the optimum conditions for hatching increases the probability that each responds only to ingestion by the appropriate host and that the infective larvae hatch in the preferred region of the alimentary tract. Some insect-parasitic nematodes also hatch inside the host. For example, infective larvae of *Mermis subnigrescens*, a parasite of grasshoppers, hatch in the intestine after the eggshell has been digested away.

Among plant-parasitic nematodes, the integration between host and parasite to ensure successful invasion has progressed furthest in the cyst forming nematodes. The sophisticated hatching mechanism of these nematodes provides one of the clearest examples of the role of host signals in synchronising host and parasite life cycles. Synchrony with the host is mediated by root diffusates from host plants which stimulate hatching; with other species of cyst nematodes that hatch freely in water, root diffusates enhance the rate of hatching (Table 6.1). The reliance on root diffusates for hatch favours persistence of the nematode in the soil in the absence of host plants and, when host plants are present, it ensures that a large number of infective larvae emerge close to the growing roots, to which the larvae are attracted. Although the requirement for diffusates to stimulate hatch is most common among species of cyst nematodes, other species, such as *Meloidogyne hapla* and *Rotylenchulus reniformis*, also hatch in response to host root diffusates.

Table 6.1. Grouping of some species of cyst nematodes into four broad categories, based on their hatching response to host root diffusates. Modified from Jones, Tylka and Perry (1998)

Group 1	Very large numbers of larvae hatching in response to host root diffusates; few hatching in water	e.g. *Globodera rostochiensis, G. pallida, Heterodera cruciferae, H. carotae, H. goettingiana, H. humuli*
Group 2	Very large numbers of larvae hatching in response to host root diffusates; moderate hatch in water.	e.g. *H. trifolii, H. galeopsidis, H. glycines*
Group 3	Very large numbers of larvae hatching in response to host root diffusates; large hatch in water	e.g. *H. schachtii, H. avenae*
Group 4	Hatching of larvae induced by diffusates only in later generations produced during the host growing season; very large hatch in water for all generations.	e.g. *H. cajani, H. sorghi*

This dependence on host root diffusates has potential for novel control strategies. Application of host root diffusates to infested soil in fields that are fallow or planted to a non-host should stimulate hatch of the target nematode, with subsequent death of the larvae by starvation in the absence of the host plant; the so-called 'suicide hatch'. This has been demonstrated with *G. rostochiensis* by Devine and Jones (2000a) who found that application of host diffusates to infested soil in the absence of potato plants gave approximately a 50% reduction in the number of viable larvae. The impetus to find alternatives to environmentally undesirable nematicides for control of economically important plant-parasitic nematodes has resulted in the acccumulation of a considerable amount of information on diffusates and the hatching factors they contain, with research focused particularly on the soybean-*Heterodera glycines* and potato-*Globodera rostochiensis/G. pallida* host-parasite interactions (Perry, 1987, 1997; Jones, Tylka and Perry, 1998). The efficacy of root diffusates is usually assessed in laboratory based hatching bioassays by determining the percentage hatch from eggs over a given period of time; the term 'active diffusates' is a convenient expression to convey that substantial hatch is induced by exposure to such diffusates. The time course of a bioassay and the total percentage hatch induced varies between species but, in general, hatch of 60-90% of larvae can be expected in active diffusates. The early work on the effects of diffusates on hatching of cyst nematodes was reviewed by Shepherd and Clarke (1971) and Clarke and Perry (1977).

The production of active diffusate declines with the onset of plant senescence (Perry, 1997) and, in several plant species, it is confined to a short period of plant growth. *Heterodera goettingiana* hatched only in diffusates from 4- and 6-week old pea plants (Perry, Clarke and Beane, 1980) and from 6- and 8-week old bean plants (Beane and Perry, 1983), whilst hatch of *H. carotae* was largely restricted to diffusates from 5- and 7-week old carrots (Greco and Brandonisio, 1986). Maximum hatching activity of potato root diffusate was reached two weeks after planting potatoes and was maintained for a further two weeks (Twomey, 1995). Hatching of larvae of *Globodera* spp. in

potato fields occurs over an 8-week period (Trudgill, Phillips and Hackett, 1996), although LaMondia and Brodie (1986) reported that most larvae which successfully invaded the host hatched within the first three weeks; the utilisation of energy reserves by late-hatching larvae (Robinson, Atkinson and Perry, 1985) may adversely affect invasion (see *Responses of the larva,* below). Diffusate from potato roots is produced along the length of the entire root, but cells near the root tip produced a more active diffusate than cells located elsewhere (Rawsthorne and Brodie, 1986). This region of the root is metabolically very active, being associated with elongation and differentiation, and is the preferred invasion site for infective larvae. Gradients of root diffusate concentration are likely to be established around the root and may aid in attracting hatched larvae of plant-endoparasitic nematodes to the invasion site (Perry, 1996). The zone of root diffusate influence is extensive; for example, hatching activity towards *G. rostochiensis* and *G. pallida* has been detected up to 80 cm from potato roots (Rawsthorne and Brodie, 1987; Malinowska, 1996). However, diffusate mobility and hatching activity are reduced in soils with high organic content. Soil moisture, temperature and microbial activity are also influential and hatching factors are rapidly inactivated at >pH 8. Fenwick (1956) considered that potato root diffusates were highly labile in the soil but this finding has since been refuted for *G. rostochiensis* (Tsutsumi, 1976; Perry, Hodges and Beane, 1981) and *H. goettingiana* (Perry, Clarke and Beane, 1980); hatching activity towards *G. rostochiensis,* for example, was detected in soil for up to 100 days after removal of the plants (Tsutsumi, 1976).

Fractionation of diffusates and macerated roots has revealed several chemicals, termed hatching factors, which elicit hatch. The first nematode hatching factor to be isolated was glycinoeclepin A, which was extracted from powdered kidney bean roots, rather than root diffusate, and induced hatch of *H. glycines* (Masamune et al., 1982; Fukuzawa et al., 1985a). The hatching factor is a pentanor-triterpene. Subsequently, two further nortriterpene hatching factors, glycinoeclepins B and C, were purified (Fukuzawa et al., 1985b). Masamune et al. (1987a) proposed that hatching factors are derivatives of an essential plant biosynthetic pathway from a cycloartane such as acerinol or cimigenol. Initial studies on potato root diffusate indicated that it contained between four and six different hatching factors that induced hatch of *G. rostochiensis* (Clarke, 1970, 1971). Later work revealed the presence of more than ten hatching factors (Devine *et al.*, 1996) and mass spectrometry distinguished them into two classes of terpenoid molecules, with molecular masses of 279 Da ('high polarity hatching factors') and 662 Da ('low polarity hatching factors') (Devine, 1994). More detailed studies on the principal hatching factors from potato root diffusate, with molecular masses of 530.5 Da, indicated that they are probably part of a common biosynthetic pathway, thus opening the possibility of using antisense technology to develop transgenic potato plants producing diffusate with reduced or no hatching activity (Devine and Jones, 2000b). A terpenoid hatching factor for *G. rostochiensis* with a molecular mass of 498 Da was purified and characterised by Mulder *et al.* (1992) and Atkinson, Fowler and Isaac (1987) reported a 437 Da molecular mass hatching factor from potato root diffusate. Although both *G. pallida* and *G. rostochiensis* hatched in response to each hatching factor tested by Byrne (1997), preferences were apparent. Individual hatching factors appear sequentially in potato root diffusate as the plant develops (Twomey, 1995) and the late production of *G. pallida*-preferred hatching factors could explain, in part, the delayed hatching of this species compared with *G. rostochiensis*.

Several empirical formulae have been given for *Globodera* hatching factors: $C_{18}H_{24}O_8$ (Marrian *et al.*, 1949), $C_{19}H_{28}O_8$ (Johnson, 1952), $C_{13}H_{12}O_3$ (Hartwell, Dahlstrom and Neal, 1959), $C_{11}H_{16}O_4$ (Clarke, 1970) and $C_{27}H_{30}O_9$ (Mulder *et al.*, 1992). Hatching factors are present only in trace amounts (for example, 1058 kg dry weight of bean roots yielded 1.25 mg of glycinoeclepin A as its bisphenacyl ester (Masamune *et al.*, 1982) and a hatching factor isolated by Devine and Jones (2000b) represented less than $2.9 \times 10^{-5}\%$ of the organic material recovered from potato root diffusate) but have very high specific activities, stimulating hatch of *H. glycines* and *G. rostochiensis* at concentrations as low as 10^{-12} g ml^{-1}; the maximum hatching activity of the hatching factors isolated by Mulder *et al.* (1992) and Devine and Jones (2000b) occurred at approximately 10^{-8} M. Total synthesis of glycinoeclepin A has been achieved (Masamune *et al.*, 1987a,b; Miwa, Okawara and Sakakibara, 1987; Murai *et al.*, 1988; Mori and Watanabe, 1989; Corey and Houpis, 1990; Watanabe and Mori, 1991) and the structure resembles that of gibberellins. Artificial compounds, with simpler partial structures of glycinoeclepin A, have been synthesised that have hatching activity towards *H. glycines* (Kraus *et al.*, 1994, 1996). Glycinoeclepin A would be prohibitively expensive for commercial use, especially as only one species of cyst nematode would be the target, but structural similarities between the A rings of glycinoeclepin A (Masamune *et al.*, 1987a,b) and the hatching factor identified by Mulder *et al.* (1992) may lead to the synthesis of an A ring-based partial structure with a

commercially more attractive broad-spectrum hatching activity. Interestingly, hatching factor stimulants have been identified in potato root diffusate. On their own, these chemicals are hatch-inactive but they can synergise the effect of natural or artificial hatching factors to increase hatch of *G. rostochiensis* (Byrne *et al.*, 1998).

Microbial Hatching Stimulants

Field observations of seasonal flushes of hatching by plant-parasitic nematodes in fallow soil have been interpreted as indicating the involvement of microbial hatching factors (Ellenby, 1963; Tsutsumi, 1976; Stelter and Sager, 1987). Carroll (1995) isolated several bacteria from the potato rhizosphere and found that the microbial hatching factors showed absolute specificity towards either *G. rostochiensis* or *G. pallida*, perhaps indicating a difference in the hatching mechanism of these two species. An isolate of the rhizobacterium *Pseudomonas fluorescens* stimulated hatch of *G. rostochiensis in vitro* and in pot trials by producing a hatch-stimulatory metabolite, 2,4-diacetylphloroglucinol (DAPG) (Cronin *et al.*, 1997a). Ionophores, which Atkinson and Ballantyne (1979) suggested were implicated in the hatch of *G. rostochiensis* (see *Changes in the lipid layer*, below), are microbial in origin (Truter, 1976). Although the very limited water solubility of ionophores indicates that plant-derived hatching factors are not ionophores, the more homogeneous distribution of bacteria within the soil profile could negate solubility limitations. Chitinase-producing bacteria are obvious candidates for use as biocontrol agents as the enzyme may degrade the eggshell and enhance hatch. In support of this hypothesis, Mercer, Greenwood and Grant (1992) demonstrated that chitinases from plant and bacterial sources increased hatch of *M. hapla*.

Only a few examples of microbial effects on hatching have been discussed here. A more extensive review of the possibilities of using microbial agents for biological control of nematodes is presented by Kerry and Hominick (this volume).

Artificial Hatching Stimulants

In *in vitro* hatching experiments with animal-parasitic nematodes that hatch in their hosts, most workers have endeavoured to mimic conditions found in the alimentary tract of the host. Although some artificial hatching agents have been reported, most of the information on so-called 'artificial hatching factors' for nematodes derives from investigations on economically important plant-parasitic nematodes, a result of the impetus to find novel control strategies to cause hatch in the absence of the host plant. Over 400 chemicals have been tested for hatching activity, primarily with *G. rostochiensis* and *H. schachtii* (Clarke and Shepherd, 1966, 1968; Shepherd and Clarke, 1971). Although many chemicals, including simple compounds such as urea and specific amino acids and sugars, induce hatch of *H. schachtii*, comparatively few are known to hatch *G. rostochiensis*. Several artificial hatching factors with broad-spectrum activity towards cyst nematodes have been identified. For example, flavianic acid causes substantial hatch of *H. cruciferae*, *H. glycines*, *H. schachtii*, *H. trifolii* and *G. tabacum*, is less active towards *H. goettingiana* and is ineffective with *H. carotae* and *G. rostochiensis* (Clarke and Shepherd, 1967). Ions such as Ba^{2+}, Zn^{2+} and La^{3+} also exhibit *in vitro* hatching activity towards several cyst nematodes (Clarke and Hennessy, 1987). However, compared with natural hatching factors, artificial hatching factors operate only at high concentrations and may act via different mechanisms (see *Eclosion*, below). Optimum concentrations for hatch of *G. rostochiensis* ranged from 50 μM (calcium ionophores) to 0.3 mM (picrolonic acid), 0.6 mM (sodium metavanadate) and 3.0 mM (anhydrotetronic acid) (Clarke and Shepherd, 1966, 1968; Atkinson and Ballantyne, 1979).

In field trials, the hatching agent, sodium metavanadate, applied to bare sandy loam at 36.7 kg ha^{-1} and rotary cultivated into the top 15 cm of soil, stimulated hatching of *G. rostochiensis* (Whitehead, 1992). Another artificial hatching agent, picrolonic acid, at 17.2 kg ha^{-1} incorporated into a sandy clay soil after potatoes also significantly increased hatch of this species (Whitehead, 1977). Banasiak *et al.* (1994) reported up to 70% hatch of *H. schachtii* after application of 1 acetoxy-2-ethylhexa-1, 3-diene to infested soil. However, the expense of the chemicals, plus the probable need to keep fields fallow for six months, would make use of these chemicals uneconomic (Whitehead, 1998).

Hatching Inhibitors

There are several reports of the inhibitory effect of plant extracts on *in vitro* hatch of plant-parasitic nematodes which may have potential for control strategies but their mode of action is unknown. Hatching inhibitors could inactivate the hatching factors, release chemicals that inhibit hatch or competitively bind to the hatching factor receptor(s) on the eggshell. Fractionation of potato root diffusate revealed the presence of hatching inhibitors which were produced earlier than the hatching factors, resulting in a net inhibition of hatching of larvae of *G. rostochiensis* immediately after planting host potatoes (Twomey, 1995). Subsequently, a rise in hatching factor:hatching inhibitor ratio in the diffusate stimulated hatch (Byrne *et al.*, 1998) at a time when the larger root mass would enhance the chances of successful invasion. Several synthetic partial structures of glycinoeclepin A inhibited *H. glycines* hatch at ppm concentrations (Kraus *et al.*, 1994) and the active component for hatching inhibitor activity appeared to be a keto diacid (Kraus *et al.*, 1996). Ellenby (1945) found that allyl isothiocyanate (the mustard oil of black mustard seed) totally inhibited hatching of *G. rostochiensis* at 500 ppm and irreversibly decreased it at 50 ppm, whilst Forrest and Farrer (1983) found that root diffusates from white mustard prevented hatch of *G. pallida* that had been previously exposed to host root diffusates. Asparagusic acid from *Asparagus* roots inhibits hatch of *G. rostochiensis* and *H. glycines* at 50 ppm (Takasugi *et al.*, 1975). DMDP (2,5-dihydroxymethyl-3, 4-dihydroxypyrrolidine), a sugar analogue

from *Derris* sp. (Birch *et al.*, 1993), and the lignans bursehernin and matairesinol from the umbellifer *Bupleurum salicifolium* (Gonzalez *et al.*, 1994) inhibited hatching of both species of *Globodera*. Matairesinol affects cyclic adenosine-3′, 5′-monophosphate (cAMP) metabolism and the hatch inhibition caused by this lignan and bursehernin may be related to their effect on respiratory metabolism (Gonzalez *et al.*, 1994).

Bacterial isolates have been shown to inhibit hatch of some species of cyst nematodes. An *Agrobacterium radiobacter* isolate inhibited hatch of *G. pallida in vitro* (Hackenberg and Sikora, 1994) and when sugarbeet seeds were inoculated with selected rhizobacterial isolates (including the *A. radiobacter* isolate) there was a reduced hatch of *H. schachtii* (Oostendorp and Sikora, 1990). In contrast to work demonstrating that bacterial chitinases enhance hatch of plant-parasitic nematodes (see *Microbial hatching stimulants*, above), Cronin *et al.* (1997b) considered that the chitinase-producing bacteria that they tested inhibited hatch of *G. rostochiensis*. This may indicate that bacteria-derived hatching inhibitors were present in the treatments. Also, hatch after treatment was assessed in picrolonic acid rather than root diffusates, so even if chitin of *G. rostochiensis* eggs was degraded, other aspects important for hatching of this species, such as direct stimulation of the larva by diffusates (see *Responses of the larva*, below), were missing.

Variations in Hatching Patterns

Hatching of intestinal animal-parasitic nematodes may be close to 100% during a single exposure to the hatching stimulus and this accords with the parasite's life cycle, where no advantage accrues by delaying eclosion once the eggs have been ingested. The hatching pattern of cyst nematodes provides an interesting contrast and illustrates further aspects of the sophisticated host-parasite interaction. Even though short exposures of 5 min to diffusate triggers hatch of *Globodera* spp. (see *Changes in the lipid layer*, below), re-exposure each week for up to five weeks is necessary to achieve substantial hatch *in vitro*. In the field, 60–80% of larvae hatch in the presence of diffusates from host crops but some larvae will remain unresponsive for several years (Turner and Evans, 1998). Presumably these larvae are in diapause and this 'carry over' population enables the species to remain viable in the field for a number of years. Zheng and Ferris (1991) identified different types of hatching response in encysted eggs of *H. schachtii*. Some larvae hatch very readily and infect any host plants present, in others hatching is delayed which reduces intraspecific competition in roots, and the remaining larvae do not hatch for a considerable period, thus increasing their chances of survival in the absence of a host.

Information on hatching of *G. rostochiensis*, *G. pallida* and *H. schachtii* in the field usually relates to the situation of a single generation during the host growing season. However, the 8-month duration of the sugar beet crop, the primary host for *H. schachtii*, enables this species to complete two generations per year in Western Europe and five in California. Other species of cyst nematodes, especially those in the tropics, complete several generations during the host growing season and research into the hatching of these species illustrates a further modification of hatching patterns. Some species of *Heterodera* extrude a proportion of their eggs into an 'eggsac', which remains attached to the cyst, as well as retaining eggs inside the cyst itself. Working with *H. glycines*, Ishibashi *et al.* (1973) were the first to note that the proportions of eggs in the cyst and eggsac vary under different environmental conditions. Under favourable conditions, *H. glycines* produced most eggs in the eggsacs and larvae from these eggs hatched in water, providing a secondary inoculum for rapid re-infestation of the host plant and permitting rapid population increase in one season. Under less favourable conditions, more eggs were retained within the cysts and a large proportion of these encysted eggs required host root diffusates (Ishibashi *et al.*, 1973) or artificial hatching factors (Thompson and Tylka, 1997) to stimulate hatch. Similar phenomena have been reported for *H. carotae* (Greco, 1981; Aubert, 1986) and *H. goettingiana* (Greco, Vito and Lamberti, 1986).

The difference between generations in hatching from cysts produced at different stages of plant growth illustrates the influence of the plant on development and subsequent hatching. Later generations of some species of polycyclic cyst nematodes show an increased dependence on root diffusates for hatch, reflecting a change of priority during the host plant growing season from rapid re-infection to survival after host senescence (Perry and Gaur, 1996). For example, hatch from cysts of *H. sacchari* extracted from young plants was not dependent on root diffusate but a percentage of larvae in cysts from older plants required root diffusate for hatch (Ibrahim *et al.*, 1993). *Heterodera sorghi* produced three types of eggs: eggs from which larvae hatched freely in soil leachate, those that required root diffusate to stimulate hatch and a large percentage from which larvae did not hatch immediately. The proportions of these three types of eggs changed with successive generations, with a trend towards increased persistence in the later generations (Gaur, Beane and Perry, 1995). For the first four generations, larvae in cysts of *H. cajani* hatched well in water with no enhancement of hatch by root diffusates but, in the fifth and sixth generations produced on senescing plants, 18–22% of the eggs required root diffusate to stimulate hatch; in the final generation, the encysted larvae contained more lipid reserves than those in eggsacs (Gaur, Perry and Beane, 1992). Thus, a large proportion of larvae from the later generations did not hatch but remained protected by the egg and cyst during the period between crops. In the research discussed above, cysts were exposed to optimum *in vitro* conditions for hatch including diffusate with maximum hatching activity. As mentioned previously, activity of diffusate declines as plants age and this is an additional factor ensuring that larvae do not hatch immediately before the next crop.

There is evidence that a similar variation in hatching between generations occurs with *Meloidogyne*. Three types of eggs are produced by *M. triticoryzae* (those that hatch in water, those that hatch in host root diffusate and those that do not hatch even in the presence of diffusate); the proportion of these three types varies with generation, with the final generation produced on senescing plants having a large proportion of unhatched larvae of the third type, which is likely to equate with diapause (Gaur, Beane and Perry, 2000).

The factors causing the change in hatching response are unknown. Compared to eggs which hatch in water, the eggs of *H. cajani*, *H. sacchari* and *H. sorghi* which are dependent on root diffusates for hatch either have a different structure or contain larvae in a modified physiological state, perhaps involving the induction of obligate quiescence (Evans and Perry, 1976). Possible changes in the feeding site with age of the host plant may be crucial to the feeding female which, in turn, may trigger biochemical changes in the larvae to enhance survival.

Effects on the Egg

The structure of the egg of the majority of species studied changes during the hatching process and, in some species, this is partially or wholly caused by the larva. However, in some parasitic species, it has been demonstrated that the host-derived hatching stimulus directly affects the eggshell. Although examination of hatching mechanisms has centred on analysis of changes in eggs, it should be remembered that, in cyst nematodes, the cyst itself has an effect on hatching. Kaul (1962) suggested that catechin-like substances from the cyst wall may be responsible for 'winter dormancy' (diapause) in *G. rostochiensis*, as their concentration within the cyst increased markedly during autumn and winter. Similarly, Okada (1972) proposed that compounds in the cyst wall of *H. glycines* were involved in inhibiting hatch under unfavourable conditions. Eggs within a cyst constitute an ecological community (Ellenby, 1956) and the changes in eggs and larvae in response to hatch stimulation may affect the subsequent behaviour of other unhatched larvae within the cyst. Release of trehalose from the perivitelline fluid (see *Importance of the perivitelline fluid*, below) may increase the osmotic stress within the cyst and temporarily inhibit the hatch of other eggs, at least until the trehalose has diffused out of the cyst. In addition, larvae which hatch first may inhibit the hatch of others within the cyst, either by releasing inhibitory metabolites or by using up essential resources within the cyst, such as oxygen. Significant spatial differences in the distribution of hatch were detected within a single cyst which may be, in part, the result of an oxygen tension gradient across the cyst (Onions, 1955). Eggs near the cyst wall hatched earlier than those near the centre of the cyst, although the distribution of hatch was not affected by proximity to the natural cyst openings.

Structure of the Egg

Knowledge of the structure of the egg is necessary to appreciate fully induced alterations to the eggshell and the subsequent changes in the perivitelline fluid surrounding the larva. These changes are associated with hatching but there is also evidence of an earlier permeability change to the eggshell during development (Matthews, 1986). The eggshell of nematodes consists of between one and five layers, depending on the species, but the most common form has three layers (Figure 6.1; Wharton, 1980). The outer vitelline layer is derived from the oolemma and it retains the unit membrane-like structure. The middle chitinous layer is often the thickest, and provides the eggshell with its structural strength. It has a chitin microfibril core (providing tensile strength) with a collagen-like protein coat (providing rigidity). The eggshell is the only structure in nematodes in which chitin has been demonstrated conclusively (Bird and Bird, 1991). The innermost lipid layer represents the main permeability barrier of the eggshell, although Matthews (1986), working with *Ancylostoma* spp., suggested that the chitinous layer may act as a sieve, preventing larger molecules reaching the lipid layer. In plant-parasitic nematodes, the inner lipid layer usually consists of two or three lipoprotein membranes (Bird and McClure, 1976; Perry, Wharton and Clarke, 1982), although in *H. schachtii* the layer is tetralaminate (Figure 6.2; Perry and Trett, 1986). In *Ascaris* spp. the lipid layer has been termed the 'ascaroside layer'. The layer contains 25% protein and 75% lipid and the lipid belongs to a unique class, called ascarosides, which consist of a series of α-glycosides each with a sugar moiety (glycone), 3,6-dideoxy-L-arabinohexose (ascarylose), that is not found in any other eukaryote group (Barrett, 1981). Ascarosides have been found only in a few species of

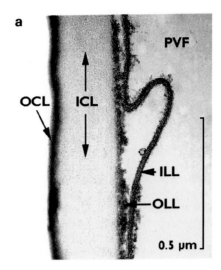

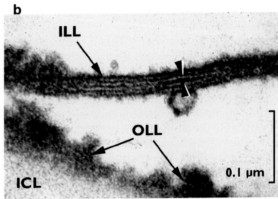

Figure 6.2. a: section through the eggshell of *Heterodera schachtii* showing the inner lipid layer (ILL) detached from the amorphus outer lipid layer (OLL); b: detail of the same inner lipid layer showing the tetralaminate structure with a mean lamina separation of 30 nm (arrowheads). For other abbreviations see legend to Figure 6.1. (From Perry and Trett, 1986).

nematodes, mainly ascarids, so in this chapter the more general term 'lipid layer' will be used throughout for the innermost layer of the eggshell.

In addition to these endogenously produced layers, the eggshells of some species of nematodes possess one or two layers secreted exogenously by the cells of the uterus. Several species of ascarids possess uterine layers, including *A. lumbricoides*, *Heterakis gallinarum* and *Ascaridia galli*, and the outer two layers of oxyurid eggshells are of uterine origin. In aquatic species, the uterine layer may form filaments which attach eggs to the substrate. As indicated by Rogers and Sommerville (1968), experiments investigating the *in vitro* hatch of *Ascaris* spp. have to be interpreted with care as some workers 'deshelled' eggs (i.e. removed the outer layers) in sodium hypochlorite before commencing hatching tests.

Changes in the Lipid Layer

One of the initial events in the hatching process of many species of nematodes is a change in eggshell permeability. As early as 1911, Looss observed that eggs of

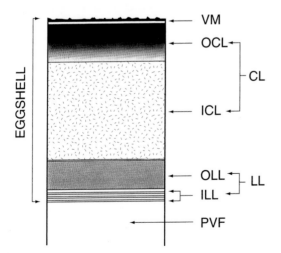

Figure 6.1. Diagrammatic reconstruction of a section through the eggshell of a generalised embryonated tylenchid egg. VM: vitelline membrane; CL: chitin layer; OCL: outer chitin layer; ICL: inner chitin layer; LL: lipid layer; OLL: outer lipid layer; ILL: inner lipid layer; PVF: perivitelline fluid (modified from Perry and Trett, 1986).

Ancylostoma duodenale in strong salt solutions collapsed more easily when ready to hatch. By determining the number of plasmolysed eggs in hypotonic solutions at various times after the start of the hatching sequence, Wilson (1958) demonstrated that eggs of *Trichostrongylus retortaeformis* became permeable to water shortly before the larvae hatched; during the pre-hatch development the eggs were completely impermeable to water. Using the same experimental method, Anya (1966) considered that hatching of *Aspiculuris tetraptera* was also associated with an increase in the permeability of the eggshell to water. Van der Gulden and van Aspert-van Erp (1976) suggested that the eggshell of *Syphacia muris* became permeable to water as an essential preliminary event that enabled hatch. Other circumstantial evidence, such as size increase of the egg, from observations on the hatching of several species of nematodes indicates a permeability change and a consequent movement of water across the eggshell into the egg.

Work by Clarke and Perry (1980) showed that the water content of unhatched larvae of *Ascaris suum*, which had not been stimulated to hatch, changed with the osmotic pressure of the medium in which the eggs were immersed, and that the effect was reversible. Thus, the eggshell is permeable to water and acts as a semipermeable membrane. It is impermeable to water-soluble molecules such as trehalose, which is present in the perivitelline fluid surrounding the unhatched larva, until the hatching sequence is initiated; then the eggshell permeability alters to allow the escape of trehalose into the external medium. The role of trehalose in the hatching process will be discussed (see *Importance of the perivitelline fluid*, below) but, first, it is necessary to examine the permeability change in more detail. The nature of the permeability change of the lipid layer has been investigated mainly in *A. suum*, *G. rostochiensis* and *H. schachtii*.

Clarke and Perry (1988) considered that the induction of permeability of the lipid layer of eggshells of *A. suum* is a Na^+-mediated process, with the Na^+ forming a complex with the lipid layer. Although the lipid layer is not permeable to Na^+ (as NaCl or NaOH) (Fairbairn and Passey, 1955), it is permeable to water (Clarke and Perry, 1980) and to CO_2 as shown by the escape of respiratory CO_2 and the incorporation within the egg of $^{14}CO_2$ from the surrounding medium (Passey and Fairbairn, 1955). Clarke and Perry (1988) suggested that free CO_2, which is only slowly hydrated (Asada, 1982), penetrates the lipid layer to occupy interstitial sites present among the ascaroside head groups. Here the CO_2 molecule is hydrated and changes from a linear, non-polar molecule into the relatively larger disc-shaped carbonic acid and its ions (Cotton, Wilkinson and Gaus, 1987); at about pH 7, the predominate species is the HCO_3^- ion. Carbonic acid and its ions are probably incompatible, because of size and charge, with the environment in which they are formed. A mutual repulsion of ascaroside and HCO_3^- results in a structural change in the lipid layer which, at about 37°C, is already on the threshold of thermally induced

changes in permeability (Barrett, 1976). As a result, a new phase is formed with the lipid layer altering from a largely impermeable structure to a permeable one with sufficiently large pores to permit the passage of trehalose and enzymes. Clarke and Perry (1988) suggested that Na^+ is involved, as this new phase may be stabilized by the formation of a $NaHCO_3^-$ ascaroside complex.

A different mechanism for eggshell permeability change, involving Ca^{2+} transport, has been suggested by Ash and Atkinson (1984) for *Nematodirus battus*. Ruthenium red and lanthanum chloride are considered to be inhibitors of biological processes involving free Ca^{2+}. Low concentrations of ruthenium red (0.005–1 mM) inhibited hatching of *N. battus* when stimulated to hatch either by 5 mM sodium fluoride or following chilling of eggs (chilled at 5°C for 8 wk, then transferred to 20°C); the inhibitory effect of lanthanum chloride was not so marked on sodium fluoride-stimulated eggs (Ash and Atkinson, 1984). Competitive binding studies between ruthenium red and Ca^{2+} revealed low- and high-affinity Ca^{2+} binding sites but their location on the eggshell is unknown (Ash and Atkinson, 1984). Ash and Atkinson (1984) considered that a Ca^{2+} transport system may operate during hatching of *N. battus* in response to both hatching stimuli. During chilling, adaptive changes may occur in the fluidity of the lipid layer which results in a change of its permeability associated with the transport of messenger cations, such as Ca^{2+}, when the ambient temperature is increased. When exposed to 5 mM sodium fluoride, the fluoride may bring about direct stimulation of the phosphorylation of proteins involved in the transport of Ca^{2+} and other ions.

By contrast, Matthews (1986) found that ruthenium red did not affect hatching of *Ancyclostoma ceylanicum* or *A. tubaeforme* and lanthanum chloride inhibited development rather than hatching of these hookworms. Ruthenium red may not have penetrated the eggshell of *Ancyclostoma* as Matthews (1986) suggested that compounds with a molecular mass greater than 400–500 Da are excluded from the eggs. Thus, piperazine citrate (642 Da) and ruthenium red (786.4 Da) did not prevent hatching and may not have reached the unhatched larva. Trehalose has a molecular mass of 378 Da and Perry and Feil (1986) showed that movement of molecules across the eggshell of *G. rostochiensis* after treatment with potato root diffusate is restricted to molecules of 400 Da or smaller; thus, trehalose can escape from the egg during the hatching process. Larger molecules, such as those isolated from root diffusates (see *Host-derived hatching stimulants*, above), may not pass across the eggshell but would bind to sites on the eggshell itself.

In *G. rostochiensis*, the changes to eggshell permeability after exposure to potato root diffusate occur rapidly even though eclosion usually does not start until at least 3 d later (Doncaster and Shepherd, 1967). In both *G. rostochiensis* (Forrest and Perry, 1980) and *G. pallida* (Perry and Beane, 1982), a 5 min exposure per week of eggs to potato root diffusate was sufficient to stimulate hatch, indicating the involvement of a receptor-ligand interaction between the eggshell lipoprotein membranes and hatching factors in potato root diffusate. The change in permeability of the lipid layer of *G. rostochiensis* is thought to involve Ca^{2+}, although there is conflicting evidence about whether a Ca^{2+}-mediated structural change or a Ca^{2+} transport system is involved. Atkinson and Ballantyne (1979) and Atkinson, Taylor and Ballantyne (1980) considered that hatching of *G. rostochiensis* involves Ca^{2+} transport. Low concentrations of lanthanum chloride and ruthenium red inhibited hatching (Atkinson and Ballantyne, 1979) and bound to eggs treated with potato root diffusate (Atkinson and Taylor, 1980). However, Clarke and Hennessy (1981) tested a range of concentrations of lanthanum chloride and ruthenium red, including those used by Atkinson and Ballantyne (1979), and came to different conclusions. They found that 0.01 mM to 10 mM lanthanum chloride *initiated* the hatching of *G. rostochiensis*; for example, 4mM lanthanum chloride in distilled water elicited nearly 40% hatch of larvae, which was equivalent to the hatch obtained by the same concentration of lanthanum chloride in root diffusate. Clarke and Hennessy (1981) confirmed that ruthenium red inhibited hatching but showed that the hatching activity of root diffusate was destroyed when ruthenium red was added to it and then removed; thus, the failure to cause hatching may be due to inactivation of the diffusate rather than interference with the hatching mechanism. Ruthenium red also inhibits movement of larvae (Clarke and Hennessy, 1981). Atkinson and Ballantyne (1979) considered that two calcium ionophores that they tested stimulated hatch and synergised diffusate-induced hatch of *G. rostochiensis*. Calcium ionophores are lipophilic compounds which sequester calcium ions and the ionophore-Ca^{2+} complex can then pass through membranes which normally control Ca^{2+} permeability. However, Clarke and Hennessy (1983) tested one of these ionophores and found that it did not stimulate hatch and nor did it synergise diffusate activity; on the contrary, it inhibited hatch. Ion transport experiments also provided no evidence of the movement of Ca^{2+} by root diffusate (Clarke and Hennessy, 1978). Metal chelating agents may be used as inhibitors of cellular Ca^{2+} transport systems but Clarke and Hennessy (1983) obtained substantial hatching of *G. rostochiensis* in root diffusate from which Ca^{2+} had been removed and which contained up to 12 mM of the calcium-chelating agent 1,2-di(2-aminoethoxyl)ethane-*N,N,N',N'*-tetra-acetic acid.

Clarke and Hennessy (1983) considered that free Ca^{2+} is not essential for hatching of larvae of *G. rostochiensis*, so the hatching mechanism may not involve the transport of Ca^{2+} through the eggshell. Instead, the initiation of hatching may be the result of changes in eggshell permeability caused by the effect of root diffusate on bound Ca^{2+}. Clarke, Cox and Shepherd (1967) showed that Ca^{2+} is a major inorganic constituent of the eggshell. In eggs of *G. rostochiensis*, hatching factors in potato root diffusate bind to or displace internal Ca^{2+}, leading to changes in lipoprotein membrane structure and permeability characteristics (Clarke and Perry, 1985a). Clarke and Perry (1985a) distinguished three types of Ca^{2+} binding sites in the *G. rostochiensis* eggshell: a) sites in the outer layers which bind Ca^{2+} but are not involved in the hatching process, b) sites on the lipid layer, from which Ca^{2+} can be displaced by hatching factors and which are associated with diffusate-stimulated hatch and c) sites on the lipid layer which bind additional Ca^{2+} ions in the presence of diffusate. Atkinson and Taylor (1983) reported a sialoglycoprotein with high affinity for Ca^{2+} which was involved in the hatching process; however, their conclusion, that the sialoglycoprotein was located on the lipid layer, was based on an experimental protocol which has been questioned (see Clarke and Perry (1985a) for full discussion). In several respects, the change in permeability of the eggshells of *H. schachtii* (Clarke and Perry, 1985b) parallels that reported by Clarke and Perry (1985a) for *G. rostochiensis*: hatching agents, by binding to or replacing membrane-bound Ca^{2+}, induced structural change and hence eggshell permeability. As well as being induced by the natural hatching stimulus, sugarbeet root diffusate, the Ca^{2+} changes in the eggshells were also caused by the artificial hatching agents, nicotinic acid, picric acid, Cd^{2+} and Zn^{2+}.

Devine *et al.* (1996) considered that the activity of hatching factors from potato root diffusate at physiological concentrations indicates the involvement of receptor-ligand interactions, which is supported by the bimodal nature of the dose response curve they found for hatching factor-induced hatch of *G. rostochiensis*; the curve had a minor peak of activity at hatching factor concentrations some 4 to 6 orders of magnitude more dilute than that which gave the major peak of hatch. Similarly, peaks of hatching activity of the potato steroidal glycoalkaloid α-chaconine occurred at 10^{-3} M and 10^{-9} M, indicating the presence of at least two classes of hatching factor receptor on the eggshell, with different affinities for a given hatching factor. At supra-optimal concentrations, suppression of hatch was observed for all hatching factors investigated by Devine *et al.* (1996); such inhibition is common in receptor-ligand interactions in hormone systems and has been reported in other plant-invertebrate interactions (Klimetzet *et al.*, 1986).

As well as α-chaconine, another glycoalkaloid, α-solanine, induces hatch of *G. rostochiensis* (Devine *et al.*, 1996). A possible mechanism whereby glycoalkaloids alter the permeability of the lipid layer was outlined by Devine *et al.* (1996). Glycoalkaloids interact strongly with sterol-containing membranes causing disruption in a glycoalkaloid- and sterol-specific manner (Keukens *et al.*, 1992). The mode of action of these glycoalkaloids is that the aglycone is inserted into the membrane bilayer, the glycoalkaloid-sterol complex forms and the membrane is rearranged by the resulting network of glycoalkaloid-sterol complexes. This rearrangement causes disruption of the bilayer, during which leakage of trehalose, for example, can occur (Keukens *et al.*, 1992). Both α-solanine and α-chaconine stimulate hatch of *Globodera* spp. but larvae of *G. rostochiensis* hatch in greater numbers than larvae of

G. pallida (Devine *et al.*, 1996; Byrne, 1997) possibly indicating evolutionary divergence between these two species for eggshell hatching factor receptors.

It is feasible that eggshell permeability change could be induced in species of *Globodera* and *Heterodera* in order to provide a novel control approach. Any agrochemicals which operate by affecting membrane permeability, such as the thiocarbamate herbicides, are potential hatching agents. Thiocarbamates, such as diallate and cycloate, stimulated hatch of *H. schachtii* directly (Kraus and Sikora, 1981, 1983; Altman and Steele, 1982) and cycloate increased the control of *H. schachtii* by the chemical aldicarb by overcoming the hatch-inhibitory effect of the nematicide (Feyaerts and Coosemans, 1992). However, Perry and Beane (1989) found that four thiocarbamate herbicides, cycloate, pebulate, vernolate and tri-allate, did not induce hatch of *G. rostochiensis* or *H. schachtii* but prevented subsequent hatch in root diffusates. These conflicting results indicate that it is probably too simplistic to expect induction of permeability change to be the only effect of thiocarbamate herbicides. As will be dicusssed below (see *Responses of the larva*), alteration of eggshell structure may not, by itself, be sufficient to cause hatch; direct stimulation of the larva appears also to be required.

Importance of the Perivitelline Fluid

The perivitelline fluid surrounding the unhatched larva contains components that play important roles in the hatching process. For example, *C. elegans* secretes a HCH-1 metalloproteinase into the perivitelline fluid prior to hatching (Hishida *et al.*, 1996) and the role of enzymes in eclosion appears to be central to the hatching of many species of nematodes (see *Involvement of enzymes*, below). Two-dimensional gel electrophoresis of perivitelline fluid from *A. suum* revealed seven major proteins, one of which was identified as a novel fatty acid binding protein, As-p18 (Mei *et al.*, 1997). Mei *et al.* (1997) suggest that As-p18 may be involved in sequestering potentially toxic fatty acids and their peroxidation products that may accumulate around the unhatched larva, or it may be involved in maintaining the permeability characteristics of the lipid layer. At least eight potential homologues of As-p18 have been identified in the *C. elegans* genome and products of the three most closely related homologues are fatty acid binding proteins (LBP-1, LBP-2 and LBP-3) which contain putative secretory signals and comprise a distinct gene class, possibly unique to nematodes (Plenefisch *et al.*, 2000). The developmental expression of *lbp*-1 was identical to that of As-p18 and Plenefisch *et al.* (2000) suggest that these, and probably other, perivitelline proteins are secreted from the epidermis before cuticle formation. Future research, linking *C. elegans* genome information with studies on parasitic nematodes, will be important in determining the roles of such secreted proteins in physiological processes, including hatching.

The majority of studies on the perivitelline fluid have focused on the function of the trehalose component in nematode hatching and survival. Much of the trehalose in nematode eggs is found in the perivitelline fluid; for example, of the total trehalose content of eggs of *Ascaris lumbricoides* and *N. battus*, over 50% and 34%, respectively, is located in the perivitelline fluid (Fairbairn and Passey, 1957; Ash and Atkinson, 1983). Cook and Bugg (1973) showed that Ca^{2+} ions and trehalose can bring about membrane stabilisation, by cross-linking of trehalose molecules with Ca^{2+} ions. Trehalose is involved in several aspects of nematode physiology (Behm, 1997) and is connected with biological membranes as a desiccation and freezing protectant by replacing water normally associated with membranes (Crowe, Hoekstra and Crowe, 1992). As a constituent of the perivitelline fluid, it has been shown to contribute to the desiccation protection of the unhatched larvae of *N. battus* (Ash and Atkinson, 1983) and *G. rostochiensis* (Perry, 1983), for example. The role of trehalose in protecting nematodes from adverse environmental conditions is discussed by Womersley, Wharton and Higa (1998) and Wharton (this volume).

In species of nematodes where the trehalose concentration of the perivitelline fluid has been estimated, it varies from 0.1 M to 0.5 M. The unhatched, unstimulated larvae are surrounded by trehalose at a concentration of 0.3 M in *H. schachtii* (Perry, Clarke and Hennessy, 1980), up to 0.328M in *N. battus* (Ash and Atkinson, 1983), 0.34 M in *G. rostochiensis* (Clarke, Perry and Hennessy, 1978) and 0.5 M in *H. goettingiana* (Perry *et al.*, 1983). The trehalose content of the perivitelline fluid in *Ascaris* eggs was estimated to be between 0.1 and 0.2 M by Clarke and Perry (1980) but 0.4 M by Fairbairn and Passey (1957). The difference is probably accounted for by the technique used by Fairbairn and Passey (1957). They derived the concentration from chemical determinations of the amount of sugar present and from the volume of the egg fluid. The latter was obtained from measurements of the free larvae and of the eggs. Larvae were measured after being hatched in 0.15 M NaCl but, as the volume of larvae alters with the osmotic pressure of the fluid in which they are immersed, measurements should ideally be taken in a medium that matches the osmotic pressure of the egg fluid.

In the perivitelline fluid, trehalose provides an osmotic stress on the unhatched larva which causes a reduction in larva water content from the normal, fully hydrated water content of approximately 72% to 66.5% in *H. goettingiana* (Perry *et al.*, 1983), 67.3% in *G. rostochiensis* (Ellenby and Perry, 1976) and 69.0% in *A. suum* (Clarke and Perry, 1980). In this incompletely hydrated state, the internal hydrostatic pressure of the nematode is likely to be insufficient to antagonise the longitudinal muscles and sustain the continual movement necessary for eclosion. Thus, the larva is quiescent and its survival is enhanced because, for example, utilisation of energy reserves is minimised. After hatch stimulation, the change in eggshell permeability permits the leakage of trehalose out of the egg. This reduces the osmotic pressure of the perivitelline fluid and the unhatched larva is gradually exposed to a new environment where the reduced osmotic pressure permits the water content and metabolic activity of the larva to increase enabling continuous movement within the egg. Leakage of trehalose during the hatching process was first detected by Fairbairn (1961) working with *Ascaris* and Clarke and Perry (1980) showed that the water content of larvae increased from 69.0% to 71.3% during this period (Figure 6.3). Martin (1994) detected trehalose efflux from *G. rostochiensis* eggs within 8 h of exposure to potato root diffusate. The increase in water content for this species reached a maximum within 24 h after stimulation for larvae in eggs freed mechanically from cysts and within 48 h for larvae in encysted eggs (Ellenby and Perry, 1976). The larva of *H. goettingiana* also shows an increase in water content before hatching but the maximum water content was not attained until 7 d after stimulation and this is correlated with a much slower rate of hatching compared with *G. rostochiensis* (Perry *et al.*, 1983). This difference may reflect variation between the species in the time taken to reduce the different initial trehalose concentrations of the perivitelline fluid; the response times of the larvae may also differ.

Nematodirus battus and *H. schachtii* provide interesting contrasts to *Ascaris*, *G. rostochiensis* and *H. goettingiana*. In experiments with *N. battus*, trehalose was detected in the hatching medium 4 h after stimulation to hatch by 100 mM sodium fluoride (Ash and Atkinson, 1984) but there was no water uptake by the third stage larva prior to hatching; the increase to the water content of 74.5% for a fully hydrated larva occurred after hatching (Perry, 1977a). Similarly, there was no water uptake by larvae of *H. schachtii* before hatching (Perry, 1977b); the unhatched, unstimulated larva contains more water than that of a comparable *G. rostochensis* larva and is sufficiently hydrated to permit movement. Perry, Clarke and Hennessy (1980) showed that the perivitelline fluid of *H. schachtii* eggs has a lower trehalose concentration (0.3 M) than that of *G. rostochiensis* eggs

and, compared with *G. rostochiensis*, the larva of *H. schachtii* is less affected by osmotic stress. These two attributes may explain why *H. schachtii* hatches readily in water. However, host root diffusates stimulate an additional hatch, indicating that the Ca^{2+}-mediated change in eggshell permeability, mentioned above, may be required for some eggs to hatch. In eggs of *H. schachtii*, lipoprotein membranes have been detected only when there was no fungal contamination (Perry and Trett, 1986) and it is possible that fungal enzymes, such as lipases, could disrupt the lipoprotein membranes thus resulting in hatch of larvae; eggs in uncontaminated cysts may still require root diffusate to alter permeability of the lipid layer. It also is possible that mechanical damage to the lipoprotein membrane may occur through movement of the unhatched larva in *H. schachtii* and in other species of nematodes where the perivitelline fluid does not induce quiescence (Perry and Trett, 1986).

Involvement of Enzymes

The involvement of enzymes to degrade the eggshell and aid eclosion has been suggested frequently but there is little direct experimental evidence to substantiate the theory. Enzymes involved in the hatching process may originate from the unhatched larva and/or may be present in the perivitelline fluid and kept inactive either by separation from their substrates by the lipid layer or by an inhibitor, such as trehalose. If trehalose suppresses enzyme activity, then trehalose release through the eggshell, facilitated by a change in permeability of the inner lipid layer, would precede activation of enzymes responsible for eroding other layers of the eggshell. In this scenario, it follows that, in species where lipase activity has been associated with hatching, lipase is more likely to be released by the larva. Rogers (1978) considered that *Ascaris* larvae released enzymes in direct response to the hatching stimulus. His hypothesis suggests that the hatching stimulus affects sulphydryl groups in a receptor in the larva, leading via non-adrenergic activity to the release of a hormone which, in turn, causes the release of enzymes which erode the eggshell. Rogers (1978) suggested that, as insect larva hormone analogues inhibit the hatch of *Nematospiroides dubius* (*Heligmosomoides polygyrus*), *Nippostrongylus brasiliensis*, *Haemonchus contortus*, *Nematodirus spathiger* and *Aphelenchus avenae*, the compounds may affect either the release or the action of the putative neurohormone.

In a minority of species, structural specialisation of the eggshell occurs at one or both poles to form an operculum, and the case for enzyme action to open the operculum is persuasive. The operculum of *Aspiculuris tetraptera* consists of a modification of the uterine and chitinous layers of the eggshell over the whole area of the operculum (Wharton, 1979a), whereas in *Syphacia obvelata* the modification occurs at the groove which delimits the operculum (Wharton, 1979b). In *S. muris*, the chitinous seal between the operculum and the eggshell is dissolved after changes in eggshell permeability (van der Gulden and van Aspert-van Erp, 1976),

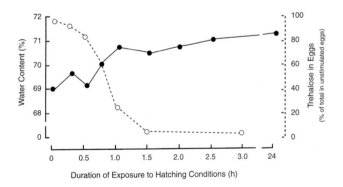

Figure 6.3. Changes in the water content of unhatched larvae (●, solid line) and in the trehalose content of eggs (○, broken line) of *Ascaris* during the hatching process. The water content of unhatched larvae was determined at set time intervals after removal of the eggs from the hatching medium (Clarke and Perry, 1980). Trehalose leakage was determined by Fairbairn (1961) after exposure of eggs to the hatching conditions.

allowing the operculum to open. The eggs of trichurid and capillarid nematodes are barrel-shaped with an opercular plug at each end. Trichurid nematodes are unusual among animal-parasitic species in that the larvae have stylets which, as with plant-parasitic species, are used to aid eclosion. When eggs of *Trichuris muris* were placed in caecal contents of mice, larvae became activated at the same time as there was an increase in the volume of the egg and larva consistent with endosmosis (Panesar and Croll, 1981). Exploratory behaviour by the larva followed by localised rubbing movements resulted in the head lying adjacent to one polar plug. The polar plug, through which the larva eventually emerged, increased in size and the stylet was used to pierce through the remaining vitelline layer; the anterior end of the larva was ejected as the internal pressure of the egg was released. The opercular plugs of eggs of *T. suis* consist of a fine network of chitin-protein microfibrils which differs in organisation from the rest of the shell (Wharton and Jenkins, 1978; Preston and Jenkins, 1985). Both plugs are partially dissolved before hatching and the larva of *T. suis* uses its stylet and emerges (Figure 6.4) in the same manner as described for *T. muris*. Monné and Hönig (1955) speculated that trichurid larvae hatch after enzymic depolymerisation of the polysaccharide plug. Wharton and Jenkins (1978) suggested that, because chitin which is bound to protein is not degradable by chitinase, the lower proportion of protein combined with the chitin-protein

arrangement of the opercular plugs makes them more susceptible to enzymic degradation than the rest of the eggshell.

Evidence for the role of enzymes in eggshell degradation derives primarily from analysis of the hatching fluid, initially from research on *Ascaris* spp. (Rogers, 1958, 1982; Fairbairn, 1961; Hinck and Ivey, 1976). The hatching fluid of *A. lumbricoides* contains a tryptic type protease, an esterase, α- and β-glycosidases, leucine aminopeptidase and a chitinase, each of which are presumed to attack different components of the eggshell prior to eclosion. The purified chitinase degrades chitin to N-acetylglucosamine. Most other chitinases only degrade chitin to di- and tri-saccharides and require a second enzyme, chitobiase, for complete breakdown (Barrett, 1981). Protease activity increased simultaneously with hatching in *A. lumbricoides* (Hinck and Ivey, 1976) and chitinase activity increased with development of eggs of *A. suum* and showed a positive correlation with the hatching stimulus (Ward and Fairbairn, 1972). Eggs of *Ascaris* have a cap at one end (Ubelaker and Allison, 1975; Kazacos and Turek, 1983) which may be a specialised region which is more susceptible to enzyme attack and, thus, plays a role in hatching.

Increase in chitinase activity as the egg matures was also found with *Heligmosomoides polygyrus* (Arnold *et al.*, 1993). The ratio of endo-/exo-chitinase activities decreases as the egg matures and Arnold *et al.* (1993) suggested that the endochitinase activity that they detected in egg fluid may soften the eggshell in strategic areas and open up otherwise inaccessible chitin chains for degradation by exochitinase. The chitinase activities were very sensitive to inhibition by allosamidin, a specific chitinase inhibitor. However, treatment of the eggs of *H. polygyrus* with 250 μM allosamidin reduced, but did not stop, hatch, perhaps because the eggshell is impermeable to exogenously applied allosamidin and the effects occur only when there is a change in eggshell permeability just prior to hatch when it would be too late to prevent hatching of the majority of larvae (Arnold *et al.*, 1993). Hatching fluid from *Haemonchus contortus* eggs contains a lipase and an aminopeptidase but no chitinase (Rogers and Brooks, 1977); chitin has not been detected in *H. contortus* eggshells (Monné and Hönig, 1955).

Tefft and Bone (1985a) considered that Zn^{2+} mediated hatch of *Heterodera glycines* and suggested that a Zn^{2+}-dependent leucine aminopeptidase, detected in the egg fluid, was involved; however, the natural hatching stimulus, soybean root diffusates, had no direct effect on the activity of this enzyme (Tefft and Bone, 1985b). The hatch of *Meloidogyne incognita* was positively correlated with lipase activity in the hatching fluid (Figure 6.5); proteinase, including collagenase, and chitinase activity also were detected and these enzymes are likely to be associated with increased flexibility of the eggshell prior to eclosion (Perry, Knox and Beane, 1992). By contrast, the eggshells of *G. rostochiensis* remain rigid throughout the hatching process and Perry, Knox and Beane (1992) were unable

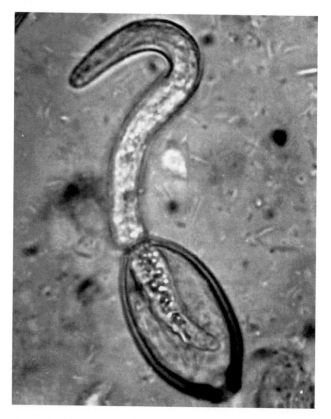

Figure 6.4. Hatching of *Trichuris suis* through a dissolved opercular plug (C. Preston and T. Jenkins, University of Portsmouth).

to detect any lipase or chitinase activity in the fluid released during the hatching of larvae (Figure 6.5). This supported and extended earlier work (Perry, Zunke and Wyss, 1989) where, although accumulation of secretory granules in the dorsal pharyngeal glands occurred, no voiding of secretions was observed indicating that the secretions were not involved in hatching (see *Responses of the larva*, below). However, the continuing rigidity of eggs of *G. rostochiensis* and some other species of plant-parasitic nematodes during the hatching process may be atypical for the Phylum Nematoda, as increased eggshell flexibility and the presumed involvement of enzymes are associated with hatching of the majority of nematode species studied.

Increase in activity of the pharyngeal glands and concomitant increased flexibility of the eggshell of several nematode species provides circumstantial evidence to support the pre-eclosion production of enzymes. In both *Xiphinema diversicaudatum* (Flegg, 1968) and *Aphelenchus avenae* (Taylor, 1962), increases in apparent secretory activity (pumping of the pharyngeal bulb and pulsation of the valve in the metacorpus, respectively) were detected prior to softening of the eggshell, although secretion through the mouth was only observed in the case of *X. diversicaudatum*. In *Necator americanus*, Croll (1974) considered that 'feeding' movements were instrumental in flushing enzymes into the egg via the anus, although this was not observed and the presence of enzymes was inferred from changes in egg volume. Hatching of *Caenorhabditis elegans* involves pharyngeal pumping and, probably,

enzymes from the pharyngeal glands and/or the intestine (Wood, 1988). About 30 min before rupture of the eggshell and hatching of *C. elegans*, pharyngeal muscles begin contracting and the intestine, initially collapsed, quickly becomes distended with ingested fluid which is recirculated through the alimentary tract by cycles of ingestion and release, presumably through the anus (Hall and Hedgecock, 1991). A protease for digesting outer layers of the eggshell has been identified genetically. The *hch*-1 gene of *C. elegans* is defined by a mutation in which the eggshell becomes soft but hatching is delayed (Hedgecock *et al.*, 1987); it appears that eggshell chitin is digested but not proteins. The *hch*-1 gene encodes a protein of the *tolloid*/bone morphogenetic protein (BMP)-1 family (Hishida *et al.*, 1996); proteins of this family contain a metalloprotease domain. Using PCR primers selected from the *hch*-1 gene sequence of *C. elegans* and genomic DNA from eggs of *H. glycines*, Bolla and Kay (1997) identified a gene in *H. glycines* with homology to *hch*-1. This indicates that similar enzymes are present in both species and similar molecular mechanisms may be used despite differences in their hatching processes. As Zn^{2+} ions mediate hatch of *H. glycines*, this nematode may regulate hatching by controlling levels or availability of enzyme co-factors.

In the absence of direct evidence for enzyme activity, alternative explanations remain a possibility. Barrett (1976) found no evidence of enzyme hydrolysis of the lipid layer of eggs of *A. suum* as it became permeable and considered that mechanical damage to the layer, caused by the activity of the unhatched larva, was responsible for the permeability change. As mentioned above, Perry and Trett (1986) suggested that mechanical damage may be a contributory factor in the hatching of *H. schachtii* and possibly other species where the perivitelline fluid does not induce larval quiescence. Wilson (1958) suggested that the inner lipid layer of the eggshell of *Trichostrongylus retortaeformis* was emulsified by the active movement of the larva, with assistance of emulsifying agents in the egg fluid. Wallace (1966) considered that this may also occur with *M. javanica*, although subsequent time-lapse and ultrastructural studies (Bird, 1968) showed no evidence of emulsification. Working with *Pratylenchus penetrans*, Thistlethwayte (1969) suggested that eggshell distortion was caused by more vigorous movements of the larva rather than increased eggshell flexibility. However, it appears that, in general, the increased activity of the unhatched larva is a *result* of eggshell permeability change, perhaps removing osmotic stress conditions (see above), rather than the *cause* of the change. Undoubtedly, increased activity will enhance the disruption of the lipid layer and speed up the process of hatching.

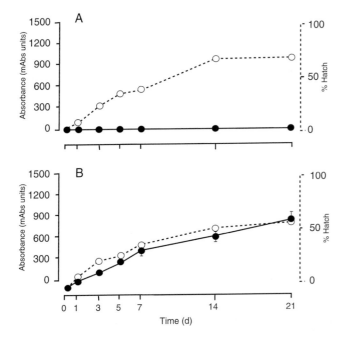

Figure 6.5. Cumulative lipase activity at pH5 (●) and percentage hatch (○) of (A) *Globodera rostochiensis* in potato root diffusate and (B) *Meloidogyne incognita* in distilled water over a three week period. Enzyme activity is expressed in absorbance units. (From Perry, Knox and Beane, 1992).

Responses of the Larva

After hatch stimulation and subsequent alteration of eggshell permeability, a series of physical and metabolic changes occur in unhatched larvae concomitant with increased hydration and commensurate with recovery from quiescence; these changes have been documented extensively in *G. rostochiensis* (Figure 6.6). When exposed to potato root diffusate, granular material in the median zone of the cuticle of unhatched larvae of *G. rostochiensis* begins to be lost while the nematode is in the egg and disappears soon after hatching (Jones, Perry and Johnston, 1993). This granular material may be directly or indirectly involved in the hatching mechanism since it remains present when larvae are artificially hatched. It also is possible that the material moves to the surface of the cuticle to form the surface coat of the hatched larva. The structure of the amphids of larvae of *G. rostochiensis* also changes during the hatching process (Jones, Perry and Johnston, 1994). The absence of secretory material and the shrunken state of the sheath cell in unhatched nematodes indicate that the amphids may not be functional during the initial phases of the hatching process. Once the water content of the larva has increased sufficiently for hatching to occur the amphids are likely to assume a functional role, not only to aid in the co-ordinated cutting of the eggshell (see *Eclosion*, below) but also to prepare for host location.

When *Ascaris* is stimulated to hatch, the oxygen quotient increases (Passey and Fairbairn, 1955) and the relatively large ATP/ADP ratio falls (Beis and Barrett, 1975). Beis and Barrett (1975) found no evidence of any general metabolic inhibitor in dormant *Ascaris* eggs, which is consistent with dormancy of larvae being maintained by a water deficit (Clarke and Perry, 1980). Similar changes occur with unhatched larvae of *G. rostochiensis* after exposure to root diffusate. As the larva of *G. rostochiensis* re-hydrates, the adenylate energy charge falls (Atkinson and Ballantyne, 1977a), oxygen consumption commences (Atkinson and Ballantyne, 1977b) and the content of cAMP rises (Atkinson, Taylor and Fowler, 1987). The changes in cAMP occurred 2.5–8 h after exposure to root diffusates and may indicate a role as a possible secondary messenger in a receptor-ligand interaction. Utilisation of endogenous lipid reserves by the larva also commences (Robinson, Atkinson and Perry, 1985), probably coinciding with the increase in locomotory activity. A decline in infectivity of second stage larvae of several species of plant-parasitic nematodes has been associated with the decline in neutral lipid reserves and the slower rate of hatch of *G. pallida* compared with *G. rostochiensis* has been positively correlated with the slower rate of lipid utilisation of *G. pallida* (Robinson, Atkinson and Perry, 1987a,b). In *Globodera* spp. C20 fatty acids predominate and Holz, Wright and Perry (1998) found that, after hatching, the degree of saturation and the percentage of monounsaturated fatty acids decreased and the percentage of polyunsaturated fatty acids increased considerably, especially in the free fatty acid fraction, where C20:1 showed an 8-fold decrease and C20:4 a 33-fold increase.

Such changes result, in part, from the recovery of the larva from a quiescent state following removal of osmotic stress. However, in *G. rostochiensis*, there is evidence that changes may be induced by direct stimulation of the larva by host root diffusates. When inactive, mechanically freed larvae of *G. rostochiensis* were placed in potato root diffusate they became active (Clarke and Hennessy, 1984) and Masamune *et al.* (1987a) demonstrated that glycinoeclepin A also stimulated movement of water-hatched larvae of *H. glycines*. Changes in the dorsal and subventral pharyngeal glands of *Globodera* spp. have been investigated in detail. Atkinson, Taylor and Fowler (1987) found that mechanically freed larvae of *G. rostochiensis* showed a partial increase in the nucleolus diameter of the dorsal pharyngeal gland when suspended in water but the larvae needed to be placed in potato root diffusate to achieve a nucleolus size similar to that of larvae that hatched after stimulation with diffusate. The nucleolus of the dorsal pharyngeal gland of unhatched larvae increased in size within 4 h of exposure to root diffusate (Perry, Zunke and Wyss, 1989) and comparisons of transcription patterns of rRNA in hatched and unhatched larvae of *G. rostochiensis* showed that the nucleolus of the dorsal pharyngeal gland cell was transcriptionally active as a result of exposure to root diffusates while the larva was in the egg (Figure 6.7; Blair *et al.*, 1999). Accumulation of secretions in the dorsal pharyngeal gland occurs before hatching but no secretory material was observed to be voided (Doncaster, 1974). Changes in the dorsal pharyngeal gland cells were interpreted by Perry, Zunke and Wyss (1989) as part of the preparation for the feeding phase occurring after hatching; the secretory granules are utilised during the establishment of the cyst nematode larva in its host (Atkinson and Harris, 1989). In the field, there are indications that the diameter of the dorsal pharyngeal gland nucleolus of unhatched larvae of *Globodera* spp. changes with the season (Holz, Riga and Atkinson, 1998) but the reasons for this are unclear; seasonal changes in protein and carbohydrate levels of eggs of *H. glycines* appeared to be primarily temperature dependent (Yen *et al.*, 1996). In contrast to the response of the dorsal pharyngeal glands, the size of the nucleoli of the subventral glands was not increased by hydration or exposure to diffusates (Perry, Zunke and Wyss, 1989) and there was no evidence of transcriptional activity in these nucleoli, even after prolonged exposure to root diffusates (Blair *et al.*, 1999). Subventral gland proteins are thought to be involved in the early events of the *G. rostochiensis*-host interaction, possibly during invasion (Smant *et al.*, 1997) and the presence of endogenous cellulases, identified in *G. rostochiensis* and *H. glycines*, probably facilitate intracellular migration through plant roots by cell wall degradation (Smant *et al.*, 1998).

Fractionation of potato root diffusate revealed that the fractions which stimulated the dorsal pharyngeal gland did not exhibit hatching activity (Atkinson, Fowler and Isaac, 1987), further evidence that potato root diffusate may induce changes in the larvae which relate to phases subsequent to hatching. The

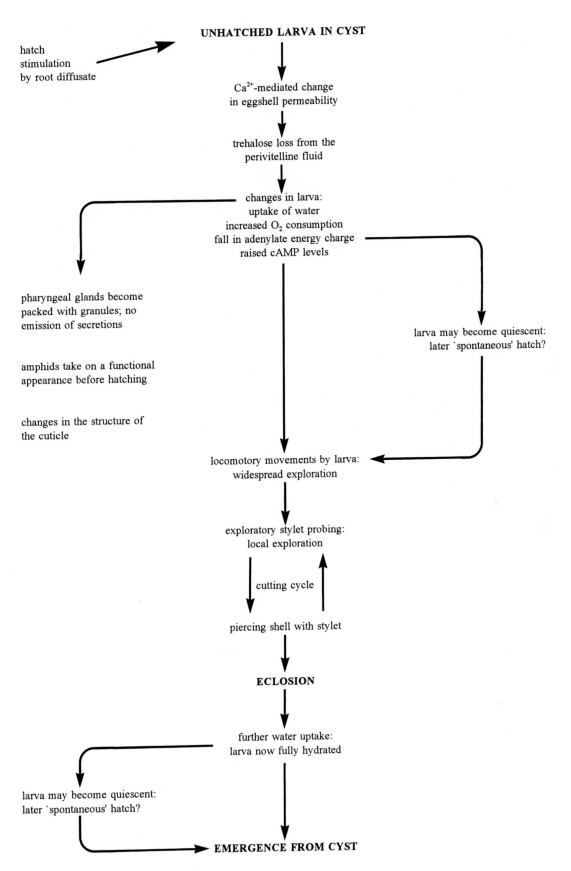

Figure 6.6. The sequence of events in the hatching process of *Globodera rostochiensis* after stimulation with potato root diffusates (from Jones, Tylka and Perry, 1998).

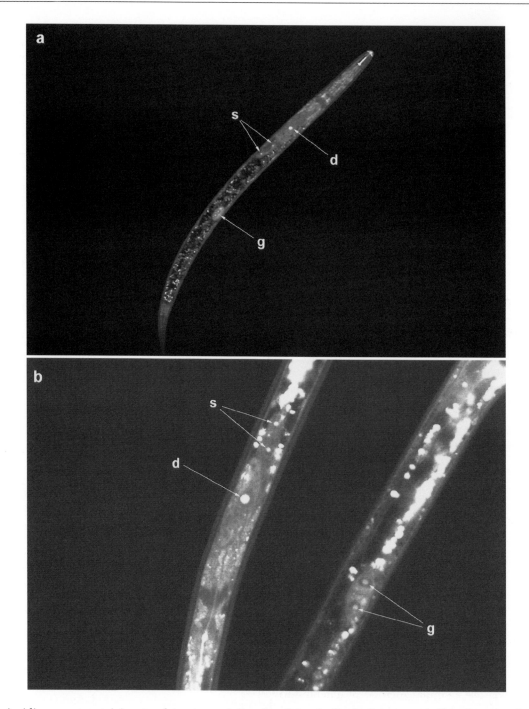

Figure 6.7. Acridine orange staining to show transcriptional activity in hatched larvae of *Globodera rostochiensis*. a: low magnification image showing the structures labelled with acridine orange including the nucleoli of the dorsal pharyngeal gland (d), the genital primordia (g) and the subventral gland cells (s); b: higher magnification image showing labelling of the same structures in more detail. Fluorescent image was captured at 525nm emission wavelength. Other visible labelling is adventitious staining of lipid droplets. (From Blair *et al.*, 1999).

fact that individual hatching factors from potato root diffusate (e.g. *α*-solanine) induce hatch of *Globodera* spp. (Devine *et al.*, 1996) indicates that changes in gland activity induced by diffusates are not essential for eclosion. Changes in gene expression do not appear to be induced directly by potato root diffusate but they seem to occur during or immediately after the hatching process (Jones *et al.*, 1997). Although this may indicate that diffusate itself may not induce all the changes in the nematode which prepare it for the parasitic phase of its life cycle, technical limitations of the differential display technique may have precluded the detection of gene expression induced by diffusate.

Eclosion

The culmination of the sequence of events in the hatching process is eclosion. The mode of exit of the larva from the egg varies markedly between species. Investigations of behaviour immediately before and during eclosion have been undertaken mainly on plant-parasitic nematodes with time-lapse ciné photomicrography providing detailed information on the hatching of *G. rostochiensis* and *D. dipsaci* (Doncaster and Shepherd, 1967; Doncaster and Seymour, 1973; Doncaster, 1974). Doncaster and Seymour (1973) defined three phases leading to eclosion of *G. rostochiensis* after stimulation with PRD. The first phase, termed 'widespread exploration', involves the larva circulating within the egg. This is followed by 'local exploration' when only the head moves and the cephalic area, where the anterior sense organs are located, is pressed and/or rubbed against the eggshell and the stylet starts to probe. These two phases have been mentioned previously (see *Involvement of enzymes*, above) as part of the hatching process of *T. muris* (Panesar and Croll, 1981). The final phase in eclosion of *G. rostochiensis*, 'stylet thrusting', causes a regular pattern of perforations in the eggshell (Figure 6.8A), each hole being close enough together for the stylet conus to break through to the preceding one resulting eventually in a slit through which the larva escapes (Figure 6.8B). Alternation of the last two phases was termed the 'cutting cycle'. The stylet thrusting behaviour shows a marked degree of sensory feedback. Doncaster and Seymour (1973) observed that, if the stylet failed to penetrate the eggshell at the first attempt, the larva persisted, with the stylet at the same angle, until it succeeded. Sometimes the slit was started near the pole and continued on one side, then the larva located the start of the slit and completed it in the opposite direction. On other occasions, stylet thrusting was started near the base of the hemispherical end of the egg on one side and continued in an arc onto the other side. The stylet thrusts were slow; one larva made 43 thrusts per min and the slit in the eggshell enlarged at 7.5 μm min^{-1}. By pressing against the eggshell, the larva causes the slit to gape and can then exit; there is no evidence that the larva moves rapidly through the slit under the influence of increased pressure within the egg. Larvae of *Ditylenchus dipsaci* use a similar method of hatching to *G. rostochiensis*, except that the stylet thrusts are rapid (120 per min), more random and only rarely penetrate the eggshell, and the larva uses its anterior end to force open the slit in the eggshell (Doncaster and Seymour, 1979). Such co-ordinated use of the stylet to produce the exit slit relies on the eggshell remaining rigid and this imposes a physical constraint on the larva of *G. rostochiensis* which, although it takes up water during early stages of the hatching sequence, only becomes fully hydrated once it has hatched (Ellenby, 1974). An increase in water content immediately after hatching has also been demonstrated in *H. schachtii* (Perry, 1977a) and *N. battus* (Perry, 1977b). In the synthetic hatching stimulant, 3 mM sodium metavana-

date, eggs of *G. rostochiensis* became soft and flexible before eclosion and, although hatch was more rapid, larvae usually made scattered perforations at both ends of the egg or made short incisions by a sawing action (Doncaster and Shepherd, 1967; Shepherd and Clarke, 1971).

Eggshells of *Ancylostoma* spp. also remain rigid during eclosion and the larva emerges through a slit in the eggshell. Larvae are not ejected under pressure, as emergence is an active process with purchase being obtained from the eggshell to enable the larva to move through the slit (Matthews, 1985). There was no change in volume of the eggs of *A. ceylanicum* (Matthews, 1986) or *Trichostrongylus colubriformis* (Wilson, 1958) during hatching. However, in eggs that are flexible immediately before the larvae hatch, eggshell permeability change has slightly different implications. Water influx could increase the volume of the perivitelline fluid and/or the water content of the unhatched larva resulting in increased internal pressure which would manifest itself as a size increase in the egg and frequently results in rapid expulsion of the larva when the eggshell is ruptured. A 15–20% increase in egg volume occurs immediately before eclosion in *Necator americanus* (Croll, 1974) and *C. elegans* (Croll, 1975). The flexible egg of *M. incognita* has about 30% free space which allows the larva to be fully hydrated by the time of eclosion (Ellenby, 1974). A firmer, stretched eggshell will assist stylet penetration and may also facilitate extension of the slit. In *M. javanica*, the anterior end of the larva projects into the flexible eggshell and stylet thrusts cause a tear through which the larva escapes. In *P. penetrans* (Thistlethwayte, 1969) and *H. avenae* (Banyer and Fisher, 1972), a single stylet thrust penetrates the eggshell, the anterior end again extending this into a tear.

There are isolated reports of nematodes using their sharply pointed tails to aid eclosion. Laughlin, Lugthart and Vargas (1974) observed that larvae of *Heterodera iri* employ the tail tip to make the initial penetration in the eggshell and emerged tail first. Silverman and Campbell (1959) stated that larvae of *Haemonchus contortus* generally emerged tail first through a hole in the eggshell made by their tail. Matthews (1985) found that this method of piercing the eggshell was occasionally used by larvae of *Ancylostoma ceylanicum* and *A. tubaeforme* but usually larvae reversed after making the slit and emerged head first. Rogers and Brooks (1977) found that larvae of *H. contortus* emerged head first and Matthews (1985) suggested that the observations of Silverman and Campbell (1959) may have been an artefact.

Summary

Co-evolution with their hosts and a close host-parasite interaction have resulted in certain species of animal- and plant-parasitic nematodes being dependent on the host to provide the stimulus for hatching of the infective larvae. Frequently, nematodes with sophisti-

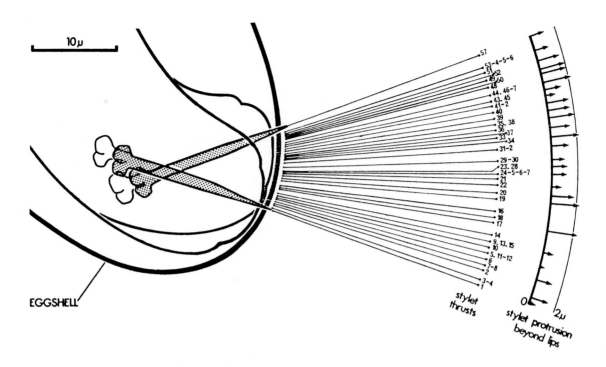

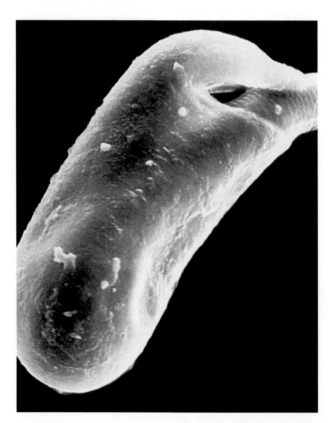

Figure 6.8. Diagram, based on analysis of ciné film, of a second-stage larva of *Globodera rostochiensis* using its stylet to perforate the eggshell and make part of the exit slit. Outlines of head and stylet (protracted and retracted) at the start and end of the sequence are shown. Within 80 sec 57 stylet thrusts were made; in 26 of these the stylet perforated the eggshell. Each arrow shows where a perforation occurred and how far the stylet then protruded beyond the lips. As the numbering of the thrusts shows, when the stylet failed to penetrate it was thrust again (up to 4 times) nearer the previous hole until it penetrated, when it usually broke into a previous perforation. After this sequence the slit was extended in the opposite direction, beginning near its original starting point. The scanning electron micrograph shows the resulting sub polar slit through which the larva can emerge. (Diagram from Doncaster and Seymour, 1973; SEM courtesy of J. Beane and R.N. Perry, IACR-Rothamsted.)

cated host-stimulated hatching mechanisms have very restricted host ranges and the hatching behaviour ensures that the nematode is able to survive in the absence of a host but will hatch when suitable hosts are available. This dependence is an obvious target for possible novel control strategies and, thus, it is unsurprising that the majority of research on hatching biology has concentrated on economically important animal- and, especially, plant-parasitic nematodes. This review of the hatching mechanisms has illustrated some general similarities but also pointed out important differences between species in detailed aspects of the responses to hatch stimulation. It is evident that there is only limited information on some aspects of the hatching process, such as the importance and origin of enzymes, and the changes in gene expression associated both with hatching and with the change from a pre-parasitic form to an infective larva; these aspects are worthy of further investigation.

Acknowledgement

The Institute of Arable Crops Research (IACR) receives grant-aided support from the Biotechnology and Biological Sciences Research Council of the UK. I am grateful to Dr Marcus Trett for help in preparing Figures 6.1, 6.2, 6.3, 6.5 and 6.8.

References

Altman, J. and Steele, A. (1982) Effect of herbicides on hatching and emergence of larvae from *Heterodera schachtii* cysts. *Phytopathology* **72**, 385.

Anya, A.O. (1966) Experimental studies on the physiology of hatching of eggs of *Aspiculuris tetraptera* Schulz (Oxyuroidea; Nematoda). *Parasitology* **56**, 733–744.

Arnold, K., Brydon, L.J., Chappell, L.H. and Gooday, G.W. (1993) Chitinolytic activities in *Heligmosomoides polygyrus* and their role in egg hatching. *Molecular and Biochemical Parasitology* **58**, 317–324.

Asada, K. (1982) Biological carboxylations. In *Organic and Bio-organic Chemistry of Carbon Dioxide*, edited by S. Inoue and N. Yamazaki, pp. 185–210. New York: Halstead Press.

Ash, C.P.J. and Atkinson, H.J. (1983) Evidence for a temperature-dependent conversion of lipid reserves to carbohydrates in quiescent eggs of the nematode *Nematodirus battus*. *Comparative Biochemistry and Physiology B* **76**, 603–610.

Ash, C.P.J. and Atkinson, H.J. (1984) *Nematodirus battus*: permeability changes, calcium binding, and phosphorylation of the eggshell during hatching. *Experimental Parasitology* **58**, 27–40.

Atkinson, H.J. and Ballantyne, A.J. (1977a) Changes in the adenine nucleotide content of cysts of *Globodera rostochiensis* associated with the hatching of juveniles. *Annals of Applied Biology* **87**, 167–174.

Atkinson, H.J. and Ballantyne, A.J. (1977b) Changes in the oxygen consumption of cysts of *Globodera rostochiensis* associated with the hatching of juveniles. *Annals of Applied Biology* **87**, 159–166.

Atkinson, H.J. and Ballantyne, A.J. (1979) Evidence for the involvement of calcium in the hatching of *Globodera rostochiensis*. *Annals of Applied Biology* **93**, 191–198.

Atkinson, H.J. and Harris, P.D. (1989) Changes in nematode antigens recognised by monoclonal antibodies during early infection of soya beans with the cyst nematode *Heterodera glycines*. *Parasitology* **98**, 479–487.

Atkinson, H.J. and Taylor, J.D. (1980) Evidence for a calcium-binding site on the eggshell of *Globodera rostochiensis* with a role in hatching. *Annals of Applied Biology* **96**, 307–315.

Atkinson, H.J. and Taylor, J.D. (1983) A calcium binding sialoglycoprotein associated with an apparent egg shell membrane of *Globodera rostochiensis*. *Annals of Applied Biology* **102**, 345–354.

Atkinson, H.J., Fowler, M. and Isaac, R.E. (1987) Partial purification of hatching activity for *Globodera rostochiensis* from potato root diffusate. *Annals of Applied Biology* **110**, 115–125.

Atkinson, H.J., Taylor, J.D. and Ballantyne, A.J. (1980) The uptake of calcium prior to the hatching of the second-stage juvenile of *Globodera rostochiensis*. *Annals of Applied Biology* **94**, 103–109.

Atkinson, H.J., Taylor, J.D. and Fowler, M. (1987) Changes in the second stage juveniles of *Globodera rostochiensis* prior to hatching in response to potato root diffusate. *Annals of Applied Biology* **110**, 105–114.

Aubert, V. (1986) Hatching of the carrot cyst nematodes. In *Cyst Nematodes*, edited by F. Lamberti and C.E. Taylor, pp. 347–348. New York: Plenum Press.

Banasiak, L., Lyr, H., Grosse, E. and Müller J. (1994) Chemical induction of larval hatching of the sugarbeet cyst nematode (*Heterodera schachtii* Schmidt) with 2-ethylex-2-en-1-al and derived compounds in soil. *Archives of Phytopathology and Plant Protection* **29**, 171–178.

Banyer, R.J. and Fisher, J.M. (1972) Motility in relation to hatching of eggs of *Heterodera avenae*. *Nematologica* **18**, 18–24.

Barrett, J. (1976) Studies on the induction of permeability in *Ascaris lumbricoides* eggs. *Parasitology* **76**, 253–262.

Barrett, J. (1981) *Biochemistry of Parasitic Helminths*, pp. 156–158. London: Macmillan.

Beane, J. and Perry, R.N. (1983) Hatching of the cyst nematode *Heterodera goettingiana* in response to root diffusate from bean (*Vicia faba*). *Nematologica* **29**, 361–363.

Behm, C.A. (1997) The role of trehalose in the physiology of nematodes. *International Journal for Parasitology* **27**, 215–229.

Beis, I. and Barrett, J. (1975) Energy metabolism in developing *Ascaris lumbricoides* eggs. II. The steady state content of intermediary metabolites. *Developmental Biology* **42**, 188–195.

Birch, A.N.E., Robertson, W.M., Geogheghan, I.E., McGavin, W.J., Alphey, T.J.W., Phillips, M.S., Fellows, L.E., Watson, A.A., Simmonds, M.S.J. and Porter, E.A. (1993) DMDP — a plant derived sugar analogue with systemic activity against plant parasitic nematodes. *Nematologica* **39**, 521–535.

Bird, A.F. (1968) Changes associated with parasitism in nematodes. III. Ultrastructure of the egg shell, larval cuticle, and contents of the subventral oesophageal glands in *Meloidogyne javanica*, with some observations on hatching. *Journal of Parasitology* **54**, 475–489.

Bird, A.F. and Bird, J. (1991) *The Structure of Nematodes*, 2nd edn, pp. 16–17. London: Academic Press.

Bird, A.F. and McClure, M.A. (1976) The tylenchid (Nematoda) egg shell: structure, composition and permeability. *Parasitology* **72**, 19–28.

Bird, A.F. and Soeffky, A. (1972) Changes in the ultrastructure of the gelatinous matrix of *Meloidogyne javanica* during dehydration. *Journal of Nematology*, **4**, 166–169.

Blair, L., Perry, R.N., Oparka, K. and Jones, J.T. (1999) Activation of transcription during the hatching process of the potato cyst nematode *Globodera rostochiensis*. *Nematology* **1**, 103–111.

Bolla, R.I. and Kay, E. (1997) A gene possibly involved in regulation of hatching of soybean cyst nematode (*Heterodera glycines*) eggs. *Second English Language International Nematology Symposium of the Russian Society of Nematologists*. Moscow, pp. 4–5.

Byrne, J. (1997) Comparative hatching behaviour of *Globodera rostochiensis* and *G. pallida*. PhD thesis, The National University of Ireland, Cork, Ireland.

Byrne, J., Twomey, U., Maher, N., Devine, K.J. and Jones, P.W. (1998) Detection of hatching inhibitors and hatching factor stimulants for golden potato cyst nematode, *Globodera rostochiensis*, in potato root leachate. *Annals of Applied Biology* **132**, 463–472.

Carroll, M. (1995) Characterisation of rhizosphere bacterial isolates which affect the hatch of potato cyst nematodes. MSc thesis, The National University of Ireland, Cork, Ireland.

Christie, M.G. (1962) On the hatching of *N. battus*, with some remarks on *N. filicollis*. *Parasitology* **52**, 297–313.

Clarke, A.J. (1970) Hatching factors and sex attractants of cyst nematodes. In *Report of Rothamsted Experimental Station for 1969*, p. 179. Harpenden: Rothamsted.

Clarke, A.J. (1971) Hatching factors and sex attractants of potato cyst nematode. In *Report of Rothamsted Experimental Station for 1970*, pp. 147–148. Harpenden: Rothamsted.

Clarke, A.J. and Hennessy, J. (1978) Hatching, osmotic stress and ion transport. In *Report of Rothamsted Experimental Station for 1977*, pp. 177–178. Harpenden: Rothamsted.

Clarke, A.J. and Hennessy, J. (1981) Calcium inhibitors and the hatching of *Globodera rostochiensis*. *Nematologica* **27**, 190–198.

Clarke, A.J. and Hennessy, J. (1983) The role of calcium in the hatching of *Globodera rostochiensis*. *Revue de Nématologie* **6**, 247–255.

Clarke, A.J. and Hennessy, J. (1984) Movement of *Globodera rostochiensis* (Wollenweber) juveniles stimulated by potato root exudate. *Nematologica* **30**, 206–212.

Clarke, A.J. and Hennessy, J. (1987) Hatching agents as stimulants of movement of *Globodera rostochiensis* juveniles. *Revue de Nématologie* **10**, 471–476.

Clarke, A.J. and Perry, R.N. (1977) Hatching of cyst-nematodes. *Nematologica* **23**, 350–368.

Clarke, A.J. and Perry, R.N. (1980) Egg-shell permeability and hatching of *Ascaris suum*. *Parasitology* **80**, 447–456.

Clarke, A.J. and Perry, R.N. (1985a) Egg-shell calcium and the hatching of *Globodera rostochiensis*. *International Journal of Parasitology* **15**, 511–516.

Clarke, A.J. and Perry, R.N. (1985b) The role of egg-shell calcium in the hatching of *Heterodera schachtii*. *Nematologica* **31**, 151–158.

Clarke, A.J. and Perry, R.N. (1988) The induction of permeability in egg-shells of *Ascaris suum* prior to hatching. *International Journal for Parasitology* **18**, 987–990.

Clarke, A.J. and Shepherd, A.M. (1966) Inorganic ions and the hatching of *Heterodera* spp. *Annals of Applied Biology* **58**, 497–508.

Clarke, A.J. and Shepherd, A.M. (1967) Flavianic acid as a hatching agent for *Heterodera cruciferae* Franklin and other cyst nematodes. *Nature* **214**, 419–420.

Clarke, A.J. and Shepherd, A.M. (1968) Hatching agents for the potato cyst-nematode, *Heterodera rostochiensis* Woll. *Annals of Applied Biology* **61**, 139–149.

Clarke, A.J., Cox, P.M. and Shepherd, A.M. (1967) Chemical composition of the eggshells of potato cyst-nematode, *Heterodera rostochiensis* Woll. *The Biochemical Journal* **104**, 1056–1060.

Clarke, A.J., Perry, R.N. and Hennessy, J. (1978) Osmotic stress and the hatching of *Globodera rostochiensis*. *Nematologica* **24**, 384–392.

Cook, W.J. and Bugg, C.E. (1973) Calcium interactions with D-glucans: crystal structure of α,α-trehalose-calcium bromide monohydrate. *Carbohydrate Research* **31**, 265–275.

Corey, E.J. and Houpis, I.N. (1990) Total synthesis of glycinoeclepin A. *Journal of the American Chemical Society* **112**, 8997–8998.

Cotton, E.A., Wilkinson, G. and Gaus, P.L. (1987) *Basic Inorganic Chemistry*, 2nd edn. New York: John Wiley.

Croll, N.A. (1974) *Necator americanus*: activity patterns in the egg and the mechanism of hatching. *Experimental Parasitology* **35**, 80–85.

Croll, N.A. (1975) Components and patterns in the behaviour of the nematode *Caenorhabditis elegans*. *Journal of Zoology* **176**, 159–176.

Cronin, D., Moënne-Loccoz, Y., Fenton, A., Dunne, C., Dowling, D.N. and O'Gara, F. (1997a) Role of 2,4-diacetylphloroglucinol in the interactions of the biocontrol pseudomonad strain F113 with the potato cyst nematode *Globodera rostochiensis*. *Applied and Environmental Microbiology* **63**, 1357–1361.

Cronin, D., Moënne-Loccoz, Y., Dunne, C. and O'Gara F. (1997b) Inhibition of egg hatch of the potato cyst nematode *Globodera rostochiensis* by chitinase-producing bacteria. *European Journal of Plant Pathology* **103**, 433–440.

Crowe, J.H., Hoekstra, F.A. and Crowe, L.M. (1992) Anhydrobiosis. *Annual Review of Physiology* **54**, 579–599.

Devine, K.J. (1994) Studies on the potato cyst nematode hatching factors of potato (*Solanum tuberosum* L.). *MSc thesis*, The National University of Ireland, Cork, Ireland.

Devine, K.J. and Jones, P.W. (2000a) Response of *Globodera rostochiensis* to exogenously applied hatching factors in soil. *Annals of Applied Biology*, **137**, 21–29.

Devine, K.J. and Jones, P.W. (2000b) Purification and partial characterisation of hatching factors for the potato cyst nematode *Globodera rostochiensis* from potato root leachate. *Nematology*, **2**, 231–236.

Devine, K.J. Byrne, J. Maher, N. and Jones, P.W. (1996) Resolution of natural hatching factors for the golden potato cyst nematode, *Globodera rostochiensis*. *Annals of Applied Biology* **129**, 323–334.

Doncaster, C.C. (1974) *Heterodera rostochiensis* (Nematoda) egg-hatch. *Institut für den Wissenschaftlichen Film, Göttingen*.

Doncaster, C.C. and Seymour, M.K. (1973) Exploration and selection of penetration site by Tylenchida. *Nematologica* **19**, 137–145.

Doncaster, C.C. and Seymour, M.K. (1979) Hatching of the stem nematode. In *Report of Rothamsted Experimental Station for 1978*, p. 186. Harpenden: Rothamsted.

Doncaster, C.C. and Shepherd, A.M. (1967) The behaviour of second-stage *Heterodera rostochiensis* larvae leading to their emergence from the eggs. *Nematologica* **13**, 476–478.

Ellenby, C. (1945) The influence of crucifers and mustard oil on the emergence of larvae of the potato root eelworm, *Heterodera rostochiensis* Wollenweber. *Annals of Applied Biology* **32**, 67–70.

Ellenby, C. (1956) The cyst of the potato root eelworm (*Heterodera rostochiensis* Wollenweber) as a hatching unit. *Annals of Applied Biology* **44**, 1–15.

Ellenby, C. (1963) Stimulation of hatching of potato root eelworm by soil leachings. *Nature* **198**, 110.

Ellenby, C. (1974) Water uptake and hatching in the potato cyst nematode, *Heterodera rostochiensis*, and the root knot nematode, *Meloidogyne incognita*. *Journal of Experimental Biology* **61**, 773–779.

Ellenby, C. and Perry, R.N. (1976) The influence of the hatching factor on the water uptake of the second stage larva of the potato cyst nematode *Heterodera rostochiensis*. *Journal of Experimental Biology* **64**, 141–147.

Evans, A.A.F. and Perry, R.N. (1976) Survival strategies in nematodes. In *The Organization of Nematodes*, edited by N.A. Croll, pp. 383–424. London: Academic Press.

Fairbairn, D. (1961) The *in vitro* hatching of *Ascaris lumbricoides* eggs. *Canadian Journal of Zoology* **39**, 153–162.

Fairbairn, D. and Passey, R.F. (1955) The lipid components in the vitelline membrane of *Ascaris lumbricoides* eggs. *Canadian Journal of Biochemistry and Physiology* **33**, 130–134.

Fairbairn, D. and Passey, R.F. (1957) Occurrence and distributions of trehalose and glycogen in the eggs and tissues of *Ascaris lumbricoides*. *Experimental Parasitology* **6**, 566–574.

Fenwick, D.W. (1956) The hatching of cyst-forming nematodes. In *Report of Rothamsted Experimental Station for 1955*, pp. 202–209. Harpenden: Rothamsted.

Feyaerts, H. and Coosemans, J. (1992) Influence of the thiocarbamate herbicide cycloate on the attraction of beet-cyst nematodes (*Heterodera schachtii* Schmidt) by their host plants. *Mededelingen van de Faculteit Landbouwetenschappen, Rijksuniversiteit Gent* **57**, 839–846.

Flegg, J.J.M. (1968) Embryogenic studies of some *Xiphinema* and *Longidorus* species. *Nematologica* **14**, 137–145.

Forrest, J.M.S. and Farrer, L.A. (1983) The response of eggs of the white potato cyst nematode *Globodera pallida* to diffusate from potato and mustard roots. *Annals of Applied Biology* **103**, 283–289.

Forrest, J.M.S. and Perry, R.N. (1980) Hatching of *Globodera pallida* eggs after brief exposure to potato root diffusate. *Nematologica* **26**, 130–132.

Fukuzawa, A., Furusaki, A., Ikura, M. and Masamune, R. (1985a) Glycinoeclepin A, a natural hatching stimulus for the soybean cyst nematode. *Journal of the Chemical Society* **4**, 222–224.

Fukuzawa, A., Matsue, H., Ikura, M. and Masamune, R. (1985b) Glycinoeclepins B and C, nortriterpenes related to glycinoeclepin A. *Tetrahedron Letters* **26**, 5539–5542.

Gaur, H.S., Beane, J. and Perry, R.N. (1995) Hatching of four successive generations of *Heterodera sorghi* in relation to the age of sorghum, *Sorghum vulgare*. *Fundamental and Applied Nematology* **18**, 599–601.

Gaur, H.S., Beane, J. and Perry, R.N. (2000) The influence of root diffusate, host age and water regimes on hatching of the root-knot nematode, *Meloidogyne triticoryzae*. *Nematology*, **2**, 191–199.

Gaur, H.S., Perry, R.N. and Beane, J. (1992) Hatching behaviour of six successive generations of the pigeon-pea cyst nematode, *Heterodera cajani*, in relation to growth and senescence of cowpea, *Vigna unguiculata*. *Nematologica* **38**, 190–202.

Gonzalez, J.A., Estevez-Braun, A., Estevez-Reyes, R. and Ravelo, A.G. (1994) Inhibition of potato cyst nematode hatch by lignans from *Bupleurum salicifolium* (Umbelliferae). *Journal of Chemical Ecology* **20**, 513–524.

Greco, N. (1981) Hatching of *Heterodera carotae* and *H. avenae*. *Nematologica* **27**, 366–371.

Greco, N. and Brandonisio, A. (1986) The biology of *Heterodera carotae*. *Nematologica* **32**, 447–460.

Greco, N., Vito, M.D. and Lamberti, F. (1986) Studies on the biology of *Heterodera goettingiana* in southern Italy. *Nematologia Mediterranea* **14**, 23–29.

Hackenberg, C. and Sikora, R.A. (1994) Influence of temperature and soil moisture on the biological control of the potato cyst nematode *Globodera pallida* using the plant-health-promoting rhizobacterium *Agrobacterium radiobacter*. *Journal of Phytopathology* **142**, 338–344.

Hall, D.H. and Hedgecock, E.M. (1991) Kinesin-related gene *unc-104* is required for axonal transport of synaptic vesicles in *C. elegans*. *Cell* **65**, 837–847.

Hartwell, W.V., Dahlstrom, R.V. and Neal, A.L. (1959) Crystallisation of a natural hatching factor for the larvae of the golden nematode. *Phytopathology* **49**, 540–541.

Hashmi, S. and Krusberg, L.R. (1995) Factors influencing emergence of juveniles from cysts of *Heterodera zeae*. *Journal of Nematology* **27**, 362–369.

Hass, D.K. and Todd, A.C. (1962) Extension of a technique for hatching ascarid eggs *in vitro*. *American Journal of Veterinary Research* **23**, 169–170.

Hedgecock, E.M., Culotti, J.G., Hall, D.H. and Stern, B.D. (1987) Genetics of the cell and axon migrations in *Caenorhabditis elegans*. *Development* **100**, 365–382.

Hinck, L.W. and Ivey, M.H. (1976) Proteinase activity in *Ascaris suum* eggs, hatching fluid, and excretions-secretions. *Journal of Parasitology* **62**, 771–774.

Hishida, R., Ishihara, T., Kondo, K. and Katsura, I. (1996) *hch-1*, a gene required for normal hatching and normal migration of a neuroblast in *C. elegans*, encodes a protein related to TOLLOID and BMP-1. *The EMBO Journal* **18**, 4111–4122.

Holz, R.A., Riga, E. and Atkinson, H.J. (1998) Seasonal changes in the dorsal pharyngeal gland nucleolus of unhatched second stage juveniles of *Globodera* spp. in Bolivia. *Journal of Nematology* **30**, 291–298.

Holz, R.A., Wright, D.J. and Perry, R.N. (1998) Changes in the lipid content and fatty acid composition of 2nd-stage juveniles of *Globodera rostochiensis* after rehydration, exposure to the hatching stimulus and hatch. *Parasitology* **116**, 183–190.

Ibrahim, S.K., Perry, R.N., Plowright, R.A. and Rowe, J. (1993) Hatching behaviour of the rice cyst nematodes *Heterodera sacchari* and *H. oryzicola* in relation to age of host plant. *Fundamental and Applied Nematology* **16**, 23–29.

Ishibashi, N., Kondo, E., Muraoka, M. and Yokoo, T. (1973) Ecological significance of dormancy in plant parasitic nematodes. I. Ecological difference between eggs in gelatinous matrix and cysts of *Heterodera glycines* Ichinohe (Tylenchida: Heteroderidae). *Applied Entomology and Zoology* **8**, 53–63.

Johnson, A.W. (1952) The eelworm problem: biological aspects. The potato eelworm hatching factor. *Chemistry and Industry* **40**, 998–999.

Jones, J.T., Perry, R.N. and Johnston, M.R.L. (1993) Changes in the ultrastructure of the cuticle of the potato cyst nematode, *Globodera rostochiensis*, during development and infection. *Fundamental and Applied Nematology* **16**, 433–445.

Jones, J.T., Perry, R.N. and Johnston, M.R.L. (1994) Changes in the ultrastructure of the amphids of the potato cyst nematode, *Globodera rostochiensis*, during development and infection. *Fundamental and Applied Nematology* **17**, 369–82.

Jones, J.T., Robertson, L., Perry, R.N. and Robertson, W.M. (1997) Changes in gene expression during stimulation and hatching of the potato cyst nematode, *Globodera rostochiensis*. *Parasitology* **114**, 309–315.

Jones, P.W., Tylka, G.L. and Perry, R.N. (1998) Hatching. In *The Physiology and Biochemistry of Free-living and Plant-parasitic Nematodes*, edited by R.N. Perry and D.J. Wright, pp. 181–212. Wallingford: CAB International.

Kaul, R. (1962) Untersuchungen über einen aus Zysten des Kartoffelnematoden (*Heterodera rostochiensis* Woll.) isolierten phenolischen Komplex. *Nematologica* **8**, 288–292.

Kazacos, K.R. and Turek, J.T. (1983) Scanning electron microscopy of the eggs of *Baylisascaris procynis*, *B. transfuga* and *Parascaris equorum*, and their comparison with *Toxacara canis* and *Ascaris suum*. *Proceedings of the Helminthological Society of Washington* **50**, 36–42.

Keukens, E.A.J., de Vrije, T., Fabrie, C.H.J.P., Demel, R.A., Jongen, W.M.F and de Kruiff, B. (1992) Dual specificity of sterol-mediated glycoalkaloid-induced membrane disruption. *Biochimica et Biophysica Acta* **1110**, 127–136.

Klimetzet, D., Kohler, J., Vite, J.P. and Kohnle, U. (1986) Dosage response to ethanol mediates host selection by 'secondary' bark beetles. *Naturwissenschaften* **73**, 270–272.

Kraus, G.A., Johnson, B., Kongsjahju, A. and Tylka, G.L. (1994) Synthesis and evaluation of compounds that affect soybean cyst nematode egg hatch. *Journal of Agricultural and Food Chemistry* **42**, 1839–1840.

Kraus, G.A., Vander Louw, S.J., Tylka, G.L. and Soh, D.H. (1996) The synthesis and testing of compounds that inhibit soybean cyst nematode egg hatch. *Journal of Agricultural and Food Chemistry* **44**, 1548–1550.

Kraus, R. and Sikora, R.A. (1981) Die Wirkungen des Herbizides Diallate auf den Befall von *Heterodera schachtii* an Zückerrüben. *Zeitschrift von Pflanzenkrankheiten und Pflanzenschutz* **88**, 210–217.

Kraus, R. and Sikora, R.A. (1983) Effects of the herbicide diallate, alone and in combination with aldicarb, on *Heterodera schachtii* population levels in sugar beets. *Zeitschrift von Pflanzenkrankheiten und Pflanzenschutz* **90**, 132–139.

LaMondia, J.A. and Brodie, B.B. (1986) The effects of potato trap crops and fallow on decline of *Globodera rostochiensis*. *Annals of Applied Biology* **108**, 347–352.

Laughlin, C.W., Lugthart, J.A. and Vargas, J.M. (1974) Observations on the emergence of *Heterodera iri* from the egg. *Journal of Nematology* **6**, 100–101.

Lee, D.L. and Atkinson, H.J. (1976) *Physiology of Nematodes*, 2nd edn, pp. 131-132. London: MacMillan.

Looss, A. (1911) The anatomy and life-history of *Agcylostoma duodenale* Dub. Part II. *Records of the Egyptian Government School of Medicine* **4** (Monograph, 446 pp.).

Malinowska, E. (1996) The range of stimulatory action of the nematode resistant cultivar Tarpan on hatching of juveniles from the cysts of golden potato cyst nematode (Rol pathotype) under field conditions. *Biuletyn Instytutu Ziemniaka* **46**,104.

Marrian, D.H., Russell, P.B., Todd, A.R. and Waring, R.S. (1949) The potato eelworm hatching factor. 3. Concentration of the factor by chromatography. Observations on the nature of eclepic acid. *Biochemical Journal* **45**, 524–528.

Martin, B. (1994) Development of a reliable assay for potato cyst nematode hatching factors. *MSc thesis*, The University of Aberdeen, UK.

Masamune, T., Anetai, M., Takasugi, M. and Katsui, N. (1982) Isolation of a natural hatching stimulus, glycinoeclepin A, for the soybean cyst nematode. *Nature* **297**, 495–496.

Masamune, T., Anetai, M., Fukuzawa, A., Takasugi, M., Matsue, H., Kabayashi, K., Ueno, S. and Katsui, N. (1987a) Glycinoeclepins, natural hatching stimuli for the soybean cyst nematode, *Heterodera glycines*. I. Isolation. *Bulletin of the Chemical Society of Japan* **60**, 981–999.

Masamune, T., Fukuzawa, A., Furusaki, A., Ikura, M., Matsue, H., Kaneko, T., Abiko, A., Sakamoto, N., Tanimoto, N. and Murai, A. (1987b) Glycinoeclepins, natural hatching stimuli for the soybean cyst nematode, *Heterodera glycines*. II. Structural elucidation. *Bulletin of the Chemical Society of Japan* **60**, 1001–1014.

Matthews, B. (1985) The influence of temperature and osmotic stress on the development and eclosion of hookworm eggs. *Journal of Helminthology* **59**, 217–224.

Matthews, B. (1986) Permeability changes in the egg-shell of hookworms during development and eclosion. *Parasitology* **93**, 547–557.

Mei, B., Kennedy, M.W., Beauchamp, J., Komuniecki, P.R. and Komuniecki, R. (1997) Secretion of a novel, developmentally regulated fatty acid-binding protein into the perivitelline fluid of the parasitic nematode, *Ascaris suum*. *The Journal of Biological Chemistry* **272**, 9933–9941.

Mercer, C.F., Greenwood, D.R. and Grant, J.L. (1992) Effect of plant and microbial chitinases on the eggs and juveniles of *Meloidigyne hapla* Chitwood (Nematoda: Tylenchida). *Nematologica* **38**, 227–236.

Miwa, A., Nii, Y., Okawara, H. and Sakakibara, M. (1987) Synthetic study on hatching stimuli for the soybean cyst nematode. *Agricultural and Biological Chemistry* **51**, 3459–3461.

Monné, L. and Hönig, G. (1955) On the properties of the egg envelopes of the parasitic nematodes *Trichuris* and *Capillaria*. *Arkiv für Zoologi* **6**, 559–562.

Mori, K. and Watanabe, H. (1989) Recent results in the synthesis of semiochemicals: synthesis of glycinoeclepin A. *Pure and Applied Chemistry* **61**, 543–546.

Mulder, J.G., Diepenhorst, P., Plieger, P. and Brüggemann-Rotgans, I.E.M. (1992) Hatching agent for the potato cyst nematode. *Patent application No. PCT/NL92/00126*.

Murai, A., Tanimoto, H., Sakamoto, H. and Masamune, T. (1988) Total synthesis of glycinoeclepin A. *Journal of the American Chemical Society* **110**, 1985–1986.

Okada, T. (1972) Hatching inhibitory factor in the cyst contents of the soybean cyst nematode *Heterodera glycines*. *Applied Entomology and Zoology* **7**, 99–102.

Onions, T.G. (1955) The distribution of hatching within the cyst of the potato-root eelworm *Heterodera rostochiensis*. *Quarterly Journal of Microscope Science* **96**, 495–513.

Oostendorp, M. and Sikora, R.A. (1990) *In vitro* interrelationships between rhizosphere bacteria and *Heterodera schachtii*. *Revue de Nématologie* **13**, 269–274.

Panesar, T.S. and Croll, N.A. (1981) The hatching process in *Trichuris muris* (Nematoda: Trichuroidea). *Canadian Journal of Zoology* **59**, 621–628.

Passey R.F. and Fairbairn, D. (1955) The respiration of *Ascaris lumbricoides* eggs. *Canadian Journal of Biochemistry and Physiology* **35**, 511–535.

Perry, R.N. (1977a) A reassessment of the variations in the water content of the larvae of *Nematodirus battus* during the hatching process. *Parasitology* **74**, 133–137.

Perry, R.N. (1977b) Water content of the second stage larva of *Heterodera schachtii* during the hatching process. *Nematologica* **23**, 431–437.

Perry, R.N. (1983) The effect of potato root diffusate on the desiccation survival of unhatched juveniles of *Globodera rostochiensis*. *Revue de Nématologie* **6**, 99–102.

Perry, R.N. (1987) Host induced hatching of phytoparasitic nematode eggs. In *Vistas on Nematology*, edited by J.A. Veech and D.W. Dickson, pp. 159–64. Hyattsville: Society of Nematologists Inc.

Perry, R.N. (1989) Dormancy and hatching of nematode eggs. *Parasitology Today* **5**, 377–383.

Perry, R.N. (1996) Chemoreception in plant parasitic nematodes. *Annual Review of Phytopathology* **34**, 181–199.

Perry, R.N. (1997) Plant signals in nematode hatching and attraction. In *Cellular and Molecular Aspects of Plant-Nematode Interactions* edited by C. Fenoll, F.M.W. Grundler and S.A. Ohl, pp. 38–50, The Netherlands: Kluwer Academic Publishers.

Perry, R.N. and Beane, J. (1982) The effect of brief exposures to potato root diffusate on the hatching of *Globodera rostochiensis*. *Revue de Nématologie* **5**, 221–224.

Perry, R.N. and Beane, J. (1989) Effects of certain herbicides on the *in vitro* hatch of *Globodera rostochiensis* and *Heterodera schachtii*. *Revue de Nématologie* **12**, 191–196.

Perry, R.N. and Clarke, A.J. (1981) Hatching mechanisms of nematodes. *Parasitology* **83**, 435–449.

Perry, R.N. and Feil, J. (1986) Observations on a novel hatching bioassay for *Globodera rostochiensis* using fluorescence microscopy. *Revue de Nématologie* **9**, 280–282.

Perry, R.N. and Gaur, H.S. (1996) Host plant influences on the hatching of cyst nematodes. *Fundamental and Applied Nematology* **19**, 505–510.

Perry, R.N. and Trett, M.W. (1986) Ultrastructure of the eggshell of *Heterodera schachtii* and *H. glycines* (Nematoda: Tylenchida). *Revue de Nématologie* **9**, 399–403.

Perry, R.N., Clarke, A.J. and Beane, J. (1980) Hatching of *Heterodera goettingiana* in vitro. *Nematologica* **26**, 493–495.

Perry, R.N., Clarke, A.J. and Hennessy, J. (1980) The influence of osmotic pressure on the hatching of *Heterodera schachtii*. *Revue de Nématologie* **3**, 3–9.

Perry, R.N., Hodges, J.A. and Beane, J. (1981) Hatching of *Globodera rostochiensis* in response to potato root diffusate persisting in soil. *Nematologica* **27**, 349–352.

Perry, R.N., Knox, D. and Beane, J. (1992) Enzymes released during hatching of *Globodera rostochiensis* and *Meloidogyne incognita*. *Fundamental and Applied Nematology* **15**, 283–288.

Perry, R.N., Wharton, D.A. and Clarke, A.J. (1982) The structure of the egg-shell of *Globodera rostochiensis* (Nematoda: Tylenchida). *International Journal for Parasitology* **12**, 481–485.

Perry, R.N., Zunke, U. and Wyss, U. (1989) Observations on the response of the dorsal and subventral oesophageal glands of *Globodera rostochiensis* to hatching stimulation. *Revue de Nématologie* **12**, 91–96.

Perry, R.N., Clarke, A.J., Hennessy, J. and Beane, J. (1983) The role of trehalose in the hatching mechanism of *Heterodera goettingiana*. *Nematologica* **29**, 324–335.

Plenefisch, J., Xiao, H., Mei, B., Geng, J., Komuniecki, P.R. and Komuniecki, R. (2000) Secretion of a novel class of iFABPs in nematodes: coordinate use of the *Ascaris/Caenorhabditis* model systems. *Molecular and Biochemical Parasitology*, **105**, 223–236.

Preston, C.M. and Jenkins, T. (1985) *Trichuris muris*: structure and formation of the polar plugs. *Zeitschrift für Parasitenkunde* **71**, 373–381.

Rawsthorne, D. and Brodie, B.B. (1986) Relationship between root growth of potato, root diffusate production, and hatching of *Globodera rostochiensis*. *Journal of Nematology* **18**, 379–384.

Rawsthorne, D. and Brodie, B.B. (1987) Movement of potato root diffusate through the soil. *Journal of Nematology* **19**, 119–122.

Robinson, M.P., Atkinson, H.J. and Perry, R.N. (1985) The effect of delayed emergence on infectivity of juveniles of the potato cyst nematode *Globodera rostochiensis*. *Nematologica* **31**, 171–178.

Robinson, M.P., Atkinson, H.J. and Perry, R.N. (1987a) The influence of soil moisture and storage time on the motility, infectivity and lipid utilization of second stage juveniles of the potato cyst nematodes *Globodera rostochiensis* and *G. pallida*. *Revue de Nématologie* **10**, 343–348.

Robinson, M.P., Atkinson, H.J. and Perry, R.N. (1987b) The influence of temperature on the hatching, activity and lipid utilization of second stage juveniles of the potato cyst nematodes *Globodera rostochiensis* and *G. pallida*. *Revue de Nématologie* **10**, 349–354.

Rogers, W.P. (1958) Physiology of the hatching of eggs of *Ascaris lumbricoides*. *Nature* **181**, 1410–1411.

Rogers, W.P. (1978) The inhibitory action of insect juvenile hormone on the hatching of nematode eggs. *Comparative Biochemistry and Physiology A* **61**, 187–190.

Rogers, W.P. (1982) Enzymes in the exsheathing fluid of nematodes and their biological significance. *International Journal for Parasitology* **12**, 495–502.

Rogers, W.P. and Brooks, F. (1977) The mechanism of hatching of eggs of *Haemonchus contortus*. *International Journal for Parasitology* **7**, 61–65.

Rogers, W.P. and Sommerville, R.I. (1968) The infectious process, and its relation to the development of early parasitic stages of nematodes. *Advances in Parasitology* **6**, 327–348.

Shepherd, A.M. and Clarke, A.J. (1971) Molting and hatching stimuli. In *Plant Parasitic Nematodes, Volume II*, edited by B.M. Zuckerman, W.F. Mai and R.A. Rohde, pp. 267–287. London: Academic Press.

Silverman, P.H. and Campbell, J.A. (1959) Studies on parasitic worms of sheep in Scotland. I. Embryonic and larval development of *Haemonchus contortus* at constant conditions. *Parasitology* **49**, 23–39.

Smant, G., Goverse, A., Stokkermans, J.P.W.G., De Boer, J.M., Pomp, H., Zilverentant, J.F., Overmars, H.A., Helder, J., Schots, A. and Bakker, L. (1997) Potato root diffusate-induced secretion of soluble, basic proteins originating from the subventral esophageal glands of potato cyst nematodes. *Phytopathology* **87**, 839–845.

Smant, G., Stokkermans, J.P.W.G., Yan, Y., de Boer, J.M., Baum, T.J., Wang, X., Hussey, R.S., Gommers, F.J., Henrissat, B., Davis, E.L., Helder, J., Schots, A. and Bakker, J. (1998) Endogenous cellulases in animals: isolation of β-1,4-endoglucanase genes from two species of plant-parasitic cyst nematodes. *Proceedings of the National Academy of Sciences* **95**, 4906–4911.

Stelter, H. and Sager, I. (1987) Larval hatching of *Globodera rostochiensis* pathotype 1 in soil leachings. *Archiv für Phytopathologie und Pflanzenschutz* **23**, 69–76.

Stoyanov, D. and Trifonova, Z. (1995) Hatching dynamics of golden potato cyst nematode larvae *Globodera rostochiensis* Woll. *Bulgarian Journal of Agricultural Science* **1**, 241–246.

Takasugi, M., Yachida, Y., Anetai, M., Masamune, T. and Kegasawa, K. (1975) Identification of asparagusic acid as a nematicide occurring naturally in the roots of asparagus. *Chemistry Letters*, 45–46.

Taylor, D.P. (1962) Effects of temperature on hatching of *Aphelenchus avenae* eggs. *Proceedings of the Helminthological Society of Washington* **29**, 52–54.

Tefft, P.M. and Bone, L.W. (1985a) Leucine aminopeptidase in eggs of the soybean cyst nematode *Heterodera glycines*. *Journal of Nematology* **17**, 270–274.

Tefft, P.M. and Bone, L.W. (1985b) Plant-induced hatching of the soybean cyst nematode *Heterodera glycines*. *Journal of Nematology* **17**, 275–279.

Thistlethwayte, B. (1969) Hatch of eggs and reproduction of *Pratylenchus penetrans* (Nematode: Tylenchida). *PhD thesis*, Cornell University, USA.

Thompson, J.M. and Tylka, G.L. (1997) Differences in hatching of *Heterodera glycines* egg-mass and encysted eggs *in vitro*. *Journal of Nematology* **29**, 315–321.

Trudgill, D.L., Phillips, M.S. and Hackett, C.A. (1996) The basis of predictive modelling for estimating yield loss and planning potato cyst nematode management. *Pesticide Science* **47**, 89–94.

Truter, M.R. (1976) Chemistry of the calcium ionophores. In *Calcium in Biological Systems*, edited by C.J. Duncan, pp. 19–40. Cambridge: Cambridge University Press.

Tsutsumi, M. (1976) Conditions for collecting the potato root diffusate and the influence on the natural hatching of potato cyst nematode. *Japanese Journal of Nematology* **6**, 10–13.

Turner, S.J. and Evans, K. (1998) The origins, global distribution and biology of potato cyst nematodes (*Globodera rostochiensis* (Woll.) and *Globodera pallida* Stone). In *Potato Cyst Nematodes: Biology, Distribution and Control*, edited by R.J. Marks and B.B. Brodie, pp. 7–26. Wallingford: CAB International.

Twomey, U. (1995) Hatching chemicals involved in the interaction between potato cyst nematodes, host and non-host plants. *PhD thesis*, The National University of Ireland, Cork, Ireland.

Ubelaker, J.E. and Allison, V.F. (1975). Scanning electron microscopy of the eggs of *Ascaris lumbricoides*, *A. suum*, *Toxocara canis* and *T. mystax*. *Journal of Parasitology* **61**, 802–807.

van der Gulden, W.J.I. and van Aspert-van Erp, A.J.M. (1976) *Syphacia muris*: water permeability of eggs and its effect on hatching. *Experimental Parasitology* **39**, 40–44.

Wallace, H.R. (1966) The influence of moisture stress on the development, hatch and survival of eggs of *Meloidogyne javanica*. *Nematologica* **12**, 57–69.

Ward, K.A. and Fairbairn, D. (1972) Chitinase in developing eggs of *Ascaris suum* (Nematoda). *Journal of Parasitology* **58**, 546–549.

Watanabe, H. and Mori, K. (1991) Triterpenoid total synthesis. Part 2. Synthesis of glycinoeclepin A, a potent hatching stimulus for the soybean cyst nematode. *Journal of the Chemical Society Perkin Transactions*, 2919–2934.

Wharton, D.A. (1979a) The structure and formation of the egg-shell of *Aspiculuris tetraptera* Schulz (Nematoda: Oxyuroidea). *Parasitology* **78**, 145–154.

Wharton, D.A. (1979b) The structure and formation of the egg-shell of *Syphacia obvelata* Rudolphi (Nematoda: Oxyuroidea). *Parasitology* **79**, 13–28.

Wharton, D.A. (1980) Nematode egg-shells. *Parasitology* **81**, 447–463.

Wharton, D.A. and Jenkins, T. (1978) Structure and chemistry of the egg-shell of a nematode (*Trichuris suis*). *Tissue and Cell* **10**, 427–440.

Whitehead, A.G. (1977) Control of potato cyst-nematode, *Globodera rostochiensis* Ro1, by picrolonic acid and trap crops. *Annals of Applied Biology* **87**, 225–227.

Whitehead, A.G. (1992) Emergence of juvenile potato cyst nematodes, *Globodera rostochiensis* and *G. pallida* and the control of *G. pallida*. *Annals of Applied Biology* **120**, 471–486.

Whitehead, A.G. (1998) *Plant Nematode Control*, pp. 170–171. Wallingford: CAB International.

Wilson, P.A.G. (1958) The effect of weak electrolyte solutions on the hatching of eggs of *Trichostrongylus retortaeformis* (Zeder) and its interpretation in terms of a proposed hatching mechanism of strongylid eggs. *Journal of Experimental Biology* **35**, 584–601.

Womersley, C.Z., Wharton, D.A. and Higa, L.M. (1998) Survival Biology. In *The Physiology and Biochemistry of Free-living and Plant-parasitic Nematodes*, edited by R.N. Perry and D.J. Wright, pp. 271–302. Wallingford: CAB International.

Wood, W.B. (1988) Embryology. In *The Nematode* Caenorhabditis elegans, edited by W.B. Wood and the community of *C. elegans* researchers, pp. 215–241. Cold Spring Harbor: Cold Spring Harbor Laboratory Press.

Yen, J.-H., Niblack, T.L., Karr, A.L. and Wiebold, W.J. (1996) Seasonal biochemical changes in eggs of *Heterodera glycines* in Missouri. *Journal of Nematology* **28**, 442–450.

Zheng, L. and Ferris, H. (1991) Four types of dormancy exhibited by eggs of *Heterodera schachtii*. *Revue de Nématologie* **14**, 419–426.

7. Cuticle, Moulting and Exsheathment

Donald L. Lee

School of Biology, The University of Leeds, Leeds LS2 9JT, UK

Introduction

The body wall of nematodes is composed of the cuticle, an underlying cellular/syncytial layer (the epidermis or hypodermis), and the longitudinally oriented somatic musculature. Classically the cytoplasmic layer underlying the cuticle has been called the hypodermis. However, Hyman (1951) and De Coninck (1965) referred to this layer as the 'epidermis' and Lee (1977) suggested that, to be comparable to many groups of animals that possess a cuticle, especially the arthropods, the term 'epidermis' is probably more appropriate than 'hypodermis'. Whilst 'hypodermis' is etymologically correct it is phylogenetically erroneous (Wright, 1987). Support for the term 'epidermis' in preference to 'hypodermis' has been forthcoming from several authors (Coomans, De Coninck and Heip, 1978; Inglis, 1983; Bird, 1984a, b; Wright, 1987; Bird and Bird, 1991; Lorenzen, 1994; Bird and Bird, 1998; Blaxter and Robertson, 1998) but the term 'hypodermis' is still extensively used in the literature. Bird and Bird (1998) stated 'There is no reason why the Nematoda should differ in this respect from all other invertebrate groups where various authorities have used the term 'epidermis' for the region beneath the cuticle'. The term 'epidermis' will therefore be used in preference to 'hypodermis' in this chapter.

The cuticle or exoskeleton of nematodes is an extracellular covering that overlies and is secreted by the hypodermis. It is invaginated at various openings such as the amphids, mouth/buccal cavity, secretory-excretory pore, vulva, cloaca and anus. It also occurs as outgrowths such as papillae, setae, spines, longitudinal ridges and lateral alae. The lumen of the pharynx is lined with cuticle but this is distinct from the cuticle of the body wall.

The cuticle of most nematodes studied consists of a three-zoned structure covered by a thin epicuticle, which in turn may be covered by a surface coat (Figure 7.1). Notwithstanding its basic three-zoned

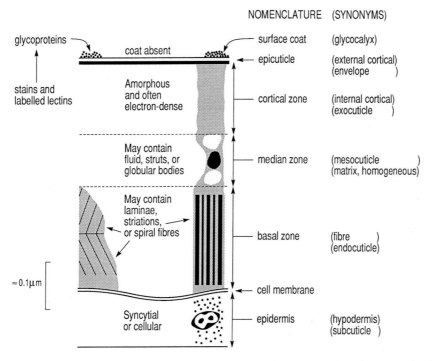

SCHEMATIC DIAGRAM OF NEMATODE CUTICLE

Figure 7.1. Nomenclature of zones or layers present in the cuticle of a typical larval nematode that is approximately 0.5 mm thick. Figure from Chapter 3 in *The Structure of Nematodes*, Second Edition by Alan F. Bird and Jean Bird, copyright © 1991 by Academic Press, reproduced by permission of the publisher.

nature, the cuticle can be a relatively simple or a very complex structure (Figures 7.2, 7.5, 7.7) that varies from one genus to another and may differ in structure, including the number of zones, throughout the life cycle of a single species and from one region of an individual to another (Chitwood and Chitwood, 1950; Malakhov, 1994).

The first observations on the structure of the nematode cuticle were made on large animal-parasitic nematodes such as *Ascaris lumbricoides*, *A. suum* and *A. megalocephala* (now known as *Parascaris equorum*). Von Siebold (1848; quoted by Chitwood and Chitwood, 1950) was the first to draw attention to the layered (zoned) nature of nematode cuticle, and these observations were developed further by several authors in the nineteenth century (see Chitwood and Chitwood, 1950). Czermak (1852) (quoted by Chitwood and Chitwood, 1950) was the first to observe surface annulations that tended to end at the lateral lines and reported on various layers in transverse and longitudinal sections, including fibrillar structures near the surface of the cuticle and fibre layers at its base that crossed at an angle of 45°. He examined these structures under polarised light and showed that many of them exhibited birefringence.

No further information on the structure of the cuticle of the large ascarids was made for more than 40 years when, at the turn of the century and coincident with development and refinement of the light microscope, several workers described the layers of these cuticles in some detail. The first and most notable of these was van Bömmel (1895) who reported the presence of nine layers in the cuticles of both *A. lumbricoides* and *A. megalocephala* (*P. equorum*). These were, from the surface inwards, outer cortical, inner cortical, fibrillar, homogeneous, boundary, three fibre layers (the middle of which is at an angle of 45° to the inner and outer layers) and a basal lamella.

Subsequent examination of sections cut through the cuticles of these ascarids, notably by Chitwood and Chitwood (1950), Bird and Deutsch (1957) and Bird (1971), have confirmed van Bömmel's nomenclature. The Chitwood's drew attention to the structures of numerous other nematode cuticles that had been sectioned and observed under the light microscope. Bird and Deutsch (1957) and Bird (1958) used phase contrast and polarised light microscopy to examine fresh frozen sections and used the transmission electron microscope to examine surfaces obtained by a layer stripping technique, and sections cut on an ultramicrotome, of cuticles of *A. lumbricoides*, *Oxyuris equi* and *Strongylus equinus*. The cuticle of *A. lumbricoides* was considered by them to be a secreted collagen and to have a thin lipid surface coating.

Crofton (1966) considered that these various layers of the ascarid cuticle could be condensed into three, namely cortex, matrix and fibre with the thin lipid coat at the surface possibly representing a fourth layer. 'Zone' was preferred to 'layer' by Wright (1968), Bird (1976), Bird and Bird (1991, 1998) and Urbancik, Bauer-Nebelsick and Ott (1996) for the three or four main

'layers' of the cuticle and will be adopted in this chapter. This is because the various regions of the cuticle sometimes merge with each other without showing any distinct boundaries and because of the complex construction of the cuticle of some marine nematodes (Bird and Bird, 1998). The three- or four-zoned cuticle is the basis for the current simplification of the classification of nematode cuticles. Thus, the cortex, matrix and fibre layers of Crofton (1966) are now the cortical, median and basal zones (Bird and Bird, 1998) with an epicuticle (Lee and Atkinson, 1976) replacing the so-called lipid membrane in Crofton's terminology (Figure 7.1). However, not all nematodes conform to the three-zoned system. For example, *Anguina tritrici* (Spiegel and Robertson, 1988) and *Cephalenchus emarginatus* (Mounport, Baujard and Martiny, 1993) have only the cortical and basal zones. A surface coat covers the epicuticle of many, if not all, nematodes (Bird and Bird 1991, 1998; Blaxter and Robertson 1998). Within this configuration there is great complexity in the cuticles of various nematodes, as has been illustrated by Malakhov (1994) (Figure 7.2). The current nomenclature of the various zones of the cuticle, together with synonymous names given to the zones, is shown in Figure 7.1.

Nematodes moult four times after formation of the first cuticle within the egg and these processes, together with exsheathment, will be considered as well as the ultrastructure, biochemistry and physiology of the various zones of the cuticle.

Cuticle Morphology

The body surface of nematodes may be smooth, and superficially annulated, especially in soil-dwelling and parasitic species. However, various species possess a variety of cuticular specialisations such as bristles, spines, hooks, warts, papillae, punctations, longitudinal striations, transverse striations, inflations, ridges, cervical alae, lateral alae and caudal alae (Hyman, 1951; Chitwood and Chitwood, 1950; Bird, 1971; Bird and Bird, 1991; Chapter 2). Various other cuticular structures, such as an inflated collar (sometimes bearing spines or hooks), cord-like thickenings called cordons, and feathery-like appendages occur at the anterior end of some nematodes. Leaf crowns and bristles also occur around the mouth of some species whilst the cuticular lining of the buccal cavity may bear tooth-like structures or cutting plates (Hyman, 1951; Chitwood and Chitwood, 1950; Chapter 2). These are all modifications, to a greater or lesser degree, of the various zones of the cuticle. Males of many families possess a copulatory bursa of varying complexity at their posterior end. This is a modification of the caudal alae and forms a thin, flexible, web-like structure supported to a greater or lesser extent by innervated rays of epidermis. The structure of the body wall cuticle of the head end often differs from that of the rest of the body.

Transverse striations and annulations

Transverse grooves, called striae, are present on the cuticle of larval and adult stages of most nematodes

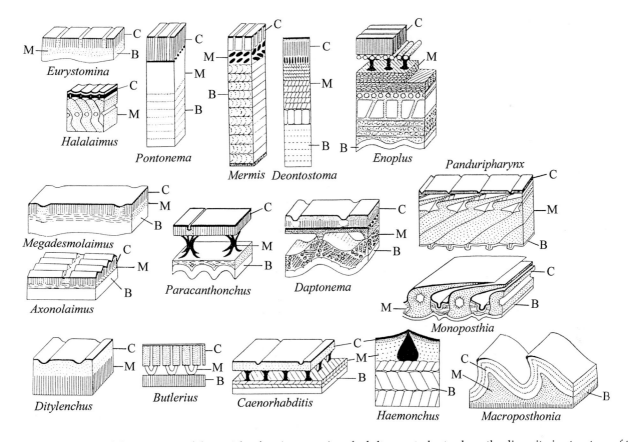

Figure 7.2. Diagrams of the structure of the cuticle of various species of adult nematodes to show the diversity in structure of the body wall cuticle. Modified with permission from Malakhov (1994) *Nematodes* and from V.V. Yushin. B = basal zone; C = cortical zone; M = median zone.

although some genera appear to lack striae and in others they are superficial or form deep grooves (Chitwood and Chitwood, 1950; Bird and Bird, 1991). Where the striae are distinct they give the nematode an annulated appearance and are referred to as annulations; some larger nematodes, such as adult *Ascaris suum*, have large annulations with many striae between the annulations (Figure 7.3). Deep annulations are characteristic of several groups of marine nematodes (e.g. desmodorids, ceramonematids, epsilonematids, draconematids) and of several groups of animal-parasitic nematodes (e.g. ascarids, oxyurids, strongylids, spirurids), and plant-parasitic nematodes (e.g. tylenchids) (Chitwood and Chitwood, 1950; De Coninck, 1965; Bird and Bird, 1991; Chapter 2). The striae do not extend across the epidermis of the body wall, although in some genera the striae extend into the median zone of the cuticle or as far as the outer epidermal membrane (e.g. *Metadasynemoides*) (Figure 7.21) (Nicholas and Stewart, 1990). The annulae are patterned by ridges that form on the surface of the epidermis during formation of the cuticle (see later). Annulations allow flexing of the body wall during locomotion and during searching movements of the anterior end.

In the marine nematode *Gonionchus australis* the body is strongly annulated, except for the smooth cephalic

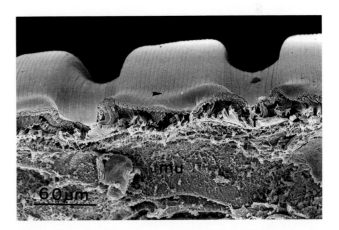

Figure 7.3. Scanning electron micrograph of a specimen of an adult *Ascaris suum* that had been fixed in formal saline and then cut longitudinally to exhibit the layers of the cuticle and body wall. Note the large annuli and the striae (arrowhead) demarking the smaller annuli of the cuticle. mu = muscle. Original.

region. In this nematode the borders of the raised annules are serrated and separated by deep striae, with an electron dense band of material encircling each annule within its crest (Nicholas and Stewart, 1997).

This presumably gives some degree of rigidity to the annule with the presumably softer grooves between the annuli allowing flexing of the body during locomotion (Figure 7.4). The body cuticle of *Desmoscolex* consists of thick cuticular rings that alternate with depressed rings, with the result that the nematode looks like a caterpillar. According to Stauffer (1924), during locomotion, waves of contraction pass from head to tail and draw together the annulations, rather than undulatory waves of contraction, with backward movement being prevented by cuticular bristles. As this is an unusual form of locomotion for nematodes it would be interesting to verify this claim. In some genera the annulations contain structural plates situated in the median layer (e.g. *Euchromadora*) (Watson, 1965a; Kulikov *et al.*, 1998). In the marine nematodes *Leptonemella* sp., *Stilbonema majum* and *Desmodora ovigera* the cuticle is deeply annulated with annules at the anterior end overlapping forwards and annules in the posterior regions overlapping backwards in a roof-tile-like fashion; the degree of overlapping decreases in the midbody region. In *Desmodora* the overlapping annuli bear numerous blunt spines along their trailing edge. These species possess a 'ring body' that is the dominant element of the median zone of the annulus and gives the nematode radial strength. *Desmodora* possesses a cushion of material below the striae of the annuli. Differences in complexity of the cuticle of several species of marine nematode are illustrated in Figures 7.5 and 7.6 (Urbancik, Bauer-Nebelsick and Ott, 1996).

Punctations and pores

Punctations and pores occur on the surface of the cuticle of many groups of free-living marine and freshwater nematodes and are usually arranged in a definite pattern. In the marine nematode, *Acanthonchus duplicatus*, the pattern of punctations occurs throughout the length of the nematode and consists of rows of transverse dots (about 3 μm in diameter with the rows about 1.5 μm apart). These dots are the surface manifestations of electron dense rods that lie within the median zone of the cuticle and appear to act as mechanical supports within it. There are also pores, that open to the surface, within the cuticle of this nematode. The pore complex consists of a depression in the surface of the cuticle, a transverse slit-like pore that opens at the surface, a duct that extends from the pore through the zones of the cuticle to the epidermis, and a collar of electron-dense material in the median zone that appears to support the pore and duct (Figure 7.7) (Wright and Hope, 1968). Adult *Mermis* spp. have a very complex cuticle with canals extending from the median zone through the cortical zone to the surface (Figure 7.8b) (Lee, 1970a). The function of these canals is not known. It is possible that they are the means by which surface coat material is passed on to the surface of the cuticle.

The trichurids (*Trichuris, Capillaria* and *Trichinella*) possess a series of pores (the bacillary band) that open into specialized cells that appear to be modified epidermal gland cells (Wright and Chan, 1973; Bird and Bird, 1991).

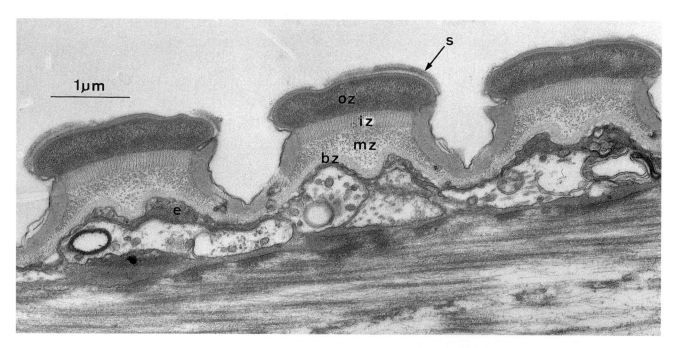

Figure 7.4. Longitudinal section through the body wall of a male *Gonionchus australis* to show the raised annules and the internal structure of the cuticle. Note the electron dense band of material (oz) and the striated layer (iz) which encircle the nematode in the crests of each annulus, and what appears to be more flexible material between the annuli. This arrangement presumably gives some degree of rigidity to the annules whilst at the same time allowing flexing of the cuticle during locomotion. bz = basal zone, e = epidermis, iz = inner striated cortical zone, mz = median zone, oz = outer cortical zone, s = surface coat. Original electron micrograph courtesy of W.L. Nicholas.

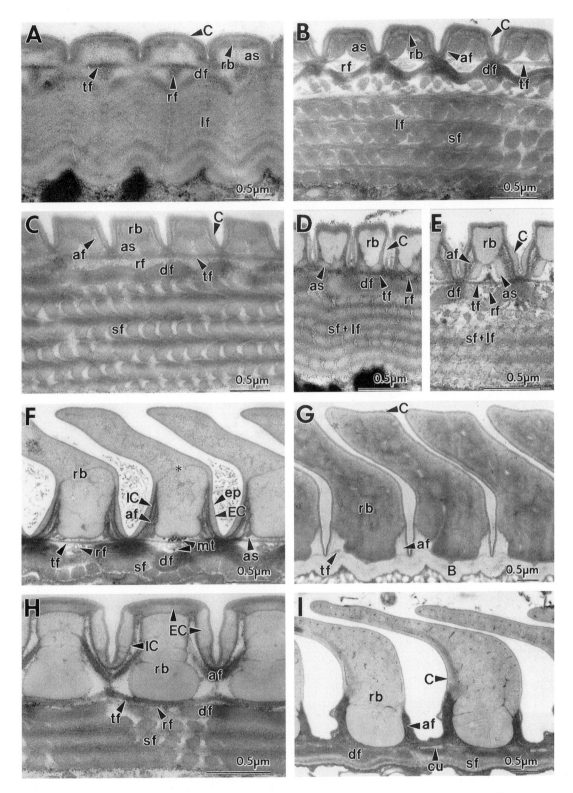

Figure 7.5. Transmission electron micrographs of mesosagittal sections through somatic cuticle of several different species of marine nematodes to show the complexity and differences in structure of the cuticles. **A,** *Eubostrichus topiarus*; **B,** *Catonema* sp.; **C,** *Laxus oneistus*; **D,** cervical region *Robbea* sp.; **E,** postcervical region *Robbea* sp.; **F,** *Leptonemella* sp. (*indicates electron-dense band); **G,** *Stilbonema majum*; **H,** *Spirinia* sp.; **I,** *Desmodora ovigera*. af = adjoining fibres, as = adjoining strands, B = basal zone, C = cortical zone, cu = cushion, df = distal fibre layer of B, EC = external cortical zone, ep = epicuticle, IC = internal cortical zone, lf = longitudinal fibres, mt = microtubule-like structures, rb = ring body, rf = ring fibres, sf = spiral fibres, tf = twin fibres. Courtesy of J.A. Ott; reproduced from Urbancik, Bauer-Nebelsick and Ott (1996) *Zoomorphologie*, **116**, with permission from Springer-Verlag Heidelberg.

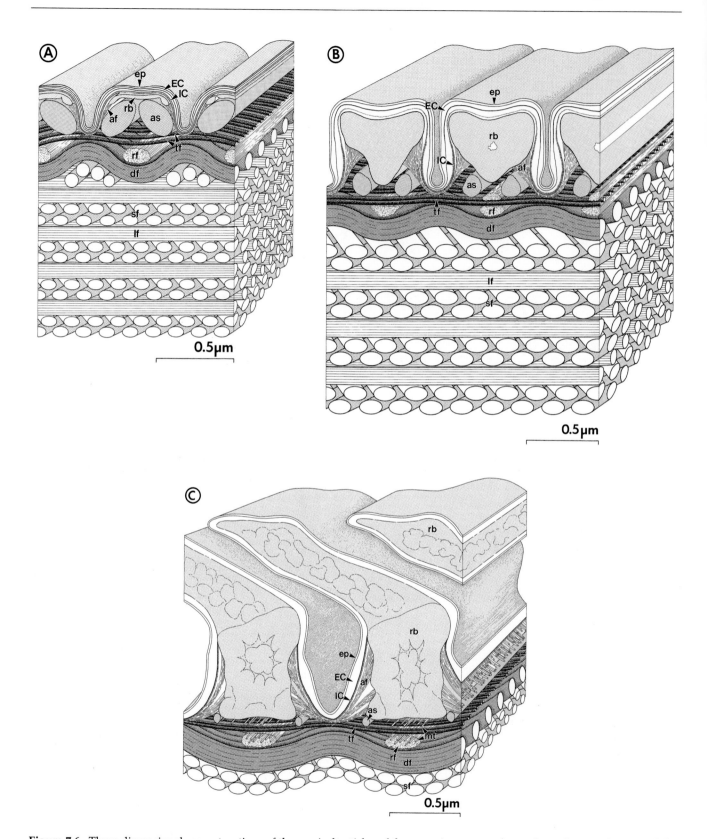

Figure 7.6. Three-dimensional reconstructions of the cervical cuticles of three marine nematodes to show the complexity and the differences in structure of the cuticles. (A) *Catanema* sp., (B) *Robbea* sp., (C) *Leptonemella* sp.. af = adjoining fibres, as = adjoining strands, B = basal zone, C = cortical zone, cu = cushion, df = distal fibre layer of B, EC = external cortical zone, ep = epicuticle, IC = internal cortical zone, lf = longitudinal fibres, mt = microtubule-like structures, rb = ring body, rf = ring fibres, sf = spiral fibres, tf = twin fibres. Courtesy of J.A. Ott, reproduced from Urbancik, Bauer-Nebelsick and Ott (1996) *Zoomorphologie*, **116**, with permission from Springer-Verlag Heidelberg.

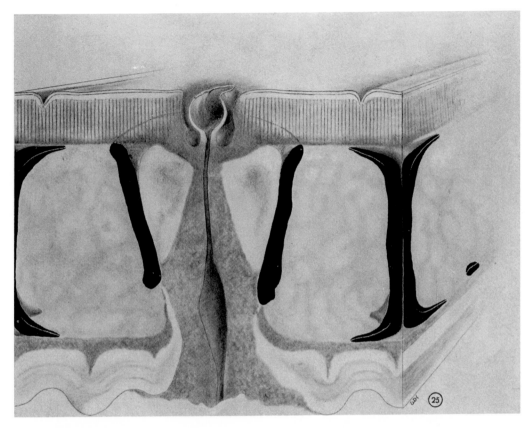

Figure 7.7. Three-dimensional reconstruction of the cuticle of *Acanthonchus duplicatus*. The dense rods with proximal and distal processes and the ring-like development of dense material around the pore complex is shown. The illustration shows a longitudinal section plane through the pore complex. Reproduced from Wright and Hope (1968) *Canadian Journal of Zoology*, **46**, with permission from NRC Press.

The amphids and phasmids are sense organs that open to the surface on the head and tail respectively (see Chapters 2 and 14). Body pores located on the lateral and median lines of the body wall of *Xiphinema americanum* have dendritic processes associated with them and are presumably sense organs (Wright and Carter, 1980).

Inflations and warts

Cuticular inflations are formed by an expansion of the median zone of the cuticle and are usually filled with fluid. The cuticle of the cephalic and cervical regions of the animal-parasitic nematode *Gongylonema* is inflated in longitudinal rows of blisters and the enoplid nematode *Prooncholaimus* has lateral and median rows of similar blisters (Chitwood and Chitwood, 1950). *Trichuris* spp. have 10–20 blister-like bulges in rows on either side of the anterior bacillary band. In living specimens these structures appear as fluid-filled blisters but they collapse during preparation for electron microscopy. The thin outermost layer of the blister is continuous with the epicuticle and cortical zone of the cuticle while the main part of the blister contains a finely anastomosing network of material (Wright, 1975a). Cephalic inflations are characteristic of many genera of animal-parasitic nematodes. In some there is complete inflation of the cephalic region, especially in trichostrongyles such as *Nippostrongylus*, *Nematodirus* and *Cooperia* (Figure 7.9; see also Chapter 2); in others, such as the oxyurid *Enterobius vermicularis*, cephalic inflations are confined to the lateral regions of the head.

Setae

Many marine and freshwater nematodes possess bristles or setae projecting from the surface of the body (see illustrations in Chitwood and Chitwood, 1950; De Coninck, 1965; Gibbons in Chapter 2). According to Maggenti (1964) the labial setae of the marine nematode, *Thoracostoma californicum*, are mechanoreceptors as they are innervated. The somatic setae of the marine nematodes *Enoplus communis* and *Chromadorina bioculata* are also mechanoreceptors as they are innervated, and arise from a small depression in the surface of the cuticle, which gives the impression that the setae are able to rock on their base if touched (Croll and Matthews (1977). Setae, apparently not connected with glands or nerves, which occur as a 'hairy' coat over the surface of the body of *Greeffiella* spp. and some other species, are thought to give traction during locomotion. However, glandular setae are also scattered over the surface of the body of *Greeffiella* (Chitwood and Chitwood, 1950; De Coninck, 1965).

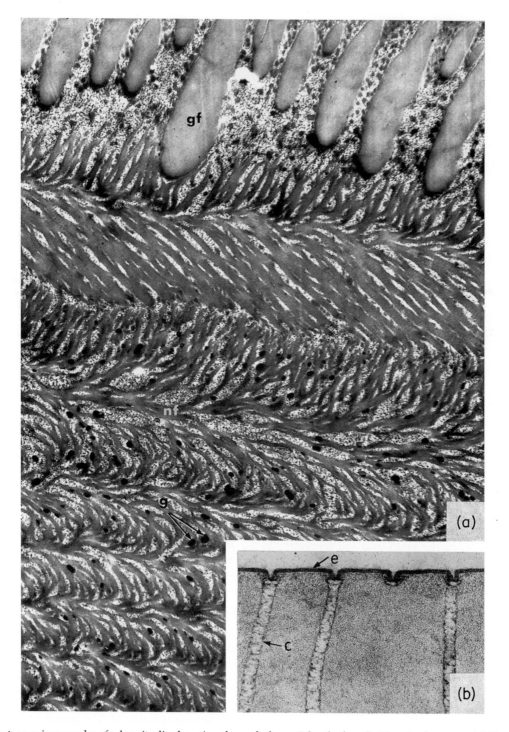

Figure 7.8. Electron micrographs of a longitudinal section through the cuticle of a female *Mermis nigrescens*. (a) The median zone containing giant fibres embedded in a slightly granular matrix, and the basal zone containing an extensive network of fibres. (b) The cortical zone pierced by canals which terminate under the striae. c = canal, e = epicuticle, g = granules, gf = giant fibre, nf = network of fibres. Reproduced from Lee (1970) *Journal of Zoology*, **161**, (with alterations to the lettering), with permission from The Zoological Society of London.

Hollow setae, connected with glands, occur on the ventral surface of the posterior end of the marine nematode *Draconema cephalatum* and other dracono-mids and are thought to have an ambulatory function (Stauffer, 1924).

Spines and hooks
Cuticular spines and hooks occur in several groups of nematodes, but are more common in animal-parasitic species than free-living ones. The spines usually take the form of elongations of the posterior margins of the

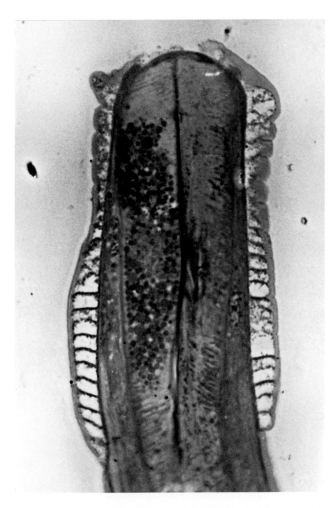

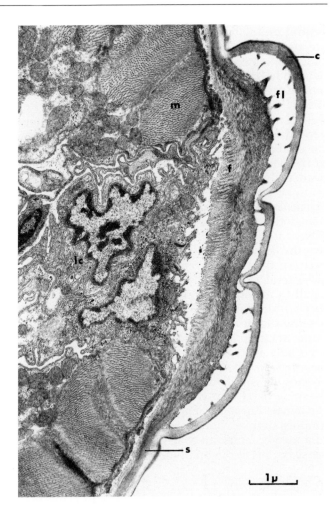

Figure 7.9. Longitudinal section through the head of an adult *Nippostrongylus brasiliensis* to show the inflated cephalic cuticle. Note the strands of material traversing the fluid-filled median zone. Note also the numerous granules in the dorsal pharyngeal gland. x 1700. Original.

Figure 7.10. Transverse section through the lateral cord area of the body wall of *Nippostrongylus brasiliensis* L3 to show the lateral ala with a small central ridge between two larger ridges. Note the change from the striated basal zone into the greatly expanded zone containing two layers of fibrils in the ala, and the fluid-filled median zone traversed by strands of material. c = cortical zone, f = fibres, fl = fluid-filled median zone, lc = lateral cord, m = muscle, s = striated layer of basal zone. Reproduced from Lee (1969) *Symposia of the British Society for Parasitology*, with permission from Blackwell Science Ltd.

annules, as in *Spinitectus* spp., which are parasitic in the alimentary tract of fish and amphibians. *Gnathostoma spinigerum*, parasitic in cats, has an inflated head end armed with circlets of backwardly pointing cuticular hooks. See chapter 2 for illustrations and more information on other species.

Alae

Alae are projections, sometimes wing-like, of the cuticle and are usually lateral or sublateral in location (Figure 7.10) (see Chapter 2). Lateral alae extend along the length of the body, are usually two to four in number, and may occur in larvae and adults of both sexes. These alae overlie the lateral epidermal cords and vary in their structure and morphology in different species and at different stages of the life cycle. Lateral alae are formed, to a greater or lesser degree, from the three major zones of the cuticle plus the epicuticle and surface coat. The lateral alae of the L1 of *Caenorhabditis elegans*, which are produced at the outer surface of lateral epidermal seam cells (Kramer,

1997), are small and bow-like in transverse section whereas the lateral alae of the dauer larva are broader and have four small ridges (Singh and Sulston, 1978). The L3 of *Nippostrongylus brasiliensis* possess lateral alae that have a smaller blunt ridge between two larger blunt ridges (Figure 7.10). The basal zone of the cuticle expands into two fibrous zones and is separated from the cortical zone of the alae by a fluid-filled zone that is traversed by strands of cuticular material (Lee, 1966a, 1969). It is probable that the fluid-filled zone dissipates strains and stresses set up in the cuticle during locomotion when the lateral alae are pressed against a firm substratum. The shape of these alae is such that they give the larva a stable base during undulatory locomotion, as nematodes lie on their sides and move by dorsoventral undulations (Lee, 1969). Fin-like lateral alae are found in several animal-parasitic genera, especially in some oxyurids such as the insect-parasitic nematode *Hammerschmidtiella diesingi* in which the L4 female and adult female have prominent lateral alae (Lee, 1958). These alae are probably used as fins during swimming with the alae acting as dorsal and ventral fins because nematodes swim on their sides. Cervical alae, when present, occur at the anterior end of the nematode (Figures 2.5, 2.6) and may give the head end of the nematode an arrowhead (*Toxocara canis, Heterakis gallinarum*) or pinhead (*Enterobius vermicularis*) appearance. The cervical alae of some nematodes (e.g. *Toxocara, Heterakis*) contain what appear to be strengthening rods of material running lengthwise along the alae. Caudal alae, when present, occur at the posterior end of male nematodes. They are lateral extensions of the cuticle and vary from small to medium lateral extensions to large copulatory bursae that may be supplied with rays, each with a sense organ at its tip. The rays of the copulatory bursa are useful in taxonomy (Durette-Desset, 1983), especially in that of trichostrongyles, such as *Haemonchus, Nematodirus* and *Trichostrongylus*. In these species the copulatory bursa, which is like a webbed grasping hand, is used to grasp the genital region of the female during copulation. Male *C. elegans* possess a rather simpler copulatory bursa that takes the form of a membranous fan containing 9 pairs of rays with sense organs at the tip of the rays.

Longitudinal ridges

Longitudinal ridges of the cuticle occur in many free-living and parasitic nematodes and there are usually several present, with some nematodes having up to 40 or more of these ridges running along most of the length of the body (e.g. *Oswaldocruzia leidyi, Gonionchus australis*). Although they occur in several groups of free-living, plant- and animal-parasitic nematodes they are particularly pronounced in adult trichostrongyles and heligmosomoids (Figures 2.3, 2.4) such as *Nippostrongylus brasiliensis* (Figures 7.11, 7.12, 7.13), *Haemonchus contortus, Trichostrongylus colubriformis, Nematodirus battus* and *Heligmosomoides polygyrus*. Durette-Desset (1971) described four types of longitudinal ridges in the Heligmosomoidea (see Chapter 2 for details). They usually extend from just behind the head region to just before the tail in unbroken ridges, except for indentations caused by superficial transverse striae (the superficial annulations). The cuticular ridges of the trichostrongyles and heligmosomoids are complex structures (Lee, 1965, 1969; Durette-Desset, 1971; Lee and Nicholls, 1983; Martin and Lee, 1983; Fok, Mizinska-Boevska and Polyakova-Krusteva, 1991) and consist of elevations of the cortex, epicuticle and surface coat supported by struts in the median zone. In adult *N. brasiliensis* rows of struts (rather like the teeth of a comb) that are suspended in the fluid-filled median layer insert at their tip into, and thus support, the longitudinal ridges (Figures 7.12, 7.13). These struts, whilst supporting the ridges, still allow flexing of the

Figure 7.11. Scanning electron micrograph of a female *Nippostrongylus brasiliensis* to show the longitudinal ridges. Note the coiled posture of the relaxed nematode. h = head. Reproduced from Lee and Biggs (1990) *Parasitology*, **101**, with permission from Cambridge University Press.

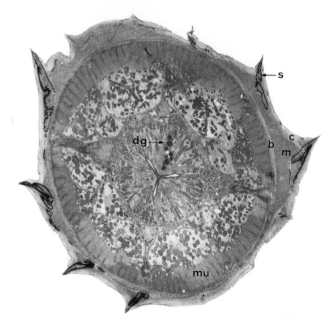

Figure 7.12. Electron micrograph of a transverse section of an adult *Nippostrongylus brasiliensis* through the pharyngeal region to show the zones of the cuticle and the struts that support the longitudinal ridges. The flocculent material in the fluid-filled median zone is probably haemoglobin. The basal zone consists of two fibre layers. b = basal zone, c = cortical zone, dg = dorsal pharyngeal gland, m = fluid-filled median zone, mu = muscle of body wall, s = strut. Reproduced by permission of John Wiley & Sons, Inc. from K.A. Wright (1991) *Microscopic Anatomy of Invertebrates. Volume IV — Aschelminthes* edited by F.W. Harrison and E.E. Ruppert, figure 15. Copy of the original electron micrograph supplied by the late K.A. Wright.

cuticle during locomotion and reduce the possibility of kinking during bending and coiling movements (Lee

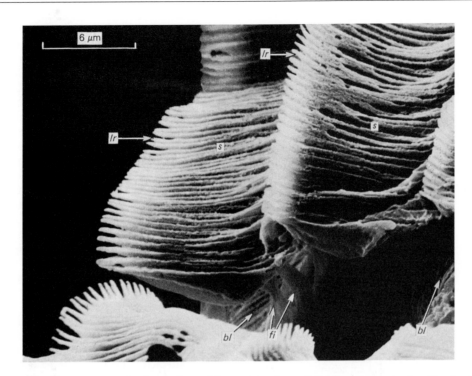

Figure 7.13. Scanning electron micrograph of a portion of the cuticle of an adult *Nippostrongylus brasiliensis* after plasma etching for 5 minutes. The cortical layer has been almost completely removed to reveal the underlying, tougher struts. The basal end of the strut is suspended in the fluid-filled median zone by fibre bundles that anchor it to the fibre layers of the basal zone. These fibres will allow limited movement of the struts in the fluid-filled median zone during movement of the nematode. *bl* = fibres of basal zone, *fi* = fibre bundles that insert onto the base of the struts and onto the fibres of the basal zone, *lr* = longitudinal ridge, *s* = struts. Reproduced from Lee and Nicholls, (1983) *Parasitology*, **86** with permission from Cambridge University Press.

and Biggs, 1990; Lee, Wright and Shivers, 1993). Deformation of the cuticle during locomotion, especially when the cuticle is pressed against the intestinal mucosa of the host, will tend to press the struts against the basal zone but the broad base of the strut will alleviate this by spreading the pressure (Lee and Nichols, 1983). Similar struts, although not usually so complex, support the ridges of other trichostrongyles, such as *H. contortus* (Figure 7.2). Adults of the marine nematode *Gonionchus australis* have numerous closely set longitudinal ridges, 30 to 70 nm high, running along the tops of the prominent annules and these appear as serrations at the anterior and posterior borders of the annules (Figure 7.14) (Nicholas and Stewart, 1997). Nematodes of the family Ceramonematidae have strongly annulated cuticles with overlapping plates and articulated longitudinal ridges that extend the length of the body. The marine nematodes *Metadasyne-moides cristatus* and *Ceramonema carinatum* have eight articulating overlapping cuticular plates per annule and the longitudinal ridges consist of vanes that arise from the plates (Figure 7.15, 7.21) (Nicholas and Stewart, 1990; Stewart and Nicholas, 1992).

Surface Coats and Accessory Layers

A surface coat (Wright, 1987) has been found on the cuticle of free-living, plant- and animal-parasitic nematodes (Figures 7.16, 7.17) and ranges in thickness from 5 nm to more than 20 nm in different species and in

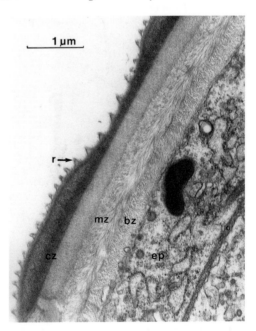

Figure 7.14. Transverse section through the cuticle of a male *Gonionchus australis* to show closely set longitudinal ridges, 30 to 70 nm high, that run along the tops of the annules and the zones of the cuticle. bz = basal zone; cz = cortical zone; ep = epidermis; mz = median zone; r = longitudinal ridge. Copy of the original electron micrograph supplied by W.L. Nicholas. Slightly modified from Figure 3D printed in Nicholas and Stewart (1997) *Journal of Nematology*, **29**. Printed with permission from The Society of Nematologists.

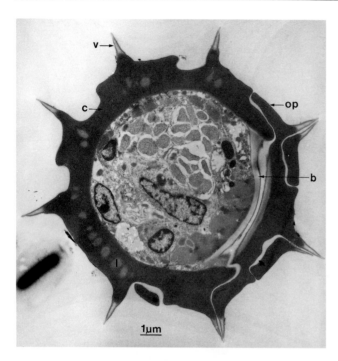

Figure 7.15. Transverse section through an adult *Ceramonema carinatum* to show longitudinal ridges of the cuticle formed by thin vanes (v) that project from each annule. The ridges divide each annule into eight overlapping plates (op) that are slightly offset and give the nematode an articulated appearance. The plates are formed from the amorphous electron-dense cortical zone (c), within which are lacunae (l). There is an electron-lucent basal zone (b) underlying the plates. Original electron micrograph courtesy of W.L. Nicholas (see Stewart and Nicholas, 1992).

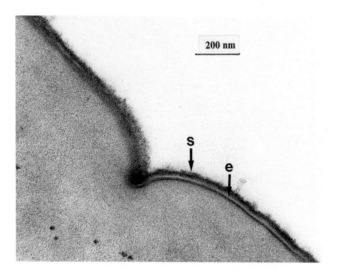

Figure 7.16. Longitudinal section through the cuticle of an adult *Necator americanus* to show the surface coat (s) and epicuticle (e). Original electron micrograph.

different stages of development. However, it has not been detected on all species examined or on all stages of development. The surface coat appears to be synthesized in the amphidial, pharyngeal and/or the secre-

tory-excretory glands and passes on to the surface of the cuticle through the amphidial pores, the mouth or the secretory-excretory pore (*Caenorhabditis elegans, Toxocara canis* (Bird, Bonig and Bacic, 1988, 1989; Nelson and Riddle, 1984; Bird and Bird, 1991; Page, Rudin and Maizels, 1992; Page, Hamilton and Maizels, 1992). Surface coat components are also passed across the cuticle (presumably from the epidermis) in some species (e.g. larval *T. canis, N. brasiliensis, Heterodera schachtii, Globodera rostochiensis*) (Aumann, Robertson and Wyss, 1991; Preston-Meek and Pritchard, 1991; Endo and Wyss, 1992; Jones, Perry and Johnston, 1993). A surface coat has been demonstrated on at least some adult animal-parasitic nematodes. Bonner, Menefee and Etges (1970) and Fok, Mizinska-Boevska and Polyakova-Krusteva (1991) demonstrated the presence of a surface coat on the cuticle of adult *Heligmosomoides polygyrus* (syn. *Nematospiroides dubius*) and *Nippostrongylus brasiliensis* by means of thorotrast or the anionic dye ruthenium red and electron microscopy. Lee, Wright and Shivers (1993) demonstrated it apparently streaming away from the surface of adult *N. brasiliensis* that had been fixed *in situ* and also on freeze-fractures of the surface. The inability to detect a surface coat should not be taken to mean that the coat is absent as particular care must be taken with some stages to preserve the coat during processing for microscopy (Blaxter, *et al.* 1992; Page *et al.*, 1992). It is therefore probable that a surface coat is present to a greater or lesser extent on the surface of all nematodes, especially as it appears to have an important role in the life of nematodes (see later).

The surface coat is composed primarily of glycoproteins, has a net negative charge and it is readily shed into the surroundings of the nematode (Vetter and Klaver-Wesseling, 1978; Philipp, Parkhouse and Ogilvie, 1980; Smith *et al.*, 1981; Philipp and Rumjaneck, 1984; Blaxter *et al.*, 1992; Page, Rudin and Maizels, 1992; Page *et al.*, 1992; Maizels, Blaxter and Selkirk, 1993; Spiegel and McClure, 1995; Gravato Nobre and Evans, 1998; Blaxter and Robertson, 1998; Joachim, Ruttkowski and Daugschies, 1999). The dynamic nature of the surface is exhibited by shedding of surface components *in vitro* by larvae of *Ancylostoma caninum* (Vetter and Klaver-Wesseling, 1978), larvae of *Haemonchus contortus* (Ashman *et al.*, 1995), the infective-stage larvae and the intestinal stages of *Trichinella spiralis* (Figure 7.18) (Philipp, Parkhouse and Ogilvie, 1980; Modha *et al.* 1999). There was a marked increase in the rate of release of surface components when *T. spiralis* was co-cultured with rat neutrophils in the presence of immune rat serum, with the neutrophils beginning to dissociate from the surface of the nematodes after 3 to 4 hours in culture. A surface-binding monoclonal antibody raised against adult *Caenorhabditis elegans* is rapidly shed after binding to the surface of the nematode (Politz and Philipp, 1992). Also, fluorescent-lectin-labelled and lectin-gold-labelled glycoproteins and some antibody complexes are rapidly shed *in vitro* from the surface of several larval animal-parasitic nematodes, such as *Toxocara canis, Nippostrongylus brasiliensis* and *Strongy-*

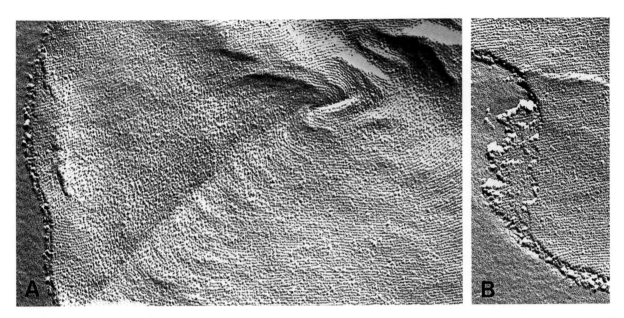

Figure 7.17. Freeze-fracture replicas of the inner surface of the accessory layer of the infective-stage larva of *Trichinella pseudospiralis*. A. Note the whorls of filaments and scattered particles of the inner surface. B. The outermost surface coat is partially hidden by the accessory layer. × 98 000. Original electron micrographs by D.L. Lee, R.R. Shivers and K.A. Wright.

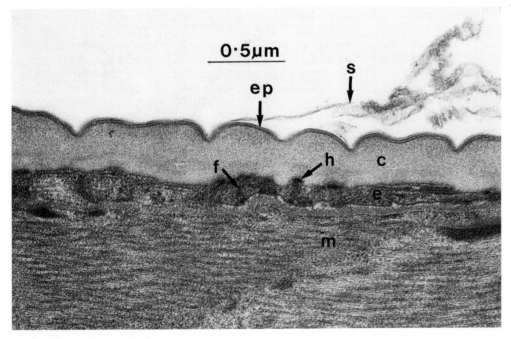

Figure 7.18. Longitudinal section through the body wall of a female *Trichinella spiralis* to show the epicuticle and the surface coat being shed from the cuticle. Note the fibres that attach the cuticle to the muscles by means of hemi-desmosomes on the outer and inner epidermal membranes. e = epidermis, ep = epicuticle, c = cuticle, f = fibres that link the cuticle to the muscle, h = hemi-desmosome, m = muscle, s = surface coat. Original electron micrograph supplied by the late K.A. Wright.

loides ratti (Smith *et al.*, 1981; Murrell and Graham,1983; Badley *et al.*, 1987; Proudfoot *et al.*, 1991; Preston-Meek and Pritchard, 1991; Page, Rudin and Maizels, 1992). Carbohydrates in the surface coat and surface access-ible proteins of the L3 of the filarial nematode *Acanthocheilonema viteae* change and increase rapidly after transmission of the L3 from the tick, *Ornithodorus*

moubata to the jird (gerbil) and there is a corresponding decrease in surface accessible lipids (Apfel, Eisenbeiss and Meyer, 1992). Also, just before moulting there is an increase in the number of detectable surface carbohy-drates, apparently linked to the initiation of ecdysis (Apfel, Eisenbeiss and Meyer, 1992). Zuckerman and Kahane (1983) found that there are stage-specific differences in the surface carbohydrates of *C. elegans*, and stage-specific differences in lectin binding occur at the surface of larval *Ascaris suum* (Hill, Fetterer and Urban, 1991) and larval *Anguina tritici* and *Meloidogyne incognita* (Spiegel and McClure, 1991). According to Hill, Fetterer and Urban (1991), excreted products are unlikely to be the source of the glycosolated surface component in the surface coat of *Ascaris* larvae as it was removed only with strong reducing agents and long-term cultured larvae do not shed the component. Hill, Fetterer and Urban suggested that the surface component of larval *Ascaris* is part of a structural component of the cuticle rather than a loosely associated surface coat. Murrell (1982) and Murrell, Graham and McGeevy (1983) also found that the surface coat of the L3 of *Strongyloides ratti* could not be dissociated from the epicuticle by a range of enzymes, detergents, and solvents. However, Preston-Meek and Pritchard (1991) have shown that if the surface coat is stripped away from the cuticle of larval *N. brasiliensis*, it is quickly replaced. A thick surface coat is present on the cuticle of infective-stage larvae of *T. canis* and anti-secretory-excretory antibody that binds to this surface coat is rapidly shed by the larvae (Smith *et al*, 1981; Page *et al.*, 1992). Similarly, antibodies bound to the surface of motile larvae of *Oesophagostomum dentatum* are shed into the culture medium in which they are kept, but this shedding is inhibited if the larvae are paralysed by the anthelmintic levamisole (Joachim, Ruttkowski and Daugschies, 1999). In the presence of complement and antibody host eosinophils adhere to, and degranulate on, the surface of *T. canis* larvae but exert no detrimental effect because the surface coat is sloughed off (Fattah *et al.*, 1986; Badley *et al.*, 1987). It would appear that this is an important mechanism that enables invading larvae of a number of animal-parasitic nematodes to evade both antibody-dependent and antibody-mediated cellular cytotoxicity *in vivo* (Ibrahim *et al.*, 1989; Edwards *et al.*, 1990; Apfel and Meyer, 1990; Smith, 1991; Page Rudin and Maizels, 1992). Page and Maizels (1992) have shown that larval *T. canis* are able to survive for extended periods of time *in vitro* and that they secrete large quantities of the secretory-excretory glycoproteins that are a component of the surface coat. The major surface glycoprotein of *Brugia malayi* is released from the surface of the adult within 5 hours of synthesis but, unlike larval *T. canis*, the glycoprotein is thought to be produced in the epidermis and transported to the surface through the cuticle (Selkirk *et al.*, 1990).

Plant-parasitic nematodes also possess a labile glycoprotein surface coat, with either the entire surface, or part of the surface, being coated with various carbohydrate moieties (Spiegel and McClure, 1991; Ibrahim, 1991; Lin and McClure, 1996; Gravato-Nobre *et al.*, 1999). Fibrillar exudates that form on the cuticle surface of feeding L2 of *Heterodera schachtii* and *H. glycines* and early L3 of *H. glycines* are formed in the epidermis and are secreted through the cuticle (Endo and Wyss, 1992; Endo, 1993). A similar process takes place in *Globodera rostochiensis* once it has been in the root for some time. Strands of material have been found apparently streaming away from the surface of the cuticle, especially from the region of the striae (Jones, Perry and Johnston, 1993). It is possible that the surface coat is more labile in some species, or at some stages in the life cycle, than others. Bird and Zuckerman (1989) found that the surface coat of the dauer larva of the plant-parasitic nematode *Anguina funesta* (called *A. agrostis* in the original paper but now regarded as *A. funesta* by A.F. Bird) could be hydrolysed by proteolytic enzymes and re-secreted over a period of 18 hour after withdrawal of the enzymes. The surface coat of *Meloidogyne incognita* L2 is shed during migration of the nematode in the root of *Arabidopsis*, leaving a trail of surface coat from the point of invasion to the feeding site in the stele, and may continue to be shed once a giant cell has been formed (Gravato-Nobre *et al.*, 1999). It is possible that carbohydrates in the surface coat of plant-parasitic nematodes may act as elicitors of a hypersensitive response by cells of the host plant (Gravato Nobre and Evans, 1998). The surface coat may also mimic components of the host and thus avoid a host reaction as antibodies raised to the surface coat of *M. incognita* specifically cross-react with phloem cells of the host plant (Bird and Wilson, 1994).

Composition of the surface coat

The surface coats of nematodes contain carbohydrate and mucin-like proteins with glycosylation of the surface coat being detected by a variety of lectin conjugates (see Kennedy, 1991; Spiegel and McClure, 1995; Blaxter and Bird, 1997; Blaxter and Robertson, 1998). Ham, Smail and Groeger (1988) used lectins to follow changes in the surface carbohydrates on *Onchocerca lienalis* larvae as they developed from microfilariae to the infective L3 in the insect vector, *Simulium ornatum*. Ham, Smail and Groeger found that none of the lectins used bound to the surface of the microfilariae but there was progressive binding to the surface of the cuticle of the L1 and L2 using Con A, lentil lectin and wheat germ agglutinin (WGA). Binding of these three lectins declined after moulting to the L3 but there was an increase in binding of peanut and *Helix pomatia* lectins. Lectin specificities indicated that, initially, mannose/glucose-type derivatives are present on the surface of the L1 and L2 but following moulting to the L3 these are progressively replaced (or overlain) with galactosamine-type derivatives. None of the L1, L2 or early L3 of *Ascaris suum* bound any of the lectins tested (Hill *et al.*, 1991). Most proteins in the surface coat of nematodes are glycoproteins, with either *N*-linked or *O*-linked sugars (Blaxter and Robertson, 1998). The surface coat of *C. elegans* and *C. briggsae* contains carbohydrate and mucin-like proteins (Zuckerman, Kahane and Himmelhoch, 1979; Zuckermann and Kahane, 1983) with one of the mucin-like proteins in *C. elegans*, a serine/threonine-rich protein, being the product of the *let-653* gene (Jones and Baillie, 1995). Each stage in the life cycle of *C. elegans* has a distinct set of surface protein molecules (Politz, Chin and Herman, 1987; Hemmer *et al.*, 1991; Blaxter, 1993; Blaxter and

Bird, 1997). The major constituent of the mucin-like surface coat of *Toxocara canis* infective-stage larvae is the TES-120 glycoprotein series and this is also serine/threonine-rich (Gems *et al.*, 1995; Gems and Maizels, 1996). The *O*-linked proteins, which are the main component, are unusually glycosylated but there is also limited *N*-linked glycosylation (Khoo *et al.*, 1991; Maizels and Page, 1990; Page and Maizels, 1992). It has been proposed that these mucin-like *O*-linkages are protected against proteolytic cleavage by enzymes secreted by the nematode and from host-derived proteases in the alimentary tract of the host as well as being involved against immune attack (Page and Maizels, 1992; Gems and Maizels, 1996). There is now considerable evidence that the surface coat of at least some nematodes contains enzymes. An *N*-glycosylated glutathione peroxidase (Gp29) (Zvelebil *et al.*, 1993 is the predominant soluble surface glycoprotein of adult *Brugia malayi* (Maizels *et al.* 1983; Maizels *et al.*, 1989; Cookson, Blaxter and Selkirk, 1992). It is synthesized in the epidermis, secreted into the cuticle and released onto the surface and into the surrounding medium (Devaney, 1988, 1991; Devaney and Jecock, 1991; Tang *et al.*, 1995). The enzyme is active against lipid peroxides but not against hydrogen peroxide (Tang *et al*, 1995) and presumably protects the epicuticle from peroxidative disruption. A CuZn superoxide dismutase secreted by, and present on, the surface of adult male and female *Brugia pahangi*, but not the microfilariae, is presumed to neutralise superoxide generated by leukocytes at the surface of the nematode, thus acting as an anti-oxidant and also as an anti-inflammatory factor, thereby contributing to survival of the parasite in its host (Tang *et al.*, 1994; Ou *et al.*, 1995). However, whilst immuno-electron microscopy localised the enzyme in the epidermis no labelling was found in the cuticle (Ou *et al.*, 1995). A cystatin-like cysteine proteinase inhibitor is also present at the surface of filarial nematodes (Lustigman *et al.*, 1992). Two glutathione *S*-tranferases (OvGST1 and OvGST2) have been detected in adults and larval stages of *Onchocerca volvulus* (Liebau, Walter and Henkle-Duhrsen, 1994a, b; Liebau *et al.*, 1994; Salinas, Braun and Taylor, 1994) but only OvGST1 has been detected in the cuticle and on its surface (Wildenburg, Liebau and Henkle-Dührsen, 1998). These enzymes may assist in counteracting damage by inhibition of the oxidative burst generated by activated leukocytes and by detoxifying secondary products of lipid peroxidation in the mammalian host (Cookson, Blaxter and Selkirk, 1992; Tang *et al.*, 1994; Wildenberg, Liebau and Henkle-Dührsen, 1998). This may be one explanation for the resistance of these nematodes to the immune response of their host and for their longevity.

The antioxidant enzyme thioredoxin (Ov-TPX-2) is present in the epidermis and cuticle of late L1, L3, post-infective larvae and adults of *Onchocerca volvulus* and is possibly a major hydrogen peroxide-detoxifying enzyme (Lu *et al.*, 1998). Thioredoxin peroxidase enzymes are also present in the epidermis of adult *Brugia malayi* but were not detected in or on the cuticle (Ghosh *et al.*, 1998).

The cuticles of the male tail and the vulval slit of hermaphrodite wild-type *C. elegans* bind the lectins WGA and SBA at low levels, and mannose or glucose residues have been detected on the cephalic region, otherwise there is little lectin binding (McClure and Zuckerman, 1982; Zuckerman and Kahane, 1983; Jannson *et al.*, 1986; Link, Ehrenfels and Wood, 1988; Link *et al*, 1992; Blaxter and Bird, 1997).

The surface coat of plant-parasitic nematodes is important not only in interactions with the host plant (see above) but also with their parasites and pathogens (e.g., nematophagous fungi) or other microorganisms (e.g., *Clavibacter*) (Bird, Bonig and Bacic, 1989; Davies and Danks, 1993; Spiegel, Mor and Sharon, 1996; Spiegel *et al.*, 1997). Attachment of fungal spores or bacteria to the surface of nematodes appears, in many instances, to be species specific, stage specific and in some nematodes, specific for certain regions of the body (e.g. the head), suggesting that the interaction may be, in part, mediated by a lectin-carbohydrate interaction (Bird, Bonig and Bacic, 1989). Adhesion of the nematode-trapping fungus *Arthrobotrys oligospora* to the cuticle of *Panagrellus redivivus* involves a carbohydrate-binding protein on the traps of the fungus binding to *N*-acetyl-galactosamine residues on the surface of the nematode (Nordbring-Hertz and Mattiasson, 1979). Bird, Bonig and Bacic (1989) found that *N*-acetyl-D-glucosamine residues in the surface coat of L2 of *Anguina agrostis, Meloidogyne javanica, M. incognita* and *M. hapla* were associated with attachment of the conidia of the fungus *Dilophospora alopecuri* and the coryneform bacterium *Clavibacter* sp. to the surface of the larvae. The surface coat of *M. incognita* L2 covers the mid-body region in a continuous layer and it binds cationised ferritin and stains with ruthenium red (Lin and McClure, 1996). Galactose and N-acetyl-D-galactosamine have been shown to occur on the surface of adult female *M. javanica* (Ibrahim, 1991). Evidence for the presence of carbohydrate-binding proteins in the surface coat of several plant-parasitic nematodes has been shown by the binding of human red blood cells to the surface (Spiegel *et al.*, 1995). Human red blood cells were found to adhere to the surface of 24 to 48-hour-old L2 of *M. javanica*, but not to 0 to 24-hour-old L2 indicating that carbohydrate-binding proteins are probably not present on the surface of the very young L2 (Sharon and Spiegel, 1996; Spiegel *et al.*, 1997). Spiegel *et al.* (1997) found that human red blood cells did not adhere to the surface coat of L2 after stripping of the coat with SDS but re-adherence of the red blood cells occurred if the L2 were then kept at 25 °C for 2 days. Red blood cells did not adhere to the surface of L2 which had been treated with SDS and then kept at 4 °C for 2 days but when these larvae were transferred to 25 °C for 2 days adhesion re-occurred indicating that replacement of the surface coat requires the nematode to be metabolically active. Spiegel *et al.* (1997) have characterised the surface proteins on the L2 of *M. javanica* by means of SDS-PAGE and labelling the glycoproteins on Western blots. ConA and WGA labelled several surface protein bands.

Available evidence shows that the surface coat has several roles in the biology of nematodes. It acts as a lubricant, it is antigenic and in animal-parasitic species the host raises antibodies against components of the surface, in some species it has enzymes that protect the nematode from host defence reactions. Various micro-organisms that attach to the cuticle of free-living stages of nematodes recognise certain components of the surface coat and may be selective in their attachment sites. The ability to change the complement of proteins in the surface coat, either during a particular stage in the life cycle or after moulting, appears to be a common feature of nematodes. Several species of animal-parasitic nematodes have been particularly successful in exploiting this, thereby enabling them to evade the defences of the host, at least for a time (Philipp, Parkhouse and Ogilvie, 1980; Philipp and Rumjaneck, 1984; Blaxter *et al.* 1992).

Accessory layer

The cuticle of isolated infective-stage larvae of *Trichinella spiralis* and *T. pseudospiralis* and of *T. spiralis in situ* within the nurse cell, is covered with an unusual accessory layer (Lee, Wright and Shivers, 1984; Wright and Hong, 1988, 1989; Gounaris, Smith and Selkirk, 1996; Modha *et al.*, 1997, Modha *et al.*, 1999). The accessory layer appears early in the development of the L1 (Despommier, 1983) and is about 15 nm thick (Lee, Wright and Shivers, 1984; Wright and Hong, 1989). This layer seems to be shed continuously *in vitro* (Modha, Kennedy and Kusel, 1995) but is substantially lost upon larval activation and host infection (Despommier, 1983; Stewart *et al.* 1987). It remains on the larva through pepsin-HCl digestion but is not replaced once the larva moults within the alimentary tract of the next host (Capo, Despommier and Silberstein, 1984). Loss of the accessory layer coincides with a change in parasite motility, increased uptake of glucose (Stewart *et al.*, 1987) and increased insertion of the fluorescent lipid probe AF18 into the surface of the cuticle (Proudfoot *et al.*, 1993; Modha, Kennedy and Kusel, 1995; Modha *et al.*, 1999). Freeze-fracturing followed by electron microscopy of the body wall has revealed the macro-molecular structure of this accessory layer and the underlying cuticle (Lee, Wright and Shivers, 1984; Wright and Hong, 1988; Gounaris, Smith and Selkirk, 1996). Three fracture planes have been described: the two most distal fracture planes revealed a non bilayer cylindrical configuration of lipid components that are reminiscent of a hexagonal-type II (H_{II}) organisation whereas the proximal plane revealed a bilayer config-uration (Gounaris, Smith and Selkirk, 1996). Modha *et al.* (1999) have shown that the shed accessory layer comprises thin multilaminate sheets and amorphous material with ridges producing a fingerprint motif, similar to that revealed by freeze-fracturing of the surface (Lee, Wright and Shivers, 1984; Gounaris, Smith and Selkirk, 1996). The outermost surface of the accessory layer contains particles, about 10 nm in diameter, as a major component within an amorphous material (Lee, Wright and Shivers, 1984). The two distal

lipid layers appear to be made up of cylinders of lipid about 6.8 nm in diameter. These cylinders run parallel to each other, occasionally forming waves or whorls out of line with their neighbours (rather like a fingerprint), and some cylinders appear as if they have been inserted into an existing pattern (Figure 7.17) (Lee, Wright and Shivers, 1984). A large number of randomly distributed particles is associated with the cylinders (Figure 7.17) and these are thought to be glycoproteins that are associated with the parasite surface via hydrophobic interactions. The accessory coat material comprises predominantly phospholipids with lysophosphatidic acid being one of the major lipid components (Gounaris, Smith and Selkirk, 1996; Modha *et al.* 1999). The function of the extensive sheets of lipid in the configuration of the accessory coat is unknown but appears to be related to the intracytoplasmic location of the larva within the nurse cell. This accessory layer is not present on the surface of adults (Figure 7.18) or on *in utero* larvae (Lee, Wright and Shivers, 1986).

The sheath found on all stages of *Hemicyliophora* (Johnson, Van Gundy and Thomson, 1970a) can also be considered to be an accessory layer but it is very different in structure from the accessory layer of *Trichinella* infective-stage larvae. Parasitic L2, 3, 4 and the adult females of *Hemicycliophora* possess a seven-layered sheath covering the five-layered cuticle whereas the males possess a four-layered sheath and a six-layered cuticle. The loose-fitting sheath appears to be attached to the underlying cuticle only at the head and tail. Examination of electron micrographs reveals what appears to be an epicuticle on the surface of both the sheath and the cuticle. The sheath is not a residual cuticle because a new sheath and a new cuticle are produced at each moult (Johnson, Van Gundy and Thomson, 1970b).

Epicuticle

The term epicuticle was proposed by Lee and Atkinson (1976) and has since been generally adopted for the trilaminar structure (Figures 7.16, 7.18, 7.19) that is present at the surface of the cuticle of all nematodes. It has been compared to a modified plasma membrane, because of its ultrastructural resemblance to a plasma membrane, and also to the extracellular envelope which is found in a wide range of organisms (Locke, 1982). It is now generally agreed that the epicuticle is not a modified cell membrane (Wright, 1987; Bird and Bird, 1991; Kennedy, 1991). Biophysical measurements of the fluidity of lipid in the epicuticle of several different species of nematode have shown that there is little mobility of lipid and this is atypical of a cell membrane (Proudfoot *et al.*, 1991; Selkirk, 1991). The outermost cell membrane of the body occurs at the epidermis/cuticle interface and not at the outer surface of the cuticle. The epicuticle allows extracellular organisation of the various components of the cuticle to occur during formation and growth of the cuticle.

The epicuticle is composed of covalently cross-linked non-collagenous proteins and lipids. When it is cleaved by freeze-fracturing two faces (P and E faces) are

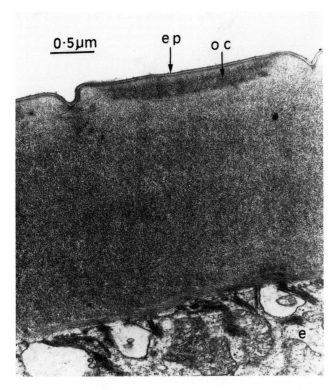

Figure 7.19. Longitudinal section through the cuticle of an adult *Heterakis gallinarum* to show the epicuticle and the band of outer cortex that encircles each annule of the nematode. ep = epicuticle, oc = outer cortex. Reproduced from Lee (1977) with permission from The Zoological Society of London

exposed. Particles resembling intramembranous particles have been found immersed in a smooth matrix on the P face, but without any particular arrangement, whilst the E face was found to possess scattered depressions in adult *Caenorhabditis elegans* (Peixoto and Desouza, 1994, 1995), adult *Nippostrongylus brasiliensis* (Lee and Bonner, 1982; Lee, Wright and Shivers, 1993), microfilariae of *Wuchereria bancrofti* (De Souza *et al.*, 1993) and *Brugia malayi* (Araujo, Souto-Padron and De Souza, 1994), and adult *Strongyloides venezuelensis* (Martinez and De Souza, 1995). The plane of fracture appears to be along the electron lucent (middle) region of the epicuticle and the particles are probably integral proteins of the P and E faces of the epicuticle (Martinez and De Souza, 1995). The external surface of the L3 of *S. venezuelensis*, as revealed by freeze-fracturing and then deep etching, showed particles and fine strands of fibrous elements and also a well organised structure with a crystalline pattern (Martinez and De Souza, 1997). It is probable that this is the surface coat/epicuticle complex. Unlike in the adult, the epicuticle of the L3 of *S. venezuelensis* did not cleave along fracture planes and this probably reflects a difference in composition between the adult and L3 (Martinez and De Souza, 1997).

Lipid components of the surface coat of nematodes have been demonstrated by lipase treatment (Spiegel,

Cohn and Spiegel, 1982). Fluorescent lipid probes have been used to study lipids in the epicuticle of several species of nematode and have led to the suggestion that the lipid is not homogeneous but may have intercalated and static phases (Kennedy *et al.*, 1987; Proudfoot *et al.*, 1991; Robertson *et al.*, 1992). Study of the lateral diffusion of fluorescent probes inserted into the surface has revealed that the probe octadecanoylaminofluorescein (AF18) is immobile in adult *Acanthocheilonema viteae*, *Brugia pahangi*, *Litomosomoides sigmodontis* (syn. *carinii*) and *Trichinella spiralis* but mobile in adult *C. elegans*. Immobility of the AF18 lipid probe in the epicuticle implies a totally rigid lipid surface. However, the non-polar lipid probe nitrobenzoxadiazolamine cholesterol (NBP-col), which detects mobile lipid, is mobile in adult *A. viteae*, *L. sigmodontis*, *B. pahangi*, *T. spiralis*, *C. elegans* and the dauer larva of *C. elegans* which suggests that in the surface of adult nematodes there is differential horizontal partitioning of the lipid probes (Proudfoot *et al.*, 1991). The infective-stage larvae of *A. viteae*, *N. brasiliensis*, *T. spiralis*, *S. ratti* and *Ostertagia ostertagi* have no affinity for either of these lipid probes but once they enter the mammalian host the biophysical properties of the epicuticle change resulting in the AF18 probe being able to insert into its surface. The dauer larva of *C. elegans*, which is similar in some ways to the infective-stage L3 of parasitic nematodes, has no affinity for the lipid probes but insertion of AF18 occurs after exposure to fresh *E. coli*, which act as a stimulus to the nematodes to begin feeding and continue development. There is the possibility that the unusual lipids of the epicuticle are waxes or ascarosides; these can exist as solids or fluids (Kennedy *et al.*, 1988) and could allow the epicuticle some flexibility in interactions with the environment while retaining resistance to chemical and physical agents.

Proteins associated with the epicuticle rapidly diffuse laterally, which suggests that the proteins are either loosely associated with ionic interactions with polar lipids, or that they may be partitioned into microdomains which contain mainly neutral lipids and are free to diffuse (Proudfoot *et al.*, 1991). A constituent of the epicuticle of the filarial nematode *Brugia pahangi* has been shown to be a 30 kDa glycoprotein. It was found to be a major antigen after labelling adults with ^{125}I and contains epitopes which are cross-reactive with other developmental stages (Philipp *et al.*, 1986; Devaney, 1991) and also with other species of filarial nematodes (Maizels *et al.*, 1983). The epicuticle also contains noncollagenous proteins called cuticlins (see below).

Work on anhydrobiotic larvae of several species of nematodes suggests that it is probable that the cuticle permeability barrier is located within the epicuticle (Bird and Bird, 1991), however, the outer membrane of the epidermis is probably very important in this respect.

Cortical Zone

The cortical zone varies in thickness from about 0.2 μm in small free-living and small larval stages of nematodes to several micrometres in large animal-parasitic species, such as *Ascaris suum*. It may consist of fine fibrils that make it appear homogeneous, may be subdivided (*Ascaris, Parascaris; Heterakis*) (Figure 7.19) (Hinz, 1963; Lee, 1977) and/or consist of perpendicularly oriented striations (*Acanthonchus, Anticoma, Enoplus, Gonionchus, Pontonema, Sphaerolaimus*) (Wright and Hope, 1968; Malakhov, 1994; Yushin and Malakhov, 1994; Tchesunov, Malakhov and Yushin, 1996; Nicholas and Stewart, 1997) (Figures 7.2, 7.4) similar to those found in the basal zone of the dauer larva of *Caenorhabditis* (Popham and Webster, 1978; Peixoto and De Souza, 1994) and the infective-stage larvae of many animal-parasitic (*Nippostrongylus*) (Figure 7.23) (Lee, 1966a) and plant-parasitic (*Globodera/Heterodera*) (Wisse and Daems, 1968) nematodes. The outer cortex, when present, is more electron dense than the inner cortex and in many species is relatively thick in the crests of the annulations as in *Gonionchus australis* (Figure 7.4) (Nicholas and Stewart, 1997). The cortical zone of adult *C. elegans* has an outer and an inner region with the outer cortex being more electron-dense than the inner cortex. This arrangement is present in the cortical zone of free-living larvae of many plant- and animal-parasitic nematodes (e.g. *Globodera, Nippostrongylus*). The outer cortex of the nematode cuticle, which contains a non-collagenous protein called cuticlin (Fujimoto and Kanaya, 1973), is thought to be tougher than the inner cortex and the median zone, and may strengthen the cuticle between the striae of the annulations. In adult *Ascaris* and *Heterakis* (Figure 7.19) the outer cortex runs as a circumferential band around each superficial annule of the cuticle. The bands are linked to each other by a thinning of the outer cortex material under each stria of the annulations in *Ascaris* and also in *Parascaris equorum* and *Heterakis gallinarum* (Figure 7.19) (Hinz, 1963; Fujimoto and Kanaya, 1973; Lee, 1977). The cuticular plates and vanes that form the longitudinal ridges of adult *Metadasynemoides* consist of the epicuticle and cortical zone with the cortex enclosing spaces traversed by a delicate lattice of fibrils (Nicholas and Stewart, 1990) (Figures 7.20, 7.21). In *Ceramonema* the vanes of the longitudinal ridges are different in appearance from the underlying cortical plate, from which they arise. In the amorphous cortical plate, discrete lacunae contain a finely granular material that is probably a system of fine fibrils (Figure 7.15) (Stewart and Nicholas, 1992).

The cortical layers of several nematodes (*Acanthonchus, Mermis*) (Wright and Hope, 1968; Lee, 1970; Malakhov, 1994) are penetrated by tubular canals that extend from the median layer, or in the case of *Acanthonchus* from the epidermis, to the surface of the cuticle (Figures 7.2, 7.7, 7.8b).

Median Zone

The median zone is usually less dense than the cortical zone and has the most varied structure, especially in many marine nematodes. It may be amorphous, vacuolated, gel-like, fluid-filled or may contain elaborate systems of plates, struts and fibres (Figure 7.2) (Watson, 1965; Shepherd, Clark and Dart, 1972; Lee and Nicholls, 1983; Stewart and Nicholas, 1992; Yushin and

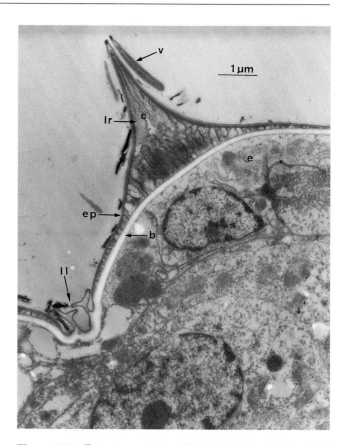

Figure 7.20. Transverse section through a ridge of an adult *Metadasynemoides cristatus* to show the vane at the tip of the ridge, the honeycomb-like structure of the cortical zone, and the electron-lucent basal zone. b = basal zone, c = cortical zone, e = epidermis, ep = epicuticle, ll = lateral line, lr = longitudinal ridge, v = vane. Copy of original electron micrograph, courtesy of W.L. Nicholas. Labelling modified from Figure 7 printed in Nicholas and Stewart (1990) *Journal of Nematology*, **22**. Printed with permission from The Society of Nematologists.

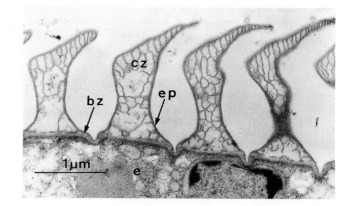

Figure 7.21. Longitudinal section through a ridge, near the centre of the ridge, of an adult *Metadasynemoides cristatus* to show that the longitudinal ridges are made up of a series of overlapping protuberances from each annule. bz = basal zone; cz = cortical zone; e = epidermis; ep = epicuticle. Original electron micrograph, courtesy of W.L. Nicholas (see Nicholas and Stewart, 1990).

Malakhov, 1994; Malakhov, 1994; Tchesunov, Mala-khov and Yushin, 1996; Urbancik, Bauer-Nebelsick and Ott, 1996; Nicholas and Stewart, 1997). The main structural component of this zone is collagen but other substances may be present, such as enzymes (Lee, 1961) or haemoglobin, as in adult *Nippostrongylus brasiliensis* (Sharpe and Lee, 1981)

Scanning electron micrographs of the body wall of adult *Ascaris suum*, after it had been exposed by means of a longitudinal cut, revealed a homogeneous gel-like material in the median zone (Figure 7.22). It is probable that a gel-filled or fluid-filled median zone will tend to dissipate shearing forces set up in the cuticle during locomotion. A number of groups of nematodes possess struts, rods or plates in what appears to be a fluid-filled median zone. (Figures 7.2, 7.13). The cuticle of adult *Caenorhabditis* contains supporting struts or strands of collagen that traverse the fluid-filled median zone (Zuckerman, Himmelhoch and Kisiel, 1973; Cox, Kusch and Edgar, 1981; Johnstone, 1994; Peixoto and De Souza, 1994, 1995). The median zone of adult *Acanthonchus duplicatus* is electron translucent (fluid-filled?) and is traversed by electron-dense rods that have 5 to 8 prong-like processes at each end and electron-dense rings that surround the cuticular pore canals (Wright and Hope, 1968) (Figure 7.7). The prongs of the rods splay out underneath the striated cortex at their distal end and splay out over the meshwork of transversely oriented filaments of the basal zone, rather as pillars supporting a roof. The dense rings that surround the pore canals presumably help to maintain the integrity of the canals during movement of the cuticular zones when the nematode moves. Similar electron dense columns, branched at

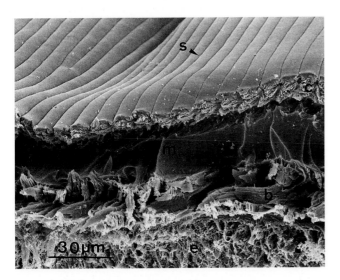

Figure 7.22. Scanning electron micrograph of a specimen of an adult *Ascaris suum* that had been fixed in formol saline and then cut longitudinally to exhibit the layers of the cuticle. Note the striae of the cuticle demarking the smaller annuli of the cuticle, the fibrous nature of the cortical zone, the gel-like appearance of the median zone, and the large fibres of the basal zone. b = fibres of basal zone, c = cortical zone, e = epidermis, m = median zone, s = striae. Original.

each end and situated in what appears to be a fluid-filled space, are also found in the median zone of *Paracanthonchus* (Figure 7.2) (Yushin and Malakhov, 1994). The median zone of *Monoposthia costata* consists of plates or scales that occupy most of the median zone in each annulus, but not between the annuli, and project anteriorly from the surface as tapering scales to end in a sharp edge (Figure 7.2) (Yushin and Malakhov, 1994).

The median zone of the plant-parasitic nematode *Tylenchorhynchus dubius* is flocculent in appearance and is possibly fluid or semi-fluid in nature (Byers and Anderson, 1972). Hatched L2 and males of *Globodera rostochiensis* possess a fluid-filled median zone traversed by small columns of material and containing some electron-dense balls of material (Wisse and Daems, 1968; Shepherd, Clark and Dart, 1972). In unhatched L2 of this nematode the median zone contains a granular material which is absent from L2 that have been experimentally hatched following exposure to potato root diffusate. After the nematode has set up its feeding site in the root, the median zone becomes less prominent. In males the median zone is similar to that of hatched L2 but contains many more of the electron-dense balls (Shepherd, Clark and Dart, 1972; Jones, Perry and Johnstone, 1993). The median zone of the cuticle of adult *Syphacia obvelata* is extensive in males, accounting for over 60% of the total cuticle thickness but in the female it accounts for only 25–30% of thickness. It appears to be fluid-filled and contains a flocculent material. Strands of striated material are present at the interfaces with the cortical and basal zones and some strands traverse the median zone (Dick and Wright, 1973). The median zone becomes expanded in the lateral alae, with strands of material traversing the zone, in the L3 of *Nippostrongylus*; this probably reduces the effect of shearing forces that may be set up in the cuticle when the nematode moves (Figure 7.10) (Lee, 1966a, 1969). In some adult nematodes the median zone of the alae contains rods of electron-dense material that run along the length of the alae. It is possible that these are stiffening rods and are important when flexing of the body occurs. The median zone of the cephalic region of many nematodes is inflated and the main component of this inflation is the fluid-filled median zone (Wright, 1975b) traversed by strands of material (Figure 7.9).

The median zone of the cuticle of the marine nematode *Euchromadora vulgaris* is formed from a series of overlapping plates that contain small canals, but these canals do not extend into the cortical zone (Watson, 1965). What appear to be toughened, more erect plates in section, and which are presumably rings of material, occur in the median zone of each annule in *E. robusta*. A process towards the base of each of these structures extends forwards and inserts into the next structure resulting in what appears to be an articulated layer embedded in a fluid-filled space in the median zone (Kulikov *et al.* 1998). The plates appear to be lying within a fluid-like matrix. A similar set of articulated plates lies in the median/basal zone of *Halalaimus leptoderma* (Figure 7.2); each of the annular plates possesses a round foramen that is part of a canal that encircles each annulus (Yushin and Malakhov, 1994). Urbancik, Bauer-Nebelsick and Ott (1996) studied the complex cuticles of several genera of marine nematodes belonging to the family Stilbonematinae and found that the median zone is the most complex zone of the cuticle. The median zone is composed of five different elements with most of the elements being located within the annulus. The principal elements are: 1. A homogeneous, ring-shaped structure which lies below the cortical zone and beneath the crest of the annulus and varies in size from a flat ring (*Catanema*) to more robust rings (*Robbea* and *Leptonemella*). 2. Bundles of fibres that link neighbouring annuli to each other and pass beneath the striae of the annuli. 3. Two transverse ring-shaped strands or fibres that encircle the annulus beneath (2) the bundles of fibres. 4. Pairs of longitudinal fibres that lie beneath the ring-shaped fibres. 5. Another set of ring fibres that lies beneath (4), encircles the annulus and lies adjacent to the fibres of the basal zone (Figures 7.6 a, b, c). Adult *Mermis nigrescens*, which lead a free-living, non-feeding existence once they have escaped from the cadaver of their insect host, also have an unusual cuticle (Lee, 1970). As mentioned earlier, the cortical zone is pierced by numerous pore canals which originate at the junction with the median layer. The median layer contains a flocculent material in which are embedded two layers of large fibres that run obliquely around the nematode in the form of two spirals, crossing each other at an angle of 110° (Figure 7.8a).

Lee (1965, 1966b), Jamuar (1966), Bonner and Weinstein (1972a, b), Lee and Nicholls (1983), and Fok, Mizinska-Boevska and Polyakova-Krusteva (1991) described rows of electron-dense struts in the fluid-filled median layer of adult *Nippostrongylus* (Figure 7.12). These struts support the longitudinal ridges of the cuticle with their pointed distal ends being inserted into the peak of the ridge and the broad base of the strut being suspended by fibrils that extend to the fibrous basal layer (Figure 7.13). Plasma etching of the cuticle revealed the comb-like appearance of these struts and demonstrated that they are more resistant than the cortex and the fluid matrix of the median zone (Figure 7.13). Similar struts are found in the median zone of other trichostrongyle and heligosomoid nematodes (Durette-Desset, 1971; Smith and Harness, 1972; Martin and Lee, 1983). Sharpe and Lee (1981) withdrew fluid from the fluid-filled median zone of the cuticle of female *N. brasiliensis* and demonstrated that it contains haemoglobin that is able to load and unload with oxygen in the living nematode. This haemoglobin probably enables the nematode to transmit oxygen from the relatively oxygen-poor mucosal surface area of the host's intestine to the muscles and other tissues of the nematode (Sharpe and Lee, 1981).

Various types of median zones, together with examples of inclusions, are illustrated in Figure 7.2 (Markhov, 1994).

Basal Zone

The main structural component of the basal zone is collagen. In the dauer larva of *C. elegans* and the pre-parasitic larvae of plant-parasitic and animal-parasitic nematodes the basal zone is a striated layer consisting of a vertically arranged lattice of rods, probably consisting of collagen, linked together by short, thin filaments, probably of lipoprotein, with a periodicity of about 20 nm and about 84 nm high (Figure 7.23) (Eckert and Schwarz, 1965; Lee, 1966a; Wisse and Daems, 1968; Popham and Webster, 1978; Bird and Bird, 1991; Jones, Perry and Johnston, 1993; Peixoto and De Souza, 1994; Martinez and De Souza, 1997). This striated layer is also found in the basal zone of some adult nematodes, such as *Capillaria hepatica* and *Trichuris myocastoris* (Wright, 1968), *Tylenchorhynchus dubius* (Byers and Anderson, 1972) and *Globodera* (syn. *Heterodera*) *rostochiensis* (Shepherd, Clark and Dart, 1972; Jones, Perry and Johnston, 1993). Larvae of *Globodera rostochiensis* lose the striated appearance of the basal layer once they have set up their feeding site in the plant root. However, males reform the striated appearance when they emerge from the root in search of a female and this lends support to the possibility that the change is associated with locomotion (Jones, Perry and Johnston, 1993). The striated layer changes into two layers of fibrils in the basal zone of the lateral alae in the cuticle of *Nippostrongylus* L3 (Figure 7.10) (Lee, 1966a) and the dauer larva of *C. elegans* (Singh and Sulston, 1978). It has been suggested that this striated layer has very close linkages between the molecules, resulting in a resistant protein, which may explain why the larva is so resistant to chemical fixatives and is not as susceptible to drying as other nematodes (Lee, 1969). Inglis (1983) suggested that this striated layer is a type of a liquid crystal. Similar striated layers are present in the cortical zone of some adult nematodes (see above). Whilst this striated basal zone is present in the dauer larva of *Caenorhabditis* it is not present in the adult nematode, this zone being of a fibrous nature with the fibrils not being arranged into distinct fibres. Several species of nematode that have been examined appear to have a

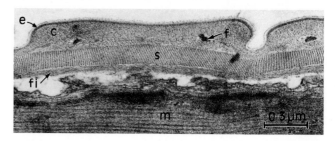

Figure 7.23. Longitudinal section through the body wall of a *Nippostrongylus brasiliensis* L3 to show the striated layer in the basal zone of the cuticle. Note the epicuticle and the two electron-dense fibres that encircle the larva in each annule. c = cortical zone, e = epicuticle, f = fibre, fi = fibrils, m = muscle, s = striated layer. Reprinted from Lee (1966a) *Parasitology*, **56** with permission from Cambridge University Press.

very simple basal layer with the median and cortical zones being the main zones of the cuticle.

The basal zone of larger nematodes consists of two or more layers of collagenous fibrils, often assembled into larger fibres, that cross each other at an angle in the cuticle of some adult free-living, plant-parasitic and animal-parasitic species. Some of the earliest descriptions are of the three fibre layers in the basal zone of *Ascaris* and of *Parascaris* (see Chitwood and Chitwood, 1950; Bird, 1957, 1958; Hinz, 1963). The crossing fibres form a trellis-like arrangement and allow a 10 to 15% elongation of adult *Ascaris* during locomotion (Harris and Crofton, 1957; see also Chapter 13). Each of the fibre layers in this nematode consists of parallel fibres which run in a spiral at about 75° to the longitudinal axis and with the middle of the three layers crossing the outer and inner layers at an angle of 135°. This allows for anisometric expansion and contraction whilst retaining the shape of the nematode as very high turgor pressures are generated when the nematode moves. The fibres are also flexible and allow coiling, such as occurs during three-dimensional locomotion in adult *Nippostrongylus brasiliensis* (Lee and Biggs, 1990). The success of the hydrostatic skeleton in larger nematodes depends on these fibres in the cuticle (Harris and Crofton, 1957). A similar lattice system of fibres has been found in most of the larger animal-parasitic nematodes and also in the basal zone of many free-living nematodes such as *Gonionchus australis* (Nicholas and Stewart, 1997). Several free-living species have a complex system of fibres in a thick basal zone. For example, the basal zone of the marine nematodes *Catanema*, *Robbea* and *Leptonemella* contains distinct layers of fibres, some of which alternate between a longitudinal arrangement and a helical arrangement, with the spiral fibres alternating at angles from 60° to 70° to the right and to the left of the body axis (Figures 7.6 a, b, c). The number of fibre layers in these genera differs between species and also between different body regions in the same species (Urbancik, Bauer-Nebelsick and Ott, 1996). Adult female *Mermis nigrescens* also possess a thick basal zone consisting of a branched network of fibres embedded in a slightly granular matrix (Figure 7.8a) (Lee, 1970). This thick network of fibres in the basal zone of *Mermis*, together with the two spirals of giant fibres in the median zone, may reflect a need for greater resistance to diametrical expansion during the spiraling locomotion of this nematode because of its great length to width ratio. The basal zone of adult female *Trichodorus* and *Paratrichodorus* species is in the form of three distinct multilaminate units or layers, presumably of a finely fibrous nature, but distinct fibres have not been observed in the basal zone (Mounport, Baujard and Martiny, 1997). When fibres of high tensile strength and elasticity are oriented in a matrix of weak tensile strength and elasticity, as in the fibre layers in nematode cuticle, then the strength of the two-phase structure is increased (Weis-Fogh, 1970; Lee, 1977).

While the cuticles of the larval stages and the male of *Globodera rostochiensis* have the basic, three-zoned plan, the cuticle of the gravid female differs significantly from this arrangement. The cuticle of the swollen, gravid, female becomes a protective cyst wall for the enclosed eggs after the female dies. The three basic zones are supplemented by two extra, fibrous layers in the gravid female, with the fibres in the fifth layer being arranged helicoidally, as in the chitin of insect cuticle, but they consist of collagen not chitin (Shepherd, Clark and Dart, 1972). A helicoidal arrangement of the fibres gives an isotropic structure, but once formed it can no longer expand (Weis-Fogh, 1970). The fibres in the innermost layer of the gravid female cuticle of *G. rostochiensis* are like this, and such an isotropic layer apparently favours a spherical structure in response to increased internal pressure as the female swells (Shepherd, Clark and Dart, 1972), unlike the system in *Ascaris* in which increased internal pressure results in lengthening of the nematode (Harris and Crofton, 1957).

Anchorage of the Cuticle

The body musculature of nematodes consists of longitudinal cells that lie beneath the epidermis; they are separated into four rows by the lateral, dorsal and ventral epidermal cords (Bird, 1984b; Bird and Bird, 1991). A basement membrane is interposed between the muscle cells and the epidermis. In the muscle cells the I-bands contain rows of periodically spaced dense bodies and these serve as attachments for thin filament anchorage. These dense bodies are continuous with the sarcolemma and are believed to be anchored by means of transmembrane components to the basement membrane that lies between the muscle cells and the epidermis (Francis and Waterston, 1991). The epidermis between the muscles and the cuticle contains bundles of tonofilament that are inserted into dense plaques, resembling hemi-desmosomes, on the inner and outer epidermal membranes. The hemi-desmosomes on the outer epidermal membrane are particularly dense and filamentous material extends from them into the basal zone of the cuticle to form attachment sites between the cuticle and the epidermis (Rosenbluth, 1967; Bartnik, Osborn and Weber, 1986; Waterston, 1988; Francis and Waterston, 1991). In *C. elegans* there are 70 of these attachment sites directly beneath each annulus but there are none beneath the striae (Francis and Waterston, 1991). The location of these fibrous bundles beneath the muscles strongly implicates them in the mechanical coupling of the muscles to the cuticle (Francis and Waterston, 1991). It is perhaps significant that the attachments between the cuticle and the epidermis are quickly reformed after detachment during moulting in *Nippostrongylus* as this enables the nematode to recommence locomotory activity and thus retain its position in the intestine of the host (Lee, 1970).

Composition of the Cortical, Median and Basal Zones

Collagens are extracellular structural proteins that have a characteristic triple-helical, rod-like structure formed by the association of three polypeptide chains. The helical portion of each protein has glycine every third residue, to give a characteristic Gly-X-Y sequence. The major proteins in the cuticle of nematodes are collagen-like proteins, found mainly in the median and basal zones and to a lesser extent in the cortical zone. Based on work on *C. elegans* it has been found that all of the collagen chains in nematode cuticle have similar domain structures and several conserved motifs. There is a long globular domain at the amino terminus that is variable in length, a central Gly-X-Y repeat domain, and a carboxyl globular domain that is variable in length (Kramer, 1997). The cuticle collagens are divided into subfamilies based on the spacing of cysteine residues that flank the Gly-X-Y domain, and collagens in the same subfamilies have more sequence similarity to each other than to collagen in other subfamilies (Kramer, 1991). Collagens of the other species of nematode that have been examined have the same conserved domain structures and sequence motifs that are found in *C. elegans* which suggests that cuticle collagen structure has been strongly conserved in the phylum Nematoda (Kramer, 1991). Another important protein component of the nematode cuticle is cuticlin, which comprises a number of different proteins and is found in the cortical zone and the epicuticle (see below).

Collagen proteins in nematode cuticle were first studied in *Ascaris* by Josse and Harrington (1964), McBride and Harrington (1967) and Fuchs and Harrington (1970). Further work on cuticle collagens of *Ascaris* has been described by, amongst others, Fujimoto (1968); Fujimoto and Kanaya (1973), Fetterer and Urban (1988), Betschart, Marti and Glaser (1990), Kingston, Wainwright and Cooper, (1989), Kingston (1991), Betschart and Wyss (1990), and Timinouni and Bazzicalupo (1997). Biochemical and molecular studies of cuticle collagens has been reported for several other nematodes, including *Haemonchus* (Shamansky *et al.*, 1989; Fetterer, 1989; Cox, 1990, 1992; Cox, Shamansky and Boisvenue, 1990), *Teladorsagia* (syn. *Ostertagia*) (Johnstone *et al.*, 1996), *Brugia* (Selkirk *et al.*, 1989; Devaney, 1991; Lewis *et al.*, 1999), *Meloidogyne* (Reddigari *et al.*, 1986; Van der Eycken *et al.*, 1994; Ray and Hussey, 1995; Ray, Wang and Hussey, 1996; Koltai *et al.*, 1997; Wang, Deom and Hussey, 1998), and especially *C. elegans* (see Cox, 1992; Johnstone, 1994, 2000; Kramer, 1998). The proteins of nematode cuticular collagens, when analyzed by SDS-PAGE, have molecular weights ranging from 30 kDa to greater than 200 kDa, with the predominant molecular weights being between 60 and 120 kDa (Cox, 1992). However, molecular studies have shown that cuticle collagen genes of those nematodes that have been studied encode approximately 30 kDa proteins which indicates that the higher molecular weights obtained by SDS-PAGE are from cross-linked aggregates of 30 kDa proteins (Cox, 1992). It would appear that the 60 to 80 kDa molecular weights represent two cross-linked polypeptide chains and the 80 to 120 kDa molecular weights represent three cross-linked collagen chains. The collagens are stabilized by interchain disulphide bonds (Fuchs and Harrington, 1970; Cox, 1992) and by tyrosine cross-links. It is thought that cuticle collagen

triple-helices assemble with one another as linear polymers and that disulphide bonds covalently cross-link monomers to one another to form a fibril and also cross-link fibrils to other fibrils within the cuticle (Cox, 1992).

Cuticle collagens are also covalently joined to one another through tyrosine-tyrosine cross-links with isotrityrosine being the predominant tyrosine cross-link in *Ascaris* and *Haemonchus* cuticle collagens (Fujimoto, Horiuchi and Hirama, 1981; Fetterer and Rhoades, 1990). A peroxidase enzyme has been implicated in the synthesis of these cross-links (Cox, 1992; Fetterer and Rhoades, 1993).

The collagen components of *C. elegans* cuticle are encoded by a multi-gene family of approximately 154 predicted cuticular collagen genes (*http://www.worms.gla.ac.uk/collagen/cecolgenes.htm*) (Johnstone, 2000). A similar sized collagen gene family has been described for *Teladorsagia* (= *Ostertagia*) *circumcincta* (Johnstone *et al.*, 1996) and possibly for *Haemonchus contortus* (Shamansky *et al.*, 1989) and *Ascaris suum* (Kingston, 1991).

The sequenced cuticle collagen genes of nematodes were grouped by Cox (1992) into 4 families, referred to as the *col-1*, *col-6*, *col-8*, and *sqt-1* families. However, the availability of the *C. elegans* genome sequence has permitted a more complete comparison of collagen-like genes (see Johnstone in *http://www.worms.gla.ac.uk/collagen/cecolgenes.htm* and Johnston (2000). The vast majority of collagen genes have been placed by Johnstone into 3 main groups and 1 sub-group, called Groups 1, 1a, 2 and 3, and into *dpy-7* and *dpy-2* groups. Group 1 and Group 1a genes are predicted to encode collagens with 1 amino acid residue between the carboxy cysteine pair. Group 2 genes have 2 amino acid residues between their carboxy cysteine pair. Group 3 genes have 3 amino acid residues between their carboxy cysteine pair. Group 1a genes are distinct from Group 1 as they have the unusual 3 cysteine residues, instead of the usual 2, in the central region between two main blocks of Gly-X-Y, but they are more closely related to Group 1 genes than to any of the other groups (Johnstone, 2000 and personal communication). In this chapter the given collagen gene names have been used. Their GENEFINDER prediction names, together with those of genes that have not been given proper gene names, can be found in *http://www.worms.gla.ac.uk/collagen/cecolgenes.htm*. Genes that have been given proper gene names are now placed in the following groups. Group1 contains *bli-1*, *col-7*, *col-8*, *col-19*, *col-35*, *col-37*, *col-38*, *col-39*, *col-41*, *sqt-1*, and *rol-6*. Group 1a contains *col-10*. Group 2 contains *bli-2*, *col-12*, *col-13*, *col-14*, *col-17*, *col-36*, *col-40*, and *rol-8 (col-6)*. Group 3 contains *col-2*, *col-3*, *col-34*, *dpy-13*, and *sqt-3*. Group *dpy-7* contains *dpy-7*. Group *dpy-2* contains *dpy-2* and *dpy-2*. The groups also contain collagen genes that have GENEFINDER prediction names but so far no other names.

Approximately 45 loci affect body shape in *C. elegans* and 17 of these have been assigned to known gene sequences with 9 of these (*bli-1*, *bli-2*, *dpy-2*, *dpy-7*, *dpy-10*, *dpy-13*, *rol-6*, *sqt-1 sqt-3*) being cuticular collagen genes. Genes encoding enzymes thought to be involved in the processing of collagen and genes involved in X-chromosome dosage compensation are among the remainder (Johnstone, 2000).

Collagen mRNA is most abundant in the epidermal cells that secrete the cuticle of *C. elegans* (Edwards and Wood, 1983; Johnstone, 2000). According to Cox and Hirsch (1985) collagen mRNA is temporarily most abundant during moulting in *C. elegans*. However, a detailed analysis of the temporal pattern of expression of a group of individual genes has generated possible contradictory data. Cuticular collagen genes are subject to spatial and temporal modes of regulation and the numbers of collagen genes that are expressed at different times are not more than at other times (Johnstone and Barry, 1996; Johnstone personal communication). Study of the temporal expression of a group of 25 genes that encode distinct cuticular collagens in *C. elegans* has found that the genes are expressed in a temporal pattern that reflects the cuticle-synthesis events (Johnstone and Barry, 1966; Johnstone, personal communication). The periods of mRNA abundance for each of the genes do not coincide with different genes being expressed at different times relative to one another within the moulting cycle. For example, *dpy-7* mRNA has a peak of abundance about 4 hours before each moult, whereas *col-12* mRNA has a peak of abundance 4 hours later at each moult. Multiple discrete temporal waves of larval collagen gene expression occur in *C. elegans* with the same ordered pattern being reiterated during each larval stage, and hence the moulting cycle (Johnstone and Barry, 1996). This ordered expression of cuticular genes has implications for the process of assembly of the cuticle. Translation can only occur when mRNA is present, thus it is probable that the observed pattern results in several waves of translation of different collagens. This could result in the restricted trimerisation of compatible collagen monomers, which could provide a mechanism for co-ordinating a critical step in cuticle formation (Johnstone, 1994). Tissue-specific expression of *dpy-7* in *C. elegans* shows that DPY-7 protein is present within the epidermal cells preceding secretion of the cuticle, then after secretion it localises to the annular furrows (striae) of the cuticle (L. McMahon and I.L. Johnstone in Johnstone, 2000).

Expression of individual cuticle collagen genes has been examined in *Ascaris suum* and found to be differentially regulated during development. The *COLA4* gene is expressed only by moulting larvae whereas the *UCOL1* gene is expressed by larvae moulting from L3 to L4 and by the adult nematode (Kingston, Wainwright and Cooper, 1989; Kingston and Pettitt, 1990). These two collagen genes of *Ascaris* probably belong in Group2. The collagen gene family of *Haemonchus contortus* is similar in size to that of *C. elegans* and the genes are similar in structure (Shamansky *et al.*, 1989). The five collagen genes of *Haemonchus* that were analyzed showed a striking similarity to members of the *C. elegans col-1* collagen subfamily (now Group 3 of Johnstone, 2000 with *col-1* =

sqt-3, Johnstone personal communication). A pair of tandemly duplicated collagen genes from *Teladorsagia* (syn. *Ostertagia*) *circumcincta*, *colost-1* and *colost-2*, appear to be the direct homologues of *col-12* and *col-13* in *C. elegans*. The interspecies comparison of these homologues indicates regions of extreme conservation and it is concluded that the gene duplication event that resulted in the creation of *col-12* and *col-13* in *C. elegans* is probably the same duplication event that generated *colost-1* and *colost-2* in *T. circumcincta*, and that this event preceded the divergence of the two species (Johnston *et al.*, 1996). Three cuticle collagen genes (*mi-col-1*, *mi-col-2* and *lemmi 5*) have been identified in *Meloidogyne incognita* (Ray and Hussey, 1995; Ray, Wang and Hussey, 1996; Van der Eycken *et al.*, 1994; Reddigari *et al.*, 1986; Wang, Deom and Hussey, 1998). None of these genes was expressed in preparasitic L2, but they were expressed in the L2 three to four days after penetration of the host's root and increasingly in the parasitic L3, L4 and adult female. Upregulation of these three collagen genes correlates with the remodelling of the L2 cuticle, which changes from a complex three-zoned to a more simple single-zoned structure following the onset of parasitism and loss of motility by the nematode in the host plant (Wang, Deom and Hussey, 1998). A polyclonal antiserum raised against a major 76 kDa collagen protein extracted from cuticles of adult female *M. incognita* was used to localise this protein collagen in the cuticles of different developmental stages. In the preparasitic L2 and adult males, which are motile and have three-zoned cuticles, the 76 kDa protein collagen was detected only in the cortical zone of the cuticle, whereas in the parasitic L2 and in the adult female the collagen was found to be uniformly distributed throughout the internal structure of the cuticle (Ray *et al.*, 1996). The collagen gene *mjcol-3* from the related plant-parasitic nematode *M. javanica* encodes a 32.4 kDa collagen protein MJCOL-3 that has a protein sequence very similar to DPY-7 of *C. elegans* (Koltai *et al.*, 1997) and probably belongs in the *dpy-7* group. This gene is also developmentally regulated with transcripts being found mainly in preparasitic developing eggs, with fewer in parasitic L3 and L4 and young females shortly after the final moult, and even fewer in females before egg-laying. It is suggested that MJCOL-3 protein may be needed to construct larval cuticle and probably does not have a major role in the synthesis of the adult female cuticle (Koltai *et al.*, 1997). Some of the genes, such as *col-12*, that are expressed in an oscillating pattern during larval development in *C. elegans* are expressed in the adult after the final moult whereas some, such as *dpy-7*, are not (Johnstone, 2000). As adult *C. elegans* and adult *Ascaris suum*, in common with many other species, especially animal-parasitic species, grow significantly in size after the final moult, it is possible that a limited set of collagen genes is expressed during the manufacture and laying down of more cuticular material in these growing adults (Johnstone, 2000).

Members of the known cuticle collagen families are thought to be fundamental, conserved aspects of nematode cuticle collagens and to be widely distributed in nematodes. For example, an antiserum that was raised against a sequence of a peptide which corresponded to a domain of the 3A3 protein of *H. contortus* which is highly conserved and diagnostic for members of the *col-1* family (now Group 3), has been shown to react with collagen proteins in eight other nematode species. These were the animal-parasitic *Ostertagia ostertagi* (L3), *Toxocara canis* (adult), *Dirofilaria immitis* (adult) and *Trichinella spiralis* (L1); free-living *C. elegans* and *Panagrellus redivivus* (mixed-stages); and entomopathogenic *Heterorhabditis bacteriophora* and *Steinernema* (syn. *Neoaplectana) carpocapsae* (dauer larvae) (Cox, Shamansky and Boisvenue, 1990; Cox, 1992).

Assembly of cuticular collagen into the multi-protein multi-layered cuticle probably begins with the formation of links between trimers with the collagen triple helix mediated by disulphide bridges and tyrosine-tyrosine bonds (see above). Collagens of the same type form triple-helices with each other (Cox, 1992). The trimers then assemble into fibrils and the fibrils become incorporated into the zones of the cuticle (Blaxter and Robertson, 1998). A cystatin-like cysteine protease inhibitor, called onchocystatin, is present in the epidermis and basal zone of the cuticle of adult female *Onchocerca volvulus* and is also expressed in the cuticles of L3 and L4 during moulting, especially around the sites of cuticle separation (Lustigman *et al.*, 1992; Lustigman, 1993). A possible role for the onchocystatin is the regulation of cysteine protease activity during moulting and loss of the cuticle (Lustigman, 1993). The median zone of the cuticle of adult *Ascaris* appears to contain a non-specific esterase (Lee, 1961), the function of which is not known but it may have a role in maintaining this gel-like layer of the cuticle.

The insoluble residue left after extraction of nematode cuticle by a mixture of strong ionic detergents and disulphide-reducing agents and which is resistant to collagenase, is called cuticulin. Cuticlin, which was first studied in *Ascaris* by Fujomoto and Kanaya (1973), is composed of a number of different proteins that are highly covalent cross-linked and have an amino acid composition and a structure, as revealed by x-ray diffraction, different from collagens. It is mainly present in the epicuticle and the cortical zone of the cuticle.

Genes encoding two cuticlin-like proteins were first isolated from *C. elegans* and were named *cut-1* and *cut-2* (Gigliotti *et al.*, 1988; Sebastiano, Lassandro and Bazzicalupo, 1991; Lassandro *et al.*, 1994). A secreted protein (CUT-1) of 423 amino acids is encoded by *cut-1* and is found in a fibrous ribbon underneath the lateral alae of the dauer larva (Sebastiano, Lassandro and Bazzicalupo, 1991; Ristoratore *et al.*, 1994). It is thought that the product encoded by the *cut-1* gene, which is only expressed in the dauer larvae, contributes to the resistance and durability of the cuticle of this stage. CUT-2 which consists of 231-amino acids is encoded by *cut-2* and contributes to the cuticles of all larval stages and is located in the cortical zone (Lassandro *et al.*, 1994; Ristoratore *et al.*, 1994; Favre *et al.*, 1995). Recombinant CUT-2 can be cross-linked *in vitro* by horse radish

peroxidase, in the presence of hydrogen peroxide, resulting in the formation of intermolecular dityrosine bridges. However, the enzymes involved in cross-linking *in vivo* are not known (Parise and Bazzicalupo, 1997). There is a suspicion that a transglutaminase-catalysed reaction may have an important role in cuticle formation. Transglutaminases are enzymes that stabilise protein structure by catalysing the formation of isopeptide bonds whose stability makes them resistant to proteolysis. This could be important in stabilising nematode cuticle and rendering it resistant to attack by host enzymes. The enzymes are present in *Brugia*, *Acanthocheilonema*, *Strongylus vulgaris*, *S. edentatus*, *Parascaris equorum* and in *Cylicocyclus insigne* (Rao *et al.*, 1991; Mehta *et al.*, 1992; Lustigman, 1993; Rao *et al.*, 1999). Timinouni and Bazzicalupo (1997) referred to CUT-1 and CUT-2 in *C. elegans* as CECUT-1 and CECUT-2.

The entomopathogenic nematode *Heterorhabditis* shows strong reactivity with anti-sera raised against CECUT-1, CECUT-2 and against the whole cuticlin residue of *C. elegans* and localisation was similar to that in *C. elegans* (Favre *et al.*, 1995). Portions of the *cut-1* and *cut-2* genes from *C. elegans* have been used to probe the genome of the plant-parasitic nematode *Meloidogyne artiella* and both genes were found to be present as single-copy genes (De Giorgi, De Luca and Lamberti, 1996; De Giorgi *et al.*, 1997). Studies on the expression pattern during development have shown that *Ma-cut-1* is transcribed at high levels in the egg and at the L1 to L2 moult but the expression rate is reduced in infective larvae which migrate in the soil; it is not expressed in the sedentary female, yet is expressed and the transcript fully processed in the motile male, all of which suggest that this gene is developmentally regulated (De Giorgi *et al.*, 1997). *cut-1*-like genes have been identified in adult *Ascaris* and one has been called *ascut-1*. This gene codes for a 385 amino acid protein called ASCUT-1 and is strongly conserved when compared to the gene for CECUT-1. As few mRNA molecules were detected it was suggested by Timinoumi and Bazzicalupo (1997) that *ascut-1*, like *cut-1* in *C. elegans*, is not functionally expressed in adults. This suggestion is strengthened by the finding that antisera against recombinant CECUT-1 protein showed similar localisation in *Ascaris* larvae within fully developed eggs, and in dauer larvae of *Heterorhabditis* and *C. elegans* (Favre *et al.*, 1998). A *cut-1*-like gene (designated *Bp-cut-1*) has been identified in the filarial nematodes *Brugia pahangi* and *B. malayi* (Lewis *et al.*, 1999). Analysis of the steady state transcript pattern throughout the mammalian stages of the life cycle has demonstrated that there are peaks of abundance prior to each moult but no signal in the adult nematode. This agrees with results from other nematodes as the transcription pattern of *cut-1* mRNAs in these other nematodes demonstrates that these homologues are transcribed to coincide with cuticle synthesis, especially at a moult (Lewis *et al.*, 1999).

In L3 and adults (both sexes) of *B. pahangi* immunolocalisation of CUT-1 epitopes showed that they were restricted to a tight band in the median zone of the

the 30-day L3 within the egg only remnants of the L1 cuticle remain, suggesting that some hydrolysis of the cuticle may have occurred (Brunaská, Dubinský and Reiterová, 1995). Cuticle formation in the L2, L3 , L4 and adult male and female of *Panagrellus silusiae* is similar for each of the moults (Samoiloff and Pasternak, 1969). Collagen synthesis is apparently discontinuous with a peak of collagen production preceding each moult (Leushner and Pasternak, 1975; Pasternak and Leushner, 1975). The inception of a moult in *Panagrellus* is marked by the appearance of fibrous material between the epidermal membrane and the old cuticle. A row of globules appears at the margin of the old and new cuticle with a fine filamentous material lying between the globules and the epidermis. The new cuticle gradually enlarges and eventually there are two structurally similar cuticles apposed to each other. The new cuticle has an annule periodicity similar to that of the old cuticle, except at the final moult into the female when the new cuticle of the female undergoes extensive folding. There is no selective destruction of the old cuticle and the old cuticle breaks off in pieces in all moults except for the final moult to the adult male when the old cuticle splits and is shed as a single piece (Samoiloff and Pasternak, 1969). In *Aporcellaimellus* a new stylet is formed at the same time as the new cuticle of the body wall. During formation of the new cuticle projections of the epidermis extend into the newly forming cuticle then later retract leaving a lacunar system in the basal zone. Differentiation of the new cuticle into distinct zones occurs once secretion of the new cuticle is completed; there is no dissolution of the old cuticle (Grootaert and Lippens, 1974). The cuticle of all parasitic stages (L2, L3, L4 and females) of *Hemicyliophora arenaria* consists of an outer seven-layered cuticular sheath and a more normal five-layered cuticle and both are formed when the nematode moults and therefore should be regarded as two parts of the same cuticle. The male has a modified four-layered sheath and a six-layered main cuticle (Johnson, Van Gundy and Thompson, 1970a). The onset of moulting in all stages of this nematode is preceded by the appearance of globular 'moulting bodies' in the epidermal cords. These bodies appear just prior to moulting and disappear shortly after completion of the moult. During moulting of the L4 female the sheath is formed first and this is then followed by formation of the new inner cuticle. The old sheath and old inner cuticle then break down and may be absorbed by the nematode (Johnson, Van Gundy and Thomson, 1970b). Moulting of *Aphelenchus avenae* and *Hirschmaniella gracilis* is a much simpler process than in *Hemicyliophora* (Johnson, Van Gundy and Thomson, 1970b) and is more like that of *C. elegans*. When new cuticle is being deposited in moulting *C. elegans*, Golgi bodies become prominent in the epidermis, particularly in the seam cells of the lateral cords. After formation of the new cuticle the old cuticle first becomes loosened from the tip of the head and seals the mouth, followed by loosening of the cuticle in the buccal cavity and around the tail. The remainder of the old cuticle also becomes

loosened but the gap between the old and new cuticles is only a fraction of a micrometre. The pharyngeal glands become active and pass secretions that probably soften and loosen the cuticular lining of the pharynx and the cuticle around the head. The pharynx then begins to contract spasmodically, the cuticular lining breaks and the posterior piece passes back into the intestine. The old cuticle of the body wall becomes inflated around the tip of the head and the nematode pulls back from it repeatedly until the remainder of the cuticular lining of the pharynx is detached and then expelled through the mouth. The nematode then escapes from the old cuticle by rolling within it and pushing against the softened old cuticle with its head until it makes a hole or a cap breaks away (Singh and Sulston, 1978). It was initially thought that *C. elegans* did not degrade the old cuticle. However, it has been shown that a low-density lipoprotein receptor LRP-1, which is a gp330/megalin-related protein, is essential for growth and development in this nematode. Megalin is thought to be a clearance receptor that maintains lipid homeostasis and regulates the activity of extracellular proteases. It appears that the *lrp-1* gene functions at the apical surface of the epidermal syncytium hyp7 in *C. elegans* and that LRP-1 is a receptor for sterols and is expressed at the surface of the epidermis. Two recessive alleles, *ku156* and *ku157* isolated in the *lrp-1* gene confer an inability to shed and degrade the old cuticle at each of the larval moults. The nematodes are able to initiate the process of moulting and synthesize a new cuticle but are unable to complete the moult. The extracellular part of LRP-1 may be required for activation of collagenase or other proteases that might be secreted during moults and partially degrade the old cuticle. Alternatively, LRP-1 may be required for proteolytic processing of procollagens or other components of the cuticle and insufficient maturation of the precursors may produce a cuticle that is resistant to degradation (Yochem *et al.*, 1999).

The cuticles of the L4 and adults of *Nippostrongylus brasiliensis* and of *Heligmosomoides polygyrus* are much more complex than the cuticle of *C. elegans* (Lee, 1965, 1970b, 1977; Lee and Nicholls, 1983; Bonner, Menefee and Etges, 1970; Bonner and Weinstein, 1972b). In these nematodes at the commencement of the moulting process the epidermis begins to retract from the old cuticle leaving a space which contains granular/filamentous material. The last areas of contact between the old cuticle and the epidermal membrane are the hemi-desmosomes that attached the cuticle to the epidermis. A finely filamentous coat then forms on the external surface of the epidermal membrane of moulting L3 and L4 of *Nippostrongylus* and develops into the new epicuticle. In moulting L4 the outer surface of the epidermis begins to be thrown into folds (plicae) that encircle the nematode and these become progressively more marked as they act as a template for the annules of the new cuticle in moulting L3 and L4 of *Nippostrongylus* (Lee, 1970b; Bonner and Weinstein, 1972b). In *Heligmosomoides* the folds appear to be narrower and form plicae (Bonner, Menefee and Etges,

1970). Small vesicles pass from the endoplasmic reticulum and coalesce with the epidermal membrane to discharge their contents into the developing cuticle in *Nippostrongylus* and at this time the basal fibre zone begins to develop (Lee, 1970b). In *Heligmosomoides* a number of large vesicles in the epidermis discharge their flocculent contents to form the fibrous layers of the new cuticle (Bonner, Menefee and Etges, 1970. As the new cuticle increases in surface area so it becomes more extensively folded, each fold forming an annule of the adult cuticle. In *Nippostrongylus* an electron dense M-shaped structure is deposited at the peak of the epidermal folds and the folds then retract leaving the M-shaped structure encircling each annule between the developing fibres of the basal zone and the epicuticle and developing cortical zone. At this stage in *Nippostrongylus* there is a large increase in the amount of rough endoplasmic reticulum in the epidermis and there are many more vesicles apparently releasing their granular/flocculent contents into the cuticle. Golgi complexes are not as prominent in the epidermis of the moulting L4 as in the moulting L3 of *Nippostrongylus* within the lungs of the mammalian host (Bonner and Weinstein, 1972a). The two fibre layers of the basal zone become more distinct and there is a rapid increase in thickness of the cuticle between the M-shaped structures and the periphery of the cuticle with the thickness varying around the nematode. In the thicker parts the material of this outer zone begins to condense down to form the electron dense struts with one strut to each annule. Eventually the struts come to lie in the fluid-filled median zone and have their pointed distal end inserted into the cortical zone where they support the longitudinal ridges of the cuticle whilst their broad proximal end is supported by bundles of cross-striated fibres that anchor the base to the fibre layers of the basal zone (Lee, 1970b; Lee and Nicholls, 1983). Smaller struts are present in the L4 and adult cuticle of *Heligmosomoides* (Bonner, Menefee and Etges, 1970) and presumably are formed in a similar manner. In *Nippostrongylus* the old cuticle is cast intact with no obvious signs of dissolution of the cuticle.

Formation of new cuticle in moulting L3 and L4 *Brugia pahangi* results in the epidermis giving rise to the new cuticle without the formation of plicae or folds of the epidermis. Vesicles within the epidermis discharge their contents into the newly forming cuticle and the contents appear to give rise to the various zones of the cuticle. The new cuticle forms numerous closely packed annulations. There is no evidence that the old cuticle is degraded and resorbed (Howells and Blainey, 1983). A cysteine proteinase has been detected in the regions of the cuticle of *Onchocerca volvulus* where separation of the old and new cuticles occurs in moulting larvae. Cysteine proteinase inhibitors did not stop the formation of a new L4 cuticle but did prevent the separation of the L3 and L4 cuticles *in vitro*, which suggests that the enzyme plays a part in this separation (Lustigman *et al.*, 1996).

Moulting in the head region of plant-parasitic nematodes is complicated because at each moult the cuticle that lines the pharynx and the cuticular components of the feeding apparatus, such as the stylet, are replaced. In larval *Labronema vulvapapillatum* and larval *Xiphinema americanum* the functional odontostyle (hollow protrusible stylet) is shed during moulting and is replaced by another ready-made odontostyle that has been stored in the pharyngeal wall. During moulting the replacement odontostyle, having moved into position, is itself replaced by a new odontostyle that is then stored in the pharyngeal wall until the next moult (Coomans and De Coninck, 1963; Grootaert and Coomans, 1980; Carter and Wright, 1984). During the early stages of moulting in *Xiphinema* the cuticle is first shed near the head end of the larva, then later near the tail. The functional odontostyle of the previous larval stage, together with the surrounding lining of the stoma, the guiding sheath and the lining of the basal portion of the spear are pulled out. The replacement odontostyle then moves into position and becomes attached to the anterior end of the basal portion of the spear, thus forming the complete spear (Coomans and De Coninck, 1963). In *Labronema* the new odontostyle is formed by two cells in the wall of the pharynx, one located posterior to the other. A deep cylindrical invagination develops in the posterior cell while the anterior cell extends a tubular sheath around the invagination. Dense granules produced by Golgi bodies in the posterior cell are exocytosed into the extracellular invagination or pocket and here they coalesce in this mould to form the next odontostyle. The shape of the odontostyle is determined by the pocket. The pocket and the sheath deepen posteriorly as formation of the odontostyle, together with polymerisation and cross-linking of the collagen, proceeds. Once fully formed the new odontostyle moves to its storage site within the pharynx where it remains until the next moult (Grootaert and Coomans, 1980; Carter and Wright, 1984). Bird (1983, 1984a) followed moulting from individual hatched L2 to the adult male and adult female in *Rotylenchus reniformis*. Individuals undergo three moults over about a week without feeding. At the commencement of moulting the stylet becomes less distinct and is eventually replaced by the stylet of the next stage. The cuticles of the previous larval stage are not cast by the nematode so that three moulted cuticles surround the young adult males and females. The inner parts of the old cuticles are partly dissolved, and are possibly resorbed, but the outer zones are retained. Presumably the old cuticles are shed when the adults begin moving in the soil. A lethargus sets in prior to moulting of *Aphelenchoides hamatus* followed by disappearance of the shaft and knobs of the stylet and cuticle in the anterior pharynx, but the conus of the stylet remains attached to the ecdysing cuticle and is withdrawn from the head of the nematode. A new conus is then formed and shaft material for the new stylet is laid down, followed by the knobs at the base of the stylet. Movement of the head of the nematode unplugs the cuticular linings of the amphids. Activity of the pharynx is associated with thinning of the old cuticle and loss of its annulations. The nematode moves actively inside the old cuticle during ecdysis and eventually breaks free from it. There is a significant decrease in volume of the nematode during the late phase of moulting followed by expansion of the moulting nematode during ecdysis. This is thought to be due to water loss that is controlled by the nematode during the late phase of the moult, followed by re-entry of water during and after ecdysis (Wright and Perry, 1991).

Exsheathment

The L3 of animal-parasitic trichostrongyle nematodes is a non-feeding, ensheathed, somewhat resistant, obligate stage in the life cycle. The cuticular sheath that encloses the infective-stage L3 is the result of an incomplete moult. The larvae have entered a form of diapause and require a stimulus from a suitable host to continue development. The sheath is the apolysed but unecdysed cuticle of the L2, the oral, anal and secretory-excretory pore orifices of which are sealed. The sheath enables trichostrongyle L3 to survive on pastures for weeks or months and in some species these ensheathed larvae become anhydrobiotic, thus enabling them to survive drying conditions. Several species of free-living nematodes (*Diplogaster* spp.) and entomophagous nematodes (*Steinernema* and *Heterorhabditis*) also produce ensheathed L3 (Bovien, 1937; Campbell and Gaugler, 1991). The term exsheathment is used to describe ecdysis of these ensheathed larvae. Exsheathment of infective-stage larvae of trichostrongyle nematodes marks the transition from a free-living to a parasitic existence and usually requires a stimulus from the host to bring about exsheathment and to continue development. The stimulus (stimuli) from a suitable region of the alimentary tract of a suitable host results in release of an exsheathing fluid by ensheathed trichostrongyle L3 (Rogers, 1962; Rogers and Sommerville, 1968; Sommerville, 1957, 1982). The exsheathment stimuli for trichostrongyle L3 are high concentrations of dissolved carbon dioxide and/or undissociated carbonic acid at 37°C with optimum conditions of pH and redox potential (Rogers, 1962; Rogers and Sommerville, 1968). The stimulation of exsheathment occurs in the region of the alimentary tract anterior to the preferred location of the adult nematode. Thus, larvae of *Haemonchus contortus* are stimulated to begin exsheathment in the rumen where the levels of carbon dioxide/undissociated carbonic acid are high and the pH of the rumen contents is about 7 to 8. They complete exsheathment and become established in the abomasum, whereas larvae of *Trichostrongylus colubriformis* are stimulated to begin exsheathment in the abomasum, where there are rather lower levels of carbon dioxide/undissociated carbonic acid and the pH is about 4 and then they complete exsheathment in the anterior small intestine of cattle (Rogers, 1962). After the ensheathed larvae are stimulated to begin exsheathment the L3 releases an 'exsheathing fluid' containing enzymes into the space between the sheath and the L3 cuticle. The exsheathing fluid contains a collagenase and a lipase (Rogers and Brooks, 1976; Sommerville, 1982). Rogers

(1982) found that a major component of the exsheathing fluid was a proteolytic enzyme that attacked the isolated sheaths but not native collagen. A 44 kD zinc metalloproteinase has been shown to be responsible for digestion of the ring region in the cuticle of the sheath surrounding the L3 of *H. contortus* (Gamble, Purcell and Fetterer, 1989). This enzyme was shown to attack the 20th posterior annulus of the sheath immediately anterior to the anterior end of the lateral alae. The annulus became indented and the cuticle immediately anterior and posterior to the indentation became swollen, holes appeared in it then the enclosed L3 used its head to force off the cap (Gamble, Lichtenfels and Purcell, 1989). The 44 kD proteinase is able to hydrolyse several substrates, such as azocoll, casein, azocasein and azoalbumin, but not native collagen and is apparently the same enzyme that was isolated and characterised by Rogers (1982). Both the secretory-excretory cells and the pharyngeal gland cells have been implicated in the release of the exsheathing fluid of trichostrongyle ensheathed larvae (Sommerville, 1982). Glycoproteins appear around the head, just underneath the sheath, of *H. contortus* L3 just before and after stimulation with carbon dioxide at 38°C (Bird, 1990). They appear to originate from the amphids and/or from the mouth and may act as a lubricant between the cuticle of the L3 and its sheath and may also be the proteinase described by Gamble, Purcell and Fetterer (1989) or possibly a proteinase similar to the cysteine proteinase described by Lustigman *et al.*, (1996) in *Onchocerca*. A leucine aminopeptidase released by the secretory-excretory gland after neurosecretory cells have been stimulated results in ecdysis during the final moult of *Pseudoterranova* (syn. *Phocanema*) *decipiens* (Davey and Kan, 1968).

Exsheathment of free-living nematodes, such as *Diplogaster* that are transported under the elytra of dung beetles from old deposits of dung to new deposits, possibly requires a stimulus from the new environment. However, methods that are effective to bring about exsheathment of ensheathed trichostrongyle L3 were found to be ineffective with ensheathed *Steinernema* and *Heterorhabditis* (Campbell and Gaugler, 1991). Movement of these nematodes can trigger exsheathment of ensheathed L3 of *Steinernema*, which indicates that the cues that trigger exsheathment of this species are not associated with host recognition. Movement did not induce exsheathment of ensheathed *Heterorhabditis* and it is suspected that a more complex set of cues are required to bring about natural exsheathment of this species (Campbell and Gaugler, 1991).

The dauer larva of *C. elegans* is not ensheathed but is a resistant L3 that differs in structure and physiology from feeding L3 that go on to produce the L4 without entering diapause. It is not an obligate stage in the life cycle but is produced in response to a deteriorating food supply and overcrowding. A dauer-inducing pheromone serves as an indicator of population density, with higher concentrations of the pheromone enhancing formation of the dauer larva and inhibiting recovery from the dauer stage. These dauer larvae do not feed but are capable of movement and their behaviour favours dispersal (Cassada and Russell, 1975; Riddle, 1988; Riddle et al., 1997; Riddle and Albert, 1997). When they encounter a suitable food signal they resume development. The ratio of pheromone to food signal influences formation and recovery from the dauer stage (Riddle, 1988; Riddle et al., 1997; Riddle and Albert, 1997). The cuticle of the moulting L2 is initially retained as a loose sheath around the fully formed dauer larva, but this is lost as the dauer larva ages (or upon washing) (Cassada and Russell, 1975) and apparently does not require a stimulus or exsheathing fluid.

Hormones and Moulting

Phylogenetic studies on the Arthropoda, Onchyophora, Tardigrada and the Nematoda have suggested that moulting arose once in evolution and that these two phyla should be grouped in a novel clade, the Ecdysozoa (Aguinaldo et al., 1997). This, together with the superficial resemblance between moulting in insects and in nematodes, notwithstanding the difference in composition of their cuticles, has raised the possibility that the respective regulatory systems may have some common features. Moulting hormones in insects occur in immature stages, in which they control moulting, and also in adults and in developing eggs. Certain moulting hormones of insects or inhibitors of these hormones have been found to have comparable effects on nematodes. However, the concentrations used were often high in an attempt to enable them to cross the cuticle and the results must therefore be treated with some caution. In insects the major ecdysteroid is 20-hydroxyecdysone and generally occurs with ecdysone (Barker and Rees, 1990), which initiates cuticle formation. The receptor for 20-hydroxyecdysone is a heterodimer of two nuclear hormone receptors, EcR and ultraspiracle. Notwithstanding the fact that no members of the *EcR* or *ultraspiracle* nuclear receptor gene classes have been found in *C. elegans* (Sluder et al., 1999), and the apparent inability of *C. elegans* to biosynthesise ecdysteroids (Barker, Chitwood and Rees, 1990; Chitwood and Feldlaufer, 1990), there is evidence that moulting and exsheathment in nematodes is under neurosecretory and endocrine control.

Ecdysteroids have been detected in the eggs, larvae and adults of several nematodes (Franke and Käuser, 1989; Barker and Rees, 1990; Barker, Chitwood and Rees, 1990).For example, ecdysone and/or its active metabolite, 20-hydroxyecdysone, or substances similar to ecdysteroids, have been detected in *Trichinella spiralis* (Hitcho and Thorsen, 1971), *Ascaris suum* (Horn, Wilkie and Thomson, 1974; Cleator et al., 1987; Fleming, 1985, 1987; O'Hanlon, et al., 1991), *Aphelenchus avenae* (Dennis, 1977), *C. elegans* (Mercer, Munn and Rees, 1988; Chitwood and Feldlaufer, 1990), *Haemonchus contortus* (Fleming, 1993), *Panagrellus redivivus* (Dennis, 1977), *Nippostrongylus brasiliensis* (Nembo et al., 1993), and in *Dirofilaria immitis* and *Brugia pahangi* (Cleator et al., 1987; Mendis et al., 1983; Mercer et al., 1989;

Barker, Chitwood and Rees, 1990). However, there is the possibility that ecdysteroids isolated from nematodes originated from the diet of the nematodes or from their hosts (Chitwood and Feldlaufer, 1990). *C. elegans* was found to contain the immunoreactive equivalent of 460 pg ecdysone per gram dry weight but the culture medium contained the immunoreactive equivalent of 68 times the quantity within the nematodes (Chitwood and Feldlaufer, 1990).

Application of ecdysteroids to different stages of various species of nematode has been found to have an effect on moulting and growth. Farnesol and farnesol methyl ether (FME) were found to delay or inhibit moulting of *Trichinella spiralis in vitro* (Meerovitch, 1965; Shanta and Meerovitch, 1970). Exposure of larval *Heterodera staccato* to high concentrations of FME and of farnesyl diethylamine resulted in supernumerary moults and failure to ecdyse (Johnson and Viglierchio, 1970). Similarly, exposure of *C. elegans* to insect juvenile hormone and mimics caused inhibition of maturation and reproduction in this nematode (Dropkin, Lower and Acedo, 1971). Synthetic juvenile hormone was found to inhibit the final moult of *Heligmosomoides polygyrus* (syn. *Nematospiroides dubius*) whereas α-ecdysone was found to stimulate the moult (Dennis, 1976). Moulting of the L1 through to the L4 stages of *C. elegans* is stimulated by 20-hydroxyecdysone at levels between 10^{-4} M and 10^{-8} M (Schmid, Franke and Koolman, 1987). Application of 10^{-5} M 20-hydroxyecdysone resulted in premature moulting of the L3 of *Dirofilaria immitis* as did the ecdysteroid agonist RH5849 but ecdysone had no effect on timing of the moult (Warbrick et al., 1993). Azadirachtin is thought to interfere with the neuroendocrine control of moulting synthesis in insects and prevented moulting of most of the larvae to the L4 stage when applied exogenously at a concentration of 10^{-5} M to L3 of *Dirofilaria immitis* (Warbrick et al., 1993).

Ecdysone and 20-hydroxyecdysone have been isolated from the reproductive tract of female *Ascaris suum* with ecdysone constituting more than 95% of the total ecdysteroids (Fleming, 1985). During cultivation *in vitro* of the L3 it was found that the levels of 20-hydroxyecdysone increased in the culture medium during ecdysis from the L3 to the L4 stage and again, but at a much lower level, when the larvae were undergoing the fourth ecdysis. Conversely, the concentration of ecdysone declined over this period. Addition of synthetic 20-hydroxyecdysone and of the 20-hydroxyecdysone fraction isolated from the female reproductive tract of *A. suum* significantly increased the numbers of L3 that moulted to L4 (Fleming, 1985). When 20-hydroxyecdysone was applied to L3 on day one in culture, after removal from lungs of rabbits 7 days after oral inoculation with embryonated eggs, and for only 24 hours, the larvae moulted earlier and grew longer. When 20-hydroxyecdysone applied later in the culture period and for longer there was no growth enhancement (Fleming, 1985). This suggests that there are time- and dose-specific effects and an endogenous role for 20-hydroxyecdysone in moulting and development in nematodes (Fleming, 1998).

Moulting of nematodes appears to be under neuro-secretory control as the final moult *in vitro* of *Pseudoterranova* (syn. *Phocanema*) *decipiens* is accompanied by a cycle of neurosecretion in some of the cells in the ventral ganglion (Davey, 1966; 1971; 1976). The cycle of neurosecretion did not control cuticle formation but did control ecdysis (Davey and Kan, 1968; Davey 1976). A synthetic preparation of insect juvenile hormone and FME had no effect on formation of the adult cuticle of *Pseudoterranova* (syn. *Phocanema*) but inhibited ecdysis and it was found that inhibition of ecdysis is linked to stimulation of the neurosecretory system of the nematode (Davey, 1971; 1976; Davey and Kan, 1968). These authors suggested that a stimulus from the final host (seal) of the nematode brings about release of an ecdysal hormone, possibly associated with neurosecretory cells in the ganglia associated with the nerve ring. This hormone then acts upon the secretory-excretory gland bringing about activation and release of leucine amino peptidase and other enzymes that are then released into the space between the old and new cuticles and these enzymes contribute to ecdysis of the old cuticle (Davey, 1976).

Proof that a circulating hormone brings about moulting in insects was initially done by ligating lepidopteran larvae and finding that a hormone secreted by the brain brought about pupation of the anterior half but not the posterior half of the larva (see Wigglesworth, 1972). Ligation of L4 of *Pseudoterranova* (syn. *Phocanema*) *decipiens* at the time of the final moult resulted in both portions producing normal cuticle, which suggested that a chemical signal produced locally in one part of the nematode is not necessary for cuticle synthesis in other regions of the nematode (Davey, 1966, 1976). However, Davey did suggest that some hormone-like influence might have been produced very much earlier. Ligatures have been used on the nematode *Aphelenchus avenae* to partition apolysis and new cuticle formation and to find the site of production and/or release of the hormones involved and the timing of their influence on moulting (Davies and Fisher, 1994). Ligatures applied to larvae at particular developmental times and positions limited the moult to the anterior or posterior position of the nematode. The ability of ligatures to partition the moult was related to the moulting cycle. Nematodes that had been ligated before receipt of the stimulus did not moult; apolysis was prevented in nematodes ligated up to 4 hours after receipt of the stimulus and was delayed when the nematodes were ligated after 4 hours. Moulting was limited by ligating one portion of the body if the ligature was applied from 4 hours after stimulation until the onset of lethargus. Nematodes ligated in the anterior third of the body moulted in the posterior two thirds; if the ligature was placed in the posterior half of the body the anterior portion moulted. This suggests that the site of production and/or release of hormone is in the anterior 40% of the nematode and that the ligatures limited movement of the moulting hormone. After the onset of lethargus the nematodes moulted on both sides of the ligature. The inability to partition moulting once the nematode has entered a period of lethargus indicates that there is a pulse of hormone release immediately prior to the onset of lethargus and that the hormone was influencing the whole of the nematode once lethargus had been initiated. High concentrations of externally applied 20-hydroxyecdysone stimulated the non-moulted portion of ligated nematodes to moult but lower concentrations did not. Extracts of *A. avenae* induced about 20% of *Musca domestica* to pupate, indicating that an ecdysteroid in the nematode had brought about pupation in the fly larva (Davies and Fisher, 1994). The fourth moult of the entomopathogenic nematode *Contortylenchus grandicolli* occurs outside the host beetle, *Ips grandicollis*, after the L4 has left the insect via the hindgut and rectum to develop into adults. The L4 is apparently stimulated to moult whilst still in the body cavity of the insect but progress in moulting is inhibited by the juvenile hormone concentration of the insect (Gibb and Fisher, 1989). It may be that insect hormones can have an effect on moulting of these entomophagous nematodes or the L4 may be using the insect hormone just as a signal to continue development.

Nuclear receptors (NR) are one of the most abundant classes of transcriptional regulators in metazoans and are involved in sexual differentiation, metabolic regulation and insect metamorphosis as well as other processes (Sluder *et al.*, 1999). They comprise a super-family of cytoplasmic/nuclear localised receptors that on ligand binding (or by phosphorylation) directly regulate the transcription of target genes. Ecdysteroids control moulting in insects by combining with an ecdysone receptor (EcR) and the ultraspiracle protein to activate a number of regulatory genes whose products both repress ongoing gene expression and also stimulate genes that are associated with the production of the next stage in the life cycle in a cascading fashion (Riddiford, 1993). Both the ecdysone receptor and ultraspiracle in insects are members of the nuclear hormone receptor (NHR) superfamily. In *C. elegans* the *daf-12* gene encodes a nuclear hormone receptor that is required to enable the nematode to form the dauer larva (Riddle and Albert, 1997). Three genes (called *chr-3*, *cnr-8* and *cnr-14*) for three nuclear hormone receptors in *C. elegans* are expressed in a developmentally regulated way (Kostrouch, Kostrouchova and Rall, 1995). As *nhr* has been adopted as the standard nomenclature for nuclear receptor genes (Hodgkin, 1997) and as a result of sequence analysis these three genes have been renamed (Sluder *et al.*, 1999). *chr-3* has been renamed *nhr-23*, *cnr-8* has become *nhr-8* and *cnr-14* has become *nhr-24*; *nhr-24* has been shown to correspond to the sex determination gene *sex-1* (Carmi, Kopczynski and Meyer, 1998; Sluder *et al.*, 1999). *nhr-23* shows homology to *DHR3*, an ecdysone-inducible gene that is involved in metamorphosis of *Drosophila* and to *MHR2* found in *Manduca*; *nhr-8* shows homology to the human early response protein NAK1; and *nhr-24* shows homology to the *Drosophila* ecdysone response gene, *E78A* (Kostrouch, Kostrouchova and Rall, 1995). *nhr-23* is expressed in the epidermis of the nematode and is most highly expressed in the adult nematode but is also expressed in the L3 and L4. *nhr-8* shows greatest expression in the L3 whilst *nhr-24* shows greatest expression in embryos, with lesser expression in the adult and other larval stages (Kostrouch, Kostrouchova and Rall, 1995). Inhibition of the gene encoding *nhr-23* results in a number of larval defects associated with abnormal functions of the epidermis, including moulting and body size. This gene is responsible for proper shedding of the L1 cuticle and it has been suggested that it might regulate the onset of moulting by activating a collagenolytic pathway or by directly regulating collagen gene expression (Kostrouchova *et al.*, 1998). Two genes, *Bp-nhr-1* and *Bp-nhr-2*, related to NRs have been isolated from *Brugia pahangi* (Moore and Devaney, 1999). *Bp-nhr-1* is related to the *C. elegans* gene *nhr-6* and is expressed at very low levels during the life cycle, with the highest level of expression being detected on post-infection L3. The low level of expression suggests that it occurs for only a very limited period or in a limited number of cells. It seems to be a member of the same gene family as *Drosophila DHR-38*, which is expressed exclusively in the epidermis of late larval, pre-pupal and pupal stages and is thought to modulate ecdysone-triggered signals in the insect (Kozlova *et al.*, 1998). *Bp-nhr-2* is more abundantly expressed in *Brugia* and has a peak of expression just before the L3 to L4 moult and seems to have a positive regulatory role at this moult but not at the L4 to adult moult. It shows the greatest similarity to the *Drosophila* receptor XR78E/F, which is thought to repress the transcriptional activity of the ecdysone receptor complex (Moore and Devaney, 1999).

Although *C. elegans* has been shown to contain ecdysteroids (see above) it does not produce them (Barker, Chitwood and Rees, 1990; Chitwood and Feldhaufer, 1990). Thus any hormone signal used in the regulation of moulting in this, and probably other nematodes, is likely to be different from the ecdysteroids of insects (Sluder *et al.*, 1999). It has been suggested that the conserved regulatory genes found in insects and in nematodes may, in nematodes, be involved in the execution of the moult rather than in reception of a hormonal signal. The cellular response cascade activated by the ecdysone receptor includes other NRs and genes related to some of these other NRs have been identified in *C. elegans* (Sluder *et al.*, 1999). Three of these genes (*daf-12*, *nhr-8* and *nhr-48*) are similar to the ecdysone responsive gene *DHR96* of *Drosophila*. *sex-1* of *C. elegans* is a close relative of the *Drosophila* ecdysone-inducible genes *E75* and *E78* but seems not to have a role in moulting (Carmi, Kopczynski and Meyer, 1998). *daf-12* mutants repeat earlier aspects of development instead of making normal progress (Antebi, Culotti and Hedgecock, 1998).

Although it appears that *C. elegans* is unable to synthesise ecdysteroids, these compounds do have various effects on moulting and development in nematodes in laboratory experiments. However, access by this nematode to these compounds in nature is probably limited or non-existent and therefore their role in moulting and development remains an enigma (Chitwood, 1999). Whilst nuclear hormone receptor genes, similar to some involved in moulting in insects, are now being identified in nematodes their role in moulting and development of nematodes remains to be clarified.

Acknowledgements

I am deeply indebted to Jean Bird, David M. Bird, Iain L. Johnstone and John T. Jones for their critical and constructive comments and suggestions on part, or all, of this chapter. Whilst these were invaluable in producing the manuscript, full responsibility for its contents, including any errors and omissions, rests entirely with the author.

References

Aguinaldo, A.M., Turbeville, J.M., Linford, L.S., Rivera, M.C., Garey, J.R., Raff, R.A. and Lake, J.A. (1997) Evidence for a clade of nematodes, arthropods and other moulting animals. *Nature*, **387**, 489–493.

Antebi, A., Culotti, J.G. and Hedgecock, E.M. (1998) *daf-12* regulates developmental age and the dauer alternative in *Caenorhabditis elegans*. *Development*, **125**, 1191–1205.

Apfel, H., Eisenbeiss, W.F. and Meyer, T.F. (1992) Changes in the surface composition after transmission of *Acanthocheilonema viteae* third stage larvae into the jird. *Molecular and Biochemical Parasitology*, **52**, 63–74.

Apfel, H. and Meyer, T.F. (1990) Active release of surface proteins: A mechanism associated with the immune escape of *Acanthocheilonema viteae* microfilariae. *Molecular and Biochemical Parasitology*, **43**, 199–210.

Araujo, A., Souto-Padron, T. and De Souza, W. (1994) An ultrastructural, cytochemical and freeze-fracture study of the surface structures of *Brugia malayi* microfilariae. *International Journal for Parasitology*, **24**, 899–907.

Ashman, K., Mather, J., Wiltshire, C., Jacobs, H.J. and Meeusen, E. (1995) Isolation of a larval surface glycoprotein from *Haemonchus contortus* and its possible role in evading host immunity. *Molecular and Biochemical Parasitology*, **70**, 175–179.

Aumann, J., Robertson, W.M. and Wyss, U. (1991) Lectin binding to cuticle exudates of sedentary *Heterodera schachtii* (Nematoda: Heteroderidae) second-stage juveniles. *Revue de Nématologie*, **14**, 113–118.

Awan, F.A. and Hominick, W.M. (1982) Observations on tanning of the potato cyst-nematode, *Globodera rostochiensis*. *Parasitology*, **85**, 61–71.

Badley, J.E., Grieve, R.B., Bowman, D.D. and Glickman, L.T. (1987) Immune-mediated adherence of eosinophils to *Toxocara canis* infective larvae: the role of excretory-secretory antigens. *Parasite Immunology*, **9**, 133–143.

Barker, G.C., Chitwood, D.J. and Rees, H.H. (1990) Ecdysteroids in helminths and annelids. *Invertebrate Reproduction and Development*, **18**, 1–11.

Barker, G.C. and Rees, H.H. (1990) Ecdysteroids in nematodes. *Parasitology Today*, **6**, 384–387.

Bartnik, E., Osborn, M. and Weber, K. (1986) Intermediate filaments in muscle and epithelial cells of nematodes. *Journal of Cell Biology*, **102**, 2033–2041.

Betschart, B., Marti, S. and Glaser, M. (1990) Antibodies against the cuticlin of *Ascaris suum* cross-react with epicuticular structures of filarial parasites. *Acta Tropica*, **47**, 331–338.

Betschart, B. and Wyss, K. (1990) Analysis of the cuticular collagens of *Ascaris suum*. *Acta Tropica*, **47**, 297–305.

Bird, A.F. (1957) Chemical composition of the nematode cuticle. Observations on individual layers and extracts from these layers in *Ascaris lumbricoides* cuticle. *Experimental Parasitology*, **6**, 383–403.

Bird, A.F. (1958) Further observations on the structure of nematode cuticle. *Parasitology*, **48**, 32–36.

Bird, A.F. (1959) Development of the root-knot nematodes *Meloidogyne javanica* (Treub) and *Meloidogyne hapla* Chitwood in the tomato. *Nematologica*, **4**, 31–42.

Bird, A.F. (1971) *The Structure of Nematodes*. 318 pp. New York and London: Academic Press.

Bird, A.F. (1976) The development and organization of skeletal structures in nematodes. In *The Organization of Nematodes*, edited by N.A. Croll, pp. 107–137. New York and London: Academic Press

Bird, A.F. (1977) Cuticle formation and moulting in the egg of *Meloidogyne javanica* (Nematoda). *Parasitology*, 74, 149–152.

Bird, A.F. (1983) Growth and moulting in nematodes: changes in the dimensions and morphology of *Rotylenchus reniformis* from start to finish of moulting. *International Journal for Parasitology*, 13, 201–206.

Bird, A.F. (1984a) Growth and moulting in nematodes: moulting and development of the hatched larva of *Rotylenchus reniformis*. *Parasitology*, 89, 107–119.

Bird, A.F. (1984b) Nematoda. In *Biology of the Integument 1. Invertebrates*, edited by J. Bereiter-Hahn, A.G. Matolsty, and K.S. Richards, pp. 212–233. Berlin: Springer-Verlag.

Bird, A.F. (1990) Vital staining of glycoprotein secreted by infective 3rd-stage larvae of *Haemonchus contortus* prior to exsheathment. *International Journal for Parasitology*, 20, 619–623.

Bird, A.F. and Bird, J. (1991) *The Structure of Nematodes*, 2nd edn., 316 pp. San Diego: Academic Press.

Bird, A.F. and Bird, J. (1998) Introduction to functional organization. In *The Physiology and Biochemistry of Free-living and Plant-parasitic Nematodes*, edited by R.N. Perry and D.J. Wright, pp. 1–24. Wallingford: CABI.

Bird, A.F., Bonig, I. and Bacic, A. (1988) A role for the excretory system in secernentean nematodes. *Journal of Nematology*, 20, 493–496.

Bird, A.F., Bonig, I. and Bacic, A. (1989) Factors affecting the adhesion of micro-organisms to the surfaces of plant-parasitic nematodes. *Parasitology*, 98, 155–164.

Bird, A.F. and Deutsch, K. (1957) The structure of the cuticle of *Ascaris lumbricoides* var. *suis*. *Parasitology* 47, 319–328.

Bird, A.F. and Rogers, G.E. (1965). Ultrastructure of the cuticle and its formation in *Meloidogyne javanica*. *Nematologica*, 11, 244–230.

Bird, A.F. and Zuckerman, B.M. (1989) Studies on the surface coat (glycocalyx) of the dauer larva of *Anguina agrostis*. *International Journal for Parasitology*, 19, 235–240.

Bird, D.M. and Wilson, M.A. (1994) Molecular and cellular dissection of giant cell function. In *Advances in Molecular Plant Nematology*, edited by F. Lamberti, C. de Giorgi and D.McK. Bird, pp. 181–195. New York: Plenum Press.

Blaxter, M.L. (1993) The cuticle surface proteins of a wild type and mutant *Caenorhabditis elegans*. *Journal of Biological Chemistry*, 268, 6600–6609.

Blaxter, M.L. and Bird, D.M. (1997) Parasites. In *C. elegans II*, edited by D. Riddle, T. Blumenthal, B. Meyer and J. Preiss, pp. 851–878. Cold Spring Harbor: Cold Spring Harbor Laboratory Press.

Blaxter, M.L., Page, A.P., Rudin, W. and Maizels, R.M. (1992) Nematode surface coats: Actively evading immunity. *Parasitology Today*, 8, 243–247.

Blaxter, M.L. and Robertson, W.M. (1998).The Cuticle. In *The Physiology and Biochemistry of Free-living and Plant-parasitic Nematodes*, edited by R.N. Perry and D.J. Wright, pp.25–48. Wallingford: CABI.

Bonner, T.P., Menefee, M.G. and Etges, F.J. (1970) Ultrastructure of cuticle formation in a parasitic nematode, *Nematospiroides dubius*. *Zeitschrift für Zellforschung und Mikroskopische Anatomie*, 104, 193–204.

Bonner, T.P. and Weinstein, P.P. (1972a) Ultrastructure of the hypodermis during cuticle formation in the third molt of the nematode *Nippostrongylus brasiliensis*. *Zeitschrift für Zellforschung und Mikroskopische Anatomie*, 126, 17–24.

Bonner, T.P. and Weinstein, P.P. (1972b) Ultrastructure of cuticle formation in the nematodes *Nippostrongylus brasiliensis* and *Nematospiroides dubius*. *Journal of Ultrastructure Research*, 40, 261–271.

Bovien, P. (1937) Some types of association between nematodes and insects. *Videnskabelige Meddelelser fra Dansk naturhistorik Forening i Kjøbenhavn*, 101, 1–114.

Brunaská, M., Dubinský, P. and Reiterová, K. (1995) *Toxocara canis*: ultrastructural aspects of larval moulting in the maturing eggs. *International Journal for Parasitology*, 25, 683–690.

Byerley, L., Cassada, R.C. and Russell, R.C. (1976) The life cycle of the nematode *Caenorhabditis elegans*. I. Wild type growth and reproduction. *Developmental Biology*, 51, 23–33

Byers, J.R. and Anderson, R.V. (1972) Ultrastructural morphology of the body wall, stoma and stomatostyle of the nematode, *Tylenchorhynchus dubius* (Bütschli, 1873). *Canadian Journal of Zoology*, 50, 457–465.

Campbell, L.R. and Gaugler, R. (1991) Mechanisms for exsheathment of entomopathogenic nematodes. *International Journal for Parasitology*, 21, 219–224.

Capo, V.A., Despommier, D.D. and Silberstein, D.S. (1984) The site of ecdysis of the L1 larva of *Trichinella spiralis*. *Journal of Parasitology*, 70, 992–994.

Carmi, I., Kopczynski, J.B. and Meyer, B.J. (1998) The nuclear hormone receptor SEX-1 is an X-chromosome signal that determines nematode sex. *Nature*, 396, 168–173.

Carter, R.F. and Wright, K.A. (1984) Formation of the odontostyle during molting of the nematode *Xiphinema americanum* (Nematoda: Dorylaimoidea). *Journal of Ultrastructure Research*, 87, 221–241.

Cassada, R.C. and Russell, R.L. (1975) The dauerlarva, a post-embryonic developmental variant of the nematode *Caenorhabditis elegans*. *Developmental Biology*, 46, 326–342.

Chitwood, B.G. and Chitwood, M.B. (1950) *An introduction to nematology*. Section I Anatomy. 2nd edn. Baltimore: B.G. Chitwood, (Monumental Printing Co.).

Chitwood, D.J. (1999) Biochemistry and function of nematode steroids. *Critical Reviews in Biochemistry and Molecular Biology*. 34, 273–284.

Chitwood, D.J. and Feldlaufer, M.F. (1990) Ecdysteroids in axenically propagated *Caenorhabditis elegans* and culture medium. *Journal of Nematology*, 22, 598–607.

Cleator, M., Delves, C.J., Howells, R.E. and Rees, H.H. (1987) Identity and tissue localisation of free and conjugated ecdysteroids in adults of *Dirofilaria immitis* and *Ascaris suum*. *Molecular and Biochemical Parasitology*, 25, 93–105.

Cookson, E., Blaxter, M.L. and Selkirk, M.E. (1992) Identification of the major soluble cuticular protein of lymphatic filarial nematode parasites as a secretory homologue of glutathione peroxidase. *Proceedings of the National Academy of Sciences of the United States of America*, 89, 5837–5841.

Coomans, A. and De Coninck, L. (1963) Observations on spear formation in *Xiphinema*. *Nematologica*, 9, 85–96.

Coomans, A., De Coninck, L.A.P. and Heip, C. (1978) Round table discussions. *Annales de la Société Royale Zoologique de Belgique*, 108, 109–113.

Costa, M., Draper, B.W. and Priess, J.R. (1997) The role of actin filaments in patterning the *Caenorhabditis elegans* cuticle. *Developmental Biology*, 184, 373–384.

Cox, G.N. (1990) Molecular biology of the cuticle collagen gene families of *Caenorhabditis elegans* and *Haemonchus contortus*. *Acta Tropica*, 47, 269–281.

Cox, G.N. (1992) Molecular and biochemical aspects of nematode collagens. *Journal of Parasitology*, 78, 1–15.

Cox, G.N. and Hirsh, D. (1985) Stage-specific patterns of collagen gene expression during development of *Caenorhabditis elegans*. *Molecular and Cellular Biology*, 5, 363–372.

Cox, G.N., Kusch, M. and Edgar, R.S. (1981) Cuticle of *Caenorhabditis elegans*: its isolation and partial characterisation. *Journal of Cell Biology*, 90, 7–17.

Cox, G.N., Shamansky, L.M. and Boisvenue, R.J. (1990) *Haemonchus contortus*: Evidence that the 3A3 collagen gene is a member of an evolutionarily conserved family of nematode collagens. *Experimental Parasitology*, 70, 175–185.

Crofton, H. D. (1966) *Nematodes*. London: Hutchinson.

Croll, N.A. and Matthews, B.E. (1977) *Biology of Nematodes*. Glasgow: Blackie.

Davey, K.G. (1965) Molting in a parasitic nematode, *Phocanema decipiens*. I. Cytological events. *Canadian Journal of Zoology*, 54, 997–1003.

Davey, K.G. (1966) Neurosecretion and molting in some parasitic nematodes. *American Zoologist*, 6, 243–249.

Davey, K.G. (1971) Molting in a parasitic nematode, *Phocanema decipiens*. VI. The mode of action of insect juvenile hormone and farnesyl methyl ether. *International Journal for Parasitology*, 1, 61–66.

Davey, K.G. (1976) Hormones in nematodes. In *The organization of nematodes*, edited by N.A. Croll, pp. 273–291. New York: Academic Press.

Davey, K.G. and Kan, S.P. (1968) Molting in a parasitic nematode, *Phocanema decipiens* — IV. Ecdysis and its control. *Canadian Journal of Zoology*, 46, 893–898.

Davies, K.A. and Fisher, J.M. (1994) On hormonal control of moulting in *Aphelenchus avenae* (Nematoda: Aphelenchida). *International Journal for Parasitology*, 24, 649–655.

Davies, K.G. and Danks, C. (1993) Carbohydrate/protein interactions between the cuticle of infective juveniles of *Meloidogyne incognita* and spores of the obligate hyperparasite *Pasteuria penetrans*. *Nematologica*, 39, 53–64.

De Coninck, L. (1965) Classe des Nématodes. In *Traité de Zoologie: Anatomie, Systématique, Biologie*. Vol. IV, Part II, edited by P.P. Grassé, pp. 3–386. Paris: Masson.

De Giorgi, C., De Luca, F. and Lamberti, F. (1996) A silent *trans*-splicing signal in the cuticlin-encoding gene of the plant-parasitic nematode *Meloidogyne artiellia*. *Gene*, **170**, 261–265.

De Giorgi, C., De Luca, F., Di Vito, M. and Lamberti, F. (1997) Modulation of expression at the level of splicing of *cut-1* RNA in the infective second-stage juvenile of the plant parasitic nematode *Meloidogyne artiellia*. *Molecular and General Genetics*, **253**, 589–598.

Dennis, R.D. (1976) Insect morphogenetic hormones and developmental mechanisms in the nematode *Nematospiroides dubius*. *Comparative Biochemistry and Physiology*, **53A**, 53–56.

Dennis, R.D.W. (1977) On ecdysone-binding proteins and ecdysone-like material in nematodes. *International Journal for Parasitology*, **7**, 181–188.

De Souza, W., Souton-Padron, T., Dreyer, G. and Andrade, L.D. (1993) *Wuchereria bancrofti*: Freeze-fracture study of the epicuticle of microfilariae. *Experimental Parasitology*, **76**, 287–290.

Despommier, D.D. (1983) Biology. In *Trichinella and Trichinellosis*, edited by W.C. Campbell, pp. 75–151. NewYork: Plenum.

Devaney, E. (1988) The biochemical and immunochemical characterisation of the 30 kilodalton surface antigen of *Brugia pahangi*. *Molecular and Biochemical Parasitology*, **27**, 83–92. London: Taylor and Francis.

Devaney, E. (1991) The surface antigens of the filarial nematode *Brugia*, and the characterization of the major 30 kDA component. In *Parasitic nematodes — antigens, membranes and genes*, edited by M.W. Kennedy, pp. 46–65.

Devaney, E. and Jecock, R. (1991) The expression of the Mr 30 000 antigen in the third stage larvae of *Brugia pahangi*. *Parasite immunology*, **13**, 75–87. London; Taylor and Francis.

Dick, T.A. and Wright, K.A. (1973) The ultrastructure of the cuticle of the nematode *Syphacia obvelata* (Rudolphi, 1802). I, The body cuticle of larvae, males and females, and observations on its development. *Canadian Journal of Zoology*, **51**, 187–196.

Dropkin, V.H., Lower, W.R. and Acedo, J. (1971) Growth inhibition of *Caenorhabditis elegans* and *Panagrellus redivivus* by selected mammalian and insect hormones. *Journal of Nematology*, **3**, 349–355.

Durette-Desset, M.-C. (1971) Essai de classification des nématodes héligmosomes. Correlations avec la paleobiogeographie des hôtes. *Mémoires du Muséum National d'Histoire Naturelle Nouvelle Série A Zoologie*, **69**, 1–126.

Durette-Desset, M.-C. (1983) Keys to genera of the superfamily Trichostrongyloidea. *CIH Keys to the Nematode Parasites of Vertebrates*, edited by R.C. Anderson and A.G. Chabaud, No. 10 pp. 1–86. Farnham Royal, Slough: Commonwealth Agricultural Bureaux.

Eckert, J. and Schwarz, R. (1965) Zur Struktur der Cuticula invasionfähiger Larven einiger Nematoden. *Zeitschrift für Parasitenkunde*, **26**, 116–142.

Edwards, M.K., Busto, P., James, E.R., Carlow, C.K.S. and Philipp, M. (1990) Antigenic and dynamic properties of the surface of *Onchocerca* microfilariae. *Tropical Medicine and Parasitology*, **41**, 174–180.

Edwards, M.K. and Wood, W.B. (1983) Location of specific messenger RNAs in *Caenorhabditis elegans* by cytological hybridisation. *Developmental Biology*, **97**, 375–390.

Ellenby, C. (1946) Nature of the cyst wall of the potato-root eelworm *Heterodera rostochiensis*, Wollenweber, and its permeability to water. *Nature*, **157**, 302.

Endo, B.Y. and Wyss, U. (1992) Ultrastructure of cuticular exudations in parasitic juvenile *Heterodera schachtii* as related to cuticle structure. *Protoplasma*, **166**, 67–77.

Endo, B.Y. (1993) Ultrastructure of cuticular exudates and related cuticular changes on juveniles in *Heterodera glycines*. *Journal of the Helminthological Society of Washington*, **60**, 76–88.

Fattah, D.I., Maizels, R.M., McLaren, D.J. and Spry, C.J.F. (1986) *Toxocara canis*: interaction of human eosinophils with the infective larvae. *Experimental Parasitology*, **61**, 421–433.

Favre, R., Hermann, R., Cermola, M., Hohenberg, H., Muller, M. and Bazzicalupo, P. (1995) Immuno-gold labelling of CUT-1, CUT-2 and cuticulin epitopes in *Caenorhabditis elegans* and *Heterorhabditis* sp. processed by high-pressure freezing and freeze-substitution. *Journal of Submicroscopic Cytology and Pathology*, **27**, 341–347.

Favre, R., Cermola, M., Nunes, C.P., Hermann, R., Muller, M. and Bazzicalupo, P. (1998) Immuno-cross-reactivity of CUT-1 and cuticlin epitopes between *Ascaris lumbricoides*, *Caenorhabditis elegans*, and *Heterorhabditis*. *Journal of Structural Biology*, **123**, 1–7.

Felix, M.A., Hill, R.J., Schwarz, H., Sternberg, P.W., Sudhaus, W. and Sommer, R.J. (1999) *Pristionchus pacificus*, a nematode with only three juvenile stages, displays major heterochronic changes relative to *Caenorhabditis elegans*. *Proceedings of the Royal Society of London*, **B, 266**, 1617–1621.

Fetterer, R.H. (1989) The cuticular proteins from free-living and parasitic stages of *Haemonchus contortus* — I. Isolation and partial characterization. *Comparative Biochemistry and Physiology*, **94B**, 383–388.

Fetterer, R.H. and Rhoades, M.L. (1990) Tyrosine-derived cross-linking amino acids in the sheath of *Haemonchus contortus* infective larvae. *Journal of Parasitology*, **76**, 619–624.

Fetterer, R.H. and Rhoades, M.L. (1993) Biochemistry of the nematode cuticle — relevance to parasitic nematodes of livestock. *Veterinary Parasitology*, **46**, 103–111.

Fetterer, R.H. and Urban, J.F. (1988) Developmental changes in cuticular proteins of *Ascaris suum*. *Comparative Biochemistry and Physiology*, **90B**, 321–327.

Fleming, M.W. (1985) *Ascaris suum*: role of ecdysteroids in molting. *Experimental Parasitology*, **60**, 207–210.

Fleming, M.W. (1987) Ecdysteroids during embryonation of eggs of *Ascaris suum*. *Comparative Biochemistry and Physiology*, **87A**, 803–805.

Fleming, M.W. (1993) Ecdysteroids during development in the ovine parasitic nematode, *Haemonchus contortus*. *Comparative Biochemistry and Physiology*, **104B**, 653–655.

Fleming, M.W. (1998) In vitro growth of swine roundworm larvae, *Ascaris suum*: cultivation techniques and endocrine regulation. *Journal of the Helminthological Society of Washington*, **65**, 69–73.

Fok, E., Mizinska-Boevska, Y. and Polyakova-Krusteva, O. (1991) *Nippostrongylus brasiliensis* — Ultrastructure of the body wall. *Parasitologia Hungarica*, **24**, 81–87.

Francis, R. and Waterston, R.H. (1991) Muscle cell attachment in *Caenorhabditis elegans*. *The Journal of Cell Biology*, **114**, 465–479.

Franke, S. and Käuser, G. (1989) Occurrence and hormonal role of ecdysteroids in non-arthropods. In *Ecdysone. From chemistry to mode of action*, edited by J. Koolman, pp. 296–307. Stuttgart: Georg Thieme Verlag.

Fuchs, S. and Harrington, W.F. (1970) Immunological properties of *Ascaris* cuticle collagen. *Biochimica et Biophysica Acta*, **221**, 119–124.

Fujimoto, D. (1968) Isolation of collagens of high hydroxyproline, hydroxylysine and carbohydrate content from muscle layer of *Ascaris lumbricoides* and pig kidney. *Biochimica et Biophysica Acta*, **168**, 537–543.

Fujimoto, D., Horiuchi, K. and Hirama, M. (1981) Isotrityrosine, a new cross-linking amino acid isolated from *Ascaris* cuticle collagen. *Biochemical and Biophysical Research Communications*, **99**, 637–643.

Fujimoto, D. and Kanaya, S. (1973) Cuticlin: a noncollagen structural protein from *Ascaris* cuticle. *Archives of Biochemistry and Biophysics*, **157**, 1–6.

Gamble, H.R., Lichtenfels, J.R. and Purcell, J.P. (1989) Light and scanning electron-microscopy of the ecdysis of *Haemonchus contortus* infective larvae. *Journal of Parasitology*, **75**, 303–307.

Gamble, H.R., Purcell, J.P. and Fetterer, R.H. (1989) Purification of a 44 kiloDalton protease which mediates the ecdysis of infective *Haemonchus contortus* larvae. *Molecular and Biochemical Parasitology*, **33**, 49–58.

Gems, D., Ferguson, C.J., Robertson, B.D., Nieves, R., Page, A.P., Blaxter, M.L. and Maizels, R.M. (1995) An abundant, trans-spliced messenger RNA from *Toxocara canis* infective larvae encodes a 26-kDa protein with homology to phosphatidylethanolamine-binding proteins. *Journal of Biological Chemistry*, **270**, 18517–18522.

Gems, D. and Maizels, R.M. (1996) An abundantly expressed mucin-like protein from *Toxocara canis* infective larvae: The precursor of the larval surface coat glycoproteins. *Proceedings of the National Academy of Sciences of the United States of America*, **93**, 1665–1670.

Ghosh, I., Eisenger, S.W., Raghavan, N. and Scott, A.L. (1998) Thioredoxin peroxidases from *Brugia malayi*. *Molecular and Biochemical Parasitology*, **91**, 207–220.

Gibb, K.S. and Fisher, J.M. (1989) Factors affecting the fourth moult of *Contortylenchus grandicolli* (Nematoda: Allantonematidae) to the free-living sexual forms. *Nematologica*, **35**, 125–128.

Gigliotti, S., Graziani, F., De Ponti, L., Rafti, F., Manzi, A., Lavorgna, G., Gargiulo, G. and Malva, C. (1988) Sex-tissue- and stage-specific expression of a vitelline membrane protein gene from region 32 of the second chromosome of *Drosophila melanogaster*. *Developmental Genetics*, **10**, 33–41.

Gounaris, K., Smith, V.P. and Selkirk, M.E. (1996) Structural composition and lipid composition of the epicuticular accessory layer of infective larvae of *Trichinella spiralis*. *Biochimica et Biophysica Acta*, **1281**, 91–100.

Gravato Nobre, M.J. and Evans, K. (1998) Plant and nematode surfaces: their structure and importance in host-parasite interactions. *Nematologica*, **44**, 103–124.

Gravato-Nobre, M.J., McClure, M.A., Dolan, L., Calder, G., Davies, K.G., Mulligan, B., Evans, K. and von Mende, N. (1999) *Meloidogyne incognita* surface antigen epitopes in infected *Arabidopsis* roots. *Journal of Nematology*, **31**, 212–223.

Grootaert, P. and Lippens, P.L. (1974) Some ultrastructural changes in cuticle and hypodermis of *Aporcellaimellus* during the first moult (Nematoda: Dorylaimoidea). *Zeitschrift für Morphologie der Tiere*, **79**, 269–282.

Grootaert, P. and Coomans, A. (1980) The formation of the feeding apparatus in dorylaims. *Nematologica*, **26**, 406–431.

Ham, P.J., Smail, A.J. and Groeger, B.K. (1988) Surface carbohydrate changes on *Onchocerca lienalis* larvae as they develop from microfilariae to the infective third-stage in *Simulium ornatum*. *Journal of Helminthology*, **62**, 195–205.

Harris, J.E. and Crofton, H.D. (1957) Structure and function in the nematodes: internal pressure and cuticular structure in *Ascaris*. *Journal of Experimental Biology*, **34**, 116–130.

Hemmer, R.M., Donkin, S.G., Chin, K.J., Grenache, D.G., Bhatt, H. and Politz, S.M. (1991) Altered expression of an L1-specific, O-linked cuticle surface glycoprotein in mutants of the nematode *Caenorhabditis elegans*. *Journal of Cell Biology*, **115**, 1237–1247.

Hill, D.E., Fetterer, R.H. and Urban, J.F. (1991) *Ascaris suum*: Stage-specific differences in lectin binding to the larval cuticle. *Experimental Parasitology*, **73**, 376–383.

Hinz, E. (1963) Elektronenmikroskopische Untersuchungen an *Parasacaris equorum*. *Protoplasma*, **56**, 202–241.

Hitcho, P.J. and Thorsen, R.E. (1971) Possible moulting and maturation controls of *Trichinella spiralis*. *Journal of Parasitology*, **57**, 787–793.

Hodgkin, J. (1997). Appendix 1: Genetics. In *C. elegans II*, edited by D.L. Riddle, T. Blumenthal, B.J. Meyer and J.R. Preiss, pp. 881–1047. Cold Spring Harbor: Cold Spring Harbor Laboratory Press.

Horn, D.H.S., Wilkie, J.S. and Thomson, J.A. (1974) Isolation of β-ecdysone (20-hydroxyecdysaone) from the parasitic nematode *Ascaris lumbricoides*. *Experientia*, **30**, 1109.

Howells, R.E. and Blainey, L.J. (1983) The moulting process and the phenomenon of intermoult growth in the filarial nematode *Brugia pahangi*. *Parasitology*, **87**, 493–505.

Hyman, L.H. (1951) *The Invertebrates: Acanthocephala, Aschelminthes, and Entoprocta. The pseudocoelomate Bilateria.* Vol. III. New York: McGraw-Hill.

Ibrahim, M.S., Tamashiro, W.K., Moraga, D.A. and Scott, A.L. (1989) Antigen shedding from the surface of the infective stage larvae of *Dirofilaria immitis*. *Parasitology*, **99**, 89–97.

Ibrahim, S.K. (1991) Distribution of carbohydrates on the cuticle of several developmental stages of *Meloidogyne javanica*. *Nematologica*, **37**, 275–284.

Inglis, W.G. (1983) The design of the nematode body wall: the ontogeny of the cuticle. *Australian Journal of Zoology*, **31**, 705–716.

Jamuar, M.P. (1966) Electron microscope studies on the body wall of the nematode *Nippostrongylus brasiliensis*. *Journal of Parasitology*, **52**, 209–232.

Jannson, H.B., Jeyaprakash, A., Coles, G.C., Marban-Mendoza, N. and Zuckerman, B.M. (1986) Fluorescent and ferritin labelling of cuticle carbohydrates of *Caenorhabditis elegans* and *Panagrellus redivivus*. *Journal of Nematology*, **18**, 570–574.

Joachim, A.M, Ruttkowski, B. and Daugschies, A. (1999) Changing surface antigen and carbohydrate patterns during the development of *Oesophagostomum dentatum*. *Parasitology*, **119**, 491–501.

Johnson, P.W., Van Gundy, S.D. and Thomson, W.W. (1970a) Cuticle ultrastructure of *Hemicycliophora arenaria*, *Aphelenchus avenae*, *Hirschmaniella gracilis* and *Hirschmaniella belli*. *Journal of Nematology*, **2**, 42–58.

Johnson, P.W., Van Gundy, S.D. and Thomson, W.W. (1970b) Cuticle formation in *Hemicycliophora arenaria*, *Aphelenchus avenae* and *Hirschmaniella gracilis*. *Journal of Nematology*, **2**, 59–79.

Johnson, R.N. and Viglierchio, D.R. (1970) *Heterodera schactii* responses to exogenous hormones. *Experimental Parasitology*, **27**, 301–309.

Johnstone, I.L. (1994) The cuticle of the nematode *Caenorhabditis elegans*: a complex collagen structure. *Bioessays*, **16**, 171–178.

Johnstone, I.L. (2000) Cuticle collagen genes — expression in *Caenorhabditis elegans*. *Trends in Genetics*, **16**, 21–27.

Johnstone, I.L. and Barry, J.D. (1996) Temporal reiteration of a precise gene expression pattern during nematode development. *The EMBO Journal*, **15**, 3633–3639.

Johnstone, I.L., Shafi, Y., Majeed, A. and Barry, J.D. (1996) Cuticular collagen genes from the parasitic nematode *Teladorsagia* (syn. *Ostertagia*) *circumcincta*. *Molecular and Biochemical Parasitology*, **80**, 103–112.

Jones, J.T., Perry, R.N. and Johnston, M.R.L. (1993) Changes in the ultrastructure of the cuticle of the potato cyst nematode, *Globodera rostochiensis*, during development and infection. *Fundamental and Applied Nematology*, **16**, 433–445.

Jones, S.J.M. and Baillie, D.L. (1995) Characterisation of the *let-653* gene in *Caenorhabditis elegans*. *Molecular and General Genetics*, **248**, 719–726.

Josse, J. and Harrington, W.F. (1964) Role of pyrrolidine residues in the structure and stabilization of collagen. *Journal of Molecular Biology*, **9**, 269–287.

Kennedy, M.W. (editor) (1991) *Parasitic nematodes — antigens, membranes and genes*. London: Taylor and Francis.

Kennedy, M.W., Foley, M., Kuo, Y.-M., Kusel, J.R. and Garland, P.B. (1987) Biophysical properties of the surface lipid of parasitic nematodes. *Molecular and Biochemical Parasitology*, **22**, 233–240.

Khoo, K.-H., Maizels, R.M., Page, A.P., Taylor, G.W., Rendell, N. and Dell, A. (1991) Characterisation of nematode glycoproteins: the major O-glycans of *Toxocara* excretory secretory antigens are methylated trisaccharides. *Glycobiology*, **1**, 163–171.

Kingston, I.B. (1991) Collagen genes in *Ascaris*. In *Parasitic nematodes — antigens, membranes and genes*, edited by M.W. Kennedy, pp.66–83. London: Taylor & Francis

Kingston, I.B. and Pettitt, J. (1990) Structure and expression of *Ascaris suum* collagen genes: A comparison with *Caenorhabditis elegans*. *Acta Tropica*, **47**, 283–287.

Kingston, I.B., Wainwright, S.M. and Cooper, D. (1989) Comparison of collagen gene sequences in *Ascaris suum* and *Caenorhabditis elegans*. *Molecular and Biochemical Parasitology*, **37**, 137–146.

Koltai, H., Chejanovsky, N., Raccah, B. and Spiegel, Y. (1997) The first isolated collagen gene of the root-knot nematode *Meloidogyne javanica* is developmentally regulated. *Gene*, **196**, 191–199.

Kostrouch, Z., Kostrouchova, M. and Rall, J.E. (1995) Steroid/thyroid hormone receptor genes in *Caenorhabditis elegans*. *Proceedings of the National Academy of Sciences of the USA*, **92**, 156–159.

Kostrouchova, M., Kause, M., Kostrouch, Z. and Rall, J.E. (1998) CHR3: a *Caenorhabditis elegans* orphan nuclear hormone receptor required for proper epidermal development and molting. *Development*, **125**, 1617–1626.

Kozlova, T., Pokolkhova, G.V., Tzertzinis, G., Sutherland, J.D., Zhimulev, I.F. and Kramer, J.M. (1997) Extracellular matrix. In: *C. elegans II*, edited by D.L. Riddle, T. Blumenthal, B. Meyer and J.R. Preiss, pp. 471–500. Cold Spring Harbor: Cold Spring Harbor Press.

Kulikov, V.V., Dashchenko, O.I., Koloss, T.V. and Yushin, V.V. (1998) A description of the free-living marine nematode *Euchromadora robusta* sp. n. (Nematoda: Chromadorida) with observations on the ultrastructure of the body cuticle. *Russian Journal of Nematology*, **6**, 103–110.

Lassandro, F., Sebastiano, M., Zei, F. and Bazzicalupo, P. (1994) The role of dityrosine formation in the cross-linking of CUT-2, the product of a second cuticlin gene of *Caenorhabditis elegans*. *Molecular and Biochemical Parasitology*, **65**, 147–159.

Lee, D.L. (1958). On the morphology of the male, female and fourth-stage larva (female) of *Hammerschmidtiella diesingi* (Hammerschmidt), a nematode parasitic in cockroaches. *Parasitology*, **48**, 433–436.

Lee, D.L. (1961). Localization of esterase in the cuticle of the nematode *Ascaris lumbricoides*. *Nature, London*, **192**, 282–283.

Lee, D.L. (1965) The cuticle of adult *Nippostrongylus brasiliensis*. *Parasitology*, **55**, 173–181.

Lee, D.L. (1966a) An electron microscope study of the body wall of the third-stage larva of *Nippostrongylus brasiliensis*. *Parasitology*, **56**, 127–135.

Lee, D.L. (1966b) The structure and composition of the helminth cuticle. *Advances in Parasitology*, **4**, 187–254.

Lee, D.L. (1969) *Nippostrongylus brasiliensis*: Some aspects of the fine structure and biology of the infective larva and the adult. In *Nippostrongylus and Toxoplasma*, pp. 3–16. *Symposia of the British Society for Parasitology*, **7**. Oxford: Blackwell Scientific Publications.

Lee, D.L. (1970a) The ultrastructure of the cuticle of adult female *Mermis nigrescens* (Nematoda). *Journal of Zoology*, **161**, 513–518.

Lee, D.L. (1970b) Moulting in nematodes: the formation of the adult cuticle during the final moult of *Nippostrongylus brasiliensis*. *Tissue & Cell*, **2**, 139–153.

Lee, D.L. (1977) The nematode epidermis and collagenous cuticle, its formation and ecdysis. In *Comparative Biology of Skin*, edited by R.I.C. Spearman. *Symposia of the Zoological Society of London*, **39**, 145–170. London: Academic Press.

Lee, D.L. and Atkinson, H.J. (1976) *Physiology of Nematodes*, 2nd edn. London: Macmillan.

Lee, D.L. and Biggs, W.D. (1990) Two- and three-dimensional locomotion of *Nippostrongylus brasiliensis*. *Parasitology*, **101**, 301–308.

Lee, D.L. and Bonner, T.P. (1982) Freeze etch studies on nematode body wall. *Parasitology*, **84**, xliv.

Lee, D.L. and Nicholls, C.D. (1983). The use of plasma etching to reveal the internal structure of *Nippostrongylus brasiliensis* (Nematoda). *Parasitology*, **86**, 477–480.

Lee, D.L., Wright, K.A. and Shivers, R.R. (1984) A freeze-fracture study of the surface of the infective-stage larva of the nematode *Trichinella*. *Tissue & Cell*, **16**, 819–828.

Lee, D.L., Wright, K.A. and Shivers, R.R. (1986) A freeze-fracture study of the body wall of adult, *in utero* larvae and infective-stage larvae of *Trichinella* (Nematoda). *Tissue & Cell*, **18**, 219–230.

Lee, D.L., Wright, K.A. and Shivers, R.R. (1993) A freeze-fracture study of the cuticle of adult *Nippostrongylus brasiliensis* (Nematoda). *Parasitology*, **107**, 545–552.

Liebau, E., Walter, R.D. and Henkle-Dührsen, K. (1994a) Isolation, sequence and expression of an *Onchocerca volvulus* glutathione S-transferase cDNA. *Molecular and Biochemical Parasitology*, **63**, 305–309.

Liebau, E., Walter, R.D. and Henkle-Dührsen, K. (1994b) *Onchocerca volvulus*: Isolation and sequence of a second glutathione S-tranferase cDNA. *Experimental Parasitology*, **79**, 68–71.

Liebau, E., Wildenburg, G., Walter, R.D. and Henkle-Dührsen, K. (1994) A novel type of glutathione s-transferase in *Onchocerca volvulus*. *Infection and Immunity*, **62**, 4762–4767.

Leushner, J.R.A. and Pasternak, J.P. (1975) Programmed synthesis of collagen during postembryonic development of the nematode *Panagrellus silusiae*. *Developmental Biology*, **47**, 68–80.

Lewis, E., Hunter, S.J., Tetley, L., Nunes, C.P., Bazzicalipo, P. and Devaney, E. (1999) *cut-1*-like genes are present in the filarial nematodes, *Brugia pahangi* and *Brugia malayi*, and, as in other nematodes, code for components of the cuticle. *Molecular and Biochemical Parasitology*, **101**, 173–183.

Lin, H.-J. and McClure, M.A. (1996) Surface coat of *Meloidogyne incognita*. *Journal of Nematology*, **28**, 216–224.

Link, C.D., Ehrenfels, C.W. and Wood, W.B. (1988) Mutant expression of male copulatory bursa surface markers on *Caenorhabditis elegans*. *Development*, **103**, 485–495.

Link, C.D., Silverman, M.A., Breen, M. and Watt, K.E. (1992) Characterisation of *Caenorhabditis elegans* lectin-binding mutants. *Genetics*, **131**, 867–881.

Locke, M. (1982) Envelopes at cell surfaces — a confused area of research of general importance. In *Parasites — their world and ours*, edited by D.F. Mettrick and S.S. Desser, pp.73–78. Amsterdam: Elsevier Biomedical Press.

Lorenzen, S. (1994) *The Phylogenetic Systematics of Freeliving Nematodes*. London: Unwin.

Lu, W., Egerton, G.L., Bianco, A.E. and Williams, S.A. (1998) Thioredoxin peroxidase from *Onchocerca volvulus*: a major hydrogen peroxide detoxifying enzyme in filarial parasites. *Molecular and Biochemical Parasitology*, **91**, 221–235.

Lustigman, S. (1993) Molting, enzymes and new targets for chemotherapy of *Onchocerca volvulus*. *Parasitology Today*, **9**, 294–297.

Lustigman, S., Brotman, B., Huima, T., Prince, A.M. and McKerrow, J.H. (1992) Molecular cloning and characterization of onchocystatin, a cysteine proteinase inhibitor of *Onchocerca volvulus*. *Journal of Biological Chemistry*, **267**, 17339–17346.

Lustigman, S., McKerrow, J.H., Shah, K., Liu, J, Huima, T., Hough, M., Brotman, B. and Liu, J. (1996) Cloning of a cysteine protease required for the molting of *Onchocerca volvulus* third stage larvae. *Journal of Biological Chemistry*, **271**, 30181–30189.

Maggenti, A.R. (1964) Morphology of somatic setae: *Thoracostoma californicum* (Nematoda: Enoplidae). *Proceedings of the Helminthological Society of Washington*, **31**, 159–166.

Maizels, R.M., Blaxter, M.L. and Selkirk, M.E. (1993) Forms and functions of nematode surfaces. *Experimental Parasitology*, **77**, 380–384.

Maizels, R.M., Gregory, W.F., Kwan-Lim, G.E. and Selkirk, M.E. (1989) Filarial surface antigens: The major 29 kilodalton glycoprotein and a novel 17–200 kilodalton complex from adult *Brugia malayi* parasites. *Molecular and Biochemical Parasitology*, **32**, 213–224.

Maizels, R.M. and Page, A.P. (1990) Surface associated glycoproteins from *Toxocara canis* larval parasites. *Acta Tropica*, **47**, 355–364.

Maizels, R.M., Partono, F., Oemijati, S., Denham, D.A. and Ogilvie, B.M. (1983) Crossreactive surface antigens on three stages of *Brugia malayi*, *B. pahangi* and *B. timor*. *Parasitology*, **87**, 249–263.

Malakhov, V. V. (1994) *Nematodes. Structure, Development, Classification, and Phylogeny*, edited by W. Duane Hope, 286 pp. Washington and London: Smithsonian Institution Press.

Martin, J. and Lee, D.L. (1983) *Nematodirus battus*: structure of the body wall of the adult. *Parasitology*, **86**, 481–488.

Martinez-Paloma, A. (1978) Ultrastructural characterization of the cuticle of *Onchocerca volvulus* microfilariae. *Journal of Parasitology*, **64**, 127–136.

Martinez, A.M.B. and De Souza, W. (1995) A quick-frozen, freeze-fracture and deep-etched study of the cuticle of adult forms of *Strongyloides venezuelensis* (Nematoda). *Parasitology*, **111**, 523–529.

Martinez, A.M.B. and De Souza, W. (1997) A freeze-fracture and deep-etch study of the cuticle and hypodermis of infective larvae of *Strongyloides venezuelensis* (Nematoda). *International Journal for Parasitology*, **27**, 289–297.

McBride, O.W. and Harrington, W.F. (1967) *Ascaris* cuticle collagen: On the disulphide cross-linkages and molecular properties of the subunits. *Biochemistry*, **6**, 1484–1498.

McClure, M.A. and Zuckerman, B.M. (1982) Localization of cuticular binding sites of concanavalin A on *Caenorhabditis elegans* and *Meloidogyne incognita*. *Journal of Nematology*, **14**, 39–44.

Meerovitch, E. (1965) Studies on the *in vitro* axenic development of *Trichinella spiralis*-II. Preliminary experiments on the effect of farnesol, cholesterol, and an insect extract. *Canadian Journal of Zoology*, **43**, 81–85.

Mehta, K., Rao, U.R., Vickery, A.C. and Fesus, L. (1992) Identification of a novel transglutaminase from the filarial parasite *Brugia malayi* and its role in growth and development. *Molecular and Biochemical Parasitology*, **53**, 1–16.

Mendis, A.H.W., Rose, M.E., Rees, H.H. and Goodwin, T.W. (1983) Ecdysteroids in adults of the nematode, *Dirofilaria immitis*. *Molecular and Biochemical Parasitology*, **9**, 209–226.

Mercer, J.G., Barker, G.C., McCall, J.W., Howells, R.E. and Rees, H.H. (1989) Studies on the biosynthesis and fate of ecdysteroids in filarial nematodes. *Tropical Medicine and Parasitology*, **40**, 429–433.

Mercer, J.G., Munn, A.E. and Rees, H.H. (1988) *Caenorhabditis elegans*: occurrence and metabolism of ecdysteroids in adults and dauer larvae. *Comparative Biochemistry and Physiology*, **90B**, 261–267.

Modha, J., Kennedy, M.W. and Kusel, J.R. (1995) A role for second messengers in the control of activation-associated modification of the surface of *Trichinella spiralis* infective larvae. *Molecular and Biochemical Parasitology*, **72**, 141–148.

Modha, J., Roberts, M.C., Kennedy, M.W. and Kusel, J.R. (1997) Induction of surface fluidity in *Trichinella spiralis* larvae during penetration of the host intestine: simulation by cyclic AMP *in vitro*. *Parasitology*, **114**, 71–77.

Modha , J., Roberts, M.C., Robertson, W.M., Sweetman, G., Powell, K.A., Kennedy, M.W. and Kusel, J.R. (1999) The surface coat of infective larvae of *Trichinella spiralis*. *Parasitology*, **118**, 509–522.

Moore, J. and Devaney, E. (1999) Cloning and characterization of two nuclear receptors from the filarial nematode *Brugia pahangi*. *Biochemical Journal*, **344**, 245–252.

Mounport, D., Baujard, P. and Martiny, B. (1993) Ultrastructural observations on the body cuticle of four species of Tylenchidae Öerley, 1880 (Nemata: Tylenchida). *Nematologica Mediterranea*, **21**, 155–159.

Mounport, D., Baujard, P. and Martiny, B. (1997) TEM observations on the body cuticle of Trichodoridae Thorne, 1935 (Nematoda: Enoplia). *Nematologica*, **43**, 253–258.

Murrell, K.D. (1982) Solubilization studies on the epicuticular antigens of *Strongyloides ratti*. *Veterinary Parasitology*, **10**, 191–203.

Murrell, K.D. and Graham, C.E. (1983) Shedding of antibody complexes by *Strongyloides ratti* (Nematoda) larvae. *Journal of Parasitology*, **69**, 70–73.

Murrel, K.D., Graham, C.E. and McGreevy, M. (1983) *Strongyloides ratti* and *Trichinella spiralis*: Net charge of epicuticle. *Experimental Parasitology*, **55**, 331–339.

Nelson, K.F. and Riddle, D.L. (1984) Functional study of the *Caenorhabditis elegans* secretory-excretory system using laser microscopy. *Journal of Experimental Zoology*, **231**, 45–56.

Nembo, B., Duie, P., Garcia, M., Breton, P., Gayral, P., Porcheron, P. and Goudeyperriere, F. (1993) Levels of ecdysteroid-like material in adults of *Nippostrongylus brasiliensis* during the intestinal phase. *Journal of Helminthology*, **67**, 305–315.

Nicholas, W.L. and Stewart, A.C. (1990) Structure of the cuticle of *Metadasynemoides cristatus* (Chromadorida, Ceramonematidae). *Journal of Nematology*, **22**, 247–261.

Nicholas, W.L. and Stewart, A.C. (1997) Ultrastructure of *Gonionchus australis* (Xyalidae, Nematoda). *Journal of Nematology*, **29**, 133–143.

Nordbring-Hertz, B. and Mattiasson, B. (1979) Action of a nematode-trapping fungus shows lectin-mediated host-microorganism interaction. *Nature, London*, **281**, 477–479.

O'Hanlon, G.M., Cleator, M., Mercer, J.G., Howells, R.E. and Rees, H.H. (1991) Metabolism and fate of ecdysteroids in the nematodes *Ascaris suum* and *Parascaris equorum*. *Molecular and Biochemical Parasitology*, **47**, 179–188.

Ou, X., Tang, L.A., McCrossan, M., Henkle-Dührsen, K. and Selkirk, M.E. (1995) *Brugia malayi*: localization and differential expression of extracellular and cytoplasmic CuZn superoxide dismutases in adults and microfilariae. *Experimental Parasitology*, **80**, 515–529.

Page, A.P., Hamilton, A.J. and Maizels, R.M. (1992) *Toxocara canis*: monoclonal antibodies to carbohydrate epitopes of secreted (TES) antigens localize to different secretion-related structures in infective larvae. *Experimental Parasitology*, **75**, 56–71.

Page, A.P. and Maizels, R.M. (1992) Biosynthesis and glycosylation of serine/threonine-rich secreted proteins from *Toxocara canis* larvae. *Parasitology*, **105**, 297–308.

Page, A.P., Rudin, W. and Maizels, R.M. (1992) Lectin binding to secretory structures, the cuticle and the surface coat of *Toxocara canis* infective larvae. *Parasitology*, **105**, 285–296.

Page, A.P., Rudin, W., Fluri, E., Blaxter, M.L. and Maizels, R.M. (1992) *Toxocara canis*: a labile antigenic surface coat overlying the epicuticle of infective larvae. *Experimental Parasitology*, **75**, 72–86.

Parise, G. and Bazzicalupo, P. (1997) Assembly of nematode cuticle: role of hydrophobic interactions in CUT-2 cross-linking. *Biochimica et Biophysica Acta*, **1337**, 295–301.

Pasternak, J.P. and Leushner, J.R.A. (1975) Programmed collagen synthesis during postembryonic development of the nematode *Panagrellus silusiae*: effect of transcription and translation inhibitors. *Journal of Experimental Zoology*, **194**, 519–528.

Peixoto, C.A. and De Souza, W. (1994) Freeze-fracture characterization of the cuticle of adult and dauer forms of *Caenorhabditis elegans*. *Parasitology Research*, **80**, 53–57.

Peixoto, C.A. and De Souza, W. (1995) Freeze-fracture and deep-etched view of the cuticle of *Caenorhabditis elegans*. *Tissue & Cell*, **27**, 561–568.

Philipp, M., Maizels, R.M., McLaren, D.J., Davies, M.W., Suswillo, R. and Denham, D.A. (1986) Expression of cross-reactive surface antigens by microfilariae and adult worms of *Brugia pahangi* during infections in cats. *Transactions of the Royal Society of Tropical Medicine and Hygiene*, **80**, 385–393.

Philipp, M., Parkhouse, R.M.E. and Ogilvie, B.M. (1980) Changing proteins on the surface of a parasitic nematode. *Nature*, **287**, 538–540.

Philipp, M. and Rumjaneck, F.D. (1984) Antigenic and dynamic properties of helminth surface structures. *Molecular and Biochemical Parasitology*, **10**, 245–268.

Platzer, A. and Platzer, E.G. (1988) Early cuticle formation in an adenophorean nematode. *International Journal for Parasitology*, **18**, 793–801.

Politz, S.M. and Philipp, M. (1992) *Caenorhabditis elegans* as a model for parasitic nematodes: a focus on the cuticle. *Parasitology Today*, **8**, 6–12.

Politz, S.M., Chin, K.J. and Herman, D.L. (1987) Genetic analysis of adult-specific surface antigenic differences between varieties of the nematode *Caenorhabditis elegans*. *Genetics*, **117**, 467–476.

Popham, J.D. and Webster, J.M. (1978) An alternative interpretation of the fine structure of the basal zone of the cuticle of the dauerlarva of the nematode *Caenorhabditis elegans* (Nematoda). *Canadian Journal of Zoology*, **56**, 1556–1563.

Preston-Meek, C.M. and Pritchard, D.I. (1991) Synthesis and replacement of nematode cuticle components. In *Parasitic nematodes — antigens, membranes and genes*, edited by M.W. Kennedy, pp. 84–94. London: Taylor and Francis.

Priess, J.R. and Hirsh, D.I. (1986) *Caenorhabditis elegans* morphogenesis — the role of the cytoskeleton in elongation of the embryo. *Developmental Biology*, **117**, 156–173.

Proudfoot, L., Kusel, J.R., Smith, H.V. and Kennedy, M.W. (1991) Biophysical properties of the nematode surface. In *Parasitic nematodes — antigens, membranes and genes*, edited by M.W. Kennedy, pp. 1–26. London: Taylor and Francis.

Proudfoot, L., Kusel, J.R., Smith, H.V., Harnett, W., Worms, M.J. and Kennedy, M.W. (1993) Rapid changes in the surface of parasitic nematodes during transition from pre- to post-parasitic forms. *Parasitology*, **107**, 107–117.

Rao, U.R., Chapman, M.R., Singh, R.N., Mehta, K. and Klei, T.R. (1999) Transglutaminase activity in equine strongyles and its potential role in growth and development. *Parasite*, **6**, 131–139.

Rao, U.R., Mehta, K., Subrahmany, D. and Vickery, A.C. (1991) *Brugia malayi* and *Acanthocheilonema viteae* — antifilarial activity of transglutaminase inhibitors *in vitro*. *Antimicrobial Agents and Chemotherapy*, **35**, 2219–2224.

Ray, C. and Hussey, R.S. (1995) Evidence for proteolytic processing of a cuticle collagen in a plant-parasitic nematode. *Molecular and Biochemical Parasitology*, **72**, 243–246.

Ray, C., Reddigari, S.R., Jansma, P.L., Allen, R. and Hussey, R.S. (1996) Immunochemical analysis of the stage-specific distribution of collagen in the cuticle of *Meloidogyne incognita*. *Fundamental and Applied Nematology*, **19**, 71–78.

Ray, C., Wang, T.Y. and Hussey, R.S. (1996) Identification and characterization of the *Meloidogyne incognita* col-1 cuticle collagen gene. *Molecular and Biochemical Parasitology*, **83**, 121–124.

Reddigari, S.R., Jansma, P.L., Premachandran, D. and Hussey, R.H. (1986) Cuticular collagenous proteins of second-stage juveniles and adult females of *Meloidogyne incognita*: Isolation and partial characterization. *Journal of Nematology*, **18**, 294–302.

Riddiford, L.M. (1993) Hormone receptors and the regulation of insect metamorphosis. *Receptor*, **3**, 203–209.

Riddle, D.L. (1988) The dauer larva. In *The Nematode Caenorhabditis elegans*, edited by W. B. Wood, pp. 393–412. Cold Spring Harbor: Cold Spring Harbor Press.

Riddle, D.L. and Albert, P.S. (1997) Genetic and environmental regulation of dauer larva development. In *C. elegans II*, edited by D.L. Riddle, T. Blumenthal, B.J. Meyer and J.R. Priess, pp. 739–768. Cold Spring Harbor: Cold Spring Harbor Press.

Riddle, D.L. and Bird, A.F. (1985) Responses of *Anguina agrostis* to detergent and anesthetic treatment. *Journal of Nematology*, **17**, 165–168 .

Riddle, D.L., Blumenthal, T. Meyer, B.J. and Priess, J.R. (1997) Introduction to *C. elegans*. In *C. elegans II*, edited by D.L. Riddle, T. Blumenthal, B.J. Meyer and J.R. Priess, pp. 1–22. Cold Spring Harbor: Cold Spring Harbor Press.

Ristoratore, F., Cermola, M., Nola, M., Bazzicalupo, P and Favre, R. (1994) Ultrastructural immuno-localization of CUT-1 and CUT-2 antigenic sites in the cuticle of the nematode *Caenorhabditis elegans*. *Journal of Submicroscopic Cytology and Pathology*, **26**, 437–443.

Robertson, W.M., Kusel, J.R., Proudfoot, L. and Prescott, A.R. (1992) The use of fluorescent probes to study the surface characteristics of *Longidorus elongatus*, *Globodera rostochiensis*, and *Anguina tritici*. *Journal of Nematology*, **24**, 614.

Rogers, W.P. (1962) *The nature of parasitism. The relationship of some metazoan parasites to their hosts*. Academic Press: New York.

Rogers, W.P. (1982) Enzymes in the exsheathing fluid of nematodes and their biological significance. *International Journal for Parasitology*, **12**, 495–502.

Rogers, W.P. and Brooks, E. (1976) Zinc as a co-factor for an enzyme involved in exsheathment in *Haemonchus contortus*. *International Journal for Parasitology*, **6**, 315–319.

Rogers, W.P. and Sommerville, R.I. (1968) The infectious process, and its relation to the development of early parasitic stages of nematodes. *Advances in Parasitology*, **6**, 327–348.

Rosenbluth, J. (1967) Obliquely striated muscle III. Contraction mechanism of *Ascaris* body muscle. *Journal of Cell Biology*, **34**, 15–33.

Salinas. G., Braun, G. and Taylor, D.W. (1994) Molecular characterization and localization of an *Onchocerca volvulus* π-class glutathione S-transferase. *Molecular and Biochemical Parasitology*, **66**, 1–9.

Samoiloff, M.R. and Pasternak, J. (1969) Nematode morphogensis: fine structure of the molting cycles in *Panagrellus silusiae* (de Man 1913) Goodey 1945. *Canadian Journal of Zoology*, **47**, 639–643.

Schmid, K., Franke, S. and Koolman, J. (1987) Detection of immunoreactive ecdysteroids in the serum of animals infected with helminths and the effect of 20-hydroxyecdysone and azadirachtin in the *Caenorhabditis elegans* model. *Zentralblatt für Bakteriologie*, **265**, 508–509.

Sebastiano, M., Lassandro, F. and Bazzicalupo, P. (1991) *cut-1* a *C. elegans* gene coding for a dauer-specific non-collagenous component of the cuticle. *Developmental Biology*, **146**, 519–530.

Selkirk, M.E. (1991) Structure and biosynthesis of cuticular proteins of lymphatic filarial parasites. In *Parasitic nematodes — antigens, membranes and genes*, edited by M.W. Kennedy, pp. 27–45. London: Taylor and Francis.

Selkirk, M.E., Gregory, W.F., Yadzdanbakhsh, M., Jenkins, R.E. and Maizels, R.M. (1990) Cuticular localisation and turnover of the major surface glycoprotein (gp29) of adult *Brugia malayi*. *Molecular and Biochemical Parasitology*, **42**, 41–44.

Selkirk, M.E., Nielsen, L., Kelly, C., Partono, F., Sayers, G. and Maizels, R.M. (1989) Identification, synthesis and immunogenicity of cuticular collagens from filarial nematodes *Brugia malayi* and *Brugia pahangi*. *Molecular and Biochemical Parasitology*, **32**, 229–246.

Shamansky, L.M., Pratt, D., Boisvenue, R.J. and Cox, G.N. (1989) Cuticle collagen genes of *Haemonchus contortus* and *Caenorhabditis elegans* are highly conserved. *Molecular and Biochemical Parasitology*, **37**, 73–86.

Shanta, C.S. and Meerovitch, E. (1970) Specific inhibition of morphogenesis in *Trichinella spiralis* by insect juvenile hormone mimics. *Canadian Journal of Zoology*, **48**, 617–620.

Sharon, E. and Spiegel, Y. (1996) Gold-conjugated reagents for labeling of carbohydrate-recognition domains and glycoconjugates on nematode surfaces. *Journal of Nematology*, **28**, 124–127.

Sharpe, M.J. and Lee, D.L. (1981) Observations on the structure and function of the haemoglobin from the cuticle of *Nippostrongylus brasiliensis*. *Parasitology*, **83**, 411–424.

Shepherd, A.M., Clark, S.A. and Dart, P.J. (1972) Cuticle structure in the genus *Heterodera*. *Nematologica*, **18**, 1–17.

Singh, R.N. and Sulston, J.E. (1978) Some observations on moulting in *Caenorhabditis elegans*. *Nematologica*, **24**, 63–71.

Sluder, A.E., Mathews, S.W., Hough, D., Yin, V.P. and Maina, C.V. (1999) The nuclear receptor superfamily has undergone extensive proliferation and diversification in nematodes. *Genome Research*, **9**, 103–120.

Smith, H.V. (1991) Immune evasion and immunopathology in *Toxocara canis* infection. In *Parasitic nematodes — antigens, membranes and genes*, edited by M.W. Kennedy, pp. 116–139. London: Taylor and Francis.

Smith, H.V., Quinn, R., Kusel, J.R. and Girdwood, R.W.A. (1981) The effect of temperature and antimetabolites on antibody-binding to the outer surface of second-stage *Toxocara canis* larvae. *Molecular and Biochemical Parasitology*, **4**, 183–193.

Smith, K. and Harness, E. (1972) The ultrastructure of the adult stage of *Trichostrongylus colubriformis* and *Haemonchus placei*. *Parasitology*, **64**, 173–179.

Sommerville, R.I. (1957) The exsheathing mechanism of nematode infective larvae. *Experimental Parasitology*, **6**, 18–30.

Sommerville, R.I. (1982) The mechanics of moulting. In *Aspects of parasitology*, edited by E. Meerovitch, pp. 407–433. Montreal: the Institute of Parasitology, McGill University.

Spiegel, Y., Cohn, E. and Spiegel, S. (1982) Characterization of sialyl and galactosyl residues on the body wall of different plant parasitic nematodes. *Journal of Nematology*, **14**, 33–39.

Spiegel, Y., Inbar, J., Kahane, I. and Sharon, E. (1995) Carbohydrate-recognition domains on the surface of phytophagous nematodes. *Experimental Parasitology*, **80**, 220–227.

Spiegel, Y., Kahane, I. Cohen, L. and Sharon, E. (1997) *Meloidogyne javanica* surface proteins: characterization and lability. *Parasitology*, **115**, 513–519.

Spiegel, Y. and McClure, M.A. (1991) Stage-specific differences in lectin binding to the surface of *Anguina tritici* and *Meloidogyne incognita*. *Journal of Nematology*, **23**, 259–263.

Spiegel, Y. and McClure, M.A. (1995) The surface coat of plant-parasitic nematodes: chemical composition, origin and biological role — a review. *Journal of Nematology*, **27**, 17–124.

Spiegel, Y., Mor, M. and Sharon, E. (1996) Attachment of *Pasteuria penetrans* endospores to the surface of *Meloidogyne javanica* juveniles. *Journal of Nematology*, **28**, 328–334.

Spiegel, Y. and Robertson, W.M. (1988) Wheat germ agglutinin binding to the outer cuticle of the plant-parasitic nematode *Anguina tritrici*. *Journal of Nematology*, **20**, 499–501.

Stauffer, H. (1924) Die Lokomotion der Nematoden. *Zoologische Jahrbücher (Sytematik)*, **49**, 1–118.

Stewart, A.C. and Nicholas, W.L. (1992) Structure of the cuticle of *Ceramonema carinatum* (Chromadorida, Ceramonenamatidae). *Journal of Nematology*, **24**, 560–570.

Stewart, G.L., Despommier, D.D., Burnham, J. and Raine, K.M. (1987) *Trichinella spiralis*: Behaviour, structure and biochemistry of larvae following exposure to components of the host enteric environment. *Experimental Parasitology*, **63**, 79–85.

Sulston, J.E., Schierenberg, E., White, J.G. and Thomson, J.N. (1983) The embryonic cell lineage of the nematode *Caenorhabditis elegans*. *Developmental Biology*, **100**, 64–119.

Tang, L., Gounaris, K., Griffiths, C. and Selkirk, M.E. (1995) Heterologous expression and enzymic properties of a selenium-independent glutathione peroxidase from the parasitic nematode *Brugia pahangi*. *Journal of Biological Chemistry*, **270**, 18313–18318.

Tang, L., Ou, X., Henkle-Dührsen, K. and Selkirk, M.E. (1994) Extracellular and cytoplasmic CuZn superoxide dismutases from *Brugia* lymphatic filarial nematode parasites. *Infection and Immunity*, **62**, 961–967.

Tchesunov, A.V., Malakhov, V.V. and Yushin, V.V. (1996) Comparative morphology and evolution of the cuticle in marine nematodes. *Russian Journal of Nematology*, **4**, 43–50.

Thust, R. (1966) Elektronenmikroskopische Untersuchungen über den Bau des larvalen Integumentes und zur Häutungsmorphologie von *Ascaris lumbricoides*. *Zoologischer Anzeiger*, **177**, 411–417.

Timinouni, M. and Bazzicalupo, P. (1997) *cut-1*-like genes of *Ascaris lumbricoides*. *Gene*, **193**, 81–87.

Urbancik, W., Bauer-Nebelsick, M. and Ott, J.A. (1996) The ultrastructure of the cuticle of Nematoda. I. The body cuticle within the Stilbonematinae (Adenophorea, Desmodoridae). *Zoomorphology*, **116**, 51–64.

van Bömmel, A. (1895). Uber Cuticular-Bildungen bei einigen Nematoden. *Arbeiten aus Wursburg Zoologischen Instituten*, **10**, 191–212.

Van der Eycken, W., de Almeida Engler, J., Van Montagu, M. and Gheysen, G. (1994) Identification and analysis of a cuticular collagen-encoding gene from the plant-parasitic nematode *Meloidogyne incognita*. *Gene*, **151**, 237–242.

Vetter, J.C. and Klaver-Wesseling, J.C. (1978) IgG antibody binding to the outer surface of infective larvae of *Ancylostoma caninum*. *Zeitschrift für Parasitenkunde*, **5**, 91–96.

Wang, T., Deom, C.M. and Hussey, R.S. (1998) Identification of a *Meloidogyne incognita* cuticle collagen gene and characterization of the developmental expression of three collagen genes in parasitic stages. *Molecular and Biochemical Parasitology*, **93**, 131–134.

Warbrick, E.V., Barker, G.C., Rees, H.H. and Howells, R.E. (1993) The effect of invertebrate hormones and potential hormone inhibitors on the third larval moult of the filarial nematode, *Dirofilaria immitis*, *in vitro*. *Parasitology*, **107**, 459–463.

Waterston, R.H. (1988) Muscle. In *The nematode Caenorhabditis elegans* edited by W.B. Wood, pp. 281–335. Cold Spring Harbor: Cold Spring Harbor Laboratory.

Watson, B.D. (1965a) The fine structure of the body-wall in a free-living nematode, *Euchromadora vulgaris*. *Quarterly Journal of Microscopical Science*, **106**, 75–81.

Watson, B.D. (1965b) The fine structure of the body-wall and the growth of the cuticle in the adult nematode *Ascaris lumbricoides*. *Quarterly Journal of Microscopical Science*, **106**, 83–91.

Weis-Fogh, T. (1970) Structure and formation of the insect cuticle. *Symposia of the Royal Entomological Society, London*, **5**, 165–185.

Wharton, D.A. and Bone, L.W. (1988) The formation of the first-stage cuticle within the egg of *Trichostrongylus colubriformis*. *Parasitology*, **97**, 459–467.

Wigglesworth, V.B. (1972) *The principles of insect physiology*. London: Chapman and Hall.

Wildenberg, G., Liebau, E. and Henkle-Dührsen, K. (1998) *Onchocerca volvulus*: Ultrastructural localization of two glutathione *S*-transferases. *Experimental Parasitology*, **88**, 34–42.

Wilson, P.A.G. (1976) Nematode growth patterns and the moulting cycle: The population growth profile. *Journal of Zoology*, **179**, 135–151.

Wisse, E. and Daems, W.T. (1968) Electron microscopic observations on second-stage larvae of the potato root eelworm *Heterodera rostochiensis*. *Journal of Ultrastructure Research*, **24**, 210–231.

Wright, K.A. (1968) The fine structure of the cuticle and interchordal hypodermis of the parasitic nematodes, *Capillaria hepatica* and *Trichuris myocastoris*. *Canadian Journal of Zoology*, **46**, 173–179.

Wright, K.A. (1975a) Cuticular inflations in whipworms, *Trichuris* spp. *International Journal for Parasitology*, **5**, 461–463.

Wright, K.A. (1975b) Cephalic sense organs of the rat hookworm, *Nippostrongylus brasiliensis* — form and function. *Canadian Journal of Zoology*, **53**, 1131–1146.

Wright, K.A. (1987) The nematode's cuticle — its surface and the epidermis: function, homology, analogy — a current consensus. *Journal of Parasitology*, **73**, 1077–1083.

Wright, K.A. and Carter, R. (1980) Cephalic sense organs and body pores of *Xiphinema americanum* (Nematoda: Dorylaimoidea). *Canadian Journal of Zoology*, **58**, 1439–1451.

Wright, K.A. and Chan, J. (1973) Sense receptors in the bacillary band of trichuroid nematodes. *Tissue and Cell*, **5**, 373–380.

Wright, K.A. and Hong. H. (1988) Characterization of the accessory layer of the cuticle of muscle larvae of *Trichinella spiralis*. *Journal of Parasitology*, **74**, 440–451.

Wright, K.A. and Hong, H. (1989) *Trichinella spiralis*: The fate of the accessory layer of the cuticle of infective larvae. *Experimental Parasitology*, **68**, 105–107.

Wright, K.A. and Hope, W.D. (1968) Elaborations of the cuticle of *Acanthonchus duplicatus* Wieser, 1959 (Nematoda: Cyatholaimidae) as revealed by light and electron microscopy. *Canadian Journal of Zoology*, **46**, 1005–1011.

Wright, K.A. and Perry, R.N. (1991) Moulting of *Aphelenchoides hamatus* with special reference to formation of the stomatostyle. *Revue de Nématologie*, **14**, 497–504.

Yochem, J., Tuck, S., Greenwood, I. and Han, M. (1999) A gp330/megalin-related protein is required in the major epidermis of *Caenorhabditis elegans* for completion of molting. *Development*, **126**, 597–606.

Yushin, V.V. and Malakhov, V.V. (1992) [Body cuticle formation in embryogenesis of a free-living marine nematode *Halichoanolaimus sonorus* (Chromadorida, Selachinematidae)]. *Zoologichesky Zhurnal*, **71**, 23–30. [In Russian].

Yushin, V.V. and Malakhov, V.V. (1994) Ultrastructure of the body cuticle of free-living marine nematodes. *Russian Journal of Nematology*, **2**, 83–98.

Zuckerman, B.M., Himmelhoch, S. and Kisiel, M. (1973). Fine structure changes in the cuticle of adult *Caenorhabditis briggsae* with age. *Nematologica*, **19**, 109–112.

Zuckerman, B.M. and Kahane, I. (1983) *Caenorhabditis elegans*: Stage specific differences in cuticle surface carbohydrates. *Journal of Nematology*, **15**, 535–538.

Zuckerman, B.M., Kahane, I. and Himmelhoch, S. (1979) *Caenorhabditis briggsae* and *C. elegans*: partial characterization of cuticle surface carbohydrates. *Experimental Parasitology*, **47**, 419–424.

Zvelebil, M.J.J., Tang, L., Cookson, E., Selkirk, M.E. and Thornton, J.M. (1993) Molecular modelling and epitope prediction of gp29 from lymphatic filariae. *Molecular and Biochemical Parasitology*, **58**, 145–154.

8. Feeding and Digestion

Edward A. Munn and Patricia D. Munn

Department of Immunology, Babraham Institute, Cambridge, CB2 4AT, UK

Introduction

Habitat and Habit

The enormous range of habitats of nematodes is paralleled by the diversity of what they feed on and the form in which they take in their food. This, in turn, is reflected in a variety of behavioural and anatomical characteristics and digestive and absorptive processes.

In general, depending on the species, food will consist of either, bacteria, algae, diatoms, protozoa, fungi, other nematodes, other invertebrates, plant tissues or animal tissues (Table 8.1). For some parasitic nematodes the larval (juvenile) stages may have completely different feeding habits to the adults. A few nematodes are omnivores. Finding food involves sensory structures and processes and associated behavioural responses. Although similar structural features associated with a particular feeding habit seem to have evolved several times, within any one group the occupation of a particular feeding niche has phylogenetic implications. It is noticeable that certain feeding behaviours are restricted to quite small groupings of nematodes. Thus parasitism of plant roots is confined to two families of dorylaimid and one suborder of tylenchid nematodes. Some of these topics are more fully considered in other chapters, but this chapter covers food, gut anatomy, aspects of feeding behaviour, digestion, nutrient uptake and defecation. For individual nematode species, the level of knowledge across these subjects ranges from the most superficial to the very detailed. Surprisingly, complete details of feeding, digestion and nutritional requirements are not known for even one species. There are other surprises, for example that blood-feeding nematodes which one might imagine only required an absorptive surface to acquire all the nutrients they need, nevertheless are dependent on digestive enzymes homologous to those in the mammalian gut.

Structure of the Gut

Overview

In general, the gut in nematodes consists of a simple tube (occasionally with diverticula) running from the anterior tip to near the posterior tip. It is composed of three distinct regions, the stomadeum, the intestine (mesenteron, or mid-gut) and the proctodeum. The stomadeum consists of the mouth (stoma), the buccal cavity and the muscular pharynx which in most

Table 8.1. Sources of food for nematodes

Food	Examples	Feeding site
Bacteria	*Caenorhabditis*	soil (and laboratory)
	Haemonchus (L1 & L2)	ruminant faeces
	Rhabditis adenobia	colleterial glands of beetles
	Tachygonetria stylosa	colon of tortoise
Diatoms	*Monohystera*	marine muds
Algae	*Halenchus fucicola*	seaweed thallus
Fungi	*Ditylenchus destructor*	potato tubers
	Aphelenchoides composticola	mushroom beds
	Aphelenchoides parietinus	lichen
	Bursaphelenchus	conifer parenchyma, roots and stems
Non-vascular plants		
	Tylenchus davainei	mosses
	Tylenchus farwicki	liverworts
	Aphelenchoides fragariae	ferns
Vascular plants		
Leaf	*Anguina tritici*	mesophyll
root tip	*Longidorus*	cortex
root	*Trichodorus*	epidermis
root	*Globodera*	vascular tissue
root	*Meloidogyne*	vascular tissue
bulb	*Hoplolaimus*	parenchyma
root, aerial tissue	*Aphelenchoides*	
Invertebrates		
Whole nematodes	*Mononchus*	soil,
haemolymph	*Sphaerularia bombi*	bumble bee haemocoel
gut contents	*Brumptaemilius*	millipede gut
fat body cells	*Physaloptera phrynsoma* (L)	ant fat body
haemolymph	*Beddingia (Deladenus) siricidicola* (adult)	woodwasp haemocoel
Vertebrates		
Blood	*Haemonchus* (L4-adult)	ruminant stomach mucosa

Table 8.1. (*continued*)

Food	Examples	Feeding site
blood	*Syngamus*	bird trachea
blood, mucosa	*Necator* (L4-adult)	man intestinal mucosa
blood	*Dirofilaria immitis*	dog ventricle and pulmonary arteries
digesta	*Ascaris* (adults)	pig, man gut lumen
digesta	*Tachygonetria robusta*	tortoise gut lumen
digesta	*Physaloptera phrynsoma* (adult)	lizard stomach
mucosa	*Nippostrongylus*	rat mucosal tissue
secretions	*Dictyocaulus*	ruminant bronchial tree
secretions	*Litomosoides*	cotton rat pleural cavity
lymph	*Wuchereria/Brugia*	man lymphatic system
lachrymal secretions	*Oxyspirura mansoni*	bird nictitating membrane
kidney	*Dioctophyme renale*	mink kidney tissue
hepatic tissue	*Capillaria hepatica*	rat liver

nematodes functions to pump food into the intestine. The stomadeum is lined with cuticle which may be modified, sometimes very intricately, into a variety of structures (plates, teeth and hollow stylets) that facilitate feeding. Sensilla and glands involved in the localisation and modification of food are present. The intestine is the principal digestive and absorptive region of the gut; the wall is one cell thick and the luminal surface is covered with microvilli. A unicellular sphincter muscle marks the beginning of the procto-deum which, like the stomadeum, is lined with cuticle, but otherwise is a simple tubular entity. Rectal glands may be present. In some adult parasitic nematodes the proctodeum degenerates so that there is no posterior opening to the gut.

Because it has long been the basis for identification and classification, the topic about which most is known for most nematodes is variation in structural organisation. This is particularly true of the stomadeum.

The Stomadeum

The stomadeum is lined with cuticle continuous with (but distinguishable from) that covering the rest of the body and which is, like the rest of the cuticle, secreted by the underlying cells. The stomadeum consists of the mouth, the buccal cavity and the pharynx.

The mouth (stoma)

The mouth is usually surrounded by cuticular pro-tuberances (lips) with one dorsal pair and two ventral pairs, some or all of which may be fused together and may be formed into elaborate prolongations. Numer-ous paired sensillae, arranged in concentric rings end in or open onto the surface of the lips and just posteriorly are the openings of the paired amphidial ducts. It is

likely that these sensory structures are involved in activities in addition to feeding. In amphids in particular (and possibly other sense organs) the nerve processes are surrounded by gland cells that release secretions. At least for some animal parasites these secretions are known to have a role in feeding.

The buccal cavity

The buccal cavity consists of five regions; from the anterior end in order these are designated the cheilo-, pro-, meso-, meta- and telo-stom (Figure 8.1). The cuticle lining each of these regions is correspondingly desig-nated the cheilo-, pro-, meso-, meta- and telo-rhabdion (Inglis, 1964). The structure of this region has been the subject of very detailed study, principally for purposes of distinguishing otherwise very closely related similar species. It must be remembered however that the differences probably reflect differences in feeding habit. Whilst some of these differences may be very subtle, others are very major. Structural differences range from variations in the diameter of the mouth (from 1–25 μm) and shape of the buccal cavity (cylindrical, cup-shaped or globular) to modifications of the rhabdions to form teeth or hooks (Figures 8.2, 8.7, 8.8), or their fusion to form a stylet (Figure 8.2) the lumen of which (in those forms in which it is hollow) is used as a passage for secretion of products of the pharyngeal glands and intake of digesta.

The pharynx

Anteriorly the pharynx is interfaced to the epidermis by way of two arcade cells, each of which is a complete toroid. The basic structural features of the pharynx (Figure 8.3) are a tri-radiate lumen, muscles with radially arranged fibrils attached between the cuticle lining the lumen and the encircling basement lamella, one or more

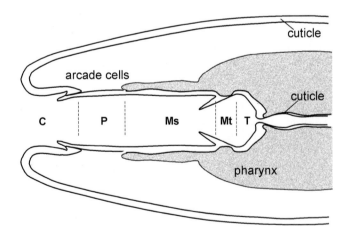

Figure 8.1. Diagram of a longitudinal section showing the five regions of the stoma; from the anterior end in order these are designated the cheilo-, pro-, meso-, meta- and telo- stom (labelled C, P, Ms, Mt and T respectively). The cuticle lining each of these regions (secreted by the underlying cells) is correspondingly designated the cheilo-, pro-, meso-, meta- and telo- rhabdion (Inglis, 1964). Diagram based on electron micrographs of *Caenorhabditis elegans* published by Wright and Thomson (1981).

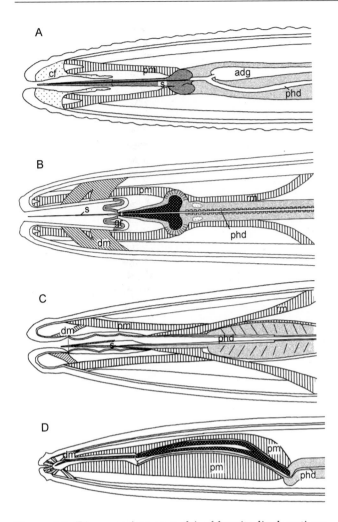

Figure 8.2. Diagrams (not to scale) of longitudinal sections to show modifications of the rhabdions to form stylets. A) *Meloidogyne*, contraction of protractor muscles (pm) attached anteriorly to a hexaradiate cuticular cephalic framework (cf) and to knobs at the base of the stylet (s) pull its tip out through the rigid entrance to the stoma. Retraction of the stylet when the muscles relax is due to elasticity in the attached pharyngeal tissue. The axial lumen of the stylet is continuous with the pharyngeal duct (phd). (adg, ampulla of dorsal gland.) B) *Xiphinema* has a hollow needle-like tip to its stylet. The lips of the stoma are pulled open by dilator muscles (dm) as the stylet protractor muscles (pm) contract. The withdrawal of the stylet is facilitated by the contraction of retractor muscles (rm). The stylet moves within a collar of highly flexible cuticle, the guide ring (gr). The stylet of *Nygolaimus* (C) is a ventrally placed spear whilst that of *Trichodorus* (D) is an elongate, ventrally curved, protrusible dorsal mural tooth, operated by large protractor muscles (pm). (Micrographs of rigid mural teeth are shown in Figures 8.7 and 8.8.) Diagrams based on Coomans (1963), Wright (1965) and Taylor *et al.* (1970).

valves, dorsal and ventral glands and neurones. The basic shape is cylindroid (Figure 8.3 A). It is encircled by a nerve ring; the regions anterior and posterior to which are designated the corpus and post-corpus respectively. Depending on the arrangement of the radial muscles, the

localised shape of the lumen and the extent of the glands, these may consist of narrowed or expanded regions designated procorpus, metacorpus, isthmus and posterior bulb (Figure 8.3 B–D). Two kinds of specialisations occur at the posterior end of the pharynx. Members of two types of parasitic nematodes (mermithids and trichurids) have a longitudinal series of large gland cells (stichocytes) that collectively form the stichosome, whilst in another group there are several caeca (diverticula) arising from a short glandular ventriculus between the posterior end of the pharynx and the intestine.

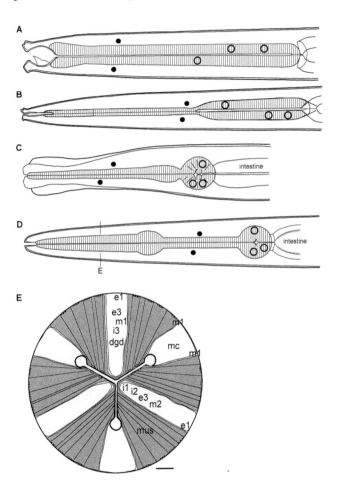

Figure 8.3. Diagrams to show the basic structural features of the pharynx. Some variations in the layout of the pharynx (striped) are illustrated by A) *Mononchus*, B) *Dorylaimus*, C) *Enterobius* and D) *Caenorhabditis*. The position of the nerve ring is shown by filled circles and the positions of glands by open circles. E) Cross-section of the pro-corpus of *Caenorhabditis* (at the position shown by the line E in D) shows the characteristic arrangement of the three pharyngeal muscles (mus) with radially arranged fibrils attached by hemidesmosomes between the cuticle lining the tri-radiate lumen and the encircling basement lamella. e1, e3, i1, i2, i3, m1 and m2 indicate the positions of epithelial cell processes, interneurones, and motor neurones respectively. mc, marginal cell; dgd, dorsal gland cell duct. The scale bar represents 1 μm. (A–D based on diagrams in Chitwood and Chitwood (1974), Maggenti (1981) and Bird and Bird (1991); E based on electron micrographs published by Albertson and Thomson, 1976.)

Cells at the posterior end of the pharynx (commonly described as a valve) mark the junction with the intestine. The pharyngeal-intestinal valve in *C. elegans* consists of six cells coupled to each other, and to the pharynx and the first intestinal cells, by desmosomes (Sulston *et al.*, 1983). Somewhat surprisingly, these cells are outside the basement membrane of the pharynx and intestine (White, 1988).

The Intestine

In almost all nematodes the intestine is a simple tube composed of a single layer of cells ensheathed by a basal lamina. In some cases, based on structural differences in the cells and thought to represent functional (secretory versus absorptive) differences, it is possible to distinguish anterior, middle and rear regions of the intestine, as in adult *Nematodirus battus* (Lee and Martin, 1980) and infective larvae of *Steinernema carpocapsae* in which the distinct anterior (ventricular) region carries bacteria into the host (Bird and Akhurst, 1983). The simple tube form is replaced in some instances (some plant- and animal-parasitic nematodes) by greatly enlarged structures described later.

The luminal surface of feeding nematodes is in the form of microvilli which even in adults may be short and sparse (as in *Muellerius capillaris*), long and numerous (as in *Ascaris suum* and *Haemonchus contortus* (Munn and Greenwood, 1984)), or somewhere in between (as in *Nematodirus battus* (Lee and Martin, 1980)). The microvilli may vary in length, from 0.8 μm in the anterior region to 0.5 μm in the pre-rectal region in *Nematodirus* for example, or be of similar length throughout the whole intestine, as in adult *H. contortus* (Figure 8.4). In general the microvilli are about 0.1 μm in diameter (the long microvilli of *Haemonchus* taper from 0.09 μm at the base to 0.06 μm at the tip) and are supported by an axial core of actin filaments (Figures 8.4, 8.5 A) that extend into the apical cytoplasm (as in *Ascaris* for example) or fuse with the cytoskeletal entity designated an endotube (Figures 8.4, 8.5 B) (Munn, 1981; Munn and Greenwood, 1984; Munn, 1983). The luminal plasma membrane, including that over the microvilli, is often coated with an amorphous glycocalyx probably composed of proteins and carbohydrates. In a few cases (all parasites in animals, infective larvae of *Trichinella spiralis* (Bruce, 1966); adult *Syphacia obvelata*, and *Haemonchus placei* (Smith and Harness, 1972), *Rhigonema* (Wright, 1991) and *H. contortus* (Munn, 1977)) regular, discrete structures visible in electron micrographs have been described coating the microvilli. The material from *H. contortus* has been shown to be a polymer of a protein, contortin (Munn, 1977). Biochemical analyses have identified a number of other proteins at the surface of the intestinal cells as described later (see page 225).

One striking difference between different genera is in the number of cells present. The range is from the dozen or so in rhabditids to the million or so in large ascarids. The latter are columnar, the former hemi-toroidal. In nematodes with small numbers of intestinal cells these sometimes form a multinucleate syncytium (as for

Figure 8.4. Electron micrograph of intestinal microvilli of an adult *Haemonchus contortus*. The long, densely packed microvilli of *Haemonchus* (and the similar arrays in other animal-parasitic nematodes such as *Ancylostoma*, *Syngamus* and *Ascaris*) provide an enormous potential digestive and absorptive surface. a, actin core of microvilli (compare Figure 8.5A); c, contortin; e, endotube; mv, microvilli. Scale bar represents 1 μm. (E.A. Munn and C.A. Greenwood, unpublished.)

example in *Ancylostoma* and *Haemonchus* (Munn, 1983)). Transport processes that depend on lateral membranes and inter-cellular spaces cannot operate in these syncytia and this may be compensated for by the observed substantial development of highly infolded basal membranes.

A few species are characterised by the possession of one, or very rarely two, diverticula usually arising from the anterior region of the intestine and extending anteriorly adjacent to the pharynx. Their function is not known, although, in one notable example to be described later, they are utilised as a storage site for bacteria.

Some, very few, nematodes have muscles from the intestine to the body wall. Members of the Dioctophymatina possess four sub-median longitudinal rows of transverse muscles, observed in *Eustrongylides ignatus* to have a role in the peristaltic movement of the intestinal contents (Chitwood and Chitwood, 1974). Similarly, in *Aspiculuris tetraptera*, muscles in the form of a network of fibrous material connecting the basement membrane with the body wall also bring about peristaltic-like movement (Lee and Anya, 1968) but it is not known why these nematodes need the muscles when so many manage without. In *C. elegans* a pair of muscles with longitudinal filaments are wrapped around the ventral surface of the posterior end of the intestine and connect it with the body wall (White, 1988). The adjacent sphincter muscle around the posterior end of the intestine usually

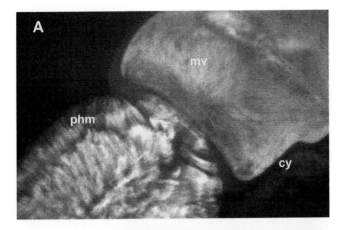

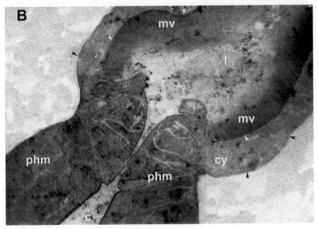

Figure 8.5. Light micrographs of the pharyngeal-intestinal junction of adult *Haemonchus contortus*. A) whole mount (confocal microscopy) showing distribution of F-actin by binding of FITC-phalloidin in the muscles of the pharynx (phm) and within the intestinal microvilli (mv); B) Section of Araldite-embedded specimen through region shown in A. c, cuticle lining the lumen of the pharynx; cy, basal cytoplasm of intestinal cell (just visible due to background fluorescence); l, lumen of intestine, mv, microvilli; phm, pharyngeal muscle. The white arrowheads label the endotube, black arrowheads show the position of the basal membrane. The scale bar represents 50 μm. (E.A. Munn, C.A. Greenwood and P. James, unpublished.)

marks its termination and is often associated with an intestino-rectal valve formed by a luminal projection of the terminal intestinal cells.

The Proctodeum

The proctodeum is a simple tube lined with cuticle. The lumen may be rounded, somewhat flattened or irregular in cross-section, but in no way matches the variations seen in the shape and structure of the stomadeum. Although continuous with the external cuticle, the cuticle lining the proctodeum is similar in appearance and staining reactions to the epicuticle (Maggenti, 1981). In some groups glands open into the proctodeum. These are particularly large in egg-laying female *Meloidogyne* in which they secrete glycoprotein

that forms a gelatinous matrix enmeshing the eggs (Dropkin and Bird, 1978). Extrusion of the matrix is aided by contractions of the depressor ani muscle. This is H-shaped and consists of two groups of fibres extending between the proctodeum and the dorsolateral body wall with the nucleus in a band forming the cross-bar of the H. In other nematodes this muscle, with others that may be present at the posterior end of the intestine, is involved in defecation.

Food and Feeding

At its simplest, food is either particulate (microorganisms and other comparatively small organisms, fragments of larger organisms) or in solution (cytosol, haemolymph, blood plasma), but in the majority of cases it is live or obtained from live plants and animals. Nematodes that feed on particles may have to separate them from a bulk medium, while those that feed on cell and body fluids may selectively avoid organelles or cells respectively. Feeding behaviour involves localisation of the food and then its ingestion. With only a few exceptions (some parasitic nematodes, bathed in nutrients) described later, the food is taken in by a pumping process due (with one known exception) to the co-ordinated activity of the pharyngeal muscles. Feeding may also involve nibbling, piercing or biting with concomitant movements of specialisations of the mouthparts. Thus it is possible to consider feeding in terms of either what the nematode feeds on (Table 8.1) or the behavioural and structural features (Table 8.2)

Table 8.2. Nematode feeding strategies

Type	Examples	Habitat/Food
Filterers	*Rhabditis*	soil, bacteria
	Caenorhabditis	soil, bacteria
	Acrobeloides	soil, bacteria
	Alaimus	soil, bacteria
Piercers	*Trichodorus*	root, epidermis
	(See Table 8.3 for hollow stylet piercers.)	
Plug feeders	*Necator*	mammal gut, mucosa, blood
	Ancylostoma	mammal gut, mucosa, blood
Engulfers	*Mononchus*	soil, nematodes
	Ironus	soil, nematodes
Absorbers	*Mermis*	invertebrates, haemolymph
	Sphaerularia	bumble bee, haemolymph
	Eudiplogaster larva	dung beetle, haemolymph
	Deladenus female	woodwasp, haemolymph

which enable it to acquire and ingest its food. Within each feeding type there can be a broad range of structural features. In the following account, the nematodes are grouped first according to their main food type and then, where appropriate, according to structural features. Grouping by food type is a little arbitrary for the so-called free-living nematodes since in many cases it is very difficult to determine exactly what a particular nematode is feeding on. The groupings used here are intended to illustrate the range of possibilities.

Microbiovores

Included in this group are nematodes that ingest bacteria, blue-green algae or other uni-cellular micro-organisms. Microbivorous nematodes live everywhere that the micro-organisms that they can eat occur. In effect, this means that they live everywhere (Alkemade, 1992; van de Velde and Coomans, 1991; Overhoff, 1993; Yeates, 1998; Schad, 1963; Bird and Akhurst, 1983; Bird and Bird, 1991). Three examples of habitats considered briefly here are soil, a vertebrate digestive tract and within a parasitised insect.

Soils of all kinds are infested with a wide range of nematodes (Yeates , 1998) . Among those feeding on micro-organisms there are bacteriovores, algivores and protozoavores. The most completely studied member of the group is the rhabditid *Caenorhabditis elegans*. In its natural habitat *C. elegans* feeds on mixed bacterial populations, but it has been very extensively cultured in the laboratory on agar plates coated with mono-cultures of *E. coli*. In addition it can be cultured in defined liquid medium and much is known about its dietary requirements and nutrition. Foraging for food by soil-dwelling bacterivorous nematodes has been studied by Young, Griffiths and Robertson (1996). In the absence of bacteria *C. elegans*, for example, moves randomly with unorientated searching behaviour, but once an appropriate signal is detected it will move along the signal gradient towards bacteria. The signals, detected by sensillae around the stoma, include CO_2, and an array of other chemicals released by the bacteria (Grewal and Wright, 1992). Some microbivorous nematodes (e.g., the rhabditid *Acrobeles* and the chromadorid *Wilsonema*) have elaborate cephalic structures (Thorne, 1925) that may aid in detecting or selecting their food.

Large numbers of micro-organisms flourish in the digestive tracts of vertebrates and it would be surprising if these did not form at least part of the diet of nematodes, such as *Heterakis*, *Enterobius* and *Syphacia*, that live there. *Tachygonetria stylosa* is representative of one group of such nematodes in that, as an adult, it lives in the colon of a vertebrate host (*Testudo*) but nonetheless feeds primarily on bacteria. The related species *T. robusta* shares the same site but feeds indiscriminately on the colon contents (Schad, 1963) and thus should perhaps be considered an omnivore (see page 222).

Larvae of *Steinernema* and its close relatives are particularly noteworthy because although they parasitise insects, they feed on bacteria (Gaugler and Kaya, 1990). They carry the spores of *Xenorhabdus* in a vesicular diverticulum arising from the intestine (Bird and Akhurst, 1983). Once the nematode has entered its host's haemocoel the bacteria are released. They multiply, digest the host tissues and then are, in turn, ingested by the nematode. Different species of *Steinernema* (and *Heterorhabditis*) carry different species of *Xenorhabdus*. The mechanisms controlling retention and release of the bacteria, or how (or indeed why) bacteria are eaten selectively are not known.

The restriction on the diet of *Tachygonetria stylosa* is attributed to the comparatively small size of its stoma (Schad, 1963). The bacteria-feeding rhabditids, in general, are characterised by having a relatively narrow stoma (1–4 μm across). Other bacterial feeders, however, e.g., *Cephalobus persegnis* and *Diplogaster* have broad stomas (some 5–7 μm diameter). Conversely, the enoplid *Alaimus thompsoni* has a narrow stoma (diameter about 1 μm) but apparently is able to ingest blue-green algae up to 17 μm in diameter (Mulk and Coomans, 1979) implying either that the stoma (and other components of the stomadeum) is able to stretch enormously or that the alga can be compressed.

With the possible exception of diplogasterids which have dentition which could rupture the prey, micro-biovores ingest prey whole and then disrupt it in the pharynx (see page 223).

Predatory Nematodes

Predatory nematodes feed on other invertebrates. Most of the known ones live in soil (Small, 1987), but some (e.g. *Mononchus* and *Tripyla*) live in freshwater and others (e.g. *Halichoanolaimus*, *Sphaerolaimus*) are marine. They may be classified as either piercers or engulfers, corresponding to the piercers and ingesters of Yeates *et al.* (1993). Piercers are equipped with a protrusible stylet with which the nematode can pierce prey such as rotifers, other nematodes and small oligochaetes, sucking out the body contents. The stylet is hollow and either occupies the whole stoma and is open to the tip (and is used like a hypodermic needle to withdraw the body contents) or placed laterally in the stoma and closed at the tip (Figure 8.2) in which case it is used like a lancet (and is called a mural tooth). The open-tip stylets of predaceous piercers in general have a comparatively wide lumen (Table 8.3); ranging from 3.5 μm in *Discolaimus* (Coomans, 1963) up to 6.0 μm for some *Aporcelaimus* species (Coomans and Van der Heiden, 1971). Piercers with a closed stylet tip ingest through the stoma which, hence, needs to be closely applied to the wound. How these piercers retain contact with their active prey does not seem to have been described, but is well known for engulfers such as *Mononchus* (Cobb, 1917; Esser, 1964) which have a large stoma able to grasp the prey (Clark, 1960). The stoma is often equipped with large teeth which may be movable as in *Ironus* (Van Der Heiden, 1974), or immovable as in *Mononchus*. *Actinolaimus* is equipped with both teeth and stylet; the latter is used first to inject a toxin and then to remove the fluid contents (Poinar, 1983). *Seinura* are also known to inject a paralysing toxin (Poinar,

Table 8.3. Representative examples of hollow stylet feeders

Type	Nematode	Stylet length (μm)	(Stylet lumen diameter), typical feeding site, comments
Predator	*Discolaimus*	25–40	(3.5 μm)
	Dorylaimus	30	(4.5 μm)
	Aporcelaimus	25–35	(6.0 μm)
Fungivores	*Aphelenchoides*	12–20	(0.2 μm)
	Tylencholaimus	15–25	0.5–1.3 μm)
	Leptonchus	20–25	(~1 μm)
Plant ecto-parasites			(most 0.1–0.3 μm)
	Tylenchulus	13	citrus roots, body penetrates to pericycle
	Tylenchorynchus	15–30	root, epidermis
	Paratylenchus	30	root, epidermis
	Xiphinema	50–200	root tip, meristem
	Longidorus	100–300	(0.2–0.7 μm) root tip, cortex
	Criconemella	30–120	peach roots, cortex
	Hemicycliophora	80	root tips, semi-sessile
	Dolichodorus	50–160	root, epidermis and cortex
	Gracilacus	50–120	root, epidermis, root hairs
Plant endo-parasites			(all 0.1–0.3 μm)
	Ditylenchus	10	stem and bulb tissue
	Pratylenchus	< 20	root, cortex, migrating
	Radopholus	14–23	root, cortex, migrating
	Meloidogyne	10–20	root, differentiating vascular tissue
	Globodera	20–40	root, xylem tracheids
	Heterodera	20–40	root, xylem parenchyma

1983). The limited data available from *in vitro* cultures indicates that of the nematodes, the bacteriovores are the principle prey of dorylaimid, diplogasterid and mononchid predatory nematodes (Bilgrami, 1992, 1993) except that *Rhabditis* and *Pelodera* frequently escape attack (Bilgrami, 1993; Grootaert, Jacques and Small, 1977).

For both piercers and engulfers, localisation of prey may result from random encounters, but it seems reasonable to suppose that like bacteria-feeding nematodes, the predators have relevant chemosensory organs. Some predatory nematodes, such as *Diplenteron colobercus* may be cultured on bacteria (Yeates, 1969), but whether this is a significant source of food in the natural environment is not known.

Parasites of Fungi and Plants

A wide range of nematodes parasitise fungi, multicellular algae, mosses and vascular plants. In common with the predatory piercers, all possess a stylet or spear with which they penetrate the wall of the host (Table 8.3) to give access to host cell contents.

Fungivores

Fungivores are ectoparasites on fungal hyphae, utilising a protrusible hollow stylet (Figure 8.2) to pierce the hyphal wall and withdraw the fluid contents. Some, for example *Aphelenchoides blastophthorus* and *A. bylurgi*, seem able to utilise a broad range of fungi equally well (Riffle, 1967; Bird *et al.*, 1989), but others such as the potato rot nematode *Ditylenchus destructor* are more restricted in their diet (Faulkner and Darling, 1961). All fungivorous nematodes have stylets with a narrow lumen (0.2 μm diameter in *Aphelenchoides* (Shepherd, 1980) and ranging from 0.5–1.3 μm in nine species of *Tylencholaimus* (Pena Santiago and Coomans, 1994)). Since all the food (cell contents) is ingested through the stylet this means in effect that the vast majority of intact cell organelles will be excluded; a phenomenon common to many of the nematodes parasitic on plants.

Parasites of plants

Nematodes parasitic on plants are restricted to members of three orders: the Aphelenchida, Tylenchida and the Dorylaimida. They all feed on essentially the same food, i.e. cell contents, but they differ in the site and type of cells they feed on; structure of the stylet or spear; the extent of associated gland cells and changes brought about in the cells fed on. In addition, nematodes parasitic on vascular plants may be grouped broadly according to whether, like fungivores, they remain outside the host or are inside and whether they move from site to site (when they are said to be migratory), or remain at one site for extended periods (sedentary).

The aphelenchids are mostly parasites of the aerial parts or shoots of plants and may feed on more than one tissue. Some, such as *Aphelenchoides ritzemabosi*, feed ectoparasitically on one host and endoparasitically in another, although this is unusual. Tylenchids that feed on the aerial parts of plants include *Anguina tritici*, an ectoparasite on young leaves of wheat seedlings and the first plant-parasitic nematode to be described (Needham, 1744). Root-parasitic tylenchids include the cyst nematodes (*Globodera, Heterodera*) and root-knot nematodes (*Meloidogyne*). Their larval (non-feeding) stages are migratory, but the parasitic adult females are sedentary and feed from sites within the roots. All the plant-parasitic dorylaimids are migratory ectoparasites; the trichodorids (e.g. *Trichodorus, Paratrichodorus*) browse from superficial epidermal cells and the longidorids (e.g. *Longidorus, Xiphinema*) feed from deeper tissues.

Three aspects will be considered briefly here; how the nematodes locate their feeding sites, how they access their food and the changes that may be induced in the host. The sedentary root parasite *Meloidogyne* will be used as the principal example. *Meloidogyne* species infect over 2,000 species of plants and are a cause of major economic losses.

The details of how *Meloidogyne* and other plant-parasitic nematodes in soil locate their hosts are not known, but two general mechanisms have been proposed for localisation of host tissue, chemo-attraction and potential gradient (these of course are not mutually exclusive) (see Chapter 15). As proposed for localisation of food by bacteriovores, detection of and movement along carbon dioxide concentration gradients have been put forward as significant components of plant root tip localisation (Pline and Dusenbery, 1987), but as with potential gradients generated by differences in redox potentials (Bird, 1959) these alone cannot account for the specificity shown by many plant-parasitic nematodes for their host or the tissue. Other, local, chemical or structural stimuli must be involved. For second-stage larval *Meloidogyne* at least, this involves contact of the lips with the surface of the epidermal cells and allows identification of the meristematic area (von Mende, 1997). The mechanisms may well involve an amphidial gland glycoprotein (M_r 32,000) since antibodies to this protein block chemoat-

traction (Stewart *et al.*, 1993; Stewart, Perry and Wright, 1993).

Localisation of host tissue for some plant-parasitic nematodes is superficially a simple matter; they are transported by insect pests. For example, larvae of *Rhadinaphelenchus cocophilus* which parasitises roots, stems and leaves of coconut and oil palms, is transmitted by the palm weevil with which it has a symbiotic relationship (Maggenti, 1981).

Once a suitable host is localised, ectoparasites will begin to feed by piercing the epidermal cells with their stylet. Endoparasites may also use their stylet to gain entrance to the plant tissue. On first contact of *Meloidogyne* and other root-knot nematodes with the growing root, the stylet is moved forwards and backwards at a rate of one to two thrusts per minute, but rarely beyond the oral aperture. Thus penetration of the wall may depend less on mechanical disruption and more on degradation by enzymes (Bird, 1968) in secretions such as observed for *Heterodera shachtii* (Wyss, 1992). Once the invasion process has begun, the stylet is protruded and the thrusts become rhythmical. The larva of *Heterodera cruciferae* does rely on mechanical disruption; it uses repeated stylet thrusts along a line to produce a slit that allows the nematode to enter the cell (Doncaster and Seymour, 1973). In *Meloidogyne* and other tylenchids, the stylet is moved by extensor (protractor) muscles attached posteriorly to its base and anteriorly to rigid cuticular infoldings and the wall muscles. To accommodate this movement, the anterior part of the pharynx at the base of the stoma has a thin-walled, folded, extendable collar region (the so-called guide or guiding ring or apparatus). In dorylaimids such as *Xiphinema*, there are also retractor muscles attached to the posterior end of the stylet and extending posteriorly to the body wall (Wright, 1983). *Meloidogyne* larvae use both intra- and inter-cellular routes to penetrate the epidermis but they then migrate between cortical cells, at first heading towards the root tip until they reach the meristem; they then turn and migrate within xylem parenchyma into the vascular cylinder (Wyss, Grundler and Münch, 1992). *Heterodera*, which has a more robust stylet, migrates intracellularly through the cortex straight into the vascular cylinder (Wyss and Zunke, 1986). Although the pharynx pumps, the larva does not take in any nutrients during invasion and migration and it appears that the stylet is used primarily as a conduit for secretions.

In the trichodorids the stylet is an elongate, ventrally curved, protrusible dorsal mural tooth. It is composed of a solid pointed tip and posteriorly, a hollow section which is part of, and is fused to, the dorsal wall of the anterior pharyngeal lumen. The two parts articulate but are not joined rigidly (Figure 8.2). In all other cases, the stylet is axial and has a narrow central lumen with an external opening just below the tip, through which secretions produced by pharyngeal glands may pass into the plant and nutrients are withdrawn. The trichodorids are able to ingest much larger components (whole mitochondria, plastids) than those with an axial stylet where the diameter of its lumen limits the size of material ingested to cytosol, some membranes and possibly small organelles. The small aperture and narrow lumen of the stylet raises the question of how clogging of the stylet is avoided. In some cells it appears that the cytoplasm gels around the end of the stylet and functions as a filter (e.g. in *Criconemella xenoplax* (Hussey, Mims and Westcott, 1992). For others, a nematode-derived entity, the feeding tube (Rumpenhorst, 1984), is formed at, and continuous with, the tip of the stylet from pharyngeal gland secretions. Feeding tubes are produced by migratory parasites, such as *Helicotylenchus* that feed for several days on a cortical cell, as well as by sedentary parasites (Jones, 1978). It is not known whether they have identical properties. Those produced by *Meloidogyne* are about 0.8–1 μm thick overall with lumen diameters in the range of 0.3–0.5 μm; the walls of the tube are 0.2–0.3 μm thick and are seen to have a para-crystalline structure when examined by electron microscopy (Hussey and Mims, 1991). The walls of tubes produced by cyst nematodes are thinner and are composed of amorphous material (Wyss, Stender and Lehmann, 1984). Aggregations of the plant cell's endoplasmic reticulum accumulate around both kinds. Based on experiments with microinjected fluorescent dextrans, the feeding tubes produced by *Heterodera* have an exclusion limit of between 20–40 kDa (Böckenhoff and Grundler, 1994). In the absence of filters, brief reversal of the flow might be involved, but unless this is a constitutive process this implies the existence of proprioceptors and associated neural control of the pumping mechanism. A second potential problem caused by the narrow lumen is how the nematodes generate sufficient force to draw the plant cell contents through the stylet which may be up to 300 μm long (Table 8.3). In part this may be overcome by the high turgor pressure which develops in feeding cells (Böckenhoff and Grundler, 1994). Lysis of the plant cell contents prior to ingestion would circumvent blockage problems and reduce flow problems. Predigestion certainly seems to be a feature of feeding by some migratory ectoparasitic nematodes (Wyss, 1987).

It is essential for sedentary nematodes that the cells they can reach remain alive and produce in abundance the nutrients that they need. To this end, sedentary feeders secrete molecules that induce structural and physiological modifications of the cells at the selected feeding site. *Meloidogyne* and other root-knot nematodes induce the formation of giant cells. Cyst nematodes such as *Heterodera* and *Globodera* induce the formation of syncytia. Migratory ectoparasitic longidorids such as *Xiphinema index* feed for a protracted period at one site in the root tip. Initial feeding on one cell induces adjacent cells to become multinucleate and metabolically highly active and they are then fed on by the parasite. Since giant cells and syncytia are wholly derived from the plant in response to chemical signals in secretions from the nematodes it may be supposed that the multinucleate food cells used by *Xiphinema* are induced directly or indirectly by similar molecules secreted by the parasite. The source

of these secretions is the large pharyngeal glands. Tylenchids such as *Meloidogyne* have three, single-cell pharyngeal glands two of which are sub-ventral, the other, dorsal. Anteriorly, each has a narrow tubular extension that terminates in an ampulla connected to the lumen of the pharynx by a cuticularized duct equipped with valves (Anderson and Byers, 1975). The dorsal gland duct opens near the base of the stylet and the sub-ventral glands open directly behind the pump chamber in the metacorpus. The sub-ventral glands, which are packed with granules in the pre-parasitic L2s, shrink as the granules are discharged during development of the parasitic stages while the dorsal gland enlarges (Bird, 1983; Hussey and Mims, 1990). These changes correlate with the change from migratory to feeding stage.

Once the feeding site is induced the second-stage larvae of sedentary endoparasites begin a repetitive cyclic feeding pattern and then moult to the third stage. Future males then cease feeding, but the future females go on feeding and develop the lemon shape characteristic of adults. More information on the feeding of plant-parasitic nematodes is provided in Chapter 9.

Parasites of Animals

Almost without exception the nematodes that are parasites of animals are endoparasites, but this unifying fact is submerged in the bewildering array of parasitic relationships that exist. First there is the enormous range of animal groups that are parasitised. Second the parasitic nematodes may actively select their hosts or be taken up passively, but either way they must locate their highly specific feeding site. There is a diverse range of these (Table 8.1) and an equally wide range of feeding strategies (Table 8.2). Finally, many of the nematodes utilise different food sources at different stages in their life histories (Table 8.1).

Following convention (e.g. Chitwood and Chitwood, 1974; Poinar, 1983), nematodes parasitic on invertebrates are considered separately from those parasitic on vertebrates. The two groups overlap to some extent in those cases where invertebrates are intermediate hosts for parasites of vertebrates.

Parasites of Invertebrates

Parasitism of invertebrates seems to have arisen independently in five groups of nematodes, the Oxyurida, Rhabdita, Tylenchida, Aphelenchida and Mermithida (Dorylaimida). There is a range of feeding types. At one end of the range are rhabditids such as *Eudiplogaster aphodii* and oxyurids such as *Thelastoma bulhōesi* that require bacteria at some stage in their life history and tylenchids such as *Parasitaphelenchus* and *Deladenus* that require fungi as well as insect tissue. At the other end of the range are, for example, mermithids such as *Mermis subnigrescens* that are wholly dependent on insect haemolymph and *Daubaylia* that feeds only on snail (*Gyraulus*) tissue (Chernin *et al.*, 1960; Poinar, 1983). The host tissue most commonly used as food is haemolymph, but secretions and, rarely, solid tissues are also utilised. The nematodes rely on two basic means of gaining entry to their invertebrate hosts; either they are eaten (and then remain in the host gut or penetrate the gut wall to reach the body cavity), or they enter actively via some external orifice (which requires the infecting stage to operate some signal recognition mechanism). For example, infective larvae of *Steinernema feltiae* infect larval houseflies through the anal aperture then migrate along the hindgut and penetrate the wall of the ileum immediately posterior to the pylorus (Renn, 1998).

The adult stage of the diplogasterid *Eudiplogaster aphodii* is a bacteriovore living in dung. Third-stage larvae become parasitic when ingested by a larval dung beetle. The nematode larvae penetrate the beetle gut wall and enter the haemocoel where they grow and accumulate nutrients that are apparently absorbed through the cuticle. They remain inside the beetle until it becomes adult when they are released and, after moulting, resume feeding on bacteria. The wholly bacteriovorous *Steinernema feltiae* that parasitises insects was described earlier (see page 216); *Heterorhabditis* is also known to carry a monoculture of symbiotic bacteria in its intestine into insect larvae as food organisms.

The mechanism by which the rhabditids penetrate the host gut wall has not been described. However, parasitic tylenchids and aphelenchids are equipped with a stylet which they use for feeding on fungi and for gaining access to their invertebrate hosts, although not for feeding thereon. *Beddingia (Deladenus)* which is parasitic on the wood-boring wasp *Sirex noctilio* is of particular interest in this context (Bedding, 1973). It will feed for several generations on fungi that the wasp introduces into its feeding tunnels. For this, the nematodes have a short stylet used for piercing and feeding. Eventually females are produced which have long stylets. These are not used for feeding but enable the nematodes to penetrate larval *Sirex* and gain access to the haemcoel. Shortly after entry into the host the nematode becomes barrel shaped, then the cuticle ruptures and the thus exposed surface of the epidermis develops microvilli, presumably for nutrient absorption. The nematodes mature and lay eggs and the resultant larvae feed on the wasp. The hosts are not killed, but carry the nematode larvae to new feeding tunnels.

Larval mermithids are also equipped with a stylet whose only purpose seems to be to enable the nematodes to enter their hosts (which include insects, isopods and echinoderms) through their body wall. The stylet disintegrates during subsequent parasitic development in which the pharynx separates from the intestine and nutrient uptake is by absorption through the body wall. In *Romanomermis culicivorax*, a parasite of mosquitoes, the cuticle is very thin and consists of three layers containing pores about 10 nm in diameter. The adjacent epidermal cells have outward facing microvilli (Poinar, 1977). Although large enough to permit uptake of nutrients such as glucose and amino acids (demonstrated in *Bradynema* (Riding, 1970) and *Mermis nigrescens* (Rutherford and Webster, 1974; Rutherford, Webster and Barlow 1977)), the role of the pores in

nutrient uptake is not known. The cuticle of *Mermis* has pores 50 nm in diameter (Lee, 1970a) and that of *Deontostoma* has pores 60 nm in diameter (Siddiqui and Viglierchio, 1977).

Sphaerularia is another parasitic nematode that does not use its gut for feeding. The adult, fertilised female *Sphaerularia bombi* is a parasite of queen bumble bees. Like its close relative *Tripius*, a parasite of flies, after entering its host the body of the young female nematode undergoes little if any increase in size, but the developing uterus everts through the vulva, carrying within it the modified intestine (and the other reproductive organs) (Poinar and van der Laan, 1972). The uterus increases enormously in size (Figure 8.6) and the surface cells become modified for absorbing nutrients from the insect haemolymph. The role, if any, of the intestine in this process is not known. The soil-dwelling larvae have a gut typical of other tylenchids with a stylet; this stage of the closely related *Prothallonema* is known to be a fungivore.

One nematode that does use a stylet to feed on an invertebrate host is the aphelenchid *Acugutturus parasiticus*. This is also remarkable because, although it feeds on haemolymph, it is an external parasite of terrestrial insects (cockroaches in the West Indies). This worm is found at various places on the outside of the insect's abdomen, wings and legs, apparently anchored in place by its extremely long stylet (Hunt, 1980).

The thelastomatids (e.g. *Thelastoma, Leidynema*) are a group of oxyurids that feed on the contents of the hind-gut of certain arthropods, such as cockroaches.

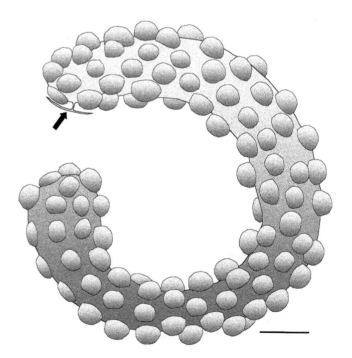

Figure 8.6. Diagram showing tubercle-covered surface of the enormously enlarged everted uterus of *Sphaerularia bombi*. The arrow points to the body of the nematode. The scale bar represents about 5 mm. Based on scanning electron micrograph in Siddiqui (1986).

Parasites of Vertebrates

The larval and adult forms of nematodes parasitic in vertebrates utilise a range of feeding strategies. There is one known example (*Probstmayria vivipara*) of a parasite that has no stage outside of its host, otherwise life histories range from simple alternation of bacterial-feeding larvae outside the host and parasitic adults, through alternation of bacterial-feeding generations with parasitic adults to utilisation of one or more intermediate hosts. Access of the infective stage to the host is by one of three routes. It may be a passive process, is the outcome of active penetration of the skin or gut surface, or is delivered, or indeed injected, by an intermediate host. Passive access depends on the host ingesting the nematode inside an egg or as a free-living larva (usually the L3) although infective-stage larvae frequently exhibit behavioural activity that facilitates their ingestion. Within the host, feeding is not indiscriminate and depends on migration to specific feeding sites. Examples of the sites occupied and the tissues fed on are listed in Table 8.1. The range will be illustrated mainly by reference to some mammalian parasites.

Ascaris has no non-parasitic feeding stage. Infection occurs when embryonated eggs consumed with contaminated food or water hatch in the jejunum. Adult *Ascaris* are facultative anaerobes in the lumen of the jejunum. Curiously, however, they do not develop there directly from ingested larvae since the first time round these penetrate the gut wall with the aid of a 'boring tooth' and then migrate in the blood stream first to the liver and then to the lungs before returning via the trachea and oesophagus to the jejunum as fourth-stage larvae. During this migration the larvae increase in size so must be feeding, but how and on what has not been described. The young adults feed on the host's ingesta (chyme) and grow rapidly (some 25 mm in the first two weeks) until they attain their mature length of 30–50 cm. Adult *Ascaris* normally move only relatively short distances up and down the host gut so must have means to establish their position; *Ascaris* has anteriorly placed sensory papillae (Ubelaker and Allison, 1972) which, by analogy with the anterior chemosensory organs of the blood-feeding parasite of birds *Syngamus trachea* (Riga *et al.*, 1995), may well have a role in this behaviour.

The first two larval stages of *Haemonchus* feed on bacteria (Table 8.1). It is assumed that they are indiscriminate feeders, but this does not seem to have been investigated. The non-feeding third-stage larva is the infective stage. It shows negative geotaxis, climbing grass stalks and thus being in a position likely to be consumed by the ruminant host. It passes through the rumen and omasum, then burrows into the abomasal mucosa where it moults to the fourth-stage larva, starts to express gut proteins of the type found in the adult and begins to feed on host blood. This is obtained by head movements lacerating the mucosa with a dorsal tooth (lancet) until a capillary is pierced. In the adult, the dorsal tooth of *Haemonchus contortus* is about 13 μm long and 3 μm wide arising from the dorsal wall of the buccal cavity. It carries a duct from the dorsal pharyngeal gland that opens on its anterior mid-ventral surface. There is a sphincter-like valve at the base of the tooth (Weise, 1977). The gland secretions include anti-coagulants (Knox, Redmond and Jones, 1993) and blood leaks from the feeding sites. Adults observed in the abomasum by means of endoscopy have been seen to move into areas of haemorrhage, which indicates that they have sense organs that are capable of detecting some constituent of the blood (Nicholls, Lee and Sharpe, 1985). *Haemonchus*, which is an obligate aerobe, achieves its maximum length (about 25 mm for the female) in about three weeks (Coadwell and Ward, 1981). This growth rate is somewhat slower than that of *Ascaris*.

Like those of *Haemonchus*, the free-living larvae of the hookworm *Necator* (and the very similar *Ancylostoma*) feed on bacteria. Further, the infective stage also shows negative geotaxis wriggling up particles of soil and herbage where, in common with the larvae of other parasitic nematodes such as *Steinernema*, it shows what has been called 'questing behaviour' (nictation) in which the anterior part of the body is raised clear of the substratum and sways. The larvae respond positively to warmth and CO_2. Unlike *Haemonchus* however, entry into the host by hookworm larvae depends on positive action. When they make contact with the skin of a mammalian host they penetrate the epidermis to reach capillaries or lymphatics in the dermis. Invasion and migration involves mechanical activity and stage-specific expression and release of hyaluronidase (Hotez *et al.*, 1992) and proteolytic enzymes. Another nematode that uses this route of entry, *Strongyloides stercoralis*, releases metalloproteases with elastase activity able to degrade dermal extracellular matrix (McKerrow *et al.*, 1990). Once established at its feeding site on the luminal surface of the mucosa of the small intestine, *Necator* utilises other proteolytic enzymes to gain access to host blood, its main food. The nematodes attach to the mucosa with stomatal plates (*Necator*) or teeth (*Ancylostoma*) and use the pharyngeal pump to suck a plug of tissue into the large globose stoma (Figure 8.7). Pharyngeal gland secretions contain proteases and hyaluronidase which break down the tissue in the plug (and the products, presumably, are ingested) and more importantly the proteases help damage capillaries so that blood is released. The pharyngeal secretions also include anti-coagulants (Capello *et al.*, 1995; Furmidge, Horn and Pritchard, 1995). The secretions are released, as in *Haemonchus*, via a duct in the dorsal tooth (McLaren, Burt and Ogilvie, 1974). Similar ducts are also present in other hookworms, such as *Bunostomum trigoncephalum* (Wilfred and Lee, 1981) a parasite of ruminants (Figure 8.8).

The whipworm *Trichuris* and its relatives also feed on intestinal mucosa, but represent yet a third method of feeding from this site. They have a tiny buccal cavity and appear to rely entirely on pre-digestion of the tissue by means of secreted digestive enzymes. Only the anterior end of the very long pharynx is muscular, the rest consisting of a thin-walled tube surrounded by

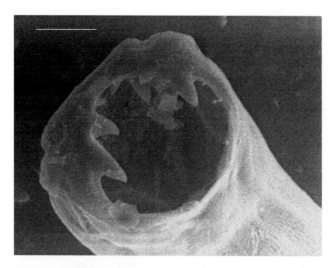

Figure 8.7. Scanning electron micrograph of the large globose stoma of *Ancylostoma caninum*. *Ancylostoma*, like *Necator* and *Bunostomum*, is a plug feeder; pumping action of the pharynx fills the stoma with mucosa which is held in place by the teeth whilst digestive proteases in pharyngeal gland secretions are released. The scale bar represents 50 µm. (Electron micrograph courtesy of F.J. Hall.)

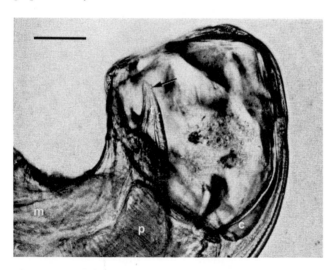

Figure 8.8. Light micrograph of the stoma of *Bunostomum trigonocephalum* showing the dorsal tooth (t) and the opening to its duct (arrow), the ventral cutting plates (c) and the dorsal muscles (m) which extend from the stoma to the pharynx (p). The scale bar represents 50 µm. (From Wilfred and Lee, 1981). Reproduced with permission from *International Journal for Parasitology*.

large unicellular glands (stichocytes). Just as many plant-parasitic nematodes induce changes in the cells they feed on, *Trichuris* induce syncytia formation, in a process which involves over-production of interferon-gamma (Artis *et al.*, 1999). *Trichinella spiralis* is also of particular interest because of its intracellular habit and the changes it induces. The infective first-stage larva lives within the cytoplasm of a striated muscle cell (Lee and Shivers, 1987), which becomes much modified to form a so-called nurse cell (Despommier, 1983) with a nexus of capillaries (Jasmer, 1995), and the other larval

and adult stages are spent within a group of enterocytes in the small intestine (Wright, 1979). The enterocytes move from the crypt, where they are formed, towards the tip of the villus. Thus by maintaining its position the parasite is presented with a supply of new cells upon which to feed. Whether the occupied enterocytes differentiate enough to begin nutrient uptake from the host digesta is not known, but the principle component of the diet of the parasite is the enterocyte contents. The filarial nematodes *Wuchereria bancrofti* and *Brugia pahangi* live in the lymphatics (Table 8.1). They have a very small mouth and long narrow pharynx. By contrast with the blood-dwelling *Dirofilaria immitis* (Lee and Miller, 1969), the intestine of the infective larva of *Brugia* is very poorly developed and even in the adult the microvilli are short and widely spaced, indicative of little or no absorptive role. Both infective larvae and adults of *Brugia* have been shown to absorb leucine, adenosine and D-glucose across the cuticle by a selective process (Chen and Howells, 1979).

Omnivores

Omnivores are nematodes that feed on different kinds of food at any given stage in their life cycle and thus are distinguishable from a number of parasitic nematodes that have very different food sources at different stages in their life histories (see Tables 8.1 and 8.3). Quite a large number of species, particularly those regarded primarily as predators, could be regarded as omnivores, but it is restricted here to nematodes that always feed indiscriminately. Despite this restriction, the group is a bit of a catch-all and can include nematodes that may also be accommodated in other groups. Thus aquatic 'Adenophorea' (refer to Chapter 1 for discussion of classification) are microbivorous, but eat any bacteria, algae and protozoa that become trapped in mucus secreted by their caudal and pharyngeal glands (Riemann and Schrage, 1978). Many 'grazers', such as *Chromadora macrolaimoids*, the stoma of which has three moveable teeth used for scraping food off substrates (Hopper and Meyers, 1967), and *Euchromadora gaulica*, which feeds on algae and fungi, can be classed as omnivores. Parasites that are non-discriminatory feeders on host gut contents may also be included here. Whilst it does not follow that ingestion means digestion, the implication is that omnivores have the means to extract nutrients from a broader range of foodstuffs than some of the more specialist feeders described above.

Non-feeding Stages

Although, strictly, the opposite of what this Chapter is about, non-feeding is a widespread phenomenon in the Nematoda and deserves at least brief mention here. As noted previously, many parasitic nematodes have non-feeding stages, when either they utilise food stored from a previous feeding stage or they enter a hypobiotic state as represented by the dauer larva. Dauer larvae are well-studied in *C. elegans*. They are regarded as a survival stage, induced by shortage of food. They are capable of movement but they do not feed nor show the prompt chemotactic response to bacteria demonstrated by L2 or adult *C. elegans*. The openings of the anterior sensillae become obscured, the buccal cavity is sealed with a cuticular plug (Albert and Riddle, 1983), there is no pharyngeal pumping, the intestinal lumen is shrunken and the microvilli are small (Popham and Webster, 1979). Recovery to the feeding state is initiated by changes in the ratio of a food signal (a neutral, carbohydrate-like substance produced by *E. coli*) and a dauer-inducing pheromone. Parallels can be drawn to the behaviour of some infective larvae of some parasitic nematodes and also to some later larval stages (for example that of *Ostertagia ostertagi* in cattle stomach mucosa) where the behaviour leads to delayed sexual maturity so that the next generation of larvae are released into environmentally favourable conditions.

Ingestion

With one known exception, ingestion depends on pumping by the pharynx of fluid or suspensions of particles of various sizes into the intestine. The exception is *Hexatylus viviparus*. This lacks a pharyngeal pump and a prerectal valve and feeding apparently depends on turgor pressure in the fungal host cells in combination with the pumping action of the rectal muscles (Doncaster and Seymour, 1975; Seymour, 1975). Most work has been done on the uptake of particles in suspension.

The process of ingestion of bacteria by rhabditids is well established (Lee, 1965; Lee and Atkinson, 1976). The mechanism (Figure 8.9) has been studied in greatest detail in *C elegans*. It involves the co-ordinated activity of the 34 pharyngeal muscle cells, their associated neurones and eighteen epithelial cells acting on parts of the cuticular lining (Albertson and Thomson, 1976). Rapid contraction of the radially orientated muscle fibres making up the procorpus and metacorpus (Figure 8.9 A, B) opens the lumen and draws liquid and bacteria in suspension into the buccal cavity (which is 3 to 4 μm wide; delta (Δ) shaped in cross-section in the larva and hexaradiate in the adult) and into the dilated lumen of the metacorpus (median bulb). At this time, the lumen of the isthmus is closed posteriorly and the grinder is dilated (Figure 8.9 B). Rapid relaxation of the procorpus and metacorpus muscles closes the lumen. In cross-section the procorpus lumen has a delta (Δ) shape when open and the characteristic Y-shape when closed but even when closed it is not fully occluded because there are dilations at the tips of the arms of the Y which form open channels (Figure 8.3 E). Liquid can be ejected forwards through these channels whilst bacteria are retained in the metacorpus which is equipped with numerous, fine finger-like projections extending into the lumen. These function as a sieve to trap ingested bacteria whilst the bulk of the associated fluid is expelled (Avery and Thomas, 1997). *C. elegans* is thus a filter feeder (Avery and Thomas, 1997). Peristaltic movement along the length of the isthmus (Figure 8.9 D, E) carries packets of bacteria from the metacorpus to

a luminal dilation within the terminal bulb. The cuticle here, called the grinder, has substantial thickenings arranged radially and longitudinally (Albertson and Thomson, 1976). As the name indicates, these disrupt the concentrated bacteria by a grinding action produced by the six muscles in the terminal bulb. (The grinding process is not complete and some bacteria remain intact.) The terminal bulb contains the cell bodies of the five pharyngeal glands. Processes from these glands are directly accessible to the lumen because they do not produce cuticle. Their secretions are thought to include digestive enzymes released by exocytosis into the grinder, perhaps under the control

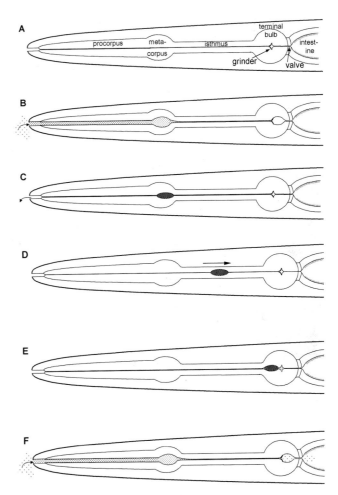

Figure 8.9. Diagrams showing mechanism of ingestion of bacteria by *C. elegans*. A, represents the resting condition; B, in response to the presence of bacteria in suspension a pumping cycle begins with contraction of muscles in the pro- and meta-corpus (and in the terminal bulb) bacteria in suspension are drawn into the corpus; C, relaxation of the muscles expels water forwards and concentrates the bacteria in the meta-corpus; D, peristalsis by sequential contraction of the muscles in the isthmus drives the packet of bacteria into the terminal bulb; F, another pump cycle draws in more bacteria through the stoma. Opening of the pharyngeal-intestinal valve allows the contents of the terminal bulb to empty into the intestine. Based on Lee, (1965); Lee and Atkinson (1976); Avery and Thomas, (1997) and White, (1988).

of motor neurones that have synaptic connections to both terminal bulb muscles and glands. Thus the pharynx ingests and concentrates the bacteria and grinds them and begins the process of digestion before pumping them into the intestine. The pharynx as a whole will function, slowly, without neuronal input. To obtain rapid pumping, one of the behavioural responses to the presence of bacteria, requires the motor neuron designated M3, the neurotransmitter for which appears to be acetylcholine. The pharynx of starved wild-type *C. elegans* will pump in the absence of bacteria under conditions in which well-fed worms do not pump. Further, starved worms respond to lower concentrations of bacteria than do fed worms (Avery and Horvitz, 1990). Whatever the pumping rate of the pharynx in *C. elegans*, the contractions of the corpus and terminal bulb are synchronized. This, however, does not seem to be an essential feature of bacterial feeding since in *Panagrellus silusiae* and *P. redivivus* the terminal bulb pumps more rapidly than the corpus (Mapes, 1965). Since *C. elegans* can be cultured on a wholly fluid diet, it is clear that it (and presumably other microbivorous nematodes) can utilise the pumping mechanism to drive liquid into the intestine.

The large size of adult *Ascaris* was exploited to enable early studies of pharyngeal pumping activity *in vitro* (Mapes, 1966). The pharynx appears to function as a simple two-stage pump in which half the pharyngeal lumen is filled with chyme at any one time. Contraction of the anterior radial muscles opens the triradiate lumen to a triangle that then clicks to a circular shape drawing in liquid as the mouth opens. The mouth closes, the anterior muscles relax and the posterior radial muscles then contract so that the posterior lumen opens and the digesta is drawn backwards. The pharynx begins to refill just before the digesta in the posterior part is fully emptied through the pharyngeal-intestinal valve into the intestine. The pumping rate is variable; sometimes it continues for several minutes at about 4 contractions per second, at other times short bursts of high frequency (up to 20 per second) occur. In *Ascaris*, the suction force generated by the pharynx is 101 kPa. This is more than enough to overcome the hydrostatic pressure which has a mean value of about 9.3 kPa and does not exceed 50 kPa. Since defecation only occurs every three or four minutes the increase in internal pressure must be due, at least in part, to the pumping activity of the pharynx.

Digestion

Nutritional Requirements

Most of the amino acids essential in the diet of mammals are, probably, also essential in the diet of nematodes. Thus *C. elegans* and *C. briggsae* require dietary histidine, leucine, isoleucine, lysine, methionine (or its precursor homocysteine), phenylalanine, threonine, tryptophan, valine and arginine (Vanfleteren, 1980), although *Meloidogyne* can synthesise labelled tryptophan from [^{14}C] acetate (Myers and Krusberg,

1965). Unless these essential amino acids are free in solution, they will need to be released from proteins and peptides.

Nematodes are also known to have a nutritional requirement for sterols. They may incorporate dietary sterols largely unchanged, as shown for *Pratylenchus agilis* feeding on *Zea mays* (Chitwood and Lusby, 1991), or may dealkylate or generate stanols from them (Chitwood, 1997). These processes are likely to occur in the intestinal cells, but probably as post-absorption events.

Water Intake

Water is the ubiquitous component of the diet of all nematodes. Whilst some, such as *C. elegans* concentrate the bacteria they feed on by filtering (Avery and Thomas, 1997) and thus appear to take steps to reduce the water intake, others such as *Heterodera schachtii* appear to take in very large amounts of water (Sijmons et al., 1991). The relative advantage or disadvantage of the level of water intake has not been established. Larger volumes taken in mean that larger volumes have to be got rid of, but if the concentration of nutrients ingested is high this of itself should not be a problem for nematodes with a high defecation rate, but otherwise would require mechanisms for disposing of the excess.

Digestive Enzymes

Digestive enzymes include endopeptidases (of which cysteine proteinases have been particularly well studied in nematodes), exopeptidases, glycosidases, lipases, esterases, phospholipases and phosphatases. The level of knowledge ranges from description of an activity to full structural characterisation. The enzymes may be secreted and released either externally or internally, or are membrane proteins at the luminal surface of the gut. It is to be expected that other proteins, such as transporters, will also be present at the luminal and basal gut surfaces and at the surface of the other absorptive organs used by nematodes.

Secreted enzymes

The secreted enzymes have two distinct functions. First, in allowing larvae of both plant- and animal-parasitic nematodes to reach their site of parasitism. This is commonly a digestive process, but is external and the products are not assimilated. The second function is in the process of digestion for assimilation and may occur externally or internally.

The larvae of a number of plant-parasitic nematodes have been shown to contain or secrete cellulases (Dropkin, 1963; Myers, 1965; Deubert and Rohde, 1971; Bird, Dowton and Hawker, 1975;) that are presumed to facilitate migration. In general, cellulases in animals are produced by associated micro-organisms, but Smant and colleagues (Smant et al., 1998) have offered evidence that the motile second-stage larvae of the cyst nematodes *Globodera rostochiensis* and *Heterodera glycines* produce β-1,4-endoglucanase endogenously. A monoclonal antibody to the cellulase cross-reacted with other species of *Globodera* and

Heterodera (de Boer *et al.*, 1999). It is possible therefore that the other nematodes with cellulase activity also produce the enzyme endogenously. At present it seems likely that the cellulase is only involved in penetration to the feeding site. To do this it needs to bind to the cellulose. A secretory cellulose-binding protein (designated Mi-cbp-1) has been cloned from *Meloidogyne incognita* (Ding *et al.*, 1998). The protein was detected in secretory granules in subventral gland cells and in stylet secretions of second-stage larvae. Based on Southern blots, the gene for Mi-cbp-1 was present in two other species of *Meloidogyne*, but not in *M. hapla*, or *H. glycines* (or *C. elegans*).

Progress of infective larvae of animal-parasitic nematodes from their site of insertion (gut or skin) to their site of feeding often involves secreted proteolytic enzymes, as has long been realised for geohelminths that gain access to internal tissues by penetrating their hosts' skin (Lewert and Lee, 1956). Among these, a 40 kDa zinc endopeptidase secreted by infective larvae of *Strongyloides stercoralis* has been characterised by McKerrow and colleagues (McKerrow *et al.*, 1990; Brindley *et al.*, 1995) and a 42 kDa protein (ASP) produced by activated *Ancylostoma* larvae, by Hawdon and colleagues (Hawdon, Jones and Hotez, 1995; Hawdon *et al.*, 1996). The *Strongyloides* peptidase has a broad specificity for elastin and other components of the dermis and facilitates rapid migration of the larvae through tissue (McKerrow *et al.*, 1990). Larval penetration of mammalian skin is blocked by the presence of peptidase inhibitors. The C-terminal half of the ASP has homology with a secretory protein in hymenopteran venom (with which it also showed immune cross-reactivity) and other insect and vertebrate secreted proteins (Hawdon *et al.*, 1996). The protein also has sequence homology with another *Ancylostoma* protein, a neutrophil inhibitory factor (Rieu *et al.*, 1994), but its role in accessing the parasite's feeding site is not known.

As noted earlier, a number of fungivores, plant- and animal-parasitic nematodes secrete digestive enzymes and then ingest the products. Few of the digestive proteins have been characterised. An endoproteinase with M_r of about 46,000 able to digest fibrinogen has been described from L4 *H. contortus* incubated *in vitro* (Gamble, Fetterer and Mansfield, 1996). A family of proteins with anti-coagulant activity secreted by the blood-feeding hookworm *Ancylostoma caninum* (Eiff, 1966; Hotez *et al.*, 1985; Capello *et al.*, 1995; Stassens *et al.*, 1996) include a set of small cysteine-rich proteins which resemble serine protease inhibitors from *Ascaris*. Thus it seems that both proteases and protease inhibitors may be secreted. Among the former, cysteine proteases are widely distributed and often expressed at high levels in the intestines of bacteriovores (e.g. *C. elegans* (Ray and McKerrow, 1992)), animal parasites (*Ascaris*, *Dirofilaria* and *Angiostrongylus* (Maki and Yanagisawa, 1986), *Haemonchus* (Knox, Redmond and Jones, 1993)) and plant parasites (*Heterodera*, *Globodera* (Lilley *et al.*, 1996)). The protease in *Heterodera* is cathepsin L-like; based on sequence analysis the enzyme in *Globodera*

pallida is cathepsin B-like (Lilley *et al.*, 1996). A cathepsin C-like enzyme, together with a phospholipase A, are secreted *in vitro* by L4 *Haemonchus contortus* (Gamble and Mansfield, 1996). Some cysteine proteases are also membrane-associated. Similarly, acid phosphatase, thought to have a role in digestion occurs, along with phosphorylcholine hydrolase, in secreted form (Fetterer and Rhoads, 2000) and as a membrane protein.

Several animal-parasitic nematodes that feed in the host gut (e.g. *Nippostrongylus*, *Trichostrongylus*, *Necator*) secrete large amounts of acetylcholinesterase (Griffiths and Pritchard, 1994; Lee, 1970b, 1996; Ogilvie *et al.*, 1973, Pritchard *et al.*, 1991). This enzyme does not have a digestive function *per se*, but may facilitate the feeding process (Foster and Lee, 1995; Lee, 1996).

Membrane-associated enzymes

These proteins have been the subject of study particularly in nematodes parasitic in animals because of their potential as vaccine components (Jasmer, Perryman and McGuire, 1996; Jasmer and McGuire, 1996; Munn, 1977, 1997; Munn *et al.*, 1993; Smith *et al.*, 1997; Smith and Munn, 1990).

The most complete studies, as far as description of activities goes, have been carried out on *Ascaris* and several disaccharidases (maltase, sucrase and trehalase), proteases and peptidases, phosphatases and lipase have been reported (Borgers, Van den Bossche and Schaper, 1970; Van den Bossche and Borgers, 1973). A number of microvillar membrane-associated proteins have been described for *Haemonchus contortus*. Quantitatively the most important integral membrane protein is H11, a glycoprotein with a mean M_r of 110,000 . This glycoprotein is only expressed in the parasitic stages and only on intestinal microvilli. It has microsomal (membrane) aminopeptidase M and microsomal aminopeptidase A activities attributable to distinct isoforms (Smith *et al.*, 1997). The structure of H11, predicted from amino acid sequence, is that of a type II integral membrane protein with a short N-terminal cytoplasmic tail, a single transmembrane region, short stalk and the remaining extracellular region folded into four domains (one containing the active site), having a total of three or four N-glycosylation sites (Smith *et al.*, 1997, Newton and Munn, 1999). Isoforms of O12, the homologue of H11 in *Ostertagia* and *Teladorsagia circumcincta*, have been cloned and characterised (D. Phillips and E.A. Munn, unpublished). As aminopeptidases these ectoenzymes would appear to have a role in the digestion of peptides generated by the action of proteases on dietary protein. A membrane protein complex designated H-gal-GP with component M_rs of about 230, 170, 42, 40 and 31,000 (Smith and Smith 1993) has aspartyl protease, neutral endopeptidase and cysteine protease activities. Two of these have been cloned (Redmond *et al.*, 1997; Longbottom *et al.*, 1997). The complex O-gal-GP, the homologue of H-gal-GP, has been purified and characterised from *O. ostertagi* and *T. circumcincta*.

Two distinct membrane associated cysteine protease fractions from *H. contortus* have been described. One is a fibrinogen-degrading complex containing 35 and 55 kDa proteins, of which the former has been cloned and sequenced and shows homology to Cathepsin B (Cox *et al.*, 1992). Second, a group of three 70 kDa cathepsin B-like, cysteine proteases with 70–90% sequence homology to each other (but distinct from the 35 kDa protein) have been obtained from detergent extracts as thiol binding proteins (Knox, Redmond and Jones, 1993). The proteases have been immunolocalised to the worm gut luminal surface. Other surface enzymes which have been identified are acid phosphatase and γ-glutamyl transferase. Acid phosphatase has been used as a marker to follow *C. elegans* intestinal cell development (Beh *et al.*, 1991).

Other intestinal membrane-associated proteins

As well as the enzymes known to be present at the luminal surface of the intestine it is expected that other proteins functioning as carriers, ion channels and transporters will be present to carry nutrients into the cells. Similar proteins to carry metabolites and ions to the other nematode tissues and receptors will be in the basal and lateral membrane. To date only two such proteins have been characterised, both in *C. elegans*. A copper trafficking pathway consisting of a copper chaperone protein, CUC-1, and a copper transporting ATPase, CUA-1, has been demonstrated in intestinal cells of adult *C. elegans* (Wakabayashi *et al.*, 1998). Receptors (ITR-1) with M_r 220 kDa, activated by inositol 1,4,5-trisphosphate, that modulate intracellular calcium signals, are strongly expressed in the intestine, (and pharynx, nerve ring, excretory cell and gonad) of *C. elegans* (Baylis *et al.*, 1999).

Two quantitatively important protein groups, contortin and *GA1*, have been described for *Haemonchus contortus* intestine. Their function is unknown, but since their expression is restricted to the luminal surface of the parasitic stages it is likely that they have a role in the acquisition of nutrients. Contortin is an extracellular, polymer (Figure 8.10). As shown by electron microscopy, on average, every microvillus in *H. contortus* has seven of the helically wound polymers (some 40 nm in diameter) attached loosely to its surface (Munn, 1977). The monomer has a M_r of 56,000. Contortin-like polymers seem to be restricted to species of *Haemonchus* (*H. contortus*, *H. placei* (Smith and Harness, 1972) and *H. longistipes*), *Ostertagia* and *Teladorsagia* (Munn, 1977) and to the ascarid *Rhigonema* (Wright, 1991), thus the function of this protein is also probably highly restricted.

The glycoprotein complex designated GA1 (for Gut Antigen 1) contains proteins with M_rs of 52 and 46,000, shown to be products of the same gene initially expressed as a polyprotein (p100). The p52 component has a GPI anchor and 47% sequence identity to p46 (Jasmer, Perryman and McGuire, 1996). The p52 and p46 components encoded by GA1 appear to have similarity to some components of the complex of glycoproteins designated P1 (with main components with M_r of 45, 49 and 53,000 under reducing conditions) purified from detergent extracts (i.e., not GPI anchored)

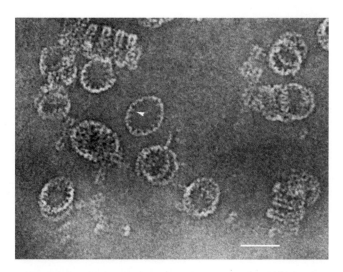

Figure 8.10. Electron micrograph of the polymeric extracellular protein contortin isolated from adult *Haemonchus contortus* and partially depolymerised by treatment with 2M-KCl. *In situ*, the polymers are up to 10 μm long. Arrowhead indicates a single Y-shaped monomer (M_r 56,000). The scale bar represents 0.5 μm. (E.A. Munn, unpublished.)

of adult *H. contortus* intestine (Munn, *et al.*, 1993). Based on deduced amino acid sequence, homologous proteins occur in *C. elegans*.

Digestion

The temporally regulated expression of digestive enzymes implies a process of digestion within the intestine, but there are many examples of parasitic nematodes in which the ingested food is apparently already in a form from which nutrients could be absorbed. As shown earlier, one parasitic nematode can live, grow and develop simply by directly absorbing insect haemolymph whilst another parasite, apparently in a position to feed on the same material can only do so after it has been converted into bacteria. Further, some mammalian parasites such as *Haemonchus* that are ingesting host blood (i.e. the distributive source of nutrients for all the host's cells) need to have digestive enzymes comparable in function to those in the host's gut in order to obtain those nutrients. The crucial role of the nematode enzymes can be inferred from the fact that specific antibodies induced by vaccination with a digestive enzyme inhibit their activity and the inhibition is highly correlated with the level of nematode mortality (Munn *et al.*, 1997). One has to suppose that crucial components, even if present in the diet, are not present in sufficient amount to meet the needs of the nematode or are not in a form in which they can be absorbed.

Absorption (Nutrient Uptake)

Under saturating conditions, the rate of absorption of the products of digestion and other nutrients will be determined by absorptive surface area and the density of any necessary transporters. Essentially nothing is known about this for those nematodes absorbing nutrients by means of a syncytial intestine or through cuticle or non-digestive tract surface. However, some generalisations can be made about absorption from the intestine.

In the intestine, absorptive surface area is a function of microvilli frequency and length. Nutrient uptake might also be expected to depend on the residence time of the digesta flow. Residence time in the intestine will be determined by input rate, length of the intestine (lumenal volume) and output rate. Input rate is dependent on the pumping rate of the pharynx, output is dependent on defecation rate. The intestine of *C. elegans* is almost completely emptied every 45 seconds, that of *Ascaris*, with a volume many hundreds of times larger, every 3 to 4 minutes. Crofton (1966) questioned the functional role of digestive enzymes expressed in the intestine on the basis that the rate of passage of digesta in many nematodes was such that there would not be time for significant digestion to occur. The fact that complete emptying does not occur may be of significance if there is a micro-environment whereby soluble products are trapped around the microvilli, whilst the bulk of the presumably indigestible food is pumped out of the intestine. Given the relatively small surface area of the *C. elegans* intestine and rapid transit of the bulk of the digesta, this is easier to envisage with nematodes with very long closely packed microvilli, such as *Haemonchus*, *Ascaris* and *Ancylostoma*, particularly if the intestinal lumen is normally collapsed so that in any cross-section the microvilli occupy a relatively large area. Mutant *C. elegans* in which the pharynx pumps food into the intestine which becomes enormously distended due to failure to defecate (see later) show the characteristics of malnourishment (Avery and Horvitz, 1990; Avery and Thomas, 1997). An explanation is that only a small proportion of the material in each food bolus pumped into the intestine is nutrient which can be assimilated. Providing that absorption rate itself is not the rate-limiting factor, then there will be advantage to the nematode to have a high through-put of digesta.

Food conversion efficiency

Data to allow rigorous comparative studies is not yet available but some nematodes seem able to utilise food far more effectively than others even within the same feeding type. Thus comparisons of the growth rate and reproductive capacity of a range of bacterivorous nematodes showed that *C. elegans*, for example, grew some 1.5 times faster when fed *E. coli* than when fed *Bacillus megaterium*. Also, *C. elegans* fed on *E. coli* grew some five times faster than e.g. *Cephalobus persegnis* and needed fifteen times less bacteria as a minimum to sustain reproduction (Venette and Ferris, 1998). As noted by Wyss (1997), trichodorids ingesting plant cell contents including organelles require very large amounts of cytoplasm to sustain reproduction and thus seem far less efficient than many of the tylenchids feeding on filtered cell fluids. As mentioned earlier, the growth rate of the gut-dwelling, animal-parasitic *H. contortus*, feeding on host blood is somewhat slower than that of *Ascaris* feeding on host gut contents.

Other Intestinal Functions

Once taken up by the absorptive surface, nutrients need to be transported directly, or after processing, to the other nematode tissues. In addition, the intestine is likely to be a major site of protein synthesis for export to other tissues and is known, in several instances, to be a food storage organ. It probably has a role in excretion in many nematodes. Two other functions are very restricted: carrying symbiont bacteria (as described earlier) and, in gut-dwelling animal-parasitic nematodes ingesting host blood, it is likely that the intestine is the major site for oxygen uptake.

Protein Synthesis

The intestinal cells are equipped with protein synthetic machinery and undoubtedly synthesise their own intracellular and extracellular enzymes and other proteins during growth and during normal turnover of cell components. In addition it is likely that these cells synthesise proteins for other tissues. This is known for sure in *C. elegans* in which it has been shown that the precursors of egg storage proteins (vitellogenins) are produced within the intestine of hermaphrodites (Kimble and Sharrock, 1983). In this respect the nematode intestine is serving the function of the fat body in insects and the liver in amphibians.

Storage

It is likely that the intestine serves as a storage organ, at least in the short term, in most nematodes. There are many reports of the presence of protein and lipid droplets and glycogen in the intestinal cell cytoplasm. Thus in infective larvae of plant-parasitic nematodes such as *Meloidogyne javanica*, stored lipid accounts for 30% of the dry weight of the nematode and occupies 43% of the area of sectioned intestinal cells (Van Gundy, Bird and Wallace, 1967). In larval mermithids, storage becomes the dominant function: at the end of the parasitic period, the intestine of larval mermithids such as *Gastromermis boophthorae* is greatly enlarged and packed with stored nutrients. It forms a cylindrical syncytium (trophosome) which, being separated from pharynx and rectum, has no openings. Subsequent transfer of the stored nutrients when they are mobilised for use by the developing adult is probably facilitated by the much infolded outer surface of the trophosome membrane (Batson, 1979).

Defecation

Defecation depends on the co-ordinated contraction of the posterior wall muscles, the anterior body wall muscles and the set of intestinal-rectal muscles (Croll and Smith, 1978). The sequence begins with the contraction of the posterior body wall muscles in all four quadrants. This causes the fluid contents of the intestine to be squeezed forwards. On relaxation of the posterior body wall muscles the intestinal contents flow backwards and collect in a bolus towards the posterior end. This action is then reinforced by contraction of the body wall muscles in the head region which pushes the pharynx backwards. As this contraction becomes maximal, the intestinal, sphincter and rectal muscles combine to open the intestinal-rectal valve and the proctodeum and expel the posterior intestinal contents (Avery and Thomas, 1997). A variation of this process occurs in *Aphelenchoides blastophorus* which has a pre-rectal as well as a rectal valve. The gut contents driven backwards by relaxation of the posterior body wall muscles are trapped in the pre-rectum when the pre-rectal valve closes and are then ejected by a separate opening of the rectal valve and proctodeum (Seymour and Doncaster, 1972). In the presence of plentiful food bacteria, hermaphrodite *C. elegans* defecate every 45 seconds. In males, defecation ceases during mating (K. Liu, quoted in Emmons (1997)); the anal sphincter surrounding the intestinal-rectal valve closes off the intestine during ejaculation and pulls it dorsally out of the way of the vas deferens (Sulston, Albertson and Thomson, 1980). Normal defecation also ceases in some mutants of *C. elegans* due to failure of the intestinal-rectal muscles to contract. These mutants continue to feed and material accumulates in the intestine until the pressure (all apparently due to the pumping of the pharynx) forces the rectal valve open and virtually the entire contents of the greatly distended intestine are explosively ejected. These mutants survive, indeed they are fertile, but they mature slowly and remain small. These are the characteristics of malnourishment (Avery and Horvitz, 1990); thus, as noted earlier, it appears that high throughput is of great importance for nutrient uptake.

Changes in the intestine of *Nematodirus battus* ascribed to host immune responses beginning about three weeks after infection include the appearance of large protein-containing crystals in the lumen (Lee and Martin, 1980). The crystals accumulate to such an extent that the intestinal–rectal junction becomes blocked and the anterior lumen distends with ingested material, apparently because, as in *C. elegans*, the nematode goes on feeding. In *N. battus*, however, the shape and size of the crystals mean that however much the pharynx pumps this blockage cannot be cleared.

Some nematodes, adult female *Meloidogyne* and mermithids for example, manage without defecation, indeed they have no rectum. Whilst this may seem a major problem, one has to assume that the diet of filtered plant cell sap does not create significant amounts of indigestible material.

Conclusion

With comparatively few exceptions nematodes are selective in what they eat, be this at the level of discriminating between one kind of bacterium or another or at the level of selecting one kind of plant

cell over another. This selectivity is based on behavioural and structural differences and is reflected in phylogenetic distinctions. The structural differences are the ones which have received the greatest attention and there is a massive literature on this subject. Interpretation of the ultrastructural observations has greatly benefited from the behavioural studies, particularly those of plant-parasitic and bacterivorous nematodes studied in the laboratory, coupled with characterisation of some of the molecules involved in different stages of the feeding processes. However, there has been no systematic study of the whole process for any one nematode and even for *C. elegans*, the nematode most intensely studied at the molecular level, the full complement of feeding-associated molecules is not known and there is as yet, very little known in detail relating diet, digestion and nutrition. For the purposes of this Chapter therefore it has been necessary to select examples both to illustrate the range of feeding types and to try to illustrate the ways in which different nematodes within each feeding type achieve their nutritional needs. Selection of examples has been part personal choice, where there was a choice, and many topics which deserve fuller treatment have had to be presented very succinctly. Nonetheless, the hope is that this chapter has conveyed some feeling for the breadth and depth of detail needed to begin to understand feeding and digestion in nematodes.

References

Albert, P.S. and Riddle, D.L. (1983) Developmental alterations in sensory neuroanatomy of the *Caenorhabditis elegans* dauer larva. *Journal of Comparative Neurology*, **219**, 435–451.

Albertson, D.G. and Thomson, J.N. (1976) The pharynx of *Caenorhabditis elegans*. *Philosophical Transactions of the Royal Society, London, B*, **275**, 299–325.

Alkemade, R., Wielemaker, A. and Hemmings, M.A. (1992) Stimulation of decomposition of *Spartina anglica* leaves by the bacterivorous marine nematode *Diplolaimelloides bruciei* (Monohysteridae). *Journal of Experimental Marine Biology and Ecology*, **159**, 267–278.

Anderson, R.V. and Byers, J.R. (1975) Ultrastructure of the esophageal procorpus in the plant parasitic nematode, *Tylenchorhynchus dubius*, and functional aspects in relation to feeding. *Canadian Journal of Zoology*, **53**, 1581–1595.

Artis, D., Potten, C.S., Else, K.J., Finkelman, F.D. and Grencis, R.K. (1999) *Trichuris muris*: host intestinal epithelial cell hyperproliferation during chronic infection is regulated by interferon-gamma. *Experimental Parasitology*, **92**, 144–153.

Avery, L. and Horvitz, H.R. (1990) Effects of starvation and neuroactive drugs on feeding in *Caenorhabditis elegans*. *Journal of Experimental Zoology*, **253**, 263–270.

Avery, L. and Thomas, J.H. (1997) Feeding and defecation. In *C. elegans II*, edited by D.L. Riddle, T. Blumenthal, B.J. Meyer and J.R. Preiss, pp. 679–716. Cold Spring Harbor: Cold Spring Harbor Laboratory Press.

Batson, B.S. (1979) Ultrastructure of the trophosome, a food storage organ in *Gastromermis boophthorae* (Nematoda: Mermithidae). *International Journal for Parasitology*, **9**, 505–514.

Baylis, H.A., Furuichi, T., Yoshikawa, F., Mikoshiba, K. and Sattelle, D.B. (1999) Inositol 1,4,5-trisphosphate receptors are strongly expressed in the nervous system, pharynx, intestine, gonad and excretory cell of *Caenorhabditis elegans* and are encoded by a single gene (itr-1). *Journal of Molecular Biology*, **294**, 467–476.

Bedding, R.A. (1973) Biology of *Deladenus siricidicola* (Neotylenchidae) an entomaphagous-mycetophagous nematode parasite in siricid wood-wasps. *Nematologica*, **18**, 482–493.

Beh, C.T., Ferrari, D.C., Chung, M.A. and McGhee, J.D. (1991) An acid phosphatase as a biochemical marker for intestinal development in the nematode *Caenorhabditis elegans*. *Developmental Biology*, **147**, 133–143

Bilgrami, A.L. (1992) Resistance and susceptibility of prey nematodes to predation and strike rate of the predators *Mononchus aquaticus, Dorylaimus stagnalis* and *Aquatides thornei*. *Fundamental and Applied Nematology*, **15**, 265–270.

Bilgrami, A.L. (1993) Analysis of the predation by *Aporcelaimus nivalis* on prey nematodes from different prey trophic categories. *Nematologica*, **39**, 356–365.

Bird, A.F. (1959) The attractiveness of roots to the plant parasitic nematodes *Meloidogyne javanica* and *M. hapla*. *Nematologica*, **4**, 322–335.

Bird, A.F. (1968) Changes associated with parasitism in nematodes, IV. Cytochemical studies on the ampulla of the dorsal oesophageal gland of *Meloidogyne javanica* and on exudations from the buccal stylet. *Journal of Parasitology*, **54**, 879–890.

Bird, A.F. (1983) Changes in the dimensions of the esophageal glands in root-knot nematodes during the onset of parasitism. *International Journal for Parasitology*, **13**, 343–348.

Bird, A.F. and Akhurst, R.J. (1983) The nature of the intestinal vesicle in nematodes of the family Steinernematidae. *International Journal for Parasitology*, **13**, 599–606.

Bird, A.F. and Bird, J. (1991) *The Structure of Nematodes*. London: Academic Press.

Bird, A.F., Bird, J., Fortuner, R. and Moen, R. (1989) Observations on *Aphelenchoides bylurgi* Massey, 1974 feeding on fungal pathogens of wheat in Australia. *Revue de Nématologie*, **12**, 27–34.

Bird, A.F., Downton, W.J.S. and Hawker, J.S. (1975) Cellulase secretion by second stage larvae of the root-knot nematode (*Meloidogyne javanica*). *Marcellia*, **38**, 165–169.

Böckenhoff, A. and Grundler, F.M.W. (1994) Studies on the nutrient uptake by the beet cyst nematode *Heterodera schachtii* by *in situ* microinjection of fluorescent probes into the feeding structures in *Arabidopsis thaliana*. *Parasitology*, **109**, 249–254.

Borgers, M., Van den Bossche, H. and Schaper, J. (1970) The ultrastructural localization of non-specific phosphatases in the intestinal epithelium of *Ascaris suum*. *Journal of Histochemistry and Cytochemistry*, **18**, 519–521.

Brindley, P.J., Gam, A.A., McKerrow, J.H. and Neva, F.A. (1995) Ss40: the zinc endopeptidase secreted by infective larvae of *Strongyloides stercoralis*. *Experimental Parasitology*, **80**, 1–7.

Bruce, R.G. (1966) The fine structure of the intestine and hind gut of the larva of *Trichinella spiralis*. *Parasitology*, **56**, 359–365.

Capello, M., Vlasuk, G.P., Bergum, P.W., Huang, S. and Hotez, P.J. (1995) *Ancylostoma caninum* anticoagulant peptide: A hookworm-derived inhibitor of human coagulation factor Xa. *Proceedings of the National Academy of Sciences, USA*, **92**, 6152–6156.

Chen, S.H. and Howells, R.E. (1979) The uptake *in vitro* of dyes, monosaccharides and amino acids by the filarial worm *Brugia pahangi*. *Parasitology*, **78**, 343–354.

Chernin, E., Michelson, E.H. and Augustine, D.L. (1960) *Daubaylia potomaca*, a nematode parasite of *Heliosoma trivolvis* transmissible to *Australorbis glabratus*. *Journal of Parasitology*, **46**, 599–607.

Chitwood, B.G. and Chitwood, M.B. (1974) *Introduction to Nematology*. Baltimore: University Park Press.

Chitwood, D.J. (1997) Biochemistry and function of nematode steroids. In *Biochemistry and Function of Sterols*, edited by E.J. Parish and W.D. Nes, pp. 169–180. Boca Raton: CRC Press.

Chitwood, D.J. and Lusby, W.R. (1991) Sterol composition of the corn root lesion nematode, *Pratylenchus agilis* and corn root cultures. *Journal of the Helminthological Society of Washington*, **58**, 43–50.

Clark, W.C. (1960) The oesophago-intestinal junction in the Mononchidae (Enoplida, Nematoda). *Nematologica*, **5**, 178–183.

Coadwell, W.J. and Ward, P.F.V. (1981) The development, composition and maintenance of experimental populations of *Haemonchus contortus* in sheep. *Parasitology*, **82**, 257–261.

Cobb, N.A. (1917) The mononchs (*Mononchus* Bastian 1866) a genus of free-living predatory nematodes. *Soil Science*, **3**, 431–486.

Coomans, A. (1963) Stoma structure in members of the Dorylaimidae. *Nematologica*, **9**, 587–601.

Coomans, A. and Van der Heiden, A. (1971) Structure and formation of the feeding apparatus in *Aporcelaimus* and *Aporcolaimellus* (Nematoda: Dorylaimida). *Zeitschrift für Morphologie der Tierre*, **70**, 103–118.

Cox, G.N., Pratt, D., Hageman, R and Boisvenue, R.J. (1992) Molecular cloning and primary sequence of a cysteine protease expressed by

Haemonchus contortus. *Molecular and Biochemical Parasitology*, **41**, 25–34.

Crofton, H.D. *Nematodes*. London: Hutchinson

Croll, N.A. and Smith, J.M. (1978) Integrated behaviour in the feeding phase of *Caenorhabditis elegans* (Nematoda). *Journal of Zoology*, **184**, 507–517.

de Boer, J.M., Yan, Y., Wang, X., Smant, G., Hussey, R.S., Davis, E.L. and Baum, T.J. (1999) Developmental expression of secretory beta-1,4-endoglucanases in the subventral esophageal glands of *Heterodera glycines*. *Molecular Plant Microbe Interactions*, **12**, 663–669.

Despommier, D.D. (1983) Biology. In *Trichinella and Trichinosis*, edited by W.C. Campbell, pp. 75–151. New York: Plenum Press.

Deubert, K.H. and Rohde, R.A. (1971) Nematode enzymes. In *Plant parasitic Nematodes: Cytogenetics, Host-Parasite Interactions and Physiology*, Vol. 2 edited by, B.M. Zuckerman, W.F. Mai and R.A. Rohde, pp. 73–90. New York: Academic Press.

Ding, X., Shields, J., Allen, R. and Hussey, R.S. (1998) A secretory cellulose-binding protein cDNA cloned from the root-knot nematode (*Meloidogyne incognita*). *Molecular Plant Microbe Interactions*, **11**, 952–959.

Doncaster, C.C. and Seymour, M.K. (1973) Exploration and selection of penetration site by Tylenchida. *Nematologica*, **19**, 137–145.

Doncaster, C.C. and Seymour, M.K. (1975) Passive ingestion in a plant nematode, *Hexatylus viviparus* (Neotylenchidae: Tylenchida). *Nematologica*, **20**, 297–307.

Dropkin, V.H. (1963) Cellulase in phytoparasitic nematodes. *Nematologica*, **9**, 444–454.

Dropkin, V.H. and Bird, A.F. (1978) An electron microscopic study of the glycogen and lipid in female *Meloidogyne incognita* (root-knot nematode). *Journal of Parasitology*, **60**, 1013–1021.

Eiff, J.A. (1966) Nature of an anti-coagulant from the cephalic glands of *Ancylostoma caninum*. *Journal of Parasitology*, **52**, 833–843.

Emmons, S.W. and Sternberg, P.W. (1997) Male development and behaviour. In *C. elegans II*, edited by D.L. Riddle, T. Blumenthal, B.J. Meyer, and J.R. Preiss, pp. 295–334. Cold Spring Harbor: Cold Spring Harbor Laboratory Press.

Esser, R.P. and Sobers, K. (1964) Natural enemies of nematodes. *Proceedings of the Soil Crop Science Society Florida*, **24**, 326–353.

Faulkner, L.R. and Darling, H.H. (1961) Pathological histology, hosts and culture of the potato rot nematode. *Phytopathology*, **51**, 778–786.

Fetterer, R.H. and Rhoads, M.L. (2000) Characterization of acid phosphatase and phosphorylcholine hydrolase in adult *Haemonchus contortus*. *Journal of Parasitology*, **86**, 1–6.

Furmidge, B.A., Horn, L.A. and Pritchard, D.I. (1995) The anti-haemostatic strategies of the human hookworm *Necator americanus*. *Parasitology*, **112**, 81–87.

Gamble, H.R., Fetterer, R.H. and Mansfield, L.S. (1996) Developmentally regulated zinc metalloproteinases from third- and fourth-stage larvae of the ovine nematode *Haemonchus contortus*. *Journal of Parasitology*, **82**, 197–202.

Gamble, H.R. and Mansfield, L.S. (1996) Characterization of excretory-secretory products from larval stages of *Haemonchus contortus* cultured *in vitro*. *Veterinary Parasitology*, **62**, 291–305.

Gaugler, R. and Kaya, H.K. (Editors) (1990) *The entomopathogenic nematodes in biological control*. Boca Baton: CRC Press.

Grewal, P.S. and Wright, D.J. (1992) Migration of *Caenorhabditis elegans* (Nematoda: Rhabditidae) larvae towards bacteria and the nature of the bacterial stimulus. *Fundamental and Applied Nematology*, **15**, 159–166.

Griffiths, G. and Pritchard, D.I. (1994) Purification and biochemical characterization of acetylcholinesterase (AChE) from the excretory/secretory products of *Trichostrongylus colubriformis*. *Parasitology*, **108**, 576–586.

Grootaert, P., Jaques, A. and Small, R.W. (1977) Prey selection in *Butlerius* sp. (Rhabditidae: Diplogasteridae). *Medelingen Faculteit Landbouwwetenschappen Rijksuniversiteit Gent*, **24**, 1559–1563.

Hawdon, J.M., Jones, B.F. and Hotez, P.J. (1995) Cloning and characterization of a cDNA encoding the catalytic subunit of a cAMP-dependent protein kinase from *Ancylostoma caninum* third-stage infective larvae. *Molecular and Biochemical Parasitology*, **69**, 127–130.

Hawdon, J.M., Jones, B.F., Hoffman, D.R. and Hotez, P.J. (1996) Cloning and characterization of *Ancylostoma*-secreted protein. A novel protein associated with the transition to parasitism by infective hookworm larvae. *Journal of Biological Chemistry*, **271**, 6672–6678.

Hopper, B.E. and Meyers, S.P. (1967) Population studies on benthic nematodes within a subtropical seagrass community. *Marine Biology*, **1**, 85–96.

Hotez, P.J., LeTrang, N., McKerrow, J.H. and Cerami, A. (1985) Isolation and characterization of a proteolytic enzyme from the adult hookworm *Ancylostoma caninum*. *Journal of Biological Chemistry*, **260**, 7343–7348.

Hotez, P.J., Narasimhan, S., Haggerty, J., Milstone, L.B., Bhopale, V., Schad, G.A. and Richards, F.F. (1992) Hyaluronidase from infective *Ancylostoma* hookworm larvae and its possible function as a virulence factor in tissue invasion and in cutaneous larva migrans. *Infection and Immunity*, **60**, 1018–1023.

Hunt, D.J. (1980) *Acugutturus parasiticus* n. gen., n. sp. a remarkable ectoparasitic aphelenchoid nematode from *Periplaneta americana* (L.) with proposal of Acugutturinae n. subf. *Systematic Parasitology*, **1**, 167–170.

Hussey, R.S. and Mims, C.W. (1990) Ultrastructure of esophageal glands and their secretory granules in the root-knot nematode *Meloidogyne incognita*. *Protoplasma*, **156**, 9–18.

Hussey, R.S. and Mims, C.W. (1991) Ultrastructure of feeding tubes in giant-cells induced in plants by the root-knot nematode *Meloidogyne incognita*. *Protoplasma*, **162**, 99–107.

Hussey, R.S., Mims, C.W. and Westcott, S.W. (1992) Ultrastructure of root cortical cells parasitized by the ring nematode *Criconemella xenoplax*. *Protoplasma*, **167**, 1–6.

Inglis, W.G. (1964) The marine Enoplida (Nematoda): A comparative study of the head. *Bulletin of the British Museum (Natural History) Zoology*, **2**, 263–376.

Jasmer, D.P. (1995) *Trichinella spiralis*: subversion of differentiated mammalian skeletal muscle cells. *Parasitology Today*, **11**, 185–188.

Jasmer, D.P., Perryman, L.E. and McGuire, T.C. (1996) *Haemonchus contortus* GA1 antigens: related, phospholipase C-sensitive, apical gut membrane proteins encoded as a polyprotein and released from the nematode during infection. *Proceedings of the National Academy of Sciences, U.S.A.*, **93**, 8642–8647.

Jasmer, D.P. and McGuire, T.C. (1996) Antigens with application toward immune control of blood-feeding parasitic nematodes. *British Veterinary Journal*, **152**, 251–268.

Jones, R.K. (1978) Histological and ultrastructural changes in cereal roots caused by feeding of *Helicotylenchus* spp. *Nematologica*, **24**, 393–397.

Kimble, J. and Sharrock, W.J. (1983) Tissue-specific synthesis of yolk proteins in *Caenorhabditis elegans*. *Developmental Biology*, **96**, 189–196.

Knox, D.P., Redmond, D.L. and Jones, D.G. (1993) Characterisation of proteinases in extracts of adult *Haemonchus contortus*, the ovine abomasal nematode. *Parasitology*, **106**, 395–404.

Lee, C.C. and Miller, J.H. (1969) Fine structure of the intestinal epithelium of *Dirofilaria immitis* and changes occurring after vermicidal treatment with caparsolate sodium. *Journal of Parasitology*, **55**, 1035–1045.

Lee, D.L. (1965) *The Physiology of Nematodes*. Edinburgh: Oliver and Boyd.

Lee, D.L. (1970a) The ultrastructure of the cuticle of adult female *Mermis nigrescens* (Nematoda). *Journal of Zoology*, **161**, 513–518.

Lee, D.L. (1970b). The fine structure of the excretory system in adult *Nippostrongylus brasiliensis* (Nematoda) and a suggested function for the 'excretory glands'. *Tissue & Cell*, **2**, 225–231.

Lee, D.L. (1996). Why do some nematode parasites of the alimentary tract secrete acetylcholinesterase? *International Journal for Parasitology*, **26**, 499–508.

Lee, D. L. and Anya, A.O. (1968). Studies on the movement, the cytology and the associated micro-organisms of the intestine of *Aspiculuris tetraptera* (Nematoda). *Journal of Zoology*, **156**, 9–14.

Lee, D. L. and Atkinson, H.J. (1976) *The Physiology of Nematodes*, 2nd edn. London: Macmillan.

Lee, D.L. and Foster, N. (1995) Gastrointestinal nematodes and host gut motility. *Helminthologia*, **32**, 107–110.

Lee, D. L. and Martin, J. (1980) The structure of the intestine of *Nematodirus battus* and changes during the course of an infection in lambs. *Parasitology*, **81**, 27–33.

Lee, D.L. and Shivers, R.R. (1987). A freeze-fracture study of muscle fibres infected with *Trichinella spiralis*. *Tissue & Cell*, **19**, 665–671.

Lewert, R.M. and Lee, C.L. (1956) Quantitative studies of the collagenase-like enzymes of cercariae of *Schistosoma mansoni* and the larvae of *Strongyloides ratti*. *Journal of Infectious Diseases*, **99**, 1–14.

Lilley, C.J., Urwin, P.E., McPherson, M.J. and Atkinson, H.J. (1996) Characterisation of intestinally active proteinases of cyst-nematodes. *Parasitology*, **113**, 415–424.

Longbottom, D., Redmond, D.L., Russell, M., Liddell, S., Smith, W.D. and Knox, D.P. (1997) Molecular cloning and characterisation of a putative aspartate proteinase associated with a gut membrane protein complex from adult *Haemonchus contortus*. *Molecular and Biochemical Parasitology*, **88**, 63–72.

Maggenti, A. (1981) *General Nematology*. New York: Springer-Verlag.

Maki, J. and Yanagisawa, T. (1986) Demonstration of carboxyl and thiol protease activities in adult *Schistosoma mansoni, Dirofilaria immitis, Angiostrongylus cantonensis* and *Ascaris suum. Journal of Helminthology*, **60**, 31–37.

Mapes, C.J. (1965) Structure and function of the nematode pharynx II. Pumping in *Panagrellus, Aplectana* and *Rhabditis. Parasitology*, **55**, 583–594.

Mapes, C.J. (1966) Structure and function in the nematode pharynx III. The pharyngeal pump of *Ascaris lumbricoides. Parasitology*, **56**, 137–149.

McKerrow, J.H., Brindley, P., Brown, M., Gam, A.A., Staunton, C. and Neva, F.A. (1990) *Strongyloides stercoralis*: identification of a protease that facilitates penetration of skin by the infective larvae. *Experimental Parasitology*, **70**, 134–143.

McLaren, D.J., Burt, J.S. and Ogilvie, B.M. (1974) The anterior glands of adult *Necator americanus* (Nematoda: Strongyloidea): II. Cytochemical and functional studies. *International Journal for Parasitology*, **4**, 39–46.

Mulk, M. M. and C, A. (1979) Three new *Alaimus*-species (Nematoda: Alaimidae) from Mount Kenya. *Nematologica*, **25**, 445–457.

Munn, E.A. (1977) A helical, polymeric extracellular protein associated with the luminal surface of *Haemonchus contortus* intestinal cells. *Tissue and Cell*, **9**, 23–34.

Munn, E.A. (1981) The endotube, a macroscopic intracellular structure from the syncytial intestine of the parasitic nematode *Haemonchus contortus. Journal of Physiology, London*, **319**, P7–P8.

Munn, E.A. (1997) Rational design of nematode vaccines: Hidden antigens. *International Journal for Parasitology*, **27**, 359–366.

Munn, E.A. and Greenwood, C.A. (1983) Endotube brush-border complexes dissected from the intestines of *Haemonchus contortus* and *Ancylostoma caninum. Parasitology*, **87**, 129–137.

Munn, E.A. and Greenwood, C.A. (1984) The occurrence of submicrovillar endotube (modified terminal web) and associated cytoskeletal structures in the intestinal epithelia of nematodes. *Philosophical Transactions of the Royal Society of London Series B-Biological Sciences*, **306**, 1–18.

Munn, E.A., Smith, T.S., Graham, M., Tavernor, A.S. and Greenwood, C.A. (1993) The potential value of integral membrane proteins in the vaccination of lambs against *Haemonchus contortus. International Journal for Parasitology*, **23**, 261–269.

Munn, E.A., Smith, T.S., Smith, H., James, F.M., Smith, F.C. and Andrews, S.J. (1997) Vaccination against *Haemonchus contortus* with denatured forms of the protective antigen H11. *Parasite Immunology*, **19**, 243–248.

Myers, R.F. (1965) Amylase, cellulase, invertase and pectinase in several free-living, mycophagous, and plant-parasitic nematodes. *Nematologica*, **11**, 441–448.

Myers, R.F. and Krusberg, L.R. (1965) Organic substances discharged by plant-parasitic nematodes. *Phytopathology*, **55**, 429–437.

Needham, T. (1744) A letter concerning certain chalky tubulous concretions, called malm; with some microscopical observations on the farina of the red lily and worms discovered in smutty corn. *Philosophical Transactions of the Royal Society, London*, **42**, 634–641.

Newton, S.E. and Munn, E.A. (1999) The development of vaccines against gastrointestinal nematode parasites, particularly *Haemonchus contortus. Parasitology Today*, **15**, 116–122.

Nicholls, C.D., Lee, D.L. and Sharpe, M.J. (1985) Scanning electron microscopy of biopsy specimens removed by a colonoscope from the abomasum of sheep infected with *Haemonchus contortus. Parasitology*, **90**, 357–363.

Ogilvie, B.M., Rothwell, T.L., Bremner, K.C., Schnitzerling, H.J., Nolan, J. and Keith, R.K. (1973) Acetylcholinesterase secretion by parasitic nematodes — I. Evidence for secretion of the enzyme by a number of species. *International Journal for Parasitology*, **3**, 589–597.

Overhoff, A., Freckman, D.W. and Virginia, R.A. (1993) Life cycle of the microbivorous Antarctic Dry Valley nematode *Scottnema lindsayae* (Timm 1971). *Polar Biology*, **13**, 151–156.

Pena Santiago, R. P. and Coomans, A. (1994) Revision of the genus *Tylencholaimus* De Man 1876. Didelphic species. *Nematologica*, **40**, 32–68.

Pline, M. and Dusenbery, D.B. (1987) Responses of the plant parasitic nematode *Meloidogyne incognita* to carbon dioxide determined by video camera-computer tracking. *Journal of Chemical Ecology*, **13**, 873–888.

Poinar, G.O. (1983) *The Natural History of Nematodes*. Englewood Cliffs: Prentice-Hall Inc.

Poinar, G.O., Jr. and van der Laan, P.A. (1972) Morphology and life history of *Sphaerularia bombi*. *Nematologica*, **18**, 239–252.

Poinar, G.O. and Hess, R. (1977) *Romanomermis culicivorax*: morphological evidence of transcuticular uptake. *Experimental Parasitology*, **42**, 27–33.

Popham, J.D. and Webster, J.M. (1979) Aspects of the fine structure of the dauer larva of the nematode *Caenorhabditis elegans*. *Canadian Journal of Zoology*, **57**, 794–800.

Pritchard, D.I., Leggett, K.V., Rogan, M.T., McKean, P.G. and Brown, A. (1991) *Necator americanus* secretory acetylcholinesterase and its purification from excretory-secretory products by affinity chromatography. *Parasite Immunology*, **13**, 187–199.

Ray, C. and McKerrow, J.H. (1992) Gut-specific and developmental expression of a *Caenorhabditis elegans* cysteine protease gene. *Molecular and Biochemical Parasitology*, **51**, 239–250.

Redmond, D.L., Knox, D.P., Newlands, G. and Smith, W.D. (1997) Molecular cloning and characterisation of a developmentally regulated putative metallopeptidase present in a host protective extract of *Haemonchus contortus*. *Molecular and Biochemical Parasitology*, **85**, 77–87.

Renn, N. (1998) Routes of penetration of the entomopathogenic nematode *Steinernema feltiae* attacking larval and adult houseflies. *Journal of Invertebrate Pathology*, **72**, 281–287.

Riding, I.L. (1970) Microvilli on the outside of a nematode. *Nature*, London, **226**, 179–180.

Riemann, F. and Schrage, M. (1978) The mucus-trap hypothesis on feeding of aquatic nematodes and implications of biodegradation and sediment texture. *Oecologia*, **34**, 75–88.

Rieu, P., Ueda, T., Haruta, I., Sharma, C.P. and Arnaout, M.A. (1994) The A-domain of beta-2 integrin CR3 (CD11b/CD18) is a receptor for the hookworm-derived neutrophil adhesion inhibitor NIF. *Journal of Cell Biology*, **127**, 2081–2091.

Riffle, J.W. (1967) Effect of an *Aphelenchoides* species on the growth of a mycorrhizal and pseudomycorrhizal fungus. *Phytopathology*, **57**, 541–544.

Riga, E., Perry, R.N., Barrett, J. and Johnston, M.R.L. (1995) Investigation of the chemosensory function of amphids of *Syngamus trachea* using electrophysiological techniques. *Parasitology*, **111**, 347–351.

Rumpenhorst, H.J. (1984) Intracellular feeding tubes associated with sedentary parasitic nematodes. *Nematologica*, **30**, 77–85.

Rutherford, T.A. and Webster, J.M. (1974) Transcuticular uptake of glucose by the entomophilic nematode, *Mermis nigrescens*. *Journal of Parasitology*, **60**, 804–808.

Rutherford, T.A., Webster, J.M. and Barlow, J.S. (1977) Physiology of nutrient uptake by the entomophilic nematode *Mermis nigrescens* (Mermithidae). *Canadian Journal of Zoology*, **55**, 1773–1778.

Schad, G.A. (1963) Niche diversification in a parasitic species flock. *Nature*, London, **198**, 404–406.

Seymour, M.K. (1975) Defaecation in a passively-feeding plant nematode, *Hexatylus viviparus* (Neotylenchidae: Tylenchida). *Nematologica*, **20**, 355–360.

Seymour, M.K. and Doncaster, C.C. (1972) Defaecation in *Aphelenchoides blastophorus* (Neotylenchidae: Tylenchida). *Nematologica*, **18**, 463–469.

Shepherd, A.M., Clark, S.A. and Hooper, D.J. (1980) Structure of the anterior alimentary tract of *Aphelenchoides blastophorus* (Nematoda: Tylenchida, Aphelenchina). *Nematologica*, **26**, 313–357.

Siddiqui, I.A. and Viglierchio, D.R. (1977) Ultrastructure of the anterior body region of the marine nematode *Deontostoma californicum*. *Journal of Nematology*, **9**, 56–82.

Siddiqui, M.R. (1986) *Tylenchida Parasites of Plants and Insects*, p. 512. Farnham Royal: Commonwealth Agricultural Bureaux.

Sijmons, P.C., Grundler, F.M.W., von Mende, N., Burrows, P.R. and Wyss, U. (1991) *Arabidopsis thaliana* as a new model for plant parasitic nematodes. *Plant Journal*, **1**, 245–254.

Small, R.W. (1987) A review of the prey of predatory soil nematodes. *Pedobiologica*, **30**, 179–206.

Smant, G., Stokkermans, J.P.W.G, Yan, Y., de Boer, J.M., Baum, T.J., Wang, X., *et al.* (1998) Endogenous cellulases in animals: Isolation of β-1,4-endoglucanase genes from two species of plant-parasitic cyst nematodes. *Proceedings of the National Academy of Sciences, USA*, **95**, 4906–4911.

Smith, K. and Harness, E. (1972) The ultrastructure of the adult stage of *Trichostrongylus colubriformis* and *Haemonchus placei*. *Parasitology*, **64**, 173–179.

Smith, T.S., Graham, M., Munn, E.A., Newton, S.E., Knox, D.P., Coadwell, W.J., *et al.* (1997) Cloning and characterization of a microsomal aminopeptidase from the intestine of the nematode *Haemonchus contortus*. *Biochimica et Biophysica Acta*, **1338**, 295–306.

Smith, T.S. and Munn, E.A. (1990) Strategies for vaccination against gastro-intestinal nematodes. *Revue Scientifique et Technique, Office International des Epizooties*, **9**, 577–595.

Smith, W.D. and Smith, S.K. (1993) Evaluation of aspects of the protection afforded to sheep with a gut membrane protein of *Haemonchus contortus*. *Research in Veterinary Science*, **55**, 1–9.

Stassens, P., Bergum, P.W., Gansemans, Y., Jespers, L., Laroche, Y., Huang, S., *et al.* (1996) Anticoagulant repertoire of the hookworm *Ancylostoma caninum*. *Proceedings of the National Academy of Sciences, USA*, **93**, 2149–2154.

Stewart, G.R., Perry, R.N., Alexander, J. and Wright, D.J. (1993) A glycoprotein specific to the amphids of *Meloidogyne* species. *Parasitology*, **106**, 405–412.

Stewart, G.R., Perry, R.N., and Wright, D.J. (1993) Studies on the amphid-specific glycoprotein gp32 in different life cycle stages of *Meloidogyne* species. *Parasitology*, **107**, 573–578.

Sulston, J.E., Albertson, D.G. and Thomson, J.N. (1980) The *Caenorhabditis elegans* male: Postembryonic development of nongonadal structures. *Developmental Biology*, **78**, 542–597.

Sulston, J.E., Schierenberg, E., White, J.G. and Thomson, J.N. (1983) The embryonic cell lineage of the nematode *Caenorhabditis elegans*. *Developmental Biology*, **100**, 64–119.

Taylor, C.E., Thomas, P.R., Robertson, W.M. and Roberts, I.M. (1970) An electron microscope study of the oesophageal region of *Longidorus elongatus*. *Nematologica*, **16**, 6–12.

Thorne, G. (1925) The genus *Acrobeles* von Linstow 1877. *Transactions of the American Microscopical Society*, **44**, 171–210.

Ubelaker, J.E. and Allison, W.F. (1972) Scanning electron microscopy of the eggs of *Ascaris lumbricoides, A. suum, Toxocara canis*, and *T. mystax*. *Journal of Parasitology*, **61**, 802–807.

Van den Bossche, H. and Borgers, M. (1973) Subcellular distribution of digestive enzymes in *Ascaris suum* intestine. *International Journal for Parasitology*, **3**, 59–65.

Van Der Heiden, A. (1974) The structure of the anterior feeding apparatus in members of the Ironidae (Nematoda: Enoplidae). *Nematologica*, **20**, 419–436.

Van de Velde, M.C. and Coomans, A. (1991) The ultrastructure of the buccal cavity of the monohysterid nematodes *Geomonhystera disjuncta* and *Diplolaimella dievengatensis*. *Revue de Nématologie*, **14**, 133–146.

Vanfleteren, J.R. (1980) Nematodes as nutritional models. In *Nematodes as Biological Models*, Vol. II, edited by B.M. Zuckerman, pp. 47–77. New York: Academic Press.

Van Gundy, S.D., Bird, A.F. and Wallace, H.R. (1967) Aging and starvation in larvae of *Meloidogyne javanica* and *Tylenchulus semipenetrans*. *Phytopathology*, **57**, 559–571.

Venette, R.C. and Ferris, H. (1998) Influence of bacterial type and density on population growth of bacterial-feeding nematodes. *Soil Biology and Biochemistry*, **30**, 949–960.

von Mende, N. (1997) Invasion and migration behaviour of sedentary nematodes. In *Cellular and Molecular aspects of Plant-Nematode Interactions*, edited by C. Fenoll, F.M.W. Grundler and S.A. Ohl, pp. 51–64. London: Kluwer Academic Publishers.

Wakabayashi, T., Nakamura, N., Sambongi, Y., Wada, Y., Oka, T. and Futai, M. (1998) Identification of the copper chaperone, CUC-1, in *Caenorhabditis elegans*: tissue specific co-expression with the copper transporting ATPase, CUA-1. *FEBS Letters*, **440**, 141–146.

Weise, R.W. (1977) A light and electron microscopic study of the dorsal buccal lancet of *Haemonchus contortus*. *Journal of Parasitology*, **63**, 854–857.

White, J. (1988) The Anatomy. In *The nematode Caenorhabditis elegans*, edited by W.B. Wood, pp. 81–122. New York: Cold Spring Harbor Laboratory.

Wilfred, M. and Lee, D.L. (1981) Observations on the buccal capsule and associated glands of adult *Bunostomum trigonocephalum* (Nematoda). *International Journal for Parasitology*, **11**, 485–492.

Wright, K.A. (1965) The histology of the oesophageal region of *Xiphinema index* Thorne and Allen, 1950, as seen with the electron microscope. *Canadian Journal of Zoology*, **43**, 689–700.

Wright, K.A. (1979) *Trichinella spiralis*: an intracellular parasite in the intestinal phase. *Journal of Parasitology*, **65**, 441–445.

Wright, K.A. (1991) Nematoda. In *Microscopic Anatomy*, Vol. 4 Aschelminthes, edited by Harrison, F.W. and Ruppert, E.E., pp. 111–195. New York: Wiley-Liss.

Wright, K.A., Carter, R.F. and Robertson, W.M. (1983) The musculature of the anterior feeding apparatus of *Xiphinema* species (Nematoda, Dorylaimoidea). *Nematologica*, **29**, 49–64.

Wright, K.A. and Thomson, J.N. (1981) The buccal capsule of *Caenorhabditis elegans* (Nematoda: Rhabditoidea): an ultrastructural study. *Canadian Journal of Zoology*, **59**, 1952–1961.

Wyss, U. (1987) Video assessment of root cell responses to dorylaimid and tylenchid nematodes. In *Vistas on Nematology* edited by J.A. Veech, and D.W. Dickson, pp. 325–351. Hyattsville: Society of Nematologists.

Wyss, U. (1992) Observations on the feeding behaviour of *Heterodera schachtii* throughout development, including events during moulting. *Fundamental and Applied Nematology*, **15**, 75–89.

Wyss, U. (1997) Root parasitic nematodes: an overview. In *Cellular and Molecular aspects of Plant-Nematode Interactions*, edited by C. Fenoll, F.M.W. Grundler and S.A. Ohl, pp. 51–64. London: Kluwer Academic Publishers.

Wyss, U., Stender, C. and Lehmann, H. (1984) Ultrastructure of feeding sites of the cyst nematode *Heterodera schachtii* Schmidt in roots of susceptible and resistant *Raphanus sativus* L. var. *oleiformis* Pers. cultivars. *Physiological Plant Pathology*, **25**, 21–37.

Wyss, U. and Zunke, U. (1986) Observations on the behaviour of second stage juveniles of *Heterodera schachtii* inside host roots. *Revue de Nématologie*, **9**, 153–165.

Wyss, U. Grundler, F.M.W. and Münch, A. (1992) The parasitic behaviour of 2nd-stage juveniles of *Meloidogyne incognita* in roots of *Arabidopsis thalina*. *Nematologica*, **38**, 98–111.

Yeates, G.W. (1969) Predation by *Monochoides potobikus* (Nematoda: Diplogasteridae) in laboratory culture. *Nematologica*, **15**, 1–9.

Yeates, G.W., Bongers, T., de Goede, R.G.M., Freckman, D.W. and Georgieva, S.S. (1993) Feeding habits in soil nematode families and genera — an outline for soil ecologists. *Journal of Nematology*, **25**, 315–331.

Yeates, G.W. (1998) Feeding in free-living soil nematodes. In *The Physiology and Biochemistry of Free-living and Plant-parasitic Nematodes*, edited by R.N. Perry and D.J. Wright, pp. 244–269. Slough: CAB International.

Young, I.M., Griffiths, B.G. and Robertson, W.M. (1996) Continuous foraging by bacterial-feeding nematodes. *Nematologica*, **42**, 378–382.

9. Feeding Behaviour of Plant-Parasitic Nematodes

Urs Wyss

Institut für Phytopathologie, Universität Kiel, Hermann-Rodewald-Straße 9, D-24118 Kiel, Germany

Keywords

Nematode vectors of plant viruses; tylenchid nematodes; feeding apparatus; parasitic strategies; associated root cell responses

Introduction

Nematodes have established themselves in nearly all possible ecological niches. Those adapted to the soil as an environment feed on many sources. Root parasitic nematodes exploit the root as their only source of nutrients and may spend their whole life cycle outside it, feeding from the surface or deeper tissues, others have evolved the capacity to invade the root and feed from cortical or stelar cells. All of them are obligate biotrophic parasites as they derive their nutrients from living cells, which they modify by pharyngeal gland secretions prior to food ingestion. In many cases food cells are transformed into highly specialised feeding structures to support nematode development and reproduction. The aim of this chapter is to summarise current knowledge of the feeding behaviour of root parasitic nematodes, first with emphasis on nematode vectors of plant viruses within the orders Triplonchida and Dorylaimida, which feed as migratory ectoparasites and second on selected examples within the Tylenchida, which have developed a highly diversified range of root parasitic strategies that culminates in sedentary endoparasitism, among which cyst and root-knot nematodes are the most economically important representatives. As cell and root tissue responses are closely related to feeding behaviour, this aspect will also be considered, however without presenting the molecular interactions in great detail. In this field of research enormous progress has been achieved in recent years and for all those interested, the book of Fenoll, Grundler and Ohl (1997) is strongly recommended. In addition, an excellent review on current approaches and progress in identifying parasitism genes in cyst and root-knot nematodes has just been published (Davis et al., 2000).

Nematode Vectors of Plant Viruses

More than forty years ago it was confirmed for the first time that a plant virus is transmitted by a root parasitic nematode. *Xiphinema index* was then shown to be the vector of grapevine fanleaf virus (Hewitt, Raski and Goheen, 1958). This finding prompted an intensive search for other virus-vector nematodes, with the result that only a few species of the nematode orders Triplonchida (family Trichodoridae) and Dorylaimida (family Longidoridae) have been found to be able to transmit soil-borne viruses (Brown, Robertson and Trudgill, 1995; Taylor and Brown, 1997).

All vectors of plant viruses are migratory ectoparasites that feed at a selected site for only a limited period of time. Trichodorids withdraw their nutrients from epidermal cells (Figure 9.1A1) which are killed after a complete feeding cycle (Figure 9.1B). Longidorid nematodes (with the main genera *Longidorus*, *Paralongidorus* and *Xiphinema*) are equipped with long stylets by means of which they are able to feed from deeper root tissues. Feeding on root tips by *Longidorus* and *Xiphinema* spp. (Figure 9.1A2,A3) transforms the tips into terminal galls that contain, among necrotic cells, modified cells (Figure 9.1C,D) as source of nutrients.

Trichodorid Nematodes

Trichodorids are plump, cigar-shaped nematodes that rarely exceed 1.5 mm in length. They are easily distinguished from other plant-parasitic nematodes by their stylet, which is a ventrally curved, elongate tooth, attached to the dorsal pharyngeal lumen wall. Trichodorids are placed into the family Trichodoridae with currently not more than 90 species within four genera (Decraemer and Baujard, 1998). Only some species of the genera *Paratrichodorus* and *Trichodorus* are vectors of soil-borne tobraviruses (Taylor and Brown, 1997). The removal of the trichodorids from the order Dorylaimida to the order Triplonchida is now generally accepted (Decraemer, 1995).

The structure and possible function of the feeding apparatus in trichodorids has been described (Robertson and Taylor, 1975; Robertson and Wyss, 1983; Taylor and Brown, 1997), but in spite of earlier ultrastructural studies (Hirumi et al., 1968; Raski, Jones and Roggen, 1969; Bird, 1971) some gaps need to be filled, especially with regard to the possible function of the secretory glands situated within the flask-shaped basal pharyngeal bulb. This bulb harbours five gland nuclei, one dorsal and two pairs in a subventral position, with the nuclei of the anterior pair being much smaller than those of the posterior pair (Figure 9.2A). The dorsal gland enters the food canal (pharyngeal lumen) just in front of the beginning of the bulbar extension, whereas

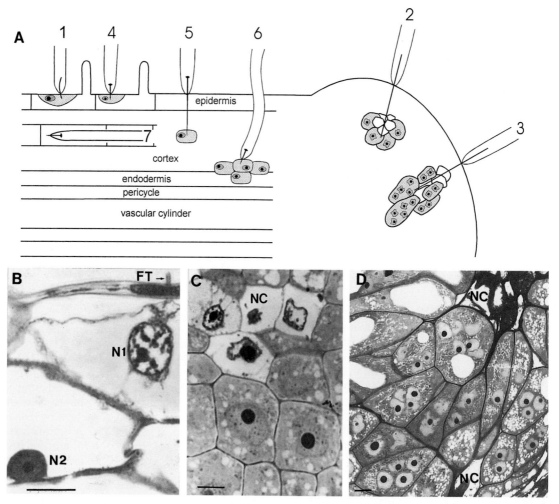

Figure 9.1. A: Schematic representation of feeding sites of some selected migratory root-parasitic nematodes and associated cell responses (B-D) shown in longitudinal sections. 1–4: Migratory ectoparasites; 1: *Trichodorus* sp.; 2: *Longidorus elongatus*; 3: *Xiphinema index*; 4: *Tylenchorhynchus dubius*; 5: Sedentary ectoparasite; *Criconemella xenoplax*; 6: Migratory ecto-endoparasite; *Helicotylenchus* sp.; 7: Migratory endoparasite; *Pratylenchus* sp. B: Epidermal cell of a *Nicotiana tabacum* root, parasitised by *Trichodorus similis*. C: Necrotic cells surrounded by hypertrophied uninucleate cells in a swollen root tip of *Apium graveolens*, parasitised by *Longidorus elongatus*. D: Necrotic cells surrounded by multinucleate cells in a root tip gall of *Ficus carica*, parasitised by *Xiphinema index*. All bars: 10 μm. Abbreviations: N1: Nucleus of parasitised epidermal cell; N2: Nucleus of unaffected subepidermal cell; NC: Necrotic cells; FT: Feeding tube.

the outlets of the two subventral pairs are presumably located near their nuclei (Decraemer and de Waele, 1981). In *T. allius* (now *P. allius*) the orifices of the dorsal and two larger subventral glands are surrounded by a sinus-like structure with radiating and branching arms (Raski *et al.*, 1969). A similar structure (Bird, 1971) was also observed around the orifice of the dorsal gland in *T. porosus* (now *P. porosus*). In both cases this fan-shaped sinus was believed to collect secretory fluids prior to their release into the food canal. The bulb also contains radial muscles in the posterior end of the bulb, about 10 μm in front of the pharyngeal-intestinal valve (Figure 9.2A). These muscles contract throughout feeding, including cell wall perforation (Wyss, 1971).

The stylet of trichodorids is composed of two parts, a protrusible solid slender and flexible anterior tooth termed the onchium (Robertson and Wyss, 1983) that

projects into the triangular food canal (Figure 9.2A1) and which is used to perforate plant cell walls. The posterior part, termed the onchiophore, is fused along its entire length with the dorsal pharyngeal lumen wall (Figure 9.2A1). A so-called 'guide ring', in reality a fold of the pharyngeal lumen wall, is formed at the site where this wall becomes first attached at the base of the onchium. When the stylet protractor muscles contract and the stylet is protruded, this fold is pulled forward (Figures 9.2B1; 9.3B), increasing the turgor pressure within the pharyngeal lumen (food canal). The stylet protractor muscles, stylet and food canal within the anterior pharynx are collectively called the oesophastome (Robertson and Wyss, 1983). The buccal cavity behind the oral aperture, called the cheilostom, is formed by an infolding of the body wall cuticle (Figure 9.2A1). At the region where the cheilostom contacts

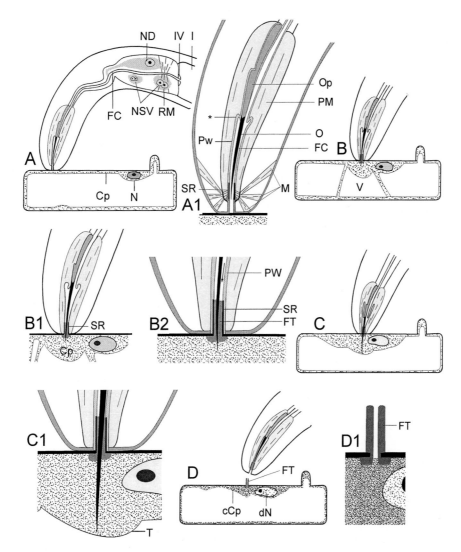

Figure 9.2. Schematic representation of a trichodorid nematode (here *Trichodorus similis*) feeding on an epidermal cell of a *Nicotiana tabacum* root. A: Exploration, with magnification of head end (A1). Asterisk indicates fusion of pharyngeal lumen wall to the posterior end of the onchium at the dorsal side. B: Injection of pharyngeal gland secretions; accumulation of cytoplasm at site of stylet penetration, with magnifications (B1; B2). Arrow in B2 indicates flow of pharyngeal gland secretions through food canal. The secretions harden into a feeding tube between adjacent strengthening rods. C: Initiation of ingestion by deep stylet thrusts into accumulated cytoplasm, with magnification (C1). D: Departure from feeding site with magnification of feeding tube left anchored in cell wall (D1). Abbreviations: Cp: Cytoplasm; cCp: coagulated cytoplasm; dN: disorganized nucleus; FC: Food canal; FT: Feeding tube; I: Intestine; IV: Pharyngeal-intestinal valve; M: Muscles at head end; N: Nucleus of epidermal root cell; ND: Nucleus of dorsal gland; NSV: Nuclei of the subventral glands; O: Onchium (anterior part of stylet); Op: Onchiophore (posterior part of stylet); PW: Pharyngeal wall; PM: Protractor muscles of stylet; RM: Radial muscles of basal bulb; SR: Strengthening rods; T: Tonoplast; V: Vacuole.

the oesophastome, there are three sets of muscles (Figure 9.2A1), first described in detail for *P. allius* (Raski *et al.*, 1969). Two of them are used to protract and retract the oesophastome, whereas the perpendicular set is thought to act on the food canal in the region of strengthening rods, possibly to break the seal between these rods and a feeding tube formed during feeding (Robertson and Wyss, 1983). The strengthening rods, recognised as refractive structures under the light microscope (Figure 9.3F), support the cuticular wall of the food canal in the anterior oesophastome. Their anterior ends are fused into the base of the cheilostome

(Figure 9.2A1). Stylet retractor muscles have not been reported, and it is thought that stylet retraction is achieved by increased turgor pressure created in the pharyngeal wall upon stylet protraction (Raski *et al.*, 1969; Robertson and Wyss, 1983).

Feeding behaviour and associated root cell responses

The feeding behaviour of *Trichodorus similis* on roots of *Brasscia rapa* var. *silvestris* (Wyss and Inst. Wiss. Film, 1972) and associated root cell responses in *Nicotiana tabacum* (Wyss and Inst. Wiss. Film, 1974), have been analysed by ciné filming and are described here briefly. Like all other

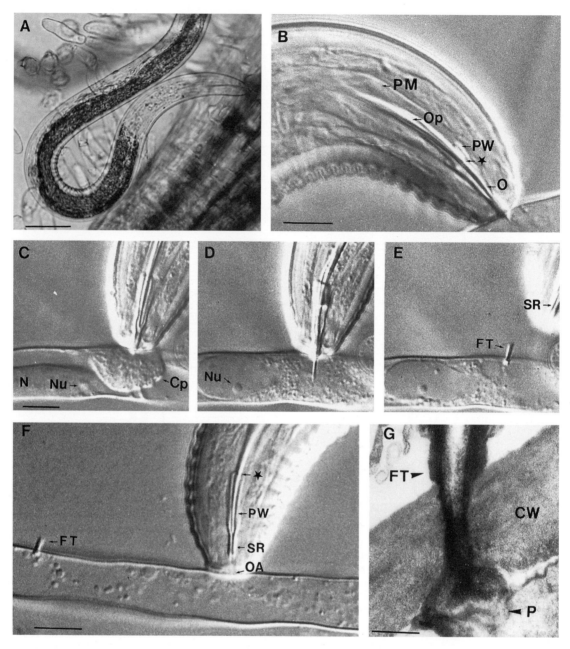

Figure 9.3. A–F: *Trichodorus similis* feeding on *Nicotiana tabacum*. A: Female feeding in root elongation region, bar: 40 μm. B-F: Feeding on root hairs, all bars: 10 μm. B: Close up of head just after cell wall perforation. Asterisk indicates fusion of pharyngeal lumen wall to dorsal side of onchium. C: During injection of pharyngeal gland secretions, accumulation of cytoplasm at site of stylet penetration, 35 sec after cell wall perforation. D: Initiation of ingestion by deep stylet thrusts, 75 sec after C. E: Departure from feeding site, 20 sec after D. F: Exploring another feeding site of a previously parasitised root hair, evident by a feeding tube. Strengthening rods of pharyngeal lumen wall still behind oral aperture. G: Transmission electron micrograph through a feeding tube left anchored in the wall of an epidermal cell, bar: 0.5 μm. Abbreviations: CW: Cell wall; Nu: Nucleolus; OA: Oral aperture; P: Basal plug of feeding tube; PW: Pharyngeal lumen wall. All other abbreviations as in Figure 9.2.

trichodorid nematodes (Wyss, 1981), *T. similis* is typically a browsing epidermal feeder, feeding on individual cells for rarely longer than just a few minutes. Preferred feeding sites are the elongation regions of rapidly growing roots (Figure 9.3A), where trichodorids tend to aggregate. Cells at the root's apex are generally only attacked for a short while when root growth stops due to gregarious feeding in the elongation region (Wyss, 1981).

A complete feeding cycle of a *T. similis* female on an individual cell lasts less than 4 minutes (Wyss, 1975) and is composed of five distinct phases: cell wall exploration, wall perforation, salivation (release of pharyngeal gland secretions), food ingestion and withdrawal from the feeding site. Prior to wall perforation, the nematode explores the surface of the wall by rubbing its lips with sensory papilla over a small area of the wall. The oesophastome is then still situated well behind the oral aperture (Figure 9.2A, A1). It is drawn forward by contraction of its protractor muscles when a suitable spot for wall perforation has been selected, so that the strengthening rods are now brought into close contact with the wall, which is thereafter perforated within less than 1 minute by continuous forceful and rapid stylet thrustings aimed at a single spot. Concurrent with each stylet protraction, the radial muscles in the basal bulb contract and thus dilate the triradiate food canal to an ellipsoid shape at the site of muscle attachment. Secretions from the pharyngeal glands are already released into the food canal during wall perforation (Wyss, Jank-Ladwig and Lehmann, 1979), probably by a mechanism of alternating high and low pressures between the food canal in the oesophastome and the site of dilation at the posterior end of the basal bulb (Wyss, 1971; 1975). As the posterior subventral pharyngeal glands appear to enter the food canal in this region (Figure 9.2A), it may be speculated that secretions from these glands could be involved in the formation of feeding tubes between the strengthening rods or in the drastic root cell responses during salivation.

After cell wall perforation, continuous stylet thrusting is maintained during salivation, which is by far the longest phase of a complete feeding cycle (Wyss, 1971). With each thrust, only the tip (2–3 μm) of the stylet is inserted through the perforated cell wall (Figure 9.3B). The cell's cytoplasm is then drawn from all directions to this site, where it accumulates into a large mass (Figures 9.2B; 9.3C). The cell's nucleus is also drawn to this aggregation; it swells and quickly loses its refractive index (Figure 9.3D) as the nucleoplasm becomes liquefied. Feeding tubes, formed by rapidly hardening pharyngeal gland secretions between the strengthening rods (Figure 9.2B2) become fully developed during salivation (Wyss et al., 1979). The secreted material also forms a small plug below the perforation hole of the cell wall and obviously functions as a one-way valve as none of the accumulated cytoplasm is ever withdrawn during salivation by the retracting stylet.

Food ingestion is initiated by a few deep stylet thrusts when most of the cell's cytoplasm has accumulated into a large mass at the site of wall perforation (Figures 9.2C; 9.3D). The main purpose of these deep thrusts is obviously to widen the feeding tube and the plug beneath the cell wall so that large proportions of the cytoplasm can now be removed concurrent with each stylet retraction. At the same time a liquefaction of the cytoplasm becomes evident, which may result from the destruction of the tonoplast during the deep stylet thrusts (Robertson and Wyss, 1983). However, this assumption is difficult to prove and it may well be that partial liquefaction of the cytoplasm occurs during salivation but does not become recognisable behind the massive aggregation. Support for this presumption is provided by the liquefaction of the nucleoplasm during salivation. Food ingestion, during which most of the accumulated cytoplasm is removed, occupies about one fifth of a complete feeding cycle. Whenever the nuclear membrane is pierced by the deep stylet thrusts, the nucleus is also removed. Departure from an emptied cell is usually accompanied with some effort as the nematodes have to detach themselves from their feeding tube (Figures 9.2D; 9.3E) that remains firmly anchored in the cell wall (Figures 9.2D1; 9.3G). Cell death soon becomes evident by the coagulation of remaining cytoplasm. However, occasionally cells survive when feeding is suddenly interrupted at the beginning of salivation. It was speculated that transmission of tobraviruses may be successful under these rather rare conditions. Another possibility may arise when subepidermal cells below emptied cells are attacked and cannot be fully exploited due to insufficient stylet length to initiate ingestion (Robertson and Wyss, 1983).

Concluding remarks

The feeding behaviour of trichodorid nematodes is unique among plant-parasitic nematodes as continuous stylet thrusting is not only required for cell wall perforation but also for the exploitation of the parasitised cell. Trichodorids have to form in addition a feeding tube for food withdrawal as their stylet is a solid tooth. All other root-parasitic nematodes use their hollow stylets for food withdrawal. Trichodorids are also the only root-parasitic nematodes capable of removing cytoplasm together with its organelles, such as mitochondria and plastids, because the lumen of the feeding tube is wide enough to permit this. Although their strategies for root cell exploitation are refined, as they first cause the cell's cytoplasm to aggregate into a large mass before they remove it by a change in feeding behaviour, they are little advanced in mode of parasitism. They require much energy to satisfy their nutritional demands and in addition need copious amounts of cytoplasm for reproduction. Cytoplasm of several hundred cells has to be withdrawn to produce just a few eggs within one week. The involvement of the five pharyngeal gland cells in feeding tube formation, cytoplasm responses and dissociation of tobraviruses from their retention sites along the cuticle of the food canal from the anterior region of the oesophastome right back to the pharyngeal-intestinal valve (Taylor and Brown, 1997) is yet unknown.

Longidorid Nematodes

Longidorids are relatively large nematodes (up to 12 mm long) and are equipped with a long hollow needle-like stylet (Figure 9.4A), by means of which they are able to feed deep within plant roots (Figure 9.4B). Only the anterior part of the stylet, termed the odontostyle (Figure 9.4A,B), is inserted into root cells. The posterior part, called the odontophore, is formed from the cuticle of the anterior pharynx and it contains, adjacent to the food canal, nerve tissues thought to have a gustatory function (Robertson, 1975). Stylet protraction is achieved by protractor muscles attached to the rear of the odontophore. The pharynx consists of a long, narrow and flexible anterior tube (Figure 9.4A) and a posterior muscular cylindrical bulb (Figure 9.4A,B), which is used to pump food through the food canal into

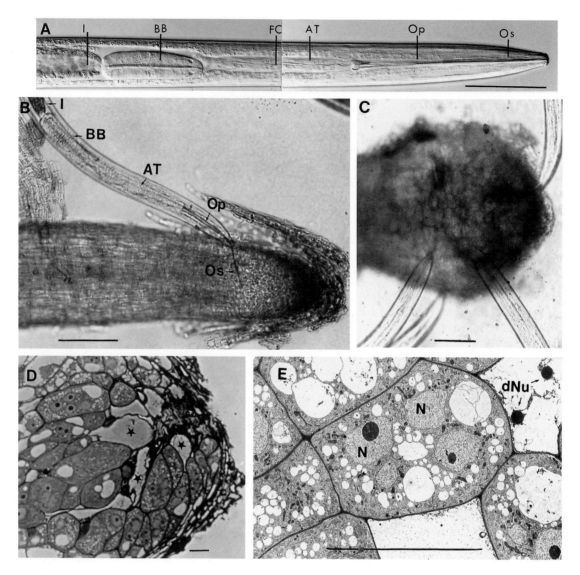

Figure 9.4. *Xiphinema index* females feeding on *Ficus carica* roots in monoxenic agar culture (b-c, bar: 100 μm and associated cell responses (d-e, bar: 10 μm). A: General overview of the feeding apparatus, bar: 100 μm. B: Female started feeding in the meristematic region of a main root. C: Four females feeding on a root tip gall. D: Light micrograph of a longitudinal section through a root tip gall with empty multinucleate cells (*) surrounded by unparasitised multinucleate cells. E: Transmission electron micrograph of a section through a multinucleate cell with four nuclei surrounded by necrotic cells with remnants of degraded nucleoli. Abbreviations: AT: anterior tube of pharynx; BB: basal pharyngeal bulb; DNU: disorganised nucleoli; FC: food canal; I: intestine; N: nucleus; OP: odontophore; OS: odontostyle.

the intestine against the turgor of the body. This bulb, generally termed the basal bulb, also harbours three pharyngeal secretory gland cells, one dorsal and two subventral, that play a decisive role in feeding/digesting processes. In this section special emphasis is placed on the structure and function of the basal bulb in *Xiphinema index*. This species transmits grapevine fanleaf virus (GFLV), the most important virus in wine growing regions, and hence *X. index* is currently one of the best examined longidorid nematodes.

Genus *Xiphinema*

The genus presently comprises about 200 species but only nine have so far proven to be vectors of nepoviruses (Taylor and Brown, 1997). Some may feed

for many hours or even several days from cells within the differentiated vascular cylinder (Wyss, 1981). Others, such as *X. index* and *X. diversicaudatum*, feed exclusively on root tips that become progressively transformed into attractive galls, containing groups of much enlarged multinucleate cells between empty cells (Figures 9.1D; 9.4D,E). *X. index* has a restricted host range, with *Vitis* spp. and fig (*Ficus carica*) being the most common hosts to support a rapid population build-up.

Feeding behaviour of X. index *and associated root cell responses*

Roots of fig plants maintained in monoxenic agar culture (Aumann, 1997) are most suitable to study in

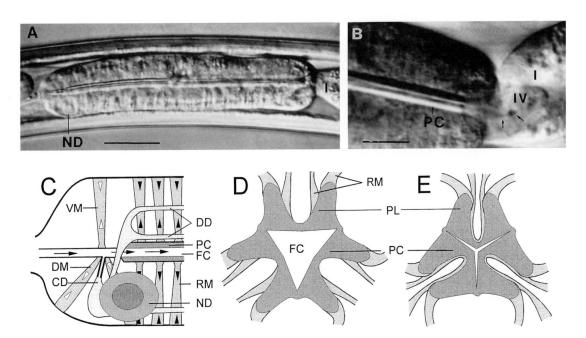

Figure 9.5. Basal pharyngeal bulb of a *Xiphinema index* female during food ingestion. A: General overview, bar: 20 μm. B: Posterior end of bulb, showing passage of ingested food (arrows) through pharyngeal-intestinal valve, bar: 10 μm. C-E: Schematic representations of parts of bulb in function. C: Most anterior part, showing flow of ingested food (arrows) through food canal. Black triangles represent contracted and white triangles relaxed muscles. D–E: Cross sections through triradiate pump chamber. D: Chamber dilated to a triangular cross section upon contraction of radial muscles attached to triangular platelets of thickened food canal cuticle. E: Chamber closed upon radial muscle relaxation. Abbreviations: CD: Collecting main duct of dorsal gland; DD: Dorsal gland ducts; DM: One of four dilator muscles that open main duct orifice of dorsal gland into food canal; FC: Food canal; I: Intestine; IV: Intestinal valve; ND: Nucleus of dorsal gland; PC: Pump chamber; PL: Platelet of thickened food canal cuticle in pump chamber; RM: Radial muscles attached to pump chamber; VM: One of three radial muscles attached to ventral side of food canal anterior to pump chamber.

detail the feeding behaviour of *X. index* and associated root cell responses. Single females usually start feeding on the main root, preferably near or in the meristematic region (Figure 9.4B), where they first explore the root surface with lip rubbings before they insert their stylets intracellularly two to three cells deep by quick successive thrusting jabs. Only thereafter they commence feeding on a column of progressively deeper cells, withdrawing food from each for a few minutes only. Upon maximal protrusion, the stylet is finally retracted and reinserted at another site, where food removal from successively deeper cells is resumed. This characteristic feeding pattern is, however, commonly interrupted by long quiescent periods during which the bulb does not pump. These periods occur at random just after the stylet has been inserted into a root cell and are invariably followed by a much prolonged period of food removal that can last more than half an hour (Wyss, 1977b; Wyss, Robertson and Trudgill, 1988).

Feeding of single females at different sites along the root tip retards root growth and the tip gradually swells into a terminal gall. Young root tip swellings contain groups of necrotic cells surrounded by slightly enlarged binucleate cells (Wyss, Lehmann and Jank-Ladwig, 1980; Bleve-Zacheo and Zacheo, 1983). A few days later these modified cells, when not attacked, have grown to large multinucleate cells by means of synchronous

mitoses without cytokinesis (Rumpenhorst and Weischer, 1978; Wyss, 1978). They predominate in root tip galls that are now strongly attractive to several females (Figure 9.4C). The multinucleate giant cells are obviously indispensable for nematode reproduction as they are not induced in resistant *Vitis* cultivars (Staudt and Weischer, 1992) and non-host plants (Wyss, 1978). In contrast to similar giant cells induced and maintained by sedentary root-knot nematodes, those induced by *X. index*, although comparatively much smaller, are eventually killed by nematode feeding. Sections through root tip galls of host plants (grape and fig) always contain empty necrotic giant cells between unparasitised multinucleate cells (Figure 9.4D, E).

Structure and function of the basal pharyngeal bulb in X. index

The basal bulb in adult *X. index* females is cylindrical in shape, it is about 100 μm long (Figure 9.5A). and contains mainly muscle, gland and nerve cells (Robertson, 1987). Right in front of the bulb the pharyngeal lumen (food canal) is still circular in cross section (Figure 9.7F), but in the pump chamber that extends along the bulb for most of its entire length (Figure 9.5A), it is triradiate and surrounded by a thickened cuticle in the form of three pairs of triangular platelets (Figure 9.5D, E). Suction for food ingestion (Figure 9.5C) is created by the contraction of six sets of radial muscles that are attached to the platelets and thus dilate the lumen of the food canal to a triangular cross section (Figure 9.5D) as proposed by Robertson and Taylor (1975) and substantiated by Towle and Doncaster (1978) and Robertson, Topham and Smith (1987) for the basal bulb of related nematodes of the genus *Longidorus*.

During food ingestion the basal bulb pumps rapidly at an average rate of three muscle contractions per second. Dilation of the pump chamber is about four times slower than its rapid closure when the radial muscles relax. The chamber then narrows from front to back, forcing food into the foregut of the intestine via a one-way pharyngeal-intestinal valve (Figure 9.5B). In longidorid nematodes, three groups of peripheral muscles are attached along the bulb to the outside of its membrane (Figure 9.6C). These muscles have been suggested to act as springs in order to conserve the volume of the bulb during pumping (Robertson *et al.*, 1987). As shown from the analysis of ciné films, the basal bulb of longidorid nematodes elongates upon contraction of the radial muscles and shortens again when the muscles relax (Wyss and Inst. Wiss. Film, 1977; Towle and Doncaster, 1978). Pumping in these nematodes is apparently driven entirely by the radial muscles (Robertson *et al.*, 1987).

Structure and function of the dorsal pharyngeal gland cell in X. index

Ciné film recordings and analysis of the feeding behaviour of *X. index* on fig roots (Wyss, 1977b; Wyss and Inst. Wiss. Film, 1977) indicated that the nematode's basal bulb harbours a complex system of ducts to be assigned to the dorsal gland, the nucleus of which is located at the anterior end of the bulb (Figure 9.5A). These ducts become dilated and are thus clearly visible when the anterior end of the bulb is suddenly stretched several seconds after odontostyle insertion into a root cell. The ducts stay dilated for a few seconds, then they narrow from back to front. After duct depletion, the bulb persists in the stretched position for a few further seconds, then it shortens and is immediately followed by a pumping action of the bulb for food ingestion.

Pumping is typically intermittent; periods of continuous bulb pulsation alternate with short pauses during which secretory fluids of the dorsal gland are rapidly flushed forward toward and through a main collecting duct that enters the food canal anterior to the bulb's pump chamber (Figure 9.5C). When pumping is resumed after these short pauses, the ducts again become dilated and refilled with secretory fluids (Wyss, *et al.*, 1988).

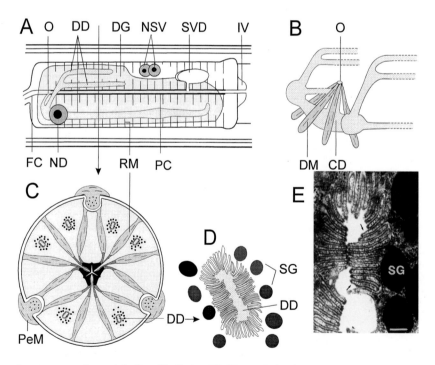

Figure 9.6. A–D: schematic representations of the basal bulb in a *Xiphinema index* female. A: Basal pharyngeal bulb, showing size of dorsal pharyngeal gland and its ducts (differentially shaded areas). B: Collecting main duct of dorsal gland ducts with orifice. C: Cross section through basal bulb (modified from a drawing by Robertson and Taylor, 1975) at level indicated by arrow. D: Enlargement of one of the six dorsal gland ducts. E: Transmission electron micrograph of a transverse section through part of one of the ducts of the dorsal pharyngeal gland, showing secretory granules close to the duct folds (arrows), bar: 0.2 μm. Abbreviations: CD: Collecting main duct of dorsal gland duct system. DD: Dorsal gland ducts (shaded); DG: Dorsal pharyngeal gland (shaded); DM: Dilator muscles that open main duct orifice of dorsal gland; FC: Food canal; IV: Pharyngeal-intestinal valve; O: Orifice of collecting main duct of dorsal gland; ND: Nucleus of dorsal gland; NSV: Nuclei of the two subventral glands. PC: Pump chamber; PeM: Peripheral muscles of basal bulb; RM: Radial muscle of basal bulb; SG: Secretory granules; SVD: Subventral gland ducts.

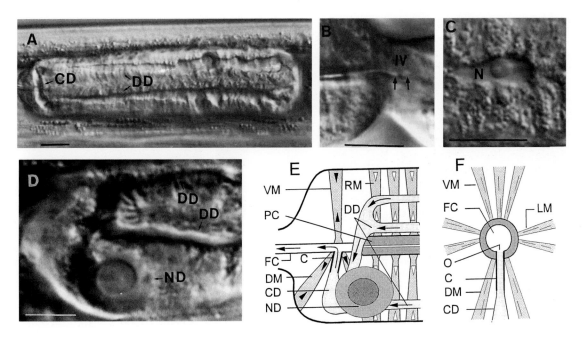

Figure 9.7. A–C: Pharyngeal glands and secretions produced in basal bulb of a *Xiphinema index* female, all bars: 10 μm. A: Dorsal view of basal bulb, showing the two main ducts of the dorsal gland and transverse collecting duct at anterior end of bulb. B: Posterior end of bulb, showing passage of secretory fluids (arrows) through pharyngeal-intestinal valve. C: Part of one of the two subventral glands with nucleus, surrounded by secretory granules. D–E: Dilation of dorsal gland ducts at anterior end of bulb, a few seconds after stylet insertion into a root cell. D: As seen under the light microscope by video contrast enhancement, bar: 5 μm. E: Schematic representation, showing flow of secretory fluids of dorsal gland to main collecting duct and from there through dilated main duct orifice into food canal. Black triangles represent contracted and white triangles relaxed muscles. F: Cross section through cuticularised basal part of main collecting duct of dorsal gland and associated muscles where it opens into food canal. Abbreviations: C: Cuticularised part of collecting main duct of dorsal gland duct system; LM: Radial muscles attached to lateral side of food canal; N: Nucleus of subventral gland O: Orifice of collecting main duct of dorsal gland into food canal. All other abbreviations as in Figures 9.5 and 9.6

An ultrastructural examination of the gland's complex duct system (Robertson and Wyss, 1979) revealed that the dorsal gland extends along the entire length on the dorsal side and half-way along the ventral side of the bulb (Figure 9.6A). Its membrane is highly convoluted to fit between and around the radial muscles (Figure 9.6C). Just anterior to the gland's nucleus, the duct system comprises a main, transverse and fan-shaped collecting duct (Figures 9.6B; 9.7A), which is fed by six longitudinal ducts. Two of these extend along the whole length of the bulb in a subdorsal position (Figures 9.6A; 9.7A) and two pairs extend along the anterior third of the bulb, each with one duct along the lateral and the other along the ventral side (Figure 9.6A, C).

The lining of the ducts has many longitudinal folds (Figure 9.6D,E) to allow extensive duct dilation. Numerous secretory granules surround these folds in feeding nematodes (Figure 9.6D,E).

Exocytosis of secretory granules in contact with the ducts is only observed to occur a few seconds after odontostyle insertion into a root cell, when the bulb is suddenly stretched and the ducts become much dilated (Wyss *et al.*, 1988; Wyss and Inst. Wiss. Film, 1988). Figures 9.7 D and E show such a moment. The main collecting duct enters the food canal just in front of the pump chamber, where the lumen is still circular in cross section (Figure 9.7F). The orifice is a longitudinal slit, and the main collecting duct that enters it is lined at its base with a thin layer of cuticle (Figure 9.7E, F) that may act as a one-way valve (Robertson and Wyss, 1979, 1983). When secretory fluids are discharged into the food canal, this slit is dilated by the contraction of four dilator muscles (Figure 9.6B) that extend radially to the periphery of the bulb (Figure 9.7E). At the site of the orifice, additional radial muscles are attached to the lateral and ventral side of the food canal (Figure 9.7F). These muscles appear to contract in unison to keep the food canal in position when the dilator muscles open the main duct orifice. Discharge of secretory fluids is most pronounced when the bulb suddenly stops pumping in order to release the contents of the re-filled ducts. Secretory fluids are then seen to be flushed forward through the ducts at great speed (Wyss, 1977b; Wyss *et al.*, 1988; Wyss and Inst. Wiss. Film, 1988). Within the food canal these fluids should be forced toward the stylet because the radial muscles attached to the pump chamber are then relaxed and the lumen within this chamber is closed (Figure 9.7E).

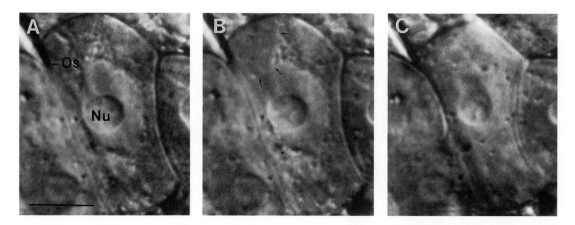

Figure 9.8. A–C: Responses of a meristematic root tip cell of *Ficus carica* to injection of dorsal gland secretions, bar: 5 μm. A: Tip of odontostyle inserted in cell, 2 sec after cell wall perforation. B: 23 sec after A and 11 sec after first injection, cytoplasm around stylet tip liquefied, marked by arrows. C: 61 sec after B, just before fourth and final injection; cytoplasm and nucleoplasm liquefied, cell partially emptied and nucleolus in process of disintegration. Abbreviations: Nu: Nucleolus of parasitised meristematic cell; Os: Tip of odontostyle.

In vivo *observations of root cell responses to dorsal gland secretions*

Before the introduction of high resolution video-enhanced contrast light microscopy (Wyss and Zunke, 1986a), it was not possible to ascertain whether the dorsal gland secretions are indeed injected into food cells and how they affect the cytoplasm of these cells. With the aid of this technique that allows observations of cell responses several layers below the root surface, it could be confirmed that the dorsal gland secretions are injected into root cells, where they are highly destructive (Wyss *et al.*, 1988; Wyss and Inst. Wiss. Film, 1988).

Whenever the stylet hits the wall of a meristematic cell, its thin wall is perforated within seconds by a few forceful jabs. The stylet tip is pushed just a few μm deep into the cytoplasm (Figure 9.8A) of the perforated cell, where it then stays immobile throughout feeding. 12–19 seconds later secretory fluids from the dorsal gland are seen to be released for the first time through the stylet tip orifice. This release coincides with the first depletion of the dorsal gland ducts. The secretions immediately liquefy the cytoplasm around the stylet tip (Figure 9.8B). After about 10 seconds, fluids from the modified cytoplasm are ingested for the first time by basal bulb pumpings. 15–40 seconds later dorsal gland secretions are injected for the second time, now however, with much greater volume and force. They further liquefy the cytoplasm and also the nucleoplasm of the nearby nucleus. After 2–3 further forceful injections, each followed by ingestion pumpings, the cell is nearly empty and the nucleolus, still retained in the nuclear envelope, has shrunk to a dense core (Figure 9.8C). The odontostyle is then pushed through the nearly emptied cell by a series of short thrusting jabs at a rate of 2–3 per second until it hits the wall of the next deeper cell, which, when perforated, suffers the same fate. In general 3–5 injections of dorsal gland secretions are required to liquefy the contents of a single meristematic

cell, which, within about two minutes, are consumed by 2–4 ingestion spells (Wyss *et al.*, 1988).

The feeding behaviour on binucleate cells is identical to that on uninucleate meristematic cells but the total time of food ingestion now lasts longer and more injections of secretory fluids from the dorsal gland (on average 12) are needed to assist depletion of these cells (Wyss *et al.*, 1988). This indicates an increased density of the cytoplasm in binucleate cells. As shown by Griffiths and Robertson (1984a), feeding by *X. diversicaudatum*, a closely related species, leads to an increased DNA content in the nuclei of these cells in strawberry root tip swellings and this is accompanied by an increase in the proportion of cytoplasm of the root tip volume (Griffiths and Robertson, 1988). Cell responses in multinucleate cells within root tip galls are difficult to observe in detail, but feeding on these cells is now much prolonged and requires numerous injections of dorsal gland secretions.

Quiescent periods and subventral gland secretions

As already mentioned, long lasting periods of apparent bulb inactivity occur at random just after odontostyle insertion into a root cell. Secretions are then seen to emanate slowly through the orifice of the odontostyle, forming a plug-like structure around the protruded stylet tip (Wyss *et al.*, 1988). Apart from a gradual increase in size of this plug, no other changes are discernible within the perforated cell. When the basal bulb finally starts pumping, the usual forceful injections of secretory fluids from the dorsal gland are obviously impeded by the presence of the plug, which has never been seen to become dislodged from the stylet tip. Degradation of the cytoplasm and nucleus within this cell is therefore considerably retarded, which explains the much prolonged ingestion duration after a quiescent period. It may be speculated that the plug-like material released during the quiescent periods originates from secretions of the two subventral glands. These gland cells occupy about 25% of the basal bulb (Robertson, 1987). The nucleus of one of these glands and its associated ducts, densely surrounded by secretory granules, is shown in Figure 9.7C. As the ducts of the subventral glands enter the lumen of the pump chamber in the posterior half of the bulb (Figure 9.6A), it is generally assumed that their secretions are passed into the intestine for digestive purposes. A steady flow of fluids through the pharyngeal-intestinal valve is clearly visible (Figure 9.7B) whenever the ducts of the dorsal gland become dilated for the first time after odontostyle insertion into a root cell, and it has been suggested that these fluids originate from the two subventral glands (Wyss et al., 1988; Wyss and Inst. Wiss. Film, 1988). However, the assumption that secretions from the subventral glands may also be released into root cells during the quiescent periods is supported by a close examination of the basal bulb during such periods. Some of the secretory granules around the ducts can then be seen to breakdown, indicating exocytosis (Wyss et al., 1988; Wyss and Inst. Wiss. Film, 1988). As no other signs of activity, including exocytosis, are clearly visible in the dorsal gland during the quiescent periods, the speculation of subventral gland involvement in this specific feeding phase appears plausible.

Concluding remarks

As shown in this section, *X. index* has developed remarkable strategies to exploit root tips of its principal host plants, grape and fig. The feeding behaviour of this species is identical to that of *X. diversicaudatum* (Trudgill, 1976; Trudgill, Robertson and Wyss, 1991), which has a broader host range (Pitcher, Siddiqi and Brown, 1974). Both species feed preferably on undifferentiated root tip cells, the contents of which are ingested within minutes by a series of dorsal gland secretions that liquefy the cytoplasm of these cells for food removal. In order to avoid a rapid depletion of food resources, both species induce the formation of large hypertrophied multinucleate cells that predominate in attractive root tip galls, the actual food sources for reproduction.

The mechanisms by which these metabolically highly active nurse cells are induced are unknown. It was originally speculated that only secretions from the dorsal gland are injected into the root and that the trigger molecules of these secretions should have a low molecular weight to enable their symplastic transport *via* plasmodesmata into adjacent yet unmodified cells to be transformed into multinucleate cells (Wyss et al., 1980). Now however, with the knowledge that the secretions of the dorsal gland are highly destructive in order to facilitate food removal, the assumption that they could play a decisive role in inducing multinucleate cells during feeding may no longer be maintained. It appears more probable that the trigger for nurse cell induction may be released during the quiescent periods.

Another question to be resolved in future research is how *Xiphinema* virus vectors, in this particular case

X. index, vector of grapevine fanleaf virus, and *X. diversicaudatum*, vector of arabis mosaic and strawberry latent ringspot virus, infect plant roots successfully with their associated nepoviruses. In both species virus particles are adsorbed in a monolayer to the cuticle of the food canal from the most anterior part of the odontophore to the posterior end of the pharangeal basal bulb (Taylor and Robertson, 1970; Raski, Maggenti and Jones, 1973). The mechanisms by which nepoviruses may become attached to their retention sites and may be dissociated from them have been discussed in recent reviews (Brown et al., 1995; Taylor and Brown, 1997). Wyss et al. (1988) and Trudgill et al. (1991) speculated that the force with which dorsal gland secretions are expelled during the brief ingestion pauses in *X.index* and *X.diversicaudatum* may be sufficient to dislodge nepoviruses from their retention sites without any specific release mechanisms. However, as the recipient host cell is then rapidly killed and emptied, this kind of virus transmission would be unsuccessful with regard to both, virus replication and cytoplasmic tubule formation by means of which nepoviruses can breach the cell wall barrier for cell-to-cell movement (Gilbertson and Lucas, 1996). The rather gentle release of dorsal gland secretions just prior to food ingestion should be equally unsuitable for a successful transmission as it initiates the liquefaction of the host cell cytoplasm. Therefore, as suggested by Wyss et al. (1988) and Trudgill et al. (1991), successful virus transmission may be associated with the quiescent periods, when host cells are killed only after considerable delay by dorsal gland secretions. Future research should prove if these quiescent periods are indeed essential for both the induction of multinucleate nurse cells and successful virus transmission and if, as assumed, the subventral gland secretions may then play a decisive role in both processes.

Genus *Longidorus*

The genus comprises about 100 species, out of which 8 have until now been identified as vectors of nepoviruses (Brown et al., 1995; Taylor and Brown, 1997). The structure of the feeding apparatus is basically similar to that of *Xiphinema* (Robertson and Wyss, 1983), with the main differences being described by Taylor and Brown (1997). However, compared to *X. index* (and *X. diversicaudatum* with an identical structure and function of the dorsal pharyngeal gland), little information is yet available on the same gland in *Longidorus*, with the exception that its duct system is much less developed than in *X. index* (Robertson and Wyss, 1983).

Longidorus spp. feed apparently exclusively on root tips, transforming them into terminal galls. Generally *Longidorus* spp. do not feed so readily on roots in agar culture as *Xiphinema* spp. However, the feeding of three species, *L. caespiticola* (Towle and Doncaster, 1978), *L. elongatus* and *L. leptocephalus* (Robertson, Trudgill and Griffiths, 1984) could be studied on roots of perennial ryegrass (*Lolium perenne*). All of them explore the surface of root tips by lip rubbings before they insert their stylets by rapid thrusting jabs deeply into the root until the odontostyle is nearly fully protracted. Root penetration is invariably followed by a period of apparent inactivity of up to 60 minutes duration during which the basal bulb does not pump and the nematodes are believed to release pharyngeal gland secretions, most probably from the dorsal gland. This salivation period is then followed by periods of continuous food ingestion that can last several hours and is only occasionally interrupted by short pauses. Calculations based on measurements of the pharyngeal bulb chamber indicated that a volume equivalent to more than 40 cells may be extracted by a single nematode feeding for 1 hour (Robertson *et al.*, 1984).

In *L. perenne*, root tip gall formation by *L. elongatus* feeding proceeds in an organised manner. Initial hypertrophy of individual uninucleate cells is followed by hyperplasia with synchronised cell divisions, which in turn is followed by secondary hypertrophy (Griffiths and Robertson, 1984b). Feeding from a single cell in root tip galls leads to the removal of contents of neighbouring cells. The walls of depleted cells adjacent to modified cells are holed, possibly as a result of local cell wall dissolution, indicating a route of food withdrawal from the modified cells (Robertson *et al.*, 1984). Similar clusters of emptied cells around modified cells have been termed lysigenous cavities in other *Longidorus* — host combinations (e.g. Zacheo and Bleve-Zacheo, 1995, Melillo *et al.*, 1997). Although modified food cells around these cavities have been described to be multinucleate in a single case (Bleve-Zacheo *et al.*, 1984), hypertrophied cells induced by *Longidorus* feeding (Figure 9.1C) do not respond by nuclear divisions without cytokinesis.

Tylenchid Nematodes

These plant-parasitic nematodes, confined to the suborder Tylenchina within the order Tylenchida, have developed a highly diversified range of root-parasitic strategies (Cohn and Spiegel, 1991; Sijmons, Atkinson and Wyss, 1994; Wyss, 1997) and they include the economically most devastating cyst and root-knot nematodes which, as sedentary endoparasites, have reached the most sophisticated level of root parasitism. In spite of this great diversity in root parasitism, all tylenchids possess basically a uniform feeding apparatus, which is here briefly described for an ectoparasite, *Tylenchorhynchus dubius*, feeding on a root hair of one of its many host plants (Figure 9.9).

The protrusible stylet has a narrow lumen throughout its length and fulfills three main functions: It is first used as a tool to pierce cell walls, then it serves to inject pharyngeal gland secretions to modify the cytoplasm of attacked cells and finally it is used to remove nutrients. Close to the tip, the lumen has a ventrally located orifice, which is usually not much larger than about 100 nm in internal diameter (Figure 9.9B). At its posterior end, the stylet bears three knobs to which protractor muscles are attached. The pharynx is divided into a nonmuscular procorpus, a muscular metacorpus

with a metacorpal bulb, containing a pump chamber, and a posterior nonmuscular glandular region. The pharyngeal lumen (food canal) is at first narrow and circular in cross section, but within the pump chamber of the metacorpal bulb it is triradiate and continues in this form right back to the intestine (Figure 9.9C).

During food ingestion the radial muscles of the metacorpal bulb contract several times per second, dilating the pump chamber to a triangular section at each contraction. Constraining muscles in front and behind the bulb (Figure 9.9A,C) stay contracted to stabilise the pulsating bulb.

The nonmuscular glandular region is composed of three secretory cells, one dorsal and two subventral. Each cell produces copious amounts of secretory proteins that are sequestered into secretory granules. The granules are then transported *via* microtubuli through a cytoplasmic extension (duct) into a terminal ampulla, where they are collected prior to the release of their contents into a valve with a complex membrane structure (Endo, 1998). The valve of the dorsal duct joins the food canal just behind the stylet knobs (Figure 9.9A), whereas the valves of the two subventral glands join the triradiate food canal just behind the pump chamber (Figure 9.9C). Due to the structure of the pump with a narrow rigid circular lumen in front and an expandable triradiate lumen behind, secretory fluids of the subventral glands were generally thought to be passed backward only into the intestine, where they may assist digestion. Now however, with the proof that subventral gland proteins can be secreted through the stylet of pre-infective and infective larvae of cyst nematodes (Smant *et al.*, 1997; Wang *et al.*, 1999), this one-way assumption can no longer be maintained. Secretory fluids from the dorsal gland are definitely injected through the stylet into root cells, where among other potential functions, they are involved in the formation of feeding tubes, which are probably indispensable for food ingestion in all sedentary endoparasites (Wyss, 1997). Additional, much more detailed information on pharyngeal gland secretions and their possible function in tylenchid nematodes is available (Jones and Robertson, 1997; Hussey and Grundler, 1998; Davis *et al.*, 2000).

Migratory Parasites

According to their modes of parasitism, root-parasitic tylenchid nematodes can be roughly divided into two main categories: migratory and sedentary parasites. Although the former include parasites that may stay for many days at a single feeding site and are thus often called sedentary (Sijmons *et al.*, 1994; Wyss, 1997) they should be placed into this category because they maintain a vermiform body shape throughout their life cycle and have not lost the capability of locomotion. Migratory root-parasitic tylenchids are generally grouped into browsing ectoparasites, sedentary ectoparasites, ecto-endoparasites and migratory endoparasites.

Browsing ectoparasites

Browsing ectoparasites are found in several subfamilies of the family Tylenchidae and in the subfamily

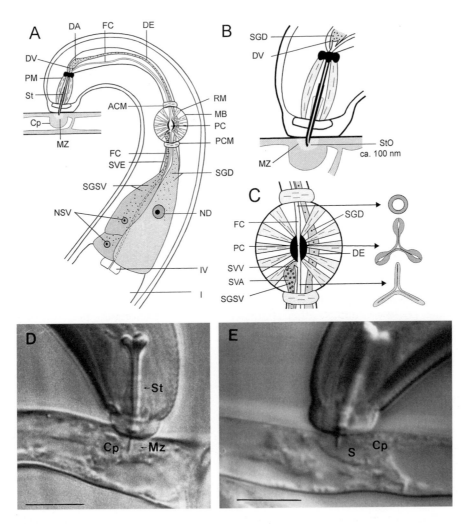

Figure 9.9. A–C: Schematic representation of the feeding apparatus of a tylenchid nematode (here *Tylenchorhynchus dubius*) feeding on a root hair. A: General view. B: Detail of head end. C: Detail of metacorpal bulb, with pump chamber closed; arrows point to cross sections through the pharyngeal lumen (food canal) within metacorpal bulb. D: Feeding on a root hair *of Brassica rapa* var. *silvestris*. E: Feeding on a root hair of *Raphanus sativus* var. *oleiformis*. bars: 10 µm. Abbreviations: ACM: Anterior constraining muscles of metacorpal bulb; Cp: Cytoplasm of root hair; DA: Dorsal gland ampulla; DE: Dorsal gland extension; DV: Dorsal gland valve; FC: Food canal; I: Intestine: IV: Pharyngeal-intestinal valve; MB: Matacorpal bulb (here pumping, pump chamber opened); MZ: Zone of modified cytoplasm around inserted stylet-tip; ND: Nucleus of dorsal gland; NSV: Nuclei of the two subventral glands; PC: Pump chamber of metacorpal bulb; PCM: Posterior constraining muscles of metacorpal bulb; PM: Stylet protractor muscles; RM: Radial muscles of metacorpal bulb; S: Secretions injected into root hair through orifice of stylet tip; SGD: Secretory granules of dorsal gland; SGSV: Secretory granules of subventral glands; St: Stylet; StO: Stylet tip orifice; SVA: Subventral gland ampulla; SVE: subventral gland extension; SVV: Subventral gland valve.

Telotylenchinae (family Belonolaimidae). A typical example is represented by *Tylenchorhynchus dubius*, as this polyphagous species feeds only for a short while on epidermal cells and its root hairs. As in browsing trichodorids, a complete feeding cycle can be divided into the same distinct phases, with the difference that most of the time is now spent on the removal of modified nutrients (Wyss, 1973; Wyss and Inst. Wiss. Film, 1973). As described in more detail by Doncaster and Seymour (1973) for tylenchid nematodes in general, exploration of the root surface is by lip rubbings accompanied by rather gentle stylet probings. Having found a suitable spot for cell wall penetration,

the stylet is vigorously and rapidly thrust in a rather irregular rhythm at a single point until the wall has been perforated within less than half a minute. The tip of the stylet then remains protruded in the cell just a few µm deep throughout salivation and ingestion. Salivation is characterised by a steady forward flow of secretory granules through the dorsal gland's extension toward the gland's ampulla and a simultaneous formation of a clear zone that surrounds the protruded stylet tip and around which the cytoplasm aggregates (Figure 9.9D). In a later study with the aid of video-enhanced contrast light microscopy, this zone was identified to represent secretory fluids (Figure 9.9E), most likely derived from the dorsal pharyngeal gland. It appears to act as a cytoplasmic filter to keep larger cell organelles at some distance from the stylet tip orifice and in addition may help to predigest some of the surrounding cytoplasm (Wyss, 1987). During ingestion, which covers about 85% of the total feeding time spent on a single cell (approximately 9 minutes), fluids are withdrawn by a rapid pumping action of the metacorpal bulb and the aggregated mass of cytoplasm shrinks in size. When the stylet is retracted, cell contents are never seen to exude through the punctured cell wall, cytoplasmic streaming is gradually reduced within the cell and finally stops when the remaining cytoplasm coagulates.

Sedentary ectoparasites

These nematodes are commonly found in members of the subfamilies Paratylenchinae (family Tylenchulidae) and Criconematinae (family Criconematidae) within the superfamily Criconematoidea. Females and larvae of these nematodes possess, amongst other characteristic diagnostic features, a very prominent metacorpal bulb with a large pump chamber but only a small glandular bulb. Within the Paratylenchinae, *Paratylenchus projectus* and *P. dianthus* were seen to feed continuously for up to one week on epidermal cells or root hairs of different plant seedlings without causing any noticeable damage (Rhoades and Linford, 1961). By far the best examined representative among this group is *Criconemella xenoplax* (Criconematinae) which pushes its long stylet between epidermal cells into an outer cortical cell (Figure 9.1A5) and then, after a secretion phase of 1–3 h duration, feeds continuously for up to 8 days from this food cell (Westcott and Hussey, 1992). Nutrients are withdrawn from a zone of modified cytoplasm in intimate contact with the stylet tip orifice, the only part of the inserted stylet not covered by callose depositions (Hussey, Mims and Westcott, 1992b). The plasmalemma of the food cell surrounds these depositions and becomes tightly appressed to the wall of the stylet orifice, creating a hole for an unimpeded uptake of nutrients (Hussey, Mims and Westcott, 1992a). Fibrillar material in an outer modified zone appears to form a barrier in the cytosol around a finely granular food removal zone in intimate contact with the stylet orifice, probably to limit movement of cell organelles toward the orifice. Plasmodesmata between the food cell and surrounding cells are modified in a unique manner to facilitate symplastic solute transport to the food cell (Hussey, Mims and Westcott, 1992a).

Ecto-endoparasites

Ecto-endoparasites are regarded as representing an intermediate step in the evolution from ectoparasitism to endoparasitism (Cohn and Spiegel, 1991). Nematodes of the subfamily Hoplolaiminae (family Hoplolaimidae), with the genera *Helicotylenchus*, *Hoplolaimus*, *Rotylenchus* and *Scutellonema* as their most important representatives, usually invade roots partially to feed on cortical or outer stelar cells. *Helicotylenchus* spp. may invade roots completely, where they feed from a single

cell for several days with alternating periods of salivation and ingestion (Jones, 1978a). The food cell, located close to the stele, is surrounded by a few metabolically active cells with dense cytoplasm but without nuclear enlargement (Figure 9.1A6). Feeding tubes, surrounded by a membranous network, were detected in the food cell (Jones, 1978b). An extensive membranous network, that appeared to originate from the plasmalemma, was also seen to surround the stylet tip and feeding tubes produced by *S. brachyurum* in its cortical food cell in potato roots (Schuerger and McClure, 1983). Stylet penetration into a cortical cell was followed by a rest period of several hours duration prior to metacorpal bulb pumping, and this sequence was repeated several times before the stylet was inserted into the next deeper cell. The stylet tip of *S. brachyurum* is surrounded by callose-like material in a similar way as in *C. xenoplax*, indicating that callose deposition may generally be one of the first responses in subepidermal root cells from which food is removed over prolonged periods of time by root-parasitic nematodes. In addition, it appears to be a common feature that the plasmalemma of the food cells becomes tightly appressed to the wall of the stylet orifice as shown in a section through the inserted stylet tip of *Hoplolaimus galeatus* (Ng and Chen, 1985).

Migratory endoparasites

Migratory endoparasites invade the roots and usually migrate intracellularly within the cortex, leaving tracks of destroyed cells behind them. They are found in members of the subfamily Pratylenchinae (family Pratylenchidae) with *Hirschmanniella*, *Pratylenchus* and *Radopholus* being the most economically important genera. *Pratylenchus* spp. have in most cases a wide host range, they can enter and leave the root at any developmental stage, and in nearly all cases their parasitism is confined to cortical cells (Figure 9.1A7). The feeding behaviour of *P. penetrans* on roots of various seedlings was studied and recorded by ciné film with the aid of video-enhanced contrast light microscopy (Zunke and Inst. Wiss. Film, 1988). Young larval stages (L2–3) prefer to feed on root hairs, whereas adults soon invade the roots. Ectoparasitic feeding on root hairs is comparable to that of *T. dubius* described above, with the difference that cell death only occurs after repeated feedings on the same hair (Zunke, 1990a). Successful root invasion by a single nematode attracts other nematodes to enter the root at this site. Migration through cortical cells is achieved by stylet thrustings first aimed to the corners of the cell walls and then over the entire wall until it ruptures. Extended food ingestion from cortical cells, which, after a short salivation period, can last many hours, leads to a loss in turgor pressure within the food cell and a gradual increase in the size of the nucleus, followed by delayed cell death after the nematode has moved away. Migration and feeding is frequently interrupted by rest periods lasting several hours, during which the nematodes become coiled inside a cell (Zunke, 1990b). In alfalfa roots, cell death inflicted by *P. penetrans* is not restricted to invaded and fed-on cells. It also extends to adjacent, uninvaded cells. In these cells tannin is deposited along the tonoplast, membrane integrity is lost, and cell organelles degenerate. This effect is most pronounced in the endodermis but can also extend into stelar cells (Townsend, Stobbs and Carter, 1989). Similarly, in roots of banana plants susceptible to the migratory endoparasite *Radopholus similis*, cell death does also spread to adjacent, uninvaded cells, and the endodermis does not inhibit nematode ingress into the vascular cylinder where vascular cells are destroyed in a similar way as in the cortex (Valette *et al.*, 1997).

Sedentary Endoparasites

Sedentary endoparasites have evolved the most advanced and spezialised strategies of root parasitism in that they induce and maintain specific permanent nurse cell structures from which they derive their food throughout development and reproduction. These feeding structures are nematode-specific, regardless of the tissue and host in which they are formed (Sijmons *et al.*, 1994). They are induced by root invasive vermiform developmental stages, which are either freshly hatched larvae (L2), e.g. species of cyst and root-knot nematodes, or adult females (e.g. species of the genera *Nacobbus, Rotylenchulus, Tylenchulus*). Soon after induction, the nematodes become sedentary due to a loss of somatic musculature. Food removal is accompanied by a continuous swelling of the body size in female nematodes, whereas nematodes destined to turn into males become vermiform again before maturity.

Nurse cell structures in the cortex

As far as known, only a few sedentary endoparasites derive their food from modified cells in the cortex of their host roots. The economically most important representative is *Tylenchulus semipenetrans* (subfamily Tylenchulinae within the family Tylenchulidae), which parasitises citrus in most areas where this crop is grown. The life cycle of this important pathogen is unusual as male larvae mature without feeding (Maggenti, 1981). Female larvae feed as migratory ectoparasites, and it is only the young adult female that enters the root to about half its body length. Six to ten cortical cells around an empty cell, in which the head of the nematode remains protruded, are transformed into metabolically highly active cells with enlarged nuclei, but which otherwise retain their normal size (Figure 9.10A). Although some information on the ultrastructure of these nurse cells is available (Himmelhoch *et al.*, 1979; B'Chir, 1988), it is not yet clear how the female derives its food from them. *Trophotylenchulus obscurus*, placed in the same subfamily as *T. semipenetrans*, is a widespread pest of coffee in West Africa. It invades the roots in the L2 stage (Vovlas, 1987), and the sedentary female feeds from a metabolically active single uninucleate cell (Figure 9.10A).

The genus *Verutus* belongs to the group of noncyst-forming heteroderid nematodes (Baldwin, 1992) within the subfamily Heteroderinae (family Heteroderidae). All members of this subfamily, including cyst-forming nematodes, enter the roots as freshly hatched L2 larvae, where they induce either a syncytium by partial wall dissolution of a group of cells or a single uninucleate giant cell, both of which are located within the vascular cylinder. *V. volvingentis* may be an exception as according to Cohn, Kaplan and Esser, (1984) it induces a syncytium within the cortex of the root (Figure 9.10A).

Nurse cell structures in the vascular cylinder

Most sedentary endoparasites migrate to the vascular cylinder of their host roots, where they select one or a few cells to become transformed into permanent feeding structures. These structures can be assigned to three groups: Uninucleate giant cells, multinucleate giant cells and syncytia. Single uninucleate giant cells are generally found in woody host plants of noncyst-forming heteroderids such as *Cryphodera utahensis* (Mundo-Ocampo and Baldwin, 1984). These nematodes cause little damage. They are considered to be ancestors of the cyst-forming nematodes and have received attention mainly for phylogenetic studies (Baldwin, 1992). Single giant cells are primarily induced in the pericycle, from where they expand into the stele (Figure 9.10A) The single nucleus of these cells is extremely enlarged and deeply invaginated to ensure an increased rate of nuclear-cytoplasmic exchange. The wall of the giant cell is considerably thickened at the head end of the nematode (Figure 9.10A); elsewhere it contains numerous pit fields with many plasmodesmata for bulk transport of solutes from adjacent cells. In ultrastructural studies of single giant cells induced by noncyst-forming Heteroderinae (Mundo-Ocampo and Baldwin, 1983a; 1983b; 1984) feeding tubes have not been recorded. However, in the first single uninucleate giant cell described, in this case for a non-heteroderid nematode, *Rotylenchulus macrodoratus*, feeding on soybean (Cohn and Mordechai, 1977), a feeding tube was shown to be attached to the nematode's stylet. The feeding tube of another, economically by far more important species, *R. reniformis*, has been described in detail (Rebois, 1980). *Rotylenchulus* is placed into the subfamily Rotylenchulinae within the family Hololaimidae, and in this respect it is interesting to note that nurse cells can differ between species of the same genus, as *R. reniformis* induces in the same host (soybean) a syncytium in the pericycle (Rebois, Madden and Eldridge, 1975; Figure 9.10A). The life cycle of *R. reniformis* is unique. Freshly hatched L2 larvae undergo three successive moults in the soil without feeding. Only the young adult females invade the roots.

Cyst nematodes of the genera *Globodera* and *Heterodera* are serious parasites of crops mainly in temperate regions of the world and their host range is usually restricted to one or few related families. Economically the most important species are *G. rostochiensis* and *G. pallida* on potatoes, *H. glycines* on soybeans, *H. schachtii* on sugar beets and *H. avenae* on cereals. All of them induce and maintain in the vascular cylinder of their host plants multinucleate syncytia (Figure 9.10A) that arise from expanding cambial cells whose protoplasts fuse after partial cell wall dissolution. The process of cell

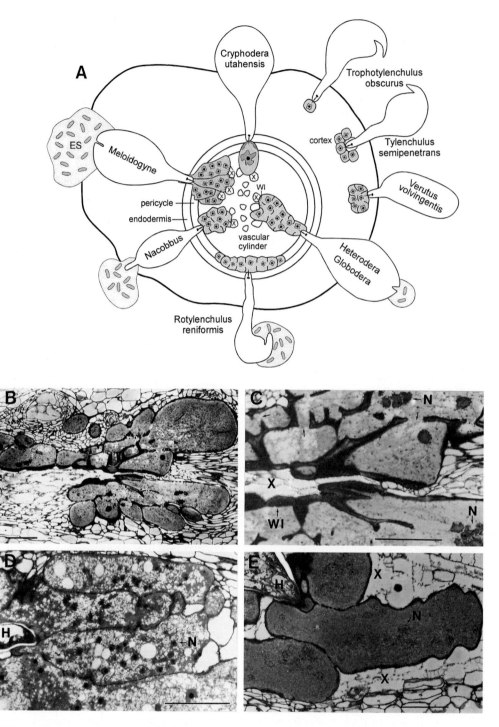

Figure 9.10. A: Schematic representation of feeding sites of some selected sedentary endoparasites and associated cell responses. B–C: Longitudinal sections through syncytium of a *Globodera rostochiensis* female in a *Lycopersicon esculentum* root. Arrows in C point to cell wall openings. D–E: Longitudinal sections through giant cells of *Meloidogyne incognita* in a *Cucumis sativus* root. D: Giant cells of a L2 larva prior to moulting. E: Giant cells of an adult female. All bars: 100 μm. Abbreviations: H: Head end of nematode; N: Nuclei of feeding structures; WI: Wall ingrowths; X: Xylem elements.

wall dissolution, first by a gradual widening of the plasmodesmata between the initial syncytial cell and adjacent cells, and later on by a temorarily and locally controlled enzymatic breakdown of intact cell walls, has been described in detail (Grundler, Sobczak and Golinowski, 1998). The nuclei of integrated cells do not divide, but concurrent with an increase in syncytial metabolism they become considerably enlarged. Hypertrophy of syncytial nuclei and nuclei of neighboring cells is caused by DNA endoreduplication, and it has been suggested that syncytium development of cyst nematodes involves several cycles of endoreduplication bypassing mitosis (De Almeida Engler *et al.*, 1999). A duplication in copy numbers of essential genes is obviously necessary to maintain the metabolically highly active state of syncytia (Goverse *et al.*, 1999). Fully developed syncytia, maintained by egg-producing females, can be composed of more than 200 integrated cells and may be considered to represent a large nurse cell unit with metabolically highly active cytoplasm (Figure 9.10B). The structure and function of syncytia induced by cyst nematodes has been extensively reviewed in the past years, most recently by Golinowski *et al.* (1997) and Hussey and Grundler (1998). Recently it could be clearly demonstrated that both induction and morphogenesis of syncytia are mediated by auxin (Goverse *et al.*, 2000). Although several cyst nematode-responsive promoters and expressed genes have been identified in the past few years (e.g. Puzio *et al.*, 1998; 1999), very little is yet known about the molecular mechanisms involved in syncytium induction. In *Arabidopsis thaliana*, expression of the cell cycle genes *cdc2a* and *cyc1At* could be detected within hours after feeding site stimulation (Niebel *et al.*, 1996). Recently it could also be demonstrated that naturally induced secretions of pre-infective *G. rostochiensis* L2 stimulate the proliferation of tobacco protoplasts in the presence of the synthetic phytohormones NAA and BAP. A low-molecular-weight peptide(s) (<3 kDa) was shown to be responsible for the observed effect (Goverse *et al.*, 1999).

Once established, syncytia of cyst nematodes are strong sinks for phloem-derived solutes (Böckenhoff *et al.*, 1996; Grundler and Böckenhoff, 1997). However, as *H.schachtii*-induced syncytia are symplastically isolated in *Arabidopsis thaliana* roots (Böckenhoff and Grundler, 1994), solutes from the phloem are specifically unloaded into the syncytia, as discussed by Grundler and Böckenhoff (1997). Increasing nutritional demands imposed by growing cyst nematodes stimulate polarised syncytial wall ingrowths adjacent to xylem vessels (Figure 9.10C). These ingrowths, lined with plasmalemma, are typical for transfer cells and are thus thought to enhance short distance solute transport between the apoplast and symplast. Syncytia are also strong sinks for water as revealed by their high turgor pressure and osmotic potential (Böckenhoff and Grundler, 1994) and hence it was concluded that wall ingrowth, adjacent to xylem vessels also function to supply the syncytia with water (Grundler and Böckenhoff, 1997).

Such wall ingrowths can be absent in syncytia of other nematodes, e.g. *Nacobbus aberrans* (Figure 9.10A), in which a high plasmodesmatal frequency is thought to ensure a symplastic solute influx (Jones and Payne, 1977). The larval stages of *Nacobbus*, the only genus in the subfamily Nacobbinae within the family Pratylenchidae, behave like *Pratylenchus* spp. in that they damage the roots of their host plants by intracellular migration (Maggenti, 1981). This is at least the case for the infective L2. The subsequent larval stages (L3 and

L4) apparently do not feed. A switch to sedentary endoparasitism occurs after the final moult, when the young females induce their typically spindle-shaped syncytia in the vascular cylinder (Souza and Baldwin, 1998). Hyperplastic responses in cortical cells are initiated concurrently, leading to the formation of galls in which the female becomes embedded.

Root galls are also formed in roots infected by root-knot nematodes (*Meloidogyne* spp.). These sedentary endoparasites belong to the subfamily Meloidogyninae within the Heteroderidae and are the most serious soil pathogens on a world wide basis. In contrast to cyst nematodes, root-knot nematodes are more frequent in areas with warm and hot climates. *Meloidogyne* spp. invariably induce and maintain multinucleate giant cells. These cells develop by the expansion of about half a dozen cambial cells within the differentiating vascular cylinder of host plants. Each cell becomes multinucleate (Figure 9.10A, D) by repeated synchronous mitoses in the absence of cytokinesis. These multiple nuclei undergo a series of endoreduplication cycles (Gheysen *et al.*, 1997), that increase their DNA content severalfold for the enhanced transcriptional and translational activity required for nematode development and reproduction. However, when mitosis is blocked by cell cycle inhibitors, further gall development is arrested, indicating that cycles of endoreduplication or other ways of of DNA amplification are insufficient to drive giant cell expansion (De Almeida Engler *et al.*, 1999). Mature giant cells, supporting female reproduction, are metabolically highly active and fully packed with cytoplasm (Figure 9.10E). Several root-knot nematode-responsive promoters and expressed genes have so far been identified (e.g. Fenoll *et al.*, 1997; Escobar *et al.*, 1999). The structure and function of *Meloidogyne*-induced giant cells has been reviewed most recently by Bleve-Zacheo and Melillo (1997).

The Life Cycle of Cyst and Root-Knot Nematodes

Cyst and root-knot nematodes are the most evolved plant-parasitic nematodes and are responsible for the vast majority of agricultural losses (Tytgat *et al.*, 2000). They invade plant roots as infective second-stage larvae (L2) that hatch from eggs retained in the protective cyst (dead female of cyst nematodes) or gelatinous egg sac produced by *Meloidogyne* females (Figure 9.11).

Cyst nematodes

In some cyst nematodes with a limited host range (e.g. the potato cyst nematodes *Globodera rostochiensis*, *G. pallida*) hatching is specifically induced by root diffusates, whereas in others with a broader host range (e.g. *Heterodera schachtii*) hatching occurs spontaneously under favourable environmental conditions. In addition to hatching stimulation, root diffusates prepare the activity of the pharyngeal glands to assist parasitism of the L2 (Perry, Zunke and Wyss, 1989; Smant *et al.*, 1997, Blair *et al.*, 1999; Goverse *et al.*, 1999) and they also play an important role in oriented searching and pre-infective exploratory behaviour. In the L2 of *H. schachtii*, for

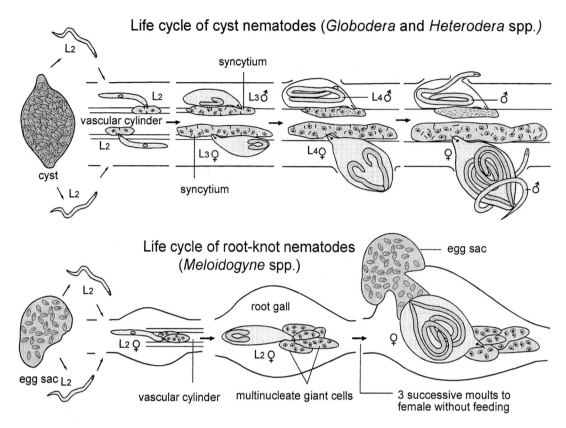

Figure 9.11. Schematic representation of the life cycles of cyst and root-knot nematodes

instance, aggregation and stylet thrusting is stimulated by diffusates of its host roots (Grundler, Schnibbe and Wyss, 1991) and this stimulation leads to oriented movement towards the attractant source (Clemens *et al.*, 1994). The L2 of cyst nematodes are equipped with a robust stylet, by means of which they invade the root and migrate within it toward the differentiating vascular cylinder, where they finally induce their syncytium. The majority of cyst nematode species reproduce by cross-fertilisation. The mechanisms involved in sex determination are not yet understood, but recent studies indicate that the invading L2 are not yet sexually determined and that sex determination is most likely controlled by the amount and quality of food available after syncytium induction (Betka, Grundler and Wyss, 1991; Grundler and Wyss, 1995; Grundler and Böckenhoff, 1997). Much more food is required by a female *H. schachtii* larva than by a male larva until moulting to the adult stage (Müller, Rehbock and Wyss, 1981), consequently the volume of syncytia maintained by males is considerably smaller (Caswell-Chen and Thomason, 1993) than that of females (Figure 9.11). In *Arabidopsis thaliana*, the model host plant for *H. schachtii* (Sijmons *et al.*, 1991; Wyss and Grundler, 1992), the syncytia associated with males are less hypertrophied and composed of more cells than those associated with females and in addition have only weakly developed cell wall ingrowths (Sobczak, Golinowski and Grundler, 1997). Male nematodes feed only until the end of the

third larval stage (L3). They become vermiform again while they moult to the L4 stage and after the last moult they emerge through the larval cuticles in search of females ready for copulation (Figure 9.11). Virgin females produce sex pheromones to attract the males (Perry and Aumann, 1998).

Root-knot nematodes

Root-knot nematodes are in most cases polyphagous. The economically most important species *M. arenaria*, *M. javanica* and *M. incognita* have an extremely wide host range, which may be facilitated by their relatively homozygous genome (Blok *et al.*, 1997) as these species, like many other *Meloidogyne* species, reproduce by mitotic parthenogenesis (Evans, 1998). Females deposit their eggs into a protective gelatinous matrix, the so-called egg sac, which is produced by six greatly enlarged rectal glands and secreted through the anus. Extrusion of the egg sac to the outside of the gall is apparently facilitated by cellulolytic and/or pectolytic enzymes (Orion and Franck, 1990). The recent detection of β-1,4-endoglucanase transcripts in *M. incognita* females (Rosso *et al.*, 1999) supports the assumption of cellulase activity during egg sac extrusion. Some protection of the eggs towards a broad spectrum of microorganisms in the soil may be provided by an antimicrobial activity of the gelatinous matrix (Orion and Kritzman, 1991). In parthenogenetic species with a wide host range, root diffusates are not involved in the hatching process of the L2 (Jones, Tylka and Perry, 1998) but they attract the L2 to the root tip of their host plants, where they generally enter the root. Inside the root the L2 migrate intercellularly to the vascular cylinder, where they become sedentary soon after giant cell induction. Differentiation of giant cells is accompanied by a pronounced galling of the surrounding root tissue while pericycle and cortex cells enlarge and divide. The L2 feed for many days from the expanding giant cells, they become saccate and finally stop feeding at the end of the L2 stage, when they enter an expanded moulting cycle (Figure 9.11), in which they moult three times in succession to adult females. The females resume feeding from the giant cells, which now function as xylem-related transfer cells and are also supplied with nutrients from the phloem (Dorhout, Gommers and Kollöffel, 1993). They are metabolically highly active to provide the females with sufficient food in order to produce many hundred eggs. Under adverse nutritional conditions, feeding L2 larvae of mitotically parthenogenetic species undergo complete or partial sex reversal and develop as males (Papadopoulou and Triantaphyllou, 1982).

Infection Processes of Cyst and Root-Knot Nematodes

Cyst nematodes

With the aid of video-enhanced contrast light microscopy (Wyss and Zunke, 1986a), the parasitic behaviour of *Heterodera schachtii*, the beet cyst nematode, could be studied in detail in thin and rather translucent roots of cruciferous host plants (Wyss, Zunke and Inst. Wiss. Film, 1986; Wyss and Zunke, 1986b; Wyss, 1992), even before the introduction of *Arabidopsis thaliana* as the most suitable model host. Freshly hatched L2 larvae invade the roots of these plants predominantly in the zone of elongation. Under the influence of stimulating root diffusates the stylet is persistently thrust with vigour at different sites of epidermal cell walls until the weakened wall finally ruptures and thus creates a hole for root invasion. Once inside the root the L2 migrate intracellularly (Figure 9.12A) through cortical cells toward the vascular cylinder. As the body of the L2 is

now enclosed by cell walls, more or less in a similar way as in the egg prior to hatching, the stylet thrusting pattern (Doncaster and Seymour, 1973) used to cut a slit into the rigid egg shell is maintained inside the root. In this way the walls of cortical and endodermal cells are cut open by highly coordinated stylet thrusts that produce a line of merging holes to form a slit, through which the L2 enter the neighbouring cell (Figure 9.12B,C).

The two subventral glands, including their ampullae, are packed with large secretory granules during intracellular migration. Based on detailed observations, it was assumed that forward progression inside the root to the final feeding site in the vascular cylinder is not assisted by secretions of these glands and occurs purely by mechanical means, i.e. by continuous accurate stylet thrusting and strong lip pressure (Wyss and Zunke, 1986b). Now, however, this assumption is no longer valid, as in both, *G. rostochiensis* and *H. glycines*, two β-1,4-endoglucanase (EGase) cDNAs were identified that are expressed specifically in the subventral pharyngeal glands of pre-infective (hatched) L2 (Smant *et al.*, 1998; Yan *et al.*, 1998). As EGase transcripts, detected by *in situ* hybridisation and immunolocalisation, prevailed predominantly in pre-infective and infective migrating L2 (De Boer *et al.*, 1999), these cellulases are thought to soften root cell walls and may thus assist intracellular migration. In *H. glycines* the gene *Hg-eng-1* encodes an EGase (HG-ENG-1) that has a catalytic domain linked to a cellulose-binding domain, while *Hg-eng-2* encodes an EGase (HG-ENG-2) that only contains a a catalytic domain. Interestingly, only HG-ENG-2-cellulase is obviously secreted into the cortical tissue of soybean roots by root invading L2 (Wang *et al.*,1999) at 24 h after inoculation of soybean roots. Recently it was shown for the first time that the L2 of cyst nematodes are also genetically equipped to produce a pectinase enzyme in the subventral glands (Popejus *et al.*, 2000).

The destructive migration behaviour changes into a subtle exploration behaviour when the L2 have reached the vascular cylinder and selected the initial cell for syncytium induction. This cell is generally called the initial syncytial cell (ISC). Once the wall of an ISC has been perforated by careful stylet thrusts, the stylet tip

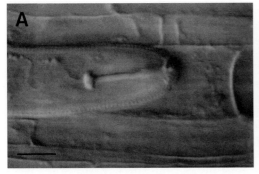
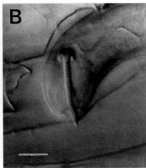
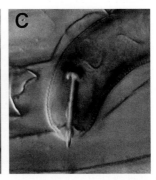

Figure 9.12. A–C: Intracellular migration of a *Heterodera schachtii* L2 larva through cortical cells of a *Brassica napus* root. A: Forward progression toward a cell wall. B: During a cycle of stylet thrustings aimed at cutting a slit in cell wall. C: Head, accompanied by stylet thrustings enters adjacent cell through cut slit. All bars: 10 μm.

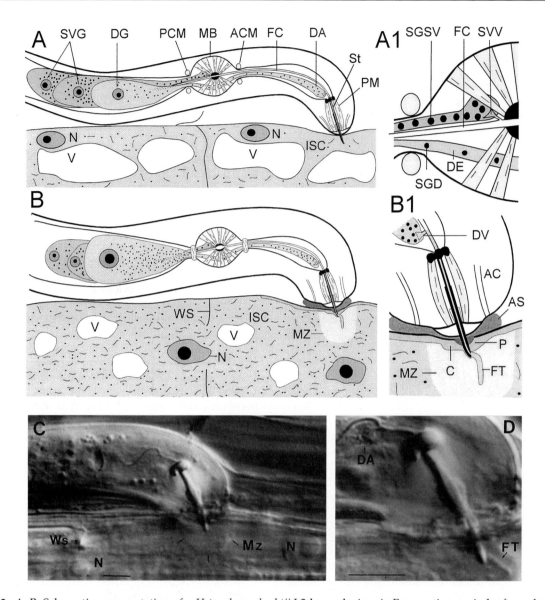

Figure 9.13. A–B: Schematic representation of a *Heterodera schachtii* L2 larva during A: Preparation period, after selection of an initial syncytial cell, showing in addition details in posterior half of metacorpal bulb (A1). B: During food withdrawal, about 36 h after A, with magnification of head end (B1). C–D: Feeding from an initial syncytial cell in a *Brassica napus* root, about one day (C) and two days (D) after induction, as observed by video-enhanced light microscopy. Abbreviations: AC: Amphidial canal; AS: Amphidial secretions; C: Callose; DG: Dorsal gland; FT: Feeding tube; ISC: Initial syncytial cell; N: Nuclei of the ISC and adjacent cell; P: Feeding plug; SVG: Subventral glands; V: Vacuoles; WS: Wall stub. All other abbreviations as in Figure 9.8.

stays protruded a few μm deep in the ISC for 6–8 hours (Figure 9.13A). During this phase, termed preparation period (Wyss, 1992), no obvious changes in the protoplast of the ISC can be recognised, and the metacorpal bulb does not pump at any time as analysed by numerous time lapse recordings.

Only two visible changes are evident during this period: A gradual decrease in the size and density of secretory granules in the subventral glands, accompanied by an increase of granules in the dorsal gland and a few sudden shrinkages of the body volume towards the end of this period, when the L2 obviously defecate. Therefore it has been suggested that secretions of the subventral glands may be used to mobilise lipid reserves, while the intestine is transformed into an

absorptive organ (Wyss, 1992). L2 that have completed this preparation period are no longer capable of leaving the root. The increase in synthetic activity of the dorsal pharyngeal gland plays most likely a decisive role in ISC initiation. The recent identification of three genes in the dorsal gland of infective *G. rostochiensis* L2 (Qin *et al.*, 2000) can be regarded as a milestone for further attempts to elucidate the precise function of this gland in the infection process.

In a transmission electronmicroscopic study it could be shown that a callose-like layer is deposited around the stylet tip and the affected area of the ISC cell wall during the preparation period. Fragments of membranes condensed in myelin bodies are scattered in the cytoplasm of the ISC and are also embedded into the newly deposited wall apposition (Grundler, Sobczak and Golinowski, 1998). Recent investigations indicate that the site of ISC selection determines the future sex in *H. schachtii*. In the differentiating vascular cylinder of *A. thaliana* roots, syncytia of females are always induced in procambial or cambial cells, depending on the developmental stage of the root (Golinowski, Grundler and Sobczak, 1996). Under conditions supporting male development, procambial cells respond by a hypersensitive reaction when selected as ISC, whereas pericycle cells do not and are thus the suitable ISCs for syncytia of males (Sobczak, Golinowski and Grundler, 1997).

First visible changes in the ISC, as seen under the light microscope by video-enhancement, become apparent a few hours after the L2 have started feeding: Cytoplasmic streaming is enhanced and the ISC's nucleus increases in size. Feeding occurs in repeated cycles, each consisting of three distinct phases (I-III). During phase I, which is by far the longest and which increases with time, nutrients are withdrawn from the ISC by a continuous rapid pumping action of the metacorpal bulb. The constraining muscles in front and behind the bulb stay contracted throughout pumping and thus they impede a forward flow of secretory granules within the duct of the dorsal gland cell (Figure 9.13B). Phase II is characterised by stylet retraction and reinsertion, and phase III by a continuous forward movement of secretory granules, especially from the dorsal gland, with the stylet tip staying inserted in the ISC. The three feeding phases, described in detail for a *H. schachtii* L2 in an early infective stage (Wyss and Zunke, 1986b) are maintained throughout development, also in the adult female stage (Wyss, 1992). These phases, including the preparation period, are most likely common to all cyst nematodes, as they have also been described for L2 and L3 larvae of *Globodera rostochiensis* feeding in tomato roots (Steinbach, 1973).

Under the light microscope, initial stages of syncytium development as revealed by partially dissolved cell walls, are visible 24 h after the L2 have commenced feeding. About 12 h later feeding tubes become distinct for the first time (Figure 9.13B). Under the transmission

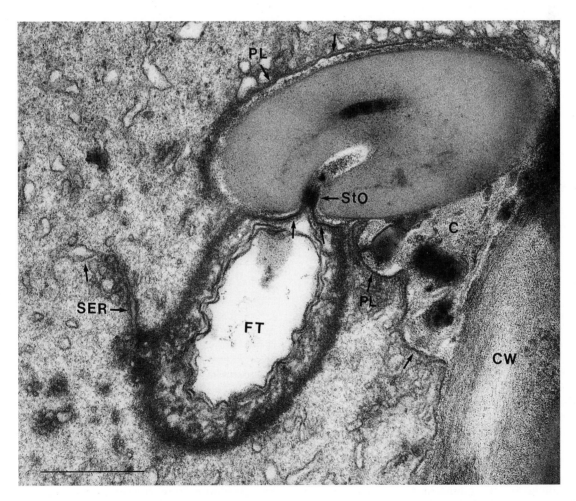

Figure 9.14. Transmission electron micrograph of a section through the stylet tip of a *Heterodera schachtii* L2 larva inserted into an initial syncytial cell of a *Raphanus sativus* var. *oleiformis* root, three days after root invasion, showing feeding tube secreted through stylet tip orifice. The plasmalemma surrounds stylet tip and callose deposition below cell wall but not feeding tube. Abbreviations: C: Callose; CW: Cell wall; FT: Feeding tube; PL: Plasmalemma; StO: Stylet tip Orifice; SER: Smooth endoplasmic reticulum. Bar: 0.5 μm.

electronmicroscope, however, tube-like material can already be detected about 6 h after the termination of the preparation period (Sobczak, 1996). *In vivo* observations show that the feeding tube is always surrounded by a zone of modified cytoplasm (Figure 9.13 B-C) that keeps larger cell organelles away (Wyss and Zunke, 1986b; Wyss *et al.*, 1986; Wyss, 1992). Electron microscopic studies confirm that this zone is free of larger organelles but rich in tubular endoplasmic reticulum (Wyss, Stender and Lehmann, 1984; Sobczak, 1996). The feeding tube stays permanently attached to the tip of the stylet during food withdrawal in phase I (Figure 9.13 B1,D). It becomes detached from the stylet tip when the stylet is retracted at the beginning of phase II. Soon afterwards the stylet is again inserted and a new feeding tube is formed during phase III, most probably by an interaction between the cytoplasm of the food cell and secretions of the dorsal gland cell.

The ultrastructure of feeding tubes, produced by *H. schachtii* larvae in *A. thaliana*, was recently examined in detail (Sobczak, 1996; Sobczak *et al.*, 1999). They consist of an osmiophilic wall and an electron translucent lumen, containing membranous structures, most probably derived from the tubular endoplasmic reticulum, which is frequently attached to the tubes. The feeding tubes are not surrounded by the plasmalemma and obviously function as a filter for a selective uptake of syncytial solutes.

Microinjection of fluorescence-labelled dextrans of different molecular weights into the syncytia showed that only dextrans of 3, 10 and 20 kDa but not of 40 and 70 kDa are ingested by the nematodes. These results suggest that only molecules of a maximum Stokes radius of 3.2 to 4.4 nm may pass from the syncytium into the feeding tubes (Böckenhoff and Grundler, 1994). This size exclusion limit (SEL) was confirmed for the engineered rice cysteine proteinase inhibitor Oc-IΔD86, a globular protein of approximately 11.2 kDa, which is ingested by *H. schachtii* and for the 28 kDa green-fluorescent protein reporter gene, which is not (Urwin *et al.*, 1997). The latter is, however, ingested by *M. incognita* females, indicating that the SEL differs between feeding tubes of cyst and root-knot nematodes. Interestingly, the SEL also differs between species of cyst nematodes, as infective larvae of *G. rostochiensis* were able to withdraw GFP with an apparent molecular weight of 32 kDa (Goverse *et al.*, 1998). Figure 9.14 shows the membranous structure of a feeding tube released through the stylet orifice of a *H. schachtii* L2, feeding from the ISC in a *Raphanus sativus* var. *oleiformis* root (Wyss *et al.*, 1984).

The plasmalemma surrounds the callose deposition around the stylet tip but not the feeding tube. Like in *C. xenoplax*, described previously, the plasmalemma is tightly appressed to the wall of the stylet orifice, creating a hole for the uptake of nutrients, which in this case are delivered through the feeding tube. In all electronmicroscopic pictures that show the inserted stylet tip of cyst nematodes, the tip is clearly surrounded by callose-like material, e.g. *H. avenae* (Bleve-Zacheo *et al.*, 1995); *H. goettingiana* (Bleve-Zacheo, Melillo and Zacheo, 1988); *H. glycines* (Endo, 1991). The question, why feeding tubes function only for a limited period of time and have to be replaced at rather regular intervals, at least in *H.schachtii* whose feeding behaviour has been recorded throughout development (Wyss, 1992), still awaits an answer. Nonfunctional feeding tubes left in the cytoplasm after stylet retraction at the beginning of phase II are eventually degraded within small vesicles into which they become enclosed (Sobczak, 1996). Reinsertion of the stylet for the production of a new feeding tube obviously requires its precise adjustment within the callose-like deposition so that the stylet orifice becomes located opposite the open pore, as can be judged from Figure 9.14. This is usually the case, but occasionally reinsertion is impeded, and repeated rather forceful stylet thrustings, lasting more than half an hour, are necessary to overcome the obstacle (Wyss, 1992).

Secretions released by the glandular sheath cell in the amphids, the primary chemosensilla at the head end of the nematodes, 'glue' the lips of the nematode to the wall of the ISC and also appear to be involved, at least partially, in the production of a feeding plug (Sobczak *et al.*, 1999) that surrounds the point where the stylet tip is inserted through the hole in the syncytial wall (Figure 9.13 B1). This plug, first described by Endo (1978) for *H.glycines*, is apparent in all ultrastructural studies of cyst nematode feeding sites and is also recognisable in living cells when viewed under the light microscope (Wyss, 1992). It is penetrated with ease upon stylet reinsertion and possibly it may help in guiding the stylet tip into the right position.

The rapid decrease in the size of the two subventral glands during the early stage of parasitism is accompanied by an increase in the size of the dorsal gland (Figure 9.13 A, B), which becomes the dominant gland throughout further parasitism. It is beyond the scope of this chapter to describe stage-specific changes in the secretory granules which have been discussed most recently by Hussey and Grundler (1998).

Root-knot nematodes

The parasitic behaviour of the infective root-knot L2 larvae has been covered in detail by von Mende (1997), and from her data it may be concluded that it does not differ between *Meloidogyne* spp. The behaviour of *Meloidogyne incognita* from root invasion until giant cell induction has been documented inside roots of *A. thaliana* with the aid of video-enhanced contrast light microscopy and time lapse studies (Wyss, Grundler and Münch, 1992). The freshly hatched L2 larvae are attracted to the tip of growing roots and usually invade in the region of elongation close to the meristematic zone (Figure 9.15A). The thin walls of epidermal and subepidermal cells are weakened and finally destroyed by continuous head rubbings and stylet movements, including occasional stylet tip protrusions followed by metacorpal bulb pumpings of a few seconds duration. This behaviour indicates that wall degrading enzymes may be involved in root invasion (Figure 9.15B-C). Support for this assumption was first provided by the

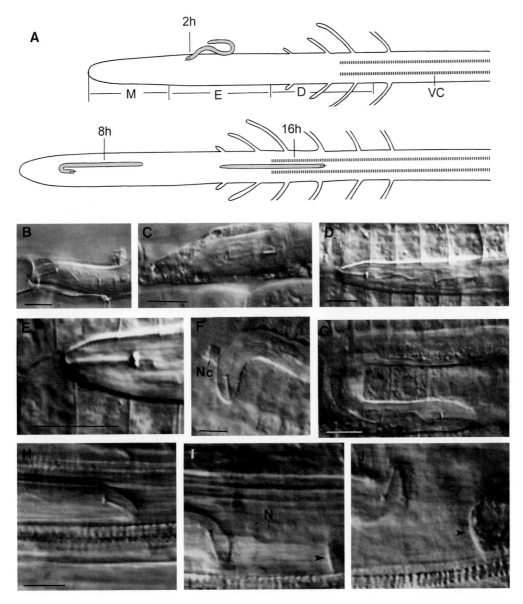

Figure 9.15. A: Schematic representation of intercellular migration of a *Meloidogyne incognita* L2 larva in an *Arabidopsis thaliana* root. Hours (h) represent average values of five continuous time lapse recordings. B–J: Intercellular migration from root invasion in elongation region to vascular cylinder and initiation of giant cells, as observed by video-enhanced light microscopy. B–C: Root invasion through necrotic epidermal cells. D: Non-destructive forward progression between meristematic cells towards root apex. E: Magnification of head. F: L2 starts turning round in root apex and then occasionally kills some cells. G: Turning round partially completed, L2 now begins to migrate towards differentiating vascular cylinder. H: Forward migration within vascular cylinder. I: L2 now sedentary, 14 h after H, numerous nuclei (some marked by arrows) close to head end. J: Same site, vascular cylinder now much expanded by multinucleate giant cells, 34 h after I. Arrowhead points to a xylem element, which compared to I (arrowhead), is now fully differentiated. All bars: 10 μm. Abbreviations: D: Differentiation region of root; E: Elongation region; M: Meristematic region; N: Some of numerous nuclei in young giant cells; VC: Vascular cylinder.

detection of a novel cellulose binding protein (MI-CBP-1), immunolocalised in the subventral glands and secreted through the stylet of pre-infective and infective L2s of *M. incognita* (Ding *et al.* 1998). One year later a β-1,4-endoglucanase encoding cDNA, named *Mi-eng-1*, was cloned from *M. incognita* L2. Transcription of the enzyme in the subventral glands of infective L2 was demonstrated by mRNA *in situ* hybridisation (Rosso *et al.*, 1999). The complete biochemical characterisation of the enzyme (MI-ENG-1) revealed no ability to bind

cellulose in spite of the presence of a cellulose-binding domain at the C-terminus (Béra-Maillet *et al.*, 2000). By dissecting the entire pharyngeal gland region from infective *M. incognita* L2, Lambert *et al.* (1999) identified a gland-specific cDNA that codes for a potentially secreted chorismate mutase (CM). It remains to be clarified if CM, a key branch-point regulatory enzyme in the shikimate pathway, does indeed play a role in successful plant parasitism as hypothesised by the authors.

Once inside the roots, the L2, equipped with only a small and rather tender stylet, are not able to migrate intracellulary to the vascular cylinder on the shortest possible route like cyst nematodes. Therefore they first orient themselves in the direction of the root-tip and migrate towards it between cortical cells without causing any damage (Figure 9.15D). The same behaviour pattern as noted during root invasion, i.e. continuous head and stylet movements, interspersed by short periods of stylet protrusion, followed by metacorpal bulb pumpings, is maintained throughout migration. No changes in the cell's protoplast become evident, when, during the short pumping periods, the stylet tip appears to be inserted into adjacent cells (Figure 9.15E). Having reached the apex of the root, the L2 start to turn around (Figure 9.15A) and then destroy some of the meristematic cells above the apical initials (Figure 9.15F). Non-destructive intercellular migration is resumed, now in the direction toward the region of root differentiation (Figure 9.15G). The mechanisms by which the root invading L2 orient themselves first towards the root tip and then in the reverse direction are unknown. Secretions on the surface of the migrating L2 may protect the nematodes from plant defense responses or may act as 'lubricating' agents during intercellular migration. It was shown that a fucosyl-bearing epitope in the surface coat of *M.incognita* L2 was shed during intercellular migration and that later on it was specifically associated with phloem sieve elements after giant cell initiation (Gravato–Nobre *et al.*, 1999).

From root invasion onwards, 14–18 h elapse until the L2 have reached the differentiating vascular cylinder (Figure 9.15A). By this time the density of secretory granules in the two subventral glands appears diminished, compared to early events of root infection when these glands are packed with granules. The reverse is true for the dorsal gland, which now contains many more granules. Based on the previously mentioned findings it can be assumed that secretions from the two subventral glands assist root invasion and intercellular migration.

When the L2 have reached the vascular cylinder, they continue to migrate between the cambial cells (Figure 9.15H) for several hours without any changes in the behavioural pattern described for root invasion and intercellular migration. Eventually forward migration ceases at the time when initial giant cells are induced. These cells become multinucleate within a few hours. About one day later, they are packed with nuclei (Figure 9.15I). Even then and also during later stages, the head of the sedentary L2 still moves in all directions (Figure 9.15J), performing the same behaviour as described for the early events of root infection. Now, however, periods of stylet tip protrusion and metacorpal pumping increase with time. In contrast to earlier phases, when it is difficult to determine whether the short periods (few seconds) of metacorpal pumping are involved in food ingestion, the L2 now withdraw nutrients from the initial giant cells. The growing larvae become saccate, and, having completed the L2 stage,

they enter the expanded moulting cycle described before. During the following adult stage, females continue feeding from the giant cells and start producing eggs. Numerous feeding tubes are now found in the cytoplasm of mature giant cells, indicating that the females continue to feed from the giant cells in the same behaviour pattern as observed during the L2 stage. Feeding tubes produced by *M. incognita* differ remarkably from those produced by cyst nematodes in that they consist of a thick, crystalline wall that encloses a smooth lumen (Hussey and Mims, 1991). A compact membrane network, formed from a rearrangement of the giant cell endomembrane system stays in intimate contact with the wall of the tubes and might thus function in synthesising and/or transporting soluble assimilates to the feeding tube for food withdrawal.

This chapter is dedicated to my wife Lea.

References

Aumann, J. (1997) Monoxenic culture of *Xiphinema index* (Nematoda: Longidoridae) on *Ficus carica*. *Nematologica Mediterranea*, **25**, 209–211.

Baldwin, J.G. (1992) Evolution of cyst and noncyst-forming Heteroderinae. *Annual Review of Phytopathology*, **30**, 271–290.

B'Chir, M.M. (1988) Organisation ultrastructurale du site trophique induit par *Tylenchulus semipenetrans* dans les racines de citrus. *Revue Nématologie*, **11**, 213–222.

Béra-Maillet., C., Arthaud, L., Abad, P. and Rosso, M.N. (2000) Biochemical characterization of MI-ENG1, a family 5 endoglucanase secreted by the root-knot nematode *Meloidogyne incognita*. *European Journal of Biochemistry*, **267**, 3255–3263.

Betka, M., Grundler, F. and Wyss, U. (1991) Influence of changes in the nurse cell system (syncytium) on the development of the cyst nematode *Heterodera schachtii*: Single amino acids. *Phytopathology*, **81**, 75–79.

Bird, G.W. (1971) Digestive system of *Trichodorus porosus*. *Journal of Nematology*, **3**, 50–57.

Blair, L., Perry, N.R., Oparka, K. and Jones, J.T. (1999) Activation of transcription during the hatching process of the potato cyst nematode *Globodera rostochiiensis*. *Nematology*, **1**, 103–111.

Bleve-Zacheo, T. and Zacheo, G. (1983) Early stage of disease in fig roots induced by *Xiphinema index*. *Nematologica Mediterranea*, **11**, 175–187.

Bleve-Zacheo, T., Andres, M., Yeves, M.F. and Zacheo, G. (1984) Development of galls induced in *Chenopodium quinoa* by *Longidorus apulus*. *Nematologica Mediterranea*, **12**, 129–139.

Bleve-Zacheo, T., Melillo, M.T. and Zacheo, G. (1988) Syncytia development in germplasm pea accessions infected with *Heterodera goettingiana*. *Nematologica Mediterranea*, **18**, 93–102.

Bleve-Zacheo, T., Melillo, M.T., Andres, M., Zacheo, G. and Romero, M.D. (1995) Ultrastructure of initial response of graminaceous roots to infection by *Heterodera avenae*. *Nematologica*, **41**, 80–97.

Bleve-Zacheo, T. and Melillo, M.T. (1997) The biology of giant cells. In *Cellular and Molecular Aspects of Plant-Nematode Interactions*, edited by C. Fenoll, F.M.W. Grundler and S.A. Ohl, pp. 65–79. Dordrecht: Kluwer Academic Publishers.

Blok, V.C., Ehwaeti, M., Fargette, M., Kumar, A., Phillips, M.S., Robertson, W.M. and Trudgill, D.L. (1997) Evolution of resistance and virulence in relation to the management of nematodes with different biology, origins and reproductive strategies. *Nematologica*, **43**, 1–13.

Böckenhoff, A. and Grundler, F.M.W. (1994) Studies on the nutrient uptake by the beet cyst nematode *Heterodera schachtii* by *in situ* microinjection of fluorescent probes into the feeding structures in *Arabidopsis thaliana*. *Parasitology*, **109**, 249–254.

Böckenhoff, A., Prior, D.A.M., Grundler, F.M.W. and Oparka, K.J. (1996) Induction of phloem unloading in *Arabidopsis thaliana* roots by the parasitic nematode *Heterodera schachtii*. *Plant Physiology*, **112**, 1421–1427.

Brown, D.J.F., Robertson, W.M. and Trudgill, D.L. (1995) Transmission of viruses by plant nematodes. *Annual Review of Phytopathology*, **33**, 223–249.

Caswell-Chen, E.P. and Thomason, I.J. (1993) Root volumes occupied by different stages of *Heterodera schachtii* in sugarbeet, *Beta vulgaris*. *Fundamental and Applied Nematology*, **16**, 39–42.

Clemens, C.D., Aumann, J., Spiegel, Y. and Wyss, U. (1994) Attractant-mediated behaviour of mobile stages of *Heterodera schachtii*. *Fundamental and Applied Nematology*, **17**, 569–574.

Cohn, E. and Mordechai, M. (1977) Uninucleate giant cell induced in soybean by the nematode *Rotylenchulus macrodoratus*. *Phytoparasitica*, **5**, 85–93.

Cohn, E., Kaplan, D.T. and Esser, R. P. (1984) Observations on the mode of parasitism and histopathology of *Meloidodera floridensis* and *Verutus volvingentis* (Heteroderidae). *Journal of Nematology*, **16**, 256–264.

Cohn, E. and Spiegel, Y. (1991) Root-nematode interactions. In *Plant Roots; The Hidden Half*, edited by Y. Waisel, A. Eshel and U. Kafkafi, pp. 789–805. New York: Marcel Dekker.

Davis, E.L., Hussey, R.S., Baum, T.J., Bakker, J., Schots, A., Rosso, M.-N. and Abad, P. (2000) Nematode parasitism genes. *Annual Review of Phytopathology*, **38**, 365–396.

De Almeida Engler, J., de Vleesschauwer, V., Burssens, S., Celenza, J.L., Jr., Inzé, D., Van Montagu, M. *et al.* (1999) Molecular markers and cell cycle inhibitors show the importance of cell cycle progression in nematode-induced galls and syncytia. *The Plant Cell*, **11**, 793–807.

De Boer J.M., Yan, Y., Wang, X., Smant, G., Hussey, R.S., Davis, E.L. and Baum, T.J. (1999) Developmental expression of secretory β-1,4-endoglucanases in the subventral esophageal glands of *Heterodera glycines*. *Molecular Plant-Microbe Interactions*, **12**, 663–669.

Decraemer, W. (1995) *The Family Trichodoridae: Stubby Root and Virus Vector Nematodes*, xvi + 360 pp. Dordrecht: Kluwer Academic Publishers.

Decraemer, W. and Baujard, P. (1998) A polytomous key for the identification of species of the family Trichodoridae Thorne, 1935 (Nematoda: Triplonchida). *Fundamental and Applied Nematology*, **21**, 37–62.

Decraemer, W. and de Waele, D. (1981) Taxonomic value of the position of oesophageal gland nuclei and of oesophageal gland overlap in the Trichodoridae (Diphtherophorina). *Nematologica*, **27**, 82–94.

Ding, X., Shields, J., Allen, R. and Hussey, R.S. (1998) A secretory cellulose-binding protein cDNA cloned from the root-knot nematode (*Meloidogyne incognita*). *Molecular Plant-Microbe Interactions*, **11**, 952–959.

Doncaster, C.C. and Seymour, M.K. (1973) Exploration and selection of penetration site by Tylenchida. *Nematologica*, **19**, 137–145.

Dorhout, R., Gommers, F.J. and Kollöffel, C. (1993) Phloem transport of carboxyfluorescein through tomato roots infected with *Meloidogyne incognita*. *Physiological and Molecular Plant Pathology*, **43**, 1–10.

Endo, B.Y. (1978) Feeding plug formation in soybean roots infected with the soybean cyst nematode. *Phytopathology*, **68**, 1022–1031.

Endo, B.Y. (1991) Ultrastructure of initial responses of susceptible and resistant soybean roots to infection by *Heterodera glycines*. *Revue de Nématologie*, **14**, 73–94.

Endo, B.Y. (1998) Atlas on ultrastructure of infective juveniles of the soybean cyst nematode, *Heterodera glycines*. U.S. Department of Agriculture. *Agriculture Handbook* No. **711**, 220 pp.

Escobar, C., De Meutter, J., Aristizábal, F.A., Sanz-Alférez, S., del Campo, F.F., Barthels, N. *et al.* (1999) Isolation of the *LEMMI9* gene and promotor analysts during a compatible plant-nematode interaction. *Molecular Plant-Microbe Interactions*, **12**, 440–449.

Evans, A.A.F. (1998) Reproductive mechanisms. In *The Physiology and Biochemistry of Free-Living and Plant-Parasitic Nematodes*, edited by R.N. Perry and D.J. Wright, pp.133–154, Wallingford: CAB International Press.

Fenoll, C., Grundler, F.M.W. and Ohl, S.A. (1997) *Cellular and Molecular Aspects of Plant-Nematode Interactions*, vii + 286 pp. Dordrecht: Kluwer Academic Publishers.

Fenoll, C., Aristizábal, F.A., Sanz-Alférez, S. and del Campo, F.F. (1997) Regulation of gene expression in feeding sites. In *Cellular and Molecular Aspects of Plant-Nematode Interactions*, edited by C. Fenoll, F.M.W. Grundler and S.A. Ohl, pp. 133–149. Dordrecht: Kluwer Academic Publishers.

Gheysen, G., de Almeida Engler, J. and Van Montagu, M. (1997) Cell cycle regulation in nematode feeding sites. In *Cellular and Molecular Aspects of Plant-Nematode Interactions*, edited by C. Fenoll, F.M.W. Grundler and S.A. Ohl, pp. 120–132. Dordrecht: Kluwer Academic Publishers.

Gilbertson, R.L. and Lucas, W.J. (1996) How do viruses traffic on the 'vascular highway'? *Trends in Plant Science*, **1**, 260–268.

Golinowski, W., Grundler, F.M.W. and Sobczak, M. (1996) Changes in the structure of *Arabidopsis thaliana* during female development of the plant-parasitic nematode *Heterodera schachtii*. *Protoplasma*, **194**, 103–116.

Golinowski, W., Sobczak, M., Kurek, W. and Grymaszewska, G. (1997) The structure of syncytia. In *Cellular and Molecular Aspects of Plant-Nematode Interactions*, edited by C. Fenoll, F.M.W. Grundler and S.A. Ohl, pp. 80–97. Dordrecht: Kluwer Academic Publishers.

Goverse, A., Biesheuvel, J., Wijers, G.J., Gommers, F.J., Bakker, J., Schots, A. and Helder, J. (1998) *In planta* monitoring of the activity of two constitutive promoters, CaMV 35S and TR2′, in developing feeding cells induced by *Globodera rostochiensis* using green fluorescent protein in combination with confocal laser scanning microscopy. *Physiological and Molecular Plant Pathology*, **52**, 275–284.

Goverse, A., van der Voort, J.R., van der Voort C.R., Kavelaars, A., Smant, G., Schots, A. *et al.* (1999) Naturally induced secretions of the potato cyst nematode co-stimulate the proliferation of both tobacco leaf protoplasts and human peripheral blood mononuclear cells. *Molecular Plant-Microbe Interactions*, **12**, 872–881.

Goverse, A., Overmars, H., Engelbertink, J., Schots, A., Bakker, J. and Helder, J. (2000) Both induction and morphogenesis of cyst nematode feeding cells are mediated by auxin. *Molecular Plant-Microbe Interactions*, **13**, 1121–1129.

Gravato-Nobre, M.J., McClure, M.A., Dolan, L., Calder, G., Davies, K.D., Mulligan, B. *et al.* (1999) *Meloidogyne incognita* surface antigen epitopes in infected *Arabidopsis* roots. *Journal of Nematology*, **31**, 212–223.

Griffiths, B.S. and Robertson, W.M. (1984a) Nuclear changes induced by the nematode *Xiphinema diversicaudatum* in root-tips of strawberry. *Histochemical Journal*, **16**, 265–273.

Griffiths, B.S. and Robertson, W.M. (1984b) Morphological and histochemical changes occurring during the life-span of root-tip galls on *Lolium perenne* induced by *Longidorus elongatus*. *Journal of Nematology*, **16**, 223–229.

Griffiths, B.S. and Robertson, W.M. (1988) A quantitative study of changes induced by *Xiphinema diversicaudatum* in root-tip galls of strawberry and ryegrass. *Nematologica*, **34**, 198–207.

Grundler, F., Schnibbe, L. and Wyss, U. (1991) *In vitro* studies on the behaviour of second-stage juveniles of *Heterodera schachtii* (Nematoda: Heteroderidae) in response to host plant root exudates. *Parasitology*, **103**, 149–155.

Grundler, F.M.W. and Wyss, U. (1995) Strategies of root parasitism by sedentary plant parasitic nematodes. In *Pathogenesis and Host Specificity in Plant Diseases. Histopathological, Biochemical, Genetic and Molecular Bases*, Vol. II, Eukaryotes, edited by K. Kohmoto, U.S. Singh and R.P. Singh, pp. 309–319. Oxford: Pergamon.

Grundler, F.M.W. and Böckenhoff, A. (1997) Physiology of nematode feeding and feeding sites. In *Cellular and Molecular Aspects of Plant-Nematode Interactions*, edited by C. Fenoll, F.M.W. Grundler and S.A. Ohl, pp. 107–119. Dordrecht: Kluwer Academic Publishers.

Grundler, F.M.W., Sobczak, M. and Golinowski, W. (1998) Formation of wall openings in root cells of *Arabidopsis thaliana* following infection by the plant-parasitic nematode *Heterodera schachtii*. *European Journal of Plant Pathology*, **104**, 545–551.

Hewitt, Wm.B., Raski, D.J. and Goheen, A.C. (1958) Nematode vector of soil-borne fanleaf virus of grapevines. *Phytopathology*, **48**, 586–595.

Himmelhoch, S., Cohn, E., Mordechai, M. and Zuckerman, B.M. (1979) Changes in fine structure of citrus root cells induced by *Tylenchulus semipenetrans*. *Nematologica*, **25**, 333–335.

Hirumi, H., Chen, T.A., Lee, K.L.and Maramorosch, K. (1968) Ultrastructure of the feeding apparatus of the nematode *Trichodorus christiei*. *Journal Ultrastructure Research*, **24**, 434–453.

Hussey, R.S. and Mims, C.W. (1991) Ultrastructure of feeding tubes formed in giant-cells induced in plants by the root-knot nematode *Meloidogyne incognita*. *Protoplasma*, **162**, 99–107.

Hussey, R.S., Mims, C.W. and Westcott, S.W. (1992a) Ultrastructure of root cortical cells parasitized by the ring nematode *Criconemella xenoplax*. *Protoplasma*, **167**, 55–65.

Hussey, R.S., Mims, C.W. and Westcott, S.W. (1992b) Immunocytochemical localization of callose in root cortical cells parasitized by the ring nematode *Criconemella xenoplax*. *Protoplasma*, **171**, 1–6.

Hussey, R.S. and Grundler, F.M.W. (1998) Nematode parasitism of plants. In *The Physiology and Biochemistry of Free-living and Plant-parasitic Nematodes*, edited by R.N. Perry and D.J. Wright, pp. 213–243, Wallingford: CAB International Press.

Jones, J.T. and Robertson, W.M. (1997) Nematode secretions. In *Cellular and Molecular Aspects of Plant-Nematode Interactions*, edited by C. Fenoll, F.M.W. Grundler and S.A. Ohl, pp. 98–106. Dordrecht: Kluwer Academic Publishers.

Jones, M.G.K. and Payne, H.L. (1977) The structure of syncytia induced by the phytoparasitic nematode *Nacobbus aberrans* in tomato roots, and the possible role of plasmodesmata in their nutrition. *Journal Cell Science*, **23**, 299–313.

Jones, P.W., Tylka, G.L. and Perry, R.N. (1998). Hatching. In *The Physiology and Biochemistry of Free-Living and Plant-Parasitic Nematodes*, edited by R.N. Perry and D.J. Wright, pp. 181–212, Wallingford: CAB International Press.

Jones, R.K. (1978a) The feeding behaviour of *Helicotylenchus* spp. on wheat roots. *Nematologica*, 24, 88–94.

Jones, R.K. (1978b) Histological and ultrastructural changes in cereal roots caused by feeding of *Helicotylenchus* spp. *Nematologica*, 24, 393–397.

Lambert, K.N., Allen, K.D. and Sussex, I.M. (1999) Cloning and characterization of an esophageal-gland-specific chorismate mutase from the phytoparasitic nematode *Meloidogyne javanica*. *Molecular Plant-Microbe Interactions*, 12, 328–326.

Maggenti, A. (1981) *General Nematology*, x + 372 pp. New York: Springer

Melillo, M.T., Lamberti, F., Choleva, B., Iovev, T. and Bleve-Zacheo, T. (1997) Cytological changes induced by the ectoparasitic nematode *Longidorus latocephalus* in tobacco roots. *Nematologia Mediterranea*, 25, 83–91.

Müller, J., Rehbock, K. and Wyss, U. (1981) Growth of *Heterodera schachtii* with remarks on amounts of food consumed. *Revue de Nématologie*, 4, 227–234.

Mundo-Ocampo, M. and Baldwin, J.G. (1983a) Host response to *Sarisodera hydrophila* Wouts and Sher, 1971. *Journal of Nematology*, 15, 259–268.

Mundo-Ocampo, M. and Baldwin, J.G. (1983b) Host response to *Meloidodera* spp. (Heteroderidae). *Journal of Nematology*, 15, 544–554.

Mundo-Ocampo, M. and Baldwin, J.G. (1984) Comparison of host response of *Cryphodera utahensis* with other Heteroderidae, and a discussion of phylogeny. *Proceedings of the Helminthological Society of Washington*, 51, 25–31.

Niebel, A., de Almeida Engler, J., Hemerly, A., Ferreira, P., Inzé, D., van Montagu, M. and Gheysen, G. (1996) Induction of *cdc2a* and *cyc1At* expression in *Arabidopsis thaliana* during early phases of nematode-induced feeding cell formation. *The Plant Journal*, 10, 1037–1043.

Ng, O.C. and Chen, T.A. (1985) The histopathology of alfalfa roots infected by *Hoplolaimus galeatus*. *Phytopathology*, 75, 297–304.

Orion, D. and Franck, A. (1990) An electron microscopy study of cell wall lysis by *Meloidogyne javanica* gelatinous matrix. *Revue de Nématologie*, 13, 105–107.

Orion, D. and Kritzman, G. (1991) Antimicrobial activity of *Meloidogyne javanica* gelatinous matrix. *Revue de Nématologie*, 14, 481–483.

Perry, R.N. and Aumann, J. (1998) Behaviour and sensory responses. In *The Physiology and Biochemistry of Free-Living and Plant-Parasitic Nematodes*, edited by R.N. Perry and D.J. Wright, pp. 75–102, Wallingford: CAB International Press.

Perry, R.N., Zunke, U. and Wyss, U. (1989) Observations on the response of the dorsal and subventral oesophageal glands of *Globodera rostochiensis* to hatching stimulation. *Revue de Nématologie*, 12, 91–96.

Papadopoulou, J. and Triantaphyllou, A.C. (1982) Sex differentiation in *Meloidogyne incognita* and anatomical evidence of sex reversal. *Journal of Nematology*, 14, 549–566.

Pitcher, R.S., Siddiqi, M.R. and Brown, D.J.F. (1974) *Xiphinema diversicaudatum*. C.I.H. Descriptions of plant-parasitic nematodes. Set 4, No. 60.

Popeijus, H., Overmars, H., Jones, J., Blok, V., Goverse, A., Helder, J. *et al.* (2000) Degradation of plant cell walls by a nematode. *Nature*, 406, 36–37.

Puzio, P.S., Cai, D., Ohl, S., Wyss, U. and Grundler, F.M.W. (1998) Isolation of regulatory DNA regions related to differentiation of nematode feeding structures in *Arabidopsis thaliana*. *Physiological and Molecular Plant Pathology*, 53, 177–193.

Puzio, P.S., Lausen, J., Almeida-Engler, J., Cai, D., Gheysen, G. and Grundler, F.M.W. (1999) Isolation of a gene from *Arabidopsis thaliana* related to nematode feeding structures. *Gene*, 239, 163–172.

Qin, L., Overmars, H., Helder, J., Popeijus, H., van der Voort, J.R., Groenink, W. *et al.* (2000) An efficient cDNA-AFLP-based strategy for the identification of putative pathogenicity factors from the potato cyst nematode *Globodera rostochiensis*. *Molecular Plant-Microbe Interactions*, 13, 830–836.

Raski, D.J., Maggenti, A.R. and Jones, N.O. (1973) Location of grapevine fanleaf and yellow mosaic virus particles in *Xiphinema index*. *Journal of Nematology*, 5, 208–211.

Raski, D.J., Jones, N.O. and Roggen, D.R. (1969) On the morphology and ultrastructure of the esophageal region of *Trichodorus allius* Jensen. *Proceedings of the Helminthological Society of Washington*, 36, 106–118.

Rebois, R.V. (1980) Ultrastructure of a feeding peg and tube associated with *Rotylenchulus reniformis* in cotton. *Nematologica*, 26, 396–405.

Rebois, R.V., Madden, P.A. and Eldridge, B.J. (1975) Some ultrastructural changes induced in resistant and susceptible soybean roots following infection by *Rotylenchulus reniformis*. *Journal of Nematology*, 7, 122–39.

Rhoades, H.L. and Linford, M.B. (1961) A study of the parasitic habit of *Paratylenchus projectus* and *P. dianthus*. *Proceedings of the Helminthological Society of Washington*, 28, 185–190.

Robertson, W.M. (1975) A possible gustatory organ associated with the odontophore in *Longidorus leptocephalus* and *Xiphinema diversicaudatum*. *Nematologica*, 21, 443–448.

Robertson, W.M. (1987) Stereology and carbohydrate histochemistry of the oesophageal bulb of *Xiphinema index* (Nematoda: Xiphinemidae). *Nematologica*, 33, 401–409.

Robertson, W.M. and Taylor, C.E. (1975) The structure and musculature of the feeding apparatus in *Longidorus* and *Xiphinema*. In *Nematode Vectors of Plant Viruses*, edited by F. Lamberti, C. E. Taylor and J. W. Seinhorst, pp. 179–194. London, New York: Plenum Press.

Robertson, W.M. and Wyss, U. (1979) Observations on the ultrastructure and function of the dorsal oesophageal gland cell in *Xiphinema index*. *Nematologica*, 25, 391–396.

Robertson, W.M. and Wyss, U. (1983) Feeding processes of virus-transmitting nematodes. In *Current Topics in Vector Research*, Vol.1, edited by K. F. Harris, pp. 271–95. New-York: Praeger Scientific.

Robertson, W.M., Trudgill, D.L. and Griffiths, B.S. (1984) Feeding of *Longidorus elongatus* and *L. leptocephalus* on root-tip galls of perennial ryegrass (*Lolium perenne*). *Nematologica*, 30, 222–229.

Robertson, W.M., Topham, P.B. and Smith, P. (1987) Observations on the action of the oesophageal pump in *Longidorus* (Nematoda). *Nematologica*, 33, 43–54.

Rumpenhorst, H.J. and Weischer, B. (1978) Histopathological and histochemical studies on grapevine roots damaged by *Xiphinema index*. *Revue de Nématologie*, 1, 217–225.

Rosso, M.N., Favery, B., Piotte, C., Arthaud, L., de Boer, J.M., Hussey, R.S. *et al.* (1999) Isolation of a cDNA encoding a β-1,4-endoglucanase in the root-knot nematode *Meloidogyne incognita* and expression analysis during plant parasitism. *Molecular Plant-Microbe Interactions*, 12, 585–591.

Schuerger, A.C. and McClure, M.A. (1983) Ultrastructural changes induced by *Scutellonema brachyurum* in potato roots. *Phytopathology*, 73, 70–81.

Sijmons, P.C., Grundler, F.M.W., von Mende, N., Burrows, P. R. and Wyss, U. (1991) *Arabidopsis thaliana* as a new model host for plant-parasitic nematodes. *Plant Journal*, 1, 245–254.

Sijmons, P.C., Atkinson, H.J. and Wyss, U. (1994) Parasitic strategies of root nematodes and associated host cell responses. *Annual Review of Phytopathology*, 32, 235–59.

Smant, G. Goverse, A., Stokkermans, J.P.W.G., de Boer, J.M., Pomp, H., Zilverentant, J.F. *et al.* (1997) Potato root diffusate-induced secretion of soluble, basic proteins originating from the subventral esophageal glands of potato cyst nematode. *Phytopathology*, 87, 839–845.

Smant, G., Stokkermans, J.P.W.G., Yan, Y., de Boer, J.M., Baum, T.J., Wang, X. *et al.* (1998) Endogenous cellulases in animals: Isolation of β-1,4-endoglucanase genes from two species of plant-parasitic cyst nematodes. *Proceedings of the National Academy of Science of the United States of America*, 95, 4906-4911.

Sobczak, M. (1996) Investigations on the structure of syncytia in roots of *Arabidopsis thaliana* induced by the beet cyst nematode *Heterodera schachtii* and its relevance to the sex of the nematode, *Ph.D. Thesis*, University of Kiel, Germany.

Sobczak, M., Golinowski, W. and Grundler, F.M.W. (1997) Changes in the structure of *Arabidopsis thaliana* roots induced during development of males of the plant parasitic nematode *Heterodera schachtii*. *European Journal of Plant Pathology*, 103, 113–124.

Sobczak, M., Golinowski, W. and Grundler, F.M.W. (1999) Ultrastructure of feeding plugs and feeding tubes formed by *Heterodera schachtii*. *Nematology*, 1, 363–377.

Souza, R.M. and Baldwin, J.G. (1998) Changes in esophageal gland activity during the life cycle of *Nacobbus abberrans* (Nemata: Pratylenchidae). *Journal of Nematology*, 30, 275–290.

Staudt, G. and Weischer, B. (1992) Resistance to transmission of grapevine fanleaf virus by *Xiphinema index* in *Vitis rotundifolia* and *Vitis munsoniana*. *Viticulture Enology Science*, 47, 56–61.

Steinbach, P. (1973) Untersuchungen über das Verhalten von Larven des Kartoffelzystenälchens (*Heterodera rostochiensis* Wollenweber, 1923) an und in Wurzeln der Wirtspflanze *Lycopersicon esculentum* Mill. III. Die Nahrungsaufnahme von Kartoffelnematodenlarven. *Biologisches Zentralblatt*, 92, 563–582.

Taylor, C.E. and Brown, D.J.F. (1997) *Nematode Vectors of Plants Viruses*, xi + 286 pp. Wallingford: CAB International.

Taylor, C.E. and Robertson, W.M. (1970) Sites of virus retention in the alimentary tract of the nematode vectors, *Xiphinema diversicaudatum* (Micol.) and *X. index* (Thorne and Allen). *Annals of Applied Biology*, **66**, 375–380.

Towle, A. and Doncaster, C.C. (1978) Feeding of *Longidorus caespiticola* on rye-grass, *Lolium perenne*. *Nematologica*, **24**, 277–285.

Townshend, J.L., Stobbs, L. and Carter, R. (1989) Ultrastructural pathology of cells affected by *Pratylenchus penetrans* in alfalfa roots. *Journal of Nematology*, **21**, 530–539.

Trudgill, D.L. (1976) Observations on the feeding of *Xiphinema diversicaudatum*. *Nematologica*, **22**, 417–423.

Trudgill, D.L., Robertson, W.M. and Wyss, U. (1991) Analysis of the feeding of *Xiphinema diversicaudatum*. *Revue de Nématologie*, **14**, 107–112.

Tytgat, T., De Meutter, J., Gheysen, G. and Coomans, A. (2000) Sedentary endoparastic nematodes as a model for other plant parasitic nematodes. *Nematology*, **2**, 113–121.

Urwin, P.E., Møller, S.G., Lilley, C.J., McPherson, M.J. and Atkinson, H.J. (1997) Continual green-fluorescent protein monitoring of cauliflower mosaic virus 35S promoter activity in nematode-induced feeding cells in *Arabidopsis thaliana*. *Molecular Plant-Microbe Interactions*, 10, 394–400.

Valette, C., Nicole, M., Sarah, J.-L., Boisseau, M., Boher, B., Fargette, M. and Geiger, J.P. (1997) Ultastructure and cytochemistry of interactions between banana and the nematode *Radopholus similis*. *Fundamental and Applied Nematology*, **20**, 65–77.

Von Mende, N. (1997) Invasion and migration behaviour of sedentary nematodes. In *Cellular and Molecular Aspects of Plant-Nematode Interactions*, edited by C. Fenoll, F.M.W. Grundler and S.A. Ohl, pp. 51–65. Dordrecht: Kluwer Academic Publishers.

Vovlas, N. (1987) Parasitism of *Trophotylenchulus obscurus* on coffee roots. *Reveu de Nématologie*, **10**, 337–342.

Wang X., Meyers, D., Yan, Y., Baum, T. Smant, G., Hussey, R. and Davis, E. (199) *In planta* localization of β-1, 4-endoglucanase secreted by *Heterodera glycines*. *Molecular Plant-Microbe Interactions*, **12**, 64–67.

Westcott, S.W. and Hussey, R.S. (1992) Feeding behavior of *Criconemella xenoplax* in monoxenic cultures. *Phytopathology*, **82**, 936–940.

Wyss, U. (1971) Der Mechanismus der Nahrungsaufname bei *Trichodorus similis*. *Nematologica*, **17**, 508–518.

Wyss, U. (1973) Feeding of *Tylenchorhynchus dubius*. *Nematologica*, 19, 125–136.

Wyss, U. (1975) Feeding of *Tricholorus, Longidorus* and *Xiphinema*. In *Nematode Vectors of Plant Viruses*, edited by F. Lamberti, C.E. Taylor and S.W. Seinhorst, pp. 203–221. London, New York: Plenum Press.

Wyss, U. (1977a) Feeding mechanisms and feeding behaviour of *Xiphinema index*. *Mededelingen van de Faculteit Landbouwwetenschappen, Universiteit Gent*, **42**, 1513–1519.

Wyss, U. (1977b) Feeding phases of *Xiphinema index* and associated processes in the feeding apparatus. *Nematologica*, **23**, 463–470.

Wyss, U. (1978) Root and cell response to feeding by *Xiphinema index*. *Nematologica*, 24, 159–166.

Wyss, U. (1981) Ectoparasitic root nematodes: Feeding behavior and plantcell responses. In *Plant Parasitic Nematodes*, Vol. III, edited by B.M. Zuckerman and R.A. Rohde, pp. 325–351. New York: Academic Press.

Wyss, U. (1987) Video assessment of root cell responses to dorylaimid and tylenchid nematodes. In *Vistas on Nematology*, edited by J.A. Veech and D.W. Dickson, pp. 211–220. Hyattsville, Maryland: Society of Nematologists.

Wyss, U. (1992) Observations on the feeding behaviour of *Heterodera schachtii* throughout development, including events during moulting. *Fundamental and Applied Nematology*, **15**, 75–89.

Wyss, U. (1997) Root parasitic nematodes: an overview. In *Cellular and Molecular Aspects of Plant-Nematode Interactions*, edited by C. Fenoll, F.M.W. Grundler and S.A. Ohl, pp. 5–22. Dordrecht: Kluwer Academic Publishers.

Wyss, U. and Grundler, F.M.W. (1992) Seminar: *Heterodera schachtii* and *Arabidopsis thaliana*, a model host-parasite interaction. *Nematologica*, **38**, 488–493.

Wyss, U. and Inst. Wiss. Film (1972) *Trichodorus similis* (Nematoda) — Saugen an Wurzeln von Sämlingen (Rübsen). Film E 1763, 12 pp. IWF Göttingen.*

Wyss, U. and Inst. Wiss. Film (1973) *Tylenchorhynchus dubius* (Nematoda) — Saugen an Wurzeln von Sämlingen (Rübsen). Film E 1902, 14 pp. IWF Göttingen.*

Wyss, U. and Inst. Wiss. Film (1974) *Trichodorus similis* (Nematoda) — Reaktion der Protoplasten von Wurzelhaaren (Nicotiana tabacum) auf den Saugvorgang. Film E 2045, 19 pp. IWF Göttingen.*

Wyss, U. and Inst. Wiss. Film (1977) *Xiphinema index* (Nematode) — Saugen an Wurzeln von Sämlingen (Feige). Film E 2375, 20 pp. IWF Göttingen.*

Wyss, U., Jank-Ladwig, R. and Lehmann, H. (1979) On the formation and ultrastructure of feeding tubes produced by trichodorid nematodes. *Nematologica*, **25**, 385–390.

Wyss, U., Lehmann, H. and Jank-Ladwig, R. (1980) Ultrastructure of modified root-tip cells in *Ficus carica*, induced by the ectoparasitic nematode *Xiphinema index*. *Journal Cell Science*, **41**, 193–208.

Wyss, U, Stender, C. and Lehmann, H. (1984) Ultrastructure of feeding sites of the cyst nematode *Heterodera schachtii* Schmidt in roots of susceptible and resistant *Raphanus sativus* L. var. *oleiformis* Pers. cultivars, *Physiological Plant Pathology*, **25**, 21–37.

Wyss, U. and Zunke, U. (1986a) The potential of high resolution video-enhanced contrast microscopy in nematological research. *Revue de Nématologie*, **9**, 91–94.

Wyss, U. and Zunke, U. (1986b) Observations on the behaviour of second stage juveniles of *Heterodera schachtii* inside host roots. *Revue de Nématologie*, **9**, 153–165.

Wyss, U., Zunke, U. and Inst. Wiss. Film (1986) *Heterodera schachtii* (Nematoda). Behaviour inside roots (rape). Film E 2904, 21 pp. IWF Göttingen.*

Wyss, U., and Inst. Wiss. Film. (1988) Responses of root-tip cells (*Ficus carica*) to feeding of the nematode *Xiphinema index*. Film D 1657, 22pp. IWF Göttingen.*

Wyss, U., Robertson, W.M. and Trudgill, D.L. (1988) Oesophageal bulb function of *Xiphinema index* and associated root cell responses, assessed by video-enhanced contrast light microscopy. *Revue de Nématologie*, **11**, 253–261.

Wyss, U., Grundler, F.M.W. and Münch, A. (1992) The parasitic behaviour of second-stage juveniles of *Meloidogyne incognita* in roots of *Arabidopsis thaliana*. *Nematologica*, **38**, 98–111.

Yan, Y., Smant, G., Stokkermans, J., Qin, L., Helder, J., Baum, T. *et al.* (1998) Genomic organization of four β-1,4-endoglucanase genes in plant-parasitic cyst nematodes and its evolutionary implications. *Gene*, **220**, 61–70.

Zacheo, G. and Bleve-Zacheo, T. (1995) Plant-nematode interactions: histological, physiological and biochemical interactions. In *Pathogenesis and Host Specificity in Plant Diseases. Histopathological, Biochemical, Genetic and Molecular Bases*, Vol. II, Eukaryotes, edited by K. Kohmoto, U.S. Singh and R.P. Singh, pp. 321–353. Oxford: Pergamon.

Zunke, U. (1990a) Ectoparasitic feeding behaviour of the root lesion nematode, *Pratylenchus penetrans*, on root hairs of different host plants, *Revue de Nématologie*, **13**, 331–337.

Zunke, U. (1990b) Observations on the invasion and endoparasitic behavior of the root lesion nematode *Pratylenchus penetrans*. *Journal of Nematology*, **22**, 309–320.

Zunke, U. and Inst. Film (1998) Behaviour of the root lesion nematode *Pratylenchus penetrans*. Film C 1676, IWF Göttingen.*

* Publications marked by an asterisk are scientific films, produced in cooperation with the Institut für den Wissenschaftlichen Film, Nonnenstieg 76, D-37075 Göttingen, Germany. They can be ordered, together with an accompanying text and commentary, from the IWF (http://www.iwf.gwdg.de/index_e.html)

10. Metabolism

Carolyn A. Behm

Division of Biochemistry & Molecular Biology, Faculty of Science, Australian National University, Canberra ACT 0200, Australia

Introduction

This chapter covers all aspects of nematode metabolism, with emphasis on the more recent developments in our understanding of this field. Energy metabolism is discussed first as it is basic to all other metabolic processes. There are large gaps in our knowledge of the details of most metabolic pathways in nematodes, many assumptions have been made and some conclusions necessarily drawn based on analogies with other organisms. The most important of these are identified in the relevant sections. Developmental aspects of metabolism are also discussed where possible or appropriate.

Although the general metabolism of free-living and plant-parasitic nematodes has been reviewed in detail recently (Barrett and Wright, 1998; Chitwood, 1998) a general review of metabolism of the animal-parasitic nematodes has not been written for some time. This chapter will therefore emphasise the more recent developments in our knowledge of the animal-parasitic nematodes. General accounts were published some years ago by Barrett (1981; 1983; 1989b) and Bryant and Behm (1989). A few more specialised areas have been reviewed more recently and these will be noted in the relevant sections.

Energy Metabolism

Most nematodes appear capable of generating energy, as ATP, via the complete oxidation of substrates, using the mitochondrial reactions of the TCA cycle and of oxidative phosphorylation via the classical electron transport chain, at least at some stages of their life cycles. The larger animal-parasitic nematodes, as well as many smaller parasitic or free-living nematodes inhabiting microaerobic or anoxic environments, also utilise energy-generating fermentative pathways that do not require oxygen as terminal electron acceptor. As nematodes do not possess an efficient circulatory system, the distribution of oxygen — where it is available in the environment — to the internal organs of the larger nematodes is not sufficient to maintain aerobic metabolism (Barrett, 1981; Barrett and Wright, 1998; Wright, 1998). These nematodes therefore appear to utilise anaerobic energy-generating pathways internally but may also utilise oxygen-based pathways, at the same time, in superficial tissues into which sufficient amounts of oxygen may diffuse (Fry and Jenkins, 1984).

Aerobic energy generation yields considerably more ATP per unit of substrate consumed than fermentative pathways. Complete oxidation of one mole of glucose as substrate generates 38 moles of ATP whereas its fermentation to lactate generates only 2 moles of ATP. Therefore, nematodes that rely on fermentative metabolism to supply most of their ATP requirements must catabolise considerable quantities of fermentable substrates such as glucose.

Aerobic Energy Generation

Aerobic energy generation in nematodes is similar to that in other eukaryotes, being sensitive to inhibition with cyanide or antimycin A, and with carbon dioxide and water being the excreted end-products (Barrett, 1981; Tielens, 1994; Kita, Hirawake and Takamiya, 1997). The major pathways are illustrated in Figures 10.1 and 10.2. Most ATP is synthesised via oxidative phosphorylation, utilising the proton motive force generated via the classical electron transport chain to drive the F_1F_0 ATP synthase complex. Substrates for mitochondrial oxidation via the TCA cycle are pyruvate, generated in the cytosol from glucose, other sugars or alanine; acetylCoA, derived from the catabolism of lipids; and possibly amino acids such as glutamine, glutamate or aspartate. Nematodes that use aerobic respiration often store large quantities of lipid for oxidation.

Anaerobic Energy Generation

In situations where insufficient oxygen is available to support aerobic metabolism, nematodes synthesise ATP via a number of fermentative pathways that generate a variety of reduced end-products such as lactate, succinate, acetate and propionate (Barrett, 1981; 1984; Saz, 1981; Köhler, 1985; Tielens and Van den Bergh, 1986; 1991) (Table 10.1). Some parasitic nematodes, such as the filarial worms living in the blood or tissues of their hosts, maintain fermentative metabolism even when sufficient oxygen would be present for aerobic metabolism, for reasons that are not clear. Carbohydrates are essential substrates for these pathways and so nematodes that respire anaerobically tend not to store significant lipid but instead accumulate considerable quantities of glycogen and, in some tissues, the disaccharide trehalose.

The generalised pathways of fermentative energy generation are illustrated in Figure 10.3. The standard reactions of glycolysis convert glucose to phosphoenolpyruvate. There follows a branchpoint at which phosphoenolpyruvate may be converted to pyruvate or

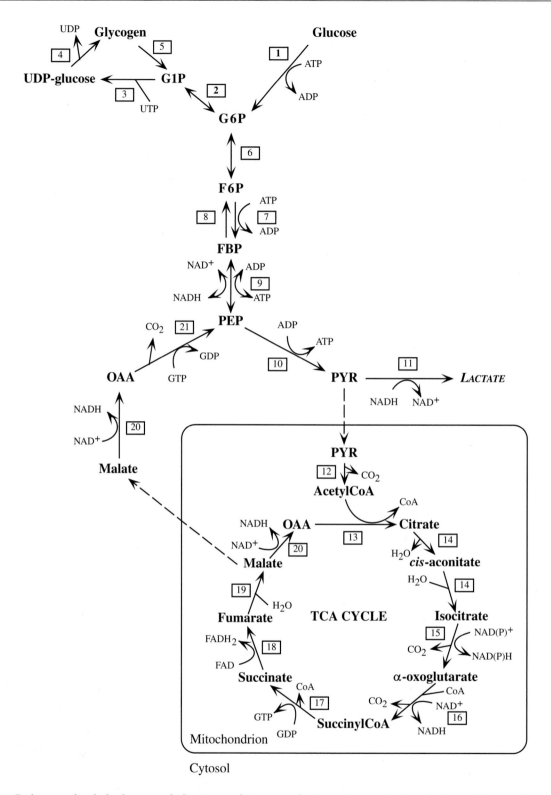

Figure 10.1. Pathways of carbohydrate metabolism in aerobic nematode tissues. Inorganic phosphate, and glycogen branching and debranching enzymes have been omitted for simplicity. Dashed lines represent transport processes. Enzymes: 1, hexokinase; 2, phosphoglucomutase; 3, UDP-glucose pyrophosphorylase; 4, glycogen synthase; 5, glycogen phosphorylase; 6, phosphoglucose isomerase; 7, phosphofructokinase; 8, fructose 1,6-bisphosphatase; 9, aldolase, triosephosphate isomerase, glyceraldehyde 3-phosphate dehydrogenase, phosphoglycerate kinase, enolase; 10, pyruvate kinase; 11, lactate dehydrogenase; 12, pyruvate dehydrogenase; 13, citrate synthase; 14, aconitase; 15, isocitrate dehydrogenase; 16, α-oxoglutarate dehydrogenase; 17, succinylCoA synthetase; 18, succinate dehydrogenase; 19, fumarase; 20, malate dehydrogenase; 21, phosphoenolpyruvate carboxykinase. Abbreviations: F6P, fructose 6-phosphate; FBP, fructose 1,6-bisphosphate; G1P, glucose 1-phosphate; G6P, glucose 6-phosphate; OAA, oxaloacetate; PEP, phosphoenolpyruvate; PYR, pyruvate; TCA, tricarboxylic acid.

Table 10.1. End-products of glucose utilisation in selected nematodes

Species	Ethanol	Lactate	Acetate	Succinate	Propionate	Other	Reference
Ascaris suum adults			+	+	+	2-methylbutyrate, 2-methylvalerate	(Saz and Vidrine, 1959; Saz and Weil, 1960; Saz and Weil, 1962)
Toxocara canis adults		+	+		+		(Learmonth et al., 1987)
Haemonchus contortus adults	+	+	+	+	+	n-propanol, CO_2	(Ward, 1974)
Trichostrongylus colubriformis adults	+		+		+	n-propanol	(Sangster and Prichard, 1985)
T. colubriformis L4s			+		+	*n*-butyrate	(Prichard and Rothwell, 1972)
Nippostrongylus brasiliensis adults		+		+			(Saz et al., 1971)
Ancylostoma caninum adults			+		+	CO_2	(Warren and Poole, 1970)
Onchocerca volvulus adults		+		+		CO_2	(Wittich and Walter, 1987)
Litomosoides sigmodontis adults		+	+			CO_2	(Wang and Saz, 1974)
Dirofilaria immitis adults		+				CO_2	(Hutchison and Turner, 1979)
Brugia pahangi adults		+					(Wang and Saz, 1974)
Acanthocheilonema viteae adults		+					(Wang and Saz, 1974)
Oesophagostomum radiatum L3s		+	+		+	CO_2	(Rew, Douvres and Saz, 1983)
Angiostrongylus cantonensis adults		+	+			alanine	(Nishina et al., 1988)
A. cantonensis eggs		+	+			alanine	(Nishina et al., 1989)
Dictyocaulus viviparus adults		+	+	+			(Learmonth et al., 1987)
Steinernema carpocapsae		+		+			(Thompson, Platzer and Lee, 1991)
Caenorhabditis elegans		+	+	+	+		(Foll et al., 1999)
C. briggsae		+				glycerol	(Liu and Rothstein, 1976)
Panagrellus redivivus	+	+	+		+	alanine, acetoin	(Clancy et al., 1992)
Aphelenchus avenae	+					glycerol	(Mendis and Evans, 1984)

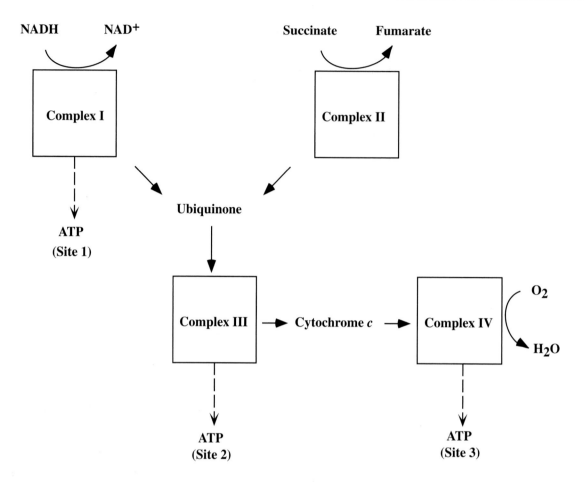

Figure 10.2. Electron transport in aerobic mitochondria. Complexes I, II, III, IV and ubiquinone are located in the inner mitochondrial membrane; cytochrome *c* is associated with complex IV but located in the mitochondrial matrix; other components are located in the matrix. The dashed line leading to (ATP) represents proton motive force-driven phosphorylation of ADP via the F_1F_O ATP synthase complex, also located on the inner mitochondrial membrane.

to oxaloacetate, depending on the relative activities of pyruvate kinase or phosphoenolpyruvate carboxykinase, respectively. Glycerol, lactate and ethanol are synthesised by reactions that take place in the cytosol whereas the synthesis of acetate, succinate, propionate and other end-products requires carbon dioxide fixation and mitochondrial pathways, some of which are TCA cycle reactions operating in reverse. The major mitochondrial substrate is malate which, on entering the mitochondrion, has two possible fates. Either malate is converted to fumarate by fumarase, a reaction normally at equilibrium, or it is oxidatively decarboxylated to pyruvate by malic enzyme. In this dismutation process conversion of malate to pyruvate provides the reducing power to drive the reduction of fumarate to succinate. Since pyruvate is subsequently oxidatively decarboxylated to acetylCoA, yielding an additional mole of NAD(P)H, the mitochondrial reactions achieve redox balance (in the absence of oxygen, and assuming no influx of extramitochondrial NAD(P)H if succinate and acetylCoA are produced in a ratio of 2:1.

Most ATP synthesis is substrate-linked but in nematodes utilising mitochondrial reactions a modified elec-

tron transport chain operates to transfer electrons from NADH to fumarate, with concomitant proton translocation driving the synthesis of ATP via the ATP synthase complex (Figure 10.4). Where required, NADPH may be converted to NADH by a transhydrogenation reaction (Goyal and Srivastava, 1990; Goyal *et al.*, 1991; Unnikrishnan and Raj, 1995). These mitochondrial reactions significantly increase the yield of ATP per mole of glucose compared with the cytosolic fermentative pathways. For example, conversion of a mole of glucose to acetate plus succinate theoretically yields approximately 3.3 net moles of ATP; further metabolism of succinate to propionate yields approximately 4.7 net moles of ATP, which is a significant increase on the 2 moles available from converting glucose to lactate alone.

Critical to the function of the anaerobic mitochondrial pathways are the enzymes phosphoenolpyruvate carboxykinase, malic enzyme, fumarate reductase, pyruvate dehydrogenase, and components of the anaerobic electron transport pathway. These essential components will be discussed in more detail using examples of nematodes where they have been studied in some depth.

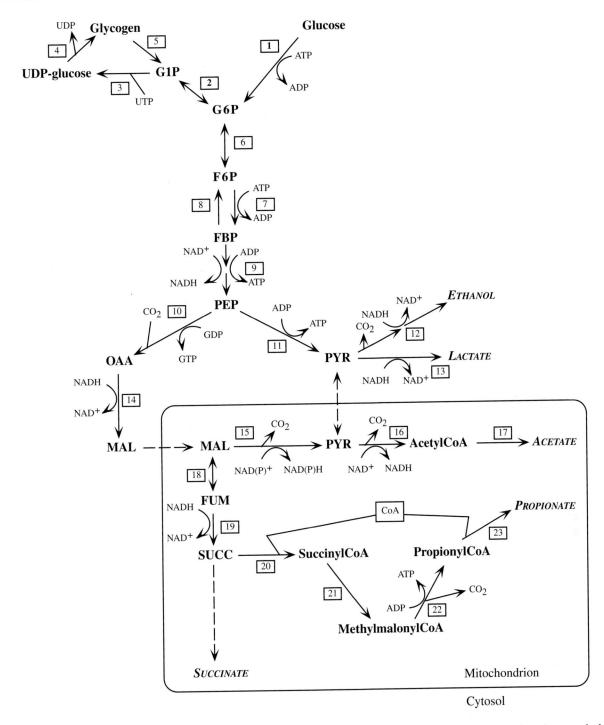

Figure 10.3. Generalised pathways of carbohydrate metabolism in anaerobic nematode tissues. Inorganic phosphate, and glycogen branching and debranching enzymes have been omitted from some reactions for simplicity. Dashed lines represent transport processes. Enzymes: 1, hexokinase; 2, phosphoglucomutase; 3, UDP-glucose pyrophosphorylase; 4, glycogen synthase; 5, glycogen phosphorylase; 6, phosphoglucose isomerase; 7, phosphofructokinase; 8, fructose 1,6-bisphosphatase; 9, aldolase, triosephosphate isomerase, glyceraldehyde 3-phosphate dehydrogenase, phosphoglycerate kinase, enolase; 10, phosphoenolpyruvate carboxykinase; 11, pyruvate kinase; 12, pyruvate decarboxylase, alcohol dehydrogenase; 13, lactate dehydrogenase; 14, malate dehydrogenase; 15, malic enzyme; 16, pyruvate dehydrogenase; 17, acylCoA transferase; 18, fumarase; 19, fumarate reductase; 20, succinylCoA synthase or acylCoA transferase; 21, methylmalonylCoA mutase; 22, propionylCoA carboxylase; 23, acylCoA transferase. Abbreviations: F6P, fructose 6-phosphate; FBP, fructose 1,6-bisphosphate; FUM, fumarate; G1P, glucose 1-phosphate; G6P, glucose 6-phosphate; MAL, malate; OAA, oxaloacetate; PEP, phosphoenolpyruvate; PYR, pyruvate; SUCC, succinate.

Note that in *A. suum* mitochondria, propionylCoA is condensed with either acetylCoA or another molecule of propionylCoA to form the enoylCoA derivatives that are reduced by enoylCoA reductase (see Figure 10.4) to form the end-products 2-methylbutyrate or 2-methylvalerate, respectively.

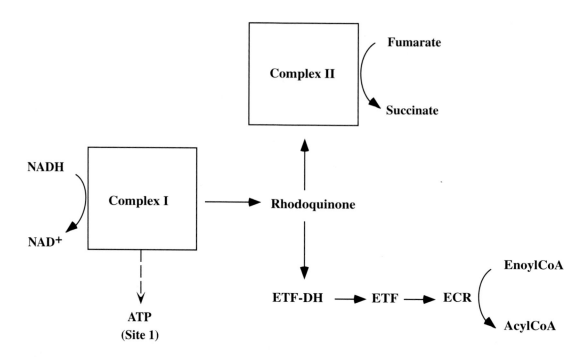

Figure 10.4. Anaerobic electron transport in *Ascaris* muscle mitochondria. ETF-DH, electron transfer flavoprotein dehydrogenase; ETF, electron transfer flavoprotein; ECR, enoylCoA reductase. Complexes I, II, rhodoquinone and ETF-DH are located in the inner mitochondrial membrane. All other components are located in the mitochondrial matrix. The dashed line leading to (ATP) represents proton motive force-driven phosphorylation of ADP via a presumptive F_1F_O ATP synthase complex, also located on the inner mitochondrial membrane.

Phosphoenolpyruvate carboxykinase

Many nematodes fix carbon dioxide and most of this fixation is due to the activity of phosphoenolpyruvate carboxykinase (PEPCK), an enzyme demonstrated in many nematodes (Barrett, 1981). In the anaerobic worms the enzyme functions to convert cytosolic phosphoenolpyruvate to oxaloacetate via fixation of carbon dioxide (Figure 10.3). A mole of GTP is synthesised in the process. Oxaloacetate is subsequently reduced to malate which enters the mitochondrion for further metabolism. In aerobic organisms PEPCK normally catalyses the reverse reaction, i.e. generation of phosphoenolpyruvate from oxaloacetate, an essential reaction of gluconeogenesis. The predicted amino acid sequences of PEPCK cDNAs derived from *Ascaris suum* and *Haemonchus contortus* possess common conserved elements that distinguish them from vertebrate PEPCKs; these may be linked to the different directions of operation of the enzyme from the different sources (Geary *et al.*, 1993). To date, only one PEPCK gene has been reported from parasitic nematodes such as *A. suum* or *H. contortus*, whereas free-living, aerobic, *Caenorhabditis elegans* has two putative genes (W05G11.6 and R11A5.4). The significance of this has yet to be determined.

PEPCK functions at an important branchpoint in glycolysis where it competes with pyruvate kinase for the same substrate, PEP. The relative activities of these enzymes may contribute to determining the flux through to pyruvate or oxaloacetate, and thus the extent to which carbon is metabolised to cytosolic products such as lactate, to pyruvate for mitochondrial oxidation, or to mitochondrial products derived from oxaloacetate via malate dismutation, such as succinate and acetate. Unlike pyruvate kinase, some isoforms of which are allosterically activated by fructose 1,6-bisphosphate (Brazier and Jaffe, 1973), PEPCK appears not to be regulated by any allosteric mechanism and its activity is determined by the concentrations of its substrates and products (Wilkes, Cornish and Mettrick, 1982; Rohrer, Saz and Nowak, 1986; Walter and Albiez, 1986).

In organisms undergoing aerobic-anaerobic transitions during development, the relative activities of PEPCK and pyruvate kinase have been used as indicators of the amount of pyruvate available for mitochondrial oxidation via the TCA cycle. For example, relative pyruvate kinase activity increased during the anaerobic to aerobic transition that occurs when the infective muscle larvae of *Trichinella spiralis* become activated upon release into the stomach of the next host (Janssen, Tetley and Kennedy, 1998).

Malic enzyme

Malic enzyme is responsible for catalysing the oxidative decarboxylation of pyruvate. It is important in the dismutation of malate in anaerobic mitochondrial metabolism as the reaction generates the intramitochondrial reducing power (NAD(P)H) to drive the fumarate reductase reaction. In most organisms, including many nematodes (e.g. filariae (Walter and Albiez, 1981)), malic enzyme utilises $NADP^+$ as the cofactor but the enzymes from *A. suum* and *Toxocara canis* are NAD^+-dependent (Fodge, Gracy and Harris,

1972; Ferreira, Andrade and Ribeiro, 1985), in common with mitochondrial malic enzymes from rapidly dividing mammalian cells and from heart and skeletal muscle. The *A. suum* enzyme exhibits sigmoidal kinetics with its substrate, malate, and its activity is stimulated by low concentrations of fumarate (Landsperger and Harris, 1976; Lai, Harris and Cook, 1992). These properties would potentially control the dismutation of malate in the mitochondria. The *Ascaris* enzyme has been crystallised (Clancy *et al.*, 1992), the reaction mechanism probed (Park *et al.*, 1984; Mallick, Harris and Cook, 1991; Weiss *et al.*, 1991) and the cDNA sequenced and expressed *in vitro* (Kulkarni, Cook and Harris, 1993; Chooback *et al.*, 1997); it possesses a putative malate binding site conserved in all malic enzymes while the ADP binding site has slight differences from the corresponding site in the NADP-dependent enzymes.

Pyruvate dehydrogenase

In anaerobic mitochondria the ratios of NADH/NAD$^+$ and acetylCoA/CoA are significantly higher than in aerobic mitochondria (Komuniecki and Komuniecki, 1995). Therefore enzymes utilising these cofactors have modifications that ensure their function under the conditions prevailing in anaerobic mitochondria. A well-studied example is the pyruvate dehydrogenase complex of *A. suum*. In aerobic mitochondria the activity of the pyruvate dehydrogenase complex is normally inhibited when intramitochondrial NADH or acetylCoA levels rise. Inhibition is achieved in two ways: by direct allosteric inhibition of the E2 and E3 components of the complex by NADH and acetylCoA, and by phosphorylation of the E1 component by pyruvate dehydrogenase kinase, another component of the complex which is activated by high levels of NADH, acetylCoA, and ATP. Pyruvate dehydrogenase from adult *A. suum* muscle is less sensitive to inhibition by NADH than the enzyme from aerobic mitochondria (Thissen *et al.*, 1986). As well, the activity of pyruvate dehydrogenase kinase is less sensitive to stimulation by NADH or acetylCoA than the corresponding enzymes from aerobic mitochondria of *C. elegans* or mammals (Chen, Komuniecki and Komuniecki, 1999). In contrast to pyruvate dehydrogenase from aerobic mitochondria, the *A. suum* enzyme is also stimulated by elevated levels of malate, the substrate for anaerobic mitochondrial pathways; this is due to allosteric activation by malate of pyruvate dehydrogenase phosphatase which activates the enzyme by dephosphorylating the E1 component (Song and Komuniecki, 1994). Two isoforms of E1 have been demonstrated in *A. suum*. One of these, E1α1, is most abundant in adult muscle whereas the other, E1αII, is abundantly expressed in the aerobic L3 (Johnson *et al.*, 1992; Huang *et al.*, 1998).

Anaerobic electron transport

In anaerobic tissues, oxygen is not available to accept electrons from cytochrome oxidase, the terminal complex (IV) of the electron transport chain (Figure 10.2). In these tissues, the electron transport chain is modified for anaerobic electron transport: the concentration of complexes III and IV is very low; ubiquinone is replaced by rhodoquinone; the composition of complex II is modified for transferring electrons to fumarate; and fumarate replaces oxygen as the terminal electron acceptor (Figure 10.4) (Tielens, 1994; Tielens and Van Hellemond, 1998). Reduction of one molecule of fumarate to succinate requires two electrons, originally derived from NADH, which are transferred to fumarate via a modified complex II (see below), plus two protons which are derived from the mitochondrial matrix. Rhodoquinone replaces ubiquinone in anaerobic nematode mitochondria (Allen, 1973; Takamiya *et al.*, 1993); the standard redox potential of rhodoquinone is sufficiently low, at –63 mV, for transferring electrons to fumarate via complex II whereas ubiquinone, at +100 mV, cannot fulfil this role (Van Hellemond *et al.*, 1995). Rhodoquinone accepts electrons from NADH via complex I and is reduced to rhodoquinol which then passes electrons to complex II.

Complex II and fumarate reductase

Fumarate reductase, which is located on the inner mitochondrial membrane, comprises the catalytic domain of complex II. The reaction it catalyses is the reverse of the TCA cycle reaction catalysed by succinate dehydrogenase, but evidence to date suggests that fumarate reductase and succinate dehydrogenase activities in nematodes may be catalysed by distinct proteins coded for by different genes (Roos and Tielens, 1994).

Complex II is comprised of four polypeptide subunits: A, commonly called F_p, which contains covalently-bound flavin, catalyses the reduction of fumarate; B, also called I_p, which possesses three Fe-S centres, in nematodes transfers electrons from rhodoquinol to the A subunit; and C and D which are small hydrophobic subunits, each of which contains a subunit of a cytochrome *b*, that anchor the complex into the inner mitochondrial membrane, and are essential for electron transfer from the quinol (Kita, 1992; Van Hellemond and Tielens, 1994; Kita, Hirawake and Takamiya, 1997; Tielens and Van Hellemond, 1998).

Fumarate reductase from *H. contortus* and *A. suum* have been studied in detail. Roos and Tielens (1994) showed in *H. contortus* that two different genes encoding isoforms of the B subunit of complex II are differentially expressed during the life cycle; one gene is expressed only in the free-living stages, which have aerobic metabolism, whereas the other is expressed throughout the life cycle, which includes the parasitic stages exhibiting fermentative metabolism. A similar situation exists in *A. suum*, where larval and adult complexes II appear to be different proteins, with different cytochrome *b* components, catalysing predominantly succinate oxidation in larvae and fumarate reduction in adult worms (Saruta *et al.*, 1995). *C. elegans*, on the other hand, exhibits only succinate dehydrogenase activity, which is consistent with aerobic metabolism throughout the life cycle. A comparative analysis of the sequences of the F_p subunit of the enzymes from *A. suum* and *C. elegans* suggests that the *Ascaris* fumarate reductase evolved from a succinate dehydrogenase present in free-living ancestors, and that it became secondarily modified to enhance fumarate reductase activity, presumably during adaptation of the parasite to low environmental oxygen levels in the gut of the host (Kuramochi *et al.*, 1994). The catalytic properties and sequence of the F_p subunit have also been analysed from the filarial nematode *Dirofilaria immitis* and shown to have highest homology to the *A. suum* F_p subunit (Kuramochi *et al.*, 1995). This, and other evidence discussed by Köhler (1991) and others (Vande Waa *et al.*, 1993), suggests that the historical dogma that filarial nematodes rely on lactate production as their primary mechanism for generating energy should be critically re-examined (see below).

The further metabolism of succinate

Many nematodes catabolise succinate further, to produce and excrete propionate as an end-product (Table 10.1; Figure 10.3). This achieves an additional yield of ATP per mole of glucose. A few gut parasites, such as *H. contortus* and *Trichostrongylus colubriformis*, metabolise propionate (or propionylCoA) further to produce n-propanol; this yields no further ATP but may act as an additional mitochondrial electron sink to maintain redox balance, and thus permit pathway function, under reducing conditions (Sangster and Prichard, 1985).

Adult worms of *A. suum* and related nematodes produce branched-chain fatty acids (2-methylbutyrate and 2-methylvalerate) as end-products of anaerobic fermentation in the body wall muscle tissue (Table 10.1; Figure 10.3). These fatty acids are excreted but also accumulate to very high concentrations (up to 100 mM) in the perienteric fluid (Sims *et al.*, 1992). The reactions of this pathway are essentially the reverse of similar reactions of the β-oxidation pathway for lipid oxidation in aerobic mitochondria and the enzymes catalysing the reactions are very similar to those catalysing β-oxidation in mammals (Suarez De Mata *et al.*, 1991; Duran *et al.*, 1993; Ma *et al.*, 1993). Electrons derived from NADH are transferred from rhodoquinol to electron-transfer flavoprotein:rhodoquinone oxidoreductase and thence via several other flavoproteins to reduce 2-methylbutyryl-CoA or 2-methylvalerylCoA. A site 1 phosphorylation of ADP contributes one mole of ATP/mole of NADH.

The function of this pathway appears to be similar to the fumarate reductase pathway, i.e. to serve as an electron sink when oxygen, and possibly fumarate, are unavailable (Ma *et al.*, 1993). Why ascarids have elaborated an additional pathway for this purpose is not yet clear. The pathway has a similar ATP yield to that derived from the reduction of fumarate to succinate. One explanation put forward for the presence of this pathway is that the products of muscle metabolism are esterified by the female reproductive system and incorporated into eggs which, being aerobic, are capable of oxidising them for energy during development (Fairbairn, 1955; Saz, 1990).

Aerobic/Anaerobic Transitions During Development

In some nematodes, developmental transitions in energy metabolism occur, most frequently from aerobic metabolism in early stages to fermentative metabolism in later stages, typically in animal parasites. For example, the eggs, L2 and L3 of *A. suum* are aerobic (Oya, Costello and Smith, 1963; Komuniecki and Vanover, 1987) but stages after the moult to L4 change to fermentative metabolism, using unsaturated organic acids as electron acceptors, and start to excrete acetate, propionate, 2-methylbutyrate and 2-methylvalerate as end-products (Rodrick *et al.*, 1982a; Komuniecki and Vanover, 1987; Vanover-Dettling and Komuniecki, 1989). These functional changes correlate with ultrastructural changes in the mitochondria (Rodrick *et al.*, 1982b), alterations in the electron transport chain (Kita, 1992; Takamiya *et al.*, 1993), in the pyruvate dehydrogenase complex (Klingbeil *et al.*, 1996; Huang *et al.*, 1998), and in expression of 2-methyl branched-chain enoylCoA reductases (Duran *et al.*, 1993; Duran *et al.*, 1998). The activities of enzymes essential for the anaerobic pathway — malic enzyme, phosphoenolpyruvate carboxykinase, and enoylCoA reductase — increase significantly in the L3 stage, prior to the critical moult, while in L4 the activities of enzymes important for aerobic metabolism — citrate synthase, isocitrate dehydrogenase and cytochrome *c* oxidase — decrease dramatically and remain low in adult worms (Vanover-Dettling and Komuniecki, 1989). Similar changes occur in *H. contortus*, which undergoes a transition from aerobic to fermentative metabolism between the L3 and adult stages (Roos and Tielens, 1994). An interesting feature of the transitions of both *A. suum* and *H. contortus* is that although the L3 stages are normally aerobic, they are already well-equipped biochemically for anaerobic metabolism — with enzymes for aerobic and anaerobic metabolism apparently coexisting in the same mitochondria (Mei, Komuniecki and Komuniecki, 1997) — even before such conditions would be encountered in a normal life cycle, an interesting example of metabolic premunition.

Carbohydrate Metabolism

Carbohydrates are essential components of nematodes as they provide an important fuel for energy metabolism, function as reserves of energy and also have structural and other roles as glycoproteins, glycolipids, proteoglycans, glycosaminoglycans (mucopolysaccharides) and chitin.

Carbohydrate Reserves

In general, the levels of free glucose in nematodes are low and most carbohydrate is stored as the polymer glycogen or the disaccharide trehalose (see Behm, 1997). Glycogen is the more common carbohydrate reserve. It is synthesised from UDP-glucose, which is derived from glycolytic glucose-6-phosphate (Figure 10.1). Glycogen may constitute from 0.1 to 25% of the dry weight of nematodes, varying with species, stage and condition. It has been possible to determine the localisation of glycogen in nematodes because it is a highly-branched, high molecular weight polymer that forms characteristic electron-dense rosettes visible in thin sections by transmission electron microscopy.

Glycogen is present in numerous tissues — epidermis, non-contractile parts of muscle cells, intestinal cells, of adults and larval stages, and in eggs (see Bird and Bird, 1991). It may accumulate to very high levels, especially in the 'anaerobic' parasitic nematodes. Glycogen stores are essential for viability in the anaerobic nematodes, except for macrofilariae which store relatively little glycogen (Barrett, 1983), but they are of variable importance in 'aerobic' worms. For example, glycogen content has been linked to survival and infectivity of facultatively aerobic, non-feeding infective L3 (dauer larvae) of *Steinernema carpocapsae* but was less critical for long-term survival in dauers of *S. feltiae*, for which lipid reserves appeared more important (Wright, Grewal and Stolinski, 1997).

Trehalose, a disaccharide of glucose, is a soluble sugar stored by many nematodes (see Behm, 1997). In most species it does not accumulate to the same levels as glycogen; values are reported in the range of 0.1–9% depending on species, stage and condition. In adults of *A. suum* the tissue distribution has been determined: trehalose was the only sugar detected in haemolymph and highest tissue concentrations were found in muscle and the male and female reproductive organs (Fairbairn and Passey, 1957). The sugar also accumulates in nematode eggs where it may reach very high concentrations — values in the range 0.1–0.5 M have been recorded (Clarke and Hennessy, 1976; Clarke, Perry and Hennessy, 1978; Perry, Clarke and Hennessy, 1980; Perry *et al.*, 1983; Weston, O'Brien and Prichard, 1984).

Catabolism

Reserve carbohydrates

The reserves of both glycogen and trehalose decline when exogenous nutrients are limiting, for example in non-feeding infective or dauer larvae, or in adult nematodes in simple medium *in vitro* (e.g. Roberts and Fairbairn, 1965; Castro and Fairbairn, 1969; Donahue *et al.*, 1981c; Bennet and Bryant, 1984; Sangster and Prichard, 1984; Wright, Grewal and Stolinski, 1997).

Glycogen is catabolised to glucose 6-phosphate by glycogen phosphorylase, debranching enzyme, and phosphoglucomutase (Figure 10.1). Phosphoglucomutase catalyses an equilibrium reaction in most organisms, including those nematodes that have been studied (Barrett, 1981; Armson and Mendis, 1995), which is consistent with its function in both the synthesis and degradation of glycogen. Glycogen phosphorylase activity, on the other hand, is under tight regulation in many organisms, including *Ascaris* muscle, and exists in two forms: relatively active phosphorylase *a* and relatively inactive phosphorylase *b*. The activity of phosphorylase *b* is regulated at two levels: (i) phosphorylation of the *b* form by phosphorylase *b* kinase activates the enzyme to the *a* form, and a phosphorylase phosphatase converts the *a* form to the inactive *b* form. Phosphorylase kinase is normally inactive until it is phosphorylated by cyclic AMP-dependent protein kinase. This mechanism of regulation is usually under hormonal control; (ii) phosphorylase *b* may also be

activated directly by binding AMP if the adenylate energy charge of the cell declines sufficiently for the concentration of AMP to increase. Phosphorylase *b* from *A. suum* muscle conforms to this pattern as it is inactive in the absence of AMP (Yacoub *et al.*, 1983), and its activity increased in perfused preparations deprived of glucose (Donahue *et al.*, 1981b) or perfused with serotonin, which increased cyclic AMP levels (Donahue *et al.*, 1981d). A cyclic-AMP-dependent protein kinase has been demonstrated, and characterised at the molecular level, from *A. suum* muscle (Donahue *et al.*, 1981a; Jung *et al.*, 1995; Treptau *et al.*, 1996). Calcium has also been implicated in control of glycogenolysis in *Ascaris* muscle (Donahue, Yacoub and Harris, 1982).

Trehalose is hydrolysed to glucose by the enzyme trehalase, which has been demonstrated in a variety of parasitic nematodes, as well as *C. elegans* (see Behm, 1997). Although neutral trehalase activity in yeast is regulated by cyclic-AMP-dependent phosphorylation (Wera *et al.*, 1999), it is not clear yet whether this occurs in nematodes. *C. elegans* possesses four genes for trehalase, all expressed, and several different biochemical activities have been demonstrated (Somerville and Behm, in preparation).

Glycolysis

The glycolytic pathway, from glucose or glucose 6-phosphate to phosphoenolpyruvate or pyruvate, is ubiquitous in nematodes and is essentially similar to the pathway in other organisms (Figure 10.1). It plays a particularly important role in those worms undergoing anaerobic metabolism because glycolysis and its various terminal pathways, discussed above, provide the major source of ATP. Filarial nematodes, in particular, appear to rely heavily on glycolysis, as they excrete lactate as the major end-product of carbohydrate catabolism when incubated aerobically, and appear not to utilise mitochondrial terminal reactions to any great extent (Barrett, 1983; Powell, Stables and Watt, 1986b; Powell, Stables and Watt, 1986a).

Glycolysis is regulated in part by the supply of substrate and in part by levels of ATP and the NAD^+/NADH ratio in the cytosol. ATP is required as substrate for both hexokinase and phosphofructokinase but phosphofructokinase, an important rate-limiting enzyme of glycolysis in *Ascaris* and probably other nematodes (Barrett and Beis, 1973; Barrett, 1981; Bennet and Bryant, 1984; Armson and Mendis, 1995), is also allosterically inhibited by elevated ATP concentrations (Hofer *et al.*, 1982b). Phosphofructokinase activity is tightly regulated in most higher eukaryotes, including nematodes. The activity of the enzymes from *A. suum* muscle and *D. immitis* is controlled by a combination of substrate (fructose 1,6-bisphosphate) concentration and allosteric effectors such as ATP (inhibitory) and AMP or fructose 2,6-bisphosphate (both stimulatory) (Hofer *et al.*, 1982b; Srinivasan *et al.*, 1988; Srinivasan, Rao and Harris, 1990; Payne *et al.*, 1991), as well as by phosphorylation of the enzyme by a cyclic-AMP-dependent protein kinase, which increases its activity (Hofer *et al.*, 1982a; Srinivasan *et al.*, 1988). The *A. suum* phosphofructokinase is also phosphorylated by a cyclic-AMP-independent protein kinase (Thalhofer et al., 1988) and several protein phosphatases have also been identified in *A. suum* muscle that reverse the phosphorylation (Daum et al., 1986; Daum et al., 1992). The protein kinases and phosphatases that act on phosphofructokinase are different from those acting on glycogen phosphorylase of *A. suum*, which permits independent control of glycolysis and glycogen catabolism. In vertebrates phosphofructokinase is not directly activated by phosphorylation.

Two other enzymes potentially important in regulating flux through glycolysis are hexokinase and pyruvate kinase (Barrett, 1981; Bennet and Bryant, 1984; Armson and Mendis, 1995). They have received less attention as phosphofructokinase activity is considered to be more critical. Hexokinase catalyses the first step in glycolysis from glucose. The enzyme has been examined from a number of animal-parasitic nematodes and a cDNA has been cloned and sequenced from *H. contortus* (Schmitt-Wrede et al., 1999). The enzymes from *A. suum* and filarial worms are inhibited by its product, glucose 6-phosphate, but at higher concentrations than required to inhibit the mammalian enzyme (Supowit and Harris, 1976; Hutchison, Turner and Oelshlegel, 1977; Barrett, 1991b). A glucokinase is present in *Angiostrongylus cantonensis* in addition to a hexokinase (Oguchi, Kanda and Akamatsu, 1979); unlike hexokinase, glucokinase is specific for glucose, for which it has a relatively low affinity, and is not inhibited by physiological concentrations of glucose 6-phosphate.

Regulation of pyruvate kinase from filarial nematodes is similar to the liver enzyme of mammals in that it is inhibited by high levels of ATP and activated by fructose 1,6-bisphosphate; these properties may be important for nematodes that produce lactate as a major metabolic end-product (Brazier and Jaffe, 1973). In mammalian liver pyruvate kinase is also regulated by phosphorylation/dephosphorylation but this has not been described for the enzyme from nematodes.

Tricarboxylic acid cycle

Many aerobic nematodes, and eggs and larvae of anaerobic nematodes, utilise the tricarboxylic acid cycle and, presumably, the standard aerobic electron transport chain (Figures 10.1 and 10.2) exclusively or facultatively for generating energy. Substrates for the cycle are acetylCoA, derived from pyruvate or lipid catabolism, and amino acids such as glutamate, glutamine and aspartate. The cycle appears to be the same as in other organisms and its presence has been demonstrated experimentally by isotope and inhibitor studies and/or demonstration of the activities of all the enzymes (Barrett, 1981; Barrett and Wright, 1998). Recent evidence suggests that tricarboxylic acid cycle activity may be more important than originally thought in some filarial nematodes (Davies and Köhler, 1990; Köhler, 1991; Vande Waa et al., 1993).

Pentose phosphate pathway

The pentose phosphate pathway is important for generating reducing power (as NADPH) in the cytosol for biosynthetic reactions such as fatty acid and nucleic acid synthesis, for providing NADPH to re-oxidise oxidised glutathione in detoxification metabolism, and for generating pentose sugars (as ribose 5-phosphate) required for nucleotide and nucleic acid synthesis. It also functions to digest pentose sugars acquired in the diet, e.g. from the digestion of nucleic acids. The substrate for the pentose phosphate pathway is glucose 6-phosphate and the products are NADPH, CO_2, fructose 6-phosphate and glyceraldehyde 3-phosphate; the latter two molecules are potentially fed back into glycolysis but whether this happens depends on the metabolic state of the cell. If cellular demand for nucleotide synthesis is high, the major product of the pathway is ribose 5-phosphate; if demand is high for NADPH, the major carbon product is glucose 6-phosphate, which can be recycled through the pentose phosphate pathway.

Operation of the pentose phosphate pathway is essential in most living organisms for nucleotide synthesis and reductive biosynthesis. Its presence is expected in all nematodes but has been positively demonstrated in only a few. It is likely that the level of activity of the pathway would vary between tissues. The first two enzymes of the pathway, glucose 6-phosphate dehydrogenase and 6-phosphogluconate dehydrogenase, have been found in all groups of nematodes — free-living, plant-parasitic, entomopathogenic, and animal-parasitic (Hutchinson and Fernando, 1974; Rathaur et al., 1982; Shih and Chen, 1982; Turner and Hutchison, 1982; Omar and Raoof, 1994; Omar, Raoof and Al Amari, 1996; Barrett and Wright, 1998). All the enzymes have been demonstrated in adult *A. suum* muscle, intestine and reproductive organs (Langer, Smith and Theodorides, 1971).

Biosynthesis

Reserve carbohydrates

Glycogen is synthesised from glucose 1-phosphate via UDP-glucose pyrophosphorylase, glycogen synthase and branching enzyme (Figure 10.1). Neither UDP-glucose pyrophosphorylase nor branching enzyme have been studied in nematodes but each is represented by a single putative gene, K08E3.5 and T04A8.7 respectively, in *C. elegans*. There is cDNA evidence that each is expressed; the putative UDP-glucose pyrophosphorylase is represented by several alternatively-spliced products (http://www.sanger.ac.uk/Projects/C_elegans/).

Glycogen synthase activity is highly regulated in most organisms, including *A. suum* and probably other nematodes, by allosteric mechanisms as well as protein phosphorylation. The active form of the enzyme, glycogen synthase I, is not phosphorylated and the phosphorylated form, glycogen synthase D, is relatively inactive unless there is an increase in the cytosolic concentration of glucose 6-phosphate, an allosteric activator of glycogen synthase D. The two forms of glycogen synthase have been demonstrated in *A. suum* muscle preparations and their relative activities were shown to change depending on the availability of exogenous glucose to the tissue (Donahue *et al.*, 1981b). Two distinct glycogen synthases have been purified from *A. suum* muscle. The first, glycogen synthase I, is very similar to the mammalian enzyme and is regulated by cyclic-AMP-dependent phosphorylation (Hannigan, Donahue and Masaracchia, 1985). The second enzyme, glycogen synthase II, is a protein-carbohydrate complex which requires glucose 6-phosphate for activity and is not phosphorylated by cyclic-AMP-dependent protein kinase (Ghosh *et al.*, 1989). As there is no sequence information for these two proteins yet it is not possible to determine whether or not they are related proteins encoded by a single gene.

Trehalose is synthesised from UDP-glucose and glucose 6-phosphate, the reactions being catalysed by trehalose 6-phosphate synthase and trehalose 6-phosphate phosphatase. Trehalose 6-phosphate synthase activity has been detected and partially characterised in a small number of parasitic and free-living nematodes (see Behm, 1997) but there is no information about regulation of activity. Two genes encoding putative trehalose 6-phosphate synthase, ZK54.2 and F19H8.1, are present in *C. elegans*; both are expressed at the mRNA level (http://www.sanger.ac.uk/Projects/C_elegans/).

The glyoxylate cycle

Unlike mammals, nematodes are capable of net synthesis of carbohydrates from 2-carbon substrates (acetylCoA). This is achieved via the glyoxylate cycle, whereby mitochondrial isocitrate (see Figure 10.1) and a molecule of acetylCoA are converted to a molecule each of succinate and malate, the reactions being catalysed by isocitrate lyase and malate synthase. Malate is then oxidised to oxaloacetate, which is converted to glucose 6-phosphate by the reverse reactions of glycolysis, a process termed glyconeogenesis. This pathway requires a source of acetylCoA, which is usually derived from β-oxidation of lipids and would only be available in aerobic tissues (see below), but which also could potentially be derived from ethanol or acetate (see Barrett and Wright, 1998).

The cycle has been shown to operate in eggs but not adult nematodes of *A. suum* (Barrett, Ward and Fairbairn, 1970), and isocitrate lyase and malate synthase have been demonstrated in a variety of free-living and animal- and plant-parasitic nematodes (Barrett, 1981; Grantham and Barrett, 1986a; Singh, Katiyar and Srivastava, 1992; Barrett and Wright, 1998). In *C. elegans* the cycle becomes active during embryogenesis and as a starvation response during larval development when nutrients are limiting (Khan and McFadden, 1980; Khan and McFadden, 1982; Liu, Thatcher and Epstein, 1997). A single gene in *C. elegans* encodes a bifunctional protein with isocitrate lyase and malate synthase activity; it is expressed during embryogenesis and in larvae and adults in the cells of the intestine and body wall muscle (Liu *et al.*, 1995). Putative operation of the cycle has been correlated with a decline in triglycerides and increase in carbohydrates in several nematode species, for example *A. suum* eggs (Barrett, Ward and

Fairbairn, 1970) and *C. elegans* embryos (Khan and McFadden, 1980).

Lipid Metabolism

Lipids are essential to all organisms, including nematodes, where they function as important fuel reserves in the aerobic worms and in eggs, as well as having structural roles in cell membranes, the egg-shell (Bird and Bird, 1991) and the cuticle (e.g. *Brugia malayi* (Smith, Selkirk and Gounaris, 1996), *Trichinella spiralis* (Modha *et al.*, 1999), for review see (Bird and Bird, 1991)), and essential functions in cell signalling (e.g. protein kinase C (Arevalo and Saz, 1991; Islas-Trejo *et al.*, 1997)) and communication (e.g. *C. elegans* pheromone (Golden and Riddle, 1982)) within worm populations. The catabolism and biosynthesis of lipids in free-living, plant-parasitic and entomopathogenic nematodes has been recently reviewed by Barrett and Wright (1998) and Chitwood (1998), so this discussion will emphasise the animal parasites and new developments.

The lipid content and composition has been analysed for many nematodes (for reviews see Lee and Atkinson, 1976; Barrett, 1981; Barrett, 1983; Frayha and Smyth, 1983; Barrett and Wright, 1998; Chitwood, 1998); total lipids vary from 1 to around 67% dry weight, depending on species, diet, season, and stage of the life cycle. The general pattern is that higher quantities of lipid are stored by the aerobic nematodes, especially the larval stages and non-feeding L3 of aerobic or anaerobic parasites, dauer larvae of free-living nematodes, and nematode eggs. Neutral lipids, particularly triacylglycerols (triglycerides), are the predominant lipid class, and these may be visible as lipid droplets that accumulate particularly in the epidermis, intestinal cells and oocytes (Bird and Bird, 1991). Other important lipid classes are the phospholipids, free fatty acids, sphingolipids and glycosphingolipids, and cholesterol and other sterols. There is great variation among nematodes in chain length and ratio of saturated to unsaturated fatty acids. Factors influencing the lipid composition of nematodes include diet and temperature, both of which influence the level of saturation of fatty acids. For example, the medium in which *Steinernema glaseri* was grown influenced both quantity and composition of lipids; there was variation in the relative amounts of phospholipids and sterols, and in fatty acid chain length and saturation (Abu Hatab, Gaugler and Ehlers, 1998). Because phospholipids and sterols are major components of cell membranes and their composition determines the fluidity of these membranes, these components are important in the adaptation of nematodes, and most other organisms, to changes in temperature. For example, lowering the growth temperature of *C. elegans* from 25°C to 15°C increased the levels of eicosapentaenoic acid both in phospholipids and in the total lipid fraction (Tanaka *et al.*, 1996; Tanaka *et al.*, 1999). This is a particular consideration for free-living nematodes, and also for animal-parasitic nematodes that undergo sudden changes in temperature on infecting or leaving their hosts.

Biosynthesis

Nematodes are able to synthesise, remodel and degrade most classes of lipids (Rothstein and Gotz, 1968; Rothstein, 1970a; Krusberg, 1972; Sarwal, Sanyal and Khera, 1989; Bankov, Timanova and Barrett, 1998), including some unusual ones not found in other organisms (e.g. a glycosphingolipid with a branched-chain sphingoid base, and 1,2-dieicosapentaenoyl-sn-glycero-3-phosphocholine in C. elegans (Chitwood et al., 1995; Tanaka et al., 1999) and a novel glycoinositolphospholipid from A. suum (Sugita et al., 1996)), but the repertoire of the parasitic nematodes is more limited (see reviews by Frayha and Smyth (1983) and Barrett (1981)). Free-living nematodes are unusual among animals in being able to synthesise unsaturated fatty acids de novo (Rothstein, 1968; Rothstein, 1970a). The information available on the synthetic processes for the different lipid classes is very limited and the nature or operation of many pathways has been surmised from other organisms.

In vertebrates fatty acids are both acquired from the diet and synthesised de novo by sequentially adding two-carbon units to an acetylCoA or propionylCoA priming group. This is catalysed in the cytosol by the fatty acid synthase complex. In nematodes de novo fatty acid synthesis occurs in free-living groups but has not been demonstrated in animal parasites. They, like the free-living nematodes are, however, able to lengthen pre-formed fatty acids by addition of acetylCoA groups, and to saturate or desaturate pre-formed saturated or mono-saturated fatty acids (e.g. B. malayi (Liu and Weller, 1989; Liu, Serhan and Weller, 1990; Liu and Weller, 1990), C. elegans (Watts and Browse, 1999)). Three new fatty acid desaturases have recently been isolated from C. elegans; in each case they are the first examples of these enzymes isolated from animals. The first, an ω^3-desaturase, inserts a double bond in the ω^3 position of a glycerolipid or 18- or 20-chain fatty acids (Spychalla, Kinney and Browse, 1997). ω^3 fatty acids comprise 17% of total fatty acids in C. elegans (Hutzell and Krusberg, 1982) and are a major component of the nematode phosphatidylcholine (Satouchi et al., 1993). Two other enzymes, fatty acid Δ^5 and Δ^6 desaturases, insert double bonds after carbons 5 and 6, respectively (Michaelson et al., 1998; Napier et al., 1998; Watts and Browse, 1999). The latter desaturases are also unique (to date) among animals as they possess an N-terminal haem-binding domain that has similarity to cytochrome b_5, which is typical of microsomal desaturases. Given that over 20% of the total fatty acids of C. elegans are Δ^5-desaturated (arachidonic and eicosapentanoic acids) (Satouchi et al., 1993; Tanaka et al., 1996), the Δ^5-desaturase is likely to play a vital role in poly-unsaturated fatty acid synthesis. What role the Δ^5 fatty acids play in C. elegans is not yet clear, but arachidonic and eicosapentaenoic acid are precursors for the eicosanoid families of signalling molecules, such as prostaglandins, leukotrienes and thromboxanes. Interestingly, the arachidonic acid concentration of the filarial nematode B. malayi is higher than in host serum and the nematodes convert arachidonic acid, of host or parasite origin, to prostaglandin E2 and prostacyclin, which are secreted and may modulate the host's immune response (Liu and Weller, 1989; Liu, Serhan and Weller, 1990; Liu and Weller, 1990; Liu, Buhlmann and Weller, 1992).

Catabolism

On a weight-for weight basis, lipids provide more stored energy than carbohydrates or proteins, because they are more highly reduced, and because, unlike glycogen molecules, lipid droplets contain no water. Many aerobic nematodes utilise stored lipids when nutrients become scarce and there is a body of literature demonstrating reduced infectivity of parasite larvae as lipid stores become depleted (Lee and Atkinson, 1976; Medica and Sukhdeo, 1997; Barrett and Wright, 1998). The major class of lipids utilised is the triacylglycerols which are digested to fatty acids plus glycerol; the fatty acids are subsequently catabolised, via mitochondrial β-oxidation, to acetylCoA which is oxidised via the TCA cycle (see below), a process yielding considerable energy as ATP.

Only nematodes respiring aerobically are able to oxidise fatty acids to release energy. This is because the NADH and $FADH_2$ released must be oxidised by the aerobic electron transport chain linked to oxygen as the terminal electron acceptor. Therefore, a functional β-oxidation pathway, which converts fatty acids to acetylCoA (plus a molecule of propionylCoA for fatty acids with odd-numbered carbon chains) is only found in aerobic nematodes, both free-living and parasitic (Barrett, 1981; Barrett and Wright, 1998).

Sterol Metabolism

Like insects, but unlike vertebrates or plants, nematodes, whether free-living or parasitic, cannot synthesise sterols de novo and must obtain them from their diet (Chitwood, 1991; Chitwood and Lusby, 1991; Chitwood, 1999). For example, when fed on bacteria, C. elegans requires supplementation of the culture medium with cholesterol or a related sterol for optimum growth (Hieb and Rothstein, 1968). This is also the case for in vitro larval development of Nippostrongylus brasiliensis (Bolla, Weinstein and Lou, 1972). In a variety of radioisotope studies, nematodes have consistently failed to synthesise sterols from acetate or mevalonate. It has not yet been possible to define precisely what is lacking in the de novo pathway as conflicting results from numerous studies in the literature reflect the fact that trace levels of sterols contaminate culture reagents and thus interfere with interpretation of the experiments (Chitwood, 1991).

Because nematode sterols are derived largely from the diet, a large diversity of sterols is found in nematodes. Animal-parasitic nematodes possess relatively high levels of cholesterol but free-living and plant-parasitic nematodes have a variety of compounds — up to 60 have been identified (Chitwood, 1991). Nematodes are capable of modifying sterols obtained from the diet, by esterification, saturation, desaturation, and, in some cases, 4α-methylation (a process unique to nematodes), and dealkylation (Chitwood and Lusby, 1991). Whether nematodes are able to degrade sterols to CO_2 is not clear (Rothstein, 1968).

Sterols are not very abundant in nematodes and it is questionable whether they are constitutively present in cell membranes at the high levels found in vertebrates (Chitwood, 1991). A freeze-fracture study of *Turbatrix aceti* showed cholesterol to be present only in one cell type and only in intracellular vesicular membranes rather than the plasma membrane (Silberkang *et al.*, 1983). Intestinal cells, in particular, showed extensive membrane disruption in larvae of *N. brasiliensis* cultured under sterol-deficient conditions (Coggins, Schaefer and Weinstein, 1985).

Steroid hormones appear to be important in developmental processes and in moulting in nematodes. For example, *daf-12*, a gene essential in the dauer developmental pathway and important in other aspects of L3 development in *C. elegans*, encodes a putative steroid or retinoid nuclear hormone receptor (Riddle and Albert, 1997; Antebi, Culotti and Hedgecock, 1998). Ecdysteroids are present in many nematodes (e.g. *H. contortus* (Fleming, 1993)), and their eggs (e.g. *Ascaris* (Fleming, 1987)), but there is no definitive evidence that they are synthesised by nematodes; whether they are involved in moulting, as has been hypothesised for some time, remains equivocal (Chitwood, 1991). Vertebrate steroid hormones such as testosterone, progesterone, oestrone and oestriol, as well as corticosteroids, have been identified in nematode extracts although their functions are obscure. Administration of hormones to nematodes in culture affected a variety of processes, including embryonation, moulting, reproduction, sex-determination, egg-laying and motility (reviewed by Chitwood, 1991). The biological significance of these studies is difficult to interpret because exogenous hormones at potentially higher concentrations than present *in vivo* are likely to cause artefacts. Function of these hormones is probably best addressed experimentally using a combination of genetic and biochemical methods: *C. elegans* possesses an estimated 270 putative nuclear hormone receptors, and many related genes have now been identified in the parasitic nematodes, which presents a daunting task to investigators (Chervitz *et al.*, 1998).

Protein and Amino Acid Metabolism

Amino acid metabolism in helminths has been reviewed in detail by Barrett (1991a). The protein fraction of an organism is comprised of water-soluble proteins which function as enzymes (e.g. triosephosphate isomerase), receptors (e.g. steroid receptors) and carrier molecules (e.g. haemoglobin), of hydrophobic proteins which act as enzymes (e.g. cytochrome P450), receptors (e.g. serotonin receptors) and carrier (e.g. chloride channels) molecules within cell membranes, and of insoluble or relatively insoluble proteins that are structural (e.g. collagens), constitute the cytoskeleton (e.g. tubulins) and contractile structures (e.g. actin), and storage molecules (e.g. yolk

proteins) within cells. Proteins constitute 50–80% of the dry weight of nematodes (Lee and Atkinson, 1976), which is relatively high compared with other organisms. Two-dimensional electrophoresis of extracts of mixed stages of *C. elegans* revealed spots on silver-stained gels representing 2000 different proteins within the pI range 3.5–9 and the molecular weight range 10–200 kDa (Bini *et al.*, 1997); it is estimated that the average eukaryotic cell expresses 5–10,000 different proteins (K.L. Williams, pers. comm.).

Proteins are synthesised within all cells from amino acids, which are obtained from the diet, by *de novo* synthesis or interconversion, or by recycling from degraded proteins. Twenty common amino acids are used by cells to synthesise the majority of proteins; as well, several minor amino acids contribute to specialised proteins, for example hydroxyproline in collagen. A variety of other amino acids are present in cells that are not used in protein synthesis, for example β-alanine which is present in pantothenic acid, taurine, which is used for osmotic purposes, and γ-aminobutyric acid, which is used as a neurotransmitter. Many proteins are post-translationally modified by glucosylation, acetylation, myristoylation, attachment of cofactors such as FAD or pyridoxal phosphate, or phosphorylation, for example.

Biosynthesis of Amino Acids

Like mammals, nematodes require essential amino acids in their diet (Barrett, 1991a). Those required by the free-living nematodes *C. elegans* and *C. briggsae* are the same as those required by mammals, i.e. histidine, isoleucine, leucine, lysine, methionine, phenylalanine, threonine, tryptophan, arginine and valine (Vanfleteren, 1980). They did not require the medium to be supplemented with alanine, aspartate, cysteine, glutamate, glutamine, glycine, proline, serine or tyrosine, which indicates that, like mammals, they are able to synthesise these amino acids from precursors.

Amino acid synthesis in nematodes and other helminths has been reviewed by Barrett (Barrett, 1991a; Walker and Barrett, 1997). Our knowledge of these processes in nematodes is rather patchy. Synthesis of alanine, aspartate and glutamate occurs by transamination of pyruvate, oxaloacetate and glutamate, respectively; the relevant aminotransferases are likely to be present in all nematodes (see below). Glutamine and asparagine would be synthesised by amidation of glutamate and aspartate, respectively. Low levels of glutamine synthetase were recorded in *Panagrellus redivivus* and *Heligmosomoides polygyrus* by Grantham and Barrett (1988), but asparagine synthetase was only detected in *P. redivivus*.

The metabolism in parasites of the sulphur amino acids, methionine and cysteine, has been reviewed in detail by Walker and Barrett (Walker and Barrett, 1997). The processes are similar to those in mammals, with some subtle differences. In most organisms methionine and cysteine can be interconverted via the methionine cycle, the intermediates and reactions of which are important in methyl group metabolism and in supplying *S*-adenosylmethionine, a molecule important in methylation reactions and in the biosynthesis of polyamines. The filarial nematodes *B. pahangi* and *D. immitis* are able to convert methionine to cysteine but are unable to convert cysteine to methionine, which must therefore be obtained from their host (Jaffe, 1980). Free-living *C. briggsae*, however, is reported to be capable of converting cysteine to methionine (Lu, Hieb and Stokstad, 1976) provided vitamin B_{12} and folate are also added to the culture. A novel type of cystathionine β-synthase, the enzyme that converts homocysteine to cystathionine (the precursor of cysteine), with catalytic properties different from the mammalian enzyme, has been identified in nematodes, both free-living and parasitic (Walker and Barrett, 1997).

Catabolism of Amino Acids

Amino acids are able to supply a significant amount of metabolic energy in aerobic organisms when they are degraded, by diverse reactions that remove the amino group to yield the carbon skeletons, to intermediates that can be oxidised via the TCA cycle — pyruvate, acetylCoA, oxaloacetate, 2-oxoglutarate, succinylCoA and fumarate (see Figure 10.1). Complete oxidation of amino acids is clearly not an important source of energy for nematodes without an active TCA cycle. For aerobic nematodes, however, amino acids may provide an additional source of energy, and this has been demonstrated for *N. brasiliensis* adults *in vitro*, where a number of amino acids supported motility and maintained glycogen and ATP levels during 48 hours' incubation (Singh *et al.*, 1992) and for *Litomosoides sigmodontis* (syn. *carinii*) *in vitro* where glutamine, but not other amino acids, maintained viability and was catabolised to CO_2, presumably via the TCA cycle (Davies and Köhler, 1990). Small amounts of amino acids were oxidised by *Brugia pahangi* adults and it was hypothesised that this oxidation could in fact be due to the metabolic activity of microfilariae and earlier developmental stages *in utero* (Srivastava, Saz and deBruyn, 1988). A variety of free-living and plant-parasitic nematodes utilise protein during starvation (Barrett and Wright, 1998). Among animal parasites the situation is variable: protein content did not decline in *Strongylus vulgaris* L3s starved for eight days at 38°C (Medica and Sukhdeo, 1997) nor was protein catabolism considered important for *Ancylostoma caninum* L3 during 20 days of starvation (Clark, 1969). But in *N. brasiliensis* exsheathed L3 starved for eight days at 25°C 2.2% of total protein was lost each day (Wilson, 1965).

Our knowledge of the enzymes and pathways used by nematodes to catabolise amino acids is rather patchy (see Barrett, 1981; Barrett, 1983; Barrett and Wright, 1998; Chitwood, 1998), except for two species. In a systematic study, remarkable in this field for its thoroughness, Grantham and Barrett (1986b; 1986a) investigated amino acid catabolism in both free-living *P. redivivus* and animal-parasitic *H. polygyrus* and concluded that the capabilities of these organisms were similar to those of mammals. Both nematodes possessed the three usual aminotransferase systems for three-carbon (pyruvate/alanine), four-carbon (oxaloa-

cetate/aspartate) and five-carbon (2-oxoglutarate/glutamate) molecules but the 2-oxoglutarate/glutamate system had highest activities. These enzymes have also been demonstrated in a variety of other nematodes and are likely to be ubiquitous. Glutamate dehydrogenase is the dominant enzyme in most organisms for oxidative deamination of amino acids. This appeared to be the case for both *P. redivivus* and *H. polygyrus*; the NAD^+-dependent enzyme, present in high activity, had regulatory properties similar to mammals. In addition, both *P. redivivus* and *H. polygyrus* had activities of L- and D-amino acid oxidases, which oxidatively deaminate amino acids such as glycine, L-aspartate, L-histidine, L-methionine, L-threonine and L-tyrosine. In contrast to *P. redivivus*, *H. polygyrus* lacked the ability to oxidise L-alanine and L-proline. The following non-oxidative deaminating enzymes were detected in both organisms: histidase, L-serine dehydratase, L-threonine dehydratase, arginase, asparaginase, glutaminase.

Ammonia is a waste product of amino acid deamination and in animal-parasitic nematodes is excreted as ammonia (Barrett, 1981; Wright, 1998). Mammals, on the other hand, excrete ammonia as the less toxic urea, a process catalysed by the urea cycle. A functional urea cycle appears to be absent in animal-parasitic nematodes and its operation in other nematode groups is also uncertain (Barrett, 1981; Barrett and Wright, 1998; Wright, 1998). Urea is excreted by many nematodes but it could be produced from arginine by the action of arginase, or from pyrimidine or purine degradation.

Nucleic Acid and Nucleotide Metabolism

DNA and RNA are, respectively, the repository of the genetic information of nematodes and the means of transcribing and translating that information. The metabolism of nucleic acids in nematodes has not been comprehensively reviewed recently although aspects of their biosynthesis in free-living nematodes was reviewed by Chitwood (1998) and earlier reviews were provided by Barrett (1981; 1983). Although it is a vital aspect of the functioning of any organism, experimental study in free-living or parasitic nematodes has not been extensive. In *C. elegans* 18% of the total complement of predicted genes is associated with DNA and RNA metabolism (Chervitz *et al.*, 1998) and the putative genes that encode the enzymes that catalyse most processes have been identified.

Unlike other macromolecules, DNA does not continuously turn over in cells but instead is replicated only before cell division; it generally resides in the nucleus (or the mitochondrion, for mtDNA) for the life of the cell. The principal enzymes catalysing DNA synthesis, the DNA polymerases, DNA topoisomerases and DNA helicase, have received little study in nematodes although their putative genes have been identified in *C. elegans* and some genomic and cDNA sequences from *B. malayi*, *O. volvulus*, *Globodera rostochiensis*, *Globodera pallida*, *Meloidogyne incognita* and *C. briggsae* are present in the databases. DNA topoisomerase I has been purified from *C. elegans* and its catalytic properties examined (Park and Koo, 1994).

When damage occurs to the macromolecule it is detected and repaired by a variety of enzymes that include DNA repair nucleases, DNA polymerase, DNA ligase, DNA glycosylases, AP endonuclease, phosphodiesterase and DNA helicase; putative genes for these enzymes have been identified in *C. elegans* but to date they have received little experimental study. As organisms age, they accumulate changes in their DNA (see, for example, a study in *C. elegans* (Klass, Nguyen and Dechavigny, 1983)), hypothesised to be caused, at least in part, by a reduced capacity to repair such changes. An age-related decline in excision repair capacity has been demonstrated in *Turbatrix aceti* (Targovnik *et al.*, 1984).

In contrast to DNA, RNA undergoes continuous synthesis and degradation as individual genes are transcribed and translated according to the requirements of the cell. RNA polymerases I, II and III have been demonstrated experimentally in *C. elegans* (Sanford, Golomb and Riddle, 1983; Sanford, Prenger and Golomb, 1985; Bird and Riddle, 1989; Dalley *et al.*, 1993; Powell-Coffman, Knight and Wood, 1996; Chervitz *et al.*, 1998; Chitwood, 1998).

The substrates for nucleic acid synthesis, the pyrimidine and purine nucleotides, are provided by biosynthetic and salvage pathways, which are discussed below. Purine or pyrimidine bases are coupled to ribose 5-phosphate to form nucleotides. Ribose 5-phosphate, a product of the pentose phosphate pathway (see above), is converted to 5'-phosphoribosyl-1-pyrophosphate (PRPP), which is transferred to the pyrimidine bases or their precursors (e.g. orotic acid), or transferred between purine or pyrimidine bases by specific phosphoribosyl transferases. Deoxyribonucleotides are formed by reduction of nucleotide diphosphates, a reaction catalysed by ribonucleotide reductase. These processes have received scant attention in nematodes.

Biosynthesis and Salvage of Pyrimidine Nucleotides

Pyrimidine nucleotides may be derived from *de novo* synthesis of pyrimidine bases followed by metabolism to nucleotides, or from salvage of preformed bases or derivatives available in the environment. *De novo* synthesis of pyrimidine nucleotides from the precursors carbon dioxide, glutamine or ammonia, and aspartate to 5'-UMP, the precursor for the synthesis of uridine, cytidine and deoxythymidine nucleotides, has been demonstrated by isotope labelling studies in only a small number of nematodes, e.g. *A. suum* ovary and *A. cantonensis* (Aoki *et al.*, 1980; So, Wong and Ko, 1992; So, Wong and Ko, 1993), but the apparent absence or low activity of *de novo* synthesis in other nematodes should be interpreted with caution as *in vitro* incubation conditions for parasites generally do not support active growth and development. Nutritional studies with *S. glaseri in vitro* demonstrated no requirement for exogenous pyrimidines, implying that the nematodes synthesise their requirements *de novo* (Jackson and Platzer, 1974).

Six enzymes catalyse *de novo* UMP synthesis: carbamoyl phosphate synthetase, aspartate transcarbamoylase, dihydro-orotase, orotate reductase, orotate phosphoribosyltransferase and orotidine 5'-phosphate decarboxylase. All five of the soluble enzymes have been detected in *N. brasiliensis* and *T. muris* (Hill *et al.*, 1981); the sixth, orotate reductase, which is a membrane protein, was not tested. Carbamoyl phosphate synthetase II, aspartate transcarbamoylase and dihydroorotase activities have been detected in *A. suum* ovary cytosolic preparations (Aoki *et al.*, 1980). In animals the activity of carbamoyl phosphate synthetase is low and highly regulated such that the reaction is considered to be potentially rate-limiting in the pyrimidine biosynthetic pathway. The activity of carbamoyl phosphate synthetase II in *Ascaris* ovary is low and its regulatory properties are consistent with an important regulatory role in this tissue (Aoki *et al.*, 1980). The enzyme has also been detected in *A. cantonensis* (see Aoki *et al.*, 1980) and free-living *P. redivivus* (Wright, 1975b).

Synthesis and interconversion of pyrimidines from preformed precursors — the salvage pathways that utilise pyrimidine bases (uracil) and other intermediates of the pathways (orotic acid, uridine, deoxyuridine, cytidine) acquired from the environment or recycled within the organism — has been demonstrated in all free-living or parasitic nematodes examined and is probably ubiquitous (see e.g. Jaffe and Doremus, 1970; Thirugnanam and Myers, 1974; Chen and Howells, 1981a; So, Wong and Ko, 1993). Parasitic nematodes do not appear to utilise thymine, cytosine, or thymidine to any great extent, however (Farland and MacInnis, 1978a; Chen and Howells, 1979; Chen and Howells, 1981a; So, Wong and Ko, 1992), and dTMP, which is essential for DNA synthesis, is only formed via methylation of dUMP, the reaction being catalysed by thymidylate synthetase (So, Wong and Ko, 1992; So, Wong and Ko, 1994; Dabrowska *et al.*, 1996). Unlike eggs or ovarian tissue of *A. suum* (Farland and MacInnis, 1978a) and microfilariae of *D. immitis* (Jaffe and Doremus, 1970), adult female *B. pahangi* are capable of incorporating exogenous thymidine into DNA *in vivo* but not *in vitro*, and activity of thymidine kinase, which catalyses the first reaction in the salvage pathway, has been detected in adult females of both *B. pahangi* and *D. immitis* (Jaffe, Comley and Chrin, 1982). Whether the *in vivo/in vitro* difference occurs in other parasitic nematodes is an open question. The salvage and interconversion reactions are catalysed by specific pyrimidine phosphoribosyl transferases and pyrimidine nucleoside phosphorylases, phosphotransferases, kinases and deaminases in a network of complex interconnecting pathways that appear to vary in detail between species.

Biosynthesis and Salvage of Purine Nucleotides

The pathway of *de novo* synthesis of purines is a series of ten reactions in which the purine ring is assembled from glycine, formate, carbon dioxide, glutamine and aspartate. The first step in the sequence is the transfer of the amido group of glutamine to PRPP and the end product of the sequence is 5'-IMP, which is then converted to adenosine and guanosine nucleotides. The formate groups are supplied by formyl-tetrahydrofolate (formyl-THF). THF is derived from folic acid, an important vitamin which is discussed below. Direct evidence of *de novo* purine synthesis, obtained by isotope labelling, has been reported for a variety of nematodes, including adult *A. cantonensis* (Wong and Ko, 1979), adult *B. pahangi* and *D. immitis* (Jaffe, 1981), adult *Metastrongylus apri* (Wong and Yeung, 1981) and the free-living *Aphelenchoides rutgersi* (Thirugnanam and Myers, 1974). Microfilariae of *D. immitis*, however, do not exhibit *de novo* purine synthesis *in vitro* (Jaffe and Doremus, 1970), and Chen and Howells (1981a) achieved enigmatic results in microfilariae, larvae and adult *B. pahangi in vitro*: radiolabelled glycine, but not formate, was incorporated into nucleic acids. As with pyrimidines, *S. glaseri in vitro* did not require exogenous purines, implying that the nematodes synthesise their requirements *de novo* (Jackson and Platzer, 1974).

Uptake or utilisation of exogenous purines (adenine, adenosine, inosine, hypoxanthine, guanine) has been demonstrated in many nematodes, for example microfilariae, infective larvae, 10 day juveniles and adults of *B. pahangi* (Chen and Howells, 1981a), adults of *D. immitis* (Chen and Howells, 1979; Chen and Howells, 1981b), adults of *M. apri* (Wong and Yeung, 1981), and *Aphelenchoides rutgersi* (Thirugnanam and Myers, 1974). Incorporation of exogenous purines into phosphorylated derivatives has been confirmed by isotope labelling studies in a variety of nematodes, including *D. immitis* microfilariae (Jaffe and Doremus, 1970) and adults of *M. apri* (Wong and Yeung, 1981). Interconversion of purines was common, with IMP being the common intermediate between the adenosine and guanosine pathways.

The cyclic nucleotides cAMP and cGMP have been demonstrated in all nematodes so far tested, as would be expected given the fundamental role they play in signal transduction. Similarly, adenylate cyclase and guanylate cyclase are likely to be present in all nematodes and are represented by numerous genes in *C. elegans* (Baude *et al.*, 1997; Berger, Hart and Kaplan, 1998).

Catabolism of Nucleic Acids

Nucleic acids are degraded by endonucleases and phosphodiesterases to yield nucleotide monophosphates. These are cleaved by nucleotidases to yield inorganic phosphate and nucleosides, which are further cleaved by nucleoside phosphorylases to yield a base plus ribose-1-phosphate. Specific phosphodiesterases catalyse the hydrolysis of cAMP or cGMP to AMP or GMP. Not all organisms carry out the full range of these reactions and intermediates may be salvaged for recycling. If not salvaged, purine or pyrimidine bases may be further degraded and excreted. The major end product of degradation of purine bases is uric acid, which may be converted to urea for excretion; those of pyrimidines are β-ureidopropionate or β-ureidoisobutyrate or their derivative amino acids, β-alanine or β-aminoisobutyrate, plus ammonia and carbon dioxide.

These processes have not been studied in any detail in nematodes. Uric acid was detected in 5-day eggs of *A. suum* and accumulated during embryonation (Far-

land and MacInnis, 1978b). Xanthine, an intermediate in the degradation of purines, was also detected. Uric acid was reported not to be an end product of adult metabolism, however (Rothstein, 1970b), and purine degradation to uric acid was insignificant in *P. redivivus* (Wright, 1975a; Wright, 1975b). Most nematodes excrete nitrogen as ammonia, with some urea, but little of this is considered likely to be derived from purines (Wright and Newall, 1976).

Vitamins

Like all animals, helminths require vitamins but careful determination of essential dietary requirements by culturing in fully-defined media has not been done for most free-living nematodes and is not possible for most parasitic nematodes. Such *in vitro* studies have been carried out on *C. briggsae*, which requires folic acid, panthothenate, thiamine, riboflavin, pyridoxine, niacinamide, biotin and cobalamin, on *A. rutgersi*, which requires folic acid, and on *S. glaseri*, which requires folic acid and biotin (Sayre, Hansen and Yarwood, 1963; Jackson and Platzer, 1974; Thirugnanam and Myers, 1974; Nicholas, 1984). The metabolism of those vitamins that have received significant attention is discussed below.

Retinol

Although nematodes do not have conventional eyes or visual pigments they utilise retinol (vitamin A). *A. suum*, *B. pahangi* and *B. malayi* absorb retinol or retinoic acid directly from the culture medium and the former two species are also capable of metabolising carotene to retinol (Barrett, 1981; Comley and Jaffe, 1983; Wolff and Scott, 1995). *A. suum* and *O. volvulus* accumulate relatively high concentrations of retinol in their perienteric fluid (Juhász and Babos, 1969; Leutskaya, 1974; Stürchler, Wyss and Hanck, 1981; Sani and Comley, 1985; Sani *et al.*, 1985; Sani and Vaid, 1988). Interestingly, the distribution of retinol, a fat-soluble vitamin, in female *A. suum* organs did not correlate with lipid content of the organs (Juhász and Babos, 1969). Retinol or its derivatives appear to be important in development, reproduction and metabolism of filarial parasites (Sani and Comley, 1985; Sani *et al.*, 1985; Sani and Vaid, 1988) and a role for retinol in glycoprotein synthesis has been suggested for *B. pahangi* (Comley and Jaffe, 1983). Excess retinol is toxic to *C. elegans*, causing a decline in fecundity after several generations of exposure to high levels; the effect is due to changes in chromosome structure which affect gene expression and chromosome pairing during meiosis (Goldstein, 1997). Synthetic and natural retinoids inhibited motility and release of microfilariae in a variety of filarial nematodes *in vitro* (Zahner *et al.*, 1989), and inhibited the L3 to L4 moult in *O. lienalis in vitro* (Lok *et al.*, 1990).

Retinol, being lipid soluble, is normally transported within organisms in a non-covalent complex with carrier proteins. It was discovered relatively recently that nematode parasites of animals secrete two new classes of high-affinity fatty acid and retinol binding proteins (for review see McDermott, Cooper and Kennedy, 1999). The first class, called nematode polyprotein allergens/antigens (NPAs), are unusual for two reasons. First, structural predictions imply that they are composed largely of α-helix, unlike fatty acid binding proteins from other organisms that are largely β-sheet structures. Second, they are initially synthesised as polyproteins which are post-translationally cleaved to multiple diverse polypeptides, each of around 14.5 KDa (see e.g. Britton et al., 1995). They are found in the perienteric fluid and are also secreted; they bind both fatty acids and retinol and have been characterised from A. suum, B. malayi, B. pahangi, D. immitis, Dictyocaulus viviparus and Ascaridia galli (McGibbon et al., 1990; Poole et al., 1992; Tweedie et al., 1993; Britton et al., 1995; Kennedy et al., 1995; Timanova et al., 1999).

The second class, which is not synthesised as a polyprotein, is predicted also to be α-helix-rich but with a different structure from the 14.5 kDa proteins; it is around 20 kDa and has highest affinity for retinol or retinoic acid (Tree et al., 1995). Ov20 from O. volvulus is expressed in the body wall of females, in microfilariae and in 3rd and 4th-stage larvae, and is secreted into the culture medium by adult females. Similar molecules were demonstrated by hybridisation in O. gibsoni, Onchocerca gutturosa, B. malayi and A. viteae (Tree et al., 1995). Genes encoding putative retinol-binding proteins with sequences similar to Ov20 have been identified in B. malayi, as well as in the plant-parasitic nematode G. pallida, and in non-parasitic C. elegans (Kennedy et al., 1997).

Although their precise functions are unclear, these hydrophobic carrier proteins are likely to be important in transferring molecules between tissues within nematodes and also in acquiring or binding molecules in their environment. These proteins are of particular interest because they are important allergens in nematode infections of mammals and could serve as potential therapeutic targets.

Vitamin B$_{12}$

Vitamin B$_{12}$, or cobalamin, is synthesised de novo by bacteria but not by animals or plants. Cobalamin must be obtained in the diet and in animals is an essential coenzyme in two reactions, those catalysed by methylmalonylCoA mutase (as deoxyadenosylcobalamin) and methionine synthetase (as methylcobalamin). The former participates in the pathway of synthesis or oxidation of propionate and the latter in the salvage of methionine. MethylmalonylCoA mutase is likely to be present in most nematodes but methionine synthase to date has only been demonstrated in the free-living C. briggsae, which has an absolute requirement for vitamin B$_{12}$ when grown on medium lacking methionine but containing homocysteine (Lu, Hieb and Stokstad, 1976; for review see Weinstein and Jaffe, 1987). A. suum converts cobalamin to adenosylcobalamin rapidly (Oya and Weinstein, 1975) and free-living larvae of N. brasiliensis accumulate large quantities of

cobalamin when feeding on bacteria in rat faeces (Weinstein, 1996).

Folic Acid

In animals folate is an essential vitamin that is metabolised to tetrahydrofolate (THF) which participates as a cofactor in one-carbon transfer reactions. Folate is reduced to dihydrofolate and then THF by two reactions catalysed by dihydrofolate reductase. THF is a very useful cofactor in intermediary metabolism as it can transfer one-carbon units at several different oxidation states (Voet, Voet and Pratt, 1999). Important biosynthetic one-carbon transfer reactions in which THF participates include those catalysed by thymidylate synthetase (as N^5N^{10}-methylene-THF) and methionine synthetase (as N^5-methyl-THF); one-carbon units are acquired through reactions catalysed by serine hydroxymethyltransferase or glycine cleavage (as N^5N^{10}-methylene-THF), and glutamate formiminotransferase (as N^5-formimino-THF).

It is likely that all nematodes require exogenous folate (for review see Weinstein and Jaffe, 1987). Dihydrofolate reductase has been demonstrated in a variety of free-living and parasitic nematodes (see Barrett, 1981). In adult B. pahangi and D. immitis tetrahydrofolate contributes carbons 2 and 8 of the purine ring during de novo purine synthesis (Jaffe, 1981), and the enzymes catalysing the interconversions of the THF cofactors have been examined in these parasitic species as well as in N. brasiliensis, S. glaseri and the free-living A. avenae (Jackson and Platzer, 1974; Platzer, 1974; Jaffe and Chrin, 1980; Jaffe, Chrin and Smith, 1980; Comley et al., 1981; Walker and Barrett, 1991).

Other Vitamins

Pyridoxine (vitamin B$_6$) is required as the cofactor pyridoxal 5'-phosphate for phosphorylases, aminotransferases and decarboxylases such as ornithine decarboxylase. The pyridoxal 5'-phosphate-dependent enzymes of N. brasiliensis were surveyed by Walker and Barrett (1991) who showed that the overall pattern of enzyme activities was similar to mammals, with some exceptions. In a study of L. sigmodontis (syn. carinii) in pyridoxal-deficient cotton rats, Beg and colleagues (1993; 1995; 1996) showed that parasite establishment, growth and embryogenesis was lower and that the activities of glycogen phosphorylase and alanine and aspartate aminotransferases were reduced. Development of Angiostrongylus costaricensis in vitro from third-stage larvae to young adults was significantly reduced by removing pyridoxine from the medium (Hata, 1994). Biotin is a coenzyme essential for carboxylases such as acylCoA carboxylase or propionylCoA carboxylase. It is a chemoattractant for C. elegans (Pierce, Morse and Lockery, 1999).

Detoxification Metabolism

Most organisms have mechanisms to protect their cells against damage from exogenous or endogenous toxic or unwanted molecules, and nematodes are no exception. Such molecules could include toxins secreted by other organisms, drugs or toxic chemicals in the environment, digestive enzymes of the host, signalling molecules, or molecules produced by the host's immune response, such as reactive oxygen or nitrogen intermediates (free radicals), proteases, antibodies, hydrolases, complement, or eosinophil granule proteins. This section will deal specifically with metabolic detoxification of exogenous chemicals and drugs, and of free radicals.

Animals metabolise foreign chemicals in three biochemical phases. In Phase I, lipophilic molecules are oxidised, reduced or hydrolysed to products that are more water-soluble and, usually, less toxic. Many Phase I reactions are catalysed by the mixed-function oxidase (MFO) system which is comprised of NADPH-cytochrome P450 reductase (a flavoprotein), cytochrome P450 and cytochrome b_5. The MFO system is inducible and capable of metabolising a wide variety of different substrates, due to the presence of a large family of inducible P450 haemoproteins, each with different substrate affinities and specificities. In animals there are currently 43 recognised genetic families of P450, each with numerous subfamilies, each in turn with numerous isoforms (Nelson, 1998). The oxidation reactions catalysed by P450 require NADPH and oxygen and include hydroxylation, epoxidation, N-oxidation, sulphooxidation, N-, S- and O-dealkylations, desulphation, dehalogenation and oxidative deamination. Other enzymes catalysing oxidation reactions of xenobiotics include alcohol dehydrogenase, aldehyde dehydrogenase, xanthine oxidase, amine oxidases and aromatases. The reductive reactions catalysed by P450 require NADH, NADH-cytochrome b_5 reductase, and cytochrome b_5. This system, which does not require oxygen, catalyses the reduction of azo, nitro and N-oxide groups. Enzymes catalysing hydrolysis reactions include a large variety of esterases.

In Phase II reactions, the products of Phase I, or other molecules already possessing suitable chemical groups, are conjugated to small polar molecules such as glucuronate, acetate, propionate, sulphate, to amino acids such as glycine or ornithine, or to glutathione, to increase hydrophilicity and facilitate excretion. These reactions are catalysed by transferases such as UDP-glucuronosyl transferases, sulphotransferases or glutathione S-transferases (GSTs). The products of these conjugation reactions may be further metabolised by Phase III reactions before excretion or storage. For example, glutathione conjugates tend to inhibit GST activity and must be removed efficiently; they may be converted to cysteine conjugates or mercapturic acids before excretion.

Detoxification metabolism in nematodes has not received extensive research despite its importance in chemical control of nematodes important in agriculture, or in chemotherapy and drug resistance in animal parasites. Recent reviews of detoxification metabolism and P450 have been written by Barrett and colleagues (Barrett, 1989a; Precious and Barrett, 1989b; Barrett, 1997; Barrett, 1998).

Phase I Metabolism

Many attempts have been made to detect cytochromes P450 and b_5 in nematodes, using either spectrophotometric, chemical or enzymatic assays, with or without a period of pre-exposure to substrates that induce expression of P450 (for reviews see Precious and Barrett, 1989a; 1989b; Barrett, 1998). For the parasitic stages of animal parasites this search has been largely unsuccessful, though P450 enzymatic activity, or the products thereof, have been demonstrated more recently in free-living larval stages of the parasites *H. polygyrus* and *H. contortus* (Kerboeuf et al., 1995; Kotze, 1997; 1999), and 80 predicted P450 genes have been identified in *C. elegans* (The *C. elegans* Sequencing Consortium, 1998), many of which are represented in the expressed sequence tag (EST) database. Of the 80 genes, 46 fall into a large grouping, termed a 'clan', containing P450 genes exclusively from *C. elegans* (Nelson, 1999). *C. elegans* possesses only a single mitochondrial P450 gene, and lacks many of the other genes for P450 molecules that normally catalyse sterol synthesis (Nelson, 1998; Nelson, pers. comm.), which could explain why *de novo* sterol biosynthesis does not occur in this nematode (see above). Sequences of genes or mRNAs putatively encoding cytochrome P450 are currently available in the public nucleotide databases for the free-living nematode *Pristionchus pacificus*, the plant parasites *G. rostochiensis* and *M. incognita*, the filarial parasite *B. malayi* and the insect parasite *Romanomermis culicivorax*, so it is clear that P450 is well represented in the phylum; whether it is important in detoxification is an open question.

The absence or low activity of P450-based oxidation in parasitic stages of animal parasites considerably limits their ability to detoxify foreign chemicals, including drugs, or to activate prodrugs by P450-linked oxidation (Precious and Barrett, 1989b; Barrett, 1997; Rothwell and Sangster, 1997), although they can catalyse oxidations not linked to P450. Nematodes are generally able to reduce or hydrolyse xenobiotics, however. Some or all of the following activities have been demonstrated in *A. suum* adults, *H. polygyrus* adults, *O. gutturosa* adults and *P. redivivus*: sulphoxidase, sulphoxide reductase, azo-reductase, nitroreductase, O-deacetylase, and N-deacetylase (for reviews see Barrett, 1981; Barrett, 1989a; Precious and Barrett, 1989b; Barrett, 1997). The inhibition profiles of the azo- and nitro-reductases from *A. suum* were different from mammals (Douch, 1975) and there is some variation in enzyme properties between species. Reduction by nematodes of lipid peroxidation-derived carbonyls is discussed below.

Phase II Metabolism

The major enzyme-dependent conjugation reactions detected in most parasitic nematodes are those catalysed by the glutathione S-transferases (GSTs) (reviewed by Brophy and Pritchard, 1994; Barrett, 1995; 1997), although metabolism of ecdysteroids by *A. suum* may involve conjugations with glucose, sulphate or phosphate (O'Hanlon et al., 1991) and N-acetyltransferase activity has been detected in A. galli (Muimo and Isaac, 1993) and Enterobius vermicularis (Chung, 1999). Two putative genes coding for UDP-glucuronosyl transferase (F56B3.7, C08B6.1), plus cognate cDNAs, have been annotated in the C. elegans genome sequence but there has been no experimental study of the enzymes. As well as catalysing conjugation to the tripeptide glutathione, GSTs also act directly as noncovalent binding proteins to sequester and transport a wide range of exogenous or endogenous hydrophobic ligands, other molecules such as haem, neurotransmitters or bile acids, and heavy metals (for reviews see Brophy and Barrett, 1990a; Barrett, 1995).

GST conjugating activity has been demonstrated in a variety of animal-parasitic nematodes (A. suum, Ascardia galli, T. canis, Necator americanus, Ancylostoma ceylanicum, H. contortus, T. colubriformis, N. brasiliensis, Cooperia oncophora, H. polygyrus, D. immitis, B. pahangi, O. gutturosa, S. feltiae) and in P. redivivus (Jaffe and Lambert, 1986; Pemberton and Barrett, 1989; Brophy, Crowley and Barrett, 1990; Brophy and Pritchard, 1992; Brophy et al., 1995a). The activities in free-living P. redivivus and entomopathogenic S. feltiae were significantly higher than those in the animal parasites (Brophy, Crowley and Barrett, 1990).

The enzyme generally occurs as a homo- or heterodimer and a complex superfamily of GST isoenzymes with distinct, but overlapping, substrate specificities is present in most organisms, including nematodes, coded for by multi-gene families (for review see Snyder and Maddison, 1997). The genomic and cDNA sequences of GSTs from C. elegans await detailed analysis but there are a large number of putative GST genes and cDNAs currently annotated in Genbank, some grouped into clusters on the same cosmids (e.g. F11G11, K08F4, F37B1). Two of these genes, gst-1 (R107.7) and R07B1.4, have been classified on the basis of their sequences into the π and σ classes, respectively. Numerous partial or complete GST genomic, mRNA or peptide sequences are also available for a variety of plant-parasitic (M. incognita, G. rostochiensis), free-living (C. elegans, P. pacificus) and animal-parasitic (N. dubius = H. polygyrus, A. suum, D. immitis, O. volvulus, B. malayi, A. galli) nematodes in which, again, more than one GST gene is expressed. It is likely that more will be found in these nematodes.

Many of the nematode enzymes tend not to group clearly into any of the major classes of GST found in mammals and/or insects (α, μ, π, σ, θ, ζ) and have characteristics of several classes (e.g. Liebau et al., 1996; Barrett, 1997). GSTs have quite diverse functions and many are expressed in a tissue-specific way. For example, GSTs with both glutathione transferase and selenium-independent glutathione peroxidase activity against alkyl hydroperoxides, a characteristic of the α class, have been detected by enzymatic activity in D. immitis (Jaffe and Lambert, 1986), H. polygyrus (Brophy et al., 1994), and T. spiralis L1s (Rojas, Rodriguez and Gomez, 1997). GST-1 from A. suum is expressed in the intestine where it may interact with exogenous molecules from the diet; the recombinant protein has peroxidase as well as transferase activity (Liebau et al., 1994a; Liebau et al., 1997). A recombinant π-class GST from O. volvulus, rOvGST2, shows low peroxidase activity and is expressed in the syncytial epidermis of adults, where it is proposed to have a 'cytosolic housekeeping' role, whereas another O. volvulus GST, rOvGST1, is expressed in the outer epidermis where it would protect the nematode against exogenous toxins (Liebau et al., 1994b; Salinas, Braun and Taylor, 1994; Liebau et al., 1996; Wildenburg, Liebau and Henkle-Dührsen, 1998). Similar localisation of immunoreactivity to rOvGST2 was observed in Onchocerca ochengi (Liebau et al., 1996). Since O. volvulus adults probably have little intestinal function (Striebel, 1988), the epidermis has apparently assumed the normal detoxification functions of the gut. Four isoforms of GST were purified from Ditylenchus myceliophagus but their properties did not conform closely to those of the mammalian α or μ classes (Persaud, Perry and Barrett, 1997). A σ-class GST, with glutathione-dependent prostaglandin-H E-isomerase activity, was purified from A. galli (Meyer et al., 1996); this enzyme may be important in suppressing host immunity. A GST with leukotriene C4 synthase activity has been purified from adult D. immitis (Weller, Longworth and Jaffe, 1989).

The enzyme in mammals and insects is inducible by exposure to potential substrates. mRNA coding for C. elegans GST-1 has been shown by differential display to be induced by paraquat, which induces oxidative stress, in adults and larvae (Tawe et al., 1998). GSTs have been implicated in drug or pesticide resistance in mammalian cancer cells and a variety of organisms but this has not yet been unequivocally demonstrated in nematodes. Elevated GST activities were correlated with cambendazole resistance in a strain of H. contortus but it is not clear whether the differences observed represent normal biological variation between strains or are directly related to drug resistance (Kawalek, Rew and Heavner, 1984).

GSTs are also important as antioxidants in many organisms, including nematodes. In parasitic nematodes selenium-independent glutathione peroxidase activity has been observed which may function to remove the toxic hydroperoxide products of lipid and nucleic acid peroxidation that are induced by free radical attack by immune cells of the host. Adult N. americanus (Brophy et al., 1995b) and the filarial worms O. volvulus and B. pahangi secrete GST into the incubation medium in vitro (Liebau et al., 1994b; Tang et al., 1996). These secreted GSTs may function just as binding proteins to sequester toxic molecules; if they were to function as transferases or peroxidases they would require glutathione, derived either from the host or secreted by the parasite, for which there is no evidence. Antibodies to T. spiralis GST have been detected in the serum of infected rabbits, and circulating T. spiralis GST has also been detected in infected rabbit serum by western blots (Rojas, Rodriguez and Gomez, 1997). Anguilicolla crassus, a parasite of eels, secretes proteins identified immunochemically as GST subunits into the medium *in vitro*; the same protein is recognised by serum from infected eels (Nielsen and Buchmann, 1997).

There is no direct evidence in nematodes of Phase III metabolism of glutathione conjugates, so it is possible that they are excreted unchanged by as yet undescribed transport processes. Although it is not known whether they participate in their metabolism, two enzymes that in other organisms catalyse the Phase III processing of glutathione conjugates, cysteine conjugate β-lyase and γ-glutamyl transpeptidase, have been identified in *N. americanus* and *H. polygyrus* (Adcock *et al.*, 1999), and *A. suum* (Dass and Donahue, 1986; Hussein and Walter, 1996), respectively.

Binding Proteins

The secretion of proteins that bind potentially toxic molecules is a common detoxification mechanism. The role of GST has been discussed above, as has the subject of binding proteins for fatty acids and retinol in nematodes. The latter proteins, some of which are secreted, bind a wide range of fatty acids and their derivatives (McDermott, Cooper and Kennedy, 1999). One of the functions of these molecules could include the sequestration of potentially toxic lipids and their peroxidation products (Mei *et al.*, 1997). The nucleophilic tripeptide glutathione also binds directly to electrophilic compounds such as free radicals and peroxides. Metallothioneins are a class of inducible small, cysteine-rich proteins that specifically bind metal ions, functioning both to sequester toxic metals such as cadmium, mercury or lead and to accumulate important trace elements such as copper or zinc. They are induced by exposure to heat shock, transition elements, ionising radiation and oxidative stress (Palmiter, 1998). In nematodes they have been investigated in detail only in *C. elegans* although cDNA sequences are currently available in Genbank for *G. rostochiensis*, *G. pallida* and *A. caninum*. *C. elegans* possesses two cadmium-inducible genes, *mtl-1* and *mtl-2*, which are expressed in intestinal cells (Freedman *et al.*, 1993; Liao and Freedman, 1998).

Antioxidants

Reactive oxygen intermediates (ROI) such as hydrogen peroxide, superoxide and hydroxyl radical are produced in all aerobic cells as a by-product of cellular reduction of oxygen. Both ROI and reactive nitrogen intermediates (RNI; nitric oxide and its derivatives nitrogen dioxide and peroxynitrite) are also produced by cells of the immune response of plant and animal hosts of parasitic nematodes, or may be generated by drug treatments. ROI and RNI damage proteins, nucleic acids and membrane lipids and may kill nematodes. Thus, whether parasitic or not, probably all nematodes are exposed to reactive oxidants and require antioxidant protection mechanisms. The major secondary toxic products generated by ROI or RNI attack on cells are reactive lipid hydroperoxides derived from unsaturated membrane lipids.

Antioxidant enzymes include catalase, superoxide dismutases, glutathione peroxidase, peroxiredoxins and ascorbate peroxidase; non-enzymatic antioxidants include vitamins E (α-tocopherol) and C (ascorbate), GSTs, glutathione, ubiquinols and albumin. Antioxidant mechanisms in parasites have been reviewed by Callahan, Crouch, and James (1988) and more recently by Selkirk and colleagues (1998) for the filarial nematodes.

Antioxidant enzymes

All nematodes that have been examined possess superoxide dismutase (SOD) (Callahan, Crouch and James, 1988; Hadas and Stankiewicz, 1998; Selkirk *et al.*, 1998), which quenches superoxide by converting it to hydrogen peroxide. Several forms of the enzyme exist in eukaryotes: MnSOD is expressed in mitochondria whereas Cu/ZnSOD is cytosolic or secreted. Some, and possibly all, nematodes upregulate expression of SOD when subjected to oxidative stress (Tawe *et al.*, 1998), or when establishing in a host (Lattemann, Matzen and Apfel, 1999). Increased expression of Cu/Zn SOD correlated with increased lifespan in *C. elegans age-1* mutants (Vanfleteren, 1993), while decreased activity of SOD correlated with decreased lifespan in *mev-1*(kn1) mutants (Matsuo, 1993). The increased longevity of the *C. elegans daf-2* mutation correlated with increased expression of *sod-3*, the mitochondrial Mn-SOD (Honda and Honda, 1999). Some animal- and plant-parasitic nematodes (e.g. adult *B. pahangi*, *B. malayi* (Tang *et al.*, 1994; Ou *et al.*, 1995a), *D. immitis* (Callahan *et al.*, 1993), *O. volvulus* (James, McLean and Perler, 1994), *Setaria cervi*, *L. carinii* (Batra, Chatterjee and Srivastava, 1992), *A. viteae* (Lattemann, Matzen and Apfel, 1999), *H. contortus* (Liddell and Knox, 1998), *N. americanus* (Taiwo *et al.*, 1999), *D. viviparus* (Britton, Knox and Kennedy, 1994), *G. rostochiensis* (Robertson, Robertson and Jones, 1999) secrete SOD into their internal fluids and also their external environment. Externally-secreted SOD may serve to detoxify host-derived superoxide, provided catalase or a peroxidase is also present to remove the resultant hydrogen peroxide. Alternatively, the hydrogen peroxide produced could attack host tissues (Brophy, Patterson and Pritchard, 1995).

Catalase, which reduces hydrogen peroxide, is also common (Callahan, Crouch and James, 1988; Lesoon, Komuniecki and Komuniecki, 1990; Molinari and Miacola, 1997; Eckelt *et al.*, 1998; Taub *et al.*, 1999), but not universal (Callahan, Crouch and James, 1991), in nematodes and was considered the major antioxidant responsible for protecting *B. malayi* adults from exposure to high levels of hydrogen peroxide *in vitro* (Ou *et al.*, 1995b). It may be secreted by *S. cervi* and *L. carinii* (Batra, Chatterjee and Srivastava, 1992). Ascorbate peroxidase activity is reported to be common in plant-parasitic nematodes (Molinari and Miacola, 1997) but there appears to be no enzymatic or sequence record of this enzyme in animal-parasitic or free-living nematodes.

Important products of lipid peroxidation are carbonyl compounds such as cytotoxic aldehydes. Nematodes have a number of mechanisms for detoxifying these molecules. In some, but not all, nematodes alk-2-enal and alk-2,4-dienal products are conjugated to glutathione in reactions possibly catalysed by GST, although they are also sufficiently reactive to bind covalently to glutathione without enzyme catalysis (Brophy and Barrett, 1990b; Brophy and Pritchard, 1992; Liebau *et al.*, 1996). Minor routes of metabolism of these products are Phase I reduction by NAD(P)H-linked activities, which have been demonstrated in *A. suum, H. contortus, H. polygyrus, N. americanus, A. ceylanicum* and *P. redivivus* (Brophy and Barrett, 1990b; Brophy and Pritchard, 1992). *P. redivivus* and *H. contortus* can also oxidise aldehydes by NAD(P)-linked activities (Brophy and Barrett, 1990b). Alkanals, on the other hand, appear not to be conjugated to glutathione in nematodes and are metabolised by NAD(P)H-linked reductions (Brophy and Barrett, 1990b; Brophy and Pritchard, 1992). The NAD(P)-linked reduction and oxidation activities were significantly higher in free-living *P. redivivus* than in *A. suum* or *H. contortus*. Glyoxalase I and II, which require glutathione and catalyse the conversion of 2-oxoaldehydes to 2-hydroxy acids, have been detected in all nematodes tested to date — *O. gutturosa* adults, *A. suum, A. galli, T. canis, H. contortus, T. colubriformis, N. brasiliensis, C. oncophora, S. feltiae* and *P. redivivus* (Pemberton and Barrett, 1989; Brophy, Crowley and Barrett, 1990)).

Glutathione peroxidases are enzymes that catalyse the glutathione-dependent reduction of both hydrogen peroxide and fatty acid peroxides (E.C. 1.11.1.9, commonly called cytosolic glutathione peroxidase), or phospholipid and other organic hydroperoxides (E.C. 1.11.1.12, commonly called phospholipid hydroperoxide glutathione peroxidase). Most glutathione peroxidases have a selenocysteine residue in the active site, and are thus termed selenium-dependent. Recently, however, selenium-independent, secreted glutathione peroxidases have also been identified; in these cysteine has replaced selenocysteine in the active site. As noted above, some nematode GSTs also have lipid hydroperoxidase activity — these, too, are selenium-independent.

Glutathione peroxidase activity has not been demonstrated in plant-parasitic nematodes (Molinari and Miacola, 1997). Activity with hydrogen peroxide as substrate is lacking in microfilariae of *Onchocerca cervicalis* (Callahan, Crouch and James, 1990), adults of *D. immitis* (Callahan, Crouch and James, 1991) and adults and microfilariae of *B. malayi* (Ou *et al.*, 1995b), but is present in a variety of other animal-parasitic nematodes (Callahan, Crouch and James, 1988; Batra, Chatterjee and Srivastava, 1990; Batra *et al.*, 1990; Batra, Chatterjee and Srivastava, 1992; Brophy and Pritchard, 1992). It is not clear in these studies whether the activities detected were catalysed by selenium-dependent or — independent enzymes. At least five putative glutathione peroxidase genes are annotated in the *C. elegans* genome sequence and a variety of cognate cDNA clones have been identified; they appear to be selenium-independent (Buettner, Harney and Berry, 1999; Gladyshev *et al.*, 1999) and thus may have low peroxidase activity. A selenium-independent, secreted enzyme with relatively low glutathione peroxidase activity is present in filarial nematodes, where it is mainly located in the matrix of the cuticle (e.g. *B. pahangi* (Cookson, Blaxter and Selkirk, 1992), *B. malayi, Wuchereria bancrofti* (Cookson, Tang and Selkirk, 1993), *D. immitis* (Tripp *et al.*, 1998)); the recombinant *B. pahangi* enzyme reduces lipid hydroperoxides but not hydrogen peroxide (Tang *et al.*, 1995; Tang *et al.*, 1996), whereas that from *D. immitis* also reduces hydrogen peroxide (Tripp *et al.*, 1998). A similar selenium-independent, apparently secreted, enzyme has been reported from genomic sequence and cDNA evidence for *C. elegans* (see Selkirk *et al.*, 1998). The filarial enzyme may function to repair membranes damaged by free radicals (Tang *et al.*, 1996; Selkirk *et al.*, 1998). At present, however, it is not clear (i) whether glutathione peroxidase activity is the primary function of these selenium-independent glycoproteins, (ii) whether glutathione is the major substrate *in vivo* and, if so, (iii) whether it originates from host or parasite, and (iv) how the oxidised glutathione that would be produced is reduced for recycling, as no recycling system is present in plasma, and glutathione reductase has not been detected in the cuticle of adult filariae (Selkirk *et al.*, 1998).

The peroxiredoxins are an important class of enzymes catalysing the reduction of hydrogen peroxide and alkyl hydroperoxides in nematodes (McGonigle, Dalton and James, 1998; Selkirk *et al.*, 1998). They have been discovered only relatively recently, primarily due to their high abundance in cDNA libraries prepared for the genome sequencing projects (Blaxter *et al.*, 1996; Lu *et al.*, 1998), and may play a critical role in detoxifying hydrogen peroxide in nematodes. Peroxidoxin cDNA sequences are now available in Genbank for a variety of animal-parasitic and free-living nematodes, including *C. elegans*. Two related classes of enzyme have been detected in nematodes, the 1-Cys peroxiredoxins, which require a small molecule such as dithiothreitol to donate electrons, and the 2-Cys peroxiredoxins (also called thioredoxin peroxidase or thiol-specific antioxidant) which accept electrons from thioredoxin.

At least one protein of each class is present in filarial nematodes. A recombinant 1-Cys peroxiredoxin from *D. immitis*, rDiPrx-1, reduces hydrogen peroxide (Chandrashekar *et al.*, 2000), but whether nematode recombinant 2-Cys peroxiredoxins reduce hydrogen peroxide directly has not been tested. The *D. immitis* 1-Cys enzyme was present in L3, L4 and adults as well as being secreted by adults and L4. A recombinant 2-Cys peroxiredoxin from *D. immitis*, nDiTPx, protected DNA from oxidative nicking in the presence of dithiothreitol; the native protein is expressed in all life cycle stages but secreted only by adults (Klimowski, Chandrashekar and Tripp, 1997). Several 2-Cys peroxiredoxins are expressed by *B. malayi*; rBm-TPx-1 exhibited antioxidant activity and the protein, which appeared to be cytosolic, was expressed at all life-cycle stages but was not exposed to the surface or secreted by adult worms (Ghosh *et al.*, 1998). *Bm*-TPx-2 possesses a putative mitochondrial import transit peptide. A recombinant 2-Cys thioredoxin peroxidase from *O. volvulus* (called OvPXN-2 or OvTPx-2 in the publications (Lu *et al.*, 1998; Schrum *et al.*, 1998) but Ov-tpx-1 in

Genbank AF029247) has antioxidant activity and the native protein is expressed in L1, L3 and adults; it is present on the surface, in epidermis, intestine and uterus, as well as being secreted into the uterine lumen where it bathes developing embryos and microfilariae (Lu *et al.*, 1998). A recombinant 1-Cys thioredoxin peroxidase has also been described from *O. volvulus* (Chandrashekar *et al.*, 1998); it has antioxidant activity and is expressed in the lateral epidermis.

Both glutathione and thioredoxin participate in detoxification reactions in their reduced form. After oxidation the reduced forms are regenerated via the action of NADPH-linked glutathione reductase and thioredoxin reductase, respectively. Both these enzymes are present in nematodes but to date have received little detailed study. Glutathione reductase from the filarial nematodes *L. carinii*, *Setaria digitata* and *O. gutturosa* are exceptionally sensitive to inhibition by organic arsenicals (Bhargava *et al.*, 1983; Müller, Walter and Fairlamb, 1995) but this was not the case with the purified enzyme from *A. suum* muscle (Komuniecki *et al.*, 1992). Significant glutathione reductase activity is present in intestine, ovaries and body wall muscle of adult *A. suum*; the enzyme purified from muscle had catalytic properties similar to the mammalian enzyme (Komuniecki *et al.*, 1992), as did the native filarial enzymes examined by Müller and colleagues (1995). The recombinant glutathione reductase from *O. volvulus* has been characterised and shown to utilise NADPH in preference to NADH (Müller *et al.*, 1997). In *C. elegans* two genes encoding putative thioredoxin reductases have been identified (Gladyshev *et al.*, 1999); one encodes a selenoprotein, as potentially is a homologue from *O. volvulus* (Buettner, Harney and Berry, 1999). The functional role of these enzymes remains to be determined.

Potential mechanisms of protection against host-derived nitric oxide have been discussed by Selkirk and colleagues (1998) but have not yet been well-studied in nematodes. By removing superoxide, SOD aids in detoxification of nitric oxide by preventing formation of peroxynitrite and thus preventing consequent lipid peroxidation and oxidation of sulphydryls. An interesting recent development is the discovery that the haemoglobin in perienteric fluid of *A. suum* could function as an enzyme, catalysing a nitric oxide 'deoxygenase' activity (Minning *et al.*, 1999; see Barrett and Brophy, 2000). This molecule, which is present in *Ascaris* adults but not in larval stages, normally binds oxygen so tightly that it would effectively function as an irreversible oxygen sink. The proposed enzymatic activity of haemoglobin would catalyse the reaction of nitric oxide, NADPH and oxygen to produce nitrate, effectively removing both nitric oxide and oxygen from the perienteric fluid. This activity could potentially remove oxygen, which is toxic to this anaerobic nematode, and also nitric oxide, which might be generated endogenously or potentially be derived from the host. The secretion of prostanoids by microfilariae of *B. malayi* has been postulated to modify the local immune response, which may include down-regulation of nitric oxide production (Liu, Serhan and Weller,

1990); NO generated by activated macrophages may play an important role in killing *B. malayi* microfilariae (Taylor *et al.*, 1996).

Chemical antioxidants

Non-protein chemical antioxidants have received little attention in nematodes although α-tocopherol has been identified as a major antioxidant in the cuticle of adults and microfilariae of *B. malayi* (Smith, Selkirk and Gounaris, 1998). Since animals are unable to synthesise α-tocopherol it is assumed to be acquired from the host. Supplementation of the culture medium of *C. elegans* with moderate levels of α-tocopherol extends the lifespan of the nematodes but higher levels decrease fecundity (Zuckerman and Geist, 1983; Harrington and Harley, 1988). *C. elegans* is able to survive and reproduce in culture in 100% oxygen only if α-tocopherol is included in the medium, and α-tocopherol also protected the worms against oxidative stress induced by paraquat (Goldstein and Modric, 1994). Inclusion of ascorbate, water-soluble antioxidant, in the culture suppressed induction of the small heat shock protein (*hsp-16-2*) in response to superoxide-generating quinones in *C. elegans*.

Conclusions and Future Prospects

We have still a lot to learn about nematode metabolism, but the future is very bright. With the availability of the complete genome sequence of *C. elegans* and the imminent availability of large collections of EST clones for nematodes of agricultural, veterinary or medical significance, and the advent of microarray technology, the probability of increasing our insight into nematode metabolism is very high, even if the funding available for the 'traditional' biochemical approaches to nematode biochemistry remains limited. It is now possible for nematologists to contribute to the field even without a laboratory: with just a suitable computer and internet access there is considerable work available just to analyse the sequence and expression data that are being generated and deposited in the public databases faster than they can be analysed.

Abbreviations

ATP, adenosine 5′-triphosphate; ADP, adenosine 5′-diphosphate; AMP, adenosine 5′-monophosphate; CoA, coenzyme A; cAMP, cyclic 3′,5′-adenosine monophosphate; DNA, deoxyribonucleic acid; dTMP, deoxythymidine 5′-monophosphate; EST, expressed sequence tag; $FADH_2$, reduced flavin adenine dinucleotide; GDP, guanosine 5′-diphosphate; GMP, guanosine 5′-monophosphate; cGMP, cyclic 3′,5′-guanosine monophosphate; GSH, reduced glutathione; GST, glutathione S-transferase; GTP, guanosine 5′-triphosphate; IMP, inosine 5′-monophosphate; mtDNA, mitochondrial DNA; NAD^+, nicotinamide adenine dinucleotide; NADH reduced nicotinamide adenine dinucleotide; $NADP^+$, nicotinamide adenine dinucleotide phosphate; NADPH reduced nicotinamide adenine dinucleotide phosphate;

NPA, nematode polyprotein allergen/antigen; PRPP, 5′-phosphoribosyl-1-pyrophosphate; RNA, ribonucleic acid; RNI, reactive nitrogen intermediate; ROI, reactive oxygen intermediate; THF, tetrahydrofolate; SOD, superoxide dismutase; UMP, uridine 5′-monophosphate.

Bibliography

Abu Hatab, M., Gaugler, R. and Ehlers, R.U. (1998) Influence of culture method on *Steinernema glaseri* lipids. *Journal of Parasitology*, **84**, 215–221.

Adcock, H.J., Brophy, P.M., Teesdale-Spittle, P.H. and Buckberry, L.D. (1999) Cysteine conjugate β-lyase activity in three species of parasitic helminth. *International Journal for Parasitology*, **29**, 543–548.

Allen, P.C. (1973) Helminths: comparison of their rhodoquinone. *Experimental Parasitology*, **34**, 211–219.

Antebi, A., Culotti, J.G. and Hedgecock, E.M. (1998) *daf-12* regulates developmental age and the dauer alternative in *Caenorhabditis elegans*. *Development*, **125**, 1191–1205.

Aoki, T., Oya, H., Mori, M. and Tatibana, M. (1980) Control of pyrimidine biosynthesis in the *Ascaris* ovary: regulatory properties of glutamine-dependent carbamoyl-phosphate synthetase and copurification of the enzyme with aspartate carbamoyltransferase and dihydroorotase. *Molecular and Biochemical Parasitology*, **1**, 55–68.

Arevalo, J. and Saz, H.J. (1991) Phospholipids and protein kinase C in acetylcholine-dependent signal transduction in *Ascaris suum*. *Molecular and Biochemical Parasitology*, **48**, 151–161.

Armson, A. and Mendis, A.H. (1995) Steady-state content of glycolytic/tricarboxylic acid-cycle intermediates, adenine nucleotide pools and the cellular redox-status in the infective (L3) larvae of (homogonic) *Strongyloides ratti*. *International Journal for Parasitology*, **25**, 197–202.

Bankov, I., Timanova, A. and Barrett, J. (1998) Sphingomyelin synthesis in helminths: a minireview. *Folia Parasitologica*, **45**, 257–260.

Barrett, J. (1981) *Biochemistry of Parasitic Helminths* London: MacMillan.

Barrett, J. (1983) Biochemistry of filarial worms. *Helminthological Abstracts*, **A52**, 1–18.

Barrett, J. (1984) The anaerobic end-products of helminths. *Parasitology*, **88**, 179–198.

Barrett, J. (1989a) Detoxification reactions in parasitic helminths. In *Comparative Biochemistry of Parasitic Helminths*, edited by E.-M. Bennet, C.A. Behm and C. Bryant, pp. 109–114. London: Chapman and Hall.

Barrett, J. (1989b) Parasitic helminths. In *Metazoan Life without Oxygen*, edited by C. Bryant, pp. 146–164. London: Chapman and Hall.

Barrett, J. (1991a) Amino acid metabolism in helminths. *Advances in Parasitology*, **30**, 39–105.

Barrett, J. (1991b) Studies on hexokinase from the filarial worms *Brugia pahangi*, *Onchocerca gutturosa* and *O. lienalis*. *Parasitology Research*, **77**, 183–184.

Barrett, J. (1995) Helminth glutathione transferases. *Helminthologia*, **32**, 125–128.

Barrett, J. (1997) Helminth detoxification mechanisms. *Journal of Helminthology*, **71**, 85–89.

Barrett, J. (1998) Cytochrome P450 in parasitic protozoa and helminths. *Comparative Biochemistry and Physiology*, **121C**, 181–183.

Barrett, J. and Beis, I. (1973) Studies on glycolysis in the muscle tissue of *Ascaris lumbricoides* (Nematoda). *Comparative Biochemistry and Physiology*, **44B**, 751–762.

Barrett, J. and Brophy, P.M. (2000) *Ascaris* haemoglobin: new tricks for an old protein. *Parasitology Today*, **16**, 90–91.

Barrett, J., Ward, C.W. and Fairbairn, D. (1970) The glyoxylate cycle and the conversion of triglycerides to carbohydrates in developing eggs of *Ascaris lumbricoides*. *Comparative Biochemistry and Physiology*, **35**, 577–586.

Barrett, J. and Wright, D.J. (1998) Intermediary metabolism. In *The Physiology and Biochemistry of Free-living and Plant-parasitic Nematodes*, edited by R.N. Perry and D.J. Wright, pp. 331–353. Wallingford: CAB International.

Batra, S., Chatterjee, R.K. and Srivastava, V.M. (1990) Antioxidant enzymes in *Acanthocheilonema viteae* and effect of antifilarial agents. *Biochemical Pharmacology*, **40**, 2363–2369.

Batra, S., Chatterjee, R.K. and Srivastava, V.M. (1992) Antioxidant system of *Litomosoides carinii* and *Setaria cervi*: effect of a macrofilaricidal agent. *Veterinary Parasitology*, **43**, 93–103.

Batra, S., Singh, S.P., Gupta, S., Katiyar, J.C. and Srivastava, V.M. (1990) Reactive oxygen intermediates metabolizing enzymes in *Ancylostoma*

ceylanicum and *Nippostrongylus brasiliensis*. *Free Radical Biology and Medicine*, **8**, 271–274.

Baude, E.J., Arora, V.K., Yu, S., Garbers, D.L. and Wedel, B.J. (1997) The cloning of a *Caenorhabditis elegans* guanylyl cyclase and the construction of a ligand-sensitive mammalian/nematode chimeric receptor. *Journal of Biological Chemistry*, **272**, 16035–16039.

Beg, M.A., Fistein, J.L., Ingram, G.A. and Storey, D.M. (1996) Activities of glycogen phosphorylase, alanine aminotransferase and aspartate aminotransferase in adult worms of *Litomosoides carinii* recovered from pyridoxine deficient cotton rats (*Sigmodon hispidus*). *Parasitology*, **112**, 227–232.

Beg, M.A., Fistein, J.L. and Storey, D.M. (1995) The host-parasite relationships in pyridoxine (vitamin B6) deficient cotton rats infected with *Litomosoides carinii* (Nematoda: Filaroidea). *Parasitology*, **111**, 111–118.

Beg, M.A. and Storey, D.M. (1993) Embryogenesis in *Litomosoides carinii* from pyridoxine deficient cotton rats. *Journal of Helminthology*, **67**, 205–212.

Behm, C.A. (1997) The role of trehalose in the physiology of nematodes. *International Journal for Parasitology*, **27**, 215–229.

Bennet, E.-M. and Bryant, C. (1984) Energy metabolism of adult *Haemonchus contortus* in vitro: a comparison of benzimidazole-susceptible and -resistant strains. *Molecular and Biochemical Parasitology*, **10**, 335–346.

Berger, A.J., Hart, A.C. and Kaplan, J.M. (1998) Gα$_s$-induced neurodegeneration in *Caenorhabditis elegans*. *Journal of Neuroscience*, **18**, 2871–2880.

Bhargava, K.K., Le Trang, N., Cerami, A. and Eaton, J.W. (1983) Effect of arsenical drugs on glutathione metabolism of *Litomosoides carinii*. *Molecular and Biochemical Parasitology*, **9**, 29–35.

Bini, L., Heid, H., Liberatori, S., Geier, G., Pallini, V. and Zwilling, R. (1997) Two-dimensional gel electrophoresis of *Caenorhabditis elegans* homogenates and identification of protein spots by microsequencing. *Electrophoresis*, **18**, 557–562.

Bird, A.F. and Bird, J. (1991) *The Structure of Nematodes*, 2nd edn, San Diego: Academic Press.

Bird, D.M. and Riddle, D.L. (1989) Molecular cloning and sequencing of *ama-1*, the gene encoding the largest subunit of *Caenorhabditis elegans* RNA polymerase II. *Molecular and Cellular Biology*, **9**, 4119–4130.

Blaxter, M.L., Raghavan, N., Ghosh, I., Guiliano, D., Lu, W., Williams, S.A., et al. (1996) Genes expressed in *Brugia malayi* infective third stage larvae. *Molecular and Biochemical Parasitology*, **77**, 77–93.

Bolla, R.I., Weinstein, P.P. and Lou, C. (1972) In vitro nutritional requirements of *Nippostrongylus brasiliensis*. I. Effects of sterols, sterol derivatives and heme compounds on the free-living stages. *Comparative Biochemistry and Physiology*, **43B**, 487–501.

Brazier, J.B. and Jaffe, J.J. (1973) Two types of pyruvate kinase in schistosomes and filariae. *Comparative Biochemistry and Physiology*, **44B**, 145–155.

Britton, C., Knox, D.P. and Kennedy, M.W. (1994) Superoxide dismutase (SOD) activity of *Dictyocaulus viviparus* and its inhibition by antibody from infected and vaccinated bovine hosts. *Parasitology*, **109**, 257–263.

Britton, C., Moore, J., Gilleard, J.S. and Kennedy, M.W. (1995) Extensive diversity in repeat unit sequences of the cDNA encoding the polyprotein antigen allergen from the bovine lungworm *Dictyocaulus viviparus*. *Molecular and Biochemical Parasitology*, **72**, 77–88.

Brophy, P.M. and Barrett, J. (1990a) Glutathione transferase in helminths. *Parasitology*, **2**, 345–349.

Brophy, P.M. and Barrett, J. (1990b) Strategies for detoxification of aldehydic products of lipid peroxidation in helminths. *Molecular and Biochemical Parasitology*, **42**, 205–211.

Brophy, P.M., Ben-Smith, A., Brown, A., Behnke, J.M. and Pritchard, D.I. (1994) Glutathione S-transferases from the gastrointestinal nematode *Heligmosomoides polygyrus* and mammalian liver compared. *Comparative Biochemistry and Physiology*, **109B**, 585–592.

Brophy, P.M., Ben-Smith, A., Brown, A., Behnke, J.M. and Pritchard, D.I. (1995a) Differential expression of glutathione S-transferase (GST) by adult *Heligmosomoides polygyrus* during primary infection in fast and slow responding hosts. *International Journal for Parasitology*, **25**, 641–645.

Brophy, P.M., Crowley, P. and Barrett, J. (1990) Relative distribution of glutathione transferase, glyoxalase I and glyoxalase II in helminths. *International Journal for Parasitology*, **20**, 259–261.

Brophy, P.M., Patterson, L.H., Brown, A. and Pritchard, D.I. (1995b) Glutathione S-transferase (GST) expression in the human hookworm *Necator americanus*: potential roles for excretory-secretory forms of GST. *Acta Tropica*, **59**, 259–263.

Brophy, P.M., Patterson, L.H. and Pritchard, D.L. (1995) Secretory nematode SOD — offensive or defensive? *International Journal for Parasitology*, **25**, 865–866.

Brophy, P.M. and Pritchard, D.I. (1992) Metabolism of lipid peroxidation products by the gastro-intestinal nematodes *Necator americanus*, *Ancylostoma ceylanicum* and *Heligmosomoides polygyrus*. *International Journal for Parasitology*, **22**, 1009–1012.

Brophy, P.M. and Pritchard, D.I. (1994) Parasitic helminth glutathione S-transferases: an update on their potential as targets for immuno- and chemotherapy. *Experimental Parasitology*, **79**, 89–96.

Bryant, C. and Behm, C.A. (1989) *Biochemical Adaptation in Parasites*, London: Chapman and Hall.

Buettner, C., Harney, J.W. and Berry, M.J. (1999) The *Caenorhabditis elegans* homologue of thioredoxin reductase contains a selenocysteine insertion sequence (SECIS) element that differs from mammalian SECIS elements but directs selenocysteine incorporation. *Journal of Biological Chemistry*, **274**, 21598–21602.

Callahan, H.L., Crouch, R.K. and James, E.R. (1988) Helminth anti-oxidant enzymes: a protective mechanism against host oxidants? *Parasitology Today*, **4**, 218–225.

Callahan, H.L., Crouch, R.K. and James, E.R. (1990) Hydrogen peroxide is the most toxic oxygen species for *Onchocerca cervicalis* microfilariae. *Parasitology*, **3**, 407–415.

Callahan, H.L., Crouch, R.K. and James, E.R. (1991) *Dirofilaria immitis* superoxide dismutase: purification and characterization. *Molecular and Biochemical Parasitology*, **49**, 245–251.

Callahan, H.L., Hazen-Martin, D., Crouch, R. and James, E.R. (1993) Immunolocalization of superoxide dismutase in *Dirofilaria immitis* adult worms. *Infection and Immunity*, **61**, 1157–1163.

Castro, G.A. and Fairbairn, D. (1969) Carbohydrates and lipids in *Trichinella spiralis* larvae and their utilization in vitro. *Journal of Parasitology*, **55**, 51–58.

Chandrashekar, R., Curtis, K.C., Lu, W.H. and Weil, G.J. (1998) Molecular cloning of an enzymatically active thioredoxin peroxidase from *Onchocerca volvulus*. *Molecular and Biochemical Parasitology*, **93**, 309–312.

Chandrashekar, R., Tsuji, N., Morales, T.H., Carmody, A.B., Ozols, V.O., Welton, J., et al. (2000) Removal of hydrogen peroxide by a 1-cysteine peroxiredoxin enzyme of the filarial parasite *Dirofilaria immitis*. *Parasitology Research*, **86**, 200–206.

Chen, S.N. and Howells, R.E. (1979) *Brugia pahangi*: uptake and incorporation of adenosine and thymidine. *Experimental Parasitology*, **47**, 209–221.

Chen, S.N. and Howells, R.E. (1981a) *Brugia pahangi*: uptake and incorporation of nucleic acid precursors by microfilariae and macrofilariae *in vitro*. *Experimental Parasitology*, **51**, 296–306.

Chen, S.N. and Howells, R.E. (1981b) The uptake in vitro of monosaccharides, disaccharide and nucleic acid precursors by adult *Dirofilaria immitis*. *Annals of Tropical Medicine and Parasitology*, **75**, 329–334.

Chen, W., Komuniecki, P.R. and Komuniecki, R. (1999) Nematode pyruvate dehydrogenase kinases: role of the C-terminus in binding to the dihydrolipoyl transacetylase core of the pyruvate dehydrogenase complex. *Biochemical Journal*, **339**, 103–109.

Chervitz, S.A., Aravind, L., Sherlock, G., Ball, C.A., Koonin, E.V., Dwight, S.S., et al. (1998) Comparison of the complete protein sets of worm and yeast: orthology and divergence. *Science*, **282**, 2022–2028.

Chitwood, D.J. (1991) Nematode sterol biochemistry. In *Physiology and Biochemistry of Sterols*, edited by G.W. Patterson, pp. 257–293. Champaign, Illinois: American Oil Chemists' Society.

Chitwood, D.J. (1998) Biosynthesis. In *The Physiology and Biochemistry of Free-living and Plant-parasitic Nematodes*, edited by R.N. Perry and D.J. Wright, pp. 303–330. Wallingford: CAB International.

Chitwood, D.J. (1999) Biochemistry and function of nematode steroids. *Critical Reviews in Biochemistry and Molecular Biology*, **34**, 273–284.

Chitwood, D.J. and Lusby, W.R. (1991) Metabolism of plant sterols by nematodes. *Lipids*, **26**, 619–627.

Chitwood, D.J., Lusby, W.R., Thompson, M.J., Kochansky, J.P. and Howarth, O.W. (1995) The glycosylceramides of the nematode *Caenorhabditis elegans* contain an unusual, branched-chain sphingoid base. *Lipids*, **30**, 567–573.

Chooback, L., Karsten, W.E., Kulkarni, G., Nalabolu, S.R., Harris, B.G. and Cook, P.F. (1997) Expression, purification, and characterization of the recombinant NAD-malic enzyme from *Ascaris suum*. *Protein Expression and Purification*, **10**, 51–54.

Chung, J.G. (1999) Purification and characterization of an arylamine N-acetyltransferase in the nematode *Enterobius vermicularis*. *Microbios*, **98**, 15–25.

Clancy, L.L., Rao, G.S., Finzel, B.C., Muchmore, S.W., Holland, D.R., Watenpaugh, K.D., et al. (1992) Crystallization of the NAD-dependent malic enzyme from the parasitic nematode *Ascaris suum*. *Journal of Molecular Biology*, **226**, 565–569.

Clark, F.E. (1969) *Ancylostoma caninum*: food reserves and changes in chemical composition with age in third stage larvae. *Experimental Parasitology*, **24**, 1–8.

Clarke, A.J. and Hennessy, J. (1976) The distribution of carbohydrates in cysts of *Heterodora rostochiensis*. *Nematologica*, **22**, 190–195.

Clarke, A.J., Perry, R.N. and Hennessy, J. (1978) Osmotic stress and the hatching of *Globodera rostochiensis*. *Nematologica*, **24**, 384–392.

Coggins, J.R., Schaefer, F.W. and Weinstein, P.P. (1985) Ultrastructural analysis of pathologic lesions in sterol-deficient *Nippostrongylus brasiliensis* larvae. *Journal of Invertebrate Pathology*, **45**, 288–297.

Comley, J.C., Jaffe, J.J., Chrin, L.R. and Smith, R.B. (1981) Synthesis of ubiquinone 9 by adult *Brugia pahangi* and *Dirofilaria immitis*: evidence against its involvement in the oxidation of 5-methyltetrahydrofolate. *Molecular and Biochemical Parasitology*, **2**, 271–283.

Comley, J.C.W. and Jaffe, J.J. (1983) The conversion of exogenous retinol and related compounds into retinyl phosphate mannose by adult *Brugia pahangi in vitro*. *Biochemical Journal*, **214**, 367–376.

Cookson, E., Blaxter, M.L. and Selkirk, M.E. (1992) Identification of the major soluble cuticular glycoprotein of lymphatic filarial nematode parasites (gp29) as a secretory homolog of glutathione peroxidase. *Proceedings of the National Academy of Sciences, U.S.A.*, **89**, 5837–5841.

Cookson, E., Tang, L. and Selkirk, M.E. (1993) Conservation of primary sequence of gp29, the major soluble cuticular glycoprotein, in three species of lymphatic filariae. *Molecular and Biochemical Parasitology*, **58**, 155–159.

Dabrowska, M., Zielinski, Z., Wranicz, M., Michalski, R., Pawelczak, K. and Rode, W. (1996) *Trichinella spiralis* thymidylate synthase: developmental pattern, isolation, molecular properties, and inhibition by substrate and cofactor analogues. *Biochemical and Biophysical Research Communications*, **228**, 440–445.

Dalley, B.K., Rogalski, T.M., Tullis, G.E., Riddle, D.L. and Golomb, M. (1993) Post-transcriptional regulation of RNA polymerase II levels in *Caenorhabditis elegans*. *Genetics*, **133**, 237–245.

Dass, P.D. and Donahue, M.J. (1986) γ-Glutamyl transpeptidase activity in *Ascaris suum*. *Molecular and Biochemical Parasitology*, **20**, 233–236.

Daum, G., Schmid, B., MacKintosh, C., Cohen, P. and Hofer, H.W. (1992) Characterization of the major phosphofructokinase-dephosphorylating protein phosphatases from *Ascaris suum* muscle. *Biochimica et Biophysica Acta*, **1122**, 23–32.

Daum, G., Thalhofer, H.P., Harris, B.G. and Hofer, H.W. (1986) Reversible activation and inactivation of phosphofructokinase from *Ascaris suum* by the action of tissue-homologous protein phosphorylating and dephosphorylating enzymes. *Biochemical and Biophysical Research Communications*, **139**, 215–221.

Davies, K.P. and Köhler, P. (1990) The role of amino acids in the energy generating pathways of *Litomosoides carinii*. *Molecular and Biochemical Parasitology*, **41**, 115–124.

Donahue, M.J., de la Houssaye, B.A., Harris, B.G. and Masaracchia, R.A. (1981a) Regulation of glycogenolysis by adenosine 3'5'-monophosphate in *Ascaris suum* muscle. *Comparative Biochemistry and Physiology*, **69B**, 693–699.

Donahue, M.J., Yacoub, N.J. and Harris, B.G. (1982) Correlation of muscle activity with glycogen metabolism in muscle of *Ascaris suum*. *American Journal of Physiology*, **242**, R514–521.

Donahue, M.J., Yacoub, N.J., Kaeini, M.R. and Harris, B.G. (1981b) Activity of enzymes regulating glycogen metabolism in perfused muscle-cuticle sections of *Ascaris suum* (Nematoda). *Journal of Parasitology*, **67**, 362–367.

Donahue, M.J., Yacoub, N.J., Kaeini, M.R., Masaracchia, R.A. and Harris, B.G. (1981c) Glycogen metabolizing enzymes during starvation and feeding of *Ascaris suum* maintained in a perfusion chamber. *Journal of Parasitology*, **67**, 505–510.

Donahue, M.J., Yacoub, N.J., Michnoff, C.A., Masaracchia, R.A. and Harris, B.G. (1981d) Serotonin (5-hydroxytryptamine): a possible regulator of glycogenolysis in perfused muscle segments of *Ascaris suum*. *Biochemical and Biophysical Research Communications*, **101**, 112–117.

Douch, P.G. (1975) The effect of flavins and enzyme inhibitors on 4-nitrobenzoic acid reductase and azo reductase of *Ascaris lumbricoides* var *suum*. *Xenobiotica*, **5**, 657–663.

Duran, E., Komuniecki, R.W., Komuniecki, P.R., Wheelock, M.J., Klingbeil, M.M., Ma, Y.C., *et al.* (1993) Characterization of cDNA clones for the 2-methyl branched-chain enoyl-CoA reductase. An enzyme involved in branched-chain fatty acid synthesis in anaerobic mitochondria of the parasitic nematode *Ascaris suum. Journal of Biological Chemistry,* **268,** 22391–22396.

Duran, E., Walker, D.J., Johnson, K.R., Komuniecki, P.R. and Komuniecki, R.W. (1998) Developmental and tissue-specific expression of 2-methyl branched-chain enoyl CoA reductase isoforms in the parasitic nematode, *Ascaris suum. Molecular and Biochemical Parasitology,* **91,** 307–318.

Eckelt, V.H.O., Liebau, E., Walter, R.D. and Henkle-Dührsen, K. (1998) Primary sequence and activity analyses of a catalase from *Ascaris suum. Molecular and Biochemical Parasitology,* **95,** 203–214.

Fairbairn, D. (1955) Embryonic and postembryonic changes in the lipids of *Ascaris lumbricoides* eggs. *Canadian Journal of Biochemistry and Physiology,* **33,** 122–129.

Fairbairn, D. and Passey, R.F. (1957) Occurrence and distribution of trehalose and glycogen in the eggs and tissues of *Ascaris lumbricoides. Experimental Parasitology,* **6,** 566–574.

Farland, W.H. and MacInnis, A.J. (1978a) In vitro thymidine kinase activity: present in *Hymenolepis diminuta* (Cestoda) and *Moniliformis dubius* (Acanthocephala), but apparently lacking in *Ascaris lumbricoides* (Nematoda). *Journal of Parasitology,* **64,** 564–565.

Farland, W.H. and MacInnis, A.J. (1978b) Purine nucleotide content of developing *Ascaris lumbricoides* eggs. *International Journal for Parasitology,* **8,** 177–186.

Ferreira, M.F.A., Andrade, C.M. and Ribeiro, L.P. (1985) Mitochondrial malic enzyme of *Toxocara canis* muscle. *Comparative Biochemistry and Physiology,* **81B,** 939–944.

Fleming, M.W. (1987) Ecdysteroids during embryonation of eggs of *Ascaris suum. Comparative Biochemistry and Physiology,* **87A,** 803–805.

Fleming, M.W. (1993) Ecdysteroids during development in the ovine parasitic nematode, *Haemonchus contortus. Comparative Biochemistry and Physiology,* **104B,** 653–655.

Fodge, D.W., Gracy, R.W. and Harris, B.G. (1972) Studies on enzymes from parasitic helminths. I. Purification and physical properties of malic enzyme from the muscle tissue of *Ascaris suum. Biochimica et Biophysica Acta,* **268,** 271–284.

Foll, R.L., Pleyers, A., Lewandovski, G.J., Wermter, C., Hegemann, V. and Paul, R.J. (1999) Anaerobiosis in the nematode *Caenorhabditis elegans. Comparative Biochemistry and Physiology,* **124B,** 269–280.

Frayha, G.J. and Smyth, J.D. (1983) Lipid metabolism in parasitic helminths. *Advances in Parasitology,* **22,** 309–387.

Freedman, J.H., Slice, L.W., Dixon, D., Fire, A. and Rubin, C.S. (1993) The novel metallothionein genes of *Caenorhabditis elegans.* Structural organization and inducible, cell-specific expression. *Journal of Biological Chemistry,* **268,** 2554–2564.

Fry, M. and Jenkins, D.C. (1984) Nematoda: aerobic respiratory pathways of adult parasitic species. *Experimental Parasitology,* **57,** 86–92.

Geary, T.G., Winterrowd, C.A., Alexander-Bowman, S.J., Favreau, M.A., Nulf, S.C. and Klein, R.D. (1993) *Ascaris suum:* cloning of a cDNA encoding phosphoenolpyruvate carboxykinase. *Experimental Parasitology,* **77,** 155–161.

Ghosh, I., Eisinger, S.W., Raghavan, N. and Scott, A.L. (1998) Thioredoxin peroxidases from *Brugia malayi. Molecular and Biochemical Parasitology,* **91,** 207–220.

Ghosh, P., Heath, A.C., Donahue, M.J. and Masaracchia, R.A. (1989) Glycogen synthesis in the obliquely striated muscle of *Ascaris suum. European Journal of Biochemistry,* **183,** 679–685.

Gladyshev, V.N., Krause, M., Xu, X.-M., Korotkov, K.V., Kryukov, G.V., Sun, Q.-A., *et al.* (1999) Selenocysteine-containing thioredoxin reductase in *C. elegans. Biochemical and Biophysical Research Communications,* **259,** 244–249.

Golden, J.W. and Riddle, D.L. (1982) A pheromone influences larval development in the nematode *Caenorhabditis elegans. Science,* **218,** 578–580.

Goldstein, P. (1997) The synaptonemal complexes of the nematode *Caenorhabditis elegans:* gametic response to retinol. *Cytobios,* **91,** 53–67.

Goldstein, P. and Modric, T. (1994) Transgenerational, ultrastructural analysis on the antioxidative effects of tocopherol on early gametogenesis in *Caenorhabditis elegans* grown in 100% oxygen. *Toxicology and Applied Pharmacology,* **124,** 212–220.

Goyal, N., Gupta, S., Katiyar, J.C. and Srivastava, V.M. (1991) NADH oxidase and fumarate reductase of *Ancylostoma ceylanicum. International Journal for Parasitology,* **21,** 673–676.

Goyal, N. and Srivastava, V.M. (1990) Mitochondrial NADH oxidase activity of *Setaria cervi. Veterinary Parasitology,* **37,** 229–236.

Grantham, B.D. and Barrett, J. (1986a) Amino acid catabolism in the nematodes *Heligmosomoides polygyrus* and *Panagrellus redivivus* 2. Metabolism of the carbon skeleton. *Parasitology,* **93,** 495–504.

Grantham, B.D. and Barrett, J. (1986b) Amino acid catabolism in the nematodes *Heligmosomoides polygyrus* and *Panagrellus redivivus* 1. Removal of the amino group. *Parasitology,* **93,** 481–493.

Grantham, B.D. and Barrett, J. (1988) Glutamine and asparagine synthesis in the nematodes *Heligmosomoides polygyrus* and *Panagrellus redivivus. Journal of Parasitology,* **74,** 1052–1053.

Hadas, E. and Stankiewicz, M. (1998) Superoxide dismutase and total antioxidant status of larvae and adults of *Trichostrongylus colubriformis, Haemonchus contortus* and *Ostertagia circumcincta. Parasitology Research,* **84,** 646–650.

Hannigan, L.L., Donahue, M.J. and Masaracchia, R.A. (1985) Comparative purification and characterization of invertebrate muscle glycogen synthase from the porcine parasite *Ascaris suum. Journal of Biological Chemistry,* **260,** 16099–16105.

Harrington, L.A. and Harley, C.B. (1988) Effect of vitamin E on lifespan and reproduction in *Caenorhabditis elegans. Mechanisms of Ageing and Development,* **43,** 71–78.

Hata, H. (1994) Essential amino acids and other essential components for development of *Angiostrongylus costaricensis* from third-stage larvae to young adults. *Journal of Parasitology,* **80,** 518–520.

Hieb, W.F. and Rothstein, M. (1968) Sterol requirement for reproduction of a free-living nematode. *Science,* **160,** 778–780.

Hill, B., Kilsby, J., Rogerson, G.W., McIntosh, R.T. and Ginger, C.D. (1981) The enzymes of pyrimidine biosynthesis in a range of parasitic protozoa and helminths. *Molecular and Biochemical Parasitology,* **2,** 123–134.

Hofer, H.W., Allen, B.J., Kaeini, M.R. and Harris, B.G. (1982a) Phosphofructokinase from *Ascaris suum.* The effect of phosphorylation on activity at near physiological conditions. *Journal of Biological Chemistry,* **257,** 3807–3810.

Hofer, H.W., Allen, B.L., Kaeini, M.R., Pette, D. and Harris, B.G. (1982b) Phosphofructokinase from *Ascaris suum.* Regulatory kinetic studies and activity near physiological conditions. *Journal of Biological Chemistry,* **257,** 3801–3806.

Honda, Y. and Honda, S. (1999) The *daf-2* gene network for longevity regulates oxidative stress resistance and Mn-superoxide dismutase gene expression in *Caenorhabditis elegans. FASEB Journal,* **13,** 1385–1393.

Huang, Y.J., Walker, D., Chen, W., Klingbeil, M. and Komuniecki, R. (1998) Expression of pyruvate dehydrogenase isoforms during the aerobic/anaerobic transition in the development of the parasitic nematode *Ascaris suum:* altered stoichiometry of phosphorylation/inactivation. *Archives of Biochemistry and Biophysics,* **352,** 263–270.

Hussein, A.S. and Walter, R.D. (1996) Purification and characterization of γ-glutamyl transpeptidase from *Ascaris suum. Molecular and Biochemical Parasitology,* **77,** 41–47.

Hutchinson, G.W. and Fernando, M.A. (1974) Enzymes of glycolysis and the pentose phosphate pathway during development of the rabbit stomach worm *Obeliscoides cuniculi. International Journal for Parasitology,* **4,** 389–395.

Hutchison, W.F. and Turner, A.C. (1979) Glycolytic end products of the adult dog heartworm, *Dirofilaria immitis. Comparative Biochemistry and Physiology,* **62B,** 71–73.

Hutchison, W.F., Turner, A.C. and Oelshlegel, F.J. (1977) Hexokinase of the adult dog heartworm, *Dirofilaria immitis. Comparative Biochemistry and Physiology,* **58B,** 131–134.

Hutzell, P.A. and Krusberg, L.R. (1982) Fatty acid compositions of *Caenorhabditis elegans* and *C. briggsae. Comparative Biochemistry and Physiology,* **73B,** 517–520.

Islas-Trejo, A., Land, M., Tcherepanova, I., Freedman, J.H. and Rubin, C.S. (1997) Structure and expression of the *Caenorhabditis elegans* protein kinase C2 gene. Origins and regulated expression of a family of Ca^{2+}-activated protein kinase C isoforms. *Journal of Biological Chemistry,* **272,** 6629–6640.

Jackson, G.J. and Platzer, E.G. (1974) Nutritional biotin and purine requirements, and the folate metabolism of *Neoaplectana glaseri. Journal of Parasitology,* **60,** 453–457.

Jaffe, J.J. (1980) Filarial folate-related metabolism is a potential target for selective inhibitors. In *The Host-Invader Interplay,* edited by H. Van den Bossche, pp. 605–614. Amsterdam: Elsevier/North Holland.

Jaffe, J.J. (1981) Involvement of tetrahydrofolate cofactors in *de novo* purine ribonucleotide synthesis by adult *Brugia pahangi* and *Dirofilaria immitis. Molecular and Biochemical Parasitology,* **2,** 259–270.

Jaffe, J.J. and Chrin, L.R. (1980) Folate metabolism in filariae: enzymes associated with 5,10-methylenetetrahydrofolate. *Journal of Parasitology*, **66**, 53–58.

Jaffe, J.J., Chrin, L.R. and Smith, R.B. (1980) Folate metabolism in filariae. Enzymes associated with 5,10-methenyltetrahydrofolate and 10-formyltetrahydrofolate. *Journal of Parasitology*, **66**, 428–433.

Jaffe, J.J., Comley, J.C. and Chrin, L.R. (1982) Thymidine kinase activity and thymidine salvage in adult *Brugia pahangi* and *Dirofilaria immitis*. *Molecular and Biochemical Parasitology*, **5**, 361–370.

Jaffe, J.J. and Doremus, H.M. (1970) Metabolic patterns of *Dirofilaria immitis* microfilariae *in vitro*. *Journal of Parasitology*, **56**, 254–260.

Jaffe, J.J. and Lambert, R.A. (1986) Glutathione *S*-transferase in adult *Dirofilaria immitis* and *Brugia pahangi*. *Molecular and Biochemical Parasitology*, **20**, 199–206.

James, E.R., McLean, D.C. and Perler, F. (1994) Molecular cloning of an *Onchocerca volvulus* extracellular Cu-Zn superoxide dismutase. *Infection and Immunity*, **62**, 713–716.

Janssen, C.S., Tetley, L. and Kennedy, M.W. (1998) Developmental activation of infective *Trichinella spiralis* larvae. *Parasitology*, **117**, 363–371.

Johnson, K.R., Komuniecki, R., Sun, Y.H. and Wheelock, M.J. (1992) Characterization of cDNA clones for the alpha subunit of pyruvate dehydrogenase from *Ascaris suum*. *Molecular and Biochemical Parasitology*, **51**, 37–48.

Juhász, S. and Babos, S. (1969) Studies of parasite metabolism. I. *In vitro* examination of the vitamin A metabolism of *Ascaris suum*. *Acta Veterinaria Hungarica*, **19**, 239–251.

Jung, S., Hoffmann, R., Rodriguez, P.H., Mutzel, R. and Hofer, H.W. (1995) The catalytic subunit of cAMP-dependent protein kinase from *Ascaris suum*. The cloning and structure of a novel subtype of protein kinase A. *European Journal of Biochemistry*, **232**, 111–117.

Kawalek, J.C., Rew, R.S. and Heavner, J. (1984) Glutathione-*S*-transferase, a possible drug-metabolizing enzyme, in *Haemonchus contortus*: comparative activity of a cambendazole-resistant and a susceptible strain. *International Journal for Parasitology*, **14**, 173–175.

Kennedy, M.W., Britton, C., Price, N.C., Kelly, S.M. and Cooper, A. (1995) The DvA-1 polyprotein of the parasitic nematode *Dictyocaulus viviparus*. A small helix-rich lipid-binding protein. *Journal of Biological Chemistry*, **270**, 19277–19281.

Kennedy, M.W., Garside, L.H., Goodrick, L.E., McDermott, L., Brass, A., Price, N.C., *et al.* (1997) The Ov20 protein of the parasitic nematode *Onchocerca volvulus*. A structurally novel class of small helix-rich retinol-binding proteins. *Journal of Biological Chemistry*, **272**, 29442–29448.

Kerboeuf, D., Soubieux, D., Guilluy, R., Brazier, J.L. and Rivière, J.L. (1995) *In vivo* metabolism of aminopyrine by the larvae of the helminth *Heligmosomoides polygyrus*. *Parasitology Research*, **81**, 302–304.

Khan, F.R. and McFadden, B.A. (1980) Embryogenesis and the glyoxylate cycle. *FEBS Letters*, **115**, 312–314.

Khan, F.R. and McFadden, B.A. (1982) *Caenorhabditis elegans*: decay of isocitrate lyase during larval development. *Experimental Parasitology*, **54**, 47–54.

Kita, K. (1992) Electron-transfer complexes of mitochondria in *Ascaris suum*. *Parasitology Today*, **8**, 155–159.

Kita, K., Hirawake, H. and Takamiya, S. (1997) Cytochromes in the respiratory chain of helminth mitochondria. *International Journal for Parasitology*, **27**, 617–630.

Klass, M., Nguyen, P.N. and Dechavigny, A. (1983) Age-correlated changes in the DNA template in the nematode *Caenorhabditis elegans*. *Mechanisms of Ageing and Development*, **22**, 253–263.

Klimowski, L., Chandrashekar, R. and Tripp, C.A. (1997) Molecular cloning, expression and enzymatic activity of a thioredoxin peroxidase from *Dirofilaria immitis*. *Molecular and Biochemical Parasitology*, **90**, 297–306.

Klingbeil, M.M., Walker, D.J., Arnette, R., Sidawy, E., Hayton, K., Komuniecki, P.R., *et al.* (1996) Identification of a novel dihydrolipoyl dehydrogenase-binding protein in the pyruvate dehydrogenase complex of the anaerobic parasitic nematode, *Ascaris suum*. *Journal of Biological Chemistry*, **271**, 5451–5457.

Köhler, P. (1985) The strategies of energy conservation in helminths. *Molecular and Biochemical Parasitology*, **17**, 1–18.

Köhler, P. (1991) The pathways of energy generation in filarial parasites. *Parasitology Today*, **7**, 21–25.

Komuniecki, P.R. and Vanover, L. (1987) Biochemical changes during the aerobic-anaerobic transition in *Ascaris suum* larvae. *Molecular and Biochemical Parasitology*, **22**, 241–248.

Komuniecki, R., Bruchhaus, I., Ilg, T., Wilson, K., Zhang, Y. and Fairlamb, A.H. (1992) Purification of glutathione reductase from muscle of the adult parasitic nematode *Ascaris suum*. *Molecular and Biochemical Parasitology*, **51**, 331–333.

Komuniecki, R. and Komuniecki, P.R. (1995) Aerobic-anaerobic transitions in energy metabolism during the development of the parasitic nematode *Ascaris suum*. In *Molecular Approaches to Parasitology*, edited by J.C. Boothroyd and R. Komuniecki, pp. 109–121. New York: Wiley-Liss.

Kotze, A.C. (1997) Cytochrome P450 monooxygenase activity in *Haemonchus contortus* (Nematoda). *International Journal for Parasitology*, **27**, 33–40.

Kotze, A.C. (1999) Peroxide-supported in-vitro cytochrome P450 activities in *Haemonchus contortus*. *International Journal for Parasitology*, **29**, 389–396.

Krusberg, L.R. (1972) Fatty acid composition of *Turbatrix aceti* and its culture medium. *Comparative Biochemistry and Physiology*, **41B**.

Kulkarni, G., Cook, P.F. and Harris, B.G. (1993) Cloning and nucleotide sequence of a full-length cDNA encoding *Ascaris suum* malic enzyme. *Archives of Biochemistry and Biophysics*, **300**, 231–237.

Kuramochi, T., Hirawake, H., Kojima, S., Takamiya, S., Furushima, R., Aoki, T., *et al.* (1994) Sequence comparison between the flavoprotein subunit of the fumarate reductase (complex II) of the anaerobic parasitic nematode, *Ascaris suum* and the succinate dehydrogenase of the aerobic, free-living nematode, *Caenorhabditis elegans*. *Molecular and Biochemical Parasitology*, **68**, 177–187.

Kuramochi, T., Kita, K., Takamiya, S., Kojima, S. and Hayasaki, M. (1995) Comparative study and cDNA cloning of the flavoprotein subunit of mitochondrial complex II (succinate-ubiquinone oxidoreductase: fumarate reductase) from the dog heartworm, *Dirofilaria immitis*. *Comparative Biochemistry and Physiology*, **111B**, 491–502.

Lai, C.J., Harris, B.G. and Cook, P.F. (1992) Mechanism of activation of the NAD-malic enzyme from *Ascaris suum* by fumarate. *Archives of Biochemistry and Biophysics*, **299**, 214–219.

Landsperger, W.J. and Harris, B.G. (1976) NAD^+-malic enzyme. Regulatory properties of the enzyme from *Ascaris suum*. *Journal of Biological Chemistry*, **251**, 3599–3602.

Langer, B.W., Smith, W.J. and Theodorides, V.J. (1971) The pentose cycle in adult *Ascaris suum*. *Journal of Parasitology*, **57**, 485–486.

Lattemann, C.T., Matzen, A. and Apfel, H. (1999) Up-regulation of extracellular copper/zinc superoxide dismutase mRNA after transmission of the filarial parasite *Acanthocheilonema viteae* in the vertebrate host *Meriones unguiculatus*. *International Journal for Parasitology*, **29**, 1437–1446.

Learmonth, M.P., Lim, M.S.L., Euerby, M.R. and Gibbons, W.A. (1987) Metabolic studies of parasitic helminths using n.m.r. spectroscopy. *Biochemical Society Transactions*, **15**, 879–880.

Lee, D.L. and Atkinson, H.J. (1976) *Physiology of Nematodes*, 2nd edn, London: Macmillan.

Lesoon, A., Komuniecki, P.R. and Komuniecki, R. (1990) Catalase activity during development of the parasitic nematode *Ascaris suum*. *Comparative Biochemistry and Physiology*, **95B**, 811–815.

Leutskaya, K.L. (1974) Content of vitamin A fractions in tissues of *Ascaris suum*. *Parazitologiia*, **8**, 83–84.

Liao, V.H.-C. and Freedman, J.H. (1998) Cadmium-regulated genes from the nematode *Caenorhabditis elegans*. Identification and cloning of new cadmium-responsive genes by differential display. *Journal of Biological Chemistry*, **273**, 31962–31970.

Liddell, S. and Knox, D.P. (1998) Extracellular and cytoplasmic Cu/Zn superoxide dismutases from *Haemonchus contortus*. *Parasitology*, **116**, 383–394.

Liebau, E., Eckelt, V.H.O., Wildenburg, G., Teesdale-Spittle, P., Brophy, P.M., Walter, R.D., *et al.* (1997) Structural and functional analysis of a glutathione *S*-transferase from *Ascaris suum*. *Biochemical Journal*, **324**, 659–666.

Liebau, E., Schönberger, O.L., Walter, R.D. and Henkle-Dührsen, K.J. (1994a) Molecular cloning and expression of a cDNA encoding glutathione *S*-transferase from *Ascaris suum*. *Molecular and Biochemical Parasitology*, **63**, 167–170.

Liebau, E., Wildenburg, G., Brophy, P.M., Walter, R.D. and Henkle-Dührsen, K. (1996) Biochemical analysis, gene structure and localization of the 24 kDa glutathione *S*-transferase from *Onchocerca volvulus*. *Molecular and Biochemical Parasitology*, **80**, 27–39.

Liebau, E., Wildenburg, G., Walter, R.D. and Henkle-Dührsen, K. (1994b) A novel type of glutathione *S*-transferase in *Onchocerca volvulus*. *Infection and Immunity*, **62**, 4762–4767.

Liu, A. and Rothstein, M. (1976) Nematode biochemistry XV. Enzyme changes related to glycerol excretion in *C. briggsae*. *Comparative Biochemistry and Physiology*, **54B**, 233–238.

Liu, F.Z., Thatcher, J.D., Barral, J.M. and Epstein, H.F. (1995) Bifunctional glyoxylate cycle protein of *Caenorhabditis elegans*: a developmentally regulated protein of intestine and muscle. *Developmental Biology*, **169**, 399–414.

Liu, F.Z., Thatcher, J.D. and Epstein, H.F. (1997) Induction of glyoxylate cycle expression in *Caenorhabditis elegans*: a fasting response throughout larval development. *Biochemistry*, **36**, 255–260.

Liu, L.X., Buhlmann, J.E. and Weller, P.F. (1992) Release of prostaglandin E2 by microfilariae of *Wuchereria bancrofti* and *Brugia malayi*. *American Journal of Tropical Medicine and Hygiene*, **46**, 520–523.

Liu, L.X., Serhan, C.N. and Weller, P.F. (1990) Intravascular filarial parasites elaborate cyclooxygenase-derived eicosanoids. *Journal of Experimental Medicine*, **172**, 993–996.

Liu, L.X. and Weller, P.F. (1989) *Brugia malayi*: microfilarial polyunsaturated fatty acid composition and synthesis. *Experimental Parasitology*, **69**, 198–203.

Liu, L.X. and Weller, P.F. (1990) Arachidonic acid metabolism in filarial parasites. *Experimental Parasitology*, **71**, 496–501.

Lok, J.B., Morris, R.A., Sani, B.P., Shealy, Y.F. and Donnelly, J.J. (1990) Synthetic and naturally occurring retinoids inhibit third- to fourth-stage larval development by *Onchocerca lienalis in vitro*. *Tropical Medicine and Parasitology*, **41**, 169–173.

Lu, N.C., Hieb, W.F. and Stokstad, E.L. (1976) Effect of vitamin B_{12} and folate on biosynthesis of methionine from homocysteine in the nematode *Caenorhabditis briggsae*. *Proceedings of the Society for Experimental Biology and Medicine*, **151**, 701–706.

Lu, W.H., Egerton, G.L., Bianco, A.E. and Williams, S.A. (1998) Thioredoxin peroxidase from *Onchocerca volvulus*: a major hydrogen peroxide detoxifying enzyme in filarial parasites. *Molecular and Biochemical Parasitology*, **91**, 221–235.

Ma, Y.C., Funk, M., Dunham, W.R. and Komuniecki, R. (1993) Purification and characterization of electron-transfer flavoprotein rhodoquinone oxidoreductase from anaerobic mitochondria of the adult parasitic nematode, *Ascaris suum*. *Journal of Biological Chemistry*, **268**, 20360–20365.

Mallick, S., Harris, B.G. and Cook, P.F. (1991) Kinetic mechanism of NAD:malic enzyme from *Ascaris suum* in the direction of reductive carboxylation. *Journal of Biological Chemistry*, **266**, 2732–2738.

Matsuo, M. (1993) Oxygen dependency of life-span in the nematode. *Comparative Biochemistry and Physiology*, **105A**, 653–658.

McDermott, L., Cooper, A. and Kennedy, M.W. (1999) Novel classes of fatty acid and retinol binding protein from nematodes. *Molecular and Cellular Biochemistry*, **192**, 69–75.

McGibbon, A.M., Christie, J.F., Kennedy, M.W. and Lee, T.D. (1990) Identification of the major *Ascaris* allergen and its purification to homogeneity by high-performance liquid chromatography. *Molecular and Biochemical Parasitology*, **39**, 163–171.

McGonigle, S., Dalton, J.P. and James, E.R. (1998) Peroxidoxins: a new antioxidant family. *Parasitology Today*, **14**, 139–145.

Medica, D.L. and Sukhdeo, M.V.K. (1997) Role of lipids in the transmission of the infective stage (L3) of *Strongylus vulgaris* (Nematoda: Strongylida). *Journal of Parasitology*, **83**, 775–779.

Mei, B., Komuniecki, R. and Komuniecki, P.R. (1997) Localization of cytochrome oxidase and the 2-methyl branched-chain enoyl CoA reductase in muscle and hypodermis of *Ascaris suum* larvae and adults. *Journal of Parasitology*, **83**, 760–763.

Mei, B.S., Kennedy, M.W., Beauchamp, J., Komuniecki, P.R. and Komuniecki, R. (1997) Secretion of a novel, developmentally regulated fatty acid-binding protein into the perivitelline fluid of the parasitic nematode, *Ascaris suum*. *Journal of Biological Chemistry*, **272**, 9933–9941.

Mendis, A.H.W. and Evans, A.A.F. (1984) Major volatile metabolites produced by two isolates of *Aphelenchus avenae* under aerobic and anaerobic conditions. *Comparative Biochemistry and Physiology*, **78B**, 737–739.

Meyer, D.J., Muimo, R., Thomas, M., Coates, D. and Isaac, R.E. (1996) Purification and characterization of prostaglandin-H E-isomerase, a σ-class glutathione S-transferase, from *Ascaridia galli*. *Biochemical Journal*, **313**, 223–227.

Michaelson, L.V., Napier, J.A., Lewis, M., Griffiths, G., Lazarus, C.M. and Stobart, A.K. (1998) Functional identification of a fatty acid Δ^5 desaturase gene from *Caenorhabditis elegans*. *FEBS Letters*, **439**, 215–218.

Minning, D.M., Gow, A.J., Bonaventura, J., Braun, R., Dewhirst, M., Goldberg, D.E., *et al.* (1999) *Ascaris* haemoglobin is a nitric oxide-activated 'deoxygenase'. *Nature*, **401**, 497–502.

Modha, J., Roberts, M.C., Robertson, W.M., Sweetman, G., Powell, K.A., Kennedy, M.W., *et al.* (1999) The surface coat of infective larvae of *Trichinella spiralis*. *Parasitology*, **118**, 509–522.

Molinari, S. and Miacola, C. (1997) Antioxidant enzymes in phytoparasitic nematodes. *Journal of Nematology*, **29**, 153–159.

Muimo, R. and Isaac, R.E. (1993) Properties of an arylalkylamine N-acetyltransferase from the nematode, *Ascaridia galli*. *Comparative Biochemistry and Physiology*, **106B**, 969–976.

Müller, S., Gilberger, T.W., Fairlamb, A.H. and Walter, R.D. (1997) Molecular characterization and expression of *Onchocerca volvulus* glutathione reductase. *Biochemical Journal*, **325**, 645–651.

Müller, S., Walter, R.D. and Fairlamb, A.H. (1995) Differential susceptibility of filarial and human erythrocyte glutathione reductase to inhibition by the trivalent organic arsenical melarsen oxide. *Molecular and Biochemical Parasitology*, **71**, 211–219.

Napier, J.A., Hey, S.J., Lacey, D.J. and Shewry, P.R. (1998) Identification of a *Caenorhabditis elegans* Δ^6-fatty-acid-desaturase by heterologous expression in *Saccharomyces cerevisiae*. *Biochemical Journal*, **330**, 611–614.

Nelson, D.R. (1998) Metazoan cytochrome P450 evolution. *Comparative Biochemistry and Physiology*, **121C**, 15–22.

Nelson, D.R. (1999) Cytochrome P450 and the individuality of species. *Archives of Biochemistry and Biophysics*, **369**, 1–10.

Nicholas, W.L. (1984) *The Biology of Free-Living Nematodes*, 2nd edn, Oxford: Clarendon Press.

Nielsen, M.E. and Buchmann, K. (1997) Glutathione-S-transferase is an important antigen in the eel nematode *Anguillicola crassus*. *Journal of Helminthology*, **71**, 319–324.

Nishina, M., Hori, E., Matsushita, K., Takahashi, M., Kato, K. and Ohsaka, A. (1988) ^{1}H- and ^{13}C-NMR spectroscopic study of the metabolites in young adult *Angiostrongylus cantonensis* maintained in vitro. *Molecular and Biochemical Parasitology*, **28**, 249–256.

Nishina, M., Hori, E., Matsushita, K., Takahashi, M., Kato, K. and Ohsaka, A. (1989) ^{13}C-NMR spectroscopic studies on the glucose metabolism of *Angiostrongylus cantonesis*' eggs with special reference to the end-products and metabolic pathway. *Physiological Chemistry and Physics and Medical NMR*, **21**, 165–170.

O'Hanlon, G.M., Cleator, M., Mercer, J.G., Howells, R.E. and Rees, H.H. (1991) Metabolism and fate of ecdysteroids in the nematodes *Ascaris suum* and *Parascaris equorum*. *Molecular and Biochemical Parasitology*, **47**, 179–187.

Oguchi, M., Kanda, T. and Akamatsu, N. (1979) Hexokinase of *Angiostrongylus cantonensis*: presence of a glucokinase. *Comparative Biochemistry and Physiology*, **63B**, 335–340.

Omar, M.S. and Raoof, A.M. (1994) Histochemical localization of key glycolytic and related enzymes in adult *Onchocerca fasciata*. *Journal of Helminthology*, **68**, 337–341.

Omar, M.S., Raoof, A.M. and Al Amari, O.M. (1996) *Onchocerca fasciata*: enzyme histochemistry and tissue distribution of various dehydrogenases in the adult female worm. *Parasitology Research*, **82**, 32–37.

Ou, X.U., Tang, L.A., McCrossan, M., Henkle-Dührsen, K. and Selkirk, M.E. (1995a) *Brugia malayi*: localisation and differential expression of extracellular and cytoplasmic CuZn superoxide dismutases in adults and microfilariae. *Experimental Parasitology*, **80**, 515–529.

Ou, X.U., Thomas, G.R., Chacon, M.R., Tang, L.A. and Selkirk, M.E. (1995b) *Brugia malayi*: differential susceptibility to and metabolism of hydrogen peroxide in adults and microfilariae. *Experimental Parasitology*, **80**, 530–540.

Oya, H., Costello, L.C. and Smith, W.N. (1963) The comparative biochemistry of developing *Ascaris* eggs. II. Changes in cytochrome *c* oxidase activity during embryonation. *Journal of Cellular and Comparative Physiology*, **62**, 287–294.

Oya, H. and Weinstein, P.P. (1975) Demonstration of cobamide coenzyme in *Ascaris suum*. *Comparative Biochemistry and Physiology*, **50B**, 435–442.

Palmiter, R.D. (1998) The elusive function of metallothioneins. *Proceedings of the National Academy of Sciences, U.S.A.*, **95**, 8428–8430.

Park, S.H., Kiick, D.M., Harris, B.G. and Cook, P.F. (1984) Kinetic mechanism in the direction of oxidative decarboxylation for NAD-malic enzyme from *Ascaris suum*. *Biochemistry*, **23**, 5446–5453.

Park, S.M. and Koo, H.S. (1994) Purification of *Caenorhabditis elegans* DNA topoisomerase I. *Biochimica et Biophysica Acta*, **1219**, 47–54.

Payne, M.A., Rao, G.S., Harris, B.G. and Cook, P.F. (1991) Fructose 2,6-bisphosphate and AMP increase the affinity of the *Ascaris suum* phosphofructokinase for fructose 6-phosphate in a process separate from the relief of ATP inhibition. *Journal of Biological Chemistry*, **266**, 8891–8896.

Pemberton, K.D. and Barrett, J. (1989) The detoxification of xenobiotic compounds by *Onchocerca gutturosa* (Nematoda: Filarioidia). *International Journal for Parasitology*, **19**, 875–878.

Perry, R.N., Clarke, A.J. and Hennessy, J. (1980) The influence of osmotic pressure on the hatching of *Heterodera schachtii*. *Revue de Nématologie*, **3**, 3–9.

Perry, R.N., Clarke, A.J., Hennessy, J. and Beane, J. (1983) The role of trehalose in the hatching mechanism of *Heterodera goettingiana*. *Nematologica*, **29**, 323–334.

Persaud, A.D., Perry, R.N. and Barrett, J. (1997) The purification and properties of glutathione *S*-transferase from *Ditylenchus myceliophagus*. *Fundamental and Applied Nematology*, **20**, 601–609.

Pierce, S.J., Morse, T.M. and Lockery, S.R. (1999) The fundamental role of pirouettes in *Caenorhabditis elegans* chemotaxis. *Journal of Neuroscience*, **19**, 9557–9569.

Platzer, E.G. (1974) Comparative biochemistry of folate metabolism in *Aphelenchus avenae* and *Nippostrongylus brasiliensis* (Nematoda). *Comparative Biochemistry and Physiology*, **49B**, 3–13.

Poole, C.B., Grandea, A.G., Maina, C.V., Jenkins, R.E., Selkirk, M.E. and McReynolds, L.A. (1992) Cloning of a cuticular antigen that contains multiple tandem repeats from the filarial parasite *Dirofilaria immitis*. *Proceedings of the National Academy of Sciences, U.S.A.*, **89**, 5986–5990.

Powell, J.W., Stables, J.N. and Watt, R.A. (1986a) An investigation of the glucose metabolism of *Brugia pahangi* and *Dipetalonema viteae* by nuclear magnetic resonance spectroscopy. *Molecular and Biochemical Parasitology*, **18**, 171–182.

Powell, J.W., Stables, J.N. and Watt, R.A. (1986b) An NMR study on the effect of glucose availability on carbohydrate metabolism in *Dipetalonema viteae* and *Brugia pahangi*. *Molecular and Biochemical Parasitology*, **19**, 265–271.

Powell-Coffman, J.A., Knight, J. and Wood, W.B. (1996) Onset of *C. elegans* gastrulation is blocked by inhibition of embryonic transcription with an RNA polymerase antisense RNA. *Developmental Biology*, **178**, 472–483.

Precious, W.Y. and Barrett, J. (1989a) The possible absence of cytochrome *P*-450 linked xenobiotic metabolism in helminths. *Biochimica et Biophysica Acta*, **992**, 215–222.

Precious, W.Y. and Barrett, J. (1989b) Xenobiotic metabolism in helminths. *Parasitology Today*, **5**, 156–160.

Prichard, R.K. and Rothwell, T.L.W. (1972) Volatile fatty acid production by parasitic fourth-stage *Trichostrongylus colubriformis*. *Journal of Parasitology*, **58**, 1161.

Rathaur, S., Anwar, N., Saxena, J.K. and Ghatak, S. (1982) *Setaria cervi*: enzymes in microfilariae and in vitro action of antifilarials. *Zeitschrift für Parasitenkunde*, **68**, 331–338.

Rew, R.S., Douvres, F.W. and Saz, H.J. (1983) *Oesophagostomum radiatum*: glucose metabolism of larvae grown *in vitro* and adults grown *in vivo*. *Experimental Parasitology*, **55**, 179–187.

Riddle, D.L. and Albert, P.S. (1997) Genetic and environmental regulation of dauer larva development. In *C. elegans II*, edited by D. Riddle, B. Meyer, J. Priess and T. Blumenthal, pp. 739–768. Cold Spring Harbor: Cold Spring Harbor Press.

Roberts, L.S. and Fairbairn, D. (1965) Metabolic studies on adult *Nippostrongylus brasiliensis* (Nematoda: Trichostrongyloidea). *Journal of Parasitology*, **51**, 129–138.

Robertson, L., Robertson, W.M. and Jones, J.T. (1999) Direct analysis of the secretions of the potato cyst nematode *Globodera rostochiensis*. *Parasitology*, **119**, 167–176.

Rodrick, G.E., Long, S.D., Sodeman, W.A. and Smith, D.L. (1982a) *Ascaris suum*: oxidative phosphorylation in mitochondria from developing eggs and adult muscle. *Experimental Parasitology*, **54**, 235–242.

Rodrick, G.E., Long, S.D., Sodeman, W.J. and Smith, D.L. (1982b) Mitochondrial ultrastructural and ATPase changes during the life cycle of *Ascaris suum*. *Memorias do Instituto Oswaldo Cruz*, **77**, 173–180.

Rohrer, S.P., Saz, H.J. and Nowak, T. (1986) Purification and characterization of phosphoenolpyruvate carboxykinase from the parasitic helminth *Ascaris suum*. *Journal of Biological Chemistry*, **261**, 13049–13055.

Rojas, J., Rodriguez, O.M. and Gomez, G.V. (1997) Immunological characteristics and localization of the *Trichinella spiralis* glutathione *S*-transferase. *Journal of Parasitology*, **83**, 630–635.

Roos, M.H. and Tielens, A.G.M. (1994) Differential expression of two succinate dehydrogenase subunit-B genes and a transition in energy metabolism during the development of the parasitic nematode *Haemonchus contortus*. *Molecular and Biochemical Parasitology*, **66**, 273–281.

Rothstein, M. (1968) Nematode biochemistry. IX. Lack of sterol biosynthesis in free-living nematodes. *Comparative Biochemistry and Physiology*, **27**, 309–317.

Rothstein, M. (1970a) Nematode biochemistry XI. Biosynthesis of fatty acids by *Caenorhabditis briggsae* and *Panagrellus redivivus*. *International Journal of Biochemistry*, **1**, 422–428.

Rothstein, M. (1970b) Nitrogen metabolism in the Aschelminthes. In *Comparative Biochemistry of Nitrogen Metabolism*, edited by J.W. Campbell, pp. 91–102. New York: Academic Press.

Rothstein, M. and Gotz, P. (1968) Biosynthesis of fatty acids in the free-living nematode *Turbatrix aceti*. *Archives of Biochemistry and Biophysics*, **126**, 131–140.

Rothwell, J. and Sangster, N. (1997) *Haemonchus contortus*: the uptake and metabolism of closantel. *International Journal for Parasitology*, **27**, 313–319.

Salinas, G., Braun, G. and Taylor, D.W. (1994) Molecular characterisation and localisation of an *Onchocerca volvulus* π-class glutathione *S*-transferase. *Molecular and Biochemical Parasitology*, **66**, 1–9.

Sanford, T., Golomb, M. and Riddle, D.L. (1983) RNA polymerase II from wild type and α-amanitin-resistant strains of *Caenorhabditis elegans*. *Journal of Biological Chemistry*, **258**, 12804–12809.

Sanford, T., Prenger, J.P. and Golomb, M. (1985) Purification and immunological analysis of RNA polymerase II from *Caenorhabditis elegans*. *Journal of Biological Chemistry*, **260**, 8064–8069.

Sangster, N.C. and Prichard, R.K. (1984) Uptake of thiabendazole and its effects on glucose uptake and carbohydrate levels in the thiabendazole-resistant and susceptible *Trichostrongylus colubriformis*. *International Journal for Parasitology*, **14**, 121–126.

Sangster, N.C. and Prichard, R.K. (1985) The contribution of a partial tricarboxylic acid cycle to volatile end-products in thiabendazole-resistant and susceptible *Trichostrongylus colubriformis*. *Molecular and Biochemical Parasitology*, **14**, 261–274.

Sani, B.P. and Comley, J.C.W. (1985) Role of retinoids and their binding proteins in filarial parasites and host tissues. *Tropenmedizin und Parasitologie*, **36** (Suppl.), 20–23.

Sani, B.P. and Vaid, A. (1988) Specific interaction of ivermectin with retinol-binding protein from filarial parasites. *Biochemical Journal*, **249**, 929–932.

Sani, B.P., Vaid, A., Comley, J.C. and Montgomery, J.A. (1985) Novel retinoid-binding proteins from filarial parasites. *Biochemical Journal*, **232**, 577–583.

Saruta, F., Kuramochi, T., Nakamura, K., Takamiya, S., Yu, Y., Aoki, T., et al. (1995) Stage-specific isoforms of complex II (succinate-ubiquinone oxidoreductase) in mitochondria from the parasitic nematode, *Ascaris suum*. *Journal of Biological Chemistry*, **270**, 928–932.

Sarwal, R., Sanyal, S.N. and Khera, S. (1989) Lipid metabolism in *Trichuris globulosa* (Nematoda). *Journal of Helminthology*, **63**, 287–297.

Satouchi, K., Hirano, K., Sakaguchi, M., Takehara, H. and Matsuura, F. (1993) Phospholipids from the free-living nematode *Caenorhabditis elegans*. *Lipids*, **28**, 837–840.

Sayre, F.W., Hansen, E.L. and Yarwood, E.A. (1963) Biochemical aspects of the nutrition of *Caenorhabditis briggsae*. *Experimental Parasitology*, **13**, 98–107.

Saz, D.K., Bonner, T.P., Karlin, M. and Saz, H.J. (1971) Biochemical observations on adult *Nippostrongylus brasiliensis*. *Journal of Parasitology*, **57**, 1159–1162.

Saz, H.J. (1981) Energy metabolisms of parasitic helminths: adaptations to parasitism. *Annual Review of Physiology*, **43**, 323–341.

Saz, H.J. (1990) Helminths — primary models for comparative biochemistry. *Parasitology Today*, **6**, 92–93.

Saz, H.J. and Vidrine, A. (1959) The mechanism of formation of succinate and propionate by *Ascaris lumbricoides* muscle. *Journal of Biological Chemistry*, **234**, 2001–2005.

Saz, H.J. and Weil, A. (1960) The mechanism of formation of α-methylbutyrate from carbohydrate by *Ascaris lumbricoides* muscle. *Journal of Biological Chemistry*, **235**, 914–918.

Saz, H.J. and Weil, A. (1962) Pathway of formation of α-methylvalerate by *Ascaris lumbricoides* muscle. *Journal of Biological Chemistry*, **237**, 2053–2056.

Schmitt-Wrede, H.P., Waldraff, A., Krücken, J., Harder, A. and Wunderlich, F. (1999) Characterization of a hexokinase encoding cDNA of the parasitic nematode *Haemonchus contortus*. *Biochimica et Biophysica Acta*, **1444**, 439–444.

Schrum, S., Bialonski, A., Marti, T. and Zipfel, P.F. (1998) Identification of a peroxidoxin protein (OvPXN-2) of the human parasitic nematode *Onchocerca volvulus* by sequential protein fractionation. *Molecular and Biochemical Parasitology*, **94**, 131–135.

Selkirk, M.E., Smith, V.P., Thomas, G.R. and Gounaris, K. (1998) Resistance of filarial nematode parasites to oxidative stress. *International Journal for Parasitology*, **28**, 1315–1332.

Shih, H.H. and Chen, S.N. (1982) Glycolytic enzymes in juvenile and adult *Angiostrongylus cantonensis*. *Southeast Asian Journal of Tropical Medicine and Public Health*, **13**, 114–119.

Silberkang, M., Havel, C.M., Friend, D.S., McCarthy, B.J. and Watson, J.A. (1983) Isoprene synthesis in isolated embryonic *Drosophila* cells. I. Sterol-deficient eukaryotic cells. *Journal of Biological Chemistry*, **258**, 8503–8511.

Sims, S.M., Magas, L.T., Barsuhn, C.L., Ho, N.F.H., Geary, T.G. and Thompson, D.P. (1992) Mechanisms of microenvironmental pH regulation in the cuticle of *Ascaris suum*. *Molecular and Biochemical Parasitology*, **53**, 135–148.

Singh, S.P., Gupta, S., Katiyar, J.C. and Srivastava, V.M.L. (1992) On the potential of amino acids to support survival and energy status of *Nippostrongylus brasiliensis in vitro*. *International Journal for Parasitology*, **22**, 131–133.

Singh, S.P., Katiyar, J.C. and Srivastava, V.M. (1992) Enzymes of the tricarboxylic acid cycle in *Ancylostoma ceylanicum* and *Nippostrongylus brasiliensis*. *Journal of Parasitology*, **78**, 24–29.

Smith, V.P., Selkirk, M.E. and Gounaris, K. (1996) Identification and composition of lipid classes in surface and somatic preparations of adult *Brugia malayi*. *Molecular and Biochemical Parasitology*, **78**, 105–116.

Smith, V.P., Selkirk, M.E. and Gounaris, K. (1998) *Brugia malayi*: resistance of cuticular lipids to oxidant-induced damage and detection of α-tocopherol in the neutral lipid fraction. *Experimental Parasitology*, **88**, 103–110.

Snyder, M.J. and Maddison, D.R. (1997) Molecular phylogeny of glutathione-*S*-transferases. *DNA and Cell Biology*, **16**, 1373–1384.

So, N., Wong, P.C.L. and Ko, R.C. (1994) *Angiostrongylus cantonensis*: characterization of thymidylate synthetase. *Experimental Parasitology*, **79**, 526–535.

So, N.N., Wong, P.C. and Ko, R.C. (1992) Precursors of pyrimidine nucleotide biosynthesis for gravid *Angiostrongylus cantonensis* (Nematoda: Metastrongyloidea). *International Journal for Parasitology*, **22**, 427–433.

So, N.N., Wong, P.C. and Ko, R.C. (1993) Pathways of pyrimidine nucleotide biosynthesis in gravid *Angiostrongylus cantonensis*. *Molecular and Biochemical Parasitology*, **60**, 45–51.

Song, H. and Komuniecki, R. (1994) Novel regulation of pyruvate dehydrogenase phosphatase purified from anaerobic muscle mitochondria of the adult parasitic nematode, *Ascaris suum*. *Journal of Biological Chemistry*, **269**, 31573–31578.

Spychalla, J.P., Kinney, A.J. and Browse, J. (1997) Identification of an animal ω-3 fatty acid desaturase by heterologous expression in *Arabidopsis*. *Proceedings of the National Academy of Sciences, U.S.A.*, **94**, 1142–1147.

Srinivasan, N.G., Rao, G.S. and Harris, B.G. (1990) Phosphofructokinase from *Dirofilaria immitis*: effect of fructose 2,6-bisphosphate and AMP on the non-phosphorylated and phosphorylated forms of the enzyme. *Molecular and Biochemical Parasitology*, **38**, 151–158.

Srinivasan, N.G., Wariso, B.A., Kulkarni, G., Rao, G.S. and Harris, B.G. (1988) Phosphofructokinase from *Dirofilaria immitis*. Stimulation of activity by phosphorylation with cyclic AMP-dependent protein kinase. *Journal of Biological Chemistry*, **263**, 3482–3485.

Srivastava, V.M.L., Saz, H.J. and deBruyn, B. (1988) Comparisons of glucose and amino acid use in adult and microfilariae of *Brugia pahangi*. *Parasitology Research*, **75**, 1–6.

Striebel, H.P. (1988) Proposed form to evaluate some histological aspects of macrofilarial morphology, its age dependent alterations and drug related changes in nodules of *Onchocerca volvulus* and *O. gibsoni*. *Tropical Medicine and Parasitology*, **39**, 367–389.

Stürchler, D., Wyss, F. and Hanck, A. (1981) Retinol, onchocerciasis and *Onchocerca volvulus*. *Transactions of the Royal Society of Tropical Medicine and Hygiene*, **75**, 617.

Suarez De Mata, Z., Lizardo, R., Diaz, F. and Saz, H.J. (1991) Propionyl-CoA condensing enzyme from *Ascaris* muscle mitochondria I. Isolation and characterization of multiple forms. *Archives of Biochemistry and Biophysics*, **285**, 158–165.

Sugita, M., Mizunoma, T., Aoki, K., Dulaney, J.T., Inagaki, F., Suzuki, M., *et al.* (1996) Structural characterization of a novel glycoinositolpho-spholipid from the parasitic nematode, *Ascaris suum*. *Biochimica et Biophysica Acta*, **1302**, 185–192.

Supowit, S.C. and Harris, B.G. (1976) *Ascaris* suum hexokinase: purification and possible function in compartmentation of glucose 6-phosphate in muscle. *Biochimica et Biophysica Acta*, **422**, 48–59.

Taiwo, F.A., Brophy, P.M., Pritchard, D.I., Brown, A., Wardlaw, A. and Patterson, L.H. (1999) Cu/Zn superoxide dismutase in excretory-secretory products of the human hookworm *Necator americanus*. An electron paramagnetic spectrometry study. *European Journal of Biochemistry*, **264**, 434–438.

Takamiya, S., Kita, K., Wang, H., Weinstein, P.P., Hiraishi, A., Oya, H., *et al.* (1993) Developmental changes in the respiratory chain of *Ascaris* mitochondria. *Biochimica et Biophysica Acta*, **1141**, 65–74.

Tanaka, T., Ikita, K., Ashida, T., Motoyama, Y., Yamaguchi, Y. and Satouchi, K. (1996) Effects of growth temperature on the fatty acid composition of the free-living nematode *Caenorhabditis elegans*. *Lipids*, **31**, 1173–1178.

Tanaka, T., Izuwa, S., Tanaka, K., Yamamoto, D., Takimoto, T., Matsuura, F., *et al.* (1999) Biosynthesis of 1,2-dieicosapentaenoyl-*sn*-glycero-3-phosphocholine in *Caenorhabditis elegans*. *European Journal of Biochemistry*, **263**, 189–195.

Tang, L., Gounaris, K., Griffiths, C.M. and Selkirk, M.E. (1995) Heterologous expression and enzymatic properties of a selenium-independent glutathione peroxidase from the parasitic nematode *Brugia pahangi*. *Journal of Biological Chemistry*, **270**, 18313–18318.

Tang, L., Ou, X., Henkle-Dührsen, K. and Selkirk, M.E. (1994) Extracellular and cytoplasmic CuZn superoxide dismutases from *Brugia* lymphatic filarial nematode parasites. *Infection and Immunity*, **62**, 961–967.

Tang, L., Smith, V.P., Gounaris, K. and Selkirk, M.E. (1996) *Brugia pahangi*: the cuticular glutathione peroxidase (gp29) protects heterologous membranes from lipid peroxidation. *Experimental Parasitology*, **82**, 329–332.

Targovnik, H.S., Locher, S.E., Hart, T.F. and Hariharan, P.V. (1984) Age-related changes in the excision repair capacity of *Turbatrix aceti*. *Mechanisms of Ageing and Development*, **27**, 73–81.

Taub, J., Lau, J.F., Ma, C., Hahn, J.H., Hoque, R., Rothblatt, J., *et al.* (1999) A cytosolic catalase is needed to extend adult lifespan in *C. elegans daf-C* and *clk-1* mutants. *Nature*, **399**, 162–166.

Tawe, W.N., Eschbach, M.L., Walter, R.D. and Henkle-Dührsen, K. (1998) Identification of stress-responsive genes in *Caenorhabditis elegans* using RT-PCR differential display. *Nucleic Acids Research*, **26**, 1621–1627.

Taylor, M.J., Cross, H.F., Mohammed, A.A., Trees, A.J. and Bianco, A.E. (1996) Susceptibility of *Brugia malayi* and *Onchocerca lienalis* microfilariae to nitric oxide and hydrogen peroxide in cell-free culture and from IFNγ-activated macrophages. *Parasitology*, **112**, 315–322.

Thalhofer, H.P., Daum, G., Harris, B.G. and Hofer, H.W. (1988) Identification of two different phosphofructokinase-phosphorylating protein kinases from *Ascaris suum* muscle. *Journal of Biological Chemistry*, **263**, 952–957.

The *C. elegans* Sequencing Consortium (1998) Genome sequence of the nematode *C. elegans*: a platform for investigating biology. *Science*, **282**, 2012–2018.

Thirugnanam, M. and Myers, R.F. (1974) Nutrient media for plant-parasitic nematodes. VI. Nucleic acids content and nucleotide synthesis in *Aphelenchoides rutgersi*. *Experimental Parasitology*, **36**, 202–209.

Thissen, J., Desai, S., McCartney, P. and Komuniecki, R. (1986) Improved purification of the pyruvate dehydrogenase complex from *Ascaris suum* body wall muscle and characterization of PDH$_a$ kinase activity. *Molecular and Biochemical Parasitology*, **21**, 129–138.

Thompson, S.N., Platzer, E.G. and Lee, R.W.-K. (1991) Bioenergetics in a parasitic nematode, *Steinernema carpocapsae*, monitored in vivo by flow NMR spectroscopy. *Parasitology Research*, **77**, 86–90.

Tielens, A.G.M. (1994) Energy generation in parasitic helminths. *Parasitology Today*, **10**, 346–352.

Tielens, A.G.M. and Van den Bergh, S.G. (1986) The (an)aerobic energy metabolism of parasitic helminths. *Molecular Physiology*, **8**, 359–369.

Tielens, A.G.M. and Van Hellemond, J.J. (1998) The electron transport chain in anaerobically functioning eukaryotes. *Biochimica et Biophysica Acta Bioenergetics*, **1365**, 71–78.

Timanova, A., Müller, S., Marti, T., Bankov, I. and Walter, R.D. (1999) *Ascaridia galli* fatty acid-binding protein, a member of the nematode polyprotein allergens family. *European Journal of Biochemistry*, **261**, 569–576.

Tree, T.I.M., Gillespie, A.J., Shepley, K.J., Blaxter, M.L., Tuan, R.S. and Bradley, J.E. (1995) Characterisation of an immunodominant glycoprotein antigen of *Onchocerca volvulus* with homologues in other filarial nematodes and *Caenorhabditis elegans*. *Molecular and Biochemical Parasitology*, **69**, 185–195.

Treptau, T., Piram, P., Cook, P.F., Rodriguez, P.H., Hoffmann, R., Jung, S., et al. (1996) Comparison of the substrate specificities of cAMP-dependent protein kinase from bovine heart and *Ascaris suum* muscle. *Biological Chemistry Hoppe Seyler*, **377**, 203–209.

Tripp, C., Frank, R.S., Selkirk, M.E., Tang, L., Grieve, M.M., Frank, G.R., et al. (1998) *Dirofilaria immitis*: molecular cloning and expression of a cDNA encoding a selenium-independent secreted glutathione peroxidase. *Experimental Parasitology*, **88**, 43–50.

Turner, A.C. and Hutchison, W.F. (1982) Oxidative decarboxylation reactions in *Dirofilaria immitis* glucose metabolism. *Comparative Biochemistry and Physiology*, **73B**, 331–334.

Tweedie, S., Paxton, W.A., Ingram, L., Maizels, R.M., McReynolds, L.A. and Selkirk, M.E. (1993) *Brugia pahangi* and *Brugia malayi*: a surface-associated glycoprotein (gp15/400) is composed of multiple tandemly repeated units and processed from a 400-kDa precursor. *Experimental Parasitology*, **76**, 156–164.

Unnikrishnan, L.S. and Raj, R.K. (1995) Transhydrogenase activities and malate dismutation linked to fumarate reductase system in the filarial parasite *Setaria digitata*. *International Journal for Parasitology*, **25**, 779–785.

Van Hellemond, J.J., Klockiewicz, M., Gaasenbeek, C.P.H., Roos, M.H. and Tielens, A.M. (1995) Rhodoquinone and complex II of the electron transport chain in anaerobically functioning eukaryotes. *Journal of Biological Chemistry*, **270**, 31065–31070.

Van Hellemond, J.J. and Tielens, A.G.M. (1994) Expression and functional properties of fumarate reductase. *Biochemical Journal*, **304**, 321–331.

Vande Waa, E.A., Foster, L.A., Deruiter, J., Guderian, R.H., Williams, J.F. and Geary, T.G. (1993) Glutamine-supported motility of adult filarial parasites in vitro and the effect of glutamine antimetabolites. *Journal of Parasitology*, **79**, 173–180.

Vanfleteren, J.R. (1980) Nematodes as nutritional models. In *Nematodes as Biological Models*, edited by B.M. Zuckerman, pp. 47–77. New York: Academic Press.

Vanfleteren, J.R. (1993) Oxidative stress and ageing in *Caenorhabditis elegans*. *Biochemical Journal*, **292**, 605–608.

Vanover-Dettling, L. and Komuniecki, P.R. (1989) Effect of gas phase on carbohydrate metabolism in *Ascaris suum* larvae. *Molecular and Biochemical Parasitology*, **36**, 29–40.

Voet, D., Voet, J.G. and Pratt, C.W. (1999) *Fundamentals of Biochemistry*, New York: John Wiley & Sons, Inc.

Walker, J. and Barrett, J. (1991) Pyridoxal 5′-phosphate dependent enzymes in the nematode *Nippostrongylus brasiliensis*. *International Journal for Parasitology*, **21**, 641–649.

Walker, J. and Barrett, J. (1997) Parasite sulphur amino acid metabolism. *International Journal for Parasitology*, **27**, 883–897.

Walter, R.D. and Albiez, E.J. (1981) Inhibition of NADP-linked malic enzyme from *Onchocerca volvulus* and *Dirofilaria immitis* by suramin. *Molecular and Biochemical Parasitology*, **4**, 53–60.

Walter, R.D. and Albiez, E.J. (1986) Phosphoenolpyruvate carboxykinase from *Onchocerca volvulus* and *O. gibsoni*. *Tropical Medicine and Parasitology*, **37**, 356–358.

Wang, E.J. and Saz, H.J. (1974) Comparative biochemical studies of *Litomosoides carinii*, *Dipetalonema viteae*, and *Brugia pahangi* adults. *Journal of Parasitology*, **60**, 316–321.

Ward, P.F.V. (1974) The metabolism of glucose by *Haemonchus contortus*, *in vitro*. *Parasitology*, **69**, 175–190.

Warren, L.G. and Poole, W.J. (1970) Biochemistry of the dog hookworm II. Nature and origin of the excreted fatty acids. *Experimental Parasitology*, **27**, 408–416.

Watts, J.L. and Browse, J. (1999) Isolation and characterization of a Δ^5-fatty acid desaturase from *Caenorhabditis elegans*. *Archives of Biochemistry and Biophysics*, **362**, 175–182.

Weinstein, P.P. (1996) Vitamin B_{12} changes in *Nippostrongylus brasiliensis* in its free-living and parasitic habitats with biochemical implications. *Journal of Parasitology*, **82**, 1–6.

Weinstein, P.P. and Jaffe, J.J. (1987) Cobalamin and folate metabolism in helminths. *Blood Reviews*, **1**, 245–253.

Weiss, P.M., Gavva, S.R., Harris, B.G., Urbauer, J.L., Cleland, W.W. and Cook, P.F. (1991) Multiple isotope effects with alternative dinucleotide substrates as a probe of the malic enzyme reaction. *Biochemistry*, **30**, 5755–5763.

Weller, P.F., Longworth, D.L. and Jaffe, J.J. (1989) Leukotriene C4 synthesis catalyzed by *Dirofilaria immitis* glutathione S-transferase. *American Journal of Tropical Medicine and Hygiene*, **40**, 171–175.

Wera, S., De Schriver, E., Geyskens, I., Nwaka, S. and Thevelein, J.M. (1999) Opposite roles of trehalase activity in heat-shock recovery and heat-shock survival in *Saccharomyces cerevisiae*. *Biochemical Journal*, **343**, 621–626.

Weston, K.M., O'Brien, R.W. and Prichard, R.K. (1984) Respiratory metabolism and thiabendazole susceptibility in developing eggs of *Haemonchus contortus*. *International Journal for Parasitology*, **14**, 159–164.

Wildenburg, G., Liebau, E. and Henkle-Dührsen, K. (1998) *Onchocerca volvulus*: ultrastructural localization of two glutathione S-transferases. *Experimental Parasitology*, **88**, 34–42.

Wilkes, J., Cornish, R.A. and Mettrick, D.F. (1982) Purification and properties of phosphoenolpyruvate carboxykinase from *Ascaris suum*. *International Journal for Parasitology*, **12**, 163–171.

Wilson, P.A.G. (1965) Changes in lipid and nitrogen content of *Nippostrongylus brasiliensis* infective larvae aged at constant temperature. *Experimental Parasitology*, **16**, 190–194.

Wittich, R.-M. and Walter, R.D. (1987) *Onchocerca volvulus*: partial glucose catabolism via fumarate and succinate. *Experimental Parasitology*, **64**, 517–518.

Wolff, K.M. and Scott, A.L. (1995) *Brugia malayi*: retinoic acid uptake and localization. *Experimental Parasitology*, **80**, 282–290.

Wong, P.C. and Ko, R.C. (1979) *De novo* purine ribonucleotide biosynthesis in adult *Angiostrongylus cantonensis* (Nematoda: Metastrongyloidea). *Comparative Biochemistry and Physiology*, **62B**, 129–132.

Wong, P.C. and Yeung, S.B. (1981) Pathways of purine ribonucleotide biosynthesis in the adult worm *Metastrongylus apri* (Nematoda: Metastrongyloidea) from pig lung. *Molecular and Biochemical Parasitology*, **2**, 285–293.

Wright, D.J. (1975a) Elimination of nitrogenous compounds by *Panagrellus redivivus*, Goodey, 1945 (Nematoda: Cephalobidae). *Comparative Biochemistry and Physiology*, **52B**, 247–253.

Wright, D.J. (1975b) Studies on nitrogen catabolism in *Panagrellus redivivus*, Goodey, 1945 (Nematoda: Cephalobidae). *Comparative Biochemistry and Physiology*, **52B**, 255–260.

Wright, D.J. (1998) Respiratory physiology, nitrogen excretion and osmotic and ionic regulation. In *The Physiology and Biochemistry of Free-living and Plant-parasitic Nematodes*, edited by R.N. Perry and D.J. Wright, pp. 103–131. Wallingford: CABI Publishing.

Wright, D.J., Grewal, P.S. and Stolinski, M. (1997) Relative importance of neutral lipids and glycogen as energy stores in dauer larvae of two entomopathogenic nematodes, *Steinernema carpocapsae* and *Steinernema feltiae*. *Comparative Biochemistry and Physiology*, **118B**, 269–273.

Wright, D.J. and Newall, D.R. (1976) Nitrogen excretion, osmotic and ionic regulation in nematodes. In *The Organization of Nematodes*, edited by N.A. Croll, pp. 163–210. London: Academic Press.

Yacoub, N.J., Allen, B.L., Payne, D.M., Masaracchia, R.A. and Harris, B.G. (1983) Purification and characterization of phosphorylase B from *Ascaris suum*. *Molecular and Biochemical Parasitology*, **9**, 297–307.

Zahner, H., Sani, B.P., Shealy, Y.F. and Nitschmann, A. (1989) Antifilarial activities of synthetic and natural retinoids in vitro. *Tropical Medicine and Parasitology*, **40**, 322–326.

Zuckerman, B.M. and Geist, M.A. (1983) Effects of vitamin E on the nematode *C. elegans*. *Age*, **6**, 1–4.

11. Excretion/Secretion, Ionic and Osmotic Regulation

David P. Thompson and Timothy G. Geary
Pharmacia and Upjohn, Inc., Kalamazoo, MI 49001, USA

Keywords

Osmotic regulation, Ionic regulation, Excretion, Electro-chemical gradients, Secretory products, *Caenorhabditis elegans*, *Ascaris suum*

Introduction

Nematodes reside in diverse environments, including fresh and salt water, soil, plant roots and various animal tissues, especially the intestinal tract of vertebrates. Indeed, a species of nematode is adapted for every environment on earth where organisms can be surrounded by at least a film of water (Wharton, 1986). Aside from moisture, each habitat poses particular challenges for nematode survival. These challenges include physicochemical stressors, such as osmotic strength, ion concentration, pH, and O_2 tension that can exceed the levels typically associated with metazoan existence. Nematodes are also subjected to hostile reactions, such as immune responses and digestive enzymes, from host animals. Considering how successful they are as a phylum of organisms, nematodes have obviously evolved ways to overcome a wide variety of environmental challenges. In this chapter, we discuss the functional, and where available, molecular biology of the processes and tissues that help maintain homeostasis in nematodes, defined here as a balance of electrochemical and osmotic (including hydrodynamic) gradients between the pseudocoelomic fluid and the external environment. Our discussion includes a consideration of the structures and tissues that establish and maintain these gradients. In addition, we distinguish processes of excretion, in which indigestible material and waste products are removed from internal tissue compartments, from those of secretion, the release of bioactive molecules that affect the environment. Nematodes secrete molecules that, among other effects, contribute to immune evasion, extracorporal digestion and localised paralysis of the host intestine. We also briefly consider the anatomical structures from which such molecules are secreted, including the tubular system, epidermal gland cells, pharyngeal glands, rectal glands and stichosomes. Detailed descriptions of each of these structures are provided in Chapter 2.

Species- and Stage-dependence

The environments of nematodes in the soil (plant-parasitic and free-living species) can fluctuate markedly in temperature and relative humidity. In contrast, aquatic nematodes enjoy an environment that may be relatively invariant, aside from gradual shifts in temperature. At the other extreme, animal-parasitic nematodes must adapt to two or more distinct and often harsh environments during their life cycle. For instance, infective larvae (embryonated eggs) of *Ascaris suum* are shed in grass or mud, where they encounter salt concentrations of only a few mM, oxygen tensions of 100–200 mm Hg, and near neutral pH. These eggs may be desiccated for long periods and even cooled to below freezing. Following ingestion, infective larvae enter the porcine stomach, a highly acidic environment that includes an array of powerful digestive enzymes, osmotic pressures from 200–500 mOsm and oxygen tension close to zero. From the stomach, they pass into the intestine, where they hatch, then travel through the liver and the highly aerobic environment of the lungs. L3 larvae then migrate to the trachea, are swallowed and return to the stomach before eventually re-entering the anaerobic environment of the small intestine, where they mature and reproduce. At each site within the host, larvae and adult stages are exposed to immune effectors, many of which are designed specifically to destroy them.

This pattern differs from that observed for other types of parasites. Filarial nematodes parasitise internal tissue compartments of their hosts and never experience existence in a free-living stage; they are transmitted by a species of insect as an intermediate host, moving from insect to mammal through the process of insect feeding. Certain plant-parasitic nematodes evoke elaborate alterations of host cell morphology and function to create tailored feeding sites. A conceptually similar phenomenon is the process of nurse cell creation by larval stages of *Trichinella spiralis* in a mammalian host muscle. Adult female *Onchocerca volvulus* elicit the construction of a nodule from the host response, in which worms can live and reproduce for up to a decade. Each of these animals is adapted to the peculiar conditions provided by its habitat. Insofar as these include markedly different pH and ionic and osmotic conditions, and vary in the abundance, quality and type of food available (which can influence excretion profiles), one should not be surprised to find that it is not easy to identify principles that apply to all nematodes. Instead, we will provide examples of the processes involved in homeostasis in different species and stages of nematodes to illustrate the kinds of complex physiology that occur in this phylum.

Processes

That nematodes are adapted to thrive in such a wide range of conditions, which often vary as the animal matures, is a testament to the effectiveness of the strategies they use to cope with their environments. In this chapter, four aspects of the worm:environment interface are considered. Nematodes, like other organisms, interact with the environment by adding material to it through the processes of excretion and secretion. The latter process carries with it the implication that the expelled molecules affect the environment in a way that enhances the fitness of the nematode. Several cell types and tissues, including the ventral gland, lateral canals, pseudocoelomocytes, intestine and epidermis, may serve both functions. It has thus been difficult to clearly distinguish products of excretion from products of secretion, and the two are often combined in the term excretory/secretory (E/S) products. Due to the size of most nematode species, it has rarely been possible to isolate and chemically characterise E/S products from individual cell types or organs. Protocols used to identify E/S products from nematodes usually involve the incubation of intact worms in defined media for some period, and measuring the appearance in the medium of specific types of molecules. These studies shed little light on the cells from which the E/S products originate, or the mechanisms underlying their extrusion. Also, some molecules generally thought of as excretory products, including H^+ and CO_2, may directly affect the nematode's environment in a way that is, for some species, essential to survival, thus satisfying one criterion for secreted products.

Nematodes, like all organisms, must regulate the flow of water and solutes (including ions) across cell membranes, including those that form the body wall. The transport processes that establish and maintain water gradients contribute to osmotic regulation. Transport processes that contribute to ionic regulation maintain ion gradients within narrow limits, are highly interdependent with each other and with water transport, and are designed to maintain homeostasis within the organism. The critical roles these processes serve are illustrated most clearly by the consequences of their inhibition by experimental treatments or by anthelmintics. For example, L4 larvae and adults of most animal-parasitic species usually swell and die when placed in distilled water. In contrast, earlier stage larvae, which are adapted for hyposmotic conditions (and thus can survive in puddles of rainwater), tolerate prolonged incubation in distilled water. Pharmacological examples of the importance of maintaining specific ionic gradients for survival are provided by salicylanilide and β-ketoamide anthelmintics, including closantel, which deplete nematodes and trematodes of ATP by abolishing proton gradients across mitochondrial membranes (Bacon et al., 1998; see Rew and Fetterer, 1986 for review). Levamisole and the macrocyclic lactone anthelmintics paralyse nematodes by opening cation and Cl^- channels, respectively, and so disrupt the normally tightly regulated concentration gradients for these inorganic ions across nerve and muscle membranes (Chapter 12). Unfortunately, few examples of anthelmintics or experimental compounds that affect ionic gradients across tissues important for excretion, secretion, or the regulation of osmotic and ionic gradients (e.g., the epidermal or intestinal membranes) have been reported.

Tissues

Details regarding the structural organisation of nematodes are provided in Chapter 2. We offer here a brief description of tissues and structures that maintain ion and water gradients across the body wall and serve in E/S processes. The physiological properties of these structures are described later in more detail.

Pseudocoelom/pseudocoelomic fluid

Nematodes are pseudocoelomates and possess a fluid-filled body cavity. The nematode pseodocoelom differs from that of a true coelom by the absence of both a mesentary (membranous lining of the body cavity) and a muscle layer surrounding the intestine. Fluid in the pseudocoelom is in constant contact with almost all of the cells in nematodes, and only one cell layer separates the pseudocoelomic fluid (PCF) from the external environment. PCF thus serves two critical roles: it is the predominant extracellular fluid compartment and also the closest thing to a circulatory system found in nematodes (Wharton, 1986; McKerrow, Huima and Lustigman, 1999). The composition of PCF is complex and probably varies considerably among nematode species. In *A. suum*, it contains inorganic and organic ions, lipid, carbohydrates and proteins (and the products of their catabolism) (von Brand, 1973) and haemoglobin (Lee and Smith, 1965).

Epidermis/cuticle complex

The pseudocoel is bounded by the epidermis/cuticle complex, which is an important site for the transport of inorganic ions, water and organic waste products. The cuticle forms as an extracellular matrix, consisting primarily of cross-linked collagen-like proteins. The structures of nematode cuticles are tremendously diverse (Howells, 1980; Maizels et al., 1993; Thompson and Geary, 1995). Among species that parasitise vertebrates, the cuticle contains several distinct layers. The outermost layer, or epicuticle, is 6–30 nm thick and appears to be trilaminate in structure at the level of electron micrographs. Considerable controversy exists about whether the epicuticle is the true limiting 'membrane' in nematodes (Wright, 1987). Several observations refute that concept. First, the epicuticle is not dissociated by treatments that usually dissolve membranes (Murrell, Graham and McGreevy, 1983; Ho et al., 1990). Second, lipophilic markers do not exhibit the lateral mobility characteristic of membranes when inserted into the surface of nematodes (Kennedy *et al.*, 1987). It is devoid of intramembranous particles that are normally associated with cell membranes (Wright, 1987). Finally, charged solutes diffuse across the cuticle at rates that indicate the presence of a microenvironment pH within aqueous pores that traverse the cuticle, and are in direct contact with the environment (Sims *et al.*, 1994).

The epidermis lies immediately beneath the cuticle, and is multicellular or syncytial, depending on the species. The outward- or cuticle-facing membrane of the epidermis is probably the true limiting membrane in nematodes, and it serves a critical role in homeostatic regulation. In adult *A. suum*, for instance, the epidermis is an electrically polarised compartment (Pax *et al.*, 1995), and both the cuticle- and muscle-facing membranes contain ion channels that conduct Cl^- and the organic anion end-products of carbohydrate metabolism (Blair *et al.*, 1999). The epidermis in several species, including *A. suum*, has been shown to contain other proteins putatively associated with nutrient and ion transport (Das and Donahue, 1986; Huntington *et al.*, 1999).

Gastrointestinal tract

The alimentary canal in most nematodes extends the full length of the organism, opening at the mouth and anus. The intestine is a tube of epithelial cells, one cell in thickness, connecting the pharynx and anus. The number of cells varies extensively by species, ranging from as few as 20 in *C. elegans* to over a million in *A. suum* (Bird and Bird, 1991). The intestine is not innervated or muscled, but is bounded by muscular structures and closed by valves at either end. In plant-parasitic and most mycophagous species, the walls of the buccal capsule are elaborated to form an extendable stylet used to inject digestive enzymes that aid in the extracorporal digestion of food (Bird, 1976). In most species, the inner lining of the intestine contains microvilli. In *A. suum*, microvilli increase the surface area for transport by almost 100-fold, and are coated by a glycocalyx similar in appearance to that of vertebrates. Some columnar cells, particularly in the anterior regions of the intestine, appear to serve glandular functions. Parts of these cells, which contain secretory material, detach from the cell and disintegrate in the lumen of the intestine (Lee and Atkinson, 1977, for review). Most evidence suggests that the intestine, when present and functional, plays a key role in nutrient absorption (Castro and Fairbairn, 1969; Schanbacher and Beames, 1973) and the elimination of some waste material (Thompson and Geary, 1995; Geary *et al.*, 1995 for review). However, *Bradynema*, which parasitises insects, lacks both a cuticle and a functional intestine. Whatever functions the gut serves in other nematodes must be unnecessary for these organisms, or else they are adopted by other tissues. For instance, the epidermis in *Bradynema* is elaborated by microvilli which greatly increase its surface area for nutrient absorption (Riding, 1970). A related phenomenon may

occur in adult female *Onchocerca volvulus*; the epidermis hypertrophies as the worm matures, coincident with the atrophy of the intestine (Franz, 1988). Finally, the intestine is not connected to the anus in certain plant-parasitic nematodes in the genus *Meloidogyne*, and serves instead as a food storage organ (Bird and Bird, 1991).

Gland cells

Glands in nematodes are composed of single cells that empty their contents directly into the digestive system. Pharyngeal (or oesophageal) glands, also referred to as dorsal or subventral glands, consist of three to five cells that secrete a broad range of digestive enzymes and other proteins into the stomodeum. In *Xiphinema*, secretions from the dorsal gland are injected into plant root cells by way of a needle-like stomatostylet. The injected material, although not fully characterised, can liquefy both the cytoplasm and nucleoplasm of the plant cell within a minute. In some nematodes, including *Globodera* and *Meloidogyne*, dorsal gland material secreted to the outside via stomatostylets is modified to form feeding tubes that aid in extracorporal digestion of host tissue. Rectal glands, which vary in number by species from zero to six, appear to secrete material into the rectum. Little is known about the biochemical nature of the contents of these secretions. In the plant-parasitic nematode, *Meloidogyne javanica*, however, the rectal glands secrete a glycoprotein-containing gelatinous material that encloses shed egg masses (Bird and Bird, 1991 for review).

Canals/tubule system

A tubular system, which is present in most nematodes, consists primarily of two lateral ducts that lie next to or within the paired lateral cords (extensions of the epidermis that also enclose the nerve cords). These ducts or tubes, and the cells that form the ampulla into which they empty before exiting the worm, were originally assigned an excretory function based on morphological evidence. More recently, however, they have been shown to play an important role in osmoregulation (Wharton and Sommerville, 1984; Nelson and Riddle, 1984), which is described in detail later in this chapter.

Uterus

The structural and functional properties of nematode reproductive systems are described in detail in Chapters 2 and 4. Nematodes possess a tubular reproductive system that includes, in females, paired germinal regions (ovaries) and oviducts that lead to a seminal receptacle, uterus and vagina. Uterine fluid from *A. suum* contains proteins and carbohydrates, but has hardly been characterised. For most species, it is not known if uterine fluid contents are secreted into the host or environment to facilitate egg hatching or early larval development, or are simply metabolic waste products that diffuse out with shed eggs. The uterus in *Heterodera* produces a gelatinous substance that encases egg masses laid in the soil (Bird and Bird, 1991).

Although the chemical composition and function(s) of this substance are unknown, the material may reduce predation by other soil organisms or prevent water loss from the eggs.

Model Organisms

Most information on the physiology of homeostatic mechanisms in nematodes comes from studies using *A. suum*, primarily because its large size permits dissection and testing of individual tissues such as the epidermis, pharynx and intestine, each of which contributes to these processes. *A. suum* provides a reasonably good model for other nematodes, based on the limited number of published studies comparing transport physiology and biochemistry among species (Hobson, 1948; Wright and Newall, 1976; Pappas and Read, 1975; Pappas, 1988; Howells, 1980; Sims *et al.*, 1994). Since the early 1990s, focus has shifted to the free-living nematode, *Caenorhabditis elegans*, as an experimental model for the study of nematode biology and genetics. This species has a very short generation time and is easy to propagate in large quantities in the laboratory. It is transparent, which facilitates visual observation of internal processes, such as pharyngeal pumping, intestinal peristalsis and defecation. The developmental lineage of every cell is known. It is possible, for instance, to capitalise on the transparency and knowledge of cell lineages to selectively eliminate specific cells using laser ablation techniques, and to infer their roles by the resulting changes in physiology or behaviour. This approach has been used, for example, to study the putative excretory-secretory tubular cell system in *C. elegans* (Nelson and Riddle, 1984). Most important, the complete genome of *C. elegans* has been sequenced and published (*C. elegans* Genome Consortium, 1998). This has provided a powerful set of tools for characterising proteins, including some that underlie physiological processes involved in homeostasis. For example, mutations in genes that encode important regulators of cellular homeostasis, such as Na^+ channels and Na^+/K^+ pumps, can be generated using functional genomics approaches developed specifically for *C. elegans*, and the phenotypic effects studied *in vitro*, sometimes all within a few days.

Excretion

As noted above, it is important to distinguish processes of excretion from those of secretion. We consider excretion in the context of this chapter to mean the removal of unusable or unnecessary material from the animal. Metazoan organisms typically excrete two distinct kinds of products. Ingested material that cannot be digested is often shed from the intestine as a solid material termed faeces. Secondly, the soluble waste products of intermediary metabolism and catabolism are typically transported via a circulatory system from the tissue of origin to an excretory organ, where they are eliminated in a fluid (e.g., urine). Nematodes possess an intestine and exhibit defecation.

However, nematodes have no recognisable kidney, and soluble waste products are apparently not concentrated prior to elimination. In addition, the late larval and adult stages of parasitic nematodes are primarily anaerobic, and oxidise completely only a small fraction of ingested carbohydrates to CO_2 prior to excretion. Instead, they produce and excrete low molecular weight organic anions, such as acetate, lactate and butyrate, which vary by species. The pseudocoelom serves as a circulatory system, transporting these molecules from the tissue of origin (mostly muscle and reproductive tissue) to, most prominently, the epidermis and intestine for excretion into the environment.

Faeces and Defecation

Although excretion through defecation is a common trait in nematodes, it is important to realise that not all species possess a functional gut. Some plant-parasitic nematodes in the genus *Meloidogyne* lack a patent connection between the gut and the anus; the gut is adapted as a food storage organ instead (Bird and Bird, 1991). The extent to which the gut is functional in adult female *O. volvulus* is an open question. It appears that this species, and perhaps other filariae as well, acquires nutrients by uptake across the cuticle in addition to (or instead of) oral ingestion (Howells, 1980; Franz, 1988). Nonetheless, most nematodes defecate, and this process represents a significant source of parasite-derived material in the environment.

Studies in *C. elegans* have revealed a great deal about the physiology of defecation in nematodes (Avery and Thomas, 1997). This process is a rhythmic, patterned behaviour that is controlled by the availability and quality of food. Transit of faecal material through the intestinal tract is rapid; the default defecation cycle in *C. elegans* is 45 sec. It has been estimated that the contents of the gut in *A. suum* are turned over every three minutes, and that defecation can propel faecal material up to 60 cm (Bird and Bird, 1991). Although it would seem a simple matter to obtain samples for analysis, it must unfortunately be acknowledged that almost nothing is known about the composition of faeces from any nematode (Harpur, 1969). Until the composition of this material is determined for several species, it is impossible to calculate the contribution of defecation to the worm:environment interaction.

Uterine Fluid

The shedding of eggs from adult female nematodes represents another process for expelling material into the environment. The expulsion of larvae from the adult is accompanied by the simultaneous release of uterine fluid. In the case of adult female *Dracunculus medinensis*, uterine fluid secreted onto host skin (i.e., from the inside) induces the formation of blisters, out of which the worm inserts her vulva and sheds eggs into the water. This is one example of an interesting bioactivity in an E/S product which, unfortunately, remains chemically undefined. Even less information is available on the composition of uterine fluid or its physiological role(s) in other species. It has been estimated that, in *A. suum*, the entire contents of the uterus (equal in volume to the PCF) are expelled every day (Fairbairn, 1957). Uterine fluid from this species was found to have a pH of 7.7, to be of lower osmolarity than PCF, and to contain a number of proteins that have not been further characterised (Fitzgerald and Foor, 1979). Whether uterine fluid in nematodes can be considered an excretory product is unknown. Further work on the composition and bioactivity of uterine fluid is clearly warranted.

End Products of Intermediary Metabolism and Catabolism

Waste products from energy generation

Nematodes, like all organisms, excrete into their environment the end-products of energy metabolism; these molecules would be toxic if allowed to accumulate to high levels in cells. Nematodes derive most of their energy from the partial or complete degradation of glucose (typically stored as glycogen or trehalose) or other sugars, including fructose (Chapter 10). Excretory products in free-living species generally remain constant throughout the life-cycle of the organism. Larval and adult *C. elegans*, for example, maintain an aerobic metabolism and excrete mainly CO_2 and glycerol (Cooper and van Gundy, 1970; 1971a). Though few data are available, plant-parasitic nematodes also are generally considered to be aerobic. In contrast, end-products excreted by animal-parasitic nematodes are highly species- and stage-specific (Table 11.1), and may include H^+, CO_2, volatile and nonvolatile fatty acids and various alcohols (Tielens and van den Berg, 1993; Komuniecki and Komuniecki, 1993; Komuniecki and Harris, 1995; see Chapter 10). Animal-parasitic species that inhabit the gastrointestinal tract of vertebrates, such as *A. suum* or *Trichostrongylus colubriformis*, typically undergo a profound metabolic conversion as they enter the nearly anaerobic environment of the gut lumen. Larval stages in the environment are typically capable of fully aerobic metabolism, with a complete and functional tricarboxylic acid cycle. Upon further development in the host, cyanide-insensitive anaerobic respiration predominates. A similar transition to anaerobic metabolism occurs in other parasitic helminths following host invasion, including the trematodes, *Schistosoma mansoni* and *Fasciola hepatica* (Tielens and van den Berg, 1993), and cestodes, including *Hymenolepis diminuta* (Behm, Bryant and Jones, 1987).

Table 11.1. Organic acid, CO_2 and nitrogen excretion in selected nematode species under aerobic or anaerobic conditions

Species	Stage	Gas phase	Major organic acids excreted	Nitrogen containing molecules excreted			
				NH_3/NH_4^+	Urea	Peptides	Amino acids
Ascaris	Adult	Aerobic	2-MV, 2-MB, AC, PR	69	7	21	+
Ascaris	Adult	Anaerobic	2-MV, 2-MB, AC, PR	71	6	18	+
Trichinella	Adult	Aerobic	AC, PR, BU, VA, SU	33	0	21	28
Haemonchus	Adult	Aerobic	CO_2, AC, PR	+	−	+	+
Haemonchus	Adult	Anaerobic	AC, LA, PR	+	−	+	+
Ditylenchus	Mixed	Aerobic	CO_2, FU	39	0	28	+
Caenorhabditis	Mixed	Aerobic	Ethanol, Glycerol	−	−	−	−
Panagrellus	Mixed	Aerobic	CO_2, LA, AC, SU, Glycerol	60	12	+	+

Expressed as % of total nitrogen excreted.
+ Present, but not quantitated
− Signifies lack of conclusive data
Abbreviations for acids: 2-MV, 2-methylvaleric; 2-MB, 2-methylbutyric; AC, acetic; PR, propionic; SU, succinic; LA, lactic; FU, fumaric; BU, butyric; VA, valeric.
Adapted from Cooper and van Gundy, 1971a; Komuniecki and Komuniecki, 1993; Sims *et al.*, 1996; Castillo and Krusberg, 1971; Wright and Newall, 1976; Rogers, 1969.

Accompanying this conversion is a marked reduction in CO_2 production and excretion and a concomitant increase in the excretion of organic acids. Migration to an anaerobic environment is not a *sine qua non* for this metabolic conversion, however. Some tissue-dwelling species, such as *O. volvulus*, reside in highly vascularised tissue compartments (i.e., the onchoceral nodule), but convert to primarily fermentative metabolism as adults (Mendis and Townson, 1985; MacKenzie *et al.*, 1989). Also, some gastrointestinal nematodes that convert to fermentative metabolism, such as *Haemonchus contortus*, reside within the mucosa of the stomach where oxygen tensions may reach 20 mM Hg, a level sufficient to maintain oxidative metabolism (Komuniecki and Harris, 1995). Finally, it is worth noting that the aerobic:anaerobic transition is found to be reversed in *T. spiralis*; infective L1 larvae reside in an anaerobic nurse cell, usually in skeletal muscle, but convert to aerobic metabolism after invading host intestinal epithelial cells. Aerobic metabolism persists through the adult stage of this intracellular parasite (Stewart, 1983).

Unfortunately, essentially nothing is known about the pathways through which CO_2 or alcohols are excreted from free-living nematodes or animal-(free-living larvae) and plant-parasitic nematodes. In vertebrates, CO_2 is typically exported from tissues via a Cl^-/HCO_3^- exchanger. None of the predicted *C. elegans* proteins has yet been assigned this function. Higher alcohols, such as ethanol, are excreted from some nematodes, including *C. elegans*; given the lipophilic nature of this molecule, it is presumably excreted by simple diffusion. Glycerol, another major excretory product, is transported out of cells by aquaporin-type channels in other organisms. Although nematodes express aquaporins (see below), no data are available on whether they are capable of transporting glycerol. In contrast to the absence of information on the mechanisms underlying CO_2 or alcohol elimination, a number of studies on organic acid excretion from adult stages of animal-parasitic nematodes have been published. The physiology of organic acid excretion in these species will be the focus of the remainder of this discussion.

In *A. suum*, like all metazoans, the H^+ concentration in cells and extracellular fluids is regulated to a level that maintains the pH close to 7.0 (Del Castillo *et al.*, 1989; Blair *et al.*, 1998). This is in the vicinity of the pH optimum for most cellular processes, including the activities of most enzymes, ion channels, contractile proteins and proteins involved in cell cycle regulation (Madshus, 1988). In the adult stage of *A. suum* and other animal-parasitic nematodes examined, the organic acids that form in cells as a result of intermediary metabolism are characterised by pK_as that range from 2.0 (e.g., lactic acid) to 4.8 (branched-chain fatty acids, such as 2-methylvaleric acid). Thus, at pH 7.0, at least 99% of the organic acids produced via carbohydrate metabolism exist in the dissociated form, i.e., as H^+ and the conjugate organic anion (e.g., lactate$^-$ or 2-methylvalerate$^-$). This point is important; non-dissociated organic acids are quite soluble in lipids,

and could thus exit tissues via simple diffusion. However, organic anions are poorly soluble in lipids and almost certainly require a protein-mediated transport process for excretion.

Sites and mechanisms of H^+ and organic anion excretion

Animal-parasitic nematodes almost invariably lower the pH of medium in which they are incubated; this phenomenon is ample evidence that these organisms excrete H^+ into their environment. The molecular mechanisms that mediate H^+ excretion from nematode cells, including the epidermis and intestine, have not been determined, though it appears that proton excretion occurs through a different process than organic anion excretion (Blair *et al.*, 1998). In vertebrates, H^+ and CO_2^-, the major end-products of carbohydrate metabolism, are excreted by a Na^+/H^+ antiporter at pH <7.0, and by an HCO_3^-/Cl^- exchanger at pH >7.0 (Weith and Brahm, 1980; Seifter and Aronson, 1986). The electrical driving force for H^+ extrusion, in the first case, is provided by the steep, inward Na^+ gradient across vertebrate cell membranes. Orthologous H^+ transporters have not yet been identified in nematodes. Other, less well-characterised transport systems, such as an electrogenic H^+-translocating ATPase in vertebrate distal nephron, and H^+ conducting cation channels in vertebrate renal brush border cells and snail neurons (Frelin *et al.*, 1988, for review) also play a role in H^+ excretion. Based on sequence homology data, a vacuolar-type H^+-ATPase is present in the H-tubular cell of *C. elegans* (Oka, Yamamoto and Futai, 1997; and see below), but the functional properties of this protein have not been determined.

The mechanisms that underlie ATP production in numerous nematode species have been delineated on the basis of *in vitro* experiments that measure accumulation of excretory products and change of pH in the incubation medium over time. While these studies have identified and biochemically characterised the enzymatic pathways of intermediary metabolism in nematodes, little information is available on the sites or mechanisms that underlie excretion of the resulting end-products from inside cells to the PCF, or from the pseudocoel to the external environment. These processes have been studied directly in only a few nematodes, including adult *A. suum* (Sims *et al.*, 1992, 1994; Robertson and Martin, 1996; Valkanov, Martin and Dixon, 1994; Valkanov and Martin, 1995; Blair *et al.*, 1998, 1999) and *H. contortus* (Sims *et al.*, 1996).

The intestine, epidermis/cuticle complex and tubular system have been suggested as surfaces for excretion of organic acids from nematodes (Thompson and Geary, 1995, for review). A role for the intestine in excretion was suggested on the basis of studies showing that each organic acid present in PCF is also present in the lumen of the intestine in adult *A. suum* (Harpur, 1969). The concentration of each organic acid was constant throughout the length of the intestine, i.e., there was no evidence of a concentrating effect as digesta moved into the caudal regions. However, indirect evidence that questions the importance of the intestine for organic acid excretion comes from more recent studies using adult *A. suum*. Neither the rate of organic acid excretion into the medium nor worm viability, based on ATP and motility levels, was affected by treatments that prevent intestinal passage of solutes (Sims *et al.*, 1992, 1994). Similarly, chemical ligation of adult *H. contortus* with 1 ηM ivermectin, which paralyses the pharynx and inhibits flow of digesta along the intestine, affected neither the rates of excretion of propionic or lactic acids nor worm viability (Geary *et al.*, 1993). These and related studies suggest that the epidermis/cuticle complex is an important site for excretion of organic acids from most nematodes.

Direct evidence that H^+ and organic anions are extruded across the cuticle/epidermis complex in nematodes was obtained in mass transport studies using isolated *A. suum* cuticle segments in 2-chamber diffusion cells (Sims *et al.*, 1992, 1994). The rate-determining barrier to transcuticular transport of organic acids depends on the presence of lipid components in the cuticle, which are probably located in the basal regions of the cuticle where collagen-like components and lipid appear to overlap (Ho *et al.*, 1990). Whether a lipid barrier also exists at the outer surface of nematodes (the epicuticle) is a matter of controversy (Kennedy *et al.*, 1987; Lee, 1966, 1972; see Wright, 1987, for review). However, there is no consistent evidence for a continuous lipid layer surrounding nematodes. The presence of an internal lipid barrier within the cuticle/epidermis complex is supported by results of *in vitro* studies showing that the pH of medium surrounding *A. suum* or *H. contortus* does not influence the rate of absorption of weak acids or bases across the cuticle. The data suggest that pH regulation occurs within the protected microenvironment of the aqueous pores that traverse the cuticle (Sims *et al.*, 1992, 1994). These findings do not demonstrate unequivocally, however, that the epicuticle, regardless of its composition, has no role in this process.

The molecular mechanisms that underlie organic acid excretion across the cuticle in nematodes have not been defined. However, anion-selective channels in the epidermis may play an important role. In adult *A. suum*, organic acid concentrations in the epidermis and muscle compartments are maintained for extended periods in culture at levels which far exceed those in the incubation medium (Blair *et al.*, 1998). The concentration and relative abundance of each acid are remarkably similar within the PCF, muscle and epidermis compartments. Voltage-clamp studies on isolated epidermal membranes reveal the existence of a large conductance, voltage-sensitive, Ca^{2+}-dependent Cl^- channel (Thompson and Geary, 1995; Blair *et al.*, 1999) that exhibits similar biophysical properties to a channel in *A. suum* muscle membrane (Valkanov and Martin, 1995; Valkanov, Martin and Dixon, 1994). The Ca^{2+}-activated Cl^- channel in *A. suum* muscle and epidermal membranes is voltage-sensitive, and open at electrical potentials recorded across the membranes of these cells. It is an inward-rectifying channel that closes as the membrane becomes depolarised. Thus, it does not appear to participate in the repolarisation of muscle membranes following acetylcholine-induced depolarisation and contraction (Valkanov, Martin and Dixon, 1994), suggesting an alternative role. Although the permeability of this channel, in both muscle and epidermal membranes, is greatest for Cl^-, larger anions, including the organic anions excreted by *A. suum*, are also conducted (Valkanov *et al.*, 1994; Valkanov and Martin, 1995; Blair *et al.*, 1999). The relationship between channel activity and the physico-chemical properties of the anions have been studied more thoroughly for the muscle channel. Permeability across the channel is inversely related to the Stoke's diameter of the anion, with the branched chain fatty acid 2-methyl butyrate (diameter = 5.62 Å) being only about 24% as permeable as Cl^- (diameter = 2.41 Å). Reducing external pH increases the probability of the channel opening at hyperpolarised membrane potentials (Robertson and Martin, 1996). Based on data for a limited set of organic anions excreted by *A. suum*, a similar trend occurs for the epidermal channel. The negative electrical potentials recorded across muscle (–30.3 mV) and epidermal membranes (–60 to –80 mV) in *A. suum* could provide a driving force for the extrusion of organic anions through the Ca^{2+}-activated Cl^- channel. This concept is supported by equilibrium potentials derived for the organic anions excreted by *A. suum*; for each there exists a net outward-directed electrical driving force of 16–27 mV across the epidermal membranes (Blair *et al.*, 1998). This driving force could supply potential energy for organic anion extrusion directly, i.e., across the anion-selective channels, or indirectly via facilitated transport processes.

Other mechanisms, less studied in nematodes, may serve equally or more important roles in organic acid excretion. Several models have been proposed for volatile fatty acid (VFA) transport in vertebrates (Titus and Ahearn, 1992). At least some transmembrane movement of VFAs is mediated by a family of H^+-monocarboxylate co-transporters, which are 12 trans-membrane domain proteins (Poole and Halestrap, 1993). Genes encoding a number of these transporters have been cloned from several organisms (Price, Jackson and Halestrap, 1998). Homologs of these genes can be identified in yeast and, importantly, in the *C. elegans* genome, which contains at least 4 of them (Price, Jackson and Halestrap, 1998). That parasitic nematodes decrease the pH of the medium in which they are kept is evidence that protons as well as organic acids are excreted; it is plausible that one or more H^+-monocarboxylate co-transporters play a key role in this process. However, no information is yet available on the expression patterns or substrate specificity of the different *C. elegans* transporters. Functional expression in a heterologous system would provide much of this information and should lead to gene knock-out experiments that could define a role for them. It is not clear if similar proteins are found in parasitic species. One confounding issue is that, unlike animal-parasitic nematodes, *C. elegans* excretes minor amounts (at most) of organic acids (see above).

Implications of the acidic cuticle pore microenvironment

By excreting organic acids across the cuticle, nematodes profoundly affect the pH of their immediate environment (Ho *et al.*, 1992, 1994; Sims *et al.*, 1994; 1996). During *in vitro* incubations, this process drives the pH of the bulk medium to values close to the aggregate pK_a of the excreted organic acids. The pH of weakly buffered media containing adult *A. suum*, for example, which excretes predominantly branched chain fatty acids with pK_as in the range of 4.76–4.88, approaches pH ~5.0 during extended incubations. Media containing species that excrete higher levels of lactic acid, such as *Brugia pahangi*, tend to become even more acidic because of the lower pK_a of this acid (MacKenzie *et al.*, 1989). This effect is most pronounced in the protected microenvironment of aqueous-filled pores that traverse the cuticle, as shown experimentally for *A. suum* (Sims *et al.*, 1992) and *H. contortus* (Sims *et al.*, 1996).

The existence of an acidic microenvironment pH within the cuticle pores has important implications for how charged molecules are transported across this structure. This includes both excreted molecules, such as organic acids and NH_4^+, and absorbed molecules, including some nutrients and drugs. The influence of cuticle pH on the excretion and absorption of charged molecules has been quantitated and mechanistically interpreted for adult *A. suum* (Ho *et al.*, 1992; Sims *et al.*, 1994) and *H. contortus* (Sims *et al.*, 1996). In these studies, the rates of uptake of several model permeants, including weak acids and weak bases, were measured as a function of pH in the medium. In some experiments, the intestine was blocked using ligatures at the head and tail (in *A. suum*) or chemically inhibited using low concentrations of ivermectin (in *H. contortus*), which paralyses the pharynx without affecting motility or ATP levels (Geary *et al.*, 1993). The rates of absorption and pattern of tissue distribution of the model weak acids, benzoic acid and p-nitrophenol, and the model weak base, aniline, were unaltered by ligation or changes in the pH or buffer capacity of the bulk medium. In the context of absorption by the nematode the critical pH controlling the ionization state of weak acids or bases is not that of the bulk medium, but rather that within the protected microenvironment of the aqueous pores in the cuticle, which is maintained at a constant level by organic anions and H^+ excreted across the epidermis/cuticle complex (Sims *et al.*, 1996).

The cuticle is also an important site for the absorption of uncharged molecules, including most anthelmintics. Accumulation of levamisole by *A. suum*, for example, can be accounted for solely by transcuticular diffusion (Verhoeven, Willemsens and Van den Bossche, 1980). Several factors influence transcuticular diffusion. For instance, absorption indices for a wide range of uncharged molecules show no obvious relationship with lipophilicity in the filarial parasites, *Brugia pahangi* and *Dipetalonema viteae* (Court *et al.*, 1988). When other physicochemical parameters, such as molecular weight, dipole moment and total energy, are considered along with logK (i.e., as an indicator of lipophilicity), it is possible to make quantitative predictions about absorption rates of uncharged molecules by these species. The absorption of drugs and other solutes by filarial nematodes is thus influenced by several physicochemical properties of the solutes and by both lipid and nonlipid components within the cuticle (Court *et al.*, 1988).

The gastro-intestinal nematodes *A. suum* and *H. contortus* have been studied to determine if the physicochemical properties of drugs can be used to predict their rates of absorption and patterns of tissue distribution (Fetterer, 1986; Ho *et al.*, 1990; Thompson *et al.*, 1993; Sims *et al.*, 1996). As in filarial nematodes, logK alone cannot be used to accurately predict how rapidly either charged or uncharged molecules are absorbed, or how extensively they accumulate in worm tissues. The relevant experiments used adult *A. suum* that were mechanically ligated to eliminate intestinal transport, or dissected into segments of body wall that contained no anatomical openings. When the isolated tissues were tested in 2-chamber diffusion cells, it was found that the collagen and lipid components of the cuticle each present a distinct barrier to diffusion of organic molecules. Penetration of the collagen barrier of the cuticle was highly dependent on the size and electrical charge of the solute. Because of the size-restriction effect of the aqueous pores, which are 15 Å in radius, the permeability of both neutral and charged solutes decreases with molecular diameter. Positively charged molecules penetrate faster than negatively charged molecules of comparable size because of the negative charge lining the aqueous pores (Ho *et al.*, 1990).

End-products of catabolism: nitrogen excretion

Nitrogen excretion in nematodes has received little attention. Most nitrogen excreted by nematodes comes from the deamination of amino acids (Rogers, 1969; Wright and Newall, 1976; Wright, 1975). From 40–90% of nitrogen is excreted as NH_3 or NH_4^+, with most of the balance eliminated as urea, amino acids and peptides (Table 11.1). Since the pH in nematode cells and PCF is maintained at 6.5–7.0, ammonia in these compartments exists in a dynamic equilibrium that greatly favors the ionised (NH_4^+) over the non-ionised (NH_3) state. Due to its charged nature, NH_4^+ transport across intestinal or epidermal membranes would require a channel or carrier protein. While there is no functional evidence that an ammonium transporter exists in nematodes, there is a homolog of the yeast ammonium permease, MEP1, in the *C. elegans* genome (Marini *et al.*, 1994). Unfortunately, no information is available about the expression or function of this putative ammonium transporter in *C. elegans*. NH_3, diffuses freely across membranes, and it is likely that most nitrogen is excreted from nematodes in this form (thus driving the conversion of additional NH_4^+ to NH_3). NH_3 excretion across the cuticle/epidermis complex would be facilitated by the acidic microenvironment pH maintained within the aqueous pores of the cuticle, which would effectively ionise NH_3 as it enters the cuticle pores. This process would ensure low concentrations of NH_3 relative to NH_4^+ in the aqueous pores, providing a driving force for continued diffusion of the uncharged species out of the worm (i.e., down its concentration gradient). Among helminths, this principal has been demonstrated experimentally only for the trematode, *Schistosoma mansoni* (Pax and Bennett, 1990). It should also be noted that nematodes contain a family of aquaporin-like genes, which may play a role in water transport (see below). At least some members of the aquaporin family can also transport urea (see Ishibashi *et al.*, 1994), and it will be interesting to determine if this route contributes to nitrogen excretion in *C. elegans* and other nematodes.

Several animal-parasitic nematodes, including *A. suum*, *Trichinella spiralis* and *Nippostrongylus muris* (*brasiliensis*), excrete low levels of short-chain aliphatic amines (Haskins and Weinstein, 1957; Weinstein, 1960; Castro, Ferguson and Gorden, 1973). Castro *et al.* (1973) suggested that excreted amines may neutralise the acidic environment of nematodes. The pK_as of these amines, however, are in the range 10.05–10.65, so it is unlikely that they contribute substantially to the buffer capacity in the pH range of gastrointestinal worms, as they would be almost completely protonated at pH values below 7.5 (Sims *et al.*, 1996). Direct biochemical evidence for nitrogen excretion is available only for the intestine and epidermis/cuticle complex of adult *A. suum* (Wright and Newall, 1976). However, it is likely that nitrogen, in some form, is also excreted from other sites in nematodes, including the pharyngeal glands and reproductive glands (Lee and Atkinson, 1977, for review). In this regard, the recent report of the functional expression of an organic cation transporter cloned from *C. elegans* provides another potential route for the excretion of nitrogen (and possibly xenobiotics; Wu *et al.*, 1999). No information is available on the expression of this gene in the nematode or on the consequences, if any, of mutations in it. The classic substrate for this transporter is tetraethylammonium, but it also transports choline and a variety of basic drugs. Its role might be revealed by knock-out mutations or the use of double-stranded RNA interference (see below).

p-Glycoproteins: A Special Case of Excretion

The accumulation of some molecules, including drugs, by nematodes may be affected by drug transporting proteins, referred to as P-glycoproteins (pgps) or multiple drug resistant proteins (mdr). Pgps actively 'pump' drugs out of some cells, thereby limiting exposure of the organism to their therapeutic actions. In vertebrates, over-expression of pgps has been linked to resistance to several drugs, most notably those used to treat solid tumors. Ivermectin is a substrate for pgp-mediated extrusion from murine kidney epithelial cells *in vitro* (Schinkel *et al.*, 1995). Mice deficient in *mdr1a*, a pgp-encoding gene, are susceptible to the toxic actions of ivermectin, which accumulates in the brains of these mutants to levels over 100-fold greater than in wild-

type mice (Schinkel *et al.*, 1994). In *C. elegans*, pgps are expressed primarily on apical membranes of cells that form the intestine and tubular system (Lincke *et al.*, 1993; Broeks *et al.*, 1995), a pattern consistent with a role in protecting the organism from toxic xenobiotics. There are no reports describing pgp expression in the epidermis. Fourteen pgp-encoding genes have been identified in *C. elegans*, some of which are probably pseudogenes (Broeks, 1997). Deletion of two of these genes leads to susceptibility to chloroquine, colchicine and to heavy metals that are not toxic to wild-type *C. elegans* (Broeks *et al.*, 1995, 1996). To this point, however, no *C. elegans* anthelmintic resistance gene has been mapped to a pgp locus.

The animal-parasitic nematode *H. contortus* contains at least four *pgp* genes (Sangster, 1994; Sangster *et al.*, 1999; Xu *et al.*, 1998). In some strains of *H. contortus*, pgp polymorphism is associated with macrocyclic lactone resistance (Xu *et al.*, 1998; Blackhall *et al.*, 1998; Sangster *et al.*, 1999). These observations, however, do not establish a causal link between pgp polymorphism and anthelmintic resistance. No definitive evidence has been reported that resistant and sensitive strains of *H. contortus* absorb these drugs at different rates, or that they accumulate in specific worm tissues in a different manner. Ascribing a role for pgps in anthelmintic resistance will require demonstrating that pgp-deficient strains accumulate drugs more rapidly, or that pgp over-expressing strains more slowly, than wild-types. In this context, a closantel-resistant strain of *H. contortus* was shown to accumulate closantel less rapidly than wild-type strains (Rothwell and Sangster, 1997). Closantel, like ivermectin, is highly lipophilic, which is a hallmark of most substrates for pgps. However, no pgp polymorphism has been associated with the closantel resistant phenotype (Kwa *et al.*, 1998; Sangster *et al.*, 1999). Because these experiments tracked variation in a limited number of restriction sites, the findings cannot rule out pgp involvement in closantel resistance.

Whether or not alterations in pgp expression or sequence have been selected as a mechanism of anthelmintic resistance, there is ample evidence to conclude that these proteins contribute to the excretion behaviour of nematodes. Their primary location in intestinal cells is consistent with a role in eliminating compounds that penetrate the cuticle, diffuse through worm tissue and accumulate in PCF. ATP-dependent transport of xenobiotics from PCF to the intestinal lumen would reduce internal drug levels and protect the organism from toxicity.

Secretion

As noted above, E/S products and processes in nematodes cannot always be clearly distinguished. Little is known about the anatomical origins of many E/S products, the mechanisms that account for their transport out of worms, or the functions they serve before or after their elimination from the parasite. Even if 'secreted' materials are defined strictly as those that originate in some type of gland or a structure otherwise specialised for secretion, and exhibit some type of defined role in enhancing nematode fitness, a substantial list of molecules exists (Table 11.2). Few have been extensively characterised. Unfortunately, identified molecules typically lack precisely defined biological activities, while interesting biological responses are caused by molecules for which a structure has not yet been determined. We consider here a few examples that illustrate the ways in which nematodes influence their environment via E/S products. It should be noted that secreted products which play a role in nutrient acquisition, such as proteases involved in extracorporal digestion and anticoagulants, are discussed in Chapter 8.

Infection and Tissue Migration

Many nematodes that parasitise plants and animals undergo a developmental stage during which migration through host tissues is required. These stages release enzymes that facilitate the journey by cleaving host structural molecules which would otherwise bar migration through the tissue (Dubremetz and McKerrow, 1995). Among animal-parasitic nematodes, an illustrative example is the release of a hyaluronidase by hookworm larvae (Hotez et al., 1992). A mucopolysaccharide known as hyaluronic acid serves as an important mediator of cell-cell attachment in mammalian dermis. Migrating hookworm larvae release a hyaluronidase, which depolymerises hyaluronic acid and facilitates movement through the skin. The activity is most pronounced in larvae of *Ancylostoma braziliense*, which infects humans and other mammals primarily by invasion through the skin. Consistent with the high levels of hyaluronidase secreted by the larvae, this species is capable of migrating through the deep layers of the epidermis and is the most common cause of cutaneous larval migrans in humans.

Plant-parasitic nematodes encounter a different set of host structural molecules, primarily cellulose. Larvae of the soy bean cyst nematode *Heterodera glycines* contain two forms of an enzyme activity ascribed to β-1,4-endoglucanase, which hydrolyses cellulose (Wang et al., 1999). The product of the *Hg-eng-2* gene was localised to the subventral gland cells in L2 larvae. By 24 hr post-infection, the encoded endoglucanase was secreted from the stylet of the larvae. Antibody staining revealed a trail of secreted enzyme deposited along the path of the invading worm, an observation consistent with a role for this enzyme in nematode migration through the root tissue.

Secreted proteases also facilitate vertebrate host invasion. Infective larvae of *Strongyloides stercoralis*, which invade through the skin, secrete a zinc metalloprotease with activity against several dermal structural proteins, including elastin (McKerrow et al., 1990; Brindley et al., 1995). The enzyme is secreted from L3 larvae in culture and is allergenic in humans. Similar proteases are found in infective stages of many other nematodes (Dubremetz and McKerrow, 1995)

Modulation of Host Immune Response

How nematodes escape host immune responses, at least temporarily, has been the subject of much research

(Maizels et al., 1993; Rifkin et al., 1996 for review). A common strategy involves the secretion of molecules that interfere with the development of humoral or cellular immunity. In some cases, this results in detectable immunosuppression. Repeating a common theme of this chapter, it is unfortunate that the most impressive biological activities are generally associated with uncharacterised parasite products, and the best characterised products have poorly defined biological functions. Although immune responses are typically specific and distinct for different parasites, it is still worthwhile to illustrate a few examples of immunomodulatory molecules secreted by nematodes.

Proteases secreted by animal-parasitic nematodes have been proposed to inactivate components of the host response in addition to pre-digesting host tissues for nutrition (Rifkin et al., 1996). A family of four cysteine proteases secreted by adult and L4 (but not L3) larvae of *H. contortus* degrade a variety of host proteins (Rhoads and Fetterer, 1995). These proteases also cleave the heavy chain of sheep IgG in the hinge region, which inactivates it. It is not known if the proteases have different substrate specificities. Whether degradation of IgG by these proteases has functional consequences for the establishment or maintenance of the infection is not known. Although a family of genes encoding cysteine proteases has been cloned from this parasite (Pratt et al., 1992), substrate specificity has not been defined for the recombinant products.

Inactivation of platelet-activating factor (PAF) is an alternative mechanism of immunomodulation. Among the many roles played by PAF in mammalian physiology is mediation of the immune response. Adult stages of the intestinal parasite *Nippostrongylus brasiliensis* maintained in culture secrete an acetylhydrolase enzyme that inactivates PAF (Grigg, Gounaris and Selkirk, 1996). The enzyme is a heterodimer. How it is secreted by *N. brasiliensis* has not been reported, nor is it evident that similar enzymes are secreted by other GI parasites. Although no studies with inhibitors of this PAF acetylhydrolase have been reported, there are good reasons to speculate that the enzyme is an important contributor to parasite establishment (Grigg et al., 1996).

Nematodes also secrete molecules that influence the responsiveness of host leukocytes. Hookworms contain a 41 kDa glycoprotein that binds with high affinity to the receptor for the integrin CD11b/CD18 on human neutrophils (Moyle et al., 1994). Based on the detection of a signal sequence encoded in the corresponding cDNA and the glycosylated nature of the protein, it is probably secreted. No further information has been published on the contribution of this neutrophil inhibitory factor (NIF) to immune evasion by *Ancylostoma caninum* (the parasite from which it was derived), nor is it known how widely similar molecules are distributed among other species of hookworms or other families of parasites. However, NIF has been found to be active as an inhibitor of neutrophil function in a variety of animal models of immune-mediated pathologies, supporting

Table 11.2. Examples of E/S products with known activities in nematodes

Secreted substance	Function	Sites of secretion	Representative species	Reference
Acetylcholinesterase	Inhibit host acetylcholine (immune evasion)	Pharyngeal glands	*Trichostrongylus colubriformis*	Ogilvie *et al.*, 1973.
Amylase	Extracorporal digestion of host (plant) starch	Pharyngeal glands	*Ditylenchus dipsaci*	Myers, 1965
Anticoagulants	Prevent host wound healing	Pharyngeal glands	*Haemonchus contortus* *Nippostrongylus brasiliensis*	Lee, 1970
Cellulase	Extracorporal digestion of host (plant) starch	Pharyngeal glands	*Meloidogyne javanica*	Bird, 1975
Chitinase	Break down eggshell in hatching process	Pharyngeal glands	*Meloidogyne javanica*	Bird, 1975
Cysteine protease	Extracorporal digestion of host protein	Pharyngeal glands	*Haemonchus contortus*	Pratt *et al.*, 1992
Esterase	Extracorporal digestion	Intestinal cells	*Ascaris suum*	Lee and Atkinson, 1977
Fatty acids (e.g., myristic, palamitic, oleic)	Egg expulsion	Ovijector of females (probably from uterus)	*Toxocara canis*	Niedfeld *et al.*, 1993
Gelatinous matrix (unknown composition)	Protect eggs from desiccation	Female uterus or rectal glands	*Meloidogyne javanica*	Bird and Bird, 1991
Glutathione-s-transferase	Anti-oxidant (immune evasion)	Pharyngeal glands	*Anguillicola crassus*	Nielson and Buchmann, 1997
Hyaluronidase	Break down host skin glycoproteins	Pharyngeal glands	*Nippostrongylus brasiliensis*	Lee, 1972
Leucine aminopeptidase	Exsheathment	Pharyngeal glands	*Haemonchus contortus* *Trichostrongylus colubriformis*	Rogers and Sommerville, 1968
Metalloprotease	Degrade host tissue during infection	Pharyngeal glands	*Haemonchus contortus* *Brugia malayi*	Gamble and Mansfield, 1996
Paralytic agents	Paralyze prey	Pharyngeal glands/stylet	*Nygolaimus predacius* *Seinure tenuicaudata*	Hechler, 1963
Pectinase	Extracorporal digestion of host (plant) pectin	Pharyngeal glands	*Ditylenchus dipsaci*	Myers, 1965
Plant cell growth regulators	Induce giant cell formation in plant hosts	Pharyngeal glands	*Heterodera xylophilus*	Bird, 1975
Prostanoids (e.g. prostaglandins, thromboxane)	Affect host endothelial cells and immune effectors	?	*Onchocerca volvulus* *Oesophagostomum dentatum*	Daugschies, 1995
Pheromones	Attract mates	Uterus or epidermis	*Globodera rostochiensis*	Green and Greet, 1966
Superoxide dismutase	Anti-oxidant used in immune evasion	Pharyngeal glands	*Onchocerca volvulus*	Henkle *et al.*, 1991
Vertebrate cell growth regulators	Induce nurse cell formation in host muscle cells	Stichosome/stichocyte	*Trichinella spiralis*	Despommier, 1993
Vasoactive intestinal polypeptide	Modulate host gut function	Pharyngeal glands	*Nippostrongylus brasiliensis*	Foster and Lee, 1996a

the concept that it is a bioactive product specifically secreted by parasites.

Modulation of Host Physiology

Parasitic nematodes modify the physiology of the host environment in ways that enhance their ability to survive and reproduce. Although a number of bioactive molecules have been identified in parasite secretions, our understanding of the *in situ* biology of these phenomena is still limited.

One of the most studied examples is the secretion of acetylcholinesterase (AChE). It has been known for some time that many animal-parasitic nematodes secrete AChE in culture (see Lee, 1996 for review). Initially, it was proposed that AChE, by hydrolysing acetylcholine (ACh), could induce local paralysis of the mammalian GI tract and so provide a more hospitable environment for intestinal parasites (ACh is an excitatory neurotransmitter for the GI musculature). Because of this activity, the secreted enzyme was labelled a 'chemical holdfast' (Lee, 1996). However, other lines of evidence argue against such a function for nematode AChE. For instance, tissue-dwelling filariae, i.e., which do not inhabit a myoactive niche, also excrete AChE (Rathaur *et al.*, 1987). Furthermore, injection of electric eel AChE into the lumen of the rat intestine does not influence muscle behaviour (Foster, Dean and Lee, 1994). An alternative role for secreted AChE is immunomodulation (Lee, 1996). ACh, acting at muscarinic cholinergic receptors in the gut, is known to mediate a variety of enterocyte secretory processes that likely influence pathways of innate immunity that contribute to worm expulsion (see Hussein *et al.*, 1999 for review). ACh enhances cellular immune responses as well, which provides a plausible explanation for the secretion of this enzyme from filariae (see Rathaur *et al.*, 1987). AChE has been localised to pharyngeal glands of the hookworm *Necator americanus* (Pritchard *et al.*, 1991), and a cDNA encoding a secreted AChE has recently been cloned from *N. brasiliensis* (Hussein *et al.*, 1999). If techniques developed for *C. elegans* can be adapted to over-express or disable parasite genes (see below), the role of secreted AChE in parasitic nematodes might be experimentally approachable.

Although secreted AChE may not affect GI muscle function in the host, nematode E/S products clearly alter intestinal motility (Foster, Dean and Lee, 1994, for review). Insight into possible effectors secreted from nematodes has come from studies on parasite proteins that are immunoreactive to antibodies against the mammalian neuropeptides, vasoactive intestinal peptide (VIP) and peptide histidine-isoleucine (PHI) (Foster and Lee, 1996a, b). Although both proteins are larger than the corresponding mammalian neuropeptides, it is possible that a conserved region of the parasite protein may mimic (or antagonise) the function of the homologous peptide. A similar activity may be found in a 30 kDa E/S protein from *T. colubriformis*, which contains a region similar to the mammalian neuropeptide valosin (Savin *et al.*, 1990); valosin influences several aspects of mammalian GI physiology. Unfortunately, no further

information on the biological activity of this nematode protein is available.

A partially purified preparation of VIP-like material from *N. brasiliensis* reduced contractility of rat intestinal segments, an effect similar to that obtained with authentic VIP (Foster and Lee, 1996a). Coadministration of antisera raised against porcine VIP abolished this response. The E/S fraction containing PHI-immunoreactive material was inactive (Foster and Lee, 1996a). Similar immunoreactive molecules are found in E/S products from a variety of other nematodes (Foster and Lee, 1996b). Immunoreactivity to both VIP and PHI has been localised to the excretory system of adult *N. brasiliensis* (Foster, 1998). Stronger evidence that parasite VIP homologs play an important role in maintaining the infection could be obtained in experiments in which a rat VIP antagonist (e.g., Gourlet *et al.*, 1997) is infused into the intestine.

Animal-parasitic nematodes also secrete non-protein factors that probably alter host physiology. Arachadonic acid metabolites (e.g., prostaglandins) have exceptionally broad activity in mammals. Prostaglandins are secreted from a number of nematodes, including filariae (Liu, Serhan and Weller, 1990; Kaiser *et al.*, 1992) and *Oesophagostomum dentatum* (Daugschies, 1995). Adult *Dirofilaria immitis* produce primarily prostaglandin D_2 (PGD_2) (Kaiser *et al.*, 1992), while microfilariae of *Brugia malayi* produce prostacyclin, PGE_2 and PGD_2 in order of descending abundance (Liu *et al.*, 1990). Histiotropic stages of *O. dentatum* produce a complex and changing mixture of prostanoids over several weeks of culture (Daugschies, 1995). A cDNA encoding an enzyme that forms arachidonic acid from di-homo-γ-linolenic acid ($\Delta 5$ fatty acid desaturase) has been cloned from *C. elegans* (Michaelson *et al.*, 1998; Watts and Browse, 1999), although no role for arachidonic acid metabolites has been defined in this free-living species.

Cyclooxygenase activities have not been characterised in nematodes, but studies with classic cyclooxygenase inhibitors, such as indomethacin, have shown that production of prostanoids by nematodes can be blocked by exposure to this drug (Liu *et al.*, 1990; Kaiser *et al.*, 1992; Daugschies, 1995). Whether long-term indomethacin treatment of the host is deleterious for nematodiases has not been conclusively shown. Since host-derived prostanoids may play a role in limiting parasite success, interpretation of the results of such a study could be complicated. There is some evidence that the activity of diethylcarbamazine against microfilariae is due, at least in part, to cyclooxygenase inhibition in the parasite and/or the host (Martin, 1997). Although it seems highly likely that parasite-derived prostanoids influence host physiology, definitive proof remains lacking.

There is convincing evidence that filarial parasites produce a variety of factors that influence host tissues. Heartworms elaborate products that alter the physiology of blood vessels and airway tissue; these effects appear to be mediated by prostanoids (Kaiser *et al.*,

1992; Collins, Williams and Kaiser, 1994). Non-prostanoid, but uncharacterised, filarial products appear to be involved in these phenomena (Mupanomunda *et al.*, 1997; Kaiser and Williams, 1998). It has recently been shown that adult filariae release nitric oxide (NO), a potent mediator of vessel tone, in culture (Kaiser, Geary and Williams, 1998), but the physiological consequences of this release for the host have not been investigated. Finally, although most of this work has been done with blood vessels and *D. immitis*, it is clear that a lymphatic system parasite, *Brugia pahangi*, produces factors that influence the physiology of lymph tissues (Kaiser, Mupanomunda and Williams, 1996). Further effort devoted to the identification and biological analysis of these products would undoubtedly be fruitful.

Modification of Host Cells

Two examples of dramatic alterations of host cells that are induced following infection of plants and mammals illustrate the remarkable biological activity associated with nematode E/S products. Following infection of plants by infective L2 larvae of the genera *Meloidogyne, Heterodera* and *Globodera*, host cells at the site of infection undergo changes that result in the development of a specialised feeding site (Sijmons, 1993; Bird, 1996; Curtis, 1996 for review). In a conceptually similar process, L1 infective larvae of the animal parasite *Trichinella spiralis* migrate to skeletal muscle tissues and elicit the formation of a nurse cell within a collagenous capsule that is surrounded by a circulatory rete (Despommier, 1993). In both situations, parasites alter gene expression programs in host cells to create a suitable environment. The consequences are different; in plant-parasitic nematodes, the feeding site provides a protective environment for the immobile female, enhancing reproductive success. In *T. spiralis*, the infective larvae reside inside the nurse cell, where it is protected from exposure to the host immune response. Upon ingestion of the muscle tissue by another host, the larvae leave the nurse cell and begin development to the adult stage in the intestinal tract of the new host.

L1 larvae of *T. spiralis* suppress the normal pattern of muscle-specific gene expression and cause arrest of the host cell cycle at the G_2/M stage (Jasmer, 1993). E/S products obtained from larvae in culture duplicate at least some of these effects when injected into muscles (Ko *et al.*, 1994). Some of these products can be localised to host nuclei, where they presumably alter gene expression patterns (Yao and Jasmer, 1998). However, very little is known about the biochemistry of the larval E/S products. Several enzyme activities have been reported from this source, but a defined function in establishing the nurse cell complex has not been assigned to any (see Mak and Ko, 1999). The profound reprogramming of skeletal muscle gene expression, which includes adaptations for collagen synthesis and angiogenesis, provide a rich source of information on basic problems in cell biology. Unraveling the phenomenon of nurse cell formation by larval *T. spiralis* will be a highly rewarding exercise.

Formation of giant cells at feeding sites in infected plant roots is also accompanied by changes in gene expression (Bird, 1996). Analysis of a subtracted cDNA library prepared from dissected giant cells has revealed several classes of transcripts, including messages involved in wound healing, signal transduction and giant cell function. Transcripts in the last category encode proteins that function in transport of nutrients, water and ions, which are thought to be elicited by the demands of the feeding parasite. Although there is little question that parasite secretory products induce these changes, none have been specifically identified. Antigenically similar products can be found in several different species of plant-parasitic nematodes (Curtis, 1996) and some are localised in root cells surrounding the feeding site. As it is technically challenging to study nematode-giant cell complexes in culture, alternative approaches may be required to define the nature of parasite secreted products that generate and maintain the feeding site.

Communication

Nematodes obtain information about other nematodes through chemical signals, or pheromones. The existence of pheromone-based communication is best demonstrated through studies on dauer larvae formation in *C. elegans* (see Riddle and Albert, 1997 for review) and on attractants released by plant- and animal-parasitic nematodes (Haseeb and Fried, 1988 for review).

Dauer larvae are adapted for persistent survival under unfavorable conditions. They arrest development in the L3 stage. Dauers exhibit reduced motility compared to non-dauer L3. The developmental pattern leading to dauer formation is normally induced after the middle of the L1 stage by exposure to chemical cues that convey the ratio of food supply to population density. Population density is communicated by release of a dauer pheromone from all stages of each worm. Although the structure of the pheromone is not yet defined, it appears to consist of a family of related molecules that have characteristics of hydroxylated fatty acids or bile acids. These molecules are stable in the environment and non-volatile. Development to the L4 stage resumes when the ratio of pheromone to food signal becomes more favorable.

Remarkable insight into a variety of fundamentally important aspects of biological signal transduction has been gained through the study of *C. elegans* mutants that are defective or constitutive for dauer formation (Riddle and Albert, 1997). Functional amphids are required for dauer formation, indicating that the pheromone receptor is probably located in these sensory structures. G-protein subunits (G_i) that are expressed in amphids are in the pheromone pathway, but the specific receptor for dauer pheromone has not yet been defined.

Parasitic nematodes exude pheromones that presumably aid in mate-finding. The phenomenon has been studied in most detail in plant-parasitic nematodes. As was the case for *C. elegans* dauer-inducing pheromone, plant-parasitic nematodes employ a mixture of molecules for this function. Although many attempts have been made at purification (Aumann *et al.*, 1998), only one compound, vanillic acid, has been proposed as a sex pheromone (for *Heterodera glycines*; see Jaffe *et al.*, 1989). How vanillic acid might be synthesised (or acquired) by nematodes is not known. The anatomical location of pheromone release also remains undefined; some evidence supports secretion from the uterus through the vulva (Green and Greet, 1972). A role for secretion across the epidermis has also been proposed (Haseeb and Fried, 1988).

Less work has been done on sex attractants from animal-parasitic nematodes, though there is good evidence for their existence (Haseeb and Fried, 1988). Animal-parasitic nematodes also release aggregating pheromones, which affect larval stages as well as adults (Huettel, 1986 for review). A receptor that mediates social behaviour was recently identified in *C. elegans* (de Bono and Bargmann, 1998). Wild-type animals typically are solitary and aggregate only in the presence of food. Loss-of-function mutations in the gene encoding this receptor cause animals to clump. In this regard, they resemble species such as *Nippostrongylus brasiliensis* and *Enoplus communis*, which aggregate into tight clumps in culture (Alphey, 1971; Lee and Atkinson, 1977, for review). The *C. elegans* receptor for this aggregating response is most similar, at the primary sequence level, to mammalian receptors for peptides in the Neuropeptide Y (NPY) family. A peptide would be an unlikely pheromone for nematodes, however, given the difficulties of environmental stability and access to internal compartments, so it is likely that this receptor transduces a downstream signal instead of responding to the pheromone. It would be interesting to discover what secreted (or excreted) molecule, if any, stimulates or inhibits aggregation in nematodes, and how important this behaviour is to the survival of parasitic species.

Ionic Regulation

Nematodes, like all metazoans, regulate the composition of ions within their cells and, to a lesser extent, extracellular fluids. Inorganic ions that contribute most importantly to homeostatic maintenance in cells include Na^+, K^+, Cl^-, Ca^{+2} and Mg^{+2}. Intracellular concentrations of these ions, particularly Ca^{+2}, are usually maintained within narrow limits. Invariably, the mechanisms that regulate inorganic ions consist of an array of integral membrane proteins, including ion channels, energy-dependent pumps (e.g., Na^+/K^+-ATPase, Ca^{2+}-ATPase, Na^+/H^+-ATPase) and energy-independent ion transporters (e.g., Na^+/H^+ antiporter, Na^+/HCO_3^- symporter, HCO_3^-/Cl^- exchanger). Control of inorganic ions in the extracellular fluids of metazoans usually involves combinations of the same transporters that regulate intracellular concentrations. These are often supplemented by additional, more complex mechanisms, including countercurrent exchangers and other coupled systems that facilitate transport across polarised epithelial or endothelial cells (Frelin *et al.*, 1988, for review).

The crucial importance of inorganic ion regulation across nematode cells is illustrated by the effects of several drugs that interfere with ion channels. The anthelmintic actions of the macrocyclic lactones (MCLs), for example, are attributed to the ability of these compounds to open Cl^- channels associated with glutamate receptors in pharyngeal and somatic muscle cell membranes (Cully *et al.*, 1996; Avery and Horvitz, 1990; Chapter 12). By increasing the permeability of these cells to Cl^-, muscle cell membranes become hyperpolarised, usually by only a few mV, and this reduces their responsiveness to excitatory nerve impulses (Parri *et al.*, 1993). This leads to flaccid paralysis of the pharynx (Geary *et al.*, 1993) and somatic muscle (Parri *et al.*, 1993) and, eventually, starvation and loss of ability by the worm to remain at its site of predilection within the host. Two other classes of anthelmintics, the imidazothiazoles and tetrahydropyrimidines, exemplified by levamisole and pyrantel, respectively, target cholinergic receptors on somatic muscle membranes. These receptors are directly coupled to channels for cations which, when opened by these drugs, leads to depolarisation and spastic paralysis of the somatic musculature (Martin *et al*, 1991; Chapters 12 and 21). The actions of several other anthelmintics are traced to their direct or indirect effects on ion channels or ion transporters, which are usually associated with nerve and muscle membranes (Rew and Fetterer, 1986; Bennett and Thompson, 1986; Geary *et al.*, 1993; Chapters 12 and 21). Unfortunately, no data are available on the effects of anthelmintics or other drugs on electrochemical gradients maintained across epidermal or intestinal cells, or on the tubular system or glandular systems in nematodes. Thus, although pharmacological experiments have provided tremendous insights to ion transport mechanisms in nematode neuromuscular systems, we have a very limited grasp of the (presumably) closely related processes that maintain ionic gradients within the pseudocoel, which forms the environment of the muscle and nerve cells.

Role of the Cuticle/Epidermis Complex in Ionic Regulation

The largest and most accessible (i.e., from the environment) structural barrier separating the inside of the nematode from the outside is the cuticle/epidermis complex. Although the cuticle forms as an extracellular secretion from the epidermis, it is possible to separate these structures by experimental procedures and study their individual contributions to the transmural transport of ions (and other solutes). The cuticle, which is 50–100 μm thick in adult *A. suum*, consists of a matrix of cross-linked collagen-like material. Aqueous pores that traverse the cuticle are negatively charged and therefore present an electrostatic barrier to the transport of large organic anions (Ho *et al.*, 1990). However, due to the large radius of these pores (15 Å in *A. suum*) compared to the sizes of relevant inorganic anions (principally Cl^-, radius = 1.2 Å), charge has little effect on inorganic anion transport across the cuticle. This is best demonstrated by data from electrophysiological studies in which electrical potential and current flux were measured across isolated body wall segments from *A. suum* in an Ussing chamber (De Mello and Tercafs, 1966; Pax *et al.*, 1995). An electrical potential of -30 mV (muscle-side negative) develops very rapidly when tissue segments containing living muscle and epidermis are examined. When muscle and epidermal tissues are mechanically scraped from this preparation, and residual lipid removed by extraction with chloroform:methanol, the transmural potential is abolished and there remains no resistive barrier to the passage of small inorganic ions (measured as electrical current) when an external voltage is imposed on the system (Pax *et al.*, 1995).

The epidermis is an anatomical syncytium in most nematodes, including *A. suum*, with apical (cuticle-facing) and basal (muscle-facing) membranes separated by a cytoplasm-filled space of variable thickness (Thompson and Geary, 1995, for review). The transmural potential is maintained, in part, by the separate contributions of the cuticle-facing and epidermis-facing membranes (Pax *et al.*, 1995; Blair *et al.*, 1998). In *A. suum*, the electrical potential within the epidermal compartment is 70 mV more negative than the external medium and 40 mV more negative than the pseudocoelomic compartment when tested in artificial media containing inorganic ion concentrations that approximate those in porcine intestinal contents. Cations do not appear to play a major role in the transmural potential. The concentrations of Na^+ and K^+ in PCF from freshly collected parasites are approximately equal to those found in swine intestinal fluid (Table 11.3), consistent with the high permeability of the body wall to these cations (Hobson, Stephenson and Beadle, 1952; Pax *et al.*, 1995). When parasites are placed in media containing elevated or reduced levels of these cations, concentrations in the PCF change fairly rapidly to approximate the outside concentrations. In contrast, the concentrations of the divalent cations Ca^{2+} and Mg^{2+} in PCF are maintained at constant levels in the face of large changes in their concentrations in the incubation medium (Hobson, Stephenson and Beadle, 1952). However, due to the low concentrations of these cations in cytoplasm or PCF, relative to Na^+, K^+, Cl^- and organic anions produced via carbohydrate metabolism, they do not contribute significantly to the transmural potential.

Chloride concentrations in *A. suum* PCF and porcine intestinal fluid are nearly equivalent, at about 60 mM

(Table 11.3) (Hobson, Stephenson and Beadle, 1952). Changing the external Cl concentration has little effect on the transmural potential in *A. suum* (Pax *et al.*, 1995), suggesting that other ions play more important roles in this process. Active, electrogenic transport of cations contributes extensively to the resting potential of vertebrate and invertebrate cells, including somatic muscle and pharyngeal cells in nematodes (Brading and Caldwell, 1971; Del Castillo, De Mello and Morales, 1964; Del Castillo and Morales, 1967; Caldwell, 1973). However, electrogenic transport of ions contributes only 36% of the transmural current in *A. suum* (DeMello and Tercafs, 1966). Most of the remaining current appears to be carried by organic anions which, at 60–80 mM (Blair *et al.*, 1998), are only slightly more abundant in the epidermal compartment than is Cl^-. Given the high concentrations of these anions in PCF relative to porcine intestinal contents (3–8 mM), it is likely that they contribute extensively to the transmural electrical potential (Pax *et al.*, 1995: Blair *et al.*, 1998; and see above).

Some of the channels and pumps that control ion concentrations and thereby maintain electrical gradients across nematode muscle and nerve membranes have been biophysically characterised in *A. suum* by *in situ* electrophysiological studies (Chapter 12). Genes that encode several homologous proteins have been cloned from *C. elegans* (*C. elegans* Genome Consortium, 1998; Bargmann, 1998, for review). A few have been functionally expressed in heterologous systems, including *Xenopus* oocytes, which has facilitated their biophysical and pharmacological characterisation (Cully *et al.*, 1996; Fleming *et al.*, 1997). By analogy with other organisms, it is likely that some of the channels and pumps that regulate ionic gradients across muscle and nerve membranes serve similar functions in epidermal and intestinal cells. As genes that encode ion transport proteins in *C. elegans* are identified through sequence homology comparisons, and definitively annotated by heterologous expression studies, it will be a straightforward experiment to determine which are also expressed in epidermal and intestinal membranes or gland cells. The highly conserved nature of specific gene regions in many ion transport proteins, such as the pore-forming domains in Na^+ channels (Jentsch, 1997) or K^+ channels (Salkoff and Jagla, 1995) has facilitated their identification in the *C. elegans* genome database (Bargmann, 1998).

In vertebrates, transport of inorganic ions and water across salivary gland cells, kidney and intestinal

Table 11.3. Inorganic and organic ionic composition, pH and osmolarity of *Ascaris suum* pseudocoelomic fluid and porcine intestinal fluid

	Na^+	K^+	Ca^{2+}	Mg^{2+}	Cl^-	Organic acids	pH	Osmolarity
Pseudocoelomic fluid	130	25	6	5	53	74	6.5	198
Porcine intestinal fluid	124	27	14	6	61	5	7.0	257

All concentrations expressed in mM.
Adapted from Baldwin and Moyle, 1946; Harpur and Popkin, 1965; Blair *et al.*, 1998.

epithelia is regulated directly by several neuroendocrine effectors, including aldosterone and antidiuretic hormone. Neurotransmitters, including adrenalin and acetylcholine, as well as hypothalamic neuropeptides that influence pituitary control of corticosteroid release also play important, albeit indirect, roles in these transport processes. However, vertebrate neuroendocrine molecules that regulate ion and water transport and have been tested in nematodes are inactive (i.e., do not affect transmural ionic or water fluxes), suggesting either different pharmacology or the absence of analogous control systems in nematodes (Pax et al., 1995). Insects, which may be more closely aligned to nematodes than previously thought (Blaxter et al., 1998), regulate water and ion flux across the Malphigian tubules, a tissue that has no counterpart in nematodes. Several neuropeptides participate in this regulation, particularly insect kinins, arginine-vasopressin immunoreactive peptides and a family of CRF-related peptides (Coast, 1996; Gade, Hoffman and Spring, 1997; Schoofs et al., 1997, for review). CRF-immunoreactive neurons have been identified in A. suum (Sithigorngul, Stretton and Cowden, 1990), but no peptide that can be assigned to this class on the basis of sequence homology has yet been identified in the C. elegans genome database (unpublished observations). This phylogenetic reach may be too long; nematodes are essentially small aquatic organisms with high surface area-to-volume ratios. The extracellular compartment of the pseudocoelom is separated from the environment by only one cell layer (intestinal epithelium or the epidermis). Therefore, remote regulation by cells far removed from the sites of ion exchange may be less critical than in larger metazoans.

Despite the lack of data supporting neural control of ionic or water transport in nematodes, several observations on nematode physiology are worthy of consideration in this context. The tubular system ('excretory ampulla') provides one example. In larvae of some species, including H. contortus, the ampulla has been observed to pump more rapidly under hyposmotic conditions. The rate of filling of the ampulla is controlled by the excretory valve (Wharton and Sommerville, 1984), which is innervated and therefore presumably under neural control. In addition, the rate of pharyngeal pumping in nematodes is controlled, in part, by serotonergic and neuropeptidergic inputs (Brownlee et al., 1995). This pumping action controls the rate at which digesta passes through the intestine, thus influencing the extent to which ions and water are allowed to enter or exit the intestinal lumen, as it does in vertebrate intestine where neurotransmitter and neuropeptide control of intestinal peristalsis is well documented. These examples illustrate how transport of ions and water could be controlled indirectly by neuronal influences on muscle groups that propel water or digesta through the tubular system or intestine. Similarly, the major muscle groups that regulate excretion, secretion and turgor pressure (pharynx, rectum and vagina vera) are innervated (Avery and Thomas, 1997; Fellowes et al., 1998, for

review). Although no examples of control of these physiological processes by neural input are yet available, an open mind must be kept to the possibility that it occurs.

Direct control of epidermal or intestinal membrane permeability by the nervous system has not been demonstrated in nematodes. However, a receptor that recognises bombesin-like neuropeptides, which serve an analogous function in some vertebrates, has recently been partially characterised (Huntington et al., 1999). In frogs, bombesin helps regulate the rate of Cl^- transport across the skin, which serves a role more closely related to the nematode cuticle/epidermis complex, perhaps, than does the vertebrate kidney. Structurally related peptides, including gastrin-releasing peptide (GRP), have also been isolated from vertebrates. Bombesin and GRP modulate several processes, in addition to ion transport, including secretion (Anastasi, Erspamer and Bucci, 1971) and mitogenesis (Cuttitia et al., 1985). Immunohistochemical studies on A. suum and Dirofilaria immitis indicate that bombesin-like immunoreactivity is localised primarily in the apical regions of the epidermis (Huntington et al., 1993), where the cuticle-facing membrane of the epidermis appears to overlap with the cuticle (Ho et al., 1990). Radioimmunoassay surveys of extracts prepared from eight nematode species indicate that a bombesin-like peptide is broadly distributed among nematodes. Competition binding assays using [^{125}I]GRP and membranes prepared from A. suum and C. elegans reveal a saturable, high-affinity binding site with a K_d of 3 nM and B_{max} of 1 fmol/mg protein (Huntington et al., 1999). Although the chemical structure of a bombesin-like peptide in nematodes has not been determined, the GRP binding data suggest the existence of a peptide that is similar in structure to the putative vertebrate homologs. A molecule from Panagrellus redivivus extract that is recognised by antibodies to GRP has a molecular weight in the 1.7 kD range, similar to that of vertebrate bombesin-like peptides (Huntington et al., 1999). Demonstrating a role for this peptide and its receptor in nematodes awaits determination of its sequence and identification of the gene that encodes its receptor.

Role of the Intestine in Ionic Regulation

Direct evidence for transport of inorganic ions by the intestine is available only for A. suum (Beames and King, 1972; Harpur and Popkin, 1973; Merz, 1977; Thompson and Geary, 1995 for review). The concentration gradients for Na^+, K^+, and Cl^- across the intestine contribute to a positive electrical potential of 10–15 mV (i.e., the lumen is depolarised relative to the surrounding PCF). This potential is highly temperature sensitive and is abolished by cooling to $20°C$. It is also reduced by agents that interfere with glycolysis (Merz, 1977). Flux measurements indicate that, among the major inorganic ions, only K^+ diffuses outward from the lumen more rapidly than it diffuses inward. Since net efflux of only K^+ would lead to hyperpolarisation of the lumen, other ions must account for the positive electrical potential recorded within this compartment. The transintestinal potential is also dependent, in part, on the presence of glucose in the lumen. Glucose is absorbed by a Na^+-coupled system (Na^+/glucose symporter; Pappas and Read, 1975), but whether this process contributes to the electrical potential has not been determined. In other organisms, a Na^+ gradient is essential for Na^+-coupled glucose transport, and this gradient is usually maintained by an Na^+/K^+ ATPase that pumps more Na^+ ions out of the cells than K^+ ions in. However, Na^+ and glucose transport across *A. suum* intestine are not affected by high levels of ouabain (Harpur and Popkin, 1973), a known inhibitor of Na^+/K^+-ATPases, indicating that if this pump regulates the concentrations of Na^+ and K^+ in *A. suum* intestine, it is pharmacologically distinct from that characterised in vertebrates as well as other invertebrates studied.

Osmotic and Volume Regulation

Fluid in the nematode pseudocoel, which is under pressure imposed by the limited ability of the cuticle to expand or contract, forms a hydrostatic skeleton that is essential to coordinated movement (Chapter 13) and to the digestive process (Lee and Atkinson, 1977, for review). Hydrostatic pressure within the pseudocoel of *A. suum* at rest is maintained at about 70 mm Hg above ambient. During contraction of the body wall musculature during locomotion, values may oscillate from 16 mm to 225 mm Hg (Harris and Crofton, 1957). Little is known about how parasitic nematodes maintain turgor pressure; the consequences of interfering with this process can only be deduced from *in vitro* experiments that are, at best, of marginal relevance to conditions that occur in the host. Puncturing the cuticle, for instance, rapidly leads to uncoordinated movement and death in *A. suum* and other nematodes *in vitro*. Loss of turgor pressure would thus likely be lethal to parasites *in vivo*. Other clues to the important links between osmotic regulation and muscle function come from studies using *C. elegans* mutants that are resistant to levamisole and appear to lack pharmacological acetylcholine receptors (Lewis *et al.*, 1989). Among the phenotypes displayed by these mutants, referred to as 'uncs' because they exhibit uncoordinated movement, is extreme sensitivity to hypotonic shock

The hydrostatic pressure within the pseudocoelom ensures that the intestine remains essentially collapsed except when occupied by food particles. This probably plays an important role in digestion by controlling the rate of passage of nutrients through the intestine. Nematodes lack splanchnic muscles, and ingested material is pumped caudally against the turgor pressure by the pharynx, which is the only muscle present in the digestive tract except for the rectal muscles that regulate defecation. Although reliable measurements of turgor pressure in nematodes other than *A. suum* are not available, the fact that most species rapidly eviscerate when the cuticle is breached suggests that high turgor pressure is a common feature in nematodes that limit the rate of water transport across their cuticle.

Osmotic and volume regulation are tightly linked processes in all metazoans. Marked fluctuations in environmental solute concentrations are encountered by most parasitic nematodes that reside in brackish water or on damp grass as larvae (Anya, 1966; Arthur and Sanborn, 1969). Adults in the mammalian gastrointestinal tract may encounter osmotic pressures in excess of 500 mOsm. Soil nematodes, such as *C. elegans*, tolerate variations in external osmotic pressure of 3–5 fold during rainy or drought conditions (Stephenson, 1942; Jones and Parrott, 1969; Wharton, 1986, for review), and the euryhaline species, *Rhabditis marine*, survives culture in salinities ranging from 0–230% sea water (Nicholas, 1984). Adult stages of parasitic nematodes are less tolerant of hyposmotic conditions than are larvae. Adult *A. suum*, for example, swell and explode within a few hours when incubated in fresh water. However, they tolerate incubations for up to 4 days in dilutions of artificial sea water that range in strength from 20–40% (Hobson, 1948). During these incubations, osmotic pressure measured in PCF is roughly equal to that of the external medium. These findings, together with results showing that inorganic ion concentrations in the PCF also follow closely changes in the environment, indicate that one mechanism used by adult nematodes to tolerate modest changes in external osmotic pressure is to adjust their internal solute concentrations. This could be accomplished by actively regulating the concentrations of organic (i.e., in addition to inorganic) solutes in their PCF and solid tissues. This concept has been studied more thoroughly in marine nematodes, where organic osmolytes such as glycerol, trehalose, amino acids and urea play a more important role than inorganic ions in osmoregulation (Yancey *et al.*, 1982; Kinne, 1993). These organic solutes are less disruptive than inorganic ions to the activities of essential enzymes, such as pyruvate dehydrogenase (Yancey *et al.*, 1982).

The most spectacular examples of osmotic stress tolerance in nematodes come from studies that examine desiccation survival (Lees, 1953; Madin and Crowe, 1975; Wharton, 1986, for review). Infective larvae of the gastrointestinal parasite *T. colubriformis* survive vacuum desiccation at 0% relative humidity for nine hours (Wharton, 1982). Some plant-parasitic and free-living species tolerate even longer periods of desiccation; encysted L2 of *Heterodera avenae* survive up to 4.5 years (Meagher, 1974), and *Anguina tritici* for 32 years (Norton, 1978). In most species, anhydrobiosis must be induced slowly in order for the worms to survive (Crowe and Madin, 1975; Madin, Crowe and Loomis, 1978). Although the mechanisms underlying anhydrobiotic survival in nematodes are unknown, several organic solutes, including trehalose (Ash and Atkinson, 1982, 1983; Crowe, Crowe and Chapman, 1984), glycerol and inositol (Womersley, 1981), have been suggested as replacement molecules for membrane-bound water. However, the observations that some species, such as *Panagrellus silusae* and *A. suum*, produce high levels of glycerol and trehalose, respectively, but are highly susceptible to desiccation (Cooper and van Gundy, 1971b; Barrett, 1981, 1982), indicate that those solutes alone do not confer tolerance to anhydrobiosis.

The ability of nematodes to withstand major osmotic changes in their environment probably depends more on the permeability of the cuticle to water than on the membrane-sparing actions of organic osmolytes. Marine nematodes are usually isosmotic with seawater and their cuticles are highly permeable to water (Lee and Atkinson, 1977; Wharton, 1986, for review). Water permeability in the marine nematode *Europlus*, for example, is 4.4 μ^3 H_2O/μ^2 body surface/hr, a value over 100-fold greater than that recorded for *Aphelenchus avenue*, a soil-dwelling species (Marks, Thomason and Castro, 1968), and over 200-fold greater than that for *C. elegans*, another soil-dweller, or larvae of the animal-parasitic species, *Nippostrongylus muris (brasiliensis)* and *Ancylostoma caninum* (Wright and Newall, 1980).

Role of the Cuticle/Epidermis Complex in Osmoregulation

The cuticle/epidermis complex, intestine and tubular system all appear to participate in osmotic and volume regulation in nematodes. However, definitive studies that delineate the relative contributions of each of these structures, or the underlying mechanisms involved, are lacking. The cuticle/epidermis complex of adult *A. suum* is highly permeable to water, as demonstrated by $[^3H]_2O$ flux studies (Wright and Newall, 1980). Water-filled pores that traverse the cuticle of *A. suum* and other species provide a highly accessible pathway for water in the environment to reach the outward-facing membrane of the epidermis, which probably forms the limiting external diffusional barrier in nematodes (Ho *et al.*, 1990).

The rates of transcuticular water flux reported for most nematodes are hundreds-fold greater than that achievable by simple diffusion across lipid membranes (Ho *et al.*, 1990). In other organisms, water transport across membranes occurs via intrinsic membrane proteins called aquaporins. These proteins have been most thoroughly studied in human red blood cells and kidney proximal tubules, but they exist in a wide range of other vertebrate tissues (Agre, Bonhivers and Borgnia, 1998, for review). Aquaporins are thought to form from 2 hemipores, each containing 3 transmembrane domains. Genes encoding some of these proteins, including that for human aquaporin-1 (AQP-1, also called CHIP28), have been expressed in *Xenopus* oocytes, which has facilitated the biophysical analysis of their transport properties. Homologous genes have been sequenced from other organisms, including bacteria (Calamita *et al.*, 1998), plants (Chrispeels and Maurel, 1994) and insects (Beuron, Le-Caherec and Guillam, 1995). Each contains the highly conserved motif Asn-Pro-Ala in the putative pore-forming region of the gene product (Agre *et al.*, 1998). Recently, a cDNA encoding a *C. elegans* aquaporin was cloned and expressed in *Xenopus* oocytes (Kuwahara *et al.*, 1998). Expression of the channel endowed oocytes with greater permeability to water and, to a lesser extent, urea, but not glycerol. This gene was expressed only in early larvae of *C. elegans* and was completely suppressed before hatching.

Other genes predicted to encode aquaporins have been found in the *C. elegans* genome database and in cDNA libraries from various parasitic nematodes, including *Toxocara canis*, *B. pahangi* and *O. volvulus* (Loukas, Hunt and Maizels, 1999). Definitive studies of the functional properties of the proteins encoded by these genes, as well as their pattern of expression in nematode tissues, are currently underway. It is highly likely that aquaporins will play a critical role in the transport of water across the cuticle/epidermis complex. Identification of the rate-limiting barrier to water diffusion across this tissue as either the epidermis or the epicuticle (or both) awaits experiments on the localisation and biophysical characterisation of water channels in these organisms.

Role of the Intestine

Water can be removed through the intestine of *A. suum* under hypotonic conditions (Wright and Newall, 1976). The mechanism by which water diffuses across the intestinal epithelium from the pseudocoel is unknown. However, active Na^+ transport occurs across the intestine of *A. suum* (Merz, 1977), and it is likely that water follows the movement of Na^+ and possibly other inorganic ions in this system. In annelid and insect gut epithelia, as well as vertebrate intestine and kidney, where these linked processes have been studied more thoroughly, epithelial Na^+ channels mediate the bulk flow of Na^+, with water following passively, probably through aquaporins. In vertebrates, epithelial Na^+ channels are regulated, in part, by vasopressin and aldosterone. The identification of a role, if any, for Na^+ channels in osmoregulation across the body wall or intestine of nematodes awaits the identification of candidate channels expressed in the tissues that form these structures. A degenerin-like Na^+ channel, which is closely related to mammalian epithelial Na^+ channels, was recently cloned from *C. elegans* (Take-Uchi *et al.*, 1998). This channel is expressed in the intestinal tract in all stages of the life cycle. Mutations in the gene encoding the channel alter the defecation rhythm, but are not reported to induce abnormalities in osmotic regulation. Further study of other *C. elegans* genes in this large family may identify one (or more) that is associated with water movement.

Under isosmotic conditions, water enters intestinal cells of *A. suum* from the lumen, then passes into intercellular spaces before entering the pseudocoelom. It is not known if water enters or exits intestinal cells through aquaporins. The rate of water transport from the gut into the pseudocoelom is higher in the presence of oxygen (Wright and Newall, 1976), suggesting a role for aerobically derived energy in this process. However, there is no reported evidence that intestinal cells in *A. suum* or other parasitic nematodes are aerobic.

The role of the intestine in osmoregulation in nematodes appears to be stage- and species-specific. The intestine is fully collapsed in some larval stages in which the mouthparts are physically occluded (Howells, 1980; Bird and Bird, 1991, for review). Based mainly on morphological evidence, the intestine is nonfunctional in some animal-parasitic nematodes, including most tissue-dwelling species examined (Howells, 1980 for review). In several filarial nematodes, water and solute transport may occur exclusively across the cuticle (Howells, 1980, for review). There are undoubtedly exceptions to this generality. For example, in L3 larvae of *Pseudoterranova decipiens*, an anisakid found in the muscle of cod, the intestine appears to play a role in osmoregulation, as the oral opening is patent and water is consumed by the oral route (Fuse, Davey and Sommerville, 1993a). Intact worms accumulate water under hyposmotic conditions. However, sacs composed of cylinders of body wall prepared without the intestine do not accumulate water under hyposmotic conditions, even though exchange of $[^3H]_2O$ between the medium and PCF occurs at a high rate. Similar results are obtained in worms mechanically ligated at the head and tail to prevent passage through the intestine. Metabolic poisons, including cyanide and dinitrophenol, abolish the ability of the body wall segments or ligated worms to osmoregulate, again suggesting a possible role for active transport of ions in this process. These results indicate that, even though the cuticle/epidermis complex provides a potential pathway for water flux, the process of regulating volume in this nematode is dependent on a patent and functional intestine (Fuse, Davey and Sommerville, 1993b).

In most nematodes, the intestine remains open and functional throughout the adult stage of the life cycle. In some filarial nematodes, however, including *O. volvulus*, the intestine appears to atrophy and eventually becomes nonfunctional as the parasite matures (Franz, 1988). Concurrent with this change is a thickening of the epidermis and an apparent increase in enzymatic activity there. These changes are most pronounced in female *O. volvulus*, which also exhibit a marked reduction in somatic muscle mass. It is not known if the apparent reduction in intestinal function precedes, or even causes, the reduction in epidermal and somatic muscle mass or *vice versa*. It is intriguing to speculate that the highly vascularised environment of the onchocerca nodule (Williams, Mackenzie and El Khalifa, 1994), coupled with the apparent loss of need for mobility in adult females, contribute to the observed atrophy of somatic muscle and hypertrophy of the epidermis. Based on these anatomical considerations, it appears likely that some transport functions previously served by the intestine are adopted by the epidermis/cuticle complex (Franz, 1988). This hypothesis could be tested by comparing transport rates of water, inorganic ions or nutrients (or their nonhydrolyzable analogs, such as 3-o-methylglucose) or excretory products (e.g., lactate) in mechanically ligated *vs* nonligated female *O. volvulus* as a function of age.

Tubular System

Nematodes possess a tubular (or glandular, in some species) system that was originally assigned an excretory function, primarily on the basis of morphological evidence (see Wharton, 1986, for review). In *A. suum*, canals that form part of this system are exposed to PCF along most of their length. These canals were shown to accumulate dyes injected into the pseudocoelom (Weinstein, 1952, 1960; Lee and Atkinson, 1977, for review). A mechanism was proposed whereby the high internal (or turgor) pressure in the nematode pseudocoel provides a driving force for filtering excretory products released from cells into the lateral canals (Harris and Crofton, 1957). Given the relatively high molecular weights and charged nature of the dyes used in these studies, however, it is difficult to conceive of a filter that would allow their passage while retaining water and other nutrients within the pseudocoel. No morphological evidence exists for endocytotic or pinocytotic mechanisms in the canals that could underlie such a process.

In *C. elegans*, the entire tubular system has been reconstructed from serial section electron micrographs (Nelson, Albert and Riddle, 1983). It consists of four cells whose nuclei are located on the ventral side of the pharynx. A *pore cell* encloses the terminus of an *excretory duct cell*, which leads to an excretory pore at the ventral midline. An *H-shaped cell* forms bilateral excretory canals that extend anteriorly and posteriorly along most of the length of the organism. These canals form numerous gap junctions with the epidermis and are in direct contact with the PCF. In addition, an *A-shaped excretory gland cell* extends bilateral processes anteriorly to cell bodies located behind the pharynx. These processes fuse with the H-shaped cell at the origin of the excretory duct. Laser ablation studies on *C. elegans* have illuminated the role of these cells in homeostatic regulation (Nelson and Riddle, 1984). If either the pore, duct or excretory cells are ablated, the nematode fills with water within 12–24 hours and dies within a few days. Ablation of the excretory gland cell results in no obvious developmental or behavioural defects. These results, which suggest a role for the tubular system in osmoregulation, are consistent with microscopic examination of the excretory duct; its rate of pulsation under hyposmotic conditions is five to six-fold higher than under isosmotic conditions (Nelson and Riddle, 1984).

Additional insights to the possible role of the tubular system in osmoregulation comes from *C. elegans* mutants. Mutants defective in the posterior migration of canal-associated neurons (CANs) exhibit arrested development and excess fluid accumulation in their pseudocoelom (Forrester and Garriga, 1997). Also, a mutation in the gene, *let-653*, which encodes a mucin-like protein, results in lethal arrest, concurrent with vacuole formation anterior to the lower pharyngeal bulb, in a position consistent with dysfunction of the tubular apparatus (Jones and Baillie, 1995).

The extent to which these physiological and genetic observations on *C. elegans* are relevant to parasitic species is largely undetermined. The existence of tubular structures in parasitic nematodes is well documented (Lee and Atkinson, 1977, for review). Recent studies suggest that these structures may be more elaborate than originally thought. For example, anterior regions of the tubular system in *O. volvulus* include paired glomerulus-like structures in the lateral cords that appear to connect to the cuticle through canals formed by projections from the basal zone of the cuticle (Strote, Bonow and Attah, 1996). Anatomically, these structures resemble organs with an osmoregulatory or excretory function. However, assigning them a function in homeostatic regulation will require additional evidence, such as localisation within their membranes of ion channels or aquaporins.

Future Directions in the Study of Homeostatic Regulation in Nematodes

The Role of *C. elegans* Genomics

The *C. elegans* genome project can resuscitate the science of nematode biology. Although *C. elegans* is not a parasite, this organism can provide unprecedented insight into the physiology of parasitic species (Bürglin, Lobos and Blaxter, 1998; Bird and Opperman, 1998; Blaxter, 1998). Now that the sequencing phase of the project is completed, attention will turn to the functional properties of the ~19,100 genes that have been sequenced. Although questions in developmental biology will continue to be of paramount interest, more research will focus on genes that encode proteins for which known orthologs in humans underlie various hereditary diseases. Approximately 36% of the predicted proteins in *C. elegans* have homologs in humans (*C. elegans* Sequencing Consortium, 1998), some of which are clearly involved in regulating the transport of ions, other solutes and water across cell membranes in the kidney and intestine. A likely consequence is that new information will emerge, perhaps unintentionally, on some of the transport processes described in this chapter. However, the opportunities for understanding how nematodes work as organisms are so great that investment must be made into research that translates the *C. elegans* genome into parasite biology. Key topics that are approachable using *C. elegans* genomics strategies, with special reference to homeostatic mechanisms, are discussed below.

Techniques

As noted above, nematode proteins that are clearly homologous to transporters known from other organisms can be readily identified in the *C. elegans* genome. Unfortunately, functional analysis of the gene products is typically lacking. Several techniques have been developed to enable the definition of the functional role of interesting proteins in *C. elegans*. Understanding what these proteins do in the free-living organism will greatly facilitate the design of studies on the conservation of those functions in plant and animal-parasitic nematodes.

Stage-and tissue-specific expression patterns. Since nematodes encounter changing environments during development, it is important to determine the life cycle stages at which genes of potential interest are expressed. An example can be found in the aquaporin gene family discussed previously. The only member of this family to be functionally expressed in a heterologous system is found only in the earliest stages of larval development (Kuwahara *et al.*, 1998). Additional members of the family can be recognised in the genomes of other nematodes (Loukas, Hunt and Maizels, 1999). Simple Northern hybrizidisation analysis, using RNA obtained from the different life cycle stages maintained in synchronous culture, can reveal whether any is expressed in adults or other larval stages. Such a study can also be combined with experiments designed to reveal tissue-specific expression of genes in asynchronous populations. *In situ* hybridisation techniques (Albertson, Fishpool and Birchall, 1995) and immunofluorescence microscopy (Miller and Shakes, 1995) can be used to monitor expression. Alternatively, upstream regulatory sequences preceding the gene of interest can be fused to a reporter gene, most commonly β-galactosidase or green fluorescent protein (Chalfie *et al.*, 1994). This construct can be injected into *C. elegans* and the location of gene expression determined in the transgenic offspring (Mello and Fire, 1995).

Loss of gene product. Elimination of the gene product of interest can help define its functional role. The process of using sequence data to analyse functional biology in this way has been termed 'reverse genetics' (Plasterk, 1995). There are two main approaches to interfering with gene expression. In the first, worm populations are subjected to random mutagenesis by activation of the Tc1 transposon, or by exposure to chemical mutagens or irradiation. Mutagenised worms can be frozen in 'libraries' and screened using a sib-selection protocol by PCR. For transposon insertions, PCR primers that recognise a region of the transposon are paired with primers designed to the gene of interest. Only pools containing animals in which the transposon has inserted in or near the target gene will produce an amplified band after PCR (Zwaal *et al.*, 1993). A similar strategy can be used after chemical mutagenesis. Pools of worms are screened with PCR primers designed to the target gene under conditions which do not permit successful amplification. Chemically-induced deletions result in a shorter distance between the primers, permitting successful amplification. Pools containing individuals with the desired deletion can thus be identified quickly via PCR screening (Jansen *et al.*, 1997).

Recently, a revolutionary method for temporarily silencing gene expression has been reported. Exposure of *C. elegans* (by injection or simply by addition to the medium) to double-stranded RNAs that encompass at least one exon of a target gene results in marked suppression of the gene product in the F_1 generation (Fire *et al.*, 1998; Tabara, Grishok and Mello, 1998). Although it is not yet known how the process works, double-stranded RNA interference (dsRNAi) promises to be an expedient way to obtain information about the function of gene products in this organism. The potential of this technology for rapid analysis of protein function cannot be overestimated.

Priority targets for functional analysis in C. elegans

Excretion-related proteins. As noted above, a number of genes that encode proteins which could mediate excretory processes have been identified in the *C. elegans* genome. Among these are the putative monocarboxylate transporters (Price, Jackson and Halestrap, 1998). Expression of these genes in heterologous systems in which substrate specificity could be established should be a high priority. If these proteins transport organic acids typically excreted by nematodes, studies to define the tissues in which they are expressed would be warranted. Disabling these genes by mutation or dsRNAi, coupled with chemical analysis of the medium, could reveal their roles in nematode physiology. Similar approaches should be supported for analysis of *C. elegans* proteins that might transport nitrogenous products, including the aquaporins (possible role in urea transport), the putative ammonium transporter (Marini *et al.*, 1994), and the putative organic cation transporters (Wu *et al.*, 1999).

Ion transport. Products of genes that encode ion channels and pumps undoubtedly contribute to homeostatic regulation in nematodes. For instance, 22 genes encoding Na^+ channels in the degenerin (e.g., *deg-1*, after 'degeneration') and mechanosensory (e.g., *mec-4* and *mec-10*; Huang and Chalfie, 1994) families have been identified in *C. elegans*. In other organisms, related channels are mechanosensory proteins that form amiloride-sensitive Na^+ channels in Na^+-resorbing epithelial cells. In vertebrates, they are targets for aldosterone in apical membranes of kidney and intestinal epithelial cells. Epithelial Na^+ channels in vertebrates are structurally unrelated to neuronal Na^+ channels (Voilley *et al.*, 1997). They form heteromultimeric proteins consisting of three or more subunits, each containing two hydrophobic transmembrane domains (which form the Na^+ pore) separated by a large extracellular loop. In *C. elegans*, these channels were originally identified in genetic screens for touch insensitivity (Chalfie and Jorgensen, 1998). Evidence for their involvement in osmoregulation in *C. elegans* comes from studies showing that dominant mutations at these genes lead to enhanced Na^+ flux, swelling and eventual degeneration of mechanosensory cells in the amphids. It would be extremely interesting to know if members of this channel family are expressed in the intestine, epidermis or tubular system in nematodes. If any are expressed at these sites, they may play a role in regulating Na^+ or water flux at the worm:environment interface. The consequences of loss-of-function mutations or dsRNAi should be determined for any Na^+ channel expressed in these tissues.

The *C. elegans* genome database contains at least 12 ATPases, including several subunits of a vacuolar-type ATPase, which transports H^+ out of cells in other organisms. This gene family was identified through the genome sequencing project (Oka, Yamamoto and Fatai, 1997, 1998). The genes are expressed in the H-shaped excretory cell, the rectum, and in two cells posterior to the anus. One subunit has been shown to be functional by its ability to complement a specific yeast mutant (Oka *et al.*, 1998). However, phenotypes associated with mutations in these genes have not been described. It should be straightforward to use dsRNAi techniques to determine if this ATPase contributes to pH regulation in nematodes, as it does in the kidney (see Oka, 1998). Na^+/K^+ ATPases are also present in *C. elegans*; mutations in one of these, *eat-6*, impair pharyngeal pumping (Davis *et al.*, 1995). These mutants show no evident alterations in somatic muscle function, suggesting that there may be additional Na^+/K^+ ATPases in these organisms. Finally, genes encoding putative Ca^{2+} ATPases have been identified in *C. elegans* (Kraev, Kraev and Carafoli, 1999). One, *mca-1*, has been functionally expressed, but no studies on tissue distribution or effects of mutations in any of them have been reported. Five candidate $Na^+/Ca^{2+}(K^+)$ exchangers are reported to be encoded in the genome (Kraev *et al.*, 1999), but no reports of biology associated with them have appeared.

Some 75–80 genes encode K^+ channels in *C. elegans* (Wei, Jegla and Salkoff, 1996; Bargmann, 1998). Among these are 20 voltage-sensitive and three inward-rectifying channels, which typically function to maintain the hyperpolarised state of cells. At least 40 genes encode 2-pore forming K^+ channels (Wei *et al.*, 1996; Bargmann, 1998). This family of K^+ channels was only recently identified in *C. elegans* (and mammals, where they are referred to as TWIK channels), and nothing is known yet about their function(s) in nematodes. However, the fact that over half of the K^+ channel-encoding genes appear to belong to this family suggests that they play diverse and important roles in nematodes. Although K^+ gradients contribute little to the transmural (Pax *et al.*, 1995) or trans-intestinal (Merz, 1977) electrical potentials, this cation is clearly important in electrical events that mediate somatic and pharyngeal muscle contraction (Brading and Caldwell, 1971; Raizen and Avery, 1994). K^+ channels may also help maintain the steep electrical gradients across the muscle- and cuticle-facing membranes of the epidermis (Pax *et al.*, 1995). In addition, indirect evidence implicates involvement of a nitric oxide sensitive K^+ channel in the hyperpolarisation of somatic muscle and subsequent flaccid paralysis associated with the FMRFamide-related peptide, SDPNFLRFamide (PF1) (Bowman *et al.*, 1995). The gene that encodes this channel has not been identified. However, this research demonstrates the existence in nematodes of G protein-coupled K^+ channels, which may be better suited for roles in homeostatic regulation of membrane potential than the usually fast-acting families of K^+ channels typically associated with the neuromusculature.

Other potential sites for K$^+$ channel involvement, in the context of structures important to homeostatic regulation in nematodes, include those found in muscles that regulate defecation and egg-laying, and channels that may be present in the tubular system. Too little is known about where most of these channels are expressed to predict their specific contributions, but the functional genomics strategies described above provide tools for investigating the distribution and functional properties of these channels. A useful example is the Shaw-type K$^+$ channel encoded by the gene *egl-36* (Johnstone *et al.*, 1997; Elkes *et al.*, 1997). Gain-of-function mutations in this gene reduce egg-laying and defecation: both processes obviously affect the way in which a nematode interacts with its environment. The apparent ease with which this *C. elegans* K$^+$ channel can be expressed in *Xenopus* oocytes (Johnstone *et al.*, 1997) may extend to other channels, which should facilitate their functional characterisation.

Finally, two genes encoding K-Cl cotransporters have been identified in *C. elegans*, one of which was functionally expressed in a heterologous system (Holtzman *et al.*, 1998). In mammals, K-Cl cotransporters typically function to balance water flux across the opposite faces of epithelial cells to maintain cell volume. It is tempting to speculate that the nematode homologs could function in osmoregulation or in regulation of turgor pressure. Tissue localisation studies and dsRNAi or gene deletion experiments could be used to test that concept.

Finding new genes in C. elegans

A little more than a third of predicted *C. elegans* proteins are considered to be homologous to known human proteins (*C. elegans* Sequencing Consortium, 1998). The number of predicted proteins with unassigned function is large. It is therefore necessary to develop strategies for searching for *C. elegans* genes encoding proteins that may be of interest for homeostatic regulation. A few examples follow.

At least six Cl$^-$ channels are found in *C. elegans* (Bargman, 1998, for review). This group includes a glutamate-gated Cl$^-$ channel that is a target for the avermectins (Cully *et al.*, 1996). Cl$^-$ appears to play only a minor role in maintaining the electrical potential across the body wall of *A. suum* (Pax *et al.*, 1995). However, as noted above, a Cl$^-$ channel, characterised in *A. suum*, may conduct the organic anions excreted as end products of carbohydrate metabolism. It is biophysically distinguishable from known vertebrate Cl$^-$ channels (Valkanov *et al.*, 1994). If this Ca^{2+}-activated Cl$^-$ channel contributes to electrochemical gradients for Cl$^-$ and organic anions in both the epidermis and somatic muscle cells of nematodes, it could serve an essential role in excretion of metabolic end-products. Identifying the gene that encodes this Cl$^-$ channel in *C. elegans*, if it exists, would permit analysis of its function and open an exciting area for research.

Other *C. elegans* Cl$^-$ channels may also play essential roles in maintaining homeostasis. In vertebrates, CLC-2 is a low conductance Cl$^-$ channel that is activated by

both swelling and acidosis. However, it is not Ca^{2+}-activated like the organic anion-conducting Cl$^-$ channel described above. Two other Cl$^-$ channels present in vertebrates are activated by swelling: CLC-3, which conducts organic anions such as acetate and gluconate, and VRAC (volume regulated anion channel). In the case of VRAC, conductance to other, uncharged organic osmolytes is greater than that to Cl$^-$ (and organic anions) by a factor of at least 6-to-1. It is not known if related channels are present in *C. elegans* or other nematodes. The ease with which the glutamate-gated Cl$^-$ channel was expressed in *Xenopus* oocytes, which greatly facilitated its characterisation (Cully, 1996), does not guarantee that all *C. elegans* Cl$^-$ channels will be functionally expressible in this or other heterologous systems (Geary, Sangster and Thompson, 1999, for review).

Bombesin-like peptides, as mentioned, are involved in diverse physiological processes in vertebrates. Although immunohistochemical experiments reveal the presence of bombesin-like peptides in various invertebrates (Huntington *et al.*, 1999), no peptide in this class has been structurally defined from an invertebrate. Pharmacological data strongly support the contention that a bombesin-like peptide is present in nematodes (Huntington *et al.*, 1999), but purification of an immunoreactive peptide to the point of sequence has not been accomplished. Antisera raised against bombesin or the related gastrin-releasing peptide (GRP) recognise a family of vertebrate peptides with minor but significant differences in primary structure (Spindel *et al.*, 1993). Attempts to locate candidate bombesin-like peptide precursors in the *C. elegans* genome by searching with short conserved peptides have so far been unsuccessful (unpublished observations). Further purification approaches might identify the responsible peptide, which could then lead to dsRNAi or knock-out experiments to demonstrate its function in *C. elegans*. Alternatively, more sophisticated bioinformatics tactics may lead to the identification of the precursor in the *C. elegans* genome.

Advances in Parasite Biology Based on *C. elegans*

Adaptations necessary for adoption of a parasitic life style almost certainly include the development of homeostatic regulatory mechanisms that are not required for life as a free-living nematode. Thus, while exploration of the physiology of *C. elegans* is a requisite first step in understanding these processes, the results must still be extended to parasites, perhaps several different species. Tools required to facilitate such experiments are sadly lacking for parasitic species, although physiological studies on *A. suum* are a notable exception. Although some progress has been made in developing genetic systems for plant-parasitic nematodes, primarily because plant hosts can be easily maintained (Bird and Opperman, 1998), it is not yet possible to passage parasitic species in axenic culture in the laboratory. Transformation systems and gene disruption techniques are not yet available. It is not yet known if dsRNAi technology can be adapted for use in parasites. Against this dreary background, which calls for urgent redress, one is left with few simple options for translating new data developed in *C. elegans* to parasites.

Most straightforward is the use of *C. elegans* genes to screen parasite cDNA libraries for homologs. It must be recognised that *C. elegans* diverged from parasitic species as much as 500 million years ago (Bird and Opperman, 1998), leaving plenty of time for sequence divergence to become problematic. Codon bias also differs remarkably among *C. elegans* and various parasites (unpublished observations). An alternative approach, of particular use when homologous genes are present in other organisms, is to use PCR techniques to amplify portions of the parasite gene from mRNA, cDNA or genomic libraries. In this case, degenerate primers can be designed to amino acid regions that are completely conserved between *C. elegans* and non-nematode organisms. Once an intact gene is in hand, it can (one hopes) be expressed in a suitable heterologous system and its function defined. It may then be possible to study that function *in situ* using parasite material. At this time, analysing the consequences of interference with the gene product on parasite viability can only be determined with an inhibitor. Such compounds can be discovered through new high-throughput screening technologies (see below), although at considerable cost.

The Biology of Nematode E/S Products

Perhaps the most crucial adaptation for parasitism is the ability to secrete products that modify the host environment to maximise fitness of the parasite. Although anaerobic metabolic pathways distinguish many parasitic species from free-living organisms, this is not a universal finding. As noted, secreted products play crucial roles in invasion, digestion of host tissues for nutrient acquisition, and defense against attack by the host immune system. As such, these kinds of secreted products may be absent from, or greatly modified in, free-living nematodes. We therefore cannot rely on the *C. elegans* genome sequence to reveal them. Instead, they must be studied in parasites. It is sobering, if not dismaying, to realise how few examples are known of secreted nematode products with defined actions on the host. Tremendous insight into the biology of parasitism will be gained from studies in which secreted products are structurally identified and assigned functions. Indeed, until such studies are done, we can claim only a dim appreciation of the molecular biology of parasitism. Although parasite secretory products are worthy of research from the most basic aspects of biology, practical benefits will also likely accrue from investment in this area. Two examples are considered below.

E/S products as model templates for pharmaceutical discovery

Bioactive molecules secreted by nematodes may have therapeutic applications. Several are attracting considerable research. These include the hookworm neutrophil inhibitory factor discussed above (Moyle *et al.*,

1994) and a small hookworm protein that is one of the most potent natural anticoagulants known (Cappello *et al.*, 1995). Other secreted molecules, such as a macrophage migration inhibitory factor found in diverse parasitic nematodes (Pennock *et al.*, 1998) may also be useful in the clinic. Nematodes secrete less well-characterised factors that suppress lymphocyte proliferation but not antigen responsiveness (Allen and MacDonald, 1998). Further work on identifying the responsible molecules is clearly warranted. In addition, as discussed above, E/S products released by filarial nematodes are capable of influencing flow in mammalian blood and lymph vessels. The molecule(s) that underlie these phenomena may have applications in hypertension and other cardiovascular diseases. E/S products from nematodes are already being evaluated for new approaches to the biological control of disease-spreading insects, including mosquitoes (intermediate host for *Plasmodium* spp.) and black flies (intermediate host for *O. volvulus*) (Williams, Mackenzie and El Khalilfa, 1994, for review).

E/S products as markers for infection

The potential that nematode E/S products hold for diagnosing and quantitating infection (i.e., by the parasites excreting them) has long been recognised. Diagnostic tests that measure circulating antigens corresponding to several helminth E/S products are already marketed. An example is Og4C3, a circulating antigen that is easily detected in patients infected with *Wuchereria bancrofti* (Chanteau *et al.*, 1994). Another example is an aspartyl protease inhibitor secreted by *Dirofilaria immitis*; antibodies against this protein provide a sensitive and specific marker for heartworm infections in cats, but not dogs (Frank *et al.*, 1998). An advantage of E/S diagnostics *per se* is that these molecules are often eliminated in fairly large quantity and will only be detected if parasites are actually present. A disadvantage is that anti-E/S product antibodies often cross-react with molecules present in multiple species of parasites. However, as genes encoding important E/S antigens are sequenced from multiple species, it is possible that regions of unique amino acids may be identified for the generation of more specific reagents.

New Targets for Nematode Control

New methods for nematode control are needed in several arenas, and different strategies may be called for in each. For instance, *plant-parasitic* nematodes are typically controlled by soil fumigants, but regulatory restrictions may soon remove the remaining effective compounds from the market. In this context, genetically engineered plants that resist nematode infection may be the best solution to the prevention of massive crop losses (see Chapter 23). For instance, transgenic plants specifically engineered for nematode resistance express inhibitors of proteases involved in nematode digestion (Urwin *et al.*, 1995, 1997). Alternative tactics will become evident when key secretory products are identified. For example, counteracting the products released by the nematode to reshape the plant root environment into a feeding site should be a specific and highly effective way to prevent reproduction.

A second arena is nematode control in livestock and companion animals. Veterinary nematode control programs typically rely on a limited set of broad-spectrum anthelmintics (Chapter 21), since multiple species of parasites are usually present simultaneously. The veterinary market for anthelmintics is large and expanding. The size of the market and the emergence of strains of parasitic nematodes resistant to existing drugs have stimulated investment by the animal health industry in the identification of anthelmintics that act by novel mechanisms. The drug discovery paradigm currently used by most pharmaceutical companies relies on mechanism-based high-throughput screens (HTS). These screens use minute quantities of specific target proteins, often expressed in a heterologous cell system that can be screened using 96-well or 384-well (and growing) microtitre plates, and require vanishingly small quantities of reagents and test compounds. This approach to screening for drugs, including anthelmintics (Geary, Thompson and Klein, 1999, for review) is driven by several factors. Principal among these is the realisation that, for cost purposes, it is essential to design selectivity into the drug discovery process as early as possible. That is, compounds that move through the screening funnel to increasingly resource-consuming stages of evaluation must first demonstrate the desired effect on the specific target protein.

The *C. elegans* genome database, coupled with functional genomics approaches that have been developed to capitalise on it (Epstein and Shakes, 1995), provide powerful tools that will facilitate mechanism-based HTS for anthelmintics (Geary *et al.*, 1999, for review). Genes that encode proteins which regulate ion flux have already been identified in *C. elegans* neurons and muscle cells by the discovery of the glutamate-gated Cl⁻ channels and acetylcholine-gated cation channels, which are targets for the macrocyclic lactones and levamisole, respectively. Whether the proteins that underlie homeostatic maintenance in nematodes are also attractive targets for anthelmintic screening remains an untested concept, but one that should now become much easier to evaluate using the *C. elegans* genome database and functional genomics techniques. Potential targets of interest already identified in *C. elegans*, in the context of homeostatic regulation in nematodes, have been discussed throughout this review.

A different situation is encountered in human medicine. Human nematodiases are usually encountered in areas where poverty and limited health care options prevail. Typically, at-risk populations suffer from a variety of more life-threatening conditions; control programs thus must be cheap and easy to administer. While a few drugs are available for human use, introduction of a new anthelmintic for this market is an extremely rare and increasingly unlikely event. In this context, vaccine-based control programs for specific human infections, particularly hookworm and the filariases, may have real advantages. Two examples of vaccine strategies that target key secretory proteins illustrate the potential. Hookworm infective larvae secrete two major antigens, termed ASP-1 and ASP-2 (Hotez *et al.*, 1996). Vaccination of mice with recombinant ASP-1 provided significant protection against subsequent challenge (reviewed in Hotez *et al.*, 1996); whether this antigen could be equally protective in humans is not yet known. Other secreted hookworm molecules, including the anticoagulant mentioned above (Cappello *et al.*, 1995), might also be useful targets for an immune response. Another example is found in *T. spiralis*. Larval stages of these parasites secrete a 43 kDa immunodominant glyco-protein. A 40-mer synthetic peptide derived from this protein induced high levels of protection against larval infection in mice (Robinson *et al.*, 1995). These examples suggest that larval secretory products might provide excellent vaccine targets for a variety of nematodes, including the filariae. Larval stages of filariae tend to develop slowly in humans, making them attractive targets for immune-based destruction before they cause pathology or reproduce. However, rational selection of vaccine candidates must await better understanding of the molecules that are secreted by these organisms and their roles in enhancing survival in the host. It must be understood that investment in such studies competes with the need for urgent attention to many other poverty-associated conditions. In addition, this research must be undertaken with the understanding that commercial success is almost certainly impossible.

Unanswered Questions

Twenty years have passed since many of the topics related to homeostatic mechanisms in nematodes were last reviewed. It is surprising, if not depressing, to realise that most of the relevant literature was written before then. This situation reflects the scant attention this topic has received since the early 1970s, when funding for helminth research became focused principally on the identification of targets for vaccines and, secondarily, targets in the neuromusculature identified by anthelmintics. By the 1980s, studies on the basic physiology of nematodes had been nearly abandoned. The recently completed *C. elegans* genome sequencing project should rekindle interest in these topics. Perhaps that interest will be inspired by the realisation that this organism is a useful model for studying some inheritable diseases in humans, including those that result from defects in genes that encode several types of ion transporters. Most new information on these transport processes in nematodes will come first from studies on *C. elegans*. Identification of orthologous genes of interest in cDNA libraries for parasitic species will be facilitated by genomics-based technologies, including use of expressed sequence tags (ESTs) to identify related genes in cDNA libraries, potentially across a broad range of organisms (Blaxter, 1998). These approaches are well suited for studying specific proteins for which a transport function can be inferred from gene structure, including ion channels and ATPases. Testing these genes or their orthologs in parasitic species can be accomplished by various genomics-based strategies (i.e., gene knockouts, dsRNA inhibition or over-expression) using *C. elegans* or by functional expression in a defined, heterologous system, such as *Xenopus* oocytes or mammalian or insect cell lines.

Other important questions related to ionic and osmotic regulation and E/S processes in nematodes may be impossible to answer using *C. elegans* genetics and functional genomics approaches. Questions that pertain to processes that occur in animal or plant-parasitic species, but are not relevant to *C. elegans* or other free-living species, provide numerous challenges. The role of the Ca^{2+}-activated Cl^- channel in *A. suum* muscle and epidermal membranes, which also conducts organic anions, is one example. Based on the biophysical data, this channel may provide an important pathway for excreting the organic acid end-products or carbohydrate metabolism in *A. suum*. It would be interesting to find out if this channel is, in fact, essential for maintaining the viability of worms, and *C. elegans* reverse genetics could provide a useful approach for this. However, *C. elegans* primarily excretes CO_2 and alcohols, and does not produce much (or any?) of the organic acids excreted in large quantities by *A. suum* and other animal-parasitic nematodes. This challenges the validity (and utility) of testing Cl^- channels in *C. elegans* for their ability to conduct organic anions. Indeed, one must question the likelihood that a true ortholog exists in that organism. Defining the precise roles of identified secretory products is perhaps the key step in understanding the biology and evolution of parasitism. Questions in this area will require renewed vigor in the use of parasitic species in technically difficult experiments.

The answers to these questions, and many others related to how nematodes maintain homeostasis, should stimulate the development of new ways to control infections of plants and animals by parasitic nematodes. The *C. elegans* genome sequencing project and related technologies in functional genomics have provided powerful tools for gaining new insights into homeostatic processes in nematodes. Realising the full potential of this opportunity will require additional basic research to reduce the sequence data to information on the functional properties of corresponding proteins, and to determine their roles and relevance *in situ* in important parasitic species. This provides the challenge and the goal for research in this area of parasitology for the coming millenium.

References

Agre, P., Bonhivers, M. and Borgnia, M.J. (1998) The aquaporins, blueprints for cellular plumbing systems. *Journal of Biological Chemistry*, **273**, 14659–14662.

Albertson, D.G., Fishpool, R.M. and Birchall, P.S. (1995) Fluorescence *in situ* hybridization for the detection of DNA and RNA. In *Caenorhabditis elegans*: modern biological analysis of an organism. *Methods in Cell Biology*, vol. 48, edited by H.F. Epstein and D.C. Shakes, pp. 340–364. San Diego: Academic Press.

Allen, J.E. and MacDonald, A.S. (1998) Profound suppression of lymphocyte proliferation mediated by the secretions of nematodes. *Parasite Immunology*, **20**, 241–247.

Alphey, T.J.W. (1971) Studies on the aggregation behaviour of *Nippostrongylus brasiliensis*. *Parasitology*, **63**, 109–117.

Anastasi, A., Erspamer, V. and Bucci, M. (1971) Isolation and structure of bombesin and alutensin, two analogous active peptides from the skin of European amphibians *Bombina* and *Alytes*. *Experientia*, **27**, 166–167.

Anya, A.O. (1966) Investigations on osmotic regulation in the parasitic nematode *Aspiculuris tetraptera* Shulz. *Parasitology*, **56**, 583–588.

Arthur, E.J. and Sanborn, R.C. (1969) Osmotic and ionic regulation in nematodes. In *Chemical Zoology*, edited by M. Florkin and B.T. Scheer, pp. 429–464. New York: Academic Press.

Ash, C.P.J. and Atkinson, H.J. (1982) The possible role of trehalose in the survival of eggs of *Nematodirus battus* during dormancy. Parasitology, **85**, liv.

Ash, C.P.J. and Atkinson, H.J. (1983) Evidence for a temperature-dependent conversion of lipid reserves to carbohydrate in quiescent eggs of the nematode, *Nematodirus battus*. *Comparative Biochemistry and Physiology*, **76B**, 603–610.

Aumann, J., Dietsche, E., Rutencrantz, S. and Ladehoff, H. (1998) Physico-chemical properties of the female sex pheromone of *Heterodera schachtii* (Nematoda: Heteroderidae). *International Journal for Parasitology*, **28**, 1691–1694.

Avery, L. and Horvitz, H.R. (1990) Effects of starvation and neuroactive drugs on feeding in *Caenorhabditis elegans*. *Journal of Experimental Zoology*, **253**, 263–270.

Avery, L. and Thomas, J.H. (1997) Feeding and defecation. In *C. elegans II*, edited by D.L. Riddle, T. Blumenthal, B.J. Meyer and J.R. Priess, pp. 679–716. Plainview, NY: Cold Spring Harbor Laboratory Press.

Bacon, J.A., Ulrich, R.G., Davis, J.P., Thomas, E.M., Johnson, S.J., Conder, G.A., Sangster, N.C., Rothwell, J.T., McCracken, R.O., Lee, B.H., Clothier, M.J., Geary, T.G. and Thompson, D.P. (1998) Comparative *in vitro* effects of closantel and selected β-ketoamide anthelmintics on a gastrointestinal nematode and vertebrate liver cells. *Journal of Veterinary Pharmacology and Therapeutics*, **21**, 190–198.

Baldwin, E. and Moyle, V. (1947) An isolated nerve-muscle preparation from *Ascaris lumbricoides*. *Journal of Experimental Biology*, **23**, 277–286.

Bargmann, C.J. (1998) Neurobiology of the *Caenorhabditis elegans* genome. *Science*, **282**, 2028–2033.

Barrett, J. (1981) Amino acid metabolism in helminths. *Advances in Parasitology*, **30**, 39–105.

Barrett, J. (1982) Metabolic responses to anabiosis in the fourth stage juveniles of *Ditylenchus dipsaci* (Nematoda). *Proceedings of the Royal Society of London*, **B216**, 159–177.

Beames, C.G. and King, G.A. (1972) Factors influencing the movement of materials across the intestine of *Ascaris*. In *Comparative Biochemistry of Parasites*, edited by H. Van den Bossche, pp. 275–282. New York: Academic Press.

Behm, C.A., Bryant, C. and Jones, A.J. (1987) Studies of glucose metabolism in *Hymenolepis diminuta* using ^{13}C nuclear magnetic resonance. *International Journal for Parasitology*, **17**, 1333–1341.

Bennett, J.L. and Thompson, D.P. (1986) Mode of action of antitrematodal agents. In *Chemotherapy of Parasitic Diseases*, edited by W.C. Campbell and R.S. Rew, pp. 427–443. New York: Plenum Press.

Beuron, F., Le-Caherec, F. and Guillam, M.T (1995) Structural analysis of a MIP family protein from the digestive tract of *Cicadella viridis*. *Journal of Biological Chemistry*, **270**, 17414–17422.

Bird, A.F. (1975) Symbiotic relationships between nematodes and plants. In *Symbiosis*, Symposia of the Society for Experimental Biology, Vol. **29**, edited by D.H. Jennings and D.L. Lee, pp. 351–371. Cambridge: Cambridge University Press.

Bird, A.F. (1976) The development and organisation of skeletal structures in nematodes. In *The Organisation of Nematodes*, edited by N.A. Croll. New York: Academic Press.

Bird, A.F. and Bird, J. (1991) *The Structure of Nematodes*, 2nd edn. San Diego: Academic Press.

Bird, D.McK. (1996) Manipulation of host gene expression by root-knot nematodes. *Journal of Parasitology*, **82**, 881–888.

Bird, D.McK. and Opperman, C.H. (1998) *Caenorhabditis elegans*: a genetic guide to parasitic nematode biology. *Journal of Nematology*, **30**, 299–308.

Blair, K.L., Barsuhn, C.L., Day, J.S., Ho, N.F.H., Geary, T.G. and Thompson, D.P. (1998) Biophysical model for organic acid excretion in *Ascaris suum. Molecular and Biochemical Parasitology*, **93**, 179–190.

Blair, K.L., Geary, T.G., Mensch, S.K., Ho, N.F.H. and Thompson, D.P. (1999) Biophysical characterization of a large conductance anion channel in hypodermal membranes of the gastrointestinal nematode, *Ascaris suum. Journal of Membrane Biology* (in press).

Blackhall, W., Liu, H.Y., Xu, M., Prichard, R.K. and Beech, R.N. (1998) Selection at a P-glycoprotein gene in ivermectin- and moxidectin-selected strains of *Haemonchus contortus. Molecular and Biochemical Parasitology*, **95**, 193–201.

Blaxter, M.L. (1998) *Caenorhabditis elegans* is a nematode. *Science*, **283**, 2041–46.

Blaxter, M.L., De Ley, P., Garey, J.R., Liu, L.X, Scheldeman, P., Vierstraete, A., Vanfleteren, J.R., Mackey, L.Y., Dorris, M., Frisse, L.M., Vida. J.T. and Thomas, W.K. (1998) A molecular evolutionary framework for the phylum Nematoda. *Nature*, **392**, 71–75.

Bowman, J.W., Winterrowd, C.A., Friedman, A.R., Thompson, D.P., Klein, R.D., Davis, J.P. Maule, A.R., Blair, K.L. and Geary, T.G. (1995) Nitric oxide mediates the inhibitory effects of SDPNFLRFamide, a nematode FMRFamide-related neuropeptide in *Ascaris suum. Journal of Neurophysiology*, **74**, 1880–1888.

Brading, A.F. and Caldwell, P.C. (1971) The resting membrane potential of the somatic muscle cells of *Ascaris lumbricoides. Journal of Physiology, London*, **217**, 605–624.

Brindley, P.J., Gam, A.A., McKerrow, J.H. and Neva, F.H. (1995) Ss40: the zinc endopeptidase secreted by infective larvae of *Strongyloides stercoralis. Experimental Parasitology*, **80**, 1–7.

Broeks, A. (1997) P-glycoproteins and multidrug resistance-associated protein function in the nematode *Caenorhabditis elegans. Ph.D. Thesis*, The University of Amsterdam, The Netherlands.

Broeks, A., Janssen, H.W.R.M., Calafat, J. and Plasterk, R.H.A. (1995) A P-glycoprotein protects *Caenorhabditis elegans* against natural toxins. *EMBO Journal*, **14**, 1858–1866.

Broeks, A., Gerrard, B., Allikmets, R., Dean, M. and Plasterk, R.H.A. (1996) Homologues of the human multidrug resistance genes *MPR* and *MDR* contribute to heavy metal resistance in the soil nematode *Caenorhabditis elegans. EMBO Journal*, **15**, 6132–6143.

Brownlee, D.J.A., Holden-Dye, L., Fairweather, I. and Walker, R.J. (1995) The action of serotonin and the nematode neuropeptide KSAYMRFamide on the pharyngeal muscle of the parasitic nematode, *Ascaris suum. Parasitology*, **111**, 379–384.

Bürglin, T.R., Lobos, E. and Blaxter, M.L. 1998. *Caenorhabditis elegans* as a model for parasitic nematodes. *International Journal for Parasitology*, **28**, 395–411.

Calamita, G., Kempf, B., Bonhivers, M., Bishai, W.R., Bremer, E. and Agre, P. (1998) Regulation of the *Escherichia coli* water channel gene *aqpZ. Proceedings of the National Academy of Science, U.S.A.*, **95**, 3627–3631.

Caldwell, P.C. (1973) Possible mechanisms for the linkage of membrane potentials to metabolism by electrogenic transport processes with special reference to *Ascaris* muscle. *Bioenergetics*, **4**, 201–208.

Cappello, M., Vlasuk, G.P., Bergum, P.W., Huang, S. and Hotez, P.J. (1995) *Ancylostoma caninum* anticoagulant peptide: a hookworm-derived inhibitor of human coagulation factor Xa. *Proceedings of the National Academy of Sciences USA*, **92**, 6152–6156.

Castro, G.A. and Fairbairn, D. (1969) Comparison of the cuticular and intestinal absorption of glucose by adult *Ascaris lumbricoides. Journal of Parasitology*, **55**, 13–16.

Castro, G.A., Ferguson, J.D. and Gorden, C.W. (1973) Amine excretion in encysted larvae and adults of *Trichinella spiralis. Comparative Biochemistry and Physiology*, **45**, 819–828.

C. elegans Genome Consortium (1998) Genome sequence of the nematode *C. elegans*. A platform for investigating biology. *Science*, **282**, 2012–2018.

Chalfie, M. and Jorgensen, E.M. (1998) *C. elegans* neuroscience: genetics to genome. *Trends in Genetics*, **14**, 506–512.

Chalfie, M., Tu, Y., Euskirchen, G., Ward, W. and Prasher, D. (1994) Green fluorescent protein as a marker for gene expression. *Science*, **263**, 802–805.

Chanteau, S., Moulia-Pelat, J.P., Glaziou, P., Nguyen, N.L., Luquiaud, P., Plichart, C., Martin, P.M.V. and Cartel, J.L. (1994) Og4C3 circulating antigen: a marker of infection and adult worm burden in *Wuchereria bancrofti* filariasis. *Journal of Infectious Diseases*, **170**, 247–250.

Chrispeels, M.J. and Maurel, C. (1994) Aquaporins: the molecular basis of facilitated water movement through living plant cells? *Plant Physiology*, **105**, 9–15.

Coast, G.M. 1996. Neuropeptides implicated in the control of diuresis in insects. *Peptides*, **17**, 327–336.

Collins, J.M., Williams, J.F. and Kaiser, L. (1994) *Dirofilaria immitis*: heartworm products contract rat trachea *in vitro. Experimental Parasitology*, **78**, 76–84.

Cooper, A.F and van Gundy, S.D. (1970) Metabolism of glycogen and neutral lipids by *Aphelenchus avenue* and *Caenorhabditis spp.* in aerobic, microaerobic and anaerobic environments. *Journal of Nematology*, **2**, 305–315.

Cooper, A.F. and van Gundy, S.D. (1971a) Ethanol production and utilization by *Aphelenchus avenue* and *Caenorhabditis spp. Journal of Nematology*, **3**, 205–214.

Cooper, A.F. and van Gundy, S.D. (1971b) Senescence, quiescence and cryptobiosis. In *Plant Parasitic Nematodes*, Vol. 2, edited by B.M. Zuckerman, W.F. Mai and R.A. Rhode. New York: Academic Press.

Court, J.P., Murgatroyd, R.C., Livingstone, D. and Rahr, E. (1988) Physicochemical characteristics of non-electrolytes and their uptake by *Brugia pahangi* and *Dipetalonema viteae. Molecular and Biochemical Parasitology*, **27**, 101–108.

Crowe, J.H. and Madin, K.A.C. (1975) Anhydrobiosis in nematodes: evaporative water loss and survival. *Journal of Experimental Zoology*, **193**, 232–234.

Crowe, J.H., Crowe, L.M. and Chapman, D. (1984) Preservation of membranes in anhydrobiotic organisms: the role of trehalose. *Science*, **223**, 701–703.

Cully, D.F., Wilkinson, H., Vassitadic, D.K., Etter, A. and Arena, J.P. (1996) Molecular biology and electrophysiology of glutamate-gated chloride channels in invertebrates. *Parasitology*, **113**, S191–S200.

Curtis, R.H.C. (1996) Identification and *in situ* and *in vitro* characterization of secreted proteins produced by plant-parasitic nematodes. *Parasitology*, **113**, 589–597.

Cuttitia, F., Carney, D.N., Mulshine, J., Moody, T.W., Fedorko, J., Fischler, A. and Minna, J.D. (1985) Bombesin-like peptides can function as autocrine growth factors in human small cell lung cancer. *Nature*, **316**, 823–826.

Das, P.D. and Donahue, M.J. (1986) γ-Glutamyl transpeptidase activity in *Ascaris suum. Molecular and Biochemical Parasitology*, **20**, 233–236.

Daugschies, A. (1995) *Oesophagostomum dentatum*: population dynamics and synthesis of prostanoids by histotropic stages cultured *in vitro. Experimental Parasitology*, **81**, 574–583.

Davis, M.W., Somerville, D., Lee, R.Y.N., Lockery, S., Avery, L. and Fambrough, D.M. (1995) Mutations in the *Caenorhabditis elegans* Na,K-ATPase I-subunit gene, *eat-6*, disrupt excitable cell function. *Journal of Neuroscience*, **15**, 8408–8418.

de Bono, M. and Bargmann, C.I. (1998) Natural variation in a neuropeptide Y receptor homolog modifies social behavior and food response in *C. elegans. Cell*, **94**, 679–689.

Del Castillo, J., De Mello, W.C. and Morales, T. (1964) Influence of some ions on the membrane potential of *Ascaris* muscle. *Journal of General Physiology*, **48**, 129–140.

Del Castillo, J. and Morales, T. (1967) The electrical and mechanical activity of the esophageal cell of *Ascaris lumbricoides. Journal of General Physiology*, **50**, 603–629.

Del Castillo, J. Rivera, A. Solarzano, S. and Sorrato, J. (1989) Some aspects of the neuromuscular system of *Ascaris. Quarterly Review of Experimental Physiology*, **74**, 1071–1087.

De Mello, W.C. and Tercafs, R.R. (1966) Ionic permeability of the 'Ascaris' body wall. *Acta Physiology of Latin America*, **16**, 121–127.

Despommier, D.D. (1993) *Trichinella spiralis* and the concept of niche. *Journal of Parasitology*, **79**, 472–482.

Dubremetz, J.F. and McKerrow, J.H. (1995) Invasion mechanisms. In *Biochemistry and Molecular Biology of Parasites*, edited by J.J. Marr and M. Müller, pp. 307–322. London: Academic Press.

Elkes, D.A., Cardozo, D.L., Madison, J. and Kaplan, J.M. (1997) EGL-36 Shaw channels regulate *C. elegans* egg-laying muscle activity. *Neuron*, **19**, 165–174.

Epstein, H.F. and Shakes, D.C. (1995) *Caenorhabditis elegans*: modern biological analysis of an organism. *Methods in Cell Biology*, vol. 48. San Diego, CA: Academic Press.

Fairbairn, D. (1957) The biochemistry of *Ascaris. Experimental Parasitology*, **6**, 491–554.

Fellowes, R.A., Maule, A.G., Marks, N., Geary, T.G., Thompson, D.P., Shaw, C. and Halton, D.W. (1998) Modulation of the motility of the vagina vera of *Ascaris suum in vitro* by FMRFamide-related peptides. *Parasitology*, **116**, 277–287.

Fetterer, R.H. (1986) Transcuticular solute movement in parasitic nematodes: relationship between non-polar solute transport and partition coefficient. *Comparative Biochemistry and Physiology*, **84A**, 461–466.

Fire, A., Xu, S., Montgomery, M.K., Kostas, S.A., Driver, S.E. and Mello, C.C. (1998) Potent and specific genetic interference by double-stranded RNA in *Caenorhabditis elegans*. *Nature*, **391**, 806–811.

Fitzgerald, L.A. and Foor, W.E. (1979) *Ascaris suum*: electrophoretic characterization of reproductive tract and perienteric fluid polypeptides, and effects of seminal and uterine fluids on spermiogenesis. *Experimental Parasitology*, **47**, 313–326.

Fleming, J.T., Squire, M.D., Barnes, T.M., Tornoe, C., Matsuda, K., Ahnn, J., Fire, A., Sulston, J.E., Barnard, E.A., Satelle, D.B. and Lewis, J.A. (1997) *Caenorhabditis elegans* levamisole resistance genes *lev-1*, *unc-29* and *unc-38* encode functional nicotinic acetylcholine receptor subunits. *Journal of Neuroscience*, **17**, 5843–5857.

Forrester, W.C. and Garriga, G. (1997) Genes necessary to *C. elegans* cell and growth migrations. *Development*, **124**, 1831–1843.

Foster, N. (1998) Immunocytochemical detection of vasoactive intestinal peptide-like and peptide histidine isoleucine-like peptides in the nervous system and excretory system of adult *Nippostrongylus brasiliensis*. *International Journal for Parasitology*, **28**, 825–829.

Foster, N., Dean, E.J. and Lee, D.L. (1994) The effect of homogenates and excretory/secretory products of *Nippostrongylus brasiliensis* and of acetylcholinesterase on the amplitude and frequency of contraction of uninfected rat intestine *in vitro*. *Parasitology*, **108**, 453–459.

Foster, N. and Lee, D.L. (1996a) A vasoactive intestinal peptide-like protein excreted/secreted by *Nippostrongylus brasiliensis* and its effect on contraction of uninfected rat intestine. *Parasitology*, **112**, 97–104.

Foster, N. and Lee, D.L. (1996b) Vasoactive intestinal polypeptide-like and peptide histidine isoleucine-like proteins excreted/secreted by *Nippostrongylus brasiliensis*, *Nematodirus battus* and *Ascaridia galli*. *Parasitology*, **113**, 287–292.

Frank, G.R., Mondesire, R.R., Brandt, K.S. and Wisnewski, N. (1998) Antibody to the *Dirofilaria immitis* aspartyl protease inhibitor homologue is a diagnostic marker for feline heartworm infections. *Journal of Parasitology*, **84**, 1231–1236.

Franz, M. (1988) The morphology of adult *Onchocerca volvulus* based on electron microscopy. *Tropical Medical Parasitology*, **39**, 359–366.

Frelin, C., Vigne, P., LaDoux, A. and Lazdunski, M. (1988) The regulation of the intracellular pH in cells from vertebrates. *European Journal of Pharmacology*, **174**, 3–14.

Fuse, M., Davey, K.G. and Sommerville, R.I. (1993a) Osmoregulation in the parasitic nematode *Pseudoterranova decipiens*. *Journal of Experimental Biology*, **175**, 127–142.

Fuse, M., Davey, K.G. and Sommerville, R.I. (1993b) Water compartments and osmoregulation in the parasitic nematode *Pseudoterranova decipiens*. *Journal of Experimental Biology*, **175**, 143–152.

Gade, G., Hoffmann, K.H. and Spring J.H. (1997) Hormonal regulation in insects: facts, gaps and future directions. *Physiological Reviews*, **77**, 963–1032.

Gamble, H.R. and Mansfield, L.S. (1996) Characterization of excretory-secretory products from larval stages of *Haemonchus contortus* cultured *in vitro*. *Veterinary Parasitology*, **62**, 291–305.

Geary, T.G., Sims, S.M., Thomas, E.M., Davis, J.P., Ho, N.F.H. and Thompson, D.P. (1993) *Haemonchus contortus*: ivermectin-induced paralysis of the pharynx. *Experimental Parasitology*, **77**, 88–96.

Geary, T.G., Blair, K.L., Ho, N.F.H., Sims, S.M. and Thompson, D.P. (1995) Biological functions of nematode surfaces. In *Molecular Approaches to Parasitology*, edited by J.C. Boothroyd and R. Komuniecki, pp. 57–76. New York: Wiley-Liss.

Geary, T.G., Sangster, N.C. and Thompson, D.P. (1999) Frontiers in anthelmintic pharmacology. *Veterinary Parasitology*, **1587**, 1–21.

Geary, T.G., Thompson, D.P. and Klein, R.D. (1999) Mechanism-based screening: discovery of the next generation of anthelmintics depends upon more basic research. *International Journal of Parasitology*, **29**, 105–112.

Green, C.D. and Greet, D.N. (1972) The location of the secretions that attract male *Heterodera schachtii* and *H. rostochiensis* to their females. *Nematologica*, **18**, 347–352.

Grigg, M.E., Gounaris, K. and Selkirk, M.E. (1996) Characterization of a platelet-activating factor acetylhydrolase secreted by the nematode parasite *Nippostrongylus brasiliensis*. *Biochemical Journal*, **317**, 541–547.

Gourlet, P., De Neef, P., Cnudde, J., Waelbroeck, M. and Robberecht, P. (1997) *In vitro* properties of a high affinity selective antagonist of the VIP1 receptor. *Peptides*, **18**, 1555–1560.

Harpur, R.P. (1969) The nematode intestine and organic acid excretion: volatile acids in *Ascaris lumbricoides* faeces. *Comparative Biochemistry and Physiology*, **28**, 865–875.

Harpur, R.P. and Popkin, J.S. (1965) Osmolarity of blood and intestinal contents in the pig, guinea pig and *Ascaris lumbricoides*. *Canadian Journal of Biochemistry*, **43**, 1157–1169.

Harpur, R.P. and Popkin, J.S. (1973) Intestinal fluid transport: studies with the gut of *Ascaris lumbricoides*. *Canadian Journal of Physiology and Pharmacology*, **51**, 79–90.

Harris, J.E. and Crofton, H.D. (1957) Structure and function of nematodes: internal pressure and the cuticular structure in *Ascaris*. *Journal of Experimental Biology*, **34**, 116–130.

Haseeb, M.A. and Fried, B. (1988) Chemical communication in helminths. *Advances in Parasitology*, **27**, 169–207.

Haskins, W.T. and Weinstein, P.P. (1957) Nitrogenous excretory products of *Trichinella spiralis* larvae. *Journal of Parasitology*, **43**, 19–24.

Hechler, H.C. (1963) Description, developmental biology and feeding habits of *Seinura tunuicaudata* (De Man), J.B. Goodey, 1960 (Nematoda: Aphelenchoididae), a nematode predator. *Proceedings of the Helminthological Society of Washington*, **30**, 182–195.

Huettel, R.N. (1986) Chemical communication in nematodes. *Journal of Nematology*, **18**, 3–8.

Ho, N.F.H., Geary, T.G., Barsuhn, C.L., Sims, S.M. and Thompson, D.P. (1992) Mechanistic studies in the transcuticular delivery of antiparasitic drugs II: ex vivo/in vitro correlation of solute transport by *Ascaris suum*. *Molecular and Biochemical Parasitology*, **52**, 1–14.

Ho, N.F.H., Geary, T.G., Raub, T.J., Barsuhn, C.L. and Thompson, D.P. (1990) Biophysical transport properties of the cuticle of *Ascaris suum*. *Molecular and Biochemical Parasitology*, **41**, 153–166.

Ho, N.F.H., Sims, S.M., Vidmar, T.J., Day, J.S., Barsuhn, C.L., Thomas, E.M., Geary, T.G. and Thompson, D.P. (1994) Theoretical perspectives on anthelmintic drug discovery: interplay of transport kinetics, physicochemical properties and in vitro activity of anthelmintic drugs. *Journal of Pharmaceutical Sciences*, **83**, 1052–1059.

Hobson, A.D. (1948) The physiology and cultivation in artificial media of nematodes parasitic in the alimentary tract of animals. *Parasitology*, **38**, 183–227.

Hobson, A.D., Stephenson, W. and Beadle, L.C. (1952) Studies on the physiology of *Ascaris lumbricoides*. 1. The relation of the total osmotic pressure, conductivity and chloride content of the body fluid to that of the external environment. *Journal of Experimental Biology*, **29**, 1–21.

Holtzman, E.J., Kumar, S., Faaland, C.A., Warner, F., Logue, P.J., Erickson, S.J., Ricken, G., Waldman, J., Kumar, S. and Dunham, P.B. (1998) Cloning, characterization, and gene organization of K-Cl cotransporter from pig and human kidney and *C. elegans*. *American Journal of Physiology*, **275**, F550–F564.

Hotez, P.J., Narasimhan, S., Haggerty, J., Milstone, L., Bhopale, V., Schad, G.A. and Richards, F.F. (1992) Hyaluronidase from infective *Ancylostoma* hookworm larvae and its possible function as a virulence factor in tissue invasion and in cutaneous larva migrans. *Infection and Immunity*, **60**, 1018–1023.

Hotez, P.J., Hawdon, J.M., Cappello, M., Jones, B.F., Ghosh, K., Volvovitz, F. and Xiao, S.H. (1996) Molecular approaches to vaccinating against hookworm disease. *Pediatric Research*, **40**, 515–521.

Howells, R.E. (1980) Filariae: dynamics of the surface. In *The Host Invader Interplay*, edited by H. Van den Bossche, pp. 69–84, Amsterdam: Elsevier/North Holland.

Huang, M. and Chalfie, M. (1994) Gene interactions affecting mechanosensory transduction in *Caenorhabditis elegans*. *Nature*, **367**, 467–470.

Huettel, R.N. (1986) Chemical communication in nematodes. *Journal of Nematology*, **18**, 3–8.

Huntington, M.K., Geary, T.G., Mackenzie, C.D. and Williams, J.F. (1999) Bombesin-like neuropeptides in nematodes. *Journal of Parasitology* (in press).

Huntington, M.K., Leykum, J., Geary, T.G., Mackenzie, C.D. and Williams, J.F. (1993) Demonstration of bombesin-like immunoreactivity and bombesin/gastrin-releasing peptide binding sites in nematodes. *Journal of Cellular Biochemistry*, **S17C**, 115.

Hussein, A.S., Chacón, M.R., Smith, A.M., Tosado-Acevedo, R. and Selkirk, M.E. (1999) Cloning, expression, and properties of a nonneuronal secreted acetylcholinesterase from the parasitic nematode *Nippostrongylus brasiliensis*. *Journal of Biological Chemistry*, **274**, 9312–9319.

Ishibashi, K., Sasaki, S., Fushimi, K., Uchida, S., Kuwahara, M., Saito, H., Furukawa, T., Nakajima, K., Yamaguchi, Y., Gojobori, T. and Marumo, F.

(1994) Molecular cloning and expression of a member of the aquaporin family with permeability to glycerol and urea in addition to water expressed at the basolateral membrane of kidney collecting ducts. *Proceedings of the National Academy of Sciences USA*, **91**, 6269–6273.

Jaffe, H., Huettel, R.N., Demilo, A.B., Hayes, D.K. and Rebois, R.V. (1989) Isolation and identification of a compound from soybean cyst nematode, *Heterodera glycines*, with sex pheromone activity. *Journal of Chemical Ecology*, **15**, 2031–2043.

Jansen, G., Hazendonk, E., Thijssen, K.L. and Plasterk, R.H. (1997) Reverse genetics by chemical mutagenesis in *Caenorhabditis elegans*. *Nature Genetics*, **17**, 119–121.

Jasmer, D.P. (1993) *Trichinella spiralis* infected skeletal muscle cells arrest in G_2/M and cease muscle gene expression. *Journal of Cell Biology*, **121**, 785–793.

Jentsch, T.J. (1997) Trinity of cation channels. *Science*, **367**, 412–413.

Johnstone, D.B., Wei, A., Butler, A., Salkoff, L. and Thomas, J.H. (1997) Behavioral defects in *C. elegans egl-36* mutants result from potassium channels shifted in voltage-dependence of activation. *Neuron*, **19**, 151–164.

Jones, F.G.W. and Parrott, D.M. (1969) The influence of soil structure and moisture on nematodes, especially *Xiphinema*, *Longidorus*, *Trichodorus* and *Heterodera* spp. *Soil Biochemistry and Soil Biology*, **1**, 153–165.

Jones, S.J.M. and Baillie, D.L. (1995) Characterization of the *let-653* gene in *Caenorhabditis elegans*. *Molecular and General Genetics*, **248**, 719–726.

Kaiser, L., Lamb, V.L., Tithof, P.K., Gage, D.A., Chamberlain, B.A., Watson, J.T. and Williams, J.F. (1992) *Dirofilaria immitis*: do filarial cyclooxygenase products depress endothelial dependent relaxation in the *in vitro* rat aorta? *Experimental Parasitology*, **75**, 153–165.

Kaiser, L., Mupanomunda, M. and Williams, J.F. (1996) *Brugia pahangi*-induced contractility of bovine mesenteric lymphatics studied in vitro: a role for filarial factors in the development of lymphedema? *American Journal of Tropical Medicine and Hygiene*, **54**, 386–390.

Kaiser, L., Geary, T.G. and Williams, J.F. (1998) *Dirofilaria immitis* and *Brugia pahangi*: filarial parasites make nitric oxide. *Experimental Parasitology*, **90**, 131–134.

Kaiser, L. and Williams, J.F. (1998) *Dirofilaria immitis*: heartworm infection converts histamine-induced constriction to endothelium-dependent relaxation in canine pulmonary artery. *Experimental Parasitology*, **88**, 146–153.

Kennedy, M.W., Foley, M., Kuo, Y.M., Kusel, J.R. and Garland, P.B. (1987) Biophysical properties of the surface lipid of parasitic nematodes. *Molecular and Biochemical Parasitology*, **22**, 233–240.

Kinne, R.K.H. (1993) The role of organic osmolytes in osmoregulation: from bacteria to mammals. *Journal of Experimental Zoology*, **265**, 346–355.

Ko, R.C., Fan, L., Lee, D.L. and Compton, H. (1994) Changes in host muscles induced by excretory/secretory products of larval *Trichinella spiralis* and *Trichinella pseudospiralis*. *Parasitology*, **108**, 195–205.

Komuniecki, R. and Harris, B.G. (1995) Carbohydrate and energy metabolism in helminths. In *Biochemistry and Molecular Biology of Parasites*, edited by J.J. Marr and M. Muller, pp. 49–66. London: Academic Press.

Komuniecki, R. and Komuniecki, P.R. (1993) *Ascaris suum*: a useful model for anaerobic mitochondrial metabolism and the aerobic-anaerobic transition in developing parasitic helminths. In *Comparative Biochemistry of Parasitic Helminths*, edited by E.M. Bennett, C. Behm and C. Bryant, pp. 1–12. London: Chapman and Hall.

Kraev, A., Kraev, N. and Carafoli, E. (1999) Identification and functional expression of the plasma membrane calcium ATPase gene family from *Caenorhabditis elegans*. *Journal of Biological Chemistry*, **274**, 4254–4258.

Kuwahara, M., Ishibashi, K., Gu, Y., Tereda, Y., Kohara, Y., Marumo, F. and Sasaki, S. (1998) A water channel of the nematode *C. elegans* and its implications for channel selectivity of MIP proteins. *American Journal of Physiology*, **275**, C1459–C1464.

Kwa, M.S.G., Okoli, M.N., Schulz-Key, H., Okongwo, P.O. and Roos, M.H. (1998) Use of P-glycoprotein gene probes to investigate anthelmintic resistance in *Haemonchus contortus* and comparison with *Onchocerca volvulus*. *International Journal of Parasitology*, **28**, 1235–1240.

Lee, D.L. (1966) The structure and composition of the helminth cuticle. *Advances in Parasitology*, **4**, 187–254.

Lee, D.L. (1970) The fine structure of the excretory system in adult *Nippostrongylus brasiliensis* (Nematoda) and a suggested function for the excretory glands. *Tissue and Cell*, **2**, 225–231.

Lee, D.L. (1972) The structure of the helminth cuticle. *Advances in Parasitology*, **10**, 347–379.

Lee, D.L. (1996) Why do some nematode parasites of the alimentary tract secrete acetylcholinesterase? *International Journal for Parasitology*, **26**, 499–508.

Lee, D.L. and Atkinson, H.J. (1977) *Physiology of Nematodes*. New York: Columbia University Press.

Lee, D.L. and Smith, M.H. (1965) Haemoglobins of parasitic animals. *Experimental Parasitology*, **16**, 392–424.

Lees, E. (1953) An investigation into method of dispersal and desiccation resistance of *Panagrellus silusae*. *Journal of Helminthology*, **27**, 95–103.

Lewis, J.A., Wu, C.H., Levine, J.H. and Berg, H. (1989) Levamisole resistant mutants of the nematode *Caenorhabditis elegans* appear to lack functional acetylcholine receptors. *Neuroscience*, **5**, 967–989.

Lincke, C.R., Broeks, A., The, I., Plasterk, R.H.A. and Borst, P. (1993) The expression of two P-glycoprotein (pgp) genes in transgenic *Caenorhabditis elegans* is confined to intestinal cells. *EMBO Journal*, **12**, 1615–1620.

Liu, L.X., Serhan, C.M. and Weller, P.F. (1990) Intravascular filarial parasites elaborate cyclooxygenase-derived eicosanoids. *Journal of Experimental Medicine*, **172**, 993–996.

Loukas, A., Hunt, P. and Maizels, R.M. (1999) Cloning and expression of an aquaporin-like gene from a parasitic nematode. *Molecular and Biochemical Parasitology*, **99**, 287–293.

Mackenzie, N.E., van de Waa, E.A., Gooley, P.R., Williams, J.F.W. and Geary, T.G. (1989) Comparison of glycolysis and glutaminolysis in *Onchocerca volvulus* and *Brugia pahangi* by ^{13}C nuclear magnetic resonance spectroscopy. *Parasitology*, **99**, 427–435.

Madin, K.A.C. and Crowe, J.H. (1975) Anhydrobiosis in nematodes: carbohydrate and lipid metabolism during dehydration. *Journal of Experimental Zoology*, **193**, 335–342.

Madin, K.A.C., Crowe, J.H. and Loomis, S.H. (1978) Metabolic transition in a nematode during induction and recovery from anhydrobiosis. In *Dry Biological Systems*, edited by J.H. Crowe and J.S. Clegg. New York: Academic Press.

Madshus, I.H. (1988) Regulation of intracellular pH in eukaryotic cells. *Biochemical Journal*, **250**, 1–8.

Maizels, R.M., Blaxter, M.L. and Selkirk, M.E. (1993) Forms and functions of nematode surfaces. *Experimental Parasitology*, **77**, 380–384.

Maizels, R.M., Bundy, D.A., Selkirk, M.E., Smith, D.M. and Anderson, R.M. (1993) Immunological modulation and evasion by helminth parasites in human populations. *Nature*, **365**, 797–805.

Mak, C-h. and Ko, R.C. (1999) Characterization of an endonuclease activity from excretory/secretory products of a parasitic nematode, *Trichinella spiralis*. *European Journal of Biochemistry*, **260**, 477–481.

Marini, A.M., Vissers, S., Urrestarazu, A. and Andre, B. (1994) Cloning and expression of the MEP1 gene encoding an ammonium transporter in *Saccharomyces cerevisiae*. *EMBO Journal*, **13**, 3456–3463.

Marks, C.F., Thomason, I.J. and Castro, C.E. (1968) Dynamics of the penetration of nematodes by water, nematocides and other substances. *Experimental Parasitology*, **22**, 321–337.

Martin, R.J. (1997) Modes of action of anthelmintic drugs. *Veterinary Journal*, **154**, 11–34.

Martin, R.J., Pennington, A.J., Duittoz, A.H., Robertson, S. and Kusel, J.R. (1991) The physiology and pharmacology of neuromuscular transmission in the nematode parasite, *Ascaris suum*. *Parasitology*, **102**, S41–S58.

McKerrow, J.H., Brindley, P., Brown, M., Gam, A.A., Staunton, C. and Neva, F.A. (1990) *Strongyloides stercoralis*: identification of a protease that facilitates penetration of skin by the infective larvae. *Experimental Parasitology*, **70**, 134–143.

McKerrow, J.H., Huima, T. and Lustigman, S. (1999) Do filarid nematodes have a vascular system? *Parasitology Today*, **15**, 123.

Meagher, J.W. (1974) Cryptobiosis of the cereal cyst nematode (*Heterodera avenae*) and effects of temperature and relative humidity on survival of eggs in storage. *Nematologica*, **20**, 323–335.

Mello, C. and Fire, A. (1995) DNA transformation. In *Caenorhabditis elegans*: modern biological analysis of an organism. *Methods in Cell Biology*, vol. 48, edited by H.F. Epstein and D.C. Shakes, pp. 452–482. San Diego: Academic Press.

Mendis, A.H.W. and Townson, S. (1985) Evidence for the occurrence of respiratory electron-transport in adult *Brugia pahangi* and *Dipetalonema viteae*. *Molecular and Biochemical Parasitology*, **14**, 337–354.

Merz, J.M. (1977) Short-circuit current and solute fluxes across the gut epithelium of *Ascaris suum*. Ph.D. Dissertation, Oklahoma State University, Stillwater, OK.

Michaelson, L.V., Napier, J.A., Lewis, M., Griffiths, G., Lazarus, C.M. and Stobart, A.K. (1998) Functional identification of a fatty acid Δ5 desaturase gene from *Caenorhabditis elegans*. *FEBS Letters*, **439**, 215–218.

Miller, D.M. and Shakes, D.C. (1995) Immunofluorescence microscopy. In *Caenorhabditis elegans*: modern biological analysis of an organism. *Methods in Cell Biology*, vol. 48, edited by H.F. Epstein and D.C. Shakes, pp. 365–394. San Diego: Academic Press.

Moyle, M., Foster, D.L., McGrath, D.E., Brown, S.M., Laroche, Y., De Meutter, J., Stanssens, P., Bogowitz, C.A., Fried, V.A., Ely, J.A., Soule, H.R. and Vlasuk, G.P. (1994) A hookworm glycoprotein that inhibits neutrophil function is a ligand of the integrin CD11b/CD18. *Journal of Biological Chemistry*, **269**, 10008–10015.

Mupanomunda, M., Williams, J.F., Mackenzie, C.D. and Kaiser, L. (1997) *Dirofilaria immitis*: heartworm infection alters pulmonary artery endothelial cell behavior. *Journal of Applied Physiology*, **82**, 389–398.

Murrell, K.D., Graham, C.E. and McGreevy, M. (1983) *Strongyloides ratti* and *Trichinella spiralis*: net charge of the epicuticle. *Experimental Parasitology*, **55**, 331–339.

Myers, R.F. (1965) Amylase, cellulase, invertase and pectinase in several free-living, mycophagous and parasitic nematodes. *Nematologica*, **11**, 441–448.

Nelson, F.K., Albert, P.S. and Riddle, D.L (1983) Fine structure of the *C. elegans* secretory-excretory system. *Journal of Ultrastructure Research*, **82**, 156–171.

Nelson, F.K. and Riddle, D.L (1984) Functional study of the *C. elegans* secretory-excretory system using laser microsurgery. *Journal of Experimental Zoology*, **231**, 45–56.

Nicholas, W.L. (1984) *The Biology of Free-living Nematodes*. Oxford: Oxford University Press.

Niedfeld, G., Pezzani, B., Minvielle, M., Basualdo Farjat, J.A. (1993) Presence of lipids in the secretory/excretory product from *Toxocara canis*. *Veterinary Parasitology*, **51**, 155–158.

Nielsen, M.E. and Buchmann, K. (1997) Glutathione-s-transferase is an important antigen in the eel nematode *Anguillicola crassus*. *Journal of Helminthology*, **71**, 319–324.

Norton, D.C. (1978) *Ecology of Plant-parasitic Nematodes*. New York: Wiley.

Ogilvie, B.M., Rothwell, T.L.W., Bremmer, K.C., Schnitzerling, H.J., Nolan, J. and Keith, R.K. (1973) Acetylcholinesterase secretion by parasitic nematodes. I. Evidence for secretion of the enzyme by a number of species. *International Journal for Parasitology*, **3**, 589–597.

Oka, T., Yamamato, R. and Futai, M. (1997) The vha genes encode proteolipids of *Caenorhabditis elegans* vacuolar-type ATPase — Gene structures and preferential expression in an H-shaped excretory cell and rectal cells. *Journal of Biological Chemistry*, **272**, 24387–24392.

Oka, T., Yamamoto, R. and Futai, M. (1998) Multiple genes for vacuolar-type ATPase proteolipids in *Caenorhabditis elegans*. A new gene, *vha-3*, has a distinct cell-specific distribution. *Journal of Biological Chemistry*, **273**, 22570–22576.

Pappas, P.W. (1988) The relative roles of the intestines and external surfaces in the nutrition of monogeneans, digeneans and nematodes. *Parasitology*, **96**, S105–S121.

Pappas, P.W. and Read, C.P. (1975) Membrane transport in helminth parasites: A review. *Experimental Parasitology*, **37**, 469–530.

Parri, H.R., Djamgoz, A.D., Holden-Dye, L. and Walker, R.J. (1993) An ion-selective microelectrode study on the effect of a high concentration of ivermectin on chloride balance in the somatic muscle bag cells of *Ascaris suum*. *Parasitology*, **106**, 421–427.

Pax, R.A. and Bennett, J.L. (1990) Studies on intrategumental pH and its regulation in adult male *Schistosoma mansoni*. *Parasitology*, **101**, 219–226.

Pax, R.A., Geary, T.G., Bennett, J.L. and Thompson, D.P. (1995) *Ascaris suum*: characterization of transmural and hypodermal potentials. *Experimental Parasitology*, **80**, 85–97.

Pennock, J.L., Behnke, J.M., Bickle, Q.D., Devaney, E., Grencis, R.K., Isaac, R.E., Joshua, G.W.P., Selkirk, M.E., Zhang, Y. and Meyer, D.J. (1998) Rapid purification and characterization of L-dopachrome-methyl ester tautomerase (macrophage-migration-inhibitory factor) from *Trichinella spiralis*, *Trichuris muris* and *Brugia pahangi*. *Biochemical Journal*, **335**, 495–498.

Plasterk, R.H. (1995) Reverse genetics: from gene sequence to mutant worm. In *Caenorhabditis elegans*: modern biological analysis of an organism. *Methods in Cell Biology*, vol. 48, edited by H.F. Epstein and D.C. Shakes, pp. 59–80. San Diego: Academic Press.

Poole, R.C. and Halestrap, A.P. (1993) Transport of lactate and other monocarboxylates across mammalian plasma membranes. *American Journal of Physiology*, **264**, 761–782.

Pratt, D., Armes, L.G., Hageman, R., Reynolds, V., Boisvenue, R.J. and Cox, G.N. (1992) Cloning and sequence comparisons of four distinct cysteine proteases expressed by *Haemonchus contortus* adult worms. *Molecular and Biochemical Parasitology*, **51**, 209–218.

Price, N.T., Jackson, V.N. and Halestrap, A.P. (1998) Cloning and sequencing of four new mammalian monocaroxylate transporter (MCT) homologues confirms the existence of a transporter family with an ancient past. *Biochemical Journal*, **329**, 321–328.

Pritchard, D.I., Leggett, K.V., Rogan, M.T., McKean, P.G. and Brown, A. (1991) *Necator americanus* secretory acetylcholinesterase and its purification from excretory-secretory products by affinity chromatography. *Parasite Immunology*, **13**, 187–199.

Raizen, D.M. and Avery, L. (1994) Electrical activity and behavior in the pharynx of *Caenorhabditis elegans*. *Neuron*, **12**, 483–495.

Rathaur, S., Robertson, B.D., Selkirk, M.E. and Maizels, R.M. (1987) Secretory acetylcholinesterases from *Brugia malayi* adult and microfilarial parasites. *Molecular and Biochemical Parasitology*, **26**, 257–265.

Rew, R.S. and Fetterer, R.H. (1986) Mode of action of antinematodal drugs. In *Chemotherapy of Parasitic Infections*, edited by W.C. Campbell and R.S. Rew, pp. 321–338. New York: Plenum Press.

Rhoads, M.L. and Fetterer, R.H. (1995) Developmentally regulated secretion of cathepsin L-like cysteine proteases by *Haemonchus contortus*. *Journal of Parasitology*, **81**, 505–512.

Riddle, D.L. and Albert, P.S. (1997) Genetic and environmental regulation of dauer larva development. In *C. elegans II*, edited by D.L. Riddle, T. Bumenthal, B.J. Meyer and J.R. Priess, pp. 739–768. Plainview, NY: Cold Spring Harbor Laboratory Press.

Riding, I. (1970) Microvilli on the outside of a nematode. *Nature*, **226**, 179–180.

Rifkin, M., Seow, H-F., Jackson, D., Brown, L. and Wood, P. (1996) Defense against the immune barrage: helminth survival strategies. *Immunology and Cell Biology*, **74**, 564–574.

Robertson, A.P. and Martin, R.J. (1996) Effects of pH on a high-conductance Ca-dependent chloride channel: a patch-clamp study in *Ascaris suum*. *Parasitology*, **113**, 191–198.

Robinson, K., Bellaby, T., Chan, W.C. and Wakelin, D. (1995) High levels of protection induced by a 40-mer synthetic peptide vaccine against the intestinal nematode parasite *Trichinella spiralis*. *Immunology*, **86**, 495–498.

Rogers, W.P. (1969) Nitrogenous components and their metabolism: Acanthocephala and Nematoda. In *Chemical Zoology*, Vol. 3, edited by M. Florkin and B.T. Scheer, pp. 379–428. New York: Academic Press.

Rogers, W.P. and Sommerville, R.I. (1968) The infectious process and its relation to the development of early parasitic stages of nematodes. *Advances in Parasitology*, **6**, 327–348.

Rothwell, J.T. and Sangster, N.C. (1997) *Haemonchus contortus*: the uptake and metabolism of closantel. *International Journal for Parasitology*, **27**, 313–319.

Salkoff, L. and Jegla, T. (1995) Surfing the DNA databases for K+ channels nets yet more diversity. *Neuron*, **15**, 489–492.

Sangster, N.C. (1994) P-glycoproteins in nematodes. *Parasitology Today*, **10**, 319–332.

Sangster, N.C., Bannon, S.C., Weiss, A.S., Nulf, N.C., Klein, R.D. and Geary, T.G. (1999) *Haemonchus contortus*: sequence heterogeneity of internucleotide binding domains from P-glycoproteins and an association with avermectin/milbemycin resistance. *Experimental Parasitology*, **91**, 250–257.

Savin, K.W., Dopheide, T.A.A., Frenkel, M.J., Wagland, B.M., Grant, W.N. and Ward, C.W. (1990) Characterization, cloning and host-protective activity of a 30-kilodalton glycoprotein secreted by the parasitic stages of *Trichostrongylus colubriformis*. *Molecular and Biochemical Parasitology*, **41**, 167–176.

Schinkel, A.H.E., Smit, J.J.M., van Tellingen, O., Beijnen, J.H., Wagenaar, E., van Deemter, L., Mol, C., van der Valk, M.A., Robanus-Maandag, E.C., te Riele, H.P.J., Berns, A.J.M. and Porst, P. (1994) Disruption of the mouse *mdr1a* P-glycoprotein gene leads to a deficiency in the blood-brain barrier and to increased sensitivity to drugs. *Cell*, **77**, 491–502.

Schinkel, A.H.E., Wagenaar, E., van Deemter, L., Mol, C. and Borst, P. (1995) Absence of the *mdr1a* P-glycoprotein in mice affects tissue distribution and pharmacokinetics of dexamethasone, digoxin and cyclosporin A. *Journal of Clinical Investigation*, **96**, 1698–1705.

Schanbacher, L.M. and Beames, C.G. (1973) Transport of 3-o-methylglucose by various regions of the intestine of *Ascaris suum*. *Journal of Parasitology*, **59**, 215.

Schoofs, L., Veelaert, D., Vanden Broeck, J. and de Loof, A. (1997) Peptides in the locusts, *Locusta migratoria* and *Schistocerca gregaria*. *Peptides*, **18**, 145–156.

Seifter, J.L. and Aronson, P.S. (1986) Properties and physiologic roles of plasma membrane sodium-hydrogen exchanger. *Journal of Clinical Investigation*, **78**, 859–864.

Sijmons, P.C. (1993) Plant-nematode interactions. *Plant Molecular Biology*, **23**, 917–931.

Sims, S.M., Ho, N.F.H., Geary, T.G., Thomas, E.M., Day, J.S., Barsuhn, C.L. and Thompson, D.P. (1996) Influence of organic acid excretion on cuticle pH and drug absorption by *Haemonchus contortus*. *International Journal for Parasitology*, **26**, 25–35.

Sims, S.M., Magas, L.T., Barsuhn, C.L., Ho, N.F.H., Geary, T.G. and Thompson, D.P. (1992) Mechanisms of microenvironment pH regulation in the cuticle of *Ascaris suum*. *Molecular and Biochemical Parasitology*, **53**, 135–148.

Sims, S.M., Ho, N.F.H., Magas, L.T., Geary, T.G., Barsuhn, C.L. and Thompson, D.P. (1994) Biophysical model of the transcuticular excretion of organic acids, cuticle pH and buffer capacity in gastrointestinal nematodes. *Journal of Drug Targeting*, **2**, 1–8.

Sithigorngul, P., Stretton, A.O.W. and Cowden, C. (1990) Neuropeptide diversity in *Ascaris*: an immunohistochemical study. *Journal of Comparative Neurology*, **284**, 389–397.

Spindel, E.R., Giladi, E., Segerson, T.P. and Nagalla, S. (1993) Bombesin-like peptides: of ligands and receptors. *Recent Progress in Hormone Research*, **48**, 365–391.

Stephenson, W. (1942) The effect of variations in osmotic pressure upon a free-living soil nematode. *Parasitology*, **34**, 253–265.

Stewart, G.L. (1983) Biochemistry. In *Trichinella and trichinosis*, edited by W.C. Campbell, pp. 153–172. New York: Plenum Press.

Strote, G., Bonow, I. and Attah, S. (1996) The ultrastructure of the anterior end of male *Onchocerca volvulus*: papillae, amphids, nerve ring and first indication of an excretory system in the adult female worm. *Parasitology*, **113**, 71–85.

Tabara, H., Grishok, A. and Mello, C.C. (1998) RNAi in *C. elegans*: soaking in the genome sequence. *Science*, **282**, 430–431.

Take-Uchi, M., Kawakami, M., Ishihara, T., Amano, T., Kondo, K. and Katsura, I. (1998) An ion channel of the degenerin/epithelial sodium channel superfamily controls the defecation rhythm in *Caenorhabditis elegans*. *Proceedings of the National Academy of Sciences USA*, **95**, 11775–11780.

Thompson, D.P. and Geary, T.G. (1995) The structure and function of helminth surfaces. In *Biochemistry and Molecular Biology of Parasites*, edited by J.J. Marr and M. Mueller, pp. 203–222. New York: Academic Press.

Thompson, D.P., Ho, N.F.H., Sims, S.M. and Geary, T.G. (1993) Mechanistic approaches to quantitate anthelmintic absorption by gastrointestinal nematodes. *Parasitology Today*, **9**, 31–35.

Tielens, A.G.M. and van den Bergh, S.G. (1993) Aerobic and anaerobic energy metabolism in the life cycle of parasitic helminths. In *Surviving Hypoxia*, edited by P.W. Houchachka, P.L. Lutz, T. Sick, M. Rosenthal and G. Van den Thillart, pp. 19–40. Boca Raton, FLA: CRC Press.

Titus, E. and Ahearn G.A. (1992) Vertebrate gastrointestinal fermentation: transport mechanisms for volatile fatty acids. *American Journal of Physiology*, **262**, R547–R553.

Urwin, P.E., Atkinson, H.J., Waller, D.A and McPherson, M.J. (1995) Engineered oryzacystatin-I expressed in transgenic hairy roots confers resistance to *Globodera pallida*. *Plant Journal*, **8**, 121–131.

Urwin, P.E., Lilley, C.J., McPherson, M.J. and Atkinson, H.J. (1997) Resistance to both cyst and root-knot nematodes conferred by transgenic *Arabidopsis* expressing a modified plant cystatin. *Plant Journal*, **12**, 455–461.

Valkanov, M. and Martin, R.J. (1995) The Cl channel in *Ascaris suum* selectively conducts dicarboxylic anion products of glucose fermentation and suggests a role in removal of waste organic anions. *Journal of Membrane Biology*, **148**, 41–49.

Valkanov, M., Martin, R.J. and Dixon, D.M. (1994) The Ca-activated chloride channel of *Ascaris suum* conducts volatile fatty acids produced by anaerobic respiration: a patch-clamp study. *Journal of Membrane Biology*, **138**, 133–141.

Verhoeven, H.L.E., Willemsens, G. and Van den Bossche, H. (1980) Uptake and distribution of levamisole in *Ascaris suum*. In *Biochemistry of Parasites and Host-Parasite Relationships*, edited by H. Van den Bossche, pp. 573–579. Amsterdam: North-Holland Biomedical Press.

Voilley, N., Galibert, A., Bassilana, F., Renard, S., Lingueglia, E., Coscoy, S., Champigny, G., Hofman, P., Lazdunski, M. and Barbry, P. (1997) The amiloride-sensitive Na$^+$ channel: from primary structure to function. *Comparative Biochemistry and Physiology Part A*, **118**, 193–200.

von Brand, T. (1973) *Biochemistry of Parasites*, 2nd edn. New York: Academic Press.

Wang, X., Meyers, D., Yan, Y., Baum, T., Smant, G., Hussey, R. and Davis, E. (1999) In planta localization of a β-1,4-endoglucanase secreted by *Heterodera glycines*. *Molecular Plant-Microbe Interactions*, **12**, 64–67.

Watts, J.L. and Browse, J. (1999) Isolation and characterization of a Δ5-fatty acid desaturase from *Caenorhabditis elegans*. *Archives of Biochemistry and Biophysics*, **362**, 175–182.

Wei, A., Jagla, T. and Salkoff, L. (1996) Eight potassium channel families revealed by the *C. elegans* genome project. *Neuropharmacology*, **35**, 805–829.

Weinstein, P.P. (1952) Regulation of water balance as a function of the excretory system of the filariform larvae of *Nippostrongylus muris* and *Ancylostoma caninum*. *Experimental Parasitology*, **1**, 363–376.

Weinstein, P.P. (1960) Excretory mechanisms and excretory products of nematodes: an appraisal. In *Host Influence on Parasite Physiology*, edited by L.A. Stauber, pp. 65–92. New Brunswick: Rutgers University Press.

Weith, J.O. and Brahm, J. (1980) Kinetics of bicarbonate exchange in human red blood cells — physiological implications. In *Membrane Transport in Erythrocytes*, edited by U.V. Lassen, H.H. Ussing and J.O. Weith, pp. 467–482. Copenhagen: Munksgaard.

Wharton, D.A. (1982) The survival of desiccation by the free-living stages of *Trichostrongylus colubriformis* (Nematoda: Trichostrongylidae). *Parasitology*, **84**, 455–462.

Wharton, D.A. (1986) *The Functional Biology of Nematodes*. Baltimore: Johns Hopkins University Press.

Wharton, D.A. and Sommerville, R.I. (1984) The structure of the excretory system of the infective larvae of *Haemonchus contortus*. *International Journal for Parasitology*, **14**, 455–462.

Williams, J.F., Mackenzie, C.D. and El Khalifa, M. (1994) Onchocerciasis and lymphatic filariasis. In *Parasitic Infections and the Immune System*, edited by F. Kierzenbaum, pp. 225–247. New York: Academic Press.

Womersley, C. (1981) Biochemical and physiological aspects of anhydrobiosis. *Comparative Biochemistry and Physiology*, **70B**, 669–678.

Wright, D.J. (1975) Studies on nitrogen catabolism in *Panagrellus redivivus* (Goodey, 1945) (Nematoda; Cephalobidae). *Comparative Biochemistry and Physiology*, **52B**, 255–260.

Wright, D.J. and Newall, D.R. (1976) Nitrogen excretion, osmotic and ionic regulation in nematodes. In *The Organization of Nematodes*, edited by N.A. Croll, pp.143–164. New York: Academic Press.

Wright, D.J. and Newall, D.R. (1980) Osmotic and ionic regulation in nematodes. In *Nematodes as Biological Models*, vol. 2, edited by B.M. Zuckerman, pp. 143–164. New York: Academic Press.

Wright, K.A. (1987) The nematode's cuticle, its surface, and the epidermis: function, analogy and homology — a current consensus. *Journal of Parasitology*, **73**, 1077–1083.

Wu, X., Fei, Y-J, Huang, W., Chancy, C., Leibach, F.H. and Ganapathy, V. (1999) Identity of the *F52F12.1* gene product in *Caenorhabditis elegans* as an organic cation transporter. *Biochimica et Biophysica Acta*, **1418**, 239–244.

Yancey, P.H., Clark, M.E., Hand, S.C., Bowlus, R.D. and Somero, G.N. (1982) Living with water stress: Evolution of osmolyte systems. *Science*, **217**, 1214–1222.

Yao, C. and Jasmer, D.P. (1998) Nuclear antigens in *Trichinella spiralis* infected muscle cells: nuclear extraction, compartmentalization and complex formation. *Molecular and Biochemical Parasitology*, **92**, 207–218.

Xu, M., Molento, M., Ribeiro, P., Beech, R. and Prichard, R. (1998) Ivermectin resistance in nematodes may be caused by alteration of P-glycoprotein homolog. *Molecular and Biochemical Parasitology*, **91**, 327–335.

Zwaal, R.R., Broeks, A., Van Meurs, J., Groenen, J.T.M. and Plasterk, R.H.A (1993) Target-selected gene inactivation in *Caenorhabditis elegans*, using a frozen transposon insertion mutant bank. *Proceedings of the National Academy of Sciences USA*, **90**, 7431–7435.

12. Neuromuscular Organisation and Control in Nematodes

Richard J. Martin[1], J. Purcell[2], A.P. Robertson[1] and M.A. Valkanov[2]
[1] Department of Biomedical Sciences, College of Veterinary Medicine, Ames, Iowa 5O011-1250, USA
[2] Department of Preclinical Veterinary Sciences, R.(D.)S.V.S., Summerhall, University of Edinburgh, Edinburgh EH9 IQH, U.K.

Introduction

Different species of parasitic nematode infect man and animals, giving rise to a range of symptoms including reduced growth, diarrhoea, and sometimes death. Some drugs used to treat these infestations interfere selectively with ion-channels in nerves or muscle cells of the nematode. Levamisole opens nicotinic acetylcholine channels in nematode somatic muscle to produce depolarisation and spastic paralysis without an effect on the host. The GABA-agonist, piperazine, and the antibiotic avermectins gate Cl channels in nerve or muscle of nematodes to produce paralysis. Quisqualate has an action on an excitatory glutamate channel. All these drugs have a selective effect on nematode parasites; their neuromuscular system is disturbed so that the parasites are no longer capable of maintaining their position in the host. In order to understand the mode of action of the anthelmintics, knowledge of the organisation of nematode neuromuscular transmission is required. The regular use of anthelmintic drugs has brought about selection of anthelmintic-resistant nematodes; the investigation of resistance involves studying the modifications of the control of neuromuscular systems associated with the resistance.

A useful nematode parasite to study is *Ascaris suum*. It is a common, large (up to 35 cm in length) nematode found in the small intestine of the pig. The large size of the parasite and the large somatic muscle cells and neurons have helped anatomical, biochemical, physiological and pharmacological studies which are more difficult on smaller nematodes. Consequently, *A. suum* has been used successfully for examining the mode of action of antinematodal drugs. Another parasite we have used is *Oesophagostomum dentatum* (Figure 12.1), an intestinal parasite 1 cm in length, that comes from the pig. The advantage of this parasite is that anthelmintic-resistant isolates are available for study and comparison with anthelmintic-sensitive isolates is possible. The results of experiments on *Ascaris* form the basis of the account presented here but we include observations on *O. dentatum*. Experiments on other smaller nematodes suggest that these observations are relevant to all nematodes. Many genetic studies have utilised the small free-living nematode, *Caenorhabditis elegans*, which is

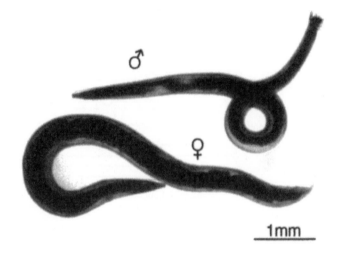

Figure 12.1. Unstained male and female adult *Oesophagostomum dentatum*

easily grown in the laboratory; so some observations on this model nematode are also included. The purpose of this chapter is to review the organisation of neuromuscular transmission in nematodes and to describe actions of antinematodal drugs which affect muscle activity.

Nematodes have Stereotyped Body Movements

Many nematode parasites have to maintain their position in the host animal (intestine, bronchi, blood vessels) by sinusoidal body movements. *A. suum* is capable of a co-ordinated wave of contraction and relaxation that is propagated anteriorly or posteriorly (Jarman, 1970; Crofton, 1971). The sinusoidal waves of nematodes are only possible in the dorso-ventral plane because the somatic muscle is divided into dorsal and ventral halves innervated by their respective major nerve cord. The more complex behaviour of the head required for feeding is under the control of the cranial ganglia and sublateral nerve cords; the behaviour required for mating is under the control of perianal ganglia (Figure 12.2A). In contrast to the active motile nematode parasites like *A. suum*, there are 'inhibited' larval stages of trichostrongylid nematodes or the adults of the filarial worm, *Onchocerca volvulus*, the nematode responsible for river blindness. These nematodes do not require co-ordinated motor activity to main-

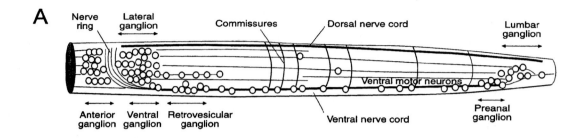

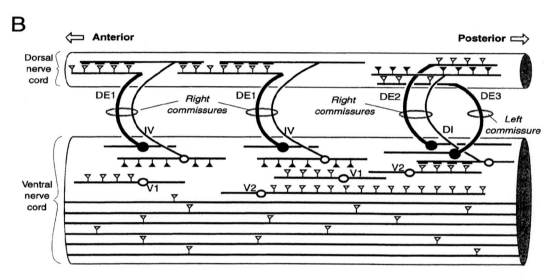

Figure 12.2. A: General organisation of the nerve cord and nerve cells of nematodes showing the dorsal and ventral nerve cords with collections of nerve cells forming ganglia. Modified from Durbin (1987). **B:** Segment of *Ascaris suum* produced by 3 right hand and 1 left hand commissures connecting the dorsal and ventral nerve cord. Each commissure contains 1 or 2 axons of a motorneuron which has its cell body in the ventral nerve cord. The ventral nerve cord contains interneurons which project between the segments.

tain their position and they are not as vulnerable to drugs affecting locomotion. Other more complex body movements are recognised in *C. elegans* in response to mechanical sensory inputs. Mechanical stimulation produced by the head meeting a solid object leads to a reversal in direction and backward movement. Mechanical stimulation of the tail region leads to forward movement. Other stereotypical motor movements include egg laying, defecation and copulation. The anatomical and physiological organisation of the nervous system is structured to permit these stereotypical body movements.

Nematode Neuroanatomy

The neuroanatomy of nematodes was studied initially by Hesse (1892) and Goldschmidt (1908; 1909), who showed that there are only about 250 neurons present in *Ascaris* and that their position and structure is completely reproducible from animal to animal. In *C. elegans* the nervous system is similar and possesses 302 neurons in each hermaphrodite animal. The neurons have been reconstructed from serial section through electron micrographs defining the morphology of each neuron. There are 118 classes of neuron and

their connections via chemical synapses or gap junction have been described (Albertson and Thomson, 1976; White *et al.*, 1976; White *et al.*, 1986). The locations of different groups of neuron cell bodies have been identified (Figure 12.2A). Each neuron can have between 1 and 30 different synaptic contacts. The entire nervous system has nearly 5000 chemical synapses, 600 gap junctions, and 2000 neuromuscular junctions. The neuromuscular synapses are symmetrical on the left and right sides and 75% reproducible from animal to animal (Bargmann and Kaplan, 1998).

In the head region of *Ascaris*, Goldschmidt (1908) described a nerve ring surrounding the pharynx and a series of associated ganglia. The ganglia, which together contain 162 neuronal nuclei, consist of a dorsal ganglion, a ventral ganglion, two lateral ganglia and the retrovesicular ganglion (Angstadt *et al.*, 1989). Two major nerve cords arise from the nerve ring dorsally and ventrally, and pass caudally as the major dorsal nerve cord and the major ventral nerve cord. The cell bodies of nearly all motorneurons are found scattered in the ventral nerve cord and total 75 in number.

Stretton *et al.* (1978) described and classified the anatomy and physiology of the motorneurons of *A. suum*. The cell bodies of motorneurons are located in the ventral nerve cord and the connections to the dorsal nerve cord are via commissures, which are single or paired (Figure 12.2B). The pattern of the commissures repeats itself down the length of the body with each repeat consisting of three paired right-hand commissures and one left-hand commissure. Each repeat then defines a segment. Each segment contains 11 motorneurons and 6 large non-segmental interneurons crossing the segments. The 11 motorneurons of each segment are separated into 7 anatomical types according to the distribution of their axon and dendrites: four types (V1, VI, V2, and DEI) are represented twice; three types (D1, D2, D3) are represented only once in each segment. The dendrites of DEI, DE2 and DE3 and VI and V2 receive their input from interneurons via *en passant* synapses. The motorneurons DI and VI receive their input from axons of the other motorneurons, from DEI, DE2, DE3 in the dorsal nerve cord and from VI and V2 in the ventral nerve cord. DI and VI are inhibitory and DEI, DE2, DE3, VI and V2 are, or are likely to be, excitatory motorneurons (Stretton *et al.*, 1978). The enzyme involved in the biosynthesis of acetylcholine, choline acetyltransferase (CAT), was localised histochemically in DEI, DE2, DE3, VI and V2 (Johnson and Stretton, 1985), a finding which is consistent with the suggestion that acetylcholine is an excitatory neurotransmitter of these motorneurons. GABA-immunoreact-ivity was observed in the inhibitory motorneurons, DI and VI (Johnson and Stretton, 1987; Sithigorngul *et al.*, 1989; Guastella *et al.*, 1991; Guastella and Stretton, 1991).

Mechanical sensory inputs (Tavernarakis and Driscoll, 1997) which initiate some reflex movements have been described in *C. elegans*. For example, a simplified description of the circuit which is involved in initiating reflex forward and reverse movement after mechanical stimulation is illustrated (Figure 12.3A). The sensory neurons known as AVM and ALM, detect stimuli on the anterior region of the body and give rise to reverse movement by inhibitory synapses onto interneurons known as PVC neurons and excitatory gap junctions onto interneurons known as AVD neurons. The AVD interneuron, in turn, projects excitatory synapses onto DA and VA motorneurons, which show reciprocal inhibition. Mechanical stimulation of the posterior region of *C. elegans* excites a PLM sensory neurons that leads to forward movement via excitatory gap junctions onto PVC interneurons and excitatory synapses onto DB and VB motorneurons.

The two nematode species, *Ascaris suum* and *C. elegans*, are well separated on an evolutionary time scale (Blaxter *et al.*, 1998) and because the nervous system of these two nematodes has significant similarities, it is suggested that the basic neuronal organisation of all nematodes is similar.

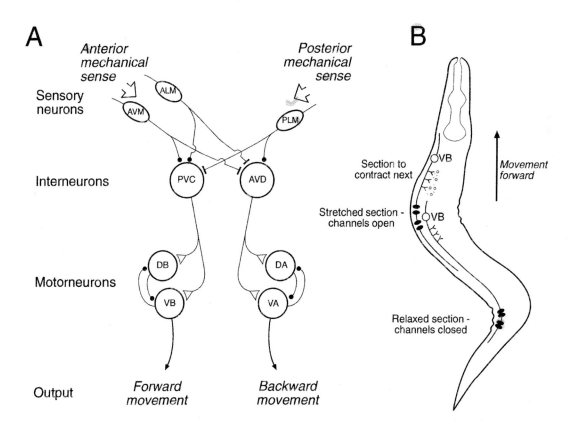

Figure 12.3. A: Nerve circuits responsible for co-ordinated movement in response to mechanical stimuli on either the anterior or posterior region of *C. elegans*. **B:** The organisation of stretch sensitive channels in ventral motor neurons of *C. elegans* which leads to contraction of appropriate regions of the body muscle and co-ordinated body movements.

Physiology of Single Neurons

The motorneurons DEI, DE2, DE3, DI, and VI have resting membrane potentials of –30 to –40 mV, like *Ascaris* muscle cells (Davis and Stretton, 1989b; Davis and Stretton, 1989a). The motorneurons are non-spiking and release transmitter tonically: hyperpolar-isation of the axons reduces release of transmitter while depolarisation increases it. The relatively short distance that these motorneurons have to conduct, and their long space-constants, allows them to be non-spiking. In contrast, the much longer interneurons found in the ventral nerve cord may spike in order to conduct the length of the adult (30 cm). Ventral inhibitory (VI) motorneurons in *A. suum* generate slow depolarising potentials in response to continuous current injection (Angstadt and Stretton, 1989). The rhythmic potentials in VI give rise to rhythmic inhibitory potentials in ventral muscle cells and may contribute to locomotory behaviour.

Another interesting aspect of the physiology of some motorneurons is the possession of stretch receptive ion-channels known as the degenerins. This is a family of channels with at least 15 members and one of them, UNC-8, is expressed in several motorneurons (Taver-narakis and Driscoll, 1997). Figure 12.3B illustrates how localised stretching of a *C. elegans* VB motorneuron, during the normal body wave associated with move-ment, leads to opening of cation-selective channels, localised release of excitatory transmitter and that then potentiates forward movement. Figure 12.3B also shows how a localised relaxed region closes the stretch-sensitive channel and reduces release of excita-tory transmitter. A similar stretch-sensitive channel, UNC-105, is present in *C. elegans* muscle cells (GarciaA-noveros *et al.*, 1998). Thus, stretch-sensitive channels in muscle cells must also be involved in generating rhythmic movement.

Motorneuron-Nerve Cell Interconnections

Over 2000 synapses from serial sections of *A. suum* nerve cords have been examined by light or electron microscopy (Stretton *et al.*, 1978). Each motorneuron has distinctive regions where it receives input from the large interneurons. In addition to these sites there are synaptic contacts between the motorneurons them-selves. This implies that motorneurons also possess acetylcholine and GABA receptors in the same way that the somatic muscle does. The majority of synaptic contacts between motorneurons are of the following types: a) excitatory axons projecting onto opposite inhibitory dendrites (ventral excitator onto dorsal inhibitor or dorsal excitator onto ventral inhibitor); and b) reciprocal synapses between excitatory and inhibitory axons in each nerve cord (ventral excitator projecting onto ventral inhibitor and dorsal excitator projecting onto dorsal inhibitor). The excitatory field in the muscle cell produced by stimulation of excitatory motorneurons is sandwiched between a cranial and a caudal inhibitory zone.

Walrond and Stretton (1985; 1985) showed that stimulation of the dorsal excitatory motorneuron, DEI, produced reciprocal inhibition in ventral muscle cells, as a result of the projection onto the ventral inhibitory motorneuron, VI. They also showed that the motor-neurons DE2 and DE3 could produce reciprocal inhibition as a result of projecting onto VIs in adjacent segments. Reciprocal inhibition in the ventral cord was suggested on anatomical grounds involving the puta-tive excitatory motorneurons VI and V2 and their projection onto the dorsal inhibitory motorneuron DI. The consequence of this synaptic organisation, with reciprocal and adjacent inhibition, is that stimulation of a single excitatory motorneuron would produce an Ω-shaped body waveform. This synaptic organisation allows propagated waves of muscle contraction.

Structure of Somatic Muscle

The anatomy of somatic muscle cells of *A. suum* was described by Schneider as early as 1866. The muscle cells have an unusual structure, with processes of the muscle (Figures 12.4A, C) going to the nervous system (Schneider, 1995; Goldschmidt, 1908; Cappe de Baillon, 1911). Figure 12.4A shows a photomicrograph of a collagenase isolated muscle cell from *A. suum*. The contractile region, known as the spindle, 'fuseau' (Cappe de Baillon, 1911), lies next to the epidermis (Figure 12.4C) and is composed of 'obliquely striated muscle' (Rosenbluth, 1965a; Rosenbluth, 1965b; Rosen-bluth, 1967; Rosenbluth, 1969). The actin and myosin filaments are arranged at an acute angle, in contrast to the 90′ angle of vertebrate skeletal muscle. This organisation allows the greater extensibility of smooth muscle but still allows the velocity of contraction to be maintained. The bag or belly, 'panse', (Cappe de Baillon, 1911) of *A. suum* (Figure 12.4A) is a large 200 μm diameter bag-shaped structure which contains the nucleus and particulate glycogen (Rosenbluth, 1965a; Rosenbluth, 1965b). It lies in the peri-enteric space and its turgidity probably provides the pressure against the collagen layers of the cuticle that allows the nematode to be sufficiently rigid to contract and move (Harris and Crofton, 1957). Bag structures are only present in large nematodes and serve to support the structure of the body by inflating against the outer collagenous cuticle like an inner tube of a tyre. Figure 12.4B shows a photomicrograph of muscle cells in a body flap preparation of the smaller parasitic nematode, *Oesopha-gostomum dentatum*; note the absence of 'bags'.

Each somatic muscle cell of *A. suum* has an average of 2.7 arms (Stretton, 1976). The arms are thin processes that pass from the base of the belly towards the syncytium that lies over the nerve cord. As the arms reach the syncytium, they break up into a number of finer processes known as 'fingers'. Tight junctions form between adjacent fingers and form a tangled complex referred to as the syncytium (Rosenbluth, 1965b) where electrical coupling between adjacent cells takes place (De Bell *et al.*, 1963).

The dorsal and ventral nerve cords are supported by an extension of the epidermis that has, in transverse section, the appearance of a chalice (wine goblet) and is known therefore as the epidermal chalice

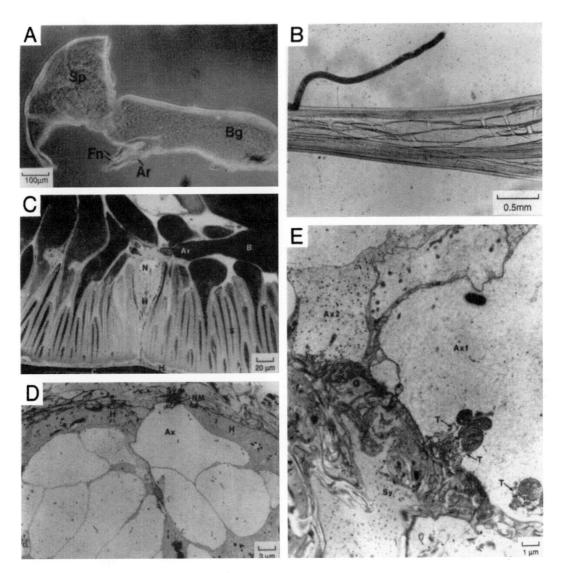

Figure 12.4. **A:** Collagenase isolated single muscle cell from an adult *Ascaris suum*. Bg: Bag region which contains the nucleus and glycogen granules. Sp: Contractile spindle region. Ar: Arm, which connects the muscle to the nerve cord. Fn: Fingers which receive the synaptic contacts from the nerve cord. **B:** A flap preparation of *Oesophagostomum dentatum*. Single muscle cells and arms can be seen. **C:** Methylene-blue-stained transverse-section through the nerve cord (N) of *Ascaris* showing the relative positions of the spindle regions (S), bag regions (B) and arms (Ar) in a section. The nerve cord is supported in the epidermal chalice which extends from the epidermis (H) found next to the cuticle (C). **D:** Transmission electron micrograph of a ventral nerve cord of *Ascaris* showing the axon (Ax) of a motor neuron forming a neuromuscular contact with the syncytium (Sy) by projecting through the epidermis. **E:** High power transmission electron micrograph of the neuromuscular junction of *Ascaris*. Two axons (Ax1 and Ax2) are shown. The axon of Ax2 contains electron dense peptide vesicles in addition to clear vesicles. Tubules (T) are also found to which the vesicles are often attached and bud off from.

(Figure 12.4C). Neuromuscular junctions are formed by 2 μm extensions of the longitudinal axons that penetrate the epidermal chalice and closely associate with the syncytium (Figure 12.4C, D). The synaptic region has clear, coated, spherical 40 nm vesicles clustered under the presynaptic membrane around tubules, dense core vesicles (80–100 nm in diameter), giant mitochondria and membrane thickenings on the presynaptic and postsynaptic membranes (Rosenbluth, 1965b). In the body region, dorsal muscle cells are connected only to the dorsal nerve cord, while the ventral muscle is connected

only to the ventral nerve cord. Contraction is therefore only possible along the dorso-ventral plane.

Pharyngeal Muscle

Pharyngeal muscle receives synaptic inputs from around 20 pharyngeal neurons. Pharyngeal muscle of nematodes forces food into its intestine. In *Ascaris* it develops from 30 cells which fuse to become one functional tube-shaped cell (Goldschmidt, 1904; Del Castillo and Morales, 1967a), which is 1 cm long and 1 mm in diameter. The triradiate lumen is closed when

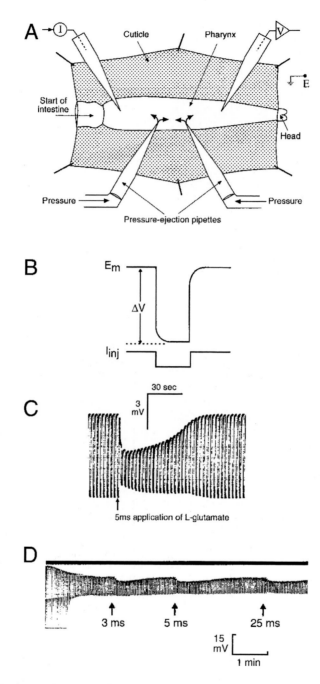

Figure 12.5. A: Diagram of the two-microelectrode current-clamp recording technique that we have used to examine electrophysiological effects of focal application of glutamate and ivermectin to the pharyngeal muscle of *Ascaris suum*. One micropipette (labelled V) is placed in the pharyngeal muscle to record the membrane potential and a second micropipette (I) is also placed in the muscle to inject hyperpolarising current pulses. **B:** Diagram of the effect of injecting a rectangular current pulse (I_{inj}) on the membrane potential (E_m) using a two microelectrode clamp. The current produces an exponentially increasing hyperpolarisation that settles after sufficient time to a change of ΔV mV. The input conductance of the pharynx can be determined from Ohms law because the current voltage relationship is linear. **C:** The trace shows that brief application of glutamate using a 5 ms 'puff' or pressure application from a micropipette filled with 0.5 M L-glutamate leads a hyperpolarisation of the membrane potential and an increase in the input conductance to 200 μS. As the ion-channels open up and carry Cl ions, the resistance of the membrane decreases and the membrane potential hyperpolarises. **D:** Effects of glutamate and ivermectin on pharyngeal membrane potential and conductance. The trace shows the membrane potential and input conductance response to a continuous application of ivermectin (horizontal bar) and 3 ms, 5 ms and 25 ms 'puffs' of 0.5 M glutamate from a micropipette.

the muscle is relaxed. Its contractile fibres are radially oriented (Mapes, 1966; Reger, 1966), so that contraction opens the lumen. When the lumen opens, food is sucked into the pharynx through a one-way valve near the mouth. When the pharyngeal muscle cell depolarises, the contractile fibres relax and elastic components that were stretched rapidly close the lumen, forcing the food in the lumen into the intestine of the *Ascaris* through a second one-way valve.

The electrophysiology of pharyngeal muscle of *A. suum* may be examined using intracellular microelectrodes (Figure 12.5). The pharyngeal muscle has a resting membrane potential of –35 mV and a positive-going action potential that reaches +30 to +50 mV and which is associated with opening of the lumen (Del Castillo and Morales, 1967a,b). The potential remains depolarised from 150 msec to several seconds with the lumen open; then the potential is suddenly returned to negative values by a unique negative-going action potential carried by a K-current to allow rapid closing of the lumen (Byerly and Masuda, 1979). The pharyngeal K-current is different from all other K-currents: the current has kinetics analogous to the classical Na-current of nerve and produces a fast negative (hyperpolarising) action potential.

Byerly and Masuda (1979) described two distinct postsynaptic potentials in the pharynx. There is a depolarising postsynaptic potential with a reversal potential of –10 mV; in addition there is a hyperpolarising postsynaptic potential with a reversal potential of –40 mV which could trigger the negative-going spike when the membrane was depolarised. We have used a two-microelectrode current-clamp technique to record changes in membrane potential and input conductance associated with the application of L-glutamate (Martin, 1996; Martin *et al.*, 1998). Glutamate has an inhibitory effect on *Ascaris* muscle, producing hyperpolarisation associated with an increase in chloride conductance (Figure 12.5B,C).

Electrophysiology of Somatic Muscle

Jarman (1959) described the presence of regular spontaneous spike-like depolarising potentials of *A. suum* body muscle which were superimposed on a resting potential of around –30 mV (Figure 12.6A). Similar membrane potentials were also observed in *Ascaridia galli* (Wann, 1987), parasitic in poultry, and *Haemonchus contortus* (Atchison *et al.*, 1992), parasitic in sheep. The ionic basis of the resting potential in *A. suum* was subsequently investigated by Del Castillo *et al.*, (1964a) and Brading and Caldwell (1971). Both studies reported that extracellular K had little effect on the potential in contrast to the effects of extracellular Cl and to a lesser extent extracellular Na. Radiolabelled ionflux experiments of Caldwell and Ellory (1968) have revealed that the permeability ratio for K, Na and Cl was 1:4:7 which again illustrates the importance of Cl and to a lesser extent Na.

The composition of peri-enteric fluid, which normally surrounds the muscle cells, is consistent with the

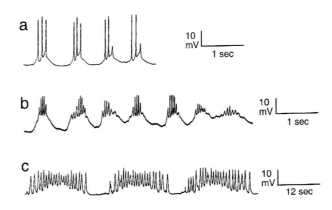

Figure 12.6. A: Spike potentials in groups of 2 or 3 sitting on a slow wave. **B:** Slow waves seen at a slower time scale with partial spike potentials sitting on top of the slow waves. **C:** Modulation waves that have a small amplitude and produce groups of slow waves.

movement of carboxylic acids across the muscle cell membrane. Hobson *et al.*, (1952a, b) found that the peri-enteric fluid contains low concentrations of Cl but high concentrations of carboxylic acid. The carboxylic acids (acetate, propionate, succinate, 2-methylbutyrate and 2-methylvalerate) are produced inside muscle cells from the anaerobic metabolism of glucose (Saz and Weil, 1962; Tsang and Saz, 1973). An investigation of the permeability properties of a 200pS conductance Ca-dependent Cl channel in the muscle membrane of *A. suum* (Thorn and Martin, 1987) found that it is permeable to these carboxylic acids (Figure 12.7) (Dixon *et al.*, 1993; Valkanov *et al.*, 1994; Valkanov and Martin, 1995). Interestingly the gating of the channel is altered by changes in pH on both sides of the membrane (Robertson and Martin, 1996): an increase in internal pH, for example, increased the probability of channel opening at hyperpolarised potentials. We have pointed out that an active proton pump in the muscle membrane would lead to an increased intracellular pH and an increase in the ionised: unionised concentration ratio of carboxylic acids. A raised internal pH favours opening of the channel and may facilitate excretion of waste organic anions. Thus, a Ca-activated Cl, channel permeable to the products of anaerobic fermentation appears to have an excretory function and to have a major influence on the membrane potential of muscle (Blair *et al.*, 1998a). A similar anion channel is also found in the epidermis (Blair *et al.*, 1998b)

Depolarising Potentials

The spontaneous depolarising potentials, described in muscle by Jarman (1959), are myogenic in origin and arise at the syncytium (De Bell *et al.*, 1963; De Bell, 1965). Three types of depolarising potential have been described (Figure 12.6). They are: (a) spikes, (b) slow waves, and (c) modulatory waves (De Bell *et al.*, 1963; Weisblat and Russel, 1976). The electrical activity seen in adjacent muscle cells is correlated (Jarman, 1959):

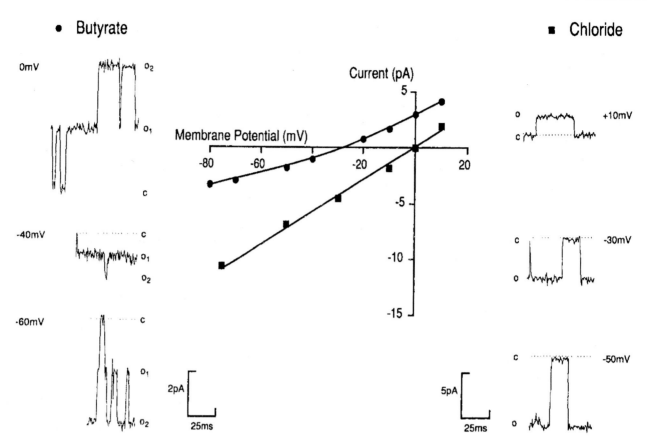

Figure 12.7. The *Ascaris* Ca-dependent CI channel is recorded in an inside-out patch. It is found frequently in patches from the bag membrane. The closed (C) state and open states (O1 and O2) are shown. On the right hand panel are channel currents at different membrane potentials in symmetrical CI solution. The I/V plot (■) is shown in the middle with the reversal potential of 0 mV. When the bath solution was changed, left hand panel, so that the CI was replaced with butyrate, the currents (●) had a reversal potential of –30 mV. Butyrate anions pass through the ion-channel and carries current as shown by the inward currents at –40 mV.

anteriorly placed cells depolarise before cells behind them; cells nearest to a nerve cord depolarise before those placed laterally. This coupling allows adjacent muscle cells to synchronise depolarisation and contraction. The coupling is consistent with the presence of tight junctions seen with the electron microscope (Rosenbluth, 1965b).

Ion-substitution experiments have been used to examine effects on the spike potentials and slow waves (Weisblat *et al.*, 1976). It was found that Ca was required to support the spike potentials and that Na and Ca were required for the slow waves. The Ca-currents responsible for the spike potentials have been recorded under a two-microelectrode voltage-clamp (Figure 12.8A) (Martin *et al.*, 1992b). Two potassium currents that facilitate repolarisation consist of a low threshold, rapidly inactivating current (an I_A-like current) and a higher threshold, non-inactivating current (a I_k-like current) (Figure 12.8B, C). K-currents are reduced by 4-aminopyridine and the anthelmintic diethylcarbamazine (Martin, 1982). The ionic basis of the slow waves and modulation waves appears to be produced by a non-selective cation current, I_{bcat} (Figure 12.8D) (Martin and Valkanov, 1996a).

Acetylcholine

Effects of Acetylcholine on *Ascaris* Muscle

Initially Baldwin and Moyle (1949) and then Norton and De Beer (1957) bath-applied acetylcholine and showed that the acetylcholine produced contraction of muscle strips. As a result, they demonstrated the presence of acetylcholine receptors on muscle. Bath-applied acetylcholine produces depolarisations and changes in spike frequency and amplitude in muscle (Del Castillo *et al.*, 1963). The receptors responsible for this are located synaptically at the syncytial region (Del Castillo *et al.*, 1963), and extrasynaptically on the bag region of the muscle (Martin, 1982). The ionic basis of electrical responses to acetylcholine has been investigated under voltage-clamp records where it was shown that acetylcholine increases the non-selective cation conductance of the membrane (Martin, 1982; Harrow and Gration, 1985): that is, acetylcholine opens ion-channels permeable to both Na and K.

Acetylcholine Single-Channel Currents

Acetylcholine-activated single-channel currents have been recorded from cell-attached and isolated inside-

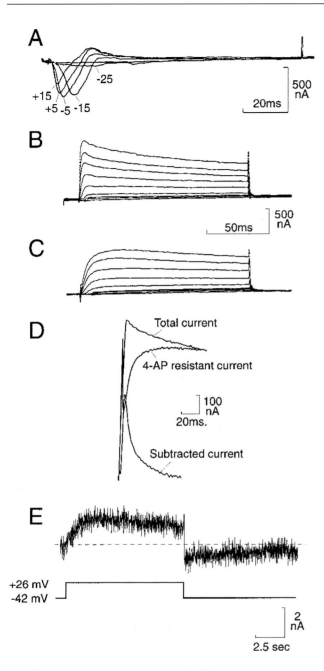

Figure 12.8. A: Voltage-activated Ca currents recorded from *Ascaris* bag membrane using a two-microelectrode voltage-clamp and barium as a charge carrier, K currents blocked with TEA. The bag was held at a membrane potential of –40 mV and stepped to the membrane potentials marked on the trace (Martin *et al.*, 1992b). **B:** Voltage-activated K-currents recorded from the *Ascaris* bag region with the two-microelectrode clamp. Calcium was omitted from the bathing solution. **C:** As **B** but in the presence of 4-aminopyridine to block the fast inactivating K current (I_k). **D:** Subtraction of **B** and **C** reveals the I_A-like, 4-aminopyridine resistant, current. This current may be responsible for repolarisation following spike potentials. **E:** slow non-selective cation current (I_{bcat}) (Martin and Valkanov, 1996c) recorded with a Kostyuk voltage-clamp. This current may underlie the slow potentials that give rise to spike potentials.

out patches using this preparation (Pennington and Martin, 1990). The channels activated by acetylcholine have at least two conductances: the larger was 40–50 pS and the smaller 25–35 pS. The average open-time was similar to that of the nicotinic channel at the frog neuromuscular junction and had a corrected mean open-time of 1.26 msec. High concentrations of acetylcholine (25–100 μm) produced a reduction in open probability and caused single-channel currents to occur in clusters with long closed-times (200 sec) between clusters; this behaviour has also been reported for the frog neuromuscular junction and described as desensitisation (Sakmann *et al.*, 1980).

Biochemistry of Acetylcholine

Cholinesterase, the enzyme inactivating acetylcholine, was first reported in *A. suum* by Bueding (1952). Lee (1962) described the distribution of this enzyme using histochemical techniques. In the head region, most of the activity is associated with the contractile region of the muscle in the extracellular matrix; enzyme activity was also observed in muscle arms near their endings on the nerve cords but not on the bag region; little or no staining was seen in the nervous tissue. No staining was seen caudal to the vulva, an observation consistent with propagation of contraction waves starting at the vulva and passing anteriorly. Johnson and Stretton (1980) have reported the presence of two types of cholinesterase, a 13 S and 5 S form, which have different distributions in the body of *Ascaris*. These two forms of acetylcholinesterase are also found in *C. elegans*, where they are products of separate genes; it is suggested that the 5 S form is involved in motor activity (Johnson and Stretton, 1980). Although cholinesterase is responsible for the breakdown of acetylcholine released from excitatory motorneurons, it is also secreted in the external environment by *A. suum* and may act on the host intestine to maintain a favourable environment for the parasite.

Cholinesterase Antagonists as Anthelmintics

Anticholinesterases potentiate the electrophysiological effects of acetylcholine in *Ascaris* (Del Castillo *et al.*, 1963). A large number of organophosphorous anticholinesterases have been tested in *A. suum* by Knowles and Casida (1966) who found that many of these compounds were toxic to the parasite and were inhibitors of *A. suum* acetylcholinesterase. The organophosphates that were not active against *Ascaris* were not active against their acetylcholinesterase.

A number of organophosphorous anticholinesterases, originally introduced as insecticides, are now used for their anthelmintic properties. They include: metriphonate (trichlorfon), dichlorvos, napthalophos, crufomate and haloxon (Roberson, 1988). The use of these compounds against nematodes shows that the cholinesterases of nematodes are different from the mammalian hosts and that they are effective target sites for anthelmintics. The role of these drugs is generally restricted to the treatment of specific roundworm infestations since the organophosphorous anthelmin-

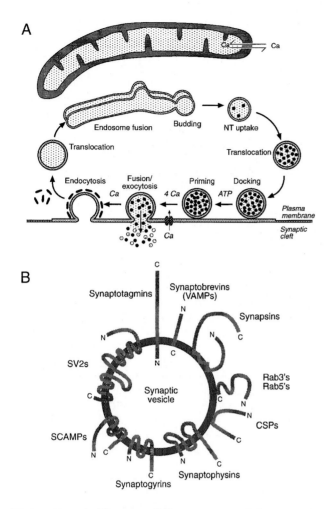

Figure 12.9. A: The nine different stages of the synaptic vesicle cycle. **B:** Proteins associated with the synaptic vesicle and required for the vesicle cycle. Additional proteins associated with the plasma membrane are also required. Diagrams modified from Sudhof (1995).

tics do not have a very broad spectrum. They have an advantage over most other anthelmintics (except ivermectin) in also controlling arthropod parasites.

Genetics of Resistance to Anticholinesterases can Involve Modification of the Synaptic Vesicle Cycle Proteins or an Acetylcholine Transporter

Figure 12.9A illustrates the synaptic vesicle cycle that governs the release of neurotransmitters, including acetylcholine. The synaptic vesicle cycle (Sudhof, 1995) is divided into 9 steps consisting of: 1) Docking (contact between the vesicle and active zone of the synaptic cleft); 2) Priming (a maturation process that makes vesicles competent for fast Ca-triggered membrane fusion); 3) Fusion/exocytosis produced by a Ca-spike; 4) Endocytosis (empty synaptic vesicles are rapidly internalized by clathrin coated pits); 5) Translocation (coated vesicle shed the clathrin, acidify and move to the interior); 6) Endosome fusion (recycling synaptic vesicles fuse with early endosomes and in *Ascaris* vesicles can be seen attached in circular arrangement to

'tubules'); 7) Budding (regeneration of synaptic vesicles); 8) Neurotransmitter uptake (transmitters are taken up into the vesicle by an active transporter); 9) Translocation (synaptic vesicles filled with transmitter move to the active zone).

A number of proteins present in the synaptic vesicle have to be present for the synaptic vesicle cycle to take place (Figure 12.9B). The proteins include synaptobrevins, synaptotagmins, SV2s, SCAMPs, synpatogyrins, synaptophysins, Rab3s, Rab5s, CSPs and synapsins. In addition synaptic plasma membrane proteins required for docking and priming that have been found include: Munc13s, neurexins, SNAP-25, syntaxins and proteins that associate with these proteins include, complexins, Munc18s, NSF, and $\alpha/\beta/\gamma$-SNAPs.

In *C. elegans* the gene *unc-17* codes for an acetylcholine transporter and a defective *UNC-17* gene mutant gives rise to anticholinesterase resistance (Sudhof, 1995). Similarly, the genes *unc-13* (which encodes Munc13), *unc-18* (which encodes Munc18s), and *syt-1* (which encodes synaptotagmin) in *C. elegans* also give rise to anticholinesterase drug resistance. The explanation for the resistance and the partial paralysis of the nematode relates to the impaired release of acetylcholine that reduces motility but makes the worm less sensitive to the effects of cholinesterase antagonism.

Pharmacology of the Acetylcholine Receptor

Low concentrations of nicotine produce contraction of body muscle strips (Toscano-Rico, 1927; Baldwin and Moyle, 1949; Natoff, 1969) like acetylcholine (Baldwin and Moyle, 1949; Natoff, 1969; Rozhova *et al.*, 1980; Grzywacz *et al.*, 1985; Onuaguluchi, 1989). The acetylcholine-induced contractions are blocked by tubocurarine but not by atropine (Baldwin and Moyle, 1949; Natoff, 1969; Rozhova *et al.*, 1980) so that the *A. suum* acetylcholine receptors have some of the pharmacological properties of vertebrate nicotinic receptors. Experiments on 'cut worm' preparations of the smaller nematode *C. elegans* following mutation led to a similar conclusion but also suggested the presence of two types of acetylcholine receptor, one of which was resistant to levamisole (Lewis *et al.*, 1980). Nicotinic-like receptors have also been observed in the sheep intestinal nematode, *H. contortus* (Atchison *et al.*, 1992), and the filarial nematode, *Dipetalonema vitiae* (Christ and Stillson, 1992).

Natoff (1969) and Rozhova *et al.*, (1980) reported that: (i) the potent ganglionic nicotinic agonist dimethylphenylpiperazinium is more potent than acetylcholine in *Ascaris*; (ii) the potent ganglionic nicotinic antagonist mecamylamine is the most potent acetylcholine antagonist in *Ascaris*, (more potent than tubocurarine); (iii) in contrast, hexamethonium, a potent ganglionic antagonist in vertebrates, had a low potency in *Ascaris*. The *Ascaris* acetylcholine receptors cannot, therefore, be classified as either ganglionic or neuromuscular and can be regarded as a separate sub-type of nicotinic receptor (e.g. the AChR$_N$).

Electrophysiological techniques have also been used to examine effects of cholinergic agonists and antagonists on muscle (Aubry *et al.*, 1970; Aceves *et al.*, 1970; Martin, 1982; Harrow and Gration, 1985; Colquhoun *et al.*, 1991; Shinozaki *et al.*, 1992). Levamisole and pyrantel are more potent agonists at the *Ascaris* muscle acetylcholine receptor than at vertebrate nicotinic receptors, where they have only weak nicotinic actions (Aubry *et al.*, 1970; Eyre, 1970). The selective action of these drugs allows them to be used as effective anthelmintics, killing the nematode parasite but not the host. The degree of selectivity will obviously affect the safety and efficacy of any nicotinic drug selected for therapeutic purposes.

In an extensive study of the pharmacology of muscle cholinoceptors in *Ascaris*, using membrane potential and input conductance responses (Colquhoun *et al.*, 1991; Colquhoun *et al.*, 1993), it was found that pilocarpine and muscarine caused a slight hyperpolarisation (up to 3 mV) rather than a depolarisation. These observations raise the possibility that in addition to the nicotinic-like receptors, muscle cells may possess muscarinic-like receptors for acetylcholine. The existence of a second type of muscle acetylcholine receptor (Martin and Valkanov, 1996b) may explain why some nematodes recover following nicotinic anthelmintic treatment (Aceves *et al.*, 1970; Coles *et al.*, 1975), and why nicotinic receptors are not essential for the survival of different strains of the nematode *C. elegans* (Lewis *et al.*, 1980; Fleming *et al.*, 1996; Fleming *et al.*, 1997).

Nicotinic Anthelmintics

The anthelmintics that act as agonists at nicotinic acetylcholine receptors of nematodes include the imidazothiazoles (levamisole and butamisole), the tetrahydropyrimidines (pyrantel, morantel and oxantel), the quaternary ammonium salts (bephenium and thenium) and the pyridines (methyridine) (Broome and Greenhall, 1961; Van den Bossche, 1985; Harrow and Gration, 1985; Roberson, 1988).

Aceves *et al.*, (1970) used intracellular recording techniques on *A. suum* muscle to observe effects of bath-applied tetramisole (a D,L racemic mixture of which levamisole is the laevo-isomer). Tetramisole produced depolarisation, an increase in spike frequency and contraction. Aubry *et al.*, (1970) described the antinematodal drug action and pharmacological properties of pyrantel, and some of its analogues, in a variety of vertebrate and helminth preparations. They showed that bath-application of pyrantel and acetylcholine produced depolarisation, and an increase in spike frequency and contracture. The effects of acetylcholine and pyrantel on *Ascaris* muscle strips were blocked by tubocurarine, suggesting that they were nicotinic. A depolarising effect of levamisole, pyrantel and morantel in *H. contortus* has also been described (Atchison *et al.*, 1992).

The two-microelectrode current-clamp and two-microelectrode voltage-clamp techniques (Martin, 1982; Harrow and Gration, 1985) have been used to

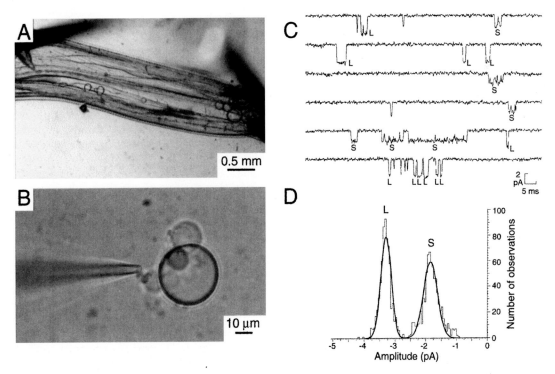

Figure 12.10. A: Muscle-flap preparation of *Oesophagostomum dentatum* showing the production of muscle membrane vesicles following collagenase treatment. **B:** A single membrane vesicle and an approaching patch-pipette viewed under phase contrast. **C:** Single channels currents activated by 10 μM levamisole in a cell-attached patch at –75 mV. Note the presence of two types of channel: the larger openings (L) and the smaller openings (S). **D:** The open channel-current histogram showing clear separation of the two channel openings (L and S).

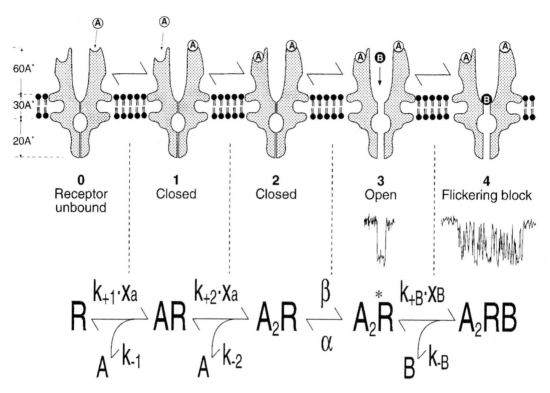

Figure 12.11. Top: Cartoon of a levamisole gated ion-channel and its opening as a result of the binding of two molecules of agonist. Two agonist molecules (A) combine sequentially with non-equivalent binding sites on the extracellular surface of the channel and permit the opening of the channel. When the channel is open a brief (a few milliseconds) current pulse of a few picoamps flows in through the channel. A large inorganic cation (B) may enter but not pass through the channel pore and produce a 'flickering' channel block as it repeatedly binds and unbinds with a block site deep in the pore.

Kinetic diagram of the same process with rate constants. The receptor (R) is in the unbound state (0) that may bind one molecule of agonist to one with a binding rate constant $K_{+1}Xa$ where Xa is the agonist concentration and the unbinding rate is K_{-1}; and the single agonist molecule state (1) is AR. A second agonist molecule may bind with a rate constant $K_{-2} Xa$ and unbind with a rate K_{-2} to the state A_2R (2) which is still closed. The double bound receptor may then open with a rate constant of β and close again with a rate constant α by returning from the open state (3) which conducts a rectangular shaped current pulse. The open channel may also become blocked if channel blocking molecule enters the channel with a rate constant of K_{+B}. X_B and an unblocking rate of K_{-B}.

examine the effects of acetylcholine, levamisole, pyrantel and morantel applied to the bag region of *A. suum* muscle. Simultaneous ionophoretic application of acetylcholine and pyrantel showed that both agonists acted on the same receptor. The relative potencies of the anthelmintics and acetylcholine are: morantel = pyrantel > levamisole = acetylcholine.

Single-Channel Currents Activated by Nicotinic Anthelmintics

Levamisole-activated channel currents have been recorded from muscle vesicle preparations (Figure 12.10C) (Martin *et al.*, 1997). The ion-channels are cation selective and have similar kinetic properties to acetylcholine-activated channels at low levamisole concentrations. There are subtypes of nicotinic receptor channel that have mean open-times in the range 0.8–2.85 msec with conductances in the range 20–45 pS. At higher levamisole concentrations a 'flickering' open channel-block is observed (Figure 12.11). The characteristic voltage-sensitive flickering seen with levamisole and other open channel-blockers (Neher and Steinbach,

1978) clearly demonstrates that anthelmintics may also act as open-channel blockers in addition to acting as agonists. In addition, levamisole, at higher concentrations (30 and 90 μm) also produced long closed-times separating clusters of openings which were interpreted as desensitisation. Single-channel currents activated by pyrantel and morantel have also been observed in vesicle preparations from *A. suum* (Robertson *et al.*, 1994; Evans and Martin, 1996).

Possible Structure of Nematode nACh Receptors

The most detailed structural studies on nicotinic receptor channels have been carried out on the nAChR derived from the *Torpedo* electric organ. It was found that the nAChR has a pentameric structure with five protein subunits (2α, 1β, 1γ and 1δ) arranged around the central ion-pore like the staves of a barrel (Figures 12.11, 12.12) (Changeaux *et al.*, 1996; Unwin, 1995). Each protein subunit is some 500 amino-acids in length with the glycosylated N-terminal forming an extracellular loop (Figure 12.12A). The α subunit, but not the β (non-α) subunits, contains two adjacent cysteine amino acids in this extracellular loop at a location referred to as 192 and 193 and is believed to contain part of the agonist binding site. Each subunit has 4 lipophillic regions (M1, M2, M3 and M4) which form α-helices that span the membrane. The M2 spanning unit forms the lining of the ion-pore with charged amino acids present so that the channel selectively binds and carries either cations or anions depending on the nature of the charged amino acids. A narrowing of the ion-channel about two-thirds of the way through the ion-channel pore restricts the movement of cations greater than 6.5 Å through the channel. Large organic ions may try and enter the channel, becoming stuck and producing a voltage-sensitive channel block (Figure 12.11).

Although the stoichiometry of vertebrate muscle nAChRs is known to be fixed, the stoichiometry of vertebrate neuronal nicotinic receptors may not be fixed (McGehee and Role, 1995). Neuronal receptors are assumed to be similar pentameric structures arising from the co-expression of 2α- and 3β-subunits but multiple stoichiometries (5α, 4α:1β, 3α:2β, 2α:3β, 1α:4β) or different arrangements around the ion-channel pore (e.g. $\alpha\alpha\beta\beta\beta$ vs. $\alpha\beta\alpha\beta\beta$) are a possibility (McGehee and Role, 1995). Families of genes for neuronal α- and β-subunits exist. The number and type of the different subunits in the neuronal nAChRs gives rise to changes in the sensitivity of the receptor to the agonist (Covernton et al., 1996), changes in the amount of calcium that can pass through the pore, and changes in the rate of desensitisation (McGehee and Role, 1995). The synthesis of nicotinic receptors is illustrated in Figures 12.13 and 12.14. Ribosomes 'read' the RNA that is translated into the subunit protein and inserted into the endoplasmic reticular membrane. Chaperone proteins (calnexin) appear to be involved in the folding of the subunit protein (Gehle et al., 1997) but the details of the mechanisms that control the stoichiometry of the nicotinic receptor are not known. There is also post-translational modification with glycosylation and phosphorylation by tryosine kinases, cAMP-dependent kinases and protein kinase C. Evidence for the glycosylation and phosphorylation of the nematode subunits is based on the presence of consensus sequences in the protein subunits (Fleming et al., 1997).

Phosphorylation of the different subunits is likely to affect opening, desensitisation and synthesis of the new nAChRs (Xie et al., 1997; Eilers et al., 1997; Khiroug et al., 1998; Hopf and Hoch, 1998). Regulatory or consensus phosphorylation sites for protein kinase C (PKC), protein kinase A (PKA) and tyrosine kinase (TK) are recognised in C. elegans. On UNC-38, there is 1 PKC site and 1 PKA site; on UNC-29 there are 3 PKC sites, one PKA site and one TK site; on LEV-1 there is only one PKC site (Figure 12.14). In Onchocerca volvulus, cDNA of a nAChR_n subunit (Ajuh et al., 1994) has identified regulatory phosphorylation sites at four places: i) three PKC sites (at Ser66, Thr289 and Ser347) and ii) one PKA site (at Arg258). Analogous phosphorylation sites are also present on nAChR_n subunits of other parasitic nematodes: HCA1 in Haemonchus contortus (Hoekstra et al., 1997), TAR1 in Trichostrongylus colubriformis (Wiley

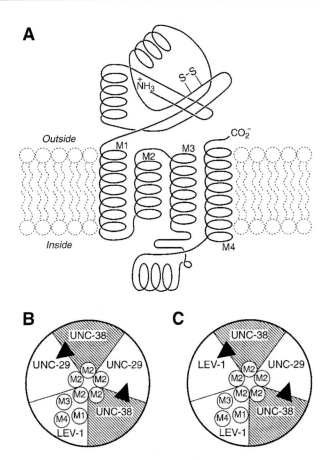

Figure 12.12. A: Diagram of an α-subunit of a nAChR showing the long extracellular N-terminal region with the cysteine bonds, the four transmembrane regions: M1, M2, M3, M4 and the cytoplasmic loop between M3 and M4. **B:** Diagram of a possible arrangement of 5 subunits comprising 2α subunits (unc-38) and 3 non-α subunits (2 unc-29 and 1 lev-1) forming the nAChR in C. elegans. The arrangement of the four transmembrane regions is such that the lining of the pore of the channel is formed by the M2 region. The binding sites of the agonist (▼) cross the interface between the α-subunit and the adjacent subunit. The two sites are equivalent in **B** but not in **C**. Thus the subunit composition will affect the binding affinity of the agonist.

et al., 1996) and ASR1 in Ascaris suum (Holden-Dye et al., Personal Communication). Phosphorylation of the levamisole receptor protein then, is one of the most effective means the parasite could have of being able to regulate and modulate the number and activity of the receptors; and so phosphorylation is likely to play a role in the development of anthelmintic resistance. The level of phosphorylation is a balance between the opposing activities of protein kinases that phosphorylate the protein and the phosphatases that dephosphorylate it. If phosphorylation enhances channel opening of levamisole receptors, phosphatases should decrease activation and depress responses to nicotinic anthelmintics.

Regulation of nicotinic receptors may also involve signalling peptides received from synaptic inputs. ARIA,

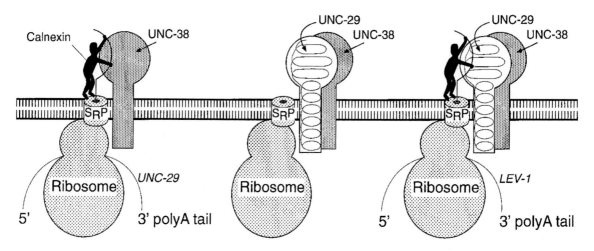

Figure 12.13. Diagram of the synthesis of the nAChR levamisole receptor. The ribosome translates the *unc-29* RNA into the UNC-29 protein with the protein being shaped and guided by the calnexin protein and passed through the membrane 4 times through the sarcoplasmic reticulum pore (SRP). Some controlling factors, possibly easy adhesion, brings the UNC-29 protein into contact with an appropriate adjacent subunit (here UNC-38). Progressively the complete nAChR receptor protein is built up. See also **Figure 12.14.**

released from motorneurons in vertebrate preparations is known (Pun *et al.*, 1997; Si *et al.*, 1997) to activate a tyrosine kinase, erbB, which leads to modification of the subunit structure of the nicotinic receptor and its biophysical properties. Analogous signalling peptides may be involved in the regulation of nematode nicotinic receptors on muscle (Figure 12.14). It is possible the level of activity of levamisole receptors, through changes in phosphorylation, is adjusted in levamisole resistant parasites. Thus, changes in amino-acid structure of the levamisole receptor may not be associated with the development of resistance.

The Presence of nAChR Subunit Genes in Nematodes Predicts Heterogeneity of Receptors

Eleven genes associated with levamisole resistance have been recognised in *C. elegans* (Fleming *et al.*, 1996). Three of these genes encoding nAChR subunits associated with strong levamisole resistance have been described in this model soil-inhabiting nematode: *unc-38*, *unc-29* (both on chromosome 1) and *lev-1* (on chromosome IV). *Unc-38* encodes an α subunit while *unc-29* and *lev-1* encode β subunits. It remains a possibility that there are additional genes (e.g. *acr-2* and *acr-3* on the X chromosome) encoding nAChR subunits that are associated with a weaker levamisole resistance.

Expression of *unc-38*, *unc-29* and *lev-1* subunits together in *Xenopus* oocytes results in levamisole-inducible currents but *unc-38* expressed alone does not produce levamisole activated currents (Fleming *et al.*, 1996). These observations suggest that the levamisole receptor cannot comprise a homooligomer of UNC-38 subunits and that channels are formed from a heterologous subunit combination, perhaps with a combination of all three subunits. Four other genes encoding subunits for the nAChR in *C. elegans* have been identified: the non-α subunits, *acr-2* and *acr-3* (on chromosome X); and the α-subunits *deg-3* and *ce21* (on

chromosome V: (Treinen and Chalfie, 1995; Ballivet *et al.*, 1996)).

Figures 12.12B and 12.12C illustrate two of the many possible arrangements of the subunits UNC-38, UNC-29, and LEV-1. The two UNC-38 subunits (the α-subunit) are fixed in position but the positions of the other subunits (the β-subunits) can vary. If a functional channel could be produced with 1–5 UNC-38 subunits, then the number of possible levamisole receptors produced by different subunit arrangements would be 3^4 (81). These receptors might be differentiated by their biophysical or pharmacological properties with the conductances being determined by the subunits making up the channel pore and the agonist binding sites being determined by the interfaces between the unc-38 subunit and adjacent units.

Heterogeneity of *O. dentatum* nAChRs

In *Oesophagostomum dentatum* the muscle nAChRs have conductances and mean open-times that show significant variation between patches and sometimes more than one conductance of nAChR is observed in one patch recording (Figures 12.10C, 12.12D) (Martin *et al.*, 1997). The variation in conductance is far wider than expected from experimental errors or open-channel noise and is greater than that produced by vertebrate muscle nAChRs (Garner *et al.*, 1984). Such variation has been used as evidence of heterogeneity of the structure and pharmacology of vertebrate neuronal nAChRs. Figure 12.15 shows a histogram of the conductances of levamisole receptors characterised by the presence of 4 peaks and a skewed distribution. We interpreted these data after fitting the sum of Gaussian distributions as indicating the occurrence of 4 different subtypes of levamisole receptor: *G25*, *G35*, *G40* and *G45*. nAChR channel currents from *Ascaris suum* with different conductances, which range between 19–50 pS and with distinguishable peaks at 24 pS and 42 pS, have

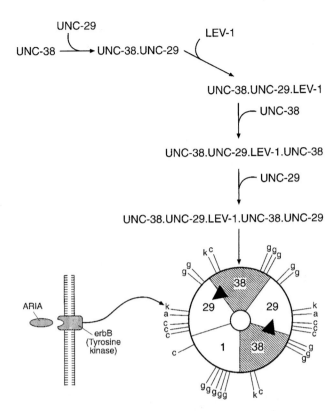

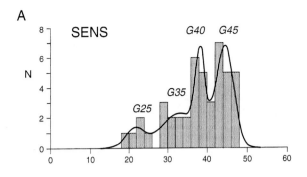

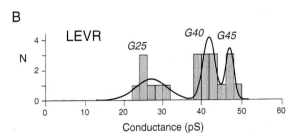

Figure 12.14. Flow diagram of a possible sequence of events whereby the 5 subunits of the nAChR levamisole receptor are added to produce the final pentameric receptor unit. This process is likely to be regulated by a phosphorylation of the receptor subunits. The UNC-38 subunit has a consensus site for protein kinase C (c) and tyrosine kinase (k). The tyrosine kinase (erbB) protein in the sarcolemma membranes may be regulated by the ARIA peptide. The UNC-29 subunit contains 3 consensus sites for protein kinase C (c), one site for protein kinase A (a) and one for tyrosine kinase (k). The lev-1 subunit contains one consensus site for protein kinase c (C). Glycosylation (g) sites are also present. Regulation or modulation of receptor activity, opening and desensitisation is likely through the physiologically regulated phosphorylation process.

Figure 12.15. A: Frequency histograms of single-channel conductances for levamisole sensitive *Oesophagostomum dentatum* parasites. Gaussian curves were fitted to each distribution using the maximum likelihood procedure. The peaks for the sensitive isolate were 21.4 ± 2.3 pS (8% area) labelled *G25*; 33.0 ± 4.8 pS (31% area) labelled *G35*; 38.1 ± 1.2 pS (19% area) labelled *G40*; and 44.3 ± 2.2 pS (42% area) labelled *G45*. The histogram shows evidence of levamisole receptor heterogeneity. **B:** The distribution of conductances observed in levamisole-resistant isolates. Note the loss of the G35 subtype.

been observed (Pennington and Martin, 1990; Robertson *et al.*, 1994) but clearer separation of subtypes was more difficult, perhaps because of variations in the source of the *Ascaris* that were derived from field infections. The function of the different subtypes of nAChR remains to be evaluated but because the pharmacology of each subtype is predicted to be slightly different we can predict changes in the proportion of receptor subtypes associated with the development of resistance to nicotinic agonists.

Recently we have been able to compare the levamisole subtypes in levamisole-sensitive and levamisole-resistant isolates of *Oesophagostomum dentatum* (Figure 12.15) (Robertson *et al.*, 1999). We found that there was a loss of the *G35* subtype with the development of resistance and that resistance was associated with a reduction in the mean proportion of time the nAChR channel spent in the open-state. We also found that the number of active receptors was less in the levamisole-resistant isolate. Quantitatively the resistance was

explained by the combination of the reduction in the number of active channels and the reduction in the average time that the channel was open. The molecular explanation for these observations remains to be determined but may involve changes in phosphorylation state of the nAChR receptor as suggested earlier.

GABA

Effects of GABA on *Ascaris* Muscle

Del Castillo *et al.* (1964b,c) demonstrated the hyperpolarising effect of bath-applied piperazine and GABA on muscle. GABA receptors are located at the syncytium (Del Castillo *et al.*, 1964b) and extrasynaptically over the surface of the rest of the cell, including the bag region (Martin, 1980). Ionophoresis of GABA onto the bag, when examined under voltage-clamp, produces a current which is explained by the opening of ion-channels permeable to Cl. The physiological function of the extrasynaptic GABA receptors is not known, however, they may be exploited in pharmacological experiments designed to look at receptor properties; they may also be activated by anthelmintics during therapy of the host animal.

The agonist profile of the *Ascaris* GABA receptor is similar to, but not identical with, that of the vertebrate GABA receptor, but the antagonist profile is very different (Martin, 1980; Wann, 1987; Holden-Dye *et al.*, 1988; Holden-Dye *et al.*, 1989; Duittoz and Martin,

1991a; Duittoz and Martin, 1991b; Martin, 1993; Martin *et al.*, 1995).

The most potent GABA$_A$ agonists are potent in *Ascaris*, so that the relative potency of GABA$_A$ agonist correlate with the relative potency of agonists at the *Ascaris* GABA receptor (r = 0.74; Holden-Dye, *et al.*, 1989). However the potency of GABA sulphonic acid derivatives (3APS and P4S) as well as muscimol and isoguvacine is less than at the vertebrate receptor (Holden-Dye *et al.*, 1988; Holden-Dye *et al.*, 1989; Table 2 Martin 1993).

In contrast, the GABA$_A$ antagonists: bicuculline, picrotoxin, securinine, pitrazepine, RU5135, are weak or inactive at the *Ascaris* muscle GABA receptor. The potency of a series of arylaminopyridazine-GABA derivatives, which act as competitive GABA$_A$ antagonists, are competitive *Ascaris* muscle GABA receptor antagonists, but the potency series differ (Duittoz and Martin, 1990a,b; 1991). A series of novel arylaminopyridazine derivatives has been synthesised and tested in *Ascaris* where it was found that NCS 281–93 was the most potent competitive GABA antagonist in *Ascaris* (Martin *et al.*, 1995).

We may summarise the pharmacology of the GABA receptor, derived from observations on *Ascaris*, by concluding that there is a separate subtype of GABA receptor present in nematodes. We may emphasise the distinctive pharmacological properties of this receptor by referring to it as the GABA$_N$ receptor (N: not antagonised by picrotoxin; N for nematodes).

Genetics of GABAergic Transmission

In *C. elegans*, 6 genes: *lin-15, unc-25, unc-30, unc-43, unc-47, unc-49*, are required for GABAergic neurons to function (Mcintire, Jorgensen and Horvitz 1993; Mcintire *et al.*, 1993; Mcintire *et al.*, 1997). *Unc-49* genes function postsynaptically and are necessary for an inhibitory effect on body muscles. A cDNA HG1 that has a high homology to the cDNA for vertebrate GABA$_A$ receptors has been recovered from *Haemonchus contortus* (Laughton *et al.*, 1994).

GABA and Piperazine Single-Channel Currents

GABA- and piperazine-activated channels have been recorded using cell-attached and outside-out patches (Martin, 1985a). The channels opened by both agonists had a mainstate conductance of 22 pS but two subconductance states were also observed. The average duration of the effective openings (bursts) produced by GABA was in the region of 32 msec while the average duration of effective openings (bursts) produced by the anthelmintic piperazine was 14 msec (Figure 12.16). High concentrations of GABA or piperazine produced desensitization, which was seen as a decrease in open probability due to the appearance of long closed-times. Piperazine is ~100 times less potent than GABA in *A. suum* (Martin, 1985b). This difference in potency may be explained by the fact that piperazine requires a higher concentration to achieve a similar opening rate to GABA, and in addition, the average duration of openings produced by piperazine is shorter.

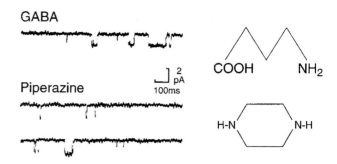

Figure 12.16. GABA and piperazine single channel currents recorded from cell attached patches of *Ascaris* bag muscle membrane. Trans-patch potential, –75 mV. GABA channel mean open time: 32 ms. Piperazine channel open-time 18 ms. GABA 3 μM in the pipette. Piperazine 500 μM in the pipette.

Glutamate

Shinozaki and Konishi (1970) first pointed out that a number of anthelmintics, including kainate, have a chemical structure similar to the excitatory amino acid glutamate. They also demonstrated the excitatory effect kainate can have on vertebrate neurons. Kainate, domoic acid and quisqualate have been used as anthelmintics in Asia for a long time and are now usually used to study vertebrate glutamate receptors. However, little is known of their action in nematodes. Davis (1998a,b) found that DE2 motorneurons show depolarising potentials and conductance changes in response to application of glutamate agonists (domoate>kainate>>glutamate>aspartate) which is similar to vertebrate kainate receptors. In the DI motorneurons, glutamate produced a hyperpolarising response that may have been mediated via an intervening inhibitory neuron. These observations indicate the presence of excitatory glutamate receptors in nematodes. The cloning of an AMPA-like glutamate receptor gene *glr-1* (Maricq *et al.*, 1995; Hart *et al.*, 1995) from *C. elegans* also indicates the presence of excitatory glutamate receptors. Mutants of *glr-1* are defective in mechanosensory behaviour and are sluggish in movement. In addition to an experiment demonstrating the presence of excitatory glutamate receptors, Davis (1998a) showed that there is also an electrogenic glutamate transporter in the epidermis. Interestingly, it was suggested that this epidermal transporter may serve as a buffer inactivating glutamate synapses from all parts of the nervous system. There are parallels with the presence of epidermal cholinesterase.

The Avermectins and GluCl Channels

The avermectins (and related milbemycins) are a group of very hydrophobic macrocyclic lactones that have broad spectrum anthelmintic (Campbell and Benz, 1984) and insecticidal properties, and are derived from *Streptomyces* micro-organisms. The group includes: ivermectin, a mixture of 80% 22,23 dihydroavermectin Bla and 20% 22,23 dihydroavermectin Blb derived from *Streptomyces avermitilis*; milbemycin D derived from

S. hygroscopocus; moxidectin derived from *S. aureolacri-mosus noncyanogenus*; and doramectin. In a number of preparations, avermectins produce a reduction in motor activity so that the parasites are excluded from the host. This immobilisation may be explained by reduced excitability of muscle or nerve produced by an increase in membrane Cl conductances.

The GluCl channels in nematodes were first recognised by expression of a glutamate-activated chloride current, sensitive to avermectins, in *Xenopus* oocytes injected with *C. elegans* RNA (Arena *et al.*, 1991; Arena *et al.*, 1992). Expression cloning then led to the recognition of two *C. elegans* channel subunits, GluClα on chromosome V and GluClβ on chromosome I that, when expressed together in *Xenopus* oocytes, produced functional ion-channels (Cully *et al.*, 1994). Subsequently, a PCR-based approach has lead to the recognition of additional GluCl related subunits from *C. elegans*, GluClX on chromosome I, and C27H5.5 on chromosome II (Cully *et al.*, 1996). Other GluCl subunits have been recognised from an ivermectin resistant *C. elegans* mutant, *avr-15*: *avr-15* encodes two alternatively spliced channel subunits present in pharyngeal muscle (Dent *et al.*, 1997). The *avr-15* encoded subunit can form a homomeric channel that is ivermectin-sensitive and glutamate-gated. The location or site of expression of the original GluClα remains to be determined. However the location of the GluClβ subunit was studied by expression of a *lacZ* reporter gene and was found distributed on the pm4 pharyngeal muscle cells of *C. elegans* (Laughton *et al.*, 1997).

The GluCl subunits are each approximately 500 amino acids in length. The presence of N-terminal cysteines in all the presently recognised GluCl subunits and hydrophobicity analysis suggests a similar motif common to all cys-loop ligand-gated channels: a big extracellular N-terminal domain carrying the ligand binding site attached to four membrane spanning α-helices with a long cytoplasmic loop between M3 and M4. Interestingly, the cytoplasmic loop has a protein kinase C phosphorylation site present in the GluClα subunit (Cully *et al.*, 1994) that may be involved in receptor desensitisation. It is assumed that 5 GluCl subunits come together, as do the subunits of the nAChRs, to produce the GluCl ion-channel, but the stoichiometric arrangement has not yet been determined. It is known that GluClα1 and GluClβ subunits may form homomeric channels as well as heteromeric ion-channels when expressed in *Xenopus* oocytes (Cully *et al.*, 1994), but the functions, locations and structures of the native/endogenous GluCl ion-channels in nematodes remain to be determined. The genetic and electrophysiological evidence suggests that in *C. elegans*, and other nematodes, a heterogeneous family of avermectin-sensitive glutamate-gated chloride channels may be involved in movement, pharyngeal pumping and egg laying (Cully *et al.*, 1996).

Avermectin sensitive sites in *Ascaris suum* have been identified on pharyngeal muscle. Avermectins produce hyperpolarisation and an increase in Cl conductance when either bath-applied or pressure-ejected onto the pharyngeal preparation and usually the response is irreversible. Figure 12.5D illustrates an experiment demonstrating the effects of glutamate and ivermectin. Glutamate and ivermectin have similar effects on expressed GluCl receptors (Cully *et al.*, 1994) and on *Ascaris suum* pharyngeal muscle where the effects were recognised with milbemycin D (Martin, 1996).

Another interesting aspect of the *Ascaris suum* pharyngeal muscle receptors is the response to aspartate as well as to glutamate. The response to aspartate is, however, biphasic with an initial depolarising response followed by the anticipated hyperpolarising response (Murray and Martin, 1997; Martin *et al.*, 1998). The initial depolarising response appears to be due to the gating of a separate aspartate-gated channel because in a number of pharyngeal preparations of *Ascaris suum*, the hyperpolarising response to glutamate and to aspartate was absent, so that only the depolarising response to aspartate was present.

The split chamber technique described by Kass *et al.*, (1980; 1982; 1984) has permitted the selective application of avermectin to the dorsal and ventral halves of *Ascaris*. It was found that avermectin blocked the DEI response to indirect stimulation but not direct stimulation: the observations were interpreted as suggesting that avermectins block transmission between interneurons (VI) in the ventral cord and the excitatory motorneurons. It was also found by these authors that the hyperpolarising response of muscle following direct stimulation of VI was blocked by avermectin, an observation consistent with the antagonism of muscle GABA receptors described above. These observations, together with the observations referred to above, show that avermectins have more than one site of action and that these include ion-channels in muscle membrane and in neuronal membranes.

A fluorescent derivative of ivermectin (4'–5, 7 dimethyl bodipy proprionylivermectin) has been prepared and injected into adult *Ascaris* where it produces dose-dependent immobilisation (Martin *et al.*, 1992a; Martin and Kusel, 1992). Fluorescent microscopy of frozen sections has revealed the distribution of the probe in the whole nematode. The probe accumulates in muscle membranes and within the nerve cord; these two sites are consistent with an action of avermectins on muscle and nerve membrane. It also accumulates under the epidermis, a site consistent with removal perhaps by a P-glycoprotein transporter (Martin and Kusel, 1992; Xu *et al.*, 1998).

Serotonin Effects and Pharmacology

Serotonin (5-HT) may play a role in regulating muscle metabolism. 5-HT can be absorbed from the host or synthesised by *Ascaris* (Martin *et al.*, 1988; Rao *et al.*, 1991); 5-HT receptors in muscle and intestine of *Ascaris* have been demonstrated using radiolabelled binding techniques (Chaudhuri and Donahue, 1989). 5-HT has also been shown to be present in two neurons, which may be neurosecretory, in the pharynx as well as cells of the posterior ventral nerve cord in males (Stretton and Johnson, 1985). 5-HT has also been shown to play an important part in egg laying in *C. elegans* (Queyroy and Verdetti, 1992).

5-HT has no observable effect on membrane potential of most somatic muscle cells of *A. suum* but appears to regulate glycogen metabolism. It does, however, produce effects on movement of the head and affects membrane potential in ventral oblique muscles of the tail of male *Ascaris* (Johnson *et al.*, 1996; Reinitz and Stretton, 1996). Injection into the body produces rapid paralysis. 5-HT reduces the amplitude of acetylcholine induced contraction. In *Ascaris*, glycogen metabolism depends on glycogen synthetase for glycogen synthesis from glucose-6-phosphate, and glycogen phosphorylase for glycogenolysis. The receptor binding studies of Chaudhuri and Donahue (1989) suggest that in *A. suum* the muscle receptor is similar to the mammalian 5-HT, receptor: ketanserin, a selective 5-HT, antagonist, that has a high affinity (K = 16.7 nm) for the muscle receptor but not the *A. suum* intestinal receptor.

Peptides

FMRFamide-like peptides have been found in the pharyngeal nerves and the four major nerve cords of *A. suum* using immunocytochemical staining (Davenport *et al.*, 1988; Cowden *et al.*, 1987; Sithigorngul *et al.*, 1989; Sithigorngul *et al.*, 1990; Sithigorngul *et al.*, 1991). A variety of antisera known to recognise FMRFamide from invertebrates have been screened and the location of some of these peptides in *A. suum* neurons has been possible.

Anti-FMRFamide antibodies recognised the largest subset of neurons and stained over half of the neurons in *A. suum*. A series of fractionations on an acid-methanol extract of around 5000 *A. suum* heads were made and assayed (Cowden *et al.*, 1989). After the final fractionation step, a single immunoreactive peptide was characterised as Lys-Asn-Glu-Phe-Ile-Arg-Phe (KNEFIRFamide). The peptide was sequenced and called AF1 (A for *Ascaris*, F for FMRFamide, and 1 for the first purified). It has been shown to be biologically active and to increase the input conductance of inhibitory motorneurons (Cowden *et al.*, 1989).

The second peptide isolated from *Ascaris* was AF2 (Cowden and Stretton, 1990) which has the amino acid sequence Lys-His-Glu-Tyr-Leu-Arg-Phe (KHEYLRFamide). In addition to these FMRFa-related peptides (FaRPs) in *Ascaris*, Smart *et al.*, (1992a) described the primary structure of a novel peptide TE-6 and the chromatographic and immunological characterisation of neuropeptide Y-like and pancreatic polypeptide-like peptides (Smart *et al.*, 1992b).

Once the sequence of the peptides is known it is possible to synthesise the peptides and to assay them for biological activity. Injection of 0.1 ml of 10^{-6} M AFI or AF2 into the anterior region of *Ascaris* blocks locomotory movement in the injection site (Stretton *et al.*, 1991). The effects on muscle-strip activity were multiple: they included transient muscle relaxation,

and rhythmic activity comprising multiple muscle contraction and relaxation events.

Intracellular records of membrane potential made from the commissural axons of motorneurons showed that effects of AF1 were to abolish slow oscillatory potentials at low concentrations (Cowden *et al.*, 1989); these effects reversed on washing. Experiments in which two intracellular micropipettes were placed in the commissural axons of motorneurons to measure input conductance changes showed that AF1 produced a large increase in input conductance in both dorsal and ventral inhibitory motorneurons. No effect was observed on one of the excitatory neurons (DEI). *Ascaris* motorneurons are non-spiking neurons which rely on their high membrane resistance to propagate electrical signals over long distances (Davis and Stretton, 1989b): an increase in input conductance reduces the space constant and removes the neuronal locomotory circuit.

Evidence of FaRPs peptides has also been reported in *C. elegans* (Rosoff *et al.*, 1992; Rosoff *et al.*, 1993; Li *et al.*, 1993; Schinkmann and Li, 1994; Nelson *et al.*, 1998b; Nelson *et al.*, 1998a; Marks *et al.*, 1998) and in *Panagrellus redivivus* (Geary *et al.*, 1992) which are free-living nematodes. Li *et al.*, (1993) cloned and sequenced a gene, flp-1, that encodes eight putative neuropeptides sequences; seven share the C-terminal sequence PNFLRFamide, including SADPNFLRFamide, SDPNFLRFamide, SQPNFLRFamide, AAADPNFLRFamide, PNFLRFamide ASGDPNFRFamide and AGSDPNFLRFamide. Geary *et al.*, (1992) have described the primary structure of two FaRPs from *Panagrellus redivivus*: PF-1 is SDPNLFLRFamide and PF2 is SADPNFLRFamide. None of these peptides is very like AF1 or AF2 in function. More FaRPs have since been isolated from nematodes (see Maule *et al.*, 1996; Davis and Stretton, 1996; Brownlee *et al.*, 1996). Furthermore, the presence of the gene *afp-1* in *A. suum* (Edison *et al.*, 1997) predicts 6 peptides sharing the-PGVLRFamide c-terminal sequence.

The function of these peptides remains to be studied in greater detail, many have effects on the neuromuscular system and affect motility at low concentrations. The successful discovery of non-peptidyl ligands for some vertebrate receptors encourages the view that FaRPs receptors may be useful target sites for future anthelmintics that have a selective action on the neuromuscular system of nematode parasites.

Nitric Oxide

Evidence that the gas, nitric oxide, is involved as a second messenger/transmitter and in neuromuscular control in nematodes has been described.

The electrophysiological effects of the FMRFamide-related neuropeptides, SDPNFLRFamide (PF1) and SADPNFLRFamide (PF2), have been examined (Bowman *et al.*, 1995) using flap preparations of *Ascaris suum*. PF1 and PF2 hyperpolarises muscle membrane and produces flaccid paralysis, independent of external Cl, in both innervated and denervated preparations. PF1 reverses spastic contractions induced by levamisole, an

Table 12.1. Selected nematode FMRFamide related peptides (FaRPs) revealing the diversity in endogenous neuropeptides

FaRP		Occurrence	References
AF1	KNEFIRF.NH$_2$	A	1
AF2	KHEYLRF.NH$_2$	A, C, H, P	1
AF3	AVPGVLRF.NH$_2$	A	1
AF4	GDVPGVLRF.NH$_2$	A	1
AF5	SGKPTFIRF.NH$_2$	A	1
AF6	FIRF.NH$_2$	A	1
AF7	AGPRFIRF.NH$_2$	A	1
AF8/PF3	KSAYMRF.NH$_2$	A, P	1
AF9	GLGPRPLRF.NH$_2$	A	1
AF10	GFGDEMSMPGVLRF.NH$_2$	A	1
AF11	SDIGISEPNFLRF.NH$_2$	A	1
AF12	FGDEMSMPGVLRF.NH$_2$	A	1
AF13	SDMPGVLRF.NH$_2$	A	2
AF14	SMPGVLRF.NH$_2$	A	2
AF15	AQTFVRF.NH$_2$	A	2
AF16	ILMRF.NH$_2$	A	2
AF17	FDRDFMHF.NH$_2$	A	2
AF??	???PNFLRF.NH$_2$	A	2
AF19	AEGLSSLPLIRF.NH$_2$	A	2
CF1/PF1	SDPNFLRF.NH$_2$	C, P	1
CF2/PF2	SADPNFLRF.NH$_2$	C, P	1
CF3	SQPNFLRF.NH$_2$	C	1
CF4	ASGDPNFLRF.NH$_2$	C	1
CF5	AAADPNFLRF.NH$_2$	C	4
CF6	PNFLRF.NH$_2$	C	1
CF7	AGSDPNFLRF.NH$_2$	C	4
	APEASPFIRF.NH$_2$	C	3
	KPSFVRF.NH$_2$	C	1
PF4	KPNFIRF.NH$_2$	C, P	1

The amino acid sequence is shown in single letter notation. A: *Ascaris suum*; C: *Caenorhabditis elegans*; H: *Haemonchus contortus*; P: *Panagrellus redivivus*.
References are as follows: 1: (Maule *et al.*, 1996); 2: (Davis and Stretton, 1996); 3: (Marks *et al.*, 1998); 4: (Nelson *et al.*, 1998a). The compounds were originally named with a letter to denote the species from which the peptide was isolated, F for FaRP, and a sequential number; thus AF1 was the first FaRP found in A. suum. As more peptides were discovered, often due to the use of different solvents for extraction (Cowden and Stretton, 1990), some were found to exist in more than one nematode species. For example, AF8 is the same as PF3; PF1 is the same as CF1; PF2 is the same as CF2.

effect blocked by pre-treatment with agents that interfere with nitric oxide (NO) synthesis (N-nitro-L-arginine), whereas sodium nitroprusside, which releases NO in solution, mimics PF1 and PF2. NO synthase activity, monitored by the conversion of [H^{-3}]arginine to [H^{-3}]citrulline, is twice as abundant in *Ascaris suum* epidermis as in muscle. These results suggest that the inhibitory effects of PF1 and PF2 on nematode somatic muscle are mediated by NO, and that the epidermis serves a role in this process analogous to that of the endothelium in vertebrate vasculature.

The distribution of NADPH diaphorase, the NO-producing enzyme, has been examined histochemically in neurons in the nervous system of *Ascaris suum* (Bascal, 1995). Positive staining was seen in the central nervous system, in selective cell bodies and fibres in the ventral ganglion, the retrovesicular ganglion, ventral and dorsal cords. Intense staining was also present in the motorneuron commissures, indicating a potential role for NO as a neurotransmitter at the neuromuscular junction. NADPH diaphorase-positive neurons were not confined to the central nervous system. Staining was also seen in the enteric nervous system, in particular the pharynx and in the peripheral nervous system innervating the sensory organs.

Other Possible Neurotransmitters

Other putative transmitters have been suggested to occur in nematodes. Dopamine has been detected biochemically in *C. elegans* (Sulston, Dew and Brenner, 1975) and the plant-parasitic nematode *Aphelenchus avenae* (Wright and Awan, 1978) and is present in eight sensory neurons in the female *Ascaris suum* (Sulston, Dew and Brenner, 1975). Catecholamine, probably dopamine, has been detected histochemically in some free-living and animal-parasitic nematodes (Sharpe and Atkinson, 1980; Sharpe *et al.*, 1980; Lee and Ko, 1991). Other possible transmitters include: adrenaline and noradrenaline (Willet, 1980), octopamine (Horvitz *et al.*, 1982) and histamine.

Acknowledgements

We are pleased to acknowledge the financial support of the Wellcome Trust. We are also particularly grateful to Professor David Halton and Dr Aaron Maule (Queen's University Belfast, Northern Ireland) and Professor Robert Walker and Dr Lindy Holden-Dye (Southampton University, UK) for reading the manuscript and for making helpful comments.

References

Aceves, J., Erliji, D. and Martinez-Marnon, R. (1970) The mechanism of the paralysing action of tetramisole on *Ascaris* somatic muscle. *British Journal of Pharmacology*, **38**, 602–607.

Ajuh, P.M., Cowell, P., Davey, M.J. and Shevde, S. (1994) Cloning of a cDNA encoding a putative nicotinic acetylcholine receptor subunit of the human filarial parasite *Onchocerca volvulus*. *Gene*, **144**, 127–129.

Albertson, D.G. and Thomson, J.N. (1976) The pharynx of *Caenorhabditis elegans*. *Philosophical Transactions of the Royal Society London, Series B*, **275**, 299–325.

Angstadt, J.D., Donmoyer, J.E. and Stretton, A.O.W. (1989) Retrovesicular ganglion of the nematode *Ascaris*. *Journal of Nematode Physiology*, **284**, 374–388.

Angstadt, J.D. and Stretton, A.O.W. (1989) Slow active potentials in ventral inhibitory motor neurons of the nematode *Ascaris*. *Journal of Comparative Physiology A: Sensory Neural and Behavioral Physiology*, **166**, 165–177.

Arena, J.P., Liu, K.K., Paress, P.S. and Cully, D.F. (1991) Avermectin-sensitive chloride currents induced by *Caenorhabditis-elegans* RNA in *Xenopus* oocytes. *Molecular Pharmacology*, **40**, 368–374.

Arena, J.P., Liu, K.K., Paress, P.S., Schaeffer, J.M. and Cully, D.F. (1992) Expression of a glutamate-activated chloride current in *Xenopus*-oocytes injected with *Caenorhabditis elegans* RNA:-evidence for modulation by avermectin. *Molecular Brain Research*, **15**, 339–348.

Atchison, W.D., Geary, T.G., Manning, B., Vandewaa, E.A. and Thompson, D.P. (1992) Comparative neuromuscular blocking actions of levamisole and pyrantel-type anthelmintics on rat and gastrointestinal nematode somatic muscle. *Toxicology and Applied Pharmacology*, **112**, 133–143.

Aubry, M.L., Cowell, P., Davey, M.J. and Shevde, S. (1970) Aspects of the pharmacology of new anthelminitics: pyrantel. *British Journal of Pharmacology*, **38**, 332–344.

Baldwin, E. and Moyle, V. (1949) A contribution to the physiology and pharmacology of *Ascaris lumbricoides* from the pig. *British Journal of Pharmacology*, **4**, 145–152.

Ballivet, M., Alliod, C., Bertrand, S. and Bertrand, D. (1996) Nicotinic acetylcholine receptors in the nematode *Caenorhabditis elegans*. *Journal of Molecular Biology*, **258**, 261–269.

Bargmann, C.I. and Kaplan, J.M. (1998) Signal transduction in the *Caenorhabditis elegans* nervous system. *Annual Reviews of Neuroscience*, **21**, 279–308.

Bascal, Z.A., Montgomery, A., Holden-Dye, L., Williams, R.G., Walker, R.J. (1995) Histochemical mapping of nadph diaphorase in the nervous-system of the parasitic nematode *Ascaris suum*. *Parasitology*, **110**, 625–637.

Blair, K.L., Barsuhn, C.L., Day, J.S., Ho, N.F.H., Geary, T.G. and Thompson, D.P. (1998a) Biophysical model for organic acid excretion in *Ascaris suum*. *Molecular and Biochemical Parasitology*, **93**, 179–190.

Blair, K.L., Geary, T.G., Mensch, S.K., Ho, N.F.H. and Thompson, D.P. (1998b) Biophysical characterization of a large conductance anion channel in hypodermal membranes of the gastrointestinal nematode, *Ascaris suum*. *Journal of Membrane Biology*, (In Press).

Blaxter, M.L., Deley, P., Garey, J.R., Liu, L.X., Scheldeman, P., Vierstraete, A., Vanfleteren, J.R., Machey, L.Y., Dorris, M., Frisse, L.M., Vida, J.T. and Thomas, W.K. (1998) A molecular evolutionary framework for the phylum Nematoda. *Nature*, **392**, 71–75.

Bowman, J.W., Winterowd, C.A., Friedman, A.R., Thompson, D.P., Klein, R.D., DAVIS, J.P., Maule, A.G., Blair, K.L., Geary, T.G. (1995) Nitric-oxide mediates the inhibitory effects of SDPNFLRFamide, a nematode FMRFamide-related peptide, in *Ascaris suum*. *Journal of Neurophysiology*, **74**, 1880–1888.

Brading, A.F. and Caldwell, P.C. (1971) The resting membrane potential of the somatic muscle cells of *Ascaris lumbricoides*. *Journal of Physiology (London)*, **217**, 605–624.

Broome, A.W.J. and Greenhall, G.H. (1961) A new anthelmintic with unusual properties. *Nature*, **189**, 59–60.

Bueding, E. (1952) Acetylcholinesterase in *Schistosoma mansoni*. *British Journal of Pharmacology*, **7**, 563–566.

Byerly, L. and Masuda, M.O. (1979) Voltage-clamp analysis of the potassium current that produces a negative-going action potential in *Ascaris* muscle. *Journal of Physiology (London)*, **288**, 263–284.

Caldwell, P.C. and Ellory, J.C. (1968) Ion movement in the somatic muscle cells of *Ascaris lumbricoides*. *Journal of Physiology (London)*, **197**, 75–76P.

Campbell, W.C. and Benz, G.W. (1984) Ivermectin: A review of efficacy and safety. *Journal of Veterinary Pharmacology and Therapeutics*, **7**, 1–16.

Cappe de Baillon, P. (1911) Etude sur les fibres musculaires d' *Ascaris*. I. Fibres pariétales. *Cellule*, **27**, 165–211.

Changeaux, J.-P., Devillers-Thiery, A. and Chemouilli, P. (1996) Acetylcholine receptor: An allosteric protein. *Science*, **225**, 1335–1345.

Chaudhuri, J. and Donahue, M.J. (1989) Serotonin receptors in the tissues of adult *Ascaris suum*. *Molecular and Biochemical Parasitology*, **35**, 191–198.

Christ, D. and Stillson, T. (1992) Effects of calcium-channel blockers on the contractility of the filariid *Acanthocheilonema viteae*. *Parasitology Research*, **78**, 489–494.

Coles, G.C., East, J.M. and Jenkins, S.N. (1975) The mechanism of action of the anthelmintic levamisole. *General Pharmacology*, **6**, 309–313.

Colquhoun, L., Holden-Dye, L. and Walker, R.J. (1991) The pharmacology of cholinoceptors on the somatic muscle-cells of the parasitic nematode *Ascaris suum*. *Journal of Experimental Biology*, **158**, 509–530.

Colquhoun, L., Holden-Dye, L. and Walker, R.J. (1993) The action of nicotinic receptor specific toxins on the somatic muscle cells of the parasitic nematode *Ascaris suum*. *Molecular Neuropharmacology*, **3**, 11–16.

Covernton, P.J.O., Kojima, H., Sivilotti, L.G., Gibb, A.J. and Colquhoun, D. (1996) Comparison of neuronal nicotinic receptors in rat sympathetic neurones with subunit pairs expressed in *Xenopus* oocytes. *Journal of Physiology (London)*, **481**, 27–34.

Cowden, C., Sithigorngul, P., Guastella, J. and Stretton, A. (1987) FMRF-amide-like peptides in *Ascaris suum*. *American Zoologist*, **27**, A 127.

Cowden, C., Stretton, A.O.W. and Davis, R.E. (1989) AF1, a sequenced bioactive neuropeptide isolated from the nematode *Ascaris suum*. *Neuron*, **2**, 1465–1473.

Cowden, C. and Stretton, A.O.W. (1990) AF2 a nematode neuropeptide. *Society for Neuroscience Abstracts*, **16**, 305.

Crofton, H.D. (1971) Form, Function and Behaviour. In *Plant Parasitic Neurotodes*. edited by B. Zuckerman, W. Mai and R. Rohds, pp. 83–113. New York: Academic Press.

Cully, D.F., Vassilatis, D.K., Liu, K.K., Paress, P.S., Vanderploeg, L.H.T. and Schaeffer, J.M. (1994) Cloning of an avermectin-sensitive glutamate-gated chloride channel from *Caenorhabditis elegans*. *Nature*, **371**, 707–711.

Cully, D.F., Wilkinson, H., Vassilitis, D.K., Etter, A. and Arena, J.P. (1996) Molecular biology and electrophysiology of glutamate-gated chloride channels of invertebrates. *Parasitology*, **114**, S191–S200.

Davenport, T.R.B., Lee, D.L. and Isaac, R.E. (1988) Immunocytochemical demonstration of a neuropeptide in *Ascaris suum* (Nematoda) using antiserum to FMRF-amide. *Parasitology*, **97**, 81–88.

Davenport, T.R.B., Eaves, L.A., Hayes, T.K., Lee, D.L. and Isaac, R.E. (1994) The detection of akh/hrth-like peptides in *Ascaridia galli* and *Ascaris suum* using an insect hyperglycemic bioassay. *Parasitology*, **108**, 479–485.

Davis, R.E. (1998a) Action of excitatory amino acids on hypodermis and the motornervous system of *Ascaris suum*: pharmacological evidence for a glutamate transporter. *Parasitology*, **116**, 487–500.

Davis, R.E. (1998b) Neurophysiology of glutamatergic signaling and anthelmintic action in *Ascaris suum*: pharmacological evidence for a kainate receptor. *Parasitology*, **116**, 471–486.

Davis, R.E. and Stretton, A.O.W. (1996) The motornervous system of *Ascaris*: electrophysiology and anatomy of the neurons and their control by neuromodulators. *Parasitology*, **113**, S199–S118.

Davis, R.E. and Stretton, A.O.W. (1989a) Signaling properties of *Ascaris* motorneurons — graded active responses, graded synaptic transmission, and tonic transmitter release. *Journal of Neuroscience*, **9**, 415–425.

Davis, R.E. and Stretton, A.O.W. (1989b) Passive membrane-properties of motorneurons and their role in long-distance signaling in the nematode *Ascaris*. *Journal of Neuroscience*, **9**, 403–414.

De Bell, J.T. (1965) A long look at neuromuscular junctions in nematodes. *Quarterly Reviews of Biology*, **40**, 233–251.

De Bell, J.T., Del Castillo, J. and Sanchez, V. (1963) Electrophysiology of the somatic muscle cells of *Ascaris lumbricoides*. *Journal of Cellular Comparative Physiology*, **62**, 159–177.

Del Castillo, J., De Mello, W.C. and Morales, T. (1963) The physiological role of acetylcholine in the neuromuscular system of *Ascaris lumbricoides*. *Archives International Physiologie et Biochimie*, **71**, 741–757.

Del Castillo, J., De Mello, W.C. and Morales, T. (1964a) Influence of some ions on the membrane potential of *Ascaris* muscle. *Journal of General Physiology*, **48**, 129–140.

Del Castillo, J., De Mello, W.C. and Morales, T. (1964b) Mechanism of the paralysing action of piperazine on *Ascaris* muscle. *British Journal of Pharmacology*, **22**, 463–477.

Del Castillo, J., De Mello, W.C. and Morales, T. (1964c) Inhibitory action of γ-aminobutyric acid (GABA) on *Ascaris* muscle. *Experientia*, **20**, 141–143.

Del Castillo, J. and Morales, T. (1967a) Extracellular action potentials recorded from the interior of the giant esophageal cell of *Ascaris lumbricoides*. *Journal of General Physiology*, **50**, 631–645.

Del Castillo, J. and Morales, T. (1967b) The electrical and mechanical activity of the esophageal cell of *Ascaris lumbricoides*. *Journal of General Physiology*, **50**, 603–630.

Dent, J.A., Davis, M.W. and Avery, L. (1997) Avr-15 encodes a chloride channel subunit that mediates inhibitory glutamatergic neurotransmission and ivermectin sensitivity in *Caenorhabditis elegans*. *Embo Journal*, **16**, 5867–5879.

Dixon, D.M., Valkanov, M. and Martin, R.J. (1993) A patch-clamp study of the ionic selectivity of the large conductance, ca-activated chloride channel in muscle vesicles prepared from *Ascaris suum*. *Journal of Membrane Biology*, **131**, 143–149.

Duittoz, A.H. and Martin, R.J. (1991a) Antagonist properties of arylaminopyridazine GABA derivatives at the *Ascaris* muscle GABA receptor. *Journal of Experimental Biology*, **159**, 149–164.

Duittoz, A.H. and Martin, R.J. (1991b) Effects of the arylaminopyridazine-GABA derivatives, sr95103 and sr95531 on the *Ascaris* muscle GABA receptor: the relative potency of the antagonists in *Ascaris* is different to that at vertebrate GABAa receptors. *Comparative Biochemistry and Physiology C: Pharmacology Toxicology & Endocrinology*, **98**, 417–422.

Edison, A.S., Messinger, L.A. and Stretton, A.O.W. (1997) Afp-1: a gene encoding multiple transcripts of a new class of FMRFamide-like neuropeptides in the nematode *Ascaris suum*. *Peptides*, **18**, 929–935.

Eilers, H., Schaeffer, E., Bickler, P.E. and Forsayeth, J.R. (1997) Functional deactivation of the major neuronal nicotinic receptor caused by nicotine and a protein kinase c-dependent mechanism. *Molecular Pharmacology*, **52**, 1105–1112.

Evans, A.M. and Martin, R.J. (1996) Activation and cooperative multi-ion block of single nicotinic-acetylcholine channel currents of *Ascaris* muscle by the tetrahydropyrimidine anthelmintic, morantel. *British Journal of Pharmacology*, **118**, 1127–1140.

Eyre, P. (1970) Some pharmacodynamic effects of the nematodes: methyridine, tetramisole and pyrantel. *Journal of Pharmacy and Pharmacology*, **22**, 26–36.

Fleming, J.T., Baylis, H.A., Satelle, D.B. and Lewis, J.A. (1996). Molecular cloning and *in vitro* expression of *C. elegans* and parasitic nematode ionotropic receptors. *Parasitology*, **113**, S175–S190.

Fleming, J.T., Squire, M.D., Barnes, T.M., Tornoe, C., Matsuda, K., Ahnn, J., Fire, A., Sulston, J.E., Barnard, E.A., Sattelle, D.B. and Lewis, J.A. (1997) *Caenorhabditis elegans* levamisole resistance genes lev-1, unc-29, and unc-38 encode functional nicotinic acetylcholine receptor subunits. *Journal of Neuroscience*, **17**, 5843–5857.

GarciaAnoveros, J.A., Liu, J.D. and Corey, D.P. (1998) The nematode degenerin UNC-105 forms ion-channels that are activated by degeneration- or hypercontraction-causing mutations. *Neuron*, **20**, 1231–1241.

Garner, P., Ogden, D.C. and Colquhoun, D. (1984) Conductances of single ion channels opened by nicotinic agonists are indistinguishable. *Nature*, **309**, 160–162.

Geary, T.G., Price, D.A., Bowman, J.W., Winterrowd, C.A., Mackenzie, C.D., Garrison, R.D., Williams, J.F. and Friedman, A.R. (1992) 2 FMRFamide-like peptides from the free-living nematode *Panagrellus redivivus*. *Peptides*, **13**, 209–214.

Gehle, V.M., Walcott, E.C., Nishizaki, T. and Sumikawa, K. (1997) N-glycosylation at the conserved sites ensures the expression of properly folded functional ACh receptors. *Molecular Brain Research*, **45**, 219–229.

Goldschmidt, R. (1904) Der Chormidialapparat Lebhaft Funktionierender Gewebszellen. *Zoologie Journal*, **21**, 41–140.

Goldschmidt, R. (1908) Das Nervensystem von Ascaris lumbricoides und Megalocephala. Ein Versuch, in den Aufbau eines einfachen Nervensystems einzudringan, Zweiter Teil. *Zeitschrift für wissenschaftliche Zoologie*, **90**, 73–136.

Goldschmidt, R. (1909) Das Nervensystem von Ascaris lumbricoides und Megalocephala. Ein Versuch, in den Aufbau eines einfachen Nervensystems einzudringen, Zweiter Teil. *Zeitschrift für wissenschaftliche Zoologie*, **92**, 396–357.

Grzywacz, M., Szkudlinsks, J. and Zandarowska, E. (1985) Pharmacological receptors of *Ascaris lumbricoides suis* L. *Wiadomosci Parazytologi*, **31**, 153–161.

Guastella, J., Johnson, C.D. and Stretton, A.O.W. (1991) GABA-immunoreactive neurons in the nematode *Ascaris*. *Journal of Comparative Neurology*, **307**, 584–597.

Guastella, J. and Stretton, A.O.W. (1991) Distribution of H^{-3} GABA uptake sites in the nematode *Ascaris*. *Journal Of Comparative Neurology*, **307**, 598–608.

Harris, J.E. and Crofton, H.D. (1957) Structure and function in the nematodes. Internal pressure and cuticular structure in *Ascaris*. *Journal of Experimental Biology*, **34**, 116–130.

Harrow, I.D. and Gration, K.A.F. (1985) Mode of action of the anthelmintics morantel, pyrantel and levamisole on muscle-cell membrane of the nematode *Ascaris-suum*. *Pesticide Science*, **16**, 662–672.

Hart, A., Sims, S. and Kaplan, J. (1995) A synaptic code for sensory modalities revealed by analysis of the *C. elegans* GLR-1 glutamate receptor. *Nature*, **378**, 82–85.

Hesse, R. (1892) Über das Nervensystem con *Ascaris lumbricoides* und *Ascaris megalocephala*. *Zeitschrift für wissenschaftliche Zoologie*, **90**, 73–136.

Hobson, A.D., Stephenson, W. and Beadle, L.C. (1952a) Studies on the physiology of *Ascaris lumbricoides*. I. The relation of total osmotic pressure, conductivity and chloride content of the body fluid to that of the external environment. *Journal of Experimental Biology*, **29**, 1–21.

Hobson, A.D., Stephenson, W. and Eden, A. (1952b) Studies on the physiology of *Ascaris* lumbricoides. II. The inorganic composition of the body fluid in relation to that of the environment. *Journal of Experimental Biology*, **29**, 22–29.

Hoekstra, R., Visser, A., Wiley, L.J., Weiss, A.S., Sangster, N.C. and Roos, M.H. (1997) Characterization of an acetylcholine receptor gene of *Haemonchus contortus* in relation to levamisole resistance. *Molecular and Biochemical Parasitology*, **84**, 179–187.

Holden-Dye, L., Hewitt, G.M., Wann, K.T., Krogsgaard-Larsen, P. and Walker, R.J. (1988) Studies involving avermectin and the 4-aminobutyric acid (GABA) receptor *of Ascaris summ* muscle. *Pesticide Science*, **24**, 231–245.

Holden-Dye, L., Krogsgaard-Larsen, P., Neilsen, L. and Walker, R.J. (1989) GABA receptors on the somatic muscle cells of the parasitic nematode, *Ascaris suum*: stereoselectivity indicates similarity to a GABA-type agonist recognition site. *British Journal of Pharmacology*, **98**, 841–850.

Hopf, C. and Hoch, W. (1998) Dimerization of the muscle-specific kinase induces tyrosine phosphorylation of acetylcholine receptors and their aggregation on the surface of myotubes. *Journal of Biological Chemistry*, **273**, 6467–6473.

Horvitz, R.H., Chalfie, M., Trent, C., Sulston, J.E. and Evans, P.D. (1982) Serotonin and octopamine in the nematode *Caenorhabditis elegans*. *Science*, **206**, 1012–1014.

Jarman, M. (1959) Electrical activity in the muscle cells of *Ascaris lumbricoides*. *Nature*, **184**, 1244.

Jarman, M. (1970) A muscle-domain hypothesis for *Ascaris* postural waves. *Parasitology*, **61**, 475–489.

Johnson, C.D., Reinitz, C.A., Sithigorngul, P. and Stretton, A.O.W. (1996) Neuronal localization of serotonin in the nematode *Ascaris suum*. *Journal of Comparative Neurology*, **367**, 352–360.

Johnson, C.D. and Stretton, A.O.W. (1980) Neural control of locomotion in *Ascaris*: anatomy, electrophysiology and biochemistry. In *Nematodes as Biological Models*, Volume 1, edited by B.M. Zuckerman, pp. 159–195. New York: Academic Press.

Johnson, C.D. and Stretton, A.O.W. (1985) Localization of cholineacetyltransferase within identified motoneurons of the nematode *Ascaris*. *Journal of Neuroscience*, **5**, 1984–1992.

Johnson, C.D. and Stretton, A.O.W. (1987) GABA-immunoreactivity in inhibitory motor neurons of the nematode *Ascaris*. *Journal of Neuroscience*, **7**, 223–235.

Kass, I.S., Wang, C.C., Waldrond, J.P. and Stretton, A.O.W. (1980) Avermectin B1a, a paralysing anthelmintic that affects interneurons and inhibitory motoneurons in *Ascaris*. *Proceedings of the National Academy of Sciences*, **77**, 6211–6215.

Kass, I.S., Larsen, D.A., Wang, C.C. and Stretton, A.O. (1982) *Ascaris suum*: differential effects of avermectin B1a on the intact animal and neuromuscular strip preparations. *Experimental Parasitology*, **54**, 166–174.

Kass, I.S., Stretton, A.O.W. and Wang, C.C. (1984) The effects of avermectin and drugs related to acetylcholine and 4-aminobutyric acid on neurotransmission in *Ascaris suum*. *Molecular and Biochemical Parasitology*, **13**, 213–225.

Khiroug, L., Sokolova, E., Giniatullin, R., Afzalov, R. and Nistri, A. (1998) Recovery from desensitization of neuronal nicotinic acetylcholine receptors of rat chromaffin cells is modulated by intracellular calcium through distinct second messengers. *Journal of Neuroscience*, **18**, 2458–2466.

Knowles, C.O. and Casida, J.E. (1966) Mode of action of organophosphates and anthelmintics. Cholinesterase inhibition in *Ascaris lumbricoides*. *Journal of Agricultural Food Chemistry*, **14**, 566.

Laughton, D.L., Amar, M., Thomas, P., Towner, P., Harris, P., Lunt, G.G. and Wolstenholme, A.J. (1994) Cloning of a putative inhibitory amino-acid receptor subunit from the parasitic nematode *Haemonchus contortus*. *Receptors & Channels*, **2**, 155–163.

Laughton, D.L., Lunt, G.G. and Wolstenholme, A.J. (1997) Reporter gene constructs suggest that the *Caenorhabditis elegans* avermectin receptor beta-subunit is expressed solely in the pharynx. *Journal of Experimental Biology*, **200**, 1509–1514.

Lee, D.L. (1962) The distribution of esterase enzymes in *Ascaris lumbricoides*. *Parasitology*, **52**, 241–260.

Lee, D.L. and Ko, R.C. (1991) Catecholaminergic neurons in *Trichinella spiralis* (Nematoda). *Parasitology Research*, **77**, 269–270.

Lewis, J.A., Wu, C.-H., Levine, J.H. and BERG, H. (1980) Levamisole-resistant mutants of the nematode *Caenorhabditis elegans* appear to lack pharmacological acetylcholine receptors. *Neuroscience*, **5**, 967–989.

Li, C., Rosoff, M. and Schinkmann, K. (1993) FMRFamide-like neuropeptides in *Caenorhabditis elegans*. *Journal of Cellular Biochemistry*, **S17C**, 98.

Mapes, C.J. (1966) Structure and function in the nematode. III. The pharyngeal pump of *Ascaris lumbricoides*. *Parasitology*, **56**, 137–149.

Maricq, A.V., Peckol, E., Driscoll, M. and Bargmann, C.I. (1995) Mechanosensory signaling in *C. elegans* mediated by the GLR-1 glutamate receptor. *Nature*, **378**, 78–81.

Marks, N.J., Maule, A.G., Geary, T.G., Thompson, D.P., Li, C., Halton, D.W. and Shaw, C. (1998) KSAYMRFamide (PF3/AF8) is present in the free-living nematode, *Caenorhabditis elegans*. *Biochemical and Biophysical Research Communications*, **248**, 422–425.

Martin, R.E., Chaudhuri, J. and Donahue, M.J. (1988) Serotonin (5-hydroxytryptamine) turnover in adult female *Ascaris suum* tissue. *Comparative Biochemistry and Physiology C–Comparative Pharmacology And Toxicology*, **91**, 307–310.

Martin, R.J. (1980) The effect of γ-aminobutyric acid in the input conductance and membrane potential of *Ascaris* muscle. *British Journal of Pharmacology*, **71**, 99–106.

Martin, R.J. (1982) Electrophysiological effects of piperazine and diethyl-carbamazine on *Ascaris suum* somatic muscle. *British Journal of Pharmacology*, **77**, 255–265.

Martin, R.J. (1985a) γ-aminobutyric acid- and piperazine-activated single channel currents from *Ascaris suum* body muscle. *British Journal of Pharmacology*, **84**, 445–461.

Martin, R.J. (1985b) Gamma-aminobutyric acid-activated and piperazine-activated single-channel currents from *Ascaris suum* body muscle. *British Journal of Pharmacology*, **84**, 445–461.

Martin, R.J., Kusel, J.R., Robertson, S.J., Minta, A. and Haugland, R.P. (1992a) Distribution of a fluorescent ivermectin probe, bodipy ivermectin, in tissues of the nematode parasite *Ascaris suum*. *Parasitology Research*, **78**, 341–348.

Martin, R.J., Thorn, P., Gration, K.A.F. and Harrow, I.D. (1992b) Voltage-activated currents in somatic muscle of the nematode parasite *Ascaris suum*. *Journal of Experimental Biology*, **173**, 75–90.

Martin, R.J. (1993) Neuromuscular transmission in nematodal parasites and antinematodal drug action. *Pharmacology and Therapeutics*, **58**, 13–50.

Martin, R.J., Sitamze, J.M., Duittoz, A.H. and Wermuth, C.G. (1995) Novel arylaminopyridazine-GABA receptor antagonists examined electrophysiologically in *Ascaris suum*. *European Journal of Pharmacology*, **276**, 9–19.

Martin, R.J. (1996) An electrophysiological preparation of *Ascaris suum* pharyngeal muscle reveals a glutamate-gated chloride channel sensitive to the avermectin analogue milbemycin D. *Parasitology*, **147**, 247–252.

Martin, R.J., Robertson, A.P., Bjorn, H. and Sangster, N.C. (1997) Heterogeneous levamisole receptors: a single-channels study of nicotinic acetylcholine receptors from *Oesophagostomum dentatum*. *European Journal of Pharmacology*, **322**, 249–257.

Martin, R.J., Murray, I., Robertson, A.P., Bjorn, H. and Sangster, N.C. (1998) Anthelmintics and ion-channels: After a puncture use a patch. *International Journal of Parasitology*, **28**, 849–862.

Martin, R.J. and Kusel, J.R. (1992) On the distribution of a fluorescent ivermectin probe (4" 5,7 dimethyl-bodipy proprionylivermectin) in *Ascaris* membranes. *Parasitology*, **104**, 549–555.

Martin, R.J., Sitamze, J-M., Duittoz, A.H. and Wermuth, C.G. (1995) Novel arylaminopyridazine-GABA receptor antagonists examined electrophysiologically in *Ascaris suum*. *European Journal of Pharmacology*, **276**, 9–19.

Martin, R.J. and Valkanov, M.A. (1996a) The Kostyuk technique for recording currents from isolated muscle bags of *Ascaris suum*. *Journal of Physiology (London)*, **491P**, 105–106P.

Martin, R.J. and Valkanov, M.A. (1996b) Effects of acetylcholine on a slow voltage-activated non-selective cation current mediated by non-nicotinic receptors on isolated *Ascaris* muscle bags. *Experimental Physiology*, **81**, 909–925.

Martin, R.J. and Valkanov, M.A. (1996c) Effects of acetylcholine on a slow voltage-activated non-selective cation-current mediated by non-nicotinic receptors on isolated *Ascaris* muscle bags. *Journal of Experimental Physiology*, **81**, 909–925.

Maule, A.G., Geary, T.G., Bowman, J.W., Shaw, C., Halton, D.W. and Thompson, D. P. (1996) The pharmacology of nematode FMRFamide-related peptides. *Parasitology Today*, **12**, 351–357.

McGehee, D.S. and Role, L.W. (1995) Physiological diversity of nictotinic acetylcholine receptors expressed by vertebrate neurons. *Annual Review of Physiology*, **57**, 521–546.

Mcintire, S.L., Jorgensen, E. and Horvitz, H.R. (1993) Genes required for GABA function in *Caenorhabditis elegans*. *Nature*, **364**, 334–337.

Mcintire, S.L., Jorgensen, E., Kaplan, J. and Horvitz, H.R. (1993) The GAGAergic nervous system of *Caenorhabditis elegans*. *Nature*, **364**, 337–341.

Mcintire, S.L., Reimer, R.J., Schuske, K., Edwards, R.H. and Jorgensen, E.M. (1997) Identification and characterization of the vesicular GABA transporter. *Nature*, **389**, 870–876.

Natoff, I.L. (1969) The pharmacology of the cholinoceptor in muscle preparations of *Ascaris lumbricoides* cat. *suum*. *British Journal of Pharmacology*, **37**, 251–257.

Neher, E. and Steinbach, J.H. (1978) Local anaesthetics transiently block currents through single acetylcholine receptor channels. *Journal of Physiology (London)*, **277**, 153–176.

Nelson, L.S., Kim, K.Y., Memmott, R.E. and Li, C. (1998a) FMRFamide-related gene family in the nematode *Caenorhabditis elegans*. *Molecular Brain Research*, **58**, 103–111.

Nelson, L.S., Rosoff, M.L. and Li, C. (1998b) Disruption of a neuropeptide gene, flp-1, causes multiple behavioral defects in *Caenorhabditis elegans*. *Science*, **281**, 1686–1690.

Norton, S. and De Beer, E.J. (1957) Investigations on the action of piperazine on *Ascaris lumbricoides*. *American Journal of Tropical Medicine*, **6**, 989–905.

Onuaguluchi, G. (1989) Some aspects of the pharmacology and physiology of the *Ascaris suum* muscle. *Archives Internationales de Pharmacodynamie et de Therapie*, **298**, 264–275.

Pennington, A.J. and Martin, R.J. (1990) A patch-clamp study of acetylcholine-activated ion channels in *Ascaris suum* muscle. *Journal of Experimental Biology*, **154**, 201–221.

Pun, S., Yang, J.F., Ng, Y.P. and Tsim, K.W.K. (1997) Ng108–15 cells express neuregulin that induces AChR alpha-subunit synthesis in cultured myotubes. *FEBS Letters*, **418**, 275–281.

Queyroy, A. and Verdetti, J. (1992) Cooperative gating of chloride channel subunits in endothelial cells. *Biochemical and Biophysical Acta Bio-Membrane*, **1108**, 159–168.

Rao, G.S.J., Cook, P.F. and Harris, B.G. (1991) Modification of the ATP inhibitory site of the *Ascaris suum* phosphofructokinase results in the stabilization of an inactive t-state. *Biochemistry*, **30**, 9998–10004.

Reger, J.F. (1966) The fine structure of fibular components and plasma membrane contract in esophageal myoepithelium of *Ascaris lumbricoides* (var. *suum*). *Journal of Ultrastructure Research*, **14**, 602–617.

Reinitz, C.A. and Stretton, A.O.W. (1996) Behavioral and cellular effects of serotonin on locomotion and male mating posture in *Ascaris suum* (nematoda). *Journal of Comparative Physiology A: Sensory Neural And Behavioral Physiology*, **178**, 655–667.

Roberson, E.L. (1988) Antinematodal Drugs. In *Veterinary Pharmacology and Therapeutics*, edited by H.B. Booth and L.E. McDonald, pp. 882–927. Ames: Iowa State University Press.

Robertson, A.P. Bjorn, H. and Martin, R.J. (1999) Resistance to levamisole resolved at the single channel level. *FASEB Journal*, **13**, 749–760.

Robertson, A.P. and Martin, R.J. (1996) Effects of pH on a high-conductance Ca-dependent chloride channel: a patch-clamp study in *Ascaris suum*. *Parasitology*, **113**, 191–198.

Robertson, S.J., Pennington, A.J., Evans, A.M. and Martin, R.J. (1994) The action of pyrantel as an agonist and an open-channel blocker at acetylcholine-receptors in isolated *Ascaris suum* muscle vesicles. *European Journal of Pharmacology*, **271**, 273–282.

Rosenbluth, J. (1965a) Ultrastructural organization of obliquely striated muscle fibers in *Ascaris lumbricoides*. *Journal of Cell Biology*, **25**, 495–515.

Rosenbluth, J. (1965b) Ultrastructure of somatic muscle cells in *Ascaris lumbricoides*. II. Intermuscular junctions, neuromuscular junctions, and glycogen stores. *Journal of Cell Biology*, **26**, 579–591.

Rosenbluth, J. (1967) Obliquely striated muscle III. Contraction mechanism of *Ascaris* body muscle. *Journal of Cell Biology*, **34**, 15–33.

Rosenbluth, J. (1969) Ultrastructure of dyads in muscle fibres of *Ascaris lumbricoides*. *Journal of Cell Biology*, **42**, 817–825.

Rosoff, M.L., Burglin, T.R. and Li, C. (1992) Alternatively spliced transcripts of the flp-1 gene encode distinct FMRFamide-like peptides in *Caenorhabditis elegans*. *Journal of Neuroscience*, **12**, 2356–2361.

Rosoff, M.L., Doble, K.E., Price, D.A. and Li, C. (1993) The flp-1 propeptide is processed into multiple, highly similar FMRFamide-like peptides in *Caenorhabditis elegans*. *Peptides*, **14**, 331–338.

Rozhova, E.K., Malyutina, T.A. and Shishov, B.A. (1980) Pharmacological characteristics if cholinoreception in somatic muscle of the nematode *Ascaris suum*. *General Pharmacology*, **11**, 141–146.

Sakmann, B., Patlak, J. and Neher, E. (1980) Single acetylcholine-activated channels show burst kinetics in presence of desensitizing concentrations of agonist. *Nature*, **286**, 71–73.

Saz, H.J. and Weil, A. (1962) A pathway of formation of α-methyl valerate by *Ascaris* lumbricoides. *Journal of Biological Chemistry*, **237**, 2053–2056.

Schinkmann, K. and Li, C. (1994) Comparison of 2 *Caenorhabditis* genes encoding FMRFamide(phe-met-arg-phe-nh2)-like peptides. *Molecular Brain Research*, **24**, 238–246.

Schneider, A. (1995) *Monograpie der Nematoden*. Berlin

Sharpe, M.J. and Atkinson, H.J. (1980) Improved visualizationof dopamineergic neurons in nematodes using the glyoxylic acid fluorescence method. *Journal of Zoology*, **190**, 273–284.

Sharpe, M.J., Atkinson, H.J., Trett, M.W. and Lee, D.L. (1980) Visualization of neurotransmitters in nematodes. In *The host-invader interplay*, edited by H. van den Bossche, pp. 713–716. Amsterdam: Elsevier/North Holland.

Shiozaki, H. and Konishi, S. (1970) Actions of several anthelmintics and insecticides on rat cortical neurons. *Brain Research*, **24**, 368–371.

Si, J.T., Miller, D.S. and Mei, L. (1997) Identification of an element required for acetylcholine receptor-inducing activity (aria)-induced expression of the acetylcholine receptor epsilon subunit gene. *Journal of Biological Chemistry*, **272**, 10367–10371.

Sithigorngul, P., Cowden, C., Guastella, J. and Stretton, A.O.W. (1989) Generation of monoclonal antibodies against a nematode peptide extract: another approach for identifying unknown neuropeptides. *Journal of Comparative Neurology*, **284**, 389–397.

Sithigorngul, P., Stretton, A.O.W. and Cowden, C. (1990) Neuropeptide diversity in *Ascaris*: an immunocytochemical study. *Journal of Comparative Neurology*, **294**, 362–376.

Sithigorngul, P., Stretton, A.O.W. and Cowden, C. (1991) A versatile dot-elisa method with fempto-mole sensitivity for detecting small peptides. *Journal of Immunological Methods*, **141**, 23–32.

Smart, D., Shaw, C., Curry, W.J., Johnston, C.F., Thim, L. and Halton, D.W. (1992a) Te-6: a novel neuropeptide from the nematode *Ascaris suum*. *Regulatory Peptides*, **40**, 252.

Smart, D., Shaw, C., Johnston, C.F., Halton, D.W., Fairweather, I. and Buchanan, K. D. (1992b) Chromatographic and immunological characterization of neuropeptide y-like and pancreatic polypeptide-like peptides from the nematode *Ascaris suum*. *Comparative Biochemistry and Physiology C: Pharmacology Toxicology & Endocrinology*, **102**, 477–481.

Stretton, A.O.W. (1976) Anatomy and development of the somatic musculature of the nematode of *Ascaris*. *Journal of Experimental Biology*, **64**, 773–788.

Stretton, A.O.W., Fishpool, R.M., Southgate, E., Donmoyer, J.E., Walrond, J.P., Moses, J.E.R. and Kass, I.S. (1978) Structure and physiological activity of the motoneurons of the nematode *Ascaris*. *Proceedings of the National Academy of Science, USA*, **75**, 3493–3497.

Stretton, A.O.W., Cowden, C., Sithigorngul, P. and Davis, R.E. (1991) Neuropeptides in the nematode *Ascaris suum*. *Parasitology*, **102**, S 107–S 116.

Stretton, A.O.W. and Johnson, C.D. (1985) GABA and 5HT immunoreactive neurons in *Ascaris*. *Society for Neuroscience Abstracts*, **11**, 626.

Sudhof, T.C. (1995) The synaptic vesicle cycle: a cascade of protein-protein interactions. *Nature*, **375**, 645–653.

Sulston, J., Dew, M. and Brenner, S. (1975) Dopamine neurons in the nematode *Caenorhabditis elegans*. *Journal of Comparative Neurology*, **163**, 215–244.

Tavemarakis, N. and Driscoll, M. (1997) Molecular modeling of mechanotransduction in the nematode *Caenorhabditis elegans*. *Annual Reviews of Physiology*, **59**, 659–689.

Thorn, P. and Martin, R.J. (1987) A high-conductance calcium-dependent chloride channel in *Ascaris suum* muscle. *Quarterly Journal of Experimental Biology*, **73**, 31–49.

Toscano Rico, J. (1927) Sur la sensitivitié de I *Ascaris* à l' action de quelques drogues. *Compte rendu des séances de la Société de Biologie, Paris*, **94**, 921–923.

Treinen, M. and Chalfie, M. (1995) A mutated acetylcholine receptor subunit causes neuronal degeneration in *C. elegans*. *Neuron*, **14**, 871–877.

Tsang, V.C. and Saz, H.J. (1973) Demonstration and function of 2-methylbutyrate racemase in *Ascaris lumbricoides*. *Journal of Comparative Biochemistry and Physiology*, **45B**, 617–623.

Unwin, N. (1995) Acetylcholine-receptor channel imaged in the open state. *Nature*, **373**, 37–43.

Valkanov, M., Martin, R.J. and Dixon, D.M. (1994) The ca-activated chloride channel of *Ascaris suum* conducts volatile fatty-acids produced by anaerobic respiration: a patch-clamp study. *Journal of Membrane Biology*, **138**, 133–141.

Valkanov, M.A. and Martin, R.J. (1995) A Cl channel in *Ascaris suum* selectivity conducts dicarboxylic anion products of glucose fermentation and suggests a role in removal of waste organic-anions. *Journal of Membrane Biology*, **148**, 41–49.

Van Den Bossche, H. (1985) Pharmacology of anthelmintics: Chemotherapy of gastrointestinal helminths. *Handbook of Experimental Parasitology*, **77**, 125–181.

Walker, R.J., Colquhoun, I. and Holden-Dye, L. (1992) Pharmacological profiles of the GABA and acetylcholine-receptors from the nematode, *Ascaris suum*. *Acta Biologica Hungarica*, **43**, 59–68.

Walrond, J.P., Kass, I.S., Stretton, A.O.W. and Donmoyer, J.E. (1985) Identification of excitatory and inhibitory motoneurons in the nematode *Ascaris* by electrophysiological techniques. *Journal of Neuroscience*, **5**, 1–8.

Walrond, J.P. and Stretton, A.O.W. (1985) Reciprocal inhibition in the motor nervous-system of the nematode *Ascaris*: direct control of ventral inhibitory motoneurons by dorsal excitatory motoneurons. *Journal of Neuroscience*, **5**, 9–15.

Wann, K.T. (1987) The electrophysiology of the somatic muscle cells of *Ascaris suum* and *Ascaridia galli*. *Parasitology*, **94**, 555–566.

Weisblat, D.A., Byerly, L. and Russel, R.L. (1976) Ionic mechanisms of electrical activity in the somatic muscle cell of the nematode *Ascaris lumbricoides*. *Journal of Comparative Physiology*, **111**, 93–113.

Weisblat, D.A. and Russel, R.L. (1976) Propagation of electrical activity in the nerve cord and muscle syncytium of the nematode *Ascaris lumbricoides*. *Journal of Comparative Physiology*, **107**, 293–307.

White, J.D., Southgate, E., Thompson, J.N. and Brenner, S. (1986) The structure of the nervous system of *Caenorhabditis elegans*. *Philosophical Transactions of the Royal Society of London, Series B*, **314**, 1–340.

White, J.G., Southgate, E., Thompson, J.N. and Brenner, S. (1976) The structure of the ventral cord of *Caenorhabditis elegans*. *Philosophical Transactions of the Royal Society. London, Series B*, **275**, 298–327.

Wiley, L.J., Weiss, A.S., Sangster, N.C. and Li, Q. (1996) Cloning and sequence analysis of the candidate nicotinic acetylcholine receptor alpha subunit gene tar-1 from *Trichostrongylus colubriformis*. *Gene*, **182**, 97–100.

Willet, J.D. (1980) Control mechanisms in nematodes. In *Nematodes as Biological Models*, edited by B.M. Zuckerman, pp. 197–225. New York: Academic Press.

Wright, D.J. and Awan, F.A. (1978) Catecholaminergic structures in the nervous system of three nematode species, with observations on related enzymes. *Journal of Zoology*, **185**, 477–489.

Xie, M.H., Yuan, J., Adams, C. and Gurney, A. (1997) Direct demonstration of musk involvement in acetylcholine receptor clustering through identification of agonist scfv. *Nature Biotechnology*, **15**, 768–771.

Xu, M., Molento, M., Blackhall, W., Ribeiro, P., Beech, R. and Prichard, R. (1998) Ivermectin resistance in nematodes may be caused by alteration of p-glycoprotein homolog. *Molecular and Biochemical Parasitology*, **91**, 327–335.

13. Locomotion

Robert McNeill Alexander

School of Biology, Leeds University, Leeds, LS2 9JT, UK

Keywords

Crawling, swimming

Introduction

Nematodes swim in water, crawl through the pore spaces of soil and move in water films on the surfaces of leaves and in the mucus of the intestinal mucosa (Wallace, 1968). This chapter discusses the mechanics of their movements in these and other environments.

Undulatory Locomotion

Kinematics

The mechanism of locomotion most used by nematodes depends on waves of bending which are made to travel backwards along the body. Travelling backwards, they drive the worm forwards, whether it is swimming through water, crawling over a solid surface or moving between solid particles such as make up soil. Similar movements are used by eels and snakes, both for swimming and for crawling (Alexander, 1992), but whereas these animals use waves of transverse bending, nematodes use dorsoventral bending. A nematode crawling on a solid surface lies on its side, so its plane of bending is parallel to the surface.

Because a swimming or crawling nematode throws its body into waves that closely resemble sine waves, it is easy and convenient to describe its motion mathematically. Doing that will enable us to describe the motion very concisely, in terms of quantities such as wavelength, wave velocity and amplitude. It will also enable us to make simple calculations that will clarify our understanding of the motion. Figure 13.1(a) represents a long slender cylindrical animal (such as a nematode) propelling itself along the x axis of Cartesian coordinates by means of backward travelling waves. The continuous wavy line represents the animal at time t and the broken one a short interval later, at time $t + \delta t$. The animal is moving to the right with velocity v_x, so its head advances a distance $v_x \cdot \delta t$ in this interval. The waves are travelling backwards relative to the ground with velocity $-v_{slip}$ (the subscript indicates that the waves are slipping over the ground), so each wave crest moves $v_{slip} \cdot \delta t$ to the left in the interval. The velocity of the waves relative to the worm's body is $v_{wave} = -(v_x + v_{slip})$. Figure 13.1(a) also shows the wavelength λ of the travelling waves and their amplitude a.

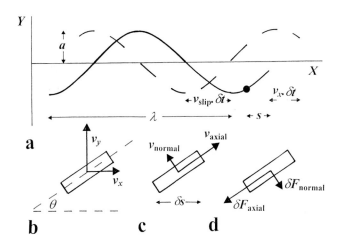

Figure 13.1. A diagram of undulatory locomotion. The animal is travelling towards the right, in a Cartesian coordinate system. Its position is shown both at time 0 (continuous outline) and at time δt (broken outline). (b), (c), (d) Diagrams of a short segment of the worm's body at the location marked by a filled circle in (a), showing (b), the longitudinal and transverse components of the velocity of this segment; (c), the axial and normal components of the velocity; and (d), the axial and normal components of the hydrodynamic force on the segment.

Gray and Lissmann (1964), in their classic account of nematode locomotion, filmed *Panagrellus*, *Turbatrix* and *Haemonchus* swimming and crawling. Their specimens were 0.5 to 1.5 mm long, with diameters 0.03 to 0.04 times body length. Whether swimming or crawling, they used waves whose ratios of amplitude to wavelength (a/λ) were about 0.2. In other respects there were many differences between swimming and crawling, as Figure 13.2 shows. Crawling (Figure 13.2(b)) was slower than swimming (Figure 13.2(a)), but the Figure overemphasises the difference because it shows the slowest crawler, among the species studied, and the fastest swimmer. *Turbatrix* crawled on agar jelly at about 0.3 mm/s and swam at 0.7 mm/s, whereas *Haemonchus* crawled at 0.05 mm/s and swam at 0.2 mm/s. These swimming speeds are similar to those of ciliate protozoans 0.03 to 1.0 mm long, and a flatworm 2 mm long, which have maximum swimming speeds of 0.5 to 1.0 mm/s (Sleigh and Blake, 1977).

Gray and Lissmann (1964) noted that the wavelengths of the travelling waves on nematodes were greater in swimming than in crawling. They give wavelengths of around 0.8 times body length for swimming, and 0.3–0.5 times body length for crawling, for all the species they studied, but they also illustrate swimming movements of *Haemonchus* in which the wavelength was a little longer than the body, which thus formed less than a complete wave.

Nematodes crawling in soil have to make their way through the narrow spaces between soil particles. To do this, the body must bend into irregular waves of which the wavelength depends on the size of the particles. This point is well illustrated by Wallace's (1958) drawings of nematodes moving through a single layer of soil particles of diameters between about 0.1 and 0.5 times the lengths of the worms. Wallace (1958) found that the worms travelled fastest when particle size was about 0.3 times body length. When the particles were smaller, many of the spaces between them were too narrow for the worm to pass through. When they were larger, some spaces were too wide for the worm to get a purchase on both sides.

In swimming, the waves always travel backwards relative to the water (Figure 13.2(a)). Gray and Lissmann (1964) found that the ratio v_x/v_{wave} (that is, the ratio of the forward speed to the velocity of the

waves relative to the worm's body) had mean values in the range 0.2 to 0.3, for the species they studied. In contrast, in crawling on some substrates the waves may remain almost stationary relative to the ground ($v_x/v_{wave} \approx 1$). Figure 13.2(b) shows an example; the worm is crawling on agar jelly, making a sinuous groove in the jelly as it moves. Similarly, the waves of a crawling snake remain stationary relative to the ground but those of a swimming snake or eel move backwards relative to the water (Alexander, 1992). Gray and Lissmann (1964) found that a little slip was usual for nematodes crawling on jelly (mean values of v_x/v_{wave} ranged from 0.77 to 0.90), and much more slip on rigid surfaces such as moist glass. They also filmed nematodes moving through suspensions of starch grains and found v_x/v_{wave} ranging from 0.4 to 0.9, presumably depending on the density of the suspension.

Nematodes cannot swim in water films of which the thickness is less than the diameter of the body, but they can crawl in them. Larvae of the chrysanthemum eelworm (*Aphelenchoides*) crawl in the water film on the outer surfaces of plants, and enter the leaves through the stomata. They can also crawl in water films on glass,

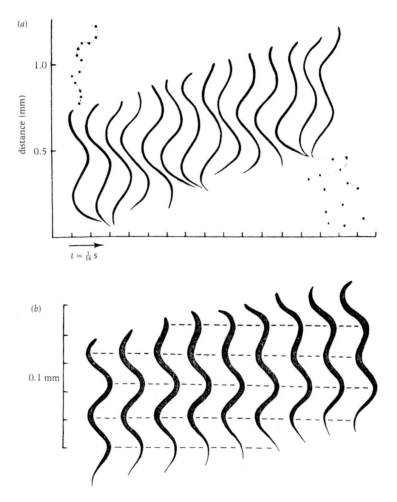

Figure 13.2. Tracings from films of nematodes. (a) Shows the position at intervals of $\frac{1}{16}$ s of *Turbatrix* swimming, and (b) shows the position at intervals of about $\frac{1}{3}$ s of *Haemonchus* crawling on the surface of an agar gel. Successive images have been displaced laterally so as not to overlap. From Gray and Lissmann (1964).

where they plainly cannot obtain a purchase by incising a groove, as nematodes crawling on agar do. Wallace (1959) showed that crawling in water films on rigid surfaces depends on surface tension. When a worm is stationary in a water film, surface tension acts symmetrically on its body. However, when it propagates waves of bending along its body, surface tension pulls at different angles on the leading and trailing edges of each wave, because advancing contact angles are greater than retreating contact angles. This asymmetry provides the resistance to backward slipping of the waves, which enables the worm to propel itself forward.

For the analysis that follows we will need to describe the travelling waves of nematodes algebraically. The coordinates at time t of the point marked on the worm of Figure 13.1(a), at a distance s from the anterior end, are

$$x = v_x t - s \qquad \text{Equation 13.1}$$

$$y = a \, \sin\,[(2\pi/\lambda)(v_{\text{wave}}t - s)] \qquad \text{Equation 13.2}$$

where a is the amplitude and λ is the wavelength. These equations assume that the waves are sinusoidal, that the worm is travelling at constant velocity, and that the wavelength and amplitude of the waves remain constant all along the body. These assumptions are not necessarily wholly realistic; the tracks of the head and tail indicated by dotted lines in Figure 13.2(a) show that in this case the amplitude increased along the length of the body.

The forward component of velocity is constant and equal to v_x for all points on the body. The transverse component v_y can be obtained by differentiating equation 13.2.

$$v_y = \mathrm{d}y/\mathrm{d}t = (2\pi v_{\text{wave}} \, a/\lambda)\cos[(2\pi/\lambda)(v_{\text{wave}}t - s)]$$

$$\text{Equation 13.3}$$

We will also need to know the angle θ between the long axis of the body at the point we are considering, and the x axis. This also is obtained from equation 13.2.

$$\tan\theta = -\mathrm{d}y/\mathrm{d}s$$
$$= -(2\pi a/\lambda)\cos[(2\pi/\lambda)(v_{\text{wave}}t - s)] = -v_y/v_{\text{wave}}$$

$$\text{Equation 13.4}$$

Reynolds Number

We are going to calculate the forces that act on a swimming nematode's body, but first we have to consider the nature of the forces. A swimming animal must exert forces on the water for two reasons. It sets the water moving, so has to exert the inertial forces needed to accelerate the water. Also, it sets up gradients of velocity in the water around it, so has to exert forces to overcome the viscosity of the water. The hydrodynamics of swimming depends on the relative importance of the inertial and viscous forces, which can be assessed by calculating a Reynolds number. If the Reynolds number is much less than one (as for swimming spermatozoa), inertial forces can be neglected, and if it is much greater than one (as for swimming snakes), viscous forces can be neglected.

The Reynolds number for an object moving in a fluid is $lv\rho/\mu$, where l and v are a length and a velocity characteristic of the motion, ρ is the density of the fluid and μ is its viscosity. In the case of water at $20°$C, the ratio μ/ρ (known as the kinematic viscosity) is 1.0×10^{-6} m^2/s. Notice the ambiguity in the definition; l and v could be defined in several different ways. One obvious choice is to make l the length of the worm and v its forward speed. Using this definition, the Reynolds number of a typical 0.8 mm (8×10^{-4} m) *Turbatrix* swimming at 0.7 mm/s (7×10^{-4} m/s; Gray and Lissmann, 1964) is $8 \times 10^{-4} \times 7 \times 10^{-4}/10^{-6} = 0.6$. Alternatively, because the pattern of flow around a long slender cylinder depends more on the diameter of the cylinder than on its length, it may seem more appropriate to choose for l the diameter of the worm (0.03 mm in the case we are considering) and for v the peak transverse speed (3.5 mm/s, calculated from the frequency and amplitude using equation 13.3). With these choices, we obtain a Reynolds number of 0.1.

When the Reynolds number is low enough for inertial forces to be neglected, the hydrodynamic forces on cylinders moving through fluids are proportional to velocity. When it is high enough for viscous forces to be neglected, these forces are proportional to velocity squared. Figure 2.1–2 of Azuma (1992) shows that these forces are reasonably nearly proportional to velocity for Reynolds numbers based on cylinder diameter up to 1.0, so we will be justified in assuming that they are proportional to diameter in our discussions of the swimming of small (e.g. 1 mm) nematodes. Conveniently for this discussion, large nematodes do not generally swim.

Dynamics

In this subsection we will calculate the forces acting on a swimming nematode and the swimming speed expected to result from waves of given wavelength, amplitude and frequency. Having done that, we will extend the analysis to describe crawling over a surface or through a suspension of particles.

We will calculate the forces on a swimming nematode by the method of Gray and Hancock (1955). Their paper was about the swimming of spermatozoa, but the approach is applicable to any slender cylindrical organism that swims by undulation at low Reynolds numbers. Gray and Lissmann (1964) used it in their discussion of nematodes. The animal is treated as a series of short cylindrical elements, each tilted at an angle θ to the x axis and moving with components of velocity v_x and v_y parallel to the x and y axes (Figure 13.1(b)). These components of velocity have a resultant which can be resolved in a different way, into a component v_{axial} along the axis of the element and a component v_{normal} at right angles to it (Figure 13.1(c)).

$$v_{\text{axial}} = v_x \cos\theta + v_y \sin\theta \qquad \text{Equation 13.5}$$

$$v_{\text{normal}} = -v_x \sin\theta + v_y \cos\theta \qquad \text{Equation 13.6}$$

The components of force on each element are calculated as if it were part of an infinitely long straight cylinder of the same diameter, moving with the same axial and normal components of velocity. This approximation has been avoided in more rigorous analyses of undulatory swimming (see Lighthill, 1976; Azuma, 1992), but will enable us to keep the mathematics in this chapter reasonably simple, and is accurate enough for our purposes. The length of the element of the worm is $\delta s \cos\theta$. If the Reynolds number is low enough (see above) the axial and normal components of force on the element (Figure 13.1(d)) are given by

$$\delta F_{\text{axial}} = C_{\text{axial}}\, v_{\text{axial}}\, \delta s / \cos\theta = C_{\text{axial}}(v_x + v_y \tan\theta)\delta s$$

$$\text{Equation 13.7}$$

$$\delta F_{\text{normal}} = C_{\text{normal}}\, v_{\text{normal}}\, \delta s / \cos\theta$$
$$= C_{\text{normal}}(-v_x \tan\theta + v_y)\delta s$$

$$\text{Equation 13.8}$$

(using equations 13.5 and 13.6). Values for the coefficients C_{axial} and C_{normal} are given by Lighthill (1976).

When the worm is swimming at constant velocity, the forward components of force on the body will balance the backward components. We can use this principle to predict the speed at which the worm will swim. Notice in Figure 13.1(d) that δF_{axial} has a backward component and δF_{normal} has a forward component. The net forward component of force on the element is thus

$$\delta F_x = \delta F_{\text{normal}} \sin\theta - \delta F_{\text{axial}} \cos\theta$$
$$= [C_{\text{normal}} \sin\theta(-v_x \tan\theta + v_y) - C_{\text{axial}}$$
$$\cos\theta(v_x + v_y \tan\theta)]\delta s$$

$$\text{Equation 13.9}$$

(using equations 13.7 and 13.8).

The force on an element of the body will fluctuate in the course of each cycle of swimming movements, but the force on a complete wavelength will remain constant and equal to zero. In other words, the integral of the right hand side of equation 13.9, over one complete wavelength, must equal zero.

$$\int_0^\lambda [C_{\text{normal}} \sin\theta(-v_x \tan\theta + v_y) -$$
$$C_{\text{axial}} \cos\theta(v_x + v_y \tan\theta)]ds = 0$$

$$\text{Equation 13.10}$$

We will assume that the amplitude is small compared to the wavelength, so that θ will always be fairly small. This enables us to make the approximations $\sin\theta \approx \tan\theta$ and

$\cos\theta \approx 1$. This is a rather rough approximation in the circumstances (notice in Figure 13.2 that parts of the body are inclined to the direction of travel at angles up to about 50°), but is adopted because it greatly simplifies the mathematics. Gray and Hancock (1955) present the mathematics for large amplitudes. With this approximation, equation 13.10 becomes

$$\int_0^\lambda [C_{\text{normal}} \tan\theta(-v_x \tan\theta + v_y) -$$
$$C_{\text{axial}}(v_x + v_y \tan\theta)]ds = 0$$

$$\text{Equation 13.11}$$

Substitution of values for v_y and $\tan\theta$ from equations 13.3 and 13.4 makes it possible to evaluate the integral. The result is remarkably simple.

$$v_x/v_{\text{wave}} =$$
$$[1 - (C_{\text{axial}}/C_{\text{normal}})]/[1 + (\lambda^2/2\pi^2 a^2)(C_{\text{axial}}/C_{\text{normal}})]$$

$$\text{Equation 13.12}$$

This equation shows that backward-travelling waves will propel the worm forward if and only if C_{axial} is less than C_{normal}. This will be the case; $(C_{\text{axial}}/C_{\text{normal}})$ will be 0.5 or a little more, depending on the ratio of worm diameter to wavelength (Lighthill, 1976).

With (diameter/wavelength) = 0.04 (a typical value for nematodes) Lighthill's (1976) equations give $(C_{\text{axial}}/C_{\text{normal}}) = 0.61$. Using this together with $a/\lambda = 0.2$ (a typical value for nematodes, see Kinematics above) equation 12 gives $(v_x/v_{\text{wave}}) = 0.22$, which is within the observed range of 0.2 to 0.3.

That analysis refers to swimming in a liquid. For a worm making its way through a dense suspension of particles, each segment of the body can move much more easily in an axial direction (along the present line of the body) than at right angles to it; it is easier to slip between the particles than to push them sideways. Consequently, $(C_{\text{axial}}/C_{\text{normal}})$ is smaller than for liquids, and equation 13.12 will predict larger values of (v_x/v_{wave}), implying less slip. The values of (v_x/v_{wave}) observed by Gray and Lissman (1964) in the densest suspensions of starch grains (see Kinematics, above) imply values of $(C_{\text{axial}}/C_{\text{normal}})$ of 0.05 or less.

For a worm crawling on a moist jelly surface, incising a channel for its body as it goes, C_{axial} may also be very small compared to C_{normal}. Equation 13.12 shows that if $(C_{\text{axial}}/C_{\text{normal}})$ becomes very small, (v_x/v_{wave}) approaches 1.0 and very little slip occurs.

Energy Cost

In this subsection we will estimate the power output required of a swimming nematode's muscles, and ask how important the energy cost of swimming is likely to be in the animal's energy budget.

The power required to move against a force is the force multiplied by the component of velocity in the direction of the force. Thus the power required to move

Locomotion

an element of the body of the nematode shown in Figure 13.1 is

$$\delta P = v_{\text{axial}}\, \delta F_{\text{axial}} + v_{\text{normal}}\, \delta F_{\text{normal}} \qquad \text{Equation 13.13}$$

which can be evaluated using equations 5 to 8. The power required to propel the whole body can then be calculated by integration, as in our calculation of force (equation 13.10). The resulting equation gives the power P required to propel a worm of length l:

$$P/l = v_x^2 C_{\text{axial}} + 6(\pi a/\lambda)^4 v_{\text{wave}}^2 C_{\text{axial}} +$$

$$2(\pi a/\lambda)^2 [(v_x^2 + v_{\text{wave}}^2)C_{\text{normal}} + 2v_x v_{\text{wave}}(C_{\text{normal}} - C_{\text{axial}})]$$

$$\text{Equation 13.14}$$

Like equation 13.12, this equation has been obtained by assuming (not entirely realistically) that the amplitude is small compared to the wavelength, so it cannot be expected to be very accurate. It will nevertheless serve for our purpose.

Consider a 1 mm worm swimming in water at 0.5 mm/s. Let $v_x/v_{\text{wave}} = 0.25$ and $a/\lambda = 0.2$, values that we have already seen to be typical. Assume that C_{axial} and C_{normal} have the values given by Lighthill (1976) for a diameter/wavelength ratio of 0.04. Then equation 13.14 gives a mechanical power requirement for swimming of 2.3×10^{-11} watts. Locomotor muscles of animals generally work with efficiencies in the range 0.03 to 0.3 (Alexander, 1999), so the metabolic power can be estimated as $0.8 - 8 \times 10^{-10}$ W.

A 1 mm nematode of diameter 1/30 times its length would have a mass of about 0.9 μg (estimated by calculating its volume and multiplying by the density of water). Its metabolic rate at $20°$C would be about 7×10^{-9} W (from the allometric equation in Peters, 1983), which is many times our estimate of the power required for swimming. It seems that, for nematodes, the energy cost of swimming is trivial. In contrast, for many larger animals, the energy cost of locomotion represents a substantial fraction of the field metabolic rate (Alexander, 1999).

Muscle Action in Undulatory Locomotion

Niebur and Erdös (1991) used a computer simulation of nematode crawling to show how the muscles may be coordinated. They represented the worm as a chain of 19 trapezoidal segments (Figure 13.3). The edges of the trapezia are non-Hookean springs, representing the elastic properties of the cuticle. In addition, longitudinal muscles in the lateral edges exert forces proportional to their states of excitation, which are controlled by stretch receptors in more posterior segments. The ratio of coefficients of force, ($C_{\text{axial}}/C_{\text{normal}}$), is assumed to be very small, as appropriate for crawling. Figure 13.3 shows a simulation which represented well the crawling of Caenorhabditis. The model is travelling towards the right. The activation levels of the muscles in each segment were controlled by stretch receptors on the same side of the

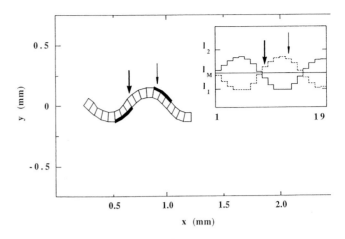

Figure 13.3. A computer simulation of a nematode crawling towards the right. Thick lines represent active muscles. The inset shows the lengths of the dorsal and ventral muscles, plotted against segment number (counting from the posterior end). From Niebur and Erdös (1991).

body, four segments further posterior; for example, muscles in the segment marked by the thin arrow are controlled by stretch receptors in the segment marked by the thick arrow. A muscle is active (indicated by a thick line) if the same side of the controlling segment is stretched above its resting length. This model shows that the crawling of nematodes could be coordinated by a very simple control system, but it is not known whether the real control system resembles the hypothetical one.

In Niebur and Erdös' (1991) model, the elastic properties of the cuticle are represented by springs arranged parallel and perpendicular to the long axis of the body. When the muscles are inactive, or when equal forces act in the dorsal and ventral muscles, these springs keep the body straight. An analysis by Alexander (1987) suggests that this may not be realistic, and that the straight posture may be unstable when the pressure in the worm exceeds ambient. If this is correct, a nematode whose dorsal and ventral muscles are symmetrically active would tend to adopt a curved or sinuous posture.

The basis for my suggestion (Alexander, 1987) is the crossed-helical arrangement of collagen fibres in the cuticle of some nematodes (see chapter 7). I considered a short length of a nematode's body as a cylinder wrapped in a layer of taut elastic fibres which formed left-handed and right-handed helices. When the worm bent its body, the cylinder became a segment of a torus, and the helices became geodesics on the surface of the torus (a torus is the shape of a wedding ring, and a geodesic is the shortest line joining two points on a curved surface). The lengths of these geodesics were calculated, and found to decrease as the worm bent its body, keeping the volume of the body unchanged. Bending would allow the stretched elastic fibres to shorten, so the straight posture would be unstable.

A crucial assumption of this analysis is that the matrix, in which the collagen fibres of the cuticle are embedded, is much less stiff than the fibres; if this were not the case, the parts of a fibre on the convex sides of bends might be stretched to different strains from the parts on the concave sides of bends, and the theory would not be applicable. The mechanical tests on nematode cuticle, that would be needed to discover whether this assumption is justified, do not seem to have been performed. There is no experimental evidence to support the prediction, that the straight posture is unstable, beyond the observation that living nematodes seldom keep their bodies straight. A further limitation of the theory is that it applies only to those nematodes that have crossed helical fibres in the cuticle.

Unusual Techniques of Locomotion

Though most nematodes crawl or swim by passing waves of bending along their bodies, a few propel themselves by other movements.

Helical Coils

The intestinal parasite *Nippostrongylus* uses a peculiar technique of locomotion by helical coils, which was studied by Lee and Biggs (1990). This is not normally used in saline, in which *Nippostrongylus* swims (rather ineffectually) by undulatory movements of the kind described above. Even this swimming is not quite like that of typical nematodes; the waves at the posterior waves of the body are asymmetrical, apparently due to failure of the dorsal muscles to shorten. The worms used similar movements to travel between sand grains in saline. However, in sodium alginate solutions of much higher viscosity than water they often moved by passing helical coils posteriorly along the body. Sometimes they used single coils (Figure 13.4(b)), sometimes several widely spaced coils (Figure 13.4(c)) and sometimes a helix of several closely spaced coils (Figure 13.4(d)). The more concentrated the alginate, the more numerous the coils were likely to be. Helical waves were also often used when moving between sand grains in alginate.

The speed of this style of swimming can be calculated very simply. Consider a worm of length l swimming at speed v_x by passing helical coils of diameter d down its body at speed (relative to its head) v_{wave}. At any instant there are n coils on the body (Figure 13.5(a)). We can model this as n rings of diameter d (circumference πd) moving backwards at velocity ($v_{wave} - v_x$) alongside a straight cylinder of length ($l - \pi nd$) which is moving forwards at velocity v_x (Figure 13.5(b)). Using the same symbols as before for the axial and normal coefficients of force, a backward hydrodynamic force ($l - \pi nd$) $C_{axial}v_x$ acts on the straight cylinder, and forward forces totalling $\pi nd C_{normal}(v_{wave} - v_x)$ on the coils. When the worm is travelling at constant velocity, the forward forces must balance the backward force, whence

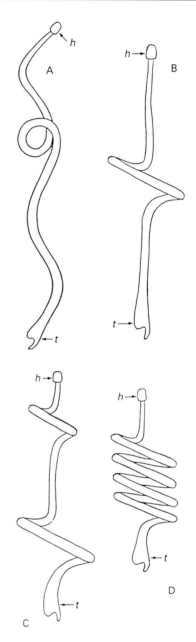

Figure 13.4. Adult *Nippostrongylus* crawling in dilute agar, using helical waves. From Lee and Biggs (1990).

$$(l - \pi nd)C_{axial}v_x = \pi nd C_{normal}(v_{wave} - v_x)$$

$$v_x/v_{wave} = [C_{normal}/C_{axial}]/[(l/\pi nd) - 1 + (C_{normal}/C_{axial})]$$

Equation 13.15

In Figure 13.4(b), ($l/\pi nd$) is about 2.2, so if (C_{axial}/C_{normal}) were about 0.5, as for water, the forward speed would be expected to be about 0.6 of the wave speed. If it were less than 0.5, as seems likely for alginate on account of its gel structure, progress would be even faster. These predictions compare favourably with undulatory swimming, but I know no data that could be used to check their accuracy.

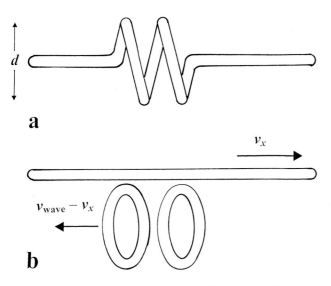

Figure 13.5. Diagrams illustrating the discussion of locomotion by helical waves.

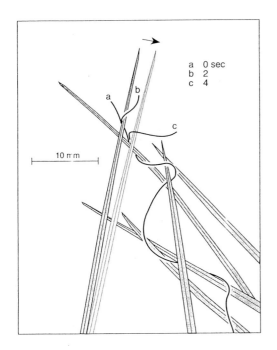

Figure 13.6. *Mermis* climbing among blades of grass. (a), (b) and (c) are successive positions of the head. From Gans and Burr (1994).

Nippostrongylus has longitudinal ridges on its cuticle, which enabled Lee and Biggs (1990) to show in scanning electron micrographs of helically coiled specimens that 360° torsion of the body occurs in each turn of the helix.

Crawling with Local Contact Only

Gans and Burr (1994) have described the locomotion of adult female *Mermis*, which are relatively more slender than most other nematodes; specimens about 120 mm long have a diameter of only 0.4 mm. Figure 13.6 shows

it climbing among blades of grass, threading its body around successive blades. The body does not form a regular sine wave, but an irregular series of bends which are propagated backwards along the body, moving the worm forwards. When the head extends some distance beyond the last support, the worm waves it around to find a new contact point. The anterior parts of the body can be bent laterally as well as dorsoventrally, facilitating these searching movements. If potential supports are widely spaced, up to half the length of the body may be extended in front of the anteriormost contact point.

Gans and Burr (1994) also observed female *Mermis* crawling on felt. They did not lie on their lateral surfaces with the median plane horizontal, as other nematodes do when they crawl, but kept the median plane vertical. Consequently, the dorsoventral bending of the body raised most of its length off the surface. Where the body contacted the felt, it was often laced under fibres at the surface of the felt, obtaining a purchase that prevented slipping as the worm moved by passing irregular waves posteriorly along its body.

Steinernema larvae also travel over the ground by reaching from one contact point to the next, but they also sometimes leap (Reed and Wallace, 1965). They prepare for a leap by bending the body into a loop, with the head attached to the posterior part of the body by surface tension in a connecting droplet of water. This attachment keeps the body bent while the muscles which will straighten it build up tension. Eventually, the head pulls free from the water droplet and the body straightens suddenly, projecting itself through the air.

Peristaltic Locomotion

The nematode *Criconemoides* crawls by passing waves of shortening forward along its body (Streu, Jenkins and Hutchinson, 1961). Note that the waves are propagated forwards, not backwards as in earthworms. Waves travelling in either direction along this worm would propel it forward, because of the ratchet-like ridges on the cuticle which allow the worm to slide forward more easily than it can slide back.

References

Alexander, R. McN. (1987) Bending of cylindrical animals with helical fibres in their skin or cuticle. *Journal of Theoretical Biology*, **124**, 97–110.

Alexander, R. McN. (1992) *Exploring Biomechanics: Animals in Motion.* New York: Scientific American Library.

Alexander, R. McN. (1999) *Energy for Animal Life.* Oxford: Oxford University Press.

Azuma, A. (1992) *The Biokinetics of Swimming and Flying.* Tokyo: Springer-Verlag.

Gans, C. and Burr, A.H.J. (1994) Unique locomotory mechanism of *Mermis nigrescens*, a large nematode that crawls over soil and climbs through vegetation. *Journal of Morphology*, **222**, 133–148.

Gray, J. and Hancock, G.J. (1955) The locomotion of sea urchin spermatozoa. *Journal of Experimental Biology*, **41**, 135–154.

Gray, J. and Lissmann, H.W. (1964) The locomotion of nematodes. *Journal of Experimental Biology*, **32**, 802–814.

Lee, D.L. and Biggs, W.D. (1990) Two- and three-dimensional locomotion of the nematode *Nippostrongylus brasiliensis. Parasitology*, **101**, 301–308.

Lighthill, J. (1976) Flagellar hydrodynamics. *SIAM Review* **18**, 161–230.

Niebur, E. and Erdös, P. (1991) Theory of the locomotion of nematodes. Dynamics of undulatory propulsion on a surface. *Biophysical Journal*, **60**, 1132–1146.

Peters, R.H. (1983) *The Ecological Implications of Body Size*. Cambridge: Cambridge University Press.

Reed, E.M. and Wallace, H.R. (1965) Leaping locomotion by an insect parasitic nematode. *Nature*, **206**, 210–211.

Sleigh, M.A. and Blake, J.R. (1977) Methods of ciliary propulsion and their size limitations. In T.J. Pedley (edit.) *Scale Effects in Animal Locomotion*. London: Academic Press.

Streu, H.T., Jenkins, W.R. and Hutchinson, M.T. (1961) *New Jersey Agricultural Experimental Station, Rutgers Bulletin*, **800**.

Wallace, H.R. (1958a) Movement of eelworms. I. The influence of pore size and moisture content of the soil on the migration of larvae of the beet eelworm, *Heterodera schachtii* Schmidt. *Annals of Applied Biology*, **46**, 74–85.

Wallace, H.R. (1958b) Movement of eelworms. II. A comparative study of the movement in soil of *Heterodera schachtii* Schmidt and of *Ditylenchus dipsaci* (Kuhn) Filipjev. *Annals of Applied Biology*, **46**, 86–94.

Wallace, H.R. (1959) The movement of eelworms in water films. *Annals of Applied Biology*, **47**, 366–370.

Wallace, H.R. (1968) The dynamics of nematode movement. *Annual Reviews in Phytopathology*, **6**, 91–114.

14. Nematode Sense Organs

John Jones

Nematology Department, Scottish Crop Research Institute, Invergowrie, Dundee, DD2 5DA, Scotland.

Keywords

Nematode, sense organ, amphid, sensillum, structure, function

Introduction

Nematodes respond to a wide variety of stimuli. These include chemical stimuli (chemoreception), touch (mechanoreception), temperature (thermoreception) and, in some cases, light (photoreception). Given the diverse array of environments occupied by nematodes it is not surprising to find that different nematodes utilise each of these senses to varying degrees. Those that are heavily dependent on one particular source of stimulus may have some sense organs developed or adapted to enhance reception of this stimulus. Detailed reviews on both the structure (McLaren, 1976a, 1976b; Wright, 1980) and function (Bargmann and Mori, 1997; Driscoll and Kaplan, 1997) of nematode sense organs have been published. Here, an attempt is made to cover both these areas and link the two together.

In almost all fields of nematode biology, studies on the free-living nematode *Caenorhabditis elegans* have advanced further than those on any other nematode species. A brief survey of the literature on nematode sense organs shows that this field is no exception. While detailed studies on the structure of sense organs from a variety of free-living and parasitic nematodes have been carried out, few are as detailed or complete as those on the sense organs of *C. elegans*. When the function or development of nematode sense organs is considered, this imbalance becomes even more extreme. The aim in this chapter has been to avoid the temptation to focus entirely on *C. elegans*. The structure and function of nematode sense organs are reviewed with as broad a brush as possible and the information from *C. elegans* is drawn upon to allow a picture of the functioning of sense organs of all nematodes to be painted.

Structure

Nematodes possess a variety of sense organs with widely varying structure. Sense organs that detect similar stimuli in the same individual can vary enormously in their architecture. Conversely, evidence is also emerging that one of the sense organs, the amphid, may allow the detection of a variety of stimuli.

In this section the structure of nematode sense organs is described and whilst focus is placed on the largest sense organs and those common to nearly all nematodes, the structure of some of the more obscure sense organs is also described.

Although widely varied in their detailed structure, nematode sense organs can be subdivided into two basic types. Cuticular sense organs detect a variety of stimuli including chemicals, mechanical stimuli, and temperature. Cuticular sense organs are composed of three basic cell types: a sheath cell, a socket cell and a variable number of dendritic processes. Cell bodies of neurons giving rise to sensory dendritic processes are located in the central nervous system (White *et al.*, 1986). Dendritic processes are formed from extensively modified cilia (Wright, 1980) and contain microtubules arranged along the length of the process. The fine structure of the microtubules may be altered in mechanoreceptors, an adaptation that reflects the role of these microtubules in detecting mechanical stimuli (see below). The sheath cell lies posterior to the socket cell and produces secretions that bathe the dendritic processes. Complex junctions are present between the sheath cell and the dendritic processes in chemoreceptors. These probably serve to isolate the space within the sense organ from the rest of the nematode body. The socket cell may produce the cuticular lining at the tip of the sensillum. In chemoreceptors it may form a channel and pore which allow exposure of the dendritic processes to the external environment, while in mechanoreceptors it may be responsible for producing a modified cuticle overlying the sense organ which enhances transmission of mechanical stimuli.

The greatest concentration of cuticular sense organs is found at the anterior end of the nematode. These usually comprise two amphids, located laterally on the nematode, six inner labial sense organs, six outer labial sense organs and four cephalic sense organs arranged in radial symmetry around the head. Cuticular sense organs are often found in abundance at the posterior end of male nematodes where they function in mating and one group of nematodes, the Secernentia, has a pair of large cuticular chemoreceptors, the phasmids, located on the tail. Cuticular sense organs are also found at other points on the body. These include mechanoreceptors such as the deirids (Ward *et al.*, 1975) and, less frequently, chemoreceptors (Wright and Chan, 1974; Wright and Carter, 1980). Many marine nematodes, often Adenophoreans, have a more complex array of cuticular sense organs, particularly chemoreceptors, than other nematodes (Wright, 1980). The structure of a variety of cuticular sense organs is considered in more detail below.

Internal sense organs show a far more diverse structure and almost always detect mechanical stimuli. Photoreceptors are considered to be internal rather than cuticular sense organs, providing the exception to the rule. A set of mechanoreceptive neurons, the touch cell receptors, run along the length of many nematodes, including *C. elegans* (Chalfie *et al.*, 1985). Internal receptors, not associated with the cuticle, are present at the anterior tip of many nematodes (Wright 1980), and probably detect external stimuli. The feeding apparatus of many nematodes is richly endowed with internal sensory structures. The pharynx of *C. elegans* contains 20 neurons, twelve of which have tips that are in close proximity to the cuticle of the lumen and may therefore have sensory roles (Albertson and Thomson, 1976). The feeding apparatus of *Xiphinema diversicaudatum* and other related nematodes contains neurons which project into the cuticle and may have a chemosensory or mechanosensory function (Robertson, 1975; Robertson, 1979.) Studies on *C. elegans* have also shown that nerve cells in muscle tissue are likely to be mechanosensitive and may play a role in generating muscle contraction patterns giving rise to the characteristic sinusoidal movement patterns of nematodes (reviewed in Driscoll and Kaplan, 1997).

Chemoreceptors

Chemoreceptors are most obviously distinguished from other sense organs in their being exposed to the external environment by a pore in the cuticle. As will be seen however, not all dendritic processes exposed to the outside world in this way function purely as chemoreceptors. The amphids are the largest nematode sense organs and function primarily as chemoreceptors, demonstrating the importance of this source of stimulus to nematodes. Other chemoreceptors include the inner labial sensillae, the phasmids and chemoreceptors associated with male reproductive structures. Each of these is considered below.

Amphids

The amphids are invariably the largest and most complex of the nematodes sense organs. Ultrastructural observations of the amphids of many nematodes have been made. Although the ultrastructure of these organs varies considerably amphids are composed, like other cuticular sense organs, of a sheath cell, a socket cell and dendritic processes. The number of dendritic processes varies from as few as 7 (e.g. *Globodera rostochiensis* — Jones, Perry and Johnston, 1994) to as many as 36 in the marine nematode *Oncholaimus vesicarius* (Burr and Burr, 1975). Some dendritic processes terminate in the amphidial canal and are exposed to the external environment via the amphidial pore, while others terminate within the sheath cell. In some (mainly plant-parasitic) nematodes, dendritic processes associated with the amphidial sheath cell leave the amphid altogether and form flattened stacks of

membranes within the cephalic framework. These accessory cilia are discussed in more detail below. Another major difference between the amphids and other sense organs is that the sheath cell is greatly enlarged and often shows very high secretory activity. The sheath cell produces secretions that bathe the dendritic processes within the receptor cavity and may also produce secretions with a function outside the nematode body (see below). In the next section the structure of the amphids is described in detail using the structure found in free-living and many plant-parasitic nematodes as a base point. Variations from this theme within these groups are then considered. Finally the structure of the amphids of animal-parasitic nematodes and marine nematodes are considered. Although the amphids of these nematodes retain the basic elements of the structure described below, variations are more widespread and much more extreme.

The structure of the amphids is summarised in Figures 14.1a–14.1d. Dendritic processes arise from sensory neurons whose cell bodies lie close to the nerve ring. In *C. elegans* all the cell bodies lie in a group posterior to the nerve ring (White *et al.*, 1986). A similar arrangement of cell bodies is found in an animal parasite *Strongyloides stercoralis*, although another group of cell bodies is found just anterior to the nerve ring in this nematode. These cell bodies give rise to sensory neurons and ultimately dendritic processes which have no homologs in *C. elegans* (Ashton and Schad, 1996; Ashton *et al.*, 1995). Few other nematodes have been studied in the detail necessary to confidently compare the positions of their sensory cell bodies to those of *C. elegans*. Moving anterior, the dendritic processes project through the base of the receptor cavity, forming tight junctions as they pass (Figure 14.1d), into a large cavity within the sheath cell known as the receptor cavity (Figure 14.1c). The tight junctions formed may serve to isolate the receptor cavity from the rest of the nematode body allowing a unique environment to be maintained and, in chemoreceptors, preventing continuity of the nematode body with the outside world. The receptor cavity contains secretions produced by the sheath cell, which shows the modifications expected of a secretory cell including the presence of a large number of secretory granules (Figure 14.1c). Golgi bodies are frequently present. In the dendritic processes at this point a circle of doublet microtubules is observed. These give rise to singlet microtubules which project to the tips of the dendritic processes (Figure 14.1b). Further anterior still, the dendritic processes become more spread apart in the receptor cavity (Figure 14.1c) and a subset of the dendritic processes continues into the amphidial canal (Figure 14.1b). This tube-like structure is formed initially by the sheath cell and, further anterior, by the socket cell. Tight junctions are formed between the sheath cell and socket cell where they join. Secretions of the sheath cell continue to bathe the dendritic processes throughout the amphidial canal and may even be seen exuding from the amphidial pore. The remaining

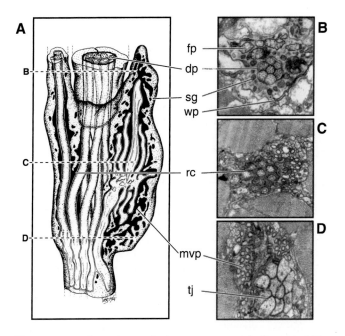

Figure 14.1. Overview of the structure of the amphid.
1a: Sketch showing features of the amphid as observed in longitudinal section. Approximate positions of the transverse sections in 1b, 1c and 1d are indicated. **1b**: Transverse section of an amphid showing dendritic processes (dp) within the amphidial canal where formed by the sheath cell. Finger-like projections (fp) and wing-like projections (wp) of dendritic processes are present within the sheath cell. Secretory activity of the sheath cell is indicated by the presence of secretory granules (sg) in this cell. **1c**: Transverse section cut at the level of the receptor cavity (rc). At this level the dendritic processes contain rings of doublet microtubules. Secretions of the sheath cell bathe the dendritic processes within the receptor cavity. Secretory granules and microvillar projections of one dendritic process are clearly visible within the sheath cell. **1d**: Transverse section cut at the base of the receptor cavity. Tight junctions (tj) formed between dendritic processes and between dendritic processes and the sheath cell are clearly visible. The microvillar projections (mvp) of one dendritic process into the sheath cell are particularly evident at this level.

dendritic processes, which do not project into the amphidial canal, give rise to more complex structures. Some project into the sheath cell and terminate as simple, finger-like dendritic processes (Figure 14.1b). Others become flattened and wider, giving rise to wing-like projections that remain within the sheath cell (Ward *et al.*, 1975, also see Figure 14.1b). One of the processes turns laterally and gives rise to microvillar processes which project both anteriorly and posteriorly into the sheath cell (Figure 14.1c). In *C. elegans* approximately 50 microvillar processes are present but in some plant-parasitic nematodes several hundred may be formed (Baldwin and Hirschmann, 1973). These microvillar processes may be tightly opposed to the membrane of the sheath cell or, in some plant-parasitic nematodes, groups of processes may be present in cavities containing secretions within the sheath cell. An intriguing anatomical difference between many plant-

parasites and *C. elegans* is that the dendritic process which gives rise to the microvilli moves into the sheath cell from the receptor cavity in plant-parasitic nematodes, whereas in *C. elegans* the process enters the sheath cell directly from the sensory nerve bundle and is never exposed to the receptor cavity (Wright, 1980).

Amphids of dorylaimid plant-parasitic nematodes, such as *X. americanum*, have a slightly different amphidial structure to that described above. The amphidial canal is broader in these nematodes and many of the sensory neurons give rise to more than one dendritic process. In *X. americanum* far fewer modified processes terminating in the sheath cell are present and no dendritic process bearing microvilli is observed (Wright and Carter, 1980). In this nematode one sensory neuron, that is not associated with the other amphidial sensory neurons, gives rise to a dendritic process which terminates at the very tip of the amphidial pore. However, this dendritic process does not pass through the receptor cavity or the sheath cell but simply turns through 90° as it reaches the tip of the nematode and enters the amphidial canal from the cephalic framework.

Tylenchid plant-parasitic nematodes, in addition to the dendritic processes described above, have four accessory cilia. The sensory neurons giving rise to these processes are associated with the neurons giving rise to the amphidial dendritic processes posterior to the amphid itself but the accessory cilia are not themselves associated with the amphid or its sheath cell. These dendritic processes give rise to extensive membranes which become swirled and folded, forming concentrically arranged stacks of membranes within the cephalic framework (Endo, 1980). An accessory cilium, from the potato cyst nematode *G. rostochiensis* is shown in Figure 14.3b. The function of these accessory cilia is unknown but Endo (1980) suggested, on the basis of structural comparisons with the amphids of *O. vesicarius*, that they may function as photoreceptors, while Wright (1980) suggested they might detect electromagnetic fields. Some plant-parasitic nematodes are known to respond to an electrical field similar to that which may be generated by roots (Robertson and Forrest, 1989) but no convincing demonstration of phototaxis by a plant-parasitic nematode has yet been made.

Amphids of animal-parasitic nematodes largely conform to the structure described above. Interspecific differences usually involve the number of dendritic processes and the structures formed by dendritic processes which project into the sheath cell. The extensive microvillar projections found in *C. elegans* and tylenchid plant-parasitic nematodes are usually absent and projections of dendritic processes into the sheath cell often arise from the base of the same processes which terminate in the amphidial canal. Sheath cells of animal-parasitic nematodes are occasionally adapted to produce anticoagulants (see below) and in these cases the sheath cell becomes greatly enlarged at the feeding stage (Jones, 1979). Ashton and Schad (1996) examined the structure of the amphids of *S. stercoralis* in detail and were able to identify, for the most part, the *C. elegans* counterparts of the processes found in the parasitic nematode indicating the degree of conservation of this structure between different nematode groups. Detailed examinations of the dog hookworm *Ancylostoma caninum* and the third stage larva of *Haemonchus contortus* also revealed a remarkable degree of similarity between the ultrastructure of the amphids of these nematodes and those of *C. elegans* (Ashton, Li and Schad, 1999). The only major difference was the absence of one of the wing cells in *A. caninum* which was replaced by an extra neuron giving rise to two dendritic processes within the sheath cell. The anterior sense organs of *O. volvulus* infective larvae and adult males have also been examined in some detail (Strote and Bonow, 1996; Strote, Bonow and Attah, 1996). Eight dendritic processes are present in the amphidial channel of the adult male while those of the infective larvae contain nine processes. In both stages processes associated with the sheath cell are present. The structure of the sense organs of a wide range of other animal parasites including *Hammerschmidtiella diesingi* (Trett and Lee, 1981) and *Nippostrongylus brasiliensis* (Trett and Lee, 1982) have been examined and reviewed in some detail by McLaren (1976a) and Wright (1980). Attempts have been made to relate differences in fine structure of the amphids of animal parasitic nematodes and *C. elegans* to differences in behaviour and the need to respond to different stimuli. Ashton *et al.* (1999) considered that a modified amphidial neuron of *S. stercoralis* may detect thermal cues related to invasion of the host and may therefore assist in controlling the transition from free-living to parasitic life cycle stages. Laser ablation studies have confirmed conservation of function of neurons which are structurally conserved in *C. elegans* and *S. stercoralis* and also indicated that the parasite specific neuron may indeed have a role in detecting thermal cues (Ashton *et al.*, 1998). These experiments demonstrate the utility of *C. elegans* as a model system for parasite studies and the way in which detailed ultrastructural information can be exploited to give ideas about biological function.

Few studies of the amphids of marine and aquatic nematodes have been carried out. The few ultrastructural studies published have confirmed the presence of the sheath cell, socket cell and dendritic processes and have shown that the basic arrangement of these cell types described above is retained. Extensive projections of dendritic processes into the sheath cell have not been described. The amphid of *O. vesicarius* is unusual in that a subset of dendritic processes form a photoreceptive organ (see below). Adults of *Gastromermis boophthorae* are aquatic organisms since they emerge from their insect hosts as larvae before moulting to the adult stage. The amphids of these nematodes show an unusual structure (Batson, 1978). The dendritic processes are held within a tight bundle by an inner channel of cuticle within the very broad amphidial canal. This inner channel is continuous with the amphidial canal itself and the significance of this structure is uncertain. The structure of the sense organs of this group of nematodes is discussed in the reviews cited above.

Other sensillae

Other structures considered to have a role in chemoreception include the inner labial sensillae, the phasmids and cuticular chemoreceptors associated with the male reproductive structures.

Six inner labial sensillae, each usually innervated by two dendritic processes, are arranged radially around the mouthparts of many nematodes. These organs are frequently, although not always (see below), exposed to the external environment and are therefore considered to function as chemoreceptors. The position of these organs has led to speculation that they may function as taste organs in some nematodes.

The inner labial sensillae are cuticular sense organs and are therefore made up of the usual sheath and socket cells in addition to the dendritic processes. The structure of a 'typical' inner labial sensillum is shown in Figures 14.2a-d. As with the amphids, dendritic processes pass through a receptor cavity formed by the sheath cell. Further anterior the canal is formed by the socket cell and ultimately by the cuticle itself. This structure has been observed in free-living nematodes (Ward *et al.*, 1975) and a wide variety of plant-parasitic forms (Baldwin and Hirschmann, 1973; Endo, 1980;

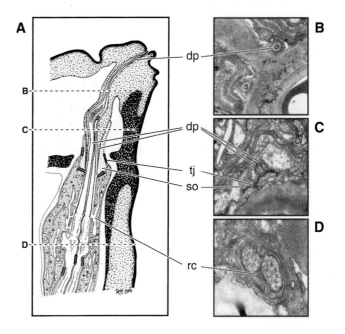

Figure 14.2. Structure of an inner labial sensillum. **2a:** Sketch showing features of the structure as observed in longitudinal section. Approximate positions of the transverse sections in 2b, 2c and 2d are indicated. **2b:** Transverse section cut at the very tip of the sense organ. Only one dendritic process (dp) is visible in the canal at this level. **2c:** Transverse section cut through the canal where formed by the socket cell (so). The canal contains two dendritic processes (dp) at this level. Note the tight junctions (tj) formed by the socket cell where its membranes meet. **2d:** Transverse section cut at the level of the receptor cavity (rc). Both dendritic processes contain a circle of doublet microtubules at this level and are bathed in secretions of the sheath cell.

Trett and Perry, 1985a). Within these groups, most variation in this structure concerns the relative length and arrangements of the two dendritic processes. However, in animal-parasitic nematodes far greater variation is observed, with the inner labial sensillae often adapted to function as mechanoreceptors (see below).

Away from the anterior tip of the nematode the most prominent chemoreceptors are the phasmids. These occur as a pair of sensillae near the tip of the tail. Phasmids are generally considered to be a feature of the Secernentea, although some examples of these organs in the Adenophorea apparently exist (Goodey, 1963). The ultrastructure of phasmids of a variety of nematodes has been described including plant-parasitic, animal-parasitic and free-living nematodes (reviewed in Bird and Bird, 1991) and the components of typical cuticular sense organs are generally present. A single dendritic process is usually present. In *Heterodera schachtii* the fine structure of the two phasmids differs with one phasmid lacking a receptor cavity and sheath cell secretions (Baldwin, 1985). The functional significance of this observation is unknown. Wang and Chen (1985) have speculated that phasmids of *Scutellonema brachyurum* may detect pheromones. However, in many species phasmids are present on female nematodes as well as males and Riga *et al.* (1996) were able to record changes in electrical activity from the anterior end of males of *Globodera* spp., suggesting a role for the anterior sense organs in detecting sex attractants. The precise role of the phasmids therefore remains uncertain.

Chemoreceptors have been reported to be present on the male sex organs of a wide variety of nematodes. Receptors may simply occur on the ventral surface of the male tail in close proximity to the reproductive structures. The copulatory bursa, a modification of the cuticle at the tail of the male that serves to position the female during mating, also carries chemoreceptors. The most detailed observations, made on *C. elegans*, showed that eighteen sense organs are present on the copulatory bursa of this species and that of these, three pairs are open to the external environment (Sulston, Albertson and Thomson, 1980). Intriguingly, these chemoreceptors do not appear to have the usual sheath and socket cells associated with the nerve processes but instead have a single structural cell. Similar observations have been made on the caudal papillae of *Acanthocheilonema (Dipetalonema) viteae*, which have a socket cell but no sheath cell (McLaren, 1972). In other nematodes however, the normal complement of supporting cells is present. The caudal papillae of *Aphelenchoides blastophthorus*, for example, do have a sheath cell and a well-developed receptor cavity is present (Clark and Shepherd, 1977).

The male reproductive structures themselves, the spicules, may also carry chemoreceptors (Lee, 1973). While ablation of spicules of *Panagrellus* spp. disrupted the ability of this nematode to locate a mate using pheromone cues (Samoiloff *et al.*, 1973), studies of mutants of *C. elegans* suggested that the cephalic chemoreceptors were responsible for detection of

pheromones (Ward *et al.*, 1975). Wright (1978) has suggested that the chemoreceptors present on spicules may sense the internal environment of the female reproductive tract as these structures are held within the male body until mating behaviour is initiated. The ultrastructure of chemoreceptors present on spicules of plant-parasitic nematodes (Clark, Shepherd and Kempton, 1973; Clark and Shepherd, 1977) and animal-parasitic nematodes (reviewed in McLaren 1976a) has been described in detail elsewhere.

Wright and Hui (1976) described a sense organ enclosed within a cuticular peg posterior to the lips of *Heterakis gallinarium*. This sense organ contained two dendritic processes, one of which was considered to be chemoreceptive while the other was suggested to have a mechanosensory role giving this organ similar properties to those described above for the inner labial sensillae of some species. Wright and Chan (1974) described a chemoreceptor on the body wall of *Capillaria hepatica*. The structure of this organ was typical of cuticular chemoreceptors. Such body wall chemoreceptors may be common in Adenophoreans but few detailed descriptions of these structures have been made.

Mechanoreceptors

Anterior receptors

The anterior end of a nematode carries up to sixteen cuticular mechanoreceptors (six outer labial sense organs, four cephalic sense organs and, in some nematodes, the six inner labial sensillae) as well as a variable number of internal receptors, not associated with cuticular modifications, which may also detect mechanical stimuli. In addition, one of the amphid neurons of *C. elegans*, which detects volatile and waterborne repellents, can also detect mechanical stimuli and direct an avoidance response (Kaplan and Horvitz, 1993). This amphid neuron accounts for almost 50% of the response to nose touch in *C. elegans* and is unique in responding to such a wide variety of stimuli.

The cuticular mechanoreceptors (cephalic and outer labial sense organs) of nematodes contain one or two dendritic processes as well as the sheath and socket cells. A receptor cavity is present and the sheath cell produces secretions which bathe the dendritic processes in this cavity. Given that these sense organs are not exposed to the external environment it seems very likely that these secretions contain material which is essential for the function of the sense organ itself. The nature and precise role of this material remains unknown. In spite of some variations (see below) it is possible to describe the general structure of a cuticular mechanoreceptor and to describe in some detail how the various components are arranged.

The structure of a cuticular mechanoreceptor is summarised diagrammatically in Figure 14.3a. Although derived from the plant-parasitic nematode *Pratylenchus penetrans* (Trett and Perry, 1985a), it can be used to describe general features of cuticular mechanoreceptors. A variable number of dendritic processes (usually

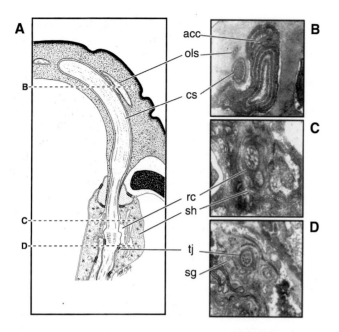

Figure 14.3. Structure of a cuticular mechanoreceptor of a nematode. **3a**: Sketch showing the arrangements of the various structures as viewed in longitudinal section. Approximate positions of the transverse sections in 3b, 3c and 3d are indicated. Note that in the plant parasitic species upon which these diagrams are based two mechanoreceptors, the cephalic sensillae (cs) and the outer labial sensillae (ols) are closely linked. **3b**: Transverse section cut at the anterior tip of the structure. The tips of the cephalic and outer labial sensillae are visible, embedded in the cuticle of the nematode. Also visible at this level are the stacks of membranes forming the accessory cilia (acc). These structures originate from nerve processes originally associated with the amphids. **3c**: Transverse section cut at the level of the receptor cavity (rc). Note that the dendritic processes are bathed in secretions of the sheath cell (sh). **3d**: Transverse section cut at the base of the receptor cavity. Tight junctions (tj) between the dendritic processes and the sheath cell are visible. The sheath cell contains secretory granules (sg).

one or two) enter the base of the socket cell forming tight junctions (Figure 14.3d) which isolate the receptor cavity from the rest of the nematode body. The receptor cavity (Figure 14.3c), filled with secretions of the sheath cell, is formed by invaginations of the inner sheath cell membrane. A ring of doublet microtubules, characteristic of sensory dendritic processes, form within the dendritic processes as they pass through the receptor cavity (Figure 14.3c). Slightly further anterior, the processes pass through the socket cell, before moving into the cuticle. Tight junctions are formed between the socket cell and the sheath cell and between the socket cell and the cuticle. The shape and length of the process tip within the cuticle varies (see below). However, in many cases the tip swells slightly and follows the contour of the cuticle for 2–3 μm (Figure 14.3a and 14.3d).

Despite these general rules, enormous variation in the fine structure of these sense organs is observed, reflecting the diverse habitats occupied by nematodes.

The tip of the mechanoreceptor may be enclosed within a short outgrowth of cuticle, deflection of which presumably triggers the mechanoreceptive dendritic process. Mechanoreceptors of this type are observed in the animal-parasitic nematodes *H. gallinarum* (Wright, 1977) and *Ascaridia galli* (Wright, 1980). The precise shape of the tip of the mechanosensitive process varies enormously. McLaren (1972) described the ultrastructure of the sense organs of *A. viteae* and found the mechanoreceptive dendritic processes to be shaped like mushrooms embedded in the cuticle. The outer labial and cephalic sense organs of *N. brasiliensis* also show unusual morphology (Wright, 1975) and these adaptations probably assist the nematode in detecting movements in different planes. In this nematode, the tip of the outer labial sense organ is orientated across the long axis of the head while the tip of the adjacent cephalic sense organ runs parallel to the longitudinal axis. The tips of both these organs are swollen and are covered in an electron dense material. In another animal-parasitic nematode, *Capillaria hepatica*, the outer labial and cephalic sense organs are exposed to the external environment and are therefore thought to have a chemoreceptive role (Wright, 1974). This is also true for *X. americanum* (Wright and Carter, 1980) but few other Adenophorean nematodes have been examined at the ultrastructural level making it difficult to assess how widespread these adaptations might be.

Touch receptors

Nematodes respond to touch along the sides of their body and this response is controlled, certainly in *C. elegans* and probably in other nematodes, by a set of touch receptor neurons which have processes that run longitudinally along the body wall. These processes are embedded below the cuticle, within the epidermis and are surrounded on their external face by a dark extracellular matrix known as the mantle. In *C. elegans*, the mantle is essential for normal function of the touch receptors. Touch cell processes are anchored to the cuticle by specialised regions which have a similar appearance to muscle attachment sites when viewed under the electron microscope (Chalfie and Sulston, 1981). The processes are filled with large numbers of unusually large microtubules that are thought to be important in the function of these cells (see below).

Other mechanoreceptors

Many other structures in nematodes may have a mechanosensory function. The inner labial sensillae often have an appearance which suggests a dual role in detecting mechanical as well as chemical stimuli. One of the dendritic processes in the inner labial sensillae of *C. elegans*, for example, is bent in a manner which suggests it is more likely to detect mechanical, rather than chemical, stimuli (Ward *et al.*, 1975). A similar structure is observed in some other soil dwelling nematodes including various *Pratylenchus* species (Trett and Perry, 1985a — see Figure 14.3a). In other nematodes, particularly animal-parasitic nematodes (e.g. *Strongyloides stercoralis* — Fine *et al.*, 1997), the inner labial sensillae do not open to the external environment and are therefore considered to function entirely as mechanoreceptors. In these nematodes the inner labial sensillae may have a role in controlling the process of penetration and migration through host tissues.

The copulatory spicules and associated reproductive structures, such as the copulatory bursa, of many male nematodes are particularly well served with mechanoreceptive sense organs. A role for these sense organs in providing tactile information during mating seems apparent. The first ultrastructural studies of these sense organs showed that the spicules of two animal-parasitic nematodes *H. gallinarium* and *N. brasiliensis* contained nerve axons and that cholinesterase activity, often associated with the presence of sensory structures, was present (Lee, 1973). Following on from this work, ultrastructural observations have been made on the reproductive structures of the males of an enormous variety of nematodes, including free-living, plant-parasitic and animal-parasitic nematodes. These studies have invariably revealed the presence of a wide array of mechanoreceptive organs.

A wide array of other, less intensively studied, mechanoreceptors are present in some nematodes including deirids, stretch receptors and receptors which control muscle contraction patterns. Some marine species have setae at various points on their body, some of which have a mechanoreceptive role. The structure of these organs is discussed in detail by McLaren (1976a).

Other Sense Organs

Thermoreception

All nematodes are sensitive to temperature, with each species having an optimum temperature for development. Responses of nematodes to thermal gradients are well documented (El-Sharif and Mai, 1969; Burman and Pye, 1980) and some nematodes display extreme sensitivity to thermal cues. *C. elegans* will move in a thermal gradient to the temperature at which it has previously been cultured, and is capable of detecting thermal gradients of less than $0.1°C$ (Hedgecock and Russell, 1975). An even more exquisite thermal sensitivity has been demonstrated for a plant-parasitic nematode *Meloidogyne incognita*. The second larval stage of this nematode can move to a preferred temperature from a higher or lower temperature (Diez and Dusenbury, 1989) and is reported to respond to changes in temperature of less than $0.001°C$ (Dusenbury, 1988). Other nematodes may have a reduced requirement for thermal sensitivity. Many parasites of mammals for example may simply need to be able to detect the dramatic shift in temperature encountered upon entering their host.

The neuron responsible for detecting thermal cues has been identified in *C. elegans* using laser ablation studies to back up initial clues from studies on mutant strains. The thermoreceptive neuron was identified as the AFD sensory neuron, which sends numerous microvillar projections in to the amphidial sheath cell (see above and Figure 14.1c and 14.1d). The finger-like projections originating from this neuron are orientated along the longitudinal axis of the nematode in *C. elegans*, although this is not always the case in other nematodes. Mutant strains, which showed normal responses to thermal cues but altered responses to chemical stimuli, retained the normal morphology of the AFD sensory neuron. However, a mutant defective in its response to thermal cues lacked the microvillar projections on this neuron (Perkins *et al.*, 1986). Further evidence linking the AFD neuron to detection of thermal cues was obtained by ablating this neuron and observing the effects on behaviour. Destruction of the neuron removed the animal's ability to respond to thermal cues, mimicking the mutant phenotypes observed (Mori and Oshima, 1995).

Neurons with a similar structure, and therefore possibly a similar functional role, are present in other nematode species. Tylenchid plant-parasitic nematodes possess a sensory neuron very similar in fine structure to the thermoreceptive AFD neuron of *C. elegans*. Since these nematodes inhabit the same environment as *C. elegans* (the soil) for at least part of their life cycles, this similarity in structure may reflect the need to detect a similar range of thermal cues. Small, soil dwelling nematodes are unable to migrate any great distance in their life cycles and it may be that thermal cues are used by the nematode to position itself at an appropriate depth in the soil. Animal-parasitic and marine forms do not usually carry such extensive microvillar structures and this may reflect a reduced requirement for detection of thermal stimuli by these nematodes.

Photoreception

Some nematodes have adaptations that allow them to detect and respond to light. Almost all nematodes possessing this characteristic are marine dwelling and approximately 50% of all known marine forms are thought to possess some structure of this type (Weiser, 1959). Photoreceptors generally take one of two forms. Specialised sensory organs, the ocelli are present in some nematodes, while in others pigment spots are present which shade photoreceptive sensory neurons present in the amphids (McLaren, 1976a). Evidence also exists suggesting that nematodes lacking ocelli and pigment spots can respond to light (Burr, 1985).

Ocelli are usually paired and associated with the pharynx. They are composed of a pigment cell, which is shaped rather like a wine glass, containing a sensory cell which harbours the photoreceptive tissues. The sides of the pigment cell contain large quantities of pigment granules which shade the sensory cell from its base and sides. The sensory cell is connected to the rest of the nervous system and, like photoreceptors from other phylogenetic groups, contains stacks of concentrically arranged lamellae (McLaren 1976a). Photoreceptors of this type have been described from a variety of aquatic nematodes including *Deontostoma californicum* (Siddiqui and Viglierchio, 1970), which were the first ultrastructural observations of any nematode photoreceptor, and a freshwater nematode *Chromadorina bioculatai* (Croll, Riding and Smith, 1972). Similar structures have also been reported from the animal-parasitic *Mermis nigrescens* (Burr and Babinszki, 1990). A broader description of photoreceptors in marine nematodes is given by McLaren (1976a).

Nematodes lacking ocelli may show adaptations of their amphids to allow photoreception. These nematodes have pigment spots which shade the amphids. Bollerup and Burr (1979) examined the distribution and chemical nature of such eyespot pigments in 10 nematode species. Pigment spots were almost always associated with the pharynx, although some pigment was also occasionally found in the epidermis. Several different substances were identified in pigment granules including melanins, haemoglobins and hemosiderins. Other pigments were also present which were not easily identifiable.

This type of photoreceptor has been described at the ultrastructural level in *O. vesicarius* (Burr and Webster, 1971; Burr and Burr 1975). In this nematode a subset of dendritic processes, presumably sensitive to light stimuli, leaves the amphids and is shaded by the pigment spot. The other dendritic processes in the amphid of this nematode respond to chemical stimuli, indicating again that nematode sense organs are capable of adaptation to allow detection of more than one source of stimulus.

Function

It is inevitable that this section draws almost exclusively on literature from the *C. elegans* community. However, some parasitic nematodes have unusually large sense organs (e.g. *Syngamus trachea*) and these nematodes have recently been used in electrophysiological studies aimed at investigating chemosensory function in nematodes (Jones, Perry and Johnston, 1991; Riga *et al.*, 1995b). Responses of the anterior nervous system to pheromones have also been recorded in *Globodera* spp. (Riga *et al.*, 1996; Perry and Aumann, 1998). The opportunities offered by bringing together molecular techniques, such as those used with *C. elegans*, and these physiological techniques remain to be fully exploited.

Chemoreception

Unravelling the molecular events surrounding the perception of odours by nematodes has been one of the many triumphs of the *C. elegans* project. *C. elegans*, until very recently, remained the only invertebrate for which olfactory receptors were identified, despite many years effort from scientists working with insects and other organisms. *C. elegans* was also the first animal for which definitive functional evidence linking a specific olfactory receptor protein to perception of a defined odorant was obtained.

The first direct evidence that the amphids and other sensillae that open to the environment functioned as chemoreceptors came from the identification of mutants of *C. elegans* which showed abnormal responses to

chemical stimulants (Lewis and Hodgkin, 1977). Electron microscopical investigations showed that many of these mutant lines had structural defects in the dendritic processes of the sensory neurons projecting into the amphids or inner labial sense organs. Other mutants had defects in the structure of the sheath cells of these sense organs. Later studies showed that many of these mutants failed to take up FITC through the sensory neurons in the same manner observed in wild-type animals, again suggesting changes in the nature of the cilia of these neurons. Laser ablation studies (Davis, Goode and Dusenbury, 1986; also see references below) in which sensory neurons were ablated and the effects on chemotaxis observed provided further evidence to support the chemosensory role of the amphids and inner labial sensillae. While molecular and genetic studies (below) have made the greatest contribution to our understanding of the function of nematode chemoreceptors, electrophysiological studies on a range of nematode species are now being used to study the physiological responses of nematodes to odorants (Jones, Perry and Johnston, 1991; Riga *et al*, 1995b; Riga *et al.*, 1996; Riga *et al.*, 1997).

A combination of laser ablation studies and genetic analysis has led to the identification of odorants detected by each of the sensory neurons of the amphids of *C. elegans* (Bargmann, Hartweig and Horvitz, 1993; reviewed in Bargmann and Mori, 1997). Sensory neurons whose dendritic processes are exposed directly to the external environment through the amphidial pore (i.e. those ending in the amphidial canal) detect water-soluble odorants including salts, the dauer pheromone and amino acids. Neurons which have dendritic processes embedded within the amphidial sheath cell tend to detect volatile odorants, with the exception of the sensory neuron bearing microvillar projections which is thought to detect thermal cues (see above). Although individual neurons can detect multiple odorants and different neurons may detect the same odorant, a general rule is that any particular neuron can only direct one type of behavioural response (attraction or avoidance). Evidence to support this was provided by an elegant set of experiments in which the receptor for diacetyl (see below), a chemical normally attractive to *C. elegans*, was expressed in a different sensory neuron which normally detects repellent chemicals. Remarkably, these transgenic animals avoided diacetyl, a complete reversal of the normal behaviour pattern (Troemel, Kimmel and Bargman, 1997).

Cellular analysis, coupled with the full wiring diagram available for *C. elegans* (White *et al.*, 1986), allows predictions to be made about how behaviour is generated. For example, two neurons which detect repellents are known to be presynaptic to two interneurons which, in turn, have output onto the motor neurons which mediate backward movement (Bargmann and Mori, 1997). *C. elegans* therefore provides a unique opportunity for the molecular and physiological mechanisms underlying complex behaviour patters to be understood.

The molecular mechanisms underlying detection of odorant molecules have also been studied in detail in *C. elegans*. A large family of genes thought to encode odorant receptor molecules have been identified (Troemel *et al.*, 1995). The odorant receptor gene family had previously been extremely difficult to identify in any invertebrate species, mainly because these molecules have no defining conserved sequence motif which could be used to isolate them. The strategy used relied on the fact that genes with related functions are frequently found in clusters in the *C. elegans* genome. Regions of sequenced DNA around genes that had potential roles in olfactory signalling pathways (below) were searched electronically for new genes that might be odorant receptors. Although few clues were available from the primary sequence it was known that the odorant receptors were likely to be a relatively large gene family and to have seven hydrophobic domains in their predicted amino acid sequence. Using these criteria over 40 candidate genes were identified. Subsequent studies showed that many of these genes were expressed specifically in sensory neurons and were therefore likely to be the receptors being sought. Receptors differed in the specificity of their expression patterns, with some being expressed only in single sensory neurons and others present in up to five sensory neurons. This linked well with cellular studies which showed that some odorants are detected by single neurons while others are detected by several different neurons.

Although the evidence is convincing that the gene family identified from *C. elegans* represents genuine odorant receptors, functional data linking a specific gene with detection of a defined odorant is available for only one gene, *odr-10*. This gene was identified using a more classical genetical approach (Sengupta *et al.*, 1996) and encodes a protein similar to those identified previously. The *odr-10* mutant phenotype is normal in its responses to all odorants tested except that animals carrying this mutation are not attracted to diacetyl. Transformation of mutant animals with the cloned *odr-10* gene restored normal behaviour, thus identifying the ODR-10 protein as the receptor for diacetyl. *odr-10* encodes a putative seven transmembrane domain receptor and, as expected, is expressed specifically in the sensory neuron associated with the response to diacetyl. The ODR-10 receptor is capable of detecting diacetyl when expressed in cultured mammalian cells (Zhang *et al.*, 1997), providing biochemical evidence to support the genetic evidence that ODR-10 is the diacetyl receptor. Intriguingly, ODR-10 also detected pyruvate and citrate, which are precursors used in diacetyl production by some bacterial species, when expressed in mammalian cells. However, no evidence for a behavioural response to these odorants mediated by ODR-10 in *C. elegans* is available.

The odorant receptors of *C. elegans* are similar to their analogues in vertebrates (Buck and Axel, 1991) in that they are members of the seven transmembrane domain superfamily of receptors. Receptors of this type are linked to G-proteins and indeed are often referred to as G-protein coupled receptors. Binding of a ligand to a G-protein coupled receptor releases a sub-unit of the G-protein, which then activates adenyl cyclase. The cyclic AMP produced by this enzyme then opens cyclic nucleotide gated channels. This may release intracellular calcium stores or may directly cause depolarisation of the sensory neuron membrane. An alternative pathway which proceeds through inisitol triphosphate is also present in olfactory tissues of other animals and it has been shown that each odorant activates only one of the two signal transduction pathways (Schandar *et al.*, 1998). Although no data is available for the secondary messenger system in nematodes, these secondary messenger cascades are conserved across vertebrates and invertebrates and it seems likely that they are used in nematodes to generate electrical activity in sensory neurons. Further evidence for this has come from experiments in which *odr-10* was expressed in cultured mammalian cell lines (Zhang *et al.*, 1997). The nematode receptor was able to activate a mammalian secondary messenger pathway, ultimately causing release of intracellular calcium stores, when the cells were exposed to diacetyl, the odorant detected by this receptor in the nematode.

In summary, the molecular mechanisms underlying odorant detection in nematodes have been thoroughly investigated and largely unravelled. Odorants bind to specific receptor molecules located in the sensory neurons which direct the appropriate behavioural response when active. Electrical response is generated by one of two secondary messenger cascades proceeding through cyclic AMP or IP$_3$. Future challenges in this area are likely to include understanding the interactions of the vast array of odorant molecules detected by nematodes with individual receptor proteins and how sensory input from a relatively large number of sources is integrated by the nervous system in order to generate appropriate behaviour patterns.

One aspect of the function of nematode chemoreceptors which has not been thoroughly addressed, even in studies on *C. elegans*, is the role of the secretions produced by the sheath cells surrounding these sense organs. Although some parasitic nematodes have adaptations which allow them to produce proteins useful in the host-parasite interaction from these secretory cells (below), several lines of evidence suggest that the sheath cell secretions are vital for chemoreception. All nematode chemoreceptors possess this secretory cell and have their sensory neurons bathed in its secretions. More importantly, mutants of *C. elegans* defective in the structure of their sheath cells show abnormal responses to odorants (Lewis and Hodgkin, 1977). What might the role of these secretions be? Some authors have suggested that the secretions serve to maintain appropriate ionic conditions around the sensory neurons (Wright, 1983; Trett and Perry, 1985b). The fact that mechanoreceptors are also bathed in secretions in their receptor cavities supports this theory but the undoubted presence of proteins in these secretions of chemoreceptors suggests a more complex role. Cholinesterase has been shown to be present in the secretions of the sense organs of a wide range of nematode species (reviewed in McLaren, 1976b, also see below) but its role in the chemoreception process remains obscure. A 32 kDa glycoprotein has been shown to be present in the amphid secretions of the root knot nematode *M. incognita* (Stewart *et al.*, 1993). This protein was present only in nematode stages such as the invasive stage larva and the adult male, which require a good chemosensory capacity, and was absent in the sedentary feeding stages, suggesting a role in chemoreception (Stewart, Perry and Wright, 1993). Further evidence to support this was provided by the fact that an antiserum against the protein blocked chemotactic behaviour (Stewart *et al*, 1993). Although nothing is known about the nature of this molecule, these experiments demonstrate that proteins in the sheath cell secretions are important for the function of the sense organs. Perhaps something can be learnt from studies of the sense organs of insects. The sensory neurons of insect sense organs are also bathed in proteinaceous secretions produced by a surrounding sheath cell. However, the much larger size of insect sensilla has meant that it has been possible to collect, identify and biochemically characterise the proteins present. The secretions of insect sense organs contain two major groups of proteins with important roles in chemoreception (Vogt, Riddiford and Prestwich, 1985). Small, abundant olfactory binding proteins (OBPs) solubilise hydrophobic odorant molecules and transport them across the hydrophilic receptor cavity to the nerve processes. The secretions of insect sense organs also contain odorant degrading enzymes which destroy odorant molecules released from receptors on sensory neurons and prevent repeated restimulation of the neuron by a single odorant molecule (Rybczynski, Reagan and Lerner, 1989; Tascayo and Prestwich, 1990; Vogt and Riddiford, 1981). Is it likely that nematode sense organs contain similar proteins in their sense organs? Nematodes respond to a variety of volatile odorants suggesting that odorant binding proteins, or functional analogues may be required. A search of the complete *C. elegans* genome sequence shows no proteins with homology to insect odorant binding proteins are present (J. Jones, unpublished results). However, these proteins are extremely diverged, even within insects, and may not be recognisable on the basis of sequence similarity if present in nematodes. The case for odorant degrading enzymes or some protein which removes 'spent' odours from the receptor cavity seems even more compelling. Studies on nematodes with sense organs large enough to allow collection of sheath cell secretions may be enlightening. Preliminary work in this area has already demonstrated the feasibility of collecting secretions in sufficient quantity for basic analysis from such nematodes (Riga *et al.*, 1995a).

Mechanoreception

Nematodes encounter a diverse range of mechanical stimuli. Stimuli from the external environment and from receptors within the body are processed by nematodes and used to regulate many behaviour patterns. In order to facilitate studies in this complex area, four basic behaviour patterns, nose touch, gentle body touch, tap and harsh body touch, have been identified for *C. elegans* and used as models to understand how mechanical stimuli are used by nematodes. The neuronal circuitry controlling these behaviour patterns has been identified and mutant strains defective in their responses to mechanical stimuli have been identified. Studies are now underway which should allow the molecular mechanisms underlying these behaviours to be discovered.

Mutant lines which are defective in their response to one type of mechanical stimulus do not necessarily respond abnormally to other mechanical stimuli. For example, some mutant strains defective in their response to gentle touch along their body wall respond normally to stimuli detected by the anterior mechanoreceptors. This suggests that different molecular mechanisms are used by nematodes in detecting and processing different mechanical stimuli. Mutations which affect ciliary structure disrupt the function of many of the anterior sense organs including the mechanoreceptors (Lewis and Hodgkin, 1977; Albert, Brown and Riddle, 1981; Perkins *et al.*, 1986), indicating the importance of these structures in the detection of mechanical stimuli. A mutation in a potential secondary messenger also affects responses to mechanical stimuli detected by the anterior sense organs and this guanylate cyclase is also required for chemoreception (cited in Driscoll and Kaplan, 1997). Other than this, little is known about the molecular events occurring during detection of mechanical stimuli by the anterior sense organs.

A detailed model for the functioning of the touch receptors running along the length of the body of *C. elegans* has been proposed (Driscoll and Kaplan, 1997). These receptors detect gentle stimuli applied to the body wall of the nematode. Detection of a stimulus at the anterior end of the nematode elicits backward movement while a gentle touch towards the posterior end of the nematode elicits forward movement. The touch receptors are presynaptic to many other neurons, including interneurons which control locomotion (Chalfie *et al.*, 1985; White *et al.*, 1986), suggesting a mechanism by which activity in a touch receptor cell may be translated into an appropriate behavioural response.

At a molecular level the proposed model suggests that the cuticle is linked to proteins in the extracellular mantle. These proteins are themselves linked to a movement sensitive ion channel. At the same time microtubules within the receptor cell may also be linked, via another protein, to the ion channel. Ion flow through the channel could then be stimulated by movement of the proteins within the extracellular mantle or by movement of the microtubules within the cells. Genetic analyses have led to the identification of a variety of proteins that may be involved in the formation of the structures described in the model above. The presence of a mechanically gated ion channel is central to the proposed model and analysis of mutant strains has led to the identification of several genes which could encode subunits of such a channel. Mutations in *mec-4, mec-10* and *mec-6* give rise to animals which do not respond normally to touch but which do not show altered structure of the touch receptors (Chalfie and Sulston, 1981). Two of the genes, *mec-4* and *mec-10*, are co-expressed in the touch receptor cells and almost nowhere else in the body (Mitani *et al.*, 1993; Huang and Chalfie, 1994). All three genes have been found to encode proteins which show similarities in their sequences to subunits of ion channels from vertebrates. Genetic analysis suggests that these three proteins interact with one another (Driscoll and Chalfie, 1991; Huang and Chalfie 1994). How is this ion channel controlled? One possibility is that proteins present in the extracellular mantle link directly to this ion channel thus allowing movement in the mantle to be translated to a change in the condition of the ion channel. Two candidates for proteins fulfilling this role exist. *mec-5* encodes a collagen expressed in the epidermis and mutations in this gene give rise to subtle changes in the biochemical properties of the mantle, indicating that the MEC-5 protein may normally be located in the mantle (Du *et al.*, 1996). *mec-9* produces two transcripts, the larger of which is restricted in its expression to the touch receptor cells and could encode a secreted protein. Mutations in *mec-9* enhance a temperature sensitive mutation of *mec-5*, suggesting an interaction between the proteins encoded by these genes which might take place in the extracellular mantle (Gu, Caldwell and Chalfie, 1996). The nature of the interactions between MEC-5 and MEC-9 and between these proteins and the ion channel is not yet fully understood.

Links between the cytoskeleton and the ion channel may also be important in mechanoreceptor function. Electron microscopical observations show that overlapping microtubules, each 10–20 μm long, run the length of the touch cells which are each approximately 400 μm in length (Chalfie and Thomson, 1979). One end of the microtubules is often linked to the membrane of the receptor cell and may therefore be associated with membrane proteins, forming a physical link between the cytoskeleton and proteins involved in generating a response to mechanical stimuli. Further evidence for the role of microtubules in mechanosensory function was obtained by exposing nematodes to very low concentrations of colchicine. This treatment disrupted the structure of touch cell microtubules and caused the nematodes to lose their sensitivity to touch stimuli (Chalfie and Thomson, 1982). Two genes, *mec-7* and *mec-12*, have been identified as encoding tubulins which are likely to form the touch cell microtubules (Savage *et al.*, 1989). Mutations in these genes give rise to touch-insensitive animals. These genes are also highly expressed in touch cell neurons (Savage *et al.*, 1994) and the presence of both proteins is required for formation of the mechanosensory microtubules (Driscoll and Kaplan, 1997).

It is thought that the link between the microtubules and the mechanosensory ion channel is an indirect one,

and that MEC-2 may function as the linking protein. The distribution of MEC-2 is dependent on normal *mec-7* and *mec-12* function, indicating a physical interaction between these proteins in touch receptor cells (Huang *et al.*, 1995), while genetic evidence suggests that MEC-2 also interacts with proteins which form the sub-units of the mechanosensory ion channel (Huang and Chalfie, 1994). Sequence features of MEC-2 also imply a role in linking the cytoskeleton to an ion channel. Domains implicated in protein-protein interactions and regions with high similarity to a human stomatin protein are present. Stomatin is known to interact with the cytoskeleton in human red blood cells and faults in this protein lead to a change in the permeability of these cells to sodium ions (Stewart, Argent and Dash, 1993). Proteins of this type may therefore be involved in linking the cytoskeleton and ion channels throughout the animal kingdom.

Unfortunately no direct information is available about the function of mechanoreceptors in any nematode other than *C. elegans*. Given that even for *C. elegans* our knowledge of how these sense organs function is far from complete, this is unlikely to change in the near future.

Thermoreception

The molecular basis of thermosensation is not well understood for any animal. However, evidence is emerging that similar mechanisms are utilised by *C. elegans* and other animals to detect thermal cues (Mori and Oshima, 1997). The *tax-2* and *tax-4* mutants are defective in their responses to a variety of stimulants, including thermal cues. The genes disrupted by these mutations have recently been identified as encoding subunits of cyclic nucleotide gated ion channels and the products have been localised to the sensory endings of the neurons mediating the sensory behaviours affected by the original mutations (Komatsu *et al.*, 1996). Although the picture is far from complete it appears that cyclic nucleotides are important secondary messengers in thermosensation as well as in other sensory behaviours. The nature of the molecules which detect thermal cues in nematodes remains unknown but it has been suggested that changes in membrane fluidity may be important, perhaps explaining the vastly increased area of membrane present on the neuron responsible for detecting thermal cues (Bargmann and Mori, 1997). However, thermoreceptive molecules which detect changes in temperature through temperature-dependent conformational changes have been identified in the bacterium *Escherichia coli* (Nara, Lee and Imae, 1991) and it is perfectly feasible that molecules which fulfil a similar function are present on the thermoreceptor neurons of nematodes.

Slightly more is known about the downstream processing of information gathered about thermal cues. The sensory neuron responsible for detecting thermal cues is presynaptic to the AIY interneuron and one of the major postsynaptic partners of this interneuron is the AIZ interneuron (White *et al.*, 1986). Laser ablation studies have shown that the AIY interneuron is responsible for thermophilic movement and the AIZ interneuron for cryophilic movement (Mori and Ohshima, 1995). These two interneurons are extensively interconnected, allowing for balancing between opposite thermal cues and are presynaptic to another interneuron, RIA, which has also been implicated in processing of thermal cues. A working model suggests that the AIZ/AIY interneurons activate/inhibit RIA, which then summates this input and stimulates the relevant behaviour pattern. However, in the absence of electrophysiological data this model remains unproven.

The molecular events underlying the detection of thermal cues remain unknown for any animal. The well defined genetics and other advantages of *C. elegans* mean that it may provide a unique opportunity for study in this area.

Photoreception

Despite the demonstration that *C. elegans* is capable of responding to light (Burr, 1985) nothing is known about the molecular or physiological mechanisms underlying phototaxis in nematodes. One reason for this is perhaps because the assays required to demonstrate phototaxis need to be incredibly complex. *C. elegans* is capable of responding to very small (0.05°C) temperature changes (Hedgecock and Russell, 1975) and this makes it extremely difficult to design experiments in which nematodes are exposed to differing light stimuli without also exposing them to thermal gradients. Thus, no mutants with altered responses to light have been found. Molecular studies on other nematodes which respond to light are currently underway (Burr *et al.*, 1997) but these are limited to molecular characterisation of the pigments found in the eyespots of these nematodes. A homeobox gene, *ceh-10* has been identified in *C. elegans* which is expressed in interneurones onto which the AFD amphidial neurone has synaptic output (Svendsen and McGhee, 1995). This gene has homology to two genes expressed in the vertebrate retina. Whether these findings tell us about the conservation of mechanisms controlling the development of photoreceptive organs or about the conservation of the control of the development of sensory structures more generally awaits further characterisation of the range of stimuli detected by the AFD neuron.

Adaptations of Sense Organs

Sense organs of nematodes are frequently adapted to allow them to fulfil another role in the nematode life cycle. The sheath cells of the amphids of parasitic nematodes occasionally become adapted to allow the production of secreted material which has a specialised role in the host-parasite interaction. This new role is not usually connected with the normal sensory function of the amphids and it is not known whether the amphids are generally capable of retaining their sensory function once adapted in this way. However, electrophysiologi-

cal studies have shown that the amphids of adult female *S. trachea*, which produce anticoagulants, also respond to chemical stimulants (Riga *et al.*, 1995b). In at least one other case the amphids revert to a sensory role in a later stage having been adapted for the feeding process in a larval stage. Specific examples are discussed below.

Anticoagulants

Many parasitic nematodes which feed on the blood of their hosts have taken advantage of the position of the amphids, near to the feeding apparatus, and the presence within the amphids of a convenient gland cell and allowed these organs to become adapted to produce anticoagulants in the feeding stages of the parasite. Anticoagulant activity has been detected in the sense organ secretions of a wide range of nematodes including *Necator americanus* (McLaren, Burt and Ogilvie, 1974), *S. trachea* (Jones, 1975; Jones, 1979) and *Ancylostoma caninum* (Thorson, 1956; Eiff, 1966). The sheath cells of the amphids of *N. americanus* and *S. trachea* are considerably enlarged in the feeding stages, presumably to allow production of very large quantities of anticaogulants. While the advantages to the parasite of subverting host haemostasis are fairly obvious the mechanisms used to achieve this have only recently been uncovered. In studies on the hookworm *A. caninum*, a family of small proteins which inhibit coagulation of host blood have been described (Stanssens *et al.*, 1996). This nematode was shown to use a mechanism distinct from those used by the host or by other blood feeding parasites (including ticks and leeches) to manipulate the mammalian blood coagulation pathway. Two proteins which inhibit the activity of blood coagulation factor Xa were produced by the parasite while a third protein inhibited the activity of a complex containing coagulation factor VIIa and tissue factor. While a detailed review of this area is beyond the scope of this article, this work illustrates the remarkable adaptations which nematode sense organs may undergo.

Feeding Plugs

The soybean cyst nematode *H. glycines*, like many endoparasitic nematodes of plants, induces remarkable changes in the roots of its host plant. After invading and migrating to a suitable site in the roots, the nematode induces the production of a large, multinucleate and highly metabolically active syncytium in the roots of its host. The nematode becomes sedentary, loses all body wall muscle and relies exclusively on the syncytium to provide all the nutrients required for development to the adult stage. Nutrients are withdrawn from the syncytium through the stylet, a nematode structure not unlike a hypodermic needle in appearance. During feeding, the stylet is repeatedly withdrawn from the syncytium as a normal part of the feeding cycle (Wyss, 1992). The stylet is also withdrawn to allow moulting to occur. In order to prevent leakage of syncytium contents and death of the syncytium, the nematode produces a feeding plug, which is thought to seal the syncytium cell wall while the stylet is retracted. This is critical since the nematode is only able to induce the formation of a single syncytium. Electron microscopical observations (Endo, 1978; 1998) suggest that the feeding plug is produced by the amphid sheath cells in *H. glycines*, although it is uncertain whether this is true for other cyst nematodes (Jones, Perry and Johnston, 1993). Whether the amphids of plant-parasitic nematodes retain their chemosensory function in this feeding stage is unknown. Feeding endoparasitic nematodes are immobile and therefore probably have no requirement for a complex chemosensory capability. A glycoprotein secreted by the amphids of a related nematode, *M. incognita*, which is thought to have a role in chemoreception is not produced by the feeding stages of this nematode (Stewart, Perry and Wright, 1993) suggesting that the chemosensory function is lost. The adult male however, reverts to the vermiform shape and locates the still immobile female for mating. In this stage it is clear that the amphid reverts to a chemosensory role.

Acetylcholinesterase

Good evidence exists for the secretion of acetylcholinesterase from the amphids and other anterior sense organs of a wide range of nematodes, including many parasitic forms (reviewed in McLaren, 1976b). Cholinesterase has been suggested to function as a 'biological holdfast', assisting some animal parasites in maintaining their position in the gut (Lee, 1969; Ogilvie and Jones, 1971). It has also been suggested that this enzyme may alter host membrane permeability, enhancing release of nutrients to the parasite (Lee, 1970). Cholinesterase in animal-parasitic nematodes is also very likely to modulate the immune response of the host, reducing inflammation in response to parasite attack (reviewed by Lee, 1996). Cholinesterase activity has been demonstrated in the sense organs of plant parasitic nematodes (Rohde, 1960) and the free-living *C. elegans* (Pertel, Paran and Mattern, 1976). This, coupled with the fact that this enzyme has been found in all the anterior chemoreceptors and in the phasmids of at least one nematode (McLaren, 1976b), suggests that this enzyme has a role in chemoreception and that some animal-parasitic nematodes have taken advantage of its presence in the amphids to enhance the efficiency of the parasitic process.

Concluding Remarks

Things have come a long way since Goldschmit (1903) first dissected *Ascaris lumbricoides* and figured that it might be useful for this nematode to have some way of finding its way around. A curious blend of literature exists on nematode sensory perception. On the one hand we have a large batch of ultrastructural papers, published shortly after electron microscopes became accessible to the research community. Another equally large group of papers exists on the function of nematode sense organs, published as the molecular

biological tools required to take advantage of the advantages offered by *C. elegans* became available. As our estimates of the number of different nematode species and the habitats in which we find them increase, it would not be surprising to find weird and wonderful additions to the ultrastructural variations found in the phylum. Since the genome of *C. elegans* has now been sequenced and interest is now focussing on determining how all these genes give rise to a functioning nematode, it is almost certain that within a few years time we will understand in enormous detail how the sense organs of at least one nematode are made and precisely how they function.

Acknowledgements

I would like to acknowledge the impact that the enthusiasm and breadth of knowledge of Dr Mike Johnston, formerly of UCW Aberystwyth but now retired, has had on my forays into the field of nematode sensory perception. I would also like to thank Dr D.L. Trudgill for helpful comments on the manuscript and Mr Ian Pitkethly for assembling the figures for this chapter. SCRI receives funding from the Scottish Office Agriculture, Environment and Fisheries Department.

References

Albert, P.S., Brown, S.J. and Riddle, D.L. (1981) Sensory control of dauer larva formation in *Caenorhabditis elegans*. *Journal of Comparative Neurology*, **198**, 435–451.

Albertson, D.G. and Thomson, J.N. (1976). The pharynx of *Caenorhabditis elegans*. *Philosphical Transactions of the Royal Society of London Series B: Biological Sciences*, **275**, 299–325.

Ashton, F.T., Bhopale, V.M., Fine, A.E. and Schad, G.A. (1995) Sensory neuroanatomy of a skin-penetrating nematode parasite *Strongyloides stercoralis*. 1. Amphidial neurones. *Journal of Comparative Neurology*, **357**, 281–295.

Ashton, F.T., Bhopale, V.M., Holt, D., Smith G. and Schad, G.A. (1998) Developmental switching in the parasitic nematode is controlled by the ASF and ASI amphidial neurons. *Journal of Parasitology*, **84**, 691–695.

Ashton, F.T., Li, J. and Schad, G.A. (1999) Chemo- and thermosensory neurons: structure and function in animal parasitic nematodes. *Veterinary Parasitology*, **84**, 297–316.

Ashton, F.T and Schad, G.A. (1996) Amphids in *Strongyloids stercoralis* and in other parasitic nematodes. *Parasitology Today*, **12**, 187–194.

Baldwin, J.G. (1985) Fine structure of the phasmid of second stage juveniles of *Heterodera schachtii* (Tylenchida: Nematoda). *Canadian Journal of Zoology*, **63**, 534–542.

Baldwin, J.G. and Hirschmann, H. (1973) Fine structure of cephalic sense organs in *Meloidogyne incognita* males. *Journal of Nematology*, **5**, 285–296.

Bargmann, C.I., Hartweig, E. and Horvitz, R. (1993) Odorant selective genes and neurones mediate olfaction in *C. elegans*. *Cell*, **74**, 515–527.

Bargmann, C.I. and Mori, I. (1997) Chemotaxis and thermotaxis. In *C. elegans II*, edited by D.L. Riddle, T. Blumenthal, B.J. Meyer and J.R. Priess, pp 717–737. Cold Spring Harbor: Cold Spring Harbor Laboratory Press.

Batson, B.S. (1978) Ultrastructure of the anterior sense organs of adult *Gastromermis boophthorae* (Nematoda: Mermithidae). *Tissue and Cell*, **10**, 51–61.

Bird, A.F. and Bird, J. (1991) *The structure of nematodes*, 2nd edn. New York: Academic Press, 316pp.

Bollerup, G. and Burr, A.H. (1979) Eyespot and other pigments in nematode esophageal muscle cells. *Canadian Journal of Zoology*, **57**, 1057–1069.

Buck, L. and Axel, R. (1991) A novel multigene family may encode odorant receptors: a molecular basis for odor recognition. *Cell*, **65**, 175–187.

Burman, M. and Pye, A.E. (1980) *Neoaplectana carpocapsae*: movements of nematode populations on a thermal gradient. *Experimental Parasitology*, **49**, 258–265.

Burr, A.H. (1985) The photomovement of *Caenorhabditis elegans*, a nematode which lacks ocelli. Proof that the response is to light not radiant heating. *Photochemistry and Photobiology*, **41**, 577–582.

Burr A.H. and Babinszki, C.P.F. (1990) Scanning motion, ocellar morphology and orientation mechanisms in the phototaxis of the nematode *Mermis nigrescens*. *Journal of Comparative Physiology*, **167A**, 257–268.

Burr A.H. and Burr C. (1975) The amphid of the nematode *Oncholaimus vesicarius*: ultrastructural evidence for a dual function as chemoreceptor and photoreceptor. *Journal of Ultrastructure Research*, **51**, 1–15.

Burr, A.H., Sidhu, P., Hunt, P., Blaxter, M. and Moens, L. (1997) Sequence and function of the eye hemoglobin of *Mermis nigrescens*, a nematode parasite of grasshoppers. *Proceedings of the 'Parasitic Helminths — from genomes to vaccines' meeting*, Edinburgh, September 1997, p20.

Burr A.H. and Webster, J.M. (1971) Morphology of the eyespot and a redescription of two pigment granules in the oesophageal muscles of a marine nematode *Oncholaimus vesicarius*. *Journal of Ultrastructure Research*, **36**, 621–632.

Chalfie, M. and Sulston, J.E. (1981) Developmental genetics of the mechanosensory neurones of *Caenorhabditis elegans*. *Developmental Biology*, **82**, 358–370.

Chalfie, M., Sulston, J.E., White, J.G., Southgate, E., Thomson, J.N. and Brenner, S. (1985) The neural cicuit for touch sensitivity in *C. elegans*. *Journal of Neuroscience*, **5**, 956–964.

Chalfie, M. and Thomson, J.N. (1979) Organisation of neuronal microtubules in the nematode *Caenorhabditis elegans*. *Journal of Cell Biology*, **82**, 278–289.

Chalfie, M. and Thomson, J.N. (1982) Structural and functional diversity in the neuronal microtubules of *C. elegans*. *Journal of Cell Biology*, **93**, 15–23.

Clark S.A. and Shepherd, A.M. (1977) Structure of the spicules and caudal sensory equipment in the male of *Aphelenchoides blastophthorus* (Nematoda: Tylenchida, Aphelenchida). *Nematologica*, **23**, 103–111.

Clark, S.A., Shepherd, A.M. and Kempton, A. (1973) Spicule structure in some *Heterodera* spp. *Nematologica*, **19**, 242–247.

Croll, N.A., Riding I.L. and Smith, J.M. (1972) A nematode photoreceptor. *Comparative Biochemistry and Physiology*, **42A**, 999–1009.

Davis, B.O., Goode, M. and Dusenbury, D.B. (1986) Laser microbeam studies of role of amphid receptors in chemosensory behaviour of nematode *Caenorhabditis elegans*. *Journal of Chemical Ecology*, **12**, 1339–1347.

Diez, J.A and Dusenbury, D.B. (1989) Preferred temperature of *Meloidogyne incognita*. *Journal of Nematology* **21**, 99–104.

Driscoll, M. and Chalfie, M. (1991) The *mec-4* gene is a member of a family of *Caenorhabditis elegans* genes that can mutate to induce neuronal degeneration. *Nature*, **349**, 588–593.

Driscoll, M. and Kaplan, J. (1997) Mechanotransduction. In *C. elegans II*, edited by D.L. Riddle, T. Blumenthal, B.J. Meyer and J.R. Priess, pp 645–677. Cold Spring Harbor: Cold Spring Harbor Laboratory Press.

Du, H., Gu, G, William, C. and Chalfie, M. (1996) Extracellular proteins needed for *C. elegans* mechanosensation. *Neuron*, **16**, 183–194.

Dusenbury, D.B. (1988) Behavioural responses of *Meloidogyne incognita* to small temperature changes. *Journal of Nematology* **20**, 351–355.

Eiff, J.A., (1966) Nature of an anticoagulant from the cephalic glands of *Ancylostoma caninum*. *Journal of Parasitology*, **52**, 833–843.

El-Sherif, M. and Mai, W.F. (1969) Thermotactic response of some plant parasitic nematodes. *Journal of Nematology*, **1**, 43–48.

Endo, B.Y. (1978) Feeding plug formation in soybean roots infected with the soybean cyst nematode. *Phytopathology*, **68**, 1022–1031.

Endo, B.Y. (1980) Ultrastructure of the anterior neurosensory organs of the larvae of the soybean cyst nematode *Heterodera glycines*. *Journal of Ultrastructure Research*, **72**, 349–366.

Endo, B.Y. (1998). *Atlas on ultrastructure of the infective juveniles of the soybean cyst nematode* Heterodera glycines. USDA Agriculture Handbook No. 711, 224 pp.

Fine, A.E., Ashton, F.T, Bhopale, V.M. and Schad, G.A. (1997) Sensory neuroanatomy of a skin-penetrating nematode parasite *Strongyloides sterocralis*, II. Labial and cephalic neurones. *Journal of Comparative Neurology*, **389**, 212–223.

Goldschmit, R. (1903) Histologische untersuchungen an nematoden, I. Die sinnesorgane von *Ascaris lumbricoides* und *A. megalocephala* Cloq. *Zoologische Jahrbuch Anatomie*, **18**, 1–57.

Goodey, J.B. (1963) *Soil and Freshwater Nematodes* 2nd edn. London:- Methuen, 544 pp.

Gu, G., Caldwell, G.A. and Chalfie, M. (1996) Genetic interactions affecting touch sensitivity in *Caenorhabditis elegans*. *Proceedings of the National Academy of Sciences of the USA*, **93**, 6577–6582.

Hedgecock, E.M. and Russell, R.L. (1975) Normal and mutant thermotaxis in the nematode *Caenorhabditis elegans*. *Proceedings of the National Academy of Sciences of the USA*, **72**, 4061–4065.

Huang, M. and Chalfie, M. (1994) Gene interactions affecting mechanosensory transduction in *Caenorhabditis elegans*. *Nature*, **367**, 467–470.

Huang, M., Gu, G., Ferguson, E.L. and Chalfie, M. (1995) A stomatin — like protein is needed for mechanosensation in *C. elegans*. *Nature*, **378**, 292–295.

Jones, G.M. (1975) The amphidial glands of *Syngamus trachea* (Nematoda). *PhD Thesis*, University of Wales.

Jones, G.M. (1979) The development of amphids and amphidial glands in adult *Syngamus trachea* (Nematoda: Syngamidae). *Journal of Morphology*, **160**, 299–322.

Jones, J.T., Perry, R.N. and Johnston, M.R.L. (1991) Electrophysiological recordings of electrical activity and responses to stimulants from *Globodera rostochiensis* and *Syngamus trachea*. *Revue de Nematologie*, **14**, 467–473.

Jones, J.T., Perry, R.N. and Johnston, M.R.L. (1993) Changes in the ultrastructure of the cuticle of the potato cyst nematode *Globodera rostochiensis* during development and infection. *Fundamental and Applied Nematology*, **16**, 433–445.

Jones, J.T., Perry, R.N. and Johnston, M.R.L. (1994) Changes in the ultrastructure of the amphids of the potato cyst nematode *Globodera rostochiensis* during development and infection. *Fundamental and Applied Nematology*, **17**, 369–382.

Kaplan, J.M. and Horvitz, H.R. (1993) A dual mechanosensory and chemosensory neurone in *Caenorhabditis elegans*. *Proceedings of the National Academy of Sciences of the USA*, **90**, 2227–2231.

Komatsu, H., Mori, I., Rhee, J.S., Akaike, N. and Ohshima, Y. (1996) Mutations in a cyclic nucleotide-gated channel lead to abnormal thermosensation and chemosensation in *C. elegans*. *Neuron*, **17**, 707–718.

Lee, D.L. (1969) Changes in adult *Nippostrongylus brasiliensis* during the development of immunity to this nematode in rats. *Parasitology*, **59**, 29–39.

Lee, D.L. (1970) The fine structure of the excretory system in adult *Nippostrongylus brasiliensis* (Nematoda) and a suggested function for the excretory glands. *Tissue and Cell*, **2**, 225–231.

Lee, D.L. (1973) Evidence for a sensory function for the copulatory spicules of nematodes. *Journal of Zoology, London*, **169**, 281–285.

Lee, D.L. (1996) Why do some nematode parasites of the alimentary tract secrete acetylcholinesterase? *International Journal for Parasitology*, **26**, 499–508.

Lewis, J.A. and Hodgkin, J.A. (1977) Specific neuroanatomical changes in chemosensory mutants of the nematode *Caenorhabditis elegans*. *Journal of Comparative Neurology*, **712**, 489–510.

McLaren, D.J. (1972) Ultrastructural and cytochemical studies on the sensory organelles and nervous system of *Dipetalonema viteae* (Nematoda: Filarioidea). *Parasitology*, **65**, 507–524.

McLaren, D.J. (1976a) Nematode Sense Organs. In *Advances in Parasitology*, edited by B. Dawes, pp 195–265. London: Academic Press.

McLaren, D.J. (1976b) Sense organs and their secretions. In *The Organisation of Nematodes*, edited by N.A. Croll, pp 139–161. London: Academic Press.

McLaren, D.J., Burt, J.S. and Ogilvie, B.M. (1974) The anterior glands of adult *Necator americanus* (Nematoda: Strongyloidea): II. Cytochemical and functional studies. *International Journal of Parasitology*, **4**, 39–46.

Mitani, S., Du, H., Hall, D.H., Driscoll, M. and Chalfie, M. (1993) Combinatorial control of touch receptor neurone expression in *Caenorhabditis elegans*. *Development*, **119**, 773–783.

Mori, I. and Ohshima, Y. (1995) Neural regulation of thermotaxis in *Caenorhabditis elegans*. *Nature*, **376**, 344–348.

Mori, I. and Ohshima, Y. (1997) Molecular neurogenetics of chemotaxis and thermotaxis in the nematode *Caenorhabditis elegans*. *Bioessays*, **19**, 1055–1064.

Nara, T., Lee, L. and Imae, Y. (1991) Thermosensing ablility of Trg and Tap chemoreceptors in *Escherichia coli*. *Journal of Bacteriology*, **173**, 1120–1124.

Ogilvie, B.M. and Jones, V.E. (1971) *Nippostrongylus brasiliensis*: A review of immunity and the host/parasite relationship in the rat. *Experimental Parasitology*, **29**, 138–177.

Perkins, L.A, Hedgecock, E.M., Thomson, J.N and Culotti, J.G. (1986) Mutant sensory cilia in the nematode *Caenorhabditis elegans*. *Developmental Biology*, **117**, 456–487.

Perry, R.N. and Aumann, J. (1998) Behaviour and Sensory Reponses. In *The Physiology and Biochemistry of Free-living and Plant-parasitic Nematodes*, edited by R.N. Perry and D.J. Wright, pp 75–102. Wallingford: CAB International.

Pertel, R., Paran, N. and Mattern, C.F.T. (1976) *Caenorhabditis elegans*: Localisation of cholinesterase associated with anterior nematode structures. *Experimental Parasitology*, **39**, 401–414.

Riga, E., Perry, R.N., Barrett, J. and Johnston, M.R.L. (1995a) Biochemical analyses on single amphidial glands, excretory — secretory gland-cells, pharyngeal glands and their secretions from the avian nematode *Syngamus trachea*. *International Journal of Parasitology*, **25**, 1151–1158.

Riga, E., Perry, R.N., Barret, J. and Johnston, M.R.L. (1995b) Investigation of the chemosensory function of amphids of *Syngamus trachea* using electrophysiological techniques. *Parasitology*, **111**, 347–351.

Riga, E., Perry, R.N., Barret, J. and Johnston, M.R.L. (1996) Electrophysiological responses of males of the potato cyst nematodes, *Globodera rostochiensis* and *G. pallida*, to their sex pheromones. *Parasitology*, **112**, 239–246.

Riga, E., Perry, R.N., Barret, J. and Johnston, M.R.L. (1997) Electrophysiological responses of males of the potato cyst nematodes, *Globodera rostochiensis* and *G. pallida*, to some chemicals. *Journal of Chemical Ecology*, **23**, 417–428.

Robertson, W.M. (1975) A possible gustatory organ associated with the odontophore in *Longidorus lectocephalus* and *Xiphinema diversicaudatum*. *Nematologica*, **21**, 443–448.

Robertson, W.M. (1979) Observations on the oesophageal nerve system of *Longidorus lectocephalus*. *Nematologica*, **25**, 245–254.

Robertson, W.M. and Forrest, J.M.S. (1989) Factors involved in host recognition by plant-parasitic nematodes. *Aspects of Applied Biology*, **22**, 129–133.

Rohde, R.A. (1960) Acetylcholinesterase in plant parasitic nematodes and an anticholinesterase from asparagus. *Proceedings of the Helminthological Society of Washington*, **27**, 121–123.

Rybczynski, R., Reagan, J. and Lerner, M.R. (1989) A pheromone degrading aldehyde oxidase in the antennae of the moth *Manduca sexta*. *Journal of Neuroscience*, **9**, 1341–1353.

Samoiloff, M.R., McNicholl, P., Cheng, R. and Balakanich, S. (1973) Regulation of nematode behaviour by physical means. *Experimental Parasitology*, **33**, 253–262.

Savage, C., Hamelin, M., Culotti, J.G., Coulson, A., Albertson, D.G. and Chalfie, M. (1989) *mec-7* is a β-tubulin gene required for the production of 15 microfilament microtubules in *Caenorhabditis elegans*. *Genes and Development*, **3**, 870–881.

Savage, C., Xue, Y.Z., Mitani, S., Hall, D., Zakhary, R. and Chalfie, M. (1994) Mutations in the *Caenorhabditis elegans* gene *mec-7*: effects on microtubule assembly and stability and on tubulin autoregulation. *Journal of Cell Science*, **107**, 2165–2175.

Schandar, M., Laugwitz, K.L., Boekhoff, I., Kroner, C., Gudermann, T., Schultz, G. and Breer, H. (1998) Odorants selectively activate distinct G protein subtypes in olfactory cilia. *Journal of Biological Chemistry*, **273**, 16669–16677.

Sengupta, P., Chou, J.H. and Bargmann, C.I. (1996) *odr-10* encodes a seven transmembrane domain olfactory receptor required for responses to the odorant diacetyl. *Cell*, **84**, 899–909.

Siddiqui, I.A. and Viglierchio, D.R. (1970) Fine structure of photoreceptors in *Deontostoma californicum*. *Journal of Nematology*, **2**, 274–276.

Stanssens, P., Bergum, P.W., Gamoermans, Y., Jespeos, L., Laroche, Y., Huang, S. *et al.*, (1996) Anticoagulant repertoire of the hookworm *Ancylostoma caninum*. *Proceedings of the National Academy of Sciences of the USA*, **93**, 2149–2154.

Stewart, G.R., Perry, R.N., Alexander, J. and Wright, D.J. (1993) A glycoprotein specific to the amphids of *Meloidogyne* species. *Parasitology*, **106**, 405–412.

Stewart, G.R., Perry, R.N. and Wright, D.J. (1993) Studies on the amphid specific glycoprotein gp32 in different life cycle stages of *Meloidogyne* species. *Parasitology*, **107**, 573–578.

Stewart, G.W., Argent, A.C. and Dash, B.C.J. (1993) Stomatin: A putative cation transport regulator in the red cell membrane. *Biochimica Biophysica Acta*, **1225**, 15–25.

Strote, G. and Bonow, I. (1996) Ultrastructural observations on the nervous system and the sensory organs of the infective stage (L3) of *Onchocerca volvulus* (Nematoda, Filarioidea). *Parasitology Research*, **79**, 213–220.

Strote, G., Bonow, I. and Attah, S. (1996) The ultrastructure of the anterior end of male *Onchocerca volvulus*: Papillae, amphids, nerve ring and first indication of an excretory system in the adult filarial worm. *Parasitology*, **113**, 71–85.

Sulston, J.E., Albertson, D.G. and Thomson, J.N. (1980) The *Caenorhabditis elegans* male: Postembryonic development of nongonadal structures. *Developmental Biology*, **78**, 542–576.

Svendsen, P.C. and McGhee, J.D. (1995). The *C. elegans* neuronally expressed homeobox gene *ceh-10* is closely related to genes expressed in the vertebrate eye. *Development*, **121**, 1253–1262.

Tascayo, M.L. and Prestwich, G.D. (1990) Aldehyde oxidases and dehydrogenases in antennae of five moth species. *Insect Biochemistry*, **20**, 691–700.

Thorsen, R.E. (1956) The effect of extracts of the amphidial glands, excretory glands and oesophagus of adults of *Ancylostoma caninum* on the coagulation of the dog's blood. *Journal of Parasitology*, **42**, 26–30.

Trett, M.W. and Lee, D.L. (1981) The cephalic sense organs of adult female *Hammerschmidtiella diesingi* (Nematoda: Oxyuroidea). *Journal of Zoology of London*, **194**, 41–52.

Trett, M.W. and Lee, D.L. (1982) A developmental study of the cephalic sense organs of the rat hookworm *Nippostrongylus brasiliensis*. *Parasitology*, **84**, R69.

Trett, M.W. and Perry, R.N. (1985a) Functional and evolutionary implications of the anterior sensory anatomy of species of root-lesion nematode (genus *Pratylenchus*). *Revue de Nématologie*, **8**, 341–355.

Trett, M.W. and Perry, R.N. (1985b) Effects of the carbamoyloxime, aldicarb, on the ultrastructure of the root lesion nematode *Pratylenchus penetrans* (Nematoda: Pratylenchidae). *Nematologica*, **31**, 321–334.

Troemel, E.R., Chou, J.H., Dwyer, N.D., Colbert, H.A. and Bargmann, C.I. (1995) Divergent seven transmembrane domain receptors are candidate chemosensory receptors in *C. elegans*. *Cell*, **83**, 207–218.

Troemel, E.R., Kimmel, B.E. and Bargmann, C.I. (1997) Reprogramming chemotaxis responses: sensory neurones define olfactory preferences in *C. elegans*. *Cell*, **91**, 161–169.

Vogt, R.G. and Riddiford, L.M. (1981) Pheromone binding and inactivation by moth antennae. *Nature*, **293**, 161–163.

Vogt, R.G., Riddiford, L.M. and Prestwich, G.D. (1985) Kinetic properties of a sex pheromone degrading enzyme: the sensillar esterase of *Antherea polyphemus*. *Proceedings of the National Academy of Sciences of the USA*, **82**, 8827–8831.

Wang, K.C. and Chen, T.A. (1985) Ultrastructure of the phasmids of *Scutellonema brachyurum*. *Journal of Nematology*, **17**, 175–186.

Ward, S., Thomson, N., White, J.G. and Brenner, S. (1975) Electron microscopical reconstruction of the anterior sensory anatomy of the nematode *Caenorhabditis elegans*. *Journal of Comparative Neurology*, **160**, 313–337.

Weiser, W. (1959) Cited in McLaren, D.J. (1976a).

White, J.G., Southgate, E., Thomson, J.N. and Brenner, S. (1986) The structure of the nervous system of the nematode *C. elegans*. *Philosophical Transactions of the Royal Society of London Series B: Biological Sciences*, **314**, 1–340.

Wright, K.A. (1974) Cephalic sense organs of the parasitic nematode *Capillaria hepatica* (Bancroft, 1893). *Canadian Journal of Zoology*, **52**, 1207–1213.

Wright, K.A. (1975) Cephalic sense organs of the rat hookworm *Nippostrongylus brasiliensis* — form and function. *Canadian Journal of Zoology*, **53**, 1131–1146.

Wright, K.A. (1977) The cephalic sense organs of the nematode *Heterakis gallinarium*. *Journal of Parasitology*, **63**, 528–539.

Wright, K.A. (1978) Structure and function of the male copulatory apparatus of the nematodes *Capillaria hepatica* and *Trichuris muris*. *Canadian Journal of Zoology*, **56**, 651–662.

Wright, K.A. (1980) Nematode sense organs. In: *Nematodes as Biological Models Volume 2*, edited by B.M Zuckerman, pp. 237–296. New York: Academic Press.

Wright, K.A. (1983) Nematode chemosensilla: form and function. *Journal of Nematology*, **15**, 151–158.

Wright, K.A. and Carter, R.F. (1980) Cephalic sense organs and body pores of *Xiphinema americanum* (Nematoda: Dorylaimoidea). *Canadian Journal of Zoology*, **58**, 1439–1451.

Wright, K.A. and Chan, J.(1974) A sense organ in the ventral anterior body wall of the nematode *Capillaria hepatica* (Bancroft, 1893). *Canadian Journal of Zoology*, **52**, 21–22.

Wright , K.A. and Hui, N.(1976) Post labial sensory structures on the cecal worm *Heterakis gallinarum. Journal of Parasitology*, **62**, 579–584.

Wyss U. (1992) Observations on the feeding behaviour of *Heterodera schachtii* throughout development, including events during moulting. *Fundamental and Applied Nematology*, **15**, 75–89.

Zhang, Y., Chou, J.H., Bradley, J., Bargmann, C.I. and Zinn, K. (1997) The *Caenorhabditis elegans* seven transmembrane protein ODR-10 functions as an odorant receptor in mammalian cells. *Proceedings of the National Academy of Sciences of the USA*, **94**, 12162–12167.

15. Behaviour

Donald L. Lee

School of Biology, University of Leeds, Leeds, LS2 9JT, UK

Introduction

Nematodes occupy a wide range of habitats and some of the early larval stages of parasitic species occupy very different habitats from those of the later larval stages and adults. It is not surprising to find, therefore, that there are differences in behaviour in the Nematoda, depending upon their ecological niche and the stage of their life cycle. Nematodes, depending upon their habitat, respond to various external stimuli, such as chemical, electrical, light, mechanical and temperature. Some of these external stimuli are important in feeding and copulation whereas others enable the nematode to locate in a suitable environment. The feeding behaviour of plant-parasitic nematodes is described by Wyss, and Ishibashi describes the behaviour of entomopathogenic nematodes, elsewhere in this volume. In this section attention will be paid to the various aspects of behaviour exhibited by free-living and parasitic nematodes with reference being made, where appropriate, to fuller descriptions in Chapters 8, 9, 14 and 20.

Behavioural Responses

Behaviour in animals is mainly the response by an animal to stimuli it receives from its environment and ranges from that which is genetically determined, as in species-specific behaviour, to behaviour that is largely acquired through learning.

Taxes and Kineses

Two of the main categories of behavioural response by nematodes to external stimuli are kineses (singular kinesis) and taxes (singular taxis) (see Croll, 1970). Kinesis is a term used to describe movement that lacks directional orientation and depends on the intensity of stimulation. The animal wanders about in an unoriented way, neither towards nor away from a stimulus. There are two types of kinesis, *orthokinesis* and *klinokinesis*. Orthokinesis is a non-directional response in which an animal's speed of movement alters with changes in the intensity of stimulation. Klinokinesis is a non-directional response in which an animal's rate of turning, or rate of change of direction, alters with changes in the intensity of stimulation. Both types of kinesis can be further divided depending upon the nature of the stimulus. Thus when the response is made to amount of contact then it is referred to as *thigmo-orthokinesis*, or *thigmokinesis*. If the response is to the amount of light then it is referred to as *photo-klinokinesis*, or *photokinesis*. These types of behaviour, with their resulting changes in speed or rate of turning, tend to make animals congregate where there are optimum conditions. Kinetic behaviour of individual nematodes can be studied by examining the tracks that they leave on the surface of an agar gel.

Taxis is a type of behaviour concerned with the directed orientation of an animal in which it moves towards (positive) or away from (negative) the source of stimulation. In *klinotaxis*, animals compare the intensities of stimulation by moving the anterior end from side to side and thus move along a concentration gradient. *Tropotaxis* involves the simultaneous comparison of the stimulation of two sense organs, for example on the head and the tail of a vermiform animal or on both sides of the body. Some authors have regarded a concentration of nematodes at one or other end of a physical or chemical gradient as a taxis solely because they have congregated there whereas they may have congregated in an area of optimum stimulation as a result of kinetic behaviour (Croll, 1970). Taxes, such as *chemotaxis*, of individual nematodes are usually studied by examining the tracks that they leave on the surface of an agar gel in the presence of a gradient of the stimulus. It is possible, by means of a suction pipette, to tether individual nematodes (*Caenorhabditis elegans, Meloidogyne incognita*) by their tails and then to record their movements. This has been done by means of photodetectors connected to multichannel recorders and, more recently, by using a video camera then studying videotapes or computer records of their behaviour in response to different concentrations of chemicals (Dusenbery, 1980a, b; 1985; Goode and Dusenbery, 1985; McCallum and Dusenbery, 1992). Populations are studied by counting the number of animals that have migrated to, or away from, the stimulus, either in thin slurries of Sephadex gel beads spread on Petri dishes (Ward, 1973) or in a counter-current apparatus (Dusenbery, 1973). In Dusenbery's apparatus the nematodes are injected into the centre of a tube in which solutions of different densities, one of which contains the chemical to be tested, flow over each other in opposite directions; and the response to the chemical is found from the proportion of the initial inoculum of nematodes recovered from each of the two solutions. Ciné time-lapse photomicrography has been successfully used to study the behaviour of plant-parasitic nematodes (Doncaster, 1962, 1966) and this technique was used in an underground laboratory to show how the plant-parasitic nematode *Trichodorus viruliferus* behaved when it attacked and fed on growing roots of apple trees in an orchard (Pitcher and Flegg, 1965; Pitcher, 1967). High-resolution video-enhanced differential interference contrast microscopy has also proved to be particularly valuable for the study of plant-parasitic nematodes (Wyss and Zunke, 1986; Grundler, Schnibbe and Wyss, 1991; Wyss, Chapter 9).

The behaviour of nematodes in a chemical gradient is regarded by many authors as a chemotaxis (Ward, 1973; Dusenbery, 1974; Jansson *et al.*, 1984; Bargmann and Horvitz, 1991; Ashton, Li and Schad, 1999) but it is not always clear if these responses are klinotactic or tropotactic (Perry and Aumann 1998). The two amphids located on the head (see Gibbons, Jones this volume) have been shown to contain chemosensory neurons (Bargmann and Horvitz, 1991; Bargmann and Mori, 1997; Riga, Perry, Barrett and Johnston, 1995). Killing cells with a laser microbeam has shown that in *C. elegans* chemosensory function is distributed among several cell types with one pair of neurons, the ASE neurons, being uniquely important for chemotaxis (Bargmann and Horvitz, 1991). Because most nematodes lie on their side then the two chemosensory amphids will lie one above the other during their sinusoidal movements. This means that in a two-dimensional environment, such as the surface of an agar gel, the amphids will probably be unable to compare simultaneously chemical stimuli from each side of the body and thus to perform tropotaxes. Most nematodes normally live in a three-dimensional environment and may be able to detect and compare simultaneously chemical stimuli from each side of the body (Perry and Aumann, 1998). However, killing of neurons in the left or the right chemosensory sensilla of *C. elegans* does not prevent chemotaxis, which suggests that the nematode does not compare chemostimulant concentrations on either side of its head (Davis, Goode and Dusenbery, 1986; Bargmann and Horvitz, 1991; Bargmann and Mori, 1997; Perry and Aumann, 1998). Lateral movements of the head during head waving or sinusoidal movement probably result in alternate sampling of the environment from either side of the head and thus klinotaxes may play an important role in orientation (Perry and Aumann, 1998). Both taxes and kineses may be used by nematodes during searching behaviour (Croll, 1967). The same stimulus may result in either a taxis or a kinesis depending upon the strength of the stimulus. Dusenbery (1980a) has suggested that klinotaxis may be the predominant mechanism for weak stimuli and klinokinesis occurs when a strong stimulus is encountered. Dusenbery (1980a), who studied the behavioural responses of *C. elegans* to controlled chemical stimulation whilst the nematode was held by the tail, showed that the frequency of change of direction of the nematode varied with the intensity of the stimulus (NaCl) in the appropriate direction and the nematode adapted quickly to a change from a strong to a weak stimulus. A change from 50 mM NaCl to none caused the nematode to pause, curl then to undergo reversal

movements which would result in the nematode stopping, backing up then setting off in a new direction. Thus klinokinesis appears to play a role in the distribution of this nematode with regard to chemical gradients. Ward (1973, 1978) has shown that the average side-to-side span of the head movement of an adult *C. elegans*, when moving in a gradient of potassium ions, was 0.15 mm with the maximum span being 0.28 mm. From the known slope of the gradient he concluded that the concentration change experienced by a worm moving at right angles to the gradient would be 3% on average and 6.2% maximum. Larvae, due to their smaller size, would experience only a 1% change in concentration from side-to-side. The nematodes sometimes changed their orientation to point more directly up the gradient after a single swing of the head, which suggests that they were able to detect the gradient in the stimulus in a single side-to-side movement of the head. Ward (1973, 1978) concluded that the nematodes must be able to detect a 1–3% change in concentration in about a second and suggested that *C. elegans* aggregated in a chemical gradient by means of chemotaxis. However, Dusenbery (1980a) has pointed out that some of the tracks published by Ward (1973) look like the spiral that would be expected for a klinokinetic response, thus both mechanisms may be used.

Responses to Chemicals

Responses by nematodes to chemical stimuli have been extensively studied in *C. elegans* and some plant-parasitic nematodes but less so in animal-parasitic nematodes. The detection of chemical stimuli by nematodes can be regarded as olfaction or taste, with the amphids being the sense organs most involved (Mori and Ohshima, 1997; Bargmann and Mori, 1997; Riga *et al.*, 1995).

Semiochemicals are defined as chemicals that cause inter- or intra-specific interactions between organisms. Allelochemicals are semiochemicals that cause a physiological or behavioural response between members of different species whereas pheromones are semiochemicals that cause physiological or behavioural responses between members of the same species (Huettel, 1986; Perry and Aumann, 1998). Allelochemicals can be subcategorised as follows: (1) an allemone brings about a negative response by the receiving organism; (2) a kairomone brings about a positive response by the receiving organism; (3) a synomone brings about a favourable response by the emitting and the receiving organisms; (4) an apneumone is a cue emitted by non-living material. Similarly pheromones can be subcategorised as: (1) a sex pheromone which causes a response between or within the sexes; (2) an epidietic pheromone which is produced by members of a species and which results in regulation of population densities, for example, the dauer-inducing pheromone of *C. elegans*; (3) an alarm pheromone produced by members of a species and which brings about a warning or protective response (Huettel, 1986; Perry and Aumann, 1998).

Responses to allelochemicals

Much work has been done on the response of *C. elegans* to allelochemicals and it has been possible to analyse the functions of individual neurons in the behaviour of the nematode (Sengupta *et al.*, 1993; Bargmann, Hartwieg and Horvitz, 1993; Bargmann and Mori, 1997; Mori and Ohshima, 1997; Troemel, Kimmel and Bargmann, 1997). Single chemosensory neurons are able to recognise several different attractive odorants and distinct sets of chemosensory neurons are able to detect high and low concentrations of a single odorant. Also, odorant responses adapt after prolonged exposure to an odorant and this is reversible (Sengupta *et al.*, 1993). In *C. elegans* the different olfactory cues elicit distinct behaviours such as attraction, avoidance, feeding, or mating and an animal's preference for an odour is defined by the sensory neurons that express a given odorant receptor molecule (Troemel, Kimmel and Bargmann, 1997). The ASH sensory neurons in *C. elegans* are able to distinguish between nose touch, hyperosmolarity and volatile repellent chemicals (Kaplan and Horvitz, 1993) and it is suggested that nose touch sensitivity and osmosensation occur via distinct signalling pathways in ASH neurons (Hart *et al.*, 1999). See Jones, Chapter 14, for more information.

Certain anions, cations, amino acids, nucleotides, basic pH, and vitamins have been shown to be water-soluble attractants for nematodes (Ward, 1973, 1978; Dusenbery, 1974, 1980a, b; Bargmann and Horvitz, 1991; Bargmann and Mori, 1997). Riddle and Bird (1985) identified the following ions as attractive to the L2 of *Rotylenchus reniformis*, the strength of attraction, with chloride ion as the strongest, was found to be: $Cl^- > Na^+ > C_2H_3O_2^- > Mg^{2+}, NH_4^+, SO_4^{2-}$. The degree of attraction was concentration dependent and thresholds of attraction varied over the 10-fold range of concentrations tested. Cyclic AMP was found to be strongly attractive to the L2 whereas AMP was only weakly attractive. These results are somewhat different from those found for *C. elegans* (Ward, 1973, 1978) and suggest that there are distinct species differences in their response to the same compounds (Riddle and Bird, 1985). Interestingly, the response of L2 of *Meloidogyne javanica* to the ions tested on *R. reniformis* did not produce a chemotactic response although they were attracted to germinated tomato seeds. Analogous assays with the dauer larvae of *Anguina agrostis* failed to detect strong orientation towards any of the attractants used in the *Rotylenchus* assay and even germinating rye grass seeds, a natural host of the nematode, did not elicit a strong response (Riddle and Bird, 1985). Tethered L2 of *M. incognita* exhibited the same patterns of behaviour as adults of *C. elegans* in response to changes in concentration of NaCl, except that the L2 of *M. incognita* were slower and less regular in behaviour. *M. incognita* initiated a reversal bout of movements in the same manner as *C. elegans* except it responded more slowly and was repelled instead of attracted (Goode and Dusenbery, 1985). AbouSetta and Duncan (1998) studied the response of *Tylenchulus*

semipenetrans and *M. javanica* in sand in a petri dish and found that *T. semipenetrans* preferred the sodium and potassium salts of acetate, formate and chloride to water whereas water was preferred to the citrates; the L2 of *M. javanica* preferred water. Sodium carbonate and sodium bicarbonate were preferred by the L2 of *M. javanica* whereas water was preferred to sodium carbonate by *T. semipenetrans*.

Several volatile organic molecules act as attractants or repellents for *M. incognita* and *C. elegans* (McCallum and Dusenbery, 1992; Bargmann, Hartwieg and Horvitz, 1993). Bargmann and Mori (1997) suggest that as volatile molecules travel quickly through diffusion and turbulence in air then they may be used by *C. elegans* for longer-range chemotaxis, whereas water-soluble molecules are used mainly for short-range chemotaxis to bacteria. Bargmann, Hartwieg and Horvitz, (1993) found that chemotaxis to volatile odorants requires different sensory neurons from chemotaxis to water-soluble attractants and suggested that *C. elegans* may have senses that correspond to smell and taste, respectively.

Using an avoidance behaviour assay for *C. elegans* it was found that the nematode moved away from toxic concentrations of cadmium and copper ions but not from nickel ions. It was suggested that avoidance of cadmium and copper ions are mediated through neural pathways including ADL, ASE, and ASH neurons and that the three sensing neurons provide increased accuracy of the sensory response and give the nematode a survival advantage in its natural habitat (Sambongi *et al.*, 1999). High osmotic strength, acid pH, copper ions, zinc ions, D-tryptophan, CO_2, extracts of garlic, camphor, menthol and α-terthienyl, are repellent to *C. elegans* (Dusenbery, 1974, 1975; Ward, 1978; Bargmann and Mori, 1997). α-Terthienyl, which is released by the roots of marigold plants (*Tagetes* sp.), kills plant-parasitic nematodes (Oostenbrink, Kuiper and s'Jacob, 1957; Uhlenbroek and Bijloo, 1958) but while it strongly repels *C. elegans* it does not kill it (Ward, 1978). Interestingly, the entomopathogenic nematode *Steinernema carpocapsae* was attracted to germinated seeds of marigold and to unsterilised pieces of root of both marigold and tomato. Aqueous extracts of marigold root and tomato root did not adversely affect the ability of *S. glaseri* to infect *Galleria mellonella*, the insect host, but 20 to 40 ppm of α-terthienyl reduced infectivity (Kanagy and Kaya, 1996). Substances in filtrates of the bacterium *Escherichia coli*, which is a suitable food source, are attractive to *C. elegans* and are released in the greatest quantity as growing cells enter the stationary phase (Ward, 1978). *C. elegans* taken from monoxenic cultures and 'starved' for 24 hours showed significantly higher levels of chemotactic response to *E. coli* or cAMP than did individuals taken from axenic or monoxenic cultures (Jansson *et al.*, 1986). Dauer larvae of *C. elegans* do not show the prompt chemotactic response to bacteria shown by well-fed L2's or adults (Albert and Riddle, 1983). Whilst many dauer larvae do not direct their movement toward the bacteria, they respond to the food signal produced by *E. coli* (a neutral, carbohydrate-like substance) by initiating the process of recovery from the dauer stage (Golden and Riddle, 1984b; Riddle, 1988). Secretions from fungal mycelia have been shown to attract *Neotylenchus linfordi* that feeds on the fungus (Klink, Dropkin and Mitchell, 1970) but the nature of the attractant is not known. Balanova and Balan (1991) found that the free-living nematode *Panagrellus redivivus* was strongly attracted to cell-free filtrates of culture media of certain yeasts and fungi, with some esters of fatty acids being most potent, and suggested that materials released by the yeasts and fungi serve as chemotactic signals in the search for food. They also found that inositol acted as a repellent. The free-living nematodes *Acrobeloides* sp. and *Mesodiplogaster lheritieri* are attracted by kairomones emitted by suitable bacterial food in culture (Anderson and Coleman, 1981). Four closely related marine bacterivorous nematodes belonging to the Monhysteridae (*Diplolaimelloides meyli*, *Diplolaimella dievengatensis*, *Monhystera* sp. and *Geomonhystera disjuncta*), which have partially overlapping microhabitat preferences associated with *Spartina anglica* decay, are potential competitors for food. It was found that they had inter specific differences in their response to three different strains of bacteria, suggesting that they had different chemotactic responses to the sources of food. This suggests that the nematodes are specialist feeders and this would allow for small differences in microhabitat choice (Moens, Verbeeck, deMaeyer, Swings and Vincx, 1999).

The fungal endozootic parasite *Harposporium anguillae* was found to attract *C. elegans* whereas secretions from three species of predacious fungi were repellent and it was suggested that these are examples of predator/parasite-prey chemical warfare (Ward, 1978). However, Balan, *et al.* (1976) showed that the predacious nematode-trapping fungus *Monacrosporium rutgeiensis* produces three different, uncharacterised, substances that attract nematodes. Calcium alginate pellets containing hyphae of the nematophagous fungi *Monacrosporium cionopagum*, *M. ellipsosporum* and *Hirsutella rhossilien*, and aqueous extracts of the pellets, have been shown to repel L2 of the plant-parasitic nematode *Meloidogyne incognita*. The fungal impregnated pellets also suppressed the invasion of cabbage roots by *M. javanica* L2 (Robinson and Jaffee, 1996). There is clearly some confusion about the attractiveness, or otherwise, of nematophagous fungi to nematodes.

Much less is known about the response of animal-parasitic nematodes to chemical gradients. Female filarial nematodes are tissue-dwelling worms which produce embryonic forms (microfilaria larvae) which are released into the tissues or blood stream of the vertebrate host. These microfilariae are taken up by blood-sucking arthopods to produce the infective stages which are then transmitted when the vector next feeds (for more information see Muller and Wakelin, 1998). Because mosquitoes, after feeding on an infected person, often contain more microfilariae than are present in the equivalent amount of blood obtained from a finger prick (O'Conner and Beatty,

1937; Croll, 1970) it was suggested that the microfilariae respond chemotactically to the saliva injected by the mosquito during feeding. However, a study of mosquitoes feeding on the North American leopard frog, *Rana spheno cephala*, indicated that there was no chemotaxis involved in the uptake of the microfilariae of *Folyella dolichoptera* as they passed through the capillaries of the webbed foot of the frog (Gordon and Lumsden, 1939). A characteristic of the lymphatic filariae is the marked periodicity of their microfilariae in the peripheral blood. Throughout most of its geographical range (Asia, Africa, Australia, South America) the microfilariae of *Wuchereria bancrofti* appear in the peripheral blood of infected people for about 1–2 hours before and after midnight then they disappear for the remainder of the 24 hour period. There is a non-periodic or slightly diurnally periodic strain in certain areas of the world (Polynesia, Philippines) and in South-East Asia there is a sun-periodic form in which the microfilariae are present in the peripheral blood throughout the 24 hour period, but numbers are elevated at night (see Muller and Wakelin, 1998). It is thought that periodicity is related to accessibility of the microfilariae to appropriate vectors with the appearance of the microfilariae in peripheral blood coinciding with feeding activity of the vector. This periodicity is related to the circadian rhythm of the host, with the microfilariae responding to physiological changes in the host. Apparently, changes in the tensions of oxygen and carbon dioxide in the pulmonary veins trigger the behavioural response of the microfilariae. Microfilariae that exhibit nocturnal periodicity accumulate in the lungs during the day, where they actively swim against the blood flow to maintain their position. When the person is asleep the difference in gas tension is decreased and the microfilariae pass out of the lungs to the peripheral circulation. Altering the daily activity patterns of an infected person can alter the periodicity of the parasite (Hawking, 1967). If a patient infected with *W. bancrofti* breathes in elevated levels of oxygen when microfilariae are circulating in the peripheral blood then the microfilariae rapidly disappear from the peripheral circulation; they return when the inhaled air contains normal levels of oxygen. Similarly, the numbers of microfilariae in the peripheral blood decrease if the carbon dioxide tension is lowered (Hawking, 1962). A number of other factors, including levels of acetylcholine and body temperature, have also been implicated in microfilarial periodicity (Croll, 1970; Muller and Wakelin, 1998). It would appear that chemicals affect periodicity and that the microfilariae respond to these chemicals. Microfilariae of *Onchocerca lienalis* have been shown to accumulate towards thoracic tissues of the black fly host, *Simulium vittatum*, to a density 4 times higher than towards abdominal tissues. Microfilariae lost their ability to differentiate thoracic and abdominal tissues following addition of thoracic attractant(s) to excised abdomens and reversed their differential response when excised thoraces were depleted of chemical cues. It would appear that the microfilariae follow unknown chemical cues in their movement towards the thoracic muscles of the black fly, the only anatomical target where their development can continue (Lehmann, Cupp and Cupp, 1995).

Skin-penetrating larvae of many animal-parasitic nematodes gain access to, and enter, the host by sensing signals that inform the larva that a potential host is near (Ashton, Li and Schad, 1999). The sense organs involved in this process are the amphids and the cuticular sensilla located on the head (see Jones, this volume; Ashton, Li and Schad, 1999). The structure of the amphidial neurons of infective-stage larvae of *Ancylostoma caninum* and *Strongyloides stercoralis*, which are parasites of warm-blooded hosts, have been shown to be very similar to those of *C. elegans* and have been named according to their homologous position in *C. elegans* (Ashton, Li and Schad, 1999). Ashton, Li and Schad (1999) propose that if positional homologies are conserved between more or less closely related species then functional homologies may also be conserved. If this is so, then the ASE-class of amphidial neurons, which are known to be chemosensory in *C. elegans* (Bargmann and Mori, 1997), are probably chemosensory in the infective-stage larvae of *A. caninum* and *S. stercoralis*. These infective-stage larvae use chemical, thermal and mechanosensory signals in host finding and in the subsequent resumption of development (Granzer and Haas, 1991; Ashton, Li and Schad, 1999). The infective larvae of *A. caninum* respond to an increase in heat, carbon dioxide and humidity by raising themselves from the substrate and waving in the air (nictation). Larvae of *A. caninum* (Granzer and Haas, 1991) and of *S. stercoralis* (Ashton, Li and Schad, 1999) show positive chemotaxis to a hydrophilic skin extract and presumably this may also attract the larvae towards the surface of the skin. Once they reach the surface of the skin the presence of serum proteins induces skin penetration. The effective component of serum is a protein and has a molecular weight of between 5000 and 30,000. Dog saliva, urine, milk, and various pure dog serum components did not stimulate penetration (Granzer and Haas, 1991). It would appear that chemo- and thermosensory neurons are involved in this process (Granzer and Haas, 1991; Ashton, Li and Schad, 1999). Ashton, Li and Schad (1999), using laser ablation techniques similar to those used to determine the function of neurons in *C. elegans*, found that when ASE-class neurons in the amphids of larval *S. stercoralis* were ablated then the ability of the larvae to migrate to skin extract was compromised. If ALD-class neurons were also ablated then very few of the larvae reached the attractant. They concluded from this that both the ASE and the ALD neurons have a chemosensory function. They also concluded that, unlike *C. elegans*, the ALD neurons do not detect a volatile substance as the larvae did not migrate to attractant applied to the lid of the container (Ashton, Li and Schad, 1999). It is suggested that resumption of development of infective-stage larvae of animal-parasitic nematodes, on entering a host, is triggered by chemical cues perceived by the amphids (Ashton, Li and Schad, 1999).

A positive correlation exists between the dispersion of the food bolus and clumped aggregations of adult *Nippostrongylus brasiliensis* in the intestine of infected rats but this behaviour is not dependent on the nature of the food (Croll, 1976; 1977a). Closed circuit television was used to record, through the wall of the small intestine, the movements of the nematode and these movements appeared to be correlated with the feeding regime of the rat thus having an effect on habitat selection (Croll, 1977a). The L3 of *N. brasiliensis* migrate to the lungs after penetrating the skin of the rat whereas the L3 of *Strongyloides ratti* apparently migrate through the lymphatic system (Wilson, 1994). The evidence suggests that this is a non-oriented migration rather than a result of chemotaxis (Croll, 1977b; Tindall and Wilson, 1990; Wilson, 1994). Endoscopy carried out on infected sheep has shown that adult *Haemonchus contortus* appear to be attracted to bleeding wounds in the abomasal wall of sheep (Lee and Nicholls, unpublished observations; Nicholls, 1987), and electrophysiological recordings of the amphids of adult *Syngamus trachea* are stimulated by serum in bird's blood (Riga *et al.*, 1995). This indicates that these blood-feeding nematodes are able to respond to the presence of blood in the environment.

Carbon dioxide can act as an allelochemical and induces a chemotactic response, either as an attractant (kairomone) or a repellent (allomone), in a range of free-living and plant-parasitic nematodes. It has been shown to attract many species of nematode over relatively long distances to sources of food: for example, the plant-parasitic nematodes *Aphelenchoides fragariae*, *Ditylenchus dipsaci*, *Heterodera schactii*, *Meloidogyne incognita*, *M. javanica* and *Pratylenchus neglectus* (Klinger, 1961, 1970; Bird, 1960; Pline and Dusenbery, 1987; Viglierchio, 1990; McCallum and Dusenbery, 1992); bacteriophagous nematodes such as *Panagrellus silusiae* (Viglierchio, 1990) and several rhabditids (Bird, 1960); and the entomopathogenic nematode *Steinernema carpocapsae* (Gaugler *et al.*, 1980). *Aphelenchoides ritzemabosi* and *A. fragariae* invade the leaves of host plants via the stomata and it has been shown that their reaction to oxygen, as a stomatal diffusion gas, was mainly negative or indifferent, whereas their reaction to carbon dioxide, the other stomatal gas, was strongly positive (Klinger, 1970). L2 of *Meloidogyne incognita* respond to a sudden increase in concentration of carbon dioxide by increasing the rate of locomotion and decreasing the frequency of changes in direction (Pline and Dusenbery, 1987). McCallum and Dusenbery (1992) used a computer tracking system to study the behaviour of infective larvae of *M. incognita* in the presence of vapour from the roots of host tomato plants and concluded that, other than carbon dioxide, there are no volatile stimuli released from host roots in effective quantities. Robinson (1995) studied the movement of vermiform stages of the plant-parasitic nematodes *M. incognita* (L2), *Rotylenchus reniformis* (mixed stages) and *Ditylenchus phyllobius* (L4), of the entomopathogenic nematode *Steinernema glaseri* (L3), and the free-living nematode *C. elegans* in a three-dimensional assay consisting of cylinders of moist sand in which was a linear gradient of carbon dioxide. *M. incognita*, *R. reniformis* and *S. glaseri* were attracted to the carbon dioxide in a linear gradient of 0.2% cm^{-1} at a mean concentration of 1.2%. There was little response from the other species. When the carbon dioxide was delivered into the sand through a syringe needle at flow rates between 2 and 130 mμ l/minute, then the optimal flow rate for attracting *M. incognita* and *R. reniformis* was 15 μl min^{-1}, and attraction of the two species from a distance of 52 mm was achieved after 29 and 40 hours, respectively. Carbon dioxide induced 96% of all *M. incognita* to move into the half of the cylinder of sand into which the carbon dioxide was delivered and more than 75% accumulated in the 9 cm^3 of sand nearest the source of the carbon dioxide (Robinson, 1995). The rate of release of carbon dioxide used in these experiments is similar to that released by <1 mg of *Saccharomyces cervisidae* (bakers' yeast) and by a young sunflower seedling and thus is very good evidence that carbon dioxide is used by some nematodes to locate food sources (Perry and Aumann, 1998). However, the concentration of the carbon dioxide in water determines if it results in positive or negative chemotaxis, at least for *Panagrellus redivivus*, as Balan and Gerber (1972) found that dilutions of 1:20,000 were repellent, those of 1: 50,000 and 1:200,000 were attractive and that 1:400,000 was without effect. The marine nematode *Adoncholaimus thalassophygas* is attracted by carbon dioxide and it is thought that this enables it to locate sites of decomposition as a source of nutrients (Riemann and Schrage, 1988).

Carbon dioxide is also an important stimulus that initiates hatching of *Ascaris lumbricoides* (see Perry, Chapter 6) and is an important component of the stimulus that initiates exsheathment of the ensheathed L3 of *Haemonchus contortus* and other trichostrongyles (Rogers, 1962; Rogers and Sommerville, 1968). Infective-stage larvae of *A. caninum*, which move from the faecal soil mixture in which they have developed, stand on their tails within their sheath and are stimulated to wave from side to side (nictate) when exposed to elevated levels of carbon dioxide, warmth and humidity. This will increase the opportunity for the larvae to make contact with a suitable host (Croll and Matthews, 1977; Ganzer and Haas, 1991; Ashton, Li and Schad, 1999).

It is well established that many plant-parasitic nematodes are attracted to the roots of host plants (Prot, 1980), presumably by allelochemicals produced by the plant as well as by other factors (Perry and Aumann, 1998). This helps to establish synchrony between the host plant and the life cycle of the nematode (Perry, 1997). However, many of their attractants are non-specific and are common to many plants and the chemotactic behaviour of infective second-stage larvae towards roots and root exudates is controversial, due to contradictory results from different studies (Grundler, Schnibbe and Wyss, 1991). The attractiveness of root exudates for plant-parasitic nematodes studied in bioassays has been described for several nematodes, e.g. *Meloidogyne javanica*, *M. hapla* (Bird, 1959), *H. glycines* (Papademetriou and Bone, 1983), *H. avenae* (Moltmann, 1990) and *H. schachtii* (Grundler, Scnibbe and Wyss, 1991). Some root exudates have a repellent effect, e.g. tomato roots appear to repel L2 of *M. incognita* (Diez and Dusenbery, 1989a). Some of the gradients around growing roots of plants are of moisture, carbon dioxide, sugars, amino acids, pH, redox potential, phenolic compounds, chelating compounds and electrical potential (Jones and Jones, 1974; Perry and Aumann, 1998). Although some of these stimuli may be specific to a given species of nematode and also to a limited region of the plant, only a few of these have so far been shown to be involved in chemotaxis (Perry and Aumann, 1998). L2 of *Heterodera schactii* have been shown to quickly aggregate on an agarose layer where root exudates of mustard (*Sinapis alba*) had been applied and they persisted there for several hours. Analysis of time-lapse video recordings revealed that this aggregation did not result from a directed orientation of the larvae towards the root exudate but reached the area by random movement. Once there they changed their random migratory behaviour into an exploratory behaviour involving stylet thrusting (Grundler, Schnibbe and Wyss, 1991). The L2 of the plant-parasitic nematode *Rotylenchus reniformis* are strongly attracted to germinated tomato seeds and this has been shown to be chemotaxis (Riddle and Bird, 1985). Although L2 of *Meloidogyne javanica* were apparently attracted to germinated seeds of tomato (see above) exudates collected from a variety of plants, including tomato, *Capsicum*, squash, water melon and soybean resulted in avoidance responses by the infective larvae of *M. incognita* (Diez and Dusenbery, 1989a). Dauer larvae of *C. elegans* do not show chemotactic behaviour in orientation assays when compared with L2 or adults, and they only begin to show chemotactic behaviour when they commence pharyngeal pumping (Albert and Riddle, 1983). It is possible that the dauer larvae of *A. agrostis*, which do not orientate to the attractants used to attract the L2 of *Rotylenchus*, are similar to *C. elegans* in this respect or possibly other environmental factors are required to initiate chemotactic behaviour (Riddle and Bird, 1985).

Electrophysiological techniques, similar to those used to record from insect receptors, have been used to obtain electrophysiological recordings from the cephalic region and from individual intact amphids of a number of plant-parasitic and animal-parasitic nematodes (Jones, Perry and Johnston, 1991; Riga et al., 1995; Riga, Perry and Barrett, 1996; Riga et al., 1966, 1997a, b). Using the electrophysiological technique and also behavioural bioassays it was found that individual males of the potato cyst nematodes, *Globodera rostochiensis* and *G. pallida* did not respond to potato root diffusate. From this it was concluded that diffusate was not involved in orientating males to roots containing females, although they did respond to sex pheromones produced by adult females (see later) (Riga, Perry and Barrett, 1966). When the response of males of both species to various chemicals was tested, responses in the form of action potentials were recorded as spike activity. Males of both species responded to 10 mM L-glutamic acid but not to the D-isomer; males of *G. pallida* responded to 10 mM γ-aminobutyric acid but not to α-aminobutyric acid whereas the reverse was true for *G. rostochiensis* (Riga *et al.*, 1997). Direct extracellular electrophysiological recordings from the amphids of *Syngamus trachea*, the gapeworm of birds, in response to bird blood serum have shown that there are at least two cell types with cell type (a) having a small amplitude and cell type (b) having a larger amplitude (Riga, Perry, Barrett and Johnston, 1995). The amphids responded with an increase in electrical activity to the blood serum, which is a feeding stimulant for the nematode, and they also responded to D-tryptophan, a known repellent for *C. elegans*. When the stimuli were removed and the nematode washed to remove traces of the serum and the D-tryptophan the amphidial responses returned to the original, resting frequency within 5–10 min (Riga, Perry, Barrett and Johnston, 1995).

Responses to pheromones

Over 30 species, but relatively few genera, of free-living, plant-parasitic and animal-parasitic nematodes have been shown to exhibit pheromone-mediated behaviour (Huettel, 1986; Haseeb and Fried, 1988). The first demonstrations of sexual attraction in nematodes were by Greet (1964) and by Beaver, Yoshida and Ash, 1964). Greet (1964) showed that males and females of *Panagrolaimus rigidus*, when separated by a cellophane barrier, moved towards members of the opposite sex but not to members of the same sex. As both males and females congregated at the barrier it would appear that both sexes must produce sex pheromones and be attracted by different specific substances. Green (1966, 1980) and Green and Plumb (1970) described the results of bioassays on the attraction of males of 10 different species of *Heterodera* (some of which are now called *Globodera*) to sedentary females. Females of these two genera secrete at least 6 different male attractants; most females attract more than one species of male and most males responded to more than one species of female. The attractants were found to be water-soluble, nonvolatile and to diffuse through dialysis membranes. Extracts of white, living, intact females of *Heterodera schactii* prepared from various solvents all showed sex pheromone activity for males in a bioassay system (Aumann and Hashem, 1993). Males of *G. rostochiensis* were attracted to a crude preparation of female *G. rostochiensis* sex pheromone but not to female sex pheromone of the closely related *G. pallida*. In contrast, males of *G. pallida* were attracted to sex pheromones from females of both species (Riga, Perry, Barrett and Johnston, 1996). It was later found that sex pheromone activity from virgin female *G. rostochiensis* was confined to two fractions of the pheromone after separation by reverse-phase high pressure liquid chromatography and subsequent testing elctrophysiologically on the cephalic region of adult males (Riga *et al.*, 1997a).

Behavioural bioassays of *Bursaphelenchus xylophilus* and *B. mucronatus* on agar showed that males of *B. xylophilus* and *B. mucronatus* were attracted to their respective females but not to the females of the other species. Neither were they attracted to chemicals released by *B. fraudulentus* or *Aphelenchoides rhyntium* (Riga and Webster, 1992). It would appear that the female sex pheromones released by these nematodes are quite distinct.

Electrophysiological recordings from the cephalic region of live, intact male *Globodera rostochiensis* have demonstrated a change of electrical activity in response to sex pheromone released by the adult female (Jones, Perry and Johnston, 1991; Riga, Perry, Barrett and Johnston, 1996). The spike frequency produced by adult males of *G. rostochiensis* and *G. pallida* increased in response to the application of their homospecific crude sex pheromone indicating that neurons in the cephalic region (amphids?) were stimulated by the pheromone from the matching female (Figure 15.1). Males of *G. pallida* responded to the application of the sex pheromone from female *G. rostochiensis* by an increase in spike activity but males of *G. rostochiensis* did not respond to female sex pheromone produced by *G. pallida* (Riga, Perry, Barrett and Johnston, 1996). These electrophysiological results confirm the bioassay results mentioned earlier. Males of *C. elegans* accumulate at hermaphrodites or at agar that has been conditioned with hermaphrodites but it is not known if they accumulate in the vicinity of hermaphrodites as a result of chemotaxis or just accumulate there as a result of kinetic movements (Emmons and Sternberg, 1997).

Sexual attraction and the sex pheromones of animal-parasitic nematodes have mainly been studied by bioassays, although there have been some *in vivo* studies and one electro-physiological study. Bonner and Etges (1967) demonstrated sex attraction in adult *Trichinella spiralis in vivo* with males being more strongly attracted to females than females to males. Beaver, Yoshida and Ash (1964) showed that *in vivo*, when mating of *Ancylostoma caninum* becomes necessary the female seeks the male to a certain extent but the male is greatly stimulated by, and attracted to, ready-to-mate females. Male *A. caninum* inserted posteriorly into the intestine of a dog moved forwards, against peristalsis, towards more anteriorly placed females which suggests that the females produced an attractant (Roche, 1966). Similar results were obtained with *Nippostrongylus brasiliensis* in the rat (Gimenez and Roche (1972) and in the mouse (Glassburg, Zalisko and Bone, 1981). Males and females of *N. brasiliensis* show mutual heterosexual attraction *in vitro*, with males being strongly attracted to females (Bone and Shorey 1977b; Roberts and Thorson, 1977; Glassburg, Zalisko and Bone, 1981). Males were attracted to as few as three females, but females were less responsive to male pheromone (Bone and Shorey, 1977a). There is an optimal concentration of female pheromone from female *N. brasiliensis* beyond which there is a reduction of movement of males towards the source of the

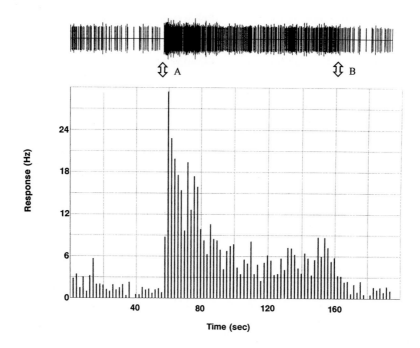

Figure 15.1. Extracellular recordings from the cephalic region of an adult male *Globodera pallida* in response to the sex pheromone from homospecific virgin females. The upper part of the figure shows actual spike events, while the histograms give the accurate total spike frequency. The arrow A indicates when the stimulus was introduced and arrow B indicates when the stimulus was removed by perfusion with artificial tap water. Reproduced with the permission of Cambridge University Press, from Riga, Perry, Barrett and Johnston (1996) Electrophysiological responses of males of the potato cyst nematodes, *Globodera rostochiensis* and *G. pallida* to their sex pheromones. *Parasitology*, **112**, figure 4.

pheromone (Bone, Shorey and Gaston, 1977, 1978). In the mouse the anterior movement of males to females was shown to be dose-dependent and was reduced or eliminated by lengthy distances from females. Males are highly responsive to female sex pheromone at 6 to 10 days post-infection whereas females become attractive to males from 4 to 10 days post-infection (Bone *et al.*, 1977a; Glassburg, Zalisko and Bone, 1981). It is interesting to note that the immune response begins to affect *N. brasiliensis* from about the tenth day post-infection, which is when the attractiveness of females to males begins to diminish. Male *N. brasiliensis* orientate to a gradient of pheromone emanating from living females *in vitro* but if they are maintained in an environment permeated with female pheromone then their ability to orientate to living females is reduced for at least 2 hours (Bone and Shorey, 1977). The sex pheromone of females of *N. brasiliensis*, released by the females into an incubating medium, brought about an almost 80% response from males within 10 minutes at a 15 females/hour dosage; the response was dosage dependent (Bone, 1980). Release of the pheromone by females kept in an incubating medium was linear for 2 hours then declined (Bone, 1982c). *In vivo* experiments indicate that there also appears to be an aggregation pheromone released by female, but not by male, *N. brasiliensis* as female excretions/secretions were attractive to both sexes whereas those of males were attractive to females only (Roberts and Thorson, 1977). The probability that an aggregation pheromone is

involved is further strengthened by the fact that the sexes become aggregated before the sex pheromone system becomes operable (Glassburg, Zalisko and Bone, 1981). Both sexes of *Aspiculuris tetraptera*, a parasite of mice, attract each other, with females being less attracted to males than males to females (Anya, 1976a,b). Males and females of *Ascaris suum* are attracted to the opposite sex from a distance of 50 cm, but not to the same sex, *in vitro* (Garcia-Rejon *et al.*, 1982). Adults have been observed responding to attractants in a three-way maze (Garcia-Rejon *et al.* (1985b). Individuals that responded positively to the attractant undulated rapidly through the bathing medium towards the inoculation zone whereas individuals that responded negatively moved more slowly. Response to sex attractant was affected by the number of pheromone-emitting individuals with the optimum response occurring when three individuals releasing pheromone were used (Garcia-Rjon *et al.*, 1985b). Males and females responded to sex pheromones at pH 7.4 but not at pH 5.5 or pH 9.0.

Characterisation of pheromones

The isolation and characterisation of sex pheromones of nematodes have been limited to a small number of species. The sex pheromone of *Pelodera teres* is thought to be water-soluble (Jones, 1966). The female sex pheromones of *G. rostochiensis* are thought to be polar compounds that probably contain several active components, are physically stable (Green 1980) and include volatile as well as non-volatile fractions (Greet, Green and Poulton, 1968). Clarke, Firth and Greet (1976) found that a non-volatile component of the pheromone is highly polar, weakly basic and soluble in water but not in most other organic solvents tested. They found that its molecular mass is about 400, that it is not a reducing sugar and is ninhydrin-negative. Bone (1986) fractionated the incubate from female *H. glycines* and found that there were aqueous and a methanol-soluble fractions, both of which attracted males in a bioassay. Two components of the sex pheromone of female *H. schactii* are volatile chemicals and have a low polarity (Aumann, Ladehoff and Rutencrantz, 1998; Aumann *et al.*, 1998). The sex pheromone from the vulval-secreted gelatinous matrices of female *H. schactii* have been analysed by gas chromatography and the pheromone was separated into two components, each with a different polarity (Aumann, Ladehoff and Rutencrantz, 1998). One component of the pheromone can be extracted from aqueous solutions with diethyl ether but most of the pheromone activity remained in the water fraction; one of the components is positively charged whereas another may be negatively charged (Aumann, Dietsche, Rutencrantz and Ladehoff, 1998). Reverse phase HPLC separation of the pheromone from virgin females of *G. rostochiensis* showed that it contained several compounds that were primarily low molecular weight, weakly basic and very polar (Riga *et al.*, 1997a). Only two of the four fractions were found to initiate a response in males (see above). The crude pheromone of *Bursaphelenchus xylophilus* is a highly polar compound with an approximate molecular weight of less than 3000 Da (Riga, 1992 in Riga *et al.*, 1997a). There are at least two components in the sex pheromone of *Panagrellus redivivus* (Balakanich and Samoiloff, 1974). Male sex attractants have been identified in *Pelodera strongyloides* (hydroxyl ion) (Stringfellow, 1974) and in *H. glycines* (vanillic acid) (Jaffee *et al.*, 1989).

Nematode sex pheromones are released into an aqueous environment and thus volatility is not a prerequisite for their efficient dissemination. However, as we have seen above, the female sex pheromones of some nematodes contain a volatile component as well as an aqueous component and both components can activate males. Greet, Green and Poulton (1968) suggested that a possible function of the volatile component in soil could be to arouse the males of plant-parasitic nematodes and to stimulate them to move at random until they come to the aqueous gradient that leads them to the female.

The female pheromone of *N. brasiliensis* appears to have two active compounds (Bone *et al.*, 1979, 1980a, b, 1982). The fraction with a molecular weight of over 500 was found in homogenates of both sexes and was attractive to both males and females. The fraction with a molecular weight of less than 400 was found only in females and attracted only males (Bone *et al.*, 1979). Gel filtration of the pheromone present in media in which females had been incubated had active elution's at K_{av} 0.64 and 1.0. The component at K_{av} 0.64 is water soluble

and is probably a peptide. The pheromone at K_{av} 1.0 is more soluble in organic solvents (Ward and Bone, 1983).

Not much is know about the site of sex pheromone production in nematodes. Cheng and Samoiloff (1972) suggested that the gonads are a likely source of pheromone production in *Panagrellus silusiae*. This has been supported by Anya (1976a, b), who suggested that the attractant is produced by the glandular cells located in the pulvilus of the female and in the caudal glands of the male of *Aspiculuris tetraptera* and by Garcia-Rejon *et al.* (1995a) who found that both male and female *Ascaris suum* were attracted to sexual organs from individuals of the opposite sex. The gelatinous matrices of female *H. schactii* have been shown to be a source of the sex pheromone that is attractive to males (Aumann, Ladehoff and Rutencrantz, 1998). Marchant (1970) reported that males of *Nematospiroides dubius* (now called *Heligmosomoides polygyrus*) responded to females by flaring their copulatory bursa in response to the presence of females but they did not respond when the vulval area of the female was treated to prevent release of secretions, thus implicating the vulval region as a release site for pheromones.

Dauer Pheromones

C. elegans forms a dauer larva instead of the normal L3 under conditions of high population density and low food availability. Chemosensory cells that respond to environmental cues of increasing population density and reduced food supply control formation of the dauer larva. This dauer larva is resistant to desiccation and harsh environments. Young larvae measure increasing population density by detecting an increase in a pheromone that the nematode produces and this, together with detection of a reducing food supply, induces formation of the dauer larva. The dauer larvae resume development in response to a decrease in pheromone concentration and an increase in food supply (Riddle, 1988; Zwaal *et al.*, 1997). At least two classes of chemosensory neurons, ADF and ASI, are implicated in the pheromone response. It is suggested that the neurons ADF and ASI are pheromone sensors that repress dauer formation in the absence of pheromone (Thomas, 1993). Mutations have been used to determine which genes are involved in dauer formation. Apparently a branched genetic pathway controls dauer formation (Malone and Thomas, 1994). Formation of the dauer larva is repressed by the chemosensory neurons ADF, ASI and ASG and the chemosensory neuron ASJ functions in parallel to induce dauer formation (Schackwitz, Inoue and Thomas (1996). The *daf-11* gene is apparently directly involved in chemosensory transduction essential for dauer formation (Vowels and Thomas, 1992). The *daf-7* gene is required for transducing environmental cues that support continuous development with plentiful food and a *daf-7* reporter construct is expressed in the ASI chemosensory neurons. The pheromone that induces dauer formation inhibits *daf-7* expression and promotes dauer formation, whereas the presence of food reactivates *daf-7* expression and promotes recovery from the dauer state. When the food/pheromone ratio is high, the level of *daf-7* mRNA peaks during the L1 stage, when the commitment to non-dauer development is made (Ren *et al.*, 1996).

The dauer-inducing pheromone of *C. elegans* is present in extremely low concentrations in liquid cultures of the nematode, making it difficult to identify the components (Golden and Riddle, 1984c). It is apparently very stable, and may be composed of a family of related hydrophobic, hydroxylated, short-chain fatty acids or bile acids having a molecular mass of less than 1000 (Golden and Riddle, 1982, 1984c; Riddle, 1988). Partial purification of the pheromone resulted in an active component being found to be methanol-soluble and ninhydrin-negative (Ohba, Nishimura and Saito, 1986).

Responses to Temperature

Nematodes are poikilothermic animals and temperature has a direct effect on their metabolic rates. This in turn can affect aspects of their behaviour. The effects of temperature on hatching (Chapter 6) and on survival (Chapter 16) of nematodes are dealt with elsewhere in this volume. Free-living, plant-parasitic and animal-parasitic nematodes appear to show sensitivity to temperature gradients resulting in oriented locomotion to, or from, the preferred temperature.

C. elegans can sense a range of environmental temperatures and will migrate along a thermal gradient towards the temperature at which it was cultivated (Hedgecock and Russell, 1975; Mori and Ohshima, 1995; Bargmann and Mori, 1997). A major thermosensory neuron, AFD, has been identified in the amphids of *C. elegans* and although both left and right AFD neurons are necessary for normal thermotaxis, killing one or other of the neurons showed that isothermal movement can be achieved without comparing information from both neurons (Mori and Ohshima, 1995). These authors have proposed that neuron AIY is responsible for thermophilic movement and that neuron AIZ is responsible for cryophilic movement. Isothermal movement may be distinct from migration to a preferred temperature. During migration the nematodes move along a thermal gradient of about $0.2°C$ mm^{-1} in a radial gradient assay, in which a relatively large increase or decrease in temperature has to be assessed over time. During isothermal movement, the nematodes usually move perpendicular to a thermal gradient, in which thermal gradients of less than $0.1°C$ have to be assessed to enable the nematode to remain in the preferred temperature (Mori and Ohshima, 1995). Temperature also modulates induction of dauer formation of *C. elegans* (Golden and Riddle, 1984 a,b). Infective larvae of *M. incognita* only feed after they locate the roots of a suitable host plant. As we have seen above, carbon dioxide from host roots attracts the larvae but it would also be advantageous for the larvae to move to an optimal depth in soil in which to search for chemical cues. The larvae are able to find out a

suitable soil depth by orienting to extremely shallow temperature gradients as they respond to temperature changes of $0.001°C$ (Dusenbery, 1988; Dusenbery, 1996) and migrate in gradients as shallow as $0.001°C/cm^{-1}$ (Pline *et al.*, 1988). They move away from extremes of temperature toward an intermediate temperature, which is their preferred temperature. Newly hatched larvae of *M. incognita* migrate to a preferred temperature that is several degrees above the temperature to which they were acclimated (Diez and Dusenbery, 1989b). Computer modelling suggests that this thermotactic response would result in the L2 migrating to the preferred temperature range in soil and thus may bring them into close proximity to plant roots and to the attractants to which they respond when host-finding (Dusenbery, 1989; Diez and Dusenbery, 1989; Dusenbery, 1996). Robinson (1994) reproduced temperature gradient fluctuations that occur naturally as a result of heating and cooling of the soil surface and introduced five ecologically different species of nematode into the system. The nematodes had been adapted to fluctuating temperatures for 20–36 hours at a simulated depth of 12.5 cm before being injected into the apparatus (acrylic tubes filled with moist sand). When heat waves were propagated horizontally (to eliminate gravitational effects) the movement of *Ditylenchus phyllbius*, *Steinernema glaseri*, and *Heterorhabditis bacteriophora* relative to the thermal surface was largely random. *Rotylenchus reniformis* and *Meloidogyne incognita* moved toward the thermal surface. When the heat waves were propagated upward or downward the responses were the same as when they were propagated horizontally, irrespective of gravity.

Infective-stage larvae of some animal-parasitic nematodes of warm-blooded hosts, such as *Strongyloides* and *Ancylostoma*, are most active around $35–40°C$ and are positively thermotactic (Lane, 1933; Croll, 1970; Croll and Smith, 1972; Ganzer and Haas, 1991; Ashton, Li and Schad, 1999). This has obvious advantages for skin-penetrating infective-stage larvae of warm-blooded animals as it will improve the probability of them making contact with, and entering, a suitable host. Ashton, Li and Schad (1999), who have reconstructed the amphidial neurons of larval *Ancylostoma caninum* and *Strongyloides stercoralis*, have suggested that the lamellar cell found in these larvae is similar to the primary thermosensor cell (finger cell or AFD neuron) found in *C. elegans* (Bargmann and Mori, 1997) and may have the same function. The infective-stage larvae of *A. caninum* respond to warmth by increased activity, including nictation. They are attracted by heat in temperature gradients as weak as $0.04°C/mm^{-1}$ (Granzer and Haas, 1991). If they make contact with a dog hair they attach to the hair and then presumably follow a heat gradient towards the surface of the skin. Larvae and adults of some animal-parasitic nematodes, whether parasitic in homeothermic animals (*Nippostrongylus brasiliensis* in the rat) or in poikilothermic animals (*Rhabdias bufonis* in the frog; *Cammallanus* sp. from turtles), continue to migrate towards the hot end of a thermal gradient until they die from thermal damage (McCue and Thorson, 1964; Croll, 1970; Croll and Matthews, 1977). In this respect they differ from the free-living and plant-parasitic nematodes that have been studied as these latter nematodes tend to aggregate at a preferred temperature (see above).

Not all animal-parasitic nematodes respond to a thermal gradient by moving along it to thermal death. The L3 of *Teladorsagia (Ostertagia) circumcincta* (a stomach worm of sheep) showed increased activity up to 25°C, reduced activity at 30°C, were inactive at 40°C and moved away from temperatures higher than 25°C (Morgan, 1928). These results were unexpected as the L3 is the infective stage of the nematode and develops at the body temperature of the sheep after ingestion. The L4 of *Pseudoterranova decipiens (syn. Terranova) decipiens*, which infects fish such as Atlantic cod (*Gadus morhua*), develops to maturity in the stomach of seals, sea lions and other Pinnipedia. The L4 moved towards the hot end of a thermal gradient until they reached 32.5°C but did not move beyond it. When placed above 32.5°C the larvae moved down the gradient until they reached 32.5°C (Ronald, 1960).

Responses to Mechanical Stimuli

The sense organs of nematodes, including those involved in mechanoreception, are described by Jones (Chapter 14). All nematodes respond to touch to a greater or lesser extent and this is very important in foraging behaviour. Free-living nematodes, including the free-living stages of plant-parasitic and animal-parasitic species, take avoiding action when they meet an obstacle, usually in the form of a reversal of forward movement then exploration of a different area (Kaplan and Horvitz, 1993). This avoidance strategy allows the nematode to escape structural traps, such as too narrow a gap between soil particles, by including random movements which enable the nematode to find a path through the particles and then to continue to react to attractant gradients by means of oriented movements (Anderson *et al.*, 1997a, b). Croll and Smith (1970) dropped tiny pins from measured heights onto specific regions of *Rhabditis* sp. and found that if the anterior half was stimulated then the nematode moved backwards and if the posterior half of the body was stimulated then the nematode moved forwards.

Touching the nematode with a fine hair, giving them a light tap, or tapping the dish in which they are contained, results in a withdrawal reflex and this has been used to study learning and habituation in *C. elegans* (Beck and Rankin, 1993; Wicks and Rankin, 1995; Staddon and Higa, 1996; Wen *et al.*, 1997; Jorgensen and Rankin, 1997; Driscoll and Kaplan, 1997). Tavernakis and Driscoll (1997) have reviewed mechanotransduction associated with this behaviour. The neural circuits for touch sensitivity in *C. elegans* have been worked out by Chalfie and Sulston (1981) and by Chalfie *et al.* (1985) and six morphologically similar neurons are responsible for sensing gentle touch to the body (Duggan, Ma and Chalfie, 1996). At least 13 genes are needed for the response to gentle touch by these six touch receptor neurons and almost all of these

touch function genes contribute to the mechanosensory apparatus (Gu, Caldwell and Chalfie, 1996). Hart *et al.* (1999) found that the ASH sensory neurons mediate responses to nose touch, as well as to hyperosmolarity and volatile repellent chemicals and that the ASH neurones are able to distinguish between these various stimuli.

Larvae of plant-parasitic nematodes orientate themselves at the plant surface prior to feeding and/or penetration (see Wyss, Chapter 9) and are able to orientate to a slit that they are making during hatching (Doncaster and Seymour, 1973; Perry, Chapter 6). Doncaster and Seymour (1973) studied the larva of *G. rostochiensis* cutting a slit to make its way out of the egg. The larva used the stylet to make perforations in the eggshell with up to 57 thrusts in 80 seconds. The positioning of the stylet was well controlled by the larva so that if a stylet failed to make a penetration adjacent to the previous penetration then the head end of the larva located the appropriate spot and continued thrusting until a hole was made. After a sequence of thrusts resulted in a slit the larva returned to the original starting point and made a slit in the opposite direction, an intact portion of shell remaining until the larva emerged from the shell. This shows very precise co-ordination of the slit-making process and presumably mechanostimulation plays an important part in this process. Similarly mechanostimulation will be an important component involved in skin penetration by infective-stage larvae when they orientate prior to, and during, penetration. Motionless infective-stage larvae of *A. caninum*, *Haemonchus contortus* and other trichostrongyles resume activity in response to vibrations.

Adult *Nippostrongylus brasiliensis* show thigmokinetic behaviour which enables them to remain amongst host intestinal villi (Lee and Atkinson, 1976).

Rheotactic behaviour (response to fluid velocity gradients) of nematodes remains unclear with most evidence suggesting that nematodes do not exhibit rheotactic behaviour (Wallace, 1963; Croll, 1970; Bone, 1985).

Responses to Electric Currents

Negative electrical potentials of about 60–100 mV occur on the root surfaces of many plants and several genera of plant-parasitic nematodes have been shown to be attracted to the cathode or anode in artificial media. Electrical potential gradients of the order of magnitude necessary to affect the orientation of nematodes do occur near plant roots thus galvanotaxis may play a role in attracting plant-parasitic nematodes to their host plants (Wallace, 1963; Croll, 1970; Croll and Matthews, 1977; Robertson and Forrest, 1989). However, there is some doubt as to how much this contributes to the complex of stimuli that attract nematodes to plants (Wallace, 1963).

Responses to Gravity

Early workers proposed that the upward migration of infective-stage larvae of animal-parasitic nematodes from dung onto the surrounding herbage was a negative geotactic response (see Croll, 1970). However,

Crofton (1954) discounted the evidence for such geotactic behaviour, concluding that the movements of infective larvae of *Trichostrongylus retortaeformis* and of *Trichonema* spp. were uninfluenced by gravity and their vertical migration was due to random movements. The plant-parasitic nematode *Aphelenchoides ritzemabosi* is parasitic upon the aerial parts of plants and migrates vertically from the soil up the stems of the plants, however, this is also thought to be the result of random movements (Barraclough and French, 1965). *Turbatrix aceti* (the vinegar eelworm) tends to swim upwards towards the surface but this is because it is tail-heavy and is not a response to gravitational forces (Peters, 1952). Larvae of *Globodera rostochienesis* tend to move downwards in soil but this may be because their centre of gravity is anterior to their mid-point (Wallace, 1963).

Responses to Light

Response of nematodes to light is controversial. Several experiments have shown the aggregation of nematodes at one or other end of a light intensity gradient, however, heat from the lamp, especially from the infra-red portion of the incident light, was not excluded in early experiments and the nematodes may have responded to a temperature gradient rather than to the light (Croll, 1970; Croll and Matthews, 1977). This is highly probable as some nematodes are extremely sensitive to temperature gradients.

Many free-living nematodes in marine and fresh-water habitats possess localised paired pigment spots, usually in the pharyngeal musculature. These are generally referred to as pigment spots, eyespots or ocelli but a positive response to light has been reported only for *Chromadorina viridis* (Croll, 1966b).

Croll (1966a, 1970) described the response of different nematodes (*Trichonema* spp. L3; *Pelodera* sp. adults; *Turbatrix aceti*, mixed stages; *Rhabdias bufonis* L1; *Anguina tritici* adults) to cold light illumination following 24 hours dark adaptation. *Pelodera* and *Turbatrix aceti* remained active in the dark and after illumination, *Anguina tritici* became mildly active for about an hour, whereas *Rhabdias bufonis* and *Trichonema* quickly reached a peak of activity, having been inactive in the dark, then activity gradually decreased until there was no movement after about 6–7 hours. When stimulated by light the L3 of *Haemonchus contortus* increased their oxygen uptake which suggests that they were responding to the light and may have a dermal light sense (Wilson, 1966).

The non-feeding female of *Mermis nigrescens* ascends from within the soil and climbs vegetation to lay her eggs after showers of rain in warm weather. She spirals up the vegetation using three-dimensional undulatory propulsion which means her head undergoes scanning movements. She possess a putative ocellus which consists of a hollow cylinder of dense haemoglobin pigmentation located in the anterior end (Ellenby, 1964; Croll, 1970; Burr and Babinski, 1990). The location of the photoreceptive nerve endings is not known. If the female is placed in the dark it ceases egg-laying but

resumes ovipositing when illuminated which tends to suggest that it is sensitive to light. It has been suggested that the chromotrope may play a role in monitoring light spectra during egg-laying (Croll, 1970). The nematode apparently carries out phototaxis during this upward migration with the anterior 2 mm continually bending or scanning. This will result in a shuttering of the light source whenever the light source is occluded from any photoreceptive components within the chromotrope and presumably the scanning motion is involved in the maintenance of orientation to light (Burr and Babinski, 1990).

Searching Behaviour

Nematodes exhibit two types of searching behaviour, namely oriented (directed) and unoriented (random) searching. Oriented searching behaviour has been clearly distinguished from unoriented searching behaviour in several free-living and plant-parasitic nematodes and in the L3 of some animal-parasitic nematodes (Croll, 1971; Croll and Smith, 1972; Croll and Blair, 1973; Samoiloff, Balakanich and Petrovich, 1974; Clemens *et al.*, 1994; Anderson *et al.*, 1997a,b). The ability of small nematodes to inscribe their tracks in the surface of agar gels as they move over it has been used to study their movement and searching behaviour. When moving forwards, the sinusoidal waves are propagated at the head and move backwards along the length of the body (see Alexander, Chapter 13) leaving a sinusoidal trail on the surface of the agar. Backward, reversal, movements are propagated at the tail and pass forward along the body. These reversal movements occur if the nematode comes up against an obstacle or initiates a change of direction, such as occurs when *C. elegans* encounters an abrupt decrease in the concentration of an attractant (Dusenbery, 1980a) or is observed moving on a homogeneous layer of agar to which has been added a monolayer of sand grains (Anderson *et al.*, 1997a, b). During unoriented, random, foraging-type behaviour in a homogeneous environment the free-living nematode *Panagrellus silusiae* is in a 'normal' behavioural state and follows a more or less linear path with the anterior and posterior of the animal sweeping through a minimal area. When the nematode is stimulated by chemical or physical means then it enters an activated state resulting in rapid changes in orientation with both ends of the nematode sweeping widely through a large area. In the presence of an attractant the nematode remains in the activated state until it senses the maximum difference in intensity of the stimulus, it then returns to the normal behavioural state and migrates to the source of the attractant (Samoiloff, Balakanich and Petrovich, 1974). The tracks tend to become straighter and more directed during oriented, non-random, movement when the nematode responds to an attractant, such as occurs when *C. elegans* is attracted to a bacterial source (Croll and Matthews, 1977; Young, Griffiths and Robertson, 1996; Anderson *et al.*, 1997a, b). Males of the plant-parasitic nematode *Heterodera schactii* exhibit oriented searching behaviour in the presence of female sex pheromone and L2 carry out oriented behaviour towards host root diffusates (see later).

The searching movements used by dauer larvae of entomopathogenic nematodes to locate hosts are described in more detail by Ishibashi (Chapter 20). Ambusher species, such as *Steinernema carpocapsae*, stand on their tails at the soil surface and wave (nictate) their bodies thus enhancing their chance of making contact with a passing insect (Campbell and Gaugler, 1993). Similar searching behaviour is exhibited by the infective stage larvae (L3) of certain species of animal-parasitic nematode, such as *N. brasiliensis*, *A. caninum* and *S. stercoralis*, which stand on their tails inside their sheath (moulted but not ecdysed cuticle of the L2) on the soil or faecal/soil mixture and wave their anterior end. It has been shown that infective-stage larvae of *A. caninum* respond to environmental and host stimuli with four behavioural phases of host-finding (Granzer and Haas, 1991). (1) Undulatory movements are initiated by warmth and by vibrations of the substratum; (2) nictation enables the larvae to contact and adhere to dog hairs and is stimulated by increases in heat radiation, carbon dioxide and humidity; (3) the larvae are attracted by weak heat gradients and by hydrophilic skin surface extracts; and (4) penetration (of agar) is stimulated by heat, dog hydrophilic skin fraction and by dog serum. Infective-stage larvae of *N. brasiliensis* begin to make active searching movements as they are brought close to rat or human skin and quickly move onto the hairs and skin if contact is made. They move over the surface of the skin until they find a suitable area of the skin, usually a crevice or wrinkle, which enables them to obtain sufficient anchorage to push into the stratum corneum. They are able to enter the skin within 5 minutes (Lee, 1972). Little is known about the searching behaviour of adult animal-parasitic nematodes. However, by means of fibre-optic endoscopy it has been possible to study the behaviour of adult *Haemonchus contortus* inside the abomasum of living sheep. The adult nematodes were seen to move actively in the mucus on the surface of the abomasum. Some that lay across the folds (*plicae abomosi*) of the wall of the abomasum were seen to be stretched across the chasm that developed when the *plicae abomosi* moved apart, before falling into the chasm, later to emerge onto the surface as they continued their journey across the mucosa. Many of the haemorrhagic foci caused by the tooth of the adult when feeding were observed directly above blood vessels in the abomasal mucosa suggesting that they may have detected the presence of the blood vessels. Occasionally they were seen to move towards, and enter, an area of haemorrhage formed when a biopsy specimen had been removed from the abomasal wall, and to remain there, presumably feeding on the blood. This suggests that the nematodes had detected and then followed a gradient to the source of the blood (Lee and Nicholls, unpublished observations; Nicholls, 1987).

Learning and Memory

Behaviour can be programmed either from within (endogenous) or from without (exogenous), with external programming being influenced by individual adjustments through experience. Memory is required for klinokinetic responses because these responses depend upon comparisons of signal strengths (Croll, 1970). Most work on learning and memory in nematodes has been done on *C. elegans* and it has been shown that it is capable of several types of non-associative learning, such as habituation, long-term retention of habituation, dishabituation and sensitisation (Rankin, Beck and Chiba, 1990; Jorgensen and Rankin, 1997; Hart *et al.*, 1999). Habituation is a decrease in response after a certain stimulus has been received repeatedly and sensitisation is an increase in response resulting from exposure to a stimulus. Both habituation and sensitisation are exhibited by nematodes.

Free-living nematodes will withdraw if they bump into an obstacle, such as sand grains (see Response to mechanostimulation above), and if repeatedly stimulated, such as being touched by a fine hair (Chalfie *et al.*, 1985), the backing response declines. Rankin, Beck and Chiba (1990) showed that motionless *C. elegans*, or those moving forward, will move backward if the side of the petri dish is tapped. Repeated taps result in the average distance the nematode moves backward decreases, thus the nematode has become habituated. Application of an electrical stimulus applied to the agar on either side of the nematode restores the normal response to a tap, i.e. dishabituation (Rankin, Beck and Chiba, 1990; Jorgensen and Rankin, 1997). Habituation is more pronounced and faster, and recovery from habituation is more rapid, if the nematode is exposed to closely spaced stimuli, and is more complete, than to widely spaced stimuli (Rankin and Broster, 1992; Broster and Rankin, 1994; Staddon and Higa, 1996). The nematode is able to record the different time intervals and these influence the behaviour of the nematode for at least 1 hour after the delivery of the last stimulus (Rankin and Broster, 1992; Jorgensen and Rankin, 1997). *C. elegans* is capable of long-term memory as it is able to retain memory of habituation for at least 24 hours. Training sessions of 60-second interstimulus intervals resulted in long-term habituation for 24 hours (Beck and Rankin, 1997). Prolonged exposure of *C. elegans* to an odorant leads to a decreased response to the odorant (olfactory adaptation) and the extent of olfactory adaptation is increased in starved animals (Colbert and Bargmann, 1997). Also, starved animals have been shown to discriminate more classes of odorants than do fed animals. The effect of olfactory adaptation together with increased olfactory discrimination will enhance the ability of starved animals to respond to other, potentially informative odorants (Colbert and Bargmann, 1997). Individuals of all ages are capable of non-associative learning but older individuals (12 days) show greater habituation than younger individuals (4 days). There is an age-related change in recovery from habituation with younger individuals recovering from habituation more quickly than older individuals (Beck and Rankin (1993).

Sensitisation to a stimulus has been shown to occur in *C. elegans*. Rankin, Beck and Chiba (1990) subjected the nematode to a single tap, then to a train of taps and then to a single tap and found that the nematode showed a larger response to the single tap following the train of taps than to the initial single tap.

Discriminative classical conditioning has been shown to occur in *C. elegans* (Jorgensen and Rankin, 1997; Wen *et al.*, 1997). Adult hermaphrodites were first deprived of food for 5 hours, and then the nematodes were conditioned by exposing them to either sodium ions or to chloride ions paired with bacteria for an hour and then exposing them to the other ion for a second hour. When they were given a choice then the conditioned nematodes displayed a preference for the ion that had been paired with the bacteria and this preference was retained for up to 7 hours after training. Similarly, if sodium or chloride was paired with a noxious stimulus, such as garlic, during conditioning then the nematodes subsequently avoided the conditioned ion in a chemotaxis assay. Comparison of the sensory abilities of L2 of *H. schactii* that had been stimulated with root exudates and those that had not been stimulated indicated the possible occurrence of a learning process in the stimulated larvae (Clemens *et al.*, 1994). L2 that had previously responded with aggregation and exploration behaviour to root exudates from mustard (*Sinapis alba*) showed a higher rate of contact with discs of filter-paper impregnated with the root exudates and had shorter tracks.

Mating Behaviour

The male accessory structures that are involved in mating vary in form from species to species. Most male nematodes possess copulatory spicules, usually a pair but sometimes one or none. They vary greatly in shape and size, especially in length, and are lodged in an invagination of the cloaca. A nerve runs through the length of the spicule and there is a pore that allows the nerve to communicate to the exterior (Lee, 1962, 1973; Bird, 1966; Clark, Shepherd and Kempton, 1973; Emmons and Sternberg, 1997). The spicules are intromittent organs as they dilate the vulva and vagina at copulation although in some genera (*Heterodera*, *Dipetalonema*) they also make a channel along which the spermatozoa can pass into the female. Males also possess several genital sensilla around the cloacal area and in some species the area of the body wall around the cloaca is expanded to form a copulatory bursa (see Gibbons, Chapter 21). The genital sensilla probably assist the male in orientating itself prior to, and during, copulation. The bursa is used to grasp the vulval region of the female during copulation. Anderson and Darling (1964) observed copulation in *Ditylenchus destructor* and found that the male made several passes around the female before the spicules and vulva became aligned. The copulatory bursa held the male and female together and the spicules were inserted deeply into the vagina.

The male then injected up to 20 sperm before disengaging. During its receptive period (about a week) the female may copulate with more than one male, although several hours elapse between matings. During copulation of *Heligmosomoides polygyrus* (*Nematospiroides dubius*) male and female lie with their anterior ends in opposite directions. The male grasps the female with its copulatory bursa prior to inserting its spicules into the vagina (Sommerville and Weinstein, 1964). The male of *Oncholaimus oxyuris* uses its genital setae and ventral tail papillae as tongs to pinch the female. If the female is unreceptive she moves away or lies still and the male moves away. If the female responds positively she aligns herself to the male, the tail of the male becomes attached to the female and the spicules of the male are used to puncture the body wall of the female. The spicules are used to form a pore into which a secretion is passed from the vas deferens followed by the sperm (Coomans, Verchuren and Vanderhaeghen, 1988; Bird and Bird, 1991). *Syngamus trachea*, the gapeworm of birds, remains in permanent copulation after pairing.

Laser ablation studies on *C. elegans* have lead to an understanding of how some of the components of the male copulatory apparatus function during mating (Chalfie and White, 1988; White, 1988). The tail of the male is composed of an elongated bursa, caudal alae which form a cuticularised fan, and the proctodeum. The fan possesses nine pairs of sensory rays which terminate in papillae and the two spicules each possess two sensory neurons (Sulston, Albertson and Thompson, 1980; Emmons and Sternberg, 1997). Other sensilla lie anterior to and posterior to the cloaca. Mating behaviour comprises a series of steps. After making contact with the hermaphrodite the male arches the posterior third of his body and apposes the ventral side of its tail to the hermaphrodite. He then backs along her body and when he approaches the end of the hermaphrodite he turns and then backs until he locates the vulva. The hook sensillum assists in locating the vulva whereas the ray sensilla contribute to the turning and arching movements of the tail during the search for the vulva (White, 1988). When the cloaca is apposed to the vulva the male inserts its spicules into the vulva and transfers sperm and seminal fluid, then moves away (Liu and Sternberg, 1995; Emmons and Sternberg, 1997). Liu and Sternberg (1995) ablated the male-specific copulatory structures and their associated neurons and identified those that are involved in copulation. The sensory rays mediate response to contact and turning; the hook, the postcloacal sensilla, and the spicules are involved in location of the vulva; the spicules also mediate insertion of the spicules and regulate transfer of the sperm.

Phoretic Behaviour

Phoresis refers to an association between two animals, usually of different species, in which one of the animals transports the other which is either permanently or temporarily sedentary. Phoretic behaviour is exhibited by the dauer larvae of several species of free-living nematode. Insects are commonly used as passive carriers when the dauer larvae attach themselves to the surface of the cuticle and are transported by the insect to a new locality. Dauer larvae of *C. elegans* can be transported this way from a deteriorating food source or crowded environment to a fresh environment. Similarly, infective-stage larvae of *Oesophagostomum* spp. and of *Ostertagia ostertagi* of cattle move from one environment to another by 'hitch-hiking' on the hairs of psychodid flies (Jacobs *et al.*, 1968; Tod, Jacobs and Dunn, 1971). The rhabditid nematode *Pelodera coarctata* is commonly found in dung where it has many generations following a normal free-living life cycle. However, when the dung begins to deteriorate dauer larvae are produced and these attach to dung beetles (*Aphodius* sp.) and remain in a dormant, immobile state until the beetle moves to freshly deposited dung when they leave the surface of the beetle and initiate a new population of the nematode in the dung (Croll and Matthews, 1977). Dauer larvae of *Pelodera strongyloides* are unusual in that they have a phoretic relationship with mammals and the relationship borders on parasitism. This species is commonly found in manure and nests of rodents where it has a free-living existence. However, dauer larvae are commonly found in the lachrymal fluid of rodents (Cliff, Anderson and Mallory, 1978) and in the hair follicles of these and other mammals (Hominick and Aston, 1981). These latter authors demonstrated that when the rodent host (*Apodemus sylvaticus*) was killed L4 emerged from the skin and moulted to adults within 24 hours at 20°C but not at 37°C. Apparently they are unable to complete their life cycle on the mammalian host. A stage in the development of parasitism is also shown by *Rhabditis pellio*. It has a free-living existence in the soil but larvae invade the tissues of earthworms, probably through the nephridial pores, and remain in the earthworm, mostly without further development, until the earthworm dies whereupon they feed on the decaying carcass and associated bacteria with several generations occurring until the carcass is consumed when they migrate into the soil (Poinar and Thomas, 1975).

Unlike the infective-stage larvae of *Haemonchus contortus*, the infective-stage larvae of *Dictyocaulus viviparus*, the lungworm of cattle, do not readily migrate, yet they become widely dispersed on grass surrounding the dung in which they have developed. The fungus *Pilobolus* sp., which grows on the dung, brings about their dispersal. The larvae, once it has reached the I3 stage, migrates to the surface of the dung, which is now heavily colonised by the fungus, and ascends the sporangiophores of the fungus. *Pilobolus* discharges its sporangia violently in response to changes in illumination and catapults the sporangia together with the larvae onto the surrounding pasture. Transport by this means is important to the nematode as it has to be ingested by a suitable host for further development and this method of transport ensures that it is dispersed away from the dung (Robinson, 1962).

References

Albert, P.S. and Riddle, D.L. (1983) Developmental alterations in the sensory neuroanatomy of the *Caenorhabditis elegans* dauer larva. *Journal of Comparative Neurology*, **219**, 461–481.

Albert, P.S. and Riddle, D.L. (1985) Responses of the plant parasitic nematodes *Rotylenchus reniformis*, *Anguina agrostis* and *Meloidogyne javanica* to chemical attractants. *Parasitology*, **91**, 185–195.

Anderson, A.R.A., Young, I.M., Sleeman, B.D., Griffiths, B.S. and Robertson, W.M. (1997a) Nematode movement along a chemical gradient in a structurally heterogeneous environment. 1. Experiment. *Fundamental and Applied Nematology*, **20**, 157–163.

Anderson, A.R.A., Sleeman, B.D., Young, I.M. and Griffiths, B.S. (1997b) Nematode movement along a chemical gradient in a structurally heterogeneous environment. 2. Theory. *Fundamental and Applied Nematology*, **20**, 165–172.

Anderson, R.V. and Coleman, D.C. (1981) Population development and interactions between two species of bacteriophagic nematodes. *Nematologica*, **27**, 6–19.

Anya, A.O. (1976a) Studies on the reproductive physiology of nematodes: The phenomenon of sexual attraction and the origin of the sex attractants in *Aspiculuris tetraptera*. *International Journal for Parasitology*, **6**, 173–177.

Anya, A.O. (1976b) Physiological aspects of reproduction in nematodes. *Advances in Parasitology*, **14**, 267–351.

Ashton, F.T., Li, J. and Schad, G.A. (1999) Chemo- and thermosensory neurons: structure and function in animal parasitic nematodes. *Veterinary Parasitology*, **84**, 297–316.

Aumann, J. (1993) Chemosensory physiology of nematodes. *Fundamental and Applied Nematology*, **16**, 193–198.

Aumann, J., Dietsche, E., Rutencrantz, S. and Ladehoff, H. (1998) Physico-chemical properties of the female sex pheromone of *Heterodera schachtii* (Nematoda: Heteroderidae). *International Journal for Parasitology*, **28**, 1691–1694.

Aumann, J., Ladehoff, H and Rutencrantz, S. (1998) Gas chromatographic characterisation of the female sex pheromone of *Heterodera schactii* (Nematoda: Heteroderidae). *Fundamental and Applied Nematology*, **21**, 119–122.

Balan, J. and Gerber, N.N. (1972) Attraction and killing of the nematode *Panagrellus redivivus* by the predaceous fungus *Arthrobotrys dactyloides*. *Nematologica*, **18**, 163–173.

Balanova, J. and Balan, J. (1991) Chemotaxis-controlled search for food by the nematode *Panagrellus redivivus*. *Biologia*, **46**, 257–263.

Balan, J., Kizkova, L., Nemec, P and Jolozsvary, A. (1976) A qualitative method for detection of nematode attracting substances and proof of production of three different attractants by the fungus *Monacrosporium rutgeriensis*. *Nematologica*, **22**, 306–311.

Baraclough, R.M. and French, N. (1965) Observations on the orientation of *Aphelenchoides ritzemabosi* (Schwartz). *Nematologica*, **11**, 199–206.

Bargmann, C.I., Hartwieg, E. and Horvitz, H.R. (1993) Odorant-selective genes and neurons mediate olfaction in *C. elegans*. *Cell*, **74**, 515–527.

Bargmann, C.I. and Horvitz, H.R. (1991) Chemosensory neurons with overlapping functions direct chemotaxis to multiple chemicals in *C. elegans*. *Neuron*, **7**, 729–742.

Bargmann, C.I and Horvitz, H.R. (1991) Chemosensory neurons with overlapping functions direct chemotaxis to multiple chemicals in *C. elegans*. *Neuron*, **7**, 729–742.

Bargmann, C.I. and Mori, I. 1997) Chemotaxis and thermotaxis. In *C. elegans II*, edited by D.L. Riddle, T. Blumenthal., B.J. Meyer and J.R. Preiss, pp. 717–737. Cold Spring Harbor: Cold Spring Harbor Laboratory Press.

Beaver, P.C., Yoshida, Y. and Ash, L.R. (1964) Mating of *Ancylostoma caninum* in relation to blood loss in the host. *Journal of Parasitology*, **50**, 286–293.

Beck, C.D.O. and Rankin, C.H. (1993) Effects of aging on habituation in the nematode *Caenorhabditis elegans*. *Behavioural Processes*, **28**, 145–163.

Beck, C.D.O. and Rankin, C.H. (1997) Long-term habituation is produced by distributed training at long ISIs and not by massed training or short ISIs in *Caenorhabditis elegans*. *Animal Learning & Behavior*, **25**, 446–457.

Bird, A.F. (1959) The attractiveness of roots to the plant parasitic nematodes *Meloidogyne javanica* and *M. hapla*. *Nematologica*, **4**, 322–335.

Bird, A.F. (1960) Additional notes on the attractiveness of roots to plant parasitic nematodes. *Nematologica*, **5**, 217.

Bird, A.F. (1962) Orientation of the larvae of *Meloidogyne javanica* relative to roots. *Nematologica*, **8**, 275–287.

Bird, A.F. (1966) Esterases in the genus *Meloidogyne*. *Nematologica*, **12**, 359–361.

Bone. L.W. (1982) *Nippostrongylus brasiliensis*: female incubation, release of pheromone, fractionation of incubates. *Experimental Parasitology*, **54**, 12–20.

Bone, L.W. (1986) Fractionation of the female's pheromone of the soybean cyst nematodes, *Heterodera glycines*. *Proceedings of the Helminthological Society of Washington*, **53**, 132–134.

Bone. L.W. and Shorey, H.H. (1977) Disruption of sex pheromone communication in a nematode. *Science*, **197**, 694–695.

Bone, L.W., Shorey, H.H. and Gaston, L.K. (1977) Sexual attraction and pheromonal dosage response of *Nippostrongylus brasiliensis*. *Journal of Parasitology*, **63**, 364–367.

Bone, L.W., Shorey, H.H. and Gaston, L.K. (1978) *Nippostrongylus brasiliensis*: Factors influencing movement of males toward a female pheromone. *Experimental Parasitology*, **44**, 100–108.

Bonner, T.P. and Etges, F.J. (1967) Chemically mediated sexual attraction in *Trichinella spiralis*. *Experimental Parasitology*, **21**, 53–60.

Broster, B.S. and Rankin, C.H. (1994) Effects of changing interstimulus-interval during habituation in *Caenorhabditis elegans*. *Behavioral Neuroscience*, **108**, 1019–1029.

Burr. A.H.J. and Babinski, C.P.F. (1990) Scanning motion, ocellar morphology and orientation mechanisms in the phototaxis of the nematode *Mermis nigrescens*. *Journal of Comparative Physiology A*, **167**, 257–268.

Campbell, J.F. and Gaugler, R. (1993) Nictation behaviour and its ecological implications in the host strategies of entomopathogenic nematodes (Heterorhabditidae and Steinernematidae). *Behaviour*, **126**, 155–169.

Chalfie, M., Sulston, J.E., White, J.G., Southgate, E., Thomson, J.N. and Brenner, S. (1985) The neural circuit for touch sensitivity in *Caenorhabditis elegans*. *Journal of Neuroscience*, **5**, 956–964.

Chalfie, M. and White, J. (1988) The nervous system. In *The nematode Caenorhabditis elegans*, edited by W.B. Wood, pp. 337–391. Cold Spring Harbor: Cold Spring Harbor Laboratory.

Cheng, R. and Samoiloff, M.R. (1972) Effects of cycloheamide and hydroyurea on mating behavior and its development in the free-living nematode *Panagrellus silusiae*. *Canadian Journal of Zoology*, **50**, 333–336.

Clarke, A.J., Firth, J.E. and Greet, D.N. (1976) The sex attractant. *Rothamsted Experimental Station, Report for 1975*, 199–200.

Clark, S.A., Shepherd, A.M. and Kempton, A. (1973) Spicule structure in some *Heterodera* spp. *Nematologica*, **18**, 242–7.

Clemens, C.D., Aumann, J., Spiegel, Y. and Wyss, U. (1994) Attractant-mediated behaviour of mobile stages of *Heterodera Schactii*. *Fundamental and Applied Nematology*, **17**, 569–574.

Cliff, G.M., Anderson, R.C. and Mallory, F.F. (1978) Dauerlarvae of *Pelodera strongyloides* (Schneider, 1860) (Nematoda: Rhabditidae) in the conjuctival sacs of lemmings. *Canadian Journal of Zoology*, **56**, 2117–2121.

Colbert, H.A. and Bargmann, C.I. (1997) Environmental signals modulate olfactory acuity, discrimination, and memory in *Caenorhabditis elegans*. *Learning & Memory*, **4**, 179–191.

Coomans, A., Verchuren, D. and Vanderhaeghen, R. (1988) The demanian system, traumatic insemination and reproductive strategy in *Oncholaimus oxyuris* Ditlevsen (Nematoda, Oncholaimina). *Zoologica Scripta*, **17**, 15–23.

Crofton, H.D. (1954) The vertical migration of infective larvae of strongyloid nematodes. *Journal of Helminthology*, **28**, 35–52.

Croll, N.A. (1966) The phototactic response and spectral sensitivity of *Chromadorina viridis* (Nematoda, Chromadorida), with a note on the nature of the paired pigment spots. *Nematologica*, **12**, 610–614.

Croll, N.A. (1967) The mechanism of orientation in nematodes. *Nematologica*, **13**, 17–22.

Croll, N.A. (1970) *The Behaviour of Nematodes*. London: Edward Arnold.

Croll, N.A. (1971) Movement patterns and photosensitivity of *Trichonema* sp. infective larvae in non-directional light. *Parasitology*, **62**, 467–478.

Croll, N.A. (1976) The location of parasites within their hosts. The influence of host feeding and diet on the dispersion of adults of *Nippostrongylus brasiliensis* in the intestine of the rat. *International Journal for Parasitology*, **6**, 441–448.

Croll, N.A. (1977a) The location of parasites within their hosts: The behaviour of *Nippostrongylus brasiliensis* in the anaesthetised rat. *International Journal for Parasitology*, **6**, 195–200.

Croll, N.A. (1977b) The location of parasites within their hosts: The behavioural component in the larval migration of *Nippostrongylus brasiliensis* in the tissues of the rat. *International Journal for Parasitology*, **6**, 201–204.

Croll, N.A. and Blair, A. (1973) Inherent movement patterns of larval nematodes, with a stochastic model to simulate movement of infective hookworm larvae. *Parasitology*, **67**, 53–66.

Croll, N.A. and Matthews, B.E. (1977) *Biology of Nematodes*. Glasgow: Blackie.

Croll, N.A. and Smith. J.M. (1970) The sensitivity and responses of *Rhabditis* sp. to peripheral mechanical stimulation. *Proceedings of the Helminthological Society of Washington*, **37**, 1–5.

Croll, N.A. and Smith, J.M. (1972) Mechanism of thermopositive behaviour in larval hookworms. *Journal of Parasitology*, **58**, 891–896.

Davis, B.O., Goode, M. and Dusenbery, D.B. (1986) Laser microbeam studies of role of amphid receptors in chemosensory behavior of nematode *Caenorhabditis elegans*. *Journal of Chemical Ecology*, **12**, 1339–1347.

Diez, J.A. and Dusenbery, D.B. (1989a) Repellent of root-knot nematodes from exudates of host roots. *Journal of Chemical Ecology*, **15**, 2445–2455.

Diez, J.A. and Dusenbery, D.B. (1989b) Preferred temperature of *Meloidogyne incognita*. *Journal of Nematology*, **21**, 99–104.

Doncaster, C.C. (1962) Nematode feeding mechanisms. 1. Observations on *Rhabditis* and *Pelodera*. *Nematologica*, **8**, 313–320.

Doncaster, C.C. (1966) Nematode feeding mechanisms. 2. Observations on *Ditylenchus destructor* and *D. myceliophagus*. *Nematologica*, **12**, 417–427.

Doncaster, C.C. and Seymour, M.K. (1973) Exploration and selection of penetration site by Tylenchida. *Nematologica*, **19**, 137–145.

Driscoll, M. and Kaplan, J. (1997) Mechanotransduction. In *C. elegans II*, edited by D.L. Riddle , T. Blumenthal, B.J. Meyer and J.R. Priess, pp. 645–678. Cold Spring Harbor: Cold Spring Harbor Laboratory Press.

Duggan, A., Ma, C. and Chalfie, M. (1996) Regulation of touch receptor differentiation by the *Caenorhabditis elegans mec-3* and *unc-86* genes. *Development*, **125**, 4107–4119.

Dusenbery, D.B. (1973) Countercurrent separation: A new method for studying behavior of small aquatic organisms. *Proceedings of the National Academy of Sciences of the United States of America*, **70**, 1349–52.

Dusenbery, D.B. (1974) Analysis of chemotaxis in the nematode *Caenorhabditis elegans* by countercurrent separation. *Journal of Experimental Zoology*, **188**, 41–48.

Dusenbery, D.B. (1980a) Responses of the nematode *Caenorhabditis elegans* to controlled chemical stimulation. *Journal of Comparative Physiology A*, **136**, 327–331.

Dusenbery, D.B. (1980b) Analysis of chemotaxis in the nematode *Caenorhabditis elegans* by countercurrent separation. *Journal of Experimental Zoology*, **188**, 41–48.

Dusenbery, D.B. (1985) Using a microcomputer and video camera to simultaneously track 25 animals. *Computers in Biology and Medicine*, **15**, 169–175.

Dusenbery, D.B. (1988) Behavioral responses of *Meloidogyne incognita* to small temperature changes. *Journal of Nematology*, **20**, 351–355.

Dusenbery, D.B. (1989) A simple animal can use a complex stimulus pattern to find a location: nematode thermotaxis in soil. *Biological Cybernetics*, **60**, 431–437.

Dusenbery, D.B. (1996) Information is where you find it. *Biological Bulletin*, **191**, 124–128.

Ellenby, C. (1964) Haemoglobin in the chromotrope of an insect parasitic nematode. *Nature*, **202**, 615–616.

Emmons, S.W. and Sternberg, P.W. (1997) Male development and mating behavior. In *C. elegans II*, edited by D.L. Riddle, T. Blumenthal, B.J. Meyer and J.R. Priess, pp. 295–334. Cold Spring Harbor: Cold Spring Harbor Press.

Garcia-Rejon, L., Sanchez-Moreno, M., Verdejo, S. and Monteoliva, M. (1982) Estudios previos sobre las feromonas sexuales en los nematodes. *Revista Ibérica de Parasitologia* (Suppl. Vol.), 307–314.

Garcia-Rejon, L., Sanchez-Moreno, M., Verdejo, S. and Monteoliva, M. (1985a) Site of pheromone production in *Ascaris suum* (Nematoda). *Canadian Journal of Zoology*, **63**, 664–665.

Garcia-Rejon, L., Sanchez-Moreno, M., Verdejo, S. and Monteoliva, M. (1985b) Some factors affecting sexual attraction in *Ascaris suum* (Nematoda).). *Canadian Journal of Zoology*, **63**, 2074–2076.

Gaugler, R., LeBeck, L., Nakagaki, B and Boush, G.M. (1980) Orientation of the entomogenous nematode *Neoaplectana carpocapse* to carbon dioxide. *Environmental Entomology*, **9**, 649–652.

Gimenez, A. and Roche, M. (1972) Influence of male and female *Nippostrongylus brasiliensis* on each other's distribution in the intestine of the rat. *Parasitology*, **64**, 305–310.

Glassburg, G.H., Zalisko, E. and Bone, L.W. (1981) In vivo pheromone activity in *Nippostrongylus brasiliensis* (Nematoda). *Journal of Parasitology*, **67**, 898–905.

Golden, J.W. and Riddle, D.L. (1982) A pheromone influences larval development in the nematode *Caenorhabditis elegans*. *Science*, **218**, 578–580.

Golden, J.W. and Riddle, D.L. (1984a) A pheromone-induced developmental switch in *Caenorhabditis elegans*: temperature-sensitive mutants reveal a wild-type temperature-dependent process. *Proceedings of the National Academy of Sciences of the United States of America*, **81**, 819–823.

Golden, J.W. and Riddle, D.L. (1984b) The *Caenorhabditis elegans* dauer larva: developmental effects of pheromone, food and temperature. *Developmental Biology*, **102**, 368–378.

Golden, J.W. and Riddle, D.L. (1984c) A *Caenorhabditis elegans* dauer-inducing pheromone and an antagonistic component of the food supply. *Journal of Chemical Ecology*, **10**, 1265–1280.

Goode, M. and Dusenbery, D.B. (1985) Behavior of tethered *Meloidogyne incognita*. *Journal of Nematology*, **17**, 460–464.

Gordon, R.M. and Lumsden, W.H.P. (1939) A study of the behaviour of the mouth parts of the mosquito when taking up blood from living tissues: together with some observations on the ingestion of microfilariae. *Annals of Tropical Medicine and Parasitology*, **23**, 259–278.

Granzer, M. and Haas, W. (1991) Host-finding and host recognition of infective *Ancylostoma caninum* larvae. *International Journal for Parasitology*, **21**, 429–440.

Green, C.D. (1966) Orientation of male *Heterodera rostochiensis* Woll. and *H. schachtii* Scm. to their females. *Annals of Applied Biology*, **58**, 327–339.

Green, C.D. (1980) Nematode sex attractants. *Helminthological Abstracts, Series B*, **49**, 81–93.

Green, C.D. and Plumb, S.C. (1970) The interrelationships of some *Heterodera* spp. indicated by the specificity of the male attractants emitted by their females. *Nematologica*, **16**, 39–46.

Greet, D.N., Green, C.D. and Poulton, M.E. (1968) Extraction, standardisation and assessment of the volatility of the sex attractants of *Heterodera rostochiensis* Woll. and *H. schachtii* Schm. *Annals of Applied Biology*, **61**, 511–519.

Grundler, F., Schnibbe, L. and Wyss, U. (1991) *In vitro* studies on the behaviour of 2nd-stage juveniles of *Heterodera schachtii* (Nematoda, Heteroderidae) in response to host plant-root exudates. *Parasitology*, **103**, 149–155.

Gu, G.Q.. Caldwell, G.A. and Chalfie, M. (1996) Genetic interactions affecting touch sensitivity in *Caenorhabditis elegans*. *Proceedings of the National Academy of Sciences of the United States of America*, **93**, 6577–6582.

Haseeb, M.A. and Fried, B. (1988) Chemical communication in helminths. *Advances in Parasitology*, **27**, 169–207.

Hawking, F. (1962) Microfilarial infestation as an instance of periodic phenomena seen in a host-parasite relationship. *Annals of the New York Academy of Sciences*, **98**, 940–953.

Hart, A.C., Kass, J., Shapiro, J.E. and Kaplan, J.M. (1999) Distinct signaling pathways mediate touch and osmosensory responses in a polymodal sensory neuron. *Journal of Neuroscience*, **19**, 1952–1958.

Hedgecock, E.M. and Russell, R.L. (1975) Normal and mutant thermotaxis in the nematode *Caenorhabditis elegans*. *Proceedings of the National Academy of Sciences of the United States of America*, **72**, 4061–4065.

Hominick, W.M. and Aston, A.J. (1981) Association between *Pelodera strongyloides* (Nematoda: Rhabditidae) and wood mice, *Apodemus sylvaticus*. *Parasitology*, **83**, 67–75.

Huettel, R.N. (1986) Chemical communicators in nematodes. *Journal of Nematology*, **18**, 3–8.

Jacobs, D.E., Tod, M.E., Dunn, A.M. and Walker, J. (1968) Farm-to-farm transmission of porcine Oesophagostomiasis. *Veterinary Record*, **82**, 57.

Jaffee, H., Huettel, R.N., DeMilo, A.B., Hayes, D.K. and Rebois, R.V. (1989) Isolation and identification of a compound from soybean cyst nematode, *Heterodera glycines*, with sex pheromone activity. *Journal of Chemical Ecology*, **15**, 2031–2043.

Jansson, H.B., Jeyaprakash, A., Damon, R.A. and Zuckerman, B.M. (1984) *Caenorhabditis elegans* and *Panagrellus redivivus*: enzyme-mediated modification of chemotaxis. *Experimental Parasitology*, **58**, 270–277.

Jansson, H.B., Jeyaprakash, A., Marban-Mendoza, N. and Zuckerman, B.M. (1986) *Caenorhabditis elegans*: comparisons of chemotactic behavior from monoxenic and axenic culture. *Experimental Parasitology*, **61**.

Jones, F.G.W. and Jones, M.G. (1974) *Pests of Field Crops* 2nd. Edn. London: Edward Arnold.

Jones, J.T., Perry, R.N. and Johnstone, M.R.L. (1991) Electrophysiological recordings of electrical activity and responses to stimulants from

Globodera rostochiensis and *Syngamus trachea*. *Revue de Nématologie*, **14**, 467–473.

Jones, T.P. (1966) Sexual attraction and copulation in *Pelodera teres*. *Nematologica*, **2**, 518–522.

Jorgensen, E.M. and Rankin, C. (1997) Neural plasticity. In *C. elegans II*, edited by D.L. Riddle, T. Blumenthal., B.J. Meyer and J.R. Preiss, pp. 769–790. Cold Spring Harbor: Cold Spring Harbor Laboratory Press.

Kanagy, J.M.N. and Kaya, H.K. (1996) The possible role of marigold roots and alpha-terthienyl in mediating host-finding by steinernematid nematodes. *Nematologica*, **42**, 220–231.

Kaplan, J.M. and Horvitz, H.R. (1993) A dual mechanosensory and chemosensory neuron in *Caenorhabditis elegans*. *Proceedings of the National Academy of Sciences of the United States of America*, **90**, 2227–2231.

Klinger, J. (1961) Anzeihungsversuche mit *Ditylenchus dipsaci* unter Berücksichtigung der Wirkung des Kohlendioxyds, des Redoxpotentials und anderer Faktoren. *Nematologica*, **6**, 69–84.

Klinger, J. (1970) The reaction of *Aphelenchoides fragariae* to slit-like microopenings and to stomatal diffusion gases. *Nematologica*, **16**, 417–422.

Klink, J.W., Dropkin, V.E. and Mitchell, J.E. (1970) Studies on the host-finding mechanisms of *Neotylenchus linfordi*. *Journal of Nematology*, **2**, 106–117.

Lane, C. (1933) The taxes of infective hookworm larvae. *Annals of Tropical Medicine and Parasitology*, **27**, 237–250.

Lee, D.L. (1962) The distribution of esterase enzymes in *Ascaris lumbricoides*. *Parasitology*, **52**, 241–260.

Lee, D.L. (1973) Evidence for a sensory function for the copulatory spicules of nematodes. *Journal of Zoology*, **169**, 281–285.

Lee, D.L. (1972) Penetration of mammalian skin by the infective larva of *Nippostrongylus brasiliensis*. *Parasitology*, **65**, 499–505.

Lee, D.L. and Atkinson, H.J. (1976) *Physiology of Nematodes*. Macmillan: London.

Lehmann, T., Cupp, S.M. and Cupp, W.E. (1995) Chemical guidance of *Onchocerca lienalis* microfilariae to the thorax of *Simulium vittatum*. *Parasitology*, **110**, 329–337.

Liu, K.S. and Sternberg, P.W. (1995) Sensory regulation of male mating-behavior in *Caenorhabditis elegans*. *Neuron*, **14**, 79–89.

Loer, C.M. and Kenyon, C.J. (1993) Serotonin-deficient mutants and male mating-behavior in the nematode *Caenorhabditis elegans*. *Journal of Neuroscience*, **13**, 5407–5417.

Marchant, H.J. (1970) Bursal response in sexually stimulated *Nematospiroides dubius* (Nematoda). *Journal of Parasitology*, **56**, 201–202.

McCallum, M.E. and Dusenbery, D.B. (1992) Computer tracking as a behavioral GC detector — Nematode responses to vapor of host roots. *Journal of Chemical Ecology*, **18**, 585–592.

McCue, J.F. and Thorson, R.E. (1964) Behaviour of parasitic stages of helminths in a thermal gradient. *Journal of Parasitology*, **50**, 67–71.

Moens, T., Verbeeck, L., deMaeyer, A., Swings, J. and Vincx, M. (1999) Selective attraction of marine bacterivorous nematodes to their bacterial food. *Marine Ecology Progress Series*, **176**, 165–178.

Moltmann, E. (1990) Kairomones in root exudates of cereals — their importance in host finding of juveniles of the cereal cyst nematode, *Heterodera avenae* (Woll.), and their characterization. *Zeitschrift für Pflanzenkrankheiten und Pflanzenschutz*, **97**, 458–469.

Morgan, D.O. (1928) On the infective larvae of *Ostertagia circumcincta* Stadelmann (1894), a stomach worm of sheep. *Journal of Helminthology*, **6**, 183–192.

Mori, I. and Ohshima, Y. (1995) Neural regulation of thermotaxis in *Caenorhabditis elegans*. *Nature*, **376**, 344–348.

Mori, I. And Ohshima, Y. (1997) Molecular neurogenetics of chemotaxis and thermotaxis in the nematode *Caenorhabditis elegans*. *Bioessays*, **19**, 1055–1064.

Muller, R. and Wakelin, D. (1998) Lymphatic filariasis. In *Topley & Wilson's Microbiology and Microbial Infections* 9th edn., edited by L. Collier, A. Balows and M. Sussma, Volume 5, *Parasitology*, edited by F.E.G. Cox, J.P. Kreier and D. Wakelin, pp. 610–619. London: Arnold.

Nicholls, C.D. (1987) Endoscopy, physiology and bacterial flora of sheep infected with abomasal nematodes. *PhD Thesis, University of Leeds*, Leeds, UK.

O'Connor, F.W. and Beatty, H.A. (1937) The abstraction by *Culex fatigans* of *Microfilaria bancrofti* from man. *Journal of Tropical Medicine and Hygiene*, **40**, 101–103.

Ohba, K., Nishimura, S. and Saito, T. (1986) Partial purification of dauer juvenile inducer of *Caenorhabditis elegans*. *Japanese Journal of Nematology*, **16**, 35–37.

Oostenbrink, M., Kuiper, K. and s'Jacob, J.J. (1957) *Tagetes* als feindpflanzen von *Pratylenchus*-arten. *Nematologica*, **2**, (Suppl.), 424–433.

Papademetriou, M.K. and Bone, L.W. (1983) Chemotaxis of larval soybean cyst nematode, *Heterodera glycines* race 3, to root leachates and ions. *Journal of Chemical Ecology*, **9**, 387–396.

Perry, R.N. (1997) Plant signals in nematode hatching and attraction. In *Cellular and Molecular Aspects of Plant-Nematode Interactions*, edited by C. Fenoll, F.M.W. Grundler and Ohl, S.A., pp. 38–50. Dordecht: Kluwer Academic Press.

Perry, R.N. and Aumann, J. (1998) Behaviour and sensory responses. In *The Physiology and Biochemistry of Free-living and Plant-parasitic Nematodes*, edited by R.N. Perry and D.J. Wright, pp. 75–102. Wallingford: CAB International Press.

Peters, B.G. (1952) Toxicity tests with vinegar eelworm. 1. Counting and culturing. *Journal of Helminthology*, **26**, 97–110.

Pitcher, R.S. and Flegg, J.J.M. (1965) Observation of root feeding by the nematode *Trichodorus viruliferus* Hooper. *Nature*, **207**, 317.

Pitcher, R.S. (1967) The host-parasite relations and ecology of *Trichodorus viruliferus* on apple roots, as observed from an underground laboratory. *Nematologica*, **13**, 547–537.

Pline, M., Diez, J.A. and Dusenbery, D.B. (1988) Extremely sensitive thermotaxis of the nematode *Meloidogyne incognita*. *Journal of Nematology*, **20**, 605–608.

Poinar, G.O. and Thomas, G.M. (1975) *Rhabditis pellio* Schneider (Nematoda) from the earthworm *Aporrectodea trapezoides* Duges (Annelida). *Journal of Nematology*, **7**, 374–379.

Prot, J.C. (1980) Migration of plant-parasitic nematodes towards plant roots. *Revue de Nématologie*, **3**, 305–318.

Rankin, C.H., Beck, C.D.O. and Chiba, C.M. (1990) *Caenorhabditis elegans*: a new model system for the study of learning and memory. *Behavioural Brain Research*, **37**, 89–92.

Rankin, C.H. and Broster, B.S. (1992) Factors affecting habituation and recovery from habituation in the nematode *Caenorhabditis elegans*. *Behavioral Neuroscience*, **106**, 239–249.

Riddle, D.L. (1988) The dauer larva. In *The nematode Caenorhabditis elegans*, edited by W.B. Wood, pp. 393–412. Cold Spring Harbor: Cold Spring Harbor Laboratory.

Riemann, F. and Schrage, M. (1988) Carbon dioxide as an attractant for the free-living marine nematode *Adoncholaimus thalassophygas*. *Marine Biology*, **98**, 81–85.

Riga, E. (1992) Multifaceted approach for differentiating isolates of *Bursaphelenchus xylophilus* and *B. mucronatus* (Nematoda), parasites of pine trees. *PhD thesis, Simon Fraser University*, Burnaby, B.C., Canada. (**not seen in original**).

Riga, E., Holdsworth, D.R., Perry, R.N., Barrett, J. and Johnston, M.R.L. (1997a) Electrophysiological analysis of the response of males of the potato cyst nematode, *Globodera rostochiensis*, to fractions of their homospecific sex pheromone. *Parasitology*, **115**, 311–316.

Riga, E., Perry, R.N., Barrett, J. and Johnston, M.R.L. (1995) Investigation of the chemosensory function of amphids of *Syngamus trachea* using electrophysiological techniques. *Parasitology*, **111**, 347–351.

Riga, E., Perry, R.N. and Barrett, J. (1996) Electrophysiological analysis of the response of males of *Globodera rostochiensis* and *G. pallida* to their female sex pheromones and to potato root diffusate. *Nematologica*, **42**, 493–498.

Riga, E., Perry, R.N., Barrett, J. and Johnston, M.R.L. (1996) Electrophysiological responses of the potato cyst nematodes, *Globodera rostochiensis* and *G. pallida*, to their sex pheromones. *Parasitology*, **112**, 239–246.

Riga, E., Perry, R.N., Barrett, J. and Johnston, M.R.L. (1997b) Electrophysiological responses of male potato cyst nematodes, *Globodera rostochiensis* and *G. pallida*, to some chemicals. *Journal of Chemical Ecology*, **23**, 417–428.

Roberts, T.M. and Thorson, R.E. (1977) Chemical attraction between adults of *Nippostrongylus brasiliensis*: description of the phenomenon and effects of host immunity. *The Journal of Parasitology*, **63**, 357–363.

Robertson, W.M. and Forrest, J.M.S. (1989) Factors involved in host recognition by plant-parasitic nematodes. *Aspects of Applied Biology*, **22**, 129–133.

Robinson, A.F. (1995) Optimal release rates for attracting *Meloidogyne incognita*, *Rotylenchus reniformis*, and other nematodes to carbon dioxide in sand. *Journal of Nematology*, **27**, 42–50.

Robinson, A.F. and Jaffe, B.A. (1996) Repulsion of *Meloidogyne incognita* by alginate pellets containing hyphae of *Monacrosporium cionopagum*, *M. ellipsosporum*, or *Hirsutella rhossiliensis*. *Journal of Nematology*, **28**, 133–147.

Robinson, J. (1962) *Pilobolus* spp. and the translation of infective larvae of *Dictyocaulus viviparus* from faeces to pasture. *Nature*, **193**, 353–354.

Roche, M. (1966) Influence of male and female *Ancylostoma caninum* on each other's distribution in the intestine of the dog. *Experimental Parasitology*, **19**, 327–331.

Rogers, W.P. (1962) *The nature of parasitism: The relationship of some metazoan parasites to their hosts*. London: Academic Press.

Rogers, W.P. and Somerville, R.I. (1968) The infectious process and its relation to the development of early parasitic stages of nematodes. *Advances in Parasitology* **6**, 327–348.

Ronald, R. (1960) The effects of physical stimuli on the larvae of *Terranova decipiens*. I. Temperature. *Canadian Journal of Zoology*, **38**, 623–642.

Sambongi, Y., Nagae, T., Liu, Y., Yoshimizu, T., Takeda, K., Wada, Y. and Futai, M. (1999) Sensing of cadmium and copper ions by externally exposed ADL, ASE, and ASH neurons elicits avoidance response in *Caenorhabditis elegans*. *Neuroreport*, **10**, 753–757.

Sengupta, P., Colbert, H.A., Kimmel, B.E., Dwyer, N. and Bargmann, C.I. (1993) The cellular and genetic basis of olfactory responses in *Caenorhabditis elegans*. *CIBA Foundation Symposia*, **179**, 235–250.

Staddon, J.E.R. and Higa, J.J. (1996) Multiple time scales in simple habituation. *Psychological Review*, **103**, 720–733.

Sulston, J.E., Albertson, D.G. and Thompson, J.N. (1980) The *Caenorhabditis elegans* male: Postembryonic development of non-gonadal structures. *Developmental Biology*, **78**, 542–576.

Sukhdeo, M.V.K. and Croll, N.A. (1981) The location of parasites within their hosts: bile and the site selection behavior of *Nematospiroides dubius*. *International Journal for Parasitology*, **11**, 157–162.

Sukhdeo, M.V.K., O'Grady, R.T. and Hsu, S.C. (1984) The site selected by the larvae of *Heligmosomoides polygyrus*. *Journal of Helminthology*, **58**, 19–23.

Sukhdeo, M.V.K. and Sukhdeo, S.C. (1994) Optimal habitat selection by helminths within the host environment. *Parasitology*, **109**, S41–S55.

Tindall, N.R. and Wilson, P.A.G. (1990) An extended proof of migration routes of immature parasites inside hosts: pathways of *Nippostrongylus brasiliensis* and *Strongyloides ratti* in the rat are mutually exclusive. *Parasitology*, **100**, 281–288.

Tod, M.E., Jacobs, D.E. and Dunn, A.M. (1971) Mechanisms for the dispersal of parasitic nematode larvae. 1. Psychodid flies as transport hosts. *Journal of Helminthology*, **45**, 133–137.

Troemel, E.R., Kimmel, B.E. and Bargmann, C.I. (1997) Reprogramming chemotaxis responses: Sensory neurones define olfactory preferences in *C. elegans*. *Cell*, **91**, 161–169.

Uhlenbroek, J.H. and Bijloo, J.D. (1958) Investigations on nematacides. I. Isolation and structure of a nematacidal principle occurring in *Tagetes* roots. *Recueil des travaux chimiques des Pay–Bas et de la Belgique*, **77**, 1004–1009.

Viglierchio, D.R. (1990) Carbon dioxide sensing by *Panagrellus silusiae* and *Ditylenchus dipsaci*. *Revue de Nématologie*, **13**, 425–432.

Wallace, H.R. (1963) *The Biology of Plant Parasitic Nematodes*. London: Edward Arnold.

Ward, S. (1973) Chemotaxis by the nematode *Caenorhabditis elegans*: identification and analysis of the response by use of mutants. *Proceedings of the National Academy of Sciences of the United States of America*, **70**, 817–821.

Ward, S. (1978) Nematode chemotaxis and chemoreceptors. In *Taxis and Behavior (Receptors and Recognition)*, Series B, Volume 5, edited by G.L. Hazelbauer, pp. 143–168. London: Chapman and Hall.

Ward, J.B. and Bone, L.W. (1983). Chromatography and isolation of the K_{av} 1.0 pheromone of female *Nippostrongylus brasiliensis* (Nematoda). *The Journal of Parasitology*, **69**, 302–306.

Ward, J.B., Nordstrom, R.M. and Bone, L.W. (1984) Chemotaxis of male *Nippostrongylus brasiliensis* (Nematoda) to some biological compounds. *Proceedings of the Helminthological Society of Washington*, **51**, 73–77.

Wen, J.Y.M., Kumar, N., Morrison, G., Rambaldini, G., Runciman, S., Rousseau, J. and van der Kooy, D. (1997) Mutations that prevent associative learning in *C. elegans*. *Behavioral Neuroscience*, **111**, 354–368.

White, J. (1988) The anatomy. In *The nematode Caenorhabditis elegans*, edited by W.B. Wood, pp. 81–122. Cold Spring Harbor: Cold Spring Harbor Laboratory.

Wicks, S.R. and Rankin, C.H. (1995) Integration of mechanosensory stimuli in *Caenorhabditis elegans*. *Journal of Neuroscience*, **15**, 2434–2444.

Wilson, P.A.G. (1966) The light sense in nematodes. *Science*, **151**, 337–338.

Wilson, P.A.G. (1994) Doubt and certainty about the pathways of invasive juvenile parasites inside hosts. *Parasitology*, **109**, S57–S67.

Wyss, U. and Zunke, U. (1986) A potential of high resolution video-enhanced contrast microscopy in nematological research. *Revue Nématologie*, **9**, 91–94.

Young, I.M., Griffiths, B.G. and Robertson, W.M. (1996) Continuous foraging by bacterial-feeding nematodes. *Nematologica*, **42**, 378–382.

Zwall, R.R., Mendel, J.E., Sternberg, P.W. and Plasterk, R.H.A. (1997) Two neuronal G proteins are involved in chemosensation of the *Caenorhabditis elegans* dauer-inducing pheromone. *Genetics*, **145**, 715–727.

16. Nematode Survival Strategies

David A. Wharton

Department of Zoology, University of Otago, P.O. Box 56, Dunedin, New Zealand

Keywords

cold, freezing, heat, desiccation, diapause, dauer larvae

Introduction

For a parasitic nematode, hosts represent oases of resource where reproduction and/or growth is possible, in an inhospitable desert where they are not. Parasites must cross this desert because their host is not immortal and the death of the host results in the death of the parasites contained within it. Thus, despite the difficulties, parasites have to face the hazards of transmission and infect new hosts to ensure the survival of the species. In this context the survival strategies of parasitic nematodes can be seen as mechanisms which either assist them to locate and infect a new host or to survive the period between hosts.

We can think of the survival strategies of free-living nematodes is a similar way. The 'oases' are areas in space and time where conditions for the nematode to grow and reproduce are favourable and the 'desert' is where they are not. Unless the nematode is able to translocate to a new habitat, temporal changes are likely to be the more important. The survival strategies of free-living nematodes are thus focussed on surviving periods when conditions such as food availability, temperature, moisture levels etc. do not allow growth and reproduction.

Responses to Adverse Conditions

Organisms may have two types of responses to adverse conditions. Resistance adaptations enable the organism to survive the stress until conditions become favourable again. Capacity adaptations enable the organism to grow and reproduce under harsh conditions (Cossins and Bowler, 1987; Precht, 1958). Although this distinction was originally proposed for adaptations to temperature stress, it could also be applied to other stresses such as anoxia, osmotic stress and desiccation.

If the conditions for the optimum growth of a nematode which lives in a harsh environment are different from those of nematodes that live in more benign environments, this would provide evidence for capacity adaptation; for example, if the optimum growth temperature of an Antarctic nematode was lower than those of its temperate relatives. Few such data are available. *Plectus antarcticus*, from the maritime Antarctic, can lay eggs at 5°C and complete its life cycle at 5°C but the generation time was much shorter at 22°C than at 10°C and lower temperatures (Caldwell, 1981). A low minimum growth temperature may indicate a capacity adaptation, although the optimal growth temperature does not. *Chiloplacus* sp., a nematode from the high arctic, had a lower temperature threshold for growth and shorter generation times than did many other nematodes at comparable low temperatures (Procter, 1984), this indicates capacity adaptation. *Panagrolaimus davidi*, from the terrestrial Antarctic, however, has a maximal intrinsic rate of natural increase at 25°C and no population growth below 6.8°C (Brown, 1993; Wharton, 1997). There is thus little evidence for capacity adaptation in this species and it appears to rely on resistance adaptations, surviving harsh conditions and growing when conditions are favourable.

Some Examples of Survival Strategies in Parasitic and Free-Living Nematodes

Any nematode species will possess a suite of adaptations which enables it to survive in the environment, or series of environments, in which it lives. This is particularly true of parasitic nematodes which may have to face the hazards of the environment outside the host. The different stages of the life cycle of a parasitic nematode may have different survival mechanisms geared to the hazards faced by each stage. The parasitic stages may have to deal with the immune responses and other features of the environment within the host (such as gut enzymes). Free-living nematodes and stages face biological hazards, such as predation, competition and pathogens. However, I will focus upon how free-living nematodes and the free-living stages of parasitic nematodes deal with the physical hazards of the environment outside a host. In this section I will consider the suite of survival mechanisms possessed by a number of parasitic and free-living nematodes.

Ascaris lumbricoides

Ascaris lumbricoides is a parasite of the small intestine of humans. It has a direct life cycle with infection of the definitive host by the chance ingestion of an egg containing an infective 2nd-stage larva (L2). Reliance on accidental ingestion means that the chance of any one egg infecting a host is extremely low and this, together with the death of eggs outside the host or their loss from the environment of the host, means that there is a very high wastage of eggs. It is often said that the high wastage rate is offset by a high fecundity. *Ascaris* can produce 200,000 eggs/female/day and may live for a year (Levine, 1968), giving a total egg production of 73 million eggs per female! Only two eggs, one destined to become a female and one a male, need to successfully infect a host in order to replace their parents, or 0.000003% of the eggs produced. The high fecundity is driven by high levels of resource availability within the host's intestine (Calow, 1983). Many parasites display high fecundities.

Another way of increasing the chances of egg transmission is to increase their chances of survival and hence the time viable eggs are retained in the host's environment. One way of doing this is to provide the egg with sufficient food to fuel embryonic and larval development and their survival until ingestion by the host. The egg of *Ascaris* is surrounded by an eggshell which provides mechanical and chemical protection and aids desiccation survival (Wharton, 1980).

The timing of egg hatching is also important for survival, since if the larva was to hatch outside the host it would presumably die. The eggs hatch in response to a specific physiological trigger, an increase in CO_2 at an appropriate temperature and pH, which indicates entry into the host's intestine (Rogers, 1960).

Globodera rostochiensis

G. rostochiensis is a parasite of potatoes and tomatoes. In many ways its life cycle and survival strategies are similar to those of *Ascaris*. After mating the female worms become gravid, with 200–300 eggs being retained within the uterus. The female then dies and her body becomes chemically hardened via a quinone-tanning process to form a protective cyst containing the eggs (Awan and Hominick, 1982; Ellenby, 1946). The eggs develop to the infective L2 which is protected by the eggshell and the cyst wall.

The infective larva within the egg is resistant to desiccation and cold and only hatches in response to some factor exuded from the roots of the potato host (Perry, 1989), although there may be some hatching in the spring in the absence of the host plant (Jones and Jones, 1974). There is evidence for a diapause in which the eggs respond to photoperiod by making hatching more likely to occur in spring, thus synchronising with the growth of the host (Hominick, 1986).

Ditylenchus dipsaci

D. dipsaci infects a variety of plant hosts. Unusually for a metazoan parasite, it can reproduce within the host and build up high levels of infection. When the host senesces the nematode accumulates as 4th-stage larvae (L4) which persist in the soil until they infect a new host. The L4 are resistant to desiccation and are capable of anhydrobiosis (Perry, 1977a).

Trichostrongyle nematodes

A number of species of trichostrongyle nematodes infect the alimentary tract of sheep and cattle, causing parasitic gastroenteritis. Most species have a similar life cycle with high fecundity, a resistant egg which hatches as a 1st-stage larva (L1). The L1 and L2 are bacterial feeders within the faeces and rapidly develop to the 3rd-stage infective larva (L3).

The L3 retains the L2 cuticle as a sheath, which completely surrounds the L3 so it is non-feeding. The L3 relies on food reserves built up by the L1 and L2 and the intestine becomes modified for food storage and loses its lumen. The L3 is resistant to a variety of environmental hazards, including desiccation and low temperatures (Allan and Wharton, 1990; Wharton and Allan, 1989). The sheath may play a role in survival by slowing down the rate of water loss when exposed to desiccation and by preventing inoculative freezing (Ellenby, 1968b; Wharton and Allan, 1989). It does not, however, prevent attack by nematophagous fungi which can trap ensheathed but not exsheathed larvae (Wharton and Murray, 1990).

When ingested by the definitive host the L3 exsheathes in response to an increase in CO_2 concentration at an appropriate temperature and pH (Rogers, 1960). This is similar to the physiological trigger initiating the hatching of *Ascaris* eggs. The region of the alimentary tract where development is completed is determined by the physiology of that region and that immediately preceding it, thus *Haemonchus contortus* begins exsheathment in the rumen and develops into the adult in the abomasum whereas *Trichostrongylus colubriformis* begins exsheathment in the abomasum and develops into the adult in the small intestine (Rogers, 1960).

Trichostrongyles may overwinter as L3 on pasture. However, in some species and strains, larvae acquired during autumn do not immediately develop into adult nematodes and are retained throughout the winter as L4. This phenomenon is referred to as hypobiosis or arrested development (Eysker, 1997; Gibbs, 1986). In addition to this winter inhibition, summer inhibition, associated with survival during dry seasons, has been reported from several parts of the world (Jacquiet, *et al.*, 1996). After winter inhibition the L4 resume development in the spring, it is thought in response to hormonal changes in the host at the start of the breeding season (Gibbs, 1986). Eggs are thus deposited on the pasture and infective larvae develop at a time when new susceptible hosts (lambs or calves) are available. Arrested development is thus a way of synchronising the life cycle of the parasite with that of the host.

Caenorhabditis elegans

C. elegans is a bacterial-feeding, free-living nematode which is widely used as a model system for investigating a variety of biological questions (Burglin, Lobos and Blaxter, 1998; Riddle, *et al.*, 1997). The nematode is hermaphroditic and, under favourable conditions, will lay about 300 eggs in its life which hatch and develop through four larval stages to the adult. The life cycle can be completed in 3 days (Wood, 1988). The short generation time of *C. elegans* means that it has a high reproductive potential (Wharton, 1986), it is thus able to rapidly exploit a transient food source. When conditions become unfavourable for growth, for example due to the exhaustion of food or at high population densities, *C. elegans* produces a special L3, called a dauer larva (Riddle, 1988).

The dauer larva is non-feeding, has a reduced epidermis and a cuticle modified by the addition of a striated layer in the basal zone. The intestine is reduced, the excretory gland is inactive and there is modification of the sense organs (Riddle, 1988). Dauer larvae have a longer life span and are more resistant to stress than normal L3. They resist high temperatures and oxygen deprivation better than adult *C. elegans* (Anderson, 1978). Their response to desiccation and to environmentally-relevant low temperatures has not, however, been investigated.

Dauer production occurs in a temperature-dependant response to an increase in the concentration of a dauer-inducing pheromone, which accumulates at high population densities, and to a decrease in a chemical signal from the bacterial food. Resumption of development is in response to an increase in food, indicating the return of conditions favourable to growth. These signals act on the L1, although this developmental decision may be reversed by stimuli acting on the L2 (Riddle and Albert, 1997). The signal appears to be mediated via the amphidial nerves and acts via the activation of genes associated with the formation of either dauer larvae or normal L3. Although the developmental pathways involved are known, the molecular mechanisms involved remain to be elucidated (Leroi and Jones, 1998; Riddle and Albert, 1997).

Panagrolaimus davidi

P. davidi is endemic to the McMurdo Sound region of Antarctica. It lives associated with the moss and algae which grow in the meltwater from glaciers and snow which forms during the brief Antarctic summer (Wharton and Brown, 1989). Its optimum growth temperature is similar to that which might be expected for a temperate species, with a maximal growth temperature of 25°C (Brown, 1993; Wharton, 1997). It can tolerate freezing and desiccation (Wharton, 1997). Its strategy appears to be a resistance adaptation, growing rapidly when conditions are favourable but lying dormant for most of the time.

Summary of Survival Strategies in Nematodes

Nematodes thus use a variety of survival strategies and a particular species is likely to display several of these. The strategies can be listed as:

1. High reproductive capacity by high fecundity (parasitic species) or short generation times (free-living species) when resources are available.
2. Survival stages: eggs, infective larvae, dauer larvae.
3. Mechanisms for synchronising the nematode's life cycle with the availability of food or hosts: diapause, arrested development, dauer larvae, infective larvae.
4. Physiological triggers which indicate entry into or the presence of a suitable host or food supply: hatching stimulus, exsheathment stimulus, resumption of development.
5. Intermediate hosts in the life cycle of parasitic species which remove the parasite from the external environment, provide additional resources for development and may facilitate the infection of the definitive host.

The strategies used by the nematodes described in this section are summarised in Table 16.1.

The variety of survival strategies employed by nematodes has led to considerable debate as to how the strategies may be classified (Antoniou, 1989; Evans, 1987; Evans and Perry, 1976; Wharton, 1986; Womersley, Wharton and Higa, 1998). I will not revisit this debate here, except to say that it may be useful to distinguish between metabolic and developmental dormancy (Wharton, 1986) or between dormancy affecting ontogenetic development and that affecting somatic development (Evans, 1987). I will attempt to define terms when we come to them.

Table 16.1. Survival strategies of some nematodes

Strategy	*A. lumbricoides*	*G. rostochiensis*	*D. dipsaci*	Trichostrongyles	*C. elegans*	*P. davidi*
High reproductive capacity	high fecundity	–	reproduction within host	high fecundity	short generation time	short generation time
Survival stage	egg	egg/cyst	L4	L3	dauer larva	all stages?
Life-cycle sychronisation	–	diapause	–	arrested development	dauer larva	–
Physiological trigger	hatching stimulus	hatching stimulus	?	exsheathment stimulus	food	–

Responses to Temperature

The responses of nematodes, and any other organism, to temperature are varied and complex. A possible, highly simplified, response is shown in Figure 16.1. There will be an optimum temperature at which life processes (growth, reproduction etc.) proceed at their fastest rate. As the temperature increases or decreases from the optimum these rates will decrease until motion becomes disoriented and normal processes disrupted (heat or cold stupor) and then cease altogether (heat or cold coma). Death may then result. The temperatures at which these transitions occur will vary from species to species and are likely to occur over a range of temperatures rather than at a single temperature. The range over which activity and normal activity can occur, and the thermal death points, may also change as a result of acclimatisation responses. I am not aware of any study in which the full range of temperature responses of a nematode has been determined.

In this section I will look at the mechanisms by which nematodes may survive high and low temperatures which could result in their death. High and low temperatures represent very different stresses. High temperatures are likely to cause irreversible damage as, for example, proteins become denatured. Proteins are not denatured by low temperatures and hence changes are potentially reversible. A major stress associated with low temperatures, however, is a change in state, i.e. freezing, of body water. Death is likely to result from high temperatures long before a change in state of body water (boiling) occurs.

Cold Tolerance in Nematodes

Nematodes which survive low temperatures in their natural habitat are said to be cold tolerant. There are a number of reports of nematodes surviving storage at ultra-low temperatures (for some examples see Antoniou, 1989), but this is of little ecological relevance.

In polar, alpine and temperate habitats free-living nematodes and the free-living stages of animal and plant-parasitic nematodes are exposed to, and must be able to survive, subzero temperatures for short or extended periods of time. Unless they are desiccated, this involves the risk of ice formation in their bodies. Nematodes are aquatic organisms and are subject to inoculative freezing when the water surrounding them freezes; unless they are able to prevent it by the presence of a structure such as an eggshell or a sheath (Wharton, 1995).

Cold tolerance strategies in invertebrates

Most research on invertebrate cold tolerance has focussed on arthropods, particularly insects (Lee and Denlinger, 1991; Sømme, 1995). Arthropods are usually considered to survive subzero temperatures by one of two strategies (Lee, 1991). Freeze-avoiding arthropods prevent freezing at subzero temperatures and super-cool (maintain their body fluids as a liquid at temperatures below their melting point) whereas freezing-tolerant arthropods survive ice formation within their bodies.

Freeze avoidance involves preventing inoculative freezing by overwintering in dry sites or by the presence of a cocoon or other structures which prevent the propagation of ice. Removal of ice nucleators within the body may include emptying the gut and the masking or inactivation of nucleators within the haemocoel. The accumulation of polyols (such as glycerol) and/or sugars depresses the melting and supercooling points and may mask or inactivate endogenous nucleators. Some insects also produce thermal hysteresis proteins (THPs) which are thought to stabilise the supercooled state by attaching to and inhibiting the growth of ice crystals (Duman, et al., 1991). Desiccation may enhance survival by increasing the osmolality of the haemolymph and the concentration of cryoprotectants (Zachariassen, 1985).

Freezing tolerant insects ensure freezing at relatively high subzero temperatures by synthesising haemolymph ice nucleating proteins (Duman, et al., 1991), although other sites such as the gut may act as nucleation sites (Block, 1995). This ensures that freezing is slow and reduces the chances of intracellular freezing. Polyols and sugars act as cryoprotectants, depressing the melting point and decreasing the amount of ice formed at any given temperature, thus reducing cellular dehydration. They may also stabilise proteins and membranes, preventing damage during freezing and desiccation (Lee, 1991). THPs are also produced by freezing tolerant insects. These may inhibit the recrystallisation which may damage cells by changes in the size of ice crystals (Knight and Duman, 1986) or prevent the migration of still-liquid salty domains (Knight, Wen and Laursen, 1995).

Cold tolerance strategies in nematodes

Some care must be taken in applying cold hardiness theory from arthropods to nematodes. Nematodes are essentially aquatic organisms and have much more of a

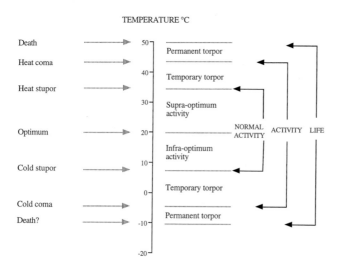

Figure 16.1. The responses to temperature in a hypothetical nematode.

problem with inoculative freezing than do the terrestrial arthropods (Wharton, 1995). Since they are living in water, ice formation in the external medium is likely to nucleate the freezing of body contents by inoculative freezing. Most cold-tolerant nematodes are thus likely to be freezing tolerant. A nematode could employ a freeze-avoiding strategy if it was free of surface water but not desiccated, a situation that is likely to be transient. To supercool in the presence of external water the nematode needs a structure, such as an eggshell or a sheath, which prevents inoculative freezing (Wharton, 1995). Another possibility is that if the nematode desiccates and can survive anhydrobiotically it would survive low temperatures without the risk of freezing, since there is no water present.

Some authors have suggested that nematodes in soil may desiccate rather than freeze when the soil freezes due to freeze-induced desiccation (Forge and Mac-Guidwin, 1992; Pickup, 1990c). This may result from ice crystal growth in the external medium concentrating salts in the non-frozen portion of the soil, raising the osmotic concentration and resulting in water loss from the nematode by osmosis, preventing the nematode from freezing. *Panagrolaimus davidi*, however, will freeze at osmolalities much higher than are likely to be attained by freeze concentration effects (Wharton and To, 1996).

A final possibility is freeze drying or 'protective dehydration mechanism' of cold hardiness (Holmstrup and Westh, 1995). This occurs because the vapour pressure of supercooled water is higher than that of ice. A supercooled animal which is surrounded by ice will therefore lose water and desiccate. This process has been demonstrated in earthworm cocoons, enchy-

traeids and springtails (Holmstrup and Sømme, 1998; Holmstrup and Westh, 1995; Worland, Gruborlajsic and Montiel, 1998). However, for it to occur in a nematode it must be surrounded by or be close to ice and yet still contain supercooled water. A nematode in contact with ice is likely to freeze by inoculative freezing. If liquid water is lost from the surface of the nematode it is likely to desiccate. For freeze drying to occur, water would have to be lost from the immediate vicinity of the nematode so there is no water in contact with its surface, there must, however, be sufficient water close to the nematode to prevent it from desiccating and that water would then have to freeze to initiate freeze drying. This sequence of events seems unlikely but if it does occur it is similar to surviving low temperatures in a state of anhydrobiosis. It is more likely to occur in a supercooled nematode which is surrounded by an eggshell or a sheath which prevents ice nucleation.

The possible strategies of cold tolerance in nematodes are summarised in Figure 16.2. It is difficult to observe nematodes in soil to determine which of these mechanisms is occurring. In the presence of water, freezing tolerance seems the most likely strategy and freeze avoidance if an eggshell or sheath is present and can prevent inoculative freezing. If the soil is dry, survival of low temperatures in a state of anhydrobiosis seems likely. There is little evidence at present for survival by freeze-induced desiccation or a protective desiccation mechanism in nematodes.

Cold tolerance in free-living nematodes

The cold tolerance of free-living nematodes has been mainly studied in Antarctic nematodes. Nematodes are

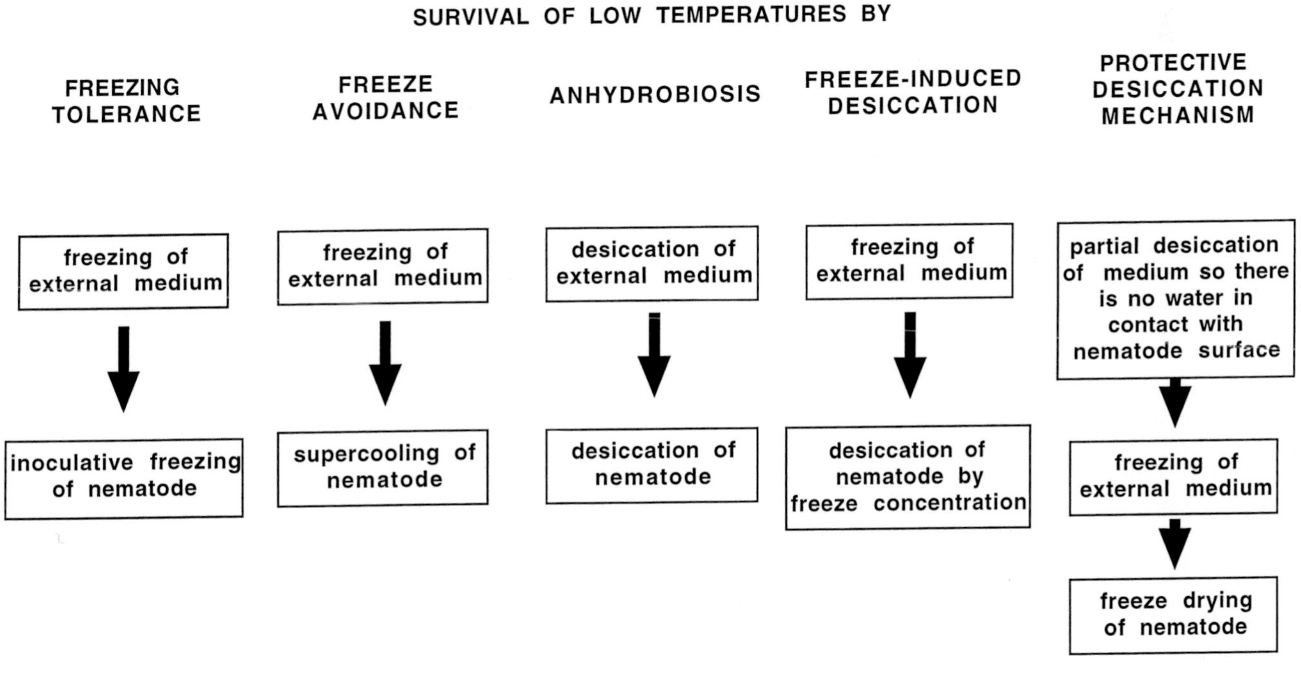

SURVIVAL OF LOW TEMPERATURES BY

FREEZING TOLERANCE	FREEZE AVOIDANCE	ANHYDROBIOSIS	FREEZE-INDUCED DESICCATION	PROTECTIVE DESICCATION MECHANISM
freezing of external medium	freezing of external medium	desiccation of external medium	freezing of external medium	partial desiccation of medium so there is no water in contact with nematode surface
inoculative freezing of nematode	supercooling of nematode	desiccation of nematode	desiccation of nematode by freeze concentration	freezing of external medium
				freeze drying of nematode

Figure 16.2. Nematode cold tolerance strategies.

found as part of the community of invertebrates which inhabit the moss and/or algae which grow in the meltwater from snow and glaciers and around the edges of bodies of freshwater which partially melt during the summer. Maslen (1979) lists 40 species from the maritime Antarctic and 10 and 22 species respectively from the continental and sub-Antarctic zones. Some of the most southerly records are from the McMurdo Sound region, which has six recorded species (Timm, 1971; Wharton and Brown, 1989). Antarctic nematodes are likely to be frozen for a substantial proportion of the year (Davey, Pickup and Block, 1992). *Panagrolaimus davidi*, from the McMurdo Sound region, has been isolated and established in culture (Wharton and Brown, 1991) and is proving a useful model for studying cold hardiness in nematodes (Wharton, 1997; Wharton, 1998).

P. *davidi* is freezing tolerant and freezes by inoculative freezing from the surrounding medium (Wharton and Brown, 1991). Inoculative freezing occurs mainly via the excretory pore (Wharton and Ferns, 1995). The degree of tolerance appears to be dependant upon the culture conditions and their thermal history, although this remains to be fully characterised. This nematode has recently been established in liquid culture, which appears to give high levels of freezing tolerance (Wharton and Block, 1997). The eggshell of P. *davidi*, however, can prevent inoculative freezing, allowing the enclosed embryo or larva to supercool and to be freeze avoiding (Wharton, 1994).

Intracellular freezing is usually thought to be fatal to cells and the ice in freezing-tolerant animals to be confined to the body cavity and extracellular spaces (Lee, 1991). This has, however, rarely been tested and survival of intracellular freezing has been described in some insect fat body cells (Lee, *et al.*, 1993). Freezing and melting in intracellular compartments has been observed in P. *davidi* using cryomicroscopy (Wharton and Ferns, 1995). Nematodes which are known to have frozen intracellularly will subsequently grow and reproduce in culture. Intracellular ice has also been demonstrated using freeze fracture (Wharton and Ferns, 1995) and freeze substitution techniques (Figure 16.3). P. *davidi* is the first intact animal in which the survival of intracellular freezing has been demonstrated and if we can understand the adaptations involved it may result in new methods for cryopreserving biological materials.

During freezing 82% of the body water of P. *davidi* is converted into ice (Wharton and Block, 1997). This is a high proportion compared with other freezing-tolerant animals and might be expected for an animal which freezes intracellularly. Freezing is rapid, with the whole body freezing within 0.21 s (Wharton and Ferns, 1995). Other freezing tolerant animals (frogs, hatchling turtles, insects) take hours or days to reach their maximum ice content. Rapid freezing may be important for surviving intracellular freezing as it would prevent the osmotic stresses which would occur if different body compartments froze at different times. Glycerol and trehalose have been identified in extracts of P. *davidi* (Wharton,

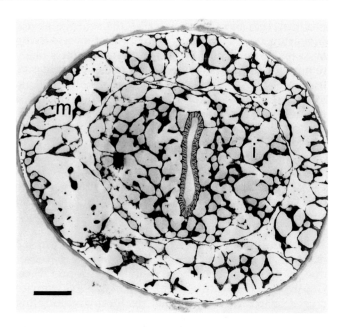

Figure 16.3. Transmission electron micrograph of a transverse section of a specimen of *Panagrolaimus davidi* after freezing at 20°C and preparation for electron microscopy by freeze substitution techniques. The white areas indicate the location of ice, which occurs throughout the cells of the body. Scale bar = 2 μm, i = intestinal cell, m = muscle cell.

unpublished results) but whether they play a role as cryoprotectants has yet to be demonstrated. The presence of THPs could not be demonstrated using differential scanning calorimetry (DSC) (Wharton and Block, 1997) or nanolitre osmometry (Wharton, unpublished results). However, recrystallisation inhibition activity, which may be the function of THPs in freezing-tolerant animals (Knight and Duman, 1986), has been demonstrated in extracts from the nematode (Ramløv, Wharton and Wilson, 1996). THPs can inhibit recrystallisation at very low concentrations (Theede, Schneppenheim and Bevess, 1976), which may not be detectable using DSC. It has also been shown, using ice nucleation spectrometry, that P. *davidi* produces a substance which inhibits ice nucleation (Wharton and Worland, 1998). This may also indicate THP activity. Work is in progress to purify and identify ice-active proteins in this nematode (Wharton, unpublished results).

Freezing tolerance has been demonstrated in nine other species of Antarctic nematodes (Wharton and Block, 1993). Supercooling points vary with season in several species of Antarctic nematode (Pickup, 1990a.b.c), which may indicate changes in the ability to tolerate freezing (Wharton, 1995).

There have been few studies on the cold tolerance of free-living nematodes from other parts of the world. *Panagrellus silusiae*, freezes by inoculative freezing from the surrounding medium but does not survive freezing (Wharton, Young and Barrett, 1984). When external water is absent the nematode will supercool and display a modest degree of cold tolerance (Wharton, et al., 1984), the supercooling point is depressed by acclimation to a lower temperature (Mabbett and Wharton, 1986). However, the opportunity for a free-living nematode to supercool in its natural habitat is likely to be limited because of inoculative freezing (Wharton, 1995). *Caenorhabditis elegans* does not appear to be cold hardy (Wharton, unpublished results). A comparison of the cold hardiness abilities and mechanisms of free-living nematodes from related genera from different habitats may explain aspects of their distribution and indicate adaptations which are specifically involved in cold tolerance. Cold-tolerant nematodes might be expected to be found in arctic, alpine and temperate habitats.

Cold tolerance in plant-parasitic nematodes

The potato-cyst nematode, *Globodera rostochiensis*, evolved in the high Andes, along with its potato host. In an alpine environment the nematode would face regular exposure to subzero temperatures and be expected to have evolved cold hardiness. The nematode has since followed the spread of its host around the World and may face subzero temperatures in many parts of its present range. *G. rostochiensis* overwinters as a second stage larva (L2) within eggs contained within a cyst, formed from the body of the female nematode. This represents a complex situation when considering the cold hardiness strategies of this nematode, since the eggshell and the cyst wall may act as barriers to ice nucleation.

G. rostochiensis cysts are tolerant of desiccation, with the enclosed larvae surviving anhydrobiotically (Evans and Perry, 1976). When desiccated the nematodes would also survive low temperatures, since there is no water to freeze; although this has not been tested. If the eggs and hatched larvae are free of surface water they are able to supercool and avoid freezing (Perry and Wharton, 1985). However, the situation where the nematode is free of surface water and yet not desiccated is likely to be transient (Wharton, 1995). Using cryomicroscopy it has been shown that the eggshell prevents inoculative freezing in water and allows the enclosed larva to supercool and survive temperatures as low as –38°C (Wharton, Perry and Beane, 1993). In contrast, hatched larvae freeze by inoculative freezing and are killed. Freeze avoidance thus appears to be the cold tolerance strategy of these larvae within the egg, although the 'protective dehydration mechanism' of cold hardiness (Holmstrup and Westh, 1995) could also occur in this situation.

The role of the eggshell in preventing inoculative freezing has been confirmed using differential scanning calorimetry (DSC). DSC thermograms of cysts during cooling show three exotherms, indicating freezing events (Wharton and Ramløv, 1995). Three endotherms, representing the melting of three compartments of different composition, occur during warming. These events can be interpreted as representing the freezing and melting of three water compartments associated with the cyst. These are: the water surrounding the cyst, the water between the cyst wall and the eggs and the water contained within the eggs (Figure 16.4). This interpretation is supported by

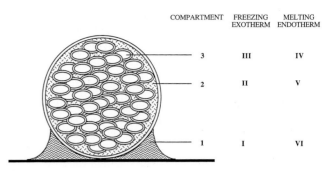

COMPARTMENT	FREEZING EXOTHERM	MELTING ENDOTHERM
3	III	IV
2	II	V
1	I	VI

Figure 16.4. Diagram showing the compartments associated with freezing and melting events in the cysts of *Globodera rostochiensis*, as detected by differential scanning calorimetry. The compartments are: (1) the water surrounding the cyst, (2) the solution between the cyst wall and the eggs, (3) the egg contents. I–VI is the order in which freezing and melting events occur. From Wharton & Ramløv (1995).

the disappearance of thermal events associated with the second compartment after the removal of eggs from the cyst and by the merging of events associated with the second and third compartments after the heating of the sample. Heating destroys the permeability barrier of the eggshell and allows mixing between these two compartments (Wharton and Ramløv, 1995). The peak of the egg exotherm (–38°C) agrees well with the mean supercooling point of eggs, determined by cryomicroscopy (–38.2°C) (Wharton, et al., 1993; Wharton and Ramløv, 1995).

The perivitelline space of the eggs of *G. rostochiensis* contains trehalose at a concentration of 0.34 mol 1^{-1} (Clarke and Hennessy, 1976). Trehalose is known to act as a cryoprotectant or antifreeze in some animals (Lee, 1991). After the heating of the sample, there is an elevation of the egg exotherm by about 18°C (Wharton and Ramløv, 1995). This may be due to loss of trehalose from the eggs following the destruction of the permeability barrier.

There are few other species of plant-parasitic nematode for which a definite statement about which cold tolerance strategy they display can be made. The survival of *Meloidogyne hapla* larvae exposed to freezing in polyethylene glycol or in soil is enhanced after prefreeze acclimation at low temperatures (Forge and MacGuidwin, 1990). The acclimation response is reversed if the nematode is returned to higher temperatures (Forge and MacGuidwin, 1992). Exposure to low water potentials also enhances freezing survival (Forge and MacGuidwin, 1992). Whether this nematode survives by freezing tolerance or freeze avoidance has not been determined. Forge and MacGuidwin (1992) suggest that in soil, nematodes may desiccate rather than freeze due to freeze concentration effects. *M. hapla* do freeze by inoculative freezing from the surrounding water (Sayre, 1964), although this might not be the case at low water potentials. Inoculative freezing of *P. davidi* will still occur at osmolalities which are much higher than those likely to result from freeze concentration effects (Wharton and To, 1996).

The survival of *Ditylenchus dipsaci* L4 is similar whether they are frozen in contact or not in contact with water (Wharton, *et al.*, 1984). This may indicate that they are freezing tolerant. Cryomicroscopy has confirmed that *D. dipsaci* L4 freeze by inoculative freezing and have some freezing tolerance (Wharton, unpublished observations). *Aphelenchoides ritzemabosi* also has some freezing tolerance ability (Asahina, 1959). There are other records of plant-parasitic nematodes surviving subzero temperatures (Antoniou, 1989; Miller, 1968), but the mechanisms by which they do so are unknown.

Cold tolerance in entomopathogenic nematodes

Entomopathogenic nematodes from the genera *Steinernema* and *Heterorhabditis* have attracted considerable interest as biological control agents for insect pests and are used commercially in many parts of the World (Gaugler and Kaya, 1990). Lack of a long-term storage method has proved a major impediment to their more widespread commercial application (Friedman, 1990). Attempts to base a storage technique on desiccation have met with limited success (Womersley, 1990). Freezing is a possible alternative storage technique and this has led to some interest in the cold tolerance strategies of these nematodes. Freezing in liquid nitrogen has been used to store small quantities (Popiel and Vasquez, 1991) but would not be economic for the storage of commercial quantities. A commercial technique is more likely to be based on the nematodes' natural capacity for cold tolerance.

Entomopathogenic nematodes have been isolated from cold temperate sites in northern Canada and Europe (Brown and Gaugler, 1996) and from alpine sites in the Swiss Alps (Steiner, 1996). They are not, however, necessarily exposed to subzero temperatures in these sites since the protection of snow cover and the soil may prevent the temperature falling below 0°C (Steiner, 1996). Attempts to isolate them from Antarctic soils have been unsuccessful (Griffin, Downes and Block, 1990).

The sheath of the infective larva of *Heterorhabditis zealandica* prevents inoculative freezing from the surrounding water and it thus avoids freezing (Wharton and Surrey, 1994). The survival of larvae is, however, lower than would be predicted if each larva which did not freeze survived. They are thus chill tolerant (Bale, 1993), suffering some prefreeze mortality at subzero temperatures above the supercooling point. Infective larvae are more cold tolerant if reared on an insect host (*Galleria mellonella*) than if grown on a Soya flour or liquid culture medium. This may be related to the better nutritional status of insect-derived larvae (Surrey, 1996). Acclimation of larvae at low temperatures (7–8°C) did not enhance their cold tolerance (Surrey, 1996).

The infective larvae of *H. bacteriophora, Steinernema feltiae, S. anomali, S. riobravis* and *S. glaseri* are freezing tolerant; surviving inoculative freezing from the surrounding water (Brown and Gaugler, 1996; Brown and Gaugler, 1998). *H. bacteriophora* and *S. feltiae* showed increased freezing tolerance after acclimation, although there was also some evidence of chilling injury in *H. bacteriophora* (Brown and Gaugler, 1996). Long-term survival was enhanced by the addition of glycerol (Brown and Gaugler, 1998).

These experiments offer some hope for the development of a long-term storage technique for commercial quantities of nematodes based on freezing or freeze drying. Entomopathogenic nematodes do have some cold tolerance ability. The survival achieved in the laboratory has, however, been modest. The lowest temperature survived is –28.5°C (Surrey, 1996) and larvae could survive –20°C for up to about 30 days after the addition of 20% glycerol (Brown and Gaugler, 1998). It may be possible to enhance the natural cold tolerance ability of the nematodes to provide a practical storage technique. Possibilities include improved acclimation regimes, adding or inducing the production of cryoprotectives and the isolation, breeding or engineering of cold-tolerant strains.

Cold tolerance in animal-parasitic nematodes

The infective L3 of trichostrongyle nematodes may persist on pasture for a year or more (Michel, 1976). L3 overwinter on pasture and are exposed to sub-zero temperatures in parts of their range. In Poland, for example, trichostrongyle L3 of a variety of species have been reported to survive temperatures as low as –28°C on pasture (Wertejuk, 1959). When frozen in water the sheath surrounding the L3 can prevent inoculative freezing, in a proportion of individuals, and allows the larva to supercool. However, a proportion do freeze and can survive freezing (Wharton and Allan, 1989). The L3 can thus survive by freezing tolerance or freeze avoidance, indicating that these two strategies are not mutually exclusive in nematodes.

Eggs and hatched L3 of *Nematodirus battus* will supercool in the absence of surface water. The degree of supercooling increased after acclimation at low temperature (Ash and Atkinson, 1986) and this was accompanied by an increase in trehalose concentration (Ash and Atkinson, 1983). Whether this nematode freezes in contact with water remains to be determined but it might be expected that the eggshell would prevent inoculative freezing.

Parasitic stages of animal parasites may be exposed to low temperatures within their hosts. Some ectothermic vertebrates, frogs, salamanders and turtles, can tolerate freezing (Storey and Storey, 1992). Parasitic nematodes of these animals would also be exposed to freezing stress. Freezing tolerance also occurs in hundreds of species of terrestrial insects, some intertidal marine invertebrates, centipedes and slugs (Storey and Storey, 1996). Little is known of the survival mechanisms of parasites within these freezing tolerant hosts. *Wetanema* sp. is a thelastomatid nematode parasitic in the hind gut of *Hemideina maori*, a New Zealand alpine weta (an orthopteran) which is the largest known freezing tolerant insect. Seasonal prevalence and intensity of *Wetanema* in *H. maori* indicate that the nematode can survive within its host over the winter, suggesting that it can survive the regular freezing of its host. In experiments *Wetanema* can survive freezing to –61°C within the host and is freezing tolerant (Tyrrell, *et al.*, 1994).

If the host is not freezing tolerant, parasitic nematodes may be exposed to freezing within the carcasses of their host after it dies. There are several reports of nematodes surviving within the frozen carcasses of their hosts (Bartlett, 1992; Gustafson, 1953; Smith, 1987; Spratt, 1972). This may be of adaptive significance if the parasite is transmitted by a host eating an infected carcass, as is the case for *Trichinella spiralis*. *Anisakis* larvae encyst within a variety of fish and are unlikely to be frozen under natural conditions before ingestion by their definitive host (cetaceans and pinnipeds). The ability of this nematode to survive a degree of freezing (Wharton & Aalders, unpublished observations; Gustafson, 1953) is thus of no adaptive significance and may reflect some general ability of nematodes to survive freezing stress. It is, however, an important phenomenon since fish and fish products which are to be consumed raw (e.g. as sushi) or lightly cooked (e.g. by cold smoking) are rendered safe for human consumption by freezing. It is therefore important to determine the degree of freezing necessary to kill the nematodes.

Heat Stress Responses in Nematodes

At high temperatures metabolism is likely to be reduced and nematodes become quiescent and exhibit first heat stupor and then heat coma (Figure 16.1). Recovery after exposure to heat coma has been reported in a few studies. At elevated temperatures the embryogenesis and hatching of the eggs of *Meloidogyne javanica* and *M. naasi* is slowed or stopped but rapidly resumes after the removal of the temperature stress (Antoniou, 1989). A cryptobiotic response to high temperatures, where recovery occurs after metabolism has ceased, has not been reported; except in the case of anhydrobiotic nematodes which can survive exposure to high temperatures in a desiccated state (Bird and Buttrose, 1974). Nematodes inhabiting high-temperature habitats must exhibit capacity adaptation with life and activity occurring at a range of temperatures higher than that of nematodes from more moderate habitats. *Aphelenchoides parientus* inhabits hot springs at a temperature of 45–51°C (Nicholas, 1984).

The ability of nematodes to survive at temperatures up to the thermal death point, and the temperature at which death occurs, may be increased by sublethal exposure to high temperatures. This is known as the heat shock response and has been demonstrated in entomopathogenic nematodes (Grewal, Gaugler and Shupe, 1996; Selvan, *et al.*, 1996) and in *C. elegans* (Snutch and Baillie, 1983); as well as a wide variety of other organisms (Parsell and Lindquist, 1994). The heat shock response is mediated by the production of heat-shock proteins (HSPs, stress proteins or molecular chaperones) which can both protect proteins against denaturation and repair damaged proteins (Parsell and Lindquist, 1994).

Heat tolerance in nematodes can be selected for (Grewal, *et al.*, 1996), is increased in nematodes from high temperature habitats (Glazer, *et al.*, 1996) and some *C. elegans* mutants have enhanced heat tolerance (Lithgow, *et al.*, 1995). This may be due to an enhancement of HSP production. A gene encoding for HSP production (hsp 70) has been identified in *C. elegans* (Dalley and Golomb, 1992). This gene has been inserted into *H. bacteriophora* and produced increased thermo-tolerance in the transgenic nematodes (Hashmi, *et al.*, 1998). Field trials have been conducted with this strain (Gaugler, Wilson and Shearer, 1997).

During the invasion of a homeothermic mammalian host a parasite is likely to experience a sudden increase in temperature, as well as other changes in its physicochemical and biological environment. The production of HSPs may enable parasites to cope with these changes (Maresca and Carratu, 1992). HSPs have been identified in various parasitic nematodes, including: *Trichinella spiralis*, *T. pseudospiralis* (Ko and Fan, 1996), *Brugia* spp. (Selkirk, *et al.*, 1989), *Haemonchus contortus* (Vanleeuwen, 1995), *Nippostrongylus brasiliensis* (Tweedie, *et al.*, 1993) and *Dirofilaria immitis* (Lillibridge, Rudin and Philipp, 1996). They are likely to be ubiquitous in nematodes, as they are in all groups of organisms (Parsell and Lindquist, 1994), and to perform a wide variety of roles.

Desiccation Stress and Anhydrobiosis

Free-living nematodes and the free-living stages of parasitic nematodes, and even the parasitic stages of some nematodes, may be faced with desiccation. All animals consist of about 76% water. Water is essential as a medium for biochemical reactions and plays an important role in the structure of biological macro-molecules and membranes. Few animals can tolerate the loss of a substantial amount of water and most die if they lose more than 15–20% of their body water (Barrett, 1991).

One way of coping with desiccation stress is to prevent water loss by seeking out moist habitats or habitats where water loss is restricted. This may be a limited option for nematodes since they are dependant upon the presence of at least a film of water for active movement. Another option may be the prevention of water loss by the presence of an impermeable cuticle or skin. Insects have a relatively impermeable cuticle but they can't be completely impermeable since they must allow the exchange of respiratory gasses and other materials. Water loss occurs via the spiracles, with the faeces and secretions, as well as by cuticular transpiration (Sømme, 1995). The permeability of some nematode eggs is very low and they can take several days to lose their enclosed water when exposed to desiccation (Wharton, 1980). Water loss by diffusion is a particularly difficult problem for a small animal like a nematode because of its large surface to volume ratio.

Despite these mechanisms for limiting water loss, nematodes will lose water when exposed to desiccation. Activity will cease as soon as the water surrounding the nematode is lost, since they are dependant upon the presence of external water for locomotion. Metabolism may, however, continue at a reduced rate and the nematode enter a state of desiccation-induced quiescence. Sufficient water may then be lost to prevent metabolism from continuing. The ultimate response to desiccation stress is to survive the almost total loss of body water and enter into a state of anhydrobiosis ('life without water') in which metabolism comes reversibly to a standstill. Measurement of metabolism in anhydrobiotic larvae of *D. dipsaci*, using oxygen uptake, heat output and the production of carbon dioxide, indicate that metabolic rates are below one tenthousandth of that of hydrated larvae (Barrett, 1982).

Desiccation Survival and Anhydrobiosis in Nematodes

Anhydrobiotic nematodes seem to be divided into two broad groups: slow-dehydration strategists and fast-dehydration strategists (Womersley, 1987). Slow-dehydration strategists need a slow rate of water loss to successfully enter anhydrobiosis. They live in habitats, such as soil and moss, which lose water slowly when exposed to desiccation and the nematodes can, therefore, rely on the dehydration characteristics of their habitat to ensure the necessary slow rate of water loss. Fast-dehydration strategists can survive immediate exposure to low relative humidity (RH). They must therefore either survive high rates of water loss or possess adaptations which slow down the rate of water loss. There is, however, likely to be a range of rates of water loss which different species can survive, rather than two distinct groups. *T. colubriformis* L3, for example, appear to survive desiccation regimes intermediate between these two groups (Allan and Wharton, 1990).

It is difficult to compare the desiccation survival of different nematodes since researchers have used different temperatures, RHs and ways of exposing nematodes to stress. Table 16.2 shows examples of studies which have tested the ability of nematodes to survive exposure to 0% RH. If they can survive this stress they can be considered to be capable of anhydrobiosis. The table includes those studies which have demonstrated survival after direct exposure to 0% RH and those which have demonstrated survival, or lack of survival, after initial slow desiccation. Studies which have failed to show survival after direct exposure to 0% RH are not included, since these nematodes may have survived if they had first been desiccated slowly. It also shows whether nematodes survive after a high rate of water loss (fast-dehydration strategists), such as immediate exposure to 0% RH, or a slow rate of water loss (slow-dehydration strategists), requiring drying at a high RH or in a substrate which produces a slow rate of water loss, before they will survive 0% RH. The longest observed period of survival whilst desiccated is also shown.

Free-living nematodes are predominantly slow-dehydration strategists, surviving the rates of water loss likely to be experienced in a soil or moss habitat

(Table 16.2). Exceptions are *Plectus* sp. and *Ditylenchus* sp. which could survive direct exposure to 0% RH and high rates of water loss (Hendriksen, 1982; Pickup and Rothery, 1991). *Ditylenchus* sp. inhabits the exposed aerial thalli of lichen whilst *Plectus* sp. was isolated from moss growing on a flat roof. These are exposed habitats where the nematodes will experience high rates of water loss. *Teratocephalus tilbrooki* will survive immediate exposure to 60% RH (Pickup and Rothery, 1991), which may indicate that it is a fast-dehydration strategist, but its survival at 0% RH was not reported.

Plant-parasitic nematodes include both fast and slow-dehydration strategists. Some fast-dehydration strategists (anguinids, *D. dipsaci*) are associated with the aerial parts of plants. Anguinids induce galls in the host inflorescence which may provide a barrier to water loss so that their first experience of desiccation may involve a slow rate of water loss (Womersley, *et al.*, 1998). They will, however, survive fast rates of water loss on subsequent rehydration and desiccation (Womersley, 1980). *D. dipsaci* once outside the host will be exposed to fast rates of water loss and appears to be one of the most resistant nematodes in this respect. *Ditylenchus* (= *Orrina*) *phyllobius* required only a brief preconditioning before their survival at a low RH (10% RH in an air stream) is enhanced (Robinson, Orr and Heintz, 1984), it may thus also be a fast-dehydration strategist. The unhatched infective larvae of many cyst nematodes could be fast-dehydration strategists since the eggshell and cyst wall may ensure the necessary slow rate of water loss.

A number of studies have exposed the free-living stages of animal-parasitic nematodes, particularly trichostrongyles, to desiccation. Most have subjected nematodes to immediate exposure to different RHs and have shown that survival at low RH does not occur under these conditions. *T. colubriformis* L3, however, will survive 0% RH if it is first dried at a high RH (Allan and Wharton, 1990). It seems likely that this is true of many infective larvae. The eggs of some animal-parasitic nematodes lose water very slowly when exposed to desiccation and this may allow the enclosed larva to survive anhydrobiotically (Wharton, 1980). Attempts to induce anhydrobiosis in entomopathogenic nematodes are aimed at developing storage methods for their commercial use as biological control agents (Womersley, 1990). This has not so far been achieved, even after desiccation at slow rates of water loss. Some species do show some degree of desiccation resistance, which holds out hope that if they are dried under the right conditions anhydrobiosis may be achieved.

Slow-dehydration strategists appear to predominate (Table 16.2) and this emphasises the importance of using conditions which produce rates of water loss which mirror those likely to be experienced by the animal in nature (Womersley, 1987). It seems likely that slow-dehydration strategists are widespread amongst free-living nematodes and the free-living stages of parasitic nematodes. Where survival at 0% RH could not be achieved, even after slow drying, these nema-

Table 16.2. Some records of anhydrobiotic survival in nematodes

Group/Species	Anhydrobiosis demonstrated	Slow or fast-rate survivors	Observed longevity whilst desiccated	References
free-living nematodes				
Aphelenchus avenae	+[1]	slow	2.2 years	Crowe and Madin, 1975; Higa and Womersley, 1993; Womersley, *et al.*, 1998
Panagrolaimus davidi	+	slow	1 day	Wharton and Barclay, 1993
Plectus sp.	+	fast	2 days	Hendriksen, 1982
Ditylenchus myceliophagus	−		10 days	Womersley and Higa, 1998
Acrobeloides nanus	+	slow	1 year	Nicholas and Stewart, 1989
Acrobeloides sp.	+	slow	1 day	Demure, *et al.*, 1979
Panagrellus silusiae P. redivivus	+	slow	1 year	Lees, 1953; Womersley, *et al.*, 1998
Ditylenchus sp.	+	fast	66 days	Pickup and Rothery, 1991
plant-parasitic nematodes				
Ditylenchus dipsaci	+	fast	23 years	Norton, 1978; Perry, 1977a; Wharton, 1996
Anguina agrostis	+	fast?[2]	4 years	Norton, 1978; Preston and Bird, 1987
A. tritici	+	fast	32 years	Norton, 1978; Womersley, 1980
Globodera rostochiensis	+	fast	8 years	Norton, 1978; Perry, 1983
Pratylenchus thornei	+	slow	1 day	Glazer and Orion, 1983
Helicotylenchus dihystera Scutellonema brachyurum	+	slow	1 day	Demure, *et al.*, 1979
S. cavenessi	+	slow	1 month	Demure, 1980
Rotylenchulus reniformis	−			Womersley and Ching, 1989
animal-parasitic nematodes				
Trichostrongylus colubriformis	+	slow	164 days	Allan and Wharton, 1990; Wharton, 1982
Heterorhabditis zealandica	−			Surrey and Wharton, 1995
Steinernema carpocapsae	−			Womersley, 1990

[1] +, survived 0%RH; −, did not survive 0%RH after slow drying
[2] survived 0%RH after desiccation at 50%RH

todes may be capable of desiccation-induced quiescence, surviving some desiccation but not anhydrobiotically. However, it is possible that survival at 0% RH could be achieved if the animal was dried at rates slower than those attained under experimental conditions. *Acrobeloides nanus* could not survive 0% RH after desiccation on an agar substrate but could after desiccation in polydextran beads, which presumably produce a slower rate of water loss (Nicholas and Stewart, 1989). It may not be necessary to show survival at 0% RH to demonstrate anhydrobiosis (the cessation of metabolism as a result of desiccation stress). Clegg

(1979) has suggested that metabolism will cease at water contents below about 20%, this is likely to occur at relative humidities much higher than 0% RH. The variation in survival between different nematodes seems to be more in the rates of water loss they can tolerate rather than in the absolute degree of desiccation stress they can survive.

Mechanisms of Anhydrobiotic Survival

Control of the rate of water loss

A slow rate of water loss is usually considered to be essential for anhydrobiosis (Evans and Perry, 1976;

Higa and Womersley, 1993). Slow-dehydration strategists rely on the water loss characteristics of their environment. Fast-dehydration strategists, however, may have adaptations which ensure the necessary slow rate of water loss. This is achieved via behavioural mechanisms, such as coiling and clumping, and by physiological mechanisms, such as a low permeability to water. The loss of water will also be affected by the vapour boundary layer above the surface of the nematode (Robinson, *et al.*, 1984). The nematode is unlikely to be able to control conductance through this layer but it could control cuticular permeability.

Low cuticular permeabilities appear to be a characteristic of nematodes exposed to harsh environments (Wharton, *et al.*, 1988). The nematode cuticle consists of several layers or zones (Bird and Bird, 1998; Blaxter and Robertson, 1998). Several of these could be involved in permeability control and the response to desiccation. The permeability barrier of the cuticle of *D. dipsaci* is heat labile and is destroyed by brief extraction with diethyl ether, suggesting that a superficial lipid layer is responsible (Wharton, *et al.*, 1988). This is most likely to be the epicuticle. This is the outer layer of the cuticle, although it may be covered by a surface coat, and consists of lipid and protein.

When *D. dipsaci* L4 are exposed to desiccation there is a marked reduction in the rate of water loss with a permeability slump occurring soon after the onset of desiccation (Figure 16.5) (Wharton, 1996). There are several mechanisms by which this change in permeability may occur. The permeability slump could result from a decrease in the depth or width of the cuticular annulations, which may represent more permeable areas of the cuticle (Wharton, 1996; Wharton and Lemmon, 1998) or the permeability of the cuticle could decrease as it dries, slowing the rate of water loss of the body (Ellenby, 1968a; Perry, 1977b). This could occur via a phase change in the lipids of the epicuticle, a change in its composition or a decrease in the thickness of the cuticle (Wharton and Lemmon, 1998). In *Aphelenchus avenae* the cuticle decreases in thickness by a factor of four after desiccation but it also becomes folded (Crowe, Lambert and Crowe, 1978). This species needs to be dried slowly before it will survive anhydrobiotically, suggesting that these changes in the cuticle are not involved in controlling water loss. Fusion of adjacent layers of epicuticle where they come into contact during coiling and clumping has been reported in *A. tritici* (Bird and Buttrose, 1974). This could provide a further barrier to water loss.

A decrease in surface area by shrinkage of the body of the nematode would also produce a decrease in the rate of water loss. A decrease in surface area does occur during desiccation in *D. dipsaci* but the reduction in the rate of water loss is much greater than could be explained by this effect. A change in cuticular permeability thus appears to be the main mechanism reducing water loss (Wharton, 1996).

The restricted permeability of the nematode eggshell ensures a slow rate of water loss when the egg is exposed to desiccation and may enable the enclosed

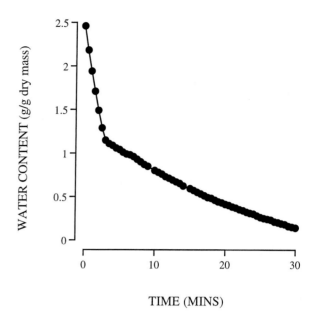

Figure 16.5. Water loss from a sample of *D. dipsaci* L4, monitored continuously on a microbalance. A permeability slump is apparent 2 min after the onset of desiccation. From Wharton (1996).

larva to survive anhydrobiotically (Wharton, 1979; Wharton, 1980). The lipid layer is the main permeability barrier of the eggshell. In some ascarid and oxyurid nematodes the lipid component has been identified as ascarosides. These are unique α-glycosides the glycone (sugar moiety) of which is ascarylose, a sugar which has not been identified from any other eucaryote (Barrett, 1981). The eggshells of oxyurid nematodes have two tertiary layers secreted by the walls of the uterus, the internal and external uterine layers, in addition to the three primary layers: the lipid, chitinous and vitelline layers. The uterine layers form complex systems of pores and spaces. These are thought to further restrict water loss from the egg by limiting gaseous exchange to the pores in the external uterine layer (Wharton, 1980). The permeability of the eggshell of *G. rostochiensis* decreases as it dries (Ellenby, 1968a). The eggs of *M. javanica* are laid in a gelatinous matrix which shrinks as it dries and may inhibit water loss from the eggs (Bird and Soeffkey, 1972).

The L3 of trichostrongyle nematodes retain the L2 cuticle as a sheath. This is an apolysed but not yet ecdysed cuticle and its retention represents a delay in the completion of the L2/L3 moult until entry into the host (Wharton, 1986). It has been proposed that the sheath plays a role in desiccation survival by slowing down the rate of water loss (Ellenby, 1968b). Exsheathed L3 of *H. contortus* can not survive desiccation (Ellenby, 1968b) but exsheathed L3 of *T. colubriformis* can (Allan and Wharton, 1990). The hydrated sheath is freely permeable to water (Allan and Wharton, 1990) and so if it is to restrict water loss during desiccation it must decrease in permeability as it dries.

Some other nematodes have sheaths which may play a role in controlling water loss during desiccation. These include the dauer larva of *Rhabditis dubius* (Bovien, 1937), the infective larvae of steinernematids and heterorhabditids (Patel and Wright, 1998; Womersley, 1990) and the adults of *Rotylenchulus reniformis* (Gaur and Perry, 1991). In viviparous nematodes, larvae may achieve a slow rate of water loss via the protection provided by the body of its mother (Ellenby, 1969; Lees, 1953). Some anguinids may produce an outer protective layer in response to desiccation (Womersley, *et al.*, 1998).

Many plant-parasitic nematodes remain within the host plant, whose tissues would provide a physical barrier to water loss. Some, such as anguinids, induce changes such as the formation of galls in the inflorescences which may further control water loss (Womersley, *et al.*, 1998). Desiccation survival of entomopathogenic nematodes may be enhanced by their retention within the cadavers of their hosts (Koppenhofer, *et al.*, 1997).

Behavioural changes may also act to reduce the rate of water loss. The major behavioural response to desiccation is coiling. This reduces the rate of water loss by reducing the surface area exposed to the air. Coiling is a widespread response to desiccation in nematodes and has been described in a variety of plant-parasitic and free-living nematodes and in the infective larvae of trichostrongyles (Wharton, 1982). In some cases, coiling appears to be essential for survival (Womersley and Ching, 1989). It is such a widespread response that some authors have considered that a coiled posture indicates that nematodes are in anhydrobiosis (Freckman, Kaplan and Van Gundy, 1977). Nematodes will, however, coil in response to stimuli other than desiccation, including increases in temperature (Wharton, 1981) and osmotic stress (Wharton, Perry and Beane, 1983). In *T. colubriformis* the main stimulus for coiling is the restriction of lateral movement that occurs in an evaporating water film (Wharton, 1981).

Cultures of some free-living and mycophagous nematodes form swarms initiated perhaps by anoxia, lack of food or the accumulation of toxins. These swarms then form clumps as they start to dry out and the nematodes within the clumps coil (Womersley, *et al.*, 1998). In nature this occurs in *D. dipsaci* and other ditylenchids where nematodes leave the plant tissue as it dies to form dense aggregations known as 'eelworm wool'. The formation of these aggregations may aid desiccation survival since the nematodes on the outside will dry first and slow the rate of water loss of those in the centre of the aggregation (Ellenby, 1968a). Drying in clumps of a sufficient size is essential for some nematodes to survive anhydrobiotically (Crowe and Madin, 1975; Higa and Womersley, 1993).

Morphological changes during desiccation

For most nematodes a slow rate of water loss is necessary in order to prevent or reduce any damage associated with the loss of water from the body and to allow biochemical changes which help the nematode to survive the desiccation stress. Several nematodes have been shown to maintain their structural integrity during anhydrobiosis, namely: *D. dipsaci* (Wharton and Barrett, 1985; Wharton and Lemmon, 1998), *A. tritici* (Bird and Buttrose, 1974) and *A. avenae* (Crowe, *et al.*, 1978). Similar studies have been made on other anhydrobiotic organisms including: rotifers (Dickson and Mercer, 1967), *Artemia* (Morris, 1968), plants (Dallavecchia, *et al.*, 1998) and cyanobacteria (Peat and Potts, 1987). Structural and membrane integrity is maintained and the cytoplasm becomes condensed. In *D. dipsaci*, changes in the muscle cells are the most noticeable (Wharton and Barrett, 1985; Wharton and Lemmon, 1998). In *P. silusiae* which were dried rapidly and did not survive desiccation, there was gross structural disruption (Wharton and Barrett, 1985). In quick-dried *A. avenae*, which do not survive the desiccation stress, structural disruption was not reported but there was less shrinkage of a number of body components than in slow-dried anhydrobiotic nematodes (Crowe, *et al.*, 1978).

The sequence of structural changes has been followed in *D. dipsaci*, using freeze substitution techniques (Wharton and Lemmon, 1998). In general changes occur in two phases: an initial rapid shrinkage during the first five minutes of desiccation, followed by a slower rate of shrinkage upon further desiccation (Figure 16.6). This pattern was observed in the cuticle, the lateral epidermal cords and the muscle cells. This mirrors the pattern of water loss from the nematode which also occurs in two phases with a permeability slump after 2–5 mins desiccation (Wharton, 1996). The contractile region of the muscle cells proved an

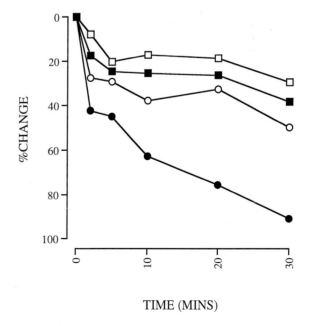

Figure 16.6. Changes in water content (●) and the thickness of the cuticle (□), muscle cells (■) and lateral epidermal cords (○) during desiccation of *D. dipsaci* at 50%RH, 20°C. Data from Wharton & Lemmon (1998).

exception to this pattern with the muscle fibres resisting packing and shrinkage until water loss became severe. The intestinal cells also change little during desiccation. They are packed with lipid droplets, which may limit the amount of shrinkage in this tissue (Wharton and Lemmon, 1998). The muscle cell mitochondria swell and then shrink, which may indicate disruption of the permeability of their outer membranes. The mitochondria of *D. dipsaci* do appear to be sensitive to desiccation stress since they do not function normally immediately after rehydration (Barrett, 1982).

Similar structural changes have been observed in other nematodes in comparisons between hydrated and anhydrobiotic specimens. Decreases in cuticle thickness occur in *A. avenae* (Crowe, *et al.*, 1978) and decreases in muscle fibre spacing in *A. tritici* and *A. avenae* (Bird and Buttrose, 1974; Crowe, *et al.*, 1978). Some changes appear to differ between different nematodes. In *A. tritici* the profiles of intestinal lipid droplets become irregular in anhydrobiotic specimens, whereas in *D. dipsaci* they remain rounded (Bird and Buttrose, 1974; Wharton and Barrett, 1985). In *A. avenae* the intestinal microvilli decrease in length and in the spacing between them. This apparently occurs by the removal of membrane material from the areas between microvilli (Crowe, *et al.*, 1978).

Some of the ultrastructural changes observed in *D. dipsaci* have been confirmed on living specimens by light microscopy, giving some confidence that they are not artefactual. Changes in length and diameter mirror ultrastructural changes and the pattern of water loss, with the largest changes occurring during the initial phase of desiccation (Wharton, 1996). Decreases in diameter are responsible for most of the decrease in volume. The decrease in diameter is mostly accounted for by a decrease in the thickness of the hyaline layer, with the intestine changing little after the initial phase of desiccation. The hyaline layer is a clear area in-between the intestine and the cuticle and consists of the muscle cells and the epidermis. A decrease in the spacing of cuticular annulations was also observed under the light microscope (Wharton, 1996).

Biochemical changes during desiccation

In addition to preventing structural damage during desiccation a slow rate of water loss may allow biochemical changes which facilitate anhydrobiotic survival. Desiccation does not result in any appreciable denaturation of metabolic enzymes in anhydrobiotic nematodes but there is no evidence that their enzymes are more stable than enzymes from other sources (Barrett, 1982). There was no increase in the frequency of DNA breaks during desiccation of *D. dipsaci* L4. This was also observed in the desiccation-susceptible nematode *Panagrellus redivivus* and thus does not indicate that DNA stability is an exclusive feature of anhydrobiotic nematodes (Barrett and Butterworth, 1985). There is evidence that many biological macromolecules have a high degree of stability (Barrett, 1991), at least if dry or frozen. Metabolite profiles from anhydrobiotic and hydrated *D. dipsaci* do not indicate extensive metabolic

reorganisation during entry into and recovery from anhydrobiosis, although there is some evidence for recovery of mitochondrial function during rehydration (Barrett, 1982). There are, however, changes in the concentrations of carbohydrates which may have adaptive significance.

In comparisons between fresh (or rehydrated) and anhydrobiotic (or desiccating) nematodes, increased levels of trehalose associated with desiccation have been demonstrated in *A. avenae* (Higa and Womersley, 1993; Madin and Crowe, 1975), *Ditylenchus myceliophagus* (Womersley and Higa, 1998), *A. tritici* (Womersley and Smith, 1981), *D. dipsaci* (Womersley and Smith, 1981) and *S. carpocapsae* (Womersley, 1990). Trehalose is also synthesised by many other desiccation-tolerant organisms including: tardigrades (Westh and Ramløv, 1991), *Artemia* (Clegg, 1965) and a variety of micro-organisms (Crowe, Hoekstra and Crowe, 1992). This disaccharide is thought to be important for anhydrobiotic survival for a number of reasons. It stabilises biological membranes by attaching to the polar head groups of membrane phospholipids and prevents them from undergoing phase changes during desiccation and rehydration. If not protected in this way membranes change to a gel state when they lose water and during rehydration reversion to a liquid crystalline state causes them to become leaky. Trehalose may also result in vitrification, the formation of a glass-like state, during desiccation (Crowe, Carpenter and Crowe, 1998). This may prevent fusion of adjacent membranes and trap tissues in the high-viscosity, low molecular mobility medium formed by the carbohydrate glass which limits a variety of deterioration processes (Franks, 1985). Other suggested roles for trehalose in anhydrobiosis include: preventing protein denaturation, prevention of oxidative damage, replacement of free and/or structural water, inhibition of browning reactions, as a free-radical scavenging agent and as an inert energy source (Higa and Womersley, 1993).

Other carbohydrates have been suggested to play a role in anhydrobiosis in nematodes. The synthesis of glycerol during desiccation of *A. avenae* (Madin and Crowe, 1975) appears to be due to the anaerobic conditions produced in large aggregates of this nematode (Higa and Womersley, 1993). Myo-inositol is produced during desiccation of *A. tritici* and *D. dipsaci* and may act as a replacement for bound water (Womersley and Smith, 1981). The addition of myo-inositol prior to desiccation improved the survival of *D. dipsaci*, *D. myceliophagus* and *A. tritici* (Womersley, 1981b). *D. myceliophagus* has been reported to accumulate ribitol (Womersley, 1987), although this was not confirmed in a subsequent report (Womersley and Higa, 1998).

Rehydration and the lag phase

Metabolism and water contents increase rapidly when anhydrobiotic nematodes are reimmersed in water (Barrett, 1982; Wharton, Barrett and Perry, 1985) but there is a period of apparent inactivity referred to as the 'lag phase' and it takes 2–3 h before activity resumes

(Wharton *et al.*, 1985). If metabolism and water contents are near normal levels during the lag phase why does it take several hours for activity to resume? In *D. dipsaci* there is an ordered series of morphological changes during the lag phase, most noticeable of which is an increase in the thickness of the muscle cells and the coalescence of small lipid droplets within the intestinal cells to form large droplets (Wharton and Barrett, 1985; Wharton, *et al.*, 1985). Coalescence of lipid droplets has also been observed in *A. agrostis* (Preston and Bird, 1987). The morphological changes occur gradually throughout the lag phase and are broadly the reverse of the changes seen during desiccation. These morphological changes suggest that the lag phase is a period of repair or restoration during which the nematode recovers from any damage sustained during desiccation. This suggestion is supported by the observation that the length of the lag phase increases with the severity of the desiccation stress experienced (Wharton and Aalders, 1999). It is the severity of the desiccation stress during dehydration, rather than the final relative humidity to which the animal is exposed which determines the length of the lag phase; again emphasising the importance of slow rates of water loss in surviving anhydrobiosis (Wharton and Aalders, 1999). A relationship between the severity of desiccation and the length of the lag phase has also been observed in *P. davidi* and the infective larvae of *T. colubriformis* (Allan and Wharton, 1990; Wharton and Barclay, 1993).

More direct evidence for repair during the lag phase comes from measurements of the cuticular permeability barrier (Preston and Bird, 1987; Wharton, *et al.*, 1988). The cuticles of *D. dipsaci* and *A. agrostis* become more permeable after desiccation but the barrier is restored during the lag phase. The restoration of the permeability barrier is disrupted by inhibitors which block post-transcriptional protein synthesis and enzyme activity, suggesting an active repair mechanism (Wharton, *et al.*, 1988). Activity does recover, however, in the presence of inhibitors of protein and RNA synthesis; which may indicate that repair of the cuticular permeability barrier is not essential for activity to resume (Barrett, 1982). *A. avenae* leaks primary amines and inorganic ions during rehydration, indicating an increase in permeability or a loss of membrane integrity (Crowe, O'Dell and Armstrong, 1979). The leakage ceases during the lag phase, indicating restoration of permeability barriers. *A. tritici* also loses ions during rehydration (Womersley, 1981a). The termination of leakage could be due to physical changes which restore permeability barriers as they take up water or to an active repair mechanism.

Changes in the muscle cells may be of particular importance during recovery of activity. In *D. dipsaci* and *T. colubriformis* the nematodes decrease in length during the lag phase before recovery of activity (Allan and Wharton, 1990; Wharton, *et al.*, 1985). This 'anomalous shrinkage' could result from a contraction of the muscle cells as they recover functionality. In *T. colubriformis* the shrinkage is accompanied by the development of birefringence under polarised light

which may reflect a change in the arrangement of muscle filaments in the contractile region of the muscle cells (Allan and Wharton, 1990).

Biochemical changes also occur during the lag phase. Trehalose is converted back into glycogen and if glycerol has formed it is converted into lipid (Madin, Crowe and Loomis, 1978). Most metabolite levels recover rapidly, although ATP does not return to normal levels for several hours. Metabolic pathways (glycolysis, the tricarboxylic acid cycle, beta-oxidation and the respiratory chain) start immediately. Mitochondria, however, are uncoupled immediately upon rehydration and hence ATP production is impaired. They show no respiratory control and no stimulation of oxygen uptake by 2.4-dinitrophenol (Barrett, 1982; Barrett, 1991). During rehydration the mitochondria swell and then shrink again suggesting changes in their functioning (Wharton and Barrett, 1985).

There is thus evidence for a process of repair or restoration during the lag phase which must occur before the nematode can recover activity. The nature of these changes are largely unknown but may include the repair of membrane damage and the restoration of the ionic gradients which are essential for nerve and muscle function.

Anhydrobiosis: has the Problem been Solved?

The widespread association between trehalose production in nematodes and other organisms and its demonstrated roles in stabilising membranes and proteins have led some to suggest that the problem has indeed been solved. Trehalose production is thought to be the principle adaptation which enables anhydrobiosis (Crowe, *et al.*, 1992). It is, however, clear that other adaptations are involved. A slow rate of water loss is essential and some nematodes have adaptations which slow down the rate of water loss. This may allow the synthesis of trehalose but it may also allow other protective mechanisms and the limitation of damage during water loss. There is also evidence for a process of repair or restoration after rehydration.

Some recent studies on the trehalose synthesis of some nematodes during desiccation throw doubt on the exclusive role of trehalose in anhydrobiosis. High levels of trehalose (10% dry weight) are synthesised by small aggregates of *A. avenae* during preconditioning at 97% RH but survival was markedly reduced during subsequent rapid desiccation upon transfer to 75%, 40% or 0% RH (Higa and Womersley, 1993). The survival of preconditioned nematodes at these RHs was enhanced by regimes of sequentially lowered RH but there was no increase in trehalose levels associated with these regimes. This indicates that processes associated with slow rates of water loss, other than the synthesis of trehalose, enhance survival.

D. myceliophagus is unable to survive 0% RH even after slow drying (Womersley and Higa, 1998). The nematodes synthesised trehalose (up to 9.3% dry weight). As fungal cultures of *D. myceliophagus* age the nematodes swarm, aggregate and coil on the sides of the culture vessel. The nematodes produce trehalose during this process (up to 16.8% dry weight) but could still not survive exposure to 0% RH (Womersley and Higa, 1998). The synthesis of high levels of trehalose did not enable these nematodes to survive anhydrobiotically at 0% RH, although they could survive 10% RH. This study did not use sequential desiccation after the initial preconditioning before exposure to low RH, the nematodes may well have survived if they had been dried more slowly (Womersley and Higa, 1998). Similar results have been obtained with *S. carpocapsae* (Womersley, 1990).

It therefore appears that trehalose is required for anhydrobiosis in nematodes but is not in itself sufficient. Other adaptations which are facilitated by a slow rate of water loss are required. These remain to be elucidated, and in that sense the problem of anhydrobiosis has not yet been solved. Some possibilities include: changes in lipid composition (Crowe, *et al.*, 1992), modification of membrane phospholipids (Womersley, 1990), the synthesis of protective antioxidants such as catalase (Gresham and Womersley, 1991; Higa and Womersley, 1993), repair or restoration process during rehydration, the prevention of structural damage during drying, the inhibition of proteases (Barrett, 1991) and the production of stress proteins. Some of the mechanisms, and possible mechanisms, involved are summarised in Figure 16.7.

Other Environmental Stresses

Osmotic Stress

Nematodes inhabit a wide range of habitats which may subject them to varying degrees of osmotic stress. In hyposmotic conditions water will enter the body and have to be removed. In hyperosmotic conditions water will be lost and stress the nematode in a fashion which may be similar to desiccation. Some nematodes have the ability to osmoregulate and some may reduce the stress by having a limited cuticular permeability (see chapter 7). The water lost under hyperosmotic stress may be sufficient to induce a state of quiescence or cryptobiosis (osmobiosis) which enables the nematode to survive the period of stress (Womersley, *et al.*, 1998). Little is known of the mechanisms involved, but we may expect some parallels with anhydrobiosis. Some types of nematode dormancy may be mediated via osmotic stress. For example, the high concentration of trehalose in the perivitelline fluid of the eggs of *G. rostochiensis* is thought to impose an osmotic stress which maintains the enclosed L2 in a state of quiescence. The dormant state is broken when the hatching stimulus results in an increase in the permeability of the eggshell, which allows the trehalose to diffuse out of the egg and water to enter, removing the osmotic stress (Jones, Tylka and Perry, 1998). A similar mechanism may operate in the eggs of other nematodes (Womersley, *et al.*, 1998).

Oxygen Stress

Free-living nematodes and plant-parasitic nematodes within the roots of plants experience low oxygen tensions under conditions where the soil is waterlogged or where there is a high rate of decomposition of organic material. The small size, the long, thin body shape, the peripheral position of nerves and muscles, and the presence of haemoglobin may enable nematodes to tolerate relatively low oxygen tensions

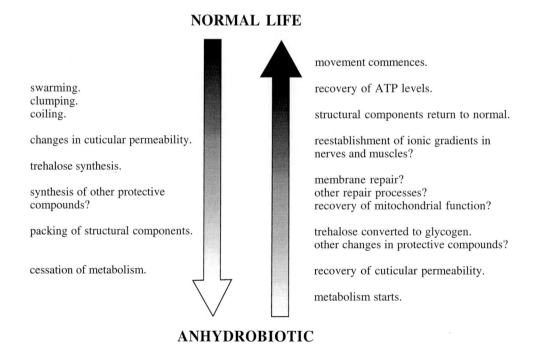

Figure 16.7. Diagram showing the possible sequence of events during entry into and recovery from anhydrobiosis in nematodes.

(Atkinson, 1976; Atkinson, 1980). Some free-living nematodes can tolerate anoxia and utilise anaerobic respiratory pathways. The time which these nematodes can survive under anaerobic conditions appears to be related to their rate of utilisation of internal glycogen reserves, since neutral lipid reserves are only catabolised under aerobic conditions (Womersley, *et al.*, 1998). Some nematodes may cease metabolism and enter a cryptobiotic state induced by low oxygen (anoxybiosis). *A. avenae* ceases metabolising under anoxia and can survive for at least 60 days (Cooper and Van Gundy, 1970). The dauer larvae of *C. elegans* can survive anoxia much longer than can the adult nematodes (Anderson, 1978).

Whilst animal-parasitic nematodes inhabiting the vertebrate intestine may have some oxygen available to them, particularly close to the mucosa, those inhabiting the gut lumen are likely to be under anaerobic conditions (Atkinson, 1980). Adult ascarids have a primarily anaerobic metabolism (Barrett, 1981).

Solar Radiation

More studies on the effects of increased UV radiation on biological systems have been initiated in recent years in response to concerns regarding the partial destruction of the earth's protective ozone layer. There are examples of such studies on nematodes (Fujiie and Yokoyama, 1998; Mills and Hartman, 1998). The tanning of the eggshell of some nematodes (Wharton, 1980) may confer some protection against UV radiation.

Interaction between Physiological Stresses

Although temperature, desiccation, osmotic and oxygen stress have been considered separately, there is likely to be a considerable interaction between them. A nematode is likely to be exposed to more than one of these stresses simultaneously. For example, freezing may induce osmotic stress via freeze concentration effects as the formation of ice raises the concentration of salts in the non-frozen portion of the medium. Being encased in ice is also likely to produce anoxic conditions since ice is an effective barrier to the diffusion of oxygen (Scholander, *et al.*, 1953). In a state of anhydrobiosis nematodes can survive extremes of temperature, as well as exposure to radiation and chemicals, which would be fatal to hydrated nematodes (Barrett, 1991). The esterase enzymes of the anhydrobiotic nematodes *Actinolaimus hintoni* and *Dorylaimus keilini* retain activity if exposed to heat when dry but the activity is destroyed if exposed when hydrated (Lee, 1964).

Timing Mechanisms

Nematodes can improve their chances of survival and/ or locating a host by synchronising their development with patterns of favourable seasons and/or the availability of hosts or suitable habitat. In cold temperate habitats this involves surviving cold winters, whilst in the tropics and hot temperate habitats it may involve surviving a dry season.

Diapause

Insects synchronise their life cycles with favourable seasons via a hormonally-regulated diapause. Hormone regulation has yet to be demonstrated in nematode diapause but the existence of a diapause has been inferred from the pattern of seasonal development and/or the response of a developmental event, such as egg hatching, to particular combinations of stimuli. Diapause may occur within every generation of the nematode as a result of endogenous factors (obligate diapause) or only in response to stimuli which indicate the onset of adverse conditions (facultative diapause). It is usually a feature of a particular stage in the life cycle. Diapause may also improve the chances of finding a mate if the development of adults is synchronised. The evidence for the existence of diapause in nematodes has been extensively reviewed (Antoniou, 1989; Evans, 1987; Evans and Perry, 1976; Jones, *et al.*, 1998; Womersley, *et al.*, 1998).

Egg diapause

The eggs of *H. avenae* and *M. naasi* show increased hatching following a period of chilling. In some populations this produces a pattern of host root invasion by infective larvae in the spring and adult maturation and egg production in the summer (Evans and Perry, 1976). Different isolates vary in their optimum temperatures for chilling and hatching and in incubation times, suggesting adaptation to local conditions (Evans, 1987; Jones, *et al.*, 1998; Womersley, *et al.*, 1998); which may allow survival of a dry season, rather than overwintering. Even within a single population there may be variation in the stimuli necessary to hatch eggs. This allows a proportion of eggs to hatch immediately and a proportion to overwinter and hatch in a second season, improving persistence of the population and reducing competition for feeding sites on hosts (Jones, *et al.*, 1998). A combination of obligate diapause, facultative diapause and quiescence enables the eggs of *G. rostochiensis* and *G. pallida* to persist in soil for many years and yet hatch rapidly in the presence of the host plant (Jones, *et al.*, 1998).

Photoperiod is a more reliable indicator of season than is temperature and evidence for a photoperiod-induced diapause has been presented for *G. rostochiensis* (Hominick, 1986). This is mediated by the effect of photoperiod on the host plant. Other factors affecting diapause include the age of the female nematode and factors involved with the senescence of the host plant (Womersley, *et al.*, 1998).

The eggs of the animal-parasitic nematode *Nematodirus battus* require a period of chilling before hatching. L3 overwinter within the eggshell and hatch *en masse* in the spring (Evans and Perry, 1976). Eggs which have developed on pasture in Britain fail to hatch if kept at 20°C but a mass hatch is elicited if they are exposed to 5°C before transfer to 20°C. Autumn conditions appear to induce a diapause which is terminated by chilling over the winter and a rise in temperature in spring. The larvae thus overwinter within the egg and hatch in spring ready to infect lambs. The response is so predictable that formulae have been produced which forecast the timing and severity of pasture contamination with hatched larvae (Thomas, 1974).

Diapause in other stages of the life cycle

The phenomenon of arrested development of trichostrongyle larvae (see section on examples of survival strategies) may be a facultative diapause induced in the parasitic L4 by the conditions experienced by the infective L3 on pasture. It may also affect the parasitic L3 and the early adult stage (Eysker, 1997). There are two distinct patterns in seasonal inhibition, associated with overwinter survival or the survival of summer or dry seasons. Low temperatures experienced by L3 on pasture during autumn trigger winter inhibition of L4 in the host but the trigger for summer inhibition is unknown. The host's immune response and the management regimes applied to domestic stock have also been implicated in the propensity of larvae to undergo arrested development, which responds rapidly to selection (Eysker, 1997). The mechanisms controlling this phenomenon are unknown but the identification of a protein which was diagnostic for inhibited *Ostertagia ostertagi* (Cross, Klesius and Williams, 1988) suggests an underlying molecular mechanism.

The infective L4 of *D. dipsaci* vary throughout the year in their ability to infect plants and a period of chilling may be necessary, indicating a diapause (Evans and Perry, 1976). A proportion of the eggs of *Meloidogyne incognita* show delayed embryonic development which has been called an 'embryonic diapause', the triggers which induce these eggs to commence development are unknown and the phenomenon might be better regarded as a life-history strategy (Evans, 1987). Diapause may also occur in the infective larvae of steinernematid nematodes, where infectivity is enhanced by a period of chilling (Fan and Hominick, 1991).

Timing Mechanisms Based on Life-History Strategies

Dauer larvae

Dauer larvae are resistant stages which may be an obligate part of the life cycle or formed in response to environmental conditions. This term has been applied to the resistant L3 of some rhabditid nematodes which are formed when environmental conditions deteriorate and which differ in their morphology, physiology and behaviour from 'normal' L3. Dauer larvae are in a state of developmental arrest and only resume development when a physiological trigger indicates the return of conditions favourable for growth; whereas normal L3 will rapidly develop to the next stage in the life cycle. This has been called a diapause but is perhaps more appropriately referred to as a life-history strategy.

Bacterial-feeding rhabditids produce dauer larvae in response to a decrease in the bacterial food, an increase in the local population of nematodes or to a decline in physical conditions, such as desiccation. For most species the physiological stimuli which induce the formation of dauer larvae are unknown. A notable exception is for *C. elegans* (see section on examples of survival strategies). The dauer larvae of some rhabditids retain the cuticle of the previous stage as sheath, have a distinct contracted shape, are resistant to environmental stresses and are non-feeding. These act as dispersal forms by attaching to insects and other invertebrates, or even vertebrates, which carry them to a fresh habitat; a relationship known as phoresis (Poinar, 1983). Contact with the phoretic host may be aided by the dauer larva mounting projections on the substrate and waving back and forth.

The infective larvae of steinernematid and heterorhabditid entomopathogenic nematodes are often referred to as dauer larvae or juveniles (Womersley, 1993). These nematodes will complete two, or even three, generations within the insect cadaver before the formation of dauer larvae which leave the corpse. Their life cycle is thus analogous to that of free-living rhabditids, with the production of L3 which develop to the next stage immediately and dauer larvae which act as survival, dispersal and infective stages. The production of dauer larvae appears to be related to the numbers of larvae present within the insect body and thus may be under the control of a pheromone-based system, similar to that controlling dauer formation in *C. elegans* (Fodor, Vecseri and Farkas, 1990).

The infective larvae (L2) of some anguinid plant-parasitic nematodes have been referred to as dauer larvae (Womersley, *et al.*, 1998) but their formation does not appear to represent an alternative pathway to the formation of a normal L2 and it perhaps best called an infective larva.

Infective larvae

Parasitic nematodes require entry into a suitable host to complete their life cycle. The stage which is infective to the host (the infective larva) is thus in a state of developmental arrest and development is resumed once physiological stimuli which indicate entry into or the proximity of a host are received. This assures that development only resumes when conditions are favourable, i.e. in a suitable host, and that development is prevented when conditions are likely to result in the death of the parasite (such as outside a host). There are numerous examples (see life cycle chapter and the section on survival strategies). The infective larvae of some species have been called dauer larvae or as being in a state of diapause but, unless the specific use of these terms is justified, they are perhaps best thought of as being in a state of developmental arrest and called infective larvae.

The basic life cycle is the same in all nematodes: consisting of the egg, four larval stages and the adult. Apart from the egg to L1, these stages are separated by the moulting of the cuticle. Each stage in the life cycle may have quite different properties. Rogers and Petronijevic (1982) have suggested that development from one stage to the next involves the suppression of the gene set associated with one stage in the life cycle and the expression of the next. The control of dauer larval formation in *C. elegans* suggests such a mechanism and parallels have been drawn between this and the formation and activation of infective larvae. In *C. elegans* transcription of the *daf-7* gene prevents the formation of dauer larvae. Noting the homology between the *daf-7* gene product and members of the transforming growth factor (TGF-β) superfamily, Rajan (1998) suggests that infective larvae act like *daf-7* defective mutants and require the *daf-7* gene product to resume development. It is suggested that this is supplied via TGF-β produced by the host and that this explains aspects of the life history and tissue specificity of some parasites (Rajan, 1998).

Conclusions

Nematodes thus possess a wide variety of survival strategies. Developmental dormancy, such as diapause and the formation of dauer or infective larvae, enables the nematode to synchronise its development with the availability of suitable habitat or conditions for growth. Metabolic dormancy, quiescence and cryptobiosis, allows survival of periods of environmental stress. These stresses include temperature extremes, desiccation, anoxia and osmotic stress. Each nematode species possesses a suite of life history and physiological adaptations which enables it to exploit its own particular habitat. The ability of some nematodes to survive environmental stress is remarkable, including the ability to survive freezing and complete desiccation. Determining how these nematodes are able to survive in these remarkable ways will assist our understanding of how life can exist in extreme environments and may facilitate the preservation of biological materials for a variety of medical and industrial purposes.

References

Allan, G.S. and Wharton, D.A. (1990) Anhydrobiosis in the infective juveniles of *Trichostrongylus colubriformis* (Nematoda: Trichostrongylidae). *International Journal for Parasitology*, **20**, 183–192.

Anderson, G.L. (1978) Responses of dauer larvae of *Caenorhabditis elegans* (Nematoda: Rhabditidae) to thermal stress and oxygen deprivation. *Canadian Journal of Zoology*, **56**, 1786–1791.

Antoniou, M. (1989) Arrested development in plant parasitic nematodes. *Helminthological Abstracts*, **B58**, 1–19.

Asahina, E. (1959) Frost-resistance in a nematode, *Aphelenchoides ritzemabosi*. *Low Temperature Science*, **B17**, 51–62.

Ash, C.P.J. and Atkinson, H.J. (1983) Evidence for a temperature-dependent conversion of lipid reserves to carbohydrate in quiescent eggs of the nematode, *Nematodirus battus*. *Comparative Biochemistry and Physiology*, **76B**, 603–610.

Ash, C.P.J. and Atkinson, H.J. (1986) *Nematodirus battus*: development of cold hardiness in dormant eggs. *Experimental Parasitology*, **62**, 24–28.

Atkinson, H.J. (1976) The respiratory physiology of nematodes. In *The Organisation of Nematodes*, edited by N.A. Croll, pp. 243–272. London & New York: Academic Press.

Atkinson, H.J. (1980) Respiration in nematodes. In *Nematodes as Biological Models*, Vol. 2, edited by B.M. Zuckerman, pp. 101–142. New York & London: Academic Press.

Awan, F.A. and Hominick, W.M. (1982) Observations on tanning of the potato cyst-nematode, *Globodera rostochiensis*. *Parasitology*, **85**, 67–71.

Bale, J.S. (1993) Classes of insect cold hardiness. *Functional Ecology*, **7**, 751–753.

Barrett, J. (1981) *Biochemistry of Parasitic Helminths*, London: Macmillan.

Barrett, J. (1982) Metabolic responses to anabiosis in the fourth stage juveniles of *Ditylenchus dipsaci* (Nematoda). *Proceedings of the Royal Society of London*, **B216**, 159–177.

Barrett, J. (1991) Anhydrobiotic nematodes. *Agricultural Zoology Reviews*, **4**, 161–176.

Barrett, J. and Butterworth, P.E. (1985) DNA stability in the anabiotic fourth-stage juveniles of *Ditylenchus dipsaci* (Nematoda). *Annals of Applied Biology*, **106**, 121–124.

Bartlett, L.M. (1992) Cold-hardiness in *Pelecitus fulicaeatrae* (Nematoda: Filarioidea), a parasite in the ankle of *Fulica americana* (Aves). *Journal of Parasitology*, **78**, 138–139.

Bird, A.F. and Bird, J. (1998) Introduction to functional organisation. In *The Physiology and Biochemistry of Free-living and Plant-parasitic Nematodes*, edited by R.N. Perry and D.J. Wright, pp. 1–24. Wallingford & New York: CABI Publishing.

Bird, A.F. and Buttrose, M.S. (1974) Ultrastructural changes in the nematode *Anguina tritici* associated with anhydrobiosis. *Journal of Ultrastructural Research*, **48**, 177–189.

Bird, A.F. and Soeffkey, A. (1972) Changes in the ultrastructure of the gelatinous matrix of *Meloidogyne javanica* during dehydration. *Journal of Nematology*, **4**, 166–169.

Blaxter, M.L. and Robertson, W.M. (1998) The cuticle. In *The Physiology and Biochemistry of Free-living and Plant-parasitic Nematodes*, edited by R.N. Perry and D.J. Wright, pp. 25–48. Wallingford & New York: CABI Publishing.

Block, W. (1995) Insects and freezing. *Science Progress*, **78**, 349–372.

Bovien, P. (1937) Some types of association between nematodes and insects. *Videnskabelige Meddelelser fra Dansk Naturhistorisk forening* **101**, 1–114.

Brown, I.M. (1993) *The Influence of Low Temperature on the Antarctic Nematode Panagrolaimus davidi*, PhD thesis, University of Otago, Dunedin, New Zealand.

Brown, I.M. and Gaugler, R. (1996) Cold tolerance of steinernematid and heterorhabditid nematodes. *Journal of Thermal Biology*, **21**, 115–121.

Brown, I.M. and Gaugler, R. (1998) Survival of steinernematid nematodes exposed to freezing. *Journal of Thermal Biology*, **23**, 75–80.

Burglin, T.R., Lobos, E. and Blaxter, M.L. (1998) *Caenorhabditis elegans* as a model for parasitic nematodes. *International Journal for Parasitology*, **28**, 395–411.

Caldwell, J.R. (1981) The Signy Island terrestrial reference sites: XIII. Population dynamics of the nematode fauna. *British Antarctic Survey Bulletin*, **54**, 33–46.

Calow, P. (1983) Pattern and paradox in parasite reproduction. *Parasitology*, **86**, 197–207.

Clarke, A.J. and Hennessy, J. (1976) The distribution of carbohydrates in cysts of *Heterodera rostochiensis*. *Nematologica*, **22**, 190–195.

Clegg, J.S. (1965) The origin of trehalose and its significance during the formation of encysted dormant embryos of *Artemia salina*. *Comparative Biochemistry and Physiology*, **14**, 135–143.

Clegg, J.S. (1979) Metabolism and the intracellular environment: the vicinal-water network model. In *Cell Associated Water*, edited by W. Drost-Hansen and J.S. Clegg, pp. 363–413. New York: Academic Press.

Cooper, A.F. and Van Gundy, S.D. (1970) Metabolism of glycogen and neutral lipids by *Aphelenchus avenae* and *Caenorhabditis* sp. in aerobic, microaerobic and anaerobic environments. *Journal of Nematology*, **2**, 305–315.

Cossins, A.R. and Bowler, K. (1987) *Temperature Biology of Animals*, London & New York: Chapman & Hall.

Cross, D.A., Klesius, P.H. and Williams, J.C. (1988) Preliminary report: immunodiagnosis of pre-type II ostertagiasis. *Veterinary Parasitology*, **27**, 151–158.

Crowe, J.H., Carpenter, J.F. and Crowe, L.M. (1998) The role of vitrification in anhydrobiosis. *Annual Review of Physiology*, **60**, 73–103.

Crowe, J.H., Hoekstra, F. and Crowe, L.M. (1992) Anhydrobiosis. *Annual Review of Physiology*, **54**, 579–599.

Crowe, J.H., Lambert, D.T. and Crowe, L.M. (1978) Ultrastructural and freeze fracture studies on anhydrobiotic nematodes. In *Dry Biological Systems*, edited by J.H. Crowe and J.S. Clegg, pp. 23–51. New York: Academic Press.

Crowe, J.H. and Madin, K.A. (1975) Anhydrobiosis in nematodes: evaporative water loss and survival. *Journal of Experimental Zoology*, **193**, 323–334.

Crowe, J.H., O'Dell, S.J. and Armstrong, D.A. (1979) Anhydrobiosis in nematodes: permeability during rehydration. *Journal of Experimental Zoology*, **207**, 431–438.

Dallavecchia, F., Elasmar, T., Calamassi, R., Rascio, N. and Vazzana, C. (1998) Morphological and ultrastructural aspects of dehydration and rehydration in leaves of *Sporobolus stapfianus*. *Plant Growth Regulation*, **24**, 219–228.

Dalley, B.K. and Golomb, M. (1992) Gene expression in the *Caenorhabditis elegans* dauer larva: developmental regulation of hsp90 and other genes. *Developmental Biology*, **151**, 80–90.

Davey, M.C., Pickup, J. and Block, W. (1992) Temperature variation and its biological significance in fellfield habitats on a maritime Antarctic island. *Antarctic Science*, **4**, 383–388.

Demure, Y. (1980) Biology of the plant-parasitic nematode, *Scutellonema cavenesii* Sher, 1964: anhydrobiosis. *Revue de Nématologie*, **3**, 283–289.

Demure, Y., Freckman, D.W. and Van Gundy, S.D. (1979) *In vitro* response of four species of nematodes to desiccation and discussion of this and related phenomena. *Revue de Nématologie*, **2**, 203–210.

Dickson, M.R. and Mercer, E.H. (1967) Fine structural changes accompanying desiccation in *Philodina roseola* (Rotifera). *Journal de Microscopie*, **6**, 331–348.

Duman, J.G., Xu, L., Neven, L.G., Thursman, D. and Wu, D.W. (1991) Hemolymph proteins involved in insect subzero-temperature tolerance: ice nucleators and antifreeze proteins. In *Insects at Low Temperatures*, edited by R.E. Lee and D.L. Denlinger, pp. 94–127. New York and London: Chapman and Hall.

Ellenby, C. (1946) Nature of the cyst wall of the potato-root eelworm *Heterodera rostochiensis*, Wollenweber, and its permeability to water. *Nature*, **157**, 302.

Ellenby, C. (1968a) Desiccation survival in the plant parasitic nematodes, *Heterodera rostochiensis* Wollenweber and *Ditylenchus dipsaci* (Kuhn) Filipjev. *Proceedings of the Royal Society of London*, **B169**, 203–213.

Ellenby, C. (1968b) Desiccation survival of the infective larva of *Haemonchus contortus*. *Journal of Experimental Biology*, **49**, 469–475.

Ellenby, C. (1969) Dormancy and survival in nematodes. *Symposium of the Society for Experimental Biology*, **23**, 83–97.

Evans, A.A.F. (1987) Diapause in nematodes as a survival strategy. In *Vistas on Nematology*, edited by J.A. Veech and D.W. Dickson, pp. 180–187. Hyattsville, Maryland: Society of Nematologists Inc.

Evans, A.A.F. and Perry, R.N. (1976) Survival strategies in nematodes. In *The Organisation of Nematodes*, edited by N.A. Croll, pp. 383–424. London & New York: Academic Press.

Eysker, M. (1997) Some aspects of inhibited development of trichostrongylids in ruminants. *Veterinary Parasitology*, **72**, 265–283.

Fan, X. and Hominick, W.M. (1991) Effects of low storage temperature on survival and infectivity of two *Steinernema* species (Nematoda: Steinernematidae). *Revue de Nématologie*, **14**, 407–412.

Fodor, A., Vecseri, G. and Farkas, T. (1990) *Caenorhabditis elegans* as a model for the study of entomopathogenic nematodes. In *Entomopathogenic Nematodes in Biological Control.*, edited by R. Gaugler and H.K. Kaya, pp. 249–265. Boca Raton: CRC Press.

Forge, T.A. and MacGuidwin, A.E. (1990) Cold hardening of *Meloidogyne hapla* second-stage juveniles. *Journal of Nematology*, **22**, 101–105.

Forge, T.A. and MacGuidwin, A.E. (1992) Effects of water potential and temperature on survival of the nematode *Meloidogyne hapla* in frozen soil. *Canadian Journal of Zoology*, **70**, 1553–1560.

Franks, F. (1985) *Biophysics and Biochemistry at Low Temperatures*, Cambridge, UK: Cambridge University Press.

Freckman, D.W., Kaplan, D.T. and Van Gundy, S.D. (1977) A comparison of techniques for extraction and study of anhydrobiotic nematodes from dry soils. *Journal of Nematology*, **9**, 176–181.

Friedman, M.J. (1990) Commercial production and development. In *Entomopathogenic Nematodes in Biological Control*, edited by R. Gaugler and H.K. Kaya, pp. 153–172. Boca Raton: CRC Press.

Fujiie, A. and Yokoyama, T. (1998) Effects of ultraviolet light on the entomopathogenic nematode, *Steinernema kushidai* and its symbiotic bacterium, *Xenorhabdus japonicus*. *Applied Entomology and Zoology*, **33**, 263–269.

Gaugler, R. and Kaya, H.K. (1990) *Entomopathogenic Nematodes in Biological Control*, Boca Raton: CRC Press.

Gaugler, R., Wilson, M. and Shearer, P. (1997) Field release and environmental fate of a transgenic entomopathogenic nematode. *Biological Control*, **9**, 75–80.

Gaur, H.S. and Perry, R.N. (1991) The role of the moulted cuticles in the desiccation survival of adults of *Rotylenchulus reniformis*. *Revue de Nématologie*, **14**, 491–496.

Gibbs, H.C. (1986) Hypobiosis in parasitic nematodes — an update. *Advances in Parasitology*, **25**, 129–174.

Glazer, I., Kozodoi, E., Hashmi, G. and Gaugler, R. (1996) Biological characteristics of the entomopathogenic nematode *Heterorhabditis* sp is-5 — a heat tolerant isolate from Israel. *Nematologica*, **42**, 481–492.

Glazer, I. and Orion, D. (1983) Studies on anhydrobiosis of *Pratylenchus thornei*. *Journal of Nematology*, **15**, 333–338.

Gresham, A. and Womersley, C. (1991) Modulation of catalase activity during the enforced induction of and revival from anhydrobiosis in nematodes. *FASEB Journal*, **5A**, 682.

Grewal, P.S., Gaugler, R. and Shupe, C. (1996) Rapid changes in thermal sensitivity of entomopathogenic nematodes in response to selection at temperature extremes. *Journal of Invertebrate Pathology*, **68**, 65–73.

Griffin, C.T., Downes, M.J. and Block, W. (1990) Tests of Antarctic soils for insect parasitic nematodes. *Antarctic Science*, **2**, 221–222.

Gustafson, P.V. (1953) The effect of freezing on encysted *Anisakis* larvae. *Journal of Parasitology*, **39**, 585–588.

Hashmi, S., Hashmi, G., Glazer, I. and Gaugler, R. (1998) Thermal response of *Heterorhabditis bacteriophora* transformed with the *Caenorhabditis elegans* Hsp70 encoding gene. *Journal of Experimental Zoology*, **281**, 164–170.

Hendriksen, N.B. (1982) Anhydrobiosis in nematodes: studies on *Plectus* sp. In *New Trends in Soil Biology*, edited by P. Lebrun, H.M. André, A. De Medts, C. Grégoire-Wibo and G. Wauthy, pp. 387–394. Louvain-la-Neurve: Dieu-Brichart.

Higa, L.M. and Womersley, C.Z. (1993) New insights into the anhydrobiotic phenomenon: the effect of trehalose content and differential rates of evaporative water loss on the survival of *Aphelenchus avenae*. *Journal of Experimental Zoology*, **267**, 120–129.

Holmstrup, M. and Sømme, L. (1998) Dehydration and cold hardiness in the arctic Collembolan *Onychiurus arcticus* Tullberg 1876. *Journal of Comparative Physiology B*, **168**, 197–203.

Holmstrup, M. and Westh, P. (1995) Effects of dehydration on water relations and survival of lumbricid earthworm egg capsules. *Journal of Comparative Physiology B*, **165**, 377–383.

Hominick, W.M. (1986) Photoperiod and diapause in the potato cyst-nematode, *Globodera rostochiensis*. *Nematologica*, **32**, 408–418.

Jacquiet, P., Cabaret, J., Dia, M.L., Cheikh, D. and Thiam, E. (1996) Adaptation to arid environment — *Haemonchus longistipes* in Dromedaries of Saharo-Sahelian areas of Mauritania. *Veterinary Parasitology*, **66**, 193–204.

Jones, F.G.W. and Jones, M.G. (1974) *Pests of field crops*, 2nd edn, New York: St. Martin's Press.

Jones, P.W., Tylka, G.L. and Perry, R.N. (1998) Hatching. In *The Physiology and Biochemistry of Free-living and Plant-parasitic Nematodes*, edited by R.N. Perry and D.J. Wright, pp. 181–212. Wallingford & New York: CABI Publishing.

Knight, C.A. and Duman, J.G. (1986) Inhibition of recrystallization of ice by insect thermal hysteresis proteins: a possible cryoprotective role. *Cryobiology*, **23**, 256–262.

Knight, C.A., Wen, D. and Laursen, R.A. (1995) Nonequilibrium antifreeze peptides and the recrystallization of ice. *Cryobiology*, **32**, 23–34.

Ko, R.C. and Fan, L. (1996) Heat shock response of *Trichinella spiralis* and *T. pseudospiralis*. *Parasitology*, **112**, 89–95.

Koppenhofer, A.M., Baur, M.E., Stock, S.P., Choo, H.Y., Chinnasri, B. and Kaya, H.K. (1997) Survival of entomopathogenic nematodes within host cadavers in dry soil. *Applied Soil Ecology*, **6**, 231–240.

Lee, D.L. (1964) Esterase enzymes in two free-living nematodes. *Proceedings of the Helminthological Society of Washington*, **31**, 285–288.

Lee, R.E. (1991) Principles of insect low temperature tolerance. In *Insects at Low Temperatures*, edited by R.E. Lee and D.L. Denlinger, pp. 17–46. New York and London: Chapman and Hall.

Lee, R.E. and Denlinger, D.L. (1991) *Insects at Low Temperatures*. New York and London: Chapman and Hall.

Lee, R.E., McGrath, J.J., Morason, R.T. and Taddeo, R.M. (1993) Survival of intracellular freezing, lipid coalescence and osmotic fragility in fat body cells of the freeze-tolerant gall fly *Eurosta solidaginis*. *Journal of Insect Physiology*, **39**, 445–450.

Lees, E. (1953) An investigation into the method of dispersal of *Panagrellus silusiae*, with particular reference to its desiccation resistance. *Journal of Helminthology*, **27**, 95–103.

Leroi, A.M. and Jones, J.T. (1998) Developmental biology. In *The Physiology and Biochemistry of Free-living and Plant-parasitic Nematodes*, edited by R.N. Perry and D.J. Wright, pp. 155–179. Wallingford & New York: CABI Publishing.

Levine, N.D. (1968) *Nematode Parasites of Domestic Animals and of Man*, Minneapolis: Burgess.

Lillibridge, C.D., Rudin, W. and Philipp, M.T. (1996) *Dirofilaria immitis* — ultrastructural localization, molecular characterization, and analysis of the expression of P27, a small heat shock protein homolog of nematodes. *Experimental Parasitology*, **83**, 30–45.

Lithgow, G.J., White, T.M., Melov, S. and Johnson, T.E. (1995) Thermotolerance and extended life-span conferred by single-gene mutations and induced by thermal stress. *Proceedings of the National Academy of Science USA*, **92**, 7540–7544.

Mabbett, K. and Wharton, D.A. (1986) Cold tolerance and acclimation in the free-living nematode, *Panagrellus redivivus*. *Revue de Nématologie*, **9**, 167–170.

Madin, K.A.C. and Crowe, J.H. (1975) Anhydrobiosis in nematodes: carbohydrate and lipid metabolism during dehydration. *Journal of Experimental Zoology*, **193**, 335–342.

Madin, K.A.C., Crowe, J.H. and Loomis, S.H. (1978) Metabolic transitions in a nematode during induction of and recovery from anhydrobiosis. In *Dry Biological Systems*, edited by J.H. Crowe and J.S. Clegg, pp. 155–174. New York: Academic Press.

Maresca, B. and Carratu, L. (1992) The biology of heat shock response in parasites. *Parasitology Today*, **8**, 260–266.

Maslen, N.R. (1979) Additions to the nematode fauna of the Antarctic region with keys to taxa. *British Antarctic Survey Bulletin*, **49**, 207–229.

Michel, J.F. (1976) The epidemiology and control of some nematode infections in grazing animals. *Advances in Parasitology*, **14**, 355–397.

Miller, P.M. (1968) The susceptibility of parasitic nematodes to sub-freezing temperatures. *Plant Disease Reporter*, **52**, 768–772.

Mills, D.K. and Hartman, P.S. (1998) Lethal consequences of simulated solar radiation on the nematode *Caenorhabditis elegans* in the presence and absence of photosensitizers. *Photochemistry and Photobiology*, **68**, 816–823.

Morris, J.E. (1968) Dehydrated cysts of *Artemia salina* prepared for electron microscopy by totally anhydrous techniques. *Journal of Ultrastructural Research*, **25**, 64–72.

Nicholas, W.L. (1984) *The Biology of Free-living Nematodes*, 2nd edn, Oxford: Clarendon Press.

Nicholas, W.L. and Stewart, A.C. (1989) Experiments on anhydrobiosis in *Acrobeloides nanus* (De Man, 1880) Anderson 1986 (Nematoda). *Nematologica*, **35**, 489–490.

Norton, D.C. (1978) *Ecology of Plant-parasitic Nematodes*, New York: Wiley and Sons.

Parsell, D.A. and Lindquist, S. (1994) Heat shock proteins and stress tolerance. In *The Biology of Heat Shock Proteins and Molecular Chaperones*, edited by R.I. Morimoto, A. Tissières and C. Georgopoulos, pp. 457–495. Cold Spring Harbor: Cold Spring Harbor Laboratory Press.

Patel, M.N. and Wright, D.J. (1998) The ultrastructure of the cuticle and sheath of infective juveniles of entomopathogenic steinernematid nematodes. *Journal of Helminthology*, **72**, 257–266.

Peat, A. and Potts, M. (1987) The ultrastructure of immobilised desiccated cells of the cyanobacterium *Nostoc commune* UTEX 584. *FEMS Microbiological Letters*, **43**, 233–237.

Perry, R.N. (1977a) Desiccation survival of larval and adult stages of the plant parasitic nematodes, *Ditylenchus dipsaci* and *D. myceliophagus*. *Parasitology*, **74**, 139–148.

Perry, R.N. (1977b) The water dynamics of stages of *Ditylenchus dipsaci* and *D. myceliophagus* during desiccation and rehydration. *Parasitology*, **75**, 45–70.

Perry, R.N. (1983) The effect of potato root diffusate on the desiccation survival of unhatched juveniles of *Globodera rostochiensis*. *Revue de Nématologie*, **6**, 33–38.

Perry, R.N. (1989) Dormancy and hatching of nematode eggs. *Parasitology Today*, **5**, 377–383.

Perry, R.N. and Wharton, D.A. (1985) Cold tolerance of hatched and unhatched second stage juveniles of *Globodera rostochiensis*. *International Journal for Parasitology*, **15**, 441–445.

Pickup, J. (1990a) Seasonal variation in the cold hardiness of three species of free-living Antarctic nematodes. *Functional Ecology*, **4**, 257–264.

Pickup, J. (1990b) Seasonal variation in the cold-hardiness of a free-living predatory nematode, *Coomansus gerlachei* (Mononchidae). *Polar Biology*, **10**, 307–315.

Pickup, J. (1990c) Strategies of cold-hardiness in three species of Antarctic dorylaimid nematodes. *Journal of Comparative Physiology B*, **160**, 167–173.

Pickup, J. and Rothery, P. (1991) Water-loss and anhydrobiotic survival in nematodes of Antarctic fellfields. *Oikos*, **61**, 379–388.

Poinar, G.O.J. (1983) *The Natural History of Nematodes*, Englewood Cliffs, New Jersey: Prentice-Hall Inc.

Popiel, I. and Vasquez, E.M. (1991) Cryopreservation of *Steinernema carpocapsae* and *Heterorhabditis bacteriophora*. *Journal of Nematology*, **23**, 432–437.

Precht, H. (1958) Concepts of the temperature adaptation of unchanging reaction systems of cold-blooded animals. In *Physiological Adaptation*, edited by C.L. Prosser, pp. 351–376. Washington: American Association for the Advancement of Science.

Preston, C.M. and Bird, A.F. (1987) Physiological and morphological changes associated with recovery from anabiosis in the dauer larva of the nematode *Anguina agrostis*. *Parasitology*, **95**, 125–133.

Procter, D.L.C. (1984) Population growth and intrinsic rate of natural increase of the high arctic nematode *Chiloplacus* sp. at low and high temperatures. *Oecologia*, **62**, 138–140.

Rajan, T.V. (1998) A hypothesis for the tissue specificity of nematode parasites. *Experimental Parasitology*, **89**, 140–142.

Ramløv, H., Wharton, D.A. and Wilson, P. (1996) Recrystallisation inhibition in a freezing tolerant Antarctic nematode, *Panagrolaimus davidi*, and an alpine weta, *Hemideina maori* (Orthoptera; Stenopelmatidae). *Cryobiology*, **33**, 607–613.

Riddle, D.L. (1988) The dauer larva. In *The Nematode Caenorhabditis elegans*, edited by W.B. Wood, pp. 393–412. Cold Spring Harbor: Cold Spring Harbor Laboratory Press.

Riddle, D.L. and Albert, P.S. (1997) Genetic and environmental regulation of dauer larva development. In *C. elegans II*, edited by D.L. Riddle, T. Blumenthal, B.J. Meyer and J.R. Priess, pp. 739–768. Cold Spring Harbor: Cold Spring Harbor Laboratory Press.

Riddle, D.L., Blumenthal, T., Meyer, B.J. and Priess, J.R. (1997) *C. elegans II*, Cold Spring Harbor: Cold Spring Harbor Laboratory Press.

Robinson, A.F., Orr, C.C. and Heintz, C.E. (1984) Some factors affecting survival of desiccation by infective juveniles of *Orrina phyllobia*. *Journal of Nematology*, **16**, 86–91.

Rogers, W.P. (1960) The physiology of the infective process of nematode parasites: the stimulus from the animal host. *Proceedings of the Royal Society of London*, **B152**, 367–386.

Rogers, W.P. and Petronijevic, T. (1982) The infective stage and the development of nematodes. In *Biology and Control of Endoparasites*, edited by L.E.A. Symons, A.D. Donald and J.K. Dineen, pp. 3–28. New York and London: Academic Press.

Scholander, P.F., Flagg, W., Hoch, R.J. and Irving, L. (1953) Climatic adaptation in arctic and tropical poikilotherms. *Physiological Zoology*, **26**, 67–92.

Selkirk, M.E., Denham, D.A., Paratono, F. and Maizels, R. (1989) Heat shock cognate 70 is a prominent immunogen in brugian filariasis. *Journal of Immunology*, **143**, 299–308.

Selvan, S., Grewal, P.S., Leustek, T. and Gaugler, R. (1996) Heat shock enhances thermotolerance of infective juvenile insect-parasitic nematodes *Heterorhabditis bacteriophora* (Rhabditida, Heterorhabditidae). *Experientia*, **52**, 727–730.

Smith, H.J. (1987) Factors affecting preconditioning of *Trichinella spiralis nativa* larvae in musculature to low temperatures. *Canadian Journal of Veterinary Research*, **51**, 169–173.

Snutch, T.P. and Baillie, D.L. (1983) Alterations in the pattern of gene expression following heat shock in the nematode *Caenorhabditis elegans*. *Canadian Journal of Biochemistry and Cell Biology*, **61**, 480–487.

Sømme, L. (1995) *Invertebrates in Hot and Cold Arid Environments*. Berlin: Springer-Verlag.

Spratt, D.M. (1972) Aspects of the life cycle of *Dirofilaria raemeri* in naturally and experimentally infected kangaroos, wallaroos and wallabies. *International Journal for Parasitology*, **2**, 139–156.

Steiner, W.A. (1996) Distribution of entomopathogenic nematodes in the Swiss Alps. *Revue Suisse de Zoologie*, **103**, 439–452.

Storey, K.B. and Storey, J.M. (1992) Natural freeze tolerance in ectothermic vertebrates. *Annual Review of Physiology*, **54**, 619–637.

Storey, K.B. and Storey, J.M. (1996) Natural freezing survival in animals. *Annual Review of Ecology and Systematics*, **27**, 365–386.

Surrey, M.R. (1996) The effect of rearing method and cool temperature acclimation on the cold tolerance of *Heterorhabditis zealandica* infective juveniles (Nematoda: Heterorhabditidae). *Cryo-Letters*, **17**, 313–320

Surrey, M.R. and Wharton, D.A. (1995) Desiccation survival of the infective larvae of the insect parasitic nematode, *Heterorhabditis zealandica* Poinar. *International Journal for Parasitology*, **25**, 749–752.

Theede, H., Schneppenheim, R. and Bevess, L. (1976) Frostschutz — Glycoproteine bei *Mytilus edulis*? *Marine Biology*, **36**, 183–189.

Thomas, R.J. (1974) The role of climate in the epidemiology of nematode parasitism in ruminants. In *The Effects of Meteorological Factors Upon Parasites*, edited by A.E.R. Taylor and R. Muller, pp. 13–32. Oxford: Blackwell.

Timm, R.W. (1971) Antarctic soil and freshwater nematodes from the McMurdo Sound region. *Proceedings of the Helminthological Society of Washington*, **38**, 42–52.

Tweedie, S., Grigg, M.E., Ingram, L. and Selkirk, M.E. (1993) The expression of a small heat shock protein homologue is developmentally regulated in *Nippostrongylus brasiliensis*. *Molecular and Biochemical Parasitology*, **61**, 149–154.

Tyrrell, C., Wharton, D.A., Ramløv, H. and Moller, H. (1994) Cold tolerance of an endoparasitic nematode within a freezing tolerant orthopteran host. *Parasitology*, **109**, 367–372.

Vanleeuwen, M.A.W. (1995) Heat-shock and stress response of the parasitic nematode *Haemonchus contortus*. *Parasitology Research*, **81**, 706–709.

Wertejuk, M. (1959) Influence of environmental temperature on the invasive larvae of gastrointestinal nematodes of sheep. *Acta Parasitologica Polonski*, **7**, 315–342.

Westh, P. and Ramløv, H. (1991) Trehalose accumulation in the tardigrade *Adorybiotus coronifer* during anhydrobiosis. *Journal of Experimental Zoology*, **258**, 303–311.

Wharton, D. (1980) Nematode egg-shells. *Parasitology*, **81**, 447–463.

Wharton, D.A. (1979) *Ascaris lumbricoides*: water loss during desiccation of embryonating eggs. *Experimental Parasitology*, **48**, 398–406.

Wharton, D.A. (1980) Studies on the function of the oxyurid egg-shell. *Parasitology*, **81**, 103–113.

Wharton, D.A. (1981) The effect of temperature on the behaviour of the infective larvae of *Trichostrongylus colubriformis*. *Parasitology*, **82**, 269–276.

Wharton, D.A. (1981) The initiation of coiling behaviour prior to desiccation in the infective larvae of *Trichostrongylus colubriformis*. *International Journal for Parasitology*, **11**, 353–357.

Wharton, D.A. (1982) Observations on the coiled posture of trichostrongyle infective larvae using a freeze-substitution method and scanning electron microscopy. *International Journal for Parasitology*, **12**, 335–343.

Wharton, D.A. (1986) *A Functional Biology of Nematodes*, London & Sydney: Croom Helm.

Wharton, D.A. (1994) Freezing avoidance in the eggs of the Antarctic nematode *Panagrolaimus davidi*. *Fundamental and Applied Nematology*, **17**, 239–243.

Wharton, D.A. (1995) Cold tolerance strategies in nematodes. *Biological Reviews*, **70**, 161–185.

Wharton, D.A. (1996) Water loss and morphological changes during desiccation of the anhydrobiotic nematode *Ditylenchus dipsaci*. *Journal of Experimental Biology*, **199**, 1085–1093.

Wharton, D.A. (1997) Survival of low temperatures by the Antarctic nematode *Panagrolaimus davidi*. In *Ecosystem Processes in Antarctic Ice-free Landscapes*, edited by W.B. Lyons, C. Howard-Williams and I. Hawes, pp. 57–60. Rotterdam: Balkema Publishers.

Wharton, D.A. (1998) Comparison of the biology and freezing tolerance of *Panagrolaimus davidi*, an Antarctic nematode, from field samples and cultures. *Nematologica*, **44**, 643–653.

Wharton, D.A. and Aalders, O. (1999) Desiccation stress and recovery in the anhydrobiotic nematode *Ditylenchus dipsaci*. (Nematoda: Anguinidae) *European Journal of Entomology*, **96**, 199–203.

Wharton, D.A. and Allan, G.S. (1989) Cold tolerance mechanisms of the free-living stages of *Trichostrongylus colubriformis* (Nematoda: Trichostrongylidae). *Journal of Experimental Biology*, **145**, 353–370.

Wharton, D.A. and Barclay, S. (1993) Anhydrobiosis in the free-living Antarctic nematode *Panagrolaimus davidi*. *Fundamental and Applied Nematology*, **16**, 17–22.

Wharton, D.A. and Barrett, J. (1985) Ultrastructural changes during recovery from anabiosis in the plant parasitic nematode, *Ditylenchus*. *Tissue & Cell*, **17**, 79–96.

Wharton, D.A., Barrett, J. and Perry, R.N. (1985) Water uptake and morphological changes during recovery from anabiosis in the plant

parasitic nematode, *Ditylenchus dipsaci*. *Journal of Zoology (London)*, **206**, 391–402.

Wharton, D.A. and Block, W. (1993) Freezing tolerance of some Antarctic nematodes. *Functional Ecology*, **7**, 578–584.

Wharton, D.A. and Block, W. (1997) Differential scanning calorimetry studies on an Antarctic nematode (*Panagrolaimus davidi*) which survives intracellular freezing. *Cryobiology*, **34**, 114–121.

Wharton, D.A. and Brown, I.M. (1989) A survey of terrestrial nematodes from the McMurdo Sound region, Antarctica. *New Zealand Journal of Zoology*, **16**, 467–470.

Wharton, D.A. and Brown, I.M. (1991) Cold tolerance mechanisms of the Antarctic nematode *Panagrolaimus davidi*. *Journal of Experimental Biology*, **155**, 629–641.

Wharton, D.A. and Ferns, D.J. (1995) Survival of intracellular freezing by the Antarctic nematode *Panagrolaimus davidi*. *Journal of Experimental Biology*, **198**, 1381–1387.

Wharton, D.A. and Lemmon, J. (1998) Ultrastructural changes during desiccation of the anhydrobiotic nematode *Ditylenchus dipsaci*. *Tissue and Cell*, **30**, 312–323.

Wharton, D.A. and Murray, D.S. (1990) Carbohydrate/lectin interactions between the nematophagous fungus, *Arthrobotrys oligospora*, and the infective juveniles of *Trichostrongylus colubriformis* (Nematoda). *Parasitology*, **101**, 101–106.

Wharton, D.A., Perry, R.N. and Beane, J. (1983) The effect of osmotic stress on behaviour and water content of the infective larvae of *Trichostrongylus colubriformis*. *International Journal for Parasitology*, **13**, 185–190.

Wharton, D.A., Perry, R.N. and Beane, J. (1993) The role of the eggshell in the cold tolerance mechanisms of the unhatched juveniles of *Globodera rostochiensis*. *Fundamental and Applied Nematology*, **16**, 425–431.

Wharton, D.A., Preston, C.M., Barrett, J. and Perry, R.N. (1988) Changes in cuticular permeability associated with recovery from anhydrobiosis in the plant parasitic nematode, *Ditylenchus dipsaci*. *Parasitology*, **97**, 317–330.

Wharton, D.A. and Ramløv, H. (1995) Differential scanning calorimetry studies on the cysts of the potato cyst nematode *Globodera rostochiensis* during freezing and melting. *Journal of Experimental Biology*, **198**, 2551–2555.

Wharton, D.A. and Surrey, M.R. (1994) Cold tolerance mechanisms of the infective larvae of the insect parasitic nematode, *Heterorhabditis zealandica* Poinar. *Cryo-letters*, **25**, 749–752.

Wharton, D.A. and To, N.B. (1996) Osmotic stress effects on the freezing tolerance of the Antarctic nematode *Panagrolaimus davidi*. *Journal of Comparative Physiology B*, **166**, 344–349.

Wharton, D.A. and Worland, M.R. (1998) Ice nucleation activity in the freezing tolerant Antarctic nematode *Panagrolaimus davidi*. *Cryobiology*, **36**, 279–286.

Wharton, D.A., Young, S.R. and Barrett, J. (1984) Cold tolerance in nematodes. *Journal of Comparative Physiology*, **B, 154**, 73–77.

Womersley, C. (1980) The effect of different periods of dehydration/rehydration periods upon the ability of second stage larvae of *Anguina tritici* to survive desiccation at 0% relative humidity. *Annals of Applied Biology*, **95**, 221–224.

Womersley, C. (1981a) The effect of dehydration and rehydration on salt loss in the second-stage larvae of *Anguina tritici*. *Parasitology*, **82**, 411–419.

Womersley, C. (1981b) The effect of myo-inositol on the ability of *Ditylenchus dipsaci*, *D. myceliophagus* and *Anguina tritici* to survive desiccation. *Comparative Biochemistry and Physiology*, **68A**, 249–252.

Womersley, C. (1987) A reevaluation of strategies employed by nematode anhydrobiotes in relation to their natural environment. In *Vistas on Nematology*, edited by J.A. Veech and D.W. Dickson, pp. 165–173. Hyattsville, Maryland: Society of Nematologists Inc.

Womersley, C.Z. (1990) Dehydration survival and anhydrobiotic survival. In *Entomopathogenic Nematodes in Biological Control*, edited by R. Gaugler and H.K. Kaya, pp. 117–137. Boca Raton: CRC Press.

Womersley, C. and Ching, C. (1989) Natural dehydration regimes as a prerequisite for the successful induction of anhydrobiosis in the nematode *Rotylenchulus reniformis*. *Journal of Experimental Biology*, **143**, 359–372.

Womersley, C. and Smith, L. (1981) Anhydrobiosis in nematodes I. The role of glycerol, myoinositol and trehalose during desiccation. *Comparative Biochemistry and Physiology*, **70B**, 579–586.

Womersley, C.Z. (1993) Factors affecting the physiological fitness and modes of survival employed by dauer juveniles and their relationship to pathology. In *Nematodes and the Biological Control of Insect Pests*, edited by

R. Bedding, R. Ackhurst and H. Kaya, pp. 79–88. East Melbourne: CSIRO Publications.

Womersley, C.Z. and Higa, L.M. (1998) Trehalose — its role in the anhydrobiotic survival of *Ditylenchus myceliophagus*. *Nematologica*, **44**, 269–291.

Womersley, C.Z., Wharton, D.A. and Higa, L.M. (1998) Survival biology. In *The Physiology and Biochemistry of Free-living and Plant-parasitic Nematodes*, edited by R.N. Perry and D.J. Wright, pp. 271–302. Wallingford & New York: CABI Publishing.

Wood, W.B. (1988) Introduction to *C. elegans* biology. In *The Nematode Caenorhabditis elegans*, edited by W.B. Wood, pp. 1–16. Cold Spring Harbour Laboratory.

Worland, M.R., Gruborlajsic, G. and Montiel, P.O. (1998) Partial desiccation induced by sub-zero temperatures as a component of the survival strategy of the Arctic Collembolan *Onychiurus arcticus* (Tullberg). *Journal of Insect Physiology*, **44**, 211–219.

Zachariassen, K.A. (1985) Physiology of cold tolerance in insects. *Physiological Reviews*, **65**, 799–832.

17. Ageing

David Gems

The Galton Laboratory, Department of Biology, University College London, 4 Stephenson Way, London NW1 2HE, England

Keywords

Ageing, nematode, evolution, *Caenorhabditis elegans*, genetics, life span

Introduction

This chapter aims to draw together the diverse elements of the study of ageing in nematodes into a single account, encompassing evolutionary biology, parasitology, nematology, gerontology and genetics. It also aims to provide an account of the biology of ageing in the model species *C. elegans* for those working on other areas of nematode biology, and information about ageing in other nematode species for *C. elegans* specialists. The tendency for separation into subdisciplines that exists in the study of nematode ageing also extends to model species studies themselves, where research on *C. elegans* has focused predominantly on the genetic specification of life span, and work on *C. briggsae* and *T. aceti* on other aspects, such as age-related changes in biochemical function and ultrastructure.

In the developing field of biological gerontology, rapid advances have recently been made in the genetics of ageing in *C. elegans*. The aim of this work is to develop an understanding of the general mechanisms determining the ageing process. Within the last 10 years the possibility of actually achieving this somewhat hubristic aim has begun to look startlingly real. The potential consequences for humanity of achieving this goal are great. In this context, knowledge of every aspect of the biology of ageing in nematodes is of added interest.

In nematology, the term ageing has been used to refer to several distinct phenomena: the loss of infectivity of developmentally arrested infective larvae of parasitic species, particularly hookworms (Croll and Matthews, 1973); the loss of infectivity over time of plant-parasitic nematodes in the soil, for example, non-feeding and/or anhydrobiotic forms; and to the age-related decline in viability in adults that eventually leads to death. This chapter deals only with ageing in the latter sense of the word.

Ageing and Life History Evolution

Life History Diversity in Nematodes

In contrast to the relative invariance in structure among the Nematoda, this phylum exhibits a remarkable degree of variation in patterns of life history. This includes a wide range of modes of sexual and asexual reproduction, including gonochorism (males and females) and parthenogenesis, which may be mitotic or meiotic. In addition there are hermaphroditic species which can reproduce either by self-fertilisation or by mating with males, and still others which exhibit pseudogamy, where eggs must be fertilised by a sperm in order to develop, but where the male pronucleus does not fuse with the female pronucleus, but instead breaks down. Pseudogamous species may be gonochoristic or hermaphroditic. Furthermore, many nematode life cycles involve developmentally arrested or quiescent stages, and these may be obligate or facultative. This high level of life history plasticity includes adult longevity.

In terms of the pattern of reproduction, life histories are sometimes divided into two sorts: semelparous life histories, where a single bout of reproduction is followed fairly rapidly by death; and iteroparous life histories, where multiple rounds of reproduction occur (Finch, 1990). Typically, longer-lived species are iteroparous, although there are exceptions, such as the 17 year cicada. Iteroparity most usefully refers to species reproducing seasonally over successive years. In the case of nematodes, species with very long-lived adults are exclusively parasitic, for example the hookworm *Necator americanus*, the patency period of which can last up to 14 years (Palmer, 1955). Here reproduction is continuous rather than intermittent, probably because host nutrients are continuously available. By contrast, the human pinworm *Enterobius vermicularis* is reproductively active for only a few weeks. Species such as these have been referred to as iteroparous and semelparous, respectively (Morand, 1996). However, in the absence of intermittent or seasonal reproduction in iteroparous species, the difference between a short-lived iteroparous species and a semelparous one is unclear.

Evolutionary Theories of Ageing

To understand the biology of ageing two sorts of questions must be answered. Firstly, why does ageing occur at all? How did it evolve? Does it have some adaptive value? Secondly, what actually happens during ageing? What mechanisms underlie it? Are there universal mechanisms of ageing or are there different primary causes of ageing in different animals groups? These two types of questions address the cause of ageing in evolutionary and mechanistic terms, respectively. While the modern evolutionary theory of ageing provides a convincing explanation for why ageing occurs, its mechanistic basis remains an unsolved puzzle.

In terms of evolution, ageing is likely to be the consequence of the inability of natural selection to eliminate mutations with effects detrimental to survival that are expressed at an age when, under natural conditions, most individuals will have already succumbed to extrinsic causes of mortality (e.g. predation, disease or starvation). By this view, ageing reflects the accumulation of late-acting deleterious mutations (Medawar, 1952), and is therefore akin to a late acting genetic disease. According to this interpretation, ageing is an entirely non-adaptive by-product of evolution, somewhat like nipples on men.

Since a given gene may affect many different aspects of the biology of an organism, some mutations may produce different pleiotropic effects at different times in the life history. In such cases, the earlier that the effects of such pleiotropic mutations are expressed, the greater their effects are likely to be on overall fitness. In theory, a pleiotropic mutation causing an early increase in reproductive fitness, but a reduction in life span, may actually increase overall fitness (Medawar, 1952; Williams, 1957; Partridge and Barton, 1993). Thus, decreases in life span potential may evolve as a consequence of antagonistic pleiotropy. As Medawar pointed out: 'A relatively small advantage conferred early in the life of an individual may outweigh a catastrophic disadvantage withheld until later.'

A related interpretation of the evolution of ageing is that it reflects the optimal partitioning of limited resources between somatic maintenance processes that assure longevity (e.g. antioxidant defence or DNA repair) on the one hand, and reproduction on the other (Kirkwood, 1977). By this view, a mutation that led to a shift in resource allocation from soma to germ line might increase overall fitness by increasing reproductive output, even though it shortened life span potential. Where extrinsic mortality rates are high, such a shortening of life span potential would have little impact on overall fitness if it affected animals at an age at which under natural conditions they would already have died from disease, or been eaten. For a detailed account of the evolutionary theories of ageing see Rose (1991), or Partridge and Barton (1993).

A key prediction of the evolutionary theory of ageing is that the maximum life span that particular species may attain under optimal conditions (e.g. in the laboratory) will reflect the life expectancy of that species in the wild, where mortality is entirely due to extrinsic causes. This is supported by the observation, in the survey of nematode life span below, that the longest-lived species are found among the parasitic nematodes. These exist under conditions of relatively low mortality, i.e. the protected environment and assured food supply of the host — assuming that the parasite possesses effective mechanisms to evade host immunity. The life span potential of parasitic nematodes that do evoke effective host immune responses

can sometimes be ascertained through the use of immunocompromised hosts, or by low level infections too small to provoke an effective host immune response.

Covariation in Nematode Life History Traits

A powerful tool for investigating the evolutionary basis of life history traits generally is the comparative method. This approach exploits the occurrence of parallel or convergent evolution. Clusters of life history traits may evolve in concert, and where this occurs independently in different taxa it is possible to identify independently evolved correlations between them. Thus, one may observe recurrent patterns of covariation of a given life history trait with particular environmental conditions and with other life history traits, and thereby understand how the trait in question has evolved to be as it is.

A number of comparative life history studies of nematodes have been carried out (Keymer et al., 1991; Skorping, Read and Keymer, 1991; Harvey and Keymer, 1991; Morand, 1996; Sorci, Morand and Hugot, 1997). Life history traits that have been examined include female size, fecundity (either daily fecundity or lifetime fecundity), duration of reproduction, time from infection to production of infective stages (prepatency, or maturation time in the case of free-living nematodes), and life expectancy. In these studies the confounding effects of phylogeny were adjusted for using contrast analysis (Harvey and Keymer, 1991). However, it is worth noting that the relationship between the major nematode taxa, using molecular phylogenetic analysis, has yet to be established definitively (Blaxter et al., 1998).

In studies of vertebrates, life history traits tend to covary along a fast-slow continuum, from small, short-lived species that have high somatic growth rates and mature early, have high fecundity and small young, to species with the opposite set of characters (Stearns, 1983; Read and Harvey, 1989). Studies of parasitic nematodes suggest a somewhat different continuum, where small species have low somatic growth rates, low fecundity and short reproductive periods (Skorping, Read and Keymer, 1991). For example, *Trichinella spiralis*, which is only a few millimetres long, starts producing progeny around five days after infection, and produces about 35 offspring per day for about two weeks. By contrast, large species have high somatic growth rates, delayed maturity, and high and extended fecundity, e.g. *Ascaris lumbricoides*, which grows to up to a third of a metre in length, and does not start to lay eggs until two months post-infection, but at a rate of up to 200,000 per day for up to a year and a half.

Certain key allometries observed among mammals are also seen among nematodes, e.g. a positive correlation of female body size with period of development and daily fecundity (Skorping, Read and Keymer, 1991; Morand, 1996). This suggests that delaying maturity leads to increased body size and therefore increased fecundity. When body size is controlled, no correlation between prepatent period

(period of development) and daily fecundity is observed, suggesting that the effect of increased prepatent period on fecundity is mediated entirely by the resulting increase in body size (Skorping, Read and Keymer, 1991).

While delaying maturity may increase fitness by increasing fecundity, it also increases generation time, thereby reducing population growth rate, and therefore may reduce evolutionary fitness. Thus, there may be a trade-off between fast generation time and high fecundity. However, given that parasitic nematode populations are typically relatively stable, population growth rate may be relatively unimportant to overall fitness (Anderson and May, 1982). In addition, there is no evidence for a trade-off between daily fecundity and patency period, consistent with high fecundity incurring no cost in terms of somatic maintenance (Keymer *et al.*, 1991; Morand, 1996).

Clearly, a key determinant of whether it is better to opt for short generation time and reduced fecundity, or longer generation time and high fecundity, is likely to be extrinsic mortality, either of larval or adult stages. Thus, whether it is better to have a small number of progeny soon, or a greater number later is likely to depend on the probability of dying while waiting to reach maturity, i.e. an increase in extrinsic mortality will favour short generation time and reduced fecundity. Factors affecting mortality rates in parasitic nematodes include intraspecific competition, host immunity and host longevity. Interestingly, comparisons of life history traits in Oxyurids (pinworms) and those of their primate hosts found a positive correlation between parasite body size and host life span, even when host body size was corrected for (Harvey and Keymer, 1991; Sorci, Morand and Hugot, 1997). However, given the great difference in magnitude of the life spans of primates and their pinworms (decades versus weeks) a direct effect of inter-species variation in primate life span on nematode life history traits does seem somewhat improbable.

In the pinworm studies, nematode body size was again positively correlated with daily fecundity and prepatent period. This suggests that reduced extrinsic mortality due to greater host longevity may result in the evolution of increased parasite body size via an evolutionary shift in the trade-off from short generation time to greater fecundity.

The role of mortality rate in driving life history evolution in nematodes was directly addressed in a comparison of life history traits in 35 nematode species, including free-living and parasitic species. Here adult mortality rate was estimated as the reciprocal of life expectancy. A significant negative correlation between adult mortality rate and prepatent period was seen among vertebrate parasitic species (Morand, 1996). A negative correlation was also seen between adult mortality rate and prepatent or maturation period in the free-living and insect- and plant-parasitic nematodes, but this did not reach statistical significance. A positive correlation between prepatent period and

body size was seen in all cases. Morand concludes that reduced adult mortality favours delayed maturity and increased fecundity.

Just how much of the variation in the prepatent period across the gastrointestinal nematode taxa can be explained in terms of trade offs between short generation time and greater fecundity has been recently addressed using an optimality model and estimates of pre-maturational mortality (Gemmill, Skorping and Read, 1999). Despite the fact that many biological details were not modelled, approximately half of the variation in the prepatent period could be explained in terms of this sort of trade off.

Given that in the Morand study covariance of life expectancy with other life history traits is examined, does this allow any conclusion to be drawn about covariation of life span potential and the rate of ageing with other life history traits? This is unclear. Morand's purpose in looking at covariance of mortality rate with other life history traits was presumably to understand how life histories evolve in response to different levels of *extrinsic* mortality. Thus, mortality rate is used as a measure of the hazardousness of the environment. Yet in the 27 nematode life expectancy estimates used, no attempt is made to distinguish between life expectancy values that reflect extrinsic mortality and those that reflect senescence. For example, the life expectancy estimate of the free-living nematode *Caenorhabditis elegans* (20 days) is clearly that of a population whose longevity is limited by senescence (Johnson and Simpson, 1985). For the purposes of comparative life history analysis one would want to distinguish life expectancy values that reflect levels of extrinsic mortality typically experienced by species, from those reflecting life span potential. Life expectancy in many parasitic nematodes clearly reflects strictly extrinsic mortality, largely inflicted by host immunity. However, there are some clear exceptions among species with effective mechanisms for evading host immunity, such as the filarial parasite *Onchocerca volvulus*, where adult life span potential is a major determinant of the length of patency (Remme, de Sole and van Oortmarssen, 1990; Plaisier *et al.*, 1991). Likewise, life expectancies of laboratory raised, free-living nematodes will reflect life span potential alone. Given a means of identifying species where life expectancy estimates give an indication of life span potential, application of comparative life history analysis (controlled for phylogenetic effects) would be informative with regard to the evolution of ageing.

Given that greater size and fecundity is a consequence of low mortality, larger, more fecund nematode species are likely to be longer-lived. This is so because reduced extrinsic mortality will allow natural selection to act on the effects of genes at later ages, resulting in the evolution of greater longevity. This is supported by the observation that as well as being smaller than parasitic nematodes on average, free-living nematodes generally age more rapidly (Morand and Sorci, 1998) (see below, Table 17.1, Table 17.2).

Comparative Biology of Ageing in Nematodes

The Problems of Observing Nematode Ageing

Ageing is one of the most difficult characteristics of animals to measure for a number of reasons. Firstly, unlike other characteristics such as size, fecundity or cell number, it is not clear what exactly ageing is, i.e. the nature of the biological processes underlying it are unknown. It is not even clear to what extent animals from different species dying of ageing are dying of the same cause. The one established universal feature of ageing is the increase in the rate of mortality with increasing age, and this is currently used as its defining characteristic (Finch, 1990). Unfortunately, measurements of age-specific mortality are almost impossible to make on most species of parasitic nematodes, and good mortality data currently exists for only a handful of free-living species. However, a number of other types of data do give at least a limited idea of nematode ageing, and to these we have to turn to know anything at all.

In the case of some free-living species there is cohort survival data giving, for example, estimates of average or maximum life span. However, several factors complicate observation of ageing, and comparisons between species of the rate of ageing. Firstly, in survival studies it is not always clear whether animals are dying as the consequence of ageing, or some other cause, e.g. non-optimal culture conditions, or bacterial infections. A second issue is that, as in other ectothermic organisms, nematode life span is dependent on temperature, being longer at lower temperatures, within the physiological temperature range (Zuckerman et al., 1971; Klass, 1977; Suzuki et al., 1978). Thirdly, life span is affected by behaviour, such as mating, which reduces life span (Honda, 1925; Suzuki et al., 1978; Gems and Riddle, 1996).

Observing ageing in parasitic nematodes is very difficult indeed. Here, estimates of life span potential and the rate of ageing can only be defined from observations of the duration of parasitic infections (the patency period). To be able to draw conclusions about the life span potential of adult parasitic nematodes, a number of factors have to be taken into account. Firstly, the length of patency can only reflect adult longevity if no reinfection occurs. Measuring patency in the absence of reinfection can be achieved either by carrying out experimental infections under conditions that avoid reinfection and periodically examining the infected host, e.g. by dissecting out the gut, or by observing the continued appearance of progeny, indicating the persistence of adults. Alternatively, in infections of humans, the course of infections can be followed in individuals who have moved away from geographical regions where reinfection may occur (Sandground, 1936). In some parasitic nematodes such measurements are not possible, either because of the occurrence of autoinfection, as in the human parasite Strongyloides stercoralis (Schad, 1989), or because not all infecting nematodes develop directly into adults, but instead arrest development for variable periods, as in the rabbit stomach worm Obeliscoides cuniculi (Watkins and Fernando, 1986).

The second factor complicating life span estimates in parasitic nematodes is that in many cases nematode life expectancy is merely a measure of the efficiency of host immunity, i.e. of extrinsic rather than intrinsic nematode mortality. However, some parasitic nematodes employ highly effective immune evasion mechanisms, such that maximum observed life spans may approximate life span potential as limited by ageing, e.g. the filarial parasite of humans Onchocerca volvulus (Remme, de Sole and van Oortmarssen, 1990; Plaisier et al., 1991). Furthermore, in some parasitic nematodes where patency period is normally limited by host immunity, some idea of life span potential may be obtained either by using immunocompromised hosts (Wakelin and Selby, 1974; Gemmill Viney and Read, 1997), or levels of infection too low to provoke a host immune response (Graham, 1938). Nonetheless, in most cases, as Caleb Finch has observed, '... there is no proof that infections of the very long lived parasitic nematodes are limited by senescence' (Finch, 1990). Thus, the period of patency is generally a mediocre indicator of longevity, telling us little about ageing, but rather giving only a lower estimate of life span potential.

Free-living Nematodes

Any comparative account of longevity and ageing among free-living nematode species is complicated by major environmental and behavioural effects on the rate of ageing. Life span may be increased by reducing temperature, food availability and reproduction. These factors are taken into account in the following survey, which is summarised in Table 17.1. Comparisons of life span values between species as shown can only give a very approximate idea of the relative differences between the longevity of species, due to differences in culture conditions, e.g. the food source, and handling methods and other conditions in different laboratories. Maximum life span is shown, since at a given temperature, where culture conditions are not highly deleterious to survival, variations in culture condition optimality are likely to cause greater variation in mean than maximum life span (Finch, 1990).

The earliest descriptions of ageing in free-living nematodes is by Maupas (1900). Estimates of life span are given for 12 species, of which nine are hermaphroditic and three parthenogenetic. A problem with Maupas' study is that while he gives the temperatures at which he measured the rate of development (which range from 12°C–27°C), it is unclear to what extent temperature varied during his life history studies. That temperature variation does occur during life span measurement is clear from his account of Cephalobus dubius, where he suggests that day to day variation in egg laying may be due to temperature variation (Maupas, 1900). Thus, the temperature given in the following description of Maupas' work on ageing are those at the start of the study. In most cases Maupas maintained his animals in drops of water to which ground meat was added as a food source.

The first species described is *Rhabditis elegans*, later renamed *Caenorhabditis elegans*. Maintained at 20°C, hermaphrodites lived 10–12 days, or 7–9 days from the onset of egg-laying. This is a very low estimate of *C. elegans* when compared to later studies (see below). Other hermaphroditic species follow. *R. caussani* is described as having the same life span as *C. elegans*. Slightly longer-lived was *R. marionis*; a single hermaphrodite studied lived 31 days, or 27.5 days from the onset of egg-laying. Similarly, *R. duthiersi* hermaphrodites lived from 20–23 days, or 16–19 days from the onset of egg-laying. In all these cases egg laying lasted about three days and was followed by a longer period of production of unfertilised eggs. *R. guignardi* hermaphrodites were cultured at 14°C, and lived for 27–28 days, or 21.5–22.5 days from the onset of egg-laying. *R. viguieri* was cultured at 22°C. Maupas describes a single hermaphrodite which died four days after the onset of egg-laying, from which he draws the somewhat shaky conclusion that this species is extremely short-lived. *R. dolichura* was grown at 23–24°C, and lived for 12–14 days or 9.3–11.3 days after the onset of egg-laying. *Diplogaster robustus* at 20°C lived 16 days, or 12.5 days after the onset of egg-laying, if internal hatching of eggs did not occur. *D. minor* at 17°C lived 16 days (from hatching), or 12 days from the onset of egg laying.

Three parthenogenetic species, one isolated in Germany, *Rhabditis schneideri*, and two from Algeria, *Cephalobus dubius* and *C. lentus* are also described. *R. schneideri* proved to be short-lived: at 12–13°C, it lived 18–19 days (approximately 15–16 days from the onset of egg-laying) and at 19–20°C, 8–9 days (or 6–7 days from the onset of egg-laying). By contrast, the two Algerian species were longer-lived. Here Maupas took the approach of picking a single egg and giving a detailed biography of the resulting worm. In the case of *C. dubius*, at 20°C this individual survived for approximately 150 days (or around 138 days from the onset of egg-laying), until finally, to Maupas' evident chagrin, it was

'...*devenu d'une inertie presque absolue, et s'éteignit d'épuisement sénile.*'

During the first four months of its life it laid 450 viable eggs, after which it continued to lay non-viable eggs. *C. lentus* was also long-lived. A single female survived 105 days (26–27°C), or 84.5 days from the onset of egg laying. It laid 315 viable eggs and was sterile during the last 40 days of life.

Life span has been examined in another parthenogenetic species in the Cephalobidae, *Acrobeloides nanus*. Longevity in *A. nanus* varies with temperature and nutrition. On full nutrition adults survived 23–27 days at 13°C and 15–21 days at 21°C. Under conditions of dietary restriction where the level of nutrients was cut by 75%, life span was extended to 21–42 days at 13°C and 12–30 days at 21°C (Sohlenius, 1973). Interestingly, under dietary restriction, the daily reproductive rate was reduced, but the life span reproductive output was

unchanged due to a lengthening of the reproductive period. Similar results were obtained for *Aphelenchus avenae* and *Tylenchus emarginatus* (Fisher, 1969; Gowen, 1970). By contrast, in *Mesodiplogaster biformis*, the rate of reproduction, the life time reproductive output and the life span were reduced when dietary restriction was imposed (Sohlenius, 1969).

Ageing has been studied extensively in the dioecious cephalobid nematode *Turbatrix aceti* (the vinegar eelworm). In one study, summed median life span of solitary males and female varied with temperature, from 104 days (211 maximum) at 15°C to 13 days (37 maximum) at 36°C (Vogel, 1974). Estimates of median life span in *T. aceti* vary between reports, from 55 days (25°C) (Vogel, 1974), to 25 days, either at 27°C (Kisiel and Zuckerman, 1972), or 30°C (Gershon, 1970), to 35–40 days (Zeelon, Gershon and Gershon, 1973). This variation may be due to strain variation or difference in culture conditions. In each case animals were cultured in liquid axenic medium (with no other organisms present), which tends to increase life span somewhat (see discussion below). In this species females and males have similar mean life spans: 73 and 66 days, respectively (Kisiel and Zuckerman, 1974). Life span was not significantly reduced in either sex by mating in axenic medium (Kisiel and Zuckerman, 1974).

Another dioecious cephalobid species is *Panagrellus redivivus*. In this species, as in most nematodes, females live longer than males, with mean life spans of 36 and 29 days, respectively (axenic medium, 25°C) (Abdulrahman and Samoiloff, 1975). The frequency of death from hypotonic shock (i.e. bursting when dropped into distilled water) also increases more rapidly with age in males than females. Mating generally reduces life span in both sexes, e.g. in females and males to 32 and 26 days, respectively, under the conditions just described. However, under nutritional conditions that slightly reduce life span in virgin animals, the effect of mating on survival is reduced or non-existent (Abdulrahman and Samoiloff, 1975). In studies of *P. redivivus* grown at 25°C on *E. coli* on agar plates, maximum life span of females and males was 20 days and 6 days, respectively (Duggal, 1978a; Duggal, 1978b). In these studies, mating shortened life span in females, but increased it in males.

The effect of mating on life span has also been examined in the dioecious species *Diplogaster aerivora* (Honda, 1925). Mating was found to reduce the mean life span of females from 52 days (range, 33–68 days) to 25 days (range, 11–54 days). Similarly, mating reduced mean male life span from 43 days (range, 15–71 days) to 33 days (range, 10–54 days) (Honda, 1925). In this study, animals were maintained in axenic liquid culture at 21°C until the end of egg laying, after which temperature varied from 22–25°C.

The tylenchid nematode *Ditylenchus triformis* has three sexes, males, females and an intersex form that behaves as a female (Hirschmann, 1962). Propagated on a fungal culture at 24–26°C, life span ranged from 31–124 days (mean, 63 days) in females, and 33–145 days (mean

Table 17.1. Variation of life span with temperature in free-living and plant-parasitic nematodes

Species	Sex[1]	Culture conditions[2]	Maximum life span (days)				Reference
			15°	20°	25°	30°	
Caenorhabditis elegans	H	Mo, S	32 (16°)	19	11 (25.5°)	–	Klass 1977
C. briggsae (G16)	H	Mo, S	–	21	–	–	Fletcher and Gems, unpublished
C. briggsae	H	Ax, L	30.0[3] (17°)	30.8 (22°)	26.2[3] (27)	–	Zuckerman *et al.* 1971
Caenorhabditis species[4]	F	Mo, S	–	38	–	–	Fletcher and Gems, unpublished
C. remanei[5]	F	Mo, S	–	58	–	–	Gems and Riddle, unpublished
Rhabditis tokai	F	Mo, S/L[6]	–	193	145	92	Suzuki *et al.* 1978
	M	Mo, S/L[6]	–	–	–	70	Suzuki *et al.* 1978
Rhabditis terracoli[7]	F	Mo, S	12	12 (22°)	–	–	Sohlenius 1968
	M	Mo, S	7	7 (22°)	–	–	Sohlenius 1968
Mesodiplogaster biformis	H	Mo, S	–	7.9[3] (21–22°)	–	–	Sohlenius 1969
Rhabditis elegans[8]	H	Ax, L	–	9[9] (12)	–	–	Maupas 1900
R. marionis	H	Ax, L	–	27.5[9] (31)	–	–	Maupas 1900
R. duthiersi	H	Ax, L	–	19[9] (23)	–	–	Maupas 1900
R. guignardi	H	Ax, L	22.5[9] (28) (14°)	–	–	–	Maupas 1900
R. viguieri	H	Ax, L	–	4[9] (6) (22°)	–	–	Maupas 1900
R. dolichura	H	Ax, L	–	–	11.3[9] (14) (23–24°)	–	Maupas 1900
R. schneideri	P	Ax, L	16[9] (19) (12–13°)	6–7[9] (9) (19–20°)	–	–	Maupas 1900
Diplogaster robustus	H	Ax, L	–	12.5[9] (16)	–	–	Maupas 1900
D. minor	H	Ax, L	12[9] (16) (17°)	–	–	–	Maupas 1900
D. aerivora	F	Ax, L	–	–	68 (21–25°)	–	Honda 1925
	M	Ax, L	–	–	71 (21–25°)	–	Honda 1925
Cephalobus dubius	P	Ax, L	–	138[9] (150)	–	–	Maupas 1900
C. lentus	P	Ax, L	–	84.5[9] (105)	–	–	Maupas 1900
Acrobeloides nanus	P	Mo, S	–	21 (21°)	–	–	Sohlenius 1973
Turbatix aceti	M, F[10]	Ax, L	211	137	123	70	Vogel 1974
	M, F	Ax, L	–	–	60.5 (27°)	–	Kisiel *et al.* 1972
Panagrellus redivivus	F	Ax, L	–	–	29.2–36.4[3]	24.5[3]	Abdulrahman and Samoiloff 1975
	M	Ax, L	–	–	21.0–29.3[3]	20.9[3]	Abdulrahman and Samoiloff 1975
Ditylenchus triformis	F	Fungal culture, S	–	–	124 (24–26°)	–	Hirschmann 1962
	M	Fungal culture, S	–	–	145 (24–26°)	–	Hirschmann 1962
Tylenchus emarginatus		Plant cell culture, S	128	66	60	–	Gowan 1970
Aphelenchus avenue	P	Fungal culture, S	–	48	37	32	Fisher 1969

[1] M, Male; F, Female; H, self-fertilising hermaphrodite; P, parthenogenetic female.
[2] S, solid medium (agar); L, liquid medium; Mo, monoxenic culture (*E. coli*); Ax, axenic culture.
[3] Mean life span.
[4] Previously *C. remanei*. Strain CB5161.
[5] *C. remanei* ssp. *vulgaris*, strain EM464. Previously known as *C. vulgaris*.
[6] Agar plates overlayed with 0.1 ml M9 buffer to reduce burrowing.
[7] Identification as *R. terracoli* in question (Sohlenius 1968)
[8] Later renamed *Caenorhabditis elegans*. Maupas' estimate of *C. elegans* life span is low compared to later estimates.
[9] From the onset of egg laying; life span including pre-adult development in parenthesis.
[10] Summed data from population of solitary animals, 40% males and 60% females.

74 days) in males. The life span of the intersex was intermediate. Females were sexually functional for about 75% of their life span, and showed retarded movement and anatomical changes in certain organ

systems later in life (Hirschmann, 1962). Another tylenchid species, *Tylenchus emarginatus*, feeds on plant root cells. Here again, life span varied with temperature, from a mean of 128 days (range 72–128 days) at 15°C, to 46 days (range 22–66 days) at 20°C, to 33 days (range 12–60 days) at 25°C (Gowen, 1970). In this study it was observed that while the rate of egg laying varied with temperature, lifetime fecundity did not. Gowen draws the unusual conclusion that longevity depends on the rate of reproduction, which is reduced at lower temperatures, or where food supply is reduced (Gowen, 1970).

The aphelenchid nematode *Aphelenchus avenae* is parthenogenetic and feeds on fungi. This species is relatively short-lived, with maximum life spans of 48, 37, 32 and 21 days at 20°C, 25°C, 30°C and 35°C, respectively (Fisher, 1969).

Among the free-living rhabditid nematodes, life span has been characterised in most detail in *Caenorhabditis elegans*. The life span of unmated hermaphrodites of this species grown on *E. coli* varies with temperature, from a mean of 9 days at 25.5°C to 35 days at 10°C (Klass, 1977). Hermaphrodite life span in the standard laboratory strain Bristol (N2) is also affected by genetic variation between different laboratory lines, with 20°C median life span varying from 11 days to 17 days (Gems and Riddle, 2000a). Dietary restriction increases life span and reduces life time fecundity in this species. Reducing bacterial concentration from 1×10^9 to 1×10^8 cells/ml increases 20°C life span from 16 to 26 days, and reduces mean brood size from 273 to 63 (Klass, 1977). Under conditions of full nutrition, hermaphrodite reproduction by self-fertilisation does not reduce life span: sterile *fer-15* and *fog-2* mutants are not longer-lived (Friedman and Johnson, 1988; Gems and Riddle, 1996), nor are sterile animals in which precursor cells of the germ-line and somatic gonad have been ablated with a laser microbeam (Kenyon *et al.*, 1993). However, mating with males reduces hermaphrodite median life span from 17 to 8 days (20°C) (Gems and Riddle, 1996). This appears to be due to the effect of copulation *per se* rather than an increase in egg production. In comparisons of groups of males and hermaphrodites, males are shorter-lived. For example, on *E. coli* lawns on agar plates male and hermaphrodite median 20°C life spans are 17 and 10 days, respectively (Gems and Riddle, 2000b). However, isolation of males increases life span up to 20 days, indicating that male-male interactions shorten life span. The biology of ageing in *C. elegans* is discussed in greater detail below.

Like *C. elegans*, *Caenorhabditis briggsae* is hermaphroditic. In axenic culture, mean life span of *C. briggsae* hermaphrodites was measured at 30 days, and 31 days, and 26 days at 17°C, 22°C and 27°C, respectively (Zuckerman *et al.*, 1971). In a limited study of the *C. briggsae* G16 strain, maintained in *E. coli* on agar plates at 20°C, life span was found to be similar to that of *C. elegans*, with median and maximum life spans of 14 and 21 days, respectively (B. Fletcher and D. Gems, unpublished results).

Another free-living rhabditid in which ageing has been examined in some detail was isolated from mud in a rice farm in the south of Japan, and designated *Rhabditis tokai* (Suzuki *et al.*, 1978). This species is unusual in that it is dioecious, but has a male: female ratio of around 1: 15. Female life span varies from 92–97 days at 20°C (maximum: 151–193 days) to 30 days at 35°C (maximum 63 days) (Suzuki *et al.*, 1978). Virgin males are shorter-lived than virgin females, with mean life spans of 38 days and 58 days, respectively (30°C). Mating with males reduces mean female life span from 92–97 days to 64 days (20°C). Fecundity in females was tested by exposure to males after 100 days at 20°C (slightly beyond average life span). 33% of females produced eggs (Suzuki *et al.*, 1978).

Another species of *Rhabditis*, isolated from soil in Sweden and identified as *R. terracoli*, proved to be much shorter lived: virgin females lived only up to 12 days, and males from 5–7 days, suprisingly at both 15°C and 22°C (Sohlenius, 1968). In another study by the same author, of hermaphrodites of the rhabditid nematode *Mesodiplogaster biformis*, a mean adult life span of 8 days (21–22°C) was observed (Sohlenius, 1969). Animals were maintained on *E. coli* lawns on agar drops. The coincidence of the unusually short life spans of these two species suggests either that free-living nematodes from high latitudes are especially short-lived (see below), or that the culture conditions employed by this investigator were not optimal for survival. Given that its life span appears to be approximately half that of *C. elegans*, *M. biformis*, which also forms dauer larvae, might be a good model organism for the study of ageing.

The observed longevities of free-living nematode species show a more than 10-fold variation at 20°C, from the short-lived *M. biformis* (mean hermaphrodite life span, 8 days) to the long-lived *R. tokai* (mean female life span, 92–97 days). Some of this variation is certainly due to variation between culture conditions: studies of the effects of culture conditions on *C. elegans* show that life span is increased in axenic liquid culture by up to 40% relative to culture on *E. coli* on agar plates (Croll, Smith and Zuckerman, 1977; Mitchell *et al.*, 1979). Furthermore, life span potential may be greatly reduced by non-optimal culture conditions. Nonetheless, it must be assumed that much of the variation in life span shown in Table 1 is due to variation in genetically determined life span potential, which has evolved due to differences in species ecology.

Why should free-living nematode species evolve life spans varying over such a wide range? According to the evolutionary theory of ageing, this variation in life span is likely to reflect the maximum age at which given species would be able to produce offspring in their normal habitat, even if they did not age at all. Natural selection will not act to remove from the population individuals carrying alleles of genes which are detrimental to survival after this maximum age.

One factor that may lead to the evolution of reduced longevity is protandrous hermaphroditism. Here, in the absence of mating with males, hermaphrodites become non-reproductive after the depletion of self-sperm. Studies of *C. elegans* suggest that the evolution of timing of the switch from sperm to egg reproduction is determined by a trade-off between the advantages of an early switch, which reduces generation time, but reduces self-brood size, and that of a later switch, which increases brood size, but increases generation time. The *tra-3(e2333)* mutation, which delays this switch and increases brood size by 50%, actually decreases the intrinsic rate of population growth, thus presumably reducing fitness (Hodgkin and Barnes, 1991).

A consequence of evolution for rapid population growth rate is a reduction in the reproductive period. In the absence of mating with males, the evolutionary theory of ageing suggests that this might result in the evolution of reduced longevity. In this context, it is interesting that the longer-lived species shown in Table 17.1 are generally dioecious or parthenogenetic rather than hermaphroditic. This issue was further addressed by comparing life spans of four species of *Caenorhabditis*: *C. elegans* and *C. briggsae*, which are hermaphroditic and *C. remanei* and *Caenorhabditis* species, which are dioecious. The longevity of females of the latter species greatly exceeds those of the hermaphrodites (Table 17.1) (D. Gems, B. Fletcher and D.L. Riddle, unpublished results).

Another possibility suggested by this survey is that nematode species from cold climates tend to be shorter-lived that those from warm climates. For example, the very short-lived *R. terracoli* and *M. biformis* were isolated from woods in Sweden, while the long-lived *R. tokai* was isolated from a rice field in Southern Japan, and *C. dubius* and *C. lentus* are from Algeria. In this context, it is worth observing that if two species, one from a cold climate and another from a hot one, have the same life expectancy in the wild, and consequently evolve similar life span potentials at their native temperature, then if longevity of these two species were compared at the same temperature, it would be expected that the species from the warm climate would be longer-lived.

Parasitic Nematodes

Compared to free-living nematodes, relatively little is known about ageing in parasitic nematodes, particularly species with invertebrate or plant hosts. Nematodes infecting higher animals can largely be divided into two groups: those that reproduce in the alimentary canal, and tissue dwelling species (although some do both). Intestinal parasitic nematodes include the strongylids (e.g. hookworms), the ascarids, oxyurids, some of the Strongyloididae, and the adenophorean trichocephalids (e.g. whip-worms). Tissue dwelling species of medical and veterinary importance are predominantly members of the filarid group (filarial worms), such as *Wuchereria bancrofti*, which causes lymphatic filariasis (elephantiasis).

Endoparasitic nematodes with mammalian hosts include some of the longest-lived nematodes known, in some cases living up to 20 years. Several factors may contribute to the evolution of such great longevity. Firstly, a low level of mortality due to a protected environment, and a constant food supply. Secondly, a relatively low level of intraspecific niche competition. This is assured in many species by a life cycle that precludes autoinfection. Thirdly, the longevity of the host species. And fourthly, the capacity to evade host immunity.

The Strongylida

The Strongylida are an order of vertebrate parasitic species, mostly intestinal, that includes the hookworms. Phylogenetic analysis using small subunit ribosomal DNA sequences indicate that the order Strongylida is most closely related to the suborder Rhabditina (Blaxter *et al.*, 1998), which includes *C. elegans*, in which the biology of ageing has been intensively studied (see below). Thus, ageing in strongylid nematodes is of special interest to those studying the biology of ageing.

Data on longevity in animal-parasitic strongylid species is largely derived from studies of experimental infections. Infections are typically relatively short-lived and in many cases are cleared relatively quickly, usually as the result of host immunity. In the case of *Ancylostoma caninum*, experimental infections of young dogs cease after 43–100 weeks (Sarles, 1929a). Egg-production in infected animals exhibits an exponential decay with time, corresponding to a constant rate of loss of adults from the gut. The period of survival before expulsion for most adult worms was estimated at only a few months. Similar results were obtained from experimental infections of dogs and cats with *Ancylostoma braziliense*: most adults survived less than two weeks in a young cat, but a few survived as long as 32 weeks, as measured by the disappearance of eggs from the faeces (Sarles, 1929b). Significantly, when the host was autopsied, the one female adult worm recovered was found to contain fertilised eggs, suggesting that infections with *A. braziliense* do not cease as the result of nematode ageing. That this is so is also suggested by the constant rate of loss of adults from the gut. *A. braziliense* survive up to 30 weeks in dogs. Thus, 100 weeks for *A. caninum* and 32 weeks for *A. braziliense* represent minimum estimates of potential life span in these species.

The same is likely to be the case for estimates of maximum life span in other strongylid parasites of animals. In *Haemonchus contortus*, the twisted wireworm of ruminants (principally sheep), the length of patency varies with the intensity of infections. Intense infections are cleared within five months (Stoll 1929, cited in Sandground 1936), in a process called self-cure. In sheep largely protected against reinfection, infections can last for more than two years (Adams and Beh, 1981). In one experimental infection of a male calf, *H. contortus* eggs had almost disappeared from the faeces after 15 months (Mayhew, 1942). One instance of an animal-parasitic strongyle where adults are not necessarily expelled from the host is *Heligmosomoides polygyrus* (*Nematospiroides dubius*). In infections of NMRI mice, adult populations showed no reduction in size over a 114 day period, although fertility did decline (Kerboeuf, 1982).

The definitive host of the strongylid nematode *Uncinaria lucasi* are young fur seals (*Callorhinus ursinus*), which are infected by larvae received post partum in their mothers' milk. Autopsies of fur seal pups performed at intervals throughout the breeding season showed that adult *U. lucasi* are cleared from the gut within three months of birth (Olsen and Lyons, 1965). The longevity of *U. stenocephala* in infections of dogs was estimated at 30 weeks (Fülleborn 1926, cited in Sarles 1929). In the metastrongylid *Oslerus (Filaroides) osleri*, which causes tracheobronchitis in dogs, tracheal nodules containing adults disappeared 5 months after the first presentation of the infected animal (Pillers, 1935). Infections of rodents by the trichostrongylid *Nippostrongylus brasiliensis* are cleared by 16 days after inoculation (Mayberry, Bristol and Villalobos, 1985), but in low level infections this nematode can survive much longer (R.M. Maizels, personal communication).

Two main approaches have been taken to estimating longevity in human parasitic species of hookworm: the survey of levels of infection among inmates of institutions such as prisons and mental hospitals, and through experimental self-infections by investigators. Mhaskar (1920) examined the prevalence of *Ancylostoma duodenale* among prisoners in Trichinopoly Jail in Madras, India, the sanitary conditions of which were believed largely to prevent reinfection (Mhaskar, 1920). Comparing prisoners whose period of stay varied from 0–5 days up to 16–17 years, the level of infection was found to decline with increasing length of incarceration. After five years infections had almost ceased, but rare animals were found even after 16 years. A similar study in Alipore Jail in Calcutta of *A. duodenale* and *Necator americanus* yielded similar results: egg output by infected prisoners was reduced by 50% within three months, by 70% within a year, by 90% within four years, and by 95% within eight years (Chandler, 1925). However, the percentage of infected individuals was only appreciably reduced after five years, with rare cases of infection being found even after 20 years of incarceration. Chandler draws the following conclusions: that the majority of hookworms are lost from the gut before reaching the limit of their life spans; that infections found after 20 or even 10 years are likely to result from low level reinfection; and that the likely limits of hookworm life span is 6–7 years. This is supported by a study from Georgia State Sanatorium in the U.S., where the percentage of hookworm infections did not drop until after six years of residence, after which they did so markedly (Willets, 1917). Sandground (1936) refers to unspecified experiments by Kendrick showing that *A. duodenale* can persist for at least seven years.

In the Mhaskar study, infections of *N. americanus* mostly ceased after about seven years, with no animals detected after 12 years. A higher estimate of longevity in *N. americanus* is given in a report of an experimental infection by Palmer (1955), presumably of himself. A plateau of egg production was reached by 11 months after infection, which lasted six years, and then gradually declined until the 14th year (Palmer, 1955) (Figure 17.1). No eggs were detected in the 15th year. The rapid reduction in the level of infection seen among the Indian prisoners is not seen in the Palmer case. This is unlikely to be due to the lower intensity of infection in the latter, whose initial worm burden was estimated at around 200 adults. Similarly, egg production was high, approximately $3–9 \times 10^3$/gram of faeces during the infection plateau (Palmer, 1955), compared to an average of approximately 1×10^3 eggs/gram in 27 new, hookworm infected prisoners at Alipore Jail, dropping to 290 eggs/gram after only a few months (Chandler, 1925). So is the maximum life span of *N. americanus* 6–7 years or 14 years? Possibly the difference between the American and Indian results may reflect differences in health and nutrition, i.e. Eddy Palmer was a more congenial host for his hookworms, while the unfortunate Indian prisoners, enduring an inferior diet, and probably more frequent gut infections and diarrhoea, would have made poorer hosts. If so then the higher estimate of maximum life span is more reliable. On the other hand, the Palmer finding involves a single infection. Furthermore, the possibility cannot be excluded that over such long periods rare instances of reinfections occurred, resulting in prolongation of the period of infection. Two other light, self-infections with *Necator* by parasitologists lasted over four years, and less than six months (Chandler and Read, 1961). Another experimental self-infection was carried out by Sandground (1936) with a species of *Trichostrongylus* originating from Zimbabwe, possibly *T. colubriformis*. No diminution of egg-laying was observed 8.5 years after infection (Sandground, 1936).

The Ascaridida

Compared to the hookworms, the typical patency periods of ascarid nematode (roundworm) infections are relatively short. The life expectancy of *Ascaris*

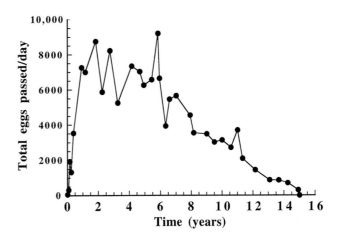

Figure 17.1. Egg output in a case of *Necator americanus* infection. Evidence from an experimental infection demonstrates the maximum fecundic life span of *N. americanus* to be at least 15 years. Data derived from Palmer 1955.

Table 17.2. Minimum estimates of adult longevity in parasitic nematode species

Species	Maximum adult life span	Class of life span estimate[1]	Host	Notes	Reference
Secernentea					
Ascaridida					
Ascaris lumbricoides	1–2 years	D	Humans		Croll *et al.* 1982
Pseudoterranova decipiens	2–3 weeks	D (C)	Seals		des Clers 1990
Toxocara canis	3–6 months	D (C)	Dogs		Lloyd 1193
Heterakis spumosa	> 10 months	D	Mouse		Winfield 1933, in Sandground 1936
Spirurida (filarids)					
Brugia malayi	8.5 years	(A)	Humans	Causes lymphatic filariasis	Wang *et al.* 1994
Brugia pahangi	7–8 years	(A)	Humans	Causes lymphatic filariasis	Wilson and Ramachandran 1971
Acanthocheilonema viteae	25 months	D	Jirds		Johnson *et al.* 1974
Dirofilaria immitis	7 years	(A)	Dogs	Heartworm	Sandground 1936
Dracunculus insignis	36, 47 weeks	D, E (males)	Ferrets	Females, males	Brandt and Eberhard 1990a, b
Litomosoides sigmodontis[2]	3 years	D	Cotton rats		Olsen 1974
Loa loa	17–20 years	(A)	Humans	Causes lymphatic filariasis	Manson-Bahr 1925; Coutelen 1935
	9 years	(A), E	Patas monkey		Orihel and Eberhard 1985
Onchocerca volvulus	13–14 years	A	Humans	Causes river blindness	Plaisier *et al.* 1991
Wuchereria bancrofti	15–16 years	(A)	Humans	Causes lymphatic filariasis	Trent 1963
Strongloididae					
Rhabdias bufonis	3, 5 days	–	FL	Males, females	Spieler and Schierenberg 1995
	3 months	D, E	Amphibians	Parthenogenetic	Goater 1991
R. fuscovensa	9 months	D, E	Snakes	Partenogenetic	Chu 1936
Strongyloides fuelleborni	9, 15 days	–	FL	Males, females	Augustine 1940
S. ratti	2 weeks	–	FL	Males, females	Gemmill *et al.* 1997
	55 weeks	B, E	Rats	Parthenogenetic	Graham 1938
S. simiae	12 days	–	FL	Females	Augustine 1940
S. stercoralis	> 2 days	–	FL	Parthenogenetic	Yamada *et al.* 1991
Strongylida (hookworms)					
Ancylostoma braziliense	> 32 weeks	D (C)	Cats		Sarles 1929b
A. caninum	100 weeks	D (C)	Dogs		Sarles 1929a

Species	Life span		Host / notes	Reference
A. duodenale	15 years	(A)	Humans	Hyman 1951
Haemonchus contortus	> 2 years	D (C)	Sheep	Adams and Beh 1981
Heligmosomoides polygyrus	114 days	A, E	Mice	Kerboeuf 1982
Necator americanus	14 years	(A)	Humans	Palmer 1955
Nippostrongylus brasiliensis	16 days	C	Rats	Mayberry *et al.* 1985
Osterus osleri (filaroides)	4–5 months	D	Dogs	Pillers 1935
Trichostrongylus	8 years	(A), E	Humans	Sandground 1936
Uncinaria lucasi	< 3 months	D	Fur seals	Olsen and Lyons 1965
Uncinaria stenocephala	30 weeks	D	Dogs	Fülleborn 1926, in Sarles 1929

Possibly *T. colubriformis* (note to *Trichostrongylus*)

Oxyuridae (pinworms)

Species	Life span		Host / notes	Reference
Enterobius vermicularis	2 months	D (C)	Human	Sandground 1936

Adenophorea

Trichocephalida (whip-worms)

Species	Life span		Host / notes	Reference
Trichinella spiralis	3 months	D (C)	Rats	Maggenti, pers.comm. in Finch 1990
	5 weeks	D (C)	Guinea pigs	McCoy 1932
Trichuris muris	10 weeks	B	Mice	Wakelin and Selby 1974

Mermithidae

Species	Life span		Host / notes	Reference
Agamermis decaudata	> 18 months	–	Insects, snails	Christie 1929

Free-living, non-feeding females

Dioctophymoideae

Species	Life span		Host / notes	Reference
Dioctophyme renale	3 years	D	Dogs	Olsen 1974

Giant kidney worm

1 The degree to which maximum reported life span of parasitic species reflects the life span potential of the species (i.e. longevity as limited to ageing) varies between species. To give an indication of the reliability of maximum reported life span as an indicator of ageing estimates are classified to five groups. (A) Species with effective host immune evasion mechanisms, where maximum life span is not limited by host immunity. (B) Species where patency is usually limited by host immunity, but where adult parasite survival data has been obtained from immunocompromised hosts or trickle infections. (C) Species where reported life span is likely to reflect host immunity alone. (D) Species where relative contribution of host immunity and nematode ageing to nematode life expectancy is unclear. In (A) and (B) species maximum reported life span is a better indicator of life span potential than in (C) and (D) species. (E) Species were maximum life span estimation was limited by the length of the observation period. Parentheses indicate tentative designation.

2 Prevously *L. carinii*.

lumbricoides, which live in the intestinal lumen of man, is of the order of a few weeks to a month (Croll *et al.*, 1982). Infections generally clear within a year (Otto, 1932; Keller, 1931), although some authors estimate *A. lumbricoides* survival at 1–2 years (Croll *et al.*, 1982). Possibly the failure to persist of *A. lumbricoides* reflects the lack of a fully effective mechanism for adult worms to anchor themselves within the gut and avoid expulsion due to the action of peristalsis (Otto, 1932). By contrast, hookworms are able to anchor themselves to the gut by burrowing their head into the intestinal mucosa. Likewise, *Trichuris* species anchor themselves to the gut wall by means of the whip-like anterior end of the body. However, *Trichuris* and *Trichinella* appear relatively short-lived (see below). Adults of the ascarid *Toxocara canis* are found in lactating bitches and pups (Lloyd, 1993). Infections in pups are cleared within 3–6 months of birth. Other estimates of life span in ascarid species are 2–3 weeks for the sealworm *Pseudoterranova decipiens,* and 279 days (life expectancy) for *Parascaris equorum* is (Mozgovoy 1953, cited in Morand, 1996). In conclusion, nothing is really known about ageing in this major group.

The Spirurida

The filarial parasitic nematodes include a number of extremely long-lived species, particularly among those infecting human beings. Estimates of the longevity of adult filariid worms in human infections come largely from studies of infected individuals after long periods during which no reinfection could occur. Since in this group autoinfection does not occur, the longevity of the infection must reflect the longevity of the adult worms.

The filariid nematode species in which ageing is best characterised is *Onchocerca volvulus*, which causes river blindness in people. In a number of areas endemic for onchocerciasis, reinfection has been prevented by eradication of the secondary host vector, the blackfly (*Simulium* species), which transmits the infective third stage larva of *O. volvulus*. In Kenya, first stage larvae, or microfilariae (mfs) were detected in human hosts 9 and 11 years after blackfly eradication, but not after 18 years

(Roberts *et al.*, 1967). Mfs themselves are able to survive for up to about 1.5 years after the death of the last adult. The most extensive studies of ageing in *O. volvulus* come from examination of infected individuals after vector eradication by the W.H.O. Onchocerciasis Programme in West Africa. Data on the changing prevalence of mfs and adults was incorporated into an epidemiological model of *O. volvulus* infection, taking into account a number of variables affecting the course of infections, including the reproductive life span of *O. volvulus*, which limits the period of infection (Remme *et al.*, 1990; Plaisier *et al.*, 1991). Good fits were obtained between the predictions of this model and skin microfilarial loads from infected populations. Model predictions and field data indicate that the average reproductive life span of *O. volvulus* is 9–11 years, with 95% of animals ceasing reproduction within 13–14 years (Plaisier *et al.*, 1991) (Figure 17.2). In this model, variation in adult life span was described in terms of probability distribution of the Weibull type. Significantly, at the end of infections, 12–14 years after vector eradication, the prevalence of infection declined at an accelerated rate (Remme, de Sole and van Oortmarssen, 1990), consistent with senescence in *O. volvulus* adult populations. An example of the change of prevalence of *O. volvulus* mfs with time since vector eradication is shown in Figure 17.3. *O. volvulus* reproductive life span has also been studied in experimental infections of chimpanzees (Duke, 1980). These gave shorter estimates of life span: mfs disappeared from the skin of four animals after 6–9 years.

O. volvulus adults (macrofilariae) are largely found in sub-dermal nodules (onchocercomata) containing between one and several dozen animals. These nodules persist after the death of the adult worms. Ageing in *O. volvulus* adults has been studied by surgical excision of nodules and isolation of worms from populations at varying intervals after vector eradication. In comparisons of populations 7–10 years after vector eradication with those in endemic areas, 30% of nodules from the former, and 90% from the latter group, respectively, contained macrofilariae (Karam *et al.*, 1987).

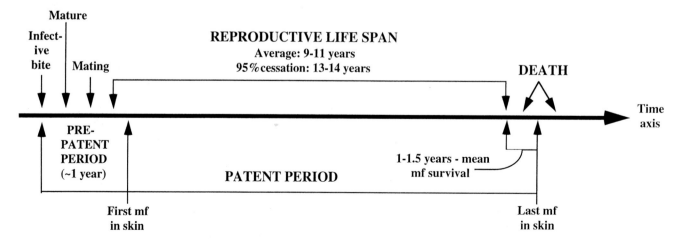

Figure 17.2. Life history of *Onchocerca volvulus* in the human host. mf = microfilariae. Adapted from Remme *et al.* 1986.

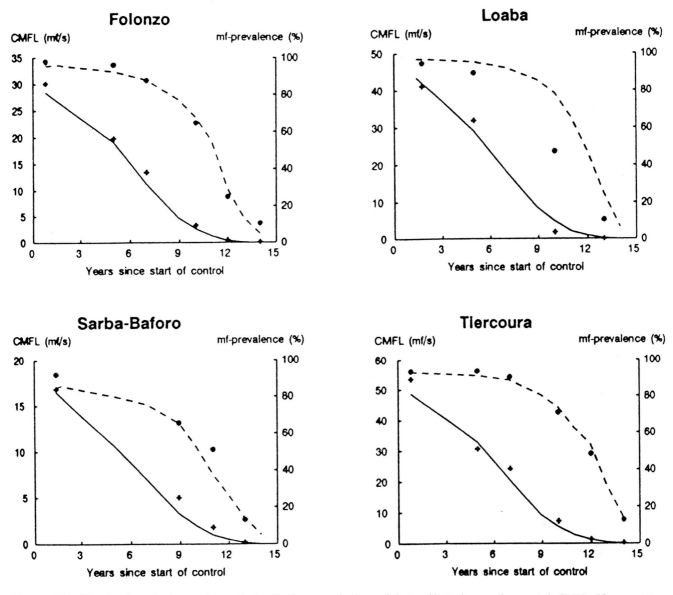

Figure 17.3. Simulated and observed trends in *Onchocera volvulus* mf (microfilariae) prevalence and CMFL (Community Microfilarial Load) in four West African villages. Symbols: filled circles = mf prevalence observed; broken line = mf prevalence simulated; + = CMFL observed; unbroken line = CMFL simulated. CMFL is expressed as the mean number of mfs per biopsy (skin snip) for all individuals over 20 years of age. The decline of mf-prevalence reflects the declining survival of adult *O. volvulus*. Reproduced with permission from Plaiser *et al.* 1991.

Increasing numbers of macrofilariae from populations 7–10 years after vector eradication exhibited signs of ageing. While young adult worms were largely transparent, and with an almost clean cuticle, older worms became increasingly opaque, and brownish in appearance (Karam, Schulz-Key and Remme, 1987), accumulating iron pigment in the body wall or the uterus (Büttner, Albiez and Parrow, 1983). Studies of older *O. volvulus* adults under the electron microscope show high densities of concentric spherules containing iron in the intestinal epithelium, uterine muscles, and to a lesser extent the uterine epithelium (Franz and Büttner, 1983). As adult females became older increasing amounts of collagenase-resistant incrustations accumulated on all regions of the cuticle apart from at the anterior end (Schulz-Key, Jean and Albiez, 1980). In addition, whitish inclusions accumulated within the muscular layer of the body wall, and in the body fluid. The final stage of these changes was body calcification and death. It is likely that these changes to some extent represent the cumulative effects of interactions with the host resulting from the sessile habit of female worms: male worms are more motile (searching for females) and retain a clean, smooth cuticle (Schulz-Key, Jean and Albiez, 1980). They are also less prone to calcification: in one survey of adult worms in an endemic area, 8% of females but only 0.5% of males were calcified (Albiez, Büttner and Schulz-Key, 1984). The motility of the anterior end of the female during mating probably explains their reduced accumulation of cuticular incrustations. That calcification is not a senescence-specific form of deterioration is demonstrated by a dramatic increase in the number of calcified worms after treatment of onchocerciasis sufferers with an anthelmintic (Wolf et al., 1980).

Of adult females it is estimated that 20% die prematurely as the consequence of calcification (Karam, Schulz-Key and Remme, 1987; Remme, de Sole and van Oortmarssen, 1990). Mean adult longevity is estimated to be 11 years for non-calcifying and 8 years for calcifying worms. The epidemiological simulation includes distinct Weibull distributions of longevity for calcified and non-calcified worms (Remme, de Sole and Oortmarssen, 1990).

The epidemiological model of onchocerciasis suggests that adult fecundity remains constant for 8 years and then slowly declines (Remme, de Sole and Oortmarssen, 1990). The change in fertility of female adult O. volvulus with age was also investigated by looking at the contents of the uterus. Among worms from endemic areas, in 21% of females the uterus was empty (Karam, Schulz-Key and Remme, 1987). By contrast, 9–10 years after blackfly eradication this had increased to 47%. The reduction in fertility to an unknown degree results from a decrease in mating: with increasing age an increasing proportion of males contained undelivered sperm, and a few very old males, with brownish bodies, were found with empty testes (Karam, Schulz-Key and Remme, 1987). While mfs were not observed in dead worms, old females which were still alive but with calcified tails were observed which contained remnant mfs. Taken together this suggests that the decrease in fecundity with age is due more to the reduced frequency of reproductive cycles, rather than decreased egg and sperm production (Karam, Schulz-Key and Remme, 1987). This suggests that the actual life span of O. volvulus does not greatly exceed its reproductive life span (Figure 17.2).

In nodules from endemic areas, females outnumber males by a factor of around 1.5 (Albiez, Büttner and Schulz-Key, 1984; Karam, Schulz-Key and Remme, 1987). Among ageing adult O. volvulus populations this ratio progressively decreases, suggesting that male life expectancy exceeds that of females (Karam, Schulz-Key and Remme, 1987). This is consistent with the observation that males exhibit a lower level of cuticular incrustations and calcification (Schulz-Key, Jean and Albiez, 1980; Albiez, Büttner and Schulz-Key, 1984). However, an alternative explanation is that males merely become less peripatetic with increasing age, and therefore appear in nodules with higher frequency.

The pattern of great longevity among filarial parasites of man continues in species which cause lymphatic filariasis (including elephantiasis), such as Loa loa, Wuchereria bancrofti and Brugia malayi. Regarding L. loa, there are numerous early reports, particularly before the availability of effective anthelmintics such as diethylcarbamazine (DEC), involving cases where Europeans or Americans returned from the tropics

with filariasis infections which endured for long periods (reviewed in Coutelen, 1935). Migrating L. loa adults have the unsettling habit of emerging from under the eyelids of their host. In one interesting case, an infected doctor extracted six virgin L. loa females from his own eyes by tying a silk thread around one end and slowly pulling them out, at intervals over a period of 10 years, the last one 15 years after leaving an area endemic for L. loa (Eveland, Yermakov and Kenny, 1975). In another case, of a man infected with both L. loa and Mansonella perstans, worms were also removed from the eyes over a 15 year period (Knabe, 1932). Earlier reports describe L. loa infections lasting 17 years (Manson-Bahr, 1925), and even 20 years (Connal 1923, cited in Coutelen, 1935). Taken together this suggests that L. loa adults can survive for at least 17–20 years, and that the longevity of this species is comparable to that of domestic dogs (maximum life span [beagles], 19 years).

Measurements of life span of adult L. loa have also been carried out by means of experimental infections of primates (Orihel and Eberhard, 1985). One African patas monkey (Erythrocebus patas) still microfilaremic after nine years, was found to contain three males and one gravid female adult L. loa.

Lower estimates of mean expected fecund life span of W. bancrofti derive from studies of infected populations in India after programmes of eradication of the mosquito vector: 10.2 years (Vanamail et al., 1990), 5.4 years (Vanamail et al., 1989), and 5.0 years (Vanamail et al., 1996). No explanation is offered by the authors of these studies for the two-fold difference in their estimates. Case histories of infected individuals have given a range of estimates of maximum life span for this species: 5 years (Bancroft, 1879; Guptavanij and Harinasuta, 1971), 6 years (Jachowski, Otto and Wharton, 1951), 7.5 years (Conn and Greenslit, 1952), and 8 years (Leeuwin, 1962; Mahoney and Aiu, 1970). In one study of U.S. ex-servicemen 15–17 years after exposure to W. bancrofti in the Pacific, increasingly severe symptoms of filariasis were observed, although mfs were not observed (Trent, 1963). Whether this is due to the presence of post-reproductive adult filarariae is unclear. Manson-Bahr estimates W. bancrofti life span as 17 years (Manson-Bahr, 1959). One interesting case history involving a woman who lived in Tahiti until 1935 is worth describing in more detail. This individual was examined again in 1975: six smears of 20 mm^3 capillary blood did not reveal mfs, but in 5 mls of venous blood, two mfs of W. bancrofti var. pacifica were found (Carme and Laigret, 1979). In the intervening years she had not visited any areas endemic for W. bancrofti. There are several alternative explanations for this unusual observation. The first, suggested by Carme and Laigret, is that W. bancrofti mf levels decline in an exponential rather than a linear fashion, such that levels of mfs drop below a concentration detectable by the standard method of examining 20 mm^3 of blood, and that the mfs can be extremely long-lived. Other possibilities are that some adults of the Tahitian strain of W. bancrofti survive to advanced ages (up to 40 years), or that the patient gave an unreliable account of her movements, and had visited areas endemic for filariasis in the interim.

In the case of *Brugia malayi*, from studies of populations in areas endemic for the periodic form after vector control, the mean fecundic life span has been estimated at 3.5 years (Krishnamoorthy *et al.*, 1991; Sabesan *et al.*, 1991). One study in which a researcher experimentally infected himself and two members of his family with periodic *B. malayi*, mfs were detected for 8.5 years (Wang *et al.*, 1994). From the report of this work it is horribly clear how much all three subjects suffered in this study. They must be admired for their self-sacrifice. In two cases of individuals in Thailand carrying *B. malayi* who left endemic areas, circulating mfs disappeared in one after seven years, while the other was still microfilaremic after seven years (Gupta-vanij and Harinasuta, 1971). Whether these infections involved the periodic or aperiodic forms is not stated. *Brugia* longevity has also been estimated by means of experimental infections of cats. Here infections of periodic and aperiodic *B. malayi* persisted approximately 4 years and 4.5 years, respectively, while in *B. pahangi* infections mfs were detectable for 7–8 years (Wilson and Ramachandran, 1971).

Another tissue-dwelling spirurid nematode is *Dracunculus insignis*, which is an animal-parasitic relative of the human guinea worm *Dracunculus medinensis*. Females of *D. insignis* in infections of ferrets are fertile 250 days post-infection (Brandt and Eberhard, 1990a), and the much smaller male worms survive up to 330 days post-infection (Brandt and Eberhard, 1990a; Brandt and Eberhard, 1990b).

Among animal-parasitic spirurid species, estimates of adult life span are, unsurprisingly, less than in human parasitic species. In infections of dogs with *Dirofilaria immitis* (dog heartworm), mfs were observed seven years after infections (Fülleborn 1929, cited in Sandground, 1936). Olsen (1974) gives an estimate of 2–5 years for the reproductive period of adult *D. immitis*.

In infections of cotton rats with *Litomosoides sigmodontis* (previously *L. carinii*) the majority of adult worms died within the first year, with a minority surviving for about three years (Olsen, 1974). Adults of another rodent-parasitic species, *Acanthocheilonema viteae*, were found to survive for up to two years in jirds (*Meriones unguiculatus*) (Johnson, Orihel and Beaver, 1974). One 25-month old female examined contained mfs in the uterine branches, and spermatozoa in the seminal vesicle, indicating that she was not senescent at this age.

The Strongyloididae and Rhabdiasidae

This group includes the genus *Strongyloides* which infects a wide range of hosts, from reptiles and amphibians to birds and mammals, and *Rhabdias*, which infects reptiles and amphibians (Speare, 1989). Human strongyloidiasis, or Cochin China diarrhoea, caused by *S. stercoralis*, can be fatal in debilitated patients, and infection with *S. fuelleborni kellyi* can prove fatal in infants (Ashford and Barnish, 1989). In terms of the biology of ageing, the Strongyloididae and Rhab-

diasidae are particularly interesting since most species include both parasitic and free-living adult forms. The scanty data that exists on longevity and ageing in the Strongyloididae and Rhabdiasidae suggest that the parasitic adult forms are much longer-lived than the free-living adult forms of the same species.

The principal laboratory model species of *Strongyloides* is *S. ratti*, in which parthenogenetic female adults dwell in the intestinal mucosa of rodents. These lay eggs which after hatching may develop through one of two routes. Homogonic (direct) development results in formation of infective filariform third stage female larvae (iL3). Heterogonic (indirect) development results in the formation of free-living rhabditid males and females (Figure 17.4). These males and females mate and give rise to filariform progeny only (Bolla and Roberts, 1968). Like *S. ratti*, in the human-parasitic species *S. stercoralis*, rhabditid adults given rise to only filariform larvae, whereas other species have been maintained through many free-living generations, e.g. *S. planiceps* (9 generations) (Yamada *et al.*, 1991), *S. simiae* (14 generations) and *S. fuelleborni* (Augustine, 1940). The life span of the parasitic adults of *S. stercoralis* is impossible to estimate owing to the occurrence of autoinfection (hyperinfection), where infective larvae reinfect the host from within the gut. By contrast, in *S. ratti* autoinfection does not occur. In infections of

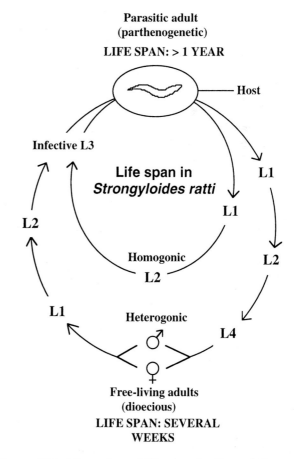

Figure 17.4. Life cycle and life span in *Strongyloides ratti*. Adapted from Viney *et al.* 1993. L denotes larval stages.

immunocompetent mice, adults are cleared from the gut within around 10 days, or in rats from 16 days (Dawkins, 1989) to 3–6 weeks (Sheldon, 1937; Gemmill, Viney and Read, 1997). However, several data suggest that the potential life span of *S. ratti* parasitic adults is much greater. Firstly, in immunodeficient hypothymic (nude) rats, infections were found still to be patent after 46 weeks (Gemmill, Viney and Read, 1997). Secondly, in rats infected with single *S. ratti* adults, a small proportion of infections were still patent after 55 weeks (Graham, 1938). Thus, the maximum life span of *S. ratti* parasitic adults appears to be at least a year. While the longevity of parasitic females of *S. stercoralis* is unknown, long-lived, non-reproductive females are thought to persist in occult infections, giving rise to recrudescing strongyloidiasis upon immunosuppression (Mansfield *et al.*, 1996).

By contrast, the maximum life span of free-living rhabditid adult *S. ratti* varies from 3–4 days at 25°C, to up to two weeks at 13°C (Gemmill, Viney and Read, 1997) (S. Harvey and M. Viney, personal communication). In the absence of mortality data it is unclear whether death in free-living adults occurs as the consequence of senescence. However, it is unlikely that they die of starvation, since they exhibit pharyngeal pumping (S. Harvey and M. Viney, personal communication). Moreover, animals show reduced movement and increased opacity with increasing age. A small proportion of mated females die as the consequence of internal hatching of larvae. Interestingly, filariform L3 can survive and retain infective ability for up to two months at 13°C (Gemmill, Viney and Read, 1997), and in a number of respects resemble the dauer larva stage of *C. elegans* (see below).

In a study of four species of *Strongyloides*, including *S. fuelleborni* and *S. simiae*, adult rhabditid males were found to live 4–9 days, and unfertilised females up to 15 days (Augustine, 1940). In another study, first and second free-living generations of rhabditid adult males and females of *S. fuelleborni* all died within four days (Hansen, Buecher and Cryan, 1969). That senescence occurs in rhabditid adults of *S. stercoralis* is suggested by the observation that after 48 hours free-living adults are sluggish and exhibit poor movement and darkly pigmented intestinal walls (Yamada *et al.*, 1991).

Parasitic adult hermaphrodites of *Rhabdias bufonis* are found in the lungs of several types of amphibians, including frogs and toads. Unlike in *Strongyloides*, larval development appears to occur exclusively via the indirect route (Spieler and Schierenberg, 1995). In this species the parasitic hermaphrodite grows to about 10-times the length of the free-living female, and lays thousands of eggs. By contrast, free-living females produce, on average, only 3 eggs; they also possess a vulva that is too small to lay eggs (Spieler and Schierenberg, 1995). Under laboratory conditions, free-living, rhabditid males start to degenerate after only one day of adulthood, after two days 80% have died, and by three days, all are dead (Spieler and Schierenberg, 1995). Similarly, free-living females die after only a few days, independent of whether they carry eggs or not. By contrast, the maximum life span of the *R. bufonis* parasitic hermaphrodite after reaching maturity is at the very least three months (Goater, 1991). Likewise, in *R. fuscovensa* var. *catanensis*, found in snakes, under conditions in which autoinfection was prevented, reproductively active adults were still present after nine months (Chu, 1936).

Overall, the available data suggests that among the Strongyloididae and Rhabdiasidae the potential life span of parasitic adults greatly exceeds that of the free-living adults, which appear to age at a rapid rate. In fact, the life span of the free-living adults is so short as to suggest that death occurs as the consequence of some degenerative process more rapid and catastrophic than ageing. It would be interesting to know whether these short-lived animals exhibit a Gompertzian exponential increase in mortality, characteristic of ageing populations (see below).

One interpretation of the dramatic difference in the pattern of ageing between parasitic and free-living adults in these two nematode groups is that the relatively low rate of extrinsic mortality of parasitic adults has resulted in the evolution of iteroparity and greater longevity, whilst high levels of extrinsic mortality among free-living adults has produced semelparity and reduced longevity. Thus, the parasitic adults resemble ascarid or strongylid nematodes in their relative longevity, while the rhabditid adults more closely resemble short-lived, free-living nematodes such as *C. elegans*. Presumably these large differences in the rate of ageing between free-living and parasitic adults of the same species result from differences in gene regulation in development and adulthood, resulting in more enduring and better maintained soma in the parasitic adults.

The parasitic Adenophorea

There is little information on longevity and ageing in Adenophorean species. Infections of the intestinal parasitic genera *Trichuris* (whip-worms) and *Trichinella* are generally cleared by host immunity within a period of weeks or months. For example, *Trichuris muris* infections are usually cleared within three weeks, although in cortisone-treated, immunocompromised hosts fertile adults persisted for 7–10 weeks after attaining sexual maturity (Wakelin and Selby, 1974). The female life span of *Trichinella spiralis* has been estimated at three months (A. Maggenti, cited in Finch, 1990). Infections with this species, of rats and guineas pigs, last 2–3 weeks and around five weeks, respectively (McCoy, 1932).

One Adenophorean species in which a survival study has been carried out is *Agamermis decaudata*, one of the Mermithidae, which parasitise insects and snails. *A. decaudata* hatch from eggs laid in the soil, and develop into infective larvae which invade grasshopper nymphs, where they mature into adults in the hemocoele (Hyman, 1951). These adults are aphagous, and survive on food stored in intestinal trophosomes, which gives them an opaque appearance (Christie, 1929). Adult females were still fecund when mated after

14 months, and virgin females were 'apparently in good condition' after 18 months (Christie, 1929). Mated females were less long-lived, and grew increasingly transparent during the egg-laying period, which was interpreted as reflecting the exhaustion of their food stores (Christie, 1929).

Nematodes as Models for the Biology of Ageing

Experimental Approaches to the Study of Ageing

Nematodes, particularly *Caenorhabditis elegans*, are currently widely used as models for the study of ageing. There are several reasons for this: they are small, morphologically and histologically simple, cheap and convenient to maintain, their life cycle is of the order of a few days, and they produce large numbers of progeny, so that it is possible to raise large numbers of animals within a short time. Most importantly, they are short-lived (2–3 weeks for *C. elegans*), exhibit distinct symptoms of senescence, and in the case of *C. elegans* are extremely well characterised (Wood, 1988; Riddle *et al.*, 1997). The presumption that underlies the use of nematodes as models for ageing in general is that 'the essential features of the ageing phenomenon are universal among metazoans' (Gershon, 1970). Establishing the validity of this presumption remains a key issue in nematode ageing studies. Yet given the present lack of understanding of ageing, even the demonstration that the primary mechanisms of ageing identified in nematodes were quite different to those of mammals would represent a major advance.

At present the primary biological determinants of ageing in animal species are unknown. This being so, the appropriate methodological course to its understanding is unclear. Historically, the use of nematodes as a model for ageing has emerged from several different traditions. The first concerted studies were an extension of the methodologies of the field of gerontology, developed to some extent in the study of human ageing. This included studies of the effect of environment on ageing, and age-related changes in behaviour, ultrastructure, enzyme function and levels of age pigment, plus the effects of drugs on the ageing process (reviewed in Zuckerman, 1980). These studies largely focused on the free-living nematodes *Caenorhabditis briggsae*, *Turbatrix aceti* and *C. elegans*, and to a lesser extent *Panagrellus redivivus*. Much of this work was carried out by Morton Rothstein, Bert Zuckerman, David Gershon and their collaborators, and was published during the 1970s.

Also in the 1970s, as the result of an initiative by the geneticist Sydney Brenner, *C. elegans* began to be used as a genetic model, and by the mid-1980s Michael Klass and, in particular, Thomas Johnson had begun taking a genetic approach to studying ageing in *C. elegans*. Subsequently over a dozen other research groups have taken this genetic approach and numerous mutations extending life span have been identified and characterised (reviewed in Kenyon, 1997) (see below).

More recently, other aspects of the study of ageing have spilled over into the *C. elegans* field: demography — the study of age-related mortality in populations (e.g. Johnson, 1990; Honda *et al.*, 1993; Brooks, Lithgow and Johnson, 1994), and to a lesser extent, the evolutionary biology of ageing (e.g. Brooks and Johnson, 1991; Gems and Riddle, 1996).

Nematode Gerontology

Environmental effects

Compared to other traits the contribution of environmental conditions to variation in life span is high relative to the contribution of the genotype. As discussed above, temperature strongly affects life span, as do culture conditions. Nematodes can be cultured axenically (in the absence of other organisms) or monoxenically (with one species of bacterium, invariably *Escherichia coli*), in liquid culture. Alternatively, animals can be cultured on agar plates on a lawn of bacteria. In axenic medium, nematodes are usually smaller and longer-lived than in monoxenic culture (Zuckerman and Himmelhoch, 1980). For example, *C. elegans* when grown on *E. coli* attain twice the volume than if grown in axenic medium, and in an early study median life span was found to be 12.3 and 17.6 days in monoxenic and axenic culture, respectively (Croll, Smith and Zuckerman, 1977). The basis for this difference remains ambiguous. One possibility is that animals are slightly starved in axenic culture, resulting in caloric restriction (CR) which extends life span (see below). If this is the case, then the life span in monoxenic medium may be viewed as the 'normal' life span. Alternatively, in monoxenic culture animals may die prematurely as the result of the harmful effects of the bacteria, perhaps due to bacterial toxins (Hansen *et al.*, 1964). If this is so, then the later stages of senescence may only be observed on axenic medium. Potentially both problems occur. Croll, Smith and Zuckerman (1977) addressed this issue by examining the change with increasing age of the mean maximum pharyngeal pumping rate and the interval between defecations in both types of media. The results suggested that in monoxenic culture animals died at physiologically younger ages. Furthermore, in *E. coli*, animals died at a more rapid rate. However, in more recent studies such a rapid die off in monoxenic culture has not been seen (e.g. Johnson and Wood, 1982; Johnson, 1990). Partly as the result of worries about bacterial toxicity, most of the work of the nematode gerontologists was carried out using liquid axenic medium.

Other workers contested the interpretations of Croll *et al.* Mitchell and co-workers observed that in axenic medium generation time is increased from 2.5 days (monoxenic) to 3 days (20°C), suggesting nutritional deficiency (Mitchell *et al.*, 1979).

One approach taken to resolve this issue has been to take animals maintained in axenic culture for a portion of their adult life and shift them to monoxenic conditions, or vice versa. If axenic conditions slow ageing by means of CR, it might be expected that axenically maintained animals shifted to monoxenic conditions late in life would be longer-lived than animals maintained throughout life in monoxenic culture. This is because at the chronological age of transfer, the axenically raised animals would be biologically younger, due to the effect of CR. By contrast, if differences in survival in axenic and monoxenic culture are due solely to the effect of bacterial toxins during old age, then the life spans of axenically maintained animals shifted to monoxenic culture would be no different to animals maintained throughout life in monoxenic culture. Several studies have taken this approach. In one, *C. elegans* maintained on live bacteria spread onto agar with no peptone (the bacterial growth substrate) were found to live longer than on plates with peptone. Furthermore, maintenance on bacteria in the presence of the antibiotic chloramphenicol increased life span still further (Hosono, Nishimoto and Kuno, 1989). Growth on *E. coli* on peptone-free medium had no effect on post-embryonic development; however, brood size was increased by approximately 30%. Significantly, it was observed that when animals were maintained on bacteria on peptone-free or antibiotic containing agar for six days (20°C) and then transferred to rich peptone plates without antibiotics, no life extension was seen. Taken together, these results indicate that actively growing *E. coli* is detrimental to survival of elderly *C. elegans* hermaphrodites. However, the reduction of brood size observed by Hosono *et al.* indicates that under these conditions, CR may still be occurring. In another study it was found that in animals transferred from axenic to monoxenic liquid culture and vice versa at late L4 or early adult stage, the survival patterns reflected the final medium, demonstrating that conditions of development do not determine differences in survival (De Cuyper and Vanfleteren, 1982). The relative effects of CR and bacterial toxicity or infection on life span in axenic and monoxenic culture remains an unresolved issue.

Behaviour

All aspects of behaviour that have been examined exhibit some degree of slowing down during ageing. These include whole body movements, such as the sinusoidal swimming motion characteristic of nematodes, mating behaviour, feeding (e.g. pharyngeal pumping and foraging movements of the head), and defecation.

Age-related decreases in overall movement in *C. elegans* have been measured in terms of decreases in frequency, which decline with age in a linear fashion in axenic culture (Croll, Smith and Zuckerman, 1977; Hosono, 1978; Bolanowski, Russell and Jacobson, 1981), and in forward and backward movement and omega turns (Duhon and Johnson, 1995). Similarly, in *P. redivivus*, there are reductions in the frequency of looping, turning and reversing behaviours (Abdulrahman and Samoiloff, 1975), and of head-swinging movements in *C. briggsae* (Zuckerman *et al.*, 1971). Ultrastructural studies of senescent nematodes have shown structural anomalies in muscle cells that may contribute to such a decline in movement. For example, interchordal hypodermal bulges appear between the body muscles and the body wall to which they are normally attached. This has been observed in *C. briggsae* (Zuckerman *et al.*, 1971) and *T. aceti* (Kisiel *et al.*, 1975).

The rate of pumping of the pharynx, reflecting the rate of ingestion of food, during ageing has been observed either by videotape analysis (Croll, Smith and Zuckerman, 1977), or by direct observation (Kenyon *et al.*, 1993; Gems *et al.*, 1998). In axenic medium at 20°C, the pharyngeal pumping rate in *C. elegans* hermaphrodites remains steady during much of adult life, declining only after 14 days (Croll, Smith and Zuckerman, 1977). By contrast, on agar plates, the rate of pharyngeal pumping declines in an approximately linear fashion with increasing age (Kenyon *et al.*, 1993; Gems *et al.*, 1998).

The defecation rate in *C. elegans* hermaphrodites in axenic medium remains steady until day eight at a mean period between defecations of approximately 70 seconds. Between day eight and nine this interval doubles to around 130 seconds, and then stays steady until day 18 (Croll, Smith and Zuckerman, 1977). On lawns of *E. coli*, the defecation rate declines rapidly from day three to day six, then stays steady until day nine, after which it declines (Bolanowski, Russell and Jacobson, 1981).

Both reproductive behaviour and the capacity to reproduce decline with age in nematodes. Most obviously, in species such as *C. elegans* which exhibit protandrous hermaphroditism, reproduction ceases upon depletion of self-sperm. However, in these cases, reproduction may resume upon mating with males. In the dioecious species *T. aceti* females mated at successively older ages take longer to begin reproduction, and produce fewer progeny (Kisiel and Zuckerman, 1974). Similar changes are seen when successively older males are mated with young females. Age-related declines in fecundity have also been observed in other species, e.g. *P. redivivus* (Duggal, 1978a; Duggal, 1978b), and *R. tokai* (Suzuki *et al.*, 1978).

Structural and physical changes

A number of changes in the structural and physical properties of nematodes occur during ageing. These include changes in the structure and permeability of the cuticle, accumulation of lipofuscin granules (see below), mitochondrial degeneration, and increases in osmotic fragility and specific gravity.

Using electron microscopy, mitochondrial degeneration has been observed during ageing in *C. briggsae* (Zuckerman *et al.*, 1971). A proportion of these organelles become electron-dense, and shrivelled.

In a manner reminiscent of senescent changes in human skin, the surface of the cuticle in young *C. briggsae* changes from smooth to wrinkled, as revealed by scanning electron microscopy (Figure 17.5) (Högger *et al.*, 1977). The biochemistry of structural changes occurring during the ageing of the cuticle, and cuticle collagens, has been investigated in *C. elegans* using spectroscopy (Davis, Anderson and Dusenbery, 1982).

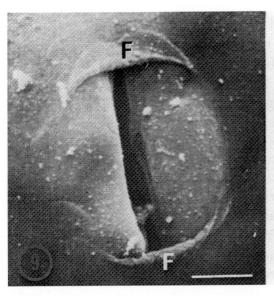
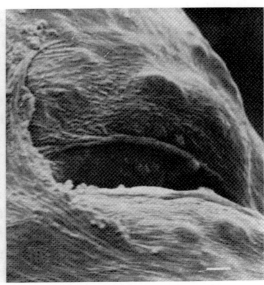

Figure 17.5. Vulva with lateral flaps (F) of *Caenorhabditis briggsae*. Left panel: young animal. Bar = 5 μm. Right panel: old animal. Bar = 2 μm. Note the shrinking of the vulval lips, and wrinkling of the cuticle surface in the old animal. Reproduced with permission from Högger *et al.* 1977.

Detergent-cleaned cuticle preparations were studied, and a fluorescence peak was seen that was identified with the pyridinoline cross-link of vertebrate collagen. This increased greatly with age, suggesting that, as in mammalian collagenous proteins, intermolecular cross-linking occurs, which is thought to cause the age-related decreases seen in elasticity and solubility. This is the result of modification of side chains of the amino acids lysine and hydroxylysine, such that ε-amino groups are converted into aldehydes, and then react together, forming cross-links via fluorescent Schiff base products. Another possibility is that reducing sugars react with the ε-amino group of lysine, which also generate fluorescent cross-links (Monnier and Cerami, 1981). These changes may cause the wrinkling of the cuticle of old nematodes.

One peculiar age-related change seen in *C. briggsae* is the appearance of balls of electron-dense material in the median layer of the cuticle (Zuckerman, Himmelhoch and Kisiel, 1973). These balls distend the median layer to up to twice its normal width, leading to bumps on the cuticular surface. Both the bumps and the balls of material that cause them can also be seen under oil immersion light microscopy. It is not known what the balls consist of.

Distal to the epicuticle is the surface coat, or glycocalyx, a thin layer of glycoprotein carrying a negative charge, as revealed by staining with ruthenium red, which stains acid mucopolysaccharide (Himmelhoch and Zuckerman, 1978), and visualisation in EM studies using cationised ferritin (Himmelhoch, Kisiel and Zuckerman, 1977; Himmelhoch and Zuckerman, 1978). In *C. briggsae* the surface coat changes with age, becoming thinner, and decreasing in charge density (Himmelhoch, Kisiel and Zuckerman, 1977). Such cuticular changes may contribute to the increase in cuticular permeability with increasing age. The permeability of the cuticles of young and old *C. briggsae* in axenic culture have been studied by means of double labelling experiments with tritiated water (Searcy, 1976). Older nematodes were twice as permeable to water as young nematodes. However, a subsequent study with *C. elegans* grown on *E. coli* lawns saw no such changes in permeability (Bolanowski, Russell and Jacobson, 1981).

Increased cuticular permeability in older nematodes may contribute to their greater sensitivity to hypotonic shock, as observed in *C. elegans* (Zuckerman and Himmelhoch, 1980) and *C. briggsae* (Zuckerman *et al.*, 1971). In *T. aceti* mating, which shortens life span, also increases female susceptibility to hypotonic shock (Kisiel *et al.*, 1975). These osmotic shock studies were carried out using axenically cultured animals; animals maintained in monoxenic liquid culture do not show the same age related increase in osmotic fragility (Zuckerman and Himmelhoch, 1980).

Another age-related change observed in *C. briggsae* is an increase in specific gravity (Zuckerman, Nelson and Kisiel, 1972), an effect not seen in *T. aceti* (Kisiel and Zuckerman, 1974). Zuckerman, Nelson and Kisiel (1972) suggest that the increase in specific gravity might accompany an increase in solute concentration in the pseudocoelom, which might, in turn, lead to increased water uptake, and increased sensitivity to osmotic shock.

Enzymes

Age-related changes in enzyme function in nematodes have been extensively and rigorously investigated (reviewed in Rothstein, 1980). Most of this work utilised *T. aceti*, for three main reasons. Firstly, for biochemical studies it is preferable to use axenic rather than monoxenic culture, in order that nematode enzyme preparations are not contaminated by those of bacteria.

T. aceti culture medium is unusually acidic (pH 3.2) which reduces the problem of bacterial contamination of axenic medium. Secondly, *T. aceti* was preferable to wild-type *C. elegans* for culture in axenic medium since in the latter there is a greater frequency of death due to internal hatching of eggs: 50–90% of animals may die this way (Mitchell *et al.*, 1979). Dauer larva formation is another problem with *C. elegans* (see below). Thirdly, it is relatively easy to prepare synchronised ageing cultures of *T. aceti* by means of repeated screening (filtration) procedures to remove progeny (Rothstein, 1980). This is partly because of the relatively large size of fully grown *T. aceti* adults (2.0 mm in length, compared to around 1.2 mm in *C. elegans*), plus the low frequency of internal hatching, which in *C. elegans* leads to contamination of ageing cultures with younger animals. More recently, the availability of sterile mutants of *C. elegans* has made the maintenance of synchronised cultures easier in this species.

The most frequently used means of preparing ageing cultures of *T. aceti* free of progeny is to block reproduction with the DNA inhibitor fluorodeoxyuridine (FUdR) (Gershon, 1970; Hieb and Rothstein, 1975; Hosono, 1978; Mitchell *et al.*, 1979). However, given the possibility that FUdR might interfere with normal ageing, it is important to verify that age-related changes in biochemistry observed in animals maintained on FUdR are also seen in ageing populations prepared by other means, e.g. filtration. By this means it has been demonstrated that age-related changes in enzyme function and protein turnover observed in FUdR-treated populations are not influenced by the presence of FUdR (Rothstein and Sharma, 1978; Sharma *et al.*, 1979).

Age-related changes have been studied in five enzymes purified from *T. aceti*: isocitrate lyase (IL) (Reiss and Rothstein, 1975), aldolase (Reznick and Gershon, 1977), phosphoglycerate kinase (PGK) (Gupta and Rothstein, 1976a), triosephosphate isomerase (TPI) (Gupta and Rothstein, 1976b), and especially enolase (Sharma, Gupta and Rothstein, 1976; Sharma and Rothstein, 1980). All but TPI show age-related changes, as measured by, for example, specific activity, heat sensitivity and immunotitration (the amount of material cross-reacting with serum raised against purified enzyme, per unit of enzyme activity).

In comparisons of enzymes purified from young and old animals, specific activity is generally reduced with age. Potentially, this may reflect the fact that old enzyme preparations consist of a) molecules the function of which are all reduced to an equal extent; b) a mixture of functional and inactive molecules; or c) a mixture of molecules with a range of activities. These possibilities can, to some degree, be distinguished by the study of the pattern of sensitivity to heat inactivation. For example, old IL exhibits a biphasic pattern such that one component is heat sensitive, and the other no different to young IL. This suggests that old IL contains a mixture of fully functional molecules, and partially active ones (those sensitive to heat). In old PGK the heat sensitivity pattern is unaltered, suggesting either that it consists of normal plus inactive molecules, or includes altered molecules whose heat sensitivity is unchanged. Old enolase exhibits a biphasic pattern where both forms have increased sensitivity to heat. By contrast, old aldolase, which also has a biphasic pattern, is actually more stable than young aldolase. This suggests that both old enolase and old aldolase preparations contain two forms of altered molecules. In conclusion, the pattern of ageing of each individual enzyme is quite individual.

What causes these changes in ageing enzymes? In a number of respects, given enzymes from young and old *T. aceti* are indistinguishable, including K_m, K_l and migratory properties in gel electrophoresis and iso-electric focusing (IEF). The latter finding indicates that the changes in old enzymes do not involve an alteration in charge. This rules out a number of possible forms of post-translational change, e.g. phosphorylation, deamidation and acylation, and also suggests that amino acid sequence changes are not involved. This finding was confirmed in *C. elegans* (Johnson and McCaffrey, 1985; Vanfleteren and De Vreese, 1994). In addition, amino acid analysis of enolase has shown that methylated derivatives of lysine and histidine and not present, therefore methylation is unlikely to cause these changes. Nor is proteolysis, as shown by the absence of changes of terminal amino acid residues, or oxidation of the SH groups of cysteine residues. By contrast, changes in the UV difference spectra and circular dichroism spectra of old enolase suggest that conformational changes occur (Sharma and Rothstein, 1978a). These spectral differences are not seen after treatment with the denaturing agent guanidine hydrochloride (6M), supporting the idea that changes in protein conformation but not sequence occur during ageing. Further support for this idea comes from the finding that enolase molecules can be unfolded and refolded using 1.25M guanidine hydrochloride. After refolding, young and old enolase appear identical, and barely distinguishable from native, old enolase by a number of criteria (Sharma and Rothstein, 1980). Furthermore, antibodies raised against partially denatured enolase (Sharma and Rothstein, 1978a) show greater affinity for old than for native, young enzyme (Sharma and Rothstein, 1978b). Thus, it appears that newly synthesised enolase has the young conformation, and with time rearranges into the slightly less active old conformational isomer, a change that does not involve covalent modification.

Why do denatured enzymes accumulate in older animals? A plausible explanation is that it reflects the slowing of protein turnover (synthesis and degradation) that occurs during ageing (Reiss and Rothstein, 1974; Rothstein, 1975; Rothstein, 1977; Sharma and Rothstein, 1978b). In *T. aceti* synthesis of both total soluble protein and purified proteins falls sharply with age (Sharma *et al.*, 1979). The rate of degradation of soluble protein has been examined by pulse chase studies. Animals were fed a pulse of [^{35}S]methionine (12 or 24 hr), which was then replaced with unlabelled methionine, and the decline in levels of labelled protein followed. In comparisons of young and old *T. aceti*, the half-life of labelled protein increased from 68 hr to 170 hr (Sharma *et al.*, 1979), and protein turnover rates in young and old *T. aceti* are estimated at 0.74%/hr and 0.3%/hr, respectively. Thus, with increasing age the life span, or 'dwell time', of individual enzyme molecules in the cell increases due to reduced rates of protein turnover, and this may lead to accumulation of conformationally altered proteins.

The question of why protein turnover decreases with age in *T. aceti* has been explored to some extent. Determinants of the rate of protein synthesis were examined using a cell-free protein synthesis system into which different elements from young and old animals are introduced. By this means it has been shown that a proportion of old ribosomes are incapable of binding the EF-1-GTP-aminoacyl-tRNA complex (Egilmez and Rothstein, 1985). In this context it is interesting that in *C. elegans* levels of the translation factor EF1-α increase with increasing age (Fabian and Johnson, 1995).What goes wrong with ageing ribosomes is not known. The cause for reduced rates of protein degradation is also unclear, although it has been established that disruption of lysosomal function does not affect protein turnover in *T. aceti* (Karey and Rothstein, 1986). In *T. aceti*, lysosomal acid proteinase levels actually increase during ageing by 30–40%. This is in contrast to *C. elegans*, where activities of three lysosomal proteases decline with age (Sarkis *et al.*, 1988). The authors of this study suggest that the increase in lysosomal protease activity seen in *T. aceti* may be artefactual. The decrease in *C. elegans* lysosomal proteases may explain the increases in levels of other lysosomal enzymes with age, e.g. glycosidases and acid phosphatases, levels of which increase up to 100-fold (Bolanowski, Jacobson and Russell, 1983).

Age pigment (lipofuscin)

A common feature of senescence in metazoans is the accumulation of fluorescent age pigment, or lipofuscin (reviewed in Porta, 1991). Age pigment largely consists of chemically highly complex aggregates of peroxidated lipid and protein. This is thought to be formed by the action of free-radicals and lipid peroxides, which enter free radical chain reactions causing membrane damage through cross-linking of lipid and protein. Its accumulation is not believed to be a primary cause of senescence, but rather its by-product. On the basis of spectrofluorometric characteristics, age pigments resembling those in mammals have been identified in *C. elegans* (Klass, 1977), *C. briggsae* (Zuckerman, Fagerson and Kisiel, 1978), and *P. redivivus* (Buecher and Hansen, 1974). Lipofuscin may be studied either by means of fluorescence microscopy of intact animals, or by spectrofluorometric analysis of aqueous homogenates or chloroform/methanol extracts. The spectral values of the latter resemble those of Schiff's base products (RN = CH-CH = CH-NH-R), which are found in mammalian lipofuscin. In *C. elegans* between days 3 and 18 the relative fluorescence per worm increases approximately 10-fold (Klass, 1977).

Ultrastructural studies have revealed the accumulation of lipofuscin granules in the intestinal epithelium of old *C. briggsae* (Epstein, Himmelhoch and Gershon, 1972), *C. elegans* (Zuckerman and Himmelhoch, 1980) and *T. aceti* (Kisiel *et al.*, 1975). Histochemical tests of the granules in *C. briggsae* show the presence of acid phosphatase, commonly found in lipofuscin granules (Epstein, Himmelhoch and Gershon, 1972). Within the 32–34 intestinal cells that contain autofluorescent lipofuscin granules, the latter appear to be located within secondary lysosomes. This was demonstrated by the capacity of these bodies to endocytose and accumulate exogenous fluorescently labelled probes, and their internal low pH, as indicated by acridine orange (Clokey and Jacobsen, 1986). The role of oxidative damage in lipofuscin accumulation has been confirmed by the demonstration that fluorescent material increases at a higher rate in *C. elegans* exposed to high levels of oxygen (Hosokawa *et al.*, 1994). Furthermore, mutations in the gene *mev-1*, which result in hypersensitivity to oxidative damage, accelerate the accumulation of fluorescent material still further (Hosokawa *et al.*, 1994).

Drugs

One aim of nematode gerontologists was to identify drugs that would slow the ageing process, either in one aspect or entirely. Such drugs would be of value as experimental tools to investigate the biology of ageing, not to mention their potential medical applications. A number of compounds were examined, including the antioxidants centrophenoxine and *N,N'*-diphenyl-1, 4-diphenylenediamine (DPPD), vitamin C, vitamin E (α-tocopherol) and its derivative α-tocopherolquinone (α-TQ), and the dental anaesthetic procaine (Novocain), which is the 'active' ingredient of the alleged anti-ageing drug Gerovital H$_3$.

The antioxidant centrophenoxine (dimethylamino *p*-chlorophenoxyacetate) has been shown to retard the accumulation of lipofuscin, and to dissolve lipofuscin granules in the brain tissue of ageing guinea pigs. Centrophenoxine also retarded lipofuscin accumulation in *C. briggsae*, but longevity and fecundity were not affected (Kisiel and Zuckerman, 1978). In 21-day-old animals treated with 6.8 mM centrophenoxine, lipofuscin levels, measured as the proportion of the cross-sectional area of electron micrographs of intestinal cells taken up by lipofuscin granules, were reduced by 42%. At higher concentrations of centrophenoxine a reduction in osmotic fragility and specific gravity in old animals was seen (Kisiel and Zuckerman, 1978). In the light of much subsequent work demonstrating the importance of free radical damage in the ageing process (e.g. Honda *et al.*, 1993; Adachi, Fujiwara and Ishii, 1998), the failure of centrophenoxine to extend life span, taken that it retards lipofuscin accumulation by reducing damage from free radicals, is surprising. Perhaps centrophenoxine blocks an intermediate step in lipofuscin formation, after free radical damage has occurred but before the formation of lipofuscin granules.

Centrophenoxine is rapidly hydrolysed to dimethylaminoethanol (DMAE) and *p*-chlorophenoxyacetic acid (PCA), so a followup study was carried out to examine their effects alone or in combination (Zuckerman and Barrett, 1978). While DMAE did not retard lipofuscin accumulation in *C. briggsae*, DMAE plus PCA did, in addition reducing the accumulation of ageing-associated electron-dense aggregates.

Vitamin E and α-TQ both increase life span in *C. briggsae*, α-TQ from 35 days to 46 days, a 31% increase (Epstein and Gershon, 1972). α-TQ also retards lipofuscin accumulation. Vitamin E also increases *C. elegans* survival by around 20%, while vitamins C and DPPD do not affect life span (Harrington and Harley, 1988). The authors of the latter study emphasised that if vitamin E extended *C. elegans* life span as the result of its antioxidant properties, then the other two antioxidants, vitamin C and DPPD ought to do so as well. They also observed that vitamin E reduced fecundity by up to 30% and delayed the timing of reproduction by up to 28%. These two facts led them to conclude that vitamin E extends life span not by virtue of its antioxidant effects, but rather due to its slight toxicity, which also causes slowed development and growth (Harrington and Harley, 1988). Similarly, low concentrations of the thymidine analogue 5-fluorodeoxyuridine result in increased life span and reduced fecundity (Gandhi *et al.*, 1980).

In other studies, vitamin E was shown to extend life span without affecting fecundity, and even increased growth in *C. elegans* (Zuckerman and Geist, 1983) and *T. aceti* (Kahn and Enesco, 1981). However, in these studies it was shown that the critical period for the effect of vitamin E on life span was during development, and animals transferred to medium without vitamin E in early adulthood exhibited the full life extension effect. Furthermore, life span was not extended in animals maintained in vitamin E from early adulthood onwards. Epstein and Gershon (1972) also found that vitamin E exerts its main effect on life span at an early age in *C. briggsae*. Thus, it seems unlikely that the effect on vitamin E on life span in these studies is due to its effects on free radical damage during ageing, but rather, a nutritional effect on growth in axenic medium.

Treatment with procaine also extends life span. In one study, mean life span was increased by over 50% by 20 mg/ml procaine (Ohba and Ishibashi, 1981). Yet, as in the case of vitamin E, strong inhibition of growth and fecundity also occurred, suggesting a possible toxic effect (Kisiel *et al.*, 1974; Ohba and Ishibashi, 1981). Castillo, Kisiel and Zuckerman, (1975) examined the effect of procaine and Gerovital H₃ on animals at low concentrations at which no growth inhibition occurred. The only effect seen was a reduction in osmotic fragility (Castillo, Kisiel and Zuckerman, 1975).

In conclusion, where antioxidants do extend life span, it is not clear that this is the consequence of retardation of free radical damage. Nonetheless, it would be interesting to know more about the mechanisms by which vitamin E and procaine extend life span.

Potentially, their effects involve the same mechanisms underlying life extension by caloric restriction and in the *clk* mutants of *C. elegans* (see below), which also affect development.

Universal mechanisms of ageing?

How far did the early gerontological studies succeed in demonstrating that the ageing process in nematodes resembles that of mammals? Clearly, they are not identical: the major age-related causes of death in man are cardiovascular disease and cancer; nematodes do not even have a cardiovascular system, and thus do not die of heart attacks or stroke. Yet certain similarities between nematode and mammalian ageing suggest the existence of some common primary mechanisms. These include the reduction in levels of movement and fecundity; the changes in specific activity and heat sensitivity of enzymes which is also seen in ageing mammalian enzymes, such as rat liver SOD and rat muscle PGK; declines in the rate of protein degradation, which are also seen in ageing rodents and humans; the accumulation of fluorescent age pigment, probably resulting from free radical damage — in old mammals lipofuscin accumulates to high levels, e.g. in cardiac muscle and many nerve cell types; the cross-linking of collagens, and the wrinkling of the cuticular surface, resembling the wrinkling of old skin; mammalian red blood cells exhibit an increase in specific gravity and osmotic fragility with age, both of which occur in ageing *C. briggsae*; and a decrease in net negative surface charge occurs during the ageing of the plasma membrane of animal cells, and the *C. briggsae* cuticular surface (nematode-mammalian ageing comparisons are reviewed in Zuckerman and Himmelhoch (1980)). Subsequent work (discussed below) has further supported the existence of mechanisms of ageing common to nematodes and other animals groups: a Gompertzian exponential increase in age-specific mortality (Johnson and Wood, 1982; Johnson, 1990); the extension of life span by caloric restriction, also seen in many other animal types, including rodents, and possibly primates; and a key role of oxidative damage in nematode ageing as reflected, for example, in the increased protein carbonyl concentration with age (Adachi, Fujiwara and Ishii, 1998). One idiosyncrasy of ageing in nematodes (or, at least, *T. aceti*) is the importance of ribosomal ageing as a determinant of declining levels of protein synthesis (Egilmez and Rothstein, 1985). In studies of *Drosophila* and rodents, this has been shown to be caused by deficiencies in soluble factors.

The Genetics of Ageing in *Caenorhabditis Elegans*

One of the greatest remaining mysteries in biology is the mechanistic basis of the genetic specification of life span (Rose, 1991; Finch, 1990; Wachter and Finch, 1997). Our best hope of understanding this complex issue is to select a very simple and rapidly ageing animal model species, study it intensively, and generate an exhaustive description of its biology, ageing included. Currently, *C. elegans* fulfils these criteria especially well. It is currently one of the only animal species for which the DNA sequence of the entire genome is known (*C. elegans* Sequencing Consortium, 1998). This consists of some 97 000 000 000 base pairs of DNA encoding over 19 000 genes. Something about the nature of the genes present, and their expression results in *C. elegans* adults that age and die within 2–3 weeks, rather than live, say, up to 20 years, as in the parasitic nematode *Loa loa*.

A further advantage of using *C. elegans* to study the genetics of ageing is that it exhibits little heterosis, or 'hybrid vigour'. The life span of hybrid progeny of two *C. elegans* isolates, N2 and BO, did not differ from that of the parental strains (Johnson and Wood, 1982). Heterosis has complicated genetic analysis of life span in other species, e.g. *Drosophila melanogaster*. A more extensive study of heterosis using seven wild isolates, whose hermaphrodite life spans were similar, confirmed this result (Johnson and Hutchinson, 1993). By contrast, life span of males in these strains varied considerably, and some hybrids exhibited heterosis, their life spans resembling that of the longer-lived parents rather than the mean of the parental life spans. This suggests that in this case heterosis is the result of partial dominance.

Two key approaches to the study of the genetics of ageing in *C. elegans* are the study of genetic polymorphisms affecting ageing, and the isolation of mutations that extend life span. A number of recent reviews describe this work, e.g. Lithgow (1996), Kenyon (1997), and Hekimi *et al.* (1998). Given the rapid pace of change in this field, the following review will certainly be out of date by the time of its publication.

Genetic polymorphisms affecting ageing

Naturally occurring genetic polymorphisms affecting the rate of ageing have been identified as follows. Two wild-type strains, *Bristol* and *Bergerac*, with similar life spans (around 18 days) were crossed. Single F2 animals were then allowed to pass through 18 rounds of self-fertilisation, resulting in recombinant inbred (RI) lines homozygous at almost all genetic loci (Johnson and Wood, 1982). The different RI lines were found to have mean life spans varying from 10–30 days. The occurrence of RI lines with increased longevity indicates the presence of genetic variation affecting the ageing rate. However, the short life spans of some RI lines might result from the presence of deleterious allelic combinations that reduce survival in a manner unrelated to the ageing process. An increase in the rate of ageing involves an increase in the exponential rise in mortality with increasing age (see below). Such increases were observed in the short-lived RI lines, indicating that their reduced life span is, in fact, the result of an accelerated rate of ageing (Johnson, 1987).

One use of RI lines is to search for life history characteristics or traits that are correlated with the rate of ageing. No covariation was observed between the rate of ageing and the rate of development (Johnson, 1987), fecundity (Brooks and Johnson, 1991; Shook, Brooks and Johnson, 1996), or the efficiency of DNA

repair mechanisms (Hartman *et al.*, 1988). The latter observation, based on a study of the sensitivity of RI lines to three DNA-damaging agents, argues against a key role of DNA damage in ageing. Using RI lines a number of quantitative trait loci (QTLs) affecting several life history traits, including life span, have been genetically mapped by means of PCR of polymorphic markers (Ebert II *et al.*, 1993; Shook, Brooks and Johnson, 1996). In the earlier study, five regions of the genome associated with longevity were identified. One of these, on chromosome *II*, mapped to the region of *sod-1*, which encodes the cytosolic form of the anti-oxidant enzyme Cu/Zn superoxide dismutase, and *age-1*, which affects life span (see below). The later study independently found three out of these five QTLs for life span (Shook, Brooks and Johnson, 1996).

Classical genetic approaches to ageing

Investigating ageing by means of the classical genetic approach, beginning with isolation of mutants with abnormal lifespan, has the advantage that it requires no reference to prior theories of ageing. These, it has been argued, have sometimes blinkered research into ageing in the past (Kenyon, 1997). The study of mutations that shorten life span has not been a favoured approach to the identification of life span-determining genes, since it is difficult to distinguish mutations that shorten life span by accelerating the rate of ageing from those which shorten life span through deleterious effects unrelated to ageing (although this may be achieved by measuring age-related mortality, as described above). By contrast, a mutation that extends life span is more likely to identify a gene involved in the ageing process. Mutations identified to date which affect ageing in *C. elegans* are listed in Table 17.3. These fall largely into two groups: those associated with dauer larva biology, and those associated with the timing of biological processes and the effects of caloric restriction (CR).

Dauer larva biology and ageing: genetics

A number of genes have been identified which affect both dauer larva development and adult longevity (reviewed in Riddle and Albert, 1997, and Kenyon, 1997). The dauer larva is a developmentally-arrested, alternative third stage larva which forms when food is scarce, and population density high (Cassada and Russell, 1975; Golden and Riddle, 1984). Higher temperatures also promote dauer larva formation. Developing larvae sense increases in population density by virtue of a constitutively released, dauer-inducing pheromone (Golden and Riddle, 1982). The word 'dauer' comes from the German *dauern*, meaning enduring: dauer larvae are stress-resistant and long-lived, with maximum life spans of around 70 days in liquid culture (Klass and Hirsh, 1976). They are also non-feeding, surviving on nutrients stored in intestinal granules, which give them a dark appearance, and possess a thickened cuticle, which renders them resistant to detergents and other environmental insults. If food levels increase and population density drops, exit from the dauer stage occurs, and animals develop

Table 17.3. Genes with effects on life span in *C. elegans*

Gene	Maximum increase in mutant mean life span, or other effects[1]	Other phenotypes[2]	Gene product[3]
age-1	190%	Daf-c, Itt, Uvr	PI 3-kinase
daf-2	200%[4]	Daf-c, Itt, Uvr	Insulin/IGF-1 receptor
daf-10	32%[4]	Daf-d	
daf-12	−25% (*m20*, 25°); enhances Age in *daf-2* class 2 alleles, partial suppression of Age in some *daf-2* class 1 alleles	Daf-d	Nuclear hormone receptor
daf-16	−29% (25°); suppresses *daf-2*, *age-1* Age	Daf-d	Forkhead TF
daf-18	−27% (25°); suppresses *age-1* Age only (*daf-2* very weakly)	Daf-d	PTEN phosphatase
daf-28	13%	Daf-c	
clk-1	40%	Slow-growing	CAT5/COQ7 homologue
clk-2, clk-3, gro-1	27%, 41%, 70%	Slow-growing	
eat-2	57%, also *eat-1*, *eat-3*, *eat-6*, *eat-13*, *eat-18*	Eat	
mev-1	−35%	Mev	Cytochrome B_{560}
rad-8	33% (16° only)	Uvs	
spe-26	65%	Fer	Actin-BP
tkr-1	100% (over-expression in transgenic lines)	Itt, Uvr	Tyrosine kinase receptor
unc-4, unc-13	50%, 80% in males only[5]	Unc	HOX TF, phorbol ester BP
unc-32	60% in males only[5]; enhances *daf-2* Age	Unc	
unc-26	47%	Unc, Eat	
unc-64	38%	Unc	Syntaxin

[1] Hermaphrodite life span, unless otherwise stated.
[2] Abbreviations: Daf-c, dauer constitutive; Daf-d, dauer defective; Eat, reduced pharyngeal pumping rate; Fer, fertilisation defective; Itt, intrinsically thermotolerant; Mev, sensitive to methyl viologen (Paraquat); Unc, uncoordinated movement; Uvr, resistant to ultraviolet radiation; Uvs, sensitive to ultraviolet radiation.
[3] Abbreviations: BP, binding protein; HOX, homeobox; TF, transcription factor.
[4] Median life span.
[5] Extension of median life span relative to solitary wild-type males.

into adults. Interestingly, the length of time spent in the dauer stage has no effect on the life span of the post-dauer adult; hence, dauer larvae are considered non-ageing (Klass and Hirsh, 1976).

Over 30 genes affecting dauer larva formation have been identified by means of mutation (reviewed in Riddle and Albert, 1997). These *daf* (dauer larva formation) genes give rise to two types of mutant phenotype. Dauer-constitutive (Daf-c) mutants form dauer larvae in non-dauer inducing conditions; dauer-defective (Daf-d) mutants are unable to form dauer larvae in dauer-inducing conditions. By studying the phenotypes of mutants bearing combinations of daf-c and daf-d mutations the *daf* genes have been ordered into a complex, branched pathway (Riddle and Albert, 1997).

Recently, it was discovered that one branch of this pathway also affects adult life span. This branch includes the daf-c genes *daf-2* and *age-1*, and the daf-d genes *daf-16* and *daf-18* (Figure 17.6). Adult life span is extended in *age-1* and *daf-2* mutants. Most *daf-2* mutations are temperature-sensitive: at the non-permissive

temperature, constitutive dauer larva formation occurs. At the permissive temperature, development into fertile adults occurs, which in the case of most *daf-2* alleles are long-lived (Kenyon *et al.*, 1993; Gems *et al.*, 1998). Similarly, all *age-1* mutants are long-lived, but only severe *age-1* alleles are Daf-c. This suggests that higher levels of wild-type *daf-2* and *age-1* gene activities are required to reduce life span than to block constitutive dauer larva formation.

The first life-extending mutation to be identified was an allele of *age-1*, and was isolated in a direct screen for mutants with increased longevity (the Age phenotype) (Klass, 1983). This loss of function allele, *age-1(hx546)*, extends maximum life span by 60–110%, but does not result in a Daf-c phenotype at temperatures up to 25°C (Friedman and Johnson, 1988a; Friedman and Johnson, 1988b). It was then established that loss of function mutations in *daf-2* can approximately double life span (Kenyon *et al.*, 1993), and furthermore, that mutations in the daf-d genes *daf-16* and *daf-18*, known to suppress the *daf-2* Daf-c phenotype, also suppress the Age phenotype (Kenyon *et al.*, 1993; Larsen, Albert and

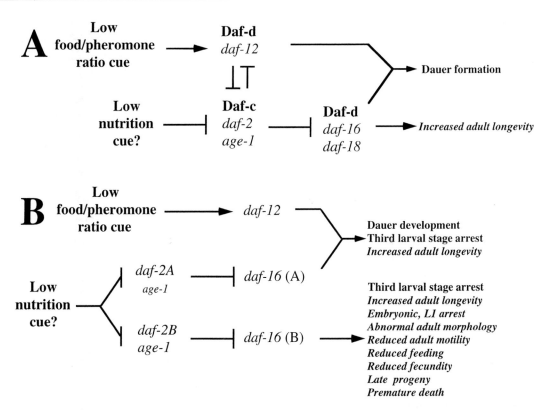

Figure 17.6. Genetic interactions of *daf-2*, *daf-12*, *daf-16*, *daf-18* and *age-1* in controlling larval development, life span and other traits. The pathway is drawn to depict wild-type gene functions that simulate (arrow) or inhibit (T bar) subsequent activities or traits. Traits only seen in *daf-2* or *age-1* mutants are in italics. A, epistasis pathway proposed by Gottlieb and Ruvkun (1994). B, Modified epistasis pathway incorporating the bifunctionality of the *daf-2* gene (Gems *et al.* 1998). *daf-2A* and *daf-2B* represent distinct functional elements within the *daf-2* gene. *daf-16* (A) and *daf-16* (B) represent the activity of *daf-16* at different times, or in different cell types, or different regulatory contexts.

Riddle, 1995; Dorman *et al.*, 1995) (Figure 17.6). Mutations in the daf-c gene *daf-23* were also found to extend life span (Larsen, Albert and Riddle, 1995). By means of complementation studies, *daf-23* was subsequently shown to be allelic with *age-1* (Malone, Inoue and Thomas, 1996; Morris, Tissenbaum and Ruvkun, 1996), although a molecular lesion in the *age-1* PI 3-kinase gene has not yet been found in the *hx546* allele. That *daf-2* and *age-1* act together to control life span is further demonstrated by the observation that the life span of *daf-2; age-1* double mutants does not exceed that of the individual mutants (Dorman *et al.*, 1995). Thus, the wild-type *daf-2* and *age-1* genes function in the adult to shorten life span. By contrast, mutations in daf-c genes in the other branches of the pathway, such as *daf-1* and *daf-7*, do not affect adult longevity (Kenyon *et al.*, 1993; Larsen, Albert and Riddle, 1995).

These findings led to two major hypotheses. Firstly, that the increased longevity of *daf-2* and *age-1* adults is the result of the heterochronic expression of a dauer larva programme of increased longevity in the adult (Kenyon *et al.*, 1993). This idea implies that the relative longevity of dauer larvae is not simply a consequence of developmental arrest, or dormancy, but rather the product of dauer-specific longevity assurance processes. Secondly, it was suggested that the wild-type *daf-2* and *age-1* genes antagonise the function of *daf-16* in wild-type adults, thereby reducing life span. According to this scheme, where *daf-2* or *age-1* action is prevented by mutation, unregulated *daf-16* activity results in an alteration of regulation of downstream genes, resulting in a slowing of the rate of ageing. The possibility that the *daf-16* gene product itself controls the ageing rate, rather than acting through downstream responder genes, seems remote.

Each of these hypotheses suggests a corresponding experimental approach. Firstly, if the heterochronic dauer longevity model is correct, then determinants of the rate of ageing may be identified as dauer larva characteristics misexpressed in the adult. Secondly, if *daf-16* regulates the dauer longevity programme, one should be able to identify the genes determining the rate of ageing among those regulated by *daf-16*.

Beside the *daf-2/age-1* branch of the genetic pathway controlling dauer development, two other branches of the pathway have been characterised. These control the detection of dauer pheromone and food by the chemosensory neurones that open into the amphid sense organs in the head, the transduction of this information through the nervous system, and its translation via a neurosecretory signal into global developmental changes (Riddle and Albert, 1997). The genetic pathways underlying this neural signalling pathway converge on the downstream daf-d gene *daf-12*. In terms of epistasis analysis, the relation between *daf-12* and *daf-2* is complex, and has been expressed in terms of mutual inhibition, as shown in Figure 13.6A (Gottlieb and Ruvkun, 1994; Larsen, Albert and Riddle, 1995). The complexity arises from the finding that interactions between *daf-2* and *daf-12* mutations vary depending on the alleles present, particularly of *daf-2*. For example, in *daf-2(e1370)*; *daf-12(m20)* animals at the non-permissive temperature (25°C), larval arrest at the L1 stage or at the L2 moult (but not as dauers) occurs (Vowels and Thomas, 1992). By contrast, in *daf-2(m41)*; *daf-12(m20)* animals, the Daf-c phenotype is fully suppressed (Larsen, Albert and Riddle, 1995). Yet in terms of the Daf-c phenotype, *daf-2(m41)* is a more severe allele than *daf-2(e1370)* (Gems *et al.*, 1998). The *daf-12(m20)* mutation has a very different effect on the *daf-2(m41)* and *daf-2(e1370)* Age phenotypes: in the case of *daf-2(m41)* the Age phenotype is partially suppressed, whereas in *daf-2(e1370)*; *daf-12* strains at 25°C, it is increased, resulting in life spans almost four-times greater than wild type (Larsen, Albert and Riddle, 1995) (Figure 17.7).

Characterisation of 15 temperature sensitive alleles of *daf-2* has shown that these differences between *daf-2(m41)* and *daf-2(e1370)* reflects the occurrence of two distinct classes of *daf-2* allele. Class 1 alleles exhibit the Daf-c, Age and Itt phenotypes alone. Class 2 alleles exhibit these traits, and many others, including morphological and behavioural abnormalities in the adult, reduced fecundity, and embryonic and L1 arrest (Gems *et al.*, 1998). The difference between the two classes suggests that in class 1 adults the effects of *daf-2(–)* on dauer life span and dauer development are fully uncoupled, whereas in class 2 *daf-2(–)* adults at higher temperatures, they are not, such that dauer behavioural and morphological traits are expressed in the adult.

While the severity of the Daf-c, Age and Itt traits are correlated with each other, the severity of this cluster of traits, and the other cluster of pleiotropic traits is not. For example, the temperature-sensitive allele with the most severe Daf-c phenotype, *daf-2(e1369)*, is a class 1 allele (Gems *et al.*, 1998). This indicates a bifunctionality in the *daf-2* gene, designated *daf-2Adaf-2B*, such that class 1 alleles are *daf-2A(–)daf-2B(+)*, and class 2 alleles *daf-2A (–)daf-2B(–)*, where '–' represents reduction or loss of function.

A comparison of *daf-2; daf-12* strains showed that when class 1 *daf-2* alleles are present, Daf-c is fully suppressed, and partial suppression of Age occurs in some cases. Where class 2 *daf-2* alleles are present, non-dauer larval arrest occurs, and the Age phenotype is enhanced at higher temperatures (Gems *et al.*, 1998). The enhancement of the Age phenotype by *daf-12* may reflect a role of this gene in the normal ageing process (Larsen, Albert and Riddle, 1995). However, an alternative explanation is that this effect is due to suppression of the life-shortening effects of some class 2-specific defects. At 15°C, where the class 2-specific defects are not seen, *daf-12* mutations have little effect on class 2 *daf-2* mutant life span (Larsen, Albert and Riddle, 1995; Gems *et al.*, 1998) (Figure 17.7). Possibly this is because in *daf-2* class 2 mutant adults, *daf-12(+)* promotes heterochronic expression of genes associated with dauer morphogenesis and behaviour. Furthermore, median life span increases relative to maximum life span in the double mutants, suggesting that premature mortality has been reduced. The bifunctional character of *daf-2* function and interactions with *daf-12* are incorporated into the epistasis pathway shown in Figure 17.6B.

Like *daf-2*, the *daf-12* gene is highly complex. Mutant alleles of *daf-12* have identified which are Daf-d, Daf-c or wild type with regard to dauer formation. Daf-c *daf-12* mutants form 'partial dauers' exhibiting incomplete radial shrinkage and do not cease feeding (Antebi,

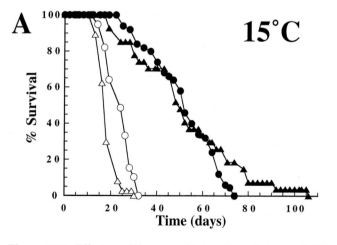

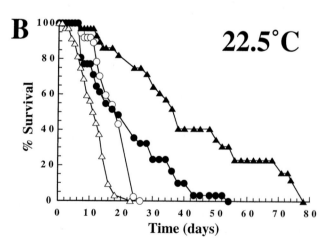

Figure 17.7. Effects on life span of interactions between *daf-2* and *daf-12* mutants at different temperatures. Survival of adult *C. elegans* hermaphrodites. Symbols: open circles = wild type (N2); open triangles = *daf-12(m20)*; filled circles = *daf-2(e979)*; filled triangles = *daf-2(e979)*; *daf-12(m20)*. *daf-2(e979)* is a severe class 2 allele. Unpublished survival curves corresponding to data in Gems *et al.* 1998.

Culotti and Hedgecock, 1998). Some alleles also exhibit heterochronic developmental defects affecting gonadal and extragonadal development.

Mechanistic basis of the age phenotype

Dauer larvae exhibit two obvious characteristics which could potentially underlie their increased longevity: reduced metabolism, and increased stress-resistance. Several lines of evidence suggest that the level of metabolic activity in dauer larvae is reduced relative to other stages. In dauer larvae the level of oxygen consumption is reduced 3.9-fold relative to L3, and the light production potential (an indicator of metabolic activity, see below) is reduced 5.6-fold (Vanfleteren and De Vreese, 1996). The level of TCA cycle activity is reduced (O'Riordan and Burnell, 1989), as are levels of high-energy phosphates such as ATP (Wadsworth and Riddle, 1989). Furthermore, RNA polymerase II-mediated transcriptional activity and mRNA levels of most genes are reduced to 11–17% of other stages (Dalley and Golomb, 1992). On agar plates dauer larvae usually adopt a characteristic resting posture resembling a question mark and remain motionless; however, if disturbed they can more rapidly.

The reduced level of metabolism in dauer larvae suggests that they are dormant, and that their relative longevity may be entirely the result of living at a slower rate. However, metabolic studies of age-1 and daf-2 adults suggest that this is not so. In young age-1 adults, oxygen consumption does not differ from that of wild-type animals of the same age (Vanfleteren and De Vreese, 1996). Oxygen consumption declines with age in both strains, but after 8.5 days of adulthood (25°C), the wild-type rate of oxygen consumption falls below that of age-1. This is what one might expect given that age-1 animals of advanced ages are presumably biologically younger than wild-type animals of the same age. Vanfleteren and DeVreese (1995, 1996) have also measured cellular metabolic activity in terms of the concentration of superoxide anions. This was done by measuring light production by the luminescent substrate lucigenin, which is oxidised by superoxide anions, in the presence of an excess of the reducing agent NADPH. The 'light production potential' or 'metabolic rate potential' so measured, reflects the level of the summed activities of all pathways that produce superoxide (e.g. mitochondrial respiration), and is initially similar in wild-type and in daf-2 and age-1 adults. Then, after 5–6 days the wild-type superoxide production rate potential drops below that of the mutants (Vanfleteren and De Vreese, 1995; Vanfleteren and De Vreese, 1996).

That increased daf-2 and age-1 adult longevity is not the result of reduced metabolic rate is also suggested by the fact that both mutants can exhibit wild-type levels of movement, fertility and rate of pharyngeal pumping. In some cases these elements are affected by mutation of daf-2 and age-1. In mutants carrying daf-2(e1370) at 20°C and above, movement levels, brood size and pumping rates are reduced (Kenyon et al., 1993, Kenyon, 1997; Gems et al., 1998). However, at 15°C daf-2(e1370)

animals behave as wild type, yet are long-lived, and class 1 alleles of daf-2 exhibit wild-type levels of movement, fertility and pharyngeal pumping even at high temperatures (Larsen, Albert and Riddle, 1995; Gems et al., 1998). The weak age-1(hx546) allele does not affect the rate of development, food uptake or behaviour (Friedman and Johnson, 1988), but does result in a 15% reduction in brood size (Lithgow et al., 1994).

Dauer larvae are also resistant to stress, in the form of thermal injury (Anderson, 1978) and oxidative damage induced by the free radical generating chemical methyl viologen (Paraquat) (Larsen, 1993). This is likely to result, at least in part, from observed increased activities of the antioxidant enzymes superoxide dismutase (SOD) (Anderson, 1982; Larsen, 1993) and catalase (Vanfleteren and De Vreese, 1995). The C. elegans genome contains at least five genes encoding SODs: sod-1 encodes a cytosolic Cu/Zn SOD (Larsen, 1993), and sod-2 and sod-3 both encode mitochondrial Mn SODs (Giglio et al., 1994; Suzuki et al., 1996; Hunter, Bannister and Hunter, 1997). Two other genes encode putative extracellular Cu/Zn SODs, the expression of one of which has been confirmed (Hunter, Bannister and Hunter, 1997). Hybridisation analysis suggests the possible existence of a third Mn SOD gene (Hunter, Bannister and Hunter, 1997). This is an unusually high number of genes encoding SOD. For example, the human and bovine genomes contain only single Mn SOD genes, although Zea mays, exceptionally, has four. The similarity of the C. elegans Mn SODs encoded by sod-2 and sod-3 suggest that these genes are the result of a relatively recent gene duplication event (Hunter, Bannister and Hunter, 1997). Dauer larvae also show a 15-fold higher level of mRNA encoding the heat shock protein Hsp90 (Dalley and Golomb, 1992).

In contrast to the case of reduced metabolism, daf-2 and age-1 adults are dauer-like in exhibiting several forms of stress-resistance. They are resistant to oxidative stress (Oxr): age-1 adults show resistance to methyl viologen (Vanfleteren, 1993), hydrogen peroxide (Larsen, 1993), and increased atmospheric oxygen (Adachi, Fujiwara and Ishii, 1998). Again, this may be due to the increased levels of SOD and catalase seen in older age-1 and daf-2 adults (Vanfleteren, 1993; Larsen, 1993; Vanfleteren and De Vreese, 1995). By contrast, SOD levels do not change during ageing in wild-type animals (Vanfleteren, 1993; Larsen, 1993), although in young but not old populations, wild-type animals are able to increase their SOD levels in response to imposed oxidative stress (Darr and Fridovich, 1995).

age-1 and daf-2 adults also exhibit increased intrinsic thermotolerance (Itt) (Lithgow et al., 1994; Lithgow et al., 1995), and age-1 adults over-accumulate the heat shock protein HSP-16 (Lithgow, 1996). However, increased intrinsic thermotolerance is not sufficient by itself to increase longevity since daf-4 and daf-7 mutant adults are Itt but not Age. By contrast, resistance to ultraviolet radiation (Uvr) is well correlated with the Age phenotype (Murakami and Johnson, 1996). It is unclear whether dauer larvae are also Uvr. On the other hand, it has been shown that exposure of wild-type animals to brief heat shock does cause a slight but statistically significant increase in life span (Lithgow *et al.*, 1995). Furthermore, in a comparison of 15 temperature-sensitive *daf-2* alleles, the Itt and Age phenotypes were positively correlated (Gems *et al.*, 1998). Both Itt and Uvr phenotypes are suppressed by mutations in *daf-16* (Lithgow *et al.*, 1995; Murakami and Johnson, 1996).

Another difference between the dauer stage, and the L3 and later stages is a relatively high level of activity of the glyoxylate cycle. This allows acetate and fatty acids to be used as sole carbon source, e.g. in gluconeogenesis. The key enzymes that modify the Krebs cycle into the glyoxylate cycle are isocitrate lyase and malate synthase. Studies of NADP isocitrate dehydrogenase, isocitrate lyase and malate synthase in larval development suggest that during non-dauer development the activity of the glyoxylate pathway is greatly reduced, and respiration via the TCA cycle increased in the transition from the L1 to the L2 stage (Wadsworth and Riddle, 1989). By contrast, in dauer development, this metabolic transition does not occur (Wadsworth and Riddle, 1989; O'Riordan and Burnell, 1990). The activity of the glyoxylate pathway is likely to reflect the importance of lipid stores in the non-feeding dauer larva.

In *daf-2(e1370)* and *age-1* adults, levels of isocitrate lyase is increased, and in *daf-2(e1370)* adults, the level of malate synthase as well (malate synthase was not examined in *age-1* adults) (Vanfleteren and De Vreese, 1995). The same study showed that the levels of isocitrate dehydrogenase was also increased, and those of acid phosphatase and alkaline phosphatase reduced in *daf-2* and *age-1* adults. Could an increase in the level of activity of the glyoxylate cycle result in retardation of ageing? One possibility is that the observed reduction in the rate of feeding in elderly worms (Kenyon *et al.*, 1993; Gems *et al.*, 1998) results in starvation and death; increased glyoxylate activity might allow more efficient use of remaining lipid resources under such conditions of starvation, thus extending life span. From the point of view of applying knowledge of mechanisms of ageing in *C. elegans* to higher animals, it is to be hoped that the glyoxylate cycle is not important, since it does not occur in animals at a higher level of organisation than helminths.

The Age mutants have also been used to test other theories of ageing. For example, it has been shown that the rate of formation of deletions in mitochondrial DNA with increasing age is reduced in *age-1* mutants (Melov *et al.*, 1994; Melov *et al.*, 1995). As often with studies of correlations of specific changes with ageing, it was not possible to demonstrate whether increased efficiency of mitochondrial DNA repair is a primary determinant of ageing, or a secondary effect.

Dauer larva biology and ageing: molecular genetics

Since 1996, the molecular identities of *daf-2*, *age-1*, *daf-16* and *daf-18* have been reported. The *daf-2* gene encodes a homologue of receptor tyrosine kinases similar to vertebrate insulin and insulin-like growth factor

(IGF-I) receptors (Kimura *et al.*, 1997). The *age-1* gene encodes the putative catalytic subunit of a lipid kinase, phosphatidylinositol 3-OH kinase (Morris, Tissenbaum and Ruvkun, 1996). A major role of PI 3-kinases in vertebrate cells is to transmit signals from cell surface receptor kinases (e.g. the insulin receptor) into the cell (Kapeller and Cantley, 1994). The *C. elegans* genome sequence also contains a gene encoding a homologue of the regulatory subunit of PI 3-kinase (Ogg and Ruvkun, 1998). In addition, *C. elegans* also possesses at least 12 genes encoding insulin-like proteins, eight of which are in two clusters of contiguous genes (Duret *et al.*, 1988; Gregoire *et al.*, 1998). Given that DAF-2 is the only protein encoded by the *C. elegans* genome resembling an insulin receptor, this suggests that the DAF-2 receptor binds multiple ligands. Of the 12 putative proteins, one is likely to be a pseudogene (Duret *et al.*, 1988). Of the others, insulin-like protein-1 (*ILP1*) is most similar to mammalian insulin, containing a putative C-peptide flanked by dibasic amino acid proteolysis sites, as found in mammalian insulin (Gregoire *et al.*, 1998). *C. elegans* also has at least three endoproteases resembling prohormone convertases (Gomez-Saladin *et al.*, 1997), consistent with the processing of the ILP1 prohormone may to remove the C-peptide. However, it is unlikely that *ILP1* encodes the ligand for the DAF-2 receptor that regulates dauer/non-dauer development, since the *ILP1* mRNA is only abundant during embryogenesis (Gregoire *et al.*, 1998). That daf mutants with affected insulin-like genes have not been identified might suggest that *C. elegans* insulin genes act in a redundant fashion. The daf-d gene *daf-16* encodes a putative transcription factor of the forkhead type (Ogg *et al.*, 1997; Lin *et al.*, 1997). In addition, the daf-d gene *daf-12* encodes a nuclear hormone receptor homologue (Yeh, 1991).

In mammalian cells, PI 3-kinases generate 3-phosphoinositides, which activate the serine/threonine kinase Akt/PKB. The *C. elegans* genome contains two genes encoding homologues of Akt/PKB, *akt-1* and *akt-2* (Paradis and Ruvkun, 1998). A gain-of-function allele of *akt-1* fully suppresses the Daf-c phenotype of severe *age-1* alleles, yet only weakly suppresses that of *daf-2(e1370)*. In neither case is the Age phenotype affected. This indicates that while *akt-1* is the major output of PI 3-kinase signalling, it is only one of several outputs of DAF-2 receptor signalling (Paradis and Ruvkun, 1998). This conforms with genetic analysis of the *daf-2* gene, which indicates its bifunctionality (Gems *et al.*, 1998; Ogg and Ruvkun, 1998). RNA interference (RNAi) studies show that *akt-1* and *akt-2* function in a redundant fashion to prevent dauer development: simultaneous inhibition of *akt-1* and *akt-2* with RNAi results in almost 100% dauer arrest, whereas inhibition of *akt-1* or *akt-2* alone has no effect (Paradis and Ruvkun, 1998). While it seems likely that these genes play a role in ageing, this remains unclear as yet.

Like *daf-16*, *daf-18* mutants are dauer defective, and suppress the *age-1* Daf-c, Age and fat accumulation traits (Gottlieb and Ruvkun, 1994; Larsen *et al.*, 1995; Ogg and Ruvkun, 1998). However, mutation of *daf-18* only weakly suppresses *daf-2* mutants (Dorman *et al.*, 1995; Larsen, Albert and Riddle, 1995; Ogg and Ruvkun, 1998). Furthermore, RNAi inhibition of *akt-1* and *akt-2* in *daf-18* mutants results in a Daf-c phenotype, indicating that *daf-18* functions upstream or in parallel to *akt-1* and *akt-2* (Ogg and Ruvkun, 1998). The *daf-18* gene encodes a homologue of a human tumour suppressor, PTEN, which has 3-phosphatase activity towards phosphatidylinositol 3,4,5-triphosphate. Thus, loss of function of DAF-18 PTEN may suppress the effect of loss of function of the AGE-1 PI 3-kinase by increasing levels of 3-phosphoinositides.

What is the cellular basis of the function of the *daf-2/age-1/daf-16* pathway? In the case of the other branches of the dauer pathway, those involved in the response to dauer pheromone and food, this is better understood (Riddle and Albert, 1997). At low pheromone:food ratios, three chemosensory neurones, ADF, ASI and ASG repress dauer larva formation via a TGF-*β*-like signalling pathway (transforming growth factor *β*)(Georgi, Albert and Riddle, 1990; Bargmann and Horvitz, 1991; Estevez *et al.*, 1993; Ren *et al.*, 1997). Dauer larva formation is promoted by the ASJ chemosensory neurones, and regulated by a cyclic nucleotide-gated ion channel (Vowels and Thomas, 1992; Thomas, Birnby and Vowels, 1993; Schackwitz Inoue and Thomas, 1996; Coburn *et al.*, 1998). The *daf-7* gene, which encodes the TGF-*β* ligand, is almost exclusively expressed in the ASI neurone (Ren *et al.*, 1997). The *daf-4* TGF-*β* receptor gene is widely expressed, including in many tissues that are remodelled during dauer development (Patterson *et al.*, 1997). This suggests that the dauer-suppressing TGF-*β* signal is released by the neurosecretory cell ASI, and received as a developmental signal throughout the organism.

The cellular basis of *daf-2/age-1/daf-16* signalling is less well understood. The cell types in which *daf-2* and *age-1* are expressed are not yet known, but *daf-16* has been shown to be widely expressed, including in neurones and the many tissues that are altered during dauer morphogenesis (Ogg *et al.*, 1997). This is consistent both with it acting in cells emitting a developmental signal, or in target tissues. The *akt-1* and *daf-18* genes are similarly widely expressed (Paradis and Ruvkun, 1998; Ogg and Ruvkun, 1998). According to one proposed model, *daf-2*, *age-1* and *daf-16*, and in addition the TGF-*β* receptor genes *daf-1* and *daf-4*, plus downstream signalling Smad transcription factor-encoding genes *daf-3*, *daf-8* and *daf-14*, all function in cells throughout the organisms to receive developmental signals, and translate them into developmental changes (Ogg *et al.*, 1997; Kimura *et al.*, 1997; Patterson *et al.*, 1997).

The question of whether *daf-2* functions in cells that principally emit, or that receive developmental signals has been addressed by studying *daf-2* mosaic animals (Apfeld and Kenyon, 1998; discussed in Wood, 1998). These were derived from *daf-2* mutants containing a wild-type copy of the *daf-2* gene on a free chromosomal duplication subject to spontaneous mitotic loss, giving rise to animals with mixtures of *daf-2*(+) and *daf-2*(−) cells. The free duplication also carries a nucleolar size marker, allowing cells which have lost *daf-2*(+) to be identified. If *daf-2* functions in cells emitting developmental signals, i.e. functions cell nonautonomously, it would be expected that *daf-2*(−) cell lineages in otherwise *daf-2*(+) animals could behave as *daf-2*(+), and vice versa. By contrast, if *daf-2* functions in cells receiving developmental signals only, then it would be expected to behave in a cell autonomous fashion, with *daf-2*(−) cell lineages in otherwise *daf-2*(+) animals exhibiting dauer-like characteristics.

The first cell division of embryogenesis in *C. elegans* gives rise to the AB cell and the P_1 cell. The AB cell gives rise to almost all the neurones; the P_1 cell gives rise to the gonad, germline, body muscle and intestine (Figure 17.8). Both cells contribute to the cuticle and pharynx. Loss of *daf-2*(+) from the P_1 cell does not cause dauer formation, whereas half of animals in

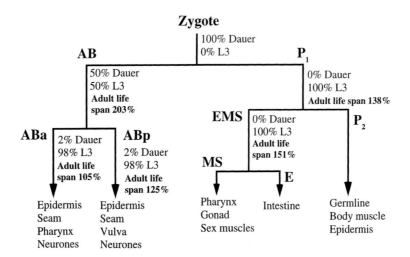

Figure 17.8. The lineal origins of the major tissues in *C. elegans*, and the phenotypes of *daf-2* genetic mosaics. Derived from Apfeld and Kenyon 1998.

which *daf-2*(+) is lost from the AB cell form dauers (Figure 17.8). Both ABa and ABp contribute to the lateral epidermal seam cells, which generate dauer-specific cuticular alae, and radial body shrinkage. ABa and ABp descendants alternate within the rows of seam cells, and if *daf-2* acted cell autonomously, one would expect an alternating pattern of dauer and non-dauer cuticle, and radial body shrinkage. Although animals with both dauer and non-dauer structures have been seen in other mutants, e.g. of *lin-14*, *lin-28* and *daf-12* (Liu and Ambros, 1989; Antebi, Culotti and Hedgecock, 1998), such animals are not seen among *daf-2* mosaics, but only dauer larvae and L3s/adults. Analysis of rare dauer larvae that formed from animals in which *daf-2*(+) had been lost from relatively small groups of cells suggested that loss of *daf-2* from at least six sets of cells could result in dauer larva formation. This indicated that *daf-2* functions cell nonautonomously, and that it functions in a semi-redundant function fashion in multiple cell lineages (Apfeld and Kenyon, 1998). Thus, *daf-2* appears to function in cells which emit a secondary signal regulating development and longevity. That *daf-2* mosaic animals either formed dauers or L3, but not a mixture of both, also suggests that secondary signalling involves some sort of consensus-generating mechanism, to ensure that all cells in an organism, whether *daf-2*(+) or *daf-2*(–) adopt the same fate, dauer or non-dauer (Apfeld and Kenyon, 1998).

In the context of this chapter, the more interesting question is, of course, whether the *daf-2* gene behaves cell autonomously or cell nonautonomously with respect to ageing. In other words, is the rate of ageing controlled at a distance by some central clock mechanism, or does each individual cell have its own clock? The life span of AB-mosaics was double that of wild type, demonstrating that ageing of individual cells can be determined cell nonautonomously. Thus, life span can be doubled, even when most of the internal organs of the worm, the musculature, gut and gonad, are *daf-2*(+) (Figure 17.8). However, in other mosaics an intermediate life span was observed: e.g. 140% of wild type in P_1 mosaics, and 125% of wild type in ABp-mosaics. Thus unlike dauer larva formation, life span is not an all or nothing decision.

The factors regulating the ligand which presumably binds to the DAF-2 receptor are unknown, but based on the similarity to the insulin receptor, one possibility is that an insulin-like signal is generated in response to a replete nutritional state. A model for the possible functioning of the *daf-2/age-1/daf-16* pathway within the context of the whole organism is shown in Figure 17.9.

Given that the DAF-2 putative protein resembles an insulin receptor, and that both dauer larva formation and life-extension by caloric restriction (CR) involve response to levels of nutrition, it has been suggested that the retardation of ageing by CR may be regulated by insulin-like signalling pathways (Kimura *et al.*, 1997). However, other findings suggest that this is unlikely (Vanfleteren and De Vreese, 1995; Lakowski

and Hekimi, 1998) (see below). An alternative possibility is that IGF-I rather than insulin has a role in the regulation of ageing in higher animals. The only mutation known to extend life span greatly in mammals is the Ames dwarf (*df*) mutation of mice, which increases mean life span by almost 70% (Brown-Borg *et al.*, 1996). Ames dwarf mice are small due to severe pituitary gland deficiency, resulting in reduced levels of growth hormone (GH). Levels of IGF-I, mediator of the action of GH on growth, are very low in Ames dwarf mice. Perhaps it is this that causes their increased longevity.

The important question remains: is it likely that homologues of *daf-2* and *age-1* function similarly to control ageing in other animal phyla? Could the relative longevity of human beings compared to other primates be the result of evolved reductions in activity of human homologues of *daf-2* and *age-1*? There are two interpretations of the role of *daf-2* and *age-1* in *C. elegans* ageing and its evolution. In the first interpretation, the ancestral role of *daf-2/age-1* was to regulate dauer development. In this view, the regulation of dauer longevity is a derived trait, suggesting that across phyla, *daf-2* and *age-1* homologues are likely to act as developmental switches, but not in regulating ageing. In the second interpretation, the evolutionary relationship between the role of *daf-2* and *age-1* in dauer larvae development and in ageing is reversed. In this case the ancestral role of *daf-2* and *age-1* was to regulate the rate of ageing, perhaps by regulating expression of stress-resistance genes. Thus, in an ancestor of *C. elegans*, down-regulation of *daf-2* and *age-1* may have extended longevity without affecting morphogenesis in larval diapause. If regulation of ageing is the ancestral function of *daf-2* and *age-1*, and regulation of dauer morphogenesis the derived trait, then it is possible that *daf-2* and *age-1* homologues regulate human ageing. If this evolutionary relationship is reversed, it is likely that the key role of *daf-2* and *age-1* in controlling ageing will be confined to nematodes, and to learn about more universal elements of ageing we must identify the genes regulated by *daf-2*, *age-1* and perhaps *daf-16*.

Clock genes affecting development and ageing

The timing of many biological processes is controlled by internal circadian clocks, such as those determining behavioural activity patterns. Is it possible that the rate of ageing is similarly controlled by a biological clock? The effects of mutations in four genes, *clk-1*, *clk-2*, *clk-3* and *gro-1* suggest that this might be so. The *clk-1*, *clk-2* and *clk-3* genes were identified in a screen for maternal effect mutations affecting the timing of development and behaviour (Wong, Boutis and Hekimi, 1995; Lakowski and Hekimi, 1996). These mutations affect the rate of embryonic and larval development, of cell division, pharyngeal pumping, defecation, egg-laying and overall movement. A fourth clock gene, *gro-1*, was identified by a mutation segregating from a wild *C. elegans* isolate producing a similar phenotype. They also cause modest increases in longevity: *clk-1*, *clk-2*, *clk-3* and *gro-1* mutants show increases in mean life

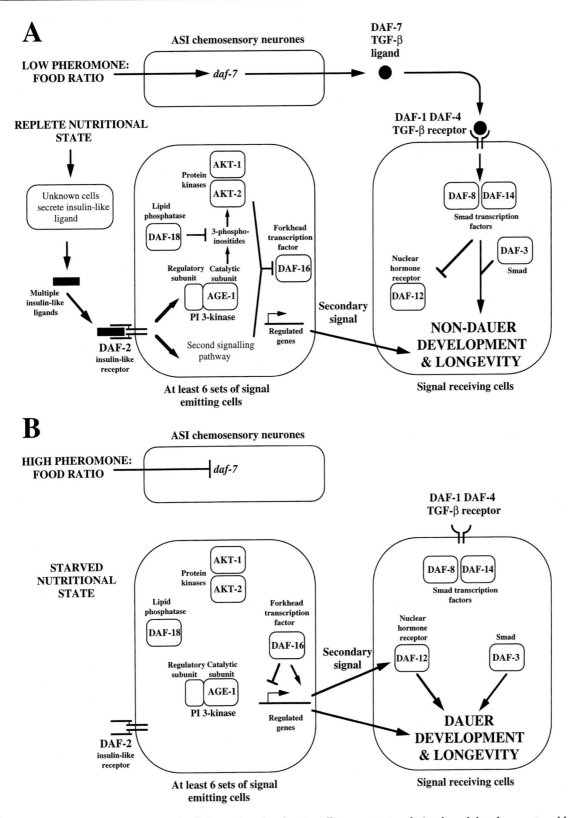

Figure 17.9. Schematic representation of cellular and molecular signalling events regulating larval development and longevity in *C. elegans*. Derived from munerous sources, including Riddle and Albert 1997, Paradis and Ruvkun 1998, Apfeld and Kenyon 1998 and Gems *et al.* 1998. Other equally plausible variants of this model may be derived from current data. As drawn, under dauer-inducing conditions, the DAF-16 transcription factor could switch off genes promoting reproductive development, or switch on genes promoting dauer formation, or both. The pathways are drawn to depict wild-type gene functions that stimulate (arrow) or inhibit (T bar) subsequent activities or traits. A, non-dauer larva inducing conditions. B, dauer larva-inducing conditions. While *daf-12* plays a key role in dauer larva development, its role in ageing is uncertain.

span of up to 40%, 27%, 41% and 70%, respectively (Lakowski and Hekimi, 1996). Part of this increase in life span is due to the reduced rate of development, in contrast to *daf-2* and *age-1* mutant adults, where the extension in life span occurs almost entirely in adulthood. Remarkably, the *clk* Age phenotype is subject to maternal rescue. Homozygous *clk-1* mutant segregants from a heterozygous parent exhibit wild-type life span. This suggests that the effects of some sort of clock-setting process occurring early in life persist throughout adulthood, or less plausibly, that maternally-derived wild-type gene product persists throughout life (Wong, Boutis and Hekimi, 1995).

Mutations in the *clk* genes also lead to a greater variation in the rates of biological processes. Thus, while they slow them down overall, in the case of *clk-1(e2519)*, in a few animals embryogenesis actually occurs faster than in the wild type (Wong, Boutis and Hekimi, 1995).

One very interesting effect of *clk-1* mutations is that they block the ability of animals to adjust their rate of development to ambient temperature. Two cell embryos were dissected from hermaphrodites grown at 15°C, 20°C or 25°C, and transferred to 20°C. In *clk-1* mutant embryos, the rate of embryogenesis depended on the temperature at which their mothers were maintained (Figure 17.10) (Wong, Boutis and Hekimi, 1995). This effect was maternally rescued. The implications of this finding are profound: the effect of temperature on the rate of biological processes, or the 'rate of living', has previously been interpreted as a passive consequence of the thermodynamic effect of temperature on biochemical processes. The properties of *clk-1* mutants suggest that in fact wild-type animals actively adjust their rate of living in response to temperature changes.

Several lines of evidence indicate that the *clk* mutant extension in life span does not involve the *daf-2/age-1/daf-16* pathway. Firstly, Clk mutants are not Daf. Secondly, *daf-2; clk-1* double mutants are much longer-lived than either individual mutant. At 25°C, *daf-2(e1370); clk-1(e2519)* animals had a mean life span 5.8-times that of N2, and 3.1-times that of *daf-2(e1370)* alone (Lakowski and Hekimi, 1996). At 18°C, the double mutant life span is 1.1-times that of *daf-2* alone. It is worth noting that this temperature-dependent enhancement effect is reminiscent of that seen in double mutants containing class 2 *daf-2* mutant alleles with *daf-12(m20)* (Larsen, Albert and Riddle, 1995; Gems *et al.*, 1998) (Figure 17.7).

The third line of evidence suggesting that the *clk* and *daf-2* Age phenotypes involve different mechanisms is that the Age phenotypes of *clk-1*, *clk-3* and *gro-1* are not suppressed by *daf-16(m26)* (Lakowski and Hekimi, 1996). However, according to Murakami and Johnson (1996), the *clk-1(e2519)* Age phenotype *is* suppressed by both *daf-16(m26)* and *daf-16(m27)*. In response to this report, a further study of *clk-1(e2519); daf-16(m26)* life span was carried out (Lakowski and Hekimi, 1998). This stressed that *daf-16(m26)* alone slightly shortens life span, and shortens the life span of *clk-1(e2519)* to a similar degree, and quite reasonably interpreted this as

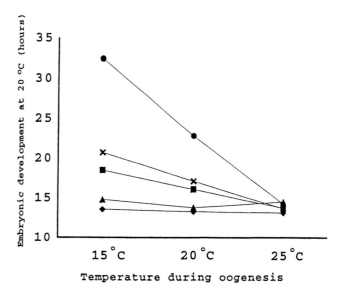

Figure 17.10. Effects of temperature shifts on the duration of embyronic development in wild-type (N2), *clk-1(e2519)*, *clk-1(qm11)*, *clk-1(qm30)* and maternally rescued *clk-1(e2519)* animals. Two-celled embryos were dissected from gravid hermaphrodites previously cultured at 15°C, 20°C and 25°C, and then incubated at 20°C for the remainder of embryogenesis. They were monitored every 30 min until they hatched. The plotted points correspond to the means. Wild type, diamond; maternally rescued *e2519* mutants, triangles; *e2519* mutants, crosses; *qm11* mutants, squares; and *qm30* mutants, circles. Reproduced with permission from Wong *et al.* 1995.

a failure of *daf-16(m26)* to suppress fully the *clk-1(e2519)* Age phenotype.

The reduced rate of biological processes in Clk mutants, and the failure of *clk-1(e2519)* embryos to adjust their rate of development to ambient temperature suggest that the *clk* mutations extend life span in a manner similar to that resulting from a reduction in temperature. Thus, their rate of living may be slowed down. If this is so, it might be expected that the metabolic rate of *clk-1* mutants might be reduced, unlike in *daf-2* and *age-1* mutants (Vanfleteren and De Vreese, 1995).

The only *clk* gene to be cloned so far, *clk-1*, encodes a protein which resembles yeast *CAT5/COQ7* (Ewbank *et al.*, 1997). Mutations in *CAT5/COQ7* result in loss of ability to grow on non-fermentable carbon sources, inability to derepress gluconeogenic enzymes, and most significantly, inability to synthesise coenzyme Q (ubiquinone; *COQ* = coenzyme Q deficient). The former traits are the consequence of the latter (Jonassen *et al.*, 1998). The Cat5p/Coq7p gene product is either itself a monoxygenase involved in coenzyme Q biosynthesis, or is a component essential for monoxygenase activity (Marbois and Clarke, 1996). Coenzyme Q is a component of the electron transfer chain in the inner mitochondrial membrane, transporting electrons from NADH dehydrogenase (complex I) and succinate dehydrogenase (complex II) to the cytochrome bc_1 complex (complex III). In the process it is interconverted between its oxidised form (ubiquinone) and its reduced form (ubiquinol). Ubiquinol also functions as a lipid soluble antioxidant. Cat5p/Coq7p is located in the inner mitochondrial membrane (Jonassen et al., 1998). However, it seems unlikely that the increased life span of clk-1 mutant is a consequence of reduced antioxidant protection by ubiquinol, which would be more likely to shorten life span. The CLK-1 protein and its homologues are functionally conserved, since clk-1 and its rat homologue are able to rescue the CAT5/COQ7 wild-type phenotype in yeast mutants (the ability to grow on glycerol) (Ewbank et al., 1997; Jonassen et al., 1996). However, it is less clear that the role of clk-1-like genes in ageing is conserved, since neither loss of function or overexpression of CAT5/COQ7 affects yeast replicative life span (Hekimi et al., 1998).

A mutation in mev-1 shortens life span and increases sensitivity to oxidative damage. The mev-1 gene encodes another element of the electron transport chain, the cytochrome B_{560} subunit of succinate dehydrogenase, which transports electrons from succinate to ubiquinone (Ishii et al., 1998) (see below). Thus, clk-1 function in the biology of ageing might be linked to that of mev-1.

Genetic analysis of caloric restriction

Reducing the caloric intake (caloric restriction, CR) of a number of animals species, including mammals, has been show to increase life span. As mentioned above, this is also true of C. elegans (Klass, 1977; Hosono, Nishimoto and Kuno, 1989). A number of eat mutants with reduced rates of feeding due to defects in pharyngeal function (the Eat phenotype), resulting in a pale, starved appearance, have recently been shown to extend life span (Lakowski and Hekimi, 1998). The greatest increases in mean life span were seen in mutants affected in eat-2 (up to 57%). This is comparable to the maximum increases in rodent life span achieved by CR. Mutants affected in eat-1, eat-3, eat-6, eat-13 and eat-18 showed weaker Age phenotypes. The magnitude of the Age trait largely corresponded with that of Eat, suggesting that Eat mutants are Age because they eat less, rather than because of any of the other pleiotropic traits seen in these Eat mutants. To exclude the possibility that the effects of other pleiotropic traits affect life span, longevity was measured in mutants affected in 15 unc genes not reported to exhibit an Eat phenotype. These were unc-1, -4, -6, -7, -9, -24, -25, -29, -30, -46, -47, -49, -79, and -80. The only gene where multiple mutant alleles were Age was unc-26. However, the unc-26 mutants were also found to be Eat.

Given that the daf-2 gene encodes a protein resembling an insulin receptor, it has been suggested that insulin signalling mediates both the daf-2 Age phenotype, and the life-extending effects of CR (Kimura et al., 1997). Two experiments employing eat-2(ad465) have recently disproved this hypothesis (Lakowski and Hekimi, 1998). First, daf-16(m26) was found not to suppress the eat-2 Age phenotype, beyond the slight reduction in life span also seen when comparing

daf-16(m26) with daf-16(+) animals. Secondly, eat-2; daf-2 double mutants lived longer than either single mutant; the presence of the eat-2 mutation increased the daf-2 mutant life span by 22%.

A further observation argues against a role of daf-2 and age-1 in CR. In axenic medium C. elegans adults exhibit a longer life span, which is likely to be due at least in part to CR. If mutation of daf-2 or age-1 increases life span by the same mechanisms as that underlying CR-mediated life extension, then one would expect that in axenic medium the differences in life span between wild-type and age-1 or daf-2 mutants would be reduced. Yet this is not the case (Vanfleteren and De Vreese, 1995).

By contrast, eat-2; clk-1 double mutants did not live longer than the single mutants. This suggests that the clk genes are involved in the effect of CR on life span (Lakowski and Hekimi, 1998). The authors put forward the hypothesis that in both clk-1 and eat-2 mutants metabolic rate is reduced, the former by direct effects on mitochondrial function, the latter by starvation.

Other genes affecting ageing

Four other genes that may interact with the daf-2/age-1/daf-16 pathway are tkr-1, daf-28, unc-64 and daf-10. The role of tkr-1 in ageing was established by the chance observation that a cosmid containing this gene extends longevity in transgenic animals (Murakami and Johnson, 1998). The tkr-1 gene encodes a putative tyrosine kinase receptor resembling mammalian platelet-derived growth factor (PDGF) and fibroblast growth factor (FGF) receptors. The mean life span of transgenic lines containing extra copies of tkr-1 is increased by 40–100% (average, 65%). These strains are also resistant to UV radiation and heat stress, but show normal rates of development and fertility. There is also evidence that tkr-1 is regulated by daf-16 (Murakami and Johnson, 1998). The authors suggest that TKR-1 may antagonise DAF-2 and AGE-1 function. However, the tkr-1 transgenic strains are not Daf-c. This study demonstrates that screening transgenic lines containing randomly selected genomic DNA sequences is potentially a powerful tool for identifying genetic determinants of ageing.

Another daf-c gene affecting ageing is daf-28. The semi-dominant mutation daf-28(sa191) results in the constitutive formation of dauer larvae which then recover (Malone, Inoue and Thomas, 1996). Mutant adult life span is extended slightly (12–13%). Its position in the epistasis pathway controlling dauer formation and life span is difficult to define.

The unc-64 gene may also have a role in regulating ageing. Sterile fer-15(b26); unc-64(e246) hermaphrodites exhibit increases in median and maximum life span of approximately 60% and 70%, respectively at 25°C (Gems and Riddle, 2000b). The fer-15 mutation, which blocks self-fertilisation, does not affect life span, and the unc-64 mutation alone, after three backcrossings with a wild-type strain, also extends life span. The e246 allele also acts as an enhancer of the Daf-c phenotype of daf-28(sa191) (Malone, Inoue and

Thomas, 1996). Thus, *unc-64* may act in the genetic pathway regulating dauer larva formation and life span.

The *unc-64* gene encodes a protein homologous to syntaxin, which is involved in synaptic vesicle fusion in neurons (Saifee *et al.*, 1998). The fact that *unc-64*, a neuronal function gene, controls adult longevity, provides direct evidence for the control of ageing by the nervous system in *C. elegans*. Neuronal control of ageing is consistent with the observation that *daf-2* controls ageing in a non-cell autonomous fashion (Apfeld and Kenyon 1998).

Mutations in the *daf-10* gene result in a dauer-defective phenotype, and chemosensory (Che) defects (Albert *et al.*, 1981). At 25°C, median life span of *daf-10(e1387)* adults was increased by 32% (Gems and Riddle, unpublished). Perhaps significantly, *daf-2(e1370); daf-10(e1387)* double mutants are sterile at 25°C (Gems and Riddle, unpublished), and a number of other mutations resulting in Daf-d and Che phenotypes cause embryonic and early larval arrest when combined with *daf-2(e1370)* (Vowels and Thomas, 1992). Thus, the effect of mutation of the *daf-10* gene on life span may be related to interactions with the *daf-2* gene.

A very different way in which mutations can lead to extension of life span is illustrated by *unc-4*, *unc-13* and *unc-32* mutants. As previously described, on agar plates interactions between wild-type males (attempted mating) reduce life span by up to 50% relative to that of solitary males (Gems and Riddle, 2000b). Surprisingly, mutations which reduce male mating behaviour increase male life span far beyond that of solitary wild-type males. For example, *unc-4(e120)*, *unc-13(e51)* and *unc-32(e189)* increase male median life span by 50–90% relative to solitary wild-type males (Gems and Riddle, 2000b). By contrast, none of these mutations extends hermaphrodite life span. These results suggest that in the absence of behaviour that shortens male life span, the potential longevity of the *C. elegans* male is 70–120% of that of the hermaphrodite.

Several mutations in the *spe-26* gene that result in a temperature-sensitive sterile phenotype have also been reported to extend life span (Van Voorhies, 1992). In this report, life spans of mated, wild-type and *spe-26* hermaphrodites are compared; the report also asserts that mating does not reduce hermaphrodite life span. However, a later, more extensive report shows that mating reduces hermaphrodite life span by up to 50% (Gems and Riddle, 1996). In another study, two mutant alleles of *spe-26* were shown to increase unmated hermaphrodite life spans by 45–46% (Murakami and Johnson, 1996). The *spe-26* gene encodes a protein resembling an actin binding protein, possibly a component of the cytoskeleton (Varkey *et al.*, 1996). The *spe-26* Age phenotype is suppressed by mutations in *daf-16*, suggesting that *spe-26*, like *daf-2* and *age-1*, acts via *daf-16* (Murakami and Johnson, 1996). Long-lived *spe-26* adults are also resistant to heat stress (Lithgow *et al.*, 1995), and UV radiation (Murakami and Johnson, 1996). Furthermore the Uvr phenotype is suppressed by mutation of *daf-16*.

Genes associated with free radical damage and ageing

The role of free radical damage in ageing has been studied in detail by means of mutations affecting *mev-1*, *rad-8* and *age-1* (Ishii *et al.*, 1990, Honda *et al.*, 1993, Adachi, Fujiwara and Ishii, 1998, Ishii *et al.*, 1998). The *mev-1(kn1)* mutation results in a four-fold increase in sensitivity to the free radical generator methyl viologen (Paraquat) (Ishii *et al.*, 1990). Mutant animals are also hypersensitive to oxygen, their longevity is reduced by about a third under normal atmospheric conditions, and levels of SOD activity are halved. The relation of oxygen concentration to ageing has been studied further by measuring age-specific mortality of wild-type and *mev-1* mutant populations under high and low oxygen concentrations (Honda *et al.*, 1993; Adachi, Fujiwara and Ishii, 1998). Interestingly, in wild-type populations, hypoxic conditions (1% oxygen) increases mean life span by 15%, and reduces the Gompertz mortality rate, regarded as a corollary of the ageing process (see below); similarly, hyperoxia reduces life span and increases the value of the Gompertz mortality rate (Honda *et al.*, 1993). The effects of oxygen concentration on the ageing rate in *mev-1* mutant populations is more pronounced. The UV-sensitive *rad-8(mn163)* mutant is also hypersensitive to hyperoxia (Honda *et al.*, 1993), and at 16°C has a mean life span 33% greater than wild type due to its retarded rate of development (Ishii *et al.*, 1994). That increased oxygen and the *mev-1(kn1)* mutation shorten life span by accelerating the ageing process is supported by the observation that fluorescent material resembling lipofuscin accumulates at an accelerated rate in *mev-1* mutant animals (Hosokawa *et al.*, 1994), as do levels of protein carbonyl (Adachi, Fujiwara and Ishii, 1998). The accumulation of protein carbonyl derivatives during ageing has been observed in many species, and may be due to metal catalysed oxidations (Stadtman, 1992). In *age-1* adults, the rate of increase in the carbonyl content with age is reduced (Adachi, Fujiwara and Ishii, 1998).

The *mev-1* gene was recently found to encode the cytochrome B_{560} subunit of succinate dehydrogenase (Ishii *et al.*, 1998). This is a component of complex II of the mitochondrial electron transport chain, and in *mev-1* mutants the transport of electrons from succinate to coenzyme Q (ubiquinone) was reduced by 80%. The authors of the cloning study suggest that the dysfunction of complex II may result in an increased level of oxidised coenzyme Q, which then reacts with molecular oxygen to form the superoxide radical, which in turn results in oxygen hypersensitivity and premature ageing (Ishii *et al.*, 1998). The relation between complex II dysfunction, and the reduced SOD activities observed in the *mev-1* mutant is unclear.

What causes ageing in C. elegans?

The work of Ishii and co-workers provides strong evidence that free radical damage is, if not the cause of ageing itself, then causally immediately distal to its primary cause. This is supported by the increased resistance to oxidative stress seen in the longest-lived

Age mutants, those affected in *age-1* and *daf-2* (Larsen, 1993; Vanfleteren, 1993). What remains to be demonstrated is that overexpression of genes determining resistance to free radical damage results in a similar Age phenotype. In this context it is notable that treatment of nematodes with antioxidant compounds does not generally extend their life span, or where it does, in a manner that suggests that it is not a consequence of their antioxidant properties (Harrington and Harley, 1988). If free radical damage were the primary cause of ageing, exogenous anti-oxidants would be expected to extend life span dramatically, assuming that they were absorbed into the body.

While the importance of free radical damage in ageing is well established, it does seem rather improbable that the difference in life span between *C. elegans* and, say, its cousin *Necator americanus* (14 year maximum life span) is due simply to the improved antioxidant defences of the latter. The free radical hypothesis of ageing may successfully describe events occurring at the molecular and cellular level during ageing, yet seems unlikely to explain the enormous variation in the ageing rates of animal species. By contrast, the evolutionary theory of ageing does provide a convincing explanation for why ageing occurs at different rates in different species, and, indeed, why it occurs at all. The evolutionary explanation is based on the inference that natural selection cannot maintain fitness at ages exceeding the life expectancy or fecundic life span of a species in the wild, where death is almost never the result of ageing, but rather, of predation, starvation *etc*. (Williams, 1957; Rose, 1991). By this view, the condition of persistence at an age beyond natural life expectancy is somewhat analogous to that of an atrophied vestigial organ, like the human appendix, or the remnants of eyes in blind cave fish. Thus, in the case of *C. elegans*, that ageing and death occurs after two weeks is likely to reflect the fact that in the wild, ancestral *C. elegans* did not survive to reproduce in significant numbers to ages greater than two weeks. This view would suppose that a hypothetical immortal variant of *C. elegans* would, in the wild, have a fecundic life span of under two weeks.

If the evolutionary theory of ageing explains *why* animal life spans are as long as they are, and the free radical hypothesis shows *how* a key component of the ageing process occurs, the question remains *how* length of life is genetically specified. The following is an attempt to explain the latter by reference to a model of ageing that incorporates the evolutionary theory of ageing, the free radical hypothesis of ageing, plus a hypothesis about the developmental genetic specification of age-specific characteristics in adults.

Let us suppose that as an animal traverses adulthood, a succession of age-specific, developmental genetic identities are expressed. The effects of genes on each successive developmental stage of adulthood is subject to natural selection — or, at advanced ages, the lack of it. If this is given, then for a species to evolve a longer life span, it must evolve fit, healthy developmental genetic identities for more advanced ages. According to

this model, at advanced ages, animals effectively start to run out of developmental programme, resulting in a gradual loss of homeostasis, and increasing fragility. Given a loss of cellular homeostasis, the first aspect of cellular function to break down is likely to be mitochondrial function, leading to increased production of free radicals, and increased free radical damage. Thus, the reason why free radical damage is so widely observed during ageing across species is that it is a natural consequence of the loss of homeostasis within cells resulting from expiration of the genetic programme specifying fit, healthy, age-specific adult developmental identities. The evolutionary theory of ageing potentially explains why this expiration occurs, and its timing. This tripartite theory of ageing is summarised schematically in Figure 17.11.

Superficially this model resembles R.G. Cutler's dysdifferentiation hypothesis for ageing. This supposed that ageing results from the non-programmed derepression of genes (Ono and Cutler, 1978; Cutler, 1982). However, Cutler's theory is essentially a guess inspired by the idea of gene regulation. The proposed model follows necessarily from the presumption that successive stages of adulthood constitute evolved, age-specific, genetically-determined identities, combined with the evolutionary theory of ageing, and the free radical hypothesis of ageing. If this model is correct, key questions are: what are the mechanisms determining the length of a developmental programme of healthy adulthood, and the rate at which it is expressed?

Is there any evidence for the occurrence of successive age-specific developmental genetic identities during adulthood in nematodes? A number of methods have been employed to look for changes in gene expression during adulthood. In one approach, animals were fed with ^{35}S-labelled *E. coli*, and proteins from 4–22 day old animals separated by two-dimensional gel electrophoresis (Johnson and McCaffrey, 1985). Over 700 newly synthesised proteins were resolved by this means, but no new major proteins were observed during the adult phase, nor did any abundant proteins cease to be made. In later studies of silver-stained cytoplasmic and nuclear protein extracts, variation in relative levels of certain proteins with age was seen (Meheus *et al.*, 1987; Vanfleteren and De Vreese, 1994). One nuclear glycoprotein which becomes more abundant with age, S-28, was purified, and the amino acid sequence of its N-terminal end ascertained (Meheus *et al.*, 1987). The sequence did not resemble that of any known protein. An alternative approach taken to look for age-related changes in gene expression was to probe replicate *C. elegans* cDNA libraries with labelled cDNA prepared from young and old nematodes (Fabian and Johnson, 1995). Nine transcripts were identified which decreased in abundance with age, two that increased slightly with age, and one that peaked in abundance in mid-adulthood. Of the nine transcripts that decreased in abundance, three encoded vitellogenins (yolk proteins), one encoded translation factor EF1-alpha, and the other five, proteins of unknown function (Fabian and Johnson, 1995).

Tripartite model of ageing

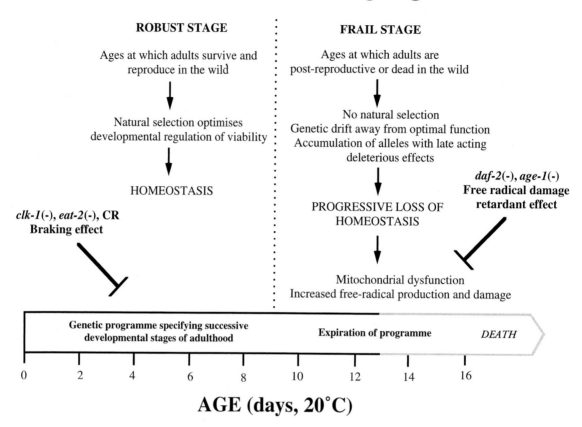

Figure 17.11. Tripartite model of ageing. See text for explanation; also Gems (2000). T bars represent inhibitory effects of mutations or environmental manipulations on different elements of ageing. Vertical dotted line depicts the approximate time of the transition from robust to frail stage. One difference between the free radical hypothesis of ageing and the tripartite model is that in the former, dyshomeostasis is viewed as the consequence of free-radical damage. In the latter, mitochondrial dysfunction is viewed as initially a consequence of a more proximal, developmentally determined dyshomeostasis. CR, caloric restriction.

Approaches such as these may not be sufficiently sensitive to detect subtle changes in transcriptional levels, for example in genes expressed at low levels or in small numbers of cells, and cannot detect spatial changes in gene expression. Furthermore, the failure of natural selection to ensure the genetic maintenance of fitness at advanced ages might result in a loss of the fine tuning of gene regulatory networks, resulting in a loss of homeostasis. Such changes might be difficult to detect at the level of protein or RNA abundance. A more sensitive and powerful approach employed in the study of gene expression and ageing in *Drosophila* is the visualisation of reporter gene expression in ageing transgenic animals. This approach has demonstrated the occurrence of complex changes in intensity and spatial distribution of gene expression during the traversal of adulthood and ageing (Helfand *et al.*, 1995; Rogina and Helfand, 1996; Rogina, Benzer and Helfand, 1997). This is consistent with the hypothesis that passage through adulthood, like development, is determined by a concerted series of changes in gene expression.

To return to the work of Ishii and co-workers: according to the proposed tripartite model, the reduction in life span resulting from mutations in *mev-1* is unlikely to represent an acceleration of the primary ageing process (the expiration of programme), but rather of the secondary process of mitochondrial dysfunction which normally ensues as the adult developmental programme ensuring full homeostasis starts to expire.

The tripartite model also suggests questions about the Age phenotypes resulting from mutation of the *clk* genes, *eat-2*, and *daf-2* or *age-1*: do they result from the elongation of the adult developmental programme ensuring full viability, or from the retardation of the secondary free radical damage element of ageing? One possibility is that in *clk* and long-lived *eat* mutants the primary, developmental ageing process is affected; and in *daf-2* and *age-1* mutants retardation of secondary ageing (free radical damage) is occurring. In the case of *clk* mutants, the rate of development is clearly affected (Wong, Boutis and Hekimi, 1995; Lakowski and Hekimi, 1996). Thus, the time taken until the expiration of the developmental programme specifying healthy adulthood, and the subsequent loss of homeostasis is delayed. In the case of *daf-2/age-1*, some evidence suggests that the Age phenotype results from a retardation of free radical damage. Consider the effect of hyperoxia on life span in wild-type, *age-1* and *mev-1* strains (Figure 17.12) (Adachi, Fujiwara and Ishii, 1998). At oxygen concentrations of 80% the life spans of the three strains are not statistically different to one another. One interpretation of this result in the light of the tripartite model is that the span of *C. elegans*

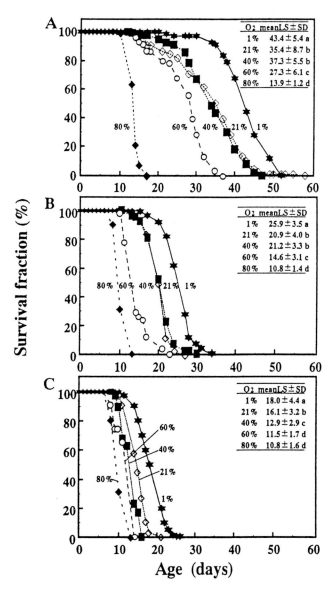

Figure 17.12. Survival curves of *age-1*, wild-type N2, and *mev-1* strains under various concentrations of oxygen continuously present later than 4 days after hatching. (A) *age-1*, (B) wild-type N2, (C) *mev-1*. Oxygen concentrations: stars, 1%; open diamonds, 21%; filled squared, 40%; open circles, 60%; and filled diamonds, 80%. Mean life span values with different letters at each strain are significantly different (*p*<0.01) according to Duncan's Multiple Range test. Reproduced with permission from Adachi *et al.* 1998.

adulthood comprises two stages: a robust stage, where a developmental genetic programme ensures full homeostasis, lasting until around day 8–10 (20°C), followed by a frail stage during which animals are increasingly susceptible to the breakdown of mitochondrial function, and the resulting downward spiral of free radical damage. The existence of robust and frail stages of adulthood is also suggested by the effects of mating with males on hermaphrodite life span. Increasing the ratio of males to hermaphrodites up to approximately 1.5 decreases hermaphrodite life span, but at higher ratios no further decrease of hermaphrodite median life span below 8–9 days is seen (Gems and Riddle, 1996). By contrast, the tripartite model predicts that the robust stage in *clk* or *eat-2* mutants would be extended.

The determinants of longevity in dauer larvae may also be considered in the context of the tripartite model. Dauer larvae maintained at 20°C in monoxenic liquid culture can retain the capacity to resume development for more than 70 days, after which they probably die of starvation, since they are not feeding and survive on stored nutrients (Klass and Hirsh, 1976). As discussed above, possible contributory factors to dauer longevity are increased stress-resistance and reduced metabolism; by contrast, in *daf-2* and *age-1* adults, while stress-resistance is increased, but metabolism is not reduced. A remaining question is whether the span of dauer larva survival involves a succession of age-specific developmental identities, or whether in the state of developmental arrest, the clock-like processes that are part of adult ageing are switched off. Possibly if dauer larvae were able to feed they would survive indefinitely. Consider the developmentally arrested L2 stage of the parasitic nematode *Toxocara canis*. Like the dauer larva, this form has a sealed buccal cavity, and is detergent-resistant (D. Gems, unpublished), but can feed trans-cuticularly. *T. canis* L2 can survive in paratenic hosts for up to nine years (Beaver, 1966).

While it cannot be said that research into ageing in nematodes so far has yielded a clear description of the mechanistic basis of ageing, it has served well to falsify a number of hypotheses about ageing in *C. elegans*. The following are some of the mechanisms that do not underlie ageing. Ageing is not caused by an 'error catastrophe' resulting from increased frequency of errors in protein synthesis, as proposed by Orgel (1963). Such a mechanism would predict the appearance of large numbers of proteins with altered primary structure during ageing, and this is not seen (Rothstein, 1980; Johnson and McCaffrey, 1985; Vanfleteren and De Vreese, 1994). Nor does it appear to be due to DNA damage (Hartman *et al.*, 1988), although DNA repair synthesis and removal of pyrimidine dimers decreases with age in *T. aceti* (Targovnik *et al.*, 1984). There is no evidence that it is linked to variation in fertility: changing fertility does not affect life span (Friedman and Johnson, 1988; Kenyon *et al.*, 1993; Gems and Riddle, 1996), nor are changes in fertility correlated with genetically specified increases in life span (Brooks and Johnson, 1991; Larsen, Albert and Riddle, 1995;

Gems *et al.*, 1998). However, while reproduction may not affect life span, it does increase the rate of metabolism: oxygen consumption and superoxide production rate potentials of reproducing hermaphrodites are increased by factors of 1.3 and 3.0, respectively, relative to non-reproducing hermaphrodites (Vanfleteren and De Vreese, 1996). A number of *ced* genes have been identified which control programmed cell death (apoptosis) in *C. elegans*. However, there is no evidence suggesting that common mechanisms control programmed cell death and ageing (Hengartner, 1997), and effects on life span by mutations in *ced* genes have not been observed (Kenyon, 1997). Ageing in *C. elegans* is also very unlikely to be due to telomere shortening, since somatic cells in adults of this species are all post-mitotic.

Demography and Ageing

Ageing has been observed in most animal species, that is to say, increasing decrepitude with increasing age leading to death. Yet when we talk of nematodes, fruit flies and mice dying as the consequence of ageing, to what extent can we say that they die as the result of the same process, given that our understanding of the basis of ageing is so poor? One feature of ageing that has been observed throughout the animal kingdom is the exponential acceleration of the mortality rate with increasing age (reviewed in Finch, 1991). This characteristic of demographic ageing was first reported in man by the actuary Benjamin Gompertz in 1825. He observed that human demographic ageing could be described by the relation between two parameters: A, the initial mortality rate, and α, an exponential function (Gompertz, 1825). These parameters are related to mortality, m at age t by the Gompertz equation:

$$m(t) = Ae^{\alpha t}$$

C. elegans populations exhibit a typical Gompertzian exponential increase in age-specific mortality (Johnson, 1987). Furthermore, genetically determined differences in life span result in changes in the Gompertz mortality rate, as in recombinant inbred lines (Johnson, 1987), short-lived *mev-1* mutants (Honda *et al.*, 1993), and long-lived *age-1* mutant populations (Johnson, 1990). In the latter case, the rate of acceleration of mortality with age was reduced by 50%. The Gompertzian mortality rate can also be increased or reduced by increasing or reducing the oxygen concentration (Honda *et al.*, 1993).

In the absence of an identifiable biological marker of ageing, demographic ageing has, to an extent, come to be viewed as a definitive characteristic of ageing. However, it remains to be demonstrated whether one could not increase the Gompertz mortality rate via mechanisms quite distinct from those determining biological ageing. This possibility may only be tested if and when the primary biological mechanisms of ageing have been established.

Concluding Remarks

Consistent with the great variety seen among the life histories of nematode species, adult nematodes exhibit a remarkable degree of variation in life span, from 2–3 days in the case of the free-living forms of *Rhabdias bufonis*, to up to 20 years in *Loa loa*. This evolutionary plasticity in the rate of ageing is mirrored by the ease with which life span can be altered in the nematode *C. elegans*, where mutations in dozens of genes and various environmental manipulations have been shown to alter life span. Currently, the ageing process in the nematode seems best to be explained by the evolutionary theory of ageing and the free radical hypothesis. Given the current fast pace of research into the genetic specification of ageing in *C. elegans*, it appears possible that we will soon achieve for the first time a clear understanding of the biological processes that determine ageing in an animal. It remains to be determined whether human ageing is capable of the same sort of plasticity as that seen in nematodes.

Acknowledgement

The author would like to thank Simon Harvey and Mark Viney for communication of unpublished data, Gordon Lithgow, Rick Maizels, Andrew Read and Mark Viney for reading parts of this chapter, and Susan Demuth and Judith Leeb for translations of French and German texts. Any errors or infelicities that remain are entirely the fault of the author.

References

Abdulrahman, M. and Samoiloff, M. (1975) Sex-specific aging in the nematode *Panagrellus redivivus*. *Canadian Journal of Zoology*, **53**, 651–656.

Adachi, H., Fujiwara, Y. and Ishii, N. (1998) Effects of oxygen on protein carbonyl and aging in *Caenorhabditis elegans* mutants with long (*age-1*) and short (*mev-1*) life spans. *Journal of Gerontology*, **53A**, B240–B244.

Adams, D.B. and Beh, K.J. (1981) Immunity acquired by sheep from an experimental infection with *Haemonchus contortus*. *International Journal of Parasitology*, **11**, 381–386.

Albert, P.S., Brown, S.J. and Riddle, D.L. (1981) Sensory control of dauer larva formation in *C. elegans*. *Journal of Comparative Neurology*, **198**, 435–451.

Albiez, E.J., Büttner, D.W. and Schulz-Key, H. (1984) Studies on nodules and adult *Onchocerca volvulus* during a nodulectomy trial in hyperendemic villages in Liberia and Upper Volta. II. Comparison of the macrofilaria population in adult nodule carriers. *Tropenmedizin und Parasitologie*, **35**, 163–166.

Anderson, G.L. (1978) Responses of dauerlarvae of *Caenorhabditis elegans* (Nematoda: Rhabditidae) to thermal stress and oxygen deprivation. *Canadian Journal of Zoology*, **56**, 1786–1791.

Anderson, G.L. (1982) Superoxide dismutase activity in dauerlarvae of *Caenorhabditis elegans* (Nematoda: Rhabditidae). *Canadian Journal of Zoology*, **60**, 288–291.

Anderson, R.M. and May, R.M. (1982) Coevolution of hosts and parasites. *Parasitology*, **85**, 411–426.

Antebi, A., Culotti, J.G. and Hedgecock, E.M. (1998) *daf-12* regulates developmental age and the dauer alternative in *Caenorhabditis elegans*. *Development*, **125**, 1191–1205.

Apfeld, J. and Kenyon, C. (1998) Cell nonautonomy of *C. elegans daf-2* function in the regulation of diapause and life span. *Cell*, **95**, 199–210.

Ashford, R.W. and Barnish, G. (1989) *Strongyloides fuelleborni* and similar parasites in animals and man. In *Strongyloidiasis: A major roundworm infection of man*, edited by D.I. Grove, pp. 271–286. London: Taylor and Francis.

Augustine, D.L. (1940) Experimental studies on the validity of species in the genus *Strongyloides*. *American Journal of Health*, **32**, 24–32.

Bancroft, J. (1879) New cases of filariasis disease. *Lancet*, 698.

Bargmann, C.I. and Horvitz, H.R. (1991) Control of larval development by chemosensory neurons in *Caenorhabditis elegans*. *Science*, 251, 1243–1246.

Beaver, P.C. (1966) Zoonoses, with particular reference to parasites of veterinary importance. In *Biology of Parasites*, edited by E.J.L. Soulsby, pp. 215–227. New York: Academic Press.

Blaxter, M.L., De Ley, P., Garey, J.R., Liu, L.X., Scheldeman, P., Vierstraete, A., Vanfleteren, J.R., Mackey, L.Y., Dorris, M., Frisse, L.M., Vida, J.T. and Thomas, W.K. (1998) A molecular evolutionary framework for the phylum Nematoda. *Nature*, 392, 71–75.

Bolanowski, M.A., Jacobson, L.A. and Russell, R.L. (1983) Quantitative measures of ageing in the nematode *Caenorhabditis elegans*: II. Lysosomal hydrolases as markers for senescence. *Mechanisms of Ageing and Development*, 21, 295–319.

Bolanowski, M.A., Russell, R.L. and Jacobson, L.A. (1981) Quantitative measures of aging in the nematode *Caenorhabditis elegans*. I. Population and longitudinal studies of two behavioural parameters. *Mechanisms of Ageing and Development*, 15, 276–270.

Bolla, R.I. and Roberts, L.S. (1968) Gametogenesis and chromosomal complement in *Strongyloides ratti* (Nematoda: Rhabdiasoidea). *Journal of Parasitology*, 54, 849–855.

Brandt, F.H. and Eberhard, M.L. (1990a) Distribution, behavior and course of patency of *Dracunculus insignis* in experimentally infected ferrets. *Journal of Parasitology*, 76, 515–518.

Brandt, F.H. and Eberhard, M.L. (1990b) *Dracunculus insignis* in ferrets: comparison of inoculation routes. *Journal of Parasitology*, 76, 93–95.

Brooks, A. and Johnson, T.E. (1991) Genetic specification of lifespan and self-fertility in recombinant-inbred strains of *Caenorhabditis elegans*. *Heredity*, 67, 19–28.

Brooks, A., Lithgow, G.J. and Johnson, T.E. (1994) Mortality rates in a genetically heterogeneous population of *Caenorhabditis elegans*. *Science*, 263, 668–671.

Brown-Borg, H.M., Borg, K.E., Meliska, C.J. and Bartke, A. (1996) Dwarf mice and the ageing process. *Nature*, 384, 33.

Buecher, E. and Hansen, E. (1974) *IRCS Library Compendium*, 2, 1595.

Büttner, D.W., Albiez, E.J. and Parrow, D. (1983) Parasitological studies on *Onchocerca volvulus* eight years after interruption of the transmission in Upper Volta. *Bulletin de la Société de Pathologie Exotique et de ses Filiales*, 76, 669–675.

Carme, B. and Laigret, J. (1979) Longevity of *Wuchereria bancrofti* var. *pacifica* and mosquito infection acquired from a patient with low level parasitemia. *American Journal of Tropical Medicine and Hygiene*, 28, 53–55.

Cassada, R.C. and Russell, R.L. (1975) The dauerlarva, a post-embryonic developmental variant of the nematode *Caenorhabditis elegans*. *Developmental Biology*, 46, 326–342.

Castillo, J.M., Kisiel, M.J. and Zuckerman, B.M. (1975) Studies on the effects of two procaine preparations on *C. elegans*. *Nematologica*, 21, 401–407.

C. elegans Consortium (1998) Genome sequence of the nematode *C. elegans*: A platform for investigating biology. *Science*, 282, 2012–2018.

Chandler, A.C. (1925) The rate of loss of hookworms in the absence of reinfections. *Indian Journal of Medical Research*, 13, 625–634.

Chandler, A.C. and Read, C.P. (1961) *Introduction to Parasitology*. New York: Wiley,.

Christie, J.R. (1929) Some observations on sex in the Mermithidae. *Journal of Experimental Zoology*, 53, 59–77.

Chu, T.C. (1936) Studies on life history of *Rhabdias fuscovenosa* var. *catanensis* (Rizzo, 1902). *Journal of Parasitology*, 22, 140–160.

Clokey, G.V. and Jacobsen, L.A. (1986) The autofluorescent 'lipofuscin' granules in the intestinal cells of *Caenorhabditis elegans* are secondary lysosomes. *Mechanisms of Ageing and Development*, 35, 79–94.

Coburn, C.M., Mori, I., Ohshima, Y. and Bargmann, C.I. (1998) A cyclic nucleotide-gated channel inhibits sensory axon outgrowth in larval and adult *Caenorhabditis elegans*: a distinct pathway for maintenance of sensory axon structure. *Development*, 125, 249–258.

Conn, H.C. and Greenslit, F.S. (1952) Filariasis residuals in veterans with report of a case of microfilaremia. *American Journal of Tropical Medicine and Hygiene*, 1, 474–476.

Coutelen, F. (1935) La longévité de la filaire *Loa loa* (Guyet, 1778) et des embryons de filaire. A propos d'un cas de filaroise diurne. *Bulletin de la Société de Pathologie Exotique*, 28, 126–134.

Croll, N.A., Anderson, R.M., Gyoarkos, T.W. and Ghadirian, E. (1982) The population biology and control of *Ascaris lumbricoides* in a rural community in Iran. *Transactions of the Royal Society of Tropical Medicine and Hygiene (London)*, 76, 187–197.

Croll, N.A. and Matthews, B.E. (1973) Activity, ageing and penetration in hookworm larvae. *Parasitology*, 66, 279–289.

Croll, N.A., Smith, J.M. and Zuckerman, B.M. (1977) The aging process of the nematode *Caenorhabditis elegans* in bacterial and axenic culture. *Experimental Aging Research*, 3, 175–189.

Cutler, R.G. (1982) The dysdifferentiative hypothesis of mammalian aging and longevity. In *The Aging Brain: Cellular and Molecular Mechanisms of Aging in the Nervous System*, edited by E. Giacobini, G. Filogamo, G. Giacobini and A. Vernadakis. New York: Raven Press.

Dalley, B.K. and Golomb, M. (1992) Gene expression in the *Caenorhabditis elegans* dauer larva: developmental regulation of Hsp90 and other genes. *Developmental Biology*, 151, 80–90.

Darr, D. and Fridovich, I. (1995) Adaptation to oxidative stress in young, but not in mature or old, *Caenorhabditis elegans*. *Free Radical Biology and Medicine*, 18, 195–201.

Davis, B.O., Anderson, G.L. and Dusenbery, D.B. (1982) Total luminescence spectroscopy of fluorescence changes during aging in *Caenorhabditis elegans*. *Biochemistry*, 21, 4089–4095.

Dawkins, H.J.S. (1989) *Strongyloides ratti* infections in rodents: value and limitations as a model for human strongyloidiasis. In *Strongyloidiasis a major roundworm infection of man*, edited by D.I. Grove, pp. 287–332. London: Taylor and Francis.

De Cuyper, C. and Vanfleteren, J.R. (1982) Nutritional alteration of life span in the nematode *Caenorhabditis elegans*. *Age (Omaha Nebraska)*, 5, 42–45.

Dorman, J.B., Albinder, B., Shroyer, T. and Kenyon, C. (1995) The *age-1* and *daf-2* genes function in a common pathway to control the lifespan of *Caenorhabditis elegans*. *Genetics*, 141, 1399–1406.

Duggal, C. (1978a) Copulatory behaviour of male *Panagrellus redivivus*. *Nematologica*, 24, 257–268.

Duggal, C. (1978b) Initiation of copulation and its effect on oocyte production and life span of adult female *Panagrellus redivivus*. *Nematologica*, 24, 269–276.

Duhon, S.A. and Johnson, T.E. (1995) Movement as an index of vitality: comparing wild-type and the *age-1* mutant of *Caenorhabditis elegans*. *Journal of Gerontology*, 50A, B254–B261.

Duke, B.O.L. (1980) Observations on *Onchocerca volvulus* in experimentally infected chimpanzees. *Tropenmedizin und Parasitologie*, 31, 41–54.

Duret, L., Guex, N., Peitsch, M.C. and Bairoch, A. (1988) New insulin-like proteins with atypical disulfide bond pattern characterized in *Caenorhabditis elegans* by comparative sequence analysis and homology modeling. *Genome Research*, 8, 348–353.

Ebert II, R.H., Cherkasova, V.A., Dennis, R.A., Wu, J.H., Ruggles, S., Perrin, T.E. and Shmookler Reis, R.J. (1993) Longevity-determining genes in *Caenorhabditis elegans*: chromosomal mapping of multiple noninteractive loci. *Genetics*, 135, 1003–1010.

Egilmez, N.K. and Rothstein, M. (1985) The effect of aging on cell-free protein synthesis in the free-living nematode *Turbatrix aceti*. *Biochimica et Biophysica Acta*, 840, 355–363.

Epstein, J. and Gershon, D. (1972) Studies on ageing in nematodes IV. The effect of anti-oxidants on cellular damage and life span. *Mechanisms of Ageing and Development*, 1, 257–264.

Epstein, J., Himmelhoch, S. and Gershon, D. (1972) Studies on aging in nematodes III. Electronmicroscopical studies on age-associated cellular damage. *Mechanisms of Ageing and Development*, 1, 245–255.

Estevez, M., Attisano, L., Wrana, J.L., Albert, P.S., Massague, J. and Riddle, D.L. (1993) The *daf-4* gene encodes a bone morphogenetic protein receptor controlling *C. elegans* dauer larva development. *Nature*, 365, 644–649.

Eveland, L.K., Yermakov, V. and Kenney, M. (1975) *Loa loa* infection without microfilaremia. *Transactions of the Royal Society of Tropical Medicine and Hygiene (London)*, 69, 354–355.

Ewbank, J.J., Barnes, T.M., Lakowski, B., Lussier, M., Bussey, H. and Hekimi, S. (1997) Structural and functional conservation of the *Caenorhabditis elegans* timing gene *clk-1*. *Science*, 275, 980–983.

Fabian, T.J. and Johnson, T.E. (1995) Identification of genes that are differentially expressed during aging in *Caenorhabditis elegans*. *Journal of Gerontology*, 50A, B245–B253.

Finch, C.E. (1991) *Longevity, Senescence and the Genome*. Chicago and London: University of Chicago Press.

Fisher, J.M. (1969) Investigations on fecundity of *Aphelenchus avenae*. *Nematologica*, 15, 22–28.

Franz, M. and Büttner, D.W. (1983) The fine structure of adult *Onchocerca volvulus*. V. The digestive tract and the reproductive system of the female worm. *Tropenmedizin und Parasitologie*, 34, 155–161.

Friedman, D.B. and Johnson, T.E. (1988a) A mutation in the *age-1* gene in *Caenorhabditis elegans* lengthens life and reduces hermaphrodite fertility. *Genetics*, **118**, 75–86.

Friedman, D.B. and Johnson, T.E. (1988b) Three mutants that extend both mean and maximum life span of the nematode, *Caenorhabditis elegans*, define the *age-1* gene. *Journal of Gerontology A*, **43**, B102–B109.

Gandhi, S., Santelli, J., Mitchell, D.G., Stiles, J.W. and Raosanadi, D. (1980) A simple method for maintaining large, aging populations of *Caenorhabditis elegans*. *Mechanisms of Ageing and Development*, **12**, 137–150.

Gemmill, A.W., Skorping, A. and Read, A.F. (1999) Optimal timing of first reproduction in parasitic nematodes. *Journal of Evolutionary Biology*, **12**, 1148–1156.

Gemmill, A.W., Viney, M.E. and Read, A.F. (1997) Host immune status determines sexuality in a parasitic nematode. *Evolution*, **51**, 393–401.

Gems, D. (2000) An integrated theory of aging in the nematode *Caenorhabditis elegans*. *Journal of Anatomy*, **In press**.

Gems, D. and Riddle, D.L. (1996) Longevity in *Caenorhabditis elegans* reduced by mating but not gamete production. *Nature*, **379**, 723–725.

Gems, D., Sutton, A.J., Sundermeyer, M.L., Albert, P.S., King, K.V., Edgley, M., Larson, P.L. and Riddle, D.L. (1998) Two pleiotropic classes of *daf-2* mutation affect larval arrest, adult behavior, reproduction and longevity in *Caenorhabditis elegans*. *Genetics*, **150**, 129–155.

Gems, D. and Riddle, D.L. (2000a) Defining wild-type life span in *Caenorhabditis elegans*. *Journal of Gerontology*, **55**, 215–219.

Gems, D. and Riddle, D.L. (2000b) Genetic, behavioral and environmental determinants of male longevity in *Caenorhabditis elegans*. *Genetics*, **154**, 1597–1610.

Georgi, L.L., Albert, P.S. and Riddle, D.L. (1990) *daf-1*, a *C. elegans* gene controlling dauer larva development, encodes a novel receptor protein kinase. *Cell*, **61**, 635–645.

Gershon, D. (1970) Studies on aging in nematodes I. The nematode as a model organism for ageing research. *Experimental Gerontology*, **5**, 7–12.

Giglio, M.-P., Hunter, T., Bannister, J.V., Bannister, W.H. and Hunter, G.J. (1994) The manganese superoxide dismutase gene of *Caenorhabditis elegans*. *Biochemistry and Molecular Biology International*, **33**, 37–40.

Goater, C.P. (1991) Experimental population dynamics for *Rhabdias bufonis* (Nematoda) in toads (*Bufo bufo*): density-dependence in primary infection. *Parasitology*, **104**, 179–187.

Golden, J.W. and Riddle, D.L. (1982) A pheromone influences larval development in the nematode *Caenorhabditis elegans*. *Science*, **218**, 578–580.

Golden, J.W. and Riddle, D.L. (1984) The *C. elegans* dauer larva: developmental effects of pheromone, food and temperature. *Developmental Biology*, **102**, 368–378.

Gomez-Saladin, E., Luebke, A.E., Wilson, D.L. and Dickerson, I.M. (1997) Isolation of a cDNA encoding a Kex2-like endoprotease with homology to furin from the nematode *Caenorhabditis elegans*. *DNA and Cell Biology*, **16**, 663–669.

Gompertz, B. (1825) Of the nature of the function expressive of the law of human mortality, and on a new mode of determining the value of life contingencies. *Philosophical Transactions of the Royal Society of London. Series A*, **57**, 513–585.

Gottlieb, S. and Ruvkun, G. (1994) *daf-2*, *daf-16* and *daf-23*: genetically interacting genes controlling dauer formation in *Caenorhabditis elegans*. *Genetics*, **137**, 107–120.

Gowen, S.R. (1970) Observations on the fecundity and longevity of *Tylenchus emarginatus* on sikta spruce seedlings at different temperatures. *Nematologica*, **16**, 267–272.

Graham, G.L. (1938) Studies on *Strongyloides*. III. The fecundity of single *S. ratti* of homogonic origin. *Journal of Parasitology*, **24**, 233–243.

Gregoire, F.M., Chomiki, N., Kachinskas, D. and Warden, C.H. (1998) Cloning and developmental regulation of a novel member of the insulin-like gene family in *Caenorhabditis elegans*. *Biochemical and Biophysical Research Communications*, **249**, 385–390.

Gupta, S.K. and Rothstein, M. (1976a) Phosphoglycerate kinase from young and old *Turbatrix aceti*. *Biochimica et Biophysica Acta*, **445**, 632–644.

Gupta, S.K. and Rothstein, M. (1976b) Triosephosphate isomerase from young and old *Turbatrix aceti*. *Arch. Biochem.*, **174**, 333–338.

Guptavanij, P. and Harinasuta, C. (1971) Spontaneous disappearance of microfilaria *Brugia malayi* and *Wuchereria bancrofti* in endemic patients living in a non-endemic area. *Southeast Asian Journal of Tropical Medicine and Public Health*, **2**, 578.

Hansen, E., Buecher, E.J. and Yarwood, E.A. (1964) Development and maturation of *Caenorhabditis briggsae* in response to growth factor. *Nematologica*, **10**, 623–630.

Hansen, E.L., Buecher, E.J. and Cryan, W.S. (1969) *Strongyloides fülleborni*: Environmental factors and free-living generations. *Experimental Parasitology*, **26**, 336–343.

Harrington, L.A. and Harley, C.B. (1988) Effect of vitamin E on lifespan and reproduction in *Caenorhabditis elegans*. *Mechanisms of Ageing and Development*, **43**, 71–78.

Hartman, P.S., Simpson, V.J., Johnson, T. and Mitchell, D. (1988) Radiation sensitivity and DNA repair in *Caenorhabditis elegans* strains with different mean life spans. *Mutation Research*, **208**, 77–82.

Harvey, P.H. and Keymer, A.E. (1991) Comparing life histories using phylogenies. *Philosophical Transactions of the Royal Society (London)*, **332**, 31–39.

Hekimi, B., Lakowski, B., Barnes, T.M. and Ewbank, J.J. (1998) Molecular genetics of life span in *C. elegans*: how much does it teach us? *Trends in Genetics*, **14**, 14–20.

Helfand, S.L., Blake, K.J., Rogina, B., Stracks, M.D., Centurion, A. and Naprta, B. (1995) Temporal patterns of gene expression in the antenna of the adult *Drosophila melanogaster*. *Genetics*, **140**, 549–555.

Hengartner, M.O. (1997) Genetic control of programmed cell death and aging in the nematode *Caenorhabditis elegans*. *Experimental Gerontology*, **32**, 363–374.

Hieb, W.F. and Rothstein, M. (1975) Aging in the free-living nematode *Turbatrix aceti*. Techniques for synchronization of large-scale axenic cultures. *Experimental Gerontology*, **10**, 145–153.

Himmelhoch, S., Kisiel, M. and Zuckerman, B. (1977) *Caenorhabditis briggsae*: electron microscope analysis of changes in negative surface charge density of the outer cuticular membrane. *Experimental Parasitology*, **41**, 118–123.

Himmelhoch, S. and Zuckerman, B.M. (1978) *Caenorhabditis briggsae*: aging and the structural turnover of the outer cuticle surface and the intestine. *Experimental Parasitology*, **45**, 208–214.

Hirschmann, H. (1962) The life cycle of *Ditylenchus triformis* (Nematoda: Tylenchida) with emphasis on post-embryonic development. *Proceedings of the Helminthological Society (Washington)*, **29**, 30–43.

Hodgkin, J. and Barnes, T.M. (1991) More is not better: brood size and population growth in a self-fertilizing nematode. *Proceedings of the Royal Society (London) B*, **246**, 19–24.

Högger, C., Estey, R., Kisiel, M. and Zuckerman, B. (1977) Surface scanning observations of changes in *Caenorhabditis elegans* during aging. *Nematologica*, **23**, 213–216.

Honda, H. (1925) Experimental and cytological studies on bisexual and hermaphroditic free-living nematodes. *Journal of Morphology*, **40**, 191–233.

Honda, S., Ishii, N., Suzuki, K. and Matsuo, M. (1993) Oxygen-dependent perturbation of life span and aging rate in the nematode. *Journal of Gerontology*, **48**, B57–B61.

Hosokawa, H., Ishii, N., Ishida, H., Ichimori, K., Nakazawa, H. and Suzuki, K. (1994) Rapid accumulation of fluorescent material with ageing in an oxygen-sensitive mutant *mev-1* of *Caenorhabditis elegans*. *Mechanisms of Ageing and Development*, **74**, 161–170.

Hosono, R. (1978) Age dependent changes in the behavior of *Caenorhabditis elegans* on attraction to *Escherichia coli*. *Experimental Gerontology*, **13**, 31–36.

Hosono, R., Nishimoto, S. and Kuno, S. (1989) Alteration of life span in the nematode *Caenorhabditis elegans* under monoxenic culture conditions. *Experimental Gerontology*, **24**, 251–264.

Hunter, T., Bannister, W.H. and Hunter, G.J. (1997) Cloning, expression, and characterization of two manganese superoxide dismutases from *Caenorhabditis elegans*. *Journal of Biological Chemistry*, **272**, 28652–28659.

Hyman, L.H. (1951) *The Invertebrates*. New York: McGraw-Hill,.

Ishii, N., Fujii, M., Hartman, P. S., Tsuda, M., Yasuda, K., Senoo-Matsuda, N., Yanase, S., Ayusawa, D. and Suzuki, K. (1998) A mutation in succinate dehydrogenase cytochrome *b* causes oxidative stress and ageing in nematodes. *Nature*, **394**, 694–697.

Ishii, N., Suzuki, N., Hartman, P.S. and Suzuki, K. (1994) The effects of temperature on the longevity of a radiation-sensitive mutant *rad-8* of the nematode *Caenorhabditis elegans*. *Journal of Gerontology*, **49**, B117–120.

Ishii, N., Takahashi, K., Tomita, S., Keino, T., Honda, S., Yoshino, K. and Suzuki, K. (1990) A methyl viologen-sensitive mutant of the nematode *Caenorhabditis elegans*. *Mutation Research*, **237**, 165–171.

Jachowski, L.A., Otto, G.F. and Wharton, J. (1951) Filariasis in American Samoa. I. Loss of microfilaria in the absence of continued reinfection. *Proceedings of the Helminthological Society (Washington)*, **18**, 25–28.

Johnson, M.H., Orihel, T.C. and Beaver, P.C. (1974) *Dipetalonema viteae* in the experimentally infected jird, *Meriones unguiculatus*. I. Insemination, development from egg to microfilaria, reinsemination, and longevity of mated and unmated worms. *Journal of Parasitology*, **60**, 302–309.

Johnson, T.E. (1987) Aging can be genetically dissected into component processes using long-lived lines of *Caenorhabditis elegans*. *Proceedings of the National Academy of Sciences, U.S.A.*, **84**, 3777–3781.

Johnson, T.E. (1990) The increased lifespan of *age-1* mutants of *Caenorhabditis elegans* results from a lowering of the Gompertz rate of aging. *Science*, **249**, 908–912.

Johnson, T.E. and Hutchinson, E.W. (1993) Absence of strong heterosis for life span and other life history traits in *Caenorhabditis elegans*. *Genetics*, **134**, 465–474.

Johnson, T.E. and McCaffrey, G. (1985) Programmed aging or error catastrophe? An examination by two-dimensional polyacrylamide gel electrophoresis. *Mechanisms of Ageing and Development*, **30**, 285–297.

Johnson, T.E. and Simpson, V.J. (1985) Aging studies in *Caenorhabditis elegans* and other nematodes. In *CRC Handbook on the Biology of Aging*, edited by V.J. Cristofalo, pp. 481–495. Boca Raton: CRC Press.

Johnson, T.E. and Wood, W.B. (1982) Genetic analysis of life-span in *Caenorhabditis elegans*. *Proceedings of the National Academy of Sciences, U.S.A.*, **79**, 6603–6607.

Jonassen, T., Marbois, B.N., Kim, L., Chin, A., Xia, Y.R., Lusis, A.J. and Clarke, C.F. (1996) Isolation and sequencing of the rat Coq7 gene and the mapping of mouse Coq7 to chromosome 7. *Archives of Biochemistry and Biophysics*, **330**, 285–289.

Jonassen, T., Proft, M., Randez-Gil, F., Schultz, J.R., Marbois, B.N., Entian, K.D. and Clarke, C.F. (1998) Yeast Clk-1 homologue (Coq7/Cat5) is a mitochondrial protein in coenzyme Q synthesis. *Journal of Biological Chemistry*, **273**, 3351–3357.

Kahn, M. and Enesco, H.E. (1981) Effect of alpha-tocopherol on the lifespan of *Turbatrix aceti*. *Age and Ageing*, **4**, 109–115.

Kapeller, R. and Cantley, L.C. (1994) Phosphatidylinositol 3-kinase. *BioEssays*, **16**, 565–576.

Karam, M., Schulz-Key, H. and Remme, J. (1987) Population dynamics of *Onchocerca volvulus* after 7 to 8 years of vector control in West Africa. *Acta Tropica*, **44**, 445–457.

Karey, K.P. and Rothstein, M. (1986) Evidence for the lack of lysosomal involvement in the age-related slowing of protein breakdown in *Turbatrix aceti*. *Mechanisms of Ageing and Development*, **35**, 169–178.

Keller, A.E. (1931) *Ascaris lumbricoides*: loss of reinfestation without treatment. *Journal of the American Medical Association*, **97**, 1299–1300.

Kenyon, C. (1997) Environmental factors and gene activities that influence life span. In *C. elegans II*, edited by D.L. Riddle, T. Blumenthal, B.J. Meyer and J.R. Priess, pp. 791–814. Plainview, N.Y.: Cold Spring Harbor Press,

Kenyon, C., Chang, J., Gensch, E., Rudener, A. and Tabtiang, R. (1993) A *C. elegans* mutant that lives twice as long as wild type. *Nature*, **366**, 461–464.

Kerboeuf, D. (1982) Egg output of *Heligmosomoides polygyrus* (*Nematospiroides dubius*) in mice infected once only. *Archives de Recherches Vétérinaire*, **13**, 69–78.

Keymer, A.E., Gregory, R.D., Harvey, P.H., Read, A.F. and Skorping, A. (1991) Parasite-host ecology: case studies in population dynamics, life history evolution and community structure. *Acta Oecologica*, **12**, 105–118.

Kimura, K.D., Tissenbaum, H.A., Liu, Y. and Ruvkun, G. (1997) *daf-2*, an insulin receptor-like gene that regulates longevity and diapause in *Caenorhabditis elegans*. *Science*, **277**, 942–946.

Kirkwood, T.B.L. (1977) Evolution of ageing. *Nature*, **270**, 301–304.

Kisiel, M. and Zuckerman, B. (1974) Studies of aging *Turbatrix aceti*. *Nematologica*, **20**, 277–282.

Kisiel, M. and Zuckerman, B.M. (1972) Effects of DNA synthesis inhibitors on *Caenorhabditis briggsae* and *Turbatrix aceti*. *Nematologica*, **18**, 373–384.

Kisiel, M.J., Castillo, J. M. and Zuckerman, B.M. (1974) Comparison of the effects of Procaine HCl and Gerovital H3 on aging parameters and development of a nematode. *Gerontologist*, **14**, 32.

Kisiel, M.J., Castillo, J.M., Zuckerman, L.S. and Zuckerman, B.M. (1975) Studies on ageing in *Turbatrix aceti*. *Mechanisms of Ageing and Development*, **4**, 81–88.

Kisiel, M.J. and Zuckerman, B.M. (1978) Effects of centrophenoxine on the nematode *Caenorhabditis briggsae*. *Age (Omaha, Nebraska)*, **1**, 17–20.

Klass, M.R. (1977) Aging in the nematode *Caenorhabditis elegans*: major biological and environmental factors influencing life span. *Mechanisms of Ageing and Development*, **6**, 413–429.

Klass, M.R. (1983) A method for the isolation of longevity mutants in the nematode *Caenorhabditis elegans* and initial results. *Mechanisms of Ageing and Development*, **22**, 279–286.

Klass, M.R. and Hirsh, D.I. (1976) Nonaging developmental variant of *C. elegans*. *Nature*, **260**, 523–525.

Knabe, K. (1932) Beitrag zur Dauer von Filarieninfecktion. *Archiv für Schiffs- und Tropenhygiene*, **36**, 496–500.

Krishnamoorthy, K., Sabesan, S., Vanamail, P. and Panicker, K.N. (1991) Influence of diethylcarbamazine on the patent period of infection in periodic *Brugia malayi*. *Indian Journal of Medical Research*, **93**, 240–244.

Lakowski, B. and Hekimi, S. (1996) Determination of life-span in *Caenorhabditis elegans* by four clock genes. *Science*, **272**, 1010–1013.

Lakowski, B. and Hekimi, S. (1998) The genetics of caloric restriction in *Caenorhabditis elegans*. *Proceedings of the National Academy of Sciences, U.S.A.*, **95**, 13091–13096.

Larsen, P.L. (1993) Aging and resistance to oxidative stress in *Caenorhabditis elegans*. *Proceedings of the National Academy of Sciences, U.S.A.*, **90**, 8905–8909.

Larsen, P.L., Albert, P.S. and Riddle, D.L. (1995) Genes that regulate both development and longevity in *Caenorhabditis elegans*. *Genetics*, **139**, 1567–1583.

Leeuwin, R.S. (1962) Microfilaraemia in Surinamese living in Amsterdam. *Tropical and Geographical Medicine*, **14**, 355–360.

Lin, K., Dorman, J.B., Rodan, A. and Kenyon, C. (1997) *daf-16*: An HNF-3/ forkhead family member that can function to double the life-span of *Caenorhabditis elegans*. *Science*, **278**, 1319–1322.

Lithgow, G.J. (1996) Invertebrate gerontology: the age mutations of *Caenorhabditis elegans*. *BioEssays*, **18**, 809–815.

Lithgow, G.J., White, T.M., Hinerfield, D.A. and Johnson, T.E. (1994) Thermotolerance of a long-lived mutant of *Caenorhabditis elegans*. *Journal of Gerontology*, **49**, B270–B276.

Lithgow, G.J., White, T.M., Melov, S. and Johnson, T.E. (1995) Thermotolerance and extended life-span conferred by single-gene mutations and induced by thermal stress. *Proceedings of the National Academy of Sciences, U.S.A.*, **92**, 7540–7544.

Liu, Z. and Ambros, V. (1989) Heterochronic genes control the stage-specific initiation and expression of the dauer developmental program in *Caenorhabditis elegans*. *Genes and Development*, **3**, 2039–2049.

Lloyd, S. (1993) *Toxocara canis*: the dog. In *Toxocara and Toxocariasis*, edited by J.W. Lewis and R.M. Maizels. Birmingham: Birbeck and Sons Limited.

Mahoney, L.E. and Aiu, P. (1970) Filariasis in Samoan immigrants to the United States. *American Journal of Tropical Medicine and Hygiene*, **19**, 629–631.

Malone, E.A., Inoue, T. and Thomas, J.H. (1996) Genetic analysis of the roles of *daf-28* and *age-1* in regulating *Caenorhabditis elegans* dauer formation. *Genetics*, **143**, 1193–1205.

Mansfield, L.S., Niamatali, S., Bhopale, V., Volk, S., Smith, G., Lok, J.B., Genta, R.M. and Schad, G.A. (1996) *Strongyloides stercoralis*: maintenance of exceedingly chronic infections. *American Journal of Tropical Medicine and Hygiene*, **55**, 617–624.

Manson-Bahr, P. (1925) On the longevity of the *Loa-loa* and some hitherto undescribed manifestations of this infection. *Archiv für Schiffs- und Tropenhygiene*, **29**, 222–224.

Manson-Bahr, P. (1959) The story of *Filaria bancrofti*. Part V. Description of *W. bancrofti* and pathology of filariasis. *Journal of Tropical Medicine and Hygiene*, **62**, 160–173.

Marbois, B.N. and Clarke, C.F. (1996) The *COQ7* gene encodes a protein in *Saccharomyces cerevisiae* necessary for ubiquinone biosynthesis. *Journal of Biological Chemistry*, **271**, 2995–3004.

Maupas, E. (1900) Modes et formes de reproduction des némadodes. *Archives de Zoologie Expérimentale et Génerale*, **8**, 463–647.

Mayberry, L.F., Bristol, J.R. and Villalobos, V.M. (1985) Intergeneric interactions between *Eimeria separata* (Apicomplexa) and *Nippostrongylus brasiliensis* (Nematoda) in the rat. *Experientia*, **41**, 689–690.

Mayhew, R.L. (1942) Preliminary note on the length of life of the stomach worm *Haemonchus contortus* in the calf. *Proceedings of the Helminthological Society (Washington)*, **9**, 28.

McCoy, O.R. (1932) Size of infection as an influence on the persistence of adult Trichinae in rats. *Science*, **75**, 364–365.

Medawar, P.B. (1952) *An Unsolved Problem Of Biology*. London: H.K. Lewis.

Meheus, L.A., Van Beeumen, J.J., Coomans, A.V. and Vanfleteren, J.R. (1987) Age-specific nuclear proteins in the nematode worm *Caenorhabditis elegans*. *Biochemical Journal*, **245**, 257–261.

Melov, S., Hertz, G.Z., Stormo, G. D. and Johnson, T.E. (1994) Detection of deletions in the mitochondrial genome of *Caenorhabditis elegans*. *Nucleic Acids Research*, **22**, 1075–1078.

Melov, S., Lithgow, G.J., Fischer, D.R., Tedesco, P.M. and Johnson, T.E. (1995) Increased frequency of deletions in the mitochondrial genome with age of *Caenorhabditis elegans*. *Nucleic Acids Research*, **23**, 1419–1425.

Mhaskar, K.S. (1920) Hookworm infection and sanitation. *Indian Journal of Medical Research*, **8**, 393–406.

Mitchell, D.H., Stiles, J.W., Santelli, J. and Sandini, D.R. (1979) Synchronous growth and aging of *Caenorhabditis elegans* in the presence of fluorodeoxyuridine. *Journal of Gerontology*, **34**, 28–36.

Monnier, V.M. and Cerami, A. (1981) Nonenzymatic browning in vivo: possible process for aging of long-lived proteins. *Science*, **211**, 491–493.

Morand, S. (1996) Life-history traits in parasitic nematodes: A comparative approach for the search for invariants. *Functional Ecology*, **10**, 210–218.

Morand, S. and Sorci, G. (1998) Determinants of life-history evolution in nematodes. *Parasitology Today*, **14**, 193–196.

Morris, J.Z., Tissenbaum, H.A. and Ruvkun, G. (1996) A phosphatidyli-nositol-3-OH kinase family member regulating longevity and diapause in *Caenorhabditis elegans*. *Nature*, **382**, 536–538.

Murakami, S. and Johnson, T.E. (1996) A genetic pathway conferring life extension and resistance to UV stress in *Caenorhabditis elegans*. *Genetics*, **143**, 1207–1218.

Murakami, S. and Johnson, T.E. (1998) Life extension and stress resistance in *Caenorhabditis elegans* modulated by the *tkr-1* gene. *Current Biology*, **8**, 1091–1094.

O'Riordan, V.B. and Burnell, A.M. (1989) Intermediary metabolism in the dauer larva of the nematode *Caenorhabditis elegans* – 1. Glycolysis, gluconeogenesis, oxidative phosphorylation and the tricarboxylic acid cycle. *Comparative Biochemistry and Physiology. B: Comparative Biochemistry*, **92B**, 233–238.

O'Riordan, V.B. and Burnell, A.M. (1990) Intermediary metabolism in the dauer larva of the nematode *Caenorhabditis elegans* – II. The glyoxylate cycle and fatty-acid oxidation. *Comparative Biochemistry and Physiology. B: Comparative Biochemistry*, **95B**, 125–130.

Ogg, S., Paradis, S., Gottlieb, S., Patterson, G.I., Lee, L., Tissenbaum, H.A. and Ruvkun, G. (1997) The Fork head transcription factor DAF-16 transduces insulin-like metabolic and longevity signals in *C. elegans*. *Nature*, **389**, 994–999.

Ogg, S. and Ruvkun, G. (1998) The *C. elegans* PTEN homolog, DAF-18, acts in the insulin receptor-like metabolic signaling pathway. *Molecular Cell*, **2**, 887–893.

Ohba, K. and Ishibashi, N. (1981) Effects of procaine on the development, longevity and fecundity of *Caenorhabditis elegans*. *Nematologica*, **27**, 275–284.

Olsen, O.W. (1974) *Animal Parasites. Their Life Cycles and Ecology*. New York: Dover Publications, Inc.

Olsen, O.W. and Lyons, E.T. (1965) Life cycle of *Uncinaria lucasi* Stiles, 1901 (Nematoda: Ancylostomatidae) of fur seals, *Callorhinus ursinus* Linn., on the Pribilof Islands, Alaska. *Journal of Parasitology*, **51**, 689–700.

Ono, T. and Cutler, R.G. (1978) Age-dependent relaxation of gene repression: increase of endogenous leukemia virus-related and globin-related RNA in brain and liver of mice. *Proceedings of the National Academy of Sciences, U.S.A.*, **75**, 4431–4435.

Orgel, L.E. (1963) The maintenance of the accuracy of protein synthesis and its relevance to aging. *Proceedings of the National Academy of Sciences, U.S.A.*, **49**, 517–521.

Orihel, T.C. and Eberhard, M.L. (1985) *Loa-loa* – Development and course of patency in experimentally infected primates. *Tropical Medicine and Parasitology*, **36**, 215–224.

Otto, G.F. (1932) Ascaris and Trichuris in the Southern United States. *Journal of Parasitology*, **18**, 200.

Palmer, E.D. (1955) Course of egg output over a 15 year period in a case of experimentally induced necatoriasis americanus, in the absence of hyperinfection. *American Journal of Tropical Medicine and Hygiene*, **4**, 756–757.

Paradis, S. and Ruvkun, G. (1998) *Caenorhabditis elegans* Akt/PKB transduces insulin receptor-like signals from AGE-1 PI3 kinase to the DAF-16 transcription factor. *Genes and Development*, **12**, 2488–2498.

Partridge, L. and Barton, N.H. (1993) Optimality, mutation and the evolution of ageing. *Nature*, **362**, 305–311.

Patterson, G.I., Koweek, A., Wong, A., Liu, Y. and Ruvkun, G. (1997) The DAF-3 Smad protein antagonizes TGF-β-related receptor signaling in the *Caenorhabditis elegans* dauer pathway. *Genes and Development* **11**, 2679–2690.

Plaisier, A.P., van Oortmarssen, G.J., Remme, J. and Habbema, J.D.F. (1991) The reproductive life span of *Onchocerca volvulus* in West African Savanna. *Acta Tropica*, **48**, 271–284.

Porta, E.A. (1991) Advances in age pigment research. *Archives of Gerontology and Geriatrics*, **12**, 303–320.

Read, A. and Harvey, P. (1989) Life history differences among eutherian radiations. *Journal of Zoology*, **219**, 329–353.

Reiss, U. and Rothstein, M. (1974) Heat-labile isozymes of isocitrate lyase from aging *Turbatrix aceti*. *Biochemical and Biophysical Research Communications*, **61**, 1012–1016.

Reiss, U. and Rothstein, M. (1975) Age-related changes in isocitrate lyase from the free-living nematode, *Turbatrix aceti*. *Journal of Biological Chemistry*, **250**, 826–830.

Remme, J., de Sole, G. and van Oortmarssen, G.J. (1990) The predicted and observed decline in onchocerciasis infection during 14 years of successful control of *Simulium* spp. in West Africa. *Bulletin of the World Health Organisation*, **68**, 331–339.

Ren, P., Lim, C.-S., Johnsen, R., Albert, P.S., Pilgrim, D. and Riddle, D.L. (1997) Control of *C. elegans* larval development by neuronal expression of a novel TGF-beta homolog. *Science*, **274**, 1389–1392.

Reznick, A.Z. and Gershon, D. (1977) Age related alterations in purified fructose 1,6-diphosphate aldolase from the nematode *Turbatrix aceti*. *Mechanisms of Ageing and Development*, **6**, 345–353.

Riddle, D.L. and Albert, P.S. (1997) Genetic and environmental regulation of dauer larva development. In *C. elegans II*, edited by D.L. Riddle, T. Blumenthal, B.J. Meyer and J.R. Priess, pp. 739–768. Plainview, N.Y.: Cold Spring Harbor Laboratory Press.

Riddle, D.L., Blumenthal, T., Meyer, B.J. and Priess, J.R. (1997) *C. elegans II*. Plainview, NY: Cold Spring Harbor Laboratory Press.

Roberts, J.M.D., Neumann, E., Göckel, C.W. and Highton, R.B. (1967) Onchocerciasis in Kenya 9, 11 and 18 years after elimination of the vector. *Bulletin of the World Health Organisation*, **37**, 195–212.

Rogina, B., Benzer, S. and Helfand, S.L. (1997) *Drosophila* drop-dead mutations accelerate the time course of age-related markers. *Proceedings of the National Academy of Sciences, U.S.A.*, **94**, 6303–6306.

Rogina, B. and Helfand, S.L. (1996) Timing of expression of a gene in the adult *Drosophila* is regulated by mechanisms independent of temperature and metabolic rate. *Genetics*, **143**, 1643–1651.

Rose, M. (1991) *Evolutionary Biology of Aging*. Oxford University Press, Oxford.

Rothstein, M. (1975) Aging and the alteration of enzymes: a review. *Mechanisms of Ageing and Development*, **4**, 325–338.

Rothstein, M. (1977) Recent developments in the age-related alteration of enzymes: a review. *Mechanisms of Ageing and Development*, **6**, 241–257.

Rothstein, M. (1980) Effects of aging on enzymes. In *Nematodes as Biological Models. Ageing and Other Model Systems*, Vol. 2, edited by B.M. Zuckerman, pp. 29–46. New York: Academic Press.

Rothstein, M. and Sharma, H.K. (1978) *Mechanisms of Ageing and Development*, **8**, 175–180.

Sabesan, S., Krishnamoorthy, K., Panicker, K.N. and Vanamail, P. (1991) The dynamics of microfilaraemia and its relation with development of disease in periodic *Brugia malayi* infection in South India. *Epidemiology and Infection*, **107**, 453–463.

Saifee, O., Wei, L.P. and Nonet, M.L. (1998) The *Caenorhabditis elegans unc-64* locus encodes a syntaxin that interacts genetically with synapto-brevin. *Molecular Biology of the Cell*, **9**, 1235–1252.

Sandground, J.H. (1936) On the potential longevity of various helminths with a record for a species of Trichostrongylus in man. *Journal of Parasitology*, **22**, 464–470.

Sarkis, G.J., Ashcom, J.D., Hawdon, J.M. and Jacobson, L.A. (1988) Decline in protease activities with age in the nematode *Caenorhabditis elegans*. *Mechanisms of Ageing and Development*, **45**, 191–201.

Sarles, M.P. (1929a) Length of life and rate of loss of dog hookworm, *Ancylostoma caninum*. *American Journal of Hygiene*, **10**, 667–682.

Sarles, M.P. (1929b) Quantitative studies on the dog and cat hookworm, *Ancylostoma brasiliense*, with special emphasis on age resistance. *American Journal of Hygiene*, **10**, 453–475.

Schackwitz, W.S., Inoue, T. and Thomas, J.H. (1996) Chemosensory neurons function in parallel to mediate a pheromone response in *C. elegans*. *Neuron*, **17**, 719–728.

Schad, G.A. (1989) Morphology and life history of *Strongyloides stercoralis*. In *Strongyloidiasis: A Major Roundworm Infection of Man*, edited by D.I. Grove. London: Taylor and Francis.

Schulz-Key, H., Jean, B. and Albiez, E.J. (1980) Investigations on female *Onchocerca volvulus* for the evaluation of drug trials. *Tropenmedizin und Parasitologie*, **31**, 34–40.

Searcy, D.G. (1976) Age-related increase of cuticle permeability in the nematode *Caenorhabditis briggsae*. *Experimental Aging Research*, **2**, 293–301.

Sharma, H.K., Gupta, S.K. and Rothstein, M. (1976) Age-related alteration of enolase in the free-living nematode, *Turbatrix aceti*. *Archives of Biochemistry and Biophysics*, **174**, 324–332.

Sharma, H.K., Prasanna, H.R., Lane, R.S. and Rothstein, M. (1979) Effect of age on enolase turnover in the free-living nematode, *Turbatrix aceti*. *Archives of Biochemistry and Biophysics*, **194**, 275–282.

Sharma, H.K. and Rothstein, M. (1978a) Age-related changes in the properties of enolase from *Turbatrix aceti*. *Biochemistry*, **17**, 2869–2876.

Sharma, H.K. and Rothstein, M. (1978b) Serological evidence for the alteration of enolase during aging. *Mechanisms of Ageing and Development*, **8**, 341–354.

Sharma, H.K. and Rothstein, M. (1980) Altered enolase in aged *Turbatrix aceti* from conformational changes in the enzyme. *Proceedings of the National Academy of Sciences, U.S.A.*, **77**, 5865–5868.

Sheldon, A.J. (1937) The rate of loss of worms (*Strongyloides ratti*) in rats. *American Journal of Hygiene*, **26**, 352–354.

Shook, D.R., Brooks, A. and Johnson, T.E. (1996) Mapping quantitative trait loci affecting life history traits in the nematode *Caenorhabditis elegans*. *Genetics*, **142**, 801–817.

Skorping, A., Read, A.F. and Keymer, A.E. (1991) Life history covariation in intestinal nematodes of mammals. *Oikos*, **60**, 365–372.

Sohlenius, B. (1968) Influence of microorganism and temperature upon some rhabditid nematodes. *Pedobiologia*, **9**, 243–253.

Sohlenius, B. (1969) Studies on the population development of *Mesodiplogaster biformis* (Nematoda, Rhabditida) in agar culture. *Pedobiologia*, **9**, 243–253.

Sohlenius, B. (1973) Growth and reproduction of a nematode. *Acrobeloides* sp. cultivated on agar. *Oikos*, **24**, 64–72.

Sorci, G., Morand, S. and Hugot, J.P. (1997) Host-parasite coevolution: comparative evidence for covariation of life history traits in primate and oxyurid parasites. *Proceedings of the Royal Society of London. Series B: Biological Sciences (London)*, **264**, 285–289.

Speare, R. (1989) Identification of species of *Strongyloides*. In *Strongyloidiasis: A major roundworm infection of man*, edited by D.I. Grove, pp. 11–83. London: Taylor and Francis.

Spieler, M. and Schierenberg, E. (1995) On the development of the alternating free-living and parasitic generations of the nematode *Rhabdias bufonis*. *Invertebrate Reproduction and Development*, **28**, 193–203.

Stadtman, E.R. (1992) Protein oxidation and aging. *Science*, **257**, 1220–1224.

Stearns, S.C. (1983) The influence of size and phylogeny on patterns of covariation among life history traits in the mammals. *Oikos*, **41**, 173–187.

Suzuki, K., Hyodo, M., Ishii, N. and Moriya, Y. (1978) Properties of a strain of free-living nematode, Rhabditidae sp.: life cycle and age-related mortality. *Experimental Gerontology*, **13**, 323–333.

Suzuki, N., Inokuma, K., Yasuda, K. and Ishii, N.(1992) Cloning, sequencing and mapping of a manganese superoxide dismutase gene of the nematode *Caenorhabditis elegans*. *DNA Research*, **3**, 171–174.

Targovnik, H.S., Locher, S.E., Hart, T.F. and Hariharan, P.V. (1984) Age-related changes in the excision repair capacity of *Turbatrix aceti*. *Mechanisms of Ageing and Development*, **27**, 73–81.

Thomas, J.H., Birnby, D.A. and Vowels, J.J. (1993) Evidence for parallel processing of sensory information controlling dauer formation in *Caenorhabditis elegans*. *Genetics*, **134**, 1105–1117.

Trent, S.C. (1963) Reevaluation of World War II veterans with filariasis acquired in the South Pacific. *American Journal of Tropical Medicine and Hygiene*, **12**, 877–887.

Van Voorhies, W.A. (1992) Production of sperm reduces nematode lifespan. *Nature*, **360**, 456–458.

Vanamail, P., Ramaiah, K.D., Pani, S.P., Das, P.K., Grenfell, B.T. and Bundy, D.A.P. (1996) Estimation of the fecund life span of *Wuchereria bancrofti* in an endemic area. *Transactions of the Royal Society for Tropical Medicine and Hygiene*, **90**, 119–121.

Vanamail, P., Subramanian, S., Das, P.K., Pani, S.P. and Rajagopalan, P.K. (1990) Estimation of fecundic life span of *Wuchereria bancrofti* from longitudinal study of human infection in an endemic area of Pondicherry (South India). *Indian Journal of Medical Research*, **91**, 293–297.

Vanamail, P., Subramanian, S., Das, P.K., Pani, S.P., Rajagopalan, P.K., Bundy, D.A.P. and Grenfell, B.T. (1989) Estimation of age-specific rates of acquisition and loss of *Wuchereria bancrofti* infection. *Transactions of the Royal Society for Tropical Medicine and Hygiene*, **83**, 689–693.

Vanfleteren, J.R. (1993) Oxidative stress and ageing in *Caenorhabditis elegans*. *Biochemical Journal*, **292**, 605–608.

Vanfleteren, J.R. and De Vreese, A. (1994) Analysis of the proteins of aging *Caenorhabditis elegans* by high resolution two-dimensional gel electrophoresis. *Electrophoresis*, **15**, 289–296.

Vanfleteren, J.R. and De Vreese, A. (1995) The gerontogenes *age-1* and *daf-2* determine metabolic rate potential in aging *Caenorhabditis elegans*. *FASEB Journal*, **9**, 1355–1361.

Vanfleteren, J.R. and De Vreese, A. (1996) Rate of aerobic metabolism and superoxide production rate potential in the nematode *Caenorhabditis elegans*. *Journal of Experimental Zoology*, **274**, 93–100.

Varkey, J.P., Muhlrad, P.J., Minniti, A.N., Do, B. and Ward, S. (1996) The *Caenorhabditis elegans spe-26* gene is necessary to form spermatids and encodes a protein similar to the actin-associated proteins kelch and scruin. *Genes and Development*, **9**, 1074–1086.

Vaupel, J.W., Johnson, T.E. and Lithgow, G.J. (1994) Rates of mortality in populations of *Caenorhabditis elegans*. *Science*, **266**, 826.

Vogel, K.G. (1974) Temperature and length of life in *Turbatrix aceti*. *Nematologica*, **29**, 361–362.

Vowels, J.J. and Thomas, J.H. (1992) Genetic analysis of chemosensory control of dauer formation in *Caenorhabditis elegans*. *Genetics*, **130**, 105–123.

Wachter, K.W. and Finch, C.E. (1997) *Between Zeus and the Salmon*. Washington D.C.: National Academy Press.

Wadsworth, W.G. and Riddle, D.L. (1989) Developmental regulation of energy metabolism in *Caenorhabditis elegans*. *Developmental Biology*, **132**, 167–173.

Wakelin, D. and Selby, G.R. (1974) The induction of immunological tolerance to the parasitic nematode *Trichuris muris* in cortisone-treated mice. *Immunology*, **26**, 1–10.

Wang, P.-Y., Zhen, T.-M., Wang, Z.-Z., Gu, Z.-F., Ren, S.-P. and Liu, L.-H. (1994) A ten-year observation on experimental infection of periodic *Brugia malayi* in man. *Journal of Tropical Medicine and Hygiene*, **97**, 269–276.

Watkins, A.R.J. and Fernando, M.A. (1986) Arrested development of the rabbit stomach worm *Obeliscoides cuniculi*: resumption of development of arrested larvae throughout the course of a single infection. *International Journal of Parasitology*, **16**, 47–54.

Willets, D.G. (1917) A statistical study of intestinal helminths. *Southern Medical Journal*, **10**, 42–49.

Williams, G.C. (1957) Pleiotropy, natural selection and the evolution of senescence. *Evolution*, **11**, 398–411.

Wilson, T. and Ramachandran, C.P. (1971) *Brugia* infections in man and animals: Long term observations on microfilaremia and estimates of the efficiency of transmission from mosquito vector to definitive host. *Annals of Tropical Medicine and Parasitology*, **65**, 525–546.

Wolf, H., Schulz-Key, H., Albiez, E.J., Geister, R. and Büttner, D.W. (1980) Analysis of enzymatically isolated adults of *Onchocerca volvulus* after treatment of patients with suramin or metrifonate. *Tropenmedizin und Parasitologie*, **31**, 143–148.

Wong, A.E., Boutis, P. and Hekimi, S. (1995) Mutations in the *clk-1* gene of *Caenorhabditis elegans* affect developmental and behavioral timing. *Genetics*, **139**, 1247–1259.

Wood, W.B. (1988) *The nematode Caenorhabditis elegans*. Plainview, N.Y.: Cold Spring Harbor Press,

Wood, W.B. (1998) Aging of *C. elegans*: mosaics and mechanisms. *Cell*, **95**, 147–150.

Yamada, M., Matsuda, S., Nakazawa, M. and Arizono, N. (1991) Species-specific differences in heterogonic development of serially transferred free-living generations of *Strongyloides planiceps* and *Strongyloides stercoralis*. *Journal of Parasitology*, **77**, 592–594.

Yeh, W.-Y. (1991) Genes acting late in the signalling pathway for *Caenorhabditis elegans* dauer larval development. *Ph.D. Thesis*, University of Missouri, Columbia, Missouri, USA.

Zeelon, P., Gershon, H. and Gershon, D. (1973) Inactive enzyme molecules in ageing organisms. Nematode fructose-1,6-diphosphate aldolase. *Biochemistry*, **12**, 1743–1750.

Zuckerman, B., Fagerson, I. and Kisiel, M. (1978) Age pigment studies on the nematode *Caenorhabditis briggsae*. *Age (Omaha Nebraska)*, **1**, 26–27.

Zuckerman, B., Himmelhoch, S. and Kisiel, M. (1973) Fine structure changes in the cuticle of adult *Caenorhabditis briggsae* with age. *Nematologica*, **19**, 109–112.

Zuckerman, B., Nelson, B. and Kisiel, M. (1972) Specific gravity increase of *Caenorhabditis elegans* with age. *Journal of Nematology*, **4**, 261–262.

Zuckerman, B.M. (1980) *Nematodes as Biological Models: Aging and Other Model Systems*, Vol. 2. New York: Academic Press.

Zuckerman, B.M. and Barrett, K.A. (1978) Effects of PCA and DMAE on the nematode *Caenorhabditis briggsae*. *Experimental Aging Research*, **4**, 133–139.

Zuckerman, B.M. and Geist, M.A. (1983) Effects of vitamin E on the nematode *Caenorhabditis elegans*. *Age (Omaha Nebraska)*, **6**, 1–4.

Zuckerman, B.M. and Himmelhoch, S. (1980) Nematodes and models to study aging. In *Nematodes as Biological Models. Ageing and Other Model Systems*, Vol. 2, edited by B.M. Zuckerman. New York: Academic Press.

Zuckerman, B.M., Himmelhoch, S., Nelson, B., Epstein, J. and Kisiel, M. (1971) Aging in *Caenorhabditis briggsae*. *Nematologica*, **17**, 478–487.

18. Immunology of Nematode Infection

Haruhiko Maruyama

Department of Medical Zoology, Nagoya City University Medical School Kawasumi-1, Mizuho, Nagoya 467-8601, Japan

Yukifumi Nawa

Department of Parasitology, Miyazaki Medical College 5200 Kiwara, Kiyotake, Miyazaki 889-1692, Japan

Keywords

cytokine, helper T cell, eosinophil, goblet cell, mast cell

Introduction

One hundred years ago, at the turn of the century, was a Golden Era of microbiology and vaccination. The causative agents of many infectious diseases were identified as bacteria and vaccination to develop specific immunity in humans and animals was becoming popular (Golub and Green, 1991). Immunology, the science of immunity and immune systems, thus started with empirical practice. People then asked: How does it work? Knowledge of animal parasites greatly increased in medicine, veterinary medicine and zoology during the two hundred years from the middle of the 17th century to the 19th century (Hoeppli, 1959; Foster, 1965), well before the work of Pasteur, Koch and their schools established the role of bacteria in disease. It was appreciated by about the 1860's that parasites were responsible for many important diseases of humans and animals. Why did parasitology not give birth to immunology? Why has immunisation against major nematode infections of humans not been achieved? While hundreds of scientific papers about parasitic nematodes are published every year, and while nearly 70,000 tons of *Ascaris* eggs are being produced annually (if one applies Stoll's estimate in 1947 to the present global epidemiological data (Stoll, 1947; Chan, 1997)), we are still unable to immunise humans against parasitic nematodes.

One of the major reasons for the great success of nematode parasites could be due to the basic attributes of these members of the phylum, e.g., a resistant cuticle and comparatively huge body size as compared to the immunocompetent cells of the host. Nematodes might appear to be beyond the capabilities of the immune system as it is difficult to imagine macrophages and lymphocytes killing a pair of 20–30 cm long adult *Ascaris lumbricoides*. Some of the most important nematode parasites of humans are actually very long-lived in the host. Chronic infections are characteristic of the gastrointestinal nematodes and tissue-dwelling filarial parasites (Behnke, Barnard and Wakelin, 1992). However, the difficulties the immune system has to overcome are not merely physical ones. Many parasitic nematodes have evolved some kind of immune evasion systems and these are currently of major interest (Maizels *et al.*, 1993). Although nematode antigens are highly immunogenic in humans and animals, and the host does respond to nematode parasites with intense cellular and humoral responses (Behnke, 1987), some reactions may not contribute to the destruction/elimination of the parasites. Nematode parasites are ancient adversaries of the immune system, which has evolved specific strategies for dealing with them (Bell, 1998).

The major players of the immune system are macrophages, T cells and B cells, which regulate functions of each other. These cells are also effectors for a number of bacterial and viral infections. In nematode infections it appears that more players are needed, especially for effector phases. When parasitic nematodes enter the body of the host they can cause remarkable and dramatic immune responses against them. Some of the responses set effector mechanisms at work, while others seem to have no protective role. Ever since the 'self-cure' phenomenon with *Haemonchus contortus*, a parasite of sheep, goats, cattle and deer, was introduced by Stoll (1929) to describe immunity to a nematode infection (Ackert 1942), we have acquired an array of experimental and clinical host-nematode systems in which immunity occurs in the host. In this chapter we will describe immune responses characteristic of nematode infections and focus on the effector functions of immune systems, which cannot be understood separately from nematode biology. The immune responses to parasitic nematodes should bring us real insight into the biology of nematodes.

Overview of Host-Parasitic Nematode Interactions

It is essential to know the basic behavioural patterns of animal-parasitic nematodes to understand the immunology of nematode infections. In the phylum Nematoda the parasitic mode of living has been employed several times independently (Blaxter *et al.*, 1998). In mammalian hosts nematodes may parasitise the eye, mouth, tongue, gastrointestinal tract, liver, lungs, muscle, connective tissue, brain, lymphatics, blood or body cavity. They can be found virtually everywhere. In spite of various life cycles, which range from the very simple to the extremely complicated, and of the phylogenetic distance between groups of parasitic nematodes, it is possible to find common motifs of nematode behaviour in the nematodes of animals and humans.

Principles of Parasitic Nematode Life Cycles

Infective larvae

All animal-parasitic nematodes are 'endoparasites', meaning parasites inside the host body. Theoretically endoparasites can be divided further into two categories. One group of parasites lives in the alimentary tract and its annex without invading the host tissue and the other enters the host body with tissue invasion. However, practically all parasitic nematodes in mammals penetrate covering epithelia of the host, namely the skin or the mucosa of the gastrointestinal tract, to enter the host body even though their final destination may be the gastrointestinal lumen. In general, parasitic nematodes enter the host when they are third-stage larvae (L3), which are thus referred to as 'infective larvae', and enter the vertebrate host orally or percutaneously. Oral infection takes place as the host drinks or ingests food that is either contaminated with eggs or larvae, or is parasitised by larvae. Larvae swallowed or hatched from eggs in the alimentary tract may immediately penetrate the epithelial layer to reach the mucosal tissue. Some species remain in the alimentary tract and others start extensive migrations (Figure 18.1).

Percutaneous infections are found in two groups of parasitic nematodes; one is the 'pioneering parasites' (Schmidt and Roberts 1989) the Rhabditida and Strongylida nematodes, in which free-living infective larvae actively penetrate the host skin, the other is the filarial worms, which are highly adapted to animal-parasitism. Members of this group use arthropods as intermediate hosts and enter the mammalian body following the bite of the infected vector. In the final host filarial worms reside in the connective tissue, lymph vessels or blood vessels, depending on the species.

Tissue migration

After entering the mammalian host many parasitic nematodes undergo tissue migration to a greater or lesser extent. In the case of *Strongyloides venezuelensis* infection in mice, third-stage larvae that have penetrated the skin, or been inoculated subcutaneously, remain in the subdermal connective tissue for a while, and then enter lymphatics or venules to reach the lungs 2 to 3 days after infection. In the lungs, they moult and break out of capillaries into the airway and move up the

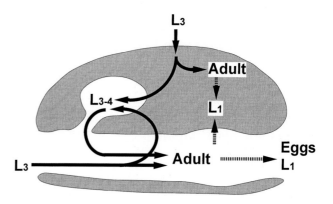

Figure 18.1. Migratory pathways of animal-parasitic nematodes.

respiratory tree to the pharynx, where they are swallowed. They pass through the stomach to the small intestine where they mature. This type of tissue migration involving the lungs is seen in various parasitic nematodes, such as *Ascaris lumbricoides* and hookworm infection in humans, *Nippostrongylus brasiliensis* in rats and *Strongyloides* spp. in humans and other mammals (Murrell, 1980; Takamure, 1995). If infective larvae enter an inappropriate host, then the larvae do not complete the migration to the final destination but wander around in the host tissue often causing severe damage to the host. This pathological condition in humans is known as 'larva migrans' (Beaver, 1969).

Interestingly some worms show reduced versions of tissue migration. For example, infective larvae of *Heligmosomoides polygyrus*, a parasite of mice, penetrate the mucosa of the anterior small intestine 24–72 hours after ingestion then invade the muscle layer of the small intestine, where they continue their development (Monroy and Enriquez, 1992). They then exit to the gut lumen and mature rapidly in the jejunum. The whipworm of humans, *Trichuris trichiurus*, also spends several days in the mucosa of the small intestine before it finally reaches the caecum where it buries its anterior end in the mucosa. The implication of these tissue migrations for the development of host immune responses is obvious. The immune system is strongly stimulated when parasites and parasitic antigens move beyond the basement membrane. Little sensitisation would occur if antigens were not brought in through the basement membrane.

Adult worms

The favoured habitats of all parasites in vertebrates are (a) the alimentary canal and its associated glands; (b) the blood stream; (c) the respiratory system; and (d) the coelom, in that order (Smyth 1994). This seems to be the case in nematode parasites as well. Intestinal nematodes contain some of the most well known or common nematodes parasitic in people. Roundworms (*Ascaris lumbricoides*), pinworms (*Enterobius vermicularis*), hookworms (*Ancylostoma*, *Necator*) and whipworms

(*Trichuris trichiurus*), are all intestinal parasites. If one combines nematodes living in the alimentary tract and the respiratory system together, as the lungs are branches of alimentary tract, then the great majority of parasitic nematodes are classified in this category. Because these organs have access to the environment, eggs or larvae of parasitic nematodes living in these organs can easily leave the host. One major group of parasitic nematodes that live and reproduce outside the alimentary tract is the filarial worms.

It has to be pointed out here that even in intestinal nematodes, adult worms usually have close contact with intestinal epithelial cells and many adult worms actually invade the epithelial layer. For example, *Strongyloides ratti* resides between intestinal epithelial cells (Dawkins, 1989), *Capillaria philipinensis* invades the intestinal mucosa repeatedly (Cross, Banzon and Singson, 1978), *Trichuris muris* is embedded deeply in the mucosal tissue (Lee and Wright, 1978), and the extraordinary *Trichinella spiralis* dwells inside the epithelial cells (Wright, 1979). *Nippostrongylus brasiliensis*, which does not undergo intestinal invasion, is tightly wedged between the intestinal villi, firmly coiling around them (Symons, 1976; Ishikawa *et al.*, 1994). As we will see later, responses that prevent continuing intimate contact of adult intestinal nematodes with the intestinal mucosa should be a basis for worm expulsion. In the expulsion of intestinal nematodes, expelled worms are not killed but are able to continue to live if implanted in a new host (Moqbel, McLaren and Wakelin, 1980; Ishikawa *et al.*, 1994).

Nematode Antigens in Infections

The immune system of the host is exposed to nematode molecules when the host is infected with a parasitic nematode. Conventionally, macromolecules such as proteins and polysaccharides of the parasite are called antigens. Antigens are substances capable of inducing a specific immune response. Because the nematode is phylogenetically distant from the host, most nematode antigens are highly immunogenic. There are two types of antigens associated with the nematode parasite; somatic antigens, and excretory/secretory (ES) antigens. Somatic antigens compose the nematode body and are fixed at external surfaces or within the parasite. ES antigens are released from the parasite during the course of infection. These antigens have various roles in parasitism and immune responses of the host.

ES antigen in parasitism

During tissue migration and establishment at the final site, nematodes release a wide variety of excretory/secretory (ES) antigens that are powerfully immunogenic (Lightowlers and Rickard, 1988). Many ES antigens have biological activities that facilitate tissue invasion and/or parasite survival. For example, hyaluronidase and metalloproteases from tissue-invading hookworm larvae (Hotez *et al.*, 1990, 1992, 1994; Hawdon *et al.*, 1995), and neutral serine protease from *Anisakis* larvae (Sakanari and McKerrow, 1990) have been reported. We also detected metalloprotease

activities in infective larvae of *Strongyloides venezuelensis* (Maruyama, unpublished observation). These enzymes are undoubtedly released to degrade host tissue. However, they may contribute to evasion of the immune response in addition to simple burrowing, as in some tissue-invading trematode parasites (Ramaswamy *et al.*, 1995; Hamajima *et al.*, 1994). Infective larvae have been shown to secrete proteins that are associated with the transition from the free-living external environment to parasitism (Hawdon *et al.*, 1996).

Adult worms also secrete various biologically active substances, *e.g.*, protease inhibitors from filarial worms and *Ascaris lumbricoides* (Lustigman *et al.*, 1992; Bennet *et al.*, 1992), phosphorylcholine and eicosanoids from filarial worms (Liu, Serhan and Weller, 1990; Lal *et al.*, 1990). These substances are implicated in inhibiting host inflammatory reactions. Hookworms, that cause blood loss during attachment to the intestinal mucosa by lacerating capillaries and ingesting blood, have evolved a family of potent anti-coagulant peptides that are specific for factor Xa (Cappello *et al.*, 1995, 1996; Stassens *et al.*, 1996). It is quite logical for blood sucking nematodes to subvert host haemostasis. In addition to anticoagulant, hookworms secrete neutrophil inhibitory factor, proteases that degrade host immunoglobulin, superoxide dismutase, and glutathione-S-transferase (reviewed by Pritchard, 1995), all of which could compromise host granulocyte attack against parasitic nematodes. Large amounts of acetylcholinesterase are present in the ES of several species of nematode, including *Ancylostoma, Necator, Nematodirus, Nippostrongylus, Oesophagostomum* and *Trichostrongylus*, and probably have an important role in immune modulation and/or reduction of inflammation in the vicinity of the nematode (reviewed by Lee, 1996). In fact, ES antigens of adult lungworm, *Dictyocaulus viviparus*, enriched with acetylcholinesterases, elicit protective immunity against a challenge infection in guinea pigs (McKeand *et al.*, 1995a, b). Nematode-derived acetylcholinesterase may contribute to parasitism by degrading host acetylcholine that stimulates histamine release from mast cells (Fantozzi *et al.*, 1978).

Protective ES antigens

ES antigens can be used to immunise animals to induce protective immunity against a challenge infection. In the intestinal nematode, *Trichinella spiralis*, protection is achieved by immunising animals with ES antigen mixtures (Robinson, Bellaby and Wakelin, 1994) and a purified ES protein preparation (Silberstein and Despommier, 1984). This protein, a 43 kD glycoprotein member of the TSL-1 family, is present in the secretions and at the cuticle surface of muscle larvae (Ortega-Pierres *et al.*, 1996). It is a stage-specific, heavily glycosylated antigen, which is involved in establishment of the nurse cell, a modified myocyte that 'nurses' intracellular *T. spiralis* muscle larvae (Vassilatis *et al.*, 1992). Immunisation of mice with synthetic peptides based on the reported sequence of this antigen induces protection against challenge infections (Robinson *et al.*, 1995).

Antigens should be protective only when they are functionally important and are inactivated by the immune system. Therefore identification of a protective antigen should add much knowledge to our understanding of the biology of parasitism. In canine hookworm, *Ancylostoma caninum*, infection in mice, a recombinant ES antigen from infective larvae, designated as *Ancylostoma*-secreted protein (ASP), has been shown to induce protective immunity (Ghosh, Hawdon and Hotez, 1996). Although physiological roles of this protein in *A. caninum* have yet to be clarified, the amino acid sequence of ASP has significant homology to hymenopteran venoms and antisera raised against the insect venom Ag5 crossreact to ASP (Hawdon *et al.*, 1996). Because hymenopteran venoms are potent allergens, ASP might have an immunomodulatory activity.

Protective somatic antigens

For immunity to work, target antigens have to be accessible to the cells and molecules of the immune system. In this regard immunisation with somatic antigens from the body surfaces of parasitic nematodes has a good chance to induce protective immunity. Beside *T. spiralis* cuticular antigen described above, immunisation with a major surface antigen on infective larvae of the sheep stomach worm, *Haemonchus contortus* , is known to elicit protective immunity to this parasite. This antigen is stage-specific, expressed on the surface of the larvae (Raleigh *et al.* 1996; Raleigh and Meeusen, 1996) and specific antibodies against it are produced by draining lymph node cells after challenge infection in immune sheep (Bowles, Brandon and Meeusen 1995). Immunised animals show significant reduction in adult worm burdens after a challenge with infective larvae. Specific antibodies against this antigen mediate killing of larvae by eosinophils *in vitro* (Rainbird, Macmillan and Meeusen, 1998).

H. contortus is an abomasal nematode that parasitises sheep, cattle and goats, causing severe anaemia and hypoalbuminaemia which can be fatal to susceptible host animals. A good deal of study has focused on gut membrane antigens in this parasite because this parasite feeds on host blood, so that circulating antibodies and cells should have good access to the gut surface of the parasite. Several antigen preparations from the integral gut membrane of *H. contortus* have in fact been shown to induce protective immunity in sheep and goats (Jasmer and McGuire, 1991; Smith, Smith and Murray, 1994; Munn *et al.*, 1997). Among these antigens, GA1 apical gut membrane proteins (Jasmer, Perryman and McGuire, 1996) are released, probably by shedding from the gut surface, and immunisation of goats with this antigen protected animals from challenge infections (Jasmer *et al.*, 1993).

These studies were started with the idea of exploiting a 'hidden antigen' strategy (Newton, 1995; Newton and Munn, 1999). This strategy has been developed for the cattle tick, *Boophilus microplus*, in which specific antibodies and cells against intestinal cell surfaces damage the ticks when they take a blood meal (Willadsen *et al.*,

1989). In the case of *H. contortus*, a hidden gut membrane antigen, H11, has been used in an attempt to immunise animals in the UK, South Africa and Australia and has resulted in a 60–90% reduction in adult worm numbers (Newton *et al.*, 1995; Smith *et al.*, 1997). The chances of producing a successful commercial vaccine based on the hidden antigen H11 seem quite good (Newton, 1995). H11 is a microsomal aminopeptidase of type II integral membrane protein (Smith *et al.*, 1997). Immune sera against H11 inhibit enzyme activities, and levels of protection and enzyme activity inhibition highly correlates (Newton and Munn, 1999). It could be that nematode enzymes are so different from those of the host, they elicit strong antibody responses. Interestingly, molecular cloning of one component of another protective integral membrane complex in *H. contortus*, *Haemonchus*-galactose containing glycoprotein (H-gal-GP), revealed it is a putative metallopeptidase (Redmond *et al.*, 1997). Nematode enzymes at the cell surfaces might be good vaccine candidates as well as the possible targets for a new anti-nematode drug.

Immunodiagnosis

Whilst nematodes produce and secrete macromolecules for their own biological purposes, these antigens may also have practical importance. ES antigens are superior targets for immunodiagnosis because they are strongly immunogenic and are released by live nematodes (Ogilvie and De Savigny, 1982). Clinically, patients who harbour migrating larvae that do not mature in the human body show quite similar symptoms, regardless of the causative species. This pathological condition, known as visceral larva migrans (VLM), is characterised by persistent eosinophilia, high IgE titre, a cough and fever or multiple nodular lesions in the lungs and/or liver (Roig *et al.*, 1992; Ishibashi *et al.*, 1992; Bartelink *et al.*, 1993). Peripheral blood eosinophilia reaches more than 50% of the total white blood cell count in some cases. Although most VLM is believed to be caused by the dog ascarid, *Toxocara canis*, other species such as *Gnathostoma doloresi* and *Ascaris suum* can cause VLM (Phills *et al.*, 1972; Seguchi *et al.*, 1995; Maruyama *et al.*, 1996). The determination of the causative species depends almost exclusively on serological tests because of the difficulty in obtaining migrating larvae of the parasites, however, intense cross-reactivities between nematode antigens have been major obstacles for diagnosis (Iglesias *et al.*, 1996; Maruyama *et al.*, 1997). In this regard, immunodiagnosis with ES antigen is recommended because it gives more specific and sensitive results than that with somatic antigens (Kennedy *et al.*, 1987a,b; Cuellar, Fenoy and Guillen, 1995).

Parasitic Nematodes Used in Experimental Studies

Examples of parasitic nematodes having been used in experimental studies are more or less limited in number. The reasons are: 1) etiological agents of human diseases are preferred, but; 2) researchers usually employ nematodes infective to mice and rats, and;

3) it is convenient to use nematodes that can be maintained in the laboratory. To understand the experimental systems the life cycles of some parasitic nematodes are described here.

Gastrointestinal nematodes with migration through the lungs

Included in this group are *Nippostrongylus brasiliensis*, *Strongyloides ratti* and *Strongyloides venezuelensis*. Basically all are native to rats and are used in experiments with mice and rats. Animals are infected with these nematodes by percutaneous or subcutaneous inoculation with infective larvae (L3). In some strains of *S. ratti*, some migrating larvae reach the airway through the brain and nasal cavity (Tada, Mimori and Nakai, 1979). In rats and mice, these nematodes are expelled within 2–4 weeks and a strong protective immunity is acquired.

Gastrointestinal nematodes with limited tissue invasion

Haemonchus contortus in sheep, goats and cattle, and *Heligmosomoides polygyrus* and *Trichuris muris* in mice are classified in this group. Animals are infected by ingesting embryonated eggs (*T. muris*) or infective larvae (*Haemonchus contortus* and *Heligmosomoides polygyrus*). After ingestion, larvae reach the abomasal mucosa (*H. contortus*) or intestinal mucosa (*H. polygyrus* and *T. muris*) where they invade the mucosal tissue to a various degree. *H. polygyrus* reaches the muscle layer, while *H. contortus* remains in the mucosa and *T. muris* remains in the epithelial surface throughout. After moulting, larvae return to the surface of the mucosa (*H. contortus*), or to the gut lumen (*H. polygyrus*), where they reach maturity and produce eggs.

Trichinella spiralis

This nematode has a unique life cycle in that the same animal serves as both definitive and intermediate host. Animals and humans acquire infection of *Trichinella spiralis* by ingesting flesh containing encysted larvae; isolated muscle larvae are used in the experimental studies. After being swallowed, larvae enter the intestinal epithelial cells, mature and then copulate. The female then gives birth to larvae (new-born larvae) that enter the circulation and distribute throughout the body. When the new-born larvae reach skeletal muscle, they penetrate the muscle fibres and parasitise the myocyte where they lie free within its cytoplasm (Lee and Shivers, 1987). The larvae modify the muscle cells vigorously, presumably through the action of secretions from the stichocytes, to form 'nurse cells' in which they live for months or years, waiting for the next host to eat them (Despommier, 1998).

Filarial nematodes

The filariae live in the blood stream, connective tissue, or body cavity of vertebrates. The ovoviviparous female discharges numerous larvae (microfilariae) into the surrounding tissue from where they may be taken into the arthropod intermediate host/vector when it feeds, e.g., mosquitoes, flies, mites, etc. Transmission from the arthropod to animals and humans occurs through the skin when the arthropod harbouring infective L3 larvae takes a blood meal. Because of the absence of ideal animal models that harbour the full developmental cycle and cause similar pathology to human infections, short-term infections in mice have been employed by researchers investigating the immunology of lymphatic filariasis and onchocerciasis (Lawrence, 1996). Experimental techniques used in experimental studies are intraperitoneal implantation of adult worms, intraperitoneal or intravenous injection of microfilariae and intraperitoneal or subcutaneous injection of infective L3 larvae. Severe combined immunodeficiency (SCID) mice can support the full development of *Brugia malayi* from L3 to microfilaria-producing adult worms (Nelson *et al.*, 1991).

T Helper Subsets and Immune Responses to Nematode Infections

Cytokines

Eosinophilia, i.e. increased eosinophilic leukocyte numbers in peripheral blood, is considered a classical manifestation of parasitic worm infections in humans as well as in animals, although it is found in many other conditions (Gutierrez, 1990). In helminthic infections, eosinophilia, mastocytosis and IgE production are said to be the worm infection-related triad (Finkelman *et al.*, 1997). Along with recent advances in molecular biology and immunology, the molecular basis for the induction of these worm-related symptoms has been clarified. Today, we are able to say that these cellular and humoral responses are largely the result of the production of a group of proteins called cytokines. Cytokines are low-molecular-weight regulatory proteins which are secreted by helper T cells and by a variety of other cells upon exposure to a number of inducing stimuli. Nematode infections are potent inducers of cytokine production. Cytokines bind to specific receptors on the target cell surfaces to elicit biological activities. Functions of a particular cytokine are determined by target cell populations that have specific membrane receptors on the cell surfaces (Kishimoto, Taga and Akira, 1994). Binding affinity for cytokines to receptors is so high that cytokines can modulate target cell activities at picomolar concentrations. A growing number of proteins have been included in the cytokines; these include interleukins, colony-stimulating factors, interferons, tumour necrosis factors and a number of growth factors (Paul and Seder, 1994).

We can see a good example of the recent development of our knowledge on cytokines and their functions in the story of eosinophils and interleukin (IL)-5. Before the molecular cloning of cDNA for IL-5 (Kinashi *et al.*, 1986), this 'factor' was designated differently by several groups according to the biological activities of their interest, e.g. T cell replacing factor (TRF), B cell growth factor-II (BCGF-II), eosinophil differentiation factor (EDF) and IgA-enhancing factor

(IgA-EF). These activities were originally detected in the culture supernatant of activated T cells (Takatsu, Tominaga and Hamaoka, 1980) and molecularly cloned thereafter. Because the broad and different activities were attributed to the same protein, TRF/BCGF-II/ EDF/IgA-EF was designated as interleukin (IL)-5 (Takatsu, 1992). It has now been established that IL-5 is produced by helper T cells, eosinophils, and mast cells. IL-5 promotes IgG and IgA production in mice, induces proliferation and differentiation of eosinophils and stimulates superoxide production by eosinophils *in vitro*. Eosinophilia induced by the intestinal nematode, *Nippostrongylus brasiliensis*, was completely suppressed by a neutralising anti-IL-5 monoclonal antibody (Coffman *et al.*, 1989; Rennick *et al.*, 1990). Later, impaired eosinophilia in nematode infections was shown in IL-5 gene-targeted (knockout) mice (Kopf *et al.*, 1996) and IL-5 receptor knockout mice (Yoshida *et al.*, 1996). Together with the finding that transgenic mice carrying cDNA for IL-5 develop spontaneous eosinophilia (Tominaga *et al.*, 1991), it has been widely accepted that the major target cells of IL-5 are eosinophils.

Similar experiments with other cytokines revealed that IgE production is dependent on IL-4 (Coffman *et al.*, 1989; Kuhn, Rajewsky and Muller, 1991) and mastocytosis on IL-3 and IL-4 (Madden *et al.*, 1991). IL-4 is a prototypic immunoregulatory cytokine (Paul, 1991) which is produced by activated T cells and mast cells. It induces immunoglobulin class switch to IgG1 and IgE, and stimulates mast cell growth. IL-3 is a haemato-poietic protein which is secreted by activated T cells and mast cells. IL-3 supports growth and differentiation of haemopoietic cells, especially mucosal mast cells, together with other cytokines such as granulocyte-macrophage colony stimulating factor (GM-CSF), IL-4 and IL-9. Mice deficient for the IL-3 gene do not show an apparent defect in haemopoiesis, however, when infected with an intestinal nematode, *Strongyloides venezuelensis*, the mice do not develop mucosal mastocytosis and worm expulsion is severely delayed (Lantz *et al.*, 1998).

In the study of immunology these kinds of 'subtraction' experiment have gained much favour. For many years, mutant mice that phenotypically lack one or several cell types have been used in subtraction experiments. Hypothymic nude mice and W/W^v mice have been used to show the importance of T cells and mast cells, respectively. SCID mice, that lack functional T and B cells, have been added to the list. After the development of monoclonal antibody techniques, *in vivo* depletion experiments became popular, in which a particular cell population that expresses a certain kind of cell surface marker is depleted. A great deal of investigation has been done with this technique to clarify the roles of, for example, CD4$^+$ T cells and CD8$^+$ T cells. Neutralising antibodies against cytokines and cytokine receptors are used as described above to determine *in vivo* functions of cytokines. Gene knockout mice have been more and more widely used. It would not be an overstatement to state that a new line of knockout mice is produced every day.

T Helper Subsets

As described above, cytokines are largely produced by activated T cells but how are these T cells stimulated to produce cytokines? In general, when an exogenous antigen is introduced into the host body, antigen-presenting cells, such as dendritic cells, macrophages and B cells, internalise the antigen, either by phago-cytosis or by endocytosis, and re-express a part of that antigen together with the MHC (major histocompat-ibility complex) molecules on their cell surface mem-brane (Germain, 1994; Watts, 1997). The helper T cells then recognise the portion of antigen associated with the MHC molecule, and are activated. T cells are activated only when properly stimulated by antigen-presenting cells (Racioppi and Germain, 1995). Activated T cells then secrete several cytokines that further stimulate the immune system and haematopoiesis resulting in the generation of acute inflammatory reactions, and hu-moral and cell-mediated immunity. Helper T cells stimulate and induce effector cells (macrophages, granulocytes, cytotoxic T cells, etc.) and promote production of effector molecules (immunoglobulins).

A key concept is that functional T helper cells can be divided into two major subsets, namely Th1 and Th2 cells, that are different in cytokine production. Th1 cells secrete IFN-γ and IL-2, but not IL-4, IL-5, IL-9 or IL-10. Th2 cells secrete IL-4, IL-5, IL-9 and IL-10, but not IFN-γ or IL-2. IL-3 is secreted by both subsets (Mosmann and Coffman 1989). Cytokines that induce the worm-related triad IL-3, IL-4 and IL-5 are all produced by Th2 cells. An interesting feature of these subsets is that they are reciprocally cross-inhibitory (Figure 18.2). For example,

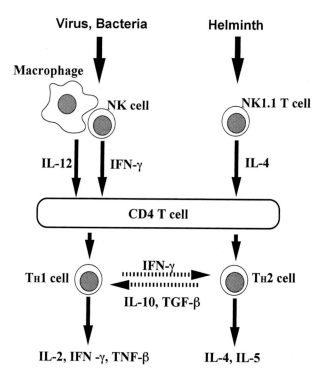

Figure 18.2. Generation of Th1 and Th2 cells and their interactions.

IL-4 secreted from Th2 cells stimulates Th2 cells and inhibits Th1 cells (Tanaka *et al.*, 1993). On the other hand, IFN-γ from Th1 cells inhibits Th2 cells. Balance or imbalance of Th1/Th2 subsets results in the production of a particular set of cytokines that characterises cellular and humoral responses in a certain infection. Although some authors describe a third Th3 subset (Letterio and Roberts, 1998), the Th1/Th2 dichotomy has been a good framework to analyse immunological regulation of infectious diseases.

Th1 and Th2 subsets were originally identified in long-term mouse T cell clones *in vitro*, however, numerous experimental and clinical studies suggest that the *in vivo* outcome of the immune responses, or severity of an infectious disease, depend on the relative levels of Th1 or Th2 activity. For example, Th1 type responses are certainly beneficial to the host in infections with intracellular parasites (reviewed by Sher and Coffman, 1992). Inbred C57BL/6 and C3H/HeN mice are resistant to infection with *Leishmania major*, whereas BALB/c mice are susceptible, developing a chronic and lethal infection. Cytokines produced by resistant C57BL/6 and C3H/HeN mice are Th1 cytokines, and susceptible BALB/c mice produce Th2 cytokines. Manipulation of cytokine production in these mice results in changes in susceptibility to the parasite. The normally resistant C3H/HeN mice developed a lethal infection when treated with anti-IFN-γ monoclonal antibody, while normally susceptible BALB/c mice became resistant when treated with anti-IL-4 antibody. These results are understandable because susceptibility of mice to infection with *L. major* is determined by the ability of macrophages to kill the parasite. Macrophages kill *L. major* by nitric oxide (NO) produced by inducible NO synthase (i NOS). Expression of i NOS is induced by IFN-γ and is suppressed by IL-4.

Obviously Th2 type response predominates in nematode infections as eosinophils, mastocytosis and IgE production are all induced by Th2 type cytokines. Indeed, elevated production of Th2 type cytokines has been shown in various nematode infections over and over again, and these reactions are said to be 'stereotypic' (Finkelman *et al.*, 1997). It would be quite natural to assume that Th2 type responses are beneficial to the host, however, the role of Th2 cytokines in host protection in nematode infections remains uncertain. There are several hypotheses concerning the roles of Th2 cytokines and related cellular and humoral responses in protection of the host (Sher and Coffman, 1992). One extreme assumption would be that all these reactions contribute to protection, and the other extreme that no Th2 reactions are protective. Alternatively, some Th2 responses are protective, but because all reactions take place as a unit some unnecessary and even pathological reactions may occur. In any case, we will have to identify truly effective mechanisms for elimination of nematodes in each case, as in NO against *L. major*. Only after we know which system is effective, can we tell unnecessary responses from indispensable ones. In any case, before examining possible effector mechanisms we will take a brief look at cytokine production in various nematode infections in humans and animals, for host responses may not be as stereotypic as they first look.

T Helper Subsets and Human Filariasis

The importance of Th1/Th2 balance would be best exemplified in immunological analyses of human lymphatic filariasis. Infection with the lymphatic dwelling nematodes *Wuchereria bancrofti* (90% of cases) and *Brugia malayi* (10% of cases) affects more than 120 million people in tropical regions of the world (Michael, Bundy and Grenfell, 1996). Because of the high prevalence and devastating pathological features, acquired immunity and immune dynamics in the pathogenesis have been extensively studied. Lymphatic filariasis is a spectral disease, showing various clinical manifestations. It has long been acknowledged that the immune responses to the parasite appear to be extremely important in determining the nature of the clinical manifestations (Nutman, 1995). One pole of the patients shows detectable microfilaremia with high parasite numbers but without any clinical manifestations of lymphatic insufficiency. At the other pole are the patients with chronic pathology of lymphatic destruction but no sign of an active infection (Maizels *et al.*, 1995). Presumably, the latter patients have cleared the parasites, but at the same time have developed pathological tissue damage due to strong immune responses. In fact, based on a large number of studies comparing these two groups of patients, one generalisation has emerged suggesting that the immune system of asymptomatic microfilaria-positive individuals is not capable of mounting specific immunity to parasite antigens, while individuals with lymphatic pathology respond vigorously to parasite antigens (Maizels *et al.*, 1995). Most microfilaremic individuals do not respond to parasite antigen in a standard T cell proliferation assay (Ottesen, Weller and Heck, 1977; Piessens *et al.*, 1980; Lammie *et al.*, 1993) or with delayed-type hypersensitivity (Weller, Ottesen and Heck, 1980). Production of IL-2 and IFN-γ by T cells stimulated with parasite antigens is largely impaired in microfilaremic individuals (Nutman, Kumaraswami and Ottesen, 1987). These depressed T cell responses are antigen-specific as responses to non-parasite antigens and mitogens are completely normal.

Specific unresponsiveness of T cells of microfilaremic individuals once led researchers to suggest the presence of parasite-derived suppressor factors, immunological tolerance, or suppressor cell populations induced by filarial nematodes (Piessens *et al.*, 1980; King *et al.*, 1992). However, it has been demonstrated that T cells of microfilaremic patients are not unresponsive as they secrete IL-4 and IL-5, but not IL-2 and IFN-γ, upon stimulation with parasite antigens (King *et al.*, 1993; Mahanty *et al.*, 1993). In addition, T cell unresponsive individuals have high titres of IgE and IgG4 subclass antibodies to parasite antigens (Maizels *et al.*, 1995). As IgE and IgG4 production is promoted by IL-4 (Wierenga *et al.*, 1990), these findings indicate a shift toward

Th2 immune responses in these asymptomatic individuals.

A number of mechanisms could account for decreased Th1 responses, however, the Th1/Th2 concept suggests active down-regulation of Th1 cells by Th2 cells in microfilaremic individuals. Among various cytokines secreted by Th2 cells, IL-10 is a major down-regulator of Th1 cells. IL-10 was originally reported as a cytokine synthesis inhibitory factor (Moore *et al.*, 1990) that suppressed IFN-γ production by T cells stimulated with antigen and macrophages (Fiorentino, Bond and Mosmann, 1989; Fiorentino *et al.*, 1991a, b). It acts on macrophages, not directly on T cells. In addition to Th2 cells, IL-10 is produced by macrophages, keratinocytes and mast cells. IL-10 also promotes proliferation of mast cells with IL-3 and IL-4. In microfilaremic individuals, IL-10 production by antigen-stimulated T cells is elevated and neutralising anti-IL10 antibodies can restore proliferation and IFN-γ production of T cells (Mahanty *et al.*, 1996, 1997; Ravichandran *et al.*, 1997). These facts indicate that IL-10 plays a role in the unresponsiveness of microfilaremic individuals.

If down-regulation of Th1 immune responses is beneficial for filarial nematodes, how do they induce this Th2 dominant state? One possibility is that parasite antigen itself induces Th2 cells. Naive T cells from unsensitised normal healthy donors secreted IFN-γ and IL-5 when cultured with microfilarial antigen of *B. malayi*. Interestingly, re-stimulation with the antigen resulted in enhanced production of IL-5 but not IFN-γ, suggesting that microfilarial antigen preferentially induced Th2 cells (Steel and Nutman, 1998). Another interesting finding is that the rodent filarial parasite, *Acanthocheilonema viteae*, secretes a 17-kDa antigen which shows amino acid homologies to cystatin C, a major cysteine protease inhibitor. This protein exhibited biological activity as a cysteine protease inhibitor, down-regulated T cell responses and up-regulated IL-10 production (Hartmann *et al.*, 1997). Hence, this filarial cystatin may be an effector molecule of immunomodulation of this filarial nematode. Recent studies indicate that inhibition of cathepsin B, a lysosomal cysteine protease in antigen-presenting cells, can induce a shift of immune responses from Th2 to Th1 in mice (Maekawa *et al.*, 1998). Filarial nematodes might be interfering with antigen processing of the host for their own survival.

In patients with chronic pathology, however, the Th1/Th2 concept may not work very well. Cells from these patients show strong immune responses to filarial antigens with production of both IL-4 and IFN-γ upon antigen stimulation. The production of IL-4 is not significantly different between microfilaremic individuals and patients with chronic pathology (King *et al.*, 1993; Ravichandran *et al.*, 1997; Mahanty *et al.*, 1997). In mice, single developmental stage infections of *Brugia malayi* have demonstrated that adult worms and L3 exclusively stimulate Th2 type responses, whereas microfilariae injected intraperitoneally induce Th1 type responses which continue during infection. However,

both IL-4 and IL-5 responses develop later (Lawrence, 1996). A complicating factor in human filariasis is the finding that the progression of filariasis towards gross pathology can be prevented by proper foot care and hygiene and antibiotic therapy (Turner and Michael, 1997). If this is true, then lymphatic pathology and Th1 responses in chronic filariasis might be the result of repeated bacterial infections. The question of whether immune responses are related to the burden of infection could be addressed with precise classification of patients in well-defined populations (Freedman 1998).

T Helper Subsets in Experimental Studies

The concept of Th1 and Th2 has stimulated much study on the analysis of immune responses in nematode infections. We focus here on early events in nematode infections that bring about such Th2 responses, because differentiation of T helper cells into either Th1 or Th2 cells is influenced by the cytokine milieu in which the initial antigen priming occurs (Rocken, Saurat and Hauser, 1992; Hsieh *et al.*, 1992; Seder *et al.*, 1992).

Continuous assay for cytokine levels in local lymph fluid was performed in rats infected with *Trichinella spiralis* (Ramaswamy, Negrao-Correa and Bell, 1996). When levels of IL-4, IL-5, TNF-α, and IFN-γ were quantified in the intestinal (afferent) and efferent thoracic duct lymph fluid during the course of infection, IL-4, IL-5 and IFN-γ were simultaneously detected in the intestinal lymph during the first 8 days after infection. IL-5 levels rose as early as 15 to 20 h and remained elevated throughout the infection. IL-4 activity appeared in intestinal lymph 60 h after infection and reached peak levels during worm expulsion. Despite the predominantly Th2 nature of the cytokine response, IFN-γ levels showed several cycles of high and low production during the course of infection. In these experiments the source of the cytokines was not identified, but the most important source of cytokines during the early infection were cells in the gut and not those in draining lymph nodes (Ramaswamy, Negrao-Correa and Bell, 1996). In mesenteric lymph node cells in mice infected with *Trichinella spiralis* or *Nippostrongylus brasiliensis*, Th1 rather than Th2 responses dominated within 2 days of worms reaching the intestine (Ishikawa *et al.*, 1998). Elevation in IFN-γ production was detectable as early as 24 hours after infection and this was abrogated by either anti-CD4 or CD8 antibody treatment. Within a few days, response to both mitogen and antigen stimulation changed to the Th2 type. Mesenteric lymph node cells, the preferred cell population for cytokine assays, might be under control of other cell populations that determine overall responses.

A role of non-T cells was suggested by experiments with *Heligmosomoides polygyrus* infection in mice (Svetic *et al.*, 1993). Chronic infection with *H. polygyrus* causes typical Th2 responses as in *T. spiralis* infections. As early as 6 hours after infection, when infective larvae are invading mucosae, mRNA for IL-5 and IL-9 were elevated in Peyer's patch cells, and IL-3 was elevated by 12 to 24 hours after infection. IL-4 was elevated only

4 to 6 days after infection. Unlike *T. spiralis* infection in rats, neither IFN-γ mRNA nor IL-2 mRNA were elevated. Interestingly, although treatment with anti-CD4 and anti-CD8 antibody markedly reduced expression of IL-3 and IL-4 mRNA 6 days after infection, early induction of IL-5 and IL-9 (6 hours after infection) was not decreased in nude mice.

We have to admit that we know little about the very early events in nematode infections. Cell populations that determine subsequent cellular responses may be $\gamma\delta$ T cells (Ferrick *et al.*, 1995) or eosinophils, as indicated in *Schistosoma mansoni* infections (Sabin and Pearce, 1995; Sabin, Kopf and Pearce, 1996). One of the major reasons why parasitology did not give birth to immunology could be that humans contract nematode infections of the same species again and again and this seems to indicate that no immunity develops to nematode parasites. However, we have a number of models in which immunity does work against nematode infections. Close examination of the effector phase should help us to comprehend the immune system and develop effective anti-nematode strategies. To know how the host rejects parasitic nematodes will provide a better understanding of the biology of these nematodes.

Effector Phase of Host Immunity Against Parasitic Nematodes

In nematode infections, eosinophils, mast cells, IgE antibodies and goblet cells have all been implicated in the elimination of the parasites. In the following sections, we will examine the roles of these responses in protection in some well-studied experimental models. Basically eosinophils play a role in destruction of tissue-invading larvae and mucosal mast cells and goblet cells are responsible for the expulsion of adult intestinal nematodes.

Eosinophils and Tissue Migrating Larvae

Recruitment and accumulation of eosinophils

There is one widely-observed principle in nematode infections; when a nematode parasite infects a non-definitive or 'abnormal' host, then the host responds to the parasite with massive eosinophilia. In an inappropriate host, infective larvae do not complete their migration to the final destination but wander around in the host tissue causing damage to the host. In humans, peripheral blood eosinophilia sometimes reaches 50–60%, which may be the only clue to the diagnosis of parasitic infections (Maruyama, Noda and Nawa, 1996; Maruyama *et al.*, 1996). In experimental models a number of nematodes, such as *Toxocara canis*, *Angiostrongylus cantonensis* and *Trichinella spiralis*, have been used to study eosinophilia in nematode infections. After infection with these nematodes, large numbers of eosinophils quickly accumulate around the migrating larvae, presumably to attack them. In mice, the demand for eosinophils is so enormous that production of eosinophils takes place not only in bone marrow but

also in the spleen and the liver with pluripotent stem cells circulating in the peripheral blood (Higa *et al.*, 1990; Maruyama *et al.*, 1991). Upon elevated demand for the mobilisation of eosinophils from bone marrow to peripheral tissues, eosinophil chemotactic substances, such as eotaxin, seem to play an important role along with IL-5. Eotaxin is a potent eosinophil chemoattractant, and a fundamental regulator of the physiological trafficking of eosinophils during healthy states (Matthews *et al.*, 1998). It is constitutively produced by various tissues such as the lungs and the intestine (Rothenberg, Luster and Leder, 1995; Collins *et al.*, 1995) and can be induced by IL-4 stimulation (Mochizuki *et al.*, 1998). Eotaxin stimulates recruitment and a rapid, selective release of eosinophil and colony-forming progenitor cells with a marked synergism with IL-5 (Mould *et al.*, 1997; Palframan *et al.*, 1998). However, the roles of eotaxin in nematode infections have not been fully examined.

Eosinophil-mediated killing of nematode larvae in vitro

Eosinophils have been demonstrated to be capable of killing nematode larvae *in vitro* in various experimental systems. In sheep, for example, eosinophils immobilise and kill *Haemonchus contortus* larvae *in vitro* in the presence of antibodies specific to a defined L3 surface antigen. Eosinophils activated *in vivo* by repeated infusion of *H. contortus* larvae are more effective. The level of larval immobilisation in the presence of antibodies is significantly increased when complement is added to cultures. The addition of IL-5 to larval cultures of inactivated eosinophils results in a significant increase in larval immobilisation (Rainbird, Macmillan and Meeusen, 1998). Similarly, eosinophils from normal volunteers with eosinophilia kill freshly collected new-born larvae of *T. spiralis* in the presence of antibodies, although eosinophils from normal individuals without eosinophilia attach to, but fail to kill, the larvae. Neither adherence nor significant mortality was observed in the absence of immune serum (Venturiello, Giambartolomei and Costantino, 1995). Rat eosinophils also kill *T. spiralis* larvae in the presence of immune sera, which is enhanced by IL-5 (Lee, 1991). These results indicate that activated eosinophils are able to kill nematode larvae with antibody-dependent cellular cytotoxic reactions *in vitro*. Ultrastructural analysis shows degranulation of adhering eosinophils onto the surface of larvae, and signs of damage to many larvae (Rainbird, Macmillan and Meeusen, 1998). Eosinophil-mediated damage of nematode larvae and its significance in protective immunity *in vivo* was demonstrated in *N. brasiliensis* infection in mice in that infective larvae are immobilized within 2 to 3 hours after penetrating the skin of mice that are immune to this nematode. The larvae become surrounded by host defence cells and bundles of collagen fibres. The cuticle is the first structure of the larva to be attacked; host defence cells may secrete a collagenase which attacks the cuticle. Disorganization of the epidermis and underlying muscle cells follows destruction of the larval cuticle (Lee, 1976).

Long standing controversy in T. spiralis *infections*

After the demonstration of eosinophil-mediating destruction of invading schistosome larvae (Butterworth *et al.*, 1975), Kazura and Grove (1978) showed that purified murine eosinophils could destroy new-born larvae of *T. spiralis* in the presence of specific antibodies. Several pieces of evidence were obtained which suggested that eosinophil-mediated killing of *T. spiralis* larvae occurred *in vivo*. In eosinophil-depleted animals that had been treated with anti-eosinophil antibodies, worm burdens were significantly higher than in control animals after challenge infections of *T. spiralis* (Mahmoud, Warren and Boros, 1973; Gleich, Loegering and Olson, 1975). Based on these experiments, the role of eosinophils in *T. spiralis* infections was accepted.

A serious doubt about the role of eosinophils *in vivo* was raised in 1992 when treatment of mice with anti-IL-5 antibody had no effect on worm burdens and fecundity of adult worms during primary as well as secondary infections. In these experiments, treatment of mice with anti-IL-5 antibody resulted in complete suppression of eosinophila in bone marrow, peripheral blood and muscular tissue around larvae. Nevertheless the number of muscle-stage larvae recovered after primary and secondary infections was unaffected by eosinophil depletion (Herndon and Kayes, 1992). Previous findings obtained with anti-eosinophil antibodies were thus thought to reflect cross-reactivity with other cell types. Moreover, no significant difference was seen in recovery of muscle-stage larva between normal and IL-5 transgenic mice that showed marked eosinophilia (Hokibara *et al.*, 1997). The odds seem to be against eosinophils as effectors of anti-*T. spiralis* immunity (Finkelman *et al.*, 1997).

There is, however, strong evidence for the presence of anti-new-born larvae immunity *in vivo* in mice and rats (Bell, Wang and Ogden, 1985; Wang and Bell, 1987, 1988), and eosinophils do kill new-born larvae *in vitro*. How can one explain all of these results together? One thing that should be considered is that when susceptibility of animals is measured by counting the number of muscle-stage larvae 3 or 4 weeks after challenge infection, the results obtained represent the summation of protective responses during the entire period (Bell, 1998). It could be possible that host animals develop a different protection mechanism during this period when eosinophils are not available. New-born larvae of *T. spiralis* enter the muscle cells quickly after discharge from the female and modify myocytes into 'nurse cells' thus ensuring long-term survival (Despommier, 1998). Any damage to nurse cells would cause loss of muscle-stage larvae. In fact, numerous mononuclear cells infiltrate adjacent to developing nurse cells in anti-IL-5 antibody-treated mice (Herndon and Kayes, 1992). Nurse cells might have been attacked by macrophages or even cytotoxic T cells when IL-5 was neutralised and eosinophils were depleted. Obviously much more has to be done to draw any conclusion about the role of eosinophils in protection against *T. spiralis* larvae.

Eosinophils in non-definitive hosts

Besides the controversial findings about *T. spiralis*, there are some instances that clearly support a protective role for eosinophils *in vivo*. *Angiostrongylus cantonensis* infections in mice represent strong evidence for eosinophil-mediated killing of nematode larvae (Yoshimura, Sugaya and Ishida, 1994). Adult worms of *A. cantonensis* reside in the pulmonary arteries of rats. Animals, including humans, are infected by ingesting intermediate hosts (land snails) or paratenic hosts that harbour infective larvae. When ingested by rats, larvae migrate to the brain through the circulation, moult twice and develop into young adult worms. The worms then migrate to the lungs where they reach maturity. During migration, rats show only slight eosinophilia in the cerebrospinal fluid. On the contrary, mice (or 'non-permissive' hosts such as humans) develop severe eosinophilia in the cerebrospinal fluid as well as in the peripheral blood after infection. In mice, larvae migrate to the brain and are killed there. Spleen cells of infected mice produce large amounts of IL-4 and IL-5 but negligible amounts of IFN-γ in response to antigen stimulation (Sugaya *et al.*, 1993, 1997a).

The role of eosinophils was clearly demonstrated when treatment with anti-IL-5 monoclonal antibody induced prolonged survival of the parasites that reached the lungs, together with complete suppression of eosinophilia (Sasaki *et al.*, 1993). In addition, IL-5 receptor alpha chain (IL-5R a)-deficient mice, that produced basal levels of eosinophils, showed impaired resistance to *A. cantonensis* (Yoshida *et al.*, 1996). In both cases, depletion of eosinophils resulted in impaired protection. Furthermore, the number of intracranial worms recovered from IL-5 transgenic mice was significantly lower than that from wild type mice (Sugaya *et al.*, 1997b). These findings indicate that eosinophils induced and activated by IL-5 are capable of eliminating *A. cantonensis* larvae *in vivo*. Electron micrographs of the infected mouse brain showed prominent eosinophil infiltration around intracranial worms. Eosinophils adhered to the worm surface and degranulated onto it (Yoshimura, Sugaya and Ishida, 1994).

Eosinophils in definitive hosts

Other cases in which nematodes mature in the intestine of the host are less striking. In *Strongyloides venezuelensis* infections in mice, subcutaneously inoculated infective larvae migrate to the lungs within 3 days and go on to the small intestine. Treatment of mice with anti-IL-5 antibody had no effect on the recovery of larvae from the lungs in the primary infection, however in the secondary infection, anti-IL-5 antibody treatment suppressed protection as well as eosinophilia (Korenaga *et al.*, 1991, 1994). Thus in infections with *S. venezuelensis*, eosinophils are implicated in protection only in secondary infections. In primary infections with *S. venezuelensis*, larvae reach the intestine before eosinophil activation and antibody production occurs. It seems that eosinophils are 'prepared' in the secondary infection. In infections with another gastrointestinal nematode, *Nippostrongylus brasiliensis*, anti-IL-5 antibody treatment had no effect on the primary infection (Finkelman *et al.*, 1997), though IL-5 transgenic mice showed resistance to migrating *N. brasiliensis* larvae in the primary infection (Shin *et al.*, 1997). In mice infected with *Heligmosomoides polygyrus*, anti-IL-5 treatment had no effect (Urban Jr *et al.*, 1991).

Eosinophils in nematode infections

Having described various examples of nematode infection, some generalisations can be made about eosinophilia and the role of eosinophils in nematode infections. 1) Eosinophils can be protective when migrating nematodes are exposed to massive amounts of activated eosinophils for an extended period of time; 2) specific antibodies, and complements as well, are necessary for eosinophils to destroy nematodes; 3) nematodes that are not adapted to the host cause severe eosinophilia and damage; 4) eosinophilia elicited by nematode infections is mostly IL-5 dependent.

In other words, the host responds to an unaccustomed nematode with massive eosinophil recruitment, but there is a much smaller response to a familiar one. If a parasite invades a suitable host it can reach the final niche without any problems. Even if a parasite enters a wrong host, in which it cannot reach maturity, massive eosinophila may be beneficial to the parasite because the only way for some species to continue further development is to be ingested, together with the present host, by a suitable definitive host. Mice infected with *A. cantonensis* show loss of Purkinje cells and spongy changes in the cerebellum due to eosinophil-derived neurotoxic substances (Yoshimura, Sugaya and Ishida, 1994). When eosinophils damage the nematodes, they release cytotoxic granular contents into surrounding tissues, and these also damage the host. These mice have a good chance of being eaten by a carnivorous animal. Tissue dwelling *T. spiralis* larvae have to be eaten by a carnivore to become adults. There was a decline in ambulatory activity, exploratory activity, running speed and in the distance travelled and an increase in the time spent immobile coincident with the release of new-born larvae into the circulation of infected mice (Rau, 1984; Rau and Putter, 1984) which means that these mice are likely to be more vulnerable to predation and thus enhance parasite transmission. Similarly, Winter *et al.*, (1994) and Harwood *et al.* (1996) showed that the diaphragm muscle of mice infected with muscle-stage larvae of *T. spiralis* or with *T. pseudospiralis* fatigued more quickly than that of uninfected mice and these infected mice would therefore be more open to predation. The racoon ascarid, *Balysascaris procyonis*, is more violent in that this nematode causes severe, sometimes fatal, dysfunction in the central nervous system in herbivores, and also in humans (Fox *et al.*, 1985). As long as nematodes are in the host tissue it would be difficult to destroy the parasites without damaging surrounding tissues, for the killing mechanism of eosinophils does not seem to be very specific for each nematode parasite.

The non-specific nature of eosinophil-mediated protection is evident when schistosome-infected mice are challenged by *Strongyloides venezuelensis* larvae. *Schistosoma mansoni* and *S. japonica* are blood flukes (trematodes) that induce tissue and blood eosinophilia in mice. When mice are inoculated with infective larvae of *S. venezuelensis* several weeks after infection with *S. mansoni* or *S. japonica*, inoculated larvae are destroyed before arriving at the lungs (Yoshida *et al.*, 1999, Maruyama *et al.*, 2000a). Eosinophilia caused by a preceded infection with one species is protective against subsequent infection with other species. It seems as if parasites make use of host immune responses for their purposes in certain situations. To defeat multicellular parasites might be a difficult task.

Goblet Cells and Expulsion of *Nippostrongylus brasiliensis*

Infection with the intestinal nematode *Nippostrongylus brasiliensis* in mice and rats is undoubtedly one of the most extensively studied host-parasite models. Immune expulsion of *N. brasiliensis* has attracted many researchers in the fields of immunology and parasitology for decades, resulting in an enormous amount of data accumulated in the literature. After percutaneous or subcutaneous inoculation, larvae migrate to the lungs, and eventually reach the small intestine, where they become established. Infection with *N. brasiliensis* generates strong T cell-dependent immune responses with eosinophilia, elevated serum IgE and intestinal mastocytosis (Befus and Bienenstock, 1979; Jarrett and Miller, 1982; Woodbury *et al.*, 1984). The intestine of mice and rats undergoing worm expulsion shows various signs of inflammation; oedema, cellular infiltration, elevated mucus secretion, etc. It seems obvious that one or several of these changes are responsible for the worm expulsion, however, it is still not clear exactly what the correlation is (Wakelin, 1994).

In general, studies on the expulsion of *N. brasiliensis* can be divided into two categories. One focuses on the immunological regulation system, especially cytokines, that controls the worm expulsion. Because of the availability of immunological reagents and knockout animals, mice are preferred for the host, and a mouse-adapted strain of *N. brasiliensis* is used. The other category deals more directly with the expulsion mechanism. Rats as well as mice have been employed as the host for this purpose. The final goal of these two approaches should be the same; identification of the effector molecules of the host and target molecules on or in *N. brasiliensis*. Although we have not identified any of the molecules responsible for worm expulsion, we are now getting close to revealing the expulsion mechanism.

Role of cytokines in N. brasiliensis *infection*

The importance of T cells in the expulsion of *N. brasiliensis* was first suggested in cortisone-treated mice (Ogilvie 1965), and more directly in nude mice (Jacobson and Reed, 1974). *In vivo* cell depletion experiments then demonstrated that it is sIg-cells in rats (Nawa and Miller, 1979) and CD4[+] cells in mice (Katona, Urban Jr. and Finkelman, 1988) that are required for the expulsion of *N. brasiliensis*. Like other nematode infections, *N. brasiliensis* elicits typical Th2 immune responses, which should induce worm expulsion in some way. How can one determine that Th2 responses are truly protective? Since Th1 and Th2 cells are cross-inhibitory, administration of cytokines that down-regulate Th2 cells would augment Th1 responses during infection. In order to test this hypothesis, mice were treated with IFN-γ and this induced a marked increase in parasite egg production and delayed expulsion of adult worms (Urban Jr. *et al.*, 1993).

The protective role of Th2 cells was further confirmed by administering IL-12 to mice infected with *N. brasiliensis* (Finkelman *et al.*, 1994). IL-12 administration *in vivo* enhanced adult worm survival and egg production during primary infections, but not secondary infections. IL-12 needed to be administered by day 4 of a primary infection to inhibit IgE and mucosal mast cell responses, and by day 6 to strongly inhibit eosinophil responses and to enhance worm survival and fecundity. Anti-IFN-γ inhibited the effects of IL-12 on parasite survival and fecundity. These observations indicate the possible protective role of Th2 responses against *N. brasiliensis*. This reminds us of the role of Th1 cells in the elimination of intracellular parasites, however, there is a big difference between intracellular parasites and the intestinal nematodes. Having shown that Th2 responses are protective, direct effects of IL-4 were examined. Mice were treated with IL-4 and infected with *N. brasiliensis*. Interestingly, administration of IL-4 restored the expulsion capacity of SCID as well as anti-CD4 antibody-treated mice that normally develop chronic infections (Urban Jr. *et al.*, 1995). Based on these experiments, it can be concluded that IL-4 can bring about worm expulsion through cells other than functional T cells. Surprisingly, however, treatment of mice with anti-IL-4 antibody or anti-IL-4 receptor antibody had no effect on the expulsion of *N. brasiliensis* from intact mice. In these mice, while IL-4 related responses, such as mastocytosis and IgE production, were completely inhibited, *N. brasiliensis* worms were expelled normally (Madden *et al.*, 1991). In addition, mice deficient for the IL-4 gene were perfectly normal in expelling adult worms of *N. brasiliensis*, although they could not mount Th2 responses (Kopf *et al.*, 1993; Lawrence *et al.*, 1996).

This apparent paradox has been resolved recently by experiments with IL-4 receptor a chain (IL-4Rα) knockout and Stat6 (signal transducer and activator of transcription 6) knockout mice. Mice deficient for the IL-4Rα gene or Stat6 genes could not expel *N. brasiliensis* (Urban Jr. *et al.*, 1998). Exogenous IL-4 administration did not improve delayed worm expulsion in these deficient mice. Conversely, neutralisation of IL-13 inhibited worm expulsion in IL-4 deficient mice, as well as in normal mice. IL-13 is a cytokine produced by activated Th2 helper cells that has various target cells (Brown *et al.*, 1989; Zurawski and de Vries, 1994). What do these facts mean? IL-4 receptor and IL-13 receptors

share the IL-4Rα chain and use Stat6 in common (Sugamura *et al.*, 1996; Takeda *et al.*, 1996; Takeda, Kishimoto and Akira, 1997) and these are the basis for the shared activities of IL-4 and IL-13 (Zurawski *et al.*, 1993). It can therefore be suggested that in IL-4 knockout mice, IL-13 produced upon infection compensated IL-4 activities and stimulated IL-4Rα/Stat6 signal transduction. Thus, it is concluded that expulsion of *N. brasiliensis* requires signalling via IL-4Rα and Stat6, and IL-13 may be more important than IL-4 as an inducer of the Stat6 signalling that leads to worm expulsion (Urban Jr. *et al.*, 1998).

It now seems certain that the critical cells for the expulsion of *N. brasiliensis* in mice are non-lymphoid cells that respond to IL-4Rα/Stat6 signal transduction. As will be described below, intestinal goblet cell mucin plays a very important role in the expulsion of *N. brasiliensis*. Recent studies suggest that impairment of worm expulsion in anti-CD4 antibody-treated mice correlates with the reduction of intestinal mucus (Khan *et al.*, 1995), and that mouse intestinal goblet cells are stimulated by Th2 cell-derived factor(s) directly or indirectly to proliferate and produce large amounts of mucin glycoprotein (Ishikawa, Wakelin and Mahida, 1997). Because IL-13R is expressed on various cell types, including epithelial cells (Zund *et al.*, 1996), intestinal epithelial stem cells and goblet cells might be directly stimulated with IL-4/IL-13 to produce large amounts of mucin.

Alterations of the terminal sugars of goblet cell mucin

Intestinal mucus has been considered to serve a protective function by excluding or trapping worms in immune animals (Wells, 1963; Miller, Huntley and Wallace, 1981) or by preventing worms from establishing in the intestine by inhibiting intimate contact with the mucosa (Miller, 1987). In *N. brasiliensis* infections in rats, goblet cell hyperplasia and worm expulsion can be transferred by immune T cells (Miller and Nawa, 1979), and worm expulsion is always associated with goblet cell hyperplasia (Nawa and Korenaga, 1983), suggesting that the effector for *N. brasiliensis* is intestinal goblet cell mucin. Recently lectin histochemical staining demonstrated that goblet cell mucin changed qualitatively as well as quantitatively during worm expulsion (Oinuma *et al.*, 1995). Goblet cell mucin in the jejunum of uninfected rats does not bind to the lectins, LFA, HPA or GSA-II, which specifically recognise sialic acid, GalNAc and GlcNAc, respectively. However, after infection with *N. brasiliensis*, goblet cell mucin bound strongly to these lectins. Such qualitative changes, or alteration of terminal sugar residues, of goblet cell mucin occurred in parallel to goblet cell hyperplasia and expulsion of worms, indicating that alteration of the terminal sugars contributes to the expulsion of *N. brasiliensis*.

The occurrence of changes in goblet cell mucin and worm expulsion was examined in a series of experiments. In *N. brasiliensis* infections in rats, adult worms in the intestine are thought to change from 'normal' to 'damaged' worms before being expelled (Ogilvie and

Jones, 1971). When adult worms of *N. brasiliensis*, obtained from donors 7 days after a primary infection, were implanted in the duodenum of naive recipient rats, implanted worms established in the second hosts and were expelled between days 10 and 14 after implantation. On the other hand, those obtained 13 days after a primary infection were expelled within 5 days after implantation in the naive recipients. Thus, 7-day-old worms were designated as 'normal', whereas 13 day-old worms as immunologically 'damaged' (Ogilvie and Jones, 1971). Lectin histochemistry revealed that in the naive recipient rats, expulsion of both 'normal' and 'damaged' worms occurred in association with quantitative and qualitative changes of goblet cell mucin. When immune rats were used as recipients, implanted 'normal' worms were expelled within 5 days, again in association with goblet cell changes (Ishikawa, Horii and Nawa, 1993). Moreover, when 'normal' worms were implanted into hypothymic nude rats, neither worm expulsion nor goblet cell changes occurred. These results clearly show that goblet cell changes are consistently associated with the immune-mediated expulsion of *N. brasiliensis*.

Quite unexpectedly, 'damaged' worms were expelled within 5 days after implantation in hypothymic rats while 'normal' worms were not expelled from hypothymic rats (Ishikawa *et al.*, 1994). One might suspect that 'damaged' worms are expelled because they are damaged. However, they were not expelled when implanted in corticosteroid-treated hypothymic rats. Histochemical examination revealed that hypothymic rats, which expelled 'damaged' worms, showed alteration in terminal sugar residues in goblet cell mucin. Moreover, 'normal' worms were rejected by euthymic as well as by hypothymic rats when goblet cell changes were induced in these rats by implantation of 'damaged' worms 3 to 5 days prior to the implantation (Ishikawa *et al.*, 1994). These findings suggest that T cells or T cell-derived factors are not necessary for the expulsion of *N. brasiliensis*, once alterations of terminal sugars of goblet cell mucin is induced.

Implantation experiments described above tell us that there are several steps which give rise to worm expulsion. First, adult *N. brasiliensis* worms elicit immune responses in the host. Second, the stimulated immune system 'damages' adult worms. Third, 'damaged' worms induce changes in mucin glycoconjugates independent of T cells. Adult worms are expelled because they cannot persist in the presence of altered terminal sugar residues. Back in the 1970's, a two-step theory was proposed for worm expulsion in which a specific antibody causes the damage to worms and, subsequently, non-specific inflammatory responses cause the expulsion (Ogilvie and Love, 1974). The results described above basically support the concept of a two-step mechanism of expulsion, with the second step mediated by T cell-independent responses.

From the point of view of *N. brasiliensis*, induction of goblet cell changes would not be desirable. An interesting finding has been obtained that suggests 'normal' *N. brasiliensis* can inhibit the induction of goblet cell changes by 'damaged' worms (Ishikawa, Horii and Nawa, 1994). When 'normal' worms were concurrently implanted with 'damaged' worms in naive recipient rats, 'damaged' worms that would have been expelled within 5 days were not expelled, and neither goblet cell hyperplasia nor sugar changes occurred. Because treatment of 'normal' worms by heating at 80°C for 1 min abolished their inhibitory effects, the inhibitory signal from the 'normal' worms was probably released as ES antigen (Uchikawa *et al.*, 1994). Since such inhibitory effects of 'normal' worms on goblet cell changes was not observed when they were implanted into immunised recipients, immune responses of the host may invalidate the inhibitory signals from 'normal' worms. The role of T cells in this suppression has not been examined.

Expulsion mechanisms for *N. brasiliensis*

What relationship could exist between these changes in goblet cell mucin in rats and cytokine responses seen in mice? In rats, immune responses to *N. brasiliensis* are dominated by Th2 cells, as in mice (Uchikawa *et al.*, 1994, 1997). It seems probable that proliferation of intestinal goblet cells is affected by Th2 type cytokine(s) as postulated in mice. Treatment of animals with antibodies against one or several Th2 cytokines may inhibit goblet cell hyperplasia after infection with *N. brasiliensis* or, on the contrary, T cell-deprived animals may undergo goblet cell hyperplasia by administrating Th2 cytokines (Figure 18.3). These kinds of experiments should clarify the roles of cytokines in goblet cell hyperplasia and mucin production. Since intestinal epithelial cells are known to respond to IFN-γ (Colgan *et al.*, 1996), there would be no reason to believe that they do not respond to Th2 cytokines. Recently trefoil factor family (TFF)-domain peptides, mainly produced by mucin-secreting epithelial cells in the gastrointestinal tract, were found to have regulatory effects on the migration of epithelial cells (Wright *et al.*, 1997; Williams and Wright, 1997). These intestinal peptides may also contribute to the mucosal defence by altering epithelial

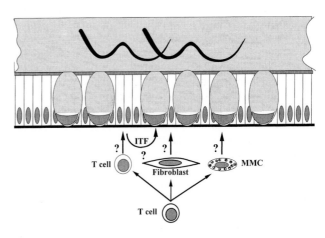

Figure 18.3. Hypothetical regulatory pathways for goblet cell changes in *Nippostrongylus brasiliensis* infection ITF: intestinal trefoil factor, MMC: mucosal mast cells.

cell functions because their expression was altered by *N. brasiliensis* infection (Tomita *et al.*, 1995).

It is uncertain whether the change in terminal sugar residues of goblet cell mucin is itself protective. Scanning electron microscopy demonstrated that when *N. brasiliensis* worms were implanted in the intestine with altered goblet cell mucin, they were trapped in the mucin and failed to migrate into the intervillous space (Ishikawa *et al.*, 1994). It could be that the goblet cell mucin binds to the surface of the worms and prevents attachment to the villous surfaces. There might be a difference in binding affinity of changed and unchanged goblet cell mucin to the surface of *N. brasiliensis*. Another interesting fact is that intestinal goblet cell mucin of mice binds to lectins such as LFA, HPA and GSA-II A. *N. brasiliensis* maintained in rats are quickly expelled as compared to the mouse-adapted strain (Khan *et al.*, 1995). *N. brasiliensis* may have changed the molecular structure of the body surface during adaptation to the mouse. Much remains to be explored in the expulsion of *N. brasiliensis*.

Expulsion of *Trichinella spiralis*

Primary infection of Trichinella spiralis

During primary infections with *T. spiralis* in mice and rats, most adult worms are expelled from the intestine in about two weeks (Alizadeh and Wakelin, 1982). The expulsion is delayed in nude mice (Ruitenberg and Elgersma, 1976) and protective immunity can be transferred to naive animals with T cells (Grencis, Riedlinger and Wakelin, 1985; Korenaga *et al.*, 1989). T cells that mediate protection are, as expected, Th2 and Th2-like cells in mice and rats respectively (Grencis, Hultner and Else, 1991; Ramaswamy, Goodman and Bell, 1994). In mice, Th2 cells have been shown to be compartmentalised to mesenteric lymph nodes (Kelly *et al.*, 1989). However, one study demonstrates that mesenteric lymph node cells of genetically resistant mice produced more IFN-γ and less IL-4 as compared to those of susceptible mice (Pond, Wassom and Hayes, 1989).

In mice, several lines of evidence indicate that mucosal mast cells are important for the expulsion of *T. spiralis*. First, there is positive correlation between resistance to *T. spiralis* and intensity of intestinal mastocytosis, and a clear association between the time at which worm expulsion takes place and peak mast cell responses in inbred mouse strains (Tuohy *et al.*, 1990). Second, W/W^v mice, known for poor mast cell responses due to aberrant stem cell factor receptor (c-kit) (Kitamura, Go and Hatanaka, 1978; Nocka *et al.*, 1990), show delayed worm expulsion and this can be restored by bone marrow reconstitution (Alizadeh and Murrell, 1984; Oku, Itayama and Kamiya, 1984). Stem cell factor (SCF) is indispensable for mast cell differentiation and maturation *in vivo* (Galli, Zsebo and Geissler, 1994). Recent experiments have demonstrated that treatment of mice with anti-SCF as well as with anti-*c-kit* antibodies causes severe impairment of worm expulsion in association with loss of intestinal masto-

cytosis (Grencis *et al.*, 1993; Donaldson *et al.*, 1996). Finally, administration of recombinant IL-3, which promotes mucosal mastocytosis (Abe and Nawa, 1988), reduces adult worm burdens and the number of muscle larvae (Korenaga, Abe and Hashiguchi, 1996). Because the effects of IL-3 are not diminished by anti-IL-4 antibody treatment, worm expulsion appears to be dependent on mucosal mast cells, but not on IgE. All of these findings suggest the importance of mucosal mast cells, however, there are still no satisfactory explanations for the expulsion mechanisms initiated by mast cells, or other cell types (Bell, 1998). Recent work (Lawrence *et al.*, 1998) has shown that pathological changes in the intestine, including mastocytosis, are not essential for worm expulsion, and emphasis has been placed on immune-mediated changes in smooth muscle function (Vallance and Collins, 1998).

In rats, intestinal mast cells do not seem to be very important in this respect. T cell populations that confer protective immunity are CD4$^+$OX22$^-$ (Korenaga *et al.* 1989), which induce antibody responses and intestinal eosinophilia in recipient rats (Wang *et al.*, 1990). On the other hand, concurrently activated CD4$^+$OX22$^+$ cells, that are not protective, induce intestinal mastocytosis with little antibody response (Wang *et al.*, 1990). During an infection with *T. spiralis*, active transport of IgE antibodies to the gut wall and into the intestinal lumen occurs in rats. Upregulation of IgE transport can be adoptively transferred with immune CD4$^+$OX22$^-$ T cells, and treatment of animals with IL-4 also induces IgE transport (Ramaswamy, Hakimi and Bell, 1994). Recently it has been found that most IgE produced as a result of infection with *T. spiralis* was found to be transported to the gut and as little as 0.02% of IgE entered the serum pool (Negrao-Correa, Adams and Bell, 1996). These findings suggest that secreted antibodies probably play an important role in mediating expulsion of *T. spiralis* in rats.

Rapid expulsion of Trichinella spiralis *in rats*

A role for antibodies has also been demonstrated in the rapid expulsion of *T. spiralis* from rats. An initial infection with *T. spiralis* confers a long-lasting immunity in rats. Immune rats are refractory to reinfection, and eliminate 90–95% of larvae of a challenge infection within an hour. This phenomenon, designated as rapid expulsion, has been studied extensively as one of the model systems for the expulsion of intestinal nematodes. Rapid expulsion affects incoming larvae and is not associated with acute inflammatory cellular responses or pathophysiological changes which occur during the expulsion of adult worms in the primary and secondary infections (Russel and Castro, 1979). It has been pointed out that mice do not express rapid expulsion as rejection of a superinfection occurs associated only with adult worm expulsion, and the response disappears quickly after the primary infection has been eliminated (Bell, 1992).

A clue to the elucidation of the rapid expulsion mechanism is that it can be adoptively transferred to naive adult rats with immune serum and thoracic duct lymphocytes (Ahmad *et al.*, 1990). Neither serum alone nor thoracic duct cells alone protect recipient animals. For the induction of rapid expulsion in recipient rats, cells have to be transferred first and immune serum next. The intestinal priming effects produced by adoptively transferred cells persist for 7 weeks, whereas immune serum has to be transferred sometime between 3 days before a challenge and 6 hours after a challenge infection (Ahmad *et al.*, 1990). Immune cells that could mediate rapid expulsion are CD4$^+$CD8$^-$ OX22$^-$ (Ahmad *et al.*, 1991). Responsible antibody classes were determined in rats transferred with immune thoracic duct lymphocytes. Serum IgE of rats infected with *T. spiralis* peaked 4 weeks after infection and declined thereafter. IgG responses on the other hand reached a peak at 4–6 weeks and maintained the same level until at least 12 weeks after infection. Antibodies which could mediate rapid expulsion were IgE in 28 day-serum, whereas they were non-IgE in 3 month-serum (Ahmad, Wang and Bell, 1991). Purified IgE from 28 day-sera recognised the TSL-1 antigen of *T. spiralis* muscle-stage larvae. While transport of large amounts of IgE into the intestinal lumen has been suggested (Ramaswamy, Hakimi and Bell, 1994), the way in which non-IgE antibodies in 3 month-serum could mediate rapid expulsion has not been clarified.

Interestingly, antibodies alone can mediate rapid expulsion in suckling rats. Appleton and McGregor (1984) reported that pups suckling an immune dam rapidly expelled *T. spiralis* muscle-stage larvae. They found that protective antibodies produced in immune dams were transferred to pups through milk and that passive transfer of antibodies alone protected pups from challenge infection. These findings were followed by the demonstration that immunisation of rat dams with ES antigen resulted in rapid expulsion in suckling rats, and that monoclonal IgG antibodies to ES antigen could mediate rapid expulsion (Appelton, Schain and McGregor, 1988). Because preincubation of muscle-stage larvae with antibodies reduced the number of worms established in the pup intestine (Carlisle, McGregor and Appelton, 1990), it is likely that pups secrete antibodies into the intestinal lumen without local T cell activation.

Expulsion mechanisms for *T. spiralis*

It appears that expulsion mechanisms for adult *T. spiralis* are different in mice and rats. Mucosal mast cells seem important in mice and antibodies in rats, however, they are not necessarily exclusive of each other. Adult *T. spiralis* in the intestine migrate through enterocytes leaving damaged epithelial cells that are sloughed off at the base of the villi (Dunn and Wright, 1985). Therefore, any host effector function that stops the worms moving would possibly result in worm expulsion, because intestinal epithelial cells are replaced every 2–5 days (Merzel and Leblond, 1969; Gordon, 1993). Obviously we have to know more about the biology of intracellular adult *T. spiralis*. In the case of rapid expulsion in rats, the mechanism appears to be

prevention of incoming larvae attaching to and invading intestinal cells (Rothwell, 1989). The role of T cells in immune adult rats is to promote immunoglobulin transport to the gut and the intestinal lumen. In this case, anything that prevents attachment and invasion of incoming larvae would cause rapid expulsion. Rejection of a superinfection in mice associated with adult worm expulsion may be due to mucosal mast cells because, as will be describe below, mucosal mast cells may be able to prevent attachment and subsequent invasion by intestinal nematodes.

Sulphated glycoconjugates and expulsion of *Strongyloides* nematodes

Experimental strongyloidiasis

Strongyloidiasis, caused by *Strongyloides stercoralis*, is a major intestinal nematode disease in humans. Free-living infective larvae penetrate the skin and migrate to the intestine through the lungs. Chronic infections are probably maintained by an autoinfection process, in which a portion of larvae develop to infective larvae in the host intestine and penetrate the lower gut mucosa or perianal skin. In an immunocompromised host, worm numbers increase enormously by this process and the disease can become fatal (Neva, 1986, Grove, 1996). Experimental studies with *S. stercoralis* have been hampered by the lack of permissive animals. Animals that can support the entire life cycle are dogs, patas monkeys, Mongolian gerbils and SCID mice (Nolan *et al.*, 1993; Rotman *et al.*, 1995). Several studies with mice inoculated with third-stage larvae indicate that mice develop immunity against challenge infections with *S. stercoralis* infective larvae. The protection mechanism is antibody-mediated killing by eosinophils, which are induced by Th2 responses (Abraham *et al.*, 1995; Rotman *et al.*, 1996, 1997).

Expulsion of *Strongyloides ratti* and *S. venezuelensis* in mice

Strongyloides ratti and *S. venezuelensis* have been used to study the expulsion mechanisms for *Strongyloides*. Expulsion mechanisms in mice for these two species seem to be the same, mucosal mast cells, especially intraepithelial ones, being critical for expulsion (Nawa *et al.*, 1994). The importance of mast cells was first demonstrated by experiments with mast cell-deficient W/W^v mice as a host (Nawa *et al.*, 1985; Khan *et al.*, 1993). When W/W^v mice were infected with *S. ratti* or *S. venezuelensis*, expulsion was delayed significantly. This delay was completely restored by bone marrow grafting, as in *T. spiralis* infections in W/W^v mice. Recently, IL-3 knockout mice and mice that have diminished expression of the IL-3 receptor α chain have been shown to have impaired worm expulsion and intestinal mastocytosis (Lantz *et al.*, 1998; Kobayashi *et al.*, 1998). Among several mouse strains tested, mice that have both W^v mutation and disrupted IL-3 gene were most severely affected (Lantz *et al.* 1998). However, defective intestinal mastocytosis and expulsion of *S. ratti* in nude mice were restored by repeated injection

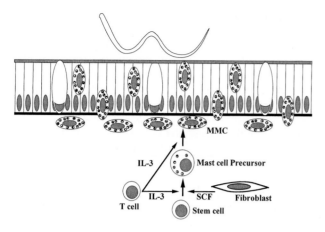

Figure 18.4. Cytokine pathways for generation of intestinal mast cells in *Strongyloides* infection.

of IL-3 *in vivo* (Abe and Nawa, 1988; Abe *et al.*, 1992). Furthermore, administration of IL-3 in normal mice induced intestinal mastocytosis and prevented *S. ratti* from establishing in the intestine (Abe *et al.*, 1993; Abe, Sugaya and Yoshimura, 1993). Based on these findings, intestinal mast cells are thought to be effector cells for the expulsion of *Strongyloides* (Figure 18.4).

Interestingly, adoptively transferred cultured mast cells, that phenotypically resemble mucosal mast cells, cannot confer expulsion capacity to W/W^v mice. In these mice only a few mucosal mast cells migrate to the intraepithelial location, whereas more than 90% of mucosal mast cells are intraepithelial in normal or W^v mice reconstituted with +/+ bone marrow cells (Abe and Nawa, 1987a, b). The critical role of intraepithelial mast cells is further suggested by *S. venezuelensis* infection in Mongolian gerbils. Mongolian gerbils cannot expel adult *S. venezuelensis*, even though they are sensitised enough to develop a strong protective immunity against tissue migrating larvae (Horii, Khan and Nawa, 1993; Khan, Horii and Nawa, 1993). In Mongolian gerbils, no mast cells migrate into the epithelial layer. Therefore it seems that mast cells have to be intraepithelial to expel *Strongyloides* nematodes.

Expulsion of Strongyloides venezuelensis *in hamsters*

Fruitful findings in parasitology are sometimes obtained from using animals other than mice and rats. When Syrian golden hamsters were infected with *S. venezuelensis* expulsion was slow and gradual as compared to that in mice and rats. Histological examination revealed goblet cell hyperplasia around the time of expulsion, which was never seen in mice and rats. Although intestinal mastocytosis was also observed, its peak did not coincide with expulsion and all mast cells remained in the lamina propria (Shi *et al.*, 1994a).

Subsequent experiments extended this finding. Four different species of hamsters were infected with *S. venezuelensis* and worm expulsion and histological

changes in the intestine were examined. Two species of hamsters, the Syrian golden hamster and the Cambell hamster, harboured *S. venezuelensis* for more than 40 days, whereas the other two species, the Chinese and Russian hamsters, expelled *S. venezuelensis* within two weeks (Shi *et al.*, 1994b). In all four species, peak responses of intestinal mastocytosis were not associated with worm expulsion, but goblet cell hyperplasia was observed around the time of worm expulsion. Curiously, the goblet cell mucin of hamsters was sulphated and the degree of sulphation was closely related to the expulsion capacity. The rapid expellers, the Chinese and Russian hamsters, had strongly sulphated goblet cell mucin whereas the slow expellers, the Syrian golden and Cambell hamsters, had moderately sulphated goblet cell mucin (Shi *et al.*, 1994b). These results suggested that sulphated mucin might be an effector for the expulsion of *S. venezuelensis* in hamsters. This hypothesis was supported by experiments in which establishment of implanted adult *S. venezuelensis* worms was inhibited in reserpine-treated rats (Ishikawa *et al.*, 1995). In these rats, intestinal goblet cell mucin was sulphated (Forstner, Roomi and Forstner, 1985).

Expulsion mechanisms for Strongyloides *nematodes*

What can one make of the findings described above? Parasitic adult *Strongyloides* are expelled by intraepithelial mast cells in mice and rats and by goblet cell mucin in hamsters and reserpine-treated rats. If it is assumed that the underlying mechanism for expulsion is the same in all of these cases then there must be something in common. One good candidate is a sulphated glycoconjugate. Mast cell granules contain highly sulphated glycosaminoglycans (Razin *et al.*, 1982; Stevens, Otsu and Austen, 1985; Stevens *et al.*, 1988) and these are released into the intestinal lumen during worm expulsion (Woodbury *et al.*, 1984; Scudamore *et al.*, 1995). Sulphated carbohydrates might be effector molecules for the expulsion of this parasite, whatever their sources.

If this is so, then how could sulphated carbohydrates expel *Strongyloides* nematodes? Parasitic adult *Strongyloides* live in between the intestinal epithelial cells, forcing a gap between them. The worms appear to exit into the lumen and re-enter the epithelial layer repeatedly (Dawkins, 1989). If they re-enter the mucosa, adult worms should adhere to the surface of intestinal cells upon invasion (Dubremetz and McKerrow, 1995). Recently we found that parasitic adult *S. venezuelensis* adhered firmly to the surface of culture plates with orally secreted adhesion substances (Maruyama and Nawa, 1997). These substances bind to red blood cells as well as to intestinal epithelial cells. Interestingly, the adhesion substances have strong binding affinity to sulphated carbohydrates (Maruyama, Nawa and Ohta, 1998). It is therefore conceivable that highly sulphated carbohydrates derived from mast cells or goblet cells may bind to adhesion substances *in vivo*, thus inhibiting attachment of the worms to intestinal epithelial cells. We have demonstrated that highly sulphated glycosaminoglycans of the type mast cell granules contain can inhibit the binding of *S. venezuelensis* adhesion substances to mucosal epithelial cells *in vitro*, and that those glycosaminoglycans block the invasion of gut mucosa by *S. venezuelensis* adults *in vivo* (Maruyama *et al.*, 2000b). Adult *Strongyloides* worms are not killed or severly damaged by the host immune system while being expelled (Maruyama *et al.*, 2000c), but they just burrow into the intestinal mucosa and sometimes come up to the surface. If they fail to re-enter the mucosa because attachment is interfered with, they would be carried down to the posterior half of the intestine where conditions are not favourable. In fact, *Strongyloides* cannot invade the intestinal mucosa when there is mastocytosis taking place (Abe *et al.*, 1993a,b; Maruyama *et al.*, 2000b). This hypothesis seems to explain why mast cells have to be in the epithelial layer. Mast cell glycosaminoglycans are covalently linked to a core peptide called serglycin to form huge molecules (Kolset and Gallagher, 1990). Such molecules would not be able to diffuse across the basement membrane.

Expulsion of Nematodes Causing Chronic Infections

Although experimental infections with *N. brasiliensis*, *T. spiralis*, and *Strongyloides* spp. have provided us with a great deal of information that leads us to the possible elucidation of the expulsion mechanisms, all of these infections 'self-cure' in the immunocompetent host. Considering that many common intestinal nematodes, such as *Ascaris lumbricoides*, pinworms (*Enterobius vermicularis*), hookworms (*Ancylostoma* and *Necator*) and whipworms (*Trichuris trichiurus*), cause chronic infections, the immunology in chronic intestinal nematode infections should be studied in addition to those expelled within weeks. Two experimental models are currently employed to investigate chronic intestinal nematode infections, namely *Trichuris muris* and *Heligmosomoides polygyrus* in mice.

Different strains of mice express varying degrees of resistance to *T. muris*. While most strains of mice rapidly expel this nematode, others fail to expel it and develop chronic infections (Else and Wakelin, 1988). Because athymic nude mice cannot bring about worm expulsion, and because depletion of CD4$^+$ T cells abrogates resistance (Koyama, Tamauchi and Ito, 1995), susceptibility to this nematode reflects altered T cell function. Comparison of various T cell parameters between susceptible and resistant strains revealed that the cytokine production patterns were markedly different. Mesenteric lymph node cells of a resistant strain produced IL-5 but not IFN-γ, and those of a susceptible strain produced IFN-γ but not IL-5 (Else and Grencis, 1991). In other words, resistant mice mounted Th2 responses to the infection. When susceptible mice, that normally develop Th1 responses, were treated with IL-4 or anti-IFN-γ antibody, the mice expelled *T. muris* (Else *et al.*, 1994). IL-4 treatment was effective even after *T. muris* started production of eggs. Susceptible mice also became resistant when Th2 responses were induced by concurrent infections with *Schistosoma mansoni* (Curry *et al.*, 1995). On the other hand, normally resistant mice became susceptible after

anti-IL-4 receptor antibody treatment (Else *et al.*, 1994) and IL-12 administration. The effects of IL-12 were abrogated by anti-IFN-γ antibody (Bancroft *et al.*, 1997). Production of Th2 cytokines was correlated to resistance, even among individuals in an inbred mouse strain (Else, Entwistle and Grencis, 1993).

Experiments with IL-4 knockout mice have demonstrated the critical role of this cytokine in resistance to *T. muris* in mice. When IL-4 deficient mice with a resistant background were infected with *T. muris*, neither Th2 responses nor worm expulsion occurred. Mesenteric lymph node cells from IL-4 knockout mice produced only a limited amount of IL-5, IL-9 and IL-13, while large amounts of IFN-γ were produced (Bancroft, McKenzie and Grencis, 1998). Interestingly, IL-13 knockout mice failed to expel *T. muris*, even though the mice produced good amounts of IL-4, IL-5 and IL-9. These findings agree with the previous report that IL-4 alone could not confer expulsion capacity to nude mice (Else *et al.*, 1994). In *T. muris* infection, therefore, large amounts of IL-13 produced by expanded Th2 cells appear to be necessary for the expression of the expulsion mechanism(s). The difference between *T. muris* and *N. brasiliensis* infections in cytokine requirements should reflect the difference in effector cells. It is becoming an important issue to identify effector cells that bring about the expulsion of *T. muris*.

Infections with *H. polygyrus* in mice are different from those with *T. muris* in that many strains of mice harbour *H. polygyrus* for months after a primary infection, although mice have raised eosinophilia and elevated IgG1 and IgE (Urban Jr. *et al.*, 1991; Svetic *et al.*, 1993). Expulsion mechanisms for this nematode may be similar to those for *N. brasiliensis* because repeated administration of large doses of IL-4 cured chronic infections in immunocompetent as well as immunodeficient mice (Urban Jr. *et al.*, 1995). The role of IL-13 or other cytokines in the expulsion of *H. polygyrus* has not yet been examined.

It has been known that adult *H. polygyrus* have immunosuppressive effects on the intestinal immunity of the host. Surgically implanted adult worms inhibited induction of protective immunity to *H. polygyrus* and other intestinal nematodes (Behnke, Hannah and Pritchard, 1983; Dehlawi, Wakelin and Behnke, 1987). Intestinal mastocytosis was suppressed in *H. polygyrus*-implanted animals (Dehlawi, Wakelin and Behnke, 1987). Adult *H. polygyrus* may induce IFN-γ production and inhibit Th2 responses.

Conclusions

Animal-parasitic nematodes stimulate the immune system of the host by invading host tissues and releasing antigens during infection. The host responds to these infections with the worm-related triad of eosinophilia, IgE production and intestinal mastocytosis, all of which are regulated by Th2 cytokines; IL-3, IL-4 and IL-5 (Figure 18.5). The mechanisms by which Th2 responses are induced by animal-parasitic nematodes

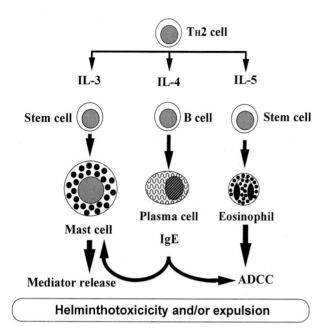

Figure 18.5. Th2 cytokine regulation on the worm-related triad ADCC: antibody-dependent cell-mediated cytotoxicity.

have not been well characterised. Some nematode antigens may directly affect antigen-presenting cells to selectively activate Th2 cells, or stimulate various cells that induce and amplify Th2 responses, e.g., $\gamma\delta$ T cells, NK T cells (Noben-Trauth *et al.*, 1997), eosinophils and B cells (Hernandez, Wang and Stadecker, 1997). The large size and multicellular nature of nematodes themselves may also contribute by presenting antigens slowly from nonphagocytable surfaces (Sher and Coffman, 1992). In any case, animal-parasitic nematodes seem to stimulate multiple pathways that lead to Th2 responses. One, or several, of these responses further induce activation and proliferation of particular cells that are important for protection. Eosinophils are implicated in the destruction of tissue-migrating larvae whilst mast cells, IgE and intestinal epithelial cells are implicated in expulsion from the intestine. Not all Th2 responses are necessary, however, to protect the host from nematode parasites.

Eosinophils are protective, especially when the infected animal is a non-definitive host for the nematodes. Production and activation of eosinophils are induced by IL-5. Massive accumulations of activated eosinophils can destroy nematode larvae in association with specific antibodies. Paradoxically, eosinophilia induced by larval nematodes may be beneficial to the nematodes in two ways. In intestinal nematodes that cause eosinophilia during tissue migration, activated eosinophils in the tissue might prevent further infection with other nematodes and ensure that first arrivals benefit from the resources of the host. When a nematode is in a non-definitive host, tissue destruction caused by massive eosinophilia could possibly increase the chance of the nematode being ingested by a definitive host. Release of chemoattractants for eosino-

phils by these nematodes may be advantageous for them. In the case of filarial nematodes, adult worms have a long life span in the host connective tissue or lymph organs where the immune system is operating. For these nematodes, immune responses of the host may not be merely undesirable but even necessary. It has been reported that natural killer cell activation promotes development and maturation of *Brugia malayi* in SCID mice (Babu *et al.*, 1998).

In the expulsion of gastrointestinal nematodes, IL-3, IL-4 and probably IL-13 are important cytokines. These cytokines stimulate intestinal epithelial cells and IgE production, induce intestinal mastocytosis and affect smooth muscle function. Effector mechanisms appear different from one nematode to another, nonetheless one general principle emerges. The host expels intestinal nematodes by inhibiting intimate contact between the parasites and the intestinal mucosa. This seems to be the case in the role of goblet cell mucin in infections with *N. brasiliensis* and of IgE in infections with *T. spiralis* in rats, of mast cell glycosaminoglycans in infections with *Strongyloides* and probably in infections with *T. spiralis* in mice as well. The host does not use a specific expulsion mechanism to a particular nematode but uses all of its available measures against them. The intestinal mucosa is an active tissue. Each crypt produces about 300 new cells a day (Potten and Loeffler, 1990) and the whole epithelium is renewed every 2–5 days. The host can expel intestinal nematodes if one of the effector functions works. However, we have little knowledge about, for example, the molecular interactions between the surface of *N. brasiliensis* and intestinal epithelial cells, and invasion mechanisms of *S. venezuelensis* and *T. spiralis*. The elucidation of molecular interactions between the animal-parasitic nematodes and the host will continue to be a rewarding research area in the future.

Animal-parasitic species occur widely in the phylum Nematoda today and probably in the past too. It is almost certain that Cambrian animals, such as *Anomalocaris* and *Opabinia*, harboured nematode parasites if they were menacing predators in a complex ecosystem (Chen, Ramskold and Zhou, 1994). Although fossils have little to tell us about how animal-parasitism evolved, we can study extant species that show primitive features as parasites. Nematodes that have free-living infective larvae in the life cycle are probably young parasites, geologically-speaking, and this can be supported by the existence of free-living adults in *Strongyloides* spp. If they are primitive as parasites, their behaviour of invading and burrowing into host tissue should also be considered as primitive. In fact, their tissue migration and intimate contact with the intestinal mucosa make them vulnerable to host protective immunity. However, the ability to burrow into animal tissue might be a prerequisite for animal-parasitism, and such behavioural changes may be induced by a slight mutation in the nematode genome (de Bono and Bargmann, 1998).

As the title of one of the earliest works on parasitology tells us (Redi, 1684), we need to observe and experiment on the living animals found in the living animals to understand these interesting creatures. There is still much to learn.

References

Abe, T. and Nawa, Y. (1987a) Reconstitution of mucosal mast cells in W/W^v mice by adoptive transfer of bone marrow-derived cultured mast cells and its effects on the protective capacity to *Strongyloides ratti*-infection. *Parasite Immunology*, **9**, 31–38.

Abe, T. and Nawa, Y. (1987b) Localization of mucosal mast cells in W/W^v mice after reconstitution with bone marrow cells or cultured mast cells, and its relation to the protective capacity to *Strongyloides ratti*-infection. *Parasite Immunology*, **9**, 477–485.

Abe, T. and Nawa, Y. (1988) Worm expulsion and mucosal mast cell response induced by repetitive IL-3 administration in *Strongyloides ratti*-infected nude mice. *Immunology*, **63**, 181–185.

Abe, T., Sugaya, H., Yoshimura, K. and Nawa, Y. (1992) Induction of the expulsion of *Strongyloides ratti* and retention of *Nippostrongylus brasiliensis* in athymic nude mice by repetitive administration of recombinant IL-3. *Immunology*, **76**, 10–14.

Abe, T., Sugaya, H., Ishida, K., Tasdemir, I. and Yoshimura, K. Khan, W.I. (1993) Intestinal protection against *Strongyloides ratti* and mastocytosis induced by administration of interleukin-3 in mice. *Immunology*, **80**, 116–121.

Abe, T., Sugaya, H. and Yoshimura, K. (1993) Different susceptibility to the IL-3 induced-protective effects between *Strongyloides ratti* and *Nippostrongylus brasiliensis* in C57BL/6 mice. *Parasite Immunology*, **15**, 643–645.

Abraham, D., Rotman, H.L., Haberstroh, H.F., Yutanawiboonchai, W., Brigandi, R.A., Leon, O. *et al.* (1995) *Strongyloides stercoralis*: protective immunity to third-stage larvae in BALB/cByJ mice. *Experimental Parasitology*, **80**, 297–307.

Ackert, J.E. (1942) Natural resistance to helminthic infections. *Journal of Parasitology*, **28**, 1–24.

Ahmad, A., Bell, R.G., Wang, C.H. and Sacuto, F.R. (1991) Characterization of the thoracic duct T-helper cells that co-mediate, with antibody, the rapid expulsion of *Trichinella spiralis* in adult rats. *Parasite Immunology*, **13**, 147–159.

Ahmad, A., Wang, C.H. and Bell, R.G. (1991) A role for IgE in intestinal immunity. Expression of rapid expulsion of *Trichinella spiralis* in rats transfused with IgE and thoracic duct lymphocytes. *Journal of Immunology*, **146**, 3563–3570.

Ahmad, A., Wang, C.H., Korenaga, M., Bell, R.G. and Adams, L.S. (1990) Synergistic interaction between immune serum and thoracic duct cells in the adoptive transfer of rapid expulsion of *Trichinella spiralis* in adult rats. *Experimental Parasitology*, **71**, 90–99.

Alizadeh, H. and Murrell, K.D. (1984) The intestinal mast cell response to *Trichinella spiralis* infection in mast cell-deficient W/W^v mice. *Journal of Parasitology*, **70**, 767–773.

Alizadeh, H. and Wakelin, D. (1982) Genetic factors controlling the intestinal mast cell response in mice infected with *Trichinella spiralis*. *Clinical and Experimental Immunology*, **49**, 331–337.

Appleton, J.A. and McGregor, D.D. (1984) Rapid expulsion of *Trichinella spiralis* in suckling rats. *Science*, **226**, 70–72.

Appleton, J.A., Schain, L.R. and McGregor, D.D. (1988) Rapid expulsion of *Trichinella spiralis* in suckling rats: mediation by monoclonal antibodies. *Immunology*, **65**, 487–924.

Ashman, K., Mather, J., Wiltshire, C., Jacobs, H.J. and Meeusen, E. (1995) Isolation of a larval surface glycoprotein from *Haemonchus contortus* and its possible role in evading host immunity. *Molecular and Biochemical Parasitology*, **70**, 175–179.

Babu, S., Porte, P., Klei, T.R., Shultz, L.D. and Rajan, T.V. (1998) Host NK cells are required for the growth of the human filarial parasite *Brugia malayi* in mice. *Journal of Immunology*, **161**, 1428–1432.

Bancroft, A.J., Else, K.J., Sypek, J.P. and Grencis, R.K. (1997) Interleukin-12 promotes a chronic intestinal nematode infection. *European Journal of Immunology*, **27**, 866–870.

Bancroft, A.J., McKenzie, A.N. and Grencis, R.K. (1998) A critical role for IL-13 in resistance to intestinal nematode infection. *Journal of Immunology*, **160**, 3453–3461.

Bartelink, A.K.M., Kortbeek, L.M., Huidekoper, H.J., Meulenbelt, J. and van Knapen, F. (1993) Acute respiratory failure due to *Toxocara* infection. *Lancet*, **342**, 1234.

Beaver, P.C. (1969) The nature of visceral larva migrans. *Journal of Parasitology*, **55**, 3–12.

Befus, A.D. and Bienenstock, J. (1979) Immunologically mediated intestinal mastocytosis in *Nippostrongylus brasiliensis*-infected rats. *Immunology*, **38**, 95–101.

Behnke, J.M. (1987) Evasion of immunity by nematode parasites causing chronic infections. *Advances in Parasitology*, **26**, 1–71.

Behnke, J.M., Barnard, C.J. and Wakelin, D. (1992) Understanding chronic nematode infections: evolutionary considerations, current hypotheses and the way forward. *International Journal for Parasitology*, **22**, 861–907.

Behnke, J.M., Hannah, J. and Pritchard, D.I. (1983) *Nematospiroides dubius* in the mouse: evidence that adult worms depress the expression of homologous immunity. *Parasite Immunology*, **5**, 397–408.

Bell, R.G. (1992) *Trichinella spiralis*: Evidence that mice do not express rapid expulsion. *Experimental Parasitology*, **74**, 417–430.

Bell, R.G. (1998) The generation and expression of immunity to *Trichinella spiralis* in laboratory rodents. *Advances in Parasitology*, **41**, 149–217.

Bell, R.G., Wang, C.H. and Ogden, R.W. (1985) *Trichinella spiralis*: nonspecific resistance and immunity to newborn larvae in inbred mice. *Experimental Parasitology*, **60**, 101–110.

Bennett, K., Levine, T., Ellis, J.S., Peanasky, R.J., Samloff, I.M., Kay, J. *et al.* (1992) Antigen processing for presentation by class II major histocompatibility complex requires cleavage by cathepsin E. *European Journal of Immunology*, **22**, 1519–1524.

Blaxter, M.L., DeLay, P., Garey, J.R., Liu, L.X., Scheldeman, P., Vierstraete, A. *et al.* (1998) A molecular evolutionary framework for the phylum Nematoda. *Nature*, **392**, 71–75.

Bowles, V.M., Brandon, M.R. and Meeusen, E. (1995) Characterization of local antibody responses to the gastrointestinal parasite *Haemonchus contortus*. *Immunology*, **84**, 669–674.

Brown, K.D., Zurawski, S.M., Mosmann, T.R. and Zurawski, G. (1989) A family of small inducible proteins secreted by leukocytes are members of a new superfamily that includes leukocyte and fibroblast-derived inflammatory agents, growth factors, and indicators of various activation processes. *Journal of Immunology*, **142**, 679–687.

Butterworth, A.E., Sturrock, R.F., Houba, V., Mahmoud, A.A., Sher, A. and Rees, P.H. (1975) Eosinophils as mediators of antibody-dependent damage to schistosomula. *Nature*, **256**, 727–729.

Cappello, M., Hawdon, J.M., Jones, B.F., Kennedy, W.P. and Hotez, P.J. (1996) *Ancylostoma caninum* anticoagulant peptide: cloning by PCR and expression of soluble, active protein in *E. coli*. *Molecular & Biochemical Parasitology*, **80**, 113–7.

Cappello, M., Vlasuk, G.P., Bergum, P.W., Huang, S. and Hotez, P.J. (1995) *Ancylostoma caninum* anticoagulant peptide: a hookworm-derived inhibitor of human coagulation factor Xa. *Proceedings of the National Academy of Sciences of the United States of America*, **92**, 6152–6156.

Carlisle, M.S., McGregor, D.D. and Appleton, J.A. (1990) The role of mucus in antibody-mediated rapid expulsion of *Trichinella spiralis* in suckling rats. *Immunology*, **70**, 126–132.

Chan, M-S. (1997) The global burden of intestinal nematode infections-Fifty years on. *Parasitology Today*, **13**, 438–443.

Chen, J-Y., Ramskold, L. and Zhou, G-Q (1994) Evidence for monophyly and arthropod affinity of Cambrian giant predators. *Science*, **264**, 1304–1308.

Coffman, R.L., Seymour, B.W., Hudak, S., Jackson, J. and Rennick, D. (1989) Antibody to interleukin-5 inhibits helminth-induced eosinophilia in mice. *Science*, **245**, 308–310.

Colgan, S.P., Morales, V.M., Madara, J.L., Polischuk, J.E., Balk, S.P. and Blumberg, R.S. (1996) IFN-γ modulates CD1d surface expression on intestinal epithelia. *American Journal of Physiology*, **271**, (Cell Physiology, **40**), C276–283.

Collins, P.D., Marleau, S., Griffiths-Johnson, D.A., Jose, P.J. and Williams, T.J. (1995) Cooperation between interleukin-5 and the chemokine eotaxin to induce eosinophil accumulation in vivo. *Journal of Experimental Medicine*, **182**, 1169–1174.

Cross, J.H., Banzon, T. and Singson, C. (1978) Further studies on *Capillaria philippinensis*: Development of the parasite in the Mongolian gerbil. *Journal of Parasitology*, **64**, 208–213.

Cuellar, C., Fenoy, S. and Guillen, J.L. (1995) Cross-reactions of sera from *Toxascaris leonina* and *Ascaris suum* infected mice with *Toxocara canis*, *Toxascaris leonina* and *Ascaris suum* antigens. *International Journal for Parasitology*, **25**, 731–739.

Curry, A.J., Else, K.J., Jones, F., Bancroft, A., Grencis, R.K. and Dunne, D.W. (1995) Evidence that cytokine-mediated immune interactions induced by *Schistosoma mansoni* alter disease outcome in mice concurrently infected with *Trichuris muris*. *Journal of Experimental Medicine*, **181**, 769–774.

Dawkins, H.J.S. (1989) *Strongyloides ratti* infections in rodents: value and limitations as a model of human strongyloidiasis. In *Strongyloidiasis: a major roundworm infection of man*, edited by D.I. Grove, pp. 287–332. Philadelphia: Taylor & Francis.

de Bono, M. and Bargmann, C.I. (1998) Natural variation in a neuropeptide Y receptor homolog modifies social behavior and food response in *C. elegans*. *Cell*, **94**, 679–689.

Dehlawi, M.S., Wakelin, D. and Behnke, J.M. (1987) Suppression of mucosal mastocytosis by infection with the intestinal nematode *Nematospiroides dubius*. *Parasite Immunology*, **9**, 187–194.

Despommier, D.D (1998) How does *Trichinella spiralis* make itself at home? *Parasitology Today*, **14**, 318–323.

Donaldson, L.E., Schmitt, E., Huntley, J.F., Newlands, G.F. and Grencis, R.K. (1996) A critical role for stem cell factor and c-kit in host protective immunity to an intestinal helminth. *International Immunology*, **8**, 559–567.

Dubremetz, J.F. and McKerrow, J.H. (1995) Invasion mechanisms. In *Biochemistry and Molecular Parasitology*, edited by J.J. Marr and M. Muller, pp. 307–322. San Diego: Academic Press.

Dunn, I.J. and Wright, K.A. (1985) Cell injury caused by *Trichinella spiralis* in the mucosal epithelium of B10A mice. *Journal of Parasitology*, **71**, 757–766.

Else, K.J. and Grencis, R.K. (1991) Cellular immune responses to the murine nematode parasite *Trichuris muris*. I. Differential cytokine production during acute or chronic infection. *Immunology*, **72**, 508–513.

Else, K.J., Entwistle, G.M. and Grencis, R.K. (1993) Correlations between worm burden and markers of Th1 and Th2 cell subset induction in an inbred strain of mouse infected with *Trichuris muris*. *Parasite Immunology*, **15**, 595–600.

Else, K.J., Finkelman, F.D., Maliszewski, C.R. and Grencis, R.K. (1994) Cytokine-mediated regulation of chronic intestinal helminth infection. *Journal of Experimental Medicine*, **179**, 347–351.

Else, K.J. and Wakelin, D. (1988) The effects of H-2 and non-H-2 genes on the expulsion of the nematode *Trichuris muris* from inbred and congenic mice. *Parasitology*, **96**, 543–550.

Fantozzi, R., Masini, E., Blandina, P., Mannaioni, P.F. and Bani-Sacchi, T. (1978) Release of histamine from rat mast cells by acetylcholine. *Nature*, **273**, 473–474.

Ferrick, D.A., Schrenzel, M.D., Mulvania, T., Hsieh B., Ferlin, W.G. and Lepper, H. (1995) Differential production of interferon-gamma and interleukin-4 in response to Th1- and Th2-stimulating pathogens by $\gamma\delta$ T cells in vivo. *Nature*, **373**, 255–257.

Finkelman, F.D., Madden, K.B., Cheever, A.W., Katona, I.M., Morris, S.C., Gately, M.K. *et al.* (1994) Effects of interleukin 12 on immune responses and host protection in mice infected with intestinal nematode parasites. *Journal of Experimental Medicine*, **179**, 1563–1572.

Finkelman, F.D., Shea-Donohue, T., Goldhill, J., Sullivan, C.A., Morris, S.C., Madden, K.B. *et al.* (1997) Cytokine regulation of host defense against parasitic gastrointestinal nematodes: lessons from studies with rodent models. *Annual Review of Immunology*, **15**, 505–533.

Fiorentino, D.F., Bond, M.W. and Mosmann, T.R. (1989) Two types of mouse T helper cell. IV. Th2 clones secrete a factor that inhibits cytokine production by Th1 clones. *Journal of Experimental Medicine*, **170**, 2081–2095.

Fiorentino, D.F., Zlotnik, A., Mosmann, T.R., Howard, M. and O'Garra, A. (1991a) IL-10 inhibits cytokine production by activated macrophages. *Journal of Immunology*, **147**, 3815–3822.

Fiorentino, D.F., Zlotnik, A., Vieira, P., Mosmann, T.R., Howard, M., Moore, K.W. *et al.* (1991b) IL-10 acts on the antigen-presenting cell to inhibit cytokine production by Th1 cells. *Journal of Immunology*, **146**, 3444–3451.

Forstner, J., Roomi, N. and Forstner, G. (1985) Intestinal glycoprotein in rats treated chronically with reserpine. In *Animal models for cystic fibrosis: The reserpine-treated rats*, edited by J.R. Martinez and G. Barbero, pp. 99–106. San Francisco: San Francisco Press.

Foster, W.D. (1965) *A History of Parasitology*, pp. 187–192. Edinburgh and London: E. &. S. Livingstone, Ltd.

Fox, A.S., Kazacos, K.R., Gould, N.S., Heydemann, P.T., Thomas, C. and Boyer, K.M. (1985) Fatal eosinophilic meningoencephalitis and visceral larva migrans caused by the raccoon ascarid *Baylisascaris procyonis*. *New England Journal of Medicine*, **312**, 1619–1623.

Freedman, D.O. (1998) Immune dynamics in the pathogenesis of human lymphatic filariasis. *Parasitology Today*, **14**, 229–234.

Galli, S.J., Zsebo, K.M. and Geissler, E.N. (1994) The kit ligand, stem cell factor. *Advances in Immunology*, **55**, 1–96.

Germain, R.N. (1994) MHC-dependent antigen processing and peptide presentation: providing ligands for T lymphocyte activation. *Cell*, **76**, 287–299.

Ghosh, K., Hawdon, J. and Hotez, P. (1996) Vaccination with alum-precipitated recombinant *Ancylostoma*-secreted protein 1 protects mice against challenge infections with infective hookworm (*Ancylostoma caninum*) larvae. *Journal of Infectious Diseases*, **174**, 1380–1303.

Gleich, G.J., Loegering, D.A. and Olson, G.M. (1975) Reactivity of rabbit antiserum to guinea pig eosinophils. *Journal of Immunology*, **115**, 950–954.

Golub, E.S. and Green, D.R. (1991) *Immunology: A Synthesis*, 2nd edn., pp. 4–8. Massachusetts: Sinauer Associates.

Gordon. J.I. (1993) Understanding gastrointestinal epithelial cell biology: lessons from mice with help from worms and flies. *Gastroenterology*, **105**, 315–324.

Grencis, R.K., Else, K.J., Huntley, J.F. and Nishikawa, S.I. (1993) The in vivo role of stem cell factor (c-*kit* ligand) on mastocytosis and host protective immunity to the intestinal nematode *Trichinella spiralis* in mice. *Parasite Immunology*, **15**, 55–59.

Grencis, R.K., Hultner, L. and Else, K.J. (1991) Host protective immunity to *Trichinella spiralis* in mice: activation of Th cell subsets and lymphokine secretion in mice expressing different response phenotypes. *Immunology*, **74**, 329–332.

Grencis, R.K., Riedlinger, J. and Wakelin, D. (1985) L3T4-positive T lymphoblasts are responsible for transfer of immunity to *Trichinella spiralis* in mice. *Immunology*, **56**, 213–218.

Grove, D.I. (1996) Human strongyloidiasis. *Advances in Parasitology*, **38**, 251–309.

Gutierrez, Y. (1990) *Diagnostic pathology of parasitic infections with clinical correlations*, pp. 1–3. Phildelphia: Lea and Febiger.

Hamajima, F., Yamamoto, M., Tsuru, S., Yamakami, K., Fujino, T., Hamajima, H. *et al.* (1994) Immunosuppression by a neutral thiol protease from parasitic helminth larvae in mice. *Parasite Immunology*, **16**, 261–73.

Hartmann, S., Kyewski, B., Sonnenburg, B. and Lucius, R. (1997) A filarial cysteine protease inhibitor down-regulates T cell proliferation and enhances interleukin-10 production. *European Journal of Immunology*, **27**, 2253–2260.

Harwood, C.L., Young, I.S., Lee, D.L. and Altringham, J.D. (1996) The effect of *Trichinella spiralis* infection on the mechanical properties of the mammalian diaphragm. *Parasitology* **113**, 535–543.

Hawdon, J.M., Jones, B.F., Hoffman, D.R. and Hotez P.J. (1996) Cloning and characterization of *Ancylostoma*-secreted protein. A novel protein associated with the transition to parasitism by infective hookworm larvae. *Journal of Biological Chemistry*, **271**, 6672–8.

Hawdon, J.M., Jones, B.F., Perregaux, M.A. and Hotez, P.J. (1995) *Ancylostoma caninum*: metalloprotease release coincides with activation of infective larvae *in vitro*. *Experimental Parasitology*, **80**, 205–11.

Hernandez, H.J., Wang, Y. and Stadecker, M.J. (1997) In infection with *Schistosoma mansoni*, B cells are required for T helper type 2 cell responses but not for granuloma formation. *Journal of Immunology*, **158**, 4832–4837.

Herndon, F.J. and Kayes, S.G. (1992) Depletion of eosinophils by anti-IL-5 monoclonal antibody treatment of mice infected with *Trichinella spiralis* does not alter parasite burden or immunologic resistance to reinfection. *Journal of Immunology*, **149**, 3642–3647.

Higa, A., Maruyama, H., Abe, T., Owhashi, M. and Nawa, Y. (1990) Effects of *Toxocara canis* infection on hemopoietic stem cells and hemopoietic factors in mice. *International Archives of Allergy and applied Immunology*, **91**, 239–243.

Hoeppli, R. (1959) *Parasites and Parasitic Infections in Early Medicine and Science*, pp. 472–486. Singapore: University of Malaya Press.

Hokibara, S., Takamoto, M., Tominaga, A., Takatsu, K. and Sugane, K. (1997) Marked eosinophilia in interleukin-5 transgenic mice fails to prevent *Trichinella spiralis* infection. *Journal of Parasitology*, **83**, 1186–1189.

Horii, Y., Khan, A.I. and Nawa, Y. (1993) Persistent infection of *Strongyloides venezuelensis* and normal expulsion of *Nippostrongylus brasiliensis* in Mongolian gerbils, *Meriones unguiculatus*, with reference to the cellular responses in the intestinal mucosa. *Parasite Immunology*, **15**, 175–179.

Hotez, P., Cappello, M., Hawdon, J., Beckers, C. and Sakanari, J. (1994) Hyaluronidases of the gastrointestinal invasive nematodes *Ancylostoma caninum* and *Anisakis simplex*: possible functions in the pathogenesis of human zoonoses. *Journal of Infectious Diseases*. **170**, 918–926.

Hotez, P., Haggerty, J., Hawdon, J., Milstone, L., Gamble, H.R., Schad, G. et al. (1990) Metalloproteases of infective *Ancylostoma* hookworm larvae and their possible functions in tissue invasion and ecdysis. *Infection & Immunity*, **58**, 3883–3892.

Hotez, P.J., Narasimhan, S., Haggerty, J., Milstone, L., Bhopale, V., Schad, G.A. and Richards, F.F. (1992) Hyaluronidase from infective *Ancylostoma* hookworm larvae and its possible function as a virulence factor in tissue invasion and in cutaneous larva migrans. *Infection & Immunity*. **60**, 1018–1023.

Hsieh, C.S., Heimberger, A.B., Gold, J.S., O'Garra, A. and Murphy, K.M. (1992) Differential regulation of T helper phenotype development by interleukins 4 and 10 in an alpha beta T-cell-receptor transgenic system. *Proceedings of the National Academy of Sciences of the United States of America*, **89**, 6065–6069.

Iglesias, R., Leiro, J., Ubeira, F.M., Santamarina, M.T., Navarrete, I. and Sanmartin, M.L. (1996) Antigenic cross-reactivity in mice between third-stage larvae of *Anisakis simplex* and other nematodes. *Parasitology Research*, **82**, 378–381.

Ishibashi, H., Shimamura, R., Hirata, Y., Kudo, J. and Onizuka, H. (1992) Hepatic granuloma in toxocaral infection: role of ultrasonography in hypereosinophilia. *Journal of Clinical Ultrasound*, **20**, 204–210.

Ishikawa, N., Goyal, P.K., Mahida, Y.R., Li, K.F. and Wakelin, D. (1998) Early cytokine responses during intestinal parasitic infections. *Immunology*, **93**, 257–263.

Jacobson, R.H. and Reed, N.D. (1974) The immune response of congenitally athymic (nude) mice to the intestinal nematode *Nippostrongylus brasiliensis*. *Proceedings of the Society for Experimental Biology and Medicine*, **147**, 667–670.

Ishikawa, N., Horii, Y., Oinuma, T., Suganuma, T. and Nawa, Y. (1994). Goblet cell mucins as the selective barrier for the intestinal helminths: T-cell independent alteration of goblet cell mucins by immunologically 'damaged' *Nippostrongylus brasiliensis* worms and its significance on the challenge infection with homologous and heterologous parasites. *Immunology*, **81**, 480–486.

Ishikawa, N., Horii, Y. and Nawa, Y. (1993). Immune-mediated alteration of the terminal sugars of goblet cell mucins in the small intestine of *Nippostrongylus brasiliensis*-infected rats. *Immunology*, **78**, 303–307.

Ishikawa, N., Horii, Y. and Nawa, Y. (1994). Inhibitory effects of concurrently present 'normal' *Nippostrongylus brasiliensis* worms on expulsion of 'damaged' worms and associated goblet cell changes in rats. *Parasite Immunology*, **16**, 329–332.

Ishikawa, N., Shi, B.-B., Khan, A. I. and Nawa, Y. (1995) Reserpine-induced sulphomucin production by goblet cells in the jejunum of rats and its significance in the establishment of intestinal helminths. *Parasite Immunology*, **17**, 581–586.

Ishikawa, N., Wakelin, D. and Mahida, Y.R. (1997) Role of T helper 2 cells in intestinal goblet cell hyperplasia in mice infected with *Trichinella spiralis*. *Gastroenterology*, **113**, 542–549.

Jarrett, E.E. and Miller, H.R. (1982) Production and activities of IgE in helminth infection. *Progress in Allergy*, **31**, 178–233.

Jasmer, D.P. and McGuire, T.C. (1991) Protective immunity to a blood-feeding nematode (*Haemonchus contortus*) induced by parasite gut antigens. *Infection and Immunity*, **59**, 4412–4417.

Jasmer, D.P., Perryman, L.E., Conder, G.A., Crow, S. and McGuire, T. (1993) Protective immunity to *Haemonchus contortus* induced by immunoaffinity isolated antigens that share a phylogenetically conserved carbohydrate gut surface epitope. *Journal of Immunology*, **151**, 5450–5460.

Jasmer, D.P., Perryman, L.E. and McGuire, T.C. (1996) *Haemonchus contortus* GA1 antigens: related, phospholipase C-sensitive, apical gut membrane proteins encoded as a polyprotein and released from the nematode during infection. *Proceedings of the National Academy of Sciences of the United States of America*, **93**, 8642–8647.

Katona, I.M., Urban, J.F. Jr. and Finkelman, F.D. (1988) The role of L3T4+ and Lyt-2+ T cells in the IgE response and immunity to *Nippostrongylus brasiliensis*. *Journal of Immunology*, **140**, 3206–3211.

Kazura, J.W. and Grove, D.I. (1978) Stage-specific antibody-dependent eosinophil-mediated destruction of *Trichinella spiralis*. *Nature*, **274**, 588–589.

Kelly, E.A., Cruz, E.S., Hauda, K.M. and Wassom, D.L. (1989) IFN-γ- and IL-5-producing cells compartmentalize to different lymphoid organs in *Trichinella spiralis*-infected mice. *Journal of Immunology*, **147**, 306–311.

Kennedy, M.W., Maizels, R.M., Meghji, M., Young, L., Qureshi, F. and Smith, H.V. (1987) Species-specific and common epitopes on the secreted and surface antigens of *Toxocara cati* and *Toxocara canis* infective larvae. *Parasite Immunology*, **9**, 407–420.

Kennedy, M.W., Qureshi, F., Haswell-Elkins, M. and Elkins, D.B. (1987) Homology and heterology between the secreted antigens of the parasitic larval stages of *Ascaris lumbricoides* and *Ascaris suum*. *Clinical and Experimental Immunology*, **67**, 20–30.

Khan, A.I., Horii, Y., Tiuria, R., Sato, Y. and Nawa, Y. (1993) Mucosal mast cells and the expulsive mechanisms of mice against *Strongyloides venezuelensis*. *International Journal for Parasitology*, **23**, 551–555.

Khan, A.I., Horii, Y. and Nawa, Y. (1993) Defective mucosal immunity and normal systemic immunity of Mongolian gerbils, *Meriones unguiculatus*, to reinfection with *Strongyloides venezuelensis*. *Parasite Immunology*, **15**, 565–571.

Khan, W.I., Abe, T., Ishikawa, N., Nawa, Y. and Yoshimura, K. (1995) Reduced amount of intestinal mucus by treatment with anti-CD4 antibody interferes with the spontaneous cure of *Nippostrongylus brasiliensis*-infection in mice. *Parasite Immunology*, **17**, 485–491.

Kinashi, T., Harada, N., Severinson, E., Tanabe, T., Sideras, P., Konishi, M. et al. (1986) Cloning of complementary DNA encoding T-cell replacing factor and identity with B-cell growth factor II. *Nature*, **324**, 70–73.

King, C.L., Kumaraswani, V., Poindexter, R.W., Kumari, S., Jayaraman, K., Alling, D.W. et al. (1992) Immunologic tolerance in lymphatic filariasis. Diminished parasite-specific T and B lymphocyte precursor frequency in the microfilaremic state. *Journal of Clinical Investigation*, **89**, 1403–1410.

King, C.L., Mahanty, S., Kumaraswami, V., Abrams, J.S., Regunathan, J., Jayaraman, K. et al. (1993) Cytokine control of parasite-specific anergy in human lymphatic filariasis. Preferential induction of a regulatory T helper type 2 lymphocyte subset. *Journal of Clinical Investigation*, **92**, 1667–1673.

Kishimoto, T., Taga, T. and Akira, S. (1994) Cytokine signal transduction. *Cell*, **76**, 253–262.

Kitamura, Y., Go, S. and Hatanaka, K. (1978) Decrease of mast cells in W/Wv mice and their increase by bone marrow transplantation. *Blood*, **52**, 447–452.

Kobayashi, T., Tsuchiya, K., Hara, T., Nakahata, T., Kurokawa, M., Ishiwata, K. et al. (1998) Intestinal mast cell response and mucosal defence against *Strongyloides venezuelensis* in interleukin-3-hyporesponsive mice. *Parasite Immunology*, **20**, 279–284.

Kolset, S.O. and Gallagher, J.T. (1990) Proteoglycans in haemopoietic cells. *Biochimica et Biophysica Acta*, **1032**, 191–211.

Kopf, M., Brombacher, F., Hodgkin, P.D., Ramsay, A.J., Milbourne, E.A., Dai, W.J. et al. (1996) IL-5-deficient mice have a developmental defect in CD5+ B-1 cells and lack eosinophilia but have normal antibody and cytotoxic T cell responses. *Immunity*, **4**, 15–24.

Kopf, M., Le Gros, G., Bachmann, M., Lamers, M.C., Bluethmann, H. and Kohler, G. (1993) Disruption of the murine IL-4 gene blocks Th2 cytokine responses. *Nature*, **362**, 245–248.

Korenaga, M., Abe, T. and Hashiguchi, Y. (1996) Injection of recombinant interleukin 3 hastens worm expulsion in mice infected with *Trichinella spiralis*. *Parasitology Research*, **82**, 108–113.

Korenaga, M., Hitoshi, Y., Yamaguchi, N., Sato, Y., Takatsu, K. and Tada, I. (1991) The role of interleukin-5 in protective immunity to *Strongyloides venezuelensis* infection in mice. *Immunology*, **72**, 502–507.

Korenaga, M., Hitoshi, Y., Takatsu, K. and Tada, I. (1994) Regulatory effect of anti-interleukin-5 monoclonal antibody on intestinal worm burden in a primary infection with *Strongyloides venezuelensis* in mice. *International Journal for Parasitology*, **24**, 951–957.

Korenaga, M., Wang, C.H., Bell, R.G., Zhu, D. and Ahmad, A. (1989) Intestinal immunity to *Trichinella spiralis* is a property of OX8− OX22− T-helper cells that are generated in the intestine. *Immunology*, **66**, 588–594.

Koyama, K., Tamauchi, H. and Ito, Y. (1995) The role of CD4+ and CD8+ T cells in protective immunity to the murine nematode parasite *Trichuris muris*. *Parasite Immunology*, **17**, 161–165.

Kuhn, R., Rajewsky, K. and Muller, W. (1991) Generation and analysis of interleukin-4 deficient mice. *Science*, **254**, 707–710.

Lal, R.B., Kumaraswami, V., Steel, C. and Nutman, T.B. (1990) Phosphocholine-containing antigens of *Brugia malayi* nonspecifically suppress lymphocyte function. *American Journal of Tropical Medicine & Hygiene*, **42**, 56–64.

Lammie, P.J., Addiss, D.G., Leonard, G., Hightower, W. and Eberhard, M.L. (1993) Heterogeneity in filarial-specific immune responsiveness among patients with lymphatic obstruction. *Journal of Infectious Diseases*, **167**, 1178–1183.

Lantz, C.S., Boesiger, J., Song, C.H., Mach, N., Kobayashi, T., Mulligan, R.C. et al. (1998) Role for interleukin-3 in mast-cell and basophil development and in immunity to parasites. *Nature*, **392**, 90-93.

Lawrence, R.A. (1996) Lymphatic filariasis: What mice can tell us. *Parasitology Today*, **12**, 267–271.

Lawrence, R.A., Gray, C.A., Osborne, J. and Maizels, R.M. (1996) *Nippostrongylus brasiliensis*: cytokine responses and nematode expulsion in normal and IL-4-deficient mice. *Experimental Parasitology*, **84**, 65–73.

Lee, D.L. (1976) Ultrastructural changes in the infective larvae of *Nippostrongylus brasiliensis* in the skin of immune mice. *Rice University Studies*, **62**, 175–182.

Lee, D.L. (1996) Why do some nematode parasite of the alimentary tract secrete acetylcholinesterase? *International Journal for Parasitology*, **26**, 499–508.

Lee, D.L. and Shivers, R.R. (1987) A freeze fracture study of muscle fibres infected with *Trichinella spiralis*. *Tissue and Cell*, **19**, 665–671.

Lee, T.D.G. (1991) Helminthotoxic responses of intestinal eosinophils to *Trichinella spiralis* newborn larvae. *Infection and Immunity*, **59**, 4405–4411.

Lee, T.D.G and Wright, K.A. (1978) The morphology of the attachment and probable feeding site of the nematode, *Trichuris muris* (Shrank 1877) Hall 1916. *Canadian Journal of Zoology*, **56**, 1889–1905.

Letterio, J.J. and Roberts, A.B. (1998) Regulation of immune responses by TGF-β. *Annual Review of Immunology*, **16**, 137–161.

Lightowlers, M.W. and Rickard, M.D. (1988) Excretory-secretory products of helminth parasites: effects of host immune responses. *Parasitology*, **96**, S123–66.

Liu, L.X., Serhan, C.N. and Weller, P.F. (1990) Intravascular filarial parasites elaborate cyclooxygenase-derived eicosanoids. *Journal of Experimental Medicine*, **172**, 993–996.

Lustigman, S., Brotman, B., Huima, T., Prince, A.M. and McKerrow, J.H. (1992) Molecular cloning and characterization of onchocystatin, a cysteine proteinase inhibitor of *Onchocerca volvulus*. *Journal of Biological Chemistry*, **267**, 17339–17346.

Madden, K.B., Urban, J.F. Jr., Ziltener, H.J., Schrader, J.W., Finkelman, F.D. and Katona, I.M. (1991) Antibodies to IL-3 and IL-4 suppress helminth-induced intestinal mastocytosis. *Journal of Immunology*, **147**, 1387–1391.

Maekawa, Y., Himeno, K., Ishikawa, H., Hisaeda, H., Sakai, T., Dainichi, T. *et al.* (1998) Switch of CD4+ T cell differentiation from Th2 to Th1 by treatment with cathepsin B inhibitor in experimental leishmaniasis. *Journal of Immunology*, **161**, 2120–2127.

Mahanty, S., King, C.L., Kumaraswami, V., Regunathan, J., Maya, A., Jayaraman, K. *et al.* (1993) IL-4- and IL-5-secreting lymphocyte populations are preferentially stimulated by parasite-derived antigens in human tissue invasive nematode infections. *Journal of Immunology*, **151**, 3704–3711.

Mahanty, S., Ravichandran, M., Raman, U., Jayaraman, K., Kumaraswami, V. and Nutman, T.B. (1997) Regulation of parasite antigen-driven immune responses by interleukin-10 (IL-10) and IL-12 in lymphatic filariasis. *Infection and Immunity*, **65**, 1742–1747.

Mahanty, S., Mollis, S.N., Ravichandran, M., Abrams, J.S., Kumaraswami, V., Jayaraman, K. *et al.* (1996) High levels of spontaneous and parasite antigen-driven interleukin-10 production are associated with antigen-specific hyporesponsiveness in human lymphatic filariasis. *Journal of Infectious Diseases*, **173**, 769–773.

Mahmoud, A.A., Warren, K.S. and Boros, D.I. (1973) Production of a rabbit antimouse eosinophil serum with no cross-reactivity to neutrophils. *Journal of Experimental Medicine*, **137**, 1526–1531.

Maizels, R.M., Bundy, D.A.P., Selkirk, M.E., Smith, D.F. and Anderson, R.M. (1993) Immunological modulation and evasion by helminth parasites in human populations. *Nature*, **365**, 797–805.

Maizels, R.M., Sartono, E., Kurniawan, A., Partono, F., Selkirk, M.E. and Yazdanbakhsh, M. (1995) T-cell activation and the balance of antibody isotypes in human lymphatic filariasis. *Parasitology Today*, **11**, 50–56.

Maruyama, H., Higa, A., Asami, M., Owhashi, M. and Nawa, Y. (1991) Hepatic eosinophilopoiesis from multipotent hemopoietic stem cells in *Toxocara canis*-infected mice. *Experimental Hematology*, **19**, 77–80.

Maruyama, H. and Nawa, Y. (1997) *Strongyloides venezuelensis*: Adhesion of adult worms to culture vessels by orally secreted mucosubstances. *Experimental Parasitology*, **85**, 10–15.

Maruyama, H., Nawa, Y., Noda, S., Mimori, T. and Choi, W-Y (1996) An outbreak of visceral larva migrans due to *Ascaris suum* in Kyushu, Japan. *Lancet*, **347**, 1766–1767.

Maruyama, H., Nawa, Y. and Ohta, N. (1998) *Strongyloides venezuelensis*: Binding of orally secreted adhesion substances to sulfated carbohydrates, *Experimental Parasitology*, **89**, 16–20.

Maruyama, H., Noda, S., Choi, W-Y., Ohta, N. and Nawa, Y. (1997) Fine binding specificities to *Ascaris suum* and *Ascaris lumbricoides* antigens of

the sera from patients of probable visceral larva migrans due to *Ascaris suum*. *Parasitology International*, **46**, 181–188.

Maruyama, H., Noda, S. and Nawa, Y. (1996) Emerging problems of parasitic diseases in Southern Kyushu, Japan. *Japanese Journal of Parasitology*, **45**, 192–200.

Maruyama, H., Osada, Y., Yoshida, A., Futakuchi, M., Kawaguchi, H., Zhang, R., Fu, J., Shirai, T., Kojima, S. and Ohta, N. (2000a) Protective mechanisms against the intestinal nematode, *Strongyloides venezuelensis*, in *Schistosoma japonicum*-infected mice. *Parasite Immunology*, **22**, 279–286.

Maruyama, H., Yabu, Y., yoshida, A., Nawa, Y. and Ohta, N. (2000b) A role of mast cell glycosaminoglycans for the immunological expulsion of intestinal nematode, *Strongyloides venezuelensis*. *Journal of Immunology*, **164**, 3749–3754.

Maruyama, H., Hatano, H., Kumagai, T., El-Malky, M., Yoshida, A., and Ohta, N. (2000c) *Strongyloides venezuelensis*: Heparin-binding adhesion substances in immunologically damaged adult worms. *Experimental Parasitology*, **95**, 170–175.

Matthews, A.N., Friend, D.S., Zimmermann, N., Sarafi, M.N., Luster, A.D., Pearlman, E. *et al.* (1998) Eotaxin is required for the baseline level of tissue eosinophils. *Proceedings of the National Academy of Sciences of the United States of America*, **95**, 6273–6278.

McKeand, J.B., Knox, D.P., Duncan, J.L. and Kennedy, M.W. (1995a) Immunisation of guinea pigs against *Dictyocaulus viviparus* using adult ES products enriched for acetylcholinesterases. *International Journal for Parasitology*, **25**, 829–37.

McKeand, J.B., Knox, D.P., Duncan, J.L. and Kennedy, M.W. (1995b) Protective immunisation of guinea pigs against *Dictyocaulus viviparus* using excretory/secretory products of adult parasites. *International Journal for Parasitology*, **25**, 95–104.

Merzel, J. and Leblond, C.P. (1969) Origin and renewal of goblet cells in the epithelium of the mouse small intestine. *American Journal of Anatomy*, **124**, 281–305.

Michael, E., Bundy, D.A. and Grenfell, B.T. (1996) Re-assessing the global prevalence and distribution of lymphatic filariasis. *Parasitology*, **112**, 409–428.

Miller, H.R.P. (1987). Gastrointestinal mucus, a medium for survival and for elimination of parasitic nematodes and protozoa. *Parasitology*, **94**, S77–S100.

Miller, H.R.P., Huntley, J.F. and Wallace, G.R. (1981) Immune exclusion and mucus trapping during the rapid expulsion of *Nippostrongylus brasiliensis* from primed rats. *Immunology*, **44**, 419–429.

Miller, H.R.P. and Nawa, Y. (1979). *Nippostrongylus brasiliensis*: Intestinal goblet-cell response in adoptively immunized rats. *Experimental Parasitology*, **47**, 81–90.

Mochizuki, M., Bartels, J., Mallet, A.I., Christophers, E. and Schroder, J.M. (1998) IL-4 induces eotaxin: a possible mechanism of selective eosinophil recruitment in helminth infection and atopy. *Journal of Immunology*, **160**, 60–68.

Monroy, F.G. and Enriquez, F.J. (1992) *Heligmosomoides polygyrus*: a model for chronic gastrointestinal helminthiasis. *Parasitology Today*, **8**, 49–54.

Moore, K.W., Vieira, P., Fiorentino, D.F., Trounstine, M.L., Khan, T.A. and Mosmann, T.R. (1990) Homology of cytokine synthesis inhibitory factor (IL-10) to the Epstein-Barr virus gene BCRFI [erratum appears in *Science*, **250**, 494] *Science*, **248**, 1230–1234.

Moqbel, R., McLaren, D.J. and Wakelin, D. (1980) *Strongyloides ratti*: reversibility of immune damage to adult worms. *Experimental Parasitology*, **49**, 153–166.

Mosmann, T.R. and Coffman R.L. (1989) Th1 and Th2 cells: Differential pattern of lymphokine secretuib lead to different functional properties. *Annual Review of Immunology*, **7**, 145–173.

Mould, A.W., Matthaei, K.I., Young, I.G. and Foster, P.S. (1997) Relationship between interleukin-5 and eotaxin in regulating blood and tissue eosinophilia in mice. *Journal of Clinical Investigation*, **99**, 1064–1071.

Munn, E.A., Smith, T.S., Smith, H., James, F.M. and Smith, F.C. (1997) Vaccination against *Haemonchus contortus* with denatured forms of the protective antigen H11. *Parasite Immunology*, **19**, 243–248.

Murrell, K.D. (1980) *Strongyloides ratti*: acquired resistance in the rat to the preintestinal migrating larvae. *Experimental Parasitology*, **50**, 417–425.

Nawa, Y., Kiyota, M., Korenaga, M. and Kotani, M. (1985) Defective protective capacity of W/Wᵛ mice against *Strongyloides ratti* infection and its reconstitution with bone marrow cells. *Parasite Immunology*, **7**, 429–438.

Nawa, Y. and Korenaga, M. (1983). Mast and goblet cell responses in the small intestine of rats concurrently infected with *Nippostrongylus brasiliensis* and *Strongyloides ratti*. *Journal of Parasitology*, **69**, 1168–1170.

Nawa, Y., Ishikawa, N., Tsuchiya, K., Horii, Y., Abe, T., Khan, A.I. *et al.* (1994) Selective effector mechanisms for the expulsion of intestinal helminths. *Parasite Immunology*, **16**, 333–338.

Negrao-Correa, D., Adams, L.S. and Bell, R.G. (1996) Intestinal transport and catabolism of IgE: a major blood-independent pathway of IgE dissemination during a *Trichinella spiralis* infection of rats. *Journal of Immunology*, **157**, 4037–4044.

Nelson, F.K., Greiner, D.L., Shultz, L.D. and Rajan, T.V. (1991) The immunodeficient scid mouse as a model for human lymphatic filariasis. *Journal of Experimental Medicine*, **173**, 659–663.

Neva, F.A. (1986) Biology and immunology of human strongyloidiasis. *Journal of Infectious Diseases*, **153**, 397–406.

Newton, S.E. (1995) Progress on vaccination against *Haemonchus contortus*. *International Journal for Parasitology*, **25**, 1281–1289.

Newton, S.E., Morrish, L.E., Martin, P.J., Montague, P.E. and Rolph, T.P. (1995) Protection against multiply drug-resistant and geographically distant strains of *Haemonchus contortus* by vaccination with H11, a gut membrane-derived protective antigen. *International Journal for Parasitology*, **25**, 511–521.

Newton, S.E. and Munn, E.A. (1999) The development of vaccines against gastrointestinal nematode parasites, particularly *Haemonchus contortus*. *Parasitology Today* **15**, 116–122.

Noben-Truth, N., Shultz, L.D., Brombacher, F., Urban, J.F Jr., Gu, H. and Paul, W.E. (1997) An interleukin 4 (IL-4)-independent pathway for CD4+ T cell IL-4 production is revealed in IL-4 receptor-deficient mice. *Proceedings of the National Academy of Sciences of the United States of America*, **94**, 10838–10843.

Nocka, K., Tan, J.C., Chiu, E., Chu, T.Y., Ray, P., Traktman, P. *et al.* (1990) Molecular basis of dominant negative and loss of function mutations at the murine c-kit/white spotting locus: W^{37}, W^v, W^{41}, and W. *EMBO Journal*, **9**, 1805–1813.

Nolan, T.J., Megyeri, Z., Bhopale, V.M. and Schad, G.A. (1993) *Strongyloides stercoralis*: the first rodent model for uncomplicated and hyperinfective strongyloidiasis, the Mongolian gerbil (*Meriones unguiculatus*). *Journal of Infectious Diseases*, **168**, 1479–1484.

Nutman, T.B. (1995) Immune responses in lymphatic dwelling filarial infections. In *Molecular Approaches to Parasitology*, edited by J.C. Boothroyd and R. Komuniecki, pp. 511–523. New York: Wiley-Liss Inc.

Nutman, T.B., Kumaraswami, V. and Ottesen, E.A. (1987) Parasite-specific anergy in human filariasis. Insights after analysis of parasite antigen-driven lymphokine production. *Journal of Clinical Investigation*, **79**, 1516–1523.

Ogilvie, B.M. (1965) Use of cortisone derivatives to inhibit resistance to *Nippostrongylus brasiliensis* and to study the fate of parasites in resistant hosts. *Parasitology*, **55**, 723–730.

Ogilvie, B.M. and Jones, V.E. (1971) *Nippostrongylus brasiliensis*: a review of immunity and host-parasite relationship in the rat. *Experimental Parasitology*, **29**, 138–177.

Ogilvie, B.M. and Love, R.J. (1974). Co-operation between antibodies and cells in immunity to a nematode parasite. *Transplantation Reviews*, **19**, 147–169.

Ogilvie, B.M. and de Savigny, D. (1982) Immune response to nematodes. In *Immunology of parasitic infections*, 2nd. edn. edited by S. Cohen and K.S. Warren, pp. 715–757. London: Blackwell Scientific Publications.

Oinuma, T., Abe, T., Nawa, Y., Kawano, J. and Suganuma, T. (1995) Glycoconjugates in rat small intestinal mucosa during infection with the intestinal nematode *Nippostrongylus brasiliensis*. In *Advances in Mucosal Immunology*, edited by J. Mestecky, pp. 975–978. New York: Plenum Press.

Oku, Y., Itayama, H. and Kamiya, M. (1984) Expulsion of *Trichinella spiralis* from the intestine of W/W^v mice reconstituted with haematopoietic and lymphopoietic cells and origin of mucosal mast cells. *Immunology*, **53**, 337–344.

Ortega-Pierres, M.G., Yepez-Mulia, L., Homan, W., Gamble, H.R., Lim, P.L., Takahashi, Y. *et al.* (1996) Workshop on a detailed characterization of *Trichinella spiralis* antigens: a platform for future studies on antigens and antibodies to this parasite. *Parasite Immunology*, **18**, 273–84.

Ottesen, E.A., Weller, P.F. and Heck, L. (1977) Specific cellular immune unresponsiveness in human filariasis. *Immunology*, **33**, 413–421.

Palframan, R.T., Collins, P.D., Williams, T.J. and Rankin, S.M. (1998) Eotaxin induces a rapid release of eosinophils and their progenitors from the bone marrow. *Blood*, **91**, 2240–2248.

Paul, W.E. (1991) Interleukin-4: a prototypic immunoregulatory lymphokine. *Blood*, **77**, 1859–1870.

Paul, W.E. and Seder, R.A. (1994) Lymphocyte responses and cytokines. *Cell*, **76**, 241–251.

Phills, J.A., Harrold, A.J., Whiteman, G.V. and Perelmutter, L. (1972) Pulmonary infiltrates, asthma and eosinophilia due to *Ascaris suum* infestation in man. *New England Journal of Medicine*, **286**, 965–970.

Piessens, W.F., McGreevy, P.B., Piessens, P.W., McGreevy, M., Koiman, I., Saroso, H.S. *et al.* (1980) Immune responses in human infections with *Brugia malayi*: Specific cellular unresponsiveness to filarial antigens. *Journal of Clinical Investigation*, **65**, 172–179.

Piessens, W.F., Ratiwayanto, S., Tuti, S., Palmieri, J.H., Piessens, P.W., Koiman, I. *et al.* (1980) Antigen-specific suppressor cells and suppressor factors in human filariasis with *Brugia malayi*. *New England Journal of Medicine*, **302**, 833–837.

Pond, L., Wassom, D.L. and Hayes, C.E. (1989) Evidence for differential induction of helper T cell subsets during *Trichinella spiralis* infection. *Journal of Immunology*, **143**, 4232–4237.

Potten, C.S. and Loeffler, M. (1990) Stem cells: attributes, cycles, spirals, pitfalls and uncertainties. Lessons for and from the crypt. *Development*, **110**, 1001–1020.

Pritchard, D.I. (1995) The survival strategies of hookworms. *Parasitology Today*, **11**, 255–259.

Racioppi, L. and Germain, R.N. (1995) Modified T-cell receptor ligands: moving beyond a strict occupancy model for T-cell activation by antigen. *Chemical Immunology*, **60**, 79–99.

Rainbird, M.A., Macmillan, D. and Meeusen, E.N. (1998) Eosinophil-mediated killing of *Haemonchus contortus* larvae: effect of eosinophil activation and role of antibody, complement and interleukin-5. *Parasite Immunology*, **20**, 93–103.

Raleigh, J.M., Brandon, M.R. and Meeusen, E. (1996) Stage-specific expression of surface molecules by the larval stages of *Haemonchus contortus*. *Parasite Immunology*, **18**, 125–132.

Raleigh, J.M. and Meeusen, E.N. (1996) Developmentally regulated expression of a *Haemonchus contortus* surface antigen. *International Journal for Parasitology*, **26**, 673–675.

Ramaswamy, K., Goodman, R.E. and Bell, R.G. (1994) Cytokine profile of protective anti-*Trichinella spiralis* CD4+ OX22− and non-protective CD4+ OX22+ thoracic duct cells in rats: secretion of IL-4 alone does not determine protective capacity. *Parasite Immunology*, **16**, 435–445.

Ramaswamy, K., Hakimi, J. and Bell, R.G. (1994) Evidence for an interleukin 4-inducible immunoglobulin E uptake and transport mechanism in the intestine. *Journal of Experimental Medicine*, **180**, 1793–1803.

Ramaswamy, K., Negrao-Correa, D. and Bell, R. (1996) Local intestinal immune responses to infections with *Trichinella spiralis*. Real-time, continuous assay of cytokines in the intestinal (afferent) and efferent thoracic duct lymph of rats. *Journal of Immunology*, **156**, 4328–4337.

Ramaswamy, K., Salafsky, B., Potluri, S., He, Y.X., Li, J.W. and Shibuya, T. (1995–96) Secretion of an anti-inflammatory factor by schistosomulae of *Schistosoma mansoni*. *Journal of Inflammation*, **46**, 13–22.

Rau, M.E. (1984) The open-field behaviour of mice infected with *Trichinella spiralis*. *Parasitology*, **86**, 311–318.

Rau, M.E. and Putter, L. (1984) Running responses of *Trichinella spiralis*-infected CD-1 mice. *Parasitology*, **89**, 579–583.

Ravichandran, M., Mahanty, S., Kumaraswami, V., Nutman, T.B. and Jayaraman, K. (1997) Elevated IL-10 mRNA expression and down-regulation of Th1-type cytokines in microfilaraemic individuals with *Wuchereria bancrofti* infection. *Parasite Immunology*, **19**, 69–77.

Razin, E., Stevens, R.L., Akiyama, F., Schmid, K. and Austen, K.F. (1982) Culture from mouse bone marrow of a subclass of mast cells possessing a distinct chondroitin sulfate proteoglycan with glycosamioglycan rich in N-acetylgalactosamine-4, 6-disulfate. *Journal of Biological Chemistry*, **257**, 7229–7236.

Redi, F. (1684) *Osservazioni intorno agli animali viventi che si trovano negli animali viventi*, title page. Firenze: Piero Matini.

Redmond, D.L., Knox, D.P., Newlands, G. and Smith, W.D. (1997) Molecular cloning of a developmentally regulated metallopeptidase present in a host protective extract of *Haemonchus contortus*. *Molecular and Biochemical Parasitology*, **85**, 77–87.

Rennick, D.M., Thompson-Snipes, L., Coffman, R.L., Seymour, B.W., Jackson, J.D. and Hudak, S. (1990) In vivo administration of antibody to interleukin-5 inhibits increased generation of eosinophils and their progenitors in bone marrow of parasitized mice. *Blood*, **76**, 312–316.

Robinson, K., Bellaby, T., Chan, W.C. and Wakelin, D. (1995) High levels of protection induced by a 40-mer synthetic peptide vaccine against the intestinal nematode parasite *Trichinella spiralis*. *Immunology*, **86**, 495–498.

Robinson, K., Bellaby, T. and Wakelin, D. (1994) Vaccination against the nematode *Trichinella spiralis* in high- and low-responder mice. Effects of different adjuvants upon protective immunity and immune responsiveness. *Immunology*, **82**, 261–267.

Rocken, M., Saurat, J.H. and Hauser, C. (1992) A common precursor for CD4+ T cells producing IL-2 or IL-4. *Journal of Immunology*, **148**, 1031–1036.

Roig, J., Romeu, J., Riera, C., Texido, A., Domingo, C. and Morera, J. (1992) Acute eosinophilic pneumonia due to toxocariasis with bronchoalveolar lavage findings. *Chest*, **102**, 294–6.

Rothenberg, M.E., Luster, A.D. and Leder, P. (1995) Murine eotaxin: an eosinophil chemoattractant inducible in endothelial cells and in interleukin 4-induced tumor suppression. *Proceedings of the National Academy of Sciences of the United States of America*, **92**, 8960–8964.

Rothwell, T.L.W. (1989) Immune expulsion of parasitic nematodes from the alimentary tract. *International Journal for Parasitology*, **19**, 139–168.

Rotman, H.L., Schnyder-Candrian, S., Scott, P., Nolan, T.J., Schad, G.A. and Abraham, D. (1997) IL-12 eliminates the Th-2 dependent protective immune response of mice to larval *Strongyloides stercoralis*. *Parasite Immunology*, **19**, 29–39.

Rotman, H.L., Yutanawiboonchai, W., Brigandi, R.A., Leon, O., Gleich, G.J., Nolan, T.J. *et al.* (1996) *Strongyloides stercoralis*: eosinophil-dependent immune-mediated killing of third stage larvae in BALB/cByJ mice. *Experimental Parasitology*, **82**, 267–278.

Rotman, H.L., Yutanawiboonchai, W., Brigandi, R.A., Leon, O., Nolan, T.J., Schad, G.A. *et al.* (1995) *Strongyloides stercoralis*: complete life cycle in SCID mice. *Experimental Parasitology*, **81**, 136–139.

Russell, D.A. and Castro, G.A. (1979) Physiological characterization of a biphasic immune response to *Trichinella* spiralis in the rat. *Journal of Infectious Diseases*, **139**, 304–312.

Ruitenberg, E.J. and Elgersma, A. (1976) Absence of intestinal mast cell response in congenitally athymic mice during *Trichinella spiralis* infection. *Nature*, **264**, 258–260.

Sabin, E.A., Kopf, M.A. and Pearce, E.J. (1996) *Schistosoma mansoni* egg-induced early IL-4 production is dependent upon IL-5 and eosinophils. *Journal of Experimental Medicine*, **184**, 1871–1878.

Sabin, E.A. and Pearce, E.J. (1995) Early IL-4 production by non-CD4+ cells at the site of antigen deposition predicts the development of a T helper 2 cell response to *Schistosoma mansoni* eggs. *Journal of Immunology*, **155**, 4844–4853.

Sakanari, J.A. and McKerrow J.H. (1990) Identification of the secreted neutral proteases from *Anisakis simplex*. *Journal of Parasitology*, **76**, 625–630.

Sasaki, O., Sugaya, H., Ishida, K. and Yoshimura, K. (1993) Ablation of eosinophils with anti-IL-5 antibody enhances the survival of intracranial worms of *Angiostrongylus cantonensis* in the mouse. *Parasite Immunology*, **15**, 349–54.

Schmidt, G.D. and Roberts, L.S. (1989) *Foundations of Parasitology*, pp. 428–433. St. Louis: Times Mirror/Mosby College Publishing.

Scudamore, C.L., Thornton, E.M., McMillan, L., Newlands, G.F. and Miller, H.R.P. (1995) Release of the mucosal mast cell granule chymase, rat mast cell protease-II, during anaphylaxis is associated with the rapid development of paracellular permeability to macromolecules in rat jejunum. *Journal of Experimental Medicine*, **182**, 1871–1881.

Seder, R.A., Paul, W.E., Davis, M.M. and Fazekas de St. Groth, B. (1992) The presence of interleukin 4 during in vitro priming determines the lymphokine-producing potential of CD4+ T cells from T cell receptor transgenic mice. *Journal of Experimental Medicine*, **176**, 1091–1098.

Seguchi, K., Matsuno, M., Kataoka, H., Kobayashi, T., Maruyama, H., Itoh, H. *et al.* (1995) A case report of colonic ileus due to eosinophilic nodular lesions caused by *Gnathostoma doloresi* infection. *American Journal of Tropical Medicine & Hygiene*, **53**, 263–266.

Sher, A. and Coffman, R.L. (1992) Regulation of immunity to parasites by T cells and T cell-derived cytokines. *Annual Review of Immunology*, **10**, 385–409.

Shi, B-B., Ishikawa, N., Khan, A.I., Tsuchiya, K., Horii, Y. and Nawa, Y. (1994a) *Strongyloides venezuelensis* infection in Syrian golden hamster, *Mesocricetus auratus*, with reference to the phenotype of intestinal mucosal mast cells. *Parasite Immunology*, **16**, 545–551.

Shi, B-B., Ishikawa, N., Itoh, H., Ide, H., Tsuchiya, K., Horii, Y., *et al.* (1994b) Goblet cell mucins of four genera of the subfamily Cricetinae with reference to the protective activity against *Strongyloides venezuelensis*. *Parasite Immunology*, **16**, 553–559.

Shin, E-H., Osada, Y., Chai, J-Y., Matsumoto, N., Takatsu, K. and Kojima, S. (1997) Protective roles of eosinophils in *Nippostrongylus brasiliensis* infection. *International Archives of Allergy and Immunology*, **114**, suppl. 1, 45–50.

Silberstein, D.S. and Despommier, D.D. (1984) Antigens from *Trichinella spiralis* that induce a protective response in the mouse. *Journal of Immunology*, **132**, 898–904.

Smith, T.S., Graham, M., Munn, E.A., Newton, S.E., Knox, D.P., Coadwell, W.J. *et al.* (1997) Cloning and characterization of a microsomal aminopeptidase from the intestine of the nematode *Haemonchus contortus*. *Biochimica et Biophysica Acta*, **1338**, 295–306.

Smith, W.D., Smith, S.K. and Murray, J.M. (1994) Protection studies with integral membrane fractions of *Haemonchus contortus*. *Parasite Immunology*, **16**, 231–241.

Smyth, J.D. (1994) *Introduction to animal parasitology*, 3rd edn, pp. 10–21. Cambridge: Cambridge University Press.

Stassens, P., Bergum, P.W., Gansemans, Y., Jespers, L., Laroche, Y., Huang, S. *et al.* (1996) Anticoagulant repertoire of the hookworm *Ancylostoma caninum*. *Proceedings of the National Academy of Sciences of the United States of America*, **93**, 2149–2154.

Steel, C. and Nutman, T.B. (1998) Helminth antigens selectively differentiate unsensitized CD45RA+ CD4+ human T cells in vitro. *Journal of Immunology*, **160**, 351–360.

Stevens, R.L., Otsu, K. and Austen, K.F. (1985) Purification and analysis of the core protein of the protease-resistant intracellular chondroitin sulfate E proteoglycan from the interleukin 3-dependent mouse mast cell. *Journal of Biological Chemistry*, **260**, 14194–14200.

Stevens, R.L., Fox, C.C., Lichtenstein, L.M. and Austen, K.F. (1988) Identification of chondroitin sulfate E proteoglycans and heparin proteoglycans in the secretory granules of human lung mast cells. *Proceedings of the National Academy of Sciences of the United States of America*, **85**, 2284–2287.

Stoll, N.R. (1927) Studies with the strongyloid nematode, *Haemonchus contortus*. I. Acquired resistance of hosts under natural reinfection conditions out-of-doors. *American Journal of Hygiene* **10**, 384–418.

Stoll, N.R. (1947) This wormy world. *Journal of Parasitology*, **33**, 1–18.

Sugamura, K., Asao, H., Kondo, M., Tanaka, N., Ishii, N., Ohbo, K. *et al.* (1996) The interleukin-2 receptor gamma chain: its role in the multiple cytokine receptor complexes and T cell development in XSCID. *Annual Review of Immunology*, **14**, 179–205.

Sugaya, H., Aoki, M., Abe, T., Ishida, K. and Yoshimura, K. (1997a) Cytokine responses in mice infected with *Angiostrongylus cantonensis*. *Parasitology Research*, **83**, 10–15.

Sugaya, H., Aoki, M., Yoshida, T., Takatsu, K. and Yoshimura, K. (1997b) Eosinophilia and intracranial worm recovery in interleukin-5 transgenic and interleukin-5 receptor alpha chain-knockout mice infected with *Angiostrongylus cantonensis*. *Parasitology Research*, **83**, 583–590.

Sugaya, H., Abe, T., Yoshimura, K. and Sasaki, O. (1993) Antigen dependent release of interleukin 5 *in vitro* from spleen cells of mice infected with *Angiostrongylus cantonensis*. *International Journal for Parasitology*, **23**, 865–986.

Svetic, A., Madden, K.B., Zhou, X.D., Lu, P., Katona, I.M., Finkelman, F.D. *et al.* (1993) A primary intestinal helminthic infection rapidly induces a gut-associated elevation of Th2-associated cytokines and IL-3. *Journal of Immunology*, **150**, 3434–3441.

Symons, L.E. (1976) Scanning electron microscopy of the jejunum of the rat infected by the nematode *Nippostrongylus brasiliensis*. *International Journal for Parasitology* **6**, 107–111.

Tada, I., Mimori, T. and Nakai, M (1979) Migration rout of *Strongyloides ratti* in albino rats. *Japanese Journal of Parasitology*, **28**, 219–227.

Takamure, A. (1995) Migration route of *Strongyloides venezuelensis* in rodents. *International Journal for Parasitology*, **25**, 907–911.

Takatsu, K. (1992) Interleukin-5. *Current Opinion in Immunology*, **4**, 299–306.

Takatsu, K., Tominaga, A. and Hamaoka, T. (1980) Antigen-induced T cell-replacing factor (TRF). I. Functional characterization of a TRF-producing helper T cell subset and genetic studies on TRF production. *Journal of Immunology*, **124**, 2414–2422.

Takeda, K., Kishimoto, T. and Akira, S. (1997) STAT6: its role in interleukin 4-mediated biological functions. *Journal of Molecular Medicine*, **75**, 317–326.

Takeda, K., Tanaka, T., Shi, W., Matsumoto, M., Minami, M., Kashiwamura, S. *et al.* (1996) Essential role of Stat6 in IL-4 signalling. *Nature*, **380**, 627–630.

Tanaka, T., Hu-Li, J., Seder, R.A. de St. Groth, B.F. and Paul, W.E. (1993) Interleukin 4 suppresses interleukin 2 and interferon γ production by naive T cells stimulated by accessory cell-dependent receptor engagement. *Proceedings of the National Academy of Sciences of the United States of America*, **90**, 5914–5918.

Tominaga, A., Takaki, S., Koyama, N., Katoh, S., Matsumoto, R., Migita, M. *et al.* (1991) Transgenic mice expressing a B cell growth and differentiation factor gene (interleukin 5) develop eosinophilia and autoantibody production. *Journal of Experimental Medicine*, **173**, 429–437.

Tomita, M., Itoh, H., Ishikawa, N., Higa, A., Ide, H., Murakumo, Y. *et al.* (1995) Molecular cloning of mouse intestinal trefoil factor and its expression during goblet cell changes. *Biochemical Journal*, **311**, 293–297.

Tuohy, M., Lammas, D.A., Wakelin, D., Huntley, J.F., Newlands, G.F. and Miller, H.R. (1990) Functional correlations between mucosal mast cell activity and immunity to *Trichinella spiralis* in high and low responder mice. *Parasite Immunology*, **12**, 675–685.

Turner, P. and Michael, E. (1997) Recent advances in the control of lymphatic filariasis. *Parasitology Today*, **11**, 410–411.

Uchikawa, R., Matsuda, S., Yamada, M., Nishida, M., Agui, T. and Arizono, N. (1997) Nematode infection induces Th2 cell-associated immune responses in LEC mutant rats with helper T cell immunodeficiency. *Parasite Immunology*, **19**, 461–468.

Uchikawa, R., Yamada, M., Matsuda, S., Kuroda, A. and Arizono, N. (1994). IgE antibody production is associated with suppressed interferon-γ levels in mesenteric lymph nodes of rats infected with the nematode *Nippostrongylus brasiliensis*. *Immunology*, **82**, 427–432.

Urban, J.F. Jr., Katona, I.M., Paul, W.E. and Finkelman, F.D. (1991) Interleukin 4 is important in protective immunity to a gastrointestinal nematode infection in mice. *Proceedings of the National Academy of Sciences of the United States of America*, **88**, 5513–5517.

Urban, J.F. Jr., Madden, K.B., Cheever, A.W., Trotta, P.P., Katona, I.M. and Finkelman, F.D. (1993) IFN inhibits inflammatory responses and protective immunity in mice infected with the nematode parasite, *Nippostrongylus brasiliensis*. *Journal of Immunology*, **151**, 7086–7094.

Urban, J.F. Jr., Maliszewski, C.R., Madden, K.B., Katona, I.M. and Finkelman, F.D. (1995) IL-4 treatment can cure established gastrointestinal nematode infections in immunocompetent and immunodeficient mice. *Journal of Immunology*, **154**, 4675–4684.

Urban, J.F. Jr., Noben-Trauth, N., Donaldson, D.D., Madden, K.B., Morris, S.C., Collins, M. *et al.* (1998) IL-13, IL-4Rα, and Stat6 are required for the expulsion of the gastrointestinal nematode parasite *Nippostrongylus brasiliensis*. *Immunity*, **8**, 255–264.

Vassilatis, D.K., Despommier, D., Misek, D.E., Polvere, R.I., Gold, A.M. and Van der Ploeg, L.H. (1992) Analysis of a 43-kDa glycoprotein from the intracellular parasitic nematode *Trichinella spiralis*. *Journal of Biological Chemistry*, **267**, 18459–18465.

Venturiello, S.M., Giambartolomei, G.H. and Costantino, S.N. (1995) Immune cytotoxic activity of human eosinophils against *Trichinella spiralis* newborn larvae. *Parasite Immunology*, **17**, 555–559.

Wakelin, D. (1994) Immunoparasitology. In *Introduction to Animal Parasitology*, 3rd edn, edited by J.D. Smyth, pp. 460–490. Cambridge: Cambridge University Press.

Wang, C.H. and Bell, R.G. (1987) *Trichinella spiralis*: intestinal expression of systemic stage-specific immunity to newborn larvae. *Parasite Immunology*, **9**, 465–475.

Wang, C.H. and Bell, R.G. (1988) Antibody-mediated *in vivo* cytotoxicity to *Trichinella spiralis* newborn larvae in immune rats. *Parasite Immunology*, **10**, 293–308.

Wang, C.H., Korenaga, M., Greenwood, A. and Bell, R.G. (1990) T-helper subset function in the gut of rats: differential stimulation of eosinophils, mucosal mast cells and antibody-forming cells by OX8⁻ OX22⁻ and OX8⁻ OX22⁺ cells. *Immunology*, **71**, 166–175.

Watts, C. (1997) Capture and processing of exogenous antigens for presentation on MHC molecules. *Annual Review of Immunology*, **15**, 821–850.

Weller, P.F., Ottesen, E.A. and Heck, L. (1980) Immediate and delayed hypersensitivity skin test responses to the *Dirofilaria immitis* filarial skin test (Sawada) antigen in *Wuchereria bancrofti* filariasis. *American Journal of Tropical Medicine and Hygiene*, **29**, 809–814.

Wells, P.D. (1963). Mucin-secreting cells in rats infected with *Nippostrongylus brasiliensis*. *Experimental Parasitology*, **14**, 15–22.

Wierenga, E.A., Snoek, M., de Groot, C., Chretien, I., Bos, J.D., Jansen, H.M. *et al.* (1990) Evidence for compartmentalization of functional subsets of CD2⁺ T lymphocytes in atopic patients. *Journal of Immunology*, **144**, 4651–4656.

Willadsen, P., Riding, G.A., McKenna, R.V., Kemp, D.H., Tellam, R.L., Nielsen, J.N. *et al.* (1989) Immunologic control of a parasitic arthropod. Identification of a protective antigen from *Boophilus microplus*. *Journal of Immunology*, **143**, 1346–1351.

Williams, G.R. and Wright, N.A. (1997) Trefoil factor family domain peptides. *Virchows Archive*, **431**, 299–304.

Winter, M.D., Ball, M.L., Altringham, J.D. and Lee, D.L. (1994) The effect of *Trichinella spiralis* and *Trichinella pseudospiralis* on the mechanical properties of mammalian diaphragm muscle. *Parasitology*, **109**, 129–134.

Woodbury, R.G., Miller, H.R., Huntley, J.F., Newlands, G.F., Palliser, A.C. and Wakelin, D. (1984) Mucosal mast cells are functionally active during spontaneous expulsion of intestinal nematode infections in rat. *Nature*, **312**, 450–452.

Wright, N.A., Hoffmann, W., Otto, W.R., Rio, M.C. and Thim, L. (1997) Rolling in the clover: trefoil factor family (TFF)-domain peptides, cell migration and cancer. *FEBS letters*, **408**, 121–123.

Wright, K.A. (1979) *Trichinella spiralis*: An intracellular parasite in the intestinal phase. *Journal of Parasitology*, **65**, 441–445.

Yoshida, A., Maruyama, H., Yabu, Y., Amano, T., Kobayakawa, T. and Ohta, N. (1999) Immune responses against protozoal and nematode infection in mice with underlying *Schistosoma mansoni* infection. *Parasitology International*, **48**, 73–79.

Yoshida, T., Ikuta, K., Sugaya, H., Maki, K., Takagi, M., Kanazawa, H. *et al.* (1996) Defective B-1 cell development and impaired immunity against *Angiostrongylus cantonensis* in IL-5R α-deficient mice. *Immunity*, **4**, 483–494.

Yoshimura, K., Sugaya, H. and Ishida, K. (1994) The role of eosinophils in *Angiostrongylus cantonensis* infection. *Parasitology Today*, **10**, 231–233.

Zund, G., Madara, J.L., Dzus, A.L., Awtrey, C.S. and Colgan, S.P. (1996) Interleukin-4 and interleukin-13 differentially regulate epithelial chloride secretion. *Journal of Biological Chemistry*, **271**, 7460–7464.

Zurawski, S.M., Vega, F. Jr., Huyghe, B. and Zurawski, G. (1993) Receptors for interleukin-13 and interleukin-4 are complex and share a novel component that functions in signal transduction. *EMBO Journal*, **12**, 2663–2670.

Zurawski, G. and de Vries, J.E. (1994) Interleukin 13, an interleukin 4-like cytokine that acts on monocytes and B cells, but not on T cells. *Immunology Today*, **15**, 19–26.

19. Biological Control

Brian R. Kerry
Entomology and Nematology Department, IACR-Rothamsted, Harpenden, Herts, AL5 2JQ, UK

William M. Hominick
CABI Bioscience, UK Centre (Egham), Bakeham Lane, Egham, Surrey, TW20 9TY, UK

Keywords

Entomogenous nematodes, nematophagous fungi and bacteria, infection processes, host recognition, population dynamics, integrated control.

Introduction

Biological control *sensu stricto* refers to the use of one living organism to control another, the latter being a pest. However, for the purposes of this chapter we have excluded the use of the host plant (see Atkinson, Chapter 23) and largely confine our comments to the use of parasites and pathogens, which has been the subject of most research on biological control both with and of nematodes. Control organisms include a wide range of vertebrates and invertebrates and microbial pathogens, particularly fungi, bacteria and viruses. Thus, the introduction of cobras to control rats in oil palm estates in Malaysia is a method of biological control, just as is the introduction of entomopathogenic fungi to control insect pests. However, while the concept is simple, its application is not. There is no unifying model for biological control, so that the basic problem is predicting in advance where biological control will succeed and the level of control that will be obtained. This is very difficult because there is usually a great diversity of potential control agents associated with any given pest, there is a number of ways in which the agents may be used, and there is a large number of complex interactions between the animals, plants, and environmental factors that can influence success. However, when biological control is successful, the environmental and economic benefits can be immense.

Nematodes that parasitise invertebrates are exploited commercially as biological control agents for insects but little is known of their ecology in soil. The greatest problems of survival of entomopathogenic nematodes (EPNs) applied to soil have been associated with the formulations used and their inherent biology, even though it is known that nematophagous fungi and bacteria may also have profound effects on the dynamics of nematodes. A wide range of organisms derive their nutrition from nematodes in soil (Stirling, 1991) and a few have been selected for their potential as biological control agents for plant-parasitic nematode species. Although several successful products based on EPNs have been developed, few commercial agents have been developed for the biological control of plant-parasitic nematodes and none is in widespread use or has provided consistent and satisfactory levels of control. In this chapter, the use of nematodes for the biological control of insects and slugs is reviewed and the natural enemies of nematodes are examined for their potential as biological control agents of plant- and animal-parasitic species.

Of the nematodes associated with insects, many have the potential to be used for the biological control of insect pests. Plant-parasitic nematodes also have associated organisms and these have the potential to control the nematodes biologically. For insect control, the basic aim is to understand the population biology of the nematodes and insects with a view to maximising the numbers of nematodes in the environment to control the insect pests. For plant-parasitic nematode control, it is also fundamental to understand the population biology of the organisms, but the aim is to minimise the numbers of nematodes. Hence, this chapter is divided into two distinct sections, the first dealing with the use of nematodes as biological control agents, the second with the use of biological control agents to control plant-parasitic nematodes. Certain common basic strategies can be employed, as outlined above, and similar biotic and abiotic factors can be limiting for both strategies. The most obvious overlap between the two concerns nematodes with the potential to control insects and which have stages that occur in the soil. These can be negatively affected by the same types of organisms that have the potential to control plant-parasitic nematodes.

Basic Strategies for Biological Control

There are four basic strategies that can be employed for biological control:

Introduction

This strategy is the classical biological control technique whereby an exotic beneficial organism is introduced into a new area and becomes permanently established. It is used most frequently against introduced pests which have no natural controlling factors; the control agents are sought in the area of origin of the pest. When successful, pest mortality is increased so that the pest population reaches an equilibrium density below the economic threshold and control continues indefinitely with little intervention required. The key components of success are the level of depression of the pest population (the final equilibrium must be below the economic threshold or the disease transmission threshold) and the subsequent stability of the interactions (the pest population must not fluctuate so widely that it crosses the economic threshold). This technique requires that pest and control agent populations persist over time at low densities. Thus, the pest must always be present because its complete elimination would lead to loss of the control agent as well.

Augmentation

Sometimes, the numbers of native natural enemies are inadequate to exert the required level of control (their numbers do not increase rapidly enough or are not large enough because of agricultural practices that restrict their numbers). In these cases, laboratory-bred individuals can be released to augment the populations of the enemies.

Inoculation

This technique can be applied when a native control agent is absent from a particular area or an introduced species cannot survive permanently. An inoculative release is made at the start of a season and, in the case of a plant pest, the agent colonises the area for the duration of the crop and prevents pest build-up. The process would need to be repeated for the next crop.

Inundation

This method refers to the mass culture of an agent, often a pathogen, for application at critical periods for short-term suppression of pest populations. Such organisms are sometimes called 'biological pesticides' in that the strategy and method of application is analogous to the use of chemical pesticides.

Total eradication of a pest is seldom feasible, so introduction, or classical biological control, is the most nearly perfect method of biological pest control and is the aim of many projects. Although classical biological control has had many successes in the management of insect pests (Waage and Greathead, 1988), there have been few cases of such successes in using nematodes to control insects or microbial agents to control plant-parasitic nematodes. Control by augmentation, inoculation and inundation requires mass culture of the control agents and, for all of these, there are issues concerning quality control and recurrent costs.

Nematodes as Biological Control Agents

Introduction

Nematode associations with invertebrates range from lethal to benign, in which the invertebrate serves as transport or shelter, and from opportunistic to obligatory. Hence, there are two terms frequently encoun-tered for nematodes associated with insects. Entomophilic (from the Greek: entomon (insect) and philic (having an affinity for or loving)) frequently refers to the more benign nematodes and entomogenous (from the Greek: genous (growing in or living on)) refers to the harmful, parasitic ones. At the extreme end of the harmful relationships are those nematodes that kill their host within hours of infection, and these have been termed entomopathogenic. As for all general terms, these are descriptive rather than definitive.

In this chapter, we focus on those entomogenous nematodes that are being used for biological control of insects. Three different groups of nematodes, the families Mermithidae (obligate parasites), Phaenopsi-tylenchidae (facultative parasites) and Steinernemati-dae and Heterorhabditidae (entomopathogenic nematodes), have been chosen because of the amount of information available for them, their success or lack of success as biological control agents, and the criteria for success that they demonstrate. We have been highly selective, with the intention to illustrate principles rather than to discuss specifics. If estimates of biodiversity are even moderately realistic, insects are the most species-rich group on the planet and nematodes are not only species-rich but also the most abundant organisms. Because only a small fraction of the insect fauna has been examined for nematode associates, the future is likely to reveal novel candidates for biological control programmes. Nevertheless, the same principles will apply.

There are a number of books and reviews that provide detailed coverage of nematodes as biological control agents. These should be consulted for details on biology, host range and safety, genetic selection and manipulation, commercial considerations, application and control, and environmental limitations. Some or all of these are covered by Gaugler (1987), Kaya (1985, 1987, 1993), Nickle (1984), Petersen (1985), Poinar (1979, 1991), and Popiel and Hominick (1992). Nickle and Welch (1984) provide an excellent review of the historical aspects of the subject.

Mermithidae

The higher order phylogeny of nematodes has always been controversial and fluid because of the morphological conservatism of the group. This is particularly so for mermithid nematodes, which are parasites of invertebrates and are highly modified for parasitism. In the host haemocoel, they absorb nutrients across the body wall and have a cellular stichosome unconnected to the intestine rather than a functional pharynx. The intestine (trophosome) serves as a food storage organ. The presence of the stichosome has meant that they are sometimes grouped with trichurids. Blaxter *et al.* (1998 and this volume) recently produced the first attempt at a phylogenetic classification of nematodes based on small subunit ribosomal DNA sequences, from 53 species chosen to represent all the major taxa including animal and plant parasites as well as free-living ones. The results indicate that convergent morphological evolution has been extensive and that present higher-level classification will need revision. They suggest that animal parasitism arose independently at least four times and plant parasitism three times. In their scheme, the vertebrate parasitic Trichocephalida (*Trichinella spiralis* and *Trichuris muris*), the insect parasitic Mermithida (*Mermis nigrescens*), the plant-parasitic Dorylaimida (Longidoridae) and the free-living Mononchida (*Mylonchulus arenicolus*) are grouped into a single clade. The mermithid is most closely related to the mononchid. In any case, Poinar, Acra and Acra (1994) described a mermithid, coiled inside the abdomen of an adult biting midge (Diptera: Ceratopogonidae), trapped in Lebanese amber. The specimen represents the oldest definite fossil nematode and was assigned to a new species, *H. libani*, in the extant genus *Heleidomermis*. This association represents the oldest known example of animal-animal internal parasitism in a terrestrial environment. The find demonstrates the antiquity of mermithid nematodes and establishes mermithid parasitism of the lower Diptera some 120–135 million years ago.

Mermithid nematodes are obligate and lethal parasites of invertebrates, particularly terrestrial and aquatic arthropods, many of which are of medical and agricultural importance. The parasites attain a length several times that of their host. In the most common type of life cycle, they gain entry to their host, usually an early instar, by hatching out of the egg and penetrating across the cuticle as second stage larvae. Once in the haemocoel, they undergo much growth, absorbing nutrients from the haemocoel fluid and storing them in the trophosome. Eventually, they occupy most of the haemocoel while the insect continues to live and feed. When they emerge from the host as post-parasitic fourth stage larvae, the insect dies within hours. The nematodes complete their development outside the host, living on their stored nutrients. Depending on the species and stage of the host penetrated, mermithids emerge from late instar larvae, pupae or even adults.

Application of mermithids for biological control has generally been by classical techniques, whereby the nematodes are released with the intention that they become established and exert acceptable levels of control thereafter. Their use as bioinsecticides was also thought possible, providing mass rearing techniques could be developed. Figure 19.1 shows the number of publications dealing with mermithids in each year from 1973 to 1996. The 1970s and early 80s were the peak years, largely due to Petersen and Willis (1972) who published procedures for mass rearing a mosquito parasite, *Romanomermis culicivorax* (first named as *Reesimermis nielseni*). For the first time, experimental material became easily available and field releases could be attempted and monitored. In the late 1970s, this parasite was briefly commercialised. However, *Bacillus thuringiensis* serotype H14 was discovered in 1977 and proved to be a potent mosquito and blackfly larvicide. It was developed quickly to the point of field use in the Onchocerciasis Control Programme in the early 1980s. Funding for work with mermithids became

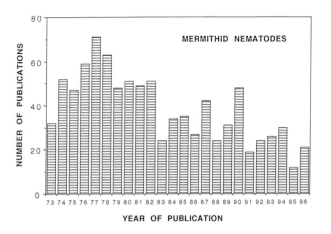

Figure 19.1. Analysis of CABPEST CD, showing the number of publications in each year from 1973 to 1996, searching for mermith*.

difficult because of the discovery of the more effective bacterial agent, which could be cheaply mass-produced and used as a bioinsecticide. This undoubtedly contributed to the reduction of mermithid publications to their traditional levels, much of which deals with systematics and the description of new taxa.

Much of the work with mermithids that relates to biological control is based on assumptions rather than data. This led Hominick and Tingley (1984) to assess the limitations of mermithids for vector control, using population biology models. They concluded that mermithid populations are controlled by such strong density dependent constraints (environmental sex determination, parasite-induced host mortality and reduced fecundity of crowded females), that they can cause only moderate long-term depressions in their host populations. Additionally, the long generation times compared to their hosts will tend to produce cycling in abundance of the insects, with periodic breakdown in control. Thus, although mermithids are lethal and there are numerous reports of insect populations with high levels of parasitism at particular times, it is unlikely that mermithids exert significant long-term, stable control over host populations. Popiel and Hominick (1992) reviewed the literature to that time and have found no reason to alter these views.

Terrestrial mermithids were identified as potential biological control agents for insect pests of the orders Orthoptera, Hemiptera, Coleoptera, Lepidoptera and Hymenoptera by Kaiser (1991) who also comprehensively reviewed their morphological and physiological characteristics, diversity, hosts, habitats, geographical distribution, taxonomy (including key to genera), life cycles and bionomics, host-parasite relationships and agricultural importance. Unfortunately, there are few quantitative data on the effects of mermithids on populations of agricultural pests. Nevertheless, the limitations that restrict the efficacy of mermithids for vector control (Hominick and Tingley, 1984) seem also to apply for control of agricultural pests. Additionally, because hosts are not killed immediately, they may cause substantial damage to crops while the parasite is developing inside them.

Most recent work has been in China and Korea. In China, field studies during 1975–1989 in Shanghai County and Xincai County of Henan Province confirmed that *Ovomermis sinensis* is an important natural enemy of the noctuid *Mythimna separata* (Chen *et al.*, 1992). The average parasitism rate was over 40%, with a maximum of 90% in 58% of observation sites over the 15 years. Feeding capacity of infected *M. separata* larvae was reduced by an average of 38%. The overwintering density of mermithid adults in early spring, the number of rainy days, precipitation and relative humidity during late April to mid-May were the main factors influencing noctuid parasitism. In May 1990, infective larvae of *Ovomermis sinensis* were released into a plot of wheat to control the first generation of the noctuids (Zhang *et al.*, 1992). When noctuid population densities were 10–20 adults/m^2, releasing the larvae at a nematode: insect ratio of 100:1, with irrigation before and after release to maintain appropriate soil humidity, controlled damage by *M. separata* for a year.

Work in Korea also supports the contention that the biology of mermithids mitigates against their use as classical biological control agents in agriculture. There, the most important biological control agent of the brown planthopper (BPH), *Nilaparvata lugens*, which is the major pest of rice in Asia, appears to be the endemic mermithid *Agamermis unka* (Choo and Kaya, 1994). While parasitism of the BPH is highly variable from place to place and from year to year, it is assumed that the mermithid exerts some control because it reduces the fecundity of the host and ultimately causes its death. Also, the mermithid has only one generation per year compared to the three to four generations of BPH. However, because the mermithid females stagger their egg production, many individuals in all BPH generations are parasitised. Augmentation of this mermithid into BPH populations is only possible on a limited scale because it is an obligate parasite and mass production technology has yet to be developed. Mermithids cannot be mass-produced *in vitro*, and *in vivo* production is labour intensive; storage and formulation are also problems yet to be solved.

Petersen (1985) listed the attributes of mermithids which might make them ideal biocontrol agents: 1) they demonstrate degrees of host specificity; 2) they are lethal, or at least sterilise their hosts; 3) they can be easily manipulated in the laboratory; 4) they have a capability for *in vivo* mass production; 5) they can be applied by standard spray equipment; 6) they are environmentally safe; 7) they have potential to exert long term control. As he pointed out, however, few species have all of these attributes. Mermithids have never attained their implied potential. Particular problems relate to their aggregated distributions, poor dispersal, long life cycles, variable effectiveness and limitations for mass production. Thus, there are two basic lessons to be learned from the long experience with mermithids. First, the ability to kill a host and

observations of high levels of infection at particular times are not necessarily an indication that an organism will be a good biological control agent. It is essential to understand the biology of the host and parasite, the complex interactions that occur between them, and the level to which the host population will be regulated. Second, unless organisms can be cultured reliably and cheaply, their potential cannot be realised.

Phaenopsitylenchidae

The most successful example of a nematode being used in a classical biological control programme is that of *Beddingia* (syn. *Deladenus*) *siricidicola* against the wood-wasp *Sirex noctilio*. The taxonomic status of *Beddingia* and other related genera is unsettled and requires further study. The nematodes concerned are facultative parasites, characterised by having both parasitic and free-living generations. For the purposes of this paper, the details of the taxonomic arguments are not relevant, but centre around the relative taxonomic importance of the free-living compared to the parasitic stages. Pending further studies, we have adopted the proposals of Remillet and Laumond (1991). Historically, the genus has also been placed in the families Neotylenchidae and Allantonematidae. There is no dispute that the nematodes belong to the Tylenchida.

Sirex noctilio is a European insect that was accidentally introduced into Australia and New Zealand and caused extensive damage to *Pinus radiata* forests. Similar examples have occurred frequently, whereby introduced pests cause large losses but cause little damage in their native habitats. This is an ideal scenario for a classical biological control programme. CSIRO scientists were sent to Europe to search for natural enemies of the wasps, and a remarkable nematode was discovered, with a life cycle showing incredible adaptations to the wasp life cycle.

When female *Sirex* oviposit into a tree, they also supply spores of a symbiotic fungus, *Amylostereum areolatum*, and a toxic mucus. *Sirex* larvae feed on the fungus, which eventually permeates the whole tree. The combination of fungus, mucus and wasps kills the tree. The nematodes are facultative parasites, which means that they have two possible life cycles. In the mycetophagous cycle, the nematodes feed and reproduce on the fungus, and can go through a number of generations in the absence of the insect. In the entomophagous cycle, the nematodes require a wood-wasp to continue development. When *Sirex* larvae are present, the fungal-feeding nematodes differentiate into morphologically distinct infective females which can no longer feed on the fungus and require an insect to continue their development. The mated infective female penetrates into the haemocoel of a *Sirex* larva and begins to grow, but her reproductive system does not develop until the host pupates. Towards the end of pupation, larval nematodes, which hatched and remained within the female, emerge into the haemocoel and migrate towards the host's reproductive organs. They enter the developing eggs of a female host, so the host is sterilised and each egg may contain up to 200 nematodes. Male hosts are not sterilised and are essentially a dead end for the parasites as there is no way for them to escape into the environment. By contrast, the nematodes escape from their female hosts during oviposition, either inside or alongside the eggs. They then feed on the associated fungus, develop into mycetophagous adults and continue through a number of generations. In this extraordinary relationship, the female wasp is not only sterilised, but also disperses the nematodes and provides them with food in a new tree. Infection levels of *Sirex* approach 100% and the population collapses. Bedding (1984a) provides details of this remarkable nematode and its successful application in Australia in this classical biological control programme. Indeed, the programme became so successful that foresters became less rigorous in maintaining the control pressure. An outbreak of *Sirex* resulted and 5 million trees were killed in Australia between 1987 and 1989. In a major response, 147,000 trees were inoculated with *B. siricidicola*. Haugen (1990), Haugen and Underdown (1990) and Haugen *et al.* (1990) detail the cause of and response to the breakdown in control and the strategy adopted to prevent its recurrence.

It is worth considering the factors that contributed to the success of this programme in more detail, because it has important general lessons for the application of biological control agents regardless of the agent (Bedding 1984a). First, it is essential to understand the biology of both the host and the parasite in great detail, so that the parasite can be manipulated and used effectively on a large scale. Here, the parasite was found to be highly host specific, so would not affect beneficials in the system. Also, it is highly adapted to its host, so much so that the host efficiently disperses the parasite and even unwittingly furnishes it with a food supply. This meant that it was not necessary to apply the parasite throughout a forest. Instead, it could be inoculated in part of the forest and the wasps acted as the dispersal agents. Second, it is important to be able to mass culture the nematodes. Even before commercialisation, this is necessary for experimentation and development of the system to the point of large scale application. *Sirex* has a 1–3 year life cycle, so *in vivo* rearing of the parasites would be impractical. The mycetophagous life cycle facilitated mass production in the absence of the insect hosts. Though the parasite lacks a resistant stage and loses viability after 8 weeks of storage, they were used soon after production by injecting them into trees infested with *Sirex*. This made formulation and long term storage unnecessary. The fungal-feeding cycle also allows the nematode to multiply not only in the environment of the host but also in the absence of host individuals, effectively increasing the numbers of infective agents.

Entomopathogenic Nematodes

Nematodes belonging to the Steinernematidae and Heterorhabditidae are referred to as entomopathogenic nematodes (EPNs). In contrast to the families described above, EPNs have been commercially developed as bioinsecticides. EPNs are attractive candidates for

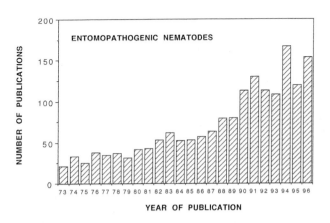

Figure 19.2. Analysis of CABPEST CD for the terms Neoaplectana or Steinernema* or Heterorhabdit* and publishing year.

commercial exploitation because several different species can be mass-produced economically, formulated and applied for control of a range of soil-dwelling insect pests. Their commercial prospects mean that there is a great deal of interest in the group, from fundamental to highly applied research. The number of annual publications for EPNs from 1973 to 1996 (Figure 19.2) provides a dramatic illustration of the increasing interest in the group, especially when compared to the work on mermithids (Figure 19.1). Thus, in 1973, similar numbers of publications were made for the two groups. Research on mermithids experienced a boom in the decade from 1973 through 1982, when publications averaged 52 per year, but has remained essentially flat at an average of 28 for the 14 years from 1983 to 1996. By contrast, the work on EPNs experienced an dramatic increase, with over 100 publications per year since 1990. The reason for this stems largely from the pioneering work of Bedding, who produced two key publications dealing with low cost production, storage and transport of EPNs (Bedding 1981, 1984b). This generated commercial opportunities which, combined with the efficacy of the nematodes as biological insecticides against important agricultural pests, stimulated the research that is reflected in the publication record.

Because there is so much interest in these nematodes, there are a number of books, special journal issues and reviews on EPNs. The book edited by Gaugler and Kaya (1990) deserves special mention because it established a foundation for work in the field. A number of books or special volumes of journals that were dedicated to EPNs followed: Abad *et al.* (1998), Bedding, Akhurst and Kaya (1993), Boemare, Nealson and Ehlers (1997), Boemare *et al.* (1996), Burnell, Ehlers and Masson (1994), Ehlers and Hokkanen (1996), Griffin, Gwynn and Masson (1995) and Simoes, Boemare and Ehlers (1998). There has also been a number of recent reviews including those of Georgis (1992), Georgis and Manweiler (1994), Hominick *et al.* (1997), Kaya and Gaugler (1993) and Popiel and Hominick (1992). Woodring and Kaya (1988) provided a handbook of techniques and Smith, Miller and Simser

(1992) produced a comprehensive bibliography with 1413 references.

Given the space available and the exceptional numbers of recent books and reviews, it seems pointless to try to provide yet another detailed review on the use of EPNs for biological control. Instead, the following is a synthesis of the present status, with some key references provided as appropriate.

Entomopathogenic Nematodes in the Phylum Nematoda

The phylogenetic analysis produced by Blaxter *et al.* (1998 and this volume) included *Steinernema carpocapsae* and *Heterorhabditis bacteriophora*. Species from both Steinernematidae and Heterorhabditidae share the same strategy of utilising specific bacterial symbionts to kill insects in which they then reproduce, but it emerged that they do not share a close ancestry. This confirms Poinar's (1993) assessment based on morphology. He concluded that both belong to the group Rhabditida, while Blaxter *et al.* (1998) show that they are more distantly related than this would imply. This is because the order Rhabditida is paraphyletic, with members in two strongly supported clades, each representing a trophically diverse assemblage. Thus, *Heterorhabditis* is grouped with *Caenorhabditis elegans* and other members of the suborder Rhabditina and is particularly associated with members of the order Strongylida, which are intestinal parasites of vertebrates. *Steinernema* is grouped with rhabditid families but also with the vertebrate-parasitic genus *Strongyloides* and the plant-parasitic orders Tylenchida and Aphelenchida. These groupings strongly support the notion that *Heterorhabditis* and *Steinernema* have independently evolved similar life strategies. Hence, it is not surprising that their bacterial symbionts belong to two different genera, *Photorhabdus* and *Xenorhabdus*, respectively.

Consideration of the systematics of the bacteria is beyond the scope of this chapter. A recent issue of the journal *Symbiosis* is devoted to papers examining various aspects of the relationships amongst EPNs and the bacterial symbionts (Boemare, Nealson and Ehlers, 1997). The isolation and characterisation of novel bioactive compounds from the bacteria is an area that will increase in importance in the future. Another useful summary is provided by Boemare, Laumond and Mauleon (1996). They cover taxonomy of the bacteria and specificity of the nematode-bacterium association. It is clear that there is great diversity amongst the bacterial symbionts of EPNs and it is proving a challenge to apply the species concept to the bacteria. In this regard, it is essential to identify the nematodes accurately in any study of the bacteria, as the nematodes provide a living environment upon which the bacterial symbionts depend and in which they are protected and evolve. As knowledge of the systematics and phylogeny increases for both groups of organisms, and molecular data become more available, a picture of evolutionary trends will emerge. It will be fascinating to compare the patterns for the bacteria within the two convergent nematode groups and to determine whether the bacteria are co-evolving with their nematode hosts, or evolving independently.

Entomopathogenic Nematode Life Cycle

The third stage larva (or dauer larva) is the only stage of the life cycle that can survive outside the host. It is the stage that is applied in control programmes. It does not feed, but searches or waits for an insect host that it can infect (see below). Once a host has been located, the infective larvae enter through the mouth, anus or spiracles. Infective larvae of *Heterorhabditis* species may also penetrate directly through the cuticle. Eventually the nematodes enter the haemocoel and release their symbiotic bacterial cells (harboured in the anterior part of the intestine) which proliferate and kill the insect through septicaemia. The bacteria establish a monoxenic culture because they produce antibiotics. The infective larvae begin to feed on the bacteria/host substrate, resume development, and mature into males and females (*Steinernema* spp.) or hermaphrodites (*Heterorhabditis* spp.). One or more generations of progeny develop, reproducing continually until the resources of the host are exhausted. As in *C. elegans*, a combination of pheromones and food stimuli influences development of early larval stages towards production of another generation within the host or towards the persistent stage (dauer larvae). These third stage larvae retain their second stage cuticle and are released from the cadaver as the insect disintegrates. This cycle may continue in the presence of sufficient insect hosts. In biological control programmes, the initial application of dauer larvae should provide adequate crop protection. However, pest population control becomes less predictable with time because a number of abiotic and biotic factors affect the reproductive rate, dispersal and persistence of the nematodes.

Entomopathogenic Nematodes and Biological Control

An OECD workshop on scientific and regulatory policy issues for the use of non-endemic EPN species was held in Malente, Germany in 1995. The papers, conclusions and recommendations have been published in a special issue of *Biocontrol Science and Technology*, edited by Ehlers and Hokkanen (1996). The workshop identified the following special features and ecological facts for EPNs:

- Natural epizootics are rare.
- The host range of EPNs in the field is limited, in contrast to the misleading broad spectrum of activity obtained experimentally in the laboratory.
- Effective control of susceptible insects by EPNs requires the application of large numbers of dauer larvae (in the region of 10^9 ha^{-1}).
- After inundative release (mass application), populations of EPNs decline rapidly to low levels that are comparable with natural densities.
- There is little dispersal of EPNs after application.

- EPNs have been widely used for many years in pest control without known detrimental effects.
- EPNs pose much less threat to the environment than chemical pesticides. They substitute for some of the broad-spectrum pesticides currently used in soil.

Based on these and other considerations, the unanimous opinion of the workshop was that EPNs should not be subject to any kind of registration. However, in some cases their use should be regulated, for example if the introduction of exotic species is contemplated. The fact that the nematodes are invertebrates (macroorganisms) associated with microorganisms may potentially complicate legal issues, as the laws for releasing macroorganisms may be different from those for microorganisms. Legislation regarding the use of EPNs in Europe, Australia, New Zealand and the USA is discussed in the OECD workshop papers.

There are a number of key factors that can influence the interactions of EPNs with their insect hosts. Considerable investment in fundamental ecological studies is required before a particular species may be taken to commercial development. The main factors that have been shown to impinge on efficacy of a particular species have been elaborated by Georgis and Manweiler (1994) under the headings: Host susceptibility; Soil type/texture/pore size/pH; Soil moisture; Temperature; Nematophagous organisms; Ultra violet light; Interaction with plants. It is important to understand environmental and physiological or behavioural limitations if EPNs are to realise their full potential. A summary with key references is provided below.

Persistence

Factors affecting persistence of EPNs have been reviewed recently by Hominick *et al.* (1996), Kaya and Koppenhofer (1996) and Smits (1996). There are few long term studies of EPNs, but it is clear that long term persistence occurs (Campbell *et al.*, 1998; Del Pino and Palomo, 1997; Gaugler *et al.*, 1992; Glazer *et al.*, 1996; Hominick and Briscoe, 1990). There may be seasonal changes and these can reflect changes in vertical distributions, particularly to avoid extreme temperatures. Such changes can move the nematodes from the sphere of activity of potential insect hosts.

The population biology of EPNs is not understood and the effects on natural insect populations are unknown. Though the nematodes are widespread, they exist naturally at low population levels, with fewer than one individual ml^{-1} of soil being the norm; epizootics are difficult to detect and hence are rarely recorded. Thus, it is unlikely that EPNs may regulate an insect population over the long term and their successful deployment in classical biocontrol programmes is therefore unlikely. It may well be that insect populations regulate the nematodes, that is, the nematodes persist and reproduce opportunistically as a susceptible host becomes available.

It is difficult to quantify the nematodes in a soil sample as results depend on the method of extraction used. Also, nematodes are very patchily distributed, both in space and time. Standard soil extraction techniques for nematodes recover all nematodes, so EPNs need to be separated and identified, a laborious procedure. Bioassays utilising a susceptible host recover only a proportion of the nematodes, and negative assays may reflect absence of nematodes or lack of infectivity of the resident population. It is known that infectivity of a population can vary or even cycle (Bohan and Hominick, 1997).

The influence of abiotic factors on EPN persistence is documented to varying degrees while the effects of biotic factors are largely unknown and will be extremely difficult to quantify. As the second half of this chapter shows, there are a large number of antagonists present in the soil, all of which have the potential to impinge on the population of EPNs. For example, Koppenhofer *et al.* (1997) observed that coastal shrubland soil in California usually contained many species of nematode-trapping fungi, which may influence the distribution of *Heterorhabditis hepialius* (syn. *H. marelatus*). They therefore quantified nematode suppression by adding combinations of fungi or single species in the form of fungal-colonized nematodes to pasteurised soil. Their results showed varying levels of suppression of *H. hepialius* as well as changes in the population densities of the fungi.

With so many unknowns and variables, it is natural that EPNs are applied inundatively, at doses around a half million m^{-1}, and so that their efficacy does not rely on their multiplication in the soil habitat. EPNs are biological insecticides and the pattern of their population change post-application is a rapid decline in the first few days, followed by a moderate decline over the next 2–6 weeks, and then a long period of recycling at a low level. That is, population density decreases to background levels within days or weeks after application. Similarly, populations of EPNs are highly aggregated, and Campbell *et al.* (1998) recently showed that uniform inoculative releases of *H. bacteriophora* tended to return to the patchy patterns of distribution typical of endemic populations.

Geographical distribution and biodiversity

Downes and Griffin (1996) and Hominick *et al.* (1996) reviewed dispersal and biodiversity. Entomopathogenic nematodes are highly aggregated in distribution and can move only centimetres under their own means. To understand geographical distribution and biodiversity, it is important to identify accurately the nematodes that are isolated. Unfortunately, this is not always done. Also, extensive and intensive surveys are labour intensive and much of the world remains to be surveyed, so our knowledge is limited.

It is intriguing that there are many species of *Steinernema* and comparatively few of *Heterorhabditis* (see Hominick *et al.*, 1997). Even when apparently unique morphological details are used to erect new heterorhabditid species, the molecular data seem to contradict these (Adams *et al.*, 1998) and the conservative *status quo* prevails. The analysis of Adams *et al.* (1998) supported the existence of three closely related sister taxa, *H. marelata* + *H. hepialius*, *H. indica* + *H. hawaiiensis* and *H. argentinensis* + *H. bacteriophora*, and they cautiously suggested that the members of each pair may be conspecific (i.e. synonymous). Hence, it would appear that there are even fewer nominal species of *Heterorhabditis* than thought, and that the ones that do exist are spread widely over the globe. Bearing in mind how little of the globe has been surveyed (Hominick *et al.*, 1996), and extrapolating from the CABI Bioscience collection which contains over 25 steinernematid species new to science, there must be large numbers of steinernematids yet to be discovered. Some steinernematids, such as *S. carpocapsae* and *S. feltiae*, are common and widespread while others appear to be more restricted in distribution. This may merely reflect sampling efforts to date and the relative prevalence of species because large numbers of samples must be taken to detect organisms that are relatively rare and usually highly aggregated.

The fact that species such as *S. feltiae*, *S. carpocapsae* and *H. bacteriophora* are essentially ubiquitous implies that dispersal for at least some species is highly efficient and probably occurs by a variety of means, including active methods by hosts and passive ones such as wind and water. Commercial activities may have complicated interpretation of natural dispersal because introductions could occur in areas where the nematodes do not occur naturally but can become established. Indeed, human activity has probably played an important part in the dispersal of EPNs. In historical times, soil was moved around the world and so were nematodes. One well-documented case concerns the potato cyst nematodes, *Globodera rostochiensis* and *G. pallida*. The potato originated in the high Andes and was introduced to Europe by the Spanish Conquistadores in the 16th century. At the same time, potato cyst nematodes were introduced on infested material. They have subsequently spread world-wide as new potato varieties emerged, the crop gained in popularity, and infested material was disseminated with no regard for quarantine measures (Sharma *et al.*, 1997). Such influences need to be appreciated as we strive to understand the biogeography of EPNs.

As data accumulate on the biodiversity and biogeography of EPNs, the information may provide biological insights that could be applied to their use as biological control agents. What are the important biological characters that allow *H. indica*, *H. bacteriophora* and *H. megidis* to disperse and establish globally? What is the gene flow between these populations? Why do steinernematids speciate more than heterorhabditids? It is known that EPNs are aggregated in distribution. Do steinernematids tend to form discrete local populations, so that the founder effect results in isolated gene pools and the evolution of new species? While acknowledging that sample sizes are small, it does appear that some species are more restricted in their distribution than others. If the reasons for such differences were known, they could perhaps be exploited for particular biological control programmes. In any case, the study of the biodiversity of EPNs is in

its infancy, and there are numerous species with unknown biologies. The application of EPNs for biological control has centered on a few species, so the potential is still essentially there to be developed.

Host and habitat specificity

Hominick *et al.* (1996), Peters (1996) and Simoes and Rosa (1996) have reviewed host and habitat specificity for EPNs. Information on their natural host range is difficult to obtain as infections are transient and hence are infrequently observed. The literature implies a broad host range, but this is based on artificial laboratory bioassays and the image of EPNs being extreme generalists is being eroded. Clear distinctions must be made between laboratory host range and field host range. The latter will involve not only the range of insects naturally parasitised, but also the range of hosts successfully controlled by inundative release of nematodes. The laboratory host range encompasses insects which are parasitised in laboratory conditions and which support propagation of the nematode. The natural host range is the least understood. However, it is critical information for biological control as a broad host range implies that the agent may pose a threat to non-target invertebrates.

Knowledge of habitat specificity is in a state similar to that of host specificity. That is, EPNs were assumed to have little or no habitat specificity. This assumption is based on results where sample sizes were inadequate or sampling strategies failed to allow for the aggregated distributions of EPNs, compounded by unreliable or inadequate identification of the nematodes isolated. As more surveys occur, providing large sample sizes, and accurate identification is made to the species level, some habitat specificity is becoming apparent. This should not be surprising, as all organisms have niche requirements that will be satisfied only in particular habitats. Similarly, the soil habitat of the nematodes has three dimensions, so specificity may extend to occurrence at particular depths in the soil. Obviously, knowledge of host and habitat specificity is fundamental for matching the best nematode to a particular insect pest control programme.

Behaviour and physiology

There are a number of studies and reviews on the behaviour of EPNs (see Ishibashi, Chapter 20; Downes and Griffin, 1996; Glazer, 1996). Two basic strategies for host finding are possible. Actively searching nematodes are termed 'cruisers,' while ones that passively wait, presumably conserving energy, until a host essentially contacts the nematode are termed 'ambushers'. Campbell and Gaugler (1997) recently challenged six EPN species with different foraging behaviours to find hosts with different levels of mobility by exposing the nematodes to restrained and unrestrained *Galleria mellonella*. *Steinernema carpocapsae* and *S. scapterisci* exhibit an ambush foraging strategy, tending to stand on their tails in a straight, non-moving posture for extended periods of time (nictation). These species were most effective at finding the mobile, unrestrained, larvae. *Heterorhabditis bacter-*

iophora and *S. glaseri* did not nictate and were most effective at finding the restrained larvae, which would be expected of a cruise forager. Another non-nictating species, *S. feltiae*, and a species which nictated infrequently, *S. riobravis*, were able to find both types of hosts, suggesting that they use an intermediate foraging strategy. The intermediate foraging strategy of *S. feltiae* may result from its tendency to raise more than 30% of its body off the substrate (body-waving) more frequently than other non-nictating species. Lewis, Campbell and Gaugler (1997) worked with *S. carpocapsae* and concluded that EPNs do not alter their foraging strategy in response to unsuccessful searching or aging. However, the question of whether these strategies are mutually exclusive remains open, and variation within a population is likely. Thus, migrators and non-migrators may occur in a population, and the tendency to migrate or respond to hosts may change over time for a particular individual.

In at least some species, infectivity may also vary, so that the population of nematodes exists as two subpopulations, one infective, the other not. The proportion that is infective can vary significantly over time (Bohan and Hominick, 1997). There is no indication of the relative importance of endogenous and exogenous cues in controlling these behaviours. However, Patel, Stolinski and Wright (1997) examined the relationship between neutral lipid utilisation and infectivity of *Steinernema carpocapsae*, *S. riobravis*, *S. feltiae* and *S. glaseri*. Neutral lipid contents of freshly harvested dauer larvae were similar in all four, measuring 31, 31, 24 and 26% dry weight, respectively. *Steinernema carpocapsae* showed a sigmoidal utilisation pattern whereas *S. riobravis* used neutral lipids at an almost constant rate. Survival of both species ranged between 120 and 135 days. By contrast, *S. feltiae* and *S. glaseri* lived beyond 450 days and had a slower rate of lipid utilization during a 260 day storage period. Oil Red O staining showed that individual dauer larvae in each population utilized lipids at different rates, even though they had the same initial lipid index. The infectivity of *S. riobravis*, *S. feltiae* and *S. glaseri* declined with lipid utilization. In contrast, *S. carpocapsae* maintained a high level of infectivity, even at relatively low lipid levels. Thus neutral lipid content is an indicator of infectivity for *S. riobravis*, *S. feltiae* and *S. glaseri* but not for *S. carpocapsae*. Wang, Kondo and Ishibashi (1997) assessed nictation of *S. carpocapsae* harvested from host cadavers of *Galleria mellonella*, on days 1, 4 and 8. The nictation rates decreased from 22% for the dauer larvae from the first day's harvest to 18% and 9% on days 4 and 8, respectively. They therefore concluded that *S. carpocapsae* has an active and an inactive host-seeking phase.

Since EPNs are found in a variety of habitats, from tropical to sub-Arctic and desert to moist, they have survival mechanisms to cope with particular environmental stresses. These include tolerance to extreme temperatures, desiccation, osmotic changes and lack of oxygen. The nematodes are also compatible with the use of most insecticides and fungicides. Indeed, these traits may be genetically enhanced. For example,

Glazer, Salame and Segal (1997) examined the feasibility of using genetic selection to enhance resistance of *Heterorhabditis bacteriophora* strain HP88 to fenamiphos, oxamyl and avermectin. Selection resulted in an 8 to 9-fold increase in resistance to fenamiphos and avermectin and a 70-fold increase in resistance to oxamyl. The enhanced resistance to oxamyl and avermectin, and to a lesser extent to fenamiphos, was stable and continued after selection was relaxed. This contrasted with some earlier attempts by other workers to select for particular traits, in which the nematodes quickly returned to the unselected state when selection was relaxed. Glazer, Salame and Segal's (1997) study was also encouraging in that no deterioration in traits relevant to biological control efficacy (including virulence, heat tolerance and reproduction potential) was observed in the selected lines as compared with the base population. Their results demonstrated that genetic selection can be used to enhance resistance of EPNs to certain environmental stresses. Such selected lines could be useful bioinsecticides in the context of integrated pest management.

The possibility that some traits could be genetically enhanced has led to concerns that continuous culture in laboratories or on an industrial scale could select for particular genotypes which might be less effective compared to the wild type. This founder effect, whereby a limited part of the gene pool exists in an isolated population which is then used to establish a long term breeding culture, is a regular occurrence in laboratories. Genetic variation in laboratory-reared biological control agents may be reduced by inbreeding and selection. Shapiro, Glazer and Segal (1997) therefore used RAPD-PCR analysis to examine genetic diversity in wild and laboratory populations of *Heterorhabditis bacteriophora*. One strain had been recently isolated from the field and the other had been reared under laboratory conditions for over 10 years. The level of within population variation detected did not differ significantly between the two strains. While this is an encouraging result, all workers should be aware of the possibility that continuous laboratory culture may result in strains of nematodes that are different from the wild type. This may also apply, even more so, to the bacterial associates. Changes may be advantageous or damaging, and cannot be predicted. Hence, it is prudent to preserve some live specimens from the wild types as soon as possible in liquid nitrogen.

Understanding survival mechanisms could be a key component in eventually selecting optimal species or strains for particular biological control programmes. Introduction of appropriate non-indigenous strains into particular habitats could be facilitated and result in increased efficacy. On the other hand, if long-term establishment is not desired, then a specific strain with a high susceptibility to a particular environmental stress could still be utilised if its other traits made it a more effective agent in the short-term. Understanding the mechanisms of survival are important for commercialisation, as these are fundamental for maximising survival during storage by appropriate formulation of the nematodes.

Commercial Development of Entomopathogenic Nematodes

The fact that EPNs do not persist to exert continuous control makes them a commercial proposition compared to agents used in a classical fashion. Indeed, commercial development has stimulated and supported substantial research. However, interpretation of ecological information requires care because of the wide utilisation and dissemination of a few species.

From a commercial point of view, an EPN strain should exhibit the following qualities:

1. It should be easy to produce on a large scale. Once a production specification has been established the strain should be robust in production i.e. the yield of dauer larvae should be consistent and the quality (lipid levels and pathogenicity) reliable.
2. The EPN strain should be suitable for the formulation available. That is, the dauer larvae should survive for the specified shelf life and remain pathogenic during the life of the product, delivered to the grower in optimum condition.
3. The EPN strain should be pathogenic over a wide temperature range. This allows application against the same pest over a wide geographic area.
4. The EPN strain should not be host specific, allowing the development of the same bioinsecticide against a range of pests, for different crops and markets (retail, horticultural, agricultural).

In direct contrast to point 4, from an environmental viewpoint, the EPN strain should demonstrate some host specificity so that it has minimal effects on non-target organisms, reducing environmental impact. Clearly, ecological facts and business priorities may conflict. Thus, while the ecological facts may indicate that each species will perform optimally in specific habitats and against a few pests, commercially it may be preferred to produce one species and, if necessary, apply larger doses to achieve control of a wider range of pest species.

While biological control offers many attractions compared to reliance on chemical pesticides, there are other methods based on traditional cultivation techniques that should not be overlooked. For example, Bellotti, Cardona and Riis (1997) reported that several species of Scarabaeidae in Colombia significantly reduce yields of beans (*Phaseolus vulgaris*), potatoes, cassava, maize, and other crops. Losses due to larval feeding on beans ranged from 40 to 60% while feeding by adults on reproductive organs in maize can prevent pollination and thereby reduce yields by more than 50%. The entomopathogenic bacterium *Bacillus popilliae*, the fungi *Beauveria bassiana* and *Metarhizium anisopliae*, and the nematode *Heterorhabditis* sp. reduced larval populations and significantly improved yields. However, the authors point out that land preparation, deep ploughing and cultivation before planting also increased yields at a lower cost than the use of entomopathogens.

Unexpected results may arise during tests, so workers should be aware of unexplained results that are worth exploring further. For example, EPNs have been observed to suppress populations or reproduction in plant-parasitic nematodes (Georgis and Kelly, 1997). Most recently, Grewal *et al.* (1997) studied the influence of *Steinernema carpocapsae* and *S. riobravis* on natural populations of plant-parasitic nematodes infesting turfgrass in Georgia and South Carolina, USA. *Steinernema riobravis* applied at 6×10^9 dauer larvae per acre provided up to complete control of *Meloidogyne* sp., *Belonolaimus longicaudatus* and *Criconemella* sp., in Georgia, although *S. carpocapsae* had no effect. *Steinernema riobravis* was as effective as fenamiphos (Nemacur 10G) at 4 weeks after treatment and more effective at 8 weeks after treatment. In South Carolina, both steinernematids applied at 10^9 dauer larvae per acre provided up to 86–100% control of the nematodes. The authors conclude that *S. riobravis* may provide effective, predictable and economical control of plant-parasitic nematodes in turfgrass. Although the mechanism causing suppression is unknown, it appears to be largely related to biostatic or repellent substances derived from the bacterial associates of the nematodes. The steinernematid is being marketed for control of plant-parasitic nematodes and other substances derived from the bacteria are being explored for insecticidal, nematicidal and fungicidal activity (Georgis and Kelly, 1997).

Biological Control of Parasitic Nematodes

Introduction

The use of biological agents to regulate parasitic nematode populations has involved the manipulation of indigenous natural enemy communities, largely through the use of soil amendments (Rodriguez-Kabana and Morgan-Jones, 1987), the fortuitous development of nematode-suppressive soils (Kerry, 1988), and the application of specific organisms (Stirling, 1991). Nematode suppressive soils do provide effective control of specific pest species but methods that rely on the manipulation of the natural enemy community or individual species from within that community have proved less successful. Biological control using a single agent to control a specific pest on a range of crops represents the most demanding situation for the exploitation of the tritrophic interaction and will only be effective if this interaction is thoroughly understood.

Successful strategies for the biological control of insect pests have been developed over the last century and an understanding of the principles of biological control is most advanced in entomology. In this discipline, biological control has usually required a detailed knowledge of the population dynamics of the pest and natural enemy. Although some mathematical models have been developed for the biological control of nematodes (see below), little is known of the quantitative relationships between nematodes and their natural enemies. In general, most knowledge of the biological control of nematodes concerns microbial agents and these are notoriously difficult to quantify in soil. Standard procedures, such as the use of the serial dilution plating of soil samples onto semi-selective media, may provide valuable information on relative changes in the abundance of culturable bacteria and fungi that are nematophagous or antagonistic to nematodes (Kerry *et al.*, 1993). However, interpretation of the data requires care, as such changes in abundance may not be correlated with changes in activity of the stages that affect nematodes. In general, there is a dearth of quantitative methods for the estimation of microbial populations in soil and several methods are required to obtain data that may be reliably interpreted.

Entomologists are also familiar with the concept of multitrophic interactions, which include hyperparasites that attack natural enemies and other organisms that may disseminate microbial agents. Insect herbivory induces the release of a range of signals from plants that may involve volatile and non-volatile compounds (Pickett *et al.*, 1992). These compounds may provide natural enemies with extremely sensitive cues to locate their hosts (Powell *et al.*, 1998). Such interactions are likely in the rhizosphere and within the root but their importance in the biological control of nematodes is unknown. The root-knot nematode, *Meloidogyne javanica*, altered the composition of tomato root exudates and resulted in *Rhizoctonia solani* changing from a rhizosphere saprophyte to a root pathogen (Van Gundy, Kirkpatrick and Golden, 1977). Also, the occurrence of certain species of *Pratylenchus* in potato roots affects the establishment of *Verticillium dahliae* (Powelson and Rowe, 1993) and the presence of specific races of soybean cyst nematodes can influence the establishment of rhizobium (Lehman, Huisingh and Barker, 1971). If such interactions have developed between nematodes and other root-colonising organisms in the rhizosphere, similar relationships have probably evolved with nematophagous organisms. Indeed, rhizobacteria may induce resistance in plants to nematode invasion through the induction of systemic cues (R. A. Sikora, *pers. comm.*) and some nematophagous fungi appear to be more abundant on nematode-infected roots (Bourne, Kerry and de Leij, 1996). Biotic factors affecting the survival and spread of the natural enemies of nematodes are poorly understood. In this review, the major components of tritrophic interactions are dealt with separately and knowledge of these interactions is brought together in a discussion of the development of management strategies.

Nematodes and Plant Interactions

The nematode target and the host plant may have profound effects on the diversity and dynamics of natural enemy communities and their effects on the regulation of nematode populations. It is essential that the biology, population dynamics and ecology of the nematode and its host are considered in order to understand the impact of natural enemies on nematode populations and to develop reliable strategies for biological control. In short, all plant-parasitic nematodes are obligate parasites and spend part of their life cycle in soil, but activity in soil may be of limited duration for those that are endoparasites of plants and for animal-parasitic nematodes. Nematodes may be migratory or sedentary endo- or ecto-parasites of plants. If migratory, parasitic natural enemies must produce an adhesive spore, or a trapping mechanism or toxin to immobilise the active host and enable infection to take place. Sedentary nematodes are exposed to parasitism by a range of relatively unspecialised bacteria and fungi. In general, female sedentary nematodes, such as root-knot nematodes (*Meloidogyne* spp.), have a fecundity that is at least an order of magnitude greater than that for migratory nematodes such as *Trichodorus* spp. Population growth rates also depend on the nematode species and the length of the life cycle, crop duration and environmental conditions, especially host species and temperature. Plant susceptibility will affect population growth and plant tolerance the proportion that must be controlled to prevent significant yield loss. As biological control agents usually act as regulators of populations they are of little value in protecting crops from virus vectors or where there is zero tolerance, such as for potato cyst nematodes on seed crops. Individual nematode species may reach population densities of 10^3 g^{-1} soil and 10^4 g^{-1} root (Whitehead, 1997) and generation times may range from 2–3 wk to more than a year. Pests that have rapid reproductive rates but survive in only small numbers in soil between susceptible crops are considered 'r strategists' and are more difficult to control with biological control agents than 'k strategists', which tend to have less fluctuating population densities (Southwood and Comins, 1976). Although such strategies form a continuum and individual species may switch between extremes depending on environmental conditions, the concepts are useful in assessing the likely impacts of natural enemies on host population dynamics. Nematodes such as *Ditylenchus dipsaci* are considered 'r strategists' whereas most cyst nematodes are 'k strategists'. However, fungal parasites of *D. dipsaci* have been reported to have a significant effect on the long-term survival of the nematode in soil (Clayden, 1985) and may affect population development (Atkinson and Dürschner-Pelz, 1995). Sedentary nematodes such as cyst and root-knot nematodes, which include the most economically important nematode pests world-wide, retain their eggs within the female (cyst nematodes only) and/or in a gelatinous matrix deposited on the root surface. A wide range of relatively unspecialised fungi, which are facultative parasites in the rhizosphere and colonise egg masses or cyst nematode females when they are exposed on the root surface, may infect these eggs. Most nematodes lay their eggs individually in soil and because these dispersed eggs are difficult to extract little is known of their natural enemies.

Interest in the biological control of animal-parasitic nematodes has been confined to killing the eggs released in faeces and the infective larvae that emerge from the egg and migrate to the pasture (Gronvold *et al.*, 1993; Waller and Faedo, 1996). Nematodes appear to be protected from parasitism within the animal host. The parasite burden, host tolerance, longevity of egg and larval stages in faeces and on pasture are greatly influenced by the nematode species and environmental conditions. Nematode control will be less effective in faeces with a large egg burden, for eggs that hatch rapidly, and if the herbivore host is intolerant of the parasites.

Natural Enemies

Natural enemies of nematodes have been identified that may be parasites/pathogens, predators, competitors or antagonists. Most have been extensively reviewed recently (Kerry, 1987; Stirling, 1991; Dickson *et al.*, 1994) and comments below relate largely to the characteristics of each group which affect their potential for exploitation as biological control agents (Figure 19.3). It must be remembered that most studies have concerned interactions between a specific natural enemy and a nematode pest. However, many of these natural enemies occur together in soil and in the rhizosphere but the measurement of the effect of the community on nematode abundance at the population or community level has rarely been attempted. Cultural methods such as crop rotation and soil amendments have been used to encourage the indigenous, antagonistic rhizosphere microflora that may reduce nematode damage and populations (Kloepper *et al.*, 1991). Most knowledge on the role of specific organisms in the regulation of nematode populations results from empirical studies and the biology and ecology of few organisms have been studied in depth. Because it is difficult to mass produce many of the predators of nematodes or manipulate their activity in soil, our understanding of their potential for nematode control is poor.

Bacterial antagonists in the soil and rhizosphere

Bacteria have long been known to produce nematicidal compounds in anaerobic (Johnston, 1957; Rodriguez-Kabana, Jordan and Hollis, 1965) and aerobic conditions (Sayre, Patrick and Thorpe, 1965) but little is known of the role of these compounds as antagonists of nematodes (Sayre and Starr, 1988). Toxins from *Streptomyces avermitilis* have been successfully commercialised as anthelmintics and there is continuing interest in screening soil and rhizosphere bacteria: approximately 8% of bacteria collected from the rhizospheres of several plant species have antagonistic activity against nematodes (Sikora, 1988). Actinomycetes have been reported from animal- (Krecek *et al.*, 1987) and plant- (Dürschner, 1984) parasitic nematodes and have been associated with nematode suppressive soils (Dicklow, Acosta and Zuckerman, 1993). Several rhizosphere bacteria, including *Agrobacterium radiobacter* (Racke and Sikora, 1992), *Pseudomonas spp.* (Oostendorp and Sikora, 1990; Speigel *et al.*, 1991), and *Bacillus subtilis* (Sikora, 1988) have been found to reduce nematode hatch or mobility and the invasion of roots. These bacteria may produce toxins or alter root exudates making the roots less attractive to nematodes (Oostendorp and Sikora, 1990). The modes of action of these bacteria are poorly understood and few active compounds have been characterised, although antagonism has been associated with the production of chitinases and collagenases in *P. chitinolytica* (Speigel *et al.*, 1991).

Microbial parasites and pathogens of active nematode stages in soil

Although a wide range of bacteria may be antagonistic to nematodes, only one genus (*Pasteuria*) has taxa which are parasites. Rickettsia-like organisms (RLOs) have been observed in second stage larvae of some cyst nematodes but their role as commensal or parasite is not clear (see Sayre and Starr, 1988). RLOs are especially associated with reproductive tissues and are transovarially transmitted but appear not to affect the development of fertilised eggs. However, treatment of *Globodera rostochiensis* second stage larvae with penicillin caused the degeneration of the RLO and an increase in nematode fecundity compared to treatment of uninfected larvae (Walsh, Shepherd and Lee, 1983). In contrast to bacteria, all the main taxonomic groups of fungi contain types that attack nematodes and those that attack the active stages are the most specialised (Dijksterhuis *et al.*, 1994).

Three gram-positive bacterial species, *Pasteuria penetrans*, *P. nishizawae* and *P. thornei*, have been distinguished in this genus because of differences in their host range and morphology. All are obligate parasites that produce spores that attach to the cuticle of nematodes. The biology and ecology of these parasites has been extensively reviewed (Sayre and Starr, 1988; Chen and Dickson, 1998). Most research has been done on *P. penetrans*, which attacks *Meloidogyne* species and may be a causal agent in root-knot nematode suppressive soils, and therefore have potential as a biological control agent (see below). The life cycle may take 20–30 days at 30°C (Stirling, 1981) but isolates of the bacterium differ in their temperature optima for spore attachment and development. The spores of *P. penetrans*

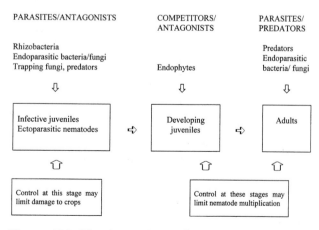

Figure 19.3. The interactions of parasites, competitors, antagonists and predators and the different types of control of different stages of plant-parasitic nematodes.

attach to the second stage larva and germinate within a few days after the larva has begun feeding within the root. Each spore gives rise to a germ tube which penetrates the body wall and establishes a dichotomously branched primary thallus within the host. This colony gives rise to daughter colonies within the pseudoceolom and eventually these fragment into enlarged quartets or doublets of cells which form sporangia, each producing single endospores. *Pasteuria penetrans* is a true parasite in that the growth and development of the nematode host appears to be unaffected until the onset of egg production, and the hyperparasite is largely confined to the reproductive system (Bird, 1986). Infected nematode females complete their development but produce few or no eggs and may contain up to 2 million spores which are released into the soil as the nematode cadaver decays. The spores are resistant to desiccation and high temperatures and may survive several years in soil (see Chen and Dickson, 1998). Spore size ranges between 1.5 and 8 μm depending on host and origin of the bacterium; second stage larvae of root-knot nematodes heavily encumbered with spores may fail to invade roots (Davies, Laird and Kerry, 1988, 1991; Stirling, Sharma and Perry, 1990). Although spores of *Pasteuria* spp. have been found adhering to c. 200 nematode taxa, it is not clear how many species are hosts in which the bacterium may complete its development.

Fungi that produce trapping mechanisms have been studied for more than a century and were used in the first commercial agents used for nematode biological control (see Kerry, 1987). Although their use against plant-parasitic nematodes has had limited success, in recent years there has been much interest in the use of selected species for the control of animal-parasitic nematodes (see below). The simplest trapping devices consist of adhesive hyphae that form part or all of the mycelium; such devices are found in two genera of the Zygomycotina, *Stylopage* and *Cystopage*. In the Deuteromycota, trapping devices are more widely distributed amongst taxa in several genera and are more specialised, being formed on only part of the mycelium. The devices consist of simple adhesive branches and knobs or the more complex two- and three-dimensional networks or non-adhesive ring structures including constricting rings (Barron, 1977). The Basidiomycotina are represented by the genus, *Nematoctonus*, in which the trapping device is a distinctive hour-glass shaped knob which produces large amounts of adhesive (Barron, 1997). Most research has been done on *Arthrobotrys oligospora* (Nordbring-Hertz, 1988; Tunlid, Jansson and Nordbring-Hertz, 1992; Dijksterhuis *et al.*, 1994) but in recent years attention has shifted to *Duddingtonia flagrans* because of its potential as a biological control agent for helminth diseases of domestic animals (Faedo, Larsen and Waller, 1997); both form adhesive networks. The trapping devices formed are specialised structures to enable parasitism and these fungi should not be considered as predators even though they are frequently referred to as such in the literature.

Nematode trapping fungi differ in their dependence on nematodes as a source of nutrition and all can develop saprophytically in soil and are readily isolated and cultured on standard laboratory media (Barron, 1977). Traps may be formed spontaneously or in response to the presence of nematodes (Jansson and Nordbring-Hertz, 1980) and trapping activity depends greatly on the nutrient status of the substrate (Cooke, 1962a,b). Different nematode trapping fungi predominate at different stages in the degradation of organic matter in soil and few individual species remain active for more than a few weeks. These fungi are widely distributed in a range of habitats and all species of nematodes that enter the traps are parasitised.

Although the extent of hyphal growth of trapping fungi in soil is not clear, the vegetative growth of endoparasitic fungi is limited mostly to the colonised hosts; hyphae extending from the nematode cadaver support the development of conidiophores. Endoparasites of nematodes in the Oomycota, such as *Catenaria anguilluluae*, produce zoospores that may be attracted to their hosts and swarm around natural openings before they encyst and give rise to a germ tube. However, the zoospores of *Haptoglossa zoospora* are mainly for dispersal and after a short swimming phase they encyst and produce an adhesive for attachment to passing nematodes. The most common endoparasite found by Barron (1969) was *Harposporium anguillulae*; this genus contains most of the nematophagous species that produce spores that are ingested by their hosts. However, plant-parasitic nematodes cannot ingest spores through the narrow lumen of their stylets, and the most studied endoparasites, such as *Drechmeria coniospora*, *Hirsutella rhossiliensis* and *Verticillium balanoides*, produce small spores that attach to the nematode cuticle. The number of spores produced differs between fungal and nematode species but >10^4 spores of *D. coniospora* were produced in a single *Panagrellus redivivus* (Dijksterhuis, Veenhuis and Harder, 1991) whereas only <10^3 spores of *V. balanoides* developed (Atkinson and Dürschner-Pelz, 1995). The spores of *H. rhossiliensis* do not attach to nematodes if they are detached from the conidiophore (McInnis and Jaffee, 1989) and, hence, the spread of infection is limited to hosts which pass close to colonised cadavers. Although, there is no saprophytic growth in soil these fungi can be cultured on laboratory media (Barron, 1982).

Some fungi and bacteria colonise plant tissue but do not cause lesions or other symptoms and are referred to as endophytes. Such organisms may be mutualistic if they protect the plant from herbivores or pathogens and parasites. *Acremonium* spp., which are common endophytes of grasses, may produce general toxins that affect grazing mammals and herbivorous insects and induce changes in roots that reduce nematode feeding and multiplication (Bernard and Gwinn, 1991). Some endophytes are non-pathogenic isolates of common plant pathogens such as *Fusarium oxysporum*, which was widespread in banana roots in East Africa and in *in vitro* tests produced metabolites that were toxic to

Radopholus similis, Meloidogyne incognita and *Pratylenchus zeae* (Hallmann and Sikora, 1994; Schuster, Sikora and Amin, 1995). Although this fungus may reduce the numbers of nematodes developing in roots, it is not clear if this effect is due to toxin production and such fungi may also compete for space in the roots, alter the physiological state of root tissue, or colonise feeding cells to the detriment of nematodes (Stiles and Glawe, 1989). Similarly, the abundance of endophytic bacterial communities may be increased in root-knot infected roots (Hallmann *et al.*, 1998) and, as some species are antagonistic to nematodes, they may have potential for biological control. Also, mycorrhizal fungi have been widely reported to improve the growth of nematode-infected plants and, in some cases, to reduce nematode infestations (Sikora and Carter, 1987; Pinochet *et al.*, 1996; Siddiqui and Mahmood, 1995). The effects of endophytes depend on the plant cultivar colonised and the species of nematode present, with the greatest reductions in nematode infestations usually occurring in roots extensively colonised by the fungus before the nematodes invade.

Parasites and pathogens of sedentary nematodes in the rhizosphere

A wide range of opportunistic soil fungi have been isolated from the eggs of cyst and root-knot nematodes (Kerry, 1988; Stirling and West, 1991) and other sedentary nematodes that lay their eggs in gelatinous matrices, such as *Tylenchulus semipenetrans* (Walter and Kaplan, 1990). In general, fungi colonising eggs and cysts are more numerous than those parasitising females (Gintis *et al.*, 1983). Some species, such as *Paecilomyces lilacinus* and *Verticillium* spp., parasitise both eggs and females and are the most studied species. Presumably, these fungi are not affected by the antimicrobial activity of the matrices produced by root-knot (Orion and Kritzman, 1991) and cyst nematodes. *Paecilomyces lilacinus* and *V. chlamydosporium* are more abundant on root-knot nematode infected roots than in the rhizospheres of healthy plants (Gaspard, Jaffee and Ferris, 1990) and the growth of the latter fungus is much affected by plant species (Bourne, Kerry and de Leij, 1996). Different isolates of these fungi differ greatly in a range of characteristics affecting their efficacy as biological control agents. Isolates of both fungal species have been collected from a wide range of cyst and root-knot nematodes and they have a worldwide distribution. Eggs that are immature are more susceptible to parasitism than those that contain second stage larvae. All species of egg parasites grow readily on artificial media and some produce resting spores. The dictyochlamydospores of *V. chlamydosporium* act as a survival stage in soil and are produced abundantly on parasitised nematodes (D. H. Crump, pers. comm.) but not on the colonised root surface. Hence, nematode parasitism may help support the long-term survival of this fungus in soil (Kerry and Crump, 1998).

Apart from the facultative parasites in the Hyphomycotina, two species of Oomycota, *Nematophthora gynophila* and an undescribed lagenidiaceous fungus,

and the chytrid *Catenaria auxiliaris* have been recorded as obligate parasites of cyst nematode females (Kerry, 1987). These fungi infect the female nematode in the rhizosphere by motile zoospores and prevent cyst formation. The species are separated by the form of the mycelium, the number of flagella on the zoospore (one in chytrids and two in oomycetes), and the morphology of the resting spore. *Nematophthora gynophila* completes its life cycle in the cereal cyst nematode within 5 days at 13°C (Kerry and Crump, 1980). The zoospores require water filled soil pores for motility and infection is limited to periods following rainfall. The obligate parasites have proved difficult or impossible to culture *in vitro* (Graff and Madelin, 1989). All three species produce thick walled resting spores that survive in soil when female nematodes are absent. *Nematophthora gynophila* is widespread in cereal cyst nematode infested soils in northern Europe and occurs together with *V. chlamydosporium*; both species are causal agents of the decline of this pest species.

Predators

Many predators have been reported to feed on both sedentary and active stages of nematodes. These have been extensively reviewed (Stirling, 1991) and include protozoa, turbellarians, tardigrades, nematodes and arthropods. The role of predators in the regulation of nematode populations is unclear but in some soils, especially in natural habitats, predator populations may be large. Research on many predator species has been limited and restricted to laboratory observations because of the difficulties of mass production. Yeates and Wardle (1996) concluded that predatory nematodes might have an important role in nutrient cycling in soil but that they were unlikely to reduce nematode herbivory and increase crop yields. Detailed studies on mites demonstrated differences in their preference for nematodes as prey but all are omnivorous and feed on both parasitic and free-living nematode species. (Walter, Hunt and Elliot, 1987). Some mite species completed their life cycle more rapidly and laid more eggs on a diet that included nematodes (Walter and Ikonen, 1989). Although it is difficult to generalise, the lack of specificity, difficulties of mass production and the inability of many arthropod predators to pursue nematodes in the pore spaces of all but the surface layers of soil make predators unlikely candidates for biological control agents. However, their impacts on nematodes in natural habitats such as the shortgrass steppe need further study (Walter, Hunt and Elliot, 1988).

Host Recognition and Infection Processes

Studies of the infection processes and host ranges of nematophagous fungi and bacteria based on light and electron microscopy and bioassays are increasingly supported by biochemical, physiological, immunological and molecular techniques. These methods have identified key compounds that may lead to the development of novel nematicides or proteins that could be expressed in transgenic plants to provide nematode resistance. For example, an ovicidal factor produced by a strain of *Pseudomonas aureofaciens* has been characterised in terms of its amino acid and nucleic acid sequences (Wechter and Kluepfel, 1997) and enzymes involved in the infection processes of egg parasitic fungi are similarly being characterised (Segers *et al.*, 1998). Several novel oligosporin antibiotics have been characterised from *A. oligospora* (Anderson, Jarman and Rickards, 1995). Also, many research groups have conducted screens of nematophagous fungi and bacteria in order to identify the structures of compounds produced *in vitro* that may have nematicidal or nematostatic properties. The role of such compounds *in vivo* is usually not known but wood rotting *Basidiomycota*, such as *Pleurotus ostreatus*, secrete droplets of a potent toxin that rapidly immobilises nematodes (Thorn and Barron, 1984) and has the structure of trans-2-decenedioic acid (Kwok *et al.*, 1992).

Infection of nematode eggs

The infection of root-knot and cyst nematode eggs by *P. lilacinus*, *V. chlamydosporium* and *V. suchlasporium* has been studied using TEM and SEM (Morgan-Jones, White and Rodriguez-Kabana, 1983; Lopez-Llorca and Claugher, 1990; Segers *et al.*, 1996). These fungi infect by producing an appressorium laterally or at the tip of hyphae growing across the egg surface, although this may not be essential for infection by *P. lilacinus* (Dunn *et al.*, 1982). The form and number of the appressoria and the extent of growth of *V. chlamydosporium* on the egg surface were affected by the nematode host and suggest that cues from the egg influence fungal morphology and development (Segers *et al.*, 1996). Typically, penetration of the egg shell is considered to be due to both enzymatic degradation and physical forces. Chitinase activity has been recorded in all egg parasitic fungi tested and Dackman, Chet and Nordbring-Hertz (1989) correlated chitinase and protease activity with the extent of parasitism of *Heterodera schactii* eggs. The eggshell of most nematodes has an outer vitelline membrane and all have chitin and lipid layers which differ in thickness (Perry and Trett, 1986). A 32 kDa protease was purified from *V. suchlasporium* (Lopez-Llorca, 1990) and its role in infection confirmed by immunolocalisation (Lopez-Llorca and Robertson, 1992). A subtilisin-like serine protease of similar size was partially characterised from the closely related *V. chlamydosporium* (Segers *et al.*, 1998) and its role in the degradation of the vitelline membrane on the surface of the eggshell and the exposure of the chitin layers was demonstrated. Bonants *et al.* (1995) partially characterised a similar serine protease in *P. lilacinus*. Vitellin and the presence of eggs induced the enzyme when *P. lilacinus* was grown on minimal media and the enzyme was repressed in the presence of glucose; immature eggs exposed to the enzyme failed to develop but it increased the hatch of more mature eggs. Different isolates of *V. chlamydosporium* produced different subtilisins and the variation in the enzyme may reflect the different niches occupied by this facultative parasite in soil (Segers *et al.*, 1998). However,

these authors reported considerable homology amongst the subtilisins produced by both nematophagous and entomopathogenic fungi.

The sterile fungus designated ARF 18, which is a member of the Ascomycotina, colonised cysts of *Heterodera glycines* from sclerotium-like structures which developed on the cyst surface but eggs were infected from an appressorium (Kim, Riggs and Kim, 1992). Inside the egg a post infection bulb is usually formed from which the egg is colonised. *Catenaria anguillulae* also infects nematode eggs after zoospores encyst on their surface and produce a germ tube that penetrates the shell (Wyss, Voss and Jansson, 1992). The lipid layer was a barrier to infection by this fungus but, once penetrated, other zoospores showed taxis towards the egg and swarmed on its surface; this observation provided supporting evidence that the lipid layer was the layer that affected eggshell permeability (Perry and Trett, 1986). The rapidity with which *C. anguillulae* killed the embryo suggested that toxins may be involved. Toxins may also be involved in the infection process of *V. chlamydosporium* (Morgan-Jones, White and Rodriguez-Kabana, 1983) and Sikora, Hiemer and Schuster (1990) suggested that this fungus was a perthophyte in nutrient rich conditions, only colonising eggs after its toxin had killed them. Although bioactive compounds have been isolated from *in vitro* cultures of *V. chlamydosporium* (Kerry, unpublished data) and *V. suchlasporium* (Lopez-Llorca, Moya and Llinares, 1993), their role in soil is not clear.

Attachment to, and penetration of, nematode cuticle

Unless toxins are involved to immobilise hosts, those microbial agents that attack active stages must be capable of adhering to the surface of the nematode in an aqueous environment. The surface coat proteins of nematodes may be altered at moulting, during development or in response to changes in the environment (Grenache *et al.*, 1996). This coat is labile and, if removed by chemical treatment, it is replaced after a recovery period in favourable conditions (Speigel *et al.*, 1997). These changes are considered to enable nematodes, especially parasitic types, to adapt to a changing environment and avoid host defences. However, such changes may also provide protection from the host recognition processes of microbial parasites in soil. Certainly, the surfaces of spores of *P. penetrans* are very variable, even those derived from a single infected female (Davies, Redden and Pearson, 1994); such variation may enable the parasite to cope with the dynamics of the surface coat of its host. The parasporal fibres are responsible for the firm attachment of spores to nematode cuticle (Sayre and Wergin 1977; Persidis *et al.*, 1991) and their host recognition processes have been studied using TEM, biochemical techniques including enzyme and lectin treatments, and immunology. The specificity of attachment has been widely reported and, in extreme cases, spores from one isolate of the bacterium may adhere to only individual populations of the host (Stirling, 1985). Differences in the proteins found on the surface coats of different root-knot nematode species may account for differences in spore attachment (Davies, Redden and Pearson, 1994). The spore has an adhesin on its surface and N-acetylglucosamine has been implicated in binding to a glycoprotein on the surface of the nematode (Davies and Danks, 1992). However, N-acetylglucosamine is not responsible for host specificity and, indeed, is present on all spores of *P. penetrans* (Davies and Redden, 1997) and also *P. ramosa*, which is a parasite of *Daphnia* and does not attach to nematodes (K.G. Davies, pers. comm.) Attachment involves hydrophobic interactions with a fibronectin-like receptor in the cuticle (Davies, Afolabi and O'Shea, 1996). The inherent specificity of *P. penetrans* means that careful selection of compatible isolates is necessary if the bacterium is applied to soil to control nematode pests that often occur as mixed populations of species. Spores of *P. penetrans* did not attach to infective larvae of several species of animal-parasitic nematodes, including entomopathogenic nematodes, or to wild type and *srf* mutants of *C. elegans* (Mendoza de Gives *et al.*, 1999).

Spores of some populations of *Pasteuria* germinate and infect the motile stages of nematodes, including the second stage larvae of cyst nematodes (Davies *et al.*, 1990; Sturhan *et al.*, 1994). However, on root-knot nematodes infection takes place in the root when the juvenile is sedentary. Plant cues may act as a signal for spore germination but these have not been demonstrated.

The attachment processes and infection of nematophagous fungi (Dijksterhuis *et al.*, 1994; Tunlid, Jansson and Nordbring-Hertz, 1992) have been observed in some detail using TEM but, apart from the much studied *A. oligospora* (Nordbring-Hertz, 1988), little is known of the biochemical and physiological events associated with parasitism. In nematode trapping fungi, the factors which cause the switch from the saprophytic to the parasitic phase remain controversial and appear in some species to be associated with a need to avoid competition in times of nutrient abundance and in others to assure long-term survival when nutrients become limiting. In Petri dish assays, nematode populations were significantly smaller in media inoculated with *A. oligospora* or *Monacrosporium cionapagum* and the competitive saprophyte, *Trichoderma* sp., than if either trapping fungus was added alone (Quinn, 1987). Parasitism begins with the induction of trap formation which, *in vitro*, may be caused by the presence of active nematodes, environmental conditions or specific peptides and amino acids. Significant numbers of traps may be formed spontaneously in some species such as *A. oligospora* and may develop directly from conidia (Dackman and Nordbring-Hertz, 1992). Jaffee, Muldoon and Tedford (1992) demonstrated that three trapping fungi produced constricting rings (*A. dactyloides*), adhesive knobs (*Monacrosporium ellipsosporum*) and two-dimensional networks (*M. cionopagum*) directly from colonised hosts where there was no competition for the rich nutrient source. They suggested that trapping fungi probably grew little in soil and were dependent on nematodes as a nutrient source and that factors influencing the induction of traps may have little relevance in soil.

Extracellular adhesins are produced on the surfaces of spores, appressoria and traps of nematophagous fungi (Tunlid, Jansson and Nordbring-Hertz, 1992) and are essential for infection (Tunlid, Jansson and Nordbring-Hertz, 1992). The adhesive layer has a fibrillar structure with residues of neutral sugars, uronic acid and proteins (Tunlid, Johansson and Nordbring-Hertz, 1991). Initial contact with the host cuticle may be followed by interactions with specific receptors, reorganisation of surface polymers to strengthen the adhesins, changes in morphology and the secretion of specific enzymes. These processes and the structures involved have been extensively reviewed (Tunlid, Jansson and Nordbring-Hertz, 1992; Dijksterhuis *et al.*, 1994). In trapping fungi, there is evidence of carbohydrate-lectin interactions in the host recognition process and a lectin has been localised in the cell wall of the trap of *A. oligospora* (Borrebaeck, Mattiasson and Nordbring-Hertz, 1984, 1985). Although different hapten sugars applied to traps of different species may inhibit attachment and infection, there is little evidence that these fungi are host specific. However, nematode species differ in the efficiency with which they are trapped by different fungi (Jaffee and Muldoon, 1995). Studies of *srf* mutants of *C. elegans* have shown that mutational changes that alter the cuticle surface affect the susceptibility of nematodes to being captured by trapping fungi (Mendoza de Gives *et al.*, 1999). Those fungi, such as *A. dactyloides* and *Dactylaria brochopaga*, that produce constricting ring traps do not rely on adhesins to facilitate infection. In response to nematode contact on the inside of the ring trap, the three cells that form the ring and whose contents are at a high osmotic pressure, rapidly imbibe water from the environment and bulge inwards to grip the nematode.

A narrow penetration tube develops from the trap cells and the nematode cuticle is breached by a combination of physical force and enzymatic degradation. A range of protease enzymes has been identified from *A. oligospora* (Tunlid and Jansson, 1991) but none is specific to the trapping device (Persson and Friman, 1993). Inside the nematode an infection bulb is formed from which trophic hyphae develop. As the conidiospores of the endophytic fungi have much less contact with the cuticle surface than the trapping structures, infection by these species is dependent on the development of an appressorium which adheres tightly to the cuticle surface and opposes the mechanical forces that occur during infection. In general, these fungi have narrower host ranges than trapping fungi but the factors affecting host specificity are poorly understood.

Suppressive Soils

Most nematode populations are regulated by the natural enemy community (Stirling, 1991) but only in suppressive soils is its impact of practical significance (Table 19.1). The suppression of nematode multiplication on intensively cropped susceptible hosts by biotic factors in soil was first demonstrated by Gair, Mathias and Harvey (1969) with the cereal cyst nematode.

Table 19.1. Characteristics of nematode-suppressive soils that affect their exploitation in the biological control of plant-parasitic nematodes

Limitations:

- pest control is often slow to establish.

- control is often specific to one nematode pest species.

- significant control has usually been achieved only in monoculture or perennial cropping systems.

- the natural enemy community is often diverse.

- the manipulation of the natural enemy community to increase natural control is often difficult to achieve with practical treatments.

Advantages:

- suppressive soils have provided the most sustainable methods of nematode management in intensive agriculture.

- suppressive soils provide a valuable source of potential biological control agents.

- soil amendments and some crop cultivars may be used to alter microbial communities in the rhizosphere to the detriment of nematode pests.

Suppressive soils contain microbial communities which have increased in size to densities that prevent nematode populations multiplying by reducing the development of larvae, fecundity of females and the survival of all stages. These natural enemy communities tend to build up relatively slowly over 4–5 years and usually under perennial crops or crops grown in monocultures. Hence, suppressive soils and the associated decline of specific nematode infestations have only been demonstrated conclusively in long term studies of >5 years. Although there is much circumstantial evidence that nematophagous bacteria and fungi may regulate nematode infestations, a large proportion of the population (>90%) must be killed to prevent population increase and few studies have demonstrated such control. For example, *Dactylella oviparasitica* may be responsible for the failure of root-knot nematodes to multiply on peaches on Lovell rootstocks in California (Stirling, KcKenry and Mankau, 1979) but, as only 20–60% of the eggs were parasitised on a single sampling occasion, the data are inconclusive. However, the amounts of parasitism were probably underestimated as all eggs within an eggmass were usually infected by the fungus and individual eggs were destroyed within 9 days at 27°C in laboratory tests (Stirling, 1979). To demonstrate that microbial agents are the cause of the decline of nematode infestations requires long-term and detailed observations.

Often, the causal agents of nematode suppression have tended to be only one or two species of nematophagous fungi. In intensive cropping systems, particular species of natural enemy appear to be selected from the community in the continued presence of a nematode host. However, in some soils several natural enemies may be involved but it is difficult to demonstrate their significance in the regulation of nematode populations. For example, a wide range of natural enemies occurs in citrus orchards infested with *Tylenchulus semipenetrans* in California (Stirling and Mankau, 1977) and with *Radopholus similis* in Florida (Walter and Kaplan, 1990). The factors which retain a diverse natural enemy community or select specific species are important for our understanding of the biological control of nematodes. Suppressive soils are induced as nematodes are usually abundant in the early stages of the monoculture and cause much damage to crops before a suppressive state is reached as a result of the build up of natural enemy densities. Suppression can be demonstrated as a biotic factor by using heat treatment and selective biocides to remove the causal agents, by introducing additional nematodes to soil and detecting no increase in multiplication of the population and by transferring the phenomenon to non-suppressive soil by inoculation (Linderman *et al.*, 1983). A number of techniques have been developed to measure the effect of the natural enemy community on nematode populations (see Kerry, 1987; Stirling, 1991) and the term 'antagonistic potential' has been used as a measure of suppressiveness (Sikora, 1992). To be of practical value the equilibrium population that establishes between the nematode pest and the natural enemy community should be below the economic threshold. Detailed research of the situations in which the natural enemy community has provided effective nematode control will lead to the identification of potential biological control agents and an understanding of the key factors and dynamics affecting their efficacy against specific nematode targets.

Pasteuria penetrans and root-knot nematodes

Mankau (1980) observed that root knot nematode populations that remained small despite the intensive cropping of susceptible vegetable crops in W. Africa contained second-stage larvae which were encumbered with spores of the bacterium *Pasteuria penetrans*. In S. Australia, Stirling and White (1982) observed that *P. penetrans* was abundant in vineyards >25 years old and there were fewer root-knot nematodes compared with nematodes densities in vineyards <10 years old where the bacterium could not be detected. Although this is only circumstantial evidence of the importance of the bacterium on nematode population regulation, further studies on the vineyard soils indicated that it was having a significant effect (Stirling, 1984; Bird and Brisbane, 1988). Presumably, the long time taken for *P. penetrans* to establish in these vineyard soils was due to the lack of cultivation operations in vineyards, which would limit the spread of inoculum in this perennial crop. There have been similar reports of large proportions of nematode populations encumbered of various species with spores on different crops throughout the world but few have been supported with data on the effects on nematode multiplication (Chen and Dickson,

1998). Even though 70% of *Belonolaimus longicaudatus* were spore-encumbered, populations on Bermudagrass were not significantly reduced (Giblin-Davis, McDaniel and Bilz, 1990). The most convincing evidence of the natural build up of *P. penetrans* and the development of suppression is provided from *M. arenaria* infested soils intensively cropped with peanuts in Florida (Minton and Sayre, 1989; Oostendorp, Dickson and Mitchell, 1991; Weibelzahl-Fulton, Dickson and Whitty, 1996).

Cyst nematode decline phenomena

Populations of the cereal cyst nematode, *Heterodera avenae*, usually decline after the fourth or fifth cereal crop to non-damaging infestations because females and eggs of the nematode are parasitised in the rhizosphere by *Nematophthora gynophila* and *Verticillium chlamydosporium* (Kerry, Crump and Mullen, 1982). The decline phenomenon is widespread in northern Europe and has been extensively studied (Kerry, 1975, 1982, 1988; Kerry and Crump, 1998). In detailed field studies, about 95% of the females and eggs were destroyed by the fungal parasites in suppressive soils (Kerry, Crump and Mullen, 1982). This control was removed by the application of formalin, which proved to be an effective fungicide but weak nematicide if applied in early spring before significant numbers of nematode eggs had hatched. Nematode densities consistently increased in soils treated with the partial soil sterilant compared to untreated soils if the fungi were abundant but there was no effect of the treatment if the fungi were inactive or absent. (Kerry, Crump and Mullen, 1980). Changes in the densities of spores of both fungi in soils were inversely correlated with the abundance of the nematode, and fungal populations required at least 3 years to reach densities which controlled the nematode on susceptible crops (Kerry and Crump, 1998). In these studies, populations of the nematode, which reached 200 eggs g^{-1} soil in some microplots, decreased to <5 eggs g^{-1} soil in 4 years despite the continuous cropping of susceptible spring barley.

Similar decline phenomena caused by the activities of nematophagous fungi have been reported in fields infested with *H. schachtii* (Crump and Kerry, 1986; Heijbroek, 1983; Thielemann and Steudel, 1973; Thomas, 1982) and *H. glycines* (Crump, Sayre and Young, 1983; Hartwig, 1982) but equilibrium populations are above the economic threshold and additional control measures are required to prevent significant yield loss. Nematode suppressive soils have developed in small plots infested with a range of cyst species, including potato cyst nematodes (Crump, 1998).

Other decline phenomena

Criconemella xenoplax is a significant problem on peach trees in the USA but populations of the nematode often collapse and *H. rhossiliensis* has been implicated in this phenomenon (Jaffee and Zehr, 1982). However, many of the data were collected from controlled laboratory tests and Jaffee and McInnes (1991) concluded that the fungus was only a weak regulator of populations of the nematode. High levels of parasitism only occurred when the nematode was abundant and populations remained at densities that caused significant damage to peach trees. In glasshouse tests, suppressive soil treated with steam at 65°C supported a nine-fold increase in nematode multiplication (McInnes, Kluepfel and Zehr, 1990) and the rhizosphere bacterium, *P. aureofaciens*, was identified as another potential agent in the decline of the nematode. The gene responsible for production of the ovicidal factor of *P. aureofaciens* is being cloned and sequenced (Wechter, Leverentz and Kluepfel, 1997). *Hirsutella rhossiliensis* occurs widely in nematode populations and may regulate populations of *H. schachtii* (Muller, 1982; Jaffee and Muldoon, 1989) and potato cyst nematodes (Velvis and Kamp, 1995) in some soils and is worthy of further investigation (Kerry and Jaffee, 1997).

Soils suppressive to nematodes have been reported in the Chinampa region of central Mexico in which *Paecilomyces marquandii* and rhizosphere bacteria are important egg parasites and antagonists respectively (Zuckerman *et al.*, 1989; Marban-Mendoza *et al.*, 1992). Plant species such as velvet bean (*Mancuna deeringiana*) and castor bean (*Ricinus communis*) are antagonistic to nematodes and have distinct rhizosphere microfloras (Kloepper *et al.*, 1991). The role of the rhizosphere microflora in the antagonism of nematodes is unclear but the rhizospheres of these plants may provide useful organisms with potential as biological control agents.

Diversity and Dynamics of Microbial Populations

Nematode populations are subject to predation and parasitism by a diverse natural enemy community and the nematophagous and antagonistic bacteria and fungi that have been isolated contain much variability within individual species. The importance of this variation in the regulation of nematode populations is unknown. The use of molecular markers to differentiate specific isolates of *V. chlamydosporium* (Arora, Hirsch and Kerry, 1996) will enable the temporal and spatial distribution of different isolates of this fungus in the rhizosphere of plants grown in suppressive soils to be determined. Such information may explain why there is such variation between isolates within the same soil in characteristics considered essential for effective biological control, and why application of a single isolate rarely provides the level of control observed in a naturally suppressive soil.

Few studies have attempted to quantify the relationships between nematodes and their natural enemies and there is a dearth of useful methods to quantify microbial agents in soil. Semi-selective media and dilution plating, including the use of most probable number analysis (Eren and Pramer, 1965; Stirling, McKenry and Mankau, 1979), have been used to quantify several nematophagous fungi but all methods have their limitations and it is not advisable to rely on one technique. For example, changes in the abundance of the nematophagous fungus, *Verticillium chlamydosporium*, were probably related to the production of conidia and not to the increases in extent of mycelial colonisation of the rhizosphere that are correlated to increased parasitism of nematode eggs (de Leij, Dennehy and Kerry, 1992). Also, changes in the activity of the fungus associated with the development of resting spores could not be detected using dilution plate techniques whereas biochemical methods based on estimating the ATP content of soil were more sensitive (Kerry *et al.*, 1993). It is therefore not possible to relate changes in abundance to changes in vegetative growth using dilution plates and methods to visualise fungal hyphae in the rhizosphere are required. The use of immunological techniques has proved successful for the detection of fungi in soil (Thornton, Dewey and Gilligan, 1994) and fungi transformed with reporter genes such as GUS (*β*-D-Glucuronidase) have been used to monitor growth in plants (Couteaudier *et al.*, 1993) and could be used to study the colonisation of nematophagous fungi in the rhizosphere.

Although the population dynamics of cereal cyst nematode in suppressive soils were first modelled by Perry (1978), there have been few attempts to develop mathematical tools to study the interactions between nematodes and their natural enemies. It appeared that the decline of *H. avenae* was temporally density dependent in that populations of the nematode were initially large and supported the build up of fungal densities that eventually regulated the population. Density dependent interactions are common between hosts and their parasites and provide a feedback mechanism for regulating populations; such interactions were reported for the obligate parasite, *P. penetrans* (Ciancio, 1995). Jaffee *et al.* (1992) modified basic models of the effects of parasites on invertebrate populations developed by Anderson and May (1981) and this group has pioneered research on quantifying such interactions for plant nematodes. In studies of *H. schachtii* and *H. rhossiliensis*, they demonstrated density dependent parasitism, host threshold densities and low transmission rates except at very high densities in soil. As a consequence, levels of parasitism change slowly, epidemics do not have explosive dynamics and natural control is unlikely to develop within a single growing season. However, transmission rates for *H. rhossiliensis* may be much lower than for other fungi as spores are only adhesive whilst still attached to the phialides close to the infected cadaver (McInnes and Jaffee, 1989). Atkinson and Dürschner-Pelz (1995) predicted that the maximum time required to replace spores that adhered to *H. schachtii* larvae and maintain the size of the reservoir in soil was 63 weeks for *H. rhossiliensis* but only 4.5 weeks for *Verticillium balanoides*. Also, *Ditylenchus dipsaci* larvae supported greater (×20) spore production of *V. balanoides* than *Panagrellus redivivus* and they suggested that this fungus responded to the pulse of *D. dipsaci* entering the soil after harvest, established an epizootic and caused the overwintering decline of this nematode. The poor production of spores in *P. redivivus* (<900 individual^{-1}) resulted in low transmission rates and no epizootic. Density dependent dynamics have also been reported for trapping fungi (Jaffee and Muldoon, 1995) and *V. chlamydosporium* (Kerry and Crump, 1998)

Table 19.2. Influence of host plant on the successful deployment of biological control agents against plant-parasitic nematodes

- Plant susceptibility affects the rate of nematode development and multiplication which may affect the efficacy of control agents.
- Plant tolerance to nematode attack affects the level of control necessary to prevent economic losses.
- Facultative nematophagous fungi and bacteria may depend on nutrients in the rhizosphere to increase their abundance on roots.
- Nematophagous fungi and bacteria are involved in signalling processes in the rhizosphere of nematode-infected plants that affect their activity as biological control agents.

and may indicate that these fungi are more dependent on nematodes than their status as facultative parasites may have suggested. Population models to study fungal colonisation in the rhizosphere (Gilligan and Kleczkowski, 1997) and biological control interactions (Kleczkowski, Bailey and Gilligan, 1996) may be adapted to examine the growth of nematophagous fungi in the rhizosphere and their impact on nematode populations.

Biological Control Strategies

The use of microbial parasites and pathogens as biological control agents for nematodes has been widely reviewed (Stirling, 1991; Kerry, 1993; Waller and Faedo, 1996; Kerry and Jaffee, 1997; Larsen *et al.*, 1997; Dickson *et al.*, 1994). A number of products have been developed but none has provided consistent or adequate control. Biological control of soil-borne diseases has had most success in situations where the target site is readily accessible and can be treated with inundative treatments and when short-term protection results in significant yield benefits (Deacon, 1991). Most plant-parasitic nematode pests present more intractable control problems and there is usually a need to protect developing root systems for several weeks. Control of animal-parasitic nematodes in faeces of cattle, sheep and horses using *Dactylella flagrans* may be more readily developed and much progress has been made in recent years. Too often the development of biological control agents has depended on empirical tests (Stirling, 1991) but careful selection and an understanding of the factors affecting the epidemiology of the agent and pest are essential for the development of successful strategies utilising biological agents. The crop plant may have a crucial effect in determining the successful deployment of biological control agents (Table 19.2).

Selection of biological control agents

There are several stages in the process of selection of biological control organisms (Kerry, 1990). Initially, selection of the type of agent will depend on whether the stage of host to be targeted is active or sedentary and on whether the strategy is to reduce the damage caused by the nematode or its multiplication (Figure 19.3). Simple, high-throughput screens are required as all groups and species of potential biological control agents have proved variable with significant differences in key factors affecting their efficacy. Nematode suppressive soils are likely to be useful sources for the isolation of potential control agents (Kerry, 1990; Dicklow *et al.*, 1993) and ecological studies supported by tests in the laboratory are required to determine the temperature, pH, moisture and nutrient conditions needed by the organism. In general, microbial agents have been isolated from parasitised hosts or, in the case of antagonists, from the rhizosphere or plant tissue. Few isolates of rhizobacteria collected at random from the rhizosphere have antagonistic activity. Simple screens using suspensions of *C. elegans* or second stage larvae of cyst and root-knot nematodes to test the nematicidal activity of metabolites from the rhizobacteria were the most efficient methods of selecting promising strains (Becker *et al.*, 1988; Oostendorp and Sikora, 1989). Although the pathogenicity of trapping fungi and egg parasites can be evaluated in tests on agar, a screen to assess growth in the rhizosphere may also be required; only isolates of *V. chlamydosporium* that extensively colonised the rhizosphere were able to control root-knot nematodes (de Leij and Kerry, 1991). The need to include plants early in the screening process will delay and increase the costs of selecting candidate organisms. Isolates are selected not only on the basis of their activity against nematodes; other factors, such as ease of production, host range, development of resting structures and growth in the rhizosphere must be considered before testing in soil. A tiered, screening process was designed for the selection of isolates of *V. chlamydosporium* collected from colonised nematodes or suppressive soils. This involved tests for growth in the rhizosphere, chlamydospore production and pathogenicity and enabled almost 90% of the isolates to be discarded before further evaluations were made in the glasshouse (Kerry, 1998).

Mass production, formulation and application of selected agents

Fewer than ten organisms have been tested for control of nematodes in the field and, as a consequence, limited efforts have been made to optimise methods for their production, formulation and application. However, much can be learnt from the development of other microbial biological control agents (see Gareth Jones, 1993). Some organisms such as *P. penetrans*, have considerable potential but the lack of suitable *in vitro* methods of mass production (Williams *et al.*, 1989; Bishop and Ellar, 1991) have limited its development. Also, *in vivo* methods (Stirling and Wachtel, 1980; Serracin *et al.*, 1994) will only be relevant for small scale farmers and gardeners. Most rhizobacteria and nematophagous fungi are able to grow on artificial media, including in liquid fermentation, but resting structures such as chlamydospores are often not produced in large numbers in submerged culture. These resting structures

often enable the agents to be readily handled and provide reasonable shelf lives without the complex formulations required for organisms applied as vegetative cells, hyphae or conidia. Chlamydospores of *V. chlamydosporium* applied to soil in aqueous suspensions enable the fungus to establish in the rhizosphere of a range of plant species and those of *D. flagrans* are essential to ensure the viability of this fungus during passage through the host gut (Larsen *et al.*, 1994). The large bulk of soil (2500t ha^{-1}) that may need to be treated for control of plant-parasitic nematodes and the large densities (10^3–10^6 g^{-1} soil) required for effective nematode control (Kerry, 1998) are likely to make broadcast treatments uneconomic and the restricted placement of inoculum essential.

Rhizobacteria and endophytes have the advantage that they may be applied as seed treatments and will proliferate and spread in the rhizosphere or within the root, protecting the plant from nematode invasion (Oostendorp and Sikora, 1989). Bacterial suspensions in 1% methyl cellulose have been used to deliver 10^6–10^8 cells seed^{-1} and such applications or bare-root dips for transplanted crops are the simplest treatments for biological control agents (Becker *et al.*, 1988). Perlite granules impregnated with suspensions of rhizobacteria or fungal conidia in 1% methyl cellulose have been applied to soil in low-pressure drip irrigation systems (Bahme *et al.*, 1988). Several granular formulations based on alginates have been used (Kerry *et al.*, 1993; Lackey, Jaffee and Muldoon, 1993; Stirling and Mani, 1995) to apply fungi that are able to colonise the soil and rhizosphere. However, most nematophagous fungi have limited abilities to colonise non-pasteurised soil and, as with obligate parasites that do not proliferate, successful establishment in the rhizosphere will require thorough mechanical incorporation.

Several nematophagous fungi have been added as active mycelia growing on colonised media that are often waste products. These applications provide an energy base to help establish the fungi in soil but they may attract competitive saprophytes that reduce proliferation of the parasite. As a consequence, nematode control is variable and any reductions in nematode populations may also be due to the soil amendment effect of the organic matter as well as any direct effect of the fungus (Stirling, 1991). In general, the combined use of a soil amendment and a biological control agent has involved large amounts of organic matter (>1 t ha^{-1}) and appears to be an inefficient method of application which could only be considered for use on small areas near to the site of production.

Nematode trapping fungi were the first to be developed as commercial products and were applied as fresh mycelium on organic substrates (Cayrol, 1983) but the control was inconsistent and the inoculum difficult to handle. The use of these fungi for the control of animal-parasitic nematodes may be more successful as inoculum can be added with the host's food. Isolates of *D. flagrans* have been selected for survival through the gut of ruminant and non-ruminant domestic animals. Viable chlamydospores are able to establish networks of traps in the faeces of sheep and cattle and significantly reduce the numbers of infective larvae of species such as *Haemonchus contortus* (Mendoza de Gives *et al.*, 1998) and *Ostertagia ostertagi* (Gronvold *et al.*, 1993). As a result, the numbers of infective larvae that contaminate the surrounding sward and infect other grazing hosts are fewer.

Integrated Control Strategies

Two approaches have been used for the exploitation of nematode natural enemies: the use of methods to increase the activity of the indigenous flora and fauna or the application of selected organisms as biological control agents. In either approach it has proved necessary to use additional control measures, as biological agents on their own rarely provide control levels that can be practically exploited (Table 19.3). In practice, the manipulation of the indigenous antagonists has been largely restricted to the use of soil amendments and crop rotation (Sikora, 1992). Chitin soil amendments have been used to increase chitinolytic activity within the soil microflora, especially actinomycetes, and have significantly reduced populations of root-knot nematodes (Rodriguez-Kabana and Morgan-Jones, 1987; Spiegel, Cohn and Chet, 1986; Spiegel, Chet and Cohn, 1987). However, the amount of material required and its cost will usually restrict the use of such materials unless suitable green manure crops can be incorporated into the cropping cycle (Schlang, Steudel and Muller, 1988).

The most promising isolates of rhizobacteria have reduced invasion by about 60–70% (Sikora, 1988) and, as a consequence, are unlikely to reduce significantly nematode multiplication (Oostendorp and Sikora, 1989). Their significance in the reduction of nematode damage to plants will depend on their ability to reduce nematode invasion for a sufficient period of time (2–3 weeks) when soil infestations are above the damage threshold. At present, these bacteria have not been evaluated at a range of nematode densities. Also, it seems unlikely that seed treatments can provide sufficient inoculum to protect roots for more than a few weeks and their use on tolerant crop cultivars may be necessary. Endophytic bacteria and fungi, especially mycorrhizae that colonise roots applied at transplanting, may provide control of a range of endoparasitic nematodes and increase the tolerance of perennial tree crops to nematode damage, but such an approach is dependent on cultivar (Pinochet *et al.*, 1996).

Those agents such as *P. penetrans*, *V. chlamydosporium*, and *P. lilacinus* that attack nematode females and eggs should be applied to non-damaging infestations of nematodes as they will not prevent initial damage to susceptible crops. Indeed, *V. chlamydosporium* is not recommended for use against root-knot nematodes on susceptible crops on which large galls are formed (Kerry, 1995a); egg masses remain embedded within large galls and escape infection as they are not exposed to the fungus, which is confined to the rhizosphere. *Verticillium chlamydosporium* should be applied to soil at the time of planting poor hosts for the nematode, so increasing the effectiveness of these plants in reducing nematode infestations before the next susceptible crop is grown in the rotation (Kerry and Bourne, 1996). Other measures successfully used in combination with biological control agents include partial soil sterilisation (B'Chir, Horrigue and Verlodt, 1983) and solarisation (Walker and Wachtel, 1988; Tzortzakakis and Gowen, 1994). Longer term control of multivoltine nematode species can be achieved if nematicides are used in combination with biological control agents such as *P. penetrans* (Brown and Nordmeyer, 1985; Tzortzakakis and Gowen, 1994). Kerry (1987) suggested that biological control agents that parasitised nematode females on resistant hosts could slow rates of selection of virulent populations. Although this hypothesis may be generally valid (van Emden, 1999), natural enemies may allow the most fit individuals to survive and increase the rates of selection of resistance breaking individuals (Gould, Kennedy and Johnson, 1991).

The use of biological control agents, even in integrated strategies, still requires the production of reliable data to demonstrate efficacy in the field and commercial development will only be achieved if successful agents can be produced cost effectively. The conditions for release of agents are becoming increasingly stringent and some countries will only allow indigenous strains to be used. There is also a need to assess the impacts of biological control agents on non-target organisms and to develop methods to monitor them after their release. Much research and development are still required before even the most studied organisms can be applied on a large scale. The development of nematode resistant transgenic crops has attracted much interest within academia and industry as the most likely approach to reduce dependence on nematicides. It remains unlikely that plant resistance (conventional or transgenic) will provide sustainable nematode control and there will

Table 19.3. Integration of biological control agents (bcas) with other management strategies

Methods to reduce nematode populations*:

- Crop rotation with poor or resistant hosts.
- Combined use with nematicides.
- Application of bcas after solarisation or steam sterilisation.

Methods to increase microbial activity:

- Soil amendments[†] and green manure crops.
- Selected plant cultivars with antagonistic rhizosphere microflora.

* Methods that reduce nematode densities in soil may reduce the efficacy of obligate biological control agents.
† Soil amendments must be carefully evaluated as they may directly reduce nematode infestations in soil and also reduce the effectiveness of some bcas by increasing competition with the general soil microflora.

be a continuing need to integrate control measures in systems that include biological agents.

Concluding Comments

The soil has abundant natural enemies to pests and diseases. There are many nematodes with the ability to kill insect pests. There are numerous organisms, including fungi and bacteria, which can kill plant-parasitic nematodes. Isolating these organisms is relatively straightforward but the question is, 'In the midst of all these natural enemies, how or why do these plant-parasitic nematodes and insects become pests?' In general, the rate of loss of entomopathogenic nematodes after application is so rapid that it is unlikely that the decrease in their numbers is due to the activities of nematophagous organisms (Kerry, 1995b). The diversity of natural enemy communities makes the rational selection of potential biological control agents a high priority. It is noteworthy that the infective stages of insect-parasitic nematodes survive in the same environment as plant-parasitic nematodes and so are potentially subjected to the same enemies. Clearly, we do not understand their population biology and their temporal and spatial dynamics sufficiently well to manipulate the organisms efficiently. On their own, they appear to be inefficient and rarely control pest populations below damage thresholds. Nevertheless, some entomophilic nematodes and nematophagous bacteria and fungi have provided long-term natural control. However, the integration of biological control agents with other management methods is usually essential for both obligate and facultative parasites and pathogens. Successful products based on entomopathogenic nematodes have been produced but there is still much to be learnt to optimise these technologies. It remains to be seen whether such technologies are relevant for the development of control agents for plant- and vertebrate-parasitic nematodes and much depends on the availability of efficient, cost-effective, mass production methods. Research on the infection processes and pathogenesis of microbial agents has already led to the identification of a number of bioactive compounds that may have application in crop protection. Studies of the biological control of insect and nematode pests has moved from a descriptive and empirical approach to one that is more mechanistic and, as a consequence, selected natural enemies are more readily manipulated. It seems likely that these agents will increase in importance in future pest control.

References

Abad, P., Burnell, A., Laumond, C., Boemare, N. and Coudert, F. (editors) (1998) *EUR 18261 COST 819 Genetic and molecular biology of entomopathogenic nematodes*. Luxembourg: Office for Official Publications of the European Communities, 163pp.

Adams, B.J., Burnell, A.M. and Powers, T.O. (1998) A phylogenetic analysis of *Heterorhabditis* (Nemata: Rhabditidae) based on internal transcribed spacer 1 DNA sequence data. *Journal of Nematology*, **30**, 22–39.

Anderson, M.G., Jarman, T.B. and Rickards, R.W. (1995) Structures and absolute configurations of antibiotics of the Oligosporon group from the nematode-trapping fungus *Arthrobotrys oligospora*. *The Journal of Antibiotics*, **48**, 391–398.

Anderson, R.M. and May, R.M. (1981) The population dynamics of microparasites and their invertebrate hosts. *Philosophical Transactions of The Royal Society, London*, **B 291**, 451–524.

Arora, D.K., Hirsch, P.R. and Kerry, B.R. (1996) PCR-based molecular discrimination of *Verticillium chlamydosporium* isolates. *Mycological Research*, **100**, 801–809.

Atkinson, H.J. and Dürschner-Pelz, U. (1995) Spore transmission and epidemiology of *Verticillium balanoides*, an endozoic fungal parasite of nematodes in soil. *Journal of Invertebrate Pathology*, **65**, 237–242.

Bahme, J.B., Schroth, M.N., Van Gundy, S.D., Weinhold, A.R. and Tolentino, D.M. (1988) Effect of inocula delivery systems on rhizobacterial colonization of underground organs of potato. *Phytopathology*, **78**, 534–542.

Barron, G.L. (1969) Isolation and maintenance of endoparasitic nematophagous hyphomycetes. *Canadian Journal of Botany*, **47**, 1899–1902.

Barron, G.L. (1977) *The Nematode-Destroying Fungi*. Guelph: Canadian Biological Publications.

Barron, G.L. (1982) Nematode destroying fungi. In *Experimental Microbiological Ecology*, edited by R.G. Burns and J.H. Slater, pp. 533–552. Oxford: Blackwell Scientific Publications.

B'Chir, M.M., Horrigue, N. and Verlodt, H. (1983) Mise au point d'une méthode de lutte intégrée associant un agent biologique et un substance chimique, pour combattre les *Meloidogyne* sous-abris plastiques en Tunisie. *Mededelingen van de Faculteit Landbouwwetenschappen Rijksuniversiteit, Gent*, **48**, 421–432.

Becker, J.O., Zavaleta-Mejia, E., Colbert, S.F., Schroth, M.N., Weinhold, A.R., Hancock, J.G. and Van Gundy, S.D. (1988) Effects of rhizobacteria on root-knot nematodes and gall formation. *Phytopathology*, **78**, 1466–1469.

Bedding, R.A. (1981) Low cost *in vitro* mass production of *Neoaplectana* and *Heterorhabditis* species (Nematoda) for field control of insect pests. *Nematologica*, **27**, 109–114.

Bedding, R.A. (1984a) Nematode parasites of Hymenoptera. In *Plant and Insect Nematodes*, edited by W.R. Nickle, pp. 755–794. New York: Marcel Dekker.

Bedding, R.A. (1984b) Large scale production, storage and transport of the insect parasitic nematodes *Neoaplectana* spp. and *Heterorhabditis* spp. *Annals of Applied Biology*, **104**, 117–120.

Bedding, R.A., Akhurst, R.J. and Kaya, H.K. (editors) (1993) *Nematodes and the Biological Control of Insect Pests*. Australia: East Melbourne, CSIRO.

Bellotti A.C., Cardona C. and Riis, L. (1997) Burrowing bugs and whitegrubs, major soil pests of food crops in Colombia. In *Soil Invertebrates in 1997. Proceedings of the 3rd Brisbane Workshop on Soil Invertebrates*, edited by P.G. Allsopp., D.J. Rogers and L.N. Robertson, pp. 130–133. Australia: Indooroopilly, Bureau of Sugar Experiment Stations.

Bernard, E.C. and Gwinn, K.D. (1991) Behaviour and reproduction of *Meloidogyne marylandi* and *Pratylenchus scribneri* in roots and rhizosphere of endophyte-infected tall fescue. *Journal of Nematology*, **23**, 520.

Bird, A.F. (1986) The influence of the actinomycete, *Pasteuria penetrans*, on the host-parasite relationship of the plant-parasitic nematode, *Meloidogyne javanica*. *Parasitology*, **93**, 571–580.

Bird, A.F. and Brisbane, P.G. (1988) The influence of *Pasteuria penetrans* in field soils on the reproduction of root-knot nematodes. *Revue de Nématologie*, **11**, 75–81.

Bishop, A.H. and Ellar, D.J. (1991) Attempts to culture *Pasteuria penetrans* in vitro. *Biocontrol Science and Technology*, **1**, 101–114.

Blaxter, M.L., De Ley, P., Garey, J.R., Liu, L.X., Scheldeman, P., Vierstraete, A., Vanfleteren, J.R., Mackey, L.Y., Dorris, M., Frisse, L.M., Vida, J.T. and Thomas, W.K. (1998) A molecular evolutionary framework for the phylum Nematoda. *Nature*, **392**, 71–75.

Boemare, N., Laumond, C. and Mauleon, H. (1996) The entomopathogenic nematode-bacterium complex: biology, life cycle and vertebrate safety. *Biocontrol Science and Technology*, **6**, 333–345.

Boemare, N., Nealson, K.H. and Ehlers, R-U. (editors) (1997) Special issue on nematode symbiosis. *Symbiosis*, **22**, 228 pp.

Boemare, N., Ehlers, R-U., Fodor, A. and Szentirmai, A. (editors) 1996. *EUR 16727-COST 819 Entomopathogenic nematodes: Symbiosis and pathogenicity of the nematode-bacterium complexes*. Luxembourg: Office for Official Publications of the European Communities. 140pp.

Bohan, D.A. and Hominick, W.M. (1997) Long-term dynamics of infectiousness within the infective stage pool of the entomopathogenic nematode *Steinernema feltiae* (Site 76 strain) Filipjev. *Parasitology*, **114**, 301–308.

Bonants, P.J.M., Fitters, P.F.L., Thijs, H., den Belder, E., Waalwijk, C. and Henfling, J.W.D.M. (1995) A basic serine protease from *Paecilomyces lilacinus* with biological activity against *Meloidogyne hapla* eggs. *Microbiology*, **141**, 775–784.

Borrebaeck, C.A.K., Mattiasson, B. and Nordbring-Hertz, B. (1984) Isolation and partial characterization of a carbohydrate-binding protein from a nematode-trapping fungus. *Journal of Bacteriology*, **159**, 53–56.

Borrebaeck, C.A.K., Mattiasson, B. and Nordbring-Hertz, B. (1985) A fungal lectin and its apparent receptors on a nematode surface. *FEMS Microbiology Letters*, **27**, 35–39.

Bourne, J.M., Kerry, B.R. and de Leij, F.A.A.M. (1996) The importance of the host plant in the interaction between root-knot nematodes (*Meloidogyne* spp.) and the nematophagous fungus, *Verticillium chlamydosporium* Goddard. *Biocontrol Science and Technology*, **6**, 539–548.

Brown, S.M. and Nordmeyer, D. (1985) Synergistic reduction in root galling by *Meloidogyne javanica* with *Pasteuria penetrans* and nematicides. *Revue de Nématologie*, **8**, 285–286.

Burnell, A.M., Ehlers, R-U. and Masson, J.P. (editors) 1994. *EUR 15681- Genetics of entomopathogenic nematode-bacterium complexes*. Luxembourg: Office for Official Publications of the European Communities, 287 pp.

Campbell, J.F. and Gaugler, R. (1997) Inter-specific variation in entomopathogenic nematode foraging strategy: dichotomy or variation along a continuum? *Fundamental and Applied Nematology*, **20**, 393–398.

Campbell, J.F., Orza, G., Yoder F., Lewis, E. and Gaugler, R. (1998). Spatial and temporal distribution of endemic and released entomopathogenic nematode populations in turfgrass. *Entomologia Experimentalis et Applicata*, **86**, 1–11.

Cayrol, J.C. (1983) Lutte biologique contre les *Meloidogyne* au moyen d'*Arthrobotrys irregularis*. *Revue de Nématologie*, **6**, 265–273.

Chen, G., Ren, H.F., Luo, L.W., Zhou, S.W., Zhang, Z.L., Lei, S.F., Chen, B.Q. and Jian, H. (1992) Infection process and natural control effect of *Ovomermis sinensis* (Nematoda: Mermithidae) to *Mythimna separata* (Lep: Noctuidae). *Wuyi Science Journal*, **9**, 249–260, In Chinese.

Chen, Z.X. and Dickson, D.W. (1998) Review of *Pasteuria penetrans*: Biology, ecology and biological control potential. *Journal of Nematology*, **30**, 313–340.

Choo, H.Y. and Kaya, H.K. (1994) Biological control of the brown planthopper by a mermithid nematode. *Korean Journal of Applied Entomology*, **33**, 207–215.

Ciancio, A. (1995) Density-dependent parasitism of *Xiphinema diversicaudatum* by *Pasteuria penetrans* in a naturally infested field. *Phytopathology*, **85**, 144–149.

Clayden, I. (1985) Factors influencing the survival of *Ditylenchus dipsaci* in soil. *PhD thesis*, University of Leeds.

Cooke, R.C. (1962a) Behaviour of nematode-trapping fungi during decomposition of organic matter in soil. *Transactions of the British Mycological Society*, **45**, 314–320.

Cooke, R.C. (1962b) The ecology of nematode-trapping fungi in the soil. *Annals of Applied Biology*, **50**, 507–513.

Couteaudier, Y., Daboussi, M-J., Eparvier, A., Langin, T. and Orcival, J. (1993) The GUS gene fusion system (*Escherichia coli β-D-Glucuronidase Gene*), a useful tool in studies of root colonization by *Fusarium oxysporum*. *Applied and Environmental Microbiology*, **59**, 1767–1773.

Crump, D.H. (1998) Biological control of potato and beet cyst nematodes. *Aspects of Applied Biology*, **52**, 383–386.

Crump, D.H. and Kerry, B.R. (1986) Studies on the population dynamics and fungal parasitism of *Heterodera schachtii* in soil from a sugar beet monoculture. *Crop Protection*, **6**, 49–55.

Crump, D.H., Sayre, R.M. and Young, L.D. (1983) Occurrence of nematophagous fungi in cyst nematode populations. *Plant Disease*, **67**, 63–64.

Dackman, C. and Nordbring-Hertz, B. (1992) Conidial traps — a new survival structure of the nematode-trapping fungus *Arthrobotrys oligospora*. *Mycological Research*, **96**, 194–198.

Dackman, C., Chet, I. and Nordbring-Hertz, B. (1989) Fungal parasitism of the cyst nematode *Heterodera schachtii*: infection and enzymatic activity. *Microbial Ecology*, **62**, 201–208.

Davies, K.G. and Danks, C. (1992) Interspecific differences in the nematode surface coat between *Meloidogyne incognita* and *M. arenaria* related to the adhesion of the bacterium *Pasteuria penetrans*. *Parasitology*, **105**, 475–480.

Davies, K.G. and Redden, M. (1997) Diversity and partial characterization of putative virulence determinants in *Pasteuria penetrans*, the hyperparasitic bacterium of root-knot nematodes (*Meloidogyne* spp.). *Journal of Applied Microbiology*, **83**, 227–235.

Davies, K.G., Afolabi, P. and O'Shea, P. (1996) Adhesion of *Pasteuria penetrans* to the cuticle of root-knot nematodes (*Meloidogyne* spp.) inhibited by fibronectin: a study of electrostatic and hydrophobic interactions. *Parasitology*, **112**, 553–559.

Davies, K.G., Kerry, B.R. and Flynn, C.A. (1988) Observations on the pathology of *Pasteuria penetrans* a parasite of root-knot nematodes. *Annals of Applied Biology*, **112**, 491–501.

Davies, K.G., Laird, V. and Kerry, B.R. (1991) The motility, development and infection of *Meloidogyne incognita* encumbered with spores of the obligate hyperparasite *Pasteuria penetrans*. *Revue de Nématologie*, **14**, 611–618.

Davies, K.G., Redden, M. and Pearson, T.K. (1994) Endospore heterogeneity in *Pasteuria penetrans* related to adhesion to plant-parasitic nematodes. *Letters in Applied Microbiology*, **19**, 370–373.

Davies, K.G., Flynn, C.A., Laird, V. and Kerry, B.R. (1990) The life-cycle, population dynamics and host specificity of a parasite of *Heterodera avenae*, similar to *Pasteuria penetrans*. *Revue de Nématologie*, **13**, 303–309.

Deacon, J.W. (1991) Significance of ecology in the development of biocontrol agents against soil-borne plant pathogens. *Biocontrol Science and Technology*, **1**, 5–20.

de Leij, F.A.A.M. and Kerry, B.R. (1991) The nematophagous fungus, *Verticillium chlamydosporium* Goddard, as a potential biological control agent for *Meloidogyne arenaria* (Neal) Chitwood. *Revue de Nématologie*, **14**, 157–164.

de Leij, F.A.A.M., Dennehy, J.A. and Kerry, B.R. (1992) The effect of temperature and nematode species on interactions between the nematophagous fungus *Verticillium chlamydosporium* and root-knot nematodes (*Meloidogyne* spp.) *Nematologica*, **38**, 65–79.

Del Pino, F.G. and Palomo, A. (1997) Temporal study of natural populations of heterorhabditid and steinernematid nematodes in horticultural crop soils. *Fundamental and Applied Nematology*, **20**, 473–480.

Dicklow, M.B., Acosta, N. and Zuckerman, B.M. (1993) A novel streptomyces species for controlling plant-parasitic nematodes. *Journal of Chemical Ecology*, **19**, 159–173.

Dickson, D.W., Oostendorp, M., Giblin-Davis, R.M. and Mitchell, D.J. (1994) Control of plant-parasitic nematodes by biological antagonists. In *Pest Management in the Subtropics Biological Control — a Florida Perspective*, edited by D. Rosen, F.D. Bennett and J.L. Capinera, pp. 575–601. UK: Intercept.

Dijksterhuis, J., Veenhuis, M. and Harder, W. (1991) Colonization and digestion of nematodes by the endoparasitic nematophagous fungus *Drechmeria coniospora*. *Mycological Research*, **95**, 873–878.

Dijksterhuis, J., Veenhuis, M., Harder, W. and Nordbring-Hertz, B. (1994) Nematophagous fungi: Physiological aspects and structure-function relationships. In *Advances in Microbial Physiology*, **36**, pp. 111–143. Academic Press.

Downes, M.J. and Griffin, C.T. (1996) Dispersal behaviour and transmission strategies of the entomopathogenic nematodes *Heterorhabditis* and *Steinernema*. *Biocontrol Science and Technology*, **6**, 347–356.

Dunn, M.T., Sayre, R.M., Carrell, A. and Wergin, W.P. (1982) Colonization of nematode eggs by *Paecilomyces lilacinus* (Thom) Samson as observed with scanning electron microscope. In *Scanning Electron Microscopy*, edited by O.M. Johari and R.M. Albrecht, pp. 1351–1357. Chicago: Sem Inc.

Dürschner, U.U. (1984) Observations on a new nematophagous actinomycete. *Proceedings of the First International Congress of Nematology*. Canada, p. 23.

Ehlers, R.U., Hokkanen, H. (editors) (1996) Special issue: OECD workshop on the introduction of non-endemic nematodes for biological control: scientific and regulatory policy issues. *Biocontrol Science and Technology*, **6**, 291–480.

Eren, J. and Pramer, D. (1965) The most probable number of nematode-trapping fungi in soil. *Soil Science*, **99**, 285.

Faedo, M., Larsen, M. and Waller, P.J. (1997) The potential of nematophagous fungi to control the free-living stages of nematode parasites of sheep: Comparison between Australian isolates of *Arthrobotrys* sp. and *Duddingtonia flagrans*. *Veterinary Parasitology*, **72**, 149–155.

Gair, R., Mathias, P.L. and Harvey, P.N. (1969) Studies of cereal nematode populations and cereal yields under continuous or intensive culture. *Annals of Applied Biology*, **63**, 503–512.

Gareth-Jones, D. (editor) (1993) *Exploitation of Microorganisms*, pp 488. London: Chapman and Hall.

Gaspard, J.T., Jaffee, B.A. and Ferris, H. (1990) Association of *Verticillium chlamydosporium* and *Paecilomyces lilacinus* with root-knot nematode infested soil. *Journal of Nematology*, **22**, 207–213.

Gaugler, R. (1987) Entomogenous nematodes and their prospects for genetic improvement. In *Biotechnology in Invertebrate Pathology and Cell Culture*, edited by K. Maramorosch, pp. 457–484. New York: Academic Press.

Gaugler, R. and Kaya, H.K. (editors) (1990). *Entomopathogenic Nematodes in Biological Control*. Florida, Boca Raton: CRC Press.

Gaugler, R., Campbell, J.F., Selvan, S. and Lewis, E.E. (1992) Large-scale inoculative releases of the entomopathogenic nematode *Steinernema glaseri*: Assessment 50 years later. *Biological Control*, **2**, 181–187.

Georgis, R. (1992) Present and future prospects for entomopathogenic nematode products. *Biocontrol Science and Technology*, **2**, 83–99.

Georgis, R. and Kelly, J. (1997) Novel pesticidal substances from the entomopathogenic nematode-bacterium complex. In *Phytochemicals for Pest Control. ACS symposium series 658*, edited by P.A. Hedin, R.M. Hollingworth, E.P. Masler., J. Miyamoto and D.G. Thompson, pp. 134–143. Washington, D.C.: American Chemical Society.

Georgis, R. and Manweiler, S.A. (1994) Entomopathogenic nematodes: A developing biological control technology. *Agricultural Zoology Reviews*, **6**, 63–94.

Giblin-Davis, R.M., McDaniel, L.L. and Bilz, F.G. (1990) Isolates of the *Pasteuria penetrans* group from phytoparasitic nematodes in bermudagrass turf. *Journal of Nematology*, **22**, 750–762.

Gilligan, C.A. and Kleczkowski, A. (1997) Population dynamics of botanical epidemics involving primary and secondary infection. *Philosophical Transactions of The Royal Society, London. B*, **353**, 591–608.

Gintis, B.O., Morgan-Jones, G. and Rodriguez-Kabana, R. (1983) Fungi associated with several developmental stages of *Heterodera glycines* from an Alabama soybean field soil. *Nematropica*, **13**, 181–200.

Glazer, I. (1996) Survival mechanisms of entomopathogenic nematodes. *Biocontrol Science and Technology*, **6**, 373–378.

Glazer, I., Salame, L. and Segal, D. (1997) Genetic enhancement of nematicide resistance in entomopathogenic nematodes. *Biocontrol Science and Technology*, **7**, 499–512.

Glazer, I., Kozodoi, E., Salame, L. and Nestel, D. (1996) Spatial and temporal occurrence of natural populations of *Heterorhabditis* spp. (Nematoda: Rhabditida) in a semiarid region. *Biological Control*, **6**, 130–136.

Gould, F., Kennedy, G.G. and Johnson, M.T. (1991) Effects of natural enemies on the rate of herbivore adaptation to resistant host plants. *Entomologia Experimentalis et Applicata*, **58**, 1–14.

Graff, N.J. and Madelin, M.F. (1989) Axenic culture of the cyst-nematode parasitising fungus, *Nematophthora gynophila*. *Journal of Invertebrate Pathology*, **53**, 301–306.

Grenache, D.G., Caldicott, I., Albert, P.S., Riddle, D.L. and Politz, S.M. (1996) Environmental induction and genetic control of surface antigen switching in the nematode *Caenorhabditis elegans*. *Proceedings of the National Academy of Sciences, USA*, **93**, 12388–12393.

Grewal, P.S., Martin, W.R., Miller, R.W. and Lewis, E.E. (1997) Suppression of plant-parasitic nematode populations in turfgrass by application of entomopathogenic nematodes. *Biocontrol Science and Technology*, **7**, 393–399.

Griffin, C.T., Gwynn, R.L. and Masson, J.P. (editors) (1995). *EUR 16269- Ecology and transmission strategies of entomopathogenic nematodes*, pp. 111. Luxembourgh: Office for Official Publications of the European Communities.

Grønvold, J., Wolstrup, J., Larsen, M., Henriksen, S.A. and Nansen, P. (1993) Biological control of *Ostertagia ostertagi* by feeding selected nematode-trapping fungi to calves. *Journal of Helminthology*, **67**, 31–36.

Hallmann, J. and Sikora, R.A. (1994) Occurrence of plant parasitic nematodes and non-pathogenic species of *Fusarium* in tomato plants in Kenya and their role as mutualistic synergists for biological control of root-knot nematodes. *International Journal of Pest Management*, **40**, 321–325.

Hallmann, J., Quadt-Hallmann, A., Rodriguez-Kabana, R. and Kloepper, J.W. (1998) Interactions between *Meloidogyne incognita* and endophytic bacteria in cotton and cucumber. *Soil Biology and Biochemistry*, **30**, 925–937.

Hartwig, E. (1982) Response of resistance and susceptible soybean cultivars to continuous cropping in areas infested with cyst nematode. *Plant Disease*, **66**, 18–20.

Haugen, D.A. (1990) Control procedures for *Sirex noctilio* in the Green Triangle: review from detection to severe outbreak (1977–1987). *Australian Forestry*, **53**, 24–32.

Haugen, D.A. and Underdown, M.G. (1990) *Sirex noctilio* control programme in response to the 1987 Green Triangle outbreak. *Australian Forestry*, **53**, 33–40.

Haugen, D.A., Bedding, R.A., Underdown, M.G. and Neumann, F.G. (1990) National strategy for control of *Sirex noctilio* in Australia. *Australian Forest Grower*, **13**, 8 pp.

Heijbroek, W. (1983) Some effects of fungal parasites on the population development of the beet cyst nematode *Heterodera schachtii* Schm. *Mededelingen Faculteit Landbouwwetenschappen Rijksuniveersiteit Gent*, **48**, 433–439.

Hominick, W.M. and Briscoe, B.R. (1990) Survey of 15 sites over 28 months for entomopathogenic nematodes (Rhabditida : Steinernematidae). *Parasitology*, **100**, 289–294.

Hominick, W.M. and Tingley, G.A. (1984) Mermithid nematodes and the control of insect vectors of human disease. *Biocontrol News and Information*, **5**, 7–20.

Hominick, W.M., Reid, A.P., Bohan, D.A. and Briscoe, B.R. (1996) Entomopathogenic nematodes: Biodiversity, geographical distribution and the Convention on Biological Diversity. *Biocontrol Science and Technology*, **6**, 317–331.

Hominick, W.M., Briscoe, B.R., del Pino, F.G., Heng, J., Hunt, D.J., Kozodoy, E., Mracek, Z., Nguyen, K.B., Reid, A.P., Spiridonov, S., Stock, P., Sturhan, D., Waturu, C. and Yoshida, M. (1997) Biosystematics of entomopathogenic nematodes: current status, protocols and definitions. *Journal of Helminthology*, **71**, 271–298.

Jaffee, B.A. and McInnis, T.M. (1991) Sampling strategies for detection of density-dependent parasitism of soil-borne nematodes by nematophagous fungi. *Revue de Nématologie*, **14**, 147–150.

Jaffee, B.A. and Muldoon, A.E. (1989) Suppression of cyst nematode by natural infestation of a nematophagous fungus. *Journal of Nematology*, **21**, 505–510.

Jaffee, B.A. and Muldoon, A.E. (1995) Numerical responses of the nematophagous fungi *Hirsutella rhossiliensis*, *Monacrosporium cionopagum*, and *M. ellipsosporum*. *Mycologia*, **87**, 643–650.

Jaffee, B.A. and Zehr, E.I. (1982) Parasitism of the nematode *Criconemella xenoplax* by the fungus *Hirsutella rhossiliensis*. *Phytopathology*, **72**, 1278–1381.

Jaffee, B.A., Muldoon, A.E. and Tedford, E.C. (1992) Trap production by nematophagous fungi growing from parasitized nematodes. *Phytopathology*, **82**, 615–620.

Jaffee, B.A., Phillips, R., Muldoon, A. and Mangel, M. (1992) Density-dependent host-pathogen dynamics in soil microcosms. *Ecology*, **73**, 495–506.

Jannsson, H.B. and Nordbring-Hertz, B. (1980) Interactions between nematophagous fungi and plant-parasitic nematodes: attraction, induction of trap formation and capture. *Nematologica*, **26**, 383–389.

Johnston, T. (1957) Further studies on microbiological reduction of nematode populations in water-saturated soils. *Phytopathology*, **47**, 525.

Kaiser, H. (1991) Terrestrial and semiterrestrial Mermithidae. In *Manual of Agricultural Nematology*, edited by W.R. Nickle, pp. 899–965. New York: Marcel Dekker.

Kaya, H.K. (1985) Entomogenous nematodes for insect control in IPM systems. In *Biological Control in Agricultural IPM Systems*, edited by M.A. Hoy, and D.C. Herzog, pp. 283–302. New York: Academic Press.

Kaya, H.K. (1987) Diseases caused by nematodes. In *Epizootiology of Insect Diseases*, edited by J.R. Fuxa and Y. Tanada, pp. 453–470. New York: John Wiley and Sons.

Kaya, H.K. (1993) Entomogenous and entomopathogenic nematodes in biological control. In *Plant Parasitic Nematodes in Temperate Agriculture*, edited by K. Evans., D.L. Trudgill and J.M. Webster, pp. 565–591. Wallingford, UK: CAB International.

Kaya, H.K. and Gaugler, R. (1993) Entomopathogenic nematodes. *Annual Review of Entomology*, **38**, 181–206.

Kaya, H.K. and Koppenhofer, A.M. (1996) Effects of microbial and other antagonistic organisms and competition on entomopathogenic nematodes. *Biocontrol Science and Technology*, **6**, 357–372.

Kerry, B.R. (1975) Fungi and the decrease of cereal cyst-nematode populations in cereal monoculture. *European Plant Protection Organisation Bulletin*, **5**, 353–361.

Kerry, B.R. (1982) The decline of *Heterodera avenae* populations. *European Plant Protection Organisation Bulletin*, **12**, 491–496.

Kerry, B.R. (1987) Biological control. In *Principles and Practice of Nematode Control in Crops*, edited by R.H. Brown and B.R. Kerry, pp. 233–263. New York: Academic Press.

Kerry, B.R. (1988) Fungal parasites of cyst nematodes. *Agriculture, Ecosystems and Environment*, **24**, 293–305.

Kerry, B.R. (1990) Selection of exploitable biological control agents for plant parasitic nematodes. *Aspects of Applied Biology*, **24**, 1–9.

Kerry, B.R. (1993) The use of microbial agents for the biological control of plant parasitic nematodes. In *Exploitation of Microorganisms*, edited by D. Gareth Jones, pp. 81–104. London: Chapman & Hall.

Kerry, B.R. (1995a) Ecological considerations for the use of the nematophagous fungus, *Verticillium chlamydosporium*, to control plant parasitic nematodes. *Canadian Journal of Botany*, 73, S65–S70.

Kerry, B.R. (1995b) The potential impact of natural enemies on the survival and efficacy of entomopathogenic nematodes. In *Ecology and Transmission Strategies of Entomopathogenic Nematodes*, edited by C.T. Griffin, L. Gwynn and J.P. Masson, pp. 7–13. Eur 16269. COST 819: Biotechnology Report.

Kerry, B.R. (1998) Progress towards biological control strategies for plant-parasitic nematodes. *Proceedings of the 1998 Brighton Conference — Pests and Diseases*. 739–746.

Kerry, B.R. and Bourne, J.M. (1996) The importance of rhizosphere interactions in the biological control of plant parasitic nematodes — a case study using *Verticillium chlamydosporium*. *Pesticide Science*, 47, 69–75.

Kerry, B.R. and Crump, D.H. (1980) Two fungi parasitic on females of cyst nematodes (*Heterodera* spp.) *Transactions of the British Mycological Society*, 74, 119–125.

Kerry, B.R. and Crump, D.H. (1998) The dynamics of the decline of the cereal cyst nematode, *Heterodera avenae*, in four soils under intensive cereal production. *Fundamental and Applied Nematology*, 21, 617–625.

Kerry, B.R. and Jaffee, B.A. (1997) Fungi as biological control agents for plant parasitic nematodes. In *The Mycota IV. Environmental and Microbial Relationships*, edited by D.T. Wicklow and B. Söderström, pp. 203–218. Berlin: Springer-Verlag.

Kerry, B.R., Crump, D.H. and Mullen, L.A. (1980) Parasitic fungi, soil moisture and multiplication of the cereal cyst nematode, *Heterodera avenae*. *Nematologica*, 26, 57–68.

Kerry, B.R., Crump, D.H. and Mullen, L.A. (1982) Studies of the cereal cyst-nematode, *Heterodera avenae* under continuous cereals, 1975–1978. II. Fungal parasitism of nematode females and eggs. *Annals of Applied Biology*, 100, 489–499.

Kerry, B.R., Kirkwood, I.A., Leij de, F.A.A.M., Barba, J., Leijdens, M.B. and Brookes, P.C. (1993) Growth and survival of *Verticillium chlamydosporium* Goddard, a parasite of nematodes in soil. *Biocontrol Science and Technology*, 3, 355–365.

Kim, D.G., Riggs, R.D. and Kim, K.S. (1992) Ultrastructure of *Heterodera glycines* parasitized by Arkansas Fungus 18. *Phytopathology*, 82, 429–433.

Kleczkowski, A., Bailey, D.J. and Gilligan, C.A. (1996) Dynamically generated variability in plant-pathogen systems with biological control. *Proceedings of The Royal Society, London B*, 263, 777–783.

Kloepper, J.W., Rodriguez-Kabana, R., McInroy, J.A. and Collins, D.J. (1991) Analysis of populations and physiological characterization of microorganisms in rhizospheres of plants with antagonistic properties to phytopathogenic nematodes. *Plant and Soil*, 136, 95–102.

Koppenhofer, A.M., Jaffee, B.A., Muldoon, A.E. and Strong, D.R. (1997) Suppression of an entomopathogenic nematode by the nematode-trapping fungi *Geniculifera paucispora* and *Monacrosporium eudermatum* as affected by the fungus *Arthrobotrys oligospora*. *Mycologia*, 89, 220–227.

Krecek, R.C., Sayre, R.M., Els, H.J., Van Niekerk, J.P. and Malan, F.S. (1987) Fine structure of a bacterial community associated with cyathostomes (Nematoda: Strongylidae) of zebras. *Proceedings of the Helminthological Society of Washington*, 54, 212.

Kwok, O.C.H., Plattner, R., Weisleder, D. and Wicklow, D.T. (1992) A nematicidal toxin from *Pleurotus ostreatus* NRRL 3526. *Journal of Chemical Ecology*, 18, 127–136.

Lackey, B.A., Muldoon, A.E. and Jaffee, B.A. (1993) Alginate pellet formulation of *Hirsutella rhossiliensis* for biological control of plant parasitic nematodes. *Biological Control*, 3, 155–160.

Larsen, M., Faedo, M. and Waller, P.J. (1994) The potential of nematophagous fungi to control the free-living stages of nematode parasites of sheep: survey for the presence of fungi in fresh faeces of grazing livestock in Australia. *Veterinary Parasitology*, 72, 479–492.

Larsen, M., Nansen, P., Grønvold, J., Wolstrup, J. and Henriksen, S.A. (1997) Biological control of gastro-intestinal nematodes — facts, future or fiction? *Veterinary Parasitology*, 72, 479–492.

Lehman, P.S., Huisingh, D. and Barker, K.R. (1971) The influence of races of *Heterodera glycines* on nodulation and nitrogen-fixing capacity of soybean. *Phytopathology*, 61, 1239–1244.

Lewis, E.E., Campbell, J.F. and Gaugler, R. (1997) The effects of ageing on the foraging behaviour of *Steinernema carpocapsae* (Rhabdita: Steinernematidae). *Nematologica*, 43, 355–362.

Linderman, R.G., Moore, L.W., Baker, K.F. and Cooksey, D.A. (1983) Strategies for detecting and characterizing systems for biological control of soilborne plant pathogens. *Plant Disease*, 67, 1058–1064.

Lopez-Llorca, L.V. (1990) Purification and properties of extracellular proteases produced by the nematophagous fungus *Verticillium suchlasporium*. *Canadian Journal of Microbiology*, 36, 530–537.

Lopez-Llorca, L.V. and Claugher, D. (1990) Appressoria of the nematophagous fungus *Verticillium suchlasporium*. *Micron and Microscopica Acta*, 21, 125–130.

Lopez-Llorca, L.V. and Robertson, W.M. (1992) Immunocytochemical localization of a 32-kDa protease from the nematophagous fungus *Verticillium suchlasporium* in infected nematode eggs. *Experimental Mycology*, 16, 261–267.

Lopez-Llorca, L.V., Moya, M. and Llinares, A. (1993) Effect of pH on growth and pigment production of nematophagous and entomogenous fungi. *Micologia Vegetazione Mediterranea*, 8, 107–112.

Mankau, R. (1980) Biological control of *Meloidogyne* populations by *Bacillus penetrans* in West Africa. *Journal of Nematology*, 12, 230.

Marban-Mendoza, N., Garcia-E, R., Bess Dicklow, M. and Zuckerman, B.M. (1992) Studies on *Paecilomyces marquandii* from nematode suppressive chinampa soils. *Journal of Chemical Ecology*, 18, 775–783.

McInnis, T.M. and Jaffee, B.A. (1989) An assay for *Hirsutella rhossiliensis* spores and the importance of phialides for nematode inoculation. *Journal of Nematology*, 21, 229–234.

McInnis, T.M., Kluepfel, D. and Zehr, E. (1990) Suppression of *Criconemella xenoplax* on peach by rhizosphere bacteria. *Second International Nematology Congress, Veldhoven, The Netherlands, Abstracts of Papers*, 106.

Mendoza de Gives, P., Davies, K.G., Morgan, M. and Behnke, J.M. (1999) Attachment tests of *Pasteuria penetrans* to the cuticle of plant and animal parasitic nematodes, free living nematodes and *srf* mutants of *Caenorhabditis elegans*. *Journal of Helminthology*, 72, (in press).

Mendoza de Gives, P., Flores Crespo, J., Herrera Rodriguez, D., Vazquez Prats, V., Liebano Hernandez, E. and Ontiveros Fernandez, G.E. (1998) Biological control of *Haemonchus contortus* infective larvae in ovine faeces by administering an oral suspension of *Duddingtonia flagrans* chlamydospores to sheep. *Journal of Helminthology*, 72, 343–347.

Minton, N.A. and Sayre, R.M. (1989) Suppressive influence of *Pasteuria penetrans* in Georgia soil on reproduction of *Meloidogyne arenaria*. *Journal of Nematology*, 21, 574–575.

Morgan-Jones, G., White, J.F. and Rodriguez-Kabana, R. (1983) Phytonematode pathology: Ultrastructural studies. 1. Parasitism of *Meloidogyne arenaria* eggs by *Verticillium chlamydosporium*. *Nematropica*, 13, 245–260.

Muller, J. (1982) The influence of fungal parasites on the population dynamics of *Heterodera schachtii* on oil radish. *Nematologica*, 28, 161.

Nickle, W.R. (editor) (1984) *Plant and Insect Nematodes*. New York: Marcel Dekker Inc.

Nickle, W.R. and Welch, H.E. (1984) History, development and importance of insect nematology. In *Plant and Insect Nematodes*, edited by W.R. Nickle, pp. 797–820. New York: Marcel Dekker Inc.

Nordbring-Hertz, B. (1988) Ecology and recognition in the nematode-nematophagous fungus system. In *Advances in Microbial Ecology*, edited by K.C. Marshall, pp. 81–114. New York: Plenum Press.

Oostendorp, M. and Sikora, R.A. (1989) Seed treatment with antagonistic rhizobacteria for the suppression of *Heterodera schachtii* early root infection of sugar beet. *Revue de Nématologie*, 12, 77–83.

Oostendorp, M. and Sikora, R.A. (1990) *In-vitro* interrelationship between rhizosphere bacteria and *Heterodera schachtii*. *Revue de Nématologie*, 13, 269–274.

Oostendorp, M., Dickson, D.W. and Mitchell, D.J. (1991) Population development of *Pasteuria penetrans* on *Meloidogyne arenaria*. *Journal of Nematology*, 23, 58–64.

Orion, D. and Kritzman, G. (1991) Antimicrobial activity of *Meloidogyne javanica* gelatinous matrix. *Revue de Nématologie*, 14, 481–483.

Patel, M.N., Stolinski, M. and Wright, D.J. (1997) Neutral lipids and the assessment of infectivity in entomopathogenic nematodes: observations on four *Steinernema* species. *Parasitology*, 114, 489–496.

Perry, J.N. (1978) A population model for the effect of parasitic fungi on numbers of the cereal cyst nematode, *Heterodera avenae*. *Journal of Applied Ecology*, 15, 781–788.

Perry, R.N. and Trett, M.W. (1986) Ultrastructure of the eggshell of *Heterodera schachtii* and *H. glycines* (Nematoda: Tylenchida). *Revue de Nématologie*, 9, 399–403.

Persidis, A., Lay, J.G., Manousis, T., Bishop, A.H. and Ellar, D.J. (1991) Characterisation of potential adhesins of the bacterium *Pasteuria penetrans* and of putative receptors on the cuticle of *Meloidogyne incognita*, a nematode host. *Journal of Cell Science*, **100**, 613–622.

Persson, Y. and Friman, E. (1993) Intracellular proteolytic activity in mycelia of *Arthrobotrys oligospora* bearing mycoparasitic or nematode trapping structure. *Experimental Mycology*, **17(3)**, 182–190.

Peters, A. (1996) The natural host range of *Steinernema* and *Heterorhabditis* spp. and their impact on insect populations. *Biocontrol Science and Technology*, **6**, 389–402.

Petersen, J.J. (1985) Nematodes as biological control agents: Part I Mermithidae. *Advances in Parasitology*, **24**, 307–346.

Petersen, J.J. and Willis, O.R. (1972) Procedures for the mass rearing of a mermithid parasite of mosquitoes. *Mosquito News*, **32**, 226–230.

Pickett, J.A., Wadhams, L.J., Woodcock, C.M. and Hardie, J. (1992) The chemical ecology of aphids. *Annual Revue of Entomology*, **37**, 67–90.

Pinochet, J., Calvet, C., Camprubí, A. and Fernández, C. (1996) Interactions between migratory endoparasitic nematodes and arbuscular mycorrhizal fungi in perennial crops: A review. *Plant and Soil*, **185**, 183–190.

Poinar, G.O. Jr. (1979) *Nematodes for Biological Control of Insects*. Florida: Boca Raton, CRC Press Inc.

Poinar, G.O. Jr. (1991) Genetic engineering of nematodes for pest control. In *Biotechnology for Biological Control of Pests and Vectors*, edited by K. Maramorosch, pp. 77–93. Florida: Boca Raton, CRC Press Inc.

Poinar, G.O. Jr. (1993) Origins and phylogenetic relationships of the entomophilic rhabditids *Heterorhabditis* and *Steinernema*. *Fundamental and Applied Nematology*, **16**, 333–338.

Poinar, G.O. Jr., Acra, A. and Acra, F. (1994) Earliest fossil nematode (Mermithidae) in cretaceous Lebanese amber. *Fundamental and Applied Nematology*, **17**, 475–477.

Popiel, I. and Hominick, W.M. (1992) Nematodes as biological control agents: Part II. *Advances in Parasitology*, **31**, 381–433.

Powell, W., Pennacchio, F., Poppy, G.M. and Tremblay, E. (1998) Strategies involved in the location of hosts by the parasitoid *Aphidius ervi* Haliday (Hymenoptera: Braconidae: Aphidiinae). *Biological Control*, **11**, 104–112.

Powelson, M.L. and Rowe, R.C. (1993) Biology and management of early dying of potatoes. *Annual Review of Phytopathology*, **31**, 111–126.

Quinn, M.A. (1987) The influence of saprophytic competition on nematode predation by nematode-trapping fungi. *Journal of Invertebrate Pathology*, **49**, 170–174.

Racke, J. and Sikora, R.A. (1992) Wirkung der pflanzengesundheitsfördernden Rhizobakterien *Agrobacterium radiobacter* und *Bacillus sphaericus* auf den *Globodera pallida* Befall der Kartoffel und das Pflanzenwachstum. *Journal of Phytopathology*, **134**, 198–208.

Remillet, M. and Laumond, C. (1991) Sphaerularioid nematodes of importance in agriculture. In *Manual of Agricultural Nematology*, edited by W.R. Nickle, pp. 967–1024. New York: Marcel Dekker.

Rodriguez-Kabana, R. and Morgan-Jones, G. (1987) Biological control of nematodes: Soil amendments and microbial antagonists. *Plant and Soil*, **100**, 237–248.

Rodriguez-Kabana, R., Jordan, J.W. and Hollis, J.P. (1965) Nematodes: Biological control in rice fields: Role of hydrogen sulphide. *Science*, **148**, 524.

Sayre, R.M. and Starr, M.P. (1988) Bacterial diseases and antagonisms of nematodes. In *Diseases of Nematodes*, edited by G.O. Poinar and H-B Jansson, pp. 69–101. Florida: CRC Press Inc.

Sayre, R.M. and Wergin, W.P. (1977) Bacterial parasite of a plant nematode: morphology and ultrastructure. *Journal of Bacteriology*, **129**, 1091–1101.

Sayre, R.M., Patrick, Z.A. and Thorpe, H.J. (1965) Identification of a selective nematicidal component in extracts of plant residues decomposing in soil. *Nematologica*, **11**, 263.

Schlang, J., Steudel, W. and Muller, J. (1988) Influence of resistant green manure crops on the population dynamics of *Heterodera schachtii* and its fungal egg parasites. *Nematologica*, **34**, 193.

Schuster, R.-P., Sikora, R.A. and Amin, N. (1995) Potential of endophytic fungi for the biological control of plant parasitic nematodes. *Mededelingen Faculteit Landbouwwetenschappen Universiteit Gent*, **60**, 1047–1052.

Segers, R., Butt, T.M., Kerry, B.R., Beckett, A. and Peberdy, J.F. (1996) The role of the proteinase VCP1 produced by the nematophagous *Verticillium chlamydosporium* in the infection process of nematode eggs. *Mycological Research*, **100**, 421–428.

Segers, R., Butt, T.M., Carder, J.H., Keen, J.N., Kerry, B.R. and Peberdy, J.F. (1998) The subtilisins of fungal pathogens of insects, nematodes and plants: distribution and variation. *Mycological Research*: **103**, 395–402.

Serracin, M., Schuerger, A.C., Dickson, D.W. and Weingartner, D.P. (1994) An alternative method for culturing *Pasteuria penetrans*. *Journal of Nematology*, **26**, 565.

Shapiro, D.I., Glazer, I. and Segal, D. (1997) Genetic diversity in wild and laboratory populations of *Heterorhabditis bacteriophora* as determined by RAPD-PCR analysis. *Fundamental and Applied Nematology*, **20**, 581–585.

Sharma, S.B., Price, N.S. and Bridge, J. (1997) The past, present and future of plant nematology in International Agricultural Research Centres. *Nematological Abstracts*, **66**, 119–142.

Siddiqui, Z.A. and Mahmood, I. (1995) Role of plant symbionts in nematode management: A Review. *Bioresource Technology*, **54b**, 217–226.

Sikora, R.A. (1988) Interrelationship between plant health promoting rhizobacteria, plant parasitic nematodes and soil microorganisms. *Mededelingen Faculteit Landbouwwetenschappen Rijksuniversiteit Gent*, **53**, 867–878.

Sikora, R.A. (1992) Management of the antagonistic potential in agricultural ecosystems for the biological control of plant parasitic nematodes. *Annual Review of Phytopathology*, **30**, 245–270.

Sikora, R.A. and Carter, W.W. (1987) Nematode interactions with fungal and bacterial plant pathogens — Fact or fantasy. In *Vistas on Nematology*, edited by J.A. Veech and D.W. Dickson, pp. 307–312. Hyattsville: Society of Nematologists, Inc.

Sikora, R.A., Hiemer, M. and Schuster, R.-P. (1990) Reflections on the complexity of fungal infection of nematode eggs and the importance of facultative perthophytic fungal pathogens in biological control of *Globodera pallida*. *Mededelingen Faculteit Landbouwwetenschappen Rijksuniveersiteit Gent*, **55**, 699–712.

Simoes, N. and Rosa, J.S. (1996) Pathogenicity and host specificity of entomopathogenic nematodes. *Biocontrol Science and Technology*, **6**, 403–412.

Simoes, N., Boemare, N. and Ehlers, R.U. (editors) (1998) *EUR 17776 COST 819: Entomopathogenic nematodes: Pathogenicity of entomopathogenic nematodes versus insect defence mechanisms: Impact on selection of virulent strains*, 271pp. Luxembourg: Office for Official Publications of the European Communities.

Smith, K.A., Miller, R.W. and Simser, D.H. (1992) Entomopathogenic nematode bibliography: Heterorhabditid and steinernematid nematodes. *Southern Cooperative Series Bulletin*, **370**, 81 pp. Arkansas: Fayetteville, Arkansas Agricultural Experiment Station.

Smits, P.H. (1996) Post-application persistence of entomopathogenic nematodes. *Biocontrol Science and Technology*, **6**, 379–387.

Southwood, T.R.E. and Comins, H.N. (1976) A synoptic population model. *Journal of Animal Ecology*, **405**, 949–965.

Spiegel, Y., Chet, I. and Cohn, E. (1987) Use of chitin for controlling plant plant-parasitic nematodes. *Plant and Soil*, **98**, 337–345.

Spiegel, Y., Cohn, E. and Chet, I. (1986) Use of chitin for controlling plant parasitic nematodes. *Plant and Soil*, **95**, 87–95.

Spiegel, Y., Kahane, I., Cohen, L. and Sharon, E. (1997) *Meloiodogyne javanica* surface proteins: characterization and lability. *Parasitology*, **115**, 513–519.

Spiegel, Y., Cohn, E., Galper, S., Sharon and Chet, I. (1991) Evaluation of a newly isolated bacterium, *Pseudomonas chitinolytica* sp. nov., for controlling the root-knot nematode *Meloidogyne javanica*. *Biocontrol Science and Technology*, **1**, 115–125.

Stiles, C.M. and Glawe, D.A. (1989) Colonisation of soybean roots by fungi isolated from cysts of *Heterodera glycines*. *Mycologia*, **81**, 797–799.

Stirling, G.R. (1979) Effect of temperature on parasitism of *Meloidogyne incognita* eggs by *Dactylella oviparasitica*. *Nematologica*, **25**, 104–110.

Stirling, G.R. (1981) Effect of temperature on infection of *Meloidogyne javanica* by *Bacillus penetrans*. *Nematologica*, **27**, 458–462.

Stirling, G.R. (1984) Biological control of *Meloidogyne javanica* with *Bacillus penetrans*. *Phytopathology*, **74**, 55–60.

Stirling, G.R. (1985) Host specificity of *Pasteuria penetrans* within the genus *Meloidogyne*. *Nematologica*, **31**, 203–209.

Stirling, G.R. (1991) *Biological Control of Plant Parasitic Nematodes: Progress, Problems and Prospects*. UK: CAB International.

Stirling, G.R. and Mankau, R. (1977) Biological control of nematode parasites of citrus by natural enemies. *Proceedings of the International Society of Citriculture*, **3**, 843–847.

Stirling, G.R. and Mani, A. (1995) The activity of nematode-trapping fungi following their encapsulation in alginate. *Nematologica*, **41**, 240–250.

Stirling, G.R. and Wachtel, M.F. (1980) Mass production of *Bacillus penetrans* for the biological control of root-knot nematodes. *Nematologica*, 26, 308–312.

Stirling, G.R. and West, L.M. (1991) Fungal parasites of root-knot nematode eggs from tropical and sub-tropical regions of Australia. *Australasian Plant Pathology*, 20, 149–154.

Stirling, G.R. and White, A.M. (1982) Distribution of a parasite of root-knot nematodes in South Australian vineyards. *Plant Disease*, 66, 52–53.

Stirling, G.R., McKenry, M.V. and Mankau, R. (1979) Biological control of root-knot nematodes (*Meloidogyne* spp.) on peach. *Phytopathology*, 69, 806–809.

Stirling, G.R., Sharma, R.D. and Perry, J. (1990) Attachment of *Pasteuria penetrans* to the root-knot nematode *Meloidogyne javanica* in soil and its effects on infectivity. *Nematologica*, 36, 246–252.

Sturhan, D., Winkelheide, R., Sayre, R.M. and Wergin, W.P. (1994) Light and electron microscopical studies of the life cycle and developmental stages of a *Pasteuria* isolate parasitising the pea cyst nematode, *Heterodera goettingiana. Fundamental and Applied Nematology*, 17, 29–42.

Thielemann, R. and Steudel, W. (1973) Nine years experience with monocultures of sugar beet in soil infested with *Heterodera schachtii* (Schmidt). *Nachrichtenblatt für den Deutschen Pflanzenshutzdienst.*, 25, 145–149.

Thomas, E. (1982) On the occurrence of parasitic fungi and other pathogens, and their influence on the development of populations of the beet cyst nematode (*Heterodera schachtii*) in the Northern Rhineland. *Gesunde Pflanzen*, 34, 162–168.

Thorn, R.G. and Barron, G.L. (1984) Carnivorous mushrooms. *Science*, 224, 76–78.

Thornton, C.R., Dewey, F.M. and Gilligan, C.A. (1994) Development of monoclonal antibody-based enzyme-linked immunosorbent assay for the detection of live propagules of *Trichoderma harzianum* in a peat-bran medium. *Soil Biology and Biochemistry*, 26, 909–920.

Tunlid, A. and Jansson, S. (1991) Proteases and their involvement in the infection and immobilisation of nematodes by the nematophagous fungus *Arthrobotrys oligospora. Applied and Environmental Microbiology*, 57, 2868–2872.

Tunlid, A., Jansson, H.-B. and Nordbring-Hertz, B. (1992) Fungal attachment to nematodes. *Mycological Research*, 96, 401–412.

Tunlid, A., Johansson, T. and Nordbring-Hertz, B. (1991) Surface polymers of the nematode-trapping fungus *Arthrobotrys stigospora. Journal of General Microbiology*, 137, 1231–1240.

Tzortzakakis, E.A. and Gowen, S.R. (1994) Evaluation of *Pasteuria penetrans* alone and in combination with oxamyl, plant resistance and solarization for control of *Meloidogyne* spp. on vegetables grown in greenhouses in Crete. *Crop Protection*, 13, 455–462.

Van Emden, H.F. (1999) Are two methods better than one? *International Journal of Pest Management*, 45, 1–2.

Van Gundy, S.D., Kirkpatrick, J.D. and Golden, J. (1977) The nature and role of metabolic leakage from root-knot nematode galls and infection by *Rhizoctonia solani. Journal of Nematology*, 9, 113–121.

Velvis, H. and Kamp, P. (1995) Infection of second stage juveniles of potato cyst nematodes by the nematophagous fungus *Hirsutella rhossiliensis* in Dutch potato fields. *Nematologica*, 41, 617–627.

Waage, J.K. and Greathead, D.J. (1988) Biological control: challenges and opportunities. *Philosophical Transactions of the Royal Society*, London, 318, 111–128.

Walker, G.E. and Wachtel, M.F. (1988) The influence of soil solarization and non-fumigant nematicides on infection of *Meloidogyne javanica* by *Pasteuria penetrans. Nematologica*, 34, 477–483.

Waller, P.J. and Faedo, M. (1996) The prospects for biological control of the free-living stages of nematode parasites of livestock. *International Journal of Parasitology*, 26, 915–925.

Walsh, J.A., Shepherd, A.M. and Lee, D.L. (1983) The distribution and effect of intracellular rickettsia-like micro-organisms infecting second-stage juveniles of the potato cyst-nematode *Globodera rostochiensis. Journal of Zoology*, 199, 395–419.

Walter, D.E. and Ikonen, E.K. (1989) Species, guilds, and functional groups: taxonomy and behaviour in nematophagous arthropods. *Journal of Nematology*, 21, 315–327.

Walter, D.E. and Kaplan, D.T. (1990) Antagonists of plant-parasitic nematodes in Florida citrus. *Journal of Nematology*, 22, 567–573.

Walter, D.E., Hunt, H.W. and Elliot, E.T. (1987) The influence of prey type on the development and reproduction of some predatory soil mites. *Pedobiologia*, 30, 419–424.

Walter, D.E., Hunt, H.W. and Elliot, E.T. (1988) Guilds or functional groups? An analysis of predatory arthropods from a shortgrass steppe soil. *Pedobiologia*, 31, 247–260.

Wang, X.D., Kondo, E. and Ishibashi, N. (1997) Infectivity of entomopathogenic nematode, *Steinernema carpocapsae*, as affected by the emergence time from the host cadaver. *Japanese Journal of Nematology*, 27, 1–6.

Wechter, P. and Kluepfel, D.A. (1997) Sequence determination and characterization of DNA fragments involved in production/expression of a nematode ovicidal factor by *Pseudomonas aureofaciens* BG33R. *Phytopathology*, 87, S116.

Wechter, W.P., Leverentz, B. and Kluepfel, D.A. (1998) Characterization of genetic loci implicated in production and exportation of a nematode ovicidal factor produced by *Pseudomonas aureofaciens* BG33R. *Phytopathology*, 88, S96.

Weibelzahl-Fulton, E., Dickson, D.W. and Whitty, E.B. (1996) Suppression of *Meloidogyne incognita* and *M. javanica* by *Pasteuria penetrans* in field soil. *Journal of Nematology*, 28, 43–49.

Whitehead, A.G. (1997) *Plant Nematode Control*. UK: CAB International.

Williams, A.B., Stirling, G.R., Hayward, A.C. and Perry, J. (1989) Properties and attempted culture of *Pasteuria penetrans*, a bacterial parasite of root-knot nematode (*Meloidogyne javanica*). *Journal of Applied Bacteriology*, 67, 145–156.

Woodring, J.L. and Kaya, H.K. (1988) Steinernematid and Heterorhabditid nematodes: A handbook of biology and techniques. *Southern Cooperative Series Bulletin*, 331, 30pp. Arkansas: Fayetteville, Arkansas Agricultural Experiment Station.

Wyss, U., Voss, B. and Jansson, H-B. (1992) *In vitro* observations on the infection of *Meloidogyne incognita* eggs by the zoosporic fungus *Catenaria anguillulae* Sorokin. *Fundamental and Applied Nematology*, 15, 133–139.

Yeates, G.W. and Wardle, D.A. (1996) Nematodes as predators and prey: relationships to biological control and soil processes. *Pedobiologia*, 40, 43–50.

Zhang, Z.L., Lei, S.F., Wang, H.S., He, J.F., Yang, Y.C. and Chen, G. (1992) Release of infective juveniles of *Ovomermis sinensis* (Nematoda: Mermithidae) to control armyworm, *Mythimna separata* (Lep.: Noctuidae) in a wheat field in 1990. *Chinese Journal of Biological Control*, 8, 148–150. In Chinese.

Zuckerman, B.M., Dicklow, M.B., Coles, G.C., Garcia-E, R. and Marban-Mendoza, N. (1989) Suppression of plant parasitic nematodes in the chinampa agricultural soils. *Journal of Chemical Ecology*, 15, 1947–1955.

20. Behaviour of Entomopathogenic Nematodes

Nobuyoshi Ishibashi

Department of Applied Biological Sciences, Saga University, Saga 840-8502, Japan

Keywords

Foraging strategy, ambusher, cruiser, infection behaviour, nictation, *Steinernema carpocapsae*, *Steinernema glaseri*

Introduction

The term 'entomopathogenic nematode' was first proposed by Gaugler and Kaya (1990) to identify insect–parasitic nematodes that kill their hosts quickly, usually in 48 hours. The 'quick-kill' by the entomopathogenic nematodes, represented by the two families Steinernematidae and Heterorhabditidae, is because they serve as vectors for the mutualistic-associated bacteria that are the true pathogens of the insects. The bacteria (*Xenorhabdus* spp. for steinernematids and *Photorhabdus* spp. for heterorhabditids) are found in the intestines of the third-stage infective nematode larvae (Kaya and Gaugler, 1993; Forst *et al.*, 1997). The infective larva (IL) infects its host by entering through natural openings (mouth, anus, spiracles) and then penetrating into the insect's haemocoel. The IL of heterorhabditid nematodes also has an anterior tooth that allows it to penetrate through soft insect cuticle directly into the haemocoel. The nematodes develop and reproduce in the insect cadaver by feeding on the bacterial cells and decayed host tissues. Under ideal conditions, the ILs begin exiting from the cadaver from 7 to 15 days after infection and search for new hosts.

Since the early 1980s, the development and use of entomopathogenic nematodes as biocontrol agents of insect pests have rapidly progressed because of their ease of mass-production, broad host range, and safety to mammals and the environment (see Kerry and Hominick, Chapter 19). The nematodes are adapted to the soil environment, and the nematodes' foraging ability makes them particularly excellent to control a number of insects in soil and cryptic habitats where they are protected from desiccation, ultraviolet light and temperature extremes. The developments in biocontrol have been concomitantly accompanied by significant advances in the taxonomy, systematics, physiology and ecology of these nematodes as well as understanding their relationships with the associated bacteria. Our insights on the behavioural ecology of these nematodes have also enhanced our understanding on how to best use them against insect pests. These nematodes, depending on species, have an 'ambush,' 'intermediate,' or 'cruise' foraging strategy. Using the nematode's behaviour as one important criterion, the appropriate nematode species can be matched against an insect pest to obtain more effective insect control. In this chapter, I will focus on the behavioural ecology of these nematodes with the emphasis on the ambush and cruise foraging strategy.

Dispersal Movement

ILs of entomopathogenic nematodes (EPNs) are motile and can search for their host. Their motility is dependent upon their foraging strategy. For example, the motility of the ambusher *Steinernema carpocapsae* IL is limited in comparison to the cruiser *S. glaseri* IL (Kaya, 1990).

A number of studies have shown that *S. carpocapsae* tends to remain where it is placed in the soil environment. However, very little research has been conducted to compare the movement of ILs and the developmental third-stage nematodes. Ishibashi and Kondo (1990) demonstrated that the motility of *S. carpocapsae* IL is nearly twice that of the third-stage developmental larva. Wavelength of the two IL species was almost the same (44 μm for developmental vs. 49 μm for the IL) when placed on 2% agar, but the amplitudes were 27 μm for the developmental larva and 17 μm for the IL (Figure 20.1). Consequently, the IL can migrate 245 μm/min on the agar surface compared

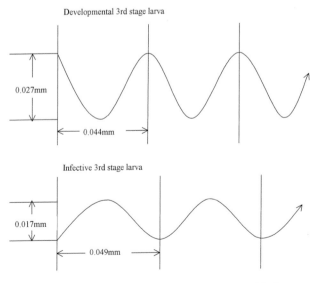

Developmental 3rd stage larva

0.027mm

0.044mm

Infective 3rd stage larva

0.017mm

0.049mm

Figure 20.1. Schematised undulatory movement of *Steinernema carpocapsae* from tracks left on a 2% water-agar film.

with 132 μm/min for the developmental larva. Undulations in physiologically balanced saline solution were 77/min for the former and 28/min for the latter on average for 4 days after incubation at 25°C (Ishibashi and Kondo, 1990). Thus, ILs are more active and faster than the developmental stage larvae (Figure 20.2). This is not only the prerequisite for the ILs to infect mobile insects but also promotes the dispersal movement.

Dispersal behaviour allows the free-living IL to locate a suitable habitat for survival and infection. This dispersal can be demonstrated in the laboratory. ILs are easily collected from the *in vitro* culture medium with blotting paper, such as used in descending chromatography (Ishibashi, unpublished). Dispersal of ILs also can be seen on the inner lid or wall of the culture vessel. To quantify this behaviour, the number of ILs ascending the inner lid of a petri dish was recorded (Figure 20.3). In the absence of a host insect, more *S. carpocapsae* ILs ascended towards the darker, inner lid of the culture vessel (e.g., petri dish) than towards the light, whereas with a host insect present, more ILs dispersed on the lid toward the light. These data indicate that ILs of *S. carpocapsae* disperse and randomly search for hosts under dark conditions; however, when hosts are present, their dispersal behaviour changes.

Ambusher and Cruiser

Ishibashi and Kondo (1986) made the seminal observation that *Steinernema feltiae* (= *S. carpocapsae*) ILs became quiescent shortly after being placed in water or soil. This behaviour has resulted in a low recovery rate from soil by the Baermann funnel method. Although some of the ILs move upward from the placement site in the soil, most remain at the placement site. On the other hand, *S. glaseri* ILs are very active, highly motile, and move out of the field of observation when viewed with a dissecting microscope. This nematode species moves downward from the soil placement site. There is a remarkable difference in behaviour between these two

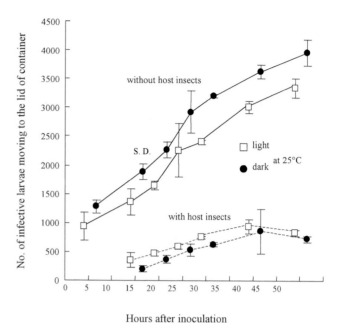

Figure 20.3. Dispersal movement of *Steinernema carpocapsae* infective larvae in the light or in the dark with or without host insects. Dispersal was recorded as the number of nematodes that moved upwards on the lid of the culture vessel.

nematode species. In addition, *S. carpocapsae* displays nictation behaviour on the substrate surface and tends to remain in a given patch, whereas *S. glaseri* actively disperse throughout the soil profile. These observations suggest that the nematodes exhibit different foraging strategies, and Kaya and Gaugler (1993) and Lewis *et al.* (1992, 1993) proposed two categories; 'ambusher' for *S. carpocapsae* and 'cruiser' for *S. glaseri* (Figure 20.4). These represent the extremes of the foraging continuum; the former represents the ambushing end of the continuum and the latter the cruising end of the continuum.

Figure 20.5 illustrates the difference between ambusher and cruiser in the route from host–finding to penetration. Ambushers have a sit-and wait strategy (*S. scapterisci* and *S. carpocapsae*) remaining near or at the soil surface, nictate, and attach to and infect mobile insects that feed at the soil-litter surface (Kaya, 1996). Cruisers have a widely foraging strategy (e.g., *S. glaseri* and *Heterorhabditis bacteriophora*) and, therefore, are highly motile, respond to chemical cues from the host, and are adapted to infect less mobile insects in the soil. There is a continuum between the two extremes with *S. feltiae* and *S. riobravis*, for example, having an intermediate foraging strategy.

Lewis (1994) defined the foraging strategies as follows. Ambushers have lower metabolic rates, are less motile and respond poorly to remote host cues, whereas cruisers have higher metabolic rates, are highly motile and are responsive to remote host cues. He further proposed five working hypotheses for explaining the foraging behaviour of these nematodes. (1) Cruise-foraging nematodes rely upon chemical cues

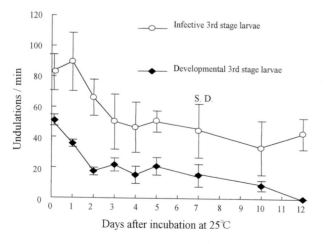

Figure 20.2. Change of the undulatory strength of *Steinernema carpocapsae* infective 3rd stage larvae and developmental 3rd stage larvae in buffered saline solution with time at 25 °C.

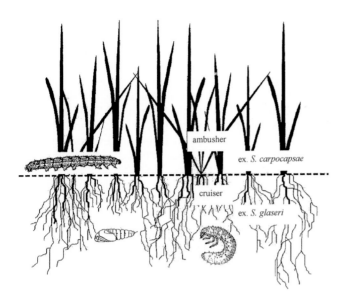

Figure 20.4. Target insects of ambusher and cruiser nematodes. Ambushers (e.g. *Steinernema carpocapsae*) sit-and-wait for passing insects at the upper soil layer. Cruisers (e.g. *Steinernema glaseri*) migrate downward in the soil and infect sedentary insects. (Original figure from Kaya, 1996).

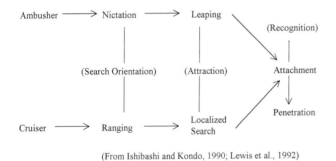

Figure 20.5. Host-finding behaviour of the two types of foragers, describing different sequences of behaviour from search orientation (nictation or ranging) to attraction (leaping or localised search), then to recognition (attachment = penetration).

for host location since their hosts are sedentary, whereas ambush foragers rely upon host movements (Lewis, Gaugler and Harrison, 1993). (2) Since the order in which cues are encountered by ambushers follows a relatively strict hierarchy, one cue may act as a 'releaser' for a response to a subsequent cue (Campbell and Gaugler, 1993). (3) Since cruise foragers (*S. glaseri* and *H. bacteriophora*) rely upon sensory input for host location, lectins should bind specifically to their amphids, whereas lectin binding to ambushers should be less specific. (4) Protandry as a mating strategy should occur among cruise foragers more so than ambush foragers. (5) If ambush foragers attach to highly mobile hosts, they should remain near the soil

surface, whereas cruisers should forage throughout the soil matrix.

Lewis' (1994) proposals have some validity, which we have confirmed, but the 4th hypothesis on protandry still remains questionable because there is no evidence that cruiser-type (*S. glaseri*) should show more protandry than ambusher-type (*S. carpocapsae*). Moreover, we need to confirm that early emerging ILs from insect cadavers should consist of more males than females. Nictating ILs of *S. carpocapsae* are supposedly composed of more males than females. Following Lewis' hypothesis, it may be that ambushers are also protandrous.

The infection behaviour of *S. carpocapsae* as an ambusher representative and *S. glaseri* as a cruiser representative will be discussed in relation to their host-finding activities, infection process, and species maintenance. In addition, we will examine the sex ratios to confirm whether protandry occurs in the ambusher and cruiser species under different conditions.

Nictation Behaviour

Ambusher-type entomopathogenic nematodes exhibit nictation behaviour. Figure 20.6 schematically illustrates the nictation behaviour from its initiation to leaping. Nictation starts when an IL standing on its tail, elevates its anterior end, and waves its head from side to side. This behaviour includes large waving, followed by small waving, pendulum waving, riding, and/or straight standing. Initially, the IL makes large waves of the head, with some individuals forming bridges with their bodies to adjacent surface projections, as if the large waving is intended for host attachment. Pendulum waving and straight standing IL occasionally provides a surface for other ILs to start nictating; other ILs crawl on these ILs to become riders. When many nematodes participate in the 'riding' formation, an aggregate is created that resembles the activity of one 'organism'. This 'organism' has a better chance of survival than a single nematode under unfavorable

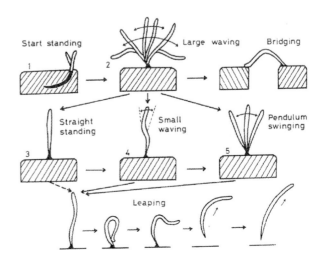

Figure 20.6. Sequential behaviour from standing to leaping of *Steinernema carpocapsae* infective larvae.

conditions such as desiccation or sunlight. That is, under these conditions ILs on the periphery of the nematode aggregation or clump will die because of the harsh environment but those within the clump have a better chance for survival (Womersley, 1990, 1993).

Small waving and pendulum waving or straight standing IL can also shift to 'leaping' or 'jumping'. When it leaps, the IL suddenly makes a large wave and leaps with a sudden reaction following loop formation (Reed and Wallace, 1965; Ishibashi and Kondo, 1990). Recently, Campbell and Kaya (1999) re-described the leaping phenomenon showing that the standing IL quickly bends the anterior half of its body until its head region makes contact with the ventral side of the body thus forming a loop. The loop appears to be held by a water film covering the IL's body. The IL has resistance to the bending of its body, so it can use its normal sinusoidal crawling behaviour to slide its body in a posterior direction. This causes the loop to become progressively smaller and the bend in its body becomes more acute. The body becomes so contorted that the cuticle on the dorsal side becomes extremely stretched and the cuticle on the ventral side kinks. This action generates sufficient force to break the surface tension forces holding the two body parts together. The body straightens out with sufficient force to break the surface tension forces holding the nematode to the substrate and propel it through the air. Leaping can be induced by a sudden change of milieu such as flashing a light or lifting the lid of the culture vessel.

Clearly, nictation behaviour is effective in assisting the ambusher ILs to attach to mobile hosts. According to Campbell and Gaugler (1993), (1) a nictating species (i.e., an ambusher) is more effective in finding mobile insect hosts than a non-nictating one (i.e., a cruiser), and (2) a nictating species tends to search along a surface rather than in a three dimensional matrix. Thus, soil moisture suitable for nictation increases the probability of high insect mortality for a mobile insect (Kondo and Ishibashi; 1985; 1986). Nictation is activated by some kinds of chemical pesticides (Ishibashi and Takii; 1993, Ishibashi, Takii and Kondo, 1994). For example, 100 to 500 ppm oxamyl enhanced the nictation level soon after treatment; nictation increased two-fold compared with nictation in the water control. On day 2, however, the nictation level precipitously declined lower than the control. Acephate and permethrin from 20 to 200 ppm kept the nictation rate at the same level as the water control throughout the 4-day study. Ishibashi and Takii (1993) suggested that nictating behaviour could be adopted as an indicator for screening pesticides to determine if they are compatible with beneficial nematodes because not all chemical pesticides enhance the nictation rate. Dichlorvos and methomyl suppressed the nictating behaviour of *S. carpocapsae* ILs even at 50 ppm and should not be applied simultaneously with nematodes used for pest suppression.

Nictation of *S. carpocapsae* ILs is a precursory behaviour to infection. High nictation rate predicts a high penetration efficiency into a host insect. We have observed a parallel relationship between nictation and infection. There was a high correlation (r = 0.720) between nictation and infection. In Figure 20.7, one dot indicates one batch population, namely, one population with a 50% nictation rate shows 50% infection on the regression line. Infectivity of nictating ILs is always higher than that of non-nictating ILs. Moreover, male ratios of penetrating ILs were 55.4%, 37.2% and 70% for nictating, non-nictating and nictating ILs, respectively. The 70% male ratio was obtained from ILs placed at the bottom of the vessel (Figure 20.8).

Another observation with nictation behaviour showed that early emerging populations of ILs from insect cadavers had a higher rate than those emerging later. In this experiment, *S. carpocapsae* ILs that emerged from insect cadavers on days 1, 4 and 8 were collected separately and observed for nictation. The later emerging ILs raised their infectivity with increasing storage time and nictation rate increased concomitantly with infectivity. Figure 20.9 shows the declining tendency of nictation rate of *S. carpocapsae* ILs emerging from insect cadavers with the delayed emergence. Figure 20.10 indicates the gradual increase in the infectivity of ILs of the day 8 emergence group; the ILs of day 1 emergence showed peak infectivity 4 days after incubation but beyond that the insect mortality rate declined. In contrast, day 8 ILs had more mortality of insects with time beyond 4 days than those of day 1 and day 4. The data showed that the later emerging ILs were not defective nematodes but they kept their infectivity longer than the early emerging ILs. Other studies have shown that the nictation rates are significantly lower in the presence of host insects than in their absence (Wang and Ishibashi, 1998), and that nictation rates are significantly higher in the presence of already infected hosts than in the non-infected (virgin) hosts (Wang and Ishibashi, 1998).

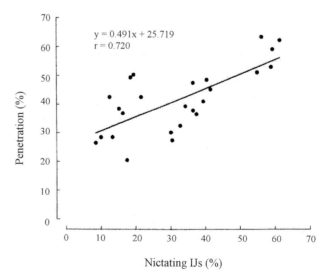

Figure 20.7. Correlation between nictation rate (X) and penetration rate (Y) into wax moth larvae by Steinernema carpocapsae infective larvae. One dot indicates one batch population. IJ's = IL's.

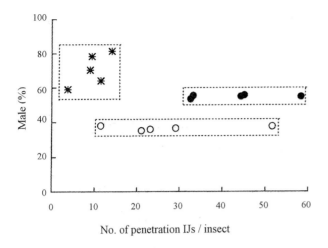

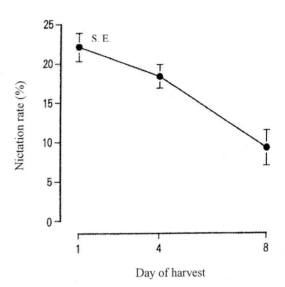

Figure 20.8. Relation between sex ratio (% male) of *Steinernema carpocapsae* and the number of nematodes invading the host insect, *Galleria mellonella*, showing a high male ratio from nictating infective larvae.
- ●: nictating ILs, being inoculated on a wax moth larva in a petri dish.
- ○: ILs collected from a nematode-water suspension, being inoculated on a wax moth larva in a petri dish.
- ★: nictating ILs, being placed at the bottom of a sandy column (5 cm- deep). The host insect, wax moth larva, was placed on the surface top of the column.
- IJ's = IL's.

Foraging Behaviour

As described above, the earlier emerging ILs from cadavers were more active in their host foraging behaviour than the later emerging ones. This staggered or 'phased' infectivity period may prevent the species from local extinction. The earlier emerging ILs were highly infectious, whereas the late emerging ILs were less infectious at the beginning, but became gradually more infectious than the earlier population with the

Figure 20.9. Showing the declining tendency of nictation rate *S. carpocapsae* ILs emerging from insect cadavers with delayed emergence.

passage of time (Figure 20.10). Fan and Hominick (1991) also demonstrated that infectivity of *S. feltiae* ILs was enhanced after being stored at 8°C with the same tendency to increase with time. Bohan and Hominick (1996) indicated that there was a density-independent infective proportion in a *S. feltiae* IL population. Dividing the reproductive sources into two or more groups has been demonstrated in plant seeds and plant-parasitic nematodes (Ishibashi, 1967) as well as in entomopathogenic nematodes. Staggering the infection period may be important for maintenance of the species in parasitic organisms (Kaya, 1996) and may be genetically programmed (Gaugler, Campbell and McGuire, 1989; Gaugler, McGuire and Campbell,

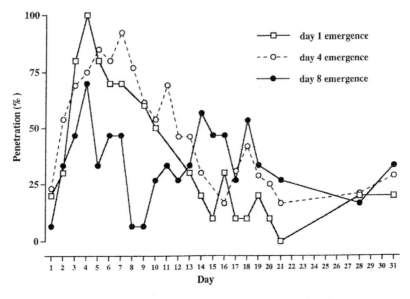

Figure 20.10. Showing gradual increase in the infectivity of *S. carpocapsae* ILs of the day 8 emergence group; the ILs of day 1 emergence shows peak infectictivity 4 days after incubation. Day 8 ILs had more mortality of insects with time beyond 4 days.

1989). However, according to Kaya (personal communication), 'phased' infectivity does not occur in all species; it occurs in some species and not others.

Grewal *et al.* (1994) suggested that entomopathogenic nematodes have some of the attributes of predators. That is, they are not parasitic but predacious because they cannot reproduce without first killing their host insects. In general, predators preclude the same brood from sharing the food. Although a predator is usually defined as the taking of a succession of living prey, the following experiments below were conducted to determine if these nematodes fit the predacious hypothesis for species maintenance as defined by Grewal *et al.* (1994).

Predacious Behaviour

In addition to a strategy adopted by steinernematid nematodes which circumvents their extinction by having active and inactive host seeking periods, another species maintenance strategy is the suppression of a subsequent invasion by a prior infection. In this study, greater wax moth larvae, *Galleria mellonella*, were injected with 50 ILs of either *S. carpocapsae* (Sc) or *S. glaseri* (Sg). At 0, 3, 6 and 9 hours after injection, the insects were exposed to ca. 100 ILs of either nematode species in a petri dish (5.5-cm-d) lined with moist filter paper. After 48 hours exposure, the insects were dissected to record the number of nematodes in the hosts. The subsequent invasion was carried out with

the following four combinations: Sc-Sc, Sc-Sg, Sg-Sc, Sg-Sg.

Figure 20.11 shows the invasion rates of *S. carpocapsae* and *S. glaseri* into greater wax moth larvae that were pre-infected by the injection of 50 ILs of either nematode species. When the insects were pre-infected with *S. carpocapsae* ILs, the subsequent invasion rates were significantly decreased from the 6-hour post-infection: 45.4, 42.6, 21.2 and 20.8% for the subsequent same *S. carpocapsae* (Figure 20.11A) and were 23.2, 22.5, 5.2 and 3.9% for *S. glaseri* (Figure 20.11B) after 0, 3, 6 and 9 hours for the subsequent inoculation, respectively. The invasion rates of *S. carpocapsae* into *G. mellonella* larvae were significantly decreased from 6 hours after prior infection with either *S. carpocapsae* or *S. glaseri*. These results concur with Glazer's findings (Glazer, 1997). Prior infection with *S. glaseri* resulted in a more drastic decrease in the subsequent invasion for the both nematode species immediately after pre-infection by injection. The invasion rates at the post-exposure time noted above were 11.0, 8.0, 6.6 and 6.0% for *S. carpocapsae* (Figure 20.11C), and 3.8, 4.4, 4.6 and 5.8% for the same time frame with *S. glaseri* (Figure 20.11D), respectively (Wang and Ishibashi, 1998). Thus, in our studies, pre-infection by *S. glaseri* drastically reduced the subsequent invasion by both nematode species immediately after the pre-infection injections. All of these values were significantly lower than those for the insects inoculated and

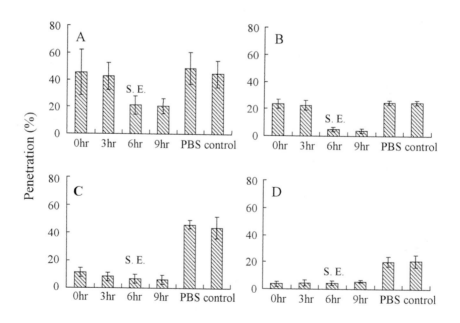

Hours after injection

Figure 20.11. Invasion of *S. carpocapsae* (Sc) and *S. glaseri* (Sg) to *Galleria mellonella* larvae injected and not injected with ca. 50 infective juveniles (ILs) into the insect haemocoel. On 0, 3, 6 and 9 hours after the injection, the insects were exposed for 48 hours to ca. 100 infective ILs.
A: Exposure to Sc after Sc injection.
B: Exposure to Sg after Sc injection.
C: Exposure to Sc after Sg injection.
D: Exposure to Sg after Sg injection.

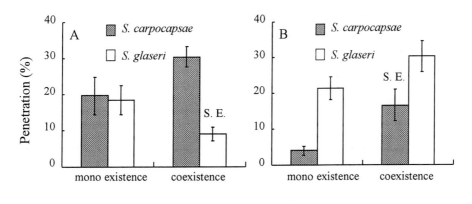

Figure 20.14. Penetration of *Steinernema carpocapsae* or *S. glaseri* infective larvae to *G. mellonella* larvae placed at 2 cm (A) or 7 cm depth (B) below the surface of sand column (7-cm-h). Small bars are standard errors of means for five replicates.

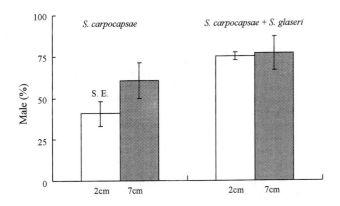

Figure 20.15. Sex ratio (% male) of *Steinernema carpocapsae* infective larvae penetrating into *Galleria mellonella* larvae placed 2 cm or 7 cm below the surface in sand column (7-cm-h) after application of *S. carpocapsae* ILs alone or with *S. glaseri* ILs on the surface of the sand. Small bars are standard errors of means for five replicates.

such as *S. glaseri*. We cannot explain why *S. carpocapsae* ILs were activated by the presence of *S. glaseri* though the former nematodes are usually more inactive or become quiescent soon after placement into a soil environment. Our results suggest that more effective control of insect pests is obtained by the combination of other biological agents with the application of *S. carpocapsae*. When we have two targets for controlling soil insect pests, e.g. sedentary insects dwelling deeper in soil and actively moving insects in the surface layers of soil, the mixed application of *S. glaseri* (cruiser) and *S. carpocapsae* (ambusher) should be recommended and can be expected to provide more effective control.

Acknowledgement

The author expresses his appreciation to Professor Harry K. Kaya of the University of California Davis, for improving the manuscript with valuable criticism.

References

Bohan, D.A. and Hominick, W.M. (1996) Investigations on the presence of an infectious proportion amongst populations of *Steinernema feltiae* (Site 76 strain) infective stages. *Parasitology*, **12**, 113–118.

Campbell, J.F. and Gaugler, R. (1993) Nictation behavior and its ecological implications in the host search strategies of entomopathogenic nematodes (Heterorhabditidae and Steinernematidae). *Behaviour*, **126**, 155–169.

Campbell, J.F. and Kaya, H.K. (1999) How and why a parasitic nematode jumps. *Nature*, **397**, 485–486.

Crankshaw, O.S. and Matthews, R.W. (1981) Sexual behavior among parasitic *Maharhyssa* wasps (Hymenoptera: Ichneumonidae). *Behaviour, Ecology, and Sociobiology*, **9**, 1–7.

Dunphy, G.B. and Webster, J.M. (1984) Interaction of *Xenorhabdus nematophilus* subsp. *nematophilus* with the haemolymph of *Galleria mellonella*. *Journal of Insect Physiology*, **30**, 883–889.

Dunphy, G.B. and Thurston, G.S. (1990) Insect Immunity. In *Entomopathogenic Nematodes in Biological Control*, edited by R. Gaugler and H.K. Kaya, pp. 301–323. Boca Raton, FL: CRC Press.

Fan, X. and Hominick, W.M. (1991) Effects of low storage temperature on survival and infectivity of 2 *Steinernema* species (Nematoda: Steinernematidae). *Review de Nématologie*, **14**, 407–412.

Forst, S., Dowds, B., Boemare, N., and Stackebrandt, E. (1997) *Xenorhabdus* and *Photorhabdus* spp.: Bugs that kill bugs. *Annual Review of Microbiology*, **51**, 47–72.

Gaugler, R. and Kaya, H.K. (1990) *Entomopathogenic Nematodes in Biological Control*, pp. 365. Boca Raton, FL: CRC Press.

Gaugler, R. Campbell, J.F. and McGuire, T. (1989) Selection for host-finding in *Steinernema feltiae*. *Journal of Invertebrate Pathology*, **54**, 363–372.

Gaugler, R., Glazer, I., Campbell, J.F. and Liran, N. (1994) Laboratory and field evaluation of an entomopathogenic nematode genetically selected for improved host-finding. *Journal of Invertebrate Pathology*, **63**, 68–73.

Gaugler, R., McGuire, T. and Campbell, J.F. (1989) Genetic variability among strains of the entomopathogenic nematode *Steinernema feltiae*. *Journal of Nematology*, **21**, 247–253.

Glazer, I. (1997) Effects of infected insects on secondary invasion of steinernematid entomopathogenic nematodes. *Parasitology*, **114**, 597–604.

Grewal, P.S., Gaugler, R., Kaya, H.K. and Wusaty, M. (1993a) Some aspects of infectivity of the entomopathogenic nematode *Steinernema scapterisci* (Nematoda: Steinernematidae). *Journal of Invertebrate Pathology*, **62**, 22–28.

Grewal, P.S., Selvan. S., Lewis, E.E. and Gaugler, R. (1993b) Males as the colonizing sex in insect parasitic nematodes. *Experientia*, **49**, 605–608.

Grewal, P.S., Lewis, E.E., Gaugler, R. and Campbell, J.F. (1994) Host finding behaviour of a predator of foraging strategy in entomopathogenic nematodes. *Parasitology*, **108**, 207–215.

Ishibashi, N. (1967) Studies on the propagation of the root-knot nematode, *Meloidogyne incognita* (Kofoid & White 1919) Chitwood, 1949. *Journal of the Central Agricultural Experiment Station*, **11**, 177–219.

Ishibashi, N. and Kondo, E. (1986) A possible quiescence of the applied entomopathogenic nematode, *Steinernema feltiae*, in soil. *Japanese Journal of Nematology*, **6**, 66–67.

Ishibashi, N. and Kondo, E. (1990) Behavior of Infective Juveniles. In *Entomopathogenic Nematodes in Biological Control*, edited by R. Gaugler and H.K. Kaya, pp. 139–150. Boca Raton, FL: CRC Press.

Ishibashi, N. and Takii, S. (1993) Effects of insecticides on movement, nictation, and infectivity of *Steinernema carpocapsae*. *Journal of Nematology*, **25**, 204–213.

Ishibashi, N., Takii, S. and Kondo, E. (1994) Infectivity of nictating juveniles of *Steinernema carpocapsae* (Rhabditida: Steinernematidae). *Japanese Journal of Nematology*, **24**, 21–29.

Ishibashi, N., Wang, X.D. and Kondo, E. (1994) *Steinernema carpocapsae*: Poststorage infectivity and sex ratio of invading infective juveniles. *Japanese Journal of Nematology*, **24**, 60–68.

Kaya, H.K. (1990) Soil ecology. In *Entomopathogenic Nematodes in Biological Control*, edited by R. Gaugler and H.K. Kaya, pp. 93–115. Boca Raton, FL: CRC Press.

Kaya, H.K. (1996) Contemporary issues in biological control with entomopathogenic nematodes. In *Biological Pest Control*, pp. 1–13. Food & Fertilizer Technology Center for the Asian and Pacific Region, Book series No.47, Taipei, Taiwan,.

Kaya, H.K. and Gaugler, R. (1993) Entomopathogenic nematodes. *Annual Review of Entomology*, **22**, 859–864.

Kondo, E. and Ishibashi, N. (1985) Effect of soil moisture on the survival and infectivity of the entomogenous nematode, *Steinernema feltiae* (DD-136). *Proceeding of Associated Plant Protection in Kyushu*, **31**, 186–190.

Kondo, E. and Ishibashi, N. (1986) Nictating behavior and infectivity of entomogenous nematodes, *Steinernema* spp., to the larvae of common cutworm, *Spodoptera litura* (Lepidoptera: Noctuidae), on the soil surface. *Applied Entomology and Zoology*, **21**, 553–560.

Lewis, E.E. and Cane, J.H. (1992) Inefficacy of courtship stridulations as a premating ethological barrier for *Ips* bark beetles (Coleoptera: Scolytidae). *Annals of Entomological Society of America*, **85**, 517–524.

Lewis, E.E., Gaugler, R. and Harrison, R. (1992) Entomopathogenic nematode host finding: response to host contact cues by cruise and ambush foragers. *Parasitology*, **105**, 109–115.

Lewis, E.E., Gaugler, R., & Harrison, R. (1993) Response of cruiser and ambusher entomopathogenic nematodes (Steinernematidae) to host volatile cues. *Canadian Journal of Zoology*, **71**, 765–769.

Lewis, E.E. (1994) Prediction of entomopathogenic nematode life history parameters based on foraging strategies. *Proceedings VIth International Colloquium on Invertebrate Pathology and Microbial Control*. Vol. 1, pp. 109–114. Montpellier, France.

Peters, A. and Ehlers, R.U. (1994) Susceptibility of leatherjackets (*Tipula paludosa* and *T. oleracea*, Tipulidae, Nematocera) to the entomopathogenic nematode *Steinernema feltiae*. *Journal of Invertebrate Pathology*, **63**, 163–171.

Reed, E.M. and Wallace, H.R. (1965) Leaping locomotion by an insect parasitic nematode. *Nature*, **206**, 210–211.

Selvan, S., Campbell, J.F., and Gaugler, R. (1993) Density-dependent effects on entomopathogenic nematodes (Heterorhabditidae and Steinernematidae) within an insect host. *Journal of Invertebrate Pathology*, **62**, 278–284.

Wang, Y. and Gaugler, R. (1995) Infection of entomopathogenic nematodes *Steinernema glaseri* and *Heterorhabditis bacteriophora* against *Popillia japonica* (Coleoptera: Scarabaeidae) larvae. *Journal of Invertebrate Pathology*, **66**, 178–184.

Wang, X.D. and Ishibashi, N. (1998) Effect of precedent infection of entomopathogenic nematodes (Steinernematidae) on the subsequent invasion of infective juveniles. *Japanese Journal of Nematology*, **28**, 8–16.

Wang, X.D. and Ishibashi, N. (1999) Infection of the entomopathogenic nematode, *Steinernema carpocapsae*, as affected by the presence of *Steinernema glaseri*. *Journal of Nematology*, **31**, 207–211.

Wang, X.D., Ishibashi, N. and Kondo, E. (1997) Infectivity of *Steinernema carpocapsae*, as affected by the emergence time from the host cadaver. *Japanese Journal of Nematology*, **27**, 1–6.

Womersley, C.Z. (1990) Dehydration survival and anhydrobiotic potential. In *Entomopathogenic Nematodes in Biological Control*, edited by R. Gaugler and H.K. Kaya, pp. 117–137. Boca Raton, FL: CRC Press.

Womersley, C.Z. (1993) Factors affecting physiological fitness and modes of survival employed by dauer juveniles and their relationship to pathogenicity. In *Nematodes and the Biological Control of Insect Pests*, edited by R. Bedding, R. Akhurst, and H. Kaya, pp. 79–95. Melbourne, Australia: CSIRO Publications.

21. Chemical Control of Animal-Parasitic Nematodes

George A. Conder

Central Research Division, Pfizer Inc., Eastern Point Road, Groton CT 06340, USA

Keywords

Anthelmintic, Endectocide, Nematicide, Nematode, Chemotherapy

Introduction

This chapter summarises the principal classes of modern broad-spectrum drugs which currently are available to control nematodes in animals and the attributes of each class, as well as briefly overviewing some narrow-spectrum drugs. Despite a very effective arsenal of nematicidal drugs, gaps still exist in our ability to deal with important nematodes, particularly filarial species. This is further complicated by the fact that resistance to the various classes of available drugs threatens control on several fronts. All references to individual drugs will use generic or non-proprietary names. A number of excellent publications are available which provide more depth on anthelmintics than is possible in this chapter (e.g. Bard, 1972; Gibson, 1975; Anonymous, 1983; Arundel, 1983; James and Gilles, 1985; Vanden Bossche, Thienpont and Janssens, 1985; Biehl, 1986; Campbell, 1986; Campbell and Rew, 1986; Coles, 1986; Prichard, 1986; Wescott, 1986; Gustafsson, Beerman and Aden Abdi, 1987; Raether, 1988; Campbell, 1989; Anonymous, 1990; Campbell, 1990; Cook, 1990; Lacey, 1990; McKellar and Scott, 1990; Courtney and Sundlof, 1991; Conder and Campbell, 1995). Drugs such as phenothiazine, disophenol, rafoxanide, the chlorinated hydrocarbons (n-butyl chloride, tetrachlorethylene), ethanolamines (bephenium, methridine), cyanine dyes (pyrvinium, dithiazanine), hygromycin B, santonin, and kainic acid have largely been replaced by safer anthelmintics with a broader spectrum of activity; hence, they have little but historical value and will not be detailed herein. No attempt will be made herein to discuss compounds used to control nematodes of plants.

Modern Broad-Spectrum Anthelmintics

The various classes of modern broad-spectrum anthelmintics used in animals are listed in Table 21.1 along with the principal drugs for each class, purported mode of action, formulations, target hosts, and spectrum.

Benzimidazoles

Benzimidazoles (Figure 21.1), with the introduction of thiabendazole in 1961 (Brown et al., 1961), were the first of the truly broad-spectrum anthelmintics. Two pro-benzimidazoles, febantel and thiophanate (Figure 21.2), also should be considered with this class. Compounds in this class are synthetically produced and have

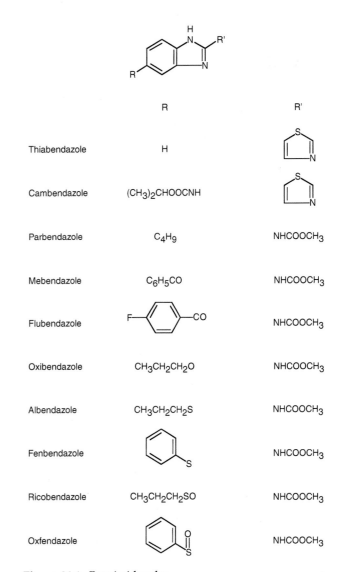

Figure 21.1. Benzimidazoles

Table 21.1. Modern broad-spectrum nematicidal drugs

Drug Class	Compounds	Mode-of-Action	Available Formulations	Principal Target Hosts	Spectrum
Benzimidazoles	Albendazole Cambendazole Febantel[a] Fenbendazole Flubendazole Mebendazole Oxfendazole Oxibendazole Parbendazole Ricobendazole Thiabendazole Thiophanate[a]	Inhibition of β-tubulin polymerisation	Oral, Pour-On, Spot-On, Injectable, Intraruminal, Sustained Release, Pulsed Release	Dog, Cat, Swine, Horse, Cattle, Sheep, Goats, Poultry, Human	Nematodes, Trematodes, Cestodes, Protozoa
Imidothiazoles	Butamisole Levamisole Tetramisole	Nicotinic agonist	Oral, Pour-On, Injectable, In-Feed, In-Water, Mineral Supplement, Sustained Release	Swine, Cattle, Sheep, Goats, Poultry, Human	Nematodes
Macrocyclic Lactones	Abamectin Doramectin Eprinomectin Ivermectin Milbemycin D Milbemycin Oxime Moxidectin Selamectin	Irreversibly opens glutamate-gated chloride ion channels	Oral, Pour-On, Spot-On, Injectable, In-Feed, Sustained-Release Bolus, Spray	Dog, Cat, Swine, Horse, Cattle, Sheep, Goats, Human	Nematodes, Arthropods
Tetrahydropyrimidines	Morantel Oxantel Pyrantel	Nicotinic agonist	Oral, In-Feed, Sustained Release	Swine, Horse, Cattle, Sheep, Goats, Human	Nematodes, Cestodes

[a] Probenzimidazoles

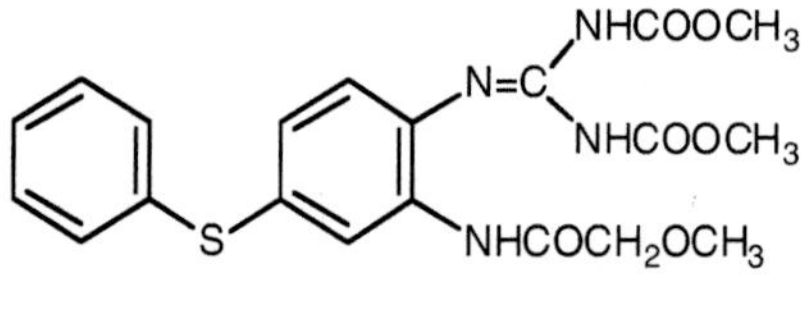

Febantel

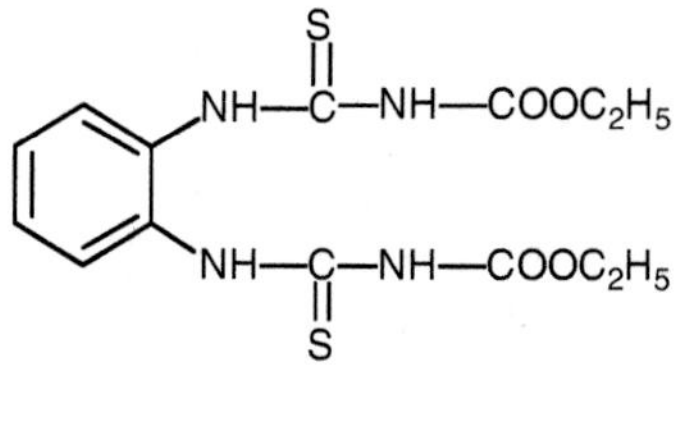

Thiophanate

Figure 21.2. Probenzimidazoles

many biological activities. Although other mechanisms have been implicated for these compounds, it appears that their effect on tubulin polymerisation is primarily responsible for their activity against nematodes, whereas secondary mechanisms generally can be related to the tubulin effects (reviewed by Lacey, 1990). This mechanism-of-action renders benzimidazoles active against protozoa, trematodes, and cestodes in addition to nematodes. Benzimidazoles also are ovicidal. An older and little used anthelmintic, phenothiazine, appears to share the benzimidazole mechanism-of-action (Rew and Fetterer, 1986) and it has been hypothesised that use of this drug may have preselected some populations of nematodes for benzimidazole resistance (Coles, 1988; Drudge *et al.*, 1990).

The mammalian toxicity of benzimidazoles is low; however, some members of the class have been shown to be teratogenic, which limits their use in pregnant animals. A variety of formulations exist, including oral, pour-on, spot-on, injectable, intraruminal, sustained-release, and pulse-release presentations. Excluding thiabendazole, doses of 5–10 mg/kg given once or in multiples in various dosing regimes are typically effective against nematodes. Current usage of this class has been limited by rapid generation of resistance in small ruminants and horses, as well as by the lack, with a single exception in each case, of pour-on (oxfendazole) and injectable (ricobendazole) formulations. This class had significant impact as the first anthelmintics with both safety and broad-spectrum activity. Notable among the activities of this class is its utility against mucosal stages of small strongyles in horses (multiple doses of fenbendazole) and muscle stages of *Trichinella spiralis*.

Imidothiazoles

This synthetic class is represented by tetramisole and levamisole (Figure 21.3), the former being a racemic mixture of D and L forms, whereas the latter is the L form. Activity resides in the L form, whereas toxicity is found with both. Hence, comparable doses of levamisole are more efficacious and safer than those of tetramisole. A third molecule, butamisole, also has been commercialised. These compounds act as agonists at an acetylcholine-gated cation channel in nematode muscle membranes (Martin *et al.*, 1996), and their receptor on nematode muscle is pharmacologically similar to mammalian nicotinic receptors (Fleming *et al.*, 1996; Martin *et al.*, 1996). The compounds have a narrow therapeutic index. They are not used in horses which are particularly sensitive (Clarkson and Beg, 1971) or in dogs and cats.

Imidothiazoles are inexpensive and have been formulated in virtually every feasible presentation (oral, pour-on, injectable, in-feed, in-water, mineral supplement, and sustained-release). Doses of 2.5–10 mg/kg are typically used. This class is useful against the majority of important nematodes but is weak against inhibited larvae. Imidothiazoles may gain some benefit from having immunostimulant effects on the host in addition to anthelmintic and ovicidal effects. Resistance is a significant problem for this class in small ruminants and swine.

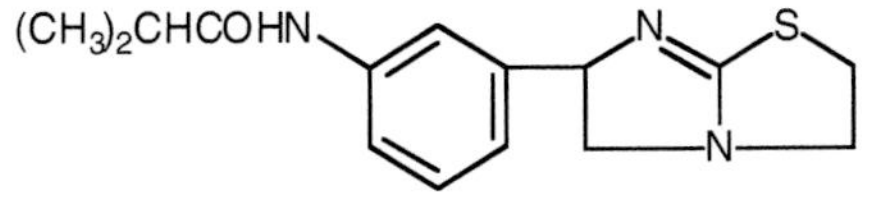

Butamisole

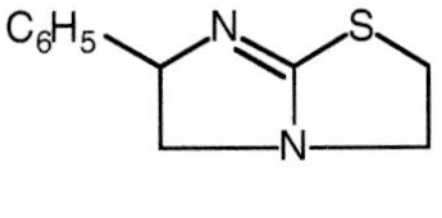

Tetramisole

Figure 21.3. Imidothiazoles

Macrocyclic Lactones

The most recent class of anthelmintics is the macrocyclic lactones. Two commercially viable subclasses exist in this class, the avermectins and the milbemycins (Figure 21.4). Macrocyclic lactones are fermentation-derived natural products. Natural avermectins are produced by *Streptomyces avermitilis*, an organism which was originally isolated from a soil isolate collected in Japan, while natural milbemycins are produced by a variety of *Streptomyces* spp. isolated throughout the world. Currently used molecules in this class include unaltered natural products (e.g. abamectin, doramectin, milbemycin D) or semi-synthetic molecules derived from the natural avermectins (eprinomectin, ivermectin, selamectin) or milbemycins (milbemycin oxime and moxidectin). In the case of doramectin, a mutational biosynthesis approach was exploited to produce a mutant *S. avermitilis* strain that serves as the producing strain. Although full synthetic approaches to some members of this class have been published (e.g. Danishefsky *et al.*, 1987a,b, 1989; Kornis *et al.*, 1991), it is currently not commercially viable to produce these molecules via total synthesis.

These molecules are believed to work by irreversibly opening glutamate-gated chloride channels (Cully *et al.*, 1994; Arena *et al.*, 1995; Molento, Wang and Prichard, 1999), a mechanism which is effective in controlling a broad spectrum of arthropod pests as well as nematodes. Hence, the term 'endectocide' is associated with compounds in this class. Selamectin, a new avermectin entry, has extended the endectocide properties of this class into companion animals for the first time, providing therapeutic and prophylactic control of fleas in addition to its nematode activity. Although there has and continues to be debate over whether both subclasses of the macrocyclic lactones utilise the same mechanism-of-action, the data suggest that the two subclasses share a common mechanism (Conder, Thompson and Johnson, 1993; Shoop *et al.*, 1993; Arena *et al.*, 1995). It has been proposed that the primary target for these drugs in nematodes is pharyngeal function, i.e., feeding, regulation of hydrostatic pressure, and/or secretion (Geary *et al.*, 1993; Brownlee, Holden-Dye and Walker, 1997). As there are beneficial species of arthropods (e.g. dung beetles) and nematodes (free-living species) which are affected by macrocyclic lactones, there have been questions raised as to the environmental impact. A recent review by the National Registry Authority of Australia (Anonymous, 1998) concluded that there is a short term impact on dung beetle populations in the microenvironment of faeces of treated animals for all members of the class but no long term or widespread impact to susceptible beneficial species or to faecal degradation. The class has an excellent safety profile in all target species, except in the dog where there is idiosyncratic toxicity in long-haired collie breeds. Selamectin exhibits an enhanced safety in these sensitive dogs, e.g., safe at ≥ 30 mg/kg ($\geq 5,000\times$ recommended minimum dose) versus 150 μg/kg ($25\times$ recommended minimum dose) for ivermectin in their respective formulations, topical and oral.

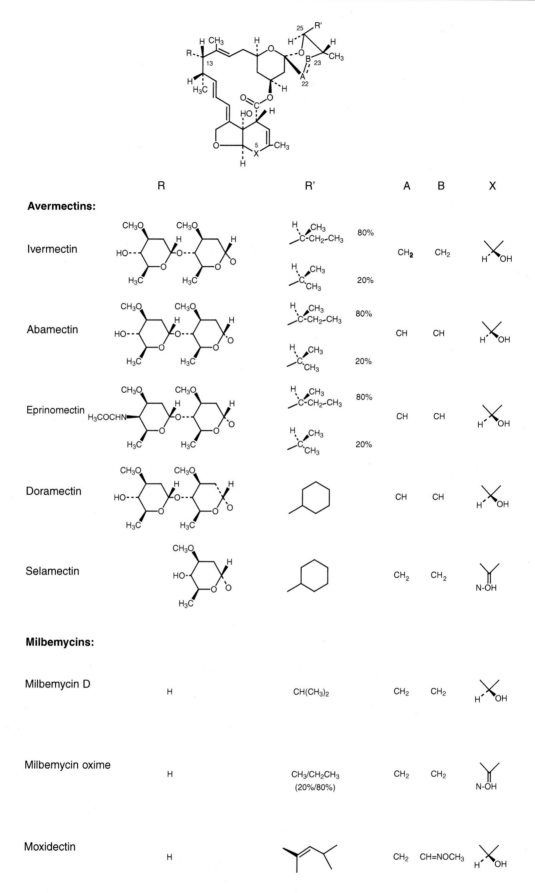

Figure 21.4. Macrocyclic lactones

The combination of broad-spectrum and persistent activity make this class an extremely important tool in not only treating existing infections, but also in reducing or preventing subsequent infection. A wide variety of delivery options exist (oral, pour-on, spot-on, injectable, in-feed, sustained-release bolus, and spray) to meet user needs. These compounds are extremely potent and are used at doses of 3–6,000 μg/kg depending on formulation and therapeutic target. Nematode resistance in small ruminants is a significant problem for the class.

Some of the particularly useful applications of this class which highlight its unique contribution to our chemotherapeutic arsenal include excellent activity against inhibited larvae in cattle and sheep, activity against the developing stages of the heartworm, *Dirofilaria immitis*, in dogs, and relatively safe microfilaricidal efficacy against *Onchocerca volvulus* in humans.

Tetrahydropyrimidines

Pyrantel, morantel, and oxantel (Figure 21.5) are the commercial members of the tetrahydropyrimidine class and are synthetically produced. With the exception of morantel, these drugs are poorly absorbed from the gut, providing high concentrations to target gut-dwelling nematodes. They are nicotinic agonists like the imidothiazoles, but they have a better therapeutic index. Unlike imidothiazoles, tetrahydropyrimidines are used in companion animals as well as farm livestock and humans.

These compounds are available as oral, in-feed, or sustained-release formulations. Doses of 6–30 mg/kg are typical. The class has good broad-spectrum activity against adult nematodes, but weak to no activity against developing and inhibited larval stages, depending on species and host. Combinations of pyrantel and oxantel have been used to enhance spectrum. In addition to nematicidal activity, pyrantel has utility against tapeworms in horses. A sustained-release bolus for cattle which provided continuous release of morantel over two months was the first product of its type, and it not only controlled nematodes in the host for an extended period of time, but it also effectively reduced pasture contamination (Jones, 1981). Resistance is a problem in small ruminants and swine, and also may be becoming a problem in horses (Herd, 1992; Chapman *et al.*, 1996; Coles, Brown and Trembath, 1999; Craven *et al.*, 1999).

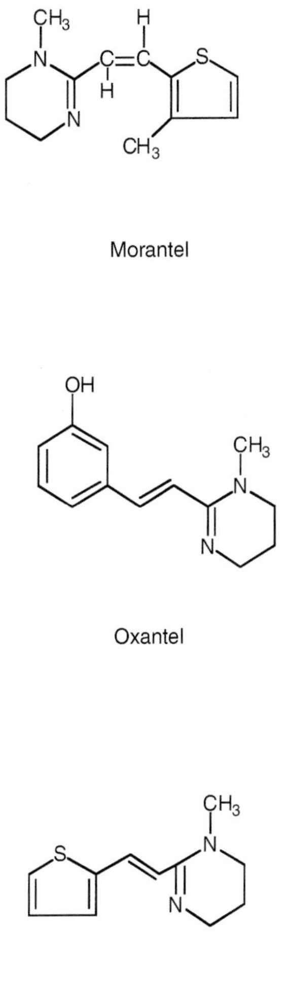

Narrow-Spectrum Nematicidal Drugs

Organophosphates

Organophosphates have only limited use today, due to both limited spectrum and toxicity. This synthetically produced class includes coumaphos, dichlorvos, haloxon, and trichlorphon (Figure 21.6). These compounds are thought to work by inhibiting acetylcholinesterase, resulting in continuous depolarisation of the postsynaptic junction and paralysis (Rew and Fetterer, 1986). This mechanism-of-action provides utility against both nematodes and arthropods. Organophosphates also

Morantel

Oxantel

Pyrantel

Figure 21.5. Tetrahydropyrimidines

inhibit mammalian acetylcholinesterase which can result in typical cholinergic signs of toxicity.

Formulations include oral, in-feed and slow-release pellet formulations. Typically a single dose of 5–100 mg/kg or lower dosages given daily for protracted periods are used. Compounds in this class have a limited spectrum and are generally weak against immature nematodes. They are used in cattle, swine, sheep, goats, and horses. Dichlorvos is particularly useful against *Trichuris suis* in swine, and until the advent of the macrocyclic lactones, trichlorphon was a mainstay in the treatment of stomach bots in horses.

Closantel

The salicylanilide closantel (Figure 21.7) is a synthetic compound with limited nematode activity. It binds very tightly to plasma proteins and has a long serum half-life. As a result of its plasma-protein binding, it is thought to be sufficiently available only to blood-feeding nematodes, and hence it is used almost exclusively to control *Haemonchus contortus* in sheep and goats where it protects therapeutically as well as

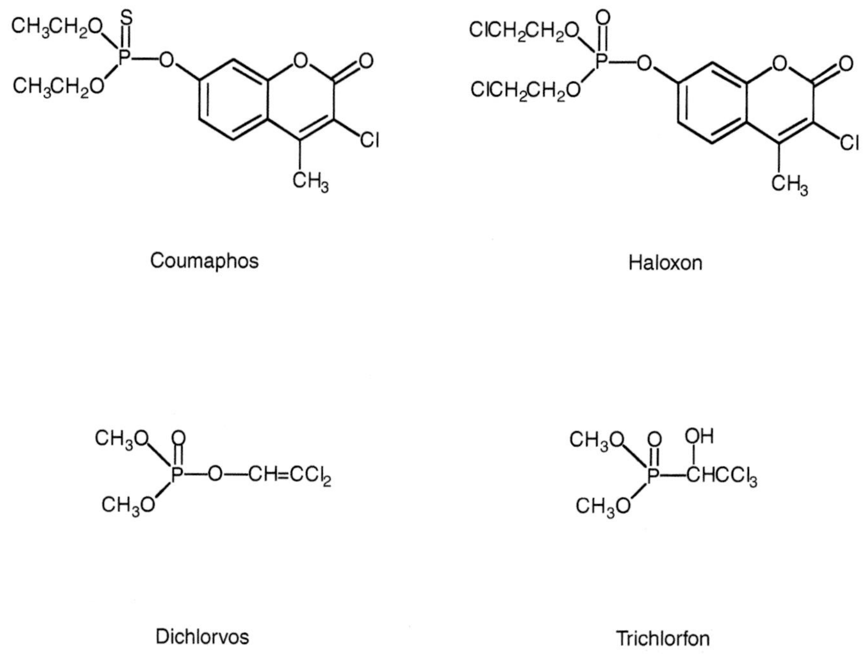

Figure 21.6. Organophosphates

prophylactically for up to a month following treatment. Closantel also has effective flukicidal activity. It is believed to work by uncoupling electron-transport-associated phosphorylation (Rew and Fetterer, 1986). Closantel is generally administered orally at a dose of 5–10 mg/kg. The long serum half-life of this compound presents a significant residue issue.

Nitroscanate

Nitroscanate (Figure 21.7) is an isothiocyanate, and this synthetic compound is used exclusively in dogs. Isothiocyanates are noted for their tapeworm activity; however, nitroscanate has utility against both nematodes and cestodes. Nitroscanate's mechanism-of-action may be related to its effect on glucose metabolism by influencing glucose transport, its synthesis into glycogen, and inhibiting its catabolism (Rew and Fetterer, 1986). The recommended dose for this compound is 50 mg/kg in a tablet formulation of micronised particles. In the micronised formulation, the compound has utility against the characteristic nematode and cestode spectrum in dogs, with the exceptions of *Trichuris vulpis* and *Echinococcus granulosus* (Rickard and Arundel, 1985). The principal problem associated with this drug is vomition which can be minimised by fasting prior to treatment followed by a small quantity of food immediately after treatment (Boray as reported in Rickard and Arundel, 1985).

Piperazine

Piperazine (Figure 21.7) is an older, synthetically produced drug. It is believed to act as a GABA-agonist, by opening chloride-ion channels which results in hyperpolarisation of muscle membrane and flaccid paralysis of the worm (Rew and Fetterer, 1986). Piperazine has a limited

spectrum and has been used primarily against ascarids and *Enterobius vermicularis*. Its use has declined significantly with the advent of newer drugs; nevertheless, piperazine remains useful in niche markets such as the poultry industry, where cost is a major consideration. A number of oral formulations exist and doses typically range from 20–200 mg/kg given once or in multiples using various dosing regimes. The compound is very safe but may on occasion cause transient nausea, vomiting, abdominal pain, diarrhoea, or ataxia.

Diethylcarbamazine

Although diethylcarbamazine (Figure 21.7) is a chemical derivative of piperazine, its mechanism-of-action remains unclear and does not appear to be the same as piperazine. Based on available data for microfilariae, diethylcarbamazine seems to interact with the parasite and/or the host immune system to render the worm susceptible to immunological attack (Rew and Fetterer, 1986), perhaps by affecting host endothelial and parasite eicosanoid production (Kanesa-thasan, Douglas and Kazura, 1991). Diethylcarbamazine has been utilised principally to control filariasis, where it remains a valuable weapon despite being an older drug. An interesting delivery approach for this drug is as a salt supplement which when used daily appears to be useful in the control of lymphatic filariasis. Diethylcarbamazine is given orally in doses of 0.5–100 mg/kg administered as a single dose or in multiples using various regimes. Side effects are similar to those for piperazine; however, diethylcarbamazine may indirectly result in an immunological reaction, known as the Mazzotti reaction in humans, resembling allergic responses, which can be cutaneous, ophthalmic, or systemic in nature, due to the compounds microfilaricidal activity.

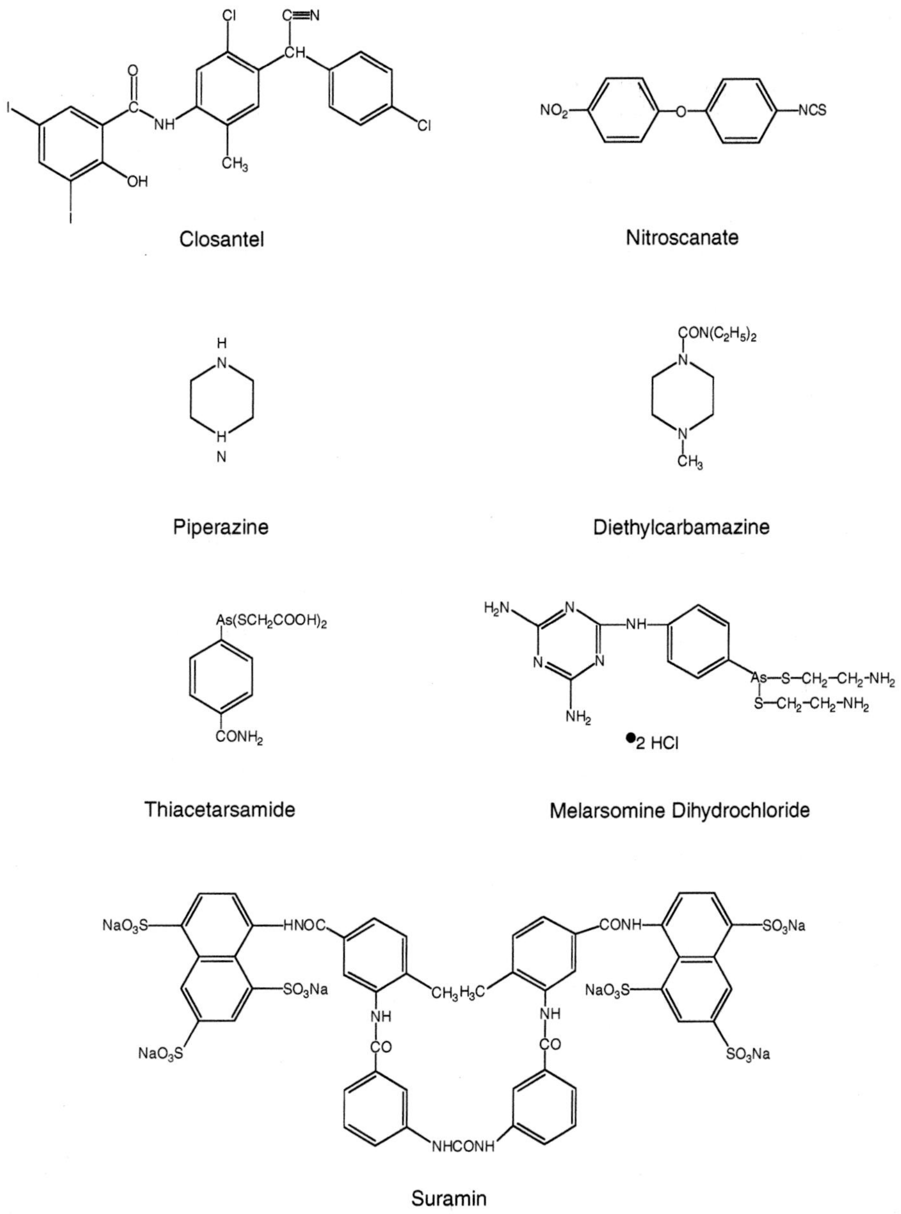

Figure 21.7. Narrow-spectrum anthelmintics

Suramin

Suramin is an old drug. It emerged serendipitously from prophylactic studies on dyes used to control trypanosomes. Suramin is a synthetic, non-metallic, symmetrical derivative of urea (Figure 21.7), which is used primarily to treat immature and adult *Onchocerca volvulus* in humans. Although the mechanism-of-action is not known, the compound is extremely reactive, exerting inhibitory effects on many enzyme systems at biologically relevant concentrations. It has strong affinity for proteins and other large molecules. Cell proliferation may be the drug's target (James and Gilles, 1985). To treat onchocerciasis, the World Health Organisation recommends the drug be administered by slow intravenous injection as a 10% solution in six incremental weekly doses of 3.3, 6.7, 10, 13.3, 16.7, and 16.7 mg/kg (Anonymous, 1990). Suramin is extremely toxic, both innately and indirectly due to the death of parasites. Severe reactions include potentially fatal collapse after first treatment, exfoliative dermatitis, heavy albuminuria, severe diarrhoea, prolonged high fever and prostration, and allergic reactions (particularly of the optic nerve and retina).

Arsenicals

Although arsenicals have a variety of nematicidal activities, they are used today for nematodes exclusively to control dog heartworm. Otto and Moren (1947) demonstrated the adulticidal effect of arsenicals against *Dirofilaria immitis* in dogs. Fifty years after this report arsenicals remain the only useful treatment for adults of this parasite, despite the obvious toxicity associated with the class. Both activity and toxicity for this class are believed to result from the binding of the arsenical to sulfhydryl groups. Thiacetarsamide (Figure 21.7) has been the drug of choice; however, melarsomine (Figure 21.7) recently has been shown to be a useful and perhaps safer alternative. Although both drugs depress vascular relaxation which can enhance embolisation of dead worms and subsequent acute pulmonary reactions in infected dogs, melarsomine has fewer vasodepressive effects (Maksimowich *et al.*, 1997). Typically, the drugs are administered intravenously (thiacetarsamide) or intramuscularly (melarsomine) at 2.2 mg/kg twice a day for two days or 2.5 mg/kg daily for two days, respectively. In severely infected animals, a single 2.5 mg/kg dosage 1 month prior to normal treatment regime is recommended.

Conclusion

In general, an impressive array of safe and effective nematicidal drugs, in a wide range of delivery options, is available to control nematode infections in humans and domestic animals. Nevertheless, new drugs are needed to control those species of nematodes which are inadequately dealt with at this time due to issues of safety and/or efficacy (e.g. filarial species), to replace existing drugs to which resistance is widespread, and to provide alternative control strategies. Although nematode vaccines, nematode-resistant breeds of production animals, biological control strategies (e.g. predacious fungi), and other innovative approaches to controlling nematodes offer promise, it will be some time before they provide a useful alternative to chemotherapy. It also is highly unlikely that non-chemotherapeutic approaches to nematode control will solve all nematode problems, and therefore, chemotherapeutic supplementation will be required. Hence, a continued effort to identify useful new nematicidal drugs is warranted and should be aggressively pursued.

Acknowledgements

The efforts of Drs. Chris Bruce and Ann Jernigan and Mr. Bernard Bishop in reviewing the manuscript are greatly appreciated.

References

Anonymous, A.A. (1983) *Anthelmintics for Cattle, Sheep, Pigs and Horses*. Ministry of Agriculture, Fisheries and Food. Booklet 2412 (83), pp. 78 Alnwick, UK: MAFF Publications.

Anonymous, A.A. (1990) *WHO Model Prescribing Information. Drugs used in parasitic disease*, pp. 126. Geneva: World Health Organization.

Anonymous, A.A. (1998) *NRA Special Review of Macrocyclic Lactones. NRA Special Review Series 98.3 by the Chemical Review Section, National Registration Authority*, pp. 79. Canberra, Australia: Commonwealth of Australia.

Arena, J.P., Liu, K.K., Paress, P.S., Frazier, E.G., Cully, D.F., Mrozik, H. *et al.* (1995) The mechanism of action of avermectins in *Caenorhabditis elegans*: Correlation between activation of glutamate-sensitive chloride current, membrane binding, and biological activity. *Journal of Parasitology*, **81**, 286–294.

Arundel, J.H. (1983) *Veterinary Anthelmintics*, pp. 90. Werribee, Australia: The University of Melbourne.

Bard, J.H. (1972) *Anthelmintic Index*, Technical Communication No. 43 of the Commonwealth Institute of Helminthology St. Albans, pp. 71. Farnham Royal, UK: Lund Humphries.

Biehl, L.G. (1986) Anthelmintics for swine. *Veterinary Clinics of North America: Food Animal Practice*, **2**, 481–487.

Brown, H.D., Matzuk, A.R., Ilves, I.R., Peterson, L.H., Harris, S.A., Sarett, L.H. *et al.* (1961) Antiparasitic drugs. IV. 2-(4'-thiazolyl)-benzimidazole, A new anthelmintic. *Journal of the American Chemical Society*, **83**, 1764–1765.

Brownlee, D.J.A., Holden-Dye, L. and Walker, R.J. (1997) Actions of the anthelmintic ivermectin on the pharyngeal muscle of the parasitic nematode, *Ascaris suum*. *Parasitology*, **115**, 553–561.

Campbell, W.C. (1986) The chemotherapy of parasitic infections. *Journal of Parasitology*, **72**, 45–61.

Campbell, W.C. (1989) *Ivermectin and Abamectin*, pp. 363. New York: Springer-Verlag.

Campbell, W.C. (1990) Benzimidazoles: Veterinary uses. *Parasitology Today*, **6**, 130–133.

Campbell, W.C. and Rew, R.S. (1986) *Chemotherapy of Parasitic Diseases*, pp. 655. New York: Plenum Press.

Chapman, M.R., French, D.D., Monahan, C.M. and Klei, T.R. (1996) Identification and characterization of a pyrantel pamoate resistant cyathostome population. *Veterinary Parasitology*, **66**, 205–212.

Clarkson, M.J. and Beg, M.K. (1971) Critical tests of levamisole as an anthelmintic in the horse. *Annals of Tropical Medicine and Parasitology*, **65**, 87–91.

Coles, G.C. (1986) Anthelmintics for small ruminants. *Veterinary Clinics of North America: Food Animal Practice*, **2**, 411–421.

Coles, G.C. (1988) Strategies for control of anthelmintic-resistant nematodes of ruminants. *Journal of the American Veterinary Medical Association*, **192**, 330–334.

Coles, G.C., Brown, S.N. and Trembath, C.M. (1999) Pyrantel-resistant large strongyles in racehorses. *Veterinary Record*, **145**, 408.

Cook, G.C. (1990) Use of benzimidazole chemotherapy in human helminthiasis: indications and efficacy. *Parasitology Today*, **6**, 133–136.

Conder, G.A. and Campbell, W.C. (1995) Chemotherapy of nematode infections of veterinary importance with special reference to drug resistance. *In Advances in Parasitology*, Vol. 35, edited by J.R. Baker, R. Muller and D. Rollinson, pp. 1–84. San Diego: Academic Press.

Conder, G.A., Thompson, D.P. and Johnson, S.S. (1993) Demonstration of co-resistance of *Haemonchus contortus* to ivermectin and moxidectin. *The Veterinary Record*, **132**, 651–652.

Courtney, C.H. and Sundlof, S.F. (1991) *Veterinary Antiparasitic Drugs*, pp. 224. Gainesville, Florida: University of Florida.

Craven, J., Bjørn, H., Barnes, E.H., Henriksen, S.A. and Nansen, P. (1999) A comparison of *in vitro* tests and a faecal egg count reduction test in detecting anthelmintic resistance in horse strongyles. *Veterinary Parasitology*, **85**, 49–59.

Cully, D.F., Vassilatis, D.K., Liu, K.K., Paress, P.S., Van der Ploeg, L.H.T., Schaeffer, J.M. *et al.* (1994) Cloning of an avermectin-sensitive glutamate-gated chloride channel from *Caenorhabditis elegans*. *Nature*, **371**, 707–711.

Danishefsky, S.J., Armistead, D.M., Wincott, F.E., Selnick, H.G. and Hungate, R. (1987a) The total synthesis of the aglycon of avermectin A (1a). *Journal of the American Chemical Society*, **109**, 8117–8119.

Danishefsky, S.J., Selnick, H.G., Armistead, D.M. and Wincott, F.E. (1987b) The total synthesis of avermectin A (1a). New protocols for the synthesis of novel 2-deoxypyranose systems and their axial glycosides. *Journal of the American Chemical Society*, **109**, 8119–8120.

Danishefsky, S.J., Armistead, D.M., Wincott, F.E., Selnick, H.G. and Hungate, R. (1989) The total synthesis of avermectin A (1a). *Journal of the American Chemical Society*, **111**, 2967–2980.

Drudge, J.H., Lyons, E.T., Tolliver, S.C. and Fallon, E.H. (1990) Phenothiazine in the origin of benzimidazole resistance in population-B equine strongyles. *Veterinary Parasitology*, **35**, 117–130.

Fleming, J.T., Baylis, H.A., Sattelle, D.B. and Lewis, J.A. (1996) Molecular cloning and *in vitro* expression of *C. elegans* and parasitic nematode ionotropic receptors. *Parasitology*, **113**, S175–S190.

Geary, T.G., Sims, S.M., Thomas, E.M., Vanover, L., Davis, J.P., Winterrowd, C.A. *et al.* (1993) *Haemonchus contortus*: Ivermectin-induced paralysis of the pharynx. *Experimental Parasitology*, **77**, 88–96.

Gibson, T.E. (1975) *Veterinary Anthelmintic Medication*, 3rd edn, Technical Communication No. 33 (3rd edition) of the Commonwealth Institute of Helminthology, St. Albans, pp. 348. London: Commonwealth Agricultural Bureaux.

Gustafsson, L.L., Beerman, B. and Aden Abdi, Y. (1987) *Handbook of Drugs for Tropical Parasitic Infections*, pp. 151. New York: Taylor & Francis.

Herd, R.P. (1992) Choosing the optimal equine anthelmintic. *Veterinary Medicine*, **87**, 231–232, 234, 236–239.

James, D.M. and Gilles, H.M. (1985) *Human Antiparasitic Drugs: Pharmacology and Usage*, pp. 289. New York: John Wiley & Sons.

Jones, R.M. (1981) A field study of the morantel sustained release bolus in the seasonal control of parasitic gastroenteritis in grazing calves. *Veterinary Parasitology*, **8**, 237–251.

Kanesa-thasan, N., Douglas, J.G. and Kazura, J.W. (1991) Diethylcarbamazine inhibits endothelial and microfilarial prostanoid metabolism in vitro. *Molecular and Biochemical Parasitology*, **49**, 11–20.

Kornis, G.I., Clothier, M.F., Nelson, S.J., Dutton, F.E. and Mizsak, S.A. (1991) Total synthesis of avermectin and milbemycin analogues. In *Synthesis and Chemistry of Agrochemicals II, ACS Series 443*, pp. 422–435.

Lacey, E. (1990) Mode of action of benzimidazoles. *Parasitology Today*, **6**, 112–115.

Maksimowich, D.S., Bell, T.G., Williams, J.F. and Kaiser, L. (1997) Effect of arsenical drugs on *in vitro* vascular responses of pulmonary artery from heartworm-infected dogs. *American Journal of Veterinary Research*, **58**, 389–393.

Martin, R.J., Valkanov, M.A., Dale, V.M.E., Robertson, A.P. and Murray, I. (1996) Electrophysiology of *Ascaris* muscle and anti-nematodal drug action. *Parasitology*, **113**, S137–S156.

McKellar, Q.A. and Scott, E.W. (1990) The benzimidazole anthelmintic agents — a review. *Journal of Veterinary Pharmacology and Therapeutics*, **13**, 223–247.

Molento, M.B., Wang, G.T. and Prichard, R.K. (1999) Decreased ivermectin and moxidectin sensitivity in *Haemonchus contortus* selected with moxidectin over 14 generations. *Veterinary Parasitology*, **86**, 77–81.

Otto, G.F. and Maren, T.H. (1947) Filaricidal activity of substituted phenyl arsenoxides. *Science*, **106**, 105–107.

Prichard, R.K. (1986) Anthelmintics for cattle. *Veterinary Clinics of North America: Food Animal Practice*, **2**, 489–501.

Raether, W. (1988) Chemotherapy and other control measures of parasitic diseases in domestic animals and man. In *Parasitology in Focus*, edited by H. Mehlhorn, pp. 819–851. Berlin: Springer-Verlag.

Rew, R.S. and Fetterer, R.H. (1986) Mode of action of antinematodal drugs. In Chemotherapy of Parasitic Diseases, edited by W.C. Campbell and R.S. Rew, pp. 321–337. New York: Plenum Press.

Rickard, M.D. and Arundel, J.H. (1985) Chemotherapy of tapeworm infections in animals. In *Chemotherapy of Gastrointestinal Helminths*, edited by H. Vanden Bossche, D. Thienpont and P.G. Janssens, pp. 557–611. New York: Springer-Verlag.

Shoop, W.L., Haines, H.W., Michael, B.F. and Eary, C.H. (1993) Mutual resistance to avermectins and milbemycins: oral activity of ivermectin and moxidectin against ivermectin-resistant and susceptible nematodes. *Veterinary Record*, **133**, 445–447.

Vanden Bossche, H., Thienpont, D. and Janssens, P.G. (1985). *Chemotherapy of Gastrointestinal Helminths*, pp. 719. New York: Springer-Verlag.

Wescott, R.B. (1986) Anthelmintics and drug resistance. *Veterinary Clinics of North America: Equine Practice*, **2**, 367–380.

22. Anthelmintic Resistance

Nick C. Sangster
Department of Veterinary Anatomy and Pathology, University of Sydney, 2006, Australia

Robert J. Dobson
CSIRO Livestock Industries, McMaster Laboratory, Blacktown, 2148, Australia

Keywords

Anthelmintic resistance; benzimidazole; imidazothiazole; avermectin; diagnosis; nematode

Introduction

Since the introduction of broad-spectrum anthelmintics these drugs have been the major means of managed control of nematode parasites. Anthelmintics have been used particularly in livestock for the control of economically important parasites. Because of heavy reliance on these drugs and their widespread use, anthelmintic-resistant parasites have been selected and anthelmintic resistance has emerged as a problem amongst parasites of livestock. The same compounds have also been used to treat parasitic infections in human populations and there are fears that resistance will develop in these parasites too. Although resistance is defined formally below, a brief definition is: the ability of helminths to survive doses of drug that would normally remove them. Resistance is genetic, that is, resistant worms carry alleles for resistance which are inherited by the next generation.

Although resistance has emerged globally in helminths of sheep and goats, at the time of writing it is only a serious economic problem in the Southern Hemisphere. For example, the cost of resistance to Australian sheep farmers has been calculated (Besier, Lyon and McQuade, 1996) as $US4 per animal which amounts to at least $US250M per annum. At the extreme, the problem in South Africa has put some sheep farmers out of business (van Wyk et al., 1989). The severity of resistance in the sheep industry has led to intensive study of this problem and for this reason many examples used in this chapter are drawn from our knowledge of anthelmintic resistance in trichostrongyloid parasites of sheep. Resistance has also arisen in parasites of horses and cattle and is an emerging international problem in both industries. If the problem of anthelmintic resistance in cattle even approaches the seriousness that exists in sheep, anthelmintic resistance will be a significant global issue. Anthelmintic resistance has been experimentally selected in nematode parasites of pigs, in particular the genus *Oesophagostomum*. Resistant strains of the free-living nematode (non-parasitic) *Caenorhabditis elegans* have been selected *in vitro*. There have been no reports of resistance in nematode parasites of humans, dogs or cats, nor has resistance been reported in plant-parasitic nematodes.

Aspects of resistance have been reviewed previously. Prichard et al. (1980) set terms and definitions for the early work. Since then reviews on occurrence (Conder and Campbell, 1995), pharmacology (Sangster and Gill, 1999), diagnosis (Lyndal-Murphy, 1992), mathematical modelling (Barnes, Dobson and Barger, 1995) and control (Herd, 1993; Waller, 1987) have been produced as well as a volume arising from a conference workshop in 1990 (Boray, Martin and Roush, 1990).

Economic issues aside, the study of anthelmintic resistance occupies a special place in our understanding of nematode biology and embraces several biological disciplines. For example, important advances have occurred in our understanding of how anthelmintics act. Anthelmintics target biological processes that must be critical to the survival of the worm and the availability of resistant isolates allows scientists to ask experimental questions on biochemical and physiological modes of drug action. The study of resistance has led to the development of mathematical models of nematode populations which have revealed information about population dynamics. In the field of genetics resistance studies have provided information on allele variation in parallel with information on population variation. Anthelmintic resistance is a science that covers many disciplines and uses many methods. Worms are recovered and identified at postmortem, eggs are recovered from faeces taken from farms, nematode development is scored under a microscope, ion channel events are measured electrophysiologically and DNA from individual worms is amplified.

The emergence of anthelmintic resistance offers an opportunity to study adaptation. Biological processes of nematodes have evolved over millennia as worms have adapted to their environments. Parasitism probably resulted from gradual adaptation to a changing environment as nematodes entered hosts so that now all of the nematodes that we study have long since evolved biological mechanisms to thrive in their host environments. Complexities such as surviving plant toxins, osmotic stress, and a hostile immune response as well as achieving developmental and reproductive imperatives are topics covered elsewhere in this volume. Our understanding of the biology of nematodes has exploded in the last few years while the worms themselves have changed little. The exception is the development of anthelmintic resistance which is a process of selection and evolution that has occurred in nematodes over less than 40 years. Although the drugs are synthetic the responses are a reflection of the adaptation of species to a changing environment. What is more, we have many of the biochemical and mathematical tools needed to study the changes that have occurred. This chapter will not only consider the practical aspects of resistance but will also consider how the study of resistance helps the study of biology and the complexity of the adaptations which have occurred and continue to occur.

Anthelmintics

Anthelmintics are drugs that remove helminth parasites. They need not kill parasites directly because dislodgment is often sufficient for loss of a favourable predilection site and elimination. For the purpose of this chapter only drugs currently available for the therapy of parasitic infections are discussed. Drugs are not used as anthelmintics against *C. elegans* but it has been a useful model and information on it has been included. Although resistance to phenothiazine, (Drudge and Elam, 1961) and piperazine (Drudge *et al.* 1988) have been reported, neither resistance has been studied beyond its original description, so are not considered further. Nevertheless, reports of these resistances emphasise the point that resistance has developed to every modern anthelmintic.

Modern anthelmintic agents can be grouped into several distinct chemical classes. Although discussed in detail elsewhere (Martin, 1997) a summary of the modes of action of each drug class is provided here and the classes are listed in Table 22.1, together with examples and abbreviations. Our information on modes of action arise from consideration of chemical structure, biochemical and physiological observations, as well as molecular evidence. In addition, the behaviour of drugs in resistant worms, including *C. elegans*, has often been used to confirm a putative mode of drug action.

Benzimidazoles (BZs) specifically bind to parasite tubulin, a process which causes the depolymerisation of microtubules. These organelles have functions in cellular transport and coordination and their loss disrupts cellular integrity. The imidazothiazole, LEV, and the related tetrahydropyrimidine anthelmintics, MOR and PYR, are cholinergic agonists with a selective pharmacology for nematode acetylcholine receptors. In nematodes LEV mimics the action of the excitatory neurotransmitter acetylcholine (ACh) and causes spastic contraction of worms. The precise action of the AM class remains unresolved. Ivermectin (IVM), for example, opens glutamate-gated Cl⁻ channels on muscles of the pharynx and, possibly, the somatic musculature. The Cl⁻ influx which follows is inhibitory to muscle contraction and is probably the cause of fatal starvation and/or paralysis which has been observed experimentally. Organophosphate compounds (OP) are acetylcholinesterase (AChE) inhibitors. In other organisms, and presumably nematodes, they cause accumulation of ACh at neuromuscular junctions. This results in over-stimulation of muscle and/or nerve receptors and causes paralysis. Closantel and rafoxanide are salicylanilides, the action of which is unresolved. Although they have been shown to uncouple oxidative phosphorylation in nematodes they also act to lower cytoplasmic pH and, consequently, glycolysis in flukes. In either case the drugs appear to inhibit energy generation in worms.

Study of Parasitic Helminths

The study of animal parasites presents special challenges. First, because they are obligate parasites, for practical purposes, they must be obtained from hosts. The cost of maintaining hosts, the ethics of euthanasia of mammals, the difficulty of *in vitro* cultivation and the poor viability of worms outside the host conspire to hamper detailed physiological studies. Such studies in sheep are feasible, but expensive. Recovery of viable parasites from cattle or horses is very costly and from humans, difficult except for the filariids. Second, even in heavy infections, quantities of parasite material are severely limited. Third, variation within isolates of resistant worms can be high. Such variation reduces the utility of molecular biology to investigate resistance because the task of finding small differences which are due to resistance is complicated by background variation. Furthermore, the genetic systems of metazoans are complex and aspects such as gene regulation are

Table 22.1. Anthelmintic classes and examples of anthelmintic drugs

Class	Benzimidazole (BZ)	Imidazothiazole/ Tetrahydropyrimidines	Avermectin/ Milbemycin (AM)*	Salicylanilide	Organophosphate Compounds
Drug examples	thiabendazole (TBZ)	levamisole (LEV)	avermectin (AVM)	closantel (CLS)	naphthalophos (NAP)
	mebendazole (MBZ)	morantel (MOR)	ivermectin (IVM)		
	oxibendazole (OBZ)	pyrantel (PYR)	moxidectin (MOX)		
	oxfendazole (OFZ)				

* Often referred to as MLs or macrocyclic lactones

poorly understood. Even the chromosomes of most nematodes cannot be separated on the basis of size. Fortunately, all of the nematodes in which anthelmintic resistance has developed have direct life cycles. This at least makes research tractable.

Several laboratory host/parasite systems have been developed. *Trichostrongylus colubriformis* in guinea pigs and *Haemonchus contortus* in jirds successfully carry their resistance phenotypes from sheep. *In vitro* techniques using free-living stages of parasites have been crucial in studying anthelmintic resistance. Assays for resistance to LEV and AMs closely reflect the *in vivo* characteristics of resistance in many cases. Nevertheless, because drug receptors are developmentally expressed and those in free-living stages could differ from those in parasitic stages, assays must be rigorously validated.

Another impediment to study is the nature and source of resistant isolates. Populations of worms are simply isolates rather than strains, a term which is used to describe well defined, inbred or clonal populations or populations with a known lineage. In general, 'field' isolates are derived from natural infections of parasites which survive treatment and are further selected in laboratory passages. Isolation involves collection of the eggs, incubation of these to infective stages and infection of worm-free hosts. At this point some steps such as selective surgical placement (Whitlock *et al.*, 1980), or narrow spectrum drug treatment are used to develop monospecific infections. Infected hosts are then treated with the recommended dose of a commercial anthelmintic, eggs collected and the process continued through sequential passages. Isolates are kept distinct by infecting each host with a single isolate.

An alternative source of isolates is through the process of the selection of 'laboratory isolates' where starting isolates are drug-susceptible and selection in penned animals only occurs. Dose rates of drug commence at levels below the recommended dose and are increased at each generation until the majority of the population survive a full recommended dose. Laboratory isolates are not the same as field isolates. For example, in using a rapid turn around of generations in order to rapidly select for resistance, researchers may unwittingly select worms with a short generation time. In another example, selection of resistant isolates in penned sheep using increasing doses of anthelmintic at each generation produces isolates with resistance phenotypes different to similar resistances recovered from the field (Gill and Lacey, 1998).

Unfortunately there are no unequivocal descriptors of resistant isolates. Typically they bear labels of percent efficacy of a certain product in a certain host, a selection history or an EC_{50} in a defined *in vitro* assay. Some isolates have been maintained for long periods in the laboratory and can be used as reference isolates. As a further benefit for referencing, third stage larvae (L3) can be stored for months at 10°C and first stage larvae (L1) (Gill and Redwin, 1995) and L3 (Coles, Simpkin

and Briscoe, 1980) cryopreserved under liquid nitrogen prior to reinfection of hosts.

Despite these difficulties, the research community have developed useful techniques and gathered considerable information. The cattle and horse industries can learn a lot from the experience in the sheep industry. With regards to management of resistance the most important factor appears to be early diagnosis. Adoption of techniques developed for sheep parasites by the cattle and horse industries is a priority.

Parasites can be studied *in vitro*, removed from their hosts, or *in vivo*, within living hosts. These descriptions are in common use and are also used in this chapter. Strictly, *in vivo* refers to studying live organisms and therefore can also refer to studies of free-living nematodes, or the free-living stages of parasites in their natural habitat.

Caenorhabditis elegans

Studies with this nematode have contributed a great deal to the understanding of nematode biology and have also provided a focus for drug resistance studies. Some studies have been driven by a desire to understand anthelmintic resistance in parasitic nematodes for which *C. elegans* acts as a model. More typically, however, studies were directed at understanding more basic aspects of biology. What sets *C. elegans* apart is the availability of laboratory tools such as laboratory propagation, the ability to clone hermaphrodites, mutational screening, lineage studies, cell ablation, physical gene mapping, transfection, gene inactivation, expression location and a comprehensive genome sequence coupled with the ability to observe behaviours and perform physiological studies. In general, resistant isolates have been selected on agar plates containing the drug of interest following ethyl methane sulphonate-induced mutagenation. They are subsequently cloned from an individual hermaphrodite and maintained on plates containing drug. Although these strains offer insights into biology and resistance the fact that they are mutagenised means that they do not offer a model of adaptation which occurs in parasite populations.

There are several experimental methods which link the studies of parasitic nematodes and *C. elegans*. DNA sequence homology between the trichostrongyloid nematode parasites and *C. elegans* is remarkably high, especially for conserved, functional genes. For example, comparison of the sequences of an ACh receptor α-subunit gene from *C. elegans* and *T. colubriformis* show a 92% amino acid similarity (Fleming *et al.*, 1997; Wiley *et al.*, 1996). Transformation of *C. elegans* with parasite genes (Grant, 1992) allows the phenotype conferred by parasite genes to be observed in the genetic background of *C. elegans*. This technique has been used to study resistance to BZs in *H. contortus* (Kwa *et al.*, 1995). On the other hand, the answer to the question of how useful *C. elegans* is as a model for parasitic nematodes (Ward, 1988) is not clear. Based on mitochondrial DNA sequence comparisons the parasitic nematodes diverged from their free-living counterparts about

80 million years ago (Okimoto *et al.*, 1992). More detailed genetic analysis shows that *C. elegans* is most closely related to strongylid nematode parasites (Blaxter, 1998; DeLey and Blaxter, Chapter 1). One would expect then that extrapolation of resistance data from *C. elegans* to parasitic nematodes requires caution. Because the selection and screening procedures are biased towards recessive characteristics and severely resistant phenotypes, many of the mutants are of no relevance to field resistance in parasites. The problem is to identify which genes, if any, are the relevant ones. Further, the biology of parasites differs fundamentally from *C. elegans* in some respects. For example, differences in feeding and reproduction indicate that the parasitic lifestyle provides different challenges. As a hermaphrodite, *C. elegans* may be a poor model for resistances in which males are more sensitive than females and there are several reports of this among the parasitic helminths (Dash, 1985). Further, the nature of drug exposure may differ significantly. The routes by which drugs reach and enter parasites are largely unknown, although the gut and the cuticle are major routes of entry. To add further complexity, some parasite species migrate through different host tissues during their development and are exposed to different drug concentrations and drug metabolites, and feed on different tissues. On the other hand, *C. elegans* has a more predictable life style and diet, as it is captive on an agar plate and fed on a lawn of *Escherichia coli*.

Definitions

What constitutes resistance has caused some controversy so clear definitions are needed. It has been defined as: '**Resistance** occurs when a greater frequency of individuals in a population of parasites, usually affected by a dose or concentration of compound, are no longer affected. Resistance is inherited.' (Prichard *et al.*, 1980). This definition needs expanding in the light of common practice and advances in our understanding. In addition to looking for a fall in efficacy at a set concentration, the definition should include the case where higher doses or concentrations of drug are required to remove or kill a certain proportion of nematodes. Because resistance is genetically based the definition may be expanded to include an increase in the proportion of worms in a population carrying a gene linked to resistance. Graphical explanations on the appearance and development of resistance are provided below.

Side resistance is the phenomenon where parasites resistant to one drug of a chemical class are also resistant to other drugs in the same class. This appears to be a universal finding. It is true for the BZs where parasites resistant to TBZ are also resistant to OFZ on the first use of the latter compound. It also holds for the imidazothiazoles/tetrahydropyrimidines and for the AMs. Some confusion has arisen when some drugs of an anthelmintic class are effective in the field while others have lost activity. For example, if two drugs act on the same target site at which resistance has arisen, a more potent drug may be active whereas a less potent compound will not. It can be shown that resistance is present in these cases. The concept is explained below.

Cross resistance describes resistance between chemical classes. As each anthelmintic group has an independent mode of action and resistance mechanisms are distinct, resistance to each appears to have developed independently. There is one exception. Cross resistance appears to be present between the imidazothiazoles and the organophosphate compounds (Sangster, 1996).

Multiple resistance and cross resistance should not be confused. Multiple resistance occurs when parasites develop resistance to several anthelmintic classes independently. For example, some isolates of *H. contortus* from sheep are resistant to all classes of drugs that are available. Apart from the case with organophosphates, just mentioned, each resistance arose separately and as a result of selection following drug treatment.

Reversion, or loss of anthelmintic resistance, occurs slowly. There are no reports of full reversion. This is probably because resistance genes persist in worm populations for many generations after the selection pressure, in the form of treatment with a specific anthelmintic class, has ceased. Resistance appears not to confer a loss of fitness. Counter selection with an alternative drug from a different class may help to accelerate reversion. However, in field situations it is likely that resistance alleles persist at low levels in the population and that once the original drug class is reintroduced, treatment selects so strongly for resistant survivors that resistance returns in a generation or two. Theoretically, single gene resistance may become fixed in a population, in which case reversion cannot occur.

The terms resistance and **clinical resistance** (or field resistance) have been used interchangeably, but they are not necessarily the same. Because of improvements in our understanding of resistance phenomena and the methods of diagnosis since 1980 a clear definition of resistance is now needed. Clinical resistance occurs when recommended doses of drug fail to remove parasites. In all cases of clinical resistance, anthelmintic resistance is present, but the reverse is not necessarily true. Measurements of the prevalence and severity of clinical resistance underestimate the prevalence of resistance.

Most surveys of clinical resistance involve treatment of infected hosts with a recommended dose of drug followed by a calculation of faecal egg count reduction compared with untreated controls. Clinical resistance, in such tests for sheep parasites, has been defined as failure of a treatment to reduce faecal egg counts by >95% (based on group arithmetic means) compared with an untreated control group. An additional criterion is that the lower 95% confidence interval of the efficacy estimate is below the 90% efficacy level. If a population meets both criteria, clinical resistance is present (Lyndal-Murphy, 1992). If it meets only one criterion, resistance is suspected (Coles *et al.*, 1992). The picture is more complex for parasites of other host species where between animal variability in infection rates, egg counts and response to treatment is higher. In cattle, horses and pigs efficacy <90% denotes clinical resistance (Coles *et al.*, 1992).

Resistance is most commonly recognised because of treatment failure. However, clinical resistance is only one form of treatment failure. Other possibilities include the following: (a) The typical signs of diarrhoea, weight loss, oedema and/or anaemia could be due to other infectious or toxic agents. (b) If parasites are present the use of an inappropriate drug will not clear them. (c) Under-dosing will leave some parasites in the host and these may continue to cause disease. Under-estimates of body weight and reduced active concentration (Monteiro *et al.*, 1998) are two possible reasons for under-dosing. (d) The more rapid metabolism of anthelmintic compounds in goats compared with sheep (Hennessy, 1994) will also result in poorer efficacy. (e) Rapid reinfection after treatment, either from incoming larvae or emergence of mucosal stages, can follow treatment and resemble treatment failure.

Resistance Factor (RF), sometimes referred to as the resistance ratio, is a measure of the degree of resistance to a particular drug of one worm isolate by comparison with another isolate, usually a known susceptible reference isolate. It is commonly estimated by dividing the ED_{50} (see below for a definition) of the suspect isolate by the ED_{50} of the susceptible reference isolate. In efficacy trials the RF of an isolate may be determined at the 95% level of drug efficacy (i.e. using the ED_{95} for both isolates).

Concept of Dose Response

Dose response describes the effect of a range of doses (or concentrations) of drug on some indicator of worm viability or activity. In the context of nematode parasites the indicator of activity may be the death of the worm, its removal from its host or its failure to develop *in vitro*. The dose is either a dose rate used *in vivo* (e.g. mg/kg) or a drug concentration present *in vitro*. Dose and concentration are used interchangeably in this section of the text. Because worms in a population vary in their susceptibility to anthelmintic drugs, there is a dose at which each individual is affected. By measuring the effect against groups of worms, which are representative of the population, at each of a range of anthelmintic concentrations, a plot of response to treatment against dose can be generated. This is termed a dose/response (D/R) curve in which increasing dose causes increasing response (e.g. kills more worms). If response is converted to logits or probits and dose to log dose, a straight line is usually achieved (referred to as LDP or LDL lines in some literature). Fits of data to these relationships are used to calculate resistance parameters (Finney, 1971; Waller *et al.*, 1985). Information on other calculations can be found in Lyndal-Murphy (1992) or Coles *et al.* (1992). For graphical purposes in the examples below the proportion of worms responding is not transformed but the dose is. This generates sigmoid curves. Because resistance causes changes in D/R curves and because they are used to describe and measure resistance, D/R

curves provide a useful way of explaining resistance phenomena. The curves used as examples below are idealised and most simulate the results of *in vitro* assays.

Two D/R curves are shown in Figure 22.1A. In curve **a** worms that are killed at low concentrations contribute to the response at the lower, left end of the curve (5% of worms are killed at log dose 2) and those which are killed at higher concentrations contribute to the right (95% of worms are killed at log dose 4). The middle point of the curve, where 50% are killed, is the effective dose 50% (ED_{50}) which equals a log dose of 3. The slope of the line indicates the heterogeneity of the population. The flatter line, **b**, reflects a population that has the same ED_{50} as **a**, but has more variability between individuals. ED_{50} is also commonly referred to as LD_{50} or EC_{50} where LD stands for Lethal Dose and EC represents Effective Concentration. ED_{95} refers to the effective dose required to kill 95% of the target population.

The relative position of the curve along the **x** axis indicates the potency of the drug. In Figure 22.1B drug **x** kills a higher proportion of worms at the same dose as drug **y** and is thus more potent. In this example **x** is 10 times more potent than **y**. The curves for more potent drugs lie to the left of those for less potent drugs.

The relative position of the line along the **x** axis also indicates the average level of susceptibility of the parasite population to a certain drug. Curve **r** (representing population r) in Figure 22.1C is more resistant than curve **s** (representing population s). Resistance is defined as an increase in the frequency of worms surviving a certain dose. So, as resistance develops, the efficacy at each dose of drug falls and the curve shifts down. For example, log dose 3 kills 70% of population s but only 5% of population r. A more common way of looking at resistance is that to achieve the same efficacy against a resistant isolate as a susceptible one, an increase in dose is required. This is seen as in a shift of the curve to the right. The ED_{50} is a measure of how far the curve is shifted to the right. For **s** and **r** the shift in ED_{50} is 1.5 on the log scale. The RF of **r** by comparison with **s** is 32 ($10^{1.5}$ or antilog of 1.5). Note that an RF based on ED_{50}s can be different from that derived from the ED_{95}, e.g. in Fig. 22.1A the RF, based on ED_{50} for population **b** is 1 (drug susceptible) and approximately 5 if determined by the ED_{95}.

In Figure 22.1C the intermediate curve is the mean position of curves **s** and **r** at each dose. It illustrates a phenomenon that sometimes occurs in worm populations where inflections in the curve indicate the presence of mixed populations of worms.

D/R curves can also be used as a model to explain the selection of resistance, although selection is covered in more detail later. Resistant worms carry alleles of genes which confer resistance. Treatment with drugs allows the most resistant worms to survive. If these survivors reproduce they pass resistance alleles on to their offspring and the frequency of resistance alleles in the population increases. The following example illustrates how selection can increase resistance levels over time. Figure 22.2A represents D/R curves for a parent (**a**) and

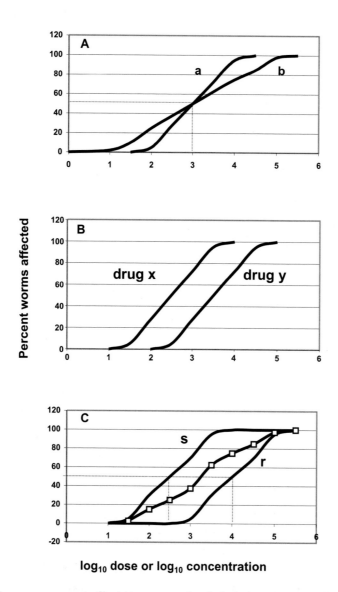

Figure 22.1. Dose/response curves to illustrate aspects of anthelmintic resistance. All are idealised for theoretical populations and drugs. A. Curves **a** and **b** represent populations with less or more heterogeneity, respectively, but the same ED_{50} ($\log ED_{50} = 3$). B. Curves showing the dose/response to the more potent drug **x** and the less potent drug **y**. C. Curves for susceptible (**s**) and resistant (**r**) populations of worms. Population **s** has an ED_{50} at log dose of 2.5 and population **r**, at log dose of 4. The resistance factor (RF) for **r** is 32 ($10^{1.5}$). The intermediate curve is the mean position of curves **s** and **r** at each dose, i.e. the expected curve for a population made up of **s** and **r** in equal proportions.

two successive generations (**b**, **c**) which have been selected by drug treatment. Several assumptions are made and listed in the figure legend. Compared with **a**, **b** is resistant and **c** is more resistant again. Although this example shows an increase in the average level of resistance, (increase in ED_{50}) the change may be small in the early stages of the development of resistance. The flatter curve in Figure 22.1A (**b**) could be resistant compared with a susceptible population **a** because part of the curve is shifted to the right. In cases of field resistance, a shift at higher doses like those shown in **b** in Figure 22.1A and apparent in Figure 22.2A can be the first sign of resistance.

A graphical representation of why some drugs of a class retain activity even when side resistance is present is shown in Figure 22.2B. Dose responses for the more potent drug **x** are shown in the upper panel and for drug **y**, below. The curves represent a susceptible isolate s and three resistant isolates, **r1, r2** and **r3**, which are shifted to the right. At log dose 5 both drugs are highly effective against s. When resistance develops to **r1**, **x** is highly effective and, as **y** is still 96% effective, clinical resistance is not yet present. (The dose intersects with a point which is still on the plateau of the D/R curve) For population r2, the efficacy of y (75% at log dose 5) indicates clinical

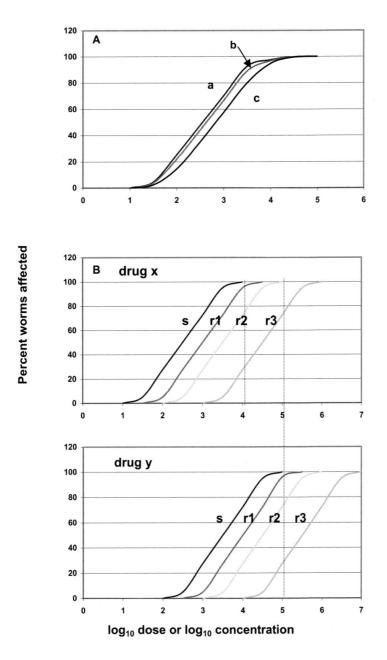

Figure 22.2. Dose/response curves illustrating the progression of resistance and a model of side resistance. All are idealised. A. Dose/response curves for a parent (**a**) and two successive generations (**b** and **c**) which have been selected by drug treatment. The following assumptions were made: one gene was involved, resistance alleles were at 5% abundance in **a**, drug was 98% effective in **a**, Hardy-Weinberg equilibrium existed, only the survivors reproduce and mating was at random. B. Curves for the more potent drug **x** are shown in the upper panel and for drug **y**, one-tenth the potency of **x**, are below. A susceptible isolate (**s**) and three isolates, **r1**, **r2** and **r3**, represent sequential steps in the development of resistance. The vertical dotted lines are used to indicate responses at log dose 4 and 5. See text for an explanation.

resistance is present to it but not to **x**. However, if a lower dose of **x** is used (log dose 4) it is less effective against **r2** than **s**. This satisfies the definition of resistance. When resistance reaches level **r3** clinical resistance is present to both drugs.

How far do curves move to the right? They are limited by several factors. *In vivo* there is a limit on how high dose

rates can be increased before the host suffers toxic drug effects. This will limit the selection pressure on resistant populations and RF values rarely exceed 20 fold. RF values derived from *in vitro* studies can exceed this as free-living stages experience different pharmacology in the absence of the host. In some cases thousand fold increases are found.

Genetics

Anthelmintic resistance is genetically based, so an understanding of genetics is essential for understanding resistance. Furthermore, genetic tools have the potential to provide sensitive and specific tests for resistance.

Selection for resistance is covered in the next section, but some consideration of the genetic processes involved is needed to understand the role of genes. For resistance to develop there must be genetic variation in the worm population and some of those variants need to code for gene products which lower susceptibility to drugs. When drug treatment is imposed individuals with those alleles survive while others die. If the survivors reproduce, the frequency of the 'resistance genes' in subsequent generations increases. Once resistance genes reach a certain level (which is undetermined) their frequency may decline but not to zero. Reversion of resistance does not generally occur.

Several possible genetic processes could lead to resistance. The simplest examples are resistances caused by a change in a gene coding for a drug receptor. Mutation, either by chance or induced, could alter the gene product to become more resistant. A change in gene expression may lead to resistance by producing more of the drug target, for example. Finally, alleles of the gene may be present at low frequency in the population. Selection favours survival of individuals carrying these alleles.

Genetic variability in some populations of trichostrongyloids is surprisingly high (based on mtDNA variation) (Blouin et al., 1992). Almost every study of putative resistance genes from H. contortus has shown that several alleles exist at each locus. For example, (Beech, Prichard and Scott, 1994) identified 13 genotypes at a β-tubulin locus (using restriction fragment length polymorphisms (RFLP)) and alleles of acetylcholine receptor genes (Hoekstra et al., 1997) and P-glycoprotein genes (Sangster et al., 1999) vary up to 83% at the nucleotide level (using sequence data). It is likely that variability in relatively small parasite populations does not limit selection for resistance. Indirect evidence for this comes from two sources. Firstly, small, isolated populations of parasites such as those maintained in penned animals can develop resistance. Second, comparisons between a worm population maintained by passage in penned sheep for 60 years and one recently recovered from the field revealed no significant difference in the degree of genetic polymorphism (Grant and Whitington, 1994).

Restriction of a population by forcing it through a bottleneck of selection should have some survival cost. Presumably resistance alleles which disadvantage worm survival would not persist. Several workers have compared the 'fitness' of worms by measuring survival and reproductive parameters in susceptible and resistant parasites. In experiments of this sort it is too difficult to control for variation or other selection pressures for any measure of genetic cost to be made.

Methods and Approaches

We have an exquisite understanding of the genetics of C. elegans. Although the power of that technology can be focussed on resistance in parasites the advances are yet to make an impact. If the level of homology between genes involved in drug action in C. elegans and those from parasites like H. contortus are any guide (Geary et al., 1992; Hoekstra et al., 1997; Delany, Laughton and Wolstenholme, 1998; Blackhall et al., 1998b), gene sequences from the C. elegans genome project will be useful for finding orthologs in parasites. What is less certain is if gene mutations which confer drug resistance in C. elegans will provide molecular probes for genes which are directly linked to field resistance in parasites.

For parasites some advances have been made in understanding the genetic mechanisms of resistance but progress has been slowed by several factors. Parasitic nematodes are difficult to obtain in quantity and cannot be cultivated in vitro in useful numbers. Further, genetic organisation of parasitic helminths is not well understood. Even the chromosomes cannot be separated on the basis of size. Perhaps the most serious problem is that parasite populations are genetically heterogeneous. If one compares populations of resistant and susceptible isolates the large majority of the differences are due to this variability which make finding differences related to resistance a very difficult task.

There are several possible approaches for describing the genetics of resistance and for identifying resistance genes. The use of Mendelian inheritance and molecular mechanisms are described below. The findings for each class of drug are covered below in a later section.

Experiments to reveal Mendelian genetics are difficult to perform in parasitic nematodes and provide crude results at best. They involve removing adult or advanced larval parasites (parent isolates) from hosts infected with a susceptible or a resistant isolate of the worm and sorting them by sex. Males of one isolate and females of another are then mixed and the crosses achieved by surgical implantation into an uninfected host. This process generates F1 eggs and, on culture, larvae which are used to infect another host. Here they interbreed to produce F2 eggs. In addition, backcrossing F1 to the susceptible parent (Martin, McKenzie and Stone, 1988) is useful in resolving modes of inheritance. Analysis of the resistance status of the parent and filial generations can be performed either by slaughter assays in sheep, or for T. colubriformis, in guinea pigs, or by in vitro assays. Sex-linked differences are generally detected by performing sex differentiation of worms during counting.

Apart from technical difficulties, the experiments suffer from the need to use mass matings of polymorphic parent isolates. Only average responses can be recorded with the result that single gene effects can look like a more complex inheritance. There is also the problem that resistance that appears as a single locus could be due to a cluster of genes. Nevertheless, the technique has been used to shed light on the number of genes likely to be involved and the type of inheritance. Selection theory suggests that selection within the phenotypic range of the population, as anthelmintics generally do, will select for polygenic resistance. Genetic studies have shown that many different patterns of resistance occur and the patterns differ between drugs and between species exposed to the same drug.

Most techniques used to search for resistance genes are based on the comparison of resistant and susceptible isolates. Differences might be sought using random methods such as random amplification of polymorphic DNA (RAPD). The statistical chance of finding a difference relevant to resistance can be improved by increasing the number of probes used and by isolating the resistance genes into a common genetic background by several generations of back-crossing the resistant isolate with a susceptible isolate and selecting with drug at each generation (Le Jambre, 1990). On the other hand, directed discovery allows a researcher to focus on a gene or gene family with a link to a putative resistance mechanism. If the mechanism of resistance is unknown this technique is little more than empirical. Table 22.2 lists some genes that have been used to investigate resistance in this fashion. Various techniques (Grant, 1994) such as RFLP (Blackhall *et al.*, 1998a) and single-strand conformation polymorphism (Blackhall *et al.*, 1998b) have been used to describe differences between isolates. Examples of genes from *C. elegans* which, when mutated, confer resistance are listed in Table 22.3. The difficulty of making rational choices of candidate genes is emphasised in the case of resistance to the OP compounds in *C. elegans* where gene products remote from the site of action can be involved. Resistance can arise from mutations in genes involved in transcription control, vesicular transport, vesicle traffic and phorbol ester receptor as well as the ACh receptor. Because of the possibility that genes remote from the site of action can code for resistance mechanisms, differences in allele frequency between resistant and susceptible isolates are insufficient evidence for confirming the role of a gene in resistance. A more rigorous approach is to measure the segregation of candidate alleles in crossbreeding studies between genetically distinct populations.

It is interesting to relate the history of how BZ resistance was linked to β-tubulin genes. In this case the mode of action of the BZ drugs was known and the mechanism of resistance known to involve a reduction in affinity of BZ drugs for tubulin from resistant worms. RFLP experiments confirmed the involvement of tubulin genes in resistance in *H. contortus* and alterations in β-tubulin sequence matched those which were linked to BZ resistance in fungi and *C. elegans*. Transfection experiments were subsequently used to confirm the role of the resistance allele. Unfortunately, we still do not fully understand the mode of action of the other classes of anthelmintics, let alone the mechanisms of resistance involved.

The Utility of Genetics

An obvious use for genetic information is in the design of highly sensitive tests for resistance. Genetics can also benefit the control of resistance in two ways. Tests have the ability to monitor the frequency of resistance genes and therefore test the success of resistance control schemes. Secondly, if a relationship between genotype and phenotype can be established then these data can be incorporated into mathematical models of resistance to improve the accuracy of their predictions. Correlations of this sort have been made for BZ resistance in *Teladorsagia (Syn. Ostertagia) circumcincta* from sheep (Elard, Cabaret and Humbert, 1999).

Resistance genetics also provides tools for studying other aspects of nematode biology. An elegant example is the breeding of resistant worms with a phenotypic marker. AM-resistance in some isolates of *H. contortus* is inherited in a dominant fashion (Le Jambre, 1993a), as is the smooth vulval morph type (Le Jambre and Royal, 1977). By cross-breeding and drug selection an isolate has been developed in which females of an AM-resistant isolate are visually distinguishable from susceptible worms. It has already been useful in efficacy studies in sheep.

Most transfection systems used in biology rely on selection of transformants which express drug resistance. Although the other processes of transfection and cloning survivors are yet to be resolved for parasites, the presence of a dominant resistance genotype in *H. contortus* could provide a tool for studying gene regulation and expression in parasitic nematodes.

Selection for Resistance

The evolution of drug resistance in a pest organism is a consequence of chemotherapy that successfully controls the disease. For parasitic nematodes this occurs when an infected host, or more commonly a herd, is treated with an anthelmintic. Anthelmintics which are, say, 95 to 99.9% effective against resident worms may leave 5 to 0.1% survivors, respectively, which are resistant to the drug. In addition to selection in the host, the extent to which resistant survivors contribute to the next generation as well as some inherent properties of the worm species affect how quickly resistance develops. Anthelmintic resistance generally develops more slowly than other types of resistances. By comparison with antibacterial resistance anthelmintic resistance is slow to develop, at least partly because of longer generation times in nematodes. Compared with insects, nematode parasites are relatively immobile so the progeny of survivors of selection are generally restricted to a farm. In contrast to insecticide use, when using anthelmintics only the organisms within the host are exposed to the selecting agent so much of the population are not selected and these dilute any resistant organisms. As with resistance in other organisms, accidental resistance can occur when resistance develops in one species of parasite, even though it was not a deliberate target of the chemotherapy.

Table 22.2. Genes investigated as markers of anthelmintic resistance in parasitic nematodes

Species	Resistance investigated	Protein	GenBank Accession number	Reference
H. contortus	BZ	β-tubulin	M76491-3	(Geary *et al.*, 1992)
H. contortus	BZ	β-tubulin	X67486	(Kwa, Veenstra and Roos, 1993)
			X67489	
H. contortus	BZ	β-tubulin	X67487	Roos, unpublished
H. contortus	BZ	β-tubulin	X67488	(Kwa, Veenstra and Roos, 1993)
T. colubriformis	BZ	β-tubulin	U39615-20	(Grant and Mascord, 1996)
			U28744	
T. (O.) circumcincta	BZ	β-tubulin	Z69258	(Elard, Comes and Humbert, 1996)
H. contortus	LEV	ACh R	U72490	(Hoekstra *et al.*, 1997)
T. colubriformis	LEV	ACh R	U56903	(Wiley *et al.*, 1996)
T. (O.) circumcincta	LEV	ACh R	Y13850	Walker *et al.*, unpublished
			Y13851	
H. contortus	AM	Glu Clα	AF034609	(Blackhall *et al.*, 1998)
H. contortus	AM	Glu Clβ	Y09796	(Delany, Laughton and Wolstenholme, 1998)
H. contortus	AM	Pgp	U94401	(Kwa *et al.*, 1998)
H. contortus	AM	Pgp	AF003908	(Xu *et al.*, 1998)
H. contortus	AM & CLS	Pgp fragments	AF055167	(Sangster *et al.*, 1999)

AChR, Acetylcholine receptor; Pgp, P-glycoprotein.

Table 22.3. Examples of anthelmintic-resistant *C. elegans* mutants

Drug group	Selective agent	Reference	Mutant	Phenotype/function in addition to drug resistance
Benzimidazole	benomyl	(Driscoll *et al.*, 1989)	*ben-1*	β-tubulin gene,
Imidothiazole THP	levamisole	(Lewis *et al.*, 1980)	*unc-38, unc-74*	α-subunit of ACh receptor
			unc-29	structural subunit of ACh receptor
			lev-1	structural subunit of ACh receptor
			unc-22	Twitcher (twitchin gene)
Avermectin/ milbemycin	AVM	(Johnson cited in Blaxter and Bird, 1997)	*avr-1 (che-3)*	Dye-filling negative (dynein gene)
			avr-5 (osm-3)	Dye-filling negative
		(Dent, Davis and Avery, 1997)	*avr-15*	Pharyngeal muscle cell Glu receptor
Organophosphate	trichlorfon	(Hosono, Sassa and Kuno, 1989)	*unc-3*	transcription factor
			unc-13	Phorbol ester receptor
			unc,17	Vesicular ACh transport
			unc-18, unc-41	Regulate vesicle traffic

Phenotypes: *avr*, AVM-resistant; *ben*, benomyl-resistant; *che*, chemotaxis abnormal; *lev*, LEV-resistant; *osm*, osmotic avoidance; *unc*, uncoordinated.

Whether, and how quickly, resistance develops depends on a multitude of interacting factors. For resistance to develop there needs to be a predisposition and enough selection pressure to allow the buildup of a significant frequency of resistance alleles (R alleles) in the population. For example, resistance will not develop when a suitable selection occurs but host immunity prevents any of the survivors producing eggs. On the other hand, an anthelmintic that is persistent in its activity can also allow resistant larvae as they are ingested to survive while susceptible phenotypes succumb. The fact that some species readily become resistant while others do not can be due to factors inherent to nematode biology, drug dynamics and the environment. Some of the factors that affect selection for resistance are considered below.

Selection pressure is a commonly used term. It defines the rate at which resistance develops and needs to consider all the factors that influence the development of resistance, both within the treated population of worms and the worms in the environment. One definition of the selective effect of a treatment is the ratio of the frequencies of resistance before and after treatment (Barnes, Dobson and Barger, 1995). A definition more relevant to the field is that of selection pressure (SP) as the per annum ratio of the resistance allele frequency in the total worm population after an anthelmintic treatment regimen (R_2) during one host management and climatic cycle to the R frequency in the worm population prior to commencement of that treatment regimen (R_1). This is given by:

$$NS_P = R_2/R_1$$

where N is the time in years of the treatment, management or climatic cycle which ever is the longer. If NS_P is > 1 resistance increases. This definition is not optimal as it is partly dependent on the R frequency. For example, if R frequency is high then S_P will tend to be low despite high levels of drug use which, if applied at a lower R frequency, would be highly selective. It also assumes that resistance is not imported over the period, N. Introduction of resistance genes by the introduction of infected stock is an effective way of introducing R alleles to a farm.

The measurement of resistance before and after a year of management or experiment can lead to the calculation of S_P. However, because resistance takes time to develop and its development depends on the interplay of many factors, mathematical models have been developed to help predict future levels of resistance. Some models have been used to evaluate the effects of management decisions on resistance (Barnes and Dobson, 1990; Gettinby, 1989; Smith, 1990; Barnes, Dobson and Barger, 1995; Smith *et al.*, 1999). They have been used below for the purpose of illustrating the impact of various factors on S_P. The diagrams that accompany the explanations are derived from the model of Barnes, Dobson and Barger (1995). In order to illustrate long term effects most simulations were run over several years and chart % R in the population over time. Simulations are performed for sheep enterprises with typical management. Generally a single parameter is altered within a diagram although in others equivalent control is administered. These examples simply provide examples for comparative purposes and do not relate to particular cases.

Worm Biology

Because anthelmintic resistance in certain worm species has arisen independently at several places around the world yet in other species susceptibility remains despite widespread usage of anthelmintics, it is likely that parasite biology is influential in driving selection for resistance. Many of the factors involved are not known, but there is evidence that parasite genetics and life history are influential. Because we generally have no control over parasite biology these factors influence the rate of development of resistance but are inherent for a parasite species, so do not influence relative S_P applied by different management practices to that species.

Population size and genetic diversity

The larger the parasite population size and the higher the genetic diversity in that population, the more likely it is that resistance alleles will be present and the more likely it is that resistance will develop.

Mendelian genetics

The method of inheritance of resistance influences how rapidly resistance develops. In the case of dominant inheritance selection for resistance is more rapid because the heterozygote is a resistant phenotype that survives treatment and resistance alleles readily build up in the population. An example of a resistance that is a dominant trait and has arisen quickly is AM resistance in *H. contortus* (Le Jambre, 1993a). In contrast, recessive traits will slow the development of resistance because heterozygotes have susceptible phenotypes and are killed by treatment. Worms carrying rare R alleles are thus removed from the population. LEV-resistance in *H. contortus* is recessive and has been slow to develop (Sangster, 1996). Results of a simulation comparing recessive and dominant inheritance are given in Figure 22.3A. Resistance due to a single gene develops quickly compared with multigenic inheritance (Barnes, Dobson and Barger, 1995).

Life history

Resistance is most common in parasites that have direct life cycles and short generation times. In such cases the progeny of resistant worms only have to infect, establish and reproduce in a host to contribute R alleles to the gene pool and the frequency of R alleles can be amplified quickly. These characteristics, which occur in the trichostrongyloid nematodes of sheep, contribute to the frequency with which resistance is found. Parasites with longer life cycles less commonly develop resistance.

It has been argued that the indirect life cycle of the human filariid nematode *Onchocerca volvulus* and the fact that the drug IVM acts against L1 (microfilaria) mitigate against the development of resistance (Shoop,

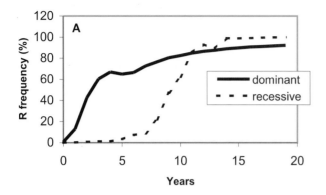

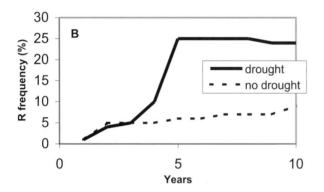

Figure 22.3. Examples of the effects of inheritance and weather on the resistance allele frequency predicted by models for anthelmintic resistance. A. Effect of the mode of inheritance. Assumes that resistance was inherited from a single gene as a dominant or recessive trait. Lambs receive 3 anthelmintic treatments per year. B. Effect of drought. Assumes 3 anthelmintic treatments per year. *Drought* shows how resistance increases under the historic weather conditions. *No drought* shows how resistance increases when the extreme dry years (3–5) were removed from the weather data file prior to simulation.

1993). Importantly, L1 are not the reproductive stage. L1 that survive IVM treatment need to survive further challenges in finding and infecting an intermediate host, developing in it then infecting and establishing in the definitive host before reproducing. Not only will these steps tend to counterselect for resistance but refugia (see below) in the intermediate host is also significant. On the other hand, Geerts, Coles and Gryseels (1997) have argued that with mass chemotherapy more resistant survivors will infect intermediate hosts and resistance may develop. It may be most likely to do so in isolated valleys where the density of intermediate hosts may be limited at certain times of the year.

Hypobiosis

Some nematode parasites, especially those of grazing animals, including horses, can undergo hypobiosis, a stage where some parasites are refractory to the effects of some drugs. Although most modern anthelmintics control hypobiotic stages, mucosal stages of horse cyathostomes are not removed by some drugs and are a potential refugium (see Sangster, 1999). The role of refugia is explored in more detail below under 'Environmental Factors' where its role is better understood. Refugia are subpopulations of worms that are not exposed to treatment. Most drugs lack consistent activity against mucosal stages of horse cyathostomes so for these drugs mucosal stages are in a refuge. If this refuge is large, and it can be up to 90% of the parasitic population, resistance will develop slowly. The BZ compounds and MOX provide two examples of drugs effective against mucosal stages. BZ compounds given daily for 5 days are effective against late L3 larvae and MOX is effective in the range 50–90% against developing L4. In both cases treatment would reduce the size of the refugium. The ultimate effect of these treatments on S_P will be influenced by the size of the mucosal refugia as well as the other factors, such as the size of the pasture refugia at the time of treatment and the efficacy of treatment.

Environmental Factors

Climatic conditions have fundamental effects on S_P through refugia. A refugium is part of the population that is not exposed to drug and is not selected. For example, worms on pasture or in an intermediate host (or hypobiotic) are in refugia. When the worms in refugia represent a large proportion of the worm population, those in the host that are exposed to treatment are in the minority. In this case the progeny of resistant survivors will be significantly diluted by susceptible worms in the refugia and S_P will be low. In Figure 22.3B, dry weather, when few worms are present in the environment, has a profound influence on resistance. The appearance of IVM resistance in *Teladorsagia* spp. in Western Australia after only two treatments with the drug illustrates the power of selection in arid areas (Swan *et al.*, 1994). On the other hand, a feature of cattle parasites is that because they are protected from climatic extremes in the dung pat and because the dung pat can be a reservoir of infective larvae for up to 12 months the level of refugia is increased which would lower S_P.

Management Factors

Frequency of treatment

An increase in treatment frequency is one of the most potent ways to increase S_P. There are experimental examples of this (Martin *et al.*, 1984; Barton, 1983) as well as evidence from surveys of drug usage coupled with resistance levels in the field (Nari *et al.*, 1996). The result of a simulation comparing 6 to 3 treatments per year is provided in Figure 22.4A. Frequent treatments provide frequent selection events. In addition, when treatments are close together and approach the pre-patent period for the parasite, there are fewer opportunities for susceptible worms to survive. In some circumstances more frequent treatment is required to achieve parasite control. Young lambs require more frequent treatment than ewes and resistance levels on pastures grazed by lambs exceeds those grazed by ewes (Barnes, Dobson and Barger, 1995).

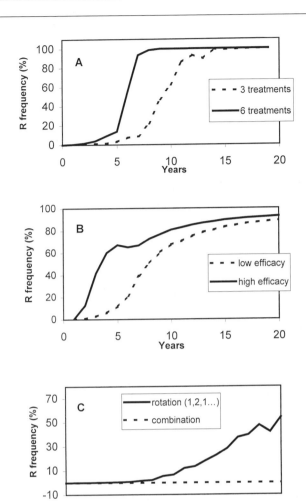

Figure 22.4. Effects of management on resistance allele frequency predicted by models for anthelmintic resistance. Assumptions were those of Barnes, Dobson and Barger (1995), unless otherwise stated: A. Effect of frequency of treatment on selection for resistance. Assumes that resistance is conferred by a single gene as a recessive trait. *3 treatments* shows the change in R-alleles for a regime where ewes and lambs receive 1 and 3 treatments per year, respectively. *6 treatments* shows the results when ewes and lambs receive 2 and 6 treatments per year, respectively. B: The effect of drug efficacy on selection for resistance. Assumes that resistance is conferred by a single gene and as a dominant trait. The low-efficacy and high-efficacy drugs allow survival of 20% and 1% of homozygous susceptible worms, respectively. C. The influence of two effective anthelmintics, used in combination or rotation. Assumes two independent genes confer resistance to the two drugs. Treatment regimes were an annual rotation regime (i.e. Year-1 use drug-1, Year-2 drug-2, Year-3 drug-1, etc.) and the combination of both drugs at each treatment. In both cases 3 anthelmintic treatments were given to lambs each year.

Efficacy

The influence of efficacy is a more complex issue. At very high efficacy there are no survivors and resistance will not develop. (This is provided that high kill is not accompanied by persistent activity, see

below.) Drugs are highly effective but do not kill every worm. Figure 22.4B models the rate of development of resistance with drugs at 99% and 80% efficacy. Resistance develops more rapidly where efficacy is high because only worms carrying resistance alleles survive and reproduce. In contrast, many of the survivors of the 80% treatment carry susceptible alleles. The principle has been supported experimentally (Sangster and Bjorn, 1995) and in simulations (Barnes, Dobson and Barger, 1995).

There is strong evidence that marginal reductions in efficacy increase S_P. For example, when goats are given the same dose rates of anthelmintic as sheep, lower drug concentrations and hence, efficacies, are achieved (Hennessy, 1994). This helps explain why resistance is detected so frequently in parasites of goats. Under-dosing with drugs would also increase S_P. For example, some anthelmintics appear to have poorer efficacy in foals compared with adult horses (Herd and Gabel, 1990). There is a genetic explanation of why underdosing selects for resistance. Worms which are heterozygous for (recessive) resistance are normally killed by full doses of drug but lower doses allow parasites carrying resistance alleles as heterozygotes to survive treatment. In future generations the frequency of resistance genes increases. Choice of dose rates is an example of how management choices can influence the importance of innate parasite factors such as Mendelian inheritance.

Drug persistence

Modelling suggests that extremely short acting drugs will reduce S_P. This is because short action only selects worms resident in the host. When long acting drugs are used, they reach peak levels in the host then fall during the elimination phase. Parasites which enter the host during the elimination phase may also be selected by a discriminating dose, rather like under-dosing. There is some experimental evidence which implicates the role of drug persistence in resistance. Controlled release capsules containing BZ drugs have been used in sheep. Although BZ resistance was present in adult worms, the drug in these devices was effective in preventing infection. With use over time, however, a higher level of resistance has developed to BZs, almost certainly as a result of persistent treatment. As concentrations fall one would predict that resistance to persistent drugs will first appear as a decline in protection period offered by the drug. This is the case for CLS resistance. At least two models indicate that persistence under some circumstances can select more strongly for resistance than short acting drugs (Dobson, Le Jambre and Gill, 1996; Smith *et al.*, 1999).

Rotations of drugs and combinations as ways of reducing selection pressure

Examples of alternating drug classes each year (rotation) and of using mixtures are shown in Figure 22.4C. Rather than relying on a single drug class, rotation is a widely used approach to slowing the rate of development of resistance. While a second drug class is being used selection pressure on the first is removed. Annual rotation of different drug groups was generally recommended at a time when the broad-spectrum anthelmintics had little persistent activity. In the light of drugs with extended activity against incoming larvae a safer option is to rotate between the drug groups each time infected hosts are treated. Combinations, if used before resistance to either drug is present, select for resistance slowly. This is because only worms simultaneously resistant to both drugs can survive, and this is very rare. Thus worms outside the host that avoid selection remain almost exclusively the source of new infection.

Laboratory Selection

It has been shown that different selection protocols can select for different resistance phenotypes and, presumably, genotypes (Gill and Lacey, 1998). Where selection is imposed by slowly increasing doses of drug at each generation, as used in laboratory selection, all the survivors of treatment have an equal chance of contributing to the next generation. This removes environmental selection for survival fitness and it would not be surprising if different alleles were selected.

Selection for resistance in *C. elegans* is a different proposition. Worms are mutagenised and then exposed to high drug levels. The entire population is screened and surviving hermaphrodites are individually cloned. Not only does resistance arise in a single generation, the bias in these selections is for single mutations and recessive inheritance. It also has the ability to isolate rare mutations. In these ways it does not resemble selection in parasites *in vivo* where it is likely that selection is for rare individuals carrying alleles which confer resistance.

Detection and Diagnosis

Detection of resistance is important both for surveillance of resistance and for parasite control. For example, early detection is essential in control schemes where we wish to switch to an alternative anthelmintic before the original compound has lost utility, and therefore have it available to return to later. Diagnosis also has a role in research where the effect of experimental control procedures on the development of resistance can be monitored. Several techniques that differ in their sensitivity and convenience are available. For example, faecal egg count reduction tests (FECRT) only detect clinical resistance, often not until the frequencies of resistance alleles reach 25% (Martin, Anderson and Jarrett, 1989). On the other hand, these tests can give a real indication of the performance of a particular drug on a particular farm. In contrast, genetic tests have the potential to detect very low levels of resistance but suffer from the need to establish a substantial understanding of resistance mechanisms and from being remote from farm realities. The middle ground is occupied by *in vitro* assays that have become useful tools for monitoring resistance. Several diagnostic techniques have been compared head to head (Johansen, 1989; Várady and Corba, 1999). A universal problem with testing for resistance is that worms

arrested in their development and residing in the mucosa can comprise a large proportion of the parasites present, yet their resistance status cannot be measured. Another problem is to identify the species of worms involved, especially when eggs or larvae are used for diagnosis. Highly fecund species may swamp less fecund ones and provide misleading results.

Treat and Slaughter Trials

The traditional and definitive test for resistance is to artificially infect worm-free hosts with infective stages of worms suspected of being resistant, treat with drug, and count the number of worms which survive at slaughter. These tests must include control groups of infected but untreated animals and of hosts infected with susceptible parasites of the same species and treated with the same drug. They are often referred to in the literature as 'drench and slaughter trials' or 'treat and slaughter trials'. Use of the recommended dose of drug can reveal clinical resistance and the use of a lower dose rate in other groups of the host can reveal information on resistance prior to the emergence of clinical resistance. Data from testing several dose rates each in a separate group considerably strengthens the utility of the data. Because dose rates are converted to logs prior to analysis, increments in dose rate should be logarithmic. A series of 0.5, 1 and 2 times the recommended dose is commonly used. As resistance is defined as the increase in % of worms surviving at a dose that is usually effective, a significantly lower efficacy of treatment for the isolate under investigation compared with a susceptible isolate run in the same trial would allow it to be classified as resistant. Significance is usually determined by analysis of variance of log transformed worm counts and these data are often supported by pre- and post-treatment egg counts from the same animals. Probit analysis (Finney, 1971; Lyndal-Murphy, 1992) can be used to calculate dose response parameters. When a sufficient number of dose rates are used log dose/response lines (like Figure 22.1) can be generated and compared in slope and position.

If properly conducted, slaughter trials provide the gold standard of diagnosis of clinical resistance. Virtually all confirmed resistant isolates have been reported on the basis of slaughter trial results. Further, adult worms which survive treatment can be unequivocally identified to species level.

Although they provide a certainty of diagnosis, slaughter trials have several drawbacks. Firstly, there is a need for a drug-susceptible reference isolate. This is routine for some species but very difficult for others. For example, susceptible isolates are often rare because of ubiquitous treatment in the field. Second, slaughter trials are expensive. This is not a serious drawback for sheep because now that we understand the relationship between *in vivo* and *in vitro* assays for anthelmintic resistance in sheep parasites (Gill *et al.*, 1995) it is only appropriate to perform slaughter trials on sheep when a novel isolation is made. On the other hand, the cost of cattle and horses means slaughter assays are rarely performed in these larger species, but the understanding of resistance in these hosts and the validation of *in vitro* assays will require slaughter trials to be conducted in the future. Slaughter experiments present an additional problem because the preparation of worm-free hosts is very expensive. Clearly, alternatives to slaughter assays are needed for the investigation of resistance in human parasites.

Experimental hosts have been useful for studying sheep trichostrongyloids. Drug efficacy in guinea pigs (Kelly *et al.*, 1981) and jirds (Conder, Thompson and Johnson, 1993) has been used to measure *in vivo* resistance in *T. colubriformis* and *H. contortus*, respectively. Although these techniques are not likely to be used for detection, they are useful experimental models because several dose rates of drug can be tested more quickly and cheaply than in sheep. Experiments reporting efficacy of IVM and MOX against AM-resistant worms in jirds (Conder, Thompson and Johnson, 1993) and sheep (Shoop *et al.*, 1993) provide examples of the use of such techniques and the good agreement of the results. Laboratory models of human hookworm infection may be a useful tool for investigating resistance in these nematodes (Richards, Behnke and Duce, 1995).

Faecal Egg Count Reduction Tests

Tests in sheep

This method is the most common method of diagnosis of clinical resistance. Recommended dose rates of drugs do not effectively reduce egg counts in resistant worm populations and this effect can be measured by a standardised faecal egg count reduction test (FECRT). Details, including calculations are provided in articles by Lyndal-Murphy (1992) and Coles *et al.* (1992). Briefly, the test involves the following steps:

1. Select a flock of sheep, 3–6 months old, not treated in the previous 6 weeks (or longer for persistent drugs), with faecal egg counts of > 200 epg.
2. Randomly allocate 10–15 animals to each drug treatment class to be tested and 10–15 controls. Use coloured marker to identify animals by group. Tests may include four drug classes plus combinations, involving 50–90 animals.
3. Weigh several of the larger animals in the flock and calculate the volume of anthelmintic to provide a recommended dose for the heaviest animal.
4. Treat each group with the dose of drug calculated in (3), leaving the controls untreated. Graze the animals together.
5. Return 14 days later, collect faecal samples from each animal (labelled by group) and perform individual faecal egg counts by a standard flotation method. Fourteen days is chosen because it is long enough for drugs to exert their effects, but not so long that newly ingested larvae will have developed to patency.
6. Calculate % reductions compared to controls, using arithmetic means, but logarithmic error calculations.

7. Anthelmintic resistance is present if efficacy is < 95% AND the lower confidence interval (95%) is < 90%.

The data can be complemented by use of culture of faeces and attribution of eggs in faeces in the same ratio as the genera identified in the L3 cultures.

This approach has the advantages of being well defined and widely adopted, using simple techniques and providing data relevant to the field resistance. Farmers can readily believe that the presence of eggs in faeces after treatment is a problem. FECRT have several disadvantages. They are insensitive. It has been estimated that at a level of 25% of resistance genes (recessive resistance) the tests will not detect resistance unequivocally (Martin, Anderson and Jarrett, 1989). Tests can only be performed when egg counts are > 200 epg in many animals in the flock. Two visits to the farm are required and farmers are often reluctant to participate in these tests, especially if sheep are due for slaughter for meat. Problems also arise in determining resistance to persistent chemicals. One way around this last problem has been to adopt protocols with lower dose rates, such as 1/3 for closantel (2.5 mg/kg). The interpretation of FECRT relies on a correlation between egg counts and worm counts. Some anthelmintics, particularly the AMs, do not kill resistant worms but, rather, suppress their egg production. Zero egg counts post-treatment are interpreted as susceptibility when in some cases worms resume egg production more than 14 days after treatment. A further potential problem with the method is that hosts may influence drug kinetics. This may be a particular problem with goats where the efficacy of many drugs is lower than in sheep.

Flocks of sheep are relatively homogeneous populations. They are usually of the same breed, many share a sire, are of a similar age and have the same grazing history. Furthermore, egg counts are a good reflection of worm burdens in most circumstances. The ruminant digestive system ensures a relatively constant passage of ingesta through the gut and eggs in faeces reflect egg production. There is still considerable variation in counts but use of group sizes of 10–15 in FECRT help to reduce variances. It has been shown that use of pre-treatment egg counts did not alter the results based on post-treatment counts alone if groups were adequately randomised. Arithmetic means are used because they have the advantages that they reflect pasture contamination, are simply calculated and are conservative estimates of resistance (Dash, Hall and Barger, 1988). The use of the two criteria (efficacy < 95% and lower confidence intervals < 90%) provides a 95% confidence of detecting clinical resistance.

FECRT in other species

Other hosts do not offer the same advantages of homogeneity of background and reliability of egg counts as measures of infection that sheep do. Cattle may offer similar within-herd consistency as sheep but egg counts do not reflect worm burdens because of variable dietary flows and the wet nature of the faeces. Although egg counts are good indicators of infection with adult parasites in horses, animals kept on properties may be quite heterogeneous. They often come from different backgrounds, are of different breeds and different ages. In order to maintain the statistical power of the test, variability would have to be offset by an increase in group size. This is generally impractical in cattle because of the additional labour required to treat more animals and to perform more egg counts. Few horse farms could provide 50 horses of similar age and management status to conduct a meaningful trial. Further, nematode parasites of horses and cattle are prone to arrested development in the host to a greater extent than parasites of sheep so only a small portion of the nematode population may be sampled.

The discriminating level for resistance based on FECRT in other species such as horses and cattle is set at an efficacy $<90\%$ (Coles et $al.$, 1992). It has been lowered from 95%, presumably, to allow for the anticipated increased variability in the data. There are additional problems in measuring resistance that places even this level in doubt, especially on horse farms. Firstly, the calculation of reduction usually relies on post-treatment counts alone. This does not account for changes in egg count between treatment and sampling which can vary between individuals in a non-homogeneous group. In order to account for these changes individual animal pre- and post-treatment egg counts can be used but this requires

individual identification of horses (see Table 22.4 for an example). The different approaches to calculation give different efficacy values and the classification of farms can shift between a resistant and a susceptible (Craven et $al.$, 1998). Secondly, there are cases where drugs are not fully effective against susceptible worms. One example is the activity of MOR against the cyathostomes. On initial release when given at 10 mg/kg it was 90–100% effective. It may be necessary to reduce the threshold for resistance to MOR and PYR in horses to 85% to account for this. Accumulation of further experimental data will help to set the appropriate methods for calculation and threshold values.

In vitro Bioassays

In vitro assays have become a popular alternative to *in vivo* assays for measuring resistance. Valid application of assays relies on co-expression of resistance traits of the target parasites in those used in the assays. *In vitro* assays, especially those using free-living stages, have several advantages over FECRT. Firstly, they avoid the difficulty of recovering parasitic stages. Second, the effects of standardised concentrations can be measured free of host influence. Third, the need for susceptible isolates to be tested simultaneously can often be dispensed with. Fourth, a single visit to the farm is required and, as only a pooled faecal sampling is required, farmers can submit samples directly. Fifth, some automation is provided by use of 96-well or 24-well format in the assays. Sixth, the assays generate dose/response data which improves sensitivity

Typically, a matrix of increasing drug concentrations are applied across a 96-well plate from left to right, and different anthelmintics are used down the rows. Duplicate wells at each concentration are usually used. Eggs or larvae are added to each well and allowed to hatch, develop or simply incubate. Where the number of surviving or developing worms need to be enumerated they are counted while viewing through a microscope. Nematodes in liquid culture are viewed through an inverted light microscope while those on agar are viewed through a stereo microscope. Viability or motility of worms can be measured either visually, following migration, or electronically and results are analysed by probit (Finney, 1971) or logit analysis (Waller et $al.$, 1985). Using these methods we can calculate ED_{50} with standard errors and derive RF values. Because survival at high concentrations may be the first indication that resistance is developing, looking for survivors in these wells increases the sensitivity of these tests over the calculation of EC_{50}. In order to simplify interpretation of the results one commercial kit is provided with wells that are colour coded. Concentrations where susceptible worms are expected to develop are coloured differently to those containing higher concentrations in which only resistant worms develop. In this case results can be calculated by rule of thumb rather than mathematically.

There are some general disadvantages of *in vitro* assays. The validity of the assay as a measure of

Table 22.4. Faecal egg count reduction test efficacy of morantel tartrate (10 mg/kg) on 6 horse properties. Figures bolded fall below 90%. % reductions were calculated by three methods:

A. {(mean post-treat. control epg–mean post-treat. epg)/mean post-treat. control epg} × 100

B. inv arcsin[mean{arcsin(SQRT (individual FECR))}]² × 100

C. 100–{(mean post-treat. epg/mean pre-treat. epg) × (mean pre-treat. control epg/mean post-treat. control epg) × 100}

(J Pook, unpublished data)

Farm	Method A	Method B	Method C
1	98.5	99.7	97
2	94.7	96.4	94.9
3	**89.8**	99.4	95.7
4	**89**	97.1	93.9
5	**74.2**	**89**	**72.6**
6	95.6	98.2	92.7

epg, eggs per gram of faeces
mean, is mean of the horses in a treatment or control group
FECR, faecal egg count reduction
Individual FECR, (pre-treatment epg-post - treatment epg)/pre-treatment epg

resistance must be confirmed under a range of conditions. There is also a requirement for trained laboratory staff and, if speciation of larvae is also expected, further skills are needed.

Várady and Corba (1999) compared several *in vitro* assays in their ability to detect BZ and LEV resistance in a parasite population. Their conclusion was that the larval development assay was both sensitive and convenient.

Egg hatch assays

The earliest *in vitro* assays involved incubating eggs freshly recovered from faeces (about 100 per well) in a range of drug concentrations (Le Jambre, 1976; Dobson *et al.*, 1986; Hunt and Taylor, 1989). The drugs inhibited hatching and hatch rate was measured by counting and comparing numbers of eggs and L1 present after overnight culture in water at around 27°C. These assays were used for the BZ group of drugs and LEV. They were simple and rapid and the ED_{50} values were compared with values for susceptible worm isolates run simultaneously. These assays had a limited utility outside the laboratory because they required fresh eggs which precluded transport of samples. Further, they were unsuitable for some drugs such as the AM class and closantel.

Larval development assays

These assays rely on drug inhibiting the *in vitro* development of eggs through to L3, a process that typically takes 5 days. Several systems use 96-well plates seeded with bacteria on which the larvae feed. Some tests rely on feeding the larvae with heat-killed *E. coli* (Coles *et al.*, 1988) while others used a defined medium such as Earles buffered salt solution with yeast extract (Hubert and Kerboeuf, 1984). In the latter case it is likely that the diet is provided by bacteria from the host faeces are transferred into the wells with the eggs. A solution of an antifungal chemical is added to the medium to suppress fungal growth. Drugs are added in liquid medium (Coles *et al.*, 1988; Taylor 1990; Hubert and Kerboeuf, 1992) or dissolved in an agar substrate prior to its setting (Lacey *et al.*, 1990) and liquid medium layered on top. Eggs (about 70 per well) are added and the plates incubated for 6 days at 25°C. After incubation, eggs and larvae, at each stage are counted and analysed as described above.

It is claimed that the use of agar to dissolve the drugs provides advantages in delivering more reliable concentrations of drug, especially lipophilic compounds, and also in facilitating transport of prepoured plates. The larval development assay (LDA) of Lacey *et al.*, (1990) has been commercialised and is used for diagnosis of field resistance in sheep and horses (Ihler and Bjorn, 1996) in several countries. It has proved to be robust and widely applicable. Standardised concentrations and good correlation with *in vivo* efficacy of the drugs in sheep (Gill *et al.*, 1995) means that a susceptible reference isolate is not required for calibration and that results can be compared between laboratories. In practice, skilled workers can differentiate the L3 of *Haemonchus* from those of *Trichostrongylus* and *Teladorsagia* directly on the plate which provides an advantage over other *in vitro* assays.

One current drawback of these tests is that there is poor discrimination between AM-resistant and susceptible populations of some species of nematode, especially *Teladorsagia* spp. However, this may be resolved by further refinement of the assay and the selection of different AM analogues for use in the plates. Another drawback is the 5–7 day delay in obtaining results. Still, this is quicker than FECRT where the delay is 2 weeks.

Typically the LDA gives resistance factors for the BZs, LEV and AMs in the range 16–60, > 50 and 5–17, respectively (Gill *et al.*, 1995). The known anthelmintic potency of several analogues had the same rank order of potency against the susceptible isolate *in vivo* confirming that, pharmacologically, the free-living stages mimicked the parasitic stages.

L3 motility

Measuring the motility of L3 in the presence of drugs is a potentially simple approach to measuring resistance. L3 are a resting stage and therefore transportable and robust. Larvae are incubated *in vitro* in the presence of drugs and then motility is measured by observation (Gill *et al.*, 1991), migration through a sieve (Sangster, Riley and Collins, 1988) or by electronic detectors (Folz *et al.*, 1987). The method is useful for detecting BZ and AM resistance. Although L3 respond in a dose-dependent manner to LEV there is no difference between resistant and susceptible isolates of *H. contortus* and of *T. colubriformis* (Sangster, Riley and Collins, 1988; Coles, Folz and Tritschler, 1989; Douch and Morum, 1994) but the method does allow differentiation between susceptible and resistant *T (O). circumcincta* (Martin and Le Jambre, 1979).

L4 (parasitic larvae) feeding assay

L3 can be cultured to parasitic L4 *in vitro* and the viability of the larvae after exposure to drug can be determined (Stringfellow, 1988). This approach has been adapted into an *in vitro* assay which can be used to detect resistance to all of the common anthelmintics including closantel (Rothwell and Sangster, 1993). The steps in this assay are to incubate (10% CO_2 atmosphere, 40°C) exsheathed L3 at the rate of about 100 per well of a 24-well plate in a medium containing sheep serum. After 4 days, drug is added at a range of concentrations across the plate. Three days later the number of worms migrating through a sieve and those that remain behind are counted and compared with results for a susceptible isolate. In Australia the assay has recently been used for a survey of closantel resistance on 300 properties up to 500 km distant from a central laboratory (Lloyd *et al.*, 1998). Although the assay provides flexibility in detecting a range of resistance phenotypes it requires more careful preparation than an LDA as well as special culture conditions.

It is interesting to explore aspects of this assay in relation to the action of closantel (Rothwell and Sangster, 1993). Although closantel affects *H. contortus* in the LDA it kills susceptible and CLS-resistant isolates at the same concentrations. This lack of discrimination may be due to a lack of the relevant site of action being present in larval stages and/or could relate to the route of entry of drug into the worm. For efficacy and for resistance to be displayed it appears that closantel must be ingested as it is by *H. contortus in vivo* with blood as it feeds. The assay described above is able to discriminate between susceptible and resistant worms because it is believed that the parasitic stages ingest closantel which is bound to ovine albumin present in the incubation medium.

Assays on parasites ex vivo

Diagnosis of resistance in parasites removed from hosts has been suggested. The motility of *Onchocerca volvulus* removed from nodules in humans has been measured in order to establish baseline data (Townson *et al.*, 1994).

Biochemically Based Assays

Biochemical assays may be those used as a measure of viability of resistant worms after drug exposure which kills susceptible worms or those where a component associated with the mechanism of resistance is measured or detected. The attraction of biochemical assays is that they have the potential to be automated.

In the study of anthelmintic resistance there are no examples of biochemical end-points for nematode viability following exposure to drugs. One may conceive of the measurement of metabolic end-products or secretions, the detection of components present in live worms, say, by the use of antibodies or the detection of intermediaries, such as ATP, which reflect high energetic state. In other fields of biology cell division and DNA replication can be tracked with dyes or isotopes, but worms do not reproduce *in vitro* so these are not useful techniques.

There are examples of tests that measure biochemical properties which are altered in resistance. Sutherland, Lee and Lewis (1998) described a two fold increase in esterase activity in several BZ-R isolates compared with susceptible *H. contortus*. Despite the relative simplicity of the colorimetric assays these workers developed, the drawback with this approach is that the changes are not linked to known resistance mechanisms and may not be generally applicable. An isotopic binding assay has been developed for BZ resistance based on changes in tubulin binding (Lacey and Snowdon, 1988). Reduced binding of BZ anthelmintics to worm tubulin is a known resistance mechanism and the assays can be performed on a pooled sample of L3. The main drawback of this method is that radiolabelled drug is required as well as the skills and equipment to use them. There are no resistance mechanisms that relate to altered enzyme activities and consequently amenable to biochemical assay.

Genetic Detection

Genetic tools potentially provide the most sensitive methods for detecting resistance, but in order to design and use gene probes, the genetic basis of resistance must be known. Further, for the test to be reliable, it must be the predominant mechanism in populations of worms, including future isolates. BZ resistance has a known mechanism in trichostrongyloid nematodes of sheep, and the tests that have been developed provide good examples. BZ resistance is linked both to a change in BZ binding to tubulin and to the presence of a predominant allele of the β-tubulin isotype 1 gene, a F200Y transition (Kwa, Veenstra and Roos, 1994) due to a single basepair difference. Although other genetic changes may contribute to resistance this allele appears to be commonly associated with resistance and is an ideal diagnostic feature. Tests have been used for pooled worm samples or single worms for the species *H. contortus* (Kwa *et al.*, 1995) and *T. circumcincta* (Elard, Comes and Humbert, 1996; Elard, Cabaret and Humbert, 1999).

Several approaches have been used. All of the assays rely on using PCR to differentially amplify tubulin DNA from susceptible and resistant worms that differ by a single base pair. The amplified products are separated by agarose gel electrophoresis and their sizes compared. Typically, template DNA is extracted by standard techniques from an individual worm or a pool of worms. There are two main approaches to primer design for PCR amplification. In the first, primers complementary to the susceptible sequence or the resistant sequence (the mismatched base pair of the point mutation is at the 3' end of the primer) are coupled with a common primer ~300 bp away. DNA is placed into two tubes and two separate reactions are performed (Kwa, Veenstra and Roos, 1994). A product in one tube but not the other indicates that homozygous alleles are present. Products in both reactions indicates heterozygous genetics. A second, more elegant and better controlled method is performed in a single tube (Elard, Cabaret and Humbert, 1999). The reaction uses four primers, two are sense and antisense primers either side of, but at different distances from, the point mutation. Within this nest, a sense primer and an antisense primer, both with 3' ends coinciding with the altered base pair, are included, one complementary to the susceptible and one to the resistant sequence. The total product, which should be present in all tubes, is the amplification control. Two differently sized products indicate heterozygous alleles are present and either product alone indicates homozygosity, the size of the fragment indicating whether susceptible or resistant alleles are present. It is conceivable that species-specific primers could also be designed by taking advantage of species specific sequences in β-tubulin and included to provide simultaneous species and resistance information. Other detection end-points may also allow automation of detection.

Use of genetic tests for other resistances awaits the elucidation of the genetics of the resistance involved. Nevertheless, these illustrate the potential for genetic tests in the future investigation of resistance. Another potential benefit of genetic studies is that if the mechanisms are conserved across parasite species they may provide diagnostic tools in parasites of horses, cattle and humans.

Pitfalls in Measuring Resistance

Resistance is a complex issue and its measurement is often indirect. Even in controlled situations results vary. Apart from the day-to-day variation in results other factors are known to influence resistance levels. Resistance varies over the course of infection (Kerboeuf and Hubert, 1987). Cold storage may also influence resistance (Hendrikx, 1988).

Measuring the Prevalence of Resistance

Resistance can be measured in several ways. One is to estimate the farm prevalence of resistance. That is, the percentage of farms on which resistance is present. In addition, on affected farms, the % of worms that are resistant (or the % of resistance genes) in a population is a measure of the population prevalence. The sampling and diagnostic techniques appropriate for measuring resistance will vary depending on whether resistance is rare or common. In areas where resistance is common, surveys are less useful than the measurement of the population prevalence on individual farms which provides useful input into local parasite control programs. In the early stages of resistance the application of sensitive tests in surveys is a logical approach. Based on the results of mathematical modelling, if we are to control resistance through detection and use of alternative chemicals before the original ones have lost their utility, we need to detect less than 1% prevalence of resistance alleles. Several examples are discussed below. In these examples we assume that only a single parasite species is present. Mixed infections, which are the norm, would reduce the ability of tests to detect resistance.

Measuring farm prevalence of resistance

In order to measure the farm prevalence, surveys need to be performed on farms selected at random from the study population of farms. Bias is common in the process of selection of properties. It may be introduced by selection of convenient properties, properties with positive egg counts, farmers willing to cooperate with sampling and a window of opportunity before the next scheduled treatment. Surveys of sheep parasites are more likely to be random as producers can be chosen from registration rolls. At the other extreme, it is more difficult to ensure randomisation in surveys of horse properties as there is a bias towards choosing better managed farms with better animal handling facilities and, possibly, atypical anthelmintic usage. On an individual farm, horses within a group usually have a more varied background and management history than

animals within a flock of sheep. Nevertheless, the survey data which are available have been painstakingly acquired and are a valuable resource.

In order to obtain random samples farms can be chosen in several ways (see Thrusfield, 1995). One approach is to take a simple random sampling from livestock rolls. Another approach is to select strata of farms and then select randomly from within each stratum so that the number of farms sampled from each stratum is proportional to the number of farms in the stratum. This approach has the advantage of representative sampling from several climatic zones or several types of farming enterprise. Nari et al., (1996) sampled 252 farms within 15 strata in an extensive survey in Uruguay. Lloyd et al., (1998) sampled two geographically distinct strata in a survey of closantel resistance in eastern Australia. These two studies provide models of good survey design.

In order to estimate the farm prevalence, the number of properties that need to be sampled depends on several factors. These are the number of farms in the sample population, the level of confidence required in the estimate, the level of accuracy required in the estimate and the expected prevalence. Table 22.5 provides some examples which illustrate that in order to detect a low prevalence, many farms need to be tested.

Measuring population prevalence

Measures of population prevalence hinge both on sampling and the type of test applied. In vitro and genetic tests are those most likely to be applied to population prevalence.

Typically, sampling is performed by taking faecal samples from a randomly selected group of animals on a farm and pooling them. Because the real value of in vitro tests is their ability to detect low levels of resistance, random sampling on a farm may not be desirable. For example, sampling from a host soon after treatment would provide a population enriched for resistant parasites and such an approach would improve the sensitivity of diagnosis. Such testing would estimate population prevalence in an animal rather than on the farm.

The second issue is the inherent sensitivity of tests. Because FECRT detect a mean response of the

Table 22.5. Number of farms that need to be sampled in order to achieve 95% confidence of measuring resistance at expected prevalences and accuracies (width of error about the prevalence estimate). Calculated by the methods in Thrusfield (1995)

Prevalence and (accuracy)	0.01 (0.005)	0.02 (0.01)	0.04 (0.02)	0.25 (0.125)
farms in study population				
5000	1184	654	343	46
1000	608	430	270	44

population, have a 95% efficacy cut off and because some drugs are highly effective against susceptible worms at recommended dose rates, large shifts in resistance can occur prior to detection (Martin *et al.*, 1989). *In vitro* assays where mean responses like an ED_{50} are calculated suffer from similar problems and may not be significantly better than FECRT. However, if assays are used to detect a shift in the dose response line to the right at any concentration, especially high concentrations, even one survivor compared with a susceptible population may indicate resistance. A single surviving worm in a well can signal that the prevalence of resistance in the population is < 4% (at 95% confidence). If duplicate wells at a single drug concentration are considered, the prevalence estimate improves to be < 2%.

Genetic tests may provide an alternative or complementary approach to the detection of resistance. They also have the potential to detect population prevalence below 1%. Assuming that a genetic test is available and that it detects a single allele (diploid and autosomal) linked to the resistant phenotype in one species of parasite, it could be applied to measure the prevalence of resistance in several ways. One would be to collect worms surviving high drug concentrations on *in vitro* assay plates and genotype them individually. Once a pattern was established genotyping may be unnecessary in every case. Another approach would be to apply genetic tests to the population of worms (e.g. eggs or larvae) recovered from a farm or animal. Because genotyping individual worms would be laborious and expensive, pooling samples of parasites would have to be considered. For example, a pool of 150 worms would have to test negative for the resistance allele to provide 95% confidence that resistance prevalence was < 1%. A useful approach would be to perform 3 PCR reactions on different pools of DNA. If no resistant alleles were detected in pooled DNA from 150, 75 and 30 worms, allele prevalences of < 1%, 1–2% and 2–5%, respectively, (at confidence level of 95%) would be present (Thrusfield, 1995). Such tests would require ideal conditions for both extraction of DNA from all of the worms and PCR-based amplification.

Once the genetic tests are available these approaches may be feasible for sheep, goat, cattle and pig parasites where the parasite species are limited in number. The horse cyathostomes present a more significant challenge. At least 10 species appear to predominate in the collections of resistant isolates (Lyons *et al.*, 1996; Chapman *et al.*, 1996). Until single species infections are achieved and species life histories are established or DNA-based techniques can be used to perform similar studies, there can be little progress. With a few modifications the LDA works well with cyathostomes (Ihler and Bjorn, 1996) and molecular phylogeny coupled with LDA will help advance our knowledge in this area.

The suitability of various tests for detecting different levels of resistance alleles in worm populations is

shown in Table 22.6. The increased sensitivity of genetic tests makes them suitable for detecting low levels of resistance, especially in the case of recessive resistance.

Resistance in *C. elegans*

Diagnosis of resistance is not relevant to *C. elegans* but measuring resistance levels are required to measure phenotypic characters of worms after selection or following transgenesis. Techniques generally involve incubating several worms in liquid culture (Simpkin and Coles, 1981) or on agar plates containing drug, at different concentrations. Assay end points can be motility or egg or worm numbers. Examples of the latter might be automated in the future by monitoring green fluorescent protein expression or accumulation of radioactive nucleotides. Several methods have been used (Kwa *et al.*, 1995; Schaeffer and Haines, 1989; Broeks *et al.*, 1995).

Occurrence of Resistance

Drugs

Anthelmintics can be grouped in classes that share a common chemistry and mechanism of action. Because nematodes develop resistance to each drug class independently and side resistance appears to be complete it is valid to record the presence of resistance in a species to a class of drugs. Such a scheme is presented in Table 22.7. The resistant parasites listed are those reported in the literature. Only cases which have been validated by experimental passage of resistant worms in hosts have been reported here. There are several cases of suspected resistance which

Table 22.6 Tests for resistance able to achieve 95% confidence of detecting resistance in populations showing different resistance gene allele frequencies and different modes of inheritance. Assumes single gene resistance and distribution of alleles in Harvey-Weinberg equilibrium. Estimates by larval development assay (LDA) assumes either calculation of EC_{50} (method a) or examination of survivors in high concentration wells (methods b and c). See text for details.

% R alleles	Recessive resistance		Dominant resistance	
	% resistant phenotype	test	% resistant phenotype	test
50	25	a-f	75	a-f
25	6	b-f	44	a-f
5	0.25	def	10	b-f
2	0.04	ef	4.0	cef
1	0.01	f	2.0	cf

a FECRT or EC_{50} on LDA.
b Single well, high concentration LDA (75 eggs/well).
c Duplicate well, high concentration LDA (75 eggs/well).
d Genetic test > 30 worms pooled.
e Genetic test > 75 worms pooled.
f Genetic test > 150 worms.

Table 22.7 Major reported resistances to anthelmintic classes (with drug examples) in parasitic helminths and *Caenorhabditis elegans*. Adapted from Conder and Campbell (1995) with additional references in the text. See text for comments on prevalence and levels of resistance.

Host	Parasite	Benzimidazole	Imidazothiazole	Avermectin/ Milbemycin	Salicylanilide	Organophosphate compounds
Sheep & goats	*Haemonchus contortus*	×	×	×[a]	×	×
	Teladorsagia spp.	×	×	×	–	
	Trichostrongylus spp	×	×	×[a]	–	
	Nematodirus spp.	×				
Pig	*Oesophagostomum* spp.	×[b]	×[a]	×[b]	–	–
Horse	Cyathostomes	×	×		–	
Cattle	*Cooperia* spp.	×		×	–	–
	Haemonchus placei	×	×			–
	Ostertagia ostertagi	×	×		–	–
	Trichostrongylus axei	×			–	–
Free-living	*Caenorhabditis elegans*[c]	×	×	×	–	×

[a] Both laboratory and field isolates have been reported. [b] Isolates are laboratory selected. [c] Selected by mutation. – indicates that for the anthelmintic class this nematode species is outside its spectrum of activity.

were either not reproducible or disappeared on passage and are probably related to one of the 'treatment failures' mentioned above. Resistances in Table 22.7 were detected as cases of clinical resistance and, as such, these reports underestimate the true level of resistance. The majority are field isolates. Others are the result of laboratory selections in penned hosts, indicating that potential for the development of resistance is present in these parasites. Confirmed cases of resistance among pig parasites have not appeared in the field except for PYR resistance.

Table 22.7 gives no indication of the prevalence of resistance either in worm populations or on farms. Farm prevalence can be measured by several methods (see diagnosis, above) but most commonly by FECRT applied to surveys. In general, resistance is reported where egg count reduction is < 95% in sheep or < 90% in horses. Other surveys are based on the results of *in vitro* assays.

Derivation of Resistant Isolates

Many cases of clinical resistance have been isolated from the field. Such isolates are further selected by anthelmintic treatment and provide worms for further laboratory experiments. Some isolates of sheep nematodes (Egerton, Suhayda and Eary, 1988; Giordano, Tritschler and Coles, 1988), and pig nematodes (Várady *et al.*, 1997) have been selected in pen-housed animals (laboratory selection) from a susceptible parent population.

Host Species

Conder and Campbell (1995) have provided a comprehensive list of reports of resistance. Readers seeking details of resistance for particular countries, drugs and parasites should consult that work. The following is a summary of current cases of resistance updated with more recent reports. Where it is available, prevalence is quoted, but prevalence is increasing with time and as surveys only measure clinical resistance the prevalence values are underestimates. In the future resistance will occur in more species, to more drugs and in more locations. At the end of each section the impact of resistance is discussed. On some properties resistance to several drugs is present. On others several genera of parasites are resistant to one drug.

Sheep and Goats

Sheep and goats carry many of the same parasite species and can cross-infect the other host species via contaminated pastures. The characteristics of resistance in these worms are retained in their new hosts. Resistance appears to develop more readily in goats than sheep. Evidence is both circumstantial, many resistances that arise in a region were observed initially in goats; and intuitive, parasites in goats are exposed to lower drug concentrations and immunity is often poorer. Both are factors that accelerate selection for resistance. When tested in the same region, resistance is usually at a higher prevalence on goat farms than sheep farms, although this could also be explained by the poorer efficacy of many drugs in goats. Goats provide a potential parasite reservoir for sheep.

Although many resistance surveys simply calculate egg count reductions in infections of mixed parasite species, some identify larvae to the genus level. In other surveys experimenters collect adult worms for speciation. These data can help us to tease out the role of individual species in sheep where multispecies infections are the norm and resistance can develop to all major species. Description of species differences are important because some species predominate in some localities and they differ in pathogenicity and resistance status. Resistance has occurred in the three genera *Haemonchus, Teladorsagia* and *Trichostrongylus* and include several species in each genus. Resistance has been described in *Nematodirus, Cooperia* and *Oesophagostomum* but are not yet problematic.

BZ resistance

BZ resistance historically appears as the first resistance in a region. This is probably because of the long term, widespread use of this class through the 1970s and 1980s. BZ resistance is the most common resistance in sheep nematodes in the Northern Hemisphere where *Teladorsagia* is the most common genus in cool climates and *Haemonchus* is a pathogen in warmer areas. In other regions, particularly the Southern Hemisphere where nematodes evolved anthelmintic resistance under strong selection pressure, resistance to BZs is widespread and occurs in virtually every country and region where it has been sought. The long history of BZ resistance in sheep parasites has provided an opportunity to study the temporal development of resistance. In a random survey of 209 properties in the UK 36–68% of properties had resistant worms (Hong *et al.*, 1992). In Australia, survey data provide a rare opportunity to explore not only the prevalence of resistance but also its species distribution and the level of resistance. The prevalence of BZ resistance (% of properties with < 95% on FECRT, range indicates prevalence in different regions) rose from 25–68% in 1986 to 81–95% in 1992 (Waller *et al.*, 1995). In 1992 in one survey 86% of properties had resistance; 32% with *Haemonchus*, 66% with *Teladorsagia* and 85% with *Trichostrongylus* infections. Further analysis of these data indicate that although efficacy was < 95% on 86% of properties, the dose failed to reduce counts by 50% on 26% of properties, suggesting that high levels of resistance were present. In random surveys in South America 37–90% of properties are affected and on many, less than 50% efficacy was achieved (Nari *et al.*, 1996; Echevarria *et al.*, 1996; Eddi *et al.*, 1996). Data from *in vitro* assays have the potential to be a more sensitive measure of resistance. Several methods are available but populations subjected to LDA were 60 fold resistant (RF > 60) compared with susceptible reference isolates (Gill *et al.*, 1995). These cases indicate a high level of resistance and are consistent with widespread failure of drug-based control.

LEV resistance

LEV resistance has only been reported occasionally in the Northern Hemisphere. There it may simply be at an earlier point in the evolution of resistance than in the Southern Hemisphere where LEV resistance in *Teladorsagia* and *Trichostrongylus* species occurs on 40–90% of sheep raising properties across Australia (Overend *et al.*, 1994; Waller *et al.*, 1995). Cases have also been reported from South America, New Zealand and South Africa. Random surveys in South American countries indicate a prevalence of 8–20% in Argentina (Eddi *et al.*, 1996)) and 84% in Brazil (Echevarria *et al.*, 1996). LEV resistance in *H. contortus* is rare as measured both by the number of recoveries from the field in several countries and in survey information in Australia where genus differentiation has been performed. It appears that the prevalence has been increasing and has reached 19–29% in Uruguay and Brazil (Nari *et al.*, 1996; Echevarria *et al.*, 1996).

AM resistance

AM-resistant parasite populations have been identified in many countries, mainly in the Southern Hemisphere (Swan *et al.*, 1994; Gopal, Pomroy and West, 1999; Watson *et al.*, 1996) although it has also arisen in goats in the USA (Craig and Miller 1990) and UK (Jackson *et al.*, 1992). *Haemonchus* and *Teladorsagia* which survive recommended doses of all commercial anthelmintics of the AM class are regularly isolated in Australia but are not sufficiently common to register in surveys. In South America prevalence is highly variable. Uruguay has a modest level (< 1%) (Nari *et al.*, 1996) and Brazil 13% (Echevarria *et al.*, 1996). A non-random survey in Paraguay revealed up to 73% prevalence (Maciel *et al.*, 1996). Several reports indicate that resistance develops very quickly. One example was that two treatments only of IVM were required to select for resistance in *Teladorsagia* in Western Australia (Swan *et al.*, 1994). Laboratory selected isolates are also available (Egerton, Suhayda and Eary 1988; Giordano, Tritschler and Coles, 1988). *In vitro* assays have been used for characterising isolates of AM-resistant nematodes. The LDA, using the IVM analogue AVMB2 give RF values in the range 2.5 to 20 for a series of isolates from South Africa and Australia (Gill *et al.*, 1995; Le Jambre *et al.*, 1995).

Resistance to other compounds

Closantel is a narrow spectrum drug used for the control of *H. contortus* in sheep. Resistance has developed in this parasite in South Africa (van Wyk, Gerber and Alves, 1982) and in Australia (Rolfe *et al.*, 1990). In affected areas of NSW (Eastern Australia) prevalence of > 25% has been estimated (Lloyd *et al.*, 1998) using an *in vitro* assay and clinical resistance is a problem in high rainfall areas.

Resistance to organophosphate compounds has been reported in *H. contortus* in south Africa (Malan *et al.*, 1990) and Australia (Green *et al.*, 1981).

Importance to industry

The inevitable consequence of these resistances has been the development of worm isolates in South Africa and Australia that have multiple anthelmintic resistance. However, their location and abundance probably reflects the geographic area and intensity of the search rather than reflecting their true prevalences. The presence of multiple resistance is disturbing because even several drugs used in combination are not fully effective against the worms. The presence of these isolates highlights the seriousness of the situation for parasite control in the Southern Hemisphere where resistance has forced several farmers out of the industry.

Pigs

In addition to a report on field resistance of *Oesophagostum* to PYR (Roepstorff, Bjorn and Nansen, 1987) several laboratory selected isolates have been described. Susceptible populations of *Oesophagostomum* spp. in pigs were used as a starting point for selection for resistance to BZs, LEV and IVM in separate lines for at least 10 generations. Worms selected with LEV survive 10 mg/kg of LEV, have RF values *in vitro* of 7.3 and share side resistance with PYR (Várady *et al.*, 1997). Resistance is not a current industry problem.

Horses

Resistance in horse parasites has occurred exclusively in the cyathostomes. Several species appear to predominate in resistant populations, the most common being *Cyathostomum catinatum*, *Cyathostomum coronatum*, *Cylicocyclus nassatus*, *Cylicostephanus longibursatus*, *Cylicostephanus calicatus*, *Cylicostephanus goldi* (Lyons *et al.*, 1996; Chapman *et al.*, 1996) although others do occur (see Conder and Campbell, 1995).

Resistance to BZs has been reported (Conder and Campbell, 1995) from Australia, Austria, Belgium, Brazil, Canada, Denmark, Germany, the Netherlands, Norway (Ihler, 1995), Slovak Republic (Várady and Corba, 1997) South Africa, Sweden, UK, Ukraine (Borgsteede, Dvojnos and Kharchenko, 1997) and USA. Diagnoses by the presence of poor egg count reductions have been supported by LDA results (Ihler and Bjorn, 1996; Pook, Power, Hodgson and Sangster, unpublished data) but RFs are difficult to establish because known susceptible populations are not available. Lyons *et al.* (1996) have recorded the efficacy of OBZ in a closed band of ponies in Kentucky between 1978 and 1991. Efficacy declined from 95% to 1% over the period and resistance in six predominant species developed at much the same rate. The prevalence of BZ resistance is difficult to ascertain because surveys are not random. It is likely that 90% of properties around the globe are affected.

MOR (and PYR) resistance is much more difficult to pin down. The problem is that even at their release on the market in the 1970's the efficacy of these compounds against the cyathostomes was in the range of 91–100% (unpublished). Clearly, the discriminating level of 90% efficacy is inappropriate as a cut off for resistance in this case. Although several studies have reported efficacies around 85–90% only one provides convincing evidence for resistance. The report of Chapman *et al.* (1996) described worm count reduction

of 25–83% and poor efficacy in a follow up study. Lyons *et al.*, (1996) also reported FECR of 72% for PYR. Surveys in USA (Woods *et al.*, 1998), Denmark (Craven *et al.*, 1998), Sweden (Nilsson, Lindholme and Christenson, 1989) Scotland (King, Love and Duncan, 1990), Norway (Ihler, 1995) and Australia (Pook, Power, Hodgson and Sangster, unpublished data) fail to provide unequivocal evidence for clinical resistance.

Although there are no reports of AM resistance at the time of writing, it is expected that resistance will emerge with time (Sangster, 1999). These drugs are the main-stay of parasite control in horses and the emergence of resistance would seriously jeopardise control of cyathostomes, particularly in foals. The factors which determine selection for resistance are more complex in the cyathostomes so the emergence of resistance is more difficult to predict. Interestingly, the importance of these parasites in equine health was not recognised until the widespread use of the AM anthelmintics effectively removed the highly pathogenic species of *Strongylus*. It is likely that BZ-resistant *T. axei* will be found in horses grazing with cattle.

The significance of resistance for the horse industry is that of the three available drug classes, the BZs have lost efficacy, that of MOR and PYR is waning. This means that the AM class is heavily used, often exclusively. Selection pressure is high and if the AM class fails the consequences for parasite control will be serious.

Cattle

It is likely that the early reports of BZ, LEV and MOR resistance in cattle parasites in Australia, USA and Europe (Conder and Campbell, 1995) were treatment failures. In particular, several reports of resistance to the older BZ compounds and MOR and LEV in *Ostertagia ostertagi* probably arose because the drugs failed to control arrested stages of the parasite. Nevertheless they are listed in Table 22.7. BZ-resistant *T. axei* have been reported from Australia and New Zealand. Because this species can parasitise several host species it is not certain in which host the resistance initially arose. Typically, it is associated with mixed grazing. *Cooperia* resistant to BZ has also been reported from New Zealand (Jackson *et al.*, 1987) and MOR-resistant *Haemonchus placei* from India (Yadav and Verma, 1997).

AMs are the most widely used anthelmintics in cattle and preserving their efficacy is central to parasite control. Therefore, a series of descriptions of IVM resistance in *Cooperia oncophora* are significant. Resistance has been reported from New Zealand (Vermunt, West and Pomroy, 1995; McKenna, 1996) and the UK (Coles, Stafford and MacKay, 1998).

Although anthelmintic resistance is not a widespread problem in cattle, several species have potential to become resistant and it is expected that the problem will escalate. There are two reasons to believe this. Firstly, the recognition of clinical resistance indicates that considerable subclinical resistance is present. Second, an increase in selection for resistance is likely to follow an anticipated rise in anthelmintic usage in cattle associated with the marketing of convenient AM preparations such as pour-ons and long acting inject-ables (McKenna, 1996).

Humans

There are no reports of resistance in nematode parasites of humans, even if the criterion of passage of resistant isolates through hosts is relaxed. On the other hand, as diagnosis of resistance is difficult, resistance may already be present. For example, treatment failure is difficult to investigate and finding worm populations to use as susceptible reference isolates will be difficult because most populations of worms have been exposed to treatment. Furthermore, such isolates would have to be maintained in experimental animals. Another complication is that there appear to be differences in drug susceptibilities between different geographic isolates and species of parasites. Some attempts have been made to develop *in vitro* tests to establish drug effect levels and potentially detect resistance if it occurs. Townson *et al.* (1994) describe an assay in which the effect of IVM on the motility of microfilaria of *O. volvulus* are measured. Richards, Behnke and Duce (1995) describe experiments which indicate that two species of human hookworm differ in their *in vitro* sensitivity to IVM by 40–50 fold.

Of the nematode parasites of humans, hookworms appear those most likely to develop resistance. They have direct life cycles and relatively short prepatent periods. Apart from selection through treatment for hookworm infection with IVM, accidental selection would also be imposed in areas where *Onchocerca* chemotherapy with IVM is being undertaken. Because the development characteristics of hookworms are similar to the trichostrongyloids of sheep an LDA may be useful in monitoring resistance. If resistance develops to a drug like IVM it could be serious, although several other compounds are available for treatment of humans.

Other Hosts

Resistance has also been reported in parasites of ostriches (Malan *et al.*, 1988) and antelope (Isaza, Courtney and Kollias, 1995).

C. elegans

A selection of genotypes and their corresponding phenotypes are listed in Table 22.3. Worms resistant to the BZ, imidazothiazole and AM classes have been characterised, although information on few of the 30 genes known to confer AM resistance have been published. Spontaneous mutations for resistance to OP compounds have also been reported.

Pharmacology of Anthelmintic Resistance

The characteristics of resistance to the different anthel-mintic classes have been studied in varying detail, mostly reflecting how long the resistance has existed. Pharmacological and genetic studies are almost exclus-ively restricted to the sheep trichostrongyloids and

C. elegans. The advance of knowledge in *C. elegans* rests on powerful genetic techniques. Progress in research to uncover the mechanisms of resistance in parasites has been slow. Part of the reason is that techniques needed for this research are difficult to develop and apply. For example, there are no methods for measuring drug efflux or influx for parasite cells. Before discussing each of the drug classes it is worth considering what mechanisms nematodes might use to develop resist-ance. Possible mechanisms have been categorised below in two ways, pharmacological/mechanistic and genetic.

Pharmacological Mechanisms

Reduced drug levels could occur through decreased uptake or increased removal (or sequestration) of anthelmintic. There are no known examples of reduced uptake in anthelmintic resistance, although reduced penetration can confer resistance in insects. Enhanced drug efflux through over-expression of worm P-glycoproteins has been hypothesised as a mechanism of AM-resistance in *H. contortus* (see below).

1. *Reduced concentration of active anthelmintic* could occur through decreased drug activation or in-creased drug metabolism. In general, any activa-tion of anthelmintics occurs in host tissues and parasite processes are not involved. Drug metabo-lism in parasites has only recently received atten-tion. Metabolites of TBZ and LEV, were not found after incubation of worms resistant to these compounds, respectively (Sangster and Prichard, 1986; Sangster, Riley and Colins, 1988). Low activities of cytochrome P450 monooxygenase and peroxygenase were found in *H. contortus* (Kotze, 1997, 1999), however, there is no evidence to link these pathways with resistance.
2. *A change in the target receptor.* This appears to be a major mechanism of BZ resistance in trichostrongy-loids where an isotype of tubulin with lower affinity for radioactive BZs predominates in resistant populations.
3. *Downstream modification* is a possible mechanism where receptor function is normal but physiological changes prevent transduction of the effect.

Genetic Mechanisms

Genetic mechanisms might affect a target receptor, a detoxifying enzyme, a protein responsible for post-translational modification or a protein with an action physiologically downstream of the target site. Some possible mechanisms are:

1. *Alteration in a gene.* For example, amino acid changes due to point mutations can alter the drug receptor affinity. Resistance could occur through allele selection in a population.
2. *Selective expression of a gene.* Several isotypes of many genes are present in the genomes of nematodes and preferential expression of one over another could occur.

3. *Increased expression of a receptor or enzyme.* This could occur through promoter changes or DNA amplification.
4. *Gene deletion.* This may cause loss of a target molecule from the organism.

Benzimidazole Resistance

Resistance to the BZs is very common in the sheep trichostrongyloids and the horse cyathostomes, but rarer amongst cattle parasites. It is the best understood resistance. The ovicidal nature of the drugs facilitated the development of simple *in vitro* assays and the abundance of the target protein, tubulin, assisted in the biochemical description of resistance.

Resistance *in vivo*

Resistance first appeared as failure of TBZ to control parasites in the field. During this time newer, more potent BZ drugs were released on the market. Some provided temporary control but, as levels of resistance increased, all of the compounds failed (see Le Jambre, Southcott and Dash, 1976). For several years oxibendazole was reported as effective against horse cyathostomes but now there is widespread resistance to this drug (Lyons *et al.*, 1996).

Controlled release capsules which deliver low doses of BZ drugs over an extended period, were marketed to control sheep trichostrongyloids. The drugs from these devices killed in-coming larvae of isolates which are resistant as adults. With continued use, a higher level of resistance has developed so that larvae of these isolates are now resistant. The recent release of the 5-day BZ treatment regime for horses to kill inhibited cyathostome larvae may also select for higher levels of resistance in these parasites.

Following treatment of infected sheep with TBZ, microtubules disappeared from gut cells of susceptible worms but not from cells of BZ-resistant worms (Sangster, Prichard and Lacey, 1985). This provides direct evidence of the mechanism of resistance to BZ drugs.

Resistance *in vitro*

Egg hatch assays (Le Jambre 1976) have been used to advance the study of BZ-resistant worms. LDA are being used to survey for resistance (Gill *et al.*, 1995). RF values of the order 10–70 are typical of resistant isolates.

Biochemistry of Resistance

Metabolism

For many years it was thought that BZs inhibited intermediary metabolism in worms. Although metabolic differences between resistant and susceptible isolates were described they were either due to the normal variation between isolates or were indirect, being a product of the interaction between BZs and tubulin (see Sangster, Prichard and Lacey, 1985).

Ligand binding

The receptor for the BZ drugs is the tubulin dimer which is comprised of α and β components which are relatively abundant in cells. Dimers polymerise into microtubules but when BZs bind to a specific high affinity-binding site on β-tubulin (Lacey, 1988) they depolymerise. Using a range of radiolabelled BZs binding experiments with worm homogenates revealed a reduction in the abundance of high affinity BZ binding sites in BZ-resistant worms compared with susceptible *T. colubriformis* (Sangster, Prichard and Lacey, 1985) and *H. contortus* (Lacey and Prichard, 1986; Lubega and Prichard, 1990). The magnitude of this reduction was dependent on the structure of the BZ tested (Lacey *et al.*, 1987), suggesting that alterations to the BZ structure could afford a drug with improved efficacy against BZ-resistant isolates. A comparative study with a number of BZ-resistant isolates showed that the reduction in high affinity [^{3}H]mebendazole binding was correlated with the known resistance status of the isolates (Lacey and Snowdon, 1988), and this became the basis for the first mechanism-based biochemical assay for the detection of anthelmintic resistance in parasitic nematodes. Binding for individual adult worms could be measured. For *T. colubriformis* RF values can be derived from comparison of binding properties. The RF (ratio of B_{max} values) for one resistant isolate was 11.2 (Russell and Lacey, 1991). BZ binding in *C. elegans* wild-type worms shows them to be relatively resistant to BZs compared with *T. colubriformis*.

Genetics

Mendelian genetics

There have been several attempts to describe the inheritance of BZ resistance. Studies of *H. contortus* generally point to polygenic inheritance which is recessive or partly dominant (Le Jambre, Royal and Martin, 1979; Herlich, Rew and Colglazier, 1981; Sangster, Redwin and Bjorn, 1998). For *T. colubriformis* resistance was incompletely recessive with a maternal effect (Martin, McKenzie and Stone, 1988).

Molecular genetics

With tubulin established as the drug target, molecular efforts were focused on β-tubulin genes of the parasite *H. contortus*. RFLP studies of DNA from several resistant and susceptible worm populations showed that the 5–10 bands hybridising to a β-tubulin isotype 1 probe present in susceptible worms were diminished to one band in resistant worms (Roos *et al.*, 1990). A similar study using an isotype 2 β-tubulin (Geary *et al.*, 1992) (clone b12–16) found that there were no differences in band number between resistant and susceptible isolates suggesting that resistance primarily concerned isotype 1.

On *in vitro* selection for resistance it was shown that BZ resistance can develop in a two step process. Alleles of isotype 1 were lost in the first step so that resistant worms contained a single allele (Kwa, Veenstra and

Roos, 1993). With further selection, isotype 2 alleles were lost (see Le Jambre, 1993b).

Upon sequencing the isotype 1 alleles from *H. contortus*, three amino acid differences were found to be related to resistance (Kwa, Veenstra and Roos, 1993). One transition, Phe to Tyr at position 200 (F200Y) (Kwa, Veenstra and Roos, 1994), was also present in the β-tubulin genes of BZ-resistant fungi and *C. elegans*. The importance of the mutation in isotype 1 was confirmed in transfection studies where the gene from BZ-resistant *H. contortus* (containing Y in the 200 position) conferred a resistant phenotype to susceptible *C. elegans*, but the susceptible *H. contortus* gene (F at 200) did not (Kwa *et al.*, 1995). Further experiments identified two sequential steps in the development of BZ-resistance and provided a PCR-based technique based on the F200Y transition capable of identifying resistant individuals. The same pattern of allele reduction has been seen in *T. colubriformis* (Grant and Mascord, 1996) and other isolates of *H. contortus* (Lubega *et al.*, 1994). The same transition has been identified in *T. circumcincta* (Elard, Comes and Humbert 1996) (see Table 22.2).

Initial identification of the F200Y transition was made in mutated fungi and *C. elegans* expressing BZ resistance and it was therefore referred to as a mutation. It is likely that the resistant isotype 1 allele was present at low levels in the susceptible worm populations prior to selection by treatment, indicating that resistance was probably due to selection of rare alleles from the population. The term mutation used in relation to parasites is probably incorrect, although a mutation may have given rise to the allelic difference prior to the introduction of BZ anthelmintics.

Mechanisms of BZ Resistance

Resistance appears to reside in expression of specific alleles of β-tubulin isotype 1 and loss of isotype 2. The presence of multiple isotypes in susceptible worm populations is probably reflected in the microheterogeneity of tubulin binding (Lacey, 1988) and in the isotype form variation observed on protein electrophoresis (Lubega and Prichard, 1991). The apparent polygenic nature of inheritance could also be explained by allele selection. Selection of alleles appears to be a universal mechanism of BZ resistance and the F200Y transition a common theme which has repeated itself at geographically distinct sites and in several species.

Resistance to controlled release capsules has developed and appears to be associated with selection at the isotype 2 locus but it is likely that other genes also modify BZ resistance too (Roos, Kwa and Grant, 1995).

Resistance in *C. elegans*

Resistant *C. elegans* show side resistance to a range of BZs (Enos and Coles, 1990) but not OBZ and parbendazole. Binding of radiolabelled BZ compounds is also reduced in resistant *C. elegans* compared with wild-type organisms (Russell and Lacey, 1991).

Woods *et al.* (1989) showed with protein electrophoresis that β-tubulin isotypes were lost in resistant worms. Twenty eight mutations all mapping to the *ben-1* locus, a β-tubulin gene, confer BZ-resistance in *C. elegans* (Driscoll *et al.*, 1989) (see Table 22.3). One of the mutations is at position 200, an F to Y transition.

Resistance to Imidazothiazoles and Tetrahydropyrimidines

Resistance to the imidazothiazoles/tetrahydropyrimidines has been restricted to the trichostrongyloid nematodes of sheep and pigs and there are some reports of PYR resistance in cyathostomes of horses. For sheep parasites, although LEV resistance is common in *T. colubriformis* and *T. circumcincta* it is rare in *H. contortus*. There are probably several reasons for this including the recessive nature of the inheritance of resistance, the high potency of LEV against *H. contortus*, and the pharmacokinetics of LEV in sheep.

LEV, the most widely used and studied drug in this class, will be considered the type compound. Most of the research into LEV resistance in parasites has been performed on sheep trichostrongyloids. In addition, there have been some comprehensive studies of LEV action and resistance in *C. elegans*. Resistance mutations in this worm fall into several categories. Phenotypically, apart from being LEV-resistant, many are uncoordinated or are 'twitchers'. Worms have a reduced contractile response to LEV and ACh and are deficient in ligand binding. Other mutations appear to disrupt muscle physiology downstream of the neuromuscular junction. Some ACh receptor subunits have been mapped, cloned, sequenced and expressed. However, mutation responsible for resistance has been described in only one gene.

Resistance *in vivo*

LEV, MOR and PYR act at the same site on the ACh receptor and share side resistance (Whitlock *et al.*, 1980; Green *et al.*, 1981). However, some interesting variations occur. There are two reports of isolates of *T. colubriformis* which have developed MOR resistance after field selection with LEV (McKenna, 1985; Waller *et al.*, 1986). On initial testing both isolates were susceptible to the recommended dose of LEV (no clinical resistance) but dose response data (Waller *et al.*, 1986) confirmed that worms were resistant to LEV. This appears to be a case where a more potent compound retained therapeutic activity even though side resistance was present.

High levels of resistance occur in *T. colubriformis*. The doses required to remove resistant worms exceed the toxic dose rate for LEV in sheep. An intriguing finding was that in one LEV-resistant isolate of *T. colubriformis* females were significantly more resistant than males (Dash, 1985) although such differences were not found in a resistant isolate of *H. contortus* (Sangster, Davis and Collins, 1991).

Model *in vivo* systems have been shown to be useful research tools. For both *T. colubriformis* in guinea pigs (Kelly *et al.*, 1981) and *H. contortus* in immunosuppressed jirds (Conder, Thompson and Johnson, 1991) resistant isolates of worms remain appropriately resistant to LEV.

In addition to the sheep parasites, PYR-resistant isolates of the pig parasites *Oesophagostomum* spp. have been recovered and characterised (Roepstorff, Bjorn and Nansen, 1987). PYR-resistant isolates were about 2 fold resistant (Bjorn *et al.*, 1989) and females were significantly more resistant than males.

LEV Resistance *in vitro*

Sangster, Riley & Collins (1988) used an egg hatch assay to investigate the cholinergic pharmacology of *T. colubriformis* and *H. contortus*. This study indicated that the cholinoreceptors displayed a pharmacology typical of other nematodes in being nicotinic and stereospecific and that LEV acted on cholinergic receptors which were altered in resistant isolates.

Larval development of several parasite species is also inhibited by LEV *in vitro* (Coles *et al.*, 1988; Lacey *et al.*, 1990). In these assays several LEV-resistant isolates contained a highly resistant sub-population. This may be an artefact of the assay because at high concentrations, such as those used in some assays, LEV displays agonist and antagonist effects on worm receptors (Robertson and Martin, 1993).

LEV also inhibited the migration of L3 of susceptible and resistant isolates of *T. colubriformis* and *H. contortus* but the potency of LEV is not altered in resistant isolates (see Detection and Diagnosis). This suggests that the receptors in these species are altered in resistant adults and L1 but are not expressed or are not essential for motility in L3. On the other hand, L3 of *T. circumcincta* do express resistance. Together these data suggest that mechanisms of anthelmintic resistance may differ between species and be developmentally expressed.

LEV Resistance and the Physiology of Parasitic Stages

In *H. contortus*, as in other nematodes, the coordinated contraction of longitudinal somatic muscle is responsible for motility. Addition of LEV (>4 μM) to adult worms in a bath causes the rapid onset of a sustained contraction (Sangster, Davis and Collins, 1991). Cholinergic compounds including MOR, PYR, DMPP, nicotine and succinylcholine cause contraction of drug susceptible worms but higher concentrations of each compound are required for the same response in resistant worms. For example, concentrations of LEV required to cause contractions were 4 and 13 fold higher for a moderately LEV-resistant and a highly resistant isolate, respectively. Concomitant resistance to eserine, an anti-acetylcholinesterase compound, suggested that these worms may also be resistant to ACh. This was tested directly in cannulated *H. contortus* (Sangster, Davis and Collins, 1991) where compounds have direct access to the neuromuscular receptors. Two LEV-resistant isolates were resistant to ACh (RF = 6). These data suggest that ACh and LEV act on a common receptor and that resistance to LEV involves an altered ACh receptor.

The patch-clamp technique has been applied to the study of LEV-resistant *Oesophagostomum* from pigs. Measurement of the biophysics of the LEV-activated cation channels from muscle has revealed a heterogeneous group of channel types. Whilst several receptor types were present in both types of worms (Martin *et al.*, 1997) other differences between resistant and susceptible worms exist:

1. Channels with lower conductance in drug-resistant worms have shorter T_{open}. (Martin *et al.*, 1998; Robertson, Bjorn and Martin, 1999).
2. The receptor type G35, was absent from resistant worms (Robertson, Bjorn and Martin, 1999).
3. There is evidence for an increased rate of desensitisation in receptors from resistant worms (Robertson, Bjorn and Martin, 1999).

It has been suggested that different combinations of receptor isotypes give rise to a variety of receptor types. A hypothesis for the development of resistance is that the proportion of isotypes alter so that those predominating in resistant worms confer a retention of function, but less current flows at the same concentration of LEV in resistant worms, meaning that contraction is less likely to occur.

Ligand Binding and LEV Resistance

Detailed binding experiments with *C. elegans* have shown that homogenates of LEV-resistant worms are deficient in binding [^{3}H]MAL compared with wild-type worms (Lewis *et al.*, 1987a; Lewis *et al.*, 1987b). A similar mechanism does not appear to occur in *H. contortus*. Binding to the receptor was specific, proportional to protein concentration and appropriately inhibited by cholinergic compounds (Sangster, Riley and Wiley, 1998). For *H. contortus* L3, two binding sites were consistently identified. The high affinity site with a K_d of 2.9×10^{-9} M and B_{max} 4×10^{-9} mol.mg protein^{-1}, was identical in susceptible and resistant isolates. On the other hand, the K_d at a second, lower affinity site of *H. contortus* ($K_d = 2$–10 μM) was higher in resistant worms. Results were consistent with there being three times as many unoccupied binding sites in resistant compared with susceptible worms. Similar findings were observed for homogenates of L1. The above data suggest that this resistant isolate of *H. contortus* could overcome the effects of LEV by having more receptors of the low affinity type available for normal transmitter function. Binding of [^{3}H]LEV was observed in another study on *H. contortus* (Moreno-Guzman *et al.*, 1998). In larvae of a highly resistant isolate, but not an isolate with intermediate resistance, reduced K_d and B_{max} were observed, but the difference was absent in adults. The K_d recorded in this study was in the μM range and may correspond to the low affinity receptor described above.

The relevance of the two binding sites found in *H. contortus* is unresolved (Sangster, Riley and Wiley, 1998). They may reflect morphologically distinct receptors such as those on nerve and muscle membranes or those with different biophysical properties

(Martin *et al.*, 1998). Another possibility is that the higher affinity receptor is desensitised and the low affinity binding sites (where the resistant and susceptible isolates differed in their binding properties) may represent the active conformation of the receptor. It also supports the notion of changes in desensitisation being involved in LEV-resistance in *Oesophagostomum*, mentioned above.

Genetics of LEV-Resistance

Mendelian genetics

Genetic crossing experiments have been performed on several LEV-resistant isolates of parasites. Martin and McKenzie (1990) described an inheritance consistent with a recessive sex-linked character due to a single gene or gene cluster for an isolate of *T. colubriformis*. There was also evidence for low level polygenic resistance on autosomes. Inheritance of resistance in a LEV-resistant *H. contortus* isolate was incompletely recessive, appeared to be due to several genes, and no evidence for sex-linkage was found (Sangster, Redwin and Bjorn, 1998).

Molecular genetics

ACh receptors are typically heteropentameric, transmembrane glycoproteins. Two members of the pentamer are α-subunits which each contain an ACh binding site. The remaining three non-α-subunits are heterogeneous and are referred to as structural subunits. Together with the α-subunits they help form the ion channel which, on binding agonists such as LEV or ACh, opens and allows passage of cations which leads to depolarisation of the cell. Some structural subunits may also contribute functionally to the binding site. Because there are differences in ligand binding in resistant worms and the binding site is predominantly on the α-subunit, it has been the target of cloning experiments. The ortholog of *C. elegans unc-38* (Fleming *et al.*, 1997) from *T. colubriformis*, *O. circumcincta* and *H. contortus* have been cloned (see Table 22.2). For all three species sequencing of the genes from resistant isolates revealed no differences that could give rise to resistance. Nor was there evidence for different allele patterns in resistant isolates. The study did reveal that *tar-1* the gene from *T. colubriformis* is likely to be located on the X chromosome (Wiley *et al.*, 1997).

Mechanisms of LEV Resistance

The weight of pharmacological evidence suggests that changes in the properties of the receptor are involved in resistance. The lack of sequence changes in the *unc38* orthologs of resistant worms raises other possibilities such as alterations in the structure of other α-subunits or of structural subunits. There is a precedent for mutation of a structural subunit in LEV-resistant *C. elegans* (Fleming *et al.*, 1997). Alternatively, selective expression of a subunit gene or allele which confers different channel properties may be involved in resistance. On the other hand modulation of the receptor may be responsible for the pharmaco-logical changes. Biochemical processes such as phosphorylation which can alter desensitisation rates of the receptor could be involved (Robertson, Bjorn and Martin, 1999).

LEV Resistance in *C. elegans*

Work on LEV-resistance in *C. elegans* predates that on parasites and many aspects of cholinergic pharmacology in this species are remarkably similar to its parasitic counterparts.

The cholinergic physiology of *C. elegans* was thoroughly investigated by Lewis *et al.* (1980a). The receptor is nicotinic in specificity and LEV acts on it directly. Interestingly, the head muscles of LEV-resistant mutants remain susceptible to LEV suggesting that their receptors have a different pharmacology. Severe loss of receptor function is not lethal but the receptor does not appear to be essential for normal behaviour and highly resistant mutants are uncoordinated in their movements. Evidence that LEV-resistant worms are also resistant to cholinergic compounds suggested that LEV receptors are a subset of the ACh receptor population. The level of resistance can be extreme, resistant isolates may survive concentrations of LEV 200,000 times that usually toxic to wild-type worms. Many resistant isolates are sensitive to hypo-osmotic shock, suggesting that this is a cholinergically controlled process.

Of 13 loci that can confer LEV resistance in *C. elegans*, 7 confer extreme resistance (Lewis *et al.*, 1980b) (see partial list in table 22.3). In excess of 250 alleles have been described and the huge majority are recessive. Some of these loci code for structural receptor genes and others may be involved in processing of the receptor subunits (Fleming *et al.*, 1997). Examples are 'twitchers' and gene products that exert their effects physiologically downstream of the LEV receptor (Lewis *et al.*, 1980a).

Detailed characterisation of the binding of [^{3}H]MAL has been performed (Lewis *et al.*, 1987a; Lewis *et al.*, 1987b). Binding is specific, saturable, displaced by cholinergic agents and most abundant in the L1 stage. In susceptible worms two binding sites are observed. The high affinity site has a K_d of 3nM and B_{max} of 2.8 fmol/mg. A second, non-saturable site has a K_d around 160 nM and represents a much smaller population of sites having a B_{max} of 0.05 fmol/mg. Resistance appears to be associated with the loss of high and low affinity sites. In three of the seven loci conferring high levels of resistance, binding is abolished and in the remaining four, binding is not activated by concentrations of the physiological inhibitor molecule mecamylamine which enhance binding in wild-type worms. Detergent-extracted membranes contain receptors in a different conformational state (Lewis and Berberich, 1992). Two sites each with increased affinity compared with the equivalent membrane-bound receptors were detected and binding was diminished in resistant worms.

Several of the genes which confer LEV resistance code for putative receptor subunits (see Table 22.3). The *unc-38* gene codes an α-subunit and *unc-29* and *lev-1* code structural subunits (Fleming *et al.*, 1993; Fleming *et al.*, 1997). In comparing sequences of the *lev-1* gene from wild-type worms with two *lev-1* resistant mutants two different mutations have been described. One results in an E to K transition and the other an L insertion in the second transmembrane region which is thought to line the ion-channel. It is conceivable that alterations in the charge of channel-lining amino acids could alter channel properties.

The study of genetic mutants of *C. elegans* has been extremely useful in describing the pharmacology and molecular biology of the LEV receptor in these worms. At present it is not known how useful this information will be for understanding resistance in parasitic nematodes. Although the pharmacological and molecular similarities between *C. elegans* and trichostrongyloid nematodes of sheep are striking, extrapolation of these similarities to resistance mechanisms is not so easy. For practical reasons highly resistant isolates of *C. elegans* were chosen for examination, yet equivalent mutations may be lethal in parasites not relevant to field resistance. For example, the uncoordinated phenotype would not be viable in natural infections and observation of hundreds of LEV-resistant *H. contortus* have failed to reveal any abnormalities in muscle contraction (unpublished observations). In the *C. elegans* genome there are about 80 neurotransmitter-gated ion channel genes and more than 40 of these code for receptor ion channels with homology to ACh receptor subunits (Bargmann, 1998). It is a large pool in which to fish for resistance homologues in parasites.

Avermectin/Milbemycin Resistance

The natural avermectins (AVMs) are a mixture of related fermentation products. In addition to having a lactone ring structure each is characterised by having one or two glycone groups. The class includes ivermectin (IVM), abamectin and doramectin. The related milbemycins share a common carbon backbone with AVM but lack the glycone groups. The available drugs include milbemycin oxime and moxidectin (MOX): these contain additional side chain variations. It is likely that milbemycins and AVM act on helminths in the same way although they do differ in potency (Shoop *et al.*, 1993) and spectrum (Sangster, 1995). Because of this similarity of action and the fact that most of the experimental work has been done with IVM or AVM these compounds will form the basis of this discussion.

Although initial reports suggested that IVM-resistant isolates were susceptible to treatment with MOX (Craig and Miller, 1990; Kieran, 1994), further experimentation and dose titration of both drugs against resistant and susceptible *H. contortus* clearly demonstrated side-resistance (Pomroy and Whelan, 1993; Shoop *et al.*, 1993; Conder, Thompson and Johnson, 1993; Leathwick, 1995; Le Jambre *et al.*, 1995). The data also support the hypothesis that these drugs have a common mode of action and mechanism of resistance. Unfortunately, resistance to AMs has occurred sooner than expected, is now established in many countries and is set to escalate world wide.

AM Resistance *in vivo*

Even for individual AMs, efficacy against different stages and species of parasites differs. For this reason they are marketed at recommended dose rates (usually 0.2 mg/kg,) well above those required to kill many stages of their target species, including the trichostrongyloid nematodes. For example, IVM is 95% effective against *H. contortus* at only 0.02 mg/kg (Egerton *et al.*, 1988). In one isolate of *H. contortus* resistant to IVM, males were more susceptible than females where the larger body size and potential for sequestering a lipophilic compound like IVM were thought to contribute to the poorer activity against female worms (Le Jambre *et al.*, 1995). When the parasites *Teladorsagia* spp. and *T. colubriformis* were subjected to dose titration assays *in vivo* RFs of 6 to 30 were found for the AM anthelmintics (Shoop *et al.*, 1993). *Cooperia oncophora* in cattle have the ability to survive a recommended dose of IVM that would be effective against susceptible parasites (Vermunt, West and Pomroy, 1995).

AM Resistance *in vitro*

Several *in vitro* assays for IVM-resistance in trichostrongyloids have been developed and have provided valuable information. AMs are thought to act on two distinct sites in worms. For *H. contortus* the less sensitive IVM site is on somatic muscle and the more sensitive site is on pharyngeal muscle. Both sites become resistant with the development of anthelmintic resistance.

Effects of IVM on somatic muscle of susceptible and resistant isolates have been assessed by measuring the effects of several analogues of IVM on the light-induced motility of L3 of four susceptible and six IVM-resistant isolates of *H. contortus* (Gill *et al.*, 1991). IVM had an EC_{50} of > 300 nM against a susceptible isolate and was several fold higher for resistant isolates. Of the analogues tested, AVM B_2 provided the largest RF for IVM-resistance (RF = 2.5–20, range for 6 isolates). L1 incubated with > 30 nM IVM assumed angular postures and twitched (Gill *et al.*, 1995). Inhibition of motility in adult *H. contortus* has also been observed at $> 10^{-8}$M IVM and paralysis was restricted to the mid body region of worms (Geary *et al.*, 1993). *In vitro* motility in an IVM-resistant isolate was less sensitive to IVM than a susceptible isolate. Martin and Pennington (1989) demonstrated that IVM (1–100 pM) opened Cl^- channels associated with somatic muscle in *Ascaris suum*.

Inhibition of larval development occurred at < 1 nM IVM. The authors (Gill *et al.*, 1995) attributed the effect to inhibition of feeding because similar concentrations inhibited feeding activity in adult *T. colubriformis* (Bottjer and Bone, 1985) and *H. contortus* (Geary *et al.*, 1993) kept *in vitro*. Compared with susceptible larvae, EC_{50}s of all of the IVM analogues tested were higher in resistant worms (Gill *et al.*, 1995). Of these analogues AVM B_2 was again the most sensitive probe for resistance (RF = 4.9–17). These data for *H. contortus*

L1 were supported in experiments using adult worms of two different IVM-resistant isolates. Concentrations of IVM 177-fold higher (Sangster, 1996) were required to inhibit feeding to the same extent in resistant compared with susceptible worms. Feeding experiments with L1 provided estimates of RF of 4.5–9 fold (Kotze, 1998).

Effects of Avermectin/Milbemycins at Two Sites

The data suggest that IVM acts on both somatic and pharyngeal muscle and that resistance of a similar magnitude occurs at both sites. The facts that L3 are paralysed and this is a non-feeding stage of the life cycle, suggest that these putative sites of action are independent. The rank order of potency of analogues is similar, but the somatic muscle site is much less sensitive. It is likely that the somatic muscle site is important in *H. contortus in vivo*. Expulsion of *H. contortus* occurs rapidly 8 hours after treatment with IVM (V. Bhardwaj, PhD Thesis, Univ. of Sydney), so drug concentrations present at 4–6 h after treatment can discriminate between resistant and susceptible worms. Following a recommended dose in sheep IVM reached blood concentrations of 20nM (Marriner, McKinnon and Bogan, 1987) and, due to its lipophilicity, is likely to reach even higher concentrations in worm tissues. These concentrations would be high enough to affect both motility and pharyngeal paralysis in susceptible worms. On the other hand they would not be likely to affect motility in resistant worms suggesting that somatic muscle is the critical site of resistance.

In vitro studies have also revealed that several phenotypes exist amongst resistance isolates. Gill and Lacey (1998) have characterised isolates using the LDA and L3 motility assay as well as the *in vitro* response to paraherquamide which is reduced in some IVM-resistant isolates of *H. contortus* (Le Jambre *et al.*, 1995). Selection history appears to alter the phenotype. Field isolates are most resistant and form a distinct group. Isolates recovered from goats form a second group and an isolate from laboratory selection comprises a third group.

Binding of IVM

Binding characteristics of IVM to homogenates of L3 from IVM-resistant and susceptible isolates of *H. contortus* have been reported. Both affinity ($K_d = 0.13$ nM) and receptor density ($B_{max} = 0.4$ pmol/mg) were similar between isolates leading to the conclusion that changes in binding at this high affinity site did not account for IVM-resistance in *H. contortus* (Rohrer *et al.*, 1994). In a second study of susceptible worms only, two binding sites were described (Gill and Lacey, 1998). The high affinity site ($K_d = 0.11$ nM) is similar to the one described in the previous report. The second site has a K_d of 8.7 nM. There is some logic for the presence of two sites. The receptor is probably a ligand-gated ion channel (Cully *et al.*, 1994) and these often exist in two states of sensitisation. On the other hand, the pharyngeal and somatic muscle receptors may differ in their binding

properties in parallel with their sensitivity to IVM. More detailed investigation of the second binding site may shed some light on resistance mechanisms.

Genetics of IVM Resistance

Mendelian genetics

Because IVM-resistance has arisen relatively recently, few genetic studies have been performed. Resistance in one isolate of IVM-resistant *H. contortus* appears to be inherited on a single gene (or closely linked cluster) which is fully dominant (Le Jambre, 1993a; Le Jambre *et al.*, 1995).

Molecular genetics

IVM is thought to act on a family of glutamate-activated Cl^- channels which have been cloned from *C. elegans* (see below). The recent cloning of 2 genes which appear to be receptor homologues from *H. contortus* (Delany, Laughton and Wolstenholme, 1998; Blackhall *et al.*, 1998b) will expedite studies on the molecular genetics of IVM resistance in this species. A β-subunit gene product has been localised by immunocytochemistry to the body nerve commissures between the nerve ring and vulva but is absent from the pharynx (Delany, Laughton and Wolstenholme, 1998). It is interesting that this broadly corresponds with the region of these worms paralysed by IVM *in vitro* (Geary *et al.*, 1993). An α-subunit gene has been used to probe for allele variability in *H. contortus* populations, two resistant and one susceptible to IVM. One allele is commonly associated with resistant worms while another with susceptible worms. Such differences are absent when a β-subunit is used as a probe (Blackhall *et al.*, 1998b). There are likely to be many glutamate receptor subunit genes in worms and their role in drug action and resistance are unknown.

P-glycoprotein efflux pumps

P-glycoproteins (Pgp) are conserved transmembrane proteins which have a role in transporting compounds, including drugs, across membranes. They were initially recognized in multidrug-resistant cancer cells where Pgp were over-expressed (Juliano and Ling, 1976). The expression of additional Pgp molecules elevates rates of drug efflux enough to lower intracellular drug concentrations and render cells resistant to a range of anticancer agents. Substrates for this transport are structurally and functionally diverse but chemical structures of several anthelmintic compounds including the BZ and AM classes and CLS designate them as potential substrates for Pgp transport in worms. IVM is a substrate for mammalian Pgp (Schinkel *et al.*, 1994; Pouliot *et al.*, 1997). It is difficult to test directly if Pgps actively remove IVM and its analogues from target sites in worms (Sangster, 1994) but genetic tests may shed light on the role of Pgp in AM resistance. There are four or more different Pgp genes in *H. contortus* (Sangster, 1994) and fragments of several genes have been used to generate Southern blots with worm DNA. The patterns fell into three categories. For several genes no polymorphisms were found between susceptible and resistant worms (Kwa et al., 1998; Sangster et al., 1999). With another, gene polymorphisms were either present (Xu et al., 1998) or absent (Sangster, et al., 1999) in different resistant isolates. A further gene (#28, Sangster et al., 1999) revealed polymorphisms between susceptible worms and two isolates of IVM-resistant worms. Until more work is done it is difficult to say if polymorphisms reflect genetic variation between worm isolates (Blackhall et al., 1998a) or relate to real differences in Pgp isotypes.

Mechanisms of IVM Resistance

There are too few data available to determine the mechanism of resistance to IVM in parasitic nematodes. Mechanisms in several insect pests have been reported (Clark et al., 1994) but the fact that more than one mechanism was described for each pest species suggests that many were induced by laboratory selection. Neither the studies on insects nor those on C. elegans are sufficiently advanced to guide us.

For H. contortus an explanation for resistance should account for how a single gene can confer similar resistances at two sites. One explanation is that genes coding the pharyngeal and somatic receptors are present in the cluster on the same chromosome. A second is that a transspliced gene (Dent, Davis and Avery, 1997) codes two tissue-specific receptor subunits. In this case a change in the common region could confer resistance through both gene products. A third explanation is that another gene modifies the response to drug. For example, drug transport or efflux via an over-expressed Pgp could account for IVM resistance by lowering drug concentrations in several worm tissues.

Studies of allelic variation which have been observed in resistant and susceptible worms (Blackhall et al., 1998a; Blackhall et al., 1998b) appear to reveal little about resistance. It is likely that one mechanism of resistance predominates in parasite populations, so polymorphisms in alleles of several genes are unlikely to be all implicated in mechanisms of resistance. On the other hand, what these studies reveal about variation is interesting. For GluCl−, for example, at least 10 alleles are present suggesting that molecular mechanisms of resistance will be difficult to elucidate.

IVM-Resistance in C. elegans

Although the pharmacology of IVM has been investigated in C. elegans, (Novak and Vanek, 1992) reports on the pharmacology of AM-resistant worms or the more than 20 genes which confer low level IVM-resistance are scant (see Blaxter and Bird, 1997). The large majority of the mutations are recessive.

IVM binds to membrane preparations from C. elegans with high affinity (K_d of 0.26 nM) at a single site (Schaeffer and Haines, 1989), presumably binding to GluCl− receptors. Several GluCl− receptor genes have been described. The α forms are the putative sites of action of IVM in this nematode (Cully et al., 1994; Dent, Davis and Avery, 1997) and the β forms a component of the receptor ion channel (Cully et al., 1994; Laughton, Lunt and Wolstenholme, 1997). At least two of the subunits (one α and one β) are expressed in the paired pharyngeal muscle cells, pm4 (Dent, Davis and Avery, 1997; Laughton, Lunt and Wolstenholme, 1997) which supports the hypothesis that IVM acts on the pharynx. In addition to the pharynx, the α form, avr-15, is also expressed in pm5 muscle cells as well as sites remote from the pharynx, such as somatic motor neurons. The avr-15 gene confers IVM sensitivity on C. elegans and expression of avr-15 in mutants (resistant to IVM) deficient in this gene restores susceptibility to IVM. In addition to avr-15, several other genes, unc-1, 7, 9 and avr-14 and 20, are associated with IVM resistance (see Table 22.3). In general, simultaneous mutations in two genes are required to confer resistance in C. elegans (Dent, Davis and Avery, 1997).

Several of the IVM-resistant C. elegans mutants have defects in dye filling of amphid neurons and may have deficiencies in avoiding adverse osmotic conditions. One gene avr-1 (Table 22.3) is a dynein which is probably involved in neuronal transport (Grant cited by Blaxter and Bird, 1997). There is no equivalent dye-filling defect in parasitic nematodes. Dye filling may reflect a defect in absorptive or transport functions related to membranes.

Several Pgp genes have been cloned from C. elegans. Two that have been studied in detail do not confer IVM-resistance when over-expressed in worms (Broeks et al., 1995). None of the genes known to confer IVM resistance in C. elegans are Pgp genes.

Resistance to Closantel

Resistance to closantel (CLS) and the related compound rafoxanide is restricted to H. contortus (and the trematode Fasciola hepatica). The anthelmintic has a 28 day protection period against infection with larvae due to its persistence in plasma of the host and resistance first appears as a reduction in the protection period afforded by the drug. This has made it difficult to diagnose resistance using in vivo assays because the recommended dose rate (7.5 mg/kg) remains effective in killing adult worms except for highly resistant isolates. In order to circumvent this problem lower dose rates which are still effective against susceptible worms have been used. In South Africa 5.0 mg/kg was 41.3 and 80.8% effective in reducing worm burdens for two isolates (van Wyk and Malan, 1988). In Australia 2.5 mg/kg reduced worm counts by 57.6 and 70.4% in two isolates (Rolfe et al., 1990).

In the L4 feeding assay resistant isolates have RFs of 2–10. Side resistance occurs with structurally related (Rothwell and Sangster, 1993) and unrelated compounds (Conder et al., 1992).

The fact that resistance could be detected in the feeding assay with parasitic stages but not larval development of free-living stages suggests that the mechanism of resistance is either restricted to parasitic stages or requires the worm's gut either as a transporter or site of action. Interestingly, C. elegans is not affected by CLS when the drug is added to liquid culture. It would be surprising if the site of action was absent from *C. elegans* which suggests that ingestion of the drug has a bearing on its activity. CLS disrupts mitochondria in the gut and other tissues of treated worms but resistant worms, similarly treated, were morphologically normal (Rothwell and Sangster, 1996).

The mechanism of resistance to CLS has been studied with two approaches. Accumulation of CLS was measured following treatment of infected sheep with radioactive CLS. Susceptible worms accumulated 4 times more isotope than resistant worms (Rothwell and Sangster, 1997). It is unknown if this reflects differences in uptake or in drug removal. Secondly, a molecular investigation of P-glycoprotein genes in *H. contortus* has been performed by comparing Southern blots of DNA from a CLS-resistant isolate with those from susceptible worms (Sangster, *et al.*, 1999). Hybridisation patterns of worm DNA with several probes from different P-glycoprotein molecules cloned from *H. contortus* were not different between the isolates. These data suggest that a Pgp-mediated process of drug efflux are unlikely to be involved with CLS resistance, but the tests were not exhaustive.

Resistance to Organophosphate Compounds

Despite its limited use in the field, resistance to the organophosphate NAP has been reported in isolates of *H. contortus*. One isolate is coincidentally resistant to LEV and BZs (Green *et al.*, 1981). Although the evidence is indirect, it is likely that use of LEV has selected for both LEV and NAP resistance. Organophosphates typically inhibit cholinesterase activity allowing ACh to accumulate at neuromuscular junctions and cause contraction and paralysis. Contraction studies revealed that worms of the Lawes isolate (Sangster, Davis and Collins, 1991) and LEV-resistant *C. elegans* (Lewis *et al.*, 1980a) were resistant to eserine, an anticholinesterase compound, as well as to LEV and ACh. In this case it is likely that a change in the ACh receptor which occurs with LEV resistance confers resistance to organophosphate compounds that acts by causing ACh accumulation.

On initial recovery of the Lawes isolate of *H. contortus* from the field, NAP (7.5 mg/kg) reduced worms counts by 74.2 to 96.8% in three experiments (Green *et al.*, 1981). After four generations of selection with a BZ and LEV the same treatment was only 22.3% effective, suggesting that LEV selection had also enhanced resistance to NAP. On the other hand there appears not to be a close correlation between the efficacy of LEV and NAP on 13 sheep farms (Cooper, *et al.*, 1996).

Organophosphate resistance has been studied in *C. elegans*. These studies provide an example of the multitude of genes which can confer resistance. In this case genes involved in ACh synthesis, transport, regulation of vesicle traffic, GTP-binding, expression of synthetic and transport genes as well as ACh receptor genes have been implicated. In addition, worms resistant to LEV are resistant to the OP eserine (Lewis *et al* 1980a) probably as a result of alterations in ACh receptor pharmacology.

Conclusions

Although we have come a long way in the study of resistance, we have not reached the stage where we can ameliorate the problem. It seems likely that parasite control will rely on anthelmintic treatment for the foreseeable future and so, selection for resistance will continue. Once resistance gene frequencies reach a few percent in a population of parasites it is a matter of time before clinical resistance develops to that drug class. Therefore strategies to reduce selection pressure and preserve the effectiveness of anthelmintics are essential. Because approaches to managing resistance has been dealt with elsewhere (Herd, 1993), only a brief overview will be provided here.

In areas where a parasite species has not developed resistance there is a chance of preventing it. The important host/parasite systems for which efficacy could be preserved are cyathostome/horse, *Ostertagia*/cattle and hookworm/human systems. Resistance to the AM class, should it develop, would have serious impacts in these cases. An approach to preventing resistance would include the following steps: develop parasite control measures which minimise drug use (e.g. limit contamination of the environment especially with eggs from worms surviving treatment), use available drugs in combinations, develop sensitive resistance detection tests to monitor resistance. Should resistance develop a switch to an alternative drug class should occur but how a decision to switch drugs might be made and advertised is one unresolved issue. The utility of low activity drugs or mosaic treatments will depend on the enterprise and the severity of the parasite on health or production. It is difficult to advise the preferential use of short acting or persistent drugs as their influence on the development of resistance is still not resolved.

Priorities

The role of mathematical models is crucial to anticipating the future of resistance. Development of such models requires detailed biological information but because these data are not available for many species, enhancement of the trichostrongyloid/sheep models are a useful way to reveal general principles. For example, further information on drug efficacy against different genotypes (heterozygotes, for example) and different life cycle stages in the host will help to refine the current models. Genetic studies are crucial too. Not only for input into models but also as a source of tools for the detection of resistance. Although *C. elegans* will provide a molecular treasure trove, the technical problems faced by nematologists will make progress in genetics difficult.

There has been much discussion on alternative control strategies for parasites. Novel chemicals would be very useful. The same can be said for antiparasite vaccines, but the technical difficulty of eliciting a protective response equivalent to that of an effective anthelmintic has eluded researchers. Nematophagous fungi also appear to offer some prospects for control. Nutritional and grazing management are also important adjuncts to control. However, the approach that seems most likely to achieve parasite control is a combination of several methods from the list above delivered in an integrated fashion.

Acknowledgements

Our thanks to Elizabeth Barnes and Andrew Kotze for critically reading the manuscript and Sharron Bannan for preparing the references.

References

Bargmann, C.I. (1998) Neurobiology of the *Caenorhabditis elegans* genome. *Science*, **282**, 2028–2033.

Barnes, E.H. and Dobson, R.J. (1990) Population dynamics of *Trichostrongylus colubriformis* in sheep: computer model to simulate grazing systems and the evolution of anthelmintic resistance. *International Journal for Parasitology*, **20**, 823–831.

Barnes, E.H., Dobson, R.J. and Barger, I.A. (1995) Worm control and anthelmintic resistance: adventures with a model. *Parasitology Today*, **11**, 56–63.

Barton, N.J. (1983) Development of anthelmintic resistance in nematodes from sheep in Australia subjected to different treatment frequencies. *International Journal for Parasitology*, **13**, 125–132.

Beech, R.N., Prichard, R.K. and Scott, M.E. (1994) Genetic variability of the beta-tubulin genes in benzimidazole-susceptible and -resistant strains of *Haemonchus contortus*. *Genetics*, **138**, 103–110.

Besier, B., Lyon, J. and McQuade, N. (1996) Drench resistance — a large economic cost. *Western Australian Journal of Agriculture*, **37**, 60–63.

Bjorn, H., Roepstorff, A., Nansen, P. and Waller, P.J. (1989) A dose-response investigation on the level of resistance to pyrantel citrate in nodular worms of pigs. *Veterinary Parasitology*, **31**, 259–267.

Blackhall, W.J., Liu, H.Y., Xu, M., Prichard, R.K. and Beech, R.N. (1998a) Selection at a P-glycoprotein gene in ivermectin- and moxidectin-selected strains of *Haemonchus contortus*. *Molecular and Biochemical Parasitology*, **95**, 193–201.

Blackhall, W.J., Pouliot, J.-F., Prichard, R.K. and Beech, R.N. (1998b) *Haemonchus contortus*: selection at a glutamate-gated chloride channel gene in ivermectin- and moxidectin-selected strains. *Experimental Parasitology*, **90**, 42–48.

Blaxter, M. (1998) *Caenorhabditis elegans* is a nematode. *Science*, **282**, 2041–2046.

Blaxter, M. and Bird, D. (1997) Parasitic nematodes. In *C. elegans II*, edited by D.L. Riddle, T. Blumenthal, B.J. Meyer and J.R. Priess, pp. 851–878. New York: Cold Spring Harbor Laboratory Press.

Blouin, M.S., Dame, J.B., Tarrant, C.A. and Courtney, C.H. (1992) Unusual population genetics of a parasitic nematode: mtDNA variation within and among populations. *Evolution*, **46**, 470–476.

Boray, J.C., Martin, P.J. and Roush, R.T. (1990) *Resistance of parasites to antiparasitic drugs*. New Jersey: MSD Agvet.

Borgsteede, F.H.M., Dvojnos, G.M. and Kharchenko, V.A. (1997) Benzimidazole resistance in cyathostomes in horses in the Ukraine. *Veterinary Parasitology*, **68**, 113–117.

Bottjer, K.P. and Bone, L.W. (1985) *Trichostrongylus colubriformis*: effect of anthelmintics on ingestion and oviposition. *International Journal for Parasitology*, **15**, 501–503.

Broeks, A., Janssen, H.W.R.M., Calafat, J. and Plasterk, R.H.A. (1995) A P-glycoprotein protects *Caenorhabditis elegans* against natural toxins. *The EMBO Journal*, **14**, 1858–1866.

Chapman, M.R., French, D.D., Monahan, C.M. and Klei, T.R. (1996) Identification and characterization of a pyrantel pamoate resistant cyathostome population. *Veterinary Parasitology*, **66**, 205–212.

Clark, J.M., Scott, J.G., Campos, F. and Bloomquist, J.R. (1994) Resistance to avermectins: extent, mechanisms, and management implications. *Annual Review of Entomology*, **40**, 1–30.

Coles, G.C., Bauer, C., Borgsteede, F.H.M., Geerts, S., Klei, T.R., Taylor, M.A. and Waller, P.J. (1992) World Association for the Advancement of Veterinary Parasitology (W.A.A.V.P.): methods for the detection of anthelmintic resistance in nematodes of veterinary importance. *Veterinary Parasitology*, **44**, 35–44.

Coles, G.C., Folz, S.D. and Tritschler, J.P. (1989) Motility response of levamisole benzimidazole resistant *Haemonchus contortus* larvae. *Veterinary Parasitology*, **31**, 253–257.

Coles, G.C., Simpkin, K.G. and Briscoe, M.G. (1980) Routine cryopreservation of ruminant nematode larvae. *Research in Veterinary Science*, **28**, 391–392.

Coles, G.C., Stafford, K.A. and MacKay, P.H.S. (1998) Ivermectin-resistant *Cooperia* species from calves on a farm in Somerset. *The Veterinary Record*, **142**, 255–256.

Coles, G.C., Tritschler II, J.P., Giordano, D.J. and Schmidt, A.L. (1988) Larval development test for detection of anthelmintic resistant nematodes. *Research in Veterinary Science*, **45**, 50–53.

Conder, G.A. and Campbell, W.C. (1995) Chemotherapy of nematode infections of veterinary importance, with special reference to drug resistance. *Advances in Parasitology*, **35**, 1–84.

Conder, G.A., Johnson, S.S., Guimond, P.M., Geary, T.G., Lee, B.L., Winterrowd, C.A., Lee, B.H. and DiRoma, P.J. (1991) Utility of a *Haemonchus contortus*/jird (*Meriones unguiculatus*) model for studying resistance to levamisole. *Journal of Parasitology*, **77**, 83–86.

Conder, G.A., Thompson, D.P. and Johnson, S.S. (1993) Demonstration of co-resistance of *Haemonchus contortus* to ivermectin and moxidectin. *The Veterinary Record*, **132**, 651–652.

Conder, G.A., Zielinski, R.J., Johnson, S.S., Kuo, M.-S.T., Cox, D.L., Marshall, V.P., Haber, C.L., DiRoma, P.J., Nelson, S.J., Conklin, R.D., Lee, B.L., Geary, T.G., Rothwell, J.T. and Sangster, N.C. (1992) Anthelmintic activity of dioxapyrrolomycin. *The Journal of Antibiotics*, **45**, 977–983.

Cooper, N.A., Rolfe, P.F., Searson, J.E. and Dawson, K.L. (1996) Naphthalophos combinations with benzimidazole or levamisole as effective anthelmintics for sheep. *Australian Veterinary Journal*, **74**, 221–224.

Craig, T.M. and Miller, D.K. (1990) Resistance by *Haemonchus contortus* to ivermectin in Angora goats. *The Veterinary Record*, **126**, 560.

Craven, J., Bjorn, H., Henriksen, S.A., Nansen, P., Larsen, M. and Lendel, S. (1998) Survey of anthelmintic resistance on Danish horse farms, using 5 different methods of calculating faecal egg count reduction. *Equine Veterinary Journal*, **30**, 289–293.

Cully, D.F., Vassilatis, D.K., Liu, K.K., Paress, P.S., van der Ploeg, L.H.T., Schaeffer, J.M. and Arena, J.P. (1994) Cloning of an avermectin-sensitive glutamate-gated chloride channel from *Caenorhabditis elegans*. *Nature*, **371**, 707–710.

Dash, K.M. (1985) Differential efficacy of levamisole and oxfendazole against resistant male and female *Trichostrongylus colubriformis*. *The Veterinary Record*, **117**, 502–503.

Dash, K.M., Hall, E. and Barger, I.A. (1988) The role of arithmetic and geometric mean worm egg counts in faecal egg count reduction tests and in monitoring strategic drenching programs in sheep. *Australian Veterinary Journal*, **65**, 66–68.

Delany, N.S., Laughton, D.L. and Wolstenholme, A.J. (1998) Cloning and localisation of an avermectin receptor-related subunit from *Haemonchus contortus*. *Molecular and Biochemical Parasitology*, **97**, 177–187.

Dent, J.A., Davis, M.W. and Avery, L. (1997) *avr-15* encodes a chloride channel subunit that mediates inhibitory glutamatergic neurotransmission and ivermectin sensitivity in *Caenorhabditis elegans*. *The EMBO Journal*, **16**, 5867–5879.

Dobson, R.J., Donald, A.D., Waller, P.J. and Snowdon, K.L. (1986) An egg-hatch assay for resistance to levamisole in trichostrongyloid nematode parasites. *Veterinary Parasitology*, **19**, 77–84.

Dobson, R.J., Le Jambre, L. and Gill, J.H. (1996) Management of anthelmintic resistance: inheritance of resistance and selection with persistent drugs. *International Journal for Parasitology*, **26**, 993–1000.

Douch, P.G. and Morum, P.E. (1994) The effects of anthelmintics on ovine larval nematode parasite migration *in vitro*. *International Journal for Parasitology*, **24**, 321–326.

Driscoll, M., Dean, E., Reilly, E., Bergholz, E. and Chalfie, M. (1989) Genetic and molecular analysis of a *Caenorhabditis elegans* beta-tubulin that conveys benzimidazole sensitivity. *The Journal of Cell Biology*, **109**, 2993–3003.

Drudge, J.H. and Elam, G. (1961) Preliminary observations on the resistance of horse strongyles to phenothiazine. *Journal of Parasitology*, **47**, 38–39.

Drudge, J.H., Lyons, E.T., Tolliver, S.C., Lowry, S.R. and Fallon, E.H. (1988) Piperazine resistance in population-B equine strongyles: a study of selection in thoroughbreds in Kentucky from 1966 through 1983. *American Journal of Veterinary Research*, **49**, 7986–994.

Echevarria, F., Borba, M.F.S., Pinheiro, A.C., Waller, P.J. and Hansen, J.W. (1996) The prevalence of anthelmintic resistance in nematode parasites of sheep in Southern Latin America: Brazil. *Veterinary Parasitology*, **62**, 199–206.

Eddi, C., Caracostantogolo, J., Pena, M., Schapiro, J., Marangunich, L., Waller, P.J. and Hansen, J.W. (1996) The prevalence of anthelmintic resistance in nematode parasites of sheep in Southern Latin America: Argentina. *Veterinary Parasitology*, **62**, 189–197.

Egerton, J.R., Suhayda, D. and Eary, C.H. (1988) Laboratory selection of *Haemonchus contortus* for resistance to ivermectin. *Journal of Parasitology*, **74**, 614–617.

Elard, L., Cabaret, J. and Humbert, J.F. (1999) PCR diagnosis of benzimidazole-susceptibility or resistance in natural populations of the small ruminant parasite, *Teladorsagia circumcinta*. *Veterinary Parasitology*, **80**, 231–237.

Elard, L., Comes, A.M. and Humbert, J.F. (1996) Sequences of β-tubulin cDNA from benzimidazole-susceptible and -resistant strains of *Teladorsagia circumcincta*, a nematode parasite of small ruminants. *Molecular and Biochemical Parasitology*, **79**, 249–253.

Enos, A. and Coles, G.C. (1990) Effect of benzimidazole drugs on tubulin in benzimidazole resistant and susceptible strains of *Caenorhabditis elegans*. *International Journal for Parasitology*, **20**, 161–167.

Finney, D.J. (1971) *Probit Analysis*, 3rd edn. Cambridge: Cambridge University Press.

Fleming, J.T., Squire, M.D., Barnes, T.M., Tornoe, C., Matsuda, K., Ahnn, J., Fire, A., Sulston, J.E., Barnard, E.A., Sattelle, D.B. and Lewis, J.A. (1997) *Caenorhabditis elegans* levamisole resistance genes *lev-1, unc-29*, and *unc-38* encode functional nicotinic acetylcholine receptor subunits. *The Journal of Neuroscience*, **17**, 5843–5857.

Fleming, J.T., Tornoe, C., Riina, H.A., Coadwell, J., Lewis, J.A. and Sattelle, D.B. (1993) Acetylcholine receptor molecules of the nematode *Caenorhabditis elegans*. *Comparative Molecular Neurobiology*, **63**, 65–80.

Folz, S.D., Pax, R.A., Thomas, E.M., Bennett, J.L., Lee, B.L. and Conder, G.A. (1987) Development and validation of an *in vitro Trichostrongylus colubriformis* motility assay. *International Journal for Parasitology*, **17**, 1441–1444.

Geary, T.G., Nulf, S.C., Favreau, M.A., Tang, L., Prichard, R.K., Hatzenbuhler, N.T., Shea, M.H., Alexander, S.J. and Klein, R.D. (1992) Three beta-tubulin cDNAs from the parasitic nematode *Haemonchus contortus*. *Molecular and Biochemical Parasitology*, **50**, 295–306.

Geary, T.G., Sims, S.M., Thomas, E.M., Vanover, L., Davis, J.P., Winterrowd, C.A., Klein, R.D., Ho, N.F.H. and Thompson, D.P. (1993) *Haemonchus contortus*: ivermectin-induced paralysis of the pharynx. *Experimental Parasitology*, **77**, 88–96.

Geerts, S., Coles, G.C. and Gryseels, B. (1997) Anthelmintic resistance in human helminths: learning from the problems with worm control in livestock. *Parasitology Today*, **13**, 149–151.

Gettinby, G. (1989) Computational veterinary parasitology with an application to chemical resistance. *Veterinary Parasitology*, **32**, 57–72.

Gill, J.H. and Lacey, E. (1998) Avermectin resistance in trichostrongyloid nematodes. *International Journal for Parasitology*, **28**, 863–877.

Gill, J.H. and Redwin, J.M. (1995) Cryopreservation of the first-stage larvae of trichostrongyloid nematode parasites. *International Journal for Parasitology*, **25**, 1421–1426.

Gill, J.H., Redwin, J.M., van Wyk, J.A. and Lacey, E. (1991) Detection of resistance to ivermectin in *Haemonchus contortus*. *International Journal for Parasitology*, **21**, 771–776.

Gill, J.H., Redwin, J.M., Van Wyk, J.A. and Lacey, E. (1995) Avermectin inhibition of larval development in *Haemonchus contortus* — effects of ivermectin resistance. *International Journal for Parasitology*, **25**, 463–470.

Giordano, D.J., Tritschler, J.P. and Coles, G.C. (1988) Selection of ivermectin-resistant *Trichostrongylus colubriformis* in lambs. *Veterinary Parasitology*, **30**, 139–148.

Gopal, R.M., Pomroy, W.E. and West, D.M. (1999) Resistance of field isolates of *Trichostrongylus colubriformis* and *Ostertagia circumcinta* to ivermectin. *International Journal for Parasitology*, **29**, 781–786.

Grant, W.N. (1992) Transformation of *Caenorhabditis elegans* with genes from parasitic nematodes. *Parasitology Today*, **8**, 344–346.

Grant, W.N. (1994) Genetic variation in parasitic nematodes and its implications. *International Journal for Parasitology*, **24**, 821–880.

Grant, W.N. and Mascord, L.J. (1996) Beta-tubulin gene polymorphism and benzamidazole resistance in *Trichostrongylus colubriformis*. *International Journal for Parasitology*, **26**, 71–77.

Grant, W.N. and Whitington, G.E. (1994) Extensive DNA polymorphism within and between two strains of *Trichostrongylus colubriformis*. *International Journal for Parasitology*, **24**, 719–725.

Green, P.E., Forsyth, B.A., Rowan, K.J. and Payne, G. (1981) The isolation of a field strain of *Haemonchus contortus* in Queensland showing multiple anthelmintic resistance. *Australian Veterinary Journal*, **57**, 79–84.

Hendrikx, W.M.L. (1988) The influence of cryopreservation on a benzimidazole-resistant isolate of *Haemonchus contortus* conditioned for inhibited development. *Parasitology Research*, **74**, 569–573.

Hennessy, D.R. (1994) The disposition of antiparasitic drugs in relation to the development of resistance by parasites of livestock. *Acta Tropica*, **56**, 125–141.

Herd, R.P. (1993) Control strategies for ruminant and equine parasites to counter resistance, encystment and ecotoxicity in the USA. *Veterinary Parasitology*, **48**, 327–336.

Herd, R.P. and Gabel, A.A. (1990) Reduced efficacy of anthelmintics in young compared with adult horses. *Equine Veterinary Journal*, **22**, 164–169.

Herlich, H., Rew, R.S. and Colglazier, M.L. (1981) Inheritance of cambendazole resistance in *Haemonchus contortus*. *American Journal of Veterinary Research*, **42**, 1342–1344.

Hoekstra, R., Visser, A., Wiley, L.J., Weiss, A.S., Sangster, N.C. and Roos, M.H. (1997) Characterization of an acetylcholine receptor gene of *Haemonchus contortus* in relation to levamisole resistance. *Molecular and Biochemical Parasitology*, **84**, 179–187.

Hong, C., Hunt, K.R., Harris, T.J., Coles, G.C., Grimshaw, W.T.R. and McMullin, P.F. (1992) A survey of benzimidazole resistant nematodes in sheep in three countries of southern England. *The Veterinary Record*, **131**, 5–7.

Hosono, R., Sassa, T. and Kuno, S. (1989) Spontaneous mutations of trichlorfon resistance in the nematode, *Caenorhabditis elegans*. *Zoological Science*, **6**, 697–708.

Hubert, J. and Kerboeuf, D. (1984) A new method for culture of larvae used in diagnosis of ruminant gastrointestinal strongylosis: comparison with fecal cultures. *Canadian Journal of Comparative Medicine*, **48**, 63–71.

Hubert, J. and Kerboeuf, D. (1992) A microlarval development assay for the detection of anthelmintic resistance in sheep nematodes. *Veterinary Record*, **130**, 442–446.

Hunt, K.R. and Taylor, M.A. (1989) Use of the egg hatch assay on sheep faecal samples for the detection of benzimidazole resistant nematodes. *The Veterinary Record*, **125**, 153–154.

Ihler, C.F. (1995) A field survey on anthelmintic resistance in equine small strongyles in Norway. *Acta Veterinaria Scandinavia*, **36**, 135–143.

Ihler, C.F. and Bjorn, H. (1996) Use of two *in vitro* methods for the detection of benzimidazole resistance in equine small strongyles (*Cyathostoma* spp.). *Veterinary Parasitology*, **65**, 117–125.

Isaza, R., Courtney, C.H. and Kollias, G.V. (1995) The prevalence of benzimidazole-resistant trichostrongyloid nematodes in antelope collections in Florida. *Journal of Zoo and Wildlife Medicine*, **26**, 260–264.

Jackson, F., Coop, R.L., Jackson, E., Scott, E.W. and Russel, A.J.F. (1992) Multiple anthelmintic resistant nematodes in goats. *The Veterinary Record*, **130**, 210–211.

Jackson, R.A., Townsend, K.G., Pyke, C. and Lance, D.M. (1987) Isolation of oxfendazole resistant *Cooperia oncophora* in cattle. *New Zealand Veterinary Journal*, **35**, 187–189.

Johansen, M.V. (1989) An evaluation of techniques used for the detection of anthelmintic resistance in nematode parasites of domestic livestock. *Veterinary Research Communications*, **13**, 455–466.

Juliano, R.L. and Ling, V. (1976) A surface glycoprotein modulating drug permeability in Chinese hamster ovary cell mutants. *Biochimica et Biophysica Acta*, **455**, 152–162.

Kelly, J.D., Sangster, N.C., Porter, C.J., Martin, I.C.A., Gunawan, M. (1981) Use of guinea pigs to assay anthelmintic resistance in ovine isolates of *Trichostrongylus colubriformis*. *Research in Veterinary Science*, **30**, 131–137.

Kerboeuf, D. and Hubert, J. (1987) Changes in response of *Haemonchus contortus* eggs to the ovicidal activity of thiabendazole during the course of infection. *Annales de Recherches Vétérinaires*, **18**, 365–370.

Kieran, P.J. (1994) Moxidectin against ivermectin-resistant nematodes — a global view. *Australian Veterinary Journal*, **71**, 18–20.

King, A.I.M., Love S. and Duncan, J.L. (1990) Field investigation of anthelmintic resistance of small strongyles in horses. *Veterinary Record*, **127**, 232–233.

Kotze, A.C. (1997) Cytochrome P450 monooxygenase activity in *Haemonchus contortus* (Nematoda). *International Journal for Parasitology*, 27, 33–40.

Kotze, A.C. (1998) Effects of macrocyclic lactones on ingestion in susceptible and resistant *Haemonchus contortus* larvae. *Journal of Parasitology*, 84, 631–635.

Kotze, A.C. (1999) Peroxide-supported *in-vitro* cytochrome P450 activities in *Haemonchus contortus*. *International Journal for Parasitology*, 29, 389–396.

Kwa, M.S.G., Kooyman, F.N., Boersema, J.H. and Roos, M.H. (1993) Effect of selection for benzimidazole resistance in *Haemonchus contortus* on beta-tubulin isotype 1 and isotype 2 genes. *Biochemical and Biophysical Research Communications*, 191, 413–419.

Kwa, M.S.G., Okoli, M.N., Schulz-Key, H., Okongkwo, P.O. and Roos, M.H. (1998) Use of P-glycoprotein gene probes to investigate anthelmintic resistance in *Haemonchus contortus* and comparison with *Onchocera volvulus*. *International Journal for Parasitology*, 28, 1235–1240.

Kwa, M.S.G., Veenstra, J.G. and Roos, M.H. (1993) Molecular characterisation of beta-tubulin genes present in benzimidazole-resistant populations of *Haemonchus contortus*. *Molecular and Biochemical Parasitology*, 60, 133–144.

Kwa, M.S.G., Veenstra, J.G. and Roos, M.H. (1994) Benzimidazole resistance in *Haemonchus contortus* is correlated with a conserved mutation at amino acid 200 in β-tubulin isotype 1. *Molecular and Biochemical Parasitology*, 63, 299–303.

Kwa, M.S.G., Veenstra, J.G.H., Van Dijk, M. and Roos, M.H. (1995) β-tubulin genes from the parasitic nematode *Haemonchus contortus* modulate drug resistance in *Caenorhabditis elegans*. *Journal of Molecular Biology*, 246, 500–510.

Lacey, E. (1988) The role of the cytoskeletal protein, tubulin, in the mode of action and mechanism of drug resistance to benzimidazoles. *International Journal for Parasitology*, 18, 885–936.

Lacey, E. and Prichard, R.K. (1986) Interactions of benzimidazoles (BZ) with tubulin from BZ-sensitive and BZ-resistant isolates of *Haemonchus contortus*. *Molecular and Biochemical Parasitology*, 19, 171–181.

Lacey, E., Redwin, J.M., Gill, J.H., Demargheriti, V.M. and Waller, P.J. (1990) A larval development assay for the simultaneous detection of broad spectrum anthelmintic resistance. In *Resistance of Parasites to Antiparasitic Drugs*, edited by J.C. Boray, P.J. Martin and R.T. Roush, pp. 177–184. New Jersey: MSD Agvet.

Lacey, E. and Snowdon, K.L. (1988) A routine diagnostic assay for the detection of benzimidazole resistance in parasitic nematodes using tritiated benzimidazole carbamates. *Veterinary Parasitology*, 27, 309–324.

Lacey, E., Snowdon, K.L., Eagleson, G.K. and Smith, E.F. (1987) Further investigations of the primary mechanisms of benzimidazole resistance in *Haemonchus contortus*. *International Journal for Parasitology*, 17, 1421–1429.

Laughton, D.L., Lunt, G.G. and Wolstenholme, A.J. (1997) Alternative splicing of *Caenorhabditis elegans* gene produces two novel inhibitory amino acid receptor subunits with identical ligand binding domains but different ion channels. *Gene*, 201, 119–125.

Laughton, D.L., Lunt, G.G. and Wolstenholme, A.J. (1997) Reporter gene constructs suggest that the *Caenorhabditis elegans* avermectin receptor beta-subunit is expressed solely in the pharynx. *Journal of Experimental Biology*, 200, 1509–1514.

Le Jambre, L.F. (1976) Egg hatch as an *in vitro* assay of thiabendazole resistance in nematodes. *Veterinary Parasitology*, 2, 385–391.

Le Jambre, L.F. (1990) Molecular biology and anthelmintic resistance in parasitic nematodes. In *Resistance of Parasites to Antiparasitic Drugs*, edited by J.C. Boray, P.J. Martin and R.T. Roush, pp. 155–164. New Jersey: MSD Agvet.

Le Jambre, L.F. (1993a) Ivermectin-resistant *Haemonchus contortus* in Australia. *Australian Veterinary Journal*, 70, 357.

Le Jambre, L.F. (1993b) Molecular variation in trichostrongylid nematodes from sheep and cattle. *Acta Tropica*, 53, 331–343.

Le Jambre, L.F., Gill, J.H., Lenane, I.J. and Lacey, E. (1995) Characterisation of an avermectin resistant strain of Australian *Haemonchus contortus*. *International Journal for Parasitology*, 25, 691–698.

Le Jambre, L.F. and Royal, W.M. (1977) Genetics of vulvar morph types in *Haemonchus contortus* from the northern tablelands of New South Wales. *International Journal for Parasitology*, 7, 481–487.

Le Jambre, L.F., Royal, W.M. and Martin, P.J. (1979) The inheritance of thiabendazole resistance in *Haemonchus contortus*. *Parasitology*, 78, 107–119.

Le Jambre, L.F., Southcott, W.H. and Dash, K.M. (1976) Resistance of selected lines of *Haemonchus contortus* to thiabendazole, morantel tartrate and levamisole. *International Journal for Parasitology*, 6, 217–222.

Leathwick (1995) A case of moxidectin failing to control resistant *Ostertagia* species in goats. *The Veterinary Record*, 136, 443–444.

Lewis, J.A. and Berberich, S. (1992) A detergent-solubilized nicotinic acetylcholine receptor of *Caenorhabditis elegans*. *Brain Research Bulletin*, 29, 667–674.

Lewis, J.A., Elmer, J.S., Skimming, J., McLafferty, S., Fleming, J. and McGee, T. (1987a) Cholinergic receptor mutants of the nematode *Caenorhabditis elegans*. *Journal of Neuroscience*, 7, 3059–3071.

Lewis, J.A., Fleming, J.T., McLafferty, S., Murphy, H. and Wu, C. (1987b) The levamisole receptor, a cholinergic receptor of the nematode *Caenorhabditis elegans*. *Molecular Pharmacology*, 31, 185–193.

Lewis, J.A., Wu, C.-H., Levine, J.H. and Berg, H. (1980a) Levamisole-resistant mutants of the nematode *Caenorhabditis* appear to lack pharmacological acetylcholine receptors. *Neuroscience*, 5, 967–989.

Lewis, J.A., Wu, C.-H., Levine, J.H. and Berg, H. (1980b) The genetics of levamisole resistance in the nematode *Caenorhabditis elegans*. *Genetics*, 95, 905–928.

Lloyd, J.B., Love, S.C.J., Fitzgibbon, C. and Davis, E.O. (1998) Closantel resistance in *Haemonchus contortus* in northern New South Wales. In *Wool and Sheepmeat Services Annual Conference*. Australia: NSW Agriculture.

Lubega, G.W., Klein, R.D., Geary, T.G. and Prichard, R.K. (1994) *Haemonchus contortus*: the role of two beta-tubulin subfamilies in the resistance to benzimidazole anthelmintics. *Biochemical Pharmacology*, 47, 1705–1715.

Lubega, G.W. and Prichard, R.K. (1990) Specific interaction of benzimidazole anthelmintics with tubulin: high-affinity binding and benzimidazole resistance in *Haemonchus contortus*. *Molecular and Biochemical Parasitology*, 38, 221–232.

Lubega, G.W. and Prichard, R.K. (1991) Beta-tubulin and benzimidazole resistance in the sheep nematode *Haemonchus contortus*. *Molecular and Biochemical Parasitology*, 47, 129–138.

Lyndal-Murphy, M. (1992) *Anthelmintic Resistance in Sheep*. In *Australian Standard Diagnostic Techniques for Animal Diseases*, Edited by L.A. Corner and T.J. Bagust, pp. 1–17. Canberra: Standing Committee on Agriculture and Resource Management.

Lyons, E.T., Tolliver, S.C., Drudge, J.H., Stamper, S., Swerczek, T.W. and Granstrom, D.E. (1996) Critical test evaluation (1977–1992) of drug efficacy against endoparasites featuring benzimidazole-resistant small strongyles (population S) in Shetland ponies. *Veterinary Parasitology*, 66, 67–73.

Maciel, S., Gimenez, A.M., Gaona, C., Waller, P.J. and Hansen, J.W. (1996) The prevalence of anthelmintic resistance in nematode parasites of sheep in Southern Latin America: Paraguay. *Veterinary Parasitology*, 62, 207–212.

Malan, F.S., Gruss, B., Roper, N.A., Ashburner, A.J. and DuPlessis, C.A. (1988) Resistance of *Libyostrongylus douglassi* in ostriches to levamisole. *Journal of the South African Veterinary Association*, 59, 202–203.

Malan, F.S., van Wyk, J.A., Gerber, H.M. and Alves, M.R. (1990) First report of organophosphate resistance in a strain of *Haemonchus contortus* in South Australia. *South African Journal of Science*, 86, 49.

Marriner, S.E., McKinnon, I. and Bogan, J.A. (1987) The pharmacokinetics of ivermectin after oral and subcutaneous administration to sheep and horses. *Journal of Veterinary Pharmacology and Therapeutics*, 10, 175–179.

Martin, P.J., Anderson, N. and Jarrett, R.G. (1989) Detecting benzimidazole resistance with faecal egg count reduction tests and *in vitro* assays. *Australian Veterinary Journal*, 66, 236–240.

Martin, P.J., Anderson, N., Lwin, T., Nelson, G. and Morgan, T.E. (1984) The association between frequency of thiabendazole treatment and the development of resistance in field isolates of *Ostertagia* spp. of sheep. *International Journal for Parasitology*, 14, 177–181.

Martin, P.J. and Le Jambre, L.F. (1979) Larval paralysis as an *in vitro* assay of levamisole and morantel tartrate resistance in *Ostertagia*. *Veterinary Science Communications*, 3, 159–164.

Martin, P.J. and McKenzie, J.A. (1990) Levamisole resistance in *Trichostrongylus colubriformis*: a sex linked recessive character. *International Journal for Parasitology*, 20, 867–872.

Martin, P.J., McKenzie, J.A. and Stone, R.A. (1988) The inheritance of thiabendazole resistance in *Trichostrongylus colubriformis*. *International Journal for Parasitology*, 18, 703–709.

Martin, R.J. (1997) Modes of action of anthelmintic drugs. *The Veterinary Journal*, **154**, 11–34.

Martin, R.J., Murray, I., Robertson, A.P., Bjorn, H. and Sangster, N. (1998) Anthelmintics and ion-channels: after a puncture, use a patch. *International Journal for Parasitology*, **28**, 849–862.

Martin, R.J. and Pennington, A.J. (1989) A patch-clamp study of effects of dihydroavermectin on *Ascaris* muscle. *British Journal of Pharmacology*, **98**, 747–756.

Martin, R.J., Robertson, A.P., Bjorn, H. and Sangster, N.C. (1997) Heterogeneous levamisole receptors: a single-channel study of nicotinic acetylcholine receptors from *Oesophagostomum dentatum*. *European Journal of Pharmacology*, **322**, 249–257.

McKenna, P.B. (1985) The efficacy of levamisole and ivermectin against a morantel resistant strain of *Trichostrongylus colubriformis*. *New Zealand Veterinary Journal*, **33**, 198–199.

McKenna, P.B. (1996) Anthelmintic resistance in cattle nematodes in New Zealand: is it increasing? *New Zealand Veterinary Journal*, **44**, 76.

Monteiro, A.M., Wanyangu, S.W., Kariuki, D.P., Bain, R., Jackson, F. and McKellar, Q.A. (1998) Pharmaceutical quality of anthelmintics sold in Kenya. *Veterinary Record*, **142**, 396–398.

Moreno-Guzman, M.J., Coles, G.C., Jimenez-Gonzalez, A., Criado-Fornelio, A., Ros-Moreno, R.M. and Rodriguez-Caabeiro, F. (1998) Levamisole binding sites in *Haemonchus contortus*. *International Journal for Parasitology*, **28**, 413–418.

Nari, A., Salles, J., Waller, P.J. and Hansen, J.W. (1996) The prevalence of anthelmintic resistance in nematode parasites of sheep in Southern Latin America: Uruguay. *Veterinary Parasitology*, **62**, 213–222.

Nilsson, O., Lindholme, A. and Christenson, D. (1989) A field evaluation of anthelmintics in horses in Sweden. *Veterinary Parasitology*, **32**, 163–171.

Novak, J. and Vanek, Z. (1992) Screening for a new generation of anthelmintic compounds: *in vitro* selection of the nematode *Caenorhabditis elegans* for ivermectin resistance. *Folia Microbiologica*, **37**, 237–238.

Okimoto, R., Macfarlane, J.L., Clary, D.O. and Wolstenholme, D.R. (1992) The mitochondrial genomes of two nematodes, *Caenorhabditis elegans* and *Ascaris suum*. *Genetics*, **130**, 471–498.

Overend, D.J., Phillips, M.L., Poulton, A.L. and Foster, C.E.D. (1994) Anthelmintic resistance in Australian sheep nematode populations. *Australian Veterinary Journal*, **71**, 117–121.

Pomroy, W.E. and Whelan, N.C. (1993) Efficacy of moxidectin against an ivermectin-resistant strain of *Ostertagia circumcincta* in young sheep. *The Veterinary Record*, **132**, 416.

Pouliot, J.-F., L'Heureux, F., Liu, Z., Prichard, R.K. and Georges, E. (1997) Reversal of P-glycoprotein-associated multidrug resistance by ivermectin. *Biochemical Pharmacology*, **53**, 17–25.

Prichard, R.K., Hall, C.A., Kelly, J.D., Martin, I.C.A. and Donald, A.D. (1980) The problem of anthelmintic resistance in nematodes. *Australian Veterinary Journal*, **56**, 239–251.

Richards, J.C., Behnke, J.M. and Duce, I.R. (1995) *In vitro* studies on the relative sensitivity to ivermectin of *Necator americanus* and *Ancylostoma ceylanicum*. *International Journal for Parasitology*, **25**, 1185–1191.

Robertson, A.P., Bjorn, H.E. and Martin, R.J. (1999) Resistance to levamisole resolved at the single-channel level. *FASEB Journal*, **13**, 749–760.

Robertson, S.J. and Martin, R.J. (1993) Levamisole-activated single-channel currents from muscle of the nematode parasite *Ascaris suum*. *British Journal of Pharmacology*, **108**, 170–178.

Roepstorff, A., Bjorn, H. and Nansen, P. (1987) Resistance of *Oesophagostomum* spp. in pigs to pyrantel citrate. *Veterinary Parasitology*, **24**, 229–239.

Rohrer, S.P., Birzin, E.T., Eary, C.H., Schaeffer, J.M. and Shoop, W.L. (1994) Ivermectin binding sites in sensitive and resistant *Haemonchus contortus*. *Journal of Parasitology*, **80**, 493–497.

Rolfe, P.F., Boray, J.C., Fitzgibbon, C., Parsons, G., Kemsley, P. and Sangster, N. (1990) Closantel resistance in *Haemonchus contortus* from sheep. *Australian Veterinary Journal*, **67**, 29–31.

Roos, M.H., Boersema, J.H., Borgsteede, F.H.M., Cornelissen, J., Taylor, M. and Ruitenberg, E.J. (1990) Molecular analysis of selection for benzimidazole resistance in the sheep parasite *Haemonchus contortus*. *Molecular and Biochemical Parasitology*, **43**, 77–88.

Roos, M.H., Kwa, M.S.G. and Grant, W.N. (1995) New genetic and practical implications of selection for anthelmintic resistance in parasitic nematodes. *Parasitology Today*, **11**, 148–150.

Rothwell, J.T. and Sangster N.C. (1993) An *in vitro* assay utilising parasitic larval *Haemonchus contortus* to detect resistance to closantel and other anthelmintics. *International Journal for Parasitology*, **23**, 573–578.

Rothwell, J.T. and Sangster N.C. (1996) The effects of closantel treatment on the ultrastructure of *Haemonchus contortus*. *International Journal for Parasitology*, **26**, 49–57.

Rothwell, J.T. and Sangster, N.C. (1997) *Haemonchus contortus*: The uptake and metabolism of closantel. *International Journal for Parasitology*, **27**, 313–319.

Russell, G.J. and Lacey, E. (1991) Temperature dependent binding of mebendazole to tubulin in benzimidazole-susceptible and -resistant strains of *Trichostrongylus colubriformis* and *Caenorhabditis elegans*. *International Journal for Parasitology*, **21**, 927–934.

Sangster, N.C. (1995) Ivermectin and moxidectin: just different names? *Australian Sheep Veterinary Society, AVA Conference Proceedings*, 144–150.

Sangster, N.C. (1994) P-glycoproteins in nematodes. *Parasitology Today*, **10**, 319–322.

Sangster, N.C. (1996) Pharmacology of anthelmintic resistance. *Parasitology*, **113**, 201–216.

Sangster, N.C. (1999) Pharmacology of anthelmintic resistance in cyathostomes: will it occur with the avermectin/milbemycins? *Veterinary Parasitology*, **85**, 189–204.

Sangster, N.C., Bannan, S.C., Weiss, A.S., Nulf, S.C., Klein, R.D. and Geary, T.G. (1999) *Haemonchus contortus*: sequence heterogeneity of internucleotide binding domains from P-glycoproteins and an association with avermectin resistance. *Experimental Parasitology*, **91**, 250–257.

Sangster, N.C. and Bjorn, H. (1995) Levamisole resistance in *Haemonchus contortus* selected at different stages of infection. *International Journal for Parasitology*, **25**, 343–348.

Sangster, N.C., Davis, C.W. and Collins, G.H. (1991) Effects of cholinergic drugs on longitudinal contraction in levamisole-susceptible and -resistant *Haemonchus contortus*. *International Journal for Parasitology*, **21**, 689–695.

Sangster, N.C. and Gill, J. (1999) Pharmacology of anthelmintic resistance. *Parasitology Today*, **15**, 141–146.

Sangster, N.C. and Prichard, R.K. (1986) Thiabendazole uptake, metabolism and excretion in thiabendazole resistant and susceptible *Trichostrongylus colubriformis*. *Journal of Parasitology*, **72**, 798–800.

Sangster, N.C., Prichard, R.K. and Lacey, E. (1985) Tubulin and benzimidazole-resistance in *Trichostrongylus colubriformis* (Nematoda). *Journal of Parasitology*, **71**, 645–651.

Sangster, N.C., Redwin, J.M. and Bjorn, H. (1998) Inheritance of levamisole and benzimidazole resistance in an isolate of *Haemonchus contortus*. *International Journal for Parasitology*, **28**, 503–510.

Sangster, N.C., Riley, F.L. and Collins, G.H. (1988) Investigation of the mechanism of levamisole resistance in trichostrongylid nematodes of sheep. *International Journal for Parasitology*, **18**, 813–818.

Sangster, N.C., Riley, F.L. and Wiley, L.J. (1998) Binding of [^{3}H]*m*-aminolevamisole to receptors in levamisole-susceptible and -resistant *Haemonchus contortus*. *International Journal for Parasitology*, **28**, 707–717.

Schaeffer, J.M. and Haines, H.W. (1989) Avermectin binding in *Caenorhabditis elegans*: a two-state model for the avermectin binding site. *Biochemical Pharmacology*, **38**, 2329–2338.

Schinkel, A.H., Smith, J.J.M., van Tellingen, O., Beijnen, J.H., Wagenaar, E., van Deemter, L., Mol, C.A.A.M., van der Valk, M.A., Robanus-Maandrag, E.C., Te Riele, H.P.J., Berns, A.J.M. and Borst, P. (1994) Disruption of the mouse *mdr1a* P-glycoprotein gene leads to a deficiency in the blood-brain barrier and to increased sensitivity to drugs. *Cell*, **77**, 491–502.

Shoop, W.L. (1993) Ivermectin resistance. *Parasitology Today*, **9**, 154–159.

Shoop, W.L., Haines, H.W., Michael, B.F. and Eary, C.H. (1993) Mutual resistance to avermectins and milbemycins: oral activity of ivermectin and moxidectin against ivermectin-resistant and susceptible nematodes. *The Veterinary Record*, **133**, 445–447.

Simpkin, K.G. and Coles, G.C. (1981) The use of *Caenorhabditis elegans* for anthelmintic screening. *Journal of Chemical Technology and Biotechnology*, **31**, 66–69.

Smith, G. (1990) Mathematical model for the evolution of anthelmintic resistance in a direct life-cycle nematode parasite. *International Journal for Parasitology*, **20**, 913–921.

Smith, G., Grenfell, B.T., Isham, V. and Cornell, S. (1999) Anthelmintic resistance revisited: under-dosing, chemoprophylactic strategies, and mating probabilities. *International Journal for Parasitology*, **29**, 77–91.

Stringfellow, F. (1988) An *in vitro* test for drug resistance in *Haemonchus contortus*. *Proceedings of the Helminthological Society of Washington*, **55**, 19–23.

Sutherland, I.A., Lee, D.L. and Lewis, D. (1998) Detection of benzimidazole resistance in trichostrongylid nematodes. *Parasitology Today*, **4**, 22–24.

Swan, N., Gardner, J.J., Besier, R.B. and Wroth, R. (1994) A field case of ivermectin resistance in *Ostertagia* of sheep. *Australian Veterinary Journal*, **71**, 302–303.

Taylor, M.A. (1990) A larval development test for the detection of anthelmintic resistance in nematodes of sheep. *Research in Veterinary Science*, **49**, 198–202.

Thrusfield, M. (1995) *Veterinary Epidemiology*, 2nd edn, pp. 183–189. Oxford: Blackwell Science.

Townson, S., Tagboto, S.K., Castro, J., Lujan, A., Awadzi, K. and Titanji, V.P.K. (1994) Comparison of the sensitivity of different geographical races of *Onchocerca volvulus* microfilariae to ivermectin: studies *in vitro. Transactions of the Royal Society of Tropical Medicine and Hygiene*, **88**, 101–106.

van Wyk, J.A., Gerber, H.M. and Alves, R.M.R. (1982) Slight resistance to the residual effect of closantel in a field strain of *Haemonchus contortus* which showed an increased resistance after one selection in the laboratory. *Onderstepoort Journal of Veterinary Research*, **49**, 257–262.

van Wyk, J.A. and Malan, F.S. (1988) Resistance of field strains of *Haemonchus contortus* to ivermectin, closantel, rafoxanide and the benzimidazoles in South Africa. *Veterinary Record*, **123**, 226–228.

van Wyk, J.A., Malan, F.S., Gerber, H.M. and Alves, R.M.R. (1989) The problem of escalating resistance of *Haemonchus contortus* to the modern anthelmintics in South Africa. *Onderstepoort Journal of Veterinary Research*, **56**, 41–49.

Várady, M., Bjorn, H., Craven, J.C. and Nansen, P. (1997) *In vitro* characterisation of lines of *Oesophagostomum dentatum* selected or not selected for resistance to pyrantel, levamisole and ivermectin. *International Journal of Parasitology*, **27**, 77–81.

Várady, M. and Corba, J. (1997) Resistance of equine small strongyles to benzimidazole in Slovak Republic. *Helminthologia*, **34**, 81–85.

Várady, M. and Corba, J. (1999) Comparison of six *in vitro* tests in determining benzimidazole and levamisole resistance in *Haemonchus contortus* and *Ostertagia circumcincta* of sheep. *Veterinary Parasitology*, **80**, **239–249.**

Vermunt, J.J., West, D.M. and Pomroy, W.E. (1995) Multiple resistance to ivermectin and oxfendazole in *Cooperia* species of cattle in New Zealand. *Veterinary Record*, **137**, 43–45.

Waller, P.J. (1987) Anthelmintic resistance and the future for roundworm control. *Veterinary Parasitology*, **25**, 177–191.

Waller, P.J., Dash, K.M., Barger, I.A., Le Jambre, L.F. and Plant, J. (1995) Anthelmintic resistance in nematode parasites in sheep: learning from the Australian experience. *The Veterinary Record*, **136**, 411–413.

Waller, P.J., Dobson, R.J., Donald, A.D., Griffiths, D.A. and Smith, E.F. (1985) Selection studies on anthelmintic resistant and susceptible populations of *Trichostrongylus colubriformis* of sheep. *International Journal for Parasitology*, **15**, 669–676.

Waller, P.J., Dobson, R.J., Obendorf, D.L. and Gillham, R.J. (1986) Resistance of *Trichostrongylus colubriformis* to levamisole and morantel: differences in relation to selection history. *Veterinary Parasitology*, **21**, 255–263.

Ward, S. (1988) *Caenorhabditis elegans*: a model for parasitic nematodes. In *The Biology of Parasitism: A Molecular and Biochemical Approach*, edited by P.T. Englund and A. Sher, pp. 503–516. New York: Liss.

Watson, T.G., Hosking, B.B., Leathwick, D.M. and McKee, P.F. (1996) Ivermectin-moxidectin side resistance by *Ostertagia* species isolated from goats and passaged to sheep. *Veterinary Record*, **138**, 472–473.

Whitlock, H.V., Sangster, N.C., Gunawan, M., Porter, C.J. and Kelly, J.D. (1980) *Trichostrongylus colubriformis* and *Ostertagia* sp. resistant to levamisole, morantel tartrate and thiabendazole: isolation into pure strain and anthelmintic titration. *Research in Veterinary Science*, **29**, 31–35.

Wiley, L.J., Ferrara, D.R., Sangster, N.C. and Weiss, A.S., (1997) The nicotinic acetylcholine receptor α-subunit is located on the X chromosome but its coding sequence is not involved in levamisole resistance in an isolate of *Trichostrongylus colubriformis*. *Molecular and Biochemical Parasitology*, **90**, 415–422.

Wiley, L.J., Weiss, A.S., Sangster, N.C. and Li, Q. (1996) Cloning and sequence analysis of the candidate nicotinic acetylcholine receptor alpha subunit gene tar-1 from *Trichostrongylus colubriformis*. *Gene*, **182**, 97–100.

Woods, R.A., Malone, K.M.B., Spence, A.M., Sigurdson, W.J. and Byard, E.H. (1989) The genetics, ultrastructure, and tubulin polypeptides of mebendazole resistant mutants of *Caenorhabditis elegans*. *Canadian Journal of Zoology*, **67**, 2422–2431.

Woods, T.F., Lane, T.J., Zeng, Q.Y. and Courtney, C.H. (1998) Anthelmintic resistance on horse farms in north central Florida. *Equine Practice*, **20**, 14–17.

Xu, M., Molento, M., Blackhall, W., Ribeiro, P., Beech, R. and Prichard, R. (1998) Ivermectin resistance in nematodes may be caused by alteration of P-glycoprotein homolog. *Molecular and Biochemical Parasitology*, **91**, 327–335.

Yadav, C.L. and Verma, S.P. (1997) Morantel resistance by *Haemonchus placei* in cattle. *Veterinary Record*, **141**, 499–500.

23. Molecular Approaches to Novel Crop Resistance Against Nematodes

Howard J. Atkinson

Centre for Plant Sciences, Leeds Institute for Plant Biotechnology & Agriculture, University of Leeds, Leeds LS2 9JT, UK

Introduction

Nematodes cause about $100 billion of losses to world agriculture annually (Sasser and Freckman, 1987). They often do not provide the clear symptoms that aid growers to identify the diseases they cause (Atkinson, 1996). One consequence of this is that the world nematicide market is only $700 m/year (Williamson, 1995). Other factors also ensure that the nematicidal market underestimates the agricultural importance of nematodes. They are the most environmentally damaging of all crop protection chemicals. This discourages some use both from grower preference and from government restriction. Environmental harm is a risk because the nematicide is incorporated into soil to ensure its distribution to the relatively immobile nematodes. Such procedures represent an environmentally unacceptable method of pesticide application. In addition, many nematicides have been shown to cause environment damage. For instance the Montreal Protocol calls for the withdrawal of the soil fumigant methyl bromide from agricultural use by 2005. It is considered a major cause of ozone layer depletion. The USA intends to eliminate its agricultural use of the compound by 2002 unless this intention is reversed as the result of lobbying by some growers (Pearce, 1997). Other halogenated hydrocarbon nematicides are also severe biohazards. Use of dibromochloropropane (DBCP) is declining and considerable safety precautions are in place in countries such as USA where some applications remain. However, investigative journalism has uncovered cases (Jenkins, 1996) where such precautions are inadequately enforced in the developing world. For example, safety equipment is often not used during nematicide application. Such misuse is highlighted by class actions in USA courts. They have been brought against chemical manufacturers by agricultural workers from Central America. The allegations arise from ill health resulting from exposure as agricultural workers (Anon, 1992). Industrial accidents have also occurred during nematicide manufacture. The most severe incident was the Bophal disaster in India but other chemical plants have had industrial accidents. Aldicarb is a widely used oxime carbamate nematicide but there are considerable concerns about the potential toxicological hazards of this compound

(Gustafson, 1993). It is detectable in groundwater after application in sandy soils and its use is closely monitored by several governments. Consequently, it has been withdrawn from many states of the USA. Nematicides are currently essential for economic production of some crops. However, they are likely to be increasingly viewed as the approach of last resort unless more environmentally benign compounds are developed.

Another main approach to nematode control is crop resistance (Cook and Evans, 1987; Atkinson, 1995). Roberts (1992) reviewed both the successes and limitations of this approach. Problems centre on availability of effective single gene resistance, the commercial viability of breeding for nematode control and its durability under challenge from different species, pathotypes or races. There seems a strong case for enhancing natural resistance using the various transgenic approaches discussed in this chapter. A main hope for the future is integration of natural resistance, transgenic approaches, cultural control, and possibly environmentally benign nematicides. The broader issues of the needs and concerns over transgenic resistance have been considered by several authors (Hobbelink, 1991; Chrispeels and Sadava, 1994; Atkinson et al., 1998).

Many of the current problems of nematode control are shown by potato cyst nematodes (PCN) (Globodera spp.) in the UK. PCN are usually controlled by integrated pest management (IPM), principally involving combinations of resistant cultivars plus chemical and cultural control. Over 40% of UK potato growers have PCN populations in their fields. Therefore growers can not afford to ignore the pest if they wish their potato cropping to remain viable. They must prevent yield loss from economically damaging populations and also prevent gradual increase to economically damaging pre-plant densities. Potato growers have shown their willingness to use effective cultivars such as those carrying the resistance gene Gro-1 (formerly termed H1) that provides full resistance to UK populations of G. rostochiensis. As a consequence G. pallida is now the prevalent species in the UK, even in regions such as E. Anglia in which it was formerly uncommon. Resistance to G. pallida has been developed for commercial use (e.g. cv Sante) but the polygenic nature slows progress of conventional plant breeding. To-date the resistance available is quantitative in nature and fails to prevent individuals of *G. pallida* from reproducing. Furthermore, individuals of *G. pallida* differ in virulence and so growing the crop gradually selects for this attribute. Even a low prevalence coupled with a widespread occurrence at the field level may ensure virulence is selected at a rate that outpaces counter measures based on new cultivar development.

Cultural practices such as rotation, when used alone, usually fail to control most PCN populations at a potato growing frequency that meets the needs of specialist growers. Declines of *G. rostochiensis* are often given as about 33% loss of eggs with viable larvae per year. This requires a 6–7 year rotation to prevent multiplication per potato crop from building an economic population. Furthermore some populations of PCN are particularly persistent. It is also probable that rotation selects for persistence.

The cloning of genes for natural resistance (see later) may speed development of agronomically acceptable cultivars and lead to new combinations of polygenic resistance. The value of this molecular approach depends upon the level of resistance and durability offered by the source of resistance. Manipulation of pathogen recognition to counter virulence with cultivars seems unlikely to be achieved for PCN within the

next decade. Therefore, there is a need for other approaches that avoid this problem. Plant defences that do not rely on natural resistance of potato plants to nematodes may avoid the issue of co-evolution of resistance and virulence. They also present distinct scientific problems and issues of public acceptability that must be addressed.

Preventing the Invasion of Plants

Ectoparasites only insert their stylets into plants and so the opportunity to disrupt their parasitism is likely to be centred on feeding (see later). However, endoparasites enter roots and so defences can also be mounted against their invasion. Cyst nematodes typically invade at the zone of root elongation and then cut their way through plant cell walls to reach the periphery of the vascular tissue where they establish as parasites (Figure 23.1 and Wyss, Chapter 9). This may be a distance of only 4 body lengths i.e. 2 mm (Robinson, Atkinson and Perry, 1988). There are two recurring natural strategies used by plants to challenge the different life processes of parasites. The first involves eliciting changes in plant cells that hinder or prevent parasitism. Examples of such responses in relation to invasion are lignification and cell death. The second strategy involves molecules that have a

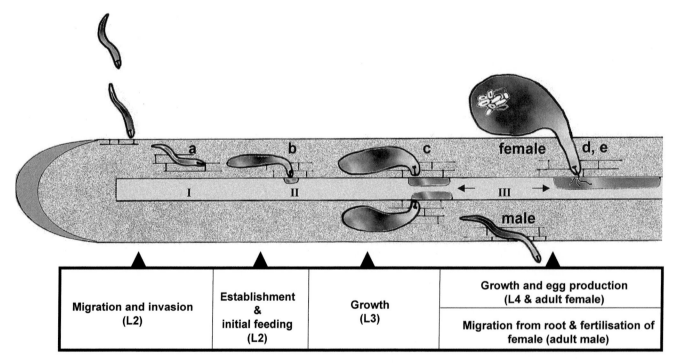

Figure 23.1. Life cycle of a cyst-nematode showing the key phases from invasion by the second-stage larvae (L2) to sexual maturity of the adults. Two principal approaches to control are expression of anti-nematode genes and destruction or attenuation of plant cells. Key opportunities for expressing anti-nematode genes occur during invasion (a) feeding site initiation, (b) early feeding of the parasite, (c) feeding and digestion of pre-adults, (d) feeding and digestion of the adult female and (e) or its reproductive development. Many of these opportunities are common to endoparasitic nematodes irrespective of their precise mode of feeding. Specific expression of genes that affect plant cells may hinder invasion (I) prevent feeding site initiation (II) or full development of the feeding site into a plant transfer system (III). Other nematodes that modify plant cells (see Wyss, Chapter 9) offer similar opportunities for engineered resistance.

direct effect on nematodes, either hindering their progress or causing mortality for those that do not leave the root.

The defences of plants to nematodes involve some pre-formed compounds but also responses that are induced during invasion or subsequent stages of parasitism. They can be sub-divided into those that depend on recognition of specific pathogens by defined resistance genes (R-genes) and more general responses against pathogens (Hunt and Ryals, 1996). One distinct general response involves a change in the level of protection provided elsewhere in the plant. This systemic acquired resistance is an inducible plant defence response in which a prior foliar pathogen interaction activates resistance in non-infected foliar tissues (Hunt and Ryals, 1996).

Nematode-resistant crops have been successfully developed by plant breeders using R-genes. There is considerable potential for extending the success of this approach now R-genes have been cloned (see later). The other responses are broadly based. Some are locally limited and may prevent the individual nematode, plus those close by, from establishing as parasites. They may involve proteins and secondary metabolites. The potential for manipulating these defences to protect against nematodes is unexplored.

Prospects for Future Manipulation of Plant Responses to Invasion

It may be possible in future to heighten plant responses to nematode invasion. There are some advantages to this approach. It may gain consumer acceptance if based on natural plant responses and it may provide a general defence against a range of nematodes. It offers a component of a stacked defence (see later). It may provide a general defence that is not species-, pathotype-or race-specific. It offers a component to a stacked defence (see later) that acts early in the parasitic life. It limits the number of nematodes that progress to damage the plant further, even if only partial protection is provided. One part of an anti-invasion defence is available. Wound-specific promoters (Hansen *et al.*, 1996) offer the opportunity of an inducible defence responding to invasion by cyst-nematodes. Therefore an anti-nematode protein that limits invasion could be delivered to the locality of attack. Some of the local responses involve secondary metabolites but manipulating them thorough changes in enzymatic pathways may prove a complex approach to nematode control. Another approach may be to amplify the natural wounding response by releasing elicitors. This would require identification of protein or peptide elicitors able to achieve this. It may prove difficult to develop mimics of those from nematodes if they prove to be glycoproteins.

Anti-Invasion Strategies Directed at the Nematode

Pre-formed Plant Defensive Compounds

The presence in the plant of pre-formed plant defensive compounds before nematode invasion requires them to lack anti-plant cell effects. A wide range of secondary metabolites has been shown to have anti-nematode effects (Huang, 1985) but there has been no attempt to manipulate their secondary pathways by transgenic means to heighten their efficacy. Therefore detailed discussion falls beyond the scope of this chapter.

Local Responses: Secondary Metabolites with Anti-nematode Activity

Some such responses involve secondary metabolites and other proteins. The toxic effects of secondary metabolites may limit the proportion of nematodes that succeed in establishing as parasites. The invasion process can initiate rapid host root responses. Auto-fluorescence of an undetermined character occurs in roots within 3 h post-invasion of potato plants by *G. rostochiensis*, probably in response to wounding during intracellular migration. This indicates the rapidity of the first response to nematode invasion or the process of wounding. This early response occurred in both compatible and incompatible interactions but was more extensive by 4 days for cv Maris Piper during an incompatible rather than a compatible interaction with *Globodera* (Robinson, Atkinson and Perry, 1988). In some cases the secondary metabolite is a phytoalexin with a known effect on the pathogen. The phytoalexin glyceollin I showed an increase from 2–4 days in roots to soybean cyst-nematode (*H. glycines*). The effect occurs at the onset of an incompatible interaction between race 1 of this nematode and cv Centennial (Huang and Barker, 1991). The pattern was similar, but the accumulation was less, during a compatible interaction with another cultivar. Serial sectioning and radio-immune assay revealed that glyceollin I accumulates from 8 h post-invasion. This occurs mainly in the locality of the head during initiation of a feeding site. At this time, the stylet is used to probe the initial feeding cell frequently and this may elicit the response strongly. Intracellular migration may not be essential for the response. It is induced by *Meloidogyne*, which primarily migrates between plant cells (Wyss, Grundler and Münch, 1992). Furthermore there is a difference between compatible and incompatible interactions between soybean cultivars and *Meloidogyne* at 2–3 days after invasion (Kaplan, Keen and Thomson, 1980). This effect is also concurrent with giant cell induction following intercellular migration by the nematode (Wyss, Chapter 9).

Nematode Secretions and the Invasion Process

Cyst and root-knot nematodes do not feed during invasion so any defensive protein must act other than by the oral route. Secretions are used by *H. glycines* shortly after invasion (Atkinson and Harris, 1989). An impressive series of experiments have established that both *H. glycines* and *G. rostochiensis* produce cellulases. The approach adopted will soon identify other secretions from nematodes. Previous work had developed procedures for raising and screening MAbs with specific reactivity to the pharyngeal glands (Atkinson *et al.*, 1988; Atkinson and Harris, 1989; Hussey, 1989b;

Goverse, Davis and Hussey, 1994). A MAb was characterised that recognises antigen(s) within the sub-ventral glands of the two species (Der Boer *et al.*, 1996; Smant *et al.*, 1998). Protein bands recognised in western blots were N' terminally sequenced. This provided information to develop a degenerative primer for the PCR approach of 3' RACE. Gene specific primers to the RACE products were then used to generate 5' RACE products. They were used to obtain full-length cDNA clones from libraries of the two nematodes. Heterologous expression in *Eshericia coli* provided fusion proteins that were shown to have cellulase activity. Polyclonal antibodies to the fusion proteins showed specific reactivity with the pharyngeal glands of the nematode. The cDNA probes were used to localise the mRNA to sub-ventral pharyngeal glands of the nematode by *in situ* hybridisation (Smant *et al.*, 1998). The presence of a leucine-rich hydrophobic region and the polyadenylated 3' cDNA ends demonstrate that the genes are of eukaryotic not bacterial origin. The genes do have close homology to the genes from bacteria (Smant *et al.*, 1998). The genes are expressed in larvae prior to establishment and in the males, which also move through roots (Smant *et al.*, 1998). The enzymes probably assist mechanical penetration of cell walls by the stylet by softening the cellulose microfibrils of plant cell walls. One of the cellulases of each nematode possesses a cellulose-binding domain as well as a catalytic domain. The polyclonal sera localised the cellulase of *H. glycines* without the cellulase-binding domain along the migration track of the invading larva. It was recognised within the nematode and was secreted through the stylet into root cortical tissue. The cellulase with an additional cellulose-binding domain was not localised outside of the nematode (Wang *et al.*, 1999). It may have a role later in invasion. For instance it may have a role in cell wall breakdown in syncytial induction or in maintaining localised cellulase activity at the site of repeated stylet insertion once a feeding site is reached. A cellulose binding protein without cellulase activity has also been cloned from *M. incognita*. It binds to cellulase and plant cell walls. It has been immuno-localised to the sub-ventral pharyngeal glands and shown to be present in stylet secretions. Its probable role in pathogenesis has yet to be established (Ding *et al.*, 1998).

Local Responses to Invasion Directed at Plant Cells

Peroxidases and Reactive Oxygen Intermediates

The sub-division of plant responses into those directed against the nematode or plant cells can not readily be made for all responses. Peroxidase activity associated with nematode invasion may produce bursts of the reactive oxygen intermediate hydrogen peroxide (H_2O_2) Their production has local effects on both the nematode and plant cells. Peroxidase isozyme patterns change both locally and systemically after pathogen challenge (Lagrimini *et al.*, 1987). The different forms

may arise by post-translational modification or phenolic modification of pre-existing peroxidase molecules associated with plant cell walls rather than activation of peroxidase genes. Peroxidases have a complex of roles beyond their basic chemical reaction of peroxidatic cycling involving H_2O_2 and a wide range of hydrogen donors. The transient increase in H_2O_2 may act as an antimicrobial agent and could have some effect on nematodes too before clearance by peroxidase. This transient increase may initiate other defensive responses thorough lipid peroxidation and cause phytoalexin release (van Huystee, 1987).

There are several additional, more complex, roles of the enzyme that may occur in response to pathogens. They include: (i) involvement in constructing intermolecular linkages in the wall matrix in wounding repair; (ii) catalysis of phenoxy radical formation in the final stages of lignification; (iii) polymerisation of phenolic molecules in the production of suberin (this involves highly anionic forms of the peroxidase); (iv) cross-linking proteins with phenolic groups (e.g. in hydroxyproline rich glycoproteins; HRGPs; extensins); (v) polysaccharide gelling, for instance by increasing differulyol bridges in pectins (van Huystee, 1987). Given the complexity of peroxidase activities, there is much to do before their importance can be defined in either general or specific local defences against nematodes.

Changes occur in peroxidase activity during a resistant response of tomato to *M. incognita* and of pea to *Heterodera goettingiana*. There are topographically distinct sites of peroxidase activity at the plant cell wall during the resistant response to *H. goetiingiana*, which presumably relates to a rapid synthesis of lignin at that time (Zacheo, Bleve-Zacheo and Melillo, 1997).

Nitric Oxide

Nitric oxide acts in the hypersensitive response to prevent pathogen spread (Delledonne *et al.*, 1998). It seems to have a complementary activity with reactive oxygen intermediates superoxide (O_2^-) and H_2O_2. This involves systemic induction of hypersensitive cell death and activation of other defence-related genes. This may provide more than one means of inducing specific defences. Reactive oxygen alone does not support a strong resistance response and nitric oxide may potentiate its activity. There is value in having complex control of hypersensitive cell death given its dramatic consequences if induced too extensively or readily.

Basic Pathogenesis-related Proteins

Pathogenesis-related proteins (PR-1-5) are polypeptides. The acidic forms have received particular attention in relation to systemic acquired resistance (see below). All but PR-4 also have basic forms which have similar amino acid sequence to their more fully studied acidic counterparts. They differ in accumulating within vacuoles. The basic PRs all show similar tissue-specificity in roots. Here they are constitutively expressed but also show increased accumulation in response to both ethylene and wounding (Linthorst, 1991). Immunoblotting using specific antibodies showed roots of potato infected with *G. pallida* had an increased presence of chitinases (Rahimi, Perry and Wright, 1998; Rahimi, Wright and Perry, 1988) but not β1-3 glucanase activity (Rahimi, Perry and Wright, 1996).

Manipulating Systemic Acquired Resistance

Pathogen invasion can lead to systemically acquired resistance (SAR) against later infection by the same or some other organism. The SAR response can be defined by expression of a specific set of genes (molecular markers) and the apparent requirement for cell death to occur (Uknes *et al.*, 1996). Knowledge of acquired resistance stems principally from work on viral/plant interactions such as Tobacco Mosaic Virus (TMV) and tobacco. Once inoculated with TMV, the same leaves and others of the same plant become resistant to further infection by TMV, by other unrelated viruses and by some fungi and bacteria. A distinction is sometimes made between acquired resistance on the same leaf (local) and elsewhere (systemic). The effects can be elicited locally or systemically (Ryals *et al.*, 1992).

The acidic pathogenesis-related proteins (PR-1-5) are characterised by stability at low pH and they are normally targeted to the apoplast of leaves. They are defined as showing concomitant *de novo* expression during the establishment of systemic acquired resistance. Some are used as reliable molecular markers of SAR (Uknes *et al.*, 1996). One standard nomenclature (van Loon, Gerritsen and Ritter, 1987) allocates each class of pathogenesis-related proteins a number (PR-1 to PR-5) plus a lower case letter for similar protein sequences (e.g. PR1a and PR1c). The number is based on relative mobility in non-denaturing conditions. Proteins sharing the same number are serologically related, have similar Mr values and share a high similarity of amino acids. In addition further proteins not accommodated within the PR nomenclature show characteristics that suggest they are defence-related proteins. Immunochemical studies show that PR-5 accumulates in extracellular pocket-like vesicles. The basic functions of PR proteins are understood. PR2 and PR3 are β1-3 glucanases and chitinases respectively and are considered to be antifungal in activity. PR4 is a thaumatin-like protein which has activity against the oomycete *Phytophora infestans*. PR-1 also accumulates in crystal idioblasts, specific leaf cells containing crystals of calcium oxalate. Wheat expresses two proteins that are homologous to basic PR-1 proteins. They were induced by both compatible and incompatible interactions involving different isolates of *Erysiphe graminis* (Powdery mildew) but not by activators of SAR. Therefore in this moncot, PR-1 is not a reliable marker of induced resistance (Molina *et al.*, 1999).

PR-proteins are induced systemically in plant leaves by nematode invasion of roots. At least 8 novel polypeptides occur in the intercellular fluid of potato leaves following establishment of *G. rostochiensis* (Hammond-Kosack, Atkinson and Bowles, 1989) or *G. pallida* (Rahimi, Perry and Wright, 1996) in the roots

of the plant. The response shows the typically slow establishment found previously for other systemic inductions of PR proteins. It was evident at 12 days and fully established by 16 days after first invasion of roots by the nematodes. This is of a similar order of response time to the response of tobacco to a viral infection (Pierpoint, Jackson and Evans, 1990; Ryals *et al.*, 1992). The Mr and isoelectric points of polypeptides induced by nematodes were similar to those reported following pathogen challenge of potato leaves in previous work. The polypeptides reported by Hammond-Kosack, Atkinson and Bowles (1989a) and by Rahimi, Perry and Wright (1996) apparently correspond to those described by White *et al.* (1987) and others to chitinase PRs and additional PRs identified previously (Kombrink, Schroder and Hahlbrock, 1988). There is no evidence that these responses affect nematodes in roots. However, root-invading nematodes may establish SAR. This may protect against subsequent challenge of leaves by other pathogens. Nematodes that colonise leaves and stems may elicit the effect and so limit subsequent challenge by further nematodes. PR-protein accumulates when sub-economic densities of *G. pallida* colonise roots (Moore, 1995). It would be intriguing to determine if sub-economic densities of *G. pallida* on plants can induce a protective SAR against leaf pathogens of the plant. Changes in chitinases occur in roots following infection by *G. pallida* (Rahinm *et al.*, 1998a, b).

These basic studies could underpin development of novel resistance in a number of ways. Possibly PR-proteins such as the chitinases (PR-3) have value in disrupting egg-shell formation through oral ingestion. Chitinases can disrupt hatch of *M. incognita* (Mercer, Greenwood and Grant, 1992). This effect may have potential, particularly against nematodes such as *Ditylenchus dipsaci* that lay eggs in the plant.

Much of the recent interest in SAR has centred on inducing it and maintaining high levels of the protective proteins throughout plant growth (Uknes *et al.*, 1996). One aim is to develop chemical sprays that induce SAR in young crop plants and subsequently provide a protective SAR throughout their growth. For instance, application of benzothiadiazole to wheat at an early stage of development induces resistance against powdery mildew, brown rust and leaf blotch (Gorlach *et al.*, 1996). The advantage of this approach centres on providing a general defence against a range of pathogens. Assuming the approach is ineffective against root parasites, this approach seems to have little potential for development as a major approach to the control of nematodes. There are a limited number of aerial nematode parasites of plants and some of them, such as those that parasitise rice (Luc, 1990), are developing world problems for which the technology would be inappropriate.

Other Proteins Involved in Systemic Responses

Cucumber, muskmelon and watermelon all show systemic induction of acidic peroxidases (c 29 kd) in the intercellular spaces of leaves following inoculation of one leaf with the fungus *Colletotrichum lagenarium*. This is correlated with the enhanced defensive ability of these plants to lignify when infected by certain pathogenic fungi. It is uncertain if the triplet defined represents three genes or is the product of a single locus undergoing differential glycosylation or interactions with phenolics (Smith and Hammerschmidt, 1988).

Infection of roots of potato plants with *Globodera rostochiensis* induced accumulation of a high abundance of the proteinase inhibitor proteins I and II in stems and leaves of the plant but not its roots (Bowles *et al.*, 1991). The effects are consistent with a similar effect restricted to stems and leaves when roots of this plant are wounded (Pena-Cortes *et al.*, 1988). Potato proteinase inhibitors I and II are well characterised serine proteinase inhibitors from non-homologous gene families and as such form two of ten known families of plant proteinase inhibitors (Ryan, 1990, 1992). They are characteristically associated with plant storage organs and considered to be active against some pathogens and insect herbivores (Ryan, 1990, 1992). There is no evidence to-date that they serve as protectants against even those nematodes that specifically attack storage organs. Further work may establish that certain PIs do have a natural protective role against at least foliar nematodes as well as insects.

The potential for manipulating peroxidases to lignify plant cells around migrating or feeding nematodes has not been explored. In contrast, proteinase inhibitors have been shown to be valuable transgenes against feeding plant-parasitic nematodes and are considered in detail below.

The Systemic Role of Reactive Oxygen Intermediates

Inoculation of *Arabidopsis* with an avirulent strain of *Pseudomonas syringae* causes a hypersensitive response at the site of inoculation. This is due to the generation of reactive oxygen intermediates, O_2^- and H_2O_2. It also causes rapid induction of secondary oxidative bursts in small groups of cells in distant tissues including leaves other than those originally challenged. The primary oxidative bursts induce the systemic responses and it seems both are required for systemic immunity (Alvarez *et al.*, 1998). The reactive oxygen intermediate O_2^- is too reactive to disperse effectively but H_2O_2 may provide a mobile signal. It sets up progressive secondary oxidative bursts that serve to amplify the initial response. This provides further signals of reactive oxygen intermediates to potentiate reiteration of the signal (Alvarez *et al.*, 1998). It is not yet known if this role for nitric oxide occurs on nematode invasion or in the root system of plants.

Nematode Elicitors of Plant Responses

R-gene responses are species specific whereas some of the other defences are general. It follows that elicitors of R-gene responses are likely to differ from those that induce a general response. Glycoproteins have been localised to the head region by lectin binding studies (Forrest and Robertson, 1986; Dürschner-Pelz and Atkinson 1988; Robertson, W.M. *et al.*, 1989) and may elicit the response. In animal-parasitic nematodes glycoproteins in the surface coat have an important role against the mammalian defences (Philipp, Parkhouse and Ogilvie, 1980; Kennedy, 1991). They may have a similar role for nematodes in soil in protecting them from bacterial (Davies and Danks, 1992) or fungal attack (Jannson, 1994). There is some evidence that this surface coat can be shed by the nematode (Badley *et al.*, 1987; Blaxter *et al.*, 1992; Dürschner-Pelz and Atkinson 1988; Lee, Wright and Shivers, 1993; Politz and Philipp, 1992), possibly in response to environmental stress, antibodies or host defence cells. Therefore they may have a similar role in protecting the nematode from certain plant defences. If this role is essential, they provide a reliable elicitor for plant responses. Other possible origins for elicitors are the mechanical damage caused by the stylet or the release of plant basic, intracellular β-1,3-glucanases resulting from stylet penetration. Cyst nematodes also release β-1-4 endoglucanases that are presumed to facilitate cell wall degradation during migration (Smant *et al.*, 1998). Their action could result in release of plant cell wall fragments that elicit the general response.

Genes for Resistance

Molecular Background to Resistance Genes (R-genes)

The active defences of plants can be sub-divided into three types. The primary response is localised to the cells in contact with the pathogen. Its main outcome is that programmed plant cell death occurs. The secondary response is induced in cells adjacent to those in contact with the pathogen. The third component is systemic acquired resistance which occurs throughout the plant (Hutcheson, 1998). However, both these latter two responses can be mounted by plants that lack the appropriate R-gene and so it is the primary response that is an essential consequence of R-gene activity.

Van der Plank (1975) characterised single dominant genes as conferring specific, vertical resistance as apposed to horizontal resistance which he envisaged as polygenic and more likely to provide durable resistance. The gene for gene hypothesis is widely viewed as providing an adequate genetic description of most differential host-pathogen interactions and polygenic resistance is not considered to be a fundamentally different type of resistance (de Wit, 1992). It envisages that for every host gene for resistance a pathogen may have a countering virulence gene. The hypothesis has been applied to nematode/plant interactions (Jones, Parrott and Perry, 1981; Fleming and Powers, 1998).

Progress is being made on understanding single gene resistance at the molecular level and presumably attention will turn to a simple, polygenic system once knowledge has improved. The molecular recognition events resulting in the primary response may depend on the type of pathogen considered. A main distinction can be anticipated between necotrophs and biotrophs. Among the latter there may also be a considerable difference depending on the locale of the pathogen. It may be within the plant cell or outside its walls. As a consequence R genes vary. At least 5 classes of R-genes are known that have different activities (Hutcheson, 1998). Pathogen recognition is often within the plant cell. In the case of viruses the avirulent product may be made by the pathogen within the plant cell. The coat protein of Tobacco Mosaic Virus seems to have a conserved central hydrophobic cavity plus surface features that vary at its periphery. This defines the elicitor and its specificity (Taraporewala and Culvert, 1997). This suggests the R-gene carries motifs or domains with a role in protein:protein interactions. Many pathogens are sited in the apoplast. In the case of the bacterium *Pseuodomonas syringeae*, a resistant response involves two distinct genes. The avirulence gene product from the bacterium is probably moved into the plant cell by a protein translocation complex (Gopalan *et al.*, 1996; Puri *et al.*, 1997). The R-gene against the fungus *Cladosporium fulvum* has a receptor for the avirulence product located outside the plant cell. It also has a membrane spanning domain and an effector system within the cell. The pathogen recognition domain of several R-genes seems to be a leucine rich repeat region. The recognition of two distinct pathotypes is switched when the extracellular domains of two similar R genes (Cf-4 and Cf-9) from tomato against *C. fulvum* are swapped. Leucine rich repeats are known to be involved in many protein:protein interactions (Kobe and Deisenhofer, 1995). Mutations in this region of a flax resistance gene (M) cause loss of specificity to rust races without any detectable new resistance specificity (Anderson *et al.*, 1997). Possibly, R genes may occur as dimers that may help provide a recognition site in the LRR regions. Known R genes can be grouped into 5 groups. They are characterised by different domains that are likely to initiate the cascade of events that follow pathogen recognition (Hutcheson, 1998). One class is represented by *pto*. It has a protein with LRR and a leucine zipper but requires a second protein *prf* to initiate the cascade of events. A second class has an LRR, a nucleotide binding site and a leucine zipper. The nucleotide binding site probably interacts with ATP. The third group differs in that a signal transduction domain replaces the leucine zipper. The remaining two groups differ in having membranes spanning domains with the leucine rich repeats outside of the cell. One has a SER-THE protein kinase domain within the cell (Song *et al.*, 1995) and the other is assumed to interact with a separate kinase (Hutcheson, 1998). *Pto* is known to interact with a Ser-Thr protein kinase and to initiate a cascade of phosphorylation of three transcriptional factors. They have the capacity to bind to specific sequences in the promoters of genes encoding PR-proteins and possibly other defence-related genes (Zhou, Tang and Martin, 1997). This example provides a clear basis for initiating a cascade of events involving post-translational protein activation but other R-genes may initiate the response through a different cascade of events.

An intracellular signal transduction pathway is triggered as the result of pathogen recognition (Dixon, Harrison and Lamb, 1994). Effects that commonly occur as part of a resistant response are callus and lignin synthesis, an oxidative burst, phytoalexin induction, salicylic acid synthesis, PR protein induction and cell death. Salicylic acid synthesis and PR protein induction are clearly part of the secondary or systemic acquired resistance response. The primary effect of the hypersensitive response may be cell death. The events may initiate programmed cell death, which is a normal response of some cells in plant development. For instance it is important in the formation of xylem. Possibly there is a class of nuclear resistance genes that initiate programmed cell death (Hutcheson, 1998). However, it has yet to be proved that cell death plays a functional role in the hypersensitive response rather than occurring as an anticlimax event after the pathogen is arrested (Gilchrist, 1998).

Programmed cell death may occur within a few hours in R-gene mediated pathogen recognition. Early events are lipid peroxidation, K^+ efflux and Ca^{2+} influx followed by a membrane depolarisation and an oxidative burst. Changes in cell morphology occur, decondensation and vacuolisation of the cytoplasm with mitochondria remaining functional. Endonucleases are induced and chromosomal cleavage may occur (Hutcheson, 1998). There are concomitant changes in gene expression but it is unsure how many are primary rather than secondary consequences of cell death process.

R-genes often occur in clusters within the genome and this is clearly a common feature among them. For instance Southern blotting suggests the *Mi* gene conferring resistance to *M. incognita* in tomato (see below) occurs in a cluster with about 6 other similar sequences (Williamson, 1998). Recombinations between these genes is one possible mechanism by which new R-genes specificity arises (Crute and Pink, 1996). If so it is possible that very distinct R-genes could emerge with resistance to nematodes.

Resistance to Nematodes

There are several examples of single gene resistance genes that are commercially effective against plant-parasitic nematodes. A single dominant R-gene (*Gro-1*; formerly *H1*) from *Solanum tuberosum* ssp. *andigena* confers resistance against many European populations of *G. rostochiensis* and is widely used commercially. It does not confer resistance to a second potato cyst-nematode, *G. pallida* (Dale and Scurrah, 1998). A gene (*Mi*) from a wild tomato species *Lycopersicon esculentum* confers resistance to several species of *Meloidogyne* (Williamson, 1998). The commercial varieties including *Mi* have been a great commercial success. Its principal limitations are: (i) an inability to control *M. hapla* which can, as a result, show increased prevalence in the field when this R-gene is used; (ii) the possibility of selecting virulent populations of *M. incognita* and; (iii) loss of resistance at high soil temperatures. Other Mi-genes are known that differ in the range of species that they recognise. There are also differences in the maximum temperature at which they are heat stable (Williamson, 1998).

Tomato has a relatively small genome and it is well mapped with many phenotypic characters. The *Mi* gene is linked to a gene (*Aps*-1) that expresses an acid phosphatase (Williamson and Colwell, 1991). Conventional plant breeders have used this isozyme of acid phosphatase as a reliable phenotypic marker for *Mi* for many years. This enhanced the rate of progress for conventional plant breeding by reducing the need to refer to nematode challenge after all crosses. Conventional plant breeding effort has accumulated a number of introgression lines with small introgressions associated with Mi. These can be identified because of a good knowledge of tomato genetics and of abundant polymorphisms between the two parent species. As a result *Mi* was localised to a region of chromosome 6. These plants have provided an essential resource from which to develop a strategy for cloning Mi. Cultivars of susceptible tomato are very similar and show little RFLP differences in this region whereas there are several differences between this species and DNA of *L. peruvianum* (Williamson, 1998). A number of RFLP markers have been obtained for the *Mi* region (Klein-Lankhorst, *et al.*, 1991; Messeguer *et al.*, 1991; Ho *et al.*, 1992). Messeguer *et al.* (1991) established the region of interest was an apparent 0.4 cM between flanking markers but suppressed segregation occurs. A more detailed mapping of this region has been developed using molecular markers. Different resistant lines have been shown to include dissimilar sizes of the introgressed *L. peruvanium* DNA segment that is present in the genome near the *Mi* gene. Initially an attempt was made to chromosome walk from a known location to *Mi* (Liharska and Williamson, 1997). However, this need was circumvented by further definition of the chromosome fragment of interest. Certain lines were identified with nematode resistance that lacked *Asp-1* but contained a DNA marker that is closer to Mi. The absence of *Asp*-1 indicates a small introgression, which was shown to be 650 kb. *Mi* was localised to a region of only 65 kb region by two approaches (Kaloshian *et al.*, 1998). A large number of F2 progeny of tomato lines was analysed with *Mi* inserts and similar studies were completed for intra-specific cross of *L. peruvanium*. Tomato DNA spanning the locus was isolated as bacterial artificial chromosome clones. A 52 kb fragment of DNA was sequenced and found to contain pseudogenes and two apparently intact genes termed Mi-1.1 and Mi-1.2. The latter alone is sufficient to confer resistance to *M. incognita* in a susceptible tomato (Milligan *et al.*, 1998).

Mi (Mi1.2) predicts a 1257 amino acid polypeptide. It contains from its N′ terminal end: a leucine zipper, a nucleotide binding site, a central hydrophobic domain and a leucine rich repeat domain. Williamson (1998) considers it most similar to *prf* gene of tomato that confers resistance to the bacterium *Pseudomonas syringea*. If so, it may require a *pto*-like protein to function. Alternatively it could be similar to another class of R-genes typified by *Rpm*1. This also provides resistance to certain races of *P. syringea* but may occur as a homodimer with no interaction with a *prf*-like gene product

(Hutcheson, 1998). Both these classes of R-genes are localised in the cytoplasm. This suggests that the R-gene recognises a protein injected into the plant cell by the nematode. This would be coincident with the onset of the HR responses occurring when feeding starts. It seems more likely that a delivery system is involved that moves the avr products from the apoplast to cytoplasm, as seems likely for *Rpm*1. Surprisingly, *Mi* did not confer resistance in tobacco when introduced as a transgene (Williamson, 1998, see earlier). This could be for many reasons but it may be that an appropriate prf-gene product or delivery system is required that tobacco lacks. The expectation is that further work will result in achieving a basis for transferring this R-gene to a range of different crops. *Mi* is unusual in also conferring resistance to the aphid *Macrosiphum euphorbiae*. This is the first anti-insect R-gene cloned and the first to confer resistance to two such distinct groups of organisms as nematodes and insects (Vos *et al.*, 1998; Rossi *et al.*, 1998).

H. schachtii

The first R-gene to be cloned that confers resistance to a nematode is *Hs1*[pro-1] from *Beta procumbens*, a wild species of beet. The approach followed was elegant but followed an approach that may not readily be adopted for other R-genes. *Hs1*[pro-1] confers resistance to *H. schachtii* (beet cyst-nematode). No commercially used beet varieties of *Beta vulgaris* carry a natural resistance gene to this nematode. Crosses of wild beet species with *B. vulgaris* provided some monosomic addition lines that had an extra wild beet chromosome. These lines showed resistance to *Heterodera schachtii* but the beet was not commercially viable because of the other additional genes that were added. Diploid genotypes were isolated from the monosomic material using irradiation to ensure efficient translocation events. By selecting a range of translocation lines, all carrying the same R-gene, it proved possible to identify and map the chromosome breakpoints and overlapping regions between lines (Jung *et al.*, 1992; Salentijn *et al.*, 1994). Molecular markers were used to identify a diploid resistant line, called A 906001, that carries the smallest wild beet translocation fragment. A yeast artificial chromosome (YAC) (Schlessinger, 1990; Albertsen *et al.*, 1990) library was generated from this line. Three YAC clones corresponding to the translocated DNA were identified and shown to correspond to a 300 kb region of *B. procumbens* DNA (Kleine *et al.*, 1995). Identification of transcripts corresponding to the R-gene was achieved by screening a root cDNA library from nematode infected A906001 plants, with the three YAC clones as probes. One cDNA, (#1832) isolated was exclusively present in the DNA of resistant lines and cosegregated with the resistance trait. It corresponded to a nematode inducible 1.6 kb transcript, and had putative protein domains that are characteristics of known resistance genes. Most significantly, the transfer of the *Hs1*[pro-1] gene to a susceptible beet and assay of hairy roots demonstrated that resistance is conferred by the 1832 cDNA (Cai *et al.*, 1997).

The protein encoded by cDNA 1832 which corresponds to $Hs1^{pro1}$ has an N-terminal signal sequence which probably leads to cytoplasm membrane localisation. There is a leucine rich region followed by a transmembrane domain and a C-terminal region. This shows no consensus features found in other R-genes (Cai *et al.*, 1997). This region may not initiate a resistance cascade directly or interact with a protein kinase or an intracellular signalling component. The extracellular leucine rich repeat domain suggests that the avirulence gene product is extracellular. This may be a secretion from the amphids or some other source or it could be a component of the nematode surface. It could also be an oral secretion that interacts with the plasma membrane during feeding.

Resistance Genes to *Globodera*

Evidence is accumulating that R-genes have conserved regions. Therefore PCR-based cloning is potentially possible using oligonucleotides to these regions. Using this approach PCR-derived clones were mapped onto the potato linkage map (Leister *et al.*, 1996). Among regions of interest was a small gene family on chromosome VII. These fragments co-segregated with *Gro1* (formerly termed H1) the dominant nematode resistance gene. No segregation was found in over 1000 plants of *Gro1* and the gene identified by the PCR-based approach. The resolution achieved is about 10 kb and R-genes in other species are typically 6 kb. It follows that *Gro1* has been localised. The approach identified genes with a nucleotide binding domain. Therefore when *Gro1* gene is characterised it may prove similar to *Mi* but different from $Hs1^{pro-1}$ in its range of domains. This information will determine variation in R-genes that are suitable for both nematode recognition and development of a resistance response.

Virulence and Avirulence

There is evidence to support the gene-for-gene hypothesis from molecular manipulation of pathogens. Mutations such as deletion or point mutation(s), or insertional inactivation of the avirulence gene lead to alteration of an avirulent to a virulent pathogen (Hammond-Kosack and Jones, 1995). An example of loss of host resistance due to single point mutations has been documented for the *avr4* gene of *Cladosporium fulvum*, leading to virulence on tomato (Joosten, Cozijnsen and Dewit, 1994). The gene-for-gene interaction envisages a plant with several R-genes, each with a recognition domain specific for a given pathogen or form of that organism. For example, tomato contains several R-genes directed against different races of the biotrophic fungus *C. fulvum* (Hammond-Kosack and Jones, 1995).

If a principal role for R-genes is pathogen recognition, it follows that a recognition domain is required in the apoplast or within the cell. The appropriate locale may depend on the nature of the plant: pathogen interaction. No avirulence gene product of a nematode has yet been characterised and shown to elicit a response. This has been achieved for other pathogens. For instance a resistant response for tomatoes carrying the *Cf-9* R-gene is elicited by *Cladosporium fulvum* isolates carrying the avirulence gene avr9. This is a cysteine rich, 28–amino acid peptide (Kooman-Gersmann *et al.*, 1997). The avirulence products could be continually present in association with the nematode surface, including secretory products, and presumably are essential to the parasite if avirulence selection is not to occur readily. Oral secretions into the plant cells would also be good candidates for plant recognition. Clearly the plant requires a reliable signal; one that is always associated with the pathogen, is reliable and cannot be readily elicited by inappropriate pathogens or self.

The Hypersensitive Response at Nematode Feeding Sites

The first indication of a resistant response for tomatoes carrying *Mi* was 12 hours after first invasion by *M. incognita*. An electron microscopic study revealed a change in electron density of the cytoplasm of plant cells surrounding the nematode. By 24 hours, dispersed electron-dense material occurred in the cell vacuoles. The cytoplasm was disorganised a further 24 hours later (Paulson and Webster, 1972). Necrosis develops around the nematode and can be visualised by light microscopy (Ho *et al.*, 1992). The resistant response of a *Raphanus sativas* cultivar to *H. schachtii* was evident from 3 days after infection. The endoplasmic reticulum in the incipient syncytium had a different appearance and cells had larger vacuoles than for cultivars supporting a compatible interaction. These changes were clearly after feeding cell induction (Wyss, Stender and Lehmann, 1984). The invasion of potato cv Maris Piper carrying the Gro-1 gene by *G. rostochiensis* (pathotype Ro1) also results in an incompatible interaction. The first difference between the susceptible and the resistant response was evident as enlargement of nuclei and cytoplasmic degeneration in cells within the vicinity of the nematode. Initiation of the feeding site resulted in cytoplasmic contents filling the initial feeding cell in a susceptible response only. Necrotic cells were evident around the syncytium formed in resistant plants within 48 hours. Two days later, other cells have been incorporated but their cytoplasmic contents had not proliferated. The cytoplasm of the syncytium was extremely electron dense and necrotic cells surrounded the syncytium. At seven days post-invasion, the syncytium in the resistant plant lacked the wall growths that were expected for transfer cell function in susceptible potato plants (Rice, Leadbeater and Stone, 1985).

Transgenic Crop Resistance Based on R-genes

There is clearly scope for the transfer of natural R-genes to related crops which may be expected to harbour suitable intracellular signalling pathways and therefore be competent to mount a resistance response. However, this may not guarantee durable plant defence as pathogens may develop resistance thereby causing breakdown of a previously effective R-gene response. An example of this problem concerns the *Mi* gene of tomato which confers resistance to *M. incognita*. Increased recent use of the R-gene in commercial tomato production in California has resulted in incidences of resistance-breaking (Kaloshian *et al.*, 1996). Populations of *M. javanica* on Crete have also been reported to break *Mi* resistance (Tzortzakakis and Gowan, 1996). Increased use of an R-gene such as *Mi* by transfer to other crops would require careful management to ensure its frequency of use in any field was not excessive. The risk is providing severe selection pressure that favours either rare individuals of a *Meloidogyne* spp. or a change in virulence of individual nematodes.

A somewhat different problem is the specificity of the R-gene. The *Gro1* resistance gene of potato confers excellent resistance against *G. rostochiensis* but cannot check the increasing problem of *G. pallida* control. Molecular characterisation of this and other R-genes may provide clues on gene modification that alters pathogen range. Altering the specificity of recognition is an important aim for such work. It is of particular interest that natural exceptions to the 'gene-for-gene' hypothesis occur. The RPM1 gene of *Arabidopsis* displays dual specificity for two unrelated avr products of *P. syringae* pathovars (Grant *et al.*, 1995). *Mi* confers resistant to both certain *Meloidogyne* spp. and the aphid *M. euphorbiae* (Vos *et al.*, 1998; Rossi *et al.*, 1998). The principal reason for the inability of a resistance gene to effect control is not due to an inability to develop a hypersensitive response. This has been demonstrated by the transfer of the $Hs1^{pro-1}$ gene to other beet species (Cai *et al.*, 1997). This suggests the critical deficiency lies in the lack of recognition of the pathogen. An R-gene product interacts with a specific pathogen-elicited molecule and triggers the plant resistance response. If an alternative pathogen has a target molecule of differing properties, or even lacks it altogether, then no resistance response can be induced by the R-gene. For R-genes containing an LRR domain it is most likely that this domain confers specificity of the host:pathogen interaction (Ori *et al.*, 1997; Parker *et al.*, 1997). The challenge, therefore, is to alter host range by selecting forms of the R-gene that interact with suitable target molecules on the new pathogen. In the longer term such an ability to extend host range should allow efficient transfer of resistance genes across species barriers. They would allow the selection of new R-genes based on *in vivo* selection against the target pathogen. In this case the nature of the Avr product would be unknown until the engineered system was studied in detail. However, the potential for molecular evolution to identify new genes rapidly is a likely important future development of novel R-genes for transgenic resistance.

Perturbing Modified Plant Cells

Raising the feeding rate possible for the nematode helps enhance both its rate of development and the daily egg production of adult females. This favours parasitic success and is usually achieved by increasing the volume of the plant cytoplasm available to support the feeding nematode. A wide range of plant cell modifications occurs to achieve this end (Sijmons, Atkinson and Wyss, 1994; Wyss, Chapter 9). One host plant species may show very different plant cell modifications according to the nematode parasite. This suggests the pathogen plays a major role in defining the plant response.

Natural resistance genes provide a good model for the design of feeding site disruption. Some of these genes invoke a hypersensitive response leading to feeding cell death. This deprives the nematode of nutrients (Wyss, Stender and Lehmann, 1984; Rice, Leadbeater and Stone, 1985; Golinowski *et al.*, 1997; see earlier). The result is that some nematodes undergo a sex switch and become males that are less damaging to the plant. Sedentary species may have lost locomotory ability before resistance is expressed and so females cannot complete development and die. The essential nature of feeding cells to many economically important nematode species makes disruption of these cells an attractive target.

Destruction of the Feeding Cell

Plant cell modification is central to successful parasitism of many plant-parasitic nematodes (Wyss, Chapter 9). The cells must provide sufficient nutrition to support the rapid growth of the nematodes. Therefore there is an opportunity to disrupt their induction and maintenance so preventing successful parasitism. One approach is to cause death of plant cells that change as required by the nematode (see below). This mimics programmed cell death which seems central to the elimination of pathogens by natural resistance genes such as *Mi* and *Gro-1* function (see above).

A plant cell can be destroyed by expressing a phytotoxic gene product specifically in the feeding cells. Proof of principle demonstration of this approach can be achieved by using barnase, a small bacterial RNAse. The utility of barnase was first demonstrated for target cell death by emasculating maize. This was achieved by expression of the RNAse from a tapetum-specific promoter (Mariani *et al.*, 1990). However, a problem with barnase is its potency, which can lead to death of non-target cells through leaky expression. This problem can be overcome by use of the inhibitor protein, barstar (Hartley, 1988). The efficacy of barstar inhibition of barnase was also demonstrated in the maize system. Barnase- and barstar-expressing plants were crossed and resulting progeny showed restoration of fertility (Mariani *et al.*, 1992). The restorer, barstar has been expressed constitutively throughout the plant from the CaMV35S promoter to avoid non-specific barnase damage in non-target cells of the plant. Only in the feeding cell should the high-level expression of barnase overcome the inhibition by barstar thereby leading to cell death.

In its simplest form targeting the feeding cell depends on identifying a gene that is expressed in feeding cells at much higher levels than in other plant cells. Several groups have identified such nematode responsive genes (see feeding-site promoters). Their coding regions are of interest for gene silencing and antisense approaches and their promoters are of value for a wide range of approaches. The promoter can be used to control expression of an introduced phytotoxic inhibitor. Some resistance to both *M. incognita* (Opperman, Taylor and Conkling, 1994) and *H. schachtii* (Ohl, van de Lee and Sijmons, 1997) has been reported using this approach.

CaMV35S is not truly constitutive in all plant cells. This generates a flaw for an approach that relies on it providing activity of a restorer throughout the plant. This promoter is down-regulated in older parts of root systems. This would lead to a failure of the restorer (e.g. barstar) to protect such regions from non-specific effector (barnase) expression. It is not clear how much of problem this will prove to be for the efficacy of such effector/inhibitor systems. Perhaps a more important issue is a regulatory one. A constitutively expressed inhibitor protein throughout a plant may not be commercially acceptable.

A development of the approach is a two component system. This involves the production of initially inactive effector molecules that are subsequently activated upon interaction with a second component. A simple example would be an inactive nuclease or proteinase containing a highly specific protease site as a first component and a protease that recognises this site as a second component. Activation would occur only when the protease is co-expressed with the inactive effector. This approach avoids the need for constitutive expression of a protector protein, and therefore limits transgene expression. A key feature of such systems is that the promoters do not need to provide expression only within feeding cells, a situation that seems increasingly unlikely to be realised. Instead it is important that the promoters should display expression patterns that do not overlap except at the feeding cell. Thus expression of one component within the root meristem would not affect these cells if the second component is not expressed in this tissue. An active effector would be limited to the feeding site where the expression patterns provided by the two promoters overlap. The feeding site would be damaged without unwanted side effects on other plant cells.

Attenuating Feeding Cells

A related approach is to allow only an impaired feeding cell to develop. The objective is to ensure that the feeding site cannot provide the quantity of nutrients required by the parasites. Female cyst-nematodes predominate under favourable conditions whereas more males occur when the nutrient supply is limiting (Grundler, Betka and Wyss, 1991; Grundler and Böckenhoff, 1997). The size of the giant cell complex varies in host/parasite interactions of *M. incognita*. It is based on the size of the giant cell complex. The nematode induces giant cells in many very different plants, including both dicots and moncots, but its success is not equal for all hosts. For instance, one population of *M. incognita* formed a large giant cell complex of *Phaseolus vulgaris* (dwarf bean) by 26 days

post-invasion. In comparison the same population induced giant cell complexes at the same time that had achieved cross-sectional areas of only 0.5× and 0.05× this size on soybean and cabbage respectively. The body volumes of the parasites were smaller in association with giant cell complexes of reduced size. Body size influences fecundity of this nematode (Atkinson *et al.*, 1996). Their total egg production by this time on cabbage, soybean and dwarf bean was zero, 64.5 ± 9.2 and 199.7 ± 23.2 eggs/females respectively (Al-Yahya, 1993). Bird (1972) showed a similar effect for *M. javanica* on cabbage and showed that some eggs eventually resulted from parasitising this host. These differences in fecundity reflect differences in the cell volume from which the parasite can feed. Therefore a defence that limits the size of the giant cells will also limit the fecundity of *Meloidogyne* spp. This nematode requires multiple generations on most crops to reach economically important levels. Attenuating feeding cells may slow development. This will reduce population build-up and prevent economic status from being achieved. The same strategy can be anticipated to be of value for single generation nematodes providing development is slowed if the fecundity of the female is severely reduced by crop harvest. This approach limits the size of the feeding cell system and not the efficiency of absorption by the nematode. Therefore the nutrients taken up per parasite are also reduced and the incremental effect per nematode on crop loss should also be reduced.

Impairing the Efficiency of Nutrient Supply by Feeding Cells

The syncytium induced by cyst nematodes (Wyss, Chapter 9) resembles a plant transfer cell. They commonly occur at sink: source sites in plants. Nitrogen sources taken up by roots are transported in the xylem as nitrates, ureides and the amides asparagine and glutamine. The ingrowths that occur between the xylem and the syncytium are probably sites for proton (H+) pumps that withdraw nitrogen sources from the xylem stream into the syncytium. Here transamination presumably ensures a plentiful and diverse amino acid pool for protein synthesis. Therefore cyst-nematodes are not merely parasitising the plant cells they modify but are using a transfer cell system that allows them to draw more widely on the resources of the plant. Parasites arranged along a root are competing for nutrients from the xylem vessels. This provides a biological basis for the competition models developed for feeding females of cyst-nematodes (Seinhorst, 1967; Phillips, Hackett and Trudgill, 1991). Therefore there may be future potential in limiting the transfer capacity of the syncytium rather than just its physical size. The same approach can be extended to other nematodes such as *Meloidogyne*, although the giant cell provides nutrients in a distinct way.

Construct of cDNA libraries (see earlier) is gradually providing an indication of those gene products that are essential for feeding cell function. For instance, the giant cell complex of *Meloidogyne* spp. is associated with expression of a water channel protein (Opperman *et al.*, 1994; Opperman, Taylor and Conkling, 1994). There is a well-established basis for suppressing the abundance of messages of genes considered to be important to the normal function of particular plant cells. A transgene was inserted into the plant that provides a mRNA transcript in the reverse, antisense direction. This resulted in a lowered level of both sense mRNA and translated protein. There are several precedents of the antisense approach in plant biotechnology (Kumria, Verma and Rajam, 1998). Antisense constructs of TobRb7 have been reported to provide partial resistance to *M. incognita* (Opperman and Conkling. 1996). Virus resistance has been achieved using untranslatable sense RNAs (Smith *et al.*, 1994). Another mechanism used with viruses is homology-dependent silencing. Both gene products are suppressed if the introduced gene provides a product with a high similarity to that normally expressed (Mueller *et al.*, 1995). Possibly a gene could be identified whose silencing would not impair most plant cells but prevent the feeding cell from supplying all the nutrients required by the nematode.

Limiting Feeding Success

Plant cell walls provide a barrier to small organisms such as nematodes. Nematode size is limited in soil by their method of locomotion and body organisation. A larger example, such as *Xiphinema*, has a body diameter that is 2–3× that of a typical plant-parasitic nematode. This limits it to a low proportion of the soil volume in a typical sandy soil (Jones, Larbey and Parrott, 1969). The stylet of nematodes overcomes the problem of the plant cell wall as a barrier. It provides a means of removing contents by piercing cell walls. This process may be aided by enzymatic activity. With the limited exception of trichodorids, the stylets of plant-parasitic nematodes are hollow. Variation in stylet length among ectoparasites allows either different plant root tissues or roots of different thickness to be attacked. Variation occurs in stylet architecture among tylenchids. This is correlated with the cell wall thickness of cell types attacked. Small size enables some nematodes such as cyst-nematodes to migrate through the plant cells to their feeding sites.

Once food is ingested it passes through the pharynx to the intestine where absorption occurs. Presumably digestive enzymes are produced in the intestine and some may also pass from the sub-ventral pharyngeal glands posteriorly to the intestine. Sedentary nematodes such as cyst nematodes with raised feeding rates have modified intestines but our understanding of nutrient absorption in these animals is inadequate.

Parasitism of plants results in considerable morphological changes for sedentary tylenchids. Locomotory ability is lost and the body wall muscles atrophy. The nematode undergoes two changes in shape. First the female swells to a saccate form and then the sub-sphere increases in diameter so that body width and length increase at about the same rate (Atkinson *et al.*, 1996). The structure of the cuticle becomes very much altered in these females (Bird and Bird, 1991). Cuticular gene expression changes can be correlated with the increase in both body length and width of sedentary females. For *Meloidogyne*, all these changes occur after the last moult and the cuticle must encompass a considerable increase in surface area. This is a very different need from that required for locomotion (Alexander, Chapter 13).

Secretions Associated with the Feeding Process

Once the parasite has completed invasion there are two important processes for those nematodes that modify plant cells. First the feeding site must be initiated. It is also likely that secretions are required to maintain it. The secretions from the unicellular gland cells of the pharynx are important in this process. Granules are budded from the cisternae of the trans-Golgi apparatus (Hussey and Mims, 1991). Those from the sub-ventral glands seem to merge (Endo, 1993). An antibody specifically recognising an epitope in the secretory granules of the subventral gland of *M. incognita* was used to isolate a cDNA clone from an expression library. The corresponding gene, *sec-1*, was cloned and the encoded protein shows slight homology to myosin heavy chains. It may play a role in the movement of secretory granules within the pharyngeal gland cells (Ray, Abbott and Hussey, 1991). Actin cables needed for such a system have not yet been identified. Microtubules in the gland extension (Endo, 1993; Hussey and Mims, 1991) probably direct granule transport as in other secretory cells (Larsen *et al.*, 1993). All three unicellular pharyngeal gland cells have neuronal process to them and these provide for regulated secretion. Each granule may contain more than one secreted protein with pre-assembly into each granule at the *trans*-Golgi bodies. Granules with different contents may also occur.

Monoclonal antibodies have been identified with specific reactivity against gland contents of *H. glycines*, *M. incognita* and *G. rostochiensis* (Atkinson *et al.*, 1988; Hussey, 1989b; Goverse, Davis and Hussey, 1994; De Boer *et al.*, 1996). The cloning of β-1-4 endoglucanase from cyst-nematodes (Smant *et al.*, 1998; see earlier) provides an example of the experimental approach that is likely soon to reveal other proteins secreted by nematodes. This approach provides a powerful basis for developing understanding of nematode/plant interactions. There is a possibility of disrupting the process by blocking the function of essential secretions. Many may be enzymes. Protein inhibitors of enzymes provide one approach. Urwin, McPherson and Atkinson (1998) provided evidence that a serine proteinase inhibitor may affect cyst nematodes without entry into the animal. The nematode is known to secrete proteinases and this may represent inhibition of such a secretion.

The first attempt to achieve resistance by expressing an antibody with reactivity against a pharyngeal gland of a nematode was unsuccessful (Baum *et al.*, 1996). Antibodies of 110 kDa that recognise secretory granules of *G. rostochiensis* have also been expressed in potato

(Stiekema *et al.*, 1997). Clearly the antibody must bind to at target of animal or plant origin in a way that disrupts parasitism. Secretions of nematodes into plants are obvious targets. Heterologous expression of protein secreted by a nematode and assays of its enzymatic activity as achieved for β1-4 endoglucanses (Smant *et al.*, 1998) provides a basis for demonstrating if an antibody blocks enzyme function before plant transformation is attempted. Targets within the intestine have also been successfully targeted in the animal-parasitic nematode *Haemonchus contortus*. An antigen prepared from one of its gut membrane glycoproteins was used to suppress the parasite by vaccinating the host (Andrews, Rolph and Munn, 1997). Sites within the plant-parasitic nematode are likely to be inaccessible to large antibodies (see later) but the value of antibodies as anti-pathogen proteins is established. However, single chain antibodies (Winter and Milstein, 1991) are typically only 26 kDa or more (Bird *et al.*, 1988) and so may be ingested from the cytosol by feeding plant-parasitic nematodes. A transgenically-expressed single chain antibody provides some protection against a virus (Taviadoraki *et al.*, 1993) and so provides a paradigm for further work. The future potential of the approach seems to depend primarily on defining an antibody that can disrupt parasitism. It is also uncertain if transgenic plants expressing even modified animal genes will be acceptable as crops.

Feeding Tubes as a Target for Disruption

Feeding tubes have been reported for a wide range of plant-parasitic nematodes including *Globodera*, *Heterodera*, *Meloidogyne*, *Rotylenchulus*, *Pratylenchus*, *Trichodorus*, *Helicotylenchus* and *Scutellonema* (see Rebois, 1980; Ruppenhorst, 1984; Wyss and Zunche, 1986; Hussey and Mims, 1991). They differ in appearance. The feeding tube of female *M. incognita* has received the most complete study. It is typically 70 μm in length and 1 μm wide with a hollow bore of about 0.35 μm and a closed distal end. Feeding tube assembly has been recorded *in planta* using high resolution video for a cyst-nematode and occurs *in vitro* under defined conditions (Cohn and Mordechai, 1977). Its wall has a different structure in cyst and root-knot nematodes (Wyss, Chapter 9). In *Meloidogyne* spp., it has a crystalline structure that is also evident for its precursor granules. They are secreted from the dorsal pharyngeal gland cell before they emerge from the stylet. The feeding tube of *Meloidogyne* is surrounded by endomembranes of the giant cell and has a similar ultrastructure to smooth endoplasmic reticulum. It may have a role in synthesising and transporting nutrients to the parasite (Hussey and Mims, 1991).

An unusual feature of feeding tubes is the suggestion from ultrastructural evidence that the plant cell membrane remains in place between the feeding tube and stylet (Hussey and Grundler, 1998). This requires all the nutrients, that may represent >50% of the feeding cell volume, to pass through the plasma membrane. This must occur in that part of the feeding cycle when ingestion occurs. Presumably it has a similar area to that of the feeding tube bore, which in *M. incognita* is only c0.01 μm^2. This seems a very high rate of flow to pass though a plasma membrane of this surface area. Possibly the membrane is highly modified by the nematode in this small region and has very different permeability characteristics to other regions of the plasma membrane. In addition, the fluid nature of lipid membrane is such that repair would be immediate once the stylet is withdrawn. Possibly, preparation for EM study ensures closure even if the membrane is temporally ruptured during inward flow of nutrients. In the latter state it may resemble a modified plasmodesmata which does allow flux between the contents of adjacent cells. The difference is that plasmodesmata have an exclusion limit of 800 Da (Leegood, 1998). There is a barrier to macromolecular uptake by nematodes but it is a higher value than characterises plasmodesmata and may occur at the feeding tube. The feeding tube of *H. schachtii* may act as a molecular sieve to exclude dextrans of 40 kDa but not 22 kDa (Böckenhoff and Grundler, 1994). Proteins of 22 kDa and 28 kDa, but not 11 kDa, are also excluded (Urwin *et al.*, 1997a; Urwin, McPherson and Atkinson, 1998). In particular, *H. schachtii* does not take up green fluorescent protein (28 kDa) from its feeding cell, in contrast to *M. incognita*.

The particular features of feeding tubes are not unique to them. Proteins that self-assemble as tubes with central bores are also known. One example is myosin although this has a reduced length and diameter relative to the feeding tube. Proteins that act as exclusion filters also occur. Lamimin is a modified collagen in the mammalian kidney that limits macromolecule uptake. One structure of particular interest in understanding the roles of the feeding tube may be the peritrophic matrix that surrounds food in the insect intestine. It is often a chitin/protein matrix but it can be mainly protein (Lehane, 1997). The exclusion limit ranges from 6 kDa to 200 kDa according to insect species. A similar layer, although probably lacking chitin, occurs in some nematodes (Borgonie *et al.*, 1995). Comparison with the peritrophic matrix suggests some possible roles for the feeding tube. These include acting as an impermeable membrane to exclude macromolecules, as a defence against ingested, pathogenic microorganisms. Another potential role for the feeding tube could be selective uptake of solutes. For instance, the kidney glomerular basal membrane hinders passage of anionically charged molecules. There is also a possibility that the feeding tube acts as a reverse osmosis membrane allowing solute uptake without substantial flow of water. The peritrophic matrix of some insects apparently allows nutrient uptake from watery sources by this mechanism (Lehane, 1997). It would provide a means for nematodes with feeding tubes to remove solutes without perturbing feeding cell volume. Solute exclusion during solvent passage may deposit a macromolecule film on the outer surface of the feeding tube. This would require its continual replacement and so have a parallel with the continuous production of the insect peritrophic matrix. This turnover may help overcome its progressive obstruction.

The feeding tube is an obvious target for disruption. It is common to many plant-parasitic nematodes and unique to them. It seems essential to the vital process of feeding and may be disrupted by means that have no adverse consequences for the plant or other organisms. The most appropriate approach for disrupting its function is likely to emerge as our basic understanding of the feeding tube functions emerge. It may involve expression in the plant of a single chain antibody that disrupts an essential function.

Inhibiting Digestive Proteinases

Work on proteinases of cyst-nematodes has dominated study of digestion by plant-parasitic nematodes at the molecular level to-date. The success achieved in developing a novel basis for resistance may be extended soon to other digestive enzymes.

Proteinases

There are four main groups of proteinases, based on their structure and mode of action (cysteine, serine, aspartic and metallo-proteinases (Rawlings and Barrett, 1993)). All four classes of proteinases occur in parasitic helminths but cysteine proteinases have received most study (Coombs and Mottram, 1997). The various roles of parasite proteinases are of particular interest because they represent good targets for development of new anthelmintics (Coombs and Mottram, 1997). The rational is that structural information gained from such studies, combined with the use of recombinant enzymes for screening inhibitors, can lead to the development of potent inhibitors with therapeutic value. The modelled structure of an elastase from the trematode *Schistosoma mansoni* has been used to design inhibitors with antiparasitic activity (Cohen *et al.*, 1991; Ring *et al.*, 1993) whilst expressed recombinant proteins have been used to identify cysteine proteinase inhibitors with activity against schistosomes (Wasilewski *et al.*, 1996).

Irrespective of the various feeding strategies adopted by nematodes there is the common need to meet the nutritional requirements of the animal. Unless nematodes can rely on a supply of free amino acids apparently available to them (Grundler, Betka and Wyss, 1991), protein in food must be digested to provide an adequate amino acid supply for protein synthesis. This aspect of metabolism is likely to be common to most plant-parasitic nematodes.

Genes encoding cysteine proteinases with cathepsin B- and L-like activities have been isolated from a range of animal-parasitic nematodes, including *Haemonchus contortus* (Pratt *et al.*, 1992), *Ostertagia ostertagi* (Pratt, Boisvenue and Cox, 1992) and *Ancylostoma caninum* (Harrop *et al.*, 1995). Localisation of cysteine proteinases to the intestine of a number of parasites supports a proposed role in digestion (Chappell and Dresden, 1986; Maki and Yanagisawa, 1986; Smith *et al.*, 1993). Cysteine proteinases are also abundant in the microbivorous nematode *Caenorhabditis elegans* (Larminie

and Johnstone, 1996; Ray and McKerrow, 1992) and the expression of at least one of the genes is restricted to the intestine, as judged by *in situ* hybridisation (Ray and McKerrow, 1992).

G. pallida has cysteine proteinase activity in homogenates of feeding females (Koritsas and Atkinson, 1994). More detailed studies have since used specific peptide substrates to localise and characterise the proteinase activity of the soybean cyst nematode *H. glycines*. Incubation of nematode cryosections with Z-Ala-Arg-Arg-4MNA localised a cysteine proteinase activity to the intestine of females. The activity was present throughout the intestine in both fusiform animals and saccate females prior to egg deposition. The cysteine proteinase activity was restricted to the regressing intestinal tissue as gravid females accumulated eggs. By using a range of specific synthetic substrates the activity was classified as cathepsin L-type (Lilley *et al.*, 1996). It was active within the intestine of *H. glycines* from late stages of the first parasitic stage to the gravid female. This suggests that the enzyme has a digestive role throughout parasitism. The cathepsin L-like activity could be completely inhibited in nematode sections by pre-incubation with the protein engineered rice cystatin Oc-IΔD86 (see later). It was unaffected by pre-treatment with either BSA or an inactive variant of Oc-I. This work also demonstrated that intestinal serine proteinase activity of *H. glycines* could be inhibited by a serine proteinase inhibitor, cowpea trypsin inhibitor, CpTI (Lilley *et al.*, 1996).

Michaud *et al.* (1996) used mildly denaturing gelatin/ polyacrylamide gel electrophoresis to establish the strength of interaction between PIs and proteolytic enzymes present in nematode homogenates of *Meloidogyne* spp. The studies revealed that *Meloidogyne hapla* females possess predominantly cysteine proteinase activity but other proteinase classes are also present in *M. incognita* and *M. javanica*.

Progress has also been made on the molecular characterisation of the proteinases of plant-parasitic nematodes. Degenerate primers designed from conserved regions of proteinase genes in other eukaryotes have previously been used to amplify fragments of proteinase genes from *C. elegans* and parasitic helminths and protozoa (Sakanari *et al.*, 1989; Eakin *et al.*, 1990; Baylis *et al.*, 1992; Rosenthal and Nelson, 1992; Heussler and Dobbelaere, 1994). They have been used to amplify gene fragments with homology to cysteine proteinases from *H. glycines*, *H. schachtii*, *G. pallida*, *M. incognita* and *M. javanica* (Lilley *et al.*, 1996: Lilley, Atkinson *et al.*, unpublished observations). The DNA fragment amplified from *H. glycines* was used to screen a cDNA library prepared from feeding females of this nematode. It resulted in two categories of clones among the large number obtained (Urwin *et al.*, 1997b). The more abundant type of clone encoded a protein of 374 amino acids (HGCP-I) with homology to cathepsin L-like cysteine proteinases. A less abundant clone encoded HGCP-II, a protein of 353 amino acids with highest homology to bovine and human cathepsin S enzymes. As with most cathepsins, the *H. glycines* cysteine proteinases seem to be synthesised as precursor molecules in a prepro-format. Each has a short signal sequence, a long pro-region and a mature proteinase of 219 amino acids. The mature protein regions of HGCP-I and HGCP-II have 63% identity although this falls to 49% if the more divergent pro-sequences are included in the comparison. Southern blot analysis suggested that *H. glycines* may have a third gene with homology to *hgcp-I* but not *hgcp-II* (Urwin *et al.*, 1997a). Screening of a *G. pallida* cDNA library with *hgcp-I* and *hgcp-II* has since led to the isolation of potato cyst nematode homologues of *hgcp-I* (C. Lilley, unpublished observations).

Serine proteinase activity has been identified from a number of parasites, including *Necator americanus* (Brown *et al.*, 1995) and *Toxocara canis* (Robertson, B.D. *et al.*, 1989), but few have been characterised at the molecular level. Genes encoding elastase (Newport *et al.*, 1988; Pierrot, Capron and Khalife, 1995) and a kallikrein-like serine proteinase (Cocude *et al.*, 1997) have been isolated from *Schistosoma mansoni*. A gene for a subtilisin-like serine proteinase is also known for *Plasmodium falciparum* (Blackman *et al.*, 1998). Currently, the serine proteinase gene of *H. glycines* is the only parasitic nematode proteinase of this class that has been characterised. It was cloned using a similar PCR-based approach to that described above for cysteine proteinases (Lilley *et al.*, 1997). Fragments of three distinct genes were obtained and each was used to screen a cDNA library resulting in the isolation of clones encoding three different serine proteinases, HGSP-I, II and III (Lilley *et al.*, 1997). HGSP-I and HGSP-III seem to be encoded by abundant transcripts in feeding female nematodes whilst only a single clone was isolated that encoded HGSP-II. All three enzymes are probably synthesised as precursor molecules in a prepro-format. The largest is HGSP-I with a mature protein of 296 amino acids and a predicted molecular mass of 38 kDa. HGSP-II and HGSP-III are predicted to be 27 and 30 kDa respectively. Homology among the three proteinases is low and mainly centred on the conserved active site regions. HGSP-I and HGSP-III also have limited homology to other known proteinases. HGSP-I is 28% identical to human plasma kallikrein whilst HGSP-III also has homology to mouse and rat kallikreins and to the achelase proteinases of *Lonomia achelous* (23% and 26% identity respectively). The protein encoded by *hgsp-II* is 41% identical to a chymotrypsin-like serine proteinase from the mollusc *Haliotis rufescens*. It is also 32% identical to bovine chymotrypsin A. Crystallographic studies of serine proteinases have led to the identification of three key residues involved in substrate binding and specificity. For HGSP-II, the residues (Ser 199, Gly 222 and Ser 232) are identical to those found in the *Haliotis* enzyme and in human, rat and porcine elastase II. These enzymes are all chymotrypsin-like proteinases that cleave substrates with bulky hydrophobic or aromatic residues at the P1 position. Results from Southern blot analysis indicate that both HGSP-II and HGSP-III are encoded

by single genes, whereas HGSP-I may belong to a small gene family with up to three members (Lilley *et al.*, 1997). A gene with homology to HGSP-III has also been identified from *G. pallida* (C. Lilley, unpublished observation).

Cysteine and serine proteinases of *H. glycines* were expressed in *E. coli* as 6xHis tagged fusion proteins. Both the mature (Lilley *et al.*, 1997; Urwin *et al.*, 1997a,b) and the pro-(C. Lilley, unpublished) forms of the enzymes were expressed and in each case insoluble protein accumulated in inclusion bodies. For HGCP-I, HGSP-I and HGSP-II, denaturation of the protein with subsequent refolding allowed confirmation of proteinase activity. However, only a low proportion of active enzyme was recovered.

The roles of the *H. glycines* proteinases *in vivo* have yet to be established. The presence of several genes for each enzyme class may suggest they have different functions. A second possibility is that they indicate digestion is more rapid or complete when several distinct proteinases function together. Animal parasites often produce a number of cysteine proteinases. Each may have a specialised role at different stages in the life cycle (Campetella *et al.*, 1992; Eakin *et al.*, 1992; Mottram *et al.*, 1997; Smith *et al.*, 1993). The abundance of the *hgcp-I* transcript suggests that this enzyme may provide the cathepsin L-like activity previously detected in the intestine of female *H. glycines* (Lilley *et al.*, 1996).

Many of the characterised serine proteinases of animal parasites have been implicated in the invasion process (Newport *et al.*, 1988; Braun-Breton *et al.*, 1992). Possibly, some of the proteinases of plant-parasitic nematodes play a role in the invasion of plant tissue. However, both the major serine proteinases of female *H. glycines* (HGSP-I and HGSP-III) seem to be digestive enzymes and are responsible for the serine proteinase activity of the intestine (Lilley *et al.*, 1996). These proteinases are probably produced within the intestinal cells but their origin in the pharyngeal glands needs to be discounted by experiment. Such an origin may allow different proteinases to be secreted and be passed posteriorly to the intestine. The unusual nature of the cyst-nematode intestine in a feeding adult female requires further study to provide understanding of the digestive and absorptive processes.

Proteinases Inhibitors

Inhibitors with activity against all four classes of proteinase occur in plants and provide one component of natural plant defence strategies (Ryan, 1990). Each is specific to one class of proteinase. Proteinase inhibitors often accumulate in aerial and certain other plant tissues in response to wounding or herbivory. They are not known to be specifically induced in roots by nematode attack. They also accumulate in seeds (Richardson, 1991) and storage organs such as potato tubers (Rodis and Hoff, 1984). Here they provide a constitutive defence for some more highly defended plant tissues and may in such locations be of value against nematodes. They are therefore present naturally in many plant foodstuffs such as rice seeds (Abe *et al.*,

1987), potato tubers (Rodis and Hoff, 1984), cowpea (Fernandes *et al.*, 1993) and also some from animal sources such as chicken egg-white (Fossum and Whitaker, 1968). Toxicological studies of certain serine proteinase inhibitors have demonstrated a lack of harmful effects in mammalian systems (Pusztai *et al.*, 1992).

The potential of PIs in novel defence strategies has been demonstrated for some insects through the transgenic expression of the serine proteinase inhibitor, cowpea trypsin inhibitor (CpTI). It retards growth of the tobacco budworm (Hilder *et al.*, 1987) but this PI provides limited protection under field conditions (Hoffmann *et al.*, 1996). Limited efficacy may occur because some mammals and several insects overcome inhibition of a single PI in their diets by inducing non-inhibited proteinases via feed-back loops to digestive proteinase synthesis (Holm, Jorgensen and Hanssen, 1991; Jongsma *et al.*, 1995; Jongsma, Stiekema and Bosch, 1996; Broadway, 1995). This presumably arises as a consequence of co-evolution between plants and seed-eating animals. PI-based plant defences tend not to be evident in roots, therefore plants transgenically expressing PIs may not suffer from problems associated with co-evolution between plants and nematode parasites. It is hoped that this will favour their durability when used as a transgenic defence against nematodes.

Results to-date do suggest that effective, proteinase inhibitor-based transgenic defences can be more readily achieved for nematodes than insects. Cowpea trypsin inhibitor expressed in transgenic potato influences the sexual fate of newly established *G. pallida* (Hepher and Atkinson, 1992; Atkinson, 1993). As a result, the population is biased towards a predominance of the much smaller and less damaging males. No reduction in fecundity of females that do establish occurs. This is consistent with biochemical analysis suggesting that cysteine proteinases predominate in females (Koritsas and Atkinson, 1994). In contrast, CpTI reduces the fecundity of females of *M. incognita* without influencing their sexual fate (Hepher and Atkinson, 1992; Urwin, McPherson and Atkinson, 1998).

As both cysteine and serine proteinases are expressed by plant-parasitic nematodes there is a possibility that proteinase switches could occur as in some insects. Further work is required to determine any responsiveness of individual cyst-nematodes to digestive proteinase inhibition and the existence of any relevant enzyme polymorphism. This is an important issue because transgenic plant resistance to nematodes using proteinase inhibitors has been achieved (see later).

Other Intestinal Targets

Endotoxins of *Bacillus thuringiensis* and the Nematode Intestine

The spore-forming bacterium *Bacillus thuringiensis* (*Bt*) is a soil micro-organism capable of killing insects. There is a very large range of isolates with different insect specificities. *Bt* is the source of two types of toxins effective against insects. One is produced in vegetative

stages and released from the bacterium. These exotoxins are encoded by the gene *vip*3a. The second group of toxins is produced in the sporulation phase and form a protein crystal that is retained within the spore. There are many such δ-endotoxins which are encoded by *cry* group of genes. They show high specificity to insects. β-exotoxins do have effects against nematodes (Devidas and Rehberger, 1992). However, most interest in control of insects via transgenic plants centres on the genes encoding variants of the δ-endotoxins (Estruch *et al.*, 1997). Each δ-endotoxin has more than one protein domain. One domain binds specifically to receptor sites in the brush border of the midgut epithelial cells. This occurs for a narrow range of host insects varying according to isolate. A conformational change occurs in the endotoxin and a pore forms in the membrane of the epithelial cell. This causes a lethal, osmotic lysis. δ-endotoxins have been reported to be ovicidal against certain animal-parasitic nematodes with a 77×10^3 fold range in LD_{50} between crystals from different isolates (Bottjer, Bone and Gill, 1985). Possibly such effects are really due to contamination of the crystal endotoxin with a highly active exotoxin.

Oral toxicity of the endotoxin occurs for some isolates against saprophagous nematodes with a pathology that involves disruption of the intestine (Borgonie *et al.*, 1996). *Bt* could be ingested by bacteriophagous nematodes when they feed and so a natural pathology may occur. In contrast, plant-parasitic nematodes ingest plant cell contents and are unlikely to ingest *Bt* in this process. If the *Bt*-receptor found in bacteriophagous nematodes also occurs in plant-parasitic nematodes, a transgenic plant expressing *Bt* might affect plant-parasitic species. Such a system may work despite the absence of a naturally existing pathology between *Bt* and plant-parasitic nematodes. The protein would require access to the nematode intestine. Many *Bt* δ-endotoxins are c130 kDa although smaller molecules are known. The protein is solubilised in the alkaline pH of the insect midgut lumen. It is also cleaved enzymatically to release an insecticidal protein. This shorter, cleaved N-terminal portion of 65–75 kDa confers effective insect resistance when expressed in plants (Fischoff *et al.*, 1987). However, a protein of even 65–75 kDa is unlikely to be ingested by *H. schachtii* (see earlier). Possibly size also prevents ingestion by plant-parasitic nematodes; exclusion and possibly the lack of appropriate receptors may also prevent *Bt*δ-endotoxins from being effective against plant-parasitic nematodes.

Lectins

Lectins bind to carbohydrates including glyco-conjugates associated with proteins. The binding affinity is altered by even minor changes in ligand structure. Some lectins are highly toxic to mammals. One example is ricin from castor bean which binds to N-acetygalactosamine and β-galactosides. In contrast pea lectin binds to α-mannosides and raw peas are not a toxic risk. Lectins may have an anti-insect role in a range of plant seeds. The efficacy of some is correlated with binding to insect midgut epithelial cells after ingestion. Anti-insect effects have been reported when lectins are expressed in transgenic plants (Hilder *et al.*, 1995).

Certain lectins do have biological activity against nematodes. Some lectins disorientate nematodes, presumably by binding to glyco-conjugates associated with chemoreceptors (Zuckerman and Jansson, 1984). Lectin-binding patterns to nematodes is complex and reflects a variable distribution of appropriate glyco-conjugates with nematode species (Dürschner-Pelz and Atkinson, 1988). It is important to be specific about the lectin, its glyco-moiety specificity, likelihood of non-target binding in the plant and any other requirements for binding in any discussion of its potential for nematode control.

Concanavalin A (mannosides) suppresses *M. incognita* when used as a soil amendment in a tomato crop (Marban-Mendoza *et al.*, 1987). To-date transgenic plants expressing two lectins have been used to challenge nematodes. Pea lectin expressed transgenically in potato did not influence either invasion or growth of *G. pallida* to the same extent as CpTI (Hepher and Atkinson, 1992). Snowdrop lectin (α-mannosides) showed a variable effect on *Pratylenchus neglectus* and no effect on *H. schachtii*. Any effect against *G. pallida* was concentration-dependent with an optimal effect at intermediate rather than high levels of expression (Burrows and De Waele, 1997). If this lectin disrupts nematodes as it does insects, by interaction with intestinal cells, its molecular size may prevent it from being ingested by *H. schachtii* and so prevent a toxic effect. The potential of lectins requires further investigation with a well defined lectin/target interaction. The promoter required will depend on the locale of the target. This could be in the apoplast if disruption of the chemoreception of an invading cyst-nematode is intended. Alternatively, it could be within plant cell cytoplasm if uptake by the animal is required.

Transgenic Resistance Based on Proteinase Inhibition

To-date the potential of serine and cysteine proteinase inhibitors have been explored. The latter are frequently termed cystatins. Their genes have been cloned from a number of plants, with the rice cystatins Oc-I and Oc-II being well characterised (Abe *et al.*, 1987; Kondo *et al.*, 1990, 1991). Demonstrations of the potential of cystatins for plant-parasitic nematode control has centred on Oc-I. This cystatin has a lower affinity for the cysteine proteinase papain than chicken egg white cystatin but it is more acceptable as a transgene for crops than a proteinase inhibitor from an animal. The affinity of a cystatin for a cysteine proteinase can be enhanced through proteinase engineering. An amino acid sequence alignment of 28 cysteine proteinase inhibitors allowed comparison of sequence features with reported inhibition values (Ki). Together with molecular modelling of Oc-I, this information led to mutagenesis experiments. Deletion of an aspartic acid residue at position 86 (Oc-IΔD86) resulted in a Ki improved 13–fold relative to native Oc-I (Urwin *et al.*, 1995). The

engineered cystatin was able to inhibit cysteine proteinase activity in cryosections of *H. glycines* (see earlier). Oc-IΔD86 was found to inhibit the growth of *G. pallida* more effectively than native Oc-I when expressed in transgenic hairy roots of tomato (Urwin *et al.*, 1995). Oc-IΔD86 was also expressed in *Arabidopsis thaliana* plants under the control of the constitutive CaMV35S promoter. A homozygous line expressing the inhibitor at 0.4% total soluble protein was used in resistance assays with both a cyst and a root-knot nematode (Urwin *et al.*, 1997b). The growth of both *H. schachtii* (Figure 23.2) and *M. incognita* on the transgenic plants was suppressed, with the cystatin having two distinct effects. Fewer female nematodes reached egg-laying size than for controls and those that did were much smaller than the controls and they had a reduced fecundity. A correlation between these effects and reduced intestinal cysteine proteinase activity for nematodes challenging the transgenic plants was established for *H. schachtii* (Urwin *et al.*, 1997b). The ability of Oc-IΔD86 to control species of three different nematode genera (*Globodera*, *Heterodera* and *Meloidogyne*; Urwin *et al.*, 1995; Urwin *et al.*, 1997c) suggests cysteine proteinase inhibitors have potential for control of a wide range of nematodes. This has two commercial advantages. First, it has the potential of offering protection against a wide range of crops such as banana, rice and potato that may be attacked by different nematodes in the same field or agricultural region. It also enhances potential for rapid adoption of transgenic resistance. It provides a generally useful approach for different transformable crops with distinct nematode problems. The cysteine proteinase genes of *H. glycines*, *H. schachtii* and *G. pallida* show a high similarity (C. Lilley and H.J. Atkinson, unpublished observations). This suggests that a single cystatin may be effective against all three cyst nematode species. However, rice cystatins Oc-I and Oc-II are reported to have distinctive effects against proteinases from

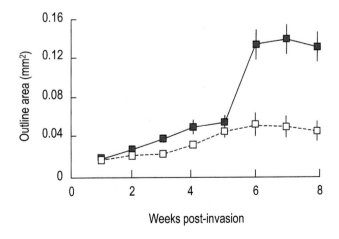

Figure 23.2. Increase in outline area as an estimate of body size of *Heterodera schachtii* developing on *Arabidopsis* expressing a cystatin (Oc-IΔD86) □ or susceptible, wild-type plants ■. Values are means ±SEM.

M. incognita, M. hapla and *M. javanica* (Michaud *et al.,* 1996). Therefore the possible advantage must be considered of either protein engineering or selecting from a panel of natural proteinase inhibitors one that has a particularly favourable Ki against the cysteine proteinase(s) of a key nematode pest. The relative important of Ki and promoters in delivering effective doses of a proteinase inhibitor to a plant-parasitic nematode also requires further study. Cowpea trypsin inhibitor (CpTI), which is a trypsin inhibitor, influences sexual fate shortly after invasion (Atkinson, 1993, Urwin, McPherson and Atkinson, 1998) whereas the cystatin reduces growth of females (Urwin *et al.,* 1995; 1997c). Therefore the two inhibitors act differently on the nematodes and have potential for use in tandem.

Additive Effects of Dual Proteinase Inhibitors

Many natural plant defences involve several components acting together. This may have two main advantages. Combinations may succeed in defending against a wide range of pathogens without need to optimise any product against a particular pathogen. In contrast, R-genes are optimised to respond through recognition to a specific pathogen. As a consequence, some R-genes can be overcome by particular pathotypes of a nematode. There are populations of *M. incognita* that overcome the Mi gene (Kaloshian *et al.,* 1996; Tzortzakakis and Gowen, 1996). Some R-genes are an exception. The resistance of *Gro-1* (H1) to *Globodera rostochiensis* has not been broken by this nematode in over 40 years of commercial use. Instead, a second species, *G. pallida,* to which it is not resistant has become prevalent. Similar switches in prevalence from *M. incognita* to *M. hapla* have occurred for tomato in the USA in response to use of Mi (Roberts, 1992).

There is a need to ensure durable transgenic plant defences. This may be achieved by copying nature and stacking several different transgenes (Boulter *et al.,* 1990; Boulter, 1993). Inclusion of both cysteine and serine proteinase inhibitors in artificial diets demonstrates a potential for synergistic toxicity towards the red flour beetle (Oppert *et al.,* 1993). The progeny of transgenic tobacco expressing CpTI crossed with tobacco expressing pea lectin showed additive efficacy of these two transgenes acting on different targets within tobacco budworm (Boulter *et al.,* 1990; Gatehouse *et al.,* 1993). The need to cross plants to study such effects can be avoided by transforming plants with tandem promoter/gene constructs. Expression of both a chitinase with either a glucanase or ribosome inactivating protein (rip) provides synergistic control of the fungus, *Rhizoctonia solani* (Jach *et al.,* 1995). Stacking can also be achieved in other ways. Bifunctional inhibitors occur for instance with efficacy against both α-amylase and serine proteinase (Wen *et al.,* 1992). In addition, multi-domain PIs are known in which active inhibitors are generated by post-translational fragmentation of a single gene product into several functional PIs (Rodis and Hoff, 1984; Waldron *et al.,* 1993; Atkinson *et al.,* 1993). The value of stacking CpTI

and snowdrop lectin has proved of limited value against *H. schachtii* or *Pratylenchus neglectus* to-date (Burrows and De Waele, 1997), primarily because the lectin seems to have little efficacy against nematodes.

Both cysteine and serine proteinase transcripts are abundant in *H. glycines* (Lilley *et al.,* 1997; Urwin *et al.,* 1997b). This suggests that a combination of inhibitors is likely to provide a more durable defence system in plants than could be achieved by expressing a single proteinase inhibitor. Co-delivery of two distinct proteinase inhibitors in *Arabidopsis* can provide additive resistance against plant-parasitic nematodes. A cysteine and a serine proteinase were joined as a translational fusion by short peptide linkers. One linker was chosen to be refractory to cleavage *in planta.* Western blot analysis of cell extracts confirmed the expected pattern of predominantly dual inhibitor. Analysis of cyst and root-knot nematodes recovered from transgenic *Arabidopsis* expressing CpTI and Oc-IδD86 as a tandem inhibitor revealed a clear additive effect over either inhibitor expressed from a single construct (Figure 23.3). Transgenic plants were

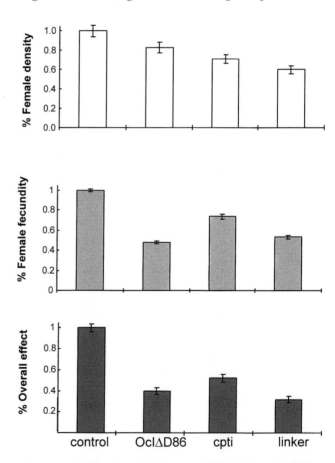

Figure 23.3. Reproductive success of *Heterodera schachtii* on four transgenic lines of *Arabidopsis* expressing a cystatin (OcIΔD86), a serine proteinase inhibitor (CpTI) or both proteinase inhibitors expressed as one protein using a peptide linker. Values for female density per plant or fecundity per individuals are expressed relative to the untransformed control. The overall effect is the product of the reduction in female density and fecundity. Values are means ±SEM.

also generated in which the two inhibitors were linked by a peptide sequence susceptible to cleavage. Cleaved inhibitors were detected *in planta* demonstrating two approaches of delivering a stacked defence (Urwin, McPherson and Atkinson, 1998). However the particular linker used requires replacement by another as the cleaved PIs being targeted to membranes by the linker fragments were unavailable for nematode uptake because of their particular method of feeding. The use of a single promoter in combination with a peptide linker strategy allows the delivery of equimolar amounts of effector proteins from a single transgene. This is advantageous in ensuring the stability of the defence within breeding programmes. This cleavable linker technology has a range of uses. It enables delivery of two or more PIs effective against the same target such as a cysteine proteinase, thereby raising the level of expression provided by a promoter. Such an approach against nematodes could either target two distinct processes, such as migration and digestion, or use specific inhibitors of two or more different targets, such as distinct proteinases required for the same process.

Promoters for Nematode control

Engineered plant resistance based on novel genes requires a promoter from a gene providing a spatial and temporal expression pattern that is appropriate for nematode control (Atkinson *et al.*, 1995). Such promoters have been discovered by several approaches. Some are known to have particular patterns of expression in plants that are appropriate for nematode control. They may be promoters of genes from other organisms such as CaMV35S from Cauliflower Mosaic virus and *Rol-1* from *Agrobacterium rhizogenes* that studies with reporter constructs have revealed to be of interest. They may also be characterised from plants during other work. Examples are the root-specific promoters *TobRB7* (Conkling *et al.*, 1990) and *Tub*-1 (Lilley and Atkinson, 1997). A third group comprises those isolated during studies of changes in plant gene expression following nematode attack. This approach has been particularly associated with identification of promoters active in the feeding sites that are associated with either cyst and/or root-knot nematodes. Conventional (Van der Eycken *et al.*, 1994, 1996) or PCR-directed cDNA libraries (Gurr *et al.*, 1991, Lambert and Williamson, 1993) have been constructed using feeding cell enriched RNA. Differential screening has revealed clones of interest (Van der Eycken *et al.*, 1994; Lambert and Williamson, 1993). A subtractive approach has been applied to enrich for differentially expressed clones in the library screens (Wilson, Bird and van der Knap, 1994).

One approach to discover promoters of interest involves *Agrobacterium*-mediated transformation of a target plant with an interposon, a promoterless β-glucuronidase (GUS) construct. Integration of the interposon within the plant genome may occur downstream from a promoter such that the GUS gene is now regulated by the plant promoter. The promoter is said to be 'tagged'. Nematode infection of tagged lines of

Arabidopsis followed by chromogenic staining for GUS activity allows visual screening of many transgenic lines. This enables co-incidence of nematode feeding sites and transcriptional activation of GUS to be revealed. Such tagged promoters can be isolated by an inverse-PCR strategy in which the regions of genomic DNA flanking the site of integration of the interposon is amplified. This DNA can be cloned into a suitable vector for further characterisation (Lindsey *et al.*, 1993). Promoter tagging has established that nematodes down-regulate some plant genes that are normally expressed in roots although up-regulation of genes has also been reported (Goddijn *et al.*, 1993). Some promoters have been identified that are induced in both the syncytia of cyst-nematodes and the giant cells of *Meloidogyne* (Atkinson *et al.*, 1994; Barthels *et al.*, 1997). Six tagged lines in *Arabidopsis* were identified with activity at feeding sites of *H. schachtii*. All but one was also active at the feeding site of *M. incognita* and in galls induced by the ectoparasites, *Xiphinema diversicaudatum* (Barthels *et al.*, 1997).

Assay of GUS *in planta* involves enzymatic activity producing a blue coloration (Figure 23.4). It is a highly sensitive approach and easy to use. Its principal disadvantage is the need for destructive sampling which ensures only periodic monitoring of the promoter is normally possible. Non-toxic substrates for GUS do help overcome this problem (Atkinson *et al.*, 1993). They have not yet been taken up for evaluation of promoters that have potential for anti-nematode constructs. One reason for this is probably the development of other reporter genes that offer distinct advantages.

Green fluorescent protein (GFP) offers a range of advantages over GUS as a reporter. Relatively non-phytotoxic forms of GFP allow continual monitoring of promoter activity by GFP expression to the sub-cellular level (Hasseloff and Amos, 1995). It has been used to monitor a constitutive promoter that shows changes in

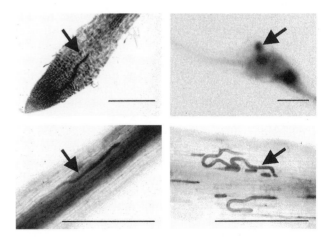

Figure 23.4. Promoter activity in transgenic roots. The promoter is controlling expression of β-glucuronidase (GUS) which provides a blue coloration (areas of dark coloration in the root tissues). Lower right image is an example of no promoter activity around the nematodes. The position of the nematode is arrowed. Scale : 500 μm.

activity when *Arabidopsis* is parasitised by *H. schachtii* (Urwin *et al.*, 1997a). It shows progressively lower GFP levels in the syncytium as parasitism develops but it also indicates that expression extends into the period of female development. The work suggests that promoter selection is an important part of optimising a transgenic defence. A constitutive promoter that responds to nematodes with down-regulation provides a less than ideal basis for resistance. Using a promoter that is induced by tetracycline, it can be shown that *de novo* expression of GFP can occur within a few minutes of induction (Urwin *et al.*, 1999). Sensitive electronic cameras and other equipment provide considerable flexibility to the researcher. The wavelength of excitation can be varied and set to any narrow-bands pass over a considerable range. Automatic continual monitoring at pre-programmed intervals can also be achieved (Urwin *et al.*, 1999).

A second reporter gene of interest is luciferase. It also has potential for studying nematode-plant interactions. Luciferase is unstable in plant cells and so could be important in monitoring transient events in nematode/plant interactions. Such work requires sensitive photon counting procedures (Millar *et al.*, 1995) and provides less precise sub-cellular localisation than GFP. New forms of GFP that are not stable *in planta* have also been protein-engineered. The lack of stability of these variants enables loss of promoter activity to be studied. It remains to be seen if luciferase is used in the future to study the response of plant promoters to nematode pathogenesis. It is likely that GFP may prove the reporter of choice when GUS is unable to meet the needs of the experimenter.

Constitutive Expression Throughout the Plant

To-date promoters shown to be of value for transgenic resistance to nematodes can be sub-divided into 4 distinct patterns of expression. Constitutive promoters are capable of provide expression of defences before and during parasitism of the plant by the nematode. The promoter CaMV35S from cauliflower mosaic virus is the best-studied example. It has been reported to be down-regulated in the syncytium of *G. pallida* (Goddijn *et al.*, 1993). The time-course of this effect in hairy roots is several weeks and so promoter activity is not fully lost even when the mature syncytium is associated with a feeding female of *H. schachtii* (Urwin *et al.*, 1997a).

GUS tends to show that CaMV35S is also more active in younger than older roots. Overall the lack of activity is some parts of the root system and older feeding cells may ensure some sites in which female nematodes can produce eggs. This may limit the overall level of resistance on the whole plant. This possibility emphasises the need for study of promoters. The aim is to select those that provide the required level of an effector gene to all nematode parasites irrespective of their location on the root system. In spite of its imperfections, CaMV35S has proved of value in demonstrating the potential of cystatins for transgenic resistance to nematodes (see earlier).

CaMV35S provides lower levels of expression in monocots than dicots. However it has been shown to have sufficient activity in rice to detect efficacy of a cystatin against *M. incognita* (Vain *et al.*, 1998). Increased constitutive expression can be obtained in monocots using other promoters such as that from the ubiquitin gene (Christensen, and Quail, 1996) in preference to CaMV35S.

Root-Specific Expression

Such promoters have commercial value because roots of most crop plants are not consumed. Therefore a defence can be developed without concern for expression in food. Several root-specific promoters of value for nematode control have been identified (Lilley and Atkinson, 1997). Ideally the optimal promoter for root nematode control would be sufficiently active to deliver the effector at an appropriate level throughout the root system. It should not be down-regulated by the feeding nematode. A gene with root-specific expression in tobacco (*TobRB7*, Yamamoto *et al.*, 1991) encoding a putative water channel has proved to be of particular interest (Opperman *et al.*, 1994). It shows substantially specific expression at root meristems in healthy plants but it also remains active in the giant cells induced by *Meloidogyne*. For other promoters, nematode directed up-regulation occurs and is superimposed on root-specific expression pattern. This increases promoter utility (Lilley and Atkinson, 1997).

Nematode Responsive Promoters Not Limited to the Syncytium

The *wun*-1 promoter (Logemann and Schell, 1989) responds to invading cyst-nematodes (Hansen *et al.*, 1996). The invading larvae of *G. pallida* induce expression that is not restricted to damaged cells in potato roots but also occurs near the invading parasite. Promoter activity is lost once the animal enters a phase of syncytial induction. The effect seems to be related to wounding of plant cells by the invading nematode. The effect is not evident when *M. incognita* invades the same roots, presumably because it moves between plant cells. A tobacco extensin gene responds to *G. pallida* in a similar manner to that described for *wun*-1. Expression is induced by the destructive intercellular migration of the invading nematode but it is not apparent in the syncytium (Niebel *et al.*, 1993).

Not all changes in plant gene expression accompanying nematode pathogenesis are part of a plant defence. The plant must also re-differentiate to ensure root function is maintained once the parasite is established. A copper diamine oxidase from *A. thaliana* has been implicated in peroxidase-mediated vascular lignification, extra-cellular cross-linking and programmed cell death (Møller and McPherson, 1998). Reporter gene studies show the promoter of this oxidase gene does not respond to root invasion by either *M. incognita* or *H. schachtii* (Møller *et al.*, 1998). Its response is limited in the case of the cyst nematode to lignification in some cell walls surrounding the syncytium. More activity is observed following establishment of *M. incognita*. It is active as the gall forms and subsequently in the vascular tissue within it. The difference in pattern reflects the greater vascular re-differentiation that the root requires to accommodate the centrally placed giant cells of *M. incognita*. The response is part of the essential dedifferentiation that the root must undergo to maintain xylem vessel continuity after *M. incognita* is established (Møller *et al.*, 1998). The syncytium of a cyst-nematode has less drastic consequences for xylem continuity and does not involve formation of new tracheary elements. Expression of the copper amine oxidase is not part of a defensive response directed at either the parasite or its feeding site.

Feeding-Site Specific Promoters

Van der Eycken *et al.* (1996) identified 10 cDNA clones of plant origin that showed differential, higher activity at the feeding sites of *M. incognita*. Two were extensins of which one may have a role in walls strengthening of large giant cells. A second cDNA evident in older feeding sites has homology to an osmoprotectant. It may have a role in protecting the giant cell from increased osmolarity associated with its function as a nutrient source for the parasite. Over 200 genes expressed at the feeding sites of *M. incognita* have been identified as a basis for understanding giant cell function (Bird, 1996).

A promising promoter identified in a GUS tagged line of *Arabidopsis* has been studied in particular detail. It showed a temporal change in proportion of the feeding cells of *H. schachtii* in which it was active. It increased from about 30 to 70% of all feeding sites between 2 and 7 days post-infection. This value declined to less than 20% of syncytia by 12 days post-infection (Puzio *et al.*, 1998).

A syncytial-specific promoter identified in this way has been used in the design of transgenic resistance to PCN (Ohl, van de Lee and Sijmons, 1997). Possibly promoters that are not active in all feeding cells for prolonged periods may prove able to provide only partial resistance.

Promoters of two cell cycle genes of *Arabidopsis* were studied using GUS as a reporter during infection by *M. incognita* and *H. schachtii*. One was *cdc2a*, which is a marker for competence of cell division in plants. It was active in giant cells which are known to undergo mitosis as their multinuclear status is achieved. The promoter was also active in syncytia and was a less expected finding. This indicates some progression in the cycle is a consequence of initiating a feeding cell. The second promoter was that from *Cyc1AT*. It is a marker of the G2 phase of the cell cycle. It was active in both giant cells but not the syncytium except at those cells that are likely to be soon incorporated into expanding syncytium. These are currently the earliest described changes in plant gene expression after initiation of a feeding cell.

The syncytia induced by cyst-nematodes are similar to transfer cells (Jones and Northcote, 1972). As such, they are within the normal repertoire of cell types for healthy uninfected plants and develop invaginated cell walls that are thought to increase solute uptake.

Apparently specific promoter activity within syncytia may therefore occur elsewhere in the plant, even if it is a rare event. The importance of such unwanted expression of the transgene depends upon the design of the defence. It is a critical matter if an effector is lethal to plant cells with damaging consequences for the plant. It is relatively unimportant at low levels if the effector is directed against nematodes and it is not a biohazard to other animals or present in harvested food.

Modifying Promoter Activity

Promoters have elements that provide different, spatial, temporal or environmentally responsive features to the expression pattern of a gene (Benfey and Chua, 1989; Benfey, Ren and Chua, 1989; Keller and Baumgartner, 1991). Rarely can these elements be recognised from sequence information. Deletion studies have been used to map elements responsible for specific expression patterns of a promoter region. This process typically involves progressive deletion from the 5'- to the 3'-end of the promoter region. In this way it is possible to separate promoter elements providing root-specific expression from those providing expression elsewhere. Deletion of the 5'-flanking region of *Tob*RB7 has demonstrated that a 300 bp promoter fragment just upstream of the coding region remains active within the giant cells induced by *M. incognita*. This truncated promoter is silenced in root meristems (Opperman *et al.*, 1994). TUB-1 also shows higher levels of activity in giant cells of *M. incognita* but deletions did not result in separating this activity from the more general, root-specific expression of this gene (C. Lilley and H.J. Atkinson, unpublished experiments).

A promoter identified by gene tagging (Puzio *et al.*, 1998) was also active in root hydathodes and stipules. Deletion studies were carried to alter its tissue specificity. One hope was probably to define a minimal promoter that provides a nematode responsive element freed from other unwanted activity. A probe for the promoter region isolated from the tagged line by inverse PCR was used to screen a genomic library of *Arabidopsis*. The isolated genomic DNA showed a number of putative regulatory elements and scaffold attachment sites were identified. Promoter deletion studies showed that different lengths of the putative promoter were required for different tissue specificity. The full length promoter was required maximum activity at all three sites. Loss of a small region just 5' of the T-DNA border eliminates promoter activity at all sites. Activity was restored sequentially to hydathodes, stipules and then the nematode feeding cells as the extent of promoter deletion was reduced. (Puzio *et al.*, 1998).

It is currently uncertain how frequently promoter modification can enhance the value of promoters for use in nematode defences.

Biosafety Issues

Criticisms of transgenic crops include ethical concerns over gene transfer, fear of ecological damage, safety of food, socio-economic and other consequences resulting from implementation and the business activities of biotechnology companies. Constructive criticism of any technology including transgenic plants is welcome. It helps prioritise scientific efforts to improve products. The risks from genes and their products in a transgenic plant can be sub-divided. Some risks are associated with genes necessary for generating transgenic plants. Others are specific to the traits being introduced.

A distinction exists in risk between limited experimental field trials of a transgenic crop and commercial release. Limited field release can be set up to investigate risk of gene transfer and can be monitored for biosafety during and after the trials. Such work involves a relatively small growing area. In contrast commercial release operates with less control over cropping and involves much greater areas for many years. The chance of a remote risk occurring increases with the crop area and time.

Risk Analysis

Authorities such as the EU treat transgenic plants as a separate category, presumably in response to political pressure. However, there are strong, logical arguments for similar risk analysis being applied to any new crop species or forms irrespective of the presence or absence of transgenes (Miller *et al.*, 1995). All transgenic experiments must meet appropriate national biosafety regulations for genetically modified organisms (GMOs). Rightly, these are likely to be more exacting for commercial release than initial field evaluation. Before the first field trial, toxicological and environmental consequences arising from release of the GMO must be considered. Detailed toxicological studies are essential before a commercial release. Environmental risks must also be considered and the problem here is to define standardised basis for identifying those short and long-term risks that deserve attention.

A crop with transgenic nematode resistance offers the real benefits of reduced pesticide use in the developed world and effective control for the developing world. Therefore they should be adopted providing a thorough risk analysis establishes the safety of an approach. A defence based on an antisense construct (Opperman, *et al.*, 1994; Opperman, Taylor and Conkling, 1994) benefits from not involving expression of a protein to control the nematode. A protein that is already consumed in plant food, for example some proteinase inhibitors, carries less concern for food safety than one from either non-food sources or one with oral toxicity against mammals. Those proteins with adverse effects on mammals, including allergenicity, are probably unsuitable for commercial use even with promoters that ensure no expression in food.

The Potential and Consequences of Transgene Escape

There are three routes for transgene escape. The first is that the transgenic plant may establish as a weed. This may involve establishment in agricultural fields, in adjacent disturbed habitats or dispersal to semi-natural or natural habitats. The second possibility is that the transgene is transferred by pollination to a non-transgenic, sexually compatible crop which may then establish in habitats as listed above. The third possibility is the transgene is transferred to a wild plant species by hybridisation (Anon, 1994).

Potato Plants as an Example of Risk

The possibility that expression of the resistance may affect non-target organisms such as herbivores needs to be evaluated to protect biodiversity. The possibility and consequences of out-crossing also needs to be considered. The USA, UK and certain other countries have developed fast track procedures for crops and certain types of transgenes that seem inherently safe. Such procedures are now allowed for registering transgenic potato plants offering pest resistance in the UK. This status partly reflects potential benefits given the urgent need to reduce use of environmentally damaging nematicides. Other points are that certain crops such as potatoes are not consumed raw and many proteins are destroyed by cooking. Its green tissues also contain toxins that are highly dangerous to mammals in comparison to any transgene products likely to be proposed for use. Safety in the UK is also favoured by clonal propagation, by the presence of male sterility in some cultivars, short-distance pollen transmission (Dale *et al.*, 1992) and by failure to hybridise with wild solanaceae. Potato plants have also proven non-invasive of semi-natural or natural habitats in four centuries of potato cultivation in Europe (Harding and Harris, 1994). It seems unlikely that an ability to resist nematodes would lead to a potato weed problem in Europe. The main economic nematodes in this continent are *Globodera* spp. There is no evidence that these nematodes prevent establishment of potatoes as a weed in fields. In addition, they are introduced animals and are restricted to cultivated fields. Therefore defence against *Globodera* spp. offers no advantage to potatoes colonising natural or semi-natural habitats.

Potato plants illustrate the issue that geographical region is important. This crop is grown in much of South America close to a range of wild relatives. The risk of transgene escape from potato to wild plants is a complex issue for this continent. It centres on several key issues. The proximity of solanaceous weeds or other potato crops needs to be considered. An ability of the cultivars to flower in the prevailing day-length is also important. Male sterility of some lines, stylar incompatibility and endosperm balance number (Hawkes, 1992) are all relevant when determining the risk of unwanted fertilisation. Some crops in Europe also grow close to wild relatives. For instance, carrot plants hybridise freely with the widely distributed wild forms of the same species.

Ecological Consequences of Transgene Escape

It is uncertain if acquisition of nematode resistance by a wild plant is of ecological significance. Some resistance strategies such as that based on proteinase inhibitor defences aim to provide general defences against all nematodes and so may have value for a range of plants. One theory for the evolution of sex in animals envisages a central importance for predators and parasites (Hamilton, 1980). We lack information on the extent that nematode resistance enhances the competitiveness of a wild plant species within a plant community. Such data is required to underpin a preliminary evaluation of the consequences of transgene escape. However, resistance to nematodes does not seem to confer considerable advantage to many wild plant species. If so, R-genes against nematodes would be expected to occur more commonly than reported.

Another distinct risk to be evaluated is the consequence on non-target animals. This can be explored using proteinase inhibitors as an example. The matters to be considered will of course vary with the transgene considered. Cystatins incorporated into artificial diets have been shown to reduce the growth and development rate of some coleopterans (Wolfson and Murdock, 1987; Hines *et al.*, 1990). Serine proteinase inhibitors reduce the growth of some lepidopteran species. Others have been shown to be able to circumvent the effects of serine proteinase inhibitors by the production of inhibitor-insensitive proteinases. A small number of protease inhibitors, including serine proteinase inhibitors from soybean, inhibit growth when fed to aphids in artificial diets (Rahbe and Febvay, 1993).

The effect of a cystatin expressed constitutively in potato has been examined for the aphid *Myzus persicae*. Addition of cystatins to artificial diet curtailed survival and suppressed their growth and fecundity. This effect was not detected for *M. persicae* feeding on transgenic potato plants in either containment or under field conditions (Cowgill, Coates and Atkinson, 1999). In these experiments the cystatin chicken egg-white was under control of the promoter CaMV35S. This provided sufficient expression in roots to control *G. pallida* without adverse effects on *M. persicae*. This is probably due to the low level of expression provided in the phloem from which the insect feeds. Therefore promoters have value in limiting non-target effects. A root-specific promoter limits expression of a cystatin for nematode control to the root system. Therefore it seems that the effect of the proteinase inhibitors on root browsers and detritus feeders requires particular study. The proteinase inhibitor concentration in tobacco fell to a very low level by day 57 of field decomposition (Donegan *et al.*, 1997). A fall in soil collembolan populations and an increase in soil nematodes were recorded in this experiment.

Study is also required for the next trophic level. The proteinases of an adult coccinellid (*Adalia bipunctata*) are inhibited when the insect is fed proteinase inhibitors. (Walker *et al.*, 1998). The pre-reproductive period, fecundity and percentage egg hatch of the predator *Perillus bioculatus* are affected when it is fed on Colorado beetle larvae spiked with a cysteine proteinase inhibitor from rice (Ashouri *et al.*, 1998). Therefore work is required to determine if herbivores ingesting a novel protein from plants can cause appreciable effects on the survival or development of their predators. Given that the nematode defence will be restricted by root-expression it would seem soil or ground-inhabiting predators of root herbivores and browsers require study.

Clearly, each transgene needs to be considered on an individual basis. Both toxicological and allegenicity information is required before commercial release can be contemplated. Risk analysis must be individual to the transgene, expression pattern, target crop and geographical region in which release is planned. Nematode resistance is unlikely to be the risky end of the spectrum of transgenes that could be contemplated for crop improvement.

Application of Transgenic Resistance to Nematodes to Developing World Needs

The green revolution helped world grain production meet world food needs for about 30 years from 1965. There is evidence of trickle down of its benefits to both urban and rural poor but there have also been negative consequences. These include increased landlessness, disruption of social systems, loss of beneficial farm practices and increased marginalisation of women. Sustainable agricultural production aims to minimise these problems but few consider it likely to meet all future needs. Increased food production is required in the developing world (Mann, 1997). Some envisage a new green revolution is required with reduced negative effects to assure future food security for all as the world population continues to grow in the next century (Conway, 1997). Approximately 26 million ha of transgenic crops were grown in the USA in 1998 and this figure will increase three fold by 2000. It is important to consider how the new technology can be adapted to meet developing world needs. The potential breadth of both crops and traits offered with the new technology ensure its potential scope is greater than the original green revolution. At best, transgenic plants involve the simple act of planting without other changes to traditional farming practices. The technology must be detached from the interests and control of biotechnology companies to achieve a clear poverty focus. (Atkinson, 1998). It must also address all real biosafety concerns.

Enabling Technology

Even critics of genetic engineering appreciate that improvements to farming will arise from the approach if the research is public-funded and for the public good. Wheat, maize and rice provide over 50% of calories to the world diet and this figure reaches 85% by adding only a further 5 species. These major crops interest agribusiness and so efficient transformation is already available for most of them. It is important that transformation of a similar standard is available in public-funded laboratories to ensure developing world needs are properly addressed. Biosafety is also centrally important to responsible technology transfer to the developing world. The paradigm must be that the biosafety standards of the donor country must be met by the recipients. There has been much effort to reach this standard in recent years.

Experience with biological control and integrated pest management has established that scientific complexity is a disadvantage in developing world implementation. Ideally new technology should not require either additional knowledge or resources from the grower before implementation can be successful. The proteinase inhibitor technology for nematode control may meet that need. The small size of nematodes ensures that many growers are unaware of the crop loss they cause. This is exacerbated in many developing world countries by the lack of extension workers to guide subsistence farmers. A technology that is effective against many different forms eliminates the need to identify them and determine if damaging levels are present in a field. A strong *prima facie* case can be made that a technology is safe. The cystatin providing the nematode resistance is already present in rice seed which is the second most consumed plant food. Furthermore the new protein can be restricted to the roots of the transgenic plant where nematodes occur. This ensures that it is not consumed in foods made from transgenic plants.

The cost of developing transgenic technology for resource-poor farmers is effectively reduced by investment by industry in first world needs. For instance the nematode-resistance technology developed with industry for European potato fields only requires research to adapt the approach before it is suitable for subsistence farmers in the Andes. The additional science needs involve transformation of potato cultivars favoured in S. America and the biosafety issues that differ between this continent and Europe. The potential benefits in countries such as Bolivia are reduced acreage for potatoes so releasing more of smallholders' land for other nutritious crops such as legumes.

Conclusions

Transgenic crop technology has the potential to provide effective and durable resistance to plant-parasitic nematodes. A successful defence strategy will have a clearly defined target as this allows selection of the most appropriate transgene. The depth of information now being accumulated about nematode proteinases will enhance future developments of a defence aimed to inhibit them. The molecular characterisation of serine proteinases opens up possibilities such as selection of alternative inhibitors or rational protein engineering of CpTI. If the aspartic proteinase genes recently isolated from cyst nematodes (C. Lilley, unpublished observations) prove to encode digestive enzymes, they offer another priority target for building an effective transgenic defence. New targets are likely to be defined as further progress is made in understanding nematode growth, development and host interaction at a biochemical and molecular level. A stacked defence could then be aimed against different aspects of parasitism, reducing the likelihood of nematodes becoming resistant to the defence. It is uncertain what range of defences will be commercially acceptable. For instance, approaches based using antibodies to target secretions

of nematodes are feasible but expressing animal genes in plants will not be acceptable to all. What technology is acceptable in USA is likely to have an important bearing on what is developed. This relates to the size of the agricultural economy of that country and to its rapid acceptance of the technology. A main concern will be ensuring the technology is not limited to major biotechnology companies. This is important if the advantages of the approach are to meet real needs of subsistence growers. A key issue will be providing sufficient durability in the field to enable science to outpace the rate of selection of virulence in nematode populations.

Dedication

This chapter is dedicated to Donald Lee for his friendship, advice and support throughout my career at the University of Leeds.

References

Abe, K., Emori, Y., Kondo, H., Suzuki, K. and Arai, S. (1987a) Molecular cloning of a cysteine proteinase inhibitor of rice (Oryzacystatin). *Journal of Biological Chemistry*, **262**, 16793–16797.

Albertsen, H.M., Abderrahim, H., Cann, H.M., Dausset, J., Le Pasilier, D. and Cohen, D. (1990) Construction and characterisation of a yeast artificial chromosome library containing seven haploid genome equivalents. *Proceedings of the National Academy of Science, USA*, **87**, 4256–4260.

Alvarez, M.E., Pennell, R.I., Mejer, P.J., Ishikawa, A., Dixon, R.A. and Lamb, C. (1998) Reactive oxygen intermediates mediate a systemic network in the establishment of plant immunity. *Cell*, **92**, 773–784.

Al-Yahya, F.A. (1993) *Aspects of the Host-Parasite Interactions of Meloidogyne spp. (Root-knot nematodes) and Crop Plants*. PhD Thesis. University of Leeds Library. 109pp.

Anderson, P.A., Lawrence, G.J., Morrish, B.C., Ayliffe, M.A., Finnegan and E.J. and Ellis, J.G. (1997) Inactivation of the flax rust resistance gene M associated with loss of a repeated unit within the leucine-rich repeat coding region. *Plant Cell*, **9**, 641–651.

Andrews, S.J., Rolph, T.P. and Munn, E.A. (1997) Duration of protective immunity against ovine haemonchosis following vaccination with the nematode gut membrane antigen H11. *Research in Veterinary Science*, **62**, 223–227.

Anon (1992) Agrow No. 171, pp. 21–22. PJB Publications LTD.

Anon (1994) *Genetically Modified Crops and their wild relatives — a UK perspective*. Research Report No. 1 Genetically modified organism Research Report. Published by Dept. of the Environment, UK, 124 pp. plus annexes.

Ashouri, A., Overney, S., Michaud, D. and Cloutier, C. (1998) Fitness and feeding are affected in the two-spotted stinkbug, *Perillus bioculatus*, by the cysteine proteinase inhibitor, oryzacystatin I. *Archives of Insect Biochemistry and Physiology*, **38**, 74–83.

Atkinson, A.H., Heath, R.L., Simpson, R.J., Clarke, A.E. and Anderson, M.A. (1993) Proteinase inhibitors in *Nicotiana alata* stigmas are derived from a precursor protein which is processed into five homogenous inhibitors. *The Plant Cell*, **5**, 203–213.

Atkinson, H.J. (1993) Opportunities for improved control of plant parasitic nematodes via plant biotechnology. In *Opportunities for Molecular Biology in Crop Production*, edited by D.J. Beadle, D.H.L. Bishop, L.G. Copping, G.K. Dixon and D.W. Holloman, pp. 257–66. British Crop Protection Council.

Atkinson, H.J. (1995) Plant-nematode interactions: molecular and genetic basis. In *Pathogenesis and Host Specificity in Plant Diseases. Histopathological, Biochemical, Genetic and Molecular Bases Diseases. Vol. II: Eukaryotes*, edited by K. Kohmoto, U.S. Singh and R.P. Singh, pp. 355–370. Oxford: Elsevier.

Atkinson, H.J. (1996) Novel defences against nematodes. *Journal of The Royal Agricultural Society*, **157**, 66–76.

Atkinson, H.J. (1998) A Robin Hood approach to transferring appropriate plant biotechnology to the developing world. *Science and Public Affairs* Winter 1998, 27–29.

Atkinson, H.J., Harris, P.D., Halk, E.L., Novitski, C. Nolan, P., Leighton-Sands, J. and Fox, P.C. (1988) Monoclonal antibodies to the soya bean cyst nematode, *Heterodera glycines*. *Annals of Applied Biology*, 112, 459–69.

Atkinson, H.J. and Harris, P.D. (1989) Changes in nematode antigens recognised by monoclonal antibodies during early infection of soya beans with the cyst nematode. *Parasitology*, 98, 479–487.

Atkinson, H.J., Blundy, K.S., Clarke, M.C., Hansen, E., Harper, G., Koritsas, V., Mcpherson, M.J., O'Reilly, D., Scollan, C., Turnbull, S.R. and Urwin, P.E. (1994) Novel defences against nematodes. In *Advances in Molecular Plant Nematology*, edited by F. Lamberti, C. De Giorgi and D. Bird, pp. 197–210. New York: Plenum Press.

Atkinson, H.J., Urwin, P.E., Hansen, E. and McPherson, M.J. (1995) Designs for engineered resistance to root-parasitic nematodes. *Trends in Biotechnology*, 13, 369–374.

Atkinson, H.J., Urwin, P.E., Clarke, M.C. and McPherson, M.J. (1996) Image analysis of the growth of *Globodera pallida* and *Meloidogyne incognita* on transgenic tomato roots expressing cystatins. *Journal of Nematology*, 28, 209–215.

Atkinson, H.J., Lilley, C.J. Urwin, P.E. and McPherson, M.J. (1998a) Engineering resistance in the potato to potato cyst nematodes. In *Potato Cyst Nematodes, Biology, Distribution and Control*, edited by R.J. Marks and B.B. Brodie, pp. 209–236. Wallingford: CAB Publishing.

Atkinson, H.J., Lilley, C.J., Urwin, P.E. and McPherson, M.J. (1998b) Engineering resistance to plant-parasitic nematodes. In *The Physiology and Biochemistry of Free-living and Plant Parasitic Nematodes*, edited by R.N. Perry and D.J. Wright, pp 381–413. Wallingford: CAB Publishing.

Badley, J.E., Grieve, R.B., Rockey, J.H. and Glickman, L. (1987) Immune-mediated adherence of eosinophils to *Toxocara canis* infective larvae: the role of excretory-secretory antigens. *Parasite Immunology*, 9, 133–43.

Barthels, N., Van Der Lee, F.M., Klap, J., Goddijn, O.J.M., Karimi, M., Puzio, P., Grundler, F.M.W., Ohl, S.A., Lindsey, K., Robertson, L., Robertson, W.M., Van Montagu, M., Gheysen, G. and Sijmons, P.C. (1997) Regulatory sequences of *Arabidopsis* drive reporter gene expression in nematode feeding structures. *Plant Cell*, 9, 2119–2134.

Baum, T.J., Hiatt, A., Parrot W.A., Pratt, L.M. and Hussey, R.S. (1996) Expression in tobacco of a functional monoclonal antibody specific to stylet secretions of the root knot nematode. *Molecular plant-microbe interactions*, 9, 382–387.

Baylis, H.A., Megson, A., Mottram, J.C. and Hall, R. (1992) Characterisation of a gene for a cysteine protease from *Theileria annulata*. *Molecular and Biochemical Parasitology*, 54, 105–108.

Benfey, P.N. and Chua, N-H. (1989) Regulated genes in transgenic plants. *Science*, 24, 174–181.

Benfey, P.N., Ren, L. and Chua, N-H. (1989) The CaMV35S enhancer contains at least two domains which can confer different developmental and tissue-specific expression patterns. *The EMBO Journal*, 8, 2195–2202.

Bird A.F. (1972) Quantitative studies on the growth of syncytia induced in plants by root knot nematodes. *International Journal for Parasitology*, 2, 157–170.

Bird, A.F. and Bird, J. (1991) *The structure of nematodes*, 2nd edn. London: Academic Press.

Bird, D.M. (1996) Manipulation of host gene expression by root-knot nematodes. *Journal of Parasitology*, 82, 881–888.

Bird, D.M. and Wilson, M.A. (1994) DNA sequence and expression analysis of root-knot nematode-elicited giant cell transcripts. *Molecular Plant-Microbe Interactions*, 7, 419–424.

Bird. R.E., Hardman, K.D., Jacobson, J.W., Johnston, S., Kaufman, B.M., Lee, S.M., Lee, T., Pope, S.H., Rioden, G.S. and Whitlow, M. (1988) Single chain antigen-binding proteins. *Science*, 242, 423–426.

Blackman, M.J., Fujioka, H., Stafford, W.H.L., Sajid, M., Clough, B., Fleck, S.L., Aikawa, M., Grainger, M. and Hackett, F. (1998) A subtilisin-like protein in secretory organelles of *Plasmodium falciparum* merozoites. *Journal of Biological Chemistry*, 273, 23398–23409.

Blaxter, M.L., Page, A.P., Rudin, W. and Maizels, R.M. (1992) Nematode surface coats: actively evading immunity. *Parasitology Today*, 8, 243–247.

Böckenhoff, A. and Grundler, F.M.W. (1994) Studies on the nutrient-uptake by the beet cyst nematode *Heterodera schachtii* by *in situ* microinjection into the feeding structures in *Arabidopsis thaliana*. *Parasitology*, 109, 249–254.

Borgonie, G., Claeys, M., Leyns, F., Arnaut, G., De Waele, D. and Coomans, A. (1996) Effect of nematicidal *Bacillus thuringiensis* strains on free-living nematodes. 1. light microscopic observations, species and biological stage specificity and identification of resistant mutants of *Caenorhabditis elegans*. *Fundamental and Applied Nematology*, 19, 391–398.

Bottjer, K.P., Bone, L.W. and Gill, S.S. (1995) Nematoda - susceptibility of the egg to *Bacillus thuringiensis* toxins. *Experimental Parasitology*, 60, 239–244.

Boulter, D. (1993) Insect pest control by copying nature using genetically engineered crops. *Phytochemistry*, 34, 1453–1466.

Boulter, D., Edwards, G.A., Gatehouse, A.M.R., Gatehouse, J.A. and Hilder, V.A. (1990) Additive protective effects of different plant-derived insect resistance genes in transgenic tobacco plants. *Crop Protection*, 9, 351–354.

Bowles, D.J., Gurr S., Scollan, S., Atkinson, H.J. and Hammond-Kosack, K. (1991) Local and systemic changes in plant gene expression. In *Biochemistry and Molecular Biology of Plant-Pathogen Interactions*, edited by C.J. Smith, pp. 225–236. Oxford: Oxford University Press.

Braun-Breton, C., Blisnick, T., Jouin, H., Barale, J.C., Rabilloud, T., Langsley, G. and Pereira Da Silva, L.H. (1992) *Plasmodium chabaudi*: p68 serine proteinase activity required for merozoite entry into mouse erythrocyte. *Proceedings of the National Academy of Sciences, USA*, 89, 9647–9651.

Broadway, R.M. (1995) Are insects resistant to plant proteinase inhibitors? *Journal of Insect Physiology*, 41, 107–116.

Brown, A., Burleigh, J.M., Billett, E.E. and Pritchard, D.I. (1995) An initial characterization of the proteolytic enzymes secreted by the adult stage of the human hookworm *Necator americanus*. *Parasitology*, 110, 555–563.

Burrows, P.R. and De Waele, D. (1997) Engineering resistance against plant parasitic nematodes using anti-nematode genes. In *Cellular and Molecular Aspects of Plant-Nematode Interactions*, edited by C. Fenoll, F.M.W. Grundler and S.A. Ohl, pp. 217–236. Dordrecht: Kluwer Academic Press.

Cai, D.G., Kleine, M., Kifle, S., Harloff, H.J., Sandal, N.N., Marcker, K.A., KleinLankhorst, R.M., Salentijn, E.M.J., Lange, W., Stiekema, W.J., Wyss, U., Grundler, F.M.W. and Jung, C. (1997) Positional cloning of a gene for nematode resistance in sugar beet. *Science*, 275, 832–834.

Campetella, O., Henriksson, J., Aslund, L., Frasch, A.C.C., Pettersson, U. and Cazzulo, J.J. (1992) The major cysteine proteinase (cruzipain) from *Trypanosoma cruzi* is encoded by multiple polymorphic tandemly organized genes located on different chromosomes. *Molecular and Biochemical Parasitology*, 50, 225–234.

Chappell, C.L. and Dresden, M.H. (1986) *Schistosoma mansoni*: proteinase activity of 'hemoglobinase' from the digestive tract of adult worms. *Experimental Parasitology*, 61, 160-167.

Chrispeels, M.J. and Sadava, D.E. (1994) *Plants, Genes and Agriculture*. Boston: Jones and Bartlett.

Christensen, A.H. and Quail, P.H. (1996) Ubiquitin promoter-based vectors for high-level expression of selectable and/or screenable marker genes in monocotyledonous plants. *Transgenic Research*, 5, 213–218.

Cocude, C., Pierrot, C., Cetr, C., Godin, C., Capron, A. and Khalife, J. (1997) Molecular characterization of a partial sequence encoding a novel *Schistosoma mansoni* serine protease. *Parasitology*, 115, 395–402.

Cohen, F.E., Gregoret, L.M., Amiri, P., Aldape, K., Railey, J. and McKerrow, J.H. (1991) Arresting tissue invasion of a parasite by protease inhibitors chosen with the aid of computer modeling. *Biochemistry*, 30, 11221–11229.

Cohn, E. and Mordechai, M. (1977) Uninucleate giant cell induced in soybean by the nematode *Rotylenchulus macrodoratus*. *Phytoparasitica*, 5, 85–93.

Conkling, M.A., Cheung, C.L., Yamamoto, Y.T. and Goodman, H.M. (1990) Isolation of transcriptionally regulated root-specific genes from tobacco. *Plant Physiology*, 93, 1203–1211.

Conway, G. (1997) *The Doubly Green Revolution: food for all in the twenty-first century*. London: Penguin Books.

Cook, R. and Evans, K. (1987) Resistance and tolerance. In *Principles and Practice of Nematode Control in Crops*, edited by R.H. Brown and B.R. Kerry, pp. 179–231. Sydney: Academic Press.

Coombs, G.H. and Mottram, J.C. (1997) Parasite proteinases and amino acid metabolism: possibilities for chemotherapeutic exploitation. *Parasitology*, 114, S61–S80.

Cowgill. S.E., Coates, D. and Atkinson, H.J. (1999) Non-target effects of proteinase inhibitors expressed in potato as an anti-nematode defence. *British Crop Protection Council*.

Crute, I.R. and Pink, D.A.C. (1996) Genetics and utilization of pathogen resistance in plants. *Plant Cell*, 8, 1747–1755.

Dale, M.F.B. and De Scurrah, M.M. (1998) Breeding for resistance to the potato cyst nematodes *Globodera rostochiensis* and *G. pallida*: strategies, mechanisms and genetic resources. In *Potato Cyst Nematodes, Biology, Distribution and Control*, edited by R.J. Marks and B.B. Brodie, pp. 167–195. Wallingford: CAB Publishing.

Dale, P.J., McPartlan, H.C., Parkinson, R., MacKay, G.R. and Scheffler, J.A. (1992) Gene dispersal from transgenic crops by pollen. In *The Biosafety Results of Filed Tests of Genetically Modified Plants and Microorganisms*, edited by W. Golsar, R. Casper and J. Landsmann, pp. 73–78. *Proceedings of the 2nd International Symposium, Biolgische Bundesanstalt fur Land-und Forstwirtschaft, Braunschweig, Germany*.

Davies, K.G. and Danks, C. (1992) Interspecific differences in the nematode surface coat between *Meloidogyne incognita* and *M. arenaria* related to the adhesion of *Pasteuria penetrans*. *Parasitology*, **105**, 475–480.

De Boer, J.M., Smant, G., Goverse, A., Davis, E.L., Overmars, H.A., Pomp, H., Van Gent-Pelzer, M., Zilverentant, J.F., Stokkermans, J.P.W.G., Hussey, R.S., Gommers, F.J., Bakker, J. and Schots, A. (1996) Secretory granule proteins from the subventral esophageal glands of the potato cyst nematode identified by monoclonal antibodies to a protein fraction from second stage juveniles. *Molecular Plant-Microbe Interactions*, **9**, 39–46.

De Giorgi, C., De Luca, F., Di Vito, M. and Lamberti, F. (1997) Modulation of expression at the level of splicing of *cut-1* RNA in the infective second-stage juvenile of the plant parasitic nematode *Meloidogyne artiellia*. *Molecular and General Genetics*, **253**, 589–598.

de Wit, P.J.G.M. (1992) Molecular characterization of gene-for-gene systems in plant-fungus interactions and the application of avirulence genes in control of plant pathogens. *Annual Revue of Phytopathology*, **30**, 391–418.

Delledonne, M., Xia, Y.J., Dixon, R.A. and Lamb, C. (1998) Nitric oxygen functions as a signal in plant disease resistance. *Nature*, **394**, 585–588.

Devidas, P. and Rehberger, L.A. (1992) The effects of exotoxin Thuringiensin from *Bacillus thuringiensis* on *Meloidogyne incognita* and *Caenorhabditis elegans*. *Plant and Soil*, **145**, 115–120.

Ding X., Shields J, Allen R. and Hussey R.S. (1998) A secretory cellulose-binding protein cDNA cloned from the root-knot nematode (*Meloidogyne incognita*). *Molecular Plant-Microbe Interactions*, **11**, 952–959.

Dixon, R.A., Harrison, M.J. and Lamb, C.J. (1994) Early events in the activation of plant defense responses. *Annual Review of Phytopathology*, **32**, 479–501.

Donegan, K.K., Seidler, R.J., Fieland, V.J., Schaller, D.L., Palm, C.J., Ganio, L.M., Cardwell, D.M. and Steinberger, Y. (1997) Decomposition of genetically engineered tobacco under field conditions: Persistence of the proteinase inhibitor I product and effects on soil microbial respiration, nematode and protozoa, and microarthropod populations. *Journal of Applied Ecology*, **34**, 767–777.

Dürschner-Pelz, U. and Atkinson, H.J. (1988) Recognition of *Ditylenchus* and other nematodes by spores of the endoparasitic fungus *Verticillium balanoides*. *Journal of Invertebrate Pathology*, **51**, 97–106.

Eakin, A.E., Bouvier, J., Sakanari, J., Criak, C.S. and McKerrow, J.H. (1990) Amplification and sequencing of genomic DNA fragments encoding cysteine proteases from protozoan parasites. *Molecular and Biochemical Parasitology*, **39**, 1–8.

Eakin, A.E., Mills, A.A., Harth, G., McKerrow, J.H. and Craik, C.S. (1992) The sequence, organization, and expression of the major cysteine protease (cruzain) from *Trypanosoma cruzi*. *Journal of Biological Chemistry*, **267**, 7411–7420.

Endo, B. (1993) Ultrastructure of subventral gland secretory granules in parasitic juveniles of the soybean cyst nematode, *Heterodera glycines*. *Journal of Heminthological Society of Washington*, **60**, 22–34.

Estruch, J.J., Carozzi, N.B., Desai, N., Duck, N.B., Warren, G.W. and Koziel, M.G. (1997) Transgenic plants: An emerging approach to pest control. *Nature Biotechnology*, **15**, 137–141.

Fernandes, K.V.S., Sabelli, P.A., Barratt, D.H.P. and Richardson, M. (1993) The resistance of cowpea seeds to bruchid beetles is not related to levels of cysteine proteinase inhibitors. *Plant Molecular Biology*, **23**, 215–219.

Fischoff, D.A., Bowdish, K.K., Palace, F.J., Marine, P.G., McCormick, S.M., Niedermeyer, J.G., Dean, D.A., Kusano-Kretzmer, K., Mayer, E.J., Rochester, D.E., Rogers, S.G. and Fralet, R.T. (1987) Insect tolerant transgenic plants. *Biotechnology*, **5**, 807–813.

Fleming, C.C. and Powers, T.O. (1998) Potato cyst nematodes: species, pathotype and virulence concepts. In *Potato Cyst Nematodes Biology, Distribution and Control*, edited by R.J. Marks and B.B. Brodie, pp 51–57. Wallingford: CAB International.

Forrest, J.M.S. and Robertson, W.M. (1986) Characterisation and localisation of saccharides on the head of four populations of the potato cyst nematode *Globodera rostochiensis* and *G. Pallida*. *Journal of Nematology*, **18**, 23–31.

Fossum, K. and Whitaker, J.R. (1968) Ficin and papain inhibitor from chicken egg white. *Archives of Biochemistry and Biophysics*, **125**, 367–375.

Gatehouse, A.M.R., Shi, Y., Powell, K.S., Brough, C., Hilder, V.A., Hamilton, W.D.O., Newell, C.A., Merryweather, A., Boulter, D. and Gatehouse, J.A. (1993) Approaches to insect resistance using transgenic plants. *Philosophical Transactions of the Royal Society of London B*, **342**, 279–286.

Gilchrist, D.G. (1998) Programmed cell death in plant disease: the purpose and promise of cellular suicide. *Annual Review of Phytopathology*, **36**, 399–414.

Goddijn, O.J.M., Lindsey, K., Van Der Lee, F.M., Klap, J.C. and Sijmons, P.C. (1993) Differential gene expression in nematode-induced feeding structures of transgenic plants harbouring promoter-gusA fusion constructs. *The Plant Journal*, **4**, 863–873.

Golinowski, W., Sobczak, M., Kurek, W. and Grymaszewska, G. (1997) The structure of syncytia. In *Cellular and Molecular Aspects of Plant-Nematode Interactions*, edited by C. Fenoll, F.M.W. Grundler and S.A. Ohl, pp 80-97. Dordrecht: Kluwer Academic Publishers.

Gopalan, S., Baurer, D.W., Afano, J.R., Loniello, A.O., He, S.Y. and Collmer, A. (1996) Expression of the *Pseudomonas synringae* avirulence protein AVrB in plant cells alleviates its dependence on the hypersensitive response and pathogenicity (Hrp) secretion system eliciting genotypic-specific hypersensitive cell death. *The Plant Cell*, **8**, 1095–1105.

Gorlach, J., Volrath, S., KnaufBeiter, G., Hengy, G., Beckhove, U., Kogel, K.H., Oostendorp, M., Staub, T., Ward, E., Kessmann, H. and Ryals, J. (1996) Benzothiadiazole, a novel class of inducers of systemic acquired resistance, activates gene expression and disease resistance in wheat. *Plant Cell*, **8**, 629–643.

Goverse, A., Davis, E.L. and Hussey, R.S. (1994) Monoclonal antibodies to the oesophageal glands and stylet secretions of *Heterodera glycines*. *Journal of Nematology*, **26**, 251–259.

Grant, M. R., Godiard, L., Straube, E., Ashfield, T., Lewald, J., Sattler, A., Innes, R. W. and Dangle, J.L. (1995) Structure of the *Arabidopsis RPM1* gene enabling dual specificity disease resistance. *Science*, **269**, 843–846.

Grundler, F., Betka, M. and Wyss, U. (1991) Influence of changes in the nurse cell system (syncytium) on sex determination and development of the cyst nematode *Heterodera schachtii*: Total amounts of proteins and amino acids. *Phytopathology*, **81**, 70-79.

Grundler, F.M.W. and Böckenhoff, A. (1997) Physiology of nematode feeding and feeding sites. In *Cellular and Molecular Aspects of Plant-Nematode Interactions*, edited by C. Fenoll, F.M.W. Grundler and S.A. Ohl, pp. 107–119. Dordrecht: Kluwer Academic Press.

Gurr, S.J., McPherson, M.J., Atkinson, H.J. and Bowles, D.J. (1992) *Plant Parasitic Nematode Control*. International Patent Application Number, PCT/GB91/01540; International Publication Number, WO 92/04453.

Gurr, S.J., McPherson, M.J., Scollan, C., Atkinson, H.J. and Bowles, D.J. (1991) Gene expression in nematode-infected plant roots. *Molecular and General Genetics*, **226**, 361–366.

Gustafson, D.I. (1993) *Pesticides in drinking water*. N. Carolina: Chappell Hill.

Hamilton, W.D. (1980) Sex versus non-sex versus parasite. *Oikos*, **35**, 282–290.

Hammond-Kosack, K., Atkinson, H.J. and Bowles, D.J. (1989a) Systemic changes in the composition of the leaf apoplast following root infection with cyst nematode *Globodera rostochiensis*. *Physiological and Molecular Plant Pathology*, **35**, 495–506.

Hammond-Kosack, K., Atkinson, H.J. and Bowles, D.J. (1989b) Local and systemic changes in gene expression in potato plants following root infection with the cyst nematode *Globodera rostochiensis*. *Physiological and Molecular Plant Pathology*, **37**, 339–354.

Hammond-Kosack, K. E. and Jones, J. D. G. (1995) Plant-Disease Resistance Genes — Unraveling How They Work. *Canadian Journal of Botany-Revue Canadienne De Botanique*, **73**, S495–S505.

Hansen, E., Harper, G., McPherson, M.J. and Atkinson, H.J. (1996) Differential expression patterns of the wound-inducible transgene *wun1–uidA* in potato roots following infection with either cyst or root-knot nematodes. *Physiological and Molecular Plant Pathology*, **48**, 161–170.

Harding, A. and Harris, P.S. (1994) Risk assessment of the release of genetically modified plants: A review. *Issued by MAFF Chief Scientists Group, London*.

Harrop, S.A., Sawangjaroen, N., Prociv, P. and Brindley, P.J. (1995) Characterization and localization of cathepsin B proteinases expressed by adult *Ancylostoma caninum* hookworms. *Molecular and Biochemical Parasitology*, **71**, 163–171.

Hartley, R.W. (1988) Barnase and barstar. Expression of its cloned inhibitor permits expression of a cloned ribonuclease. *Journal of Molecular Biology*, **202**, 913–915.

Hasseloff, J. and Amos, B. (1995) GFP in plants. *Trends in Genetics*, **11**, 328–329.

Hawkes, J.G. (1992) Biosystematics of the Potato. In *The Potato Crop*, edited by P.M. Harris, pp 13–64. London: Chapman and Hall.

Hepher, A. and Atkinson, H.J. (1992) *Nematode Control with Proteinase Inhibitors*. European Patent Publication Number 0 502 730 A1.

Heussler, V.T. and Dobbelaere, D.A.E. (1994) Cloning of a protease gene family of *Fasciola hepatica* by the polymerase chain reaction. *Molecular and Biochemical Parasitology*, **64**, 11–23.

Hilder, V.A., Gatehouse, A.M.R., Sheerman, S.E., Barker, R.F. and Boulter, D. (1987) A novel mechanism of insect resistance engineered into tobacco. *Nature*, **330**, 160-163.

Hilder, V.A., Powell, K.S., Gatehouse, A.M.R., Gatehouse, J.A., Shi, Y., Hamilton, W.D.O., Merryweather, A., Newell, C.A., Timans, J.C., Peumans, W.J., Vandamme, E. and Boulter, D. (1995) Expression of snowdrop lectin in transgenic tobacco plants results in added protection against aphids. *Transgenic Research*, **4**, 18–25.

Hines, M.E., Nielssen, S.S., Shade, R.E. and Pomperoy, M.A. (1990) The effect of 2 proteinase inhibitors, E-64 and the Bowman-Birk inhibitor, on the developmental time and mortality of *Acanthoscelides obtectus*. *Entomologia Experimentalis et Applicata*, **57**, 201–207.

Ho, J.Y., Weide, R., Ma, H.M., Wordragen, M.F., Lambert, K.N., Koornneef, M., Zabel, P. and Williamson, V.M. (1992) The root-knot nematode resistance gene Mi in tomato: construction of a molecular linkage map and identification of dominant cDNA markers in resistant genotypes. *The Plant Journal*, **2**, 971–982.

Hobbelink, H. (1991) *Biotechnology and the Future of World Agriculture*. London : Zed Books Ltd.

Hoffmann, M.P., Zalom, F.G., Wilson, L.T., Smilanick, J.M., Malyj, L.D., Kiser, J., Hilder, V.A. and Barnes, W.M. (1992) Field evaluation of transgenic tobacco containing genes encoding *Bacillus thuringiensis* δ endotoxin or cowpea trypsin inhibitor: efficacy against *Helicoverpera zea* (Lepidoptera: Noctuidae). *Journal of Economic Entomology*, **85**, 2516–2522.

Holm, H., Jorgensen, A. and Hanssen, L.E. (1991) Raw soy and purified proteinase inhibitors induce the appearance of inhibitor-resistant trypsin and chymotrypsin activities in Wistar rat duodenal juice. *Journal of Nutrition*, **121**, 532–538.

Huang, J.S. (1985) Mechanisms of resistance to root-knot nematodes. In *An advanced treatise on Meloidogyne* volume 1. *Biology and Control*, edited by J.N. Sasser and C.C. Carter, pp. 165–174. Raleigh, North Carolina: State University Graphics.

Huang, J.S. and Barker, K.R. (1991) Glyceollin 1 in soybean-cyst nematode interactions. 1. Spatial and temporal distribution in roots of resistant and susceptible soybeans. *Plant Physiology*, **96**, 1302–1307.

Hunt, M.D. and Ryals, J.A. (1996) Systemic acquired resistance signal transduction. *Critical Reviews in Plant Sciences*, **15**, 583–606.

Hussey, R.S. (1989a) Disease-inducing secretions of plant parasitic nematodes. *Annual Review of Phytopathology*, **27**, 123–141.

Hussey, R.S. (1989b) Monoclonal antibodies to secretory granules in esophageal glands of *Meloidogyne* species. *Journal of Nematology*, **21**, 392–398.

Hussey, R.S. and Grundler, F.M.W. (1998) Nematode parasitism in plants. In *The Physiology and Biochemistry of Free-living and Plant-Parasitic Nematodes*, edited by R.N. Perry and D.J. Wright, pp. 213–243. Wallingford: CAB Publishing.

Hussey, R.S. and Mims, C.W. (1991) Ultrastructure of feeding tubes formed in giant cells induced in plants by the root-knot nematode *Meloidogyne incognita*. *Protoplasma*, **162**, 99–107.

Hutcheson, S.W. (1998) Current concepts in active defense in plants. *Annual Review of Phytopathology*, **36**, 59–90.

Jach, G., Gornhardt, B., Mundy, J., Logemann, J., Pinsdorf, E., Leah, R., Schell, J. and Maas, C. (1995) Enhanced quantitative resistance against fungal disease by combinatorial expression of different barley anti-fungal proteins in transgenic tobacco. *The Plant Journal*, **8**, 97–109.

Jannson, H.B. (1994) Adhesion of conidia of *Drechmeria coniospora* to *Caenorhabditis elegans* wild type and mutants. *Journal of Nematology*, **26**, 430-435.

Jenkins, J. (1996) *File on 4*: Transcript of broadcast on Radio 4, 26/3/1996. Programme number 96VY3012NHO. London: British Broadcasting Corporation.

Jones, F.G.W., Larbey, D.W. and Parrott, D.M. (1969) The influence of soil structure and moisture on nematodes especially *Xiphinema, Longidorus, Trichodorus* and *Heterodera* spp. *Soil Biology and Biochemistry*, **1**, 153–165.

Jones, D.A., Thomas, C.M., Hammond-Kosack, K.E., Balintkurti, P.J. and Jones, J.D.G. (1994) Isolation of the Tomato Cf-9 Gene For Resistance to Cladosporium-Fulvum By Transposon Tagging. *Science*, **266**, 789–793.

Jones, F.G.W., Parrott, D.M. and Perry, J.N. (1981) The gene-for-gene relationship and its significance for potato cyst nematodes and their Solanaceous hosts. In *Plant Parasitic Nematodes* Vol. III, edited by B.M. Zuckerman and R.A. Rohde, pp. 255–279. New York: Academic Press.

Jones, M.G.K. and Northcote, D.H. (1972) Nematode-induced syncytium — a multinucleate transfer cell. *Journal of Cell Science*, **10**, 789–809.

Jongsma, M., Jung. A., Bakker, P.L., Peters, J., Bosch, D. and Stiekema, W.J. (1995) Adaption of *Spodoptera exigua* larvae to plant proteinase inhibitors by induction of gut proteinase activity insensitive to inhibition. *Proceedings of the National Academy of Sciences, USA*, **92**, 8041–8045.

Jongsma, M.A., Stiekema, W.J. and Bosch, D. (1996) Combating inhibitor-insensitive proteases of insect pests. *Trends in Biotechnology*, **14**, 331–333.

Joosten, M., Cozijnsen, T.J. and Dewit, P. (1994) Host-Resistance to a fungal tomato pathogen lost by a single base-pair change in an avirulence gene. *Nature*, **367**, 384–386.

Jung, C., Koch, R., Fischer, F., Brandes, A., Wricke, G. and Herrmann, R.G. (1992) DNA markers closely linked to nematode resistance genes in sugar beet (*Reta vulgaris* L.) using chromosome additions and translocations originating from wild beets of the *Procumbentes* species. *Molecular and General Genetics*, **232**, 271–278.

Kaloshian, I., Williamson, V.M., Miyao, G., Lawn, D.A. and Westerdahl, B.B. (1996) 'Resistance-breaking' nematodes in California tomatoes. *Californian Agriculture*, **50**, 18–19.

Kaloshian, I., Yahoobi, J., Liharska, T., Hontelez, J., Hanson, D., Hogan, P., Jesse, T., Wijbrandi, J., Simons, G., Vos, P., Zabel, P. and Williamson, V.M. (1998) Genetic and physical localisation of the root-knot nematode resistance locus Mi in tomato. *Molecular and General Genetics*, **257**, 376–385.

Kaplan, D.J., Keen, N.T. and Thomson, I.J. (1980) Association of glyceollin with the incompatible response of soybean roots to *Meloidogyne incognita*. *Physiological Plant Pathology*, **16**, 309–318.

Keller, B. and Baumgartner, C. (1991) Vascular-specific expression of the bean GRP 1.8 gene is negatively regulated. *The Plant Cell*, **3**, 1051–1061.

Kennedy, M.W. (1991) *Parasitic nematodes: antigens, membranes and genes*. London: Taylor and Francis.

Kleine, M., Cai, D., Eibl, C., Herrmann, R.G. and Jung, C. (1995) Physical mapping and cloning of a translocation in sugar beet (*Reta vulgaris* L.) carrying a gene for nematode (*Heterodera schachtii*) resistance from *B. procumbens*. *Theoretical and Applied Genetics*, **90**, 399–406.

Kleine, M., Cai, D., Klein-Lankhorst, R.M., Sandal, N.N., Salentijn, E.M.J., Harloff, H., Kifle, S., Marcker, K.A., Stiekema, W.J. and Jung, C. (1997) Breeding for nematode resistance in sugarbeet: A molecular approach. In *Cellular and molecular Aspects of Plant-nematode Interactions*, edited by C. Fenoll, F.M.W. Grundler and S.A. Ohl, pp. 176–190. Dordrecht: Kluwer Academic Press.

Klein-Lankhorst, R.P. Rietveld, B., Machiels, R., Verkerk, R., Weide, C., Gebhardt, M., Koornneef and P. Zabel. (1991) RFLP markers linked to the root knot nematode resistance gene *Mi* in tomato. *Theoretical and Applied Genetics*, **81**, 661–667.

Kobe, B. and Deisenhofer, J. (1995) Proteins with leucine-rich repeats. *Current Opinion in Structural Biology*, **5**, 409–416.

Kombrink, E., Schroder, M. and Hahlbrock, K. (1988) Several pathogen-esis-related proteins are β-1,3-glucanases and chitinases. *Proceedings of the National Academy of Sciences, USA*, **85**, 782–786.

Kondo, H., Abe, K., Emori, Y. and Arai, S. (1991) Gene organisation of oryzacystatin-II, a new cystatin superfamily member of plant origin, is closely related to that of oryzacystatin-I but different from those of animal cystatins, *FEBS Letters*, **278**, 87–90.

Kondo, H., Abe, K., Nishimura, I., Watanabe, H., Emori, Y. and Arai, S. (1990) Two distinct cystatin species in rice seeds with different specificities against cysteine proteinases. *Journal of Biological Chemistry*, **265**, 15832–15837.

Kooman-Gersmann, M., Vogelsang, R., Hoogendijk, E.R.M. and de Wit, P.J.G.M. (1997) Assignment of amino acid residues of the AVR9 peptide of *Cladosporium fulvum* that determine elicitor activity. *Molecular Plant Microbe Interactions*, **10**, 821–829.

Koritsas, V.M. and Atkinson, H.J. (1994) Proteinases of females of the phytoparasite *Globodera pallida* (potato cyst nematode). *Parasitology*, **109**, 357–365.

Kumria, R., Verma, R. and Rajam, M.V. (1998) Potential applications of antisense RNA technology in plants. *Current Science*, **74**, 35–41.

Lagrimini, L.M., Burkhart, W., Moyer, M. and Rothstein, S. (1987) Molecular cloning of complementing DNA encoding the lignin-forming peroxidase from tobacco: Molecular analysis and tissue specific expression. *Proceedings of the National Academy of Sciences, USA*, **84**, 7542–7546.

Lambert, K.N. and Williamson, V.M. (1993) cDNA library construction from small amounts of RNA using paramagnetic beads and PCR. *Nucleic Acids Research*, **3**, 775–776.

Larminie, C.G.C. and Johnstone, I.L. (1996) Isolation and characterization of four developmentally regulated cathepsin B-like cysteine protease genes from the nematode *Caenorhabditis elegans*. *DNA and Cell Biology*, **15**, 75–82.

Lee, D.L., Wright, K.A. and Shivers, R.R. (1993) A freeze-fracture study of the cuticle of adult *Nippostrongylus brasiliensis* (Nematoda). *Parasitology*, **107**, 545–552.

Lehane, M.J. (1997) Peritrophic matrix structure and function. *Annual Review of Entomology*, **42**, 525–550.

Leister, D., Ballvora, A., Salamini, F. and Gebhardt, C. (1996) A pcr-based approach for isolating pathogen resistance genes from potato with potential for wide application in plants. *Nature Genetics*, **14**, 421–429.

Liharska, T.B. and Williamson, V.M. (1997) Resistance to root-knot nematode in tomato: towards the molecular cloning of the Mi-1 locus. In *Cellular and Molecular Aspects of Plant Nematode Interactions*, edited by C. Fenoll, F.M.W. Grundler and S.A. Ohl, pp. 191–200. Dordrecht: Kluwer Academic Press.

Lilley, C.J., Urwin, P.E., McPherson, M.J. and Atkinson, H.J. (1996) Characterisation of intestinally active proteinases of cyst-nematodes. *Parasitology*, **113**, 415–424.

Lilley C.J. and Atkinson, H.J. (1997) Promoters for control of root-feeding nematodes. UK patent application number 9524395.2

Lilley, C.J., Urwin, P.E., Atkinson, H.J. and McPherson, M.J. (1997) Characterisation of cDNAs encoding serine proteinases from the soybean cyst nematode *Heterodera glycines*. *Molecular and Biochemical Parasitology*, **89**, 195–207.

Lindsey, K., Wei, W., Clarke, M.C., McArdle, H.F., Rooke, L.M. and Topping, J.F. (1993) Tagging genomic sequences that direct gene expression by activation of a promoter trap in plants. *Transgenic Research*, **2**, 33–47.

Linthorst, H.J.M. (1991) Pathogenesis-related proteins of plants. *Critical Reviews in Plant Science*, **10**, 123–150.

Logemann, J. and Schell, J. (1989) Nucleotide sequence and regulated expression of a wound-inducible potato gene. *Molecular and General Genetics*, **219**, 81–88.

Maki, J. and Yanagisawa, T. (1986) Demonstration of carboxyl and thiol protease activities in adult *Schistosoma mansoni*, *Dirofilaria immitis*, *Angiostrongylus cantonensis* and *Ascaris suum*. *Journal of Helminthology*, **60**, 31–7.

Mann, C. (1997) Reseeding the Green Revolution. *Science*, **277**, 1038–1043.

Marban-Mendoza, N., Jeyaprakash, A., Jansson, H.-B., Damon, Jr., R.A. and Zuckerman, B.M. (1987) Control of root-knot nematodes on tomato by lectins. *Journal of Nematology*, **19**, 331–335.

Mariani, C., Beuckeleer, M., Truettner, J., Leemans, J. and Goldberg, R.B. (1990) Induction of male sterility in plants by a chimaeric ribonuclease gene. *Nature*, **347**, 737–741.

Mariani, C., Gossele, V., Beuckeleer, M., De Block, M., Goldberg, R.B., De Greef, W. and Leemans, J. (1992) A chimaeric ribonuclease-inhibitor gene restores fertility to male sterile plants. *Nature*, **357**, 384–387.

Mercer, C.F., Greenwood, D.R. and Grant, J.L. (1992) Effect of plant and microbial chitinases on the eggs and juveniles of *Meloidogyne hapla* Chitwood. *Nematologia*, **37**, 227–236.

Messeguer, R., Ganal, M., deViceute, M.C., Young N.D., Bolkan, H. and Tanksley, S.D. (1991) High Resolution RFLP map around the root knot nematode resistance gene (*Mi*) in tomato. *Theoretical and Applied Genetics*, **82**, 529–536.

Michaud, D., Cantin, L., Bonade-Bottino, M., Jouanin, L. and Vrain, T.C. (1996) Identification of stable plant cystatin/nematode proteinase complexes using mildly denaturing gelatin/polyacrylamide gel electrophoresis. *Electrophoresis*, **17**, 1373–1379.

Miller, H.I., Altman, D.W., Barton, J.H. and Huttner, S.L. (1995) An algorithm for the oversight of field trials in economically developing countries. *Bio-technology*, **13**, 955–959.

Milligan, S.B., Bodeau, J., Yaghoobi, J., Kaloshian, I., Zabel, P. and Williamson, V.M. (1998) The root-knot nematode resistance gene Mi from tomato is a member of the leucine zipper, nucleotide binding, leucine-rich repeat family of plant genes. *Plant Cell*, **10**, 1307–1319.

Molina, A., Gorlach, J., Volrath, S. and Ryals J. (1999) Wheat genes encoding two types of PR-1 proteins are pathogen inducible, but do not respond to activators of systemic acquired resistance. *Molecular Plant-Microbe Interactions*, **12**, 53–58.

Møller, S.G. and McPherson, M.J. (1998) Developmental expression and biochemical analysis of the *Arabidopsis atao1* gene encoding an H_2O_2-generating diamine oxidase. *The Plant Journal*, **13**, 781–791.

Møller, S.G., Urwin, P.E., Atkinson, H.J. and McPherson, M.J. (1998) Nematode-induced expression of atao1, a gene encoding an extracellular diamine oxidase associated with developing vascular tissue. *Physiological and Molecular Plant Pathology*, **53**, 73–79.

Moore, H. (1995) A new approach to the control of potato cyst nematodes in ware and seed potatoes. PhD Thesis, University of Leeds Library.

Mottram, J.C., Frame, M.J., Brooks, D.R., Tetley, L., Hutchison, J.E., Souza, A.E. and Coombs, G.H. (1997) The multiple cpb cysteine proteinase genes of *Leishmania mexicana* encode isoenzymes that differ in their stage-regulation and substrate preferences. *Journal of Biological Chemistry*, **272**, 14285–14293.

Mueller, E., Gilbert, J., Davenport, G., Brigneti, G. and Baulcombe, D.C. (1995) Homology dependent resistance-Transgenic virus-resistance in plants related to homology-dependent gene silencing. *The Plant Journal*, **7**, 1001–1003.

Newport, G.R., McKerrow, J.H., Hedstrom, R., Petit, M., Mcgarrigle, L., Barr, P.J. and Agabian, N. (1988) Cloning of the proteinase that facilitates infection by schistosome parasites. *Journal of Biological Chemistry*, **263**, 13179–13184.

Niebel, A., Engler, J.D., Tiré, C., Engler, G., Van Montagu, M. and Gheysen, G. (1993) Induction patterns of an extensin gene in tobacco upon nematode infection. *Plant Cell*, **5**, 1697–1710.

Niebel, A., Engler, J.D., Hemerly, A.. Ferreira, P., Inze, D., Van Montagu, M. and Gheysen, G. (1996) Induction of cdc2a and cyc1At expression in *Arabidopsis thaliana* during early phases of nematode-induced feeding cell formation. *Plant Journal*, **10**, 1037–1043.

Ohl S. A., van de Lee, F.M. and Sijmons, P.C. (1997) Anti-feeding structure approaches to nematode resistance. In *Cellular and molecular aspects of plant-nematode interactions*, edited by C. Fenoll, S. Ohl and F. Grundler, pp 250-261. Dordrecht: Kluwer Academic Press.

Opperman, C.H. and Conkling, M. A. (1996) Analysis of transgenic root knot nematode resistant tobacco. *Abstract. Third International Nematology Congress, Guadeloupe.*

Opperman, C.H., Acedo, G.N., Saravitz, D.M., Skantar, A.M., Song, W., Taylor, C.G. and Conkling M.A. (1994) Bioengineering resistance to sedentary endoparasitic nematodes. In *Advances in Molecular Plant Nematology*, edited by F. Lamberti, C., De Giorgi and D. McK. Bird, pp. 221–232. New York: Plenum Press.

Opperman, C.H., Taylor, C.G. and Conkling, M.A. (1994) Root-knot nematode-directed expression of a plant root-specific gene. *Science*, **263**, 221–223.

Oppert, B., Morgan, T.D., Culbertson, C. and Kramer, K.J. (1993) Dietary mixtures of cysteine and serine proteinase inhibitors exhibit synergistic toxicity toward the red flour beetle, *Tribolium castaneum*. *Comparative Biochemistry and Physiology*, **105C**, 379–385.

Ori, N., Eshed, Y., Paran, I., Presting, G., Aviv, D., Tanksley, S., Zamir, D. and Fluhr, R. (1997) The I2C family from the wilt disease resistance locus I2 belongs to the nucleotide binding, leucine-rich repeat superfamily of plant resistance genes. *Plant Cell*, **9**, 521–532.

Parker, J.E., Coleman, M.J., Szabo, V., Frost, L.N., Schmidt, R., vanderBiezen, E.A., Moores, T., Dean, C., Daniels, M.J. and Jones, J.D.G. (1997) The *Arabidopsis* downy mildew resistance gene RPP5 shares similarity to the toll and interleukin-1 receptors with N and L6. *Plant Cell*, **9**, 879–894.

Paulson, R.E. and Webster, J.M. (1972) Ultrastructure of the hypersensitive reaction in roots of tomato, *Lycopersicon esculentum* L., to infection by the root knot nematode *Meloidogyne incognita*. *Physiological Plant Pathology*, **2**, 227–234.

Pearce, F. (1997) Promising the Earth. *New Scientist*, **155**, 4.

Pena-Cortes, H., Sanchez-Serrano, J., Rocha-Sosa, M., and Willmitzer, L. (1988) Systemic induction of proteinase-inhibitor-II gene expression in potato plants by wounding. *Planta*, **174**, 84–9.

Philipp, M, Parkhouse, R.M.E. and Ogilvie, B.M. (1980) Changing proteins on the surface of a parasitic nematode. *Nature*, **257**, 538–540.

Phillips, M.S. Hackett, C.A. and Trudgill, D.L. (1991) The relationship between the initial and final populations densities of the potato cyst nematode *Globodera pallida* for partially resistant potatoes. *Journal of Applied Ecology*, **28**, 109–119.

Pierpoint, W.S., Jackson, P.J., and Evans, R.M. (1990) The presence of a thaumatin-like protein, a chitinase and a glucanase among the pathogenesis-related proteins of potato (*Solanum tuberosum*). *Physiological Molecular Plant Pathology*, **36**, 325–338.

Pierrot, C., Capron, A. and Khalife, J. (1995) Cloning and characterization of two genes encoding *Schistosoma mansoni* elastase. *Molecular and Biochemical Parasitology*, **75**, 113–117.

Politz, S.M. and Philipp, M. (1992) *Caenorhabditis elegans* as a model for parasitic nematodes: a focus on the cuticle. *Parasitology Today*, **8**, 6–12.

Pratt, D., Armes, L.G., Hageman, R., Reynolds, V., Boisvenue, R.J. and Cox, G.N. (1992) Cloning and sequence comparisons of four distinct cysteine proteases expressed in *Haemonchus contortus* adult worms. *Molecular and Biochemical Parasitology*, **51**, 209–18.

Pratt, D., Boisvenue, R.J. and Cox, G.N. (1992) Isolation of putative cysteine proteinase genes from *Ostertagia ostertagi*. *Molecular and Biochemical Parasitology*, **56**, 39–48.

Puri, N., Jenner. C., Bennett, M., Stewart, R., Mansfield, J., Lyons, N. and Taylor, J. (1997) Expression of avPphB, an avirulence gene from *Pseudomonas syringae* pv *phaseolicola* and the delivery of signals causing hypersensitive reaction in bean. *Molecular Plant-Microbe Interactions*, **10**, 247–256.

Pusztai, A., Grant, G., Brown, D.J., Stewart, C. and Bardocz, S. (1992) Nutritional evaluation of the trypsin (EC 3.4.21.4) inhibitor from cowpea (*Vigna unguiculata* Walp.) *British Journal of Nutrition*, **68**, 783–791.

Puzio, P.S., Cai, D., Ohl, S., Wyss, U. and Grundler, F.M.W. (1998) Isolation of regulatory DNA regions related to differentiation of nematode feeding structures in *Arabidopsis thaliana*. *Physiological and Molecular Plant Pathology*, **53**, 177–193.

Rahbe, Y. and Febvay, G. (1993) Protein toxicity to aphids. An *in vitro* test on *Acyrohosiphon pisum*. *Entomologia Experimentalis et Applicta*, **67**, 149–160.

Rahimi, S., Perry, R.N. and Wright, D.J. (1996) Identification of pathogenesis-related proteins induced in leaves of potato plants infected with potato cyst nematodes *Globodera* species. *Physiological and Molecular Plant Pathology*, **49**, 49–59.

Rahimi, S., Perry, R.N. and Wright, D.J. (1998) Detection of chitinases in potato roots following infection with the potato cyst-nematodes, *Globodera rostochiensis* and *G. pallida*. *Nematologica*, **44**, 181–193.

Rahimi, S., Wright, D.J. and Perry, R.N. (1998) Identification and localisation of chitinases induced in the roots of potato plants infected with the potato cyst-nematode *Globodera pallida*. *Fundamental and Applied Nematology*, **21**, 705–713.

Rawlings, N.D. and Barrett, A.J. (1993) Evolutionary families of peptidases. *The Biochemical Journal*, **290**, 205–218.

Ray, C. and McKerrow, J.H. (1992) Gut-specific and developmental expression of a *Caenorhabditis elegans* cystein protease gene. *Molecular and Biochemical Parasitology*, **51**, 239–250.

Ray, C., Abbott, A.G. and Hussey, R.S. (1994) Trans-splicing of a *Meloidogyne incognita* mRNA encoding a putative esophageal gland protein. *Molecular and Biochemical Parasitology*, **68**, 93–101.

Rebois, R.V. (1980) Ultrastructure of a feeding peg and tube associated with *Rotylenchulus reniformis* in cotton. *Nematologica*, **26**, 396–405.

Rice, S.L., Leadbeater, B.S.C and Stone, A.R. (1985) Changes in cell structure in roots of resistant potatoes parasitised by potato cyst-nematodes. I. Potatoes with resistance gene H1 derived from *Solanum tuberosum* ssp. andigena, *Physiological Plant Pathology*, **27**, 219–234.

Richardson, M. (1991) Seed storage proteins: the enzyme inhibitors. *Methods in Plant Biochemistry*, **5**, 259–305.

Ring, M.K., Sun, E., McKerrow, J.H., Lee, G.K., Rosenthal, P.J., Kuntz, I.D. and Cohen, F.E. (1993) Structure-based inhibitor design by using protein models for the development of antiparasitic agents. *Proceedings of the National Academy of Sciences, USA*, **90**, 3583–3587.

Roberts, P.A. (1992) Current status of the availability, development and use of host plant resistance to nematodes. *Journal of Nematology*, **24**, 213–227.

Robertson, B.D., Bianco, A.E., McKerrow, J.H. and Maizels, R.M. (1989) *Toxocara canis*: Proteolytic enzymes secreted by the infective larvae *in vitro*. *Experimental Parasitology*, **69**, 30-36.

Robertson, W.M., Spiegel, Y., Yansson, H.B., Marban-Mendoza, N. and Zuckerman, B.M. (1989) Surface carbohydrates on plant parasitic nematodes. *Nematologica*, **35**, 180-86.

Robinson, M.P., Atkinson, H.J. and Perry, R.N. (1988) The association and partial characterisation of a fluorescent hypersensitive response of potato roots to the cyst nematodes, *Globodera rostochienis* and *G. pallida*. *Revue Nematology*, **11**, 99–107.

Rodis, P. and Hoff, J.E. (1984) Naturally occurring protein crystals in the potato. *Plant Physiology*, **74**, 907–911.

Rossi, M., Goggin, F.L., Milligan, S.B., Kaloshian, I., Ullman, D.E. and Williamson, V.M. (1998) The nematode resistance gene Mi of tomato confers resistance against potato aphid. *Proceedings of the National Academy of Sciences, USA*, **95**, 9750-9754.

Ruppenhorst, H.J. (1984) Intracellular feeding tubes associated with sedentary plant parasitic nematodes. *Nematologica*, **30**, 77–85.

Ryals, J., Ward, E., Ahl-Goy, P. and Metraux, J.P. (1992) Systemic acquired resistance: an inducible defence mechanism in plants. In *Society for Experimental Biology Seminar Series 49: Inducible Plant Proteins*, edited by J.L. Way, pp. 205–229. Cambridge: Cambridge University Press.

Ryan, C.A. (1990) Protease inhibitors in plants: Genes for improving defenses against insects and pathogens. *Annual Review of Phytopathology*, **28**, 425–49.

Ryan, C.A. (1992) The search for the proteinase inhibitor-inducing factor, PIIF. *Plant Molecular Biology*, **19**, 123–133.

Sakanari, J.A., Staunton, C.E., Eakin, A.E., Craik, C.S. and McKerrow, J.H. (1989) Serine proteases from nematode and protozoan parasites: Isolation of sequence homologs using generic molecular probes. *Proceedings of the National Academy of Sciences, USA*, **86**, 4863–67.

Salentijn, E.M.J., Sandal, N.N., Klein-Lankhorst, R., Lange, W., De Bock, T.S.M., Marcker, K.A. and Stiekema, W.J. (1994) Long-range organization of a satellite DNA family flanking the beet cyst nematode resistance locus (*Hsl*) on chromosome-l of *R. patellaris* and *R. procumbens*. *Theoretical and Applied Genetics*, **89**, 459–466.

Sasser, J.N. and Freckman, D.W. (1987) A world perspective on nematology: the role of the society. In *Vistas on Nematology*, edited by J.A. Veech and D.W. Dickerson, pp. 7–14. Hyatsville: Society of Nematologists.

Schlessinger, D. (1990) Yeast artificial chromosomes: tools for mapping and analysis of complex genomes. *Trends in Genetics*, **6**, 248–258.

Seinhorst, J.W. (1967) The relationships between population increase and population density in plant parasitic nematodes. II sedentary nematodes. *Nematologica*, **13**, 157–171.

Sijmons, P.C., Atkinson, H.J. and Wyss, U. (1994) Parasitic strategies of root nematodes and associated host cell responses. *Annual Review of Phytopathology*, **32**, 235–259.

Smant, G., Stokkermans, J.P.W.G., Yan, Y., De Boer, J.M., Baum, T.J., Wang, X., Hussey, R.S., Gommers, F.J., Henrissat, B., Davis, E.L., Helder, J., Schots, A. and Bakker, J. (1998) Endogenous cellulases in animals: Isolation of β-1,4–endoglucanase genes from two species of plant-parasitic cyst nematodes. *Proceedings of the National Academy of Sciences, USA*, **95**, 4906–4911.

Smith, A.M., Dowd, A.J., McGongle, S., Keegan, P.S., Brennan, G., Trudgett, A. and Dalton, J.P. (1993) Purification of a cathepsin L-like proteinase secreted by adult *Fasciola hepatica*. *Molecular and Biochemical Parasitology*, **62**, 1–8.

Smith, H.A., Swaney, S.L., Parks, T.D., Wernsman, E.A., and Dougherty, W.G. (1994) Transgenic plant-virus resistance mediated by untranslatable sense RNAs-expression, regulation and fate of nonessential RNAs. *Plant Cell*, **6**, 1441–1453.

Smith, J.A. and Hammerschmidt, R. (1988) Comparative study of acidic peroxidases associated with induced resistance in cucumber, musk lemon and water-melon. *Physiological and Molecular Plant Pathology*, **33**, 255–61.

Song, W.Y., Wang, G.L., Chen, L.L., Kim, H.S., Pi, L.Y., Holsten, T., Gardner, J., Wang, B., Zhai, W.X., Zhu, L.H., Fauquet, C. and Ronald, P. (1995) A receptor kinase-like protein encoded by the rice disease resistance gene, Xa21. *Science*, **270**, 1804–1806.

Stiekema, W.J., Bosch, D., Wilmink, A., De Boer, J., Schouten, A., Rossien, J., Goverse, A., Smant, G., Stokkermans, J., Gommers, F., Schots, A. and

Bakker, J. (1997) Towards plantibody-mediated resistance against nematodes. In *Cellular and molecular aspects of plant-nematode interactions*, edited by C. Fenoll, S. Ohl and F. Grundler, pp. 262–271. Dordrecht: Kluwer Academic Press.

Taraporewala, Z.F. and Culvert, J.N. (1997) Structural and functional conservation of the tobamovirus coat protein elicitor active site. *Molecular Plant-Microbe Interactions*, **10**, 597–604.

Taviadoraki, P., Benvenuto, E., Trinca, S., De Martinis, D., Cattaneo, A. and Galeffi, P. (1993) Transgenic plants expressing a functional single-chain Fv antibody are specifically protected from virus attack. *Nature*, **366**, 469–472.

Tzortzakakis, E.A. and Gowan, S.R. (1996). Occurrence of a resistance-breaking pathotype of *Meloidogyne javanica* on tomatoes in Crete. *Fundamental and Applied Nematology*, **19**, 960-967.

Uknes, S., Vernooij, B., Morris, S., Chandler, D., Steiner, H.Y., Specker, N., Hunt, M., Neuenschwander, U, Lawton, K., Starrett, M., Friedrich, L., Weymann, K., Negrotto, D., Gorlach, J., Lanahan, M., Salmeron, J., Ward, E., Kessmann, H. and Ryals J. (1996) Reduction in risk for growers. Methods for the development of disease-resistant crops. *New Phytologist*, **133**, 3–10.

Urwin, P.E., Atkinson, H.J., Waller, D.A. and McPherson, M.J. (1995) Engineered oryzacystatin-I expressed in transgenic hairy roots confers resistance to *Globodera pallida*. *The Plant Journal*, **8**, 121–131.

Urwin, P.E., Lilley, C.J., McPherson, M.J. and Atkinson, H.J. (1997a) Characterisation of two cDNAs encoding cysteine proteinases from the soybean cyst nematode *Heterodera glycines*. *Parasitology*, **114**, 605–613.

Urwin, P.E., Lilley, C.J., McPherson, M.J. and Atkinson, H.J. (1997b) Resistance to both cyst- and root-knot nematodes conferred by transgenic *Arabidopsis* expressing a modified plant cystatin. *The Plant Journal*, **12**, 455–461.

Urwin, P.E., McPherson, M.J. and Atkinson, H.J. (1998) Enhanced transgenic plant resistance to nematodes by dual proteinase inhibitor constructs. *Planta*, **204**, 472–479.

Urwin, P.E., Møller, S.G., Lilley, C.J., Atkinson, H.J. and McPherson, M.J. (1997c) Continual GFP monitoring of CaMV35S promoter activity in nematode induced feeding cells in *Arabidopsis thaliana*. *Molecular Plant Pathogen Interactions*, **10**, 394–400.

Urwin, P.E., Møller, S.G., Blumsom, J.K. and Atkinson, H.J. (1999) Continual monitoring of promoter activity in plants. *Methods in Enzymology*, **302**, 316–328.

Vain, P., Worland, B., Clarke, M.C., Richard, G., Beavis, M., Lin, H., Kohli, A., Leech, M., Snape, J., Christou, P. and Atkinson, H.J. (1998) Expression if an engineered cysteine proteinase inhibitor (Oryzacystatin-IδD86) for nematode resistance in transgenic rice plants. *Theoretical and Applied Genetics*, **96**, 266–271.

Van der Eycken, W., De Almeida Engler, J., Van Montagu, M. and Gheysen, G. (1994) Identification and analysis of a cuticular collagen-encoding gene from the plant-parasitic nematode *Meloidogyne incognita*. *Gene*, **151**, 237–242.

Van der Eycken, W., Engler, J.D., Inze, D., Van Montagu, M. and Gheysen, G. (1996) Molecular study of root-knot nematode-induced feeding sites. *The Plant Journal*, **9**, 45–54.

Van der Plank, J.E. (1975) *Principles of plant infection*. New York: Academic Press.

van Huystee, R.B., (1987) Some molecular aspects of plant peroxidase biosynthetic studies. *Annual Revue of Plant Physiology*, **38**, 205–219.

van Loon, L.C., Gerritsen, Y.A.M., and Ritter, C.E. (1987), Identification, purification and characterization of pathogenesis-related proteins from virus-infected Samsun NN tobacco leaves. *Plant Molecular Biology*, **9**, 593–609.

Vos, P., Simons, G., Jesse, T., Wijbrandi, J., Heinen, L., Hogers, R., Frijters, A., Groenendijk, J., Diergaarde, P., Reijans, M., FierensOnstenk, J., deBoth, M., Peleman, J., Liharska, T., Hontelez, J. and Zabeau, M. (1998) The tomato Mi-1 gene confers resistance to both root-knot nematodes and potato aphids. *Nature Biotechnology*, **16**, 1365–1369.

Waldron, C., Wegrich, L.M., Owens Merlo, P.A. and Walsh, T.A. (1993) Characterisation of a genomic sequence coding for potato multicystatin, an eight-domain cysteine proteinase inhibitor. *Plant Molecular Biology*, **23**, 801–812.

Walker, A.J., Ford, L., Majerus, M.E.N., Geoghegan, I.E., Birch, N., Gatehouse, J.S.A. and Gatehouse, A.M.R. (1998) Characterisation of the mid-gut digestive proteinase activity of the two-spot ladybird, *Adalia bipunctata*, and its sensitivity to proteinase inhibitors. *Insect Biochemistry and Molecular Biology*, **28**, 173–180.

Wang, X., Meyers, D., Yan, Y., Baum, T., Smant, G., Hussey, R. and Davis, E. (1999) In Planta localisation of a β 1–4–endoglucnases secreted by *Heterodera glycines*. *Molecular Plant-Microbe Interactions*, **12**, 64–67.

Wasilewski, M.M., Lim, K.C., Phillips, J. and McKerrow, J.H. (1996) Cysteine protease inhibitors block schistosome hemoglobin degradation *in vitro* and decrease worm burden and egg production *in vivo*. *Molecular and Biochemical Parasitology*, **81**, 179–189.

Wen, L., Haung, J.K., Zen, K.C., Johnston, B.H., Muthukrishnan S., MacKay, V., Manney, T.R., Manney, M. and Reeck, G.R. (1992) Nucleotide sequence of a cDNA clone that encodes the maize inhibitor of trypsin and activated Hageman factor. *Plant Molecular Biology*, **18**, 813–814.

White, R.F., Rybicki, E.P., Von Wechmar, M.B., Dekker, J.L. and Antoniw, J.F. (1987) Detection of PR-1 type proteins in Amaranthaceae, Chenopodiaceae, Graminae, and Solanaceae by immunoblotting. *Journal of General Virology*, **68**, 2043–2048.

Williamson, F.A. (1995) Products and opportunities in nematode control. *AGROW Reports* DS 108.

Williamson, V.M. (1998) Root-knot nematode resistance genes in tomato and their potential for future use. *Annual Review of Phytopathology*, **36**, 277–293.

Williamson, V.M. and Colwell, G. (1991) Acid phosphatase-1 from nematode resistant tomato: isolation and characterisation of the gene. *Plant Physiology*, **97**, 129–146.

Wilson, M.A., Bird, D.McK. and van der Knap, E. (1994) A comprehensive subtractive cDNA cloning approach to identify nematode-induced transcripts in tomato. *Phytopathology*, **84**, 299–303.

Winter, G. and Milstein, C. (1991) Man-made antibodies. *Nature*, **349**, 293–299.

Wolfson, J.L. and Murdock, L.L. (1987) Suppression of larval Colorado Potato Beetle growth and development by digestive proteinase inhibitors. *Entomologia Experimentalis et Applicata*, **44**, 235–240.

Wyss, U., Stender, C. and Lehmann, H. (1984) Ultrastructure of feeding sites of the cyst nematode *Heterodera schachtii* Schmidt in roots of susceptible and resistant *Raphanus sativus* L. var. oleiformis Pers. Cultivars. *Physiological Plant Pathology*, **9**, 153–165.

Wyss, U., Grundler, F.M.W. and Münch, A. (1992) The parasitic behaviour of second-stage juveniles of *Meloidogyne incognita* in roots of *Arabidopsis thaliana*. *Nematologica*, **38**, 98–111.

Wyss, U. and Zunche, U. (1986) Observations on the behaviour of second stage juveniles of *Heterodera schachtii* inside host roots. *Revue de Nématologie*, **9**, 153–165.

Yamamoto, Y.T., Taylor, C.G., Acedo, G.N., Cheng, C-L. and Conkling, M.A. (1991) Characterisation of cis-acting sequences regulating root-specific gene expression in tobacco. *Plant Cell*, **3**, 371–382.

Zacheo, G., Bleve-Zacheo, T. and Melillo, M.T. (1997) Biochemistry of plant defence responses to nematode infection In *Cellular and Molecular Aspects of Plant Nematode Interactions*, edited by C. Fenoll, F.M.W. Grundler and S.A. Ohl, pp. 201–213. Dordrecht: Kluwer Academic Press.

Zhou, J., Tang, X. and Martin, G.B. (1997) The Pto kinase conferring resistance to tomato bacterial speck disease interacts with proteins that bind a cis-element of pathogenesis-related genes. *The EMBO Journal*, **16**, 3207–3218.

Zuckerman, B.M. and Jansson, H.B. (1984) Nematode chemotaxis and possible mechanisms of host/prey interactions. *Annual Review of Phytopathology*, **22**, 95–113.

24. Epidemiology and Control of Nematode Infection and Disease in Humans

Donald A.P. Bundy, Helen L. Guyatt and Edwin Michael
The Wellcome Trust Centre for the Epidemiology of Infectious Disease, University of Oxford, South Parks Road, Oxford, OX1 3FY, UK

Introduction

The epidemiology of nematode infections is remarkably well understood. This is because nematodes were amongst the first recognised pathogens of humans, because there have been several major efforts to achieve control, and because there is now a well developed body of ecological theory to help explain their population dynamics. The long history of people and their worms is explored in detail in other volumes (see, for example, Bundy and Michael, 1996; Cox, 1996; Grove, 1990) and elsewhere in this text, and will not be dealt with here. Similarly there are excellent sources describing the extraordinary efforts of the major control programmes which have been conducted on a scale only exceeded by global vaccination programmes. The Rockefeller hookworm eradication campaign, for example, was started in the southern United States at the turn of this century, extended globally in 1913, and only terminated in the 1950s (Bundy, 1990). The observations by Cort, Styles, Stoll and others laid the foundations for epidemiological understanding and provided a data resource which remains useful today. The extensive literature generated by the forty years of operation of the Japanese national control programme — which reduced the prevalence of intestinal nematode infection from 73% in 1949 to <1% in 1990 (WHO, 1996a) — has also provided an important basis for epidemiological understanding, and the full value of these data is only now beginning to be recognised (Kunni, 1992).

Interpretation of this wealth of empirical data has been assisted by the creation of ecological theory, based firmly on empirical evidence but expressed in the language of mathematics, which allows insight into the typically counter-intuitive world of population dynamics. While there have been a number of individuals who have played a crucial role in the development of a theory of helminth dynamics — notably including MacDonald, Morris and Bartlett — the conceptual basis of most current understanding can be traced to the work over the last 25 years of Roy Anderson and Robert May (see their 1991 textbook for a comprehensive treatment of the mathematical and statistical theory).

In this chapter we specifically avoid any direct reference to mathematics, but instead use the theoretical approaches to aid understanding of empirical observation, and hopefully show the value of ecological theory in explaining epidemiology. In the first part of the chapter we examine the population dynamics of helminths and how this relates to patterns of infection and disease. We then explore the use of dynamic models to gain an understanding of the process, costs and effectiveness of different control strategies. The chapter finishes with a brief look forward into the future for disease control.

Population Dynamics of Infection and Disease

Understanding why the size of worm burdens differs between individuals is the starting point for understanding the patterns of helminth infection and disease in communities (Bundy and Medley, 1992). The intensity of infection is a central determinant of the severity of morbidity: the anaemia of hookworm infection, the colitis of trichuriasis and the growth stunting of ascariasis are all related to the number of worms harboured (Stephenson, 1987; Cooper and Bundy, 1988a,b). Intensity also determines the net rate of transmission within a community: the density of infective stages in stool is a function of worm burden, and targeting treatment at the most intensely infected age groups has a disproportionate effect on rates of reinfection (Bundy et al., 1990; Thein-Hlaing et al., 1991). What remains unclear is what determines intensity, both at an individual level and within communities.

Determinants of Worm Burden

Worm burden is a function of the rate of exposure to infective stages, the rate of successful establishment in the face of host protective responses, and the rate of adult parasite mortality (Warren, 1973; Bradley and McCullough, 1974). On first reading it appears that the processes involved are self-evident, but in practice the interpretation of these relationships tends to vary with experience and, hence, intuition. To the health worker it is clear that the level of infection is related to sanitation, hygienic behaviour and socio-economic circumstances. To the laboratory worker it seems equally clear that the human immune response plays a decisive role. It is the relative contribution of these processes that is uncertain.

The level of infection in an individual host is determined by the balance between parasite input

599

(= parasite establishment) and output (= parasite death). This system is reduced to its simplest form by making two assumptions. First, that establishment occurs at a constant rate: the number of parasites that establish per week (say) is constant regardless of the number of parasites already living within the host. Second, that death of parasites occurs at a constant *per capita* rate: the chance of an individual parasite dying is unaltered by the number of parasites with which it shares its host. The second assumption produces a linear relationship between the total death rate (the number of parasites dying per week, say) and parasite burden. The rate of change of the parasite population is then simply the difference between the rates of establishment and death. If the establishment rate is greater than the death rate then the rate of change will be negative, and the parasite population will decrease. Thus, beginning from a situation of zero parasites, the population will grow, but at an ever decreasing rate, until the point is reached where the growth rate becomes (effectively) zero and the parasite population is at stable equilibrium. At this point establishment and death still occur, but they balance each other, so that the parasite population remains constant in size, if not in its composition (Anderson, 1986).

Infection and Host Age

Nematodes are amongst the most ubiquitous of all human infectious pathogens (Table 24.1). Patterns of infection prevalence with age are generally rather similar amongst the major nematode species, exhibiting a rise in childhood to a relatively stable asymptote in adulthood. Maximum prevalence is usually attained before 5 years of age for *Ascaris lumbricoides*, *Trichuris trichiura* and *Enterobius vermicularis*, and in young adults with hookworm infection. In *A. lumbricoides* infection there is usually a slight decline in prevalence during adulthood, but this is less common with the other major nematode species.

There is no simple correspondence between prevalence and intensity, which has the consequence that the observed age-prevalence profiles provide little indication of the underlying profiles of age-intensity. For most

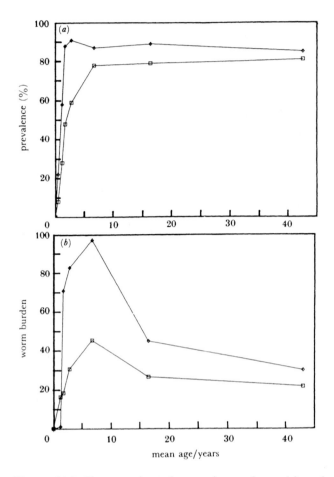

Figure 24.1. The age dependency of prevalence (a) and intensity (b) of *T. trichiura* (□) and *A. lumbricoides* (◆) infection. Prevalence rises monotonically in childhood and attains a stable asymptote in adulthood. Intensity profiles are convex, attaining a maximum value in the 5–10 year age-class and declining in adulthood (*A. lumbricoides* worm burdens x7) (modified from Bundy *et al.* 1987b).

Table 24.1. Global Numbers of the major human nematode infections (in millions)[a]

Helminth	Sub-saharan Africa	Latin America[b]	Middle Eastern Crescent	India	China	Other Asia and Islands	Total
Ascaris lumbricoides	105	171	96	188	410	303	1273
Trichuris trichiura	88	147	64	134	220	249	902
Hookworm[c]	138	130	95	306	367	242	1277
Onchocerca volvulus	17.5	0.14	0.03	–	–	–	17.7
Wuchereria bancrofti[d]	50.2	0.4	0.34	45.5	5.5	13.15	115.1
Brugia malayi[d]	–	–	–	2.6	4.2	6.2	12.9

[a] Regions as defined for the Global Burden of Disease Study. Estimates of prevalence are for both sexes combined. Regional population data on which the geohelminths, *O. volvulus*, *W. bancrofti* and *B. malayi* estimates are based are from the World Bank.
[b] Includes the Caribbean nations.
[c] Both *Necator Americanus* and *Ancylostoma duodenale* combined.
[d] Denotes both infection and disease cases. (modified from PCD, 1997)

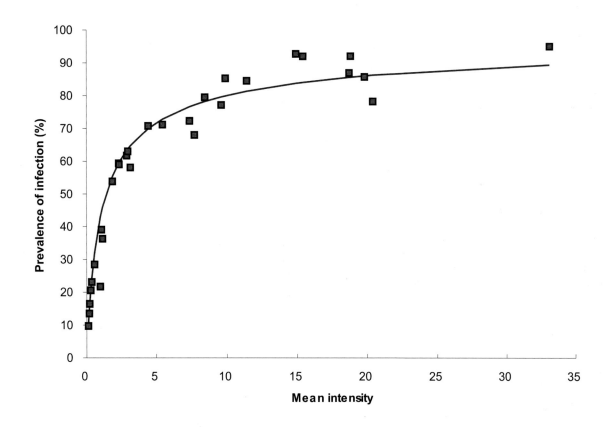

Figure 24.2. The relationship between prevalence and mean worm burden of *Ascaris lumbricoides* infection. Each data point corresponds to a single locality which assessed both the prevalence of infection and the mean intensity by anthelmintic expulsion of worms. Prevalence and mean worm burden have a non-linear association which is well described by the negative binomial probability distribution. The line represents the best maximum-likelihood fits for the negative. (Modified from Guyatt *et al.*, 1990).

nematode species the initial rise in intensity with age closely mirrors that of prevalence but occurs at a slightly slower rate. Maximum intensity occurs at a host age which is parasite species-specific and dependent on parasite longevity, but independent of local transmission rates (Anderson, 1986). For *A. lumbricoides* and *T. trichiura*, maximum worm burdens occur in human populations at 5–10 years of age, and for hookworms at 20–25 years.

Important differences in the age-intensity profiles of these species become apparent after peak intensity has been attained. *A. lumbricoides* and *T. trichiura* both exhibit a marked decline in intensity to a low level which then persists throughout adulthood. Age-profiles based on egg density in stool has suggested that there was considerable variation in the patterns seen in hookworm (Behnke, 1987; Bundy, 1990), but it now appears, from studies where burdens have been enumerated by anthelmintic expulsion, that the intensity attains a stable asymptote, or rises marginally, in adulthood (Pritchard *et al.*, 1990; Bradley *et al.*, 1992).

Thus, for those species with convex age-intensity profiles, but asymptotic age-prevalence profiles, a similar proportion of children and adults are infected, but the adults have substantially smaller worm burdens. With hookworm infection, where both prevalence and intensity are asymptotic, more adults are infected and they have larger worm burdens.

Studies of reinfection with *A. lumbricoides* (Cabrera, 1981) and *T. trichiura* (Bundy *et al.*, 1988; Bundy and Cooper, 1988a) demonstrate that the rate of reacquisition of infection after treatment is age dependent, with children reacquiring worm burdens at a higher rate than adults. This may suggest that age-related changes in establishment rather than death of adult worms determine age-related changes in intensity. However, establishment will be affected by age-related changes in exposure, survival of immature stages or both. Since the rate of establishment reflects a balance between exposure and resistance to infection, these results imply, for the species other than hookworm, that either children are more exposed to infection or adults are less susceptible, or some combination of both processes.

Note that such longitudinal reinfection studies, as against cross-sectional studies, allow the separation of the age — and time — dependent components of infection.

Infection in the Host Population

The distribution of helminths among hosts is typically observed to be overdispersed. The fact that most hosts harbour few or no worms, while the majority of the worm population is found in the minority of hosts, has many consequences, both in terms of the population biology of the parasites and the clinical consequences for the host (Anderson and Medley, 1985; Guyatt and Bundy, 1991).

Simple ecological theory shows that density dependence must be operating at some point in the parasite life-cycle, and it is probable that it is the parasitic stages within the host that are regulated the most intensely by, for example, acquired immunity (Anderson and May, 1985a). Using simulation models to compare different methods of heterogeneity generation and density dependence, Quinnell, Medley and Keymer (1990)

showed that the specific parasite stage which attracts regulation will alter both the demographic characteristics of helminth infections (such as parasite age-distributions within hosts) and the patterns of parasite population change in response to perturbation. It has been shown that the *per capita* egg production rate is reduced as worm burden increases for all the major geohelminths (Anderson and Schad, 1986; Elkins *et al.*, 1986; Bundy, 1986); however, demonstration and detection of the ecological consequences of such density dependence is fraught with empirical and statistical difficulties (Keymer and Slater, 1987).

One of the more important empirical observations, and one that may help explain the mechanisms of over-dispersion, arose from studies of reinfection which indicated that individuals tend to be predisposed to a high or low intensity of infection. Predisposition has been demonstrated for human infection with *A. lumbricoides* (Elkins *et al*, 1986; Haswell-Elkins *et al.*, 1987; Holland *et al.*, 1989), *T. trichiura* (Bundy *et al.*, 1987a; Bundy and Cooper, 1988b), and hookworm (Schad and Anderson, 1985; Bradley and Chandiwana, 1990).

There is also a trend for more individuals in a family group to have intense (or light) infection than one would expect by chance (Forrester *et al.*, 1988), and the mean intensity of infection in a family unit exhibits predisposition in the same way as infection in an individual (Forrester *et al.*, 1990). Covariance between initial and reinfection mean worm burdens is significantly stronger if the data are analysed using natural family units rather than randomly assigned groups of the same individuals (Chan *et al.*, 1992). *A. lumbricoides* and *T. trichiura* expulsion data from families were analysed to seek correlates between the age-standardised worm burdens of parents and between the burdens of parents and their children (Chan, 1991).

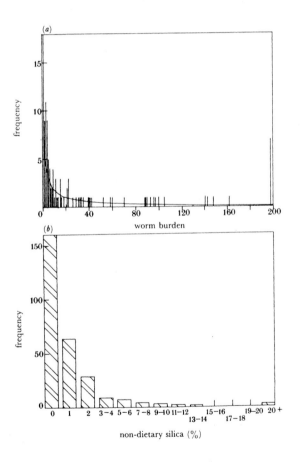

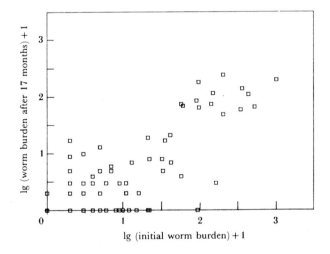

Figure 24.3. *(a)* Frequency distribution of *T. trichiura* infection intensity. The distribution is markedly overdispersed (data from Cooper and Bundy, 1988b). *(b)* Frequency distribution of geophagia in children as assessed by levels of silica in stool which exceed normative dietary amounts. The distribution of this exposure parameter is overdispersed and resembles that shown in *(a)* above.

Figure 24.4. Predisposition to *T. trichiura* infection. A comparison between initial worm burden (determined by anthelmintic expulsion) and the worm burden acquired during 17 months of re-exposure to infection (data from Bundy, 1986).

Despite analysis of burdens from the same families over three cycles of anthelmintic intervention, no consistent trend was detected. This suggests that if there is a genetic basis for predisposition its effects tend to be overwhelmed by other household-specific factors, such as the peridomiciliary environment and family-specific practices and behaviours. It is, perhaps evidence that the balance of influence tends towards ecology rather than immunology.

Analysis of human hookworm (Anderson and May, 1985a) and *Schistosoma haematobium* infections (Woolhouse *et al.*, 1991) has predicted that acquired immunity, engendered by accumulated past experience of infection and acting to moderate the rate of parasite establishment, will cause age-infection profiles to become more convex in areas of higher transmission, resulting in larger and earlier peaks of intensity: the so-called 'peak shift' phenomenon (Anderson and May, 1985a; Woolhouse *et al.*, 1991; Fulford *et al.*, 1992). Reductions in transmission, due to mass chemotherapy for example, will therefore act to diminish the peak intensity. However, they may also delay the onset of immunity and move the peak of intensity into an older age class (Anderson and May, 1985a).

These studies assume that the observed age-infection profile is a reflection of the dynamic interaction between exposure and acquired herd immunity (Anderson and May, 1985a; Woolhouse *et al.*, 1991). Yet the parasitic infections upon which these studies are based allow no direct estimates of exposure, which is estimated retrospectively from the levels of infection achieved, although attempts to examine the impact of water contact patterns on age patterns of infection have been made for schistosomiasis (Hagan, 1992; Woolhouse, 1994). Parallel estimates of community exposure rates and age-infection profiles are, however, readily available for the mosquito-borne lymphatic filariasis (Southgate, 1992), and this allows not only the examination of the role of herd immunity in observed age-infection patterns, but also the direct examination of the proposed relationship between exposure and the generation of immunity to the parasite.

Analysis of available data on anopheline and culicine transmitted *W. bancrofti* infection shows that observed infection patterns are consistent with the occurrence and operation of an effective herd immunity in endemic human communities (Michael and Bundy, 1998). The effect of this immunity is greater and occurs earlier in

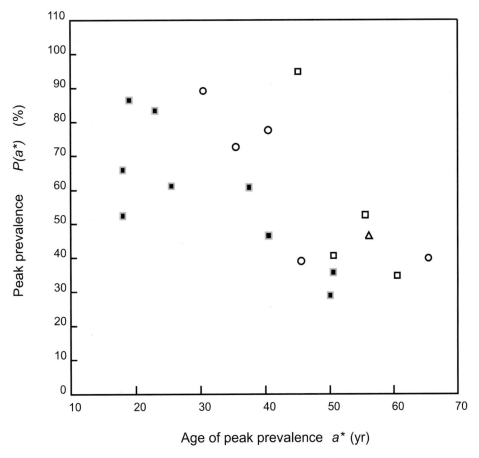

Figure 24.5. Observed relationship between peak prevalence of infection, $P(a^*)$, and age and peak of prevalence, a^*, for human *W. bancrofti* infections. Symbols differentiate the data by the type of mosquito species implicated in transmission at each study site as follows: filled squares, culicine transmission; open circles, anopheline transmission; open square, mixed culicine and anopheline transmission; open triangles, *Aedes* transmission. The depicted relationship between $P(a^*)$ and a^* is in the direction predicted by the model incorporating acquired immunity.

areas with higher rates of mosquito biting. This provides an explicit demonstration that human acquired immunity to macroparasites is correlated with infection exposure (Anderson and May, 1985a; Crombie and Anderson, 1985).

These results have important implications for the control of nematodes, including filarial nematodes, by mass chemotherapy. They imply that if control is implemented at a level less than that required to eradicate the parasite it has the potential to significantly reduce the level of herd immunity in a population and raise the infection burden in the older age classes above the level pertaining before control, especially in areas of previously high transmission (Anderson and May, 1985a). This is not an argument against control, but is a strong implication that it may be prudent to carefully define control components (e.g. target age groups and coverage), perhaps with the aid of model predictions, if helminth control is to be successfully implemented on a large scale.

Patterns of Morbidity and Infection

Our understanding of the impact of intestinal nematode infections on the health of humans has advanced in recent years. It is becoming increasingly clear that the effects of infection are worse than they were assumed to be in the past, and there is increasing evidence that chemotherapy (i.e. treatment of the infection) can reverse most of these effects.

Hookworms have long been recognised as an important cause of iron-deficiency anaemia. The feeding behaviour of hookworms results in mechanical blood loss into the gut and the *per capita* loss of blood is constant. Despite this, the severity of the resultant iron-deficiency anaemia is not linearly related to the worm burden, as it is governed by many other factors, including the iron status and general nutritional status of the host, the quality and quantity of iron sources in the diet, the absorptive properties in the gut, ulcers, gut lesions, and other infections which may adversely influence iron status (Crompton and Whitehead, 1993). Since the largest worm burdens in hookworm infections are generally seen in adults, hookworm-related anaemia has been viewed principally as a problem of the older population, especially adolescent girls and women of child-bearing age. However, there is now evidence from intervention studies in Zanzibar and mainland Tanzania that hookworm infections also contribute significantly to anaemia in schoolchildren (Stoltzfus *et al.*, 1997; Beasley *et al.*, 1999).

In contrast to hookworm, *A. lumbricoides* and *T. trichiura* were widely held to cause little ill-health in the host, except for the rare infections which led to serious complications such as intestinal obstruction (in ascariasis) or rectal prolapse (in trichuriasis), complications generally associated with a very heavy worm burden. However, intervention studies have shown that infection with as few as 10–15 *A. lumbricoides* is associated with reversible deficits in growth and physical fitness in school age children (Stephenson *et al.*, 1993a; Stephenson *et al.*, 1993b; Thein-Hlaing, 1993). Similarly *T. trichiura* has been

recognised as causing Trichuris Dysentery Syndrome in heavily infected children, and growth retardation and anaemia in children with less heavy infections (Cooper and Bundy, 1988a).

Fortunately, it appears that many of the growth and nutritional deficits caused by these helminth infections are reversible with the use of anthelmintics; in fact many of the studies that provide the clearest causative evidence for health deficits in intestinal nematode infections are intervention studies. For example, children with Trichuris Dysentery Syndrome have been shown to exhibit 'catch-up growth' after treatment of intense infections with *T. trichiura* (Cooper *et al.*, 1995). In addition, a recent study in northern India has shown that albendazole given every six months to children aged 1–4 years led to an extra weight gain of one kg over a period of 2 years, a 35% difference (Awasthi *et al.*, 1997).

The global burden of disease attributable to the major geohelminths has been estimated using the Disability Adjusted Life Year (DALY) metric (World Bank, 1993; Chan *et al.*, 1994a; Bundy *et al.*, 1998). These estimates indicate that intestinal nematodes are a leading contributor to the burden of disease in school children.

Effects of Infection on Intellectual Development

Intestinal helminth infections can have a detrimental effect on cognition and educational achievement in children (Sternberg, Grigorenko and Nokes, 1997). Many of the sequelae of helminth infection such as undernutrition, and iron deficiency anaemia, are associated with deficits in cognitive functioning (Simeon and Grantham-McGregor, 1990; Watkins and Pollitt, 1997). Low height for age (stunting) has been associated with detriments in cognitive function, in mental development, in behaviour and in educational achievement; it is a striking feature of intense trichuriasis (Cooper and Bundy, 1988b) and a not uncommon consequence of ascariasis (Thein-Hlaing, 1993).

Early studies provided correlational evidence that children infected with hookworm and to a lesser extent, *A. lumbricoides*, suffered detrimental effects on educational achievement but did not separate the effects of infection from those of confounding variables such as socio-economic status (World Bank, 1993). The possibility of a causal association between helminth infection and education or cognitive function has been addressed by a few intervention studies. The effect of Trichuris Dysentery Syndrome on the mental development of Jamaican schoolchildren was examined in a case-controlled study (Callendar *et al.*, 1992). Significant improvement was seen after one year, in nutritional status and in the locomotor subscale of the Griffiths test of mental development. In a double-blind placebo controlled trial involving children with moderate-heavy loads of *T. trichiura*, expulsion of worms led to a significant improvement in tests of auditory short-term memory and a highly significant improvement in the scanning and retrieval of long-term memory (Nokes *et al.*, 1992).

Thus the implied evidence for an effect of helminth infection on cognitive function is persuasive, but the evidence from correlational and intervention studies still leaves many uncertainties concerning the extent and nature of the effect. Even small effects will have major practical implications for child development since the peak of infection intensity, and presumably impact on cognitive ability coincides with the age when children are in school. For most children in low income countries this will be their only opportunity for formal education, and the opportunity may be compromised by ill-health.

Infection and Mortality

Although geohelminth infections rarely cause death the large number of infections which occur globally has the effect that even these rare events are of significance. The WHO attributes 65,000 deaths a year to hookworm infections, 60,000 deaths a year to *A. lumbricoides*, and 10,000 deaths a year to trichuriasis (WHO, 1997); the means by which these estimates are derived are not described. Using a mathematical model (Chan *et al.*, 1994) and whatever little empirical data are available, it has been estimated that the acute complications of *A. lumbricoides* infections (mainly intestinal obstruction and biliary complications) could result in approximately 10,000 deaths each year, mostly affecting children below the age of 10 years (de Silva, Chan and Bundy, 1998). *T. trichiura* may cause a similar number of deaths through Trichuris Dysentery Syndrome or intussusception and hookworm probably contributes to a larger number of deaths than do ascariasis and trichuriasis, but there is a general lack of empirical data. It is safe to conclude, however, that the burden of ill health attributable to geohelminths is related less to mortality than to the insidious effects on physical and mental growth.

Mathematical Models of the Population Dynamics of Nematodes

A number of epidemiological models of nematode infections, based on differential equation systems, have been developed (reviewed by Anderson and May, 1991; Woolhouse, 1991), although the computational implementation of these models presents difficulties due to the importance of the non-normal parasite distributions in determining the dynamics of the system (Anderson and Medley, 1985; Medley *et al.*, 1993). An alternative modelling approach, using stochastic microsimulations has been used to model some helminths, but not yet for intestinal nematodes (Plaiser *et al.*, 1990; Habbema *et al.*, 1992). The empirical manner with which many epidemiological and environmental processes are incorporated and the large number of local parameters required as inputs suggests that complex simulation models may have limited applicability when used outside of the originally intended locality or time-scale.

A computer-based epidemiological model to assess community chemotherapy programmes for intestinal helminths was developed by Medley *et al.* (1993), based on the differential equation models of Anderson (1980). This model includes a dynamically changing worm burden distribution which more accurately reflects the effect of control measures and improves the estimation of morbidity. An age-structured version of this model was developed for *A. lumbricoides* and *T. trichiura* (Chan *et al.*, 1994b) in order to examine the effects of host age on transmission dynamics. For these species, the highest intensity of infection is generally observed in school-age children who may also contribute relatively more to transmission due to insanitary behaviour patterns.

In these deterministic models the helminth life-cycle is conceptualised as consisting of a population of mature worms in human hosts and a population of free-living infective stages in the environment. The host population becomes infected by contact with the infective stages. The model divides the host population into children and adults (Figure 24.6) with the important assumed difference that children both acquire infections and contaminate the environment at a higher rate than adults. For both host groups, there is an overdispersed worm burden distribution such that most hosts harbour few worms and a few hosts have consistently higher worm burdens. These differential rates of infection acquisition are preserved during reinfection after treatment, as has been observed (Bundy *et al.*, 1988). Mass chemotherapy is simulated as instantaneous killing of worms. The coverage and frequency of treatment can be varied and treatment can be given to the whole population or to children only. The morbidity due to helminth infection is assumed to be associated with the worm burden by using a threshold worm burden above which morbidity occurs (Guyatt and Bundy, 1991). The changes in mean worm burden, infection prevalence and prevalence of morbidity for both groups can be followed over several rounds of treatment.

Dynamic Behaviour and Model Validation

The results of a simulated age-targeted treatment programme using the model are shown in Figure 24.7. In this simulation, children were treated during years 1–5 and adults were not treated at all. The figure shows the initial morbidity levels during the first year, the effect of the 5 annual treatments, and then the rebound in morbidity levels over the next 5 years. The morbidity (estimated as the prevalence of infections with at least 20 worms) is plotted over time separately for adults and children. When the treatment occurs, morbidity drops instantaneously in the children, but then rises again as they become reinfected. After the last treatment occurs, the morbidity slowly rebounds to the pre-control level after about 5 years. After each treatment, morbidity levels do not increase immediately and may even decrease initially. This initial decrease is because of the overall reduction in transmission caused by the intervention. The figure also shows that a significant reduction in morbidity for the adults can result from treatment of children only.

The benefit of the control programme in terms of morbidity reduction is estimated as the area between the morbidity curve and the equilibrium level. The

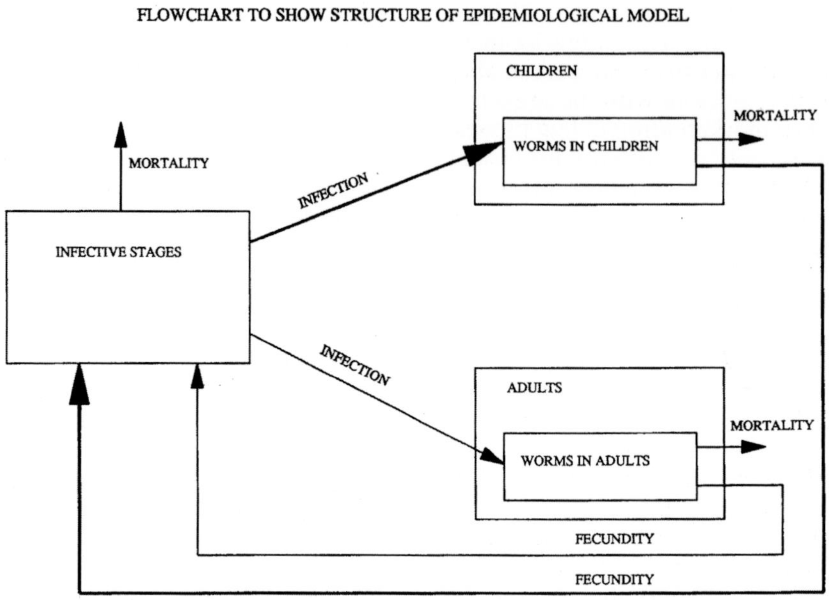

FLOWCHART TO SHOW STRUCTURE OF EPIDEMIOLOGICAL MODEL

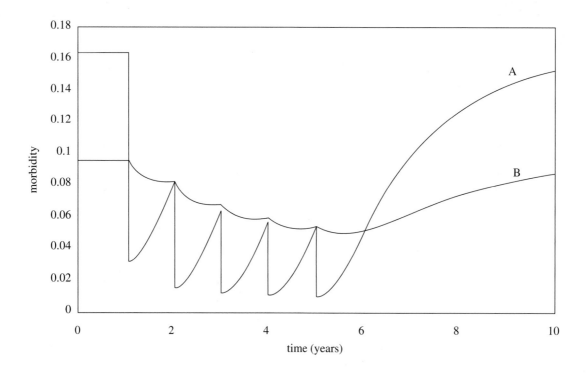

Figure 24.6. Flowchart to show structure of a 2-age-group model. The host population is divided into children and adults with higher rates of infection acquisition and contamination for children.

Figure 24.7. An example of the results of a simulated chemotherapy programme. The prevalence of morbidity (estimated as the prevalence of infections with more than 20 worms) is plotted against time. A shows child morbidity and B adult morbidity. Children are treated at a coverage of 80% with a drug of efficacy 90% yearly from years 1 to 5. Adults are not treated. The parameter values are, $W_C = 10$, $W_A = 7$, $R_{OC} = 1.5$, $\pi_C = 0.25$, $\mu = 1$, k = 0.543.

benefit in terms of morbidity reduction is therefore greater in the children than in the adults. However, since the total population benefit is the sum of the benefits of each group, weighted by the population in each group, and the population of adults is larger than that of the children, the benefit to untreated adults may be a significant proportion of the total community benefit of the chemotherapy programme. The benefits of a child targeted chemotherapy programme will therefore depend on the relative contributions of adults and children to both transmission dynamics and community morbidity.

The model has been validated using data from an *A. lumbricoides* control programme in Myanmar (Thein-Hlaing *et al.*, 1991), in which children under 15 years old were treated every 3 months over a period of 2 years, but adults were not treated. For model validation, parameter values were taken from the original data and the reproductive number (R_o) was assumed to be equal in adults and children. Using a correction to exclude male-worm-only , and therefore non-fecund infections (from Guyatt, 1992; Guyatt and Bundy, 1993), the model gave a very good fit to the worm burden data which were determined directly by anthelmintic expulsion. Although the assumption that the parasite population reproductive number is equal in adults and children gives a good fit to the data, the possibility that the reproductive numbers are different cannot be excluded.

The model was also validated using data from a *T. trichiura* control programme in 11,500 people on the island of Montserrat, West Indies (Bundy *et al.*, 1990 and Bundy unpublished data). At least 90% of children

aged 2–15 years were treated every 4 months for more than 2 years, and although treatment was offered to adults, coverage in adults was less than 4% for any treatment and it was assumed in the model that adults were not treated. After alternate treatment cycles, infection prevalence was estimated separately for target and non-target age groups in a random age-stratified sample of the population. Parameter values were again taken from the original data and it was assumed that the ratio of reproductive numbers in adults and children was 0.5. The results (Figure 24.8) demonstrate that the model is able to reflect the decline in prevalence over time in both the target and untreated groups; a result which implies that children are an important source of infection to adults, and that reduction in infection levels in the whole population can be achieved by treating the children alone.

Estimating the Cost and Cost-Effectiveness of Nematode Control

Since the resources available for health care are finite, consideration must be given not only to the impact of a control programme, but also to the costs involved. A number of economic evaluation techniques have been used to aid the allocation of scarce resources to helminth control. The techniques differ in the outcome measure assessed, and the choice of technique depends on the aims of the analysis (Bundy and Guyatt, 1992). *Cost-benefit analysis* values the outcome in monetary terms and is concerned with the broad question of whether control of a specific helminth infection is

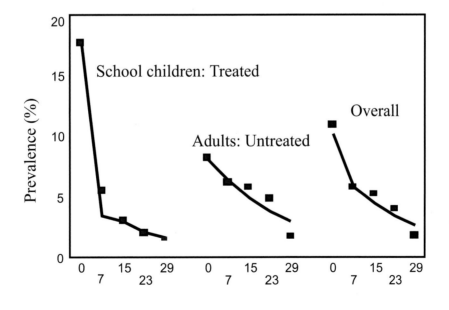

Time of Survey (months)

Figure 24.8. Validation of model with field data for *T. trichiura* infection from the island of Montserrat, West Indies. Data have been obtained from a study by Bundy *et al.* (1990) in which children were treated every 4 months throughout the study period. Parameters used for the models are the same as reported in the field study and are: $W_C = 0.3$, $W_A = 0.1$, $R_{OC} = 2$, $R_{OA} = 1$, $\pi_C = 0.22$, $\mu = 1$, $k = 0.3$. The graph shows prevalence of infection over time for children and adults: (■) data points; (—) model output.

worthwhile in comparison with expenditure on other projects (health or otherwise); this is difficult to apply to helminth control because of the complexities associated with valuing any health benefits (but see Stephenson *et al.*, 1990). *Cost-effectiveness analysis* is concerned with identifying the most efficient option among a range of alternative strategies. Cost-effectiveness analysis can be used to facilitate control programme design, once it is accepted that control of a given nematode infection is socially worthwhile (Evans and Guyatt, 1995). *Cost-utility analysis* addresses the question of whether intervention of any form, even the most efficient, is a cost-effective use of resources, taking into account the competing demands of other health interventions. The health benefit outcome measure in the example given in this chapter is the Disability Adjusted Life Year (DALY) metric (Murray and Lopez, 1994).

Most cost-effectiveness analyses have focused on strategies for delivering chemotherapy, as this is generally considered to be the most cost-effective approach to controlling intestinal nematode infection (Bundy and de Silva, 1998). The basis for these strategies is the control of morbidity, rather than infection, and includes targeting treatment at high-risk groups, such as children, and reducing intervals between treatment cycles (Bundy, 1988). Cost-effectiveness analysis has been used to assess the relative merits of these and other treatment strategies (Guyatt *et al.*, 1993; 1995).

The examples given in this chapter address a number of delivery issues including the age of the target population, mass treatment versus treatment based on diagnosis (screening), variations in the frequency of drug delivery, the coverage, drug efficacy and drug price. Although detailed empirical estimates of the costs and effectiveness of control (e.g. by Savioli *et al.*, 1989) have considerable value in indicating to the health planner what can be achieved in practice, since only one study can be managed at one site, evaluating the cost-effectiveness of different strategies in different sites leads to problems of interpretation since the control approach is only one amongst many variables. Even when comparative studies using similar and adjacent sites are undertaken there are still problems, not least being the considerable expense and the long time frame required (Bundy and Guyatt, 1992). The advantage of the modelling approach to cost-effectiveness analysis is that such analytical frameworks can permit predictions to be made in different endemic settings, and sensitivity analysis of key variables can be undertaken (Guyatt and Tanner, 1996).

Estimating Costs Using a Static Model of Nematode Control

Prescott (1987) developed a generalised static economic framework for investigating the optimal strategy for chemotherapy which has been used to explore the cost-effectiveness of six strategies of drug delivery for the combined control of intestinal nematodes and schistosomiasis (Warren *et al.*, 1993) (see Figure 24.9).

As was demonstrated by the earlier model (Prescott, 1987), compliance behaviour was critical in determining the effectiveness of screening approaches. Based on empirical cost data, the screening options — in which individual diagnosis preceded treatment — were also shown to be more costly than the mass treatment options. The cost disadvantage of the screening options occurred because the slightly lower treatment costs achieved by only treating infected or heavily-infected individuals are offset by the very high costs of the screening process itself. This disadvantage was most acute for the screening for heavy infection option, because of the lower technician productivity in assessing specimens for intensity rather than mere presence of infection. Sensitivity analysis was undertaken to explore how the ranking of the different strategies would compare if screening compliance was increased to 100%, and if the prevalence of infection was highest in adults, as would be the case for hookworm infection. Under both these scenarios, mass population chemotherapy was still shown to dominate in terms of cost-effectiveness. When screening compliance was increased to 100%, the screening options achieved the same effectiveness as the corresponding mass treatment options, but the costs increased because of the higher volume of screening. The major consequence of increasing the prevalence of infection to 100% in adults was to lower the effectiveness and increase the cost of all the screening options. The general conclusion of the analysis was that if there is a budget constraint, mass chemotherapy options are preferred; child-targeted treatment at lower budget levels and population treatment at higher budget levels.

Estimating Costs Using a Dynamic Model of Nematode Control

In the application of mass chemotherapy, one important realm of indecision lies in the frequency of treatment, which depends in part on the intensity of transmission in that endemic locality (Anderson and May, 1985b). Guyatt *et al.* (1993) incorporated cost analysis into the dynamic framework of Medley *et al.* (1993) to assess the cost-effectiveness of alternative strategies of mass chemotherapy which varied in the frequency of treatment of *Ascaris lumbricoides* infection in high and low transmission areas. The effectiveness of a 5 year programme, with treatment at intervals of 4 months, 6 months, 1 year and 2 years, was assessed over 10 years. Since the control programme component is assumed to last for 5 years, it is obvious that the less frequent the treatments, the fewer treatments that are given.

One of the most important qualitative results from this analysis was that measuring effectiveness and cost-effectiveness in terms of the reduction in prevalence of infection gives conclusions which conflict with assessment in terms of reduction in intensity. This is an important observation given that most control programmes are evaluated in terms only of the reduction in prevalence of infection. Figure 24.10 illustrates the relationship between cost and effectiveness for the four

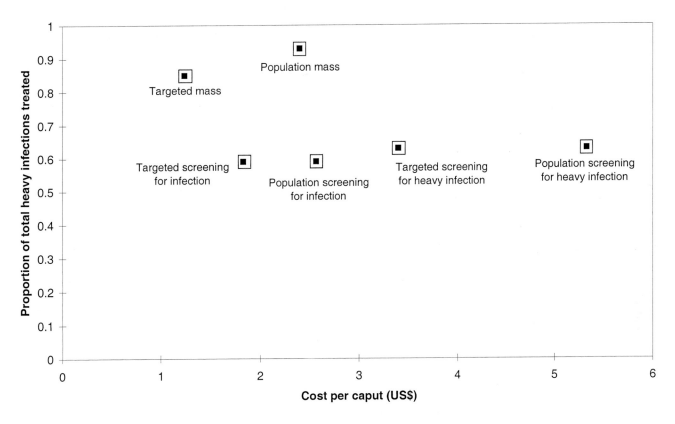

Figure 24.9. The costs and effectiveness (proportion of total heavy infections treated) of six strategies for the delivery of anthelmintics. (modified from Warren *et al.*, 1993).

treatment options in the high and low transmission areas, in terms of reduction in infection and in heavy infection. Reduction in heavy infection was maximised in the high transmission area and when treating frequently (every 4 months). Maximal reduction in infection prevalence, in contrast, was observed in the low transmission area. Since, in this model, the costs are independent of endemicity, the cost-effectiveness ratios for heavy infection reduction are consistently lower in the high transmission area than in the low transmission area (see Table 24.2), suggesting it is more cost-effective to intervene in the high transmission area. The cost-effectiveness ratios also indicate that the most cost-effective intervention in terms of heavy infection reduction is to treat every 2 years.

Although treating every 2 years minimises the cost per heavy infection case prevented per person, treating every year provides an extra gain in effectiveness, but at an extra cost. The extra cost required to obtain an extra unit gain in effectiveness, by treating more frequently, is expressed in terms of incremental cost-effectiveness ratios (see Table 24.2). In pictorial terms, these values can be understood by examining the gradient of a line joining any two alternative strategies in Figure 24.10. The steeper the line, the more efficient the more costly

alternative, as any increase in cost returns a high increase in effectiveness. For instance, treating every year rather than every 2 years requires an extra cost of US$0.83 per extra gain in heavy infection cases prevented per person in the high transmission area, and an extra US$138.62 in the low transmission area (see Table 24.2). It is evident that treating more frequently requires less cost input to achieve a gain in disease prevalence reduction in a high transmission area than in a low transmission area. The incremental cost-effectiveness ratios for disease reduction also reveal diminishing marginal returns; that is, as the frequency of treatment is increased, a higher cost investment is required per extra gain in effectiveness (for heavy infection reduction) (see Table 24.2). Studies have suggested that the interval between treatments for *A. lumbricoides* infection should be relatively short, of the order of 3 to 4 months (Cabrera *et al.*, 1975; Arfaa and Ghadirian, 1977) depending on the infection rate of the endemic area (Morishita, 1972). These estimates, however, were based on the rate of rebound in infection prevalence after treatment. Previous studies have shown that prevalence recovers more rapidly than intensity and thus that the frequency of treatment required to maintain low levels of intensity is typically

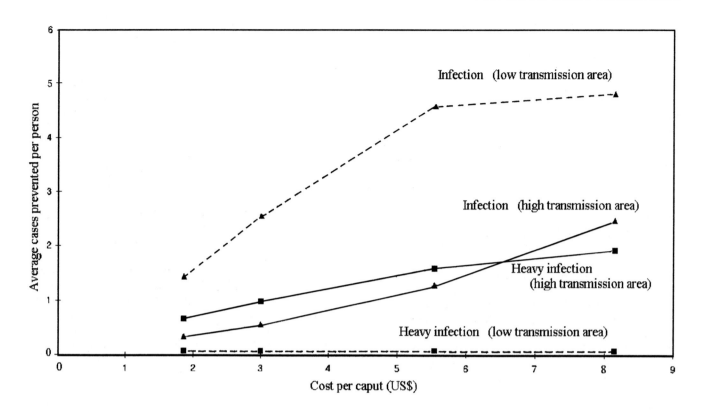

Figure 24.10. The relationship between effectiveness (average infection and heavy infection cases prevented per person) and costs per caput at increasing frequency of treatment directed at *Ascaris lumbricoides* in a high and low transmission area (modified from Guyatt *et al.*, 1993).

Table 24.2. Cost-effectiveness ratios and incremental cost-effectiveness ratios for delivering anthelmintic treatment directed at *Ascaris lumbricoides* at different frequencies in a high and low transmission area (modified from Guyatt *et al.*, 1993)

	High transmission area		Low transmission area	
	Heavy infection	Infection	Heavy infection	Infection
Cost (US$) per case prevented				
2 yearly treatment	2.82	5.65	27.88	1.30
yearly treatment	3.05	5.44	40.97	1.18
6 monthly treatment	3.48	4.34	72.01	1.21
4 monthly treatment	4.23	3.30	104.72	1.69
Extra cost per case prevented				
1 years instead of 2 years	0.83	5.15	138.62	0.95
6 months versus 1 year	4.19	3.50	722.38	1.25
4 months versus 6 months	8.02	2.16	17000	12.71

much less than that required to minimise prevalence (Anderson and Medley, 1985). The analyses by Guyatt *et al.* (1993) indicate further that relatively long intervals between treatment offers a cost-effective approach to morbidity reduction, and that measuring effectiveness in terms of infection prevalence reduction can lead to the identification of options that do not optimise morbidity control.

The Effect of Population Age-Structure on the Costs of Control

The cost-effectiveness of age-targeted treatment has also been investigated for different delivery options using the age-structured dynamic model described in the previous section. The first delivery option considered was the target group (Guyatt et al., 1995). Although child-targeted treatment can never be more effective than population treatment (see above), it is less costly and may prove to be more cost-effective since this target group has the highest intensity of infection, and is therefore most likely to suffer disease and to be responsible for a larger proportion of the transmission stages (Anderson and May, 1979). The cost-effectiveness analysis demonstrated that child-targeted treatment was more cost-effective than population treatment in terms of reduction in heavy infection. For example, in the high transmission area, the cost per heavy infection case prevented was US$ 1.78 for child-targeted treatment, compared to US$ 2.45 for population treatment. The main reason for this difference was that the intensity of infection was highest in the child age classes, and therefore treating children benefits the population as a whole since the rate of transmission is greatly reduced. In this analysis, the unit cost of treating children was assumed to be identical to that for adults. In practice, the costs are likely to be much smaller for children, since they are more easily accessible through schools (Partnership for Child Development, 1997; World Bank, 1993). Including such a differential cost advantage would further favour child-targeted treatment (Bundy and Guyatt, 1996).

With the exception of drug efficacy, the coverage and frequency of treatment are often the only aspects of a delivery programme that can be controlled. Theoretical studies have investigated a control criterion for chemotherapy, which corresponds to the threshold coverage per time period required to eradicate infection or to control infection at a given level (Anderson and May, 1985b; Anderson and Medley, 1985). There have been few attempts, however, to dissect the interaction between coverage and frequency of treatment, or to investigate the implications for cost-effectiveness. Guyatt et al. (1995) examined three different frequencies of treatment (every 6 months, every year and every 2 years) and three levels of coverage of children (50%, 70% and 90%) over a 5 year treatment period. The analysis demonstrated that of the nine treatment options, the most cost-effective in terms of the reduction in heavy infection was to treat every 2 years at a coverage of 90% (see Table 24.3). This suggests that it is more cost-effective to extend the coverage of an existing cycle than to increase the number of cycles. The main reasons for this are that it is cheaper to treat more people at one visit than to make more visits (due to the relatively lower costs of drugs versus delivery), and that the dynamics are such that the lower the infection levels achieved, the slower the return to equilibrium levels. Medley et al. (1993) showed that increasing the coverage of treatment was increasingly beneficial to the untreated portion of the population because of reduced infection rates overall.

The most cost-effective option for child-targeted treatment was shown to be 2 yearly treatment of 90% of children. However, it is possible, in practice, that a 2 yearly treatment programme is chosen, but only 70% coverage is achieved. In this situation, one may consider trying to increase coverage to 90%. This may involve extra costs to motivate the community (e.g. through health education) and the staff (e.g. through incentives). In this instance, one would be interested in determining the maximum amount of money that could be spent achieving a 90% coverage, which would be more cost-effective than leaving the coverage at 70%. This can be calculated by determining how much more the costs of the 90% coverage programme can be increased, such that this option still has a lower cost-effectiveness ratio than leaving coverage at 70%. The analyses suggest that, in addition to the extra treatment

Table 24.3. The cost-effectiveness of different combinations of frequency of treatment and coverage for child-targeted treatment directed at *Ascaris lumbricoides* in a high transmission area (note: cost and effectiveness are rounded up to the nearest thousand, but the cost-effectiveness ratios are calculated directly from the original data). (modified from Guyatt *et al.*, 1995)

| | Coverage | Frequency of treatment | | |
		every 6 months	every year	every 2 years
Cost (US$)	0.5	145	72	44
	0.7	194	97	58
	0.9	243	122	73
Number of heavy infection cases prevented	0.5	52	32	20
	0.7	70	50	33
	0.9	83	68	49
Cost per heavy infection case prevented (US$)	0.5	2.80	2.25	2.16
	0.7	2.77	1.95	1.78
	0.9	2.92	1.78	1.48

costs, it would be possible to invest up to US$ 14,967 over the 5 year programme (approximately 25% of the initial investment) on increasing coverage from 70% to 90%, and this would still be the more cost-effective option.

Cost Utility: the Cost of Gaining a Disability-Adjusted Life-Year

Nematodes have their main effects on morbidity rather than mortality, so the impact of treatment cannot be determined in terms of commonly used indicators such as the number of lives saved. However, a recent study sponsored by the World Bank, has lead to the development of a technique which allows the impact of disease and control programmes on both morbidity and mortality to be quantified and combined into one measure. The effectiveness measure is termed the disability-adjusted life year (DALY), and a number of health interventions, including school-targeted anthelmintic chemotherapy, have been assessed in terms of the cost per DALY gained (Jamison, 1993). DALYs translate disabilities into years of healthy life lost by giving each disease state a disability weight ranging from 0 (healthy) to 1 (death). Additional complexities in the formulation of DALYs include an age-weighting function whereby life at different ages is valued unequally (weighting function peaks at age 25), the long term consequences of a disability caused by a risk factor this year are counted back to the present year (requires incidence data), and disability experienced in future years is discounted. Further details on the calculation of DALYs can be found in Murray and Lopez (1994).

The first published attempt at assessing the cost per disability-adjusted life year gained for anthelmintic chemotherapy was undertaken by Warren *et al.* (1993) using a simple static framework. The total cost per child for a 10 year programme (albendazole, praziquantel and delivery costs) was assumed to be between US$8–US$18. The estimated gains in DALYs arose from a reduction in morbidity and mortality in children during the intervention, post-intervention health benefits for the target group and indirect health benefits for the families of the target group during the intervention. The reduction in morbidity was assessed as functions of the proportion of children with high or mild-moderate infections, and the disability-weights for the morbidity associated with high or mild-moderate infections. The mortality prevented was determined from the estimated number of deaths, and the life expectancy at death. It was estimated that the programme would cost between US$ 6–12 (moderate effect estimates) or US$ 15–33 (low effect estimates) per DALY gained.

The model by Warren *et al.* (1993) was essentially static. A more recent cost-utility analysis in terms of cost per DALYs gained for school-based delivery of anthelmintics has been undertaken using a population dynamic model in relation to *A. lumbricoides* infection (Chan, 1997). The model is an extension of the Chan *et al.* (1994b) age-structured model for intestinal nematodes described before, with three rather than two age-groups: pre-school children, school age children and adults. Morbidity is classified into 4 types, defined by group affected, duration of disability and the number of worms associated with the condition. A low threshold worm burden is associated with reversible growth faltering in children and/or reduced physical fitness in children and adults (Type A), and permanent growth retardation in 3% of children (Type B). A high threshold burden is associated with clinically overt acute illness of short duration (Type C) and acute complications in 70% of those with Type C (Type D). Mortality was assumed to occur in 5% of Type D cases. The low and high threshold worm burdens were age-dependent, and each morbidity type was associated with assumed disability weights. Chan (1997) predicted that 70% of the total DALY loss in a community with a high prevalence of infection could be averted by treating 60% of school-aged children every year for 10 years, at a cost per DALY averted of US$8.

Both these estimates would suggest that helminth control is an exceptionally good 'buy' in public health terms (World Bank, 1993).

Epidemiology and the Future of Control

The empirical results and analyses described above have led to a reassessment of the optimal strategies for controlling disease due to nematodes. These experiences have also led to renewed interest in implementing control. The focus in control today is on carefully targeting control at those most in need, and ensuring that all those in the target group receive treatment.

Control of Intestinal Nematodes

Although intestinal nematodes can infect all members of a population, it is clear that there are specific groups who are at greater risk of heavy infection than others, and that some are more vulnerable than others to the harmful effects of chronic infections. They are: school age children; pre-school children, especially those nearing school age; and women of child-bearing age, including adolescent girls. Infrastructures exist whereby treatment can be delivered to each of these three groups, and there are many examples of practical control programmes which have successfully delivered treatments at very low cost.

Children of school age

School children harbour the most intense infections with some of the commonest worms and are thus the age-group most at risk of morbidity and, simultaneously, the major contributors to transmission. Theory and practice indicate that targeting treatment at school children reduces infection levels in the community as a whole (WHO, 1996a; Bundy, 1995). As we have seen above, the impact of intervention programmes in terms of averting DALY losses has been estimated using a population dynamic model and has shown that a sustained chemotherapy programme where only the schoolchildren are treated, could result in aversion of 70% of the DALY loss in the entire population (Chan, 1997).

The fact that children assemble daily in one place, their school, provides an opportunity to deliver mass treatments through an existing infrastructure. The experience of programmes in which teachers have treated children with anthelmintic drugs indicates that school-based programmes are feasible and practicable (Partnership for Child Development, 1997). In many developing societies, there are more teachers than health workers and more schools than clinics; the education sector provides an already developed and supported infrastructure that might usefully supplement the existing primary health care system. There have been concerns that the education systems and teachers are already too stretched to take on additional tasks, and that teachers and the community would resent the education sector playing a role in health. Carefully monitored school based health and nutrition programmes in Ghana, Tanzania, India and Indonesia have now shown that the education sector is capable of delivering a simple health package (health education, anthelmintics, sometimes with micronutrients) to large numbers of school children (50,000 to 3 million) without the creation of specific infrastructures (Partnership for Child Development, 1997). Furthermore, they have shown that teachers perceive this role in health as an acceptable, even welcome extension of their overall role in the community, and that both students and parents concur with this view. Analyses of costs tend to confirm the prediction that such education sector delivery methods are associated with small financial cost (Partnership for Child Development, 1998a). Cost analyses of the programmes in Ghana and Tanzania indicate that the cost of delivering albendazole was US$ 0.03 per child per annum while the cost of the treatment itself can be as low as US$ 0.04. For the India programme in Gujarat (2.83 million children), the cost of locally purchased ferrous sulphate, albendazole and vitamin A, delivered as a standard dose twice a year, was US$ 0.50 per child (Partnership for Child Development, 1997).

Children under five years of age

Although more research on the benefits of treatment in the pre-school age group is required, a recently concluded consultative meeting sponsored by UNICEF recommended that where the prevalence of infection is high, programmes which currently deliver health care to children aged between 1 and school age should consider adding periodic anthelmintic treatments to the health care currently provided (UNICEF, 1997). Several opportunities for delivery of anthelmintic therapy were identified including: immunization programmes, vitamin A capsule distribution programmes, Maternal-Child Health clinics, the Integrated Management of Childhood Illness programme, and visits to homes by Community Health Workers.

Adolescent girls and women of child-bearing age

The contribution of hookworm infection to anaemia is such that all women of childbearing age could benefit from periodic treatment in areas where these worms are endemic, although treatment is not recommended in the first trimester of pregnancy. There are three critical periods to consider for the intervention to improve or restore iron status in women. The first is around puberty, in preparation for the years of reproduction and greatest economic activity (Partnership for Child Development, 1998b). It is possible that deworming during the pubertal growth spurt could yield a height benefit that a girl would carry with her throughout her life, thus reducing her risk of complications during childbirth, and increasing her physical capacity for work (WHO, 1996b).

The second critical intervention time is around reproductive events, during pregnancy or during the post-partum period. Intervention in pregnancy could favourably influence infant birth outcomes and prevent maternal morbidity and mortality. For example, in Sri Lanka, because hookworm infection is widespread, and maternal anaemia is a major problem, it has been national policy since 1994 to treat all pregnant mothers with a single dose of mebendazole at the first ante-natal clinic visit after completion of the first trimester (WHO, 1996b). Anthelmintic chemotherapy at delivery or immediately post-partum might help reduce the vertical transmission of *A. duodenale*.

Thirdly, if the programme objective is either to improve work efficiency and the economic productivity of women, or to improve their sense of well-being, quality of life and caring capacity, then the goal should be to alleviate iron deficiency and the hookworm load of women throughout their adulthood. Where transmission of hookworm is intense, contacting women only once per pregnancy will probably not achieve this goal, especially in societies where family spacing is fairly successful; other strategies such as involving Community Health Workers, or the delivery of treatments through the work place, such as factories or plantations, may have to be considered.

Control of Lymphatic Filariasis

The control of lymphatic filariasis has been pursued in a number of countries over the last several decades using Diethylcarbamazine (DEC) chemotherapy and control of the larvae and adults of the mosquito vector (Sasa, 1976). There have been a number of successes in achieving control by these means, for example, in Japan and Polynesia; most famously through the use of DEC fortified common salt in China (Gelband, 1994).

In the 1990s there has been a significant revival of interest in controlling filariasis by mass chemotherapy. A number of factors have helped this process. An important contributor was the recognition that single dose DEC, as against the 12 days of treatment originally recommended by WHO, was sufficient to achieve a significant reduction in microfilaraemia (WHO, 1994). This observation has led to a more thorough evaluation of chemotherapy in general, and to the availability of single dose Ivermectin as a safe alternative to DEC for use in Africa where the latter is contraindicated because of cross-reactivity with onchocerciasis and loaiasis

(Ottesen, 1990). More recent studies suggest that the efficacy of both DEC and ivermectin can be enhanced by concomitant administration of albendazole (Ismail *et al.*, 1998).

These technical advances in treatment, as well as advances in diagnosis (More and Copeman, 1990; Weil *et al.*, 1987), have made the delivery of treatment logistically easier. But there has also been a major change in the perception of the causation of the disease which has made mass treatment a plausible means of control. The traditional perception of the relationship between infection and the many disease syndromes of filariasis was that the disease state was determined by host-specific immunological responses which themselves were largely under host genetic control (see Ottesen (1990) for a recent description). This raised serious concerns about the potential benefit, and indeed potential harm, of mass treatment. A series of studies of the population biology of lymphatic filariasis (Das *et al.*, 1990; Srividya *et al.*, 1991) showed that the age profiles and population distribution of infection were not dissimilar from those of nematodes in general, while population dynamic analyses argues strongly that the dynamics of transmission, and probably disease, were of the general form common to helminths (Bundy, Grenfell and Rajagopalan, 1991; Michael, Grenfell and Bundy, 1998; Michael and Bundy, 1998).

Field studies have now confirmed that *Wuchereria bancrofti* infection patterns respond to mass chemotherapy in the manner predicted by dynamic theory, with a reduction in microfilaraemia being associated with a reduction in transmission of infection in the vector population (Bockarie *et al.*, 1998; Das *et al.*, 1990). Mathematical models of these patterns have now been developed using microsimulation (Plaiser *et al.*, 1998) and deterministic (Chan *et al.*, 1998) approaches.

Conclusion

The improved understanding of the epidemiology of nematode infections, the recognition of the importance of infection for human development, and the remarkable cost-effectiveness of control methods have led to the growing involvement of countries and agencies in programmes that are either aimed primarily at helminth control or that incorporate helminth control as part of an integrated health package (Savioli, Bundy and Tomkins, 1992).

The WHO, in association with its collaborating centres, has taken the lead in the technical design of control programmes (WHO, 1996a). In 1996 the WHO issued guidelines on strengthening interventions to prevent helminth infection through schools, as an entry point for the development of 'health promoting schools' under its global school health initiative (WHO, 1997). The World Bank's World Development Report entitled 'Investing in Health' (World Bank, 1993), identified school health programmes, including deworming, as one of five priority public health measures that had the potential to produce substantial gains in health at a modest cost, and the World Bank's International School Health Initiative is now assisting the incorporation of this approach into government programmes supported by the World Bank. UNICEF is launching a programme to promote deworming in populations at risk (Hall *et al.*, 1997). Numerous international and national non-governmental organisations are including deworming in their health promotion strategies (Bundy *et al.*, 1998). There is growing interest in incorporating deworming in early child development and maternal and child health programmes, but school based delivery remains the most popular and cost-effective approach to anthelmintic distribution.

The WHO, in partnership with SmithKline Beecham, Merck and Co., the World Bank and others, is currently developing a programme for the global elimination of lymphatic filariasis as a public health problem. This programme is based around the use of chemotherapy to modify the transmission dynamics of infection, and so reduce the incidence of infection and hence disease.

These new initiatives are a reflection, in part, of the availability of safe, simple, low cost and effective treatments. But they also reflect the very considerable understanding of the epidemiology of nematode infections that has resulted from the ecological approach to describing transmission dynamics.

References

Anderson, R.M. (1980) The dynamics and control of direct life-cycle helminth parasites. *Lecture Notes in Biomathematics*, **39**, 278–322.

Anderson, R.M. (1986) The population dynamics and epidemiology of intestinal nematode infections. *Transactions of the Royal Society of Tropical Medicine and Hygiene*, **80**, 686–696.

Anderson, R.M. and May, R.M. (1979) Population biology of infectious diseases: Part 1. *Nature*, **280**, 361–367.

Anderson, R.M. and May, R.M. (1985a) Herd immunity to helminth infections: implications for parasite control. *Nature*, **315**, 493–496.

Anderson, R.M. and May, R.M. (1985b) Helminth infections of humans: mathematical models, populations dynamics and control. *Advances in Parasitology*, **24**, 1–101.

Anderson, R.M. and May, R.M. (1991) *Infectious Diseases of Humans: Dynamics and Control*. Oxford: Oxford University Press.

Anderson, R.M. and Medley, G.F. (1985) Community control of helminth infections of man by mass and selective chemotherapy. *Parasitology*, **90**, 629–660.

Anderson, R.M. and Schad, G.A. (1986) Hookworm burdens and faecal egg counts: an analysis of the biological basis of variation. *Transactions of the Royal Society of Tropical Medicine and Hygiene*, **79**, 812–825.

Arfaa, F. and Ghadirian, E. (1977) Epidemiology and mass-treatment of ascariasis in six rural communities in central Iran. *American Journal of Tropical Medicine and Hygiene*, **26**, 866–871.

Awasthi, S., Peto, R., Bundy, D.A.P., Kumar Pandy, V. and Fletcher R.H. (1997) Improvement in nutritional status among preschool children in Lucknow, India. A randomised trial of albendazole. King George's Medical College, Lucknow, India; University of Oxford, UK; and Harvard Medical School, USA. *International Clinical Epidemiology Network Annual Meeting, Mexico, February 1997*.

Beasley, N.M.R., Tomkins, A.M., Hall, A., Kihamia, C.M., Lorri, W., Nduma, B., Issae, W., Nokes, C. and Bundy, D.A.P. (1999) The impact of population level deworming on the haemoglobin levels in Tanga, Tanzania. *Tropical Medicine and International Health*, **11**, 744–750.

Behnke, J.M. (1987) Do hookworms elicit protective immunity in man? *Parasitology Today*, **3**, 200–206.

Bockarie, M.J., Alexander, N.D.E., Hyun, P., Dimber, Z., Bockarie, F., Ibam, E., Alpers, M.P. and Kazura, J.W. (1998) Randomised community-based trial of annual single-dose diethylcarbamazine with or without ivermectin against *Wuchereria bancrofti* infection in human beings and mosquitoes. *Lancet*, **351**, 162–168.

Bradley, D.J. and McCullough, F.S. (1974) Egg output stability and epidemiology of *Schistosoma haematobium*. II. An analysis of the epidemiology of endemic *S. haematobium*. *Transactions of the Royal Society of Tropical Medicine and Hygiene*, **80**, 706–718.

Bradley, M. and Chandiwana, S.K. (1990) Age-dependency in predisposition to hookworm infection in the Burma valley area of Zimbabwe. *Transactions of the Royal Society of Tropical Medicine and Hygiene*, **84**, 826–828.

Bradley, M., Chandiwana, S.K., Medley, G.F. and Bundy, D.A.P. (1992) The epidemiology and population biology of *Necator americanus* in a rural community on Zimbabwe. *Transactions of the Royal Society of Tropical Medicine and Hygiene*, **86**, 73–76.

Bundy, D.A.P. (1986) Epidemiological aspects of *Trichuris* and Trichuriasis in Caribbean communities. *Transactions of the Royal Society of Tropical Medicine and Hygiene*, **80**, 706–718.

Bundy, D.A.P. (1988) Population ecology of intestinal helminth infections in human communities. *Philosophical Transactions of the Royal Society of London B*, **321**, 405–420.

Bundy, D.A.P. (1990) Is the hookworm just another geohelminth. In *Hookworm Disease. Current Status and New Directions*, edited by G.A. Schad and K.S. Warren, pp. 147–164. London: Taylor & Francis.

Bundy, D.A.P. (1995) Epidemiology and transmission of intestinal helminths. In *Enteric infection 2. Intestinal helminths*, edited by M.J.G. Farthing, G.T. Keusch, D. Wakelin, pp. 5–24. London: Chapman & Hall Medical.

Bundy, D.A.P., Chan, M.S., Medley, G.F., Jamison, D., de Silva, N.R. and Savioli, L. (1998) Intestinal nematode infections. In *Health Priorities and Burden of Disease Analysis: Methods and Appliances from Global, National and Sub-national Studies*. Cambridge: Harvard University Press (on behalf of World Health Organization and World Bank).

Bundy, D.A.P., McGuire, J., Hall, A. and Dolan, C. (1998a) School based health and nutrition programmes: a survey of donor and agency support. Sub-Committee on Nutrition News: No. 16, July 1998.

Bundy, D.A.P. and Cooper, E.S. (1988a) The evidence for predisposition to trichuriasis in humans: comparison of institutional and community studies. *Annals of Tropical Medicine and Parasitology*, **82**, 251–256.

Bundy, D.A.P. and Cooper, E.S. (1988b) Human *Trichuris* and trichuriasis. *Advances in Parasitology*, **28**, 107–173.

Bundy, D.A.P. Cooper, E.S., Thompson, D.E., Didier, J.M. and Simmons, I. (1988) Effect of age and initial status on the rate of infection with *Trichuris trichiura* after treatment. *Parasitology*, **97**, 469–476.

Bundy, D.A.P., Cooper, E.S., Thompson, D.E., Didier, J.M., Anderson, R.M. and Simmons, I. (1987a) Predisposition to *Trichuris trichiura* infections in humans. *Epidemiology and Infection*, **98**, 65–71.

Bundy, D.A.P., Cooper, E.S., Thompson, D.E., Didier, J.M. and Simmons, I. (1987b) Epidemiology and population dynamics of *Ascaris lumbricoides* and *Trichuris trichiura* infection in the same community. *Transactions of the Royal Society of Tropical Medicine and Hygiene*, **81**, 987–993.

Bundy, D.A.P. and de Silva, N.R. (1998). Can we deworm this Wormy World. *British Medical Bulletin* **54**, 421–432.

Bundy, D.A.P., Grenfell, B.T. and Rajagopalan, P.K. (1991) Immunoepidemiology of lymphatic filariasis: the relationship between infection and disease. In *Immunoparasitology Today* (ed. C. Ash & R.B. Gallagher), pp. A71–A75. Cambridge: Elsevier Trends Journals.

Bundy, D.A.P. and Guyatt, H.L. (1992) Cost analysis of schistosomiasis. *Transactions of the Royal Society of Tropical Medicine and Hygiene*, **86**, 646–648.

Bundy, D.A.P. and Guyatt, H.L. (1996) Schools for Health: focus on health education and the school-age child. *Parasitology Today*, **12**, 1–16.

Bundy, D.A.P., Kan, S.P. and Rose, R. (1988) Age-related prevalence, intensity and frequency distribution of gastrointestinal helminth infection in urban slum children from Kuala Lumpur, Malaysia. *Transactions of the Royal Society of Tropical Medicine and Hygiene*, **82**, 298–294.

Bundy, D.A.P. and Medley, G. (1992) Immuno-epidemiology of human geohelminthiasis: ecological and immunological determinants of worm burden. *Parasitology*, **104**, S105–S109.

Bundy, D.A.P. and Michael, E. (1996) Trichinosis. In *Illustrated History of Tropical Diseases*, edited by F.E.G. Cox, pp. 310–317. London :The Wellcome Trust.

Bundy, D.A.P., Wong, M.S., Lewis, L.L. and Horton, J. (1990) Control of geohelminths by delivery of targeted chemotherapy through schools. *Transactions of the Royal Society of Tropical Medicine and Hygiene*, **84**, 115–120.

Cabrera, B.D. (1981) Reinfection and infection rate studies of soil-transmitted helminthiases in Juban, Sorsogon. In *Collected Papers on the Control of Soil-Transmitted Helminthiases*, Vol. 1, pp. 181–191. Tokyo: Asian Parasitic Control Organisation.

Cabrera, B.D., Aramulo, P.V. III and Portillo, G.P. (1975) Ascariasis control and/or eradication in a rural community in the Philippines. *South East Asian Journal of Tropical Medicine and Public Health*, **6**, 510–518.

Callendar, J.E.M., McGregor, S.M., Walker, S. and Cooper, E.S. (1992) *Trichuris* infection and mental development in children. *Lancet*, **339**, 181.

Chan, L.L. (1991) The Epidemiology of Intestinal Nematode Infection in Urban Malaysia. *Ph.D. Thesis*, Faculty of Science, University of London.

Chan, L.L., Kan, S.P. and Bundy, D.A.P. (1992) The effect of repeated chemotherapy on age-related predisposition to *Ascaris lumbricoides* and *Trichuris trichiura*. *Parasitology*, **104**, 371–377.

Chan M.S. (1997) The global burden of intestinal nematode infections: 50 years on. *Parasitology Today* **13**, 438–443.

Chan, M.S., Medley, G.F., Jamison, D. and Bundy, D.A.P. (1994a) The evaluation of potential global morbidity due to intestinal nematode infections. *Parasitology*, **109**, 373–387.

Chan, M.S., Medley, G.F., Bundy, D.A.P. and Guyatt, H.L. (1994b) The development and validation of an age structured model for the evaluation of disease control strategies for intestinal helminths. *Parasitology*, **109**, 389–396.

Chan, M.S., Srividya, A., Norman, R.A., Pani, S.P., Ramaiah, K.D., Vanamail, P., Michael, E., Das, P.K. and Bundy, D.A.P. (1998) EPIFIL: A dynamic model of infection and disease in lymphatic filariasis. *American Journal of Tropical Medicine and Hygiene*, **59**, 606–614.

Cooper, E.S. and Bundy, D.A.P. (1988a) *Trichuris* is not trivial. *Parasitology Today*, **4**, 301–306.

Cooper, E.S. and Bundy, D.A.P. (1988b) Trichuriasis. In *Intestinal Helminthic Infections*, edited by Z.S. Pawlowski. Bailliére's Clinical Tropical Medicine and Communicable Disease, Vol. 2, pp. 629–643. London :Bailliére Tindall.

Cooper, E.S., Duff, E.M.W., Howell, S. and Bundy, D.A.P. (1995) 'Catch-up' growth velocities after treatment for *Trichuris* dysentery syndrome. *Transactions of the Royal Society of Tropical Medicine and Hygiene*, **89**, 653.

Cox, F.E.G. (1996) Brucellosis. In *Illustrated History of Tropical Diseases*, edited by F.E.G. Cox, pp. 50–59. London :The Wellcome Trust.

Crombie, J.A. and Anderson, R.M. (1985) Population dynamics of *Schistosoma mansoni* in mice repeatedly exposed to infection. *Nature*, **315**, 491.

Crompton, D.W.T. and Whitehead, R.R. (1993) Hookworm infections and human iron metabolism. *Parasitology*, **107**, S137–S145.

Das, P.K., Manoharan, A., Srividya, A., Grenfell, B.T., Bundy, D.A.P. and Vanamail, P. (1990) Frequency distribution of *Wuchereria bancrofti* microfilariae in human populations and its relationship with age and sex. *Parasitology*, **101**, 429–434.

de Silva, N.R., Chan, M.S. and Bundy, D.A.P. (1998) Morbidity and mortality due to ascariasis: re-estimation and sensitivity analysis of global numbers at risk. *Tropical Medicine and International Health*, **2**, 517–528.

Elkins, D.B., Haswell-Elkins, M. and Anderson, R.M. (1986) The epidemiology and control of intestinal helminths in the Pulicat Lake region of Southern India. I. Study design and pre and post-treatment observations on *Ascaris lumbricoides* infection. *Transactions of the Royal Society of Tropical Medicine and Hygiene*, **80**, 774–792.

Evans, D. and Guyatt, H.L. (1995) The cost-effectiveness of mass drug therapy for intestinal parasites. *PharmacoEconomics*, **8**, 14–22.

Forrester, J.E., Scott, M.E., Bundy, D.A.P. and Golden, M.H.N. (1988) Clustering of *Ascaris lumbricoides* and *Trichuris trichuria* infections within households. *Transactions of the Royal Society of Tropical Medicine and Hygiene*, **82**, 282–288.

Forrester, J.E., Scott, M.E., Bundy, D.A.P. and Golden, M.H.N. (1990) Predisposition of individuals and families in Mexico to heavy infection with *Ascaris lumbricoides* and *Trichuris trichiura*. *Transactions of the Royal Society of Tropical Medicine and Hygiene*, **84**, 272–276.

Fulford, A.J.C., Butterworth, A.E., Sturrock, R.F., Ouma, J.H. (1992) On the use of age-intensity data to detect immunity to parasitic infections, with special reference to *Schistosoma mansoni* in Kenya. *Parasitology*, **105**, 219–227.

Gelband, H. (1994) Diethylcarbamazine salt in the control of lymphatic filariasis. *American Journal of Tropical Medicine and Hygiene*, **50**, 655–662.

Grove, D.I. (1990) *A history of human helminthology*. Wallingford :CAB International.

Guyatt, H.L. (1992). Parasite population biology and the design and evaluation of helminth control programmes. *Ph.D thesis*, University of London.

Guyatt, H.L. and Bundy, D.A.P. (1991) Estimating prevalence of community morbidity due to intestinal helminths: prevalence of infection as an indicator of the prevalence of disease. *Transactions of the Royal Society of Tropical Medicine and Hygiene*, **85**, 778–782.

Guyatt, H.L. and Bundy, D.A.P. (1993). Estimation of intestinal nematode prevalence: influence of parasite mating patterns. *Parasitology*, **107**, 99–106.

Guyatt, H.L., Bundy, D.A.P., Medley, G.F. and Grenfell, B.T. (1990) The relationship between the frequency distribution of *Ascaris lumbricoides* and the prevalence and intensity of infection in human communities. *Parasitology*, **101**, 139–143.

Guyatt, H.L., Bundy, D.A.P. and Evans, D. (1993) A population approach to the cost-effectiveness analysis of mass anthelmintic treatment: effects of treatment frequency on *Ascaris* infection. *Transactions of the Royal Society of Tropical Medicine and Hygiene*, **87**, 570–575.

Guyatt, H.L., Chan, M.S., Medley, G.F., and Bundy, D.A.P. (1995) Control of *Ascaris* infection by chemotherapy: which is the most cost-effective option? *Transactions of the Royal Society of Tropical Medicine and Hygiene*, **89**, 16–20.

Guyatt, H.L. and Tanner, M. (1996) Different approaches to modelling the cost-effectiveness of schistosomiasis control. *American Journal of Tropical Medicine and Hygiene*, **55**, 159–164.

Habbema, J.D.F., Alley, E.S., Plaisier, A.P., Van Oortmarssen, G.J. and Remme, J.H.F. (1992) Epidemiological modelling for onchocerciasis control. *Parasitology Today*, **8**, 99–103.

Hagan, P. (1992) Reinfection, exposure and immunity in human schistosomiasis. *Parasitology Today*, **8**, 12–16.

Hall, A., Orinda, V., Bundy, D.A.P. and Broun, D. (1997) Promoting Child Health through Helminth Control: A Way Forward. *Parasitology Today*, **13**, 411–413.

Haswell-Elkins, M.R., Elkins, D.B. and Anderson, R.M. (1987) Evidence for predisposition in humans to infections with *Ascaris*, hookworm, *Enterobius* and *Trichuris* in a South Indian fishing community. *Parasitology*, **95**, 323–327.

Holland, C.V., Asaolu, S.O., Crompton, D.W.T., Stoddart, R.C., MacDonald, R. and Torimiro, S.E.A. (1989) The epidemiology of *Ascaris lumbricoides* and other soil-transmitted helminths in primary school children from Ile-Ife, Nigeria. *Parasitology*, **99**, 275–285.

Ismail, M.M., Jayakody, R.L., Weil, G.J., Nirmalan, N., Jayasinghe, K.S., Abeyewickrema, W., Rezvi-Sheriff, M.H., Rajaratnam, H.N., Amarasekera, N., de-Silva, D.C., Michalski, M.L. and Dissanaike, A.S. (1998) Efficacy of single dose combinations of albendazole, ivermectin and diethylcarbamazine for the treatment of bancroftian filariasis. *Transactions of the Royal Society of Tropical Medicine and Hygiene*, **92**, 94–97.

Jamison, D.T. (1993) Disease control priorities in developing countries: an overview. In *Disease Control Priorities in Developing Countries*, edited by D.T. Jamison, W.H. Mosley, A.R. Measham, and J.L. Bobadilla, pp. 3–34. Oxford: Oxford University Press.

Keymer, A. and Slater, A.F.G. (1987) Helminth fecundity: density dependence or statistical illusion? *Parasitology Today*, **3**, 56–58.

Kunni, C. (1992) *It all started from worms: the 45 year record of Japan's Post-World War II National Health and Family Planning Movement*. Tokyo: Hoken Kaikan Foundationm.

Medley, G.F. Guyatt, H.L. and Bundy, D.A.P. (1993) A quantitative framework for evaluating the effect of community treatment on the morbidity due to ascariasis. *Parasitology*, **106**, 211–221.

Michael, E., Grenfell, B.T. and Bundy, D.A.P. (1994) The association between microfilaraemia and disease in lymphatic filariasis. *Proceedings of the Royal Society of London B*, **256**, 33–44.

Michael, E. and Bundy, D.A.P. (1998) Herd immunity to filarial infection is a function of vector biting rate. *Proceedings of the Royal Society*, **265**, 855–860.

Michael, E., Grenfell, B.T., Isham, V.S., Denham, D.A. and Bundy, D.A.P. (1998) Modelling variability in lymphatic filariasis: macrifilarial dynamics in the *Brugia pahangi* — cat model. *Proceedings of the Royal Society*, **265**, 155–165.

More, S.J. and Copeman, D.B. (1990) A highly specific and sensitive monoclonal antibody-based ELISA for the detection of circulating antigen in Bancroftian filariasis. *Tropical Medicine and Parasitolology*, **41**, 403–406.

Morishita, K. (1972) Studies on epidemiological aspects of ascariasis in Japan and basic knowledge concerning its control. *Progress of Medical Parasitology in Japan*, **4**, 1–153.

Murray, C.J.L. and Lopez, A.D. (1994) eds. *Global Comparative Assessments in the Health Sector: Disease Burden, Expenditures and Intervention Packages*. Geneva: World Health Organization.

Nokes, C., Grantham-McGregor, S.M., Sawyer, A.W., Cooper, E.S., Robinson, B.A. and Bundy, D.A.P. (1992) Moderate to heavy infections of *Trichuris trichiura* affect cognitive function in Jamaican school children. *Parasitology*, **104**, 539–547.

Ottesen, E.A. (1990) Activity of ivermectin in human parasitic infections other than onchocerciasis. *Current Opinion in Infections Diseases*, **3**, 834–837.

Partnership for Child Development (1997) Better health, nutrition and education for the school-aged child. *Transactions of the Royal Society of Tropical Medicine and Hygiene* **91**, 1–2.

Partnership for Child Development (1998a) Cost of school based drug treatment in Tanzania. *Health Policy and Planning* **13**, 384–396.

Partnership for Child Development (1998b) The health and nutritional status of school children in Africa: evidence from school-based programmes in Ghana and Tanzania. *Transaction of the Royal Society of Tropical Medicine and Hygiene*, **92**, 254–261.

Plaisier, A.P., Van Oortmarssen, G.J., Habbema, J.D.F., Remme, J. and Alley, E.S. (1990) ONCHOSIM: A model and computer simulation programme for the transmission and control of onchocerciasis. *Computer Methods and Programmes in Biomedicine*, **31**, 43–56.

Plaisier, A.P., Subramanian, S., Das, P.K., Souza, W., Lapa, T. and Furtado, A.F. (1998) The LYMFASIM simulation programme for modelling lymphatic filariasis and its control. *Methods of Information in Medicine*, **37**, 97–108.

Prescott, N.M. (1987) The economics of schistosomiasis chemotherapy. *Parasitology Today*, **3**, 21–25.

Pritchard, D.I., Quinnell, R.J., Slater, A.F.G., McKean, P.G., Dale, D.D.S., Raiko, A. and Keymer, A.E. (1990) Epidemiology and immunology of *Necator americanus* infection in a community in Papua New Guinea: humoral responses to excretory-secretory and cuticular collagen antigens. *Parasitology*, **100**, 317–326.

Quinnell, R.J., Medley, G.F. and Keymer, A.E. (1990) The regulation of gastrointestinal helminth populations. *Philosophical Transactions of the Royal Society of London* **B**, **330**, 191–201.

Sasa, M. (1976) *Human filariasis — a Global Survey of Epidemiology and Control. Baltimore, Maryland*. Tokyo: University Park Press.

Savioli, L., Dixon, H., Kisumku, U.M. and Mott, K. (1989) Control of morbidity due to *Schistosoma haematobium* on Pemba Island: selective population chemotherapy of schoolchildren with haematuria to identify high-risk communities. *Transactions of the Royal Society of Tropical Medicine and Hygiene*, **83**, 805–810.

Savioli, L., Bundy, D.A.P. and Tomkins, A.M. (1992) Intestinal parasitic infections: a soluble public health problem. *Transaction of the Royal Society of Tropical Medicine and Hygiene*, **86**, 353–354.

Schad, G.A. and Anderson, R.M. (1985) Predisposition to hookworm infection in humans. *Science*, **228**, 1537–1540.

Simeon, D.T. and Grantham-McGregor, S.M. (1990) Nutritional deficiency and children's behaviour and mental development. *Nutritional Research Review*, **3**, 1–24.

Southgate, B.A. (1992) Intensity and efficiency of transmission and the development of microfilaraemia and disease — their relationship in lymphatic filariasis. *Journal of Tropical Medicine and Hygiene*, **95**, 1–12.

Srividya, A., Das, P.K. Subramanian, S.P. Ramaiah, K.D., Grenfell, B.T., Michael, E. and Bundy, D.A.P. (1996) Past exposure and the dynamics of lymphatic filariasis infection in young children. *Epidemiology and Infection*, **117**, 195–201.

Srividya, A., Pani, S.P., Rajagopalan, P.K., Bundy, D.A.P. and Grenfell, B.T. (1991) The dynamics of infection and disease in Bancroftian filariasis. *Transactions of the Royal Society of Tropical Medicine and Hygiene*, **85**, 255–259.

Stephenson, L.S. (1987) *Impact of Helminth Infections on Human Nutrition: Schistosomes and Soil Transmitted Helminths*. London: Taylor & Francis.

Stephenson, L.S., Crompton, D.W., Latham, M.C., Schulpen, T.W., Nesheim, M.C., Jansen, A.A. (1980) Relationships between *Ascaris* infection and growth of malnourished preschool children in Kenya. *American Journal of Clinical Nutrition* **33**(5), 1165–72.

Stephenson, L.S., Latham, M.C., Kinoti, S.N., Kurz, K.M. and Brigham, H. (1990) Improvements in physical fitness on Kenyan schoolboys infected with hookworm, *Trichuris trichiura* and *Ascaris lumbricoroides* following a single dose of albendazole. *Transactions of the Royal Society of Tropical Medicine and Hygiene*, **84**, 277–282.

Stephenson, L., Latham, M., Adams, E., Kinoti, S. and Pertet, A. (1993a) Weight gain of Kenyan school children infected with hookworm, *Trichuris trichiura* and *Ascaris lumbricoides* is improved following once- or twice-yearly treatment with albendazole. *Journal of Nutrition*, **123**, 656–665.

Stephenson, L., Latham, M., Adams, E., Kinoti, S. and Pertet, A. (1993b) Physical fitness, growth and appetite of Kenyan schoolboys with hookworm, *Trichuris trichiura* and *Ascaris lumbricoides* infections are improved four months after a single dose of albendazole. *Journal of Nutrition*, **123**, 1036–1046.

Sternberg, R.T., Grigorenko, E.L. and Nokes, C. (1997) Effects of children's ill health on cognitive development. In *Early Child Development*, edited by M.E. Young. Washington DC: World Bank.

Stoltzfus, R.J., Chwaya, H.M., Tielsch, J.M., Schulze, K.J., Albonico, M. and Savioli, L. (1997) Epidemiology of iron deficiency anaemia in Zanzibari schoolchildren: the importance of hookworms. *American Journal of Clinical Nutrition*, **65**, 153–159.

Thein-Hlaing (1993) Ascariasis and childhood malnutrition. *Parasitology*, **107**, S125–S136.

Thein-Hlaing, Than-Saw and Myat-Lay-Kyin (1991). The impact of three-monthly age-targeted chemotherapy on *Ascaris lumbricoides* infection. *Transactions of the Royal Society of Tropical Medicine and Hygiene*, **85**, 519–522.

UNICEF (1997) Promoting child development through helminth control programmes. Report of a workshop 24–25 February 1997. New York: UNICEF.

Warren, K.S. (1973) Regulation of the prevalence and intensity of schistosomiasis in man: immunology or ecology? *Journal of Infectious Diseases*, **127**, 595–609.

Warren, K.S., Bundy, D.A.P., Anderson, R.M., Davis, A.R., Henderson, D.A., Jamison, D.T., Prescott, N. and Senft, A. (1993) Helminth infections. In *Disease Control Priorities in Developing Countries*, edited by D.T. Jamison, W.H. Mosley, A.R. Measham and J.L. Bobadilla, pp 131–60. Oxford: Oxford University Press.

Watkins, W.E. and Pollitt, E. (1997) 'Stupidity or worms': do intestinal worms impair mental performance? *Psychology Bulletin*, **121**, 171–191.

Weil, G.J., Jain, D.C., Santhanam, S., Malhotra, A., Kumar, H., Sethumadhavan, K.V.P., Liftis, F. and Ghosh, T.K. (1987) A monoclonal antibody-based enzyme-immunoassay for detecting parasite antigenemia in Bancroftian filariasis. *Journal of Infectious Diseases*, **157**, 350–355.

WHO (1994) Lymphatic filariasis infection and disease: control strategies. TDR/CTD/PENANG/94.1 (mimeogr)

WHO (1996b) Report of the WHO Informal Consultation on hookworm infection and anaemia in girls and women. WHO/CTD/SIP/96.1 Geneva: WHO.

WHO (1996a) Report of the WHO Informal Consultation on the use of chemotherapy for the control of morbidity due to soil-transmitted nematodes in humans. WHO/CTD/SIP/96.2. Geneva: WHO.

WHO (1997) *The World Health Report 1997: Conquering Suffering. Enriching Humanity*. Geneva: WHO.

Woolhouse, M.E.J. (1991) On the application of mathematical models of schistosome dynamics. 1. Natural transmission. *Acta Tropica*, **49**, 241–270.

Woolhouse, M.E.J. (1994) Immunoepidemiology of human schistosomes — taking the theory into the field. *Parasitology Today*, **10**, 196–202.

Woolhouse, M.E.J., Taylor, P., Matanhire, D. and Chandiwana, S.K. (1991) Acquired immunity and Epidemiology of *Schistosoma haematobium*. *Nature*, **351**, 757–759.

World Bank (1993) *World Development Report: Investing in Health*. Oxford University Press, Oxford.

Index